IB BIOLOGY

DIPLOMA PROGRAMME

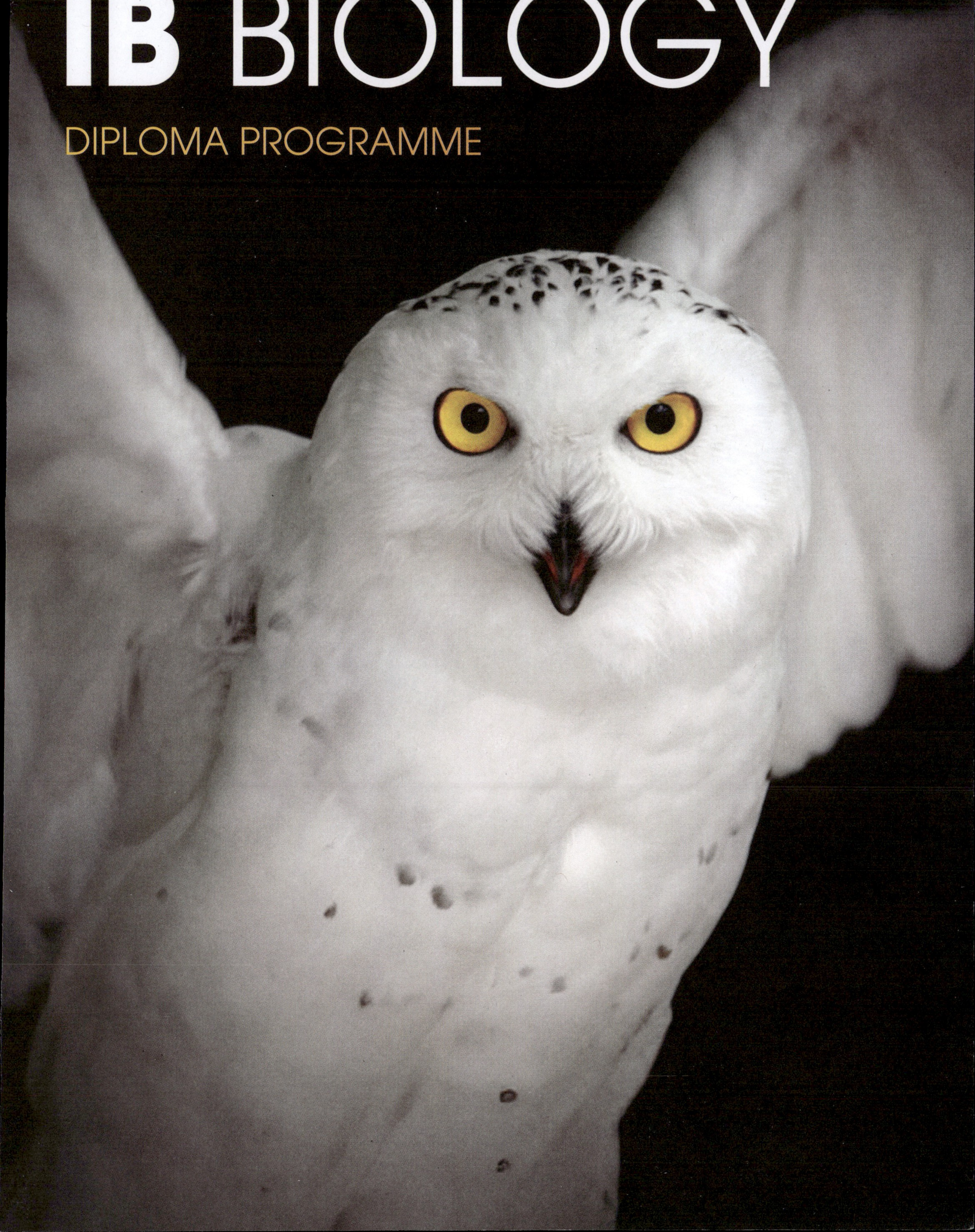

Third edition 2024

Second Printing

ISBN 978-1-99-101410-8

Copyright ©2024 Richard Allan
First published 2024
by BIOZONE International Ltd

Printed by Thomson Press using paper sourced using sustainable methods.

This product has been developed independently of the International Baccalaureate Organization. It is not endorsed.

Acknowledgements

BIOZONE wishes to thank and acknowledge the team for their efforts and contributions to the production of this title.

Cover Photograph

Photo: iStock - Adam Bennie

The snowy owl (*Bubo scandiacus*) is a migratory bird inhabiting the Arctic tundra in warmer months, and migrating south to North America, Europe, and Asia in winter. It is diurnal, hunting its primary food source (lemmings) day and night. Snowy owls breed on the Arctic tundra and are highly territorial, defending the nest vigorously against much larger animals (including wolves). Its magnificent plumage ranges from snowy white (in older males), to white with dark bars and spots (in females and juveniles).

BIOZONE International Ltd.
32 Somerset Street, Frankton,
Hamilton 3204, New Zealand

PH: +64 7 856 8104
FAX: +64 7 856 9243
Email: sales@biozone.com

www.BIOZONE.com

IB BIOLOGY

About the Authors

Jillian Mellanby *Editor*

Jill began her science career with a degree in biochemistry and, after some time working in research institutes, became a science teacher, working in the UK and New Zealand. She spent many years as managing editor of a suite of science journals and has also written science articles for a public audience. She joined BIOZONE in late 2021.

Kent Pryor *Author*

Kent has a BSc from Massey University majoring in zoology and ecology and taught secondary school biology and chemistry for 9 years before joining BIOZONE as an author in 2009.

Sarah Gaze *Author*

Sarah has 16 years of experience as a Science and Chemistry teacher, recently completing M.Ed. (1st class hons) with a focus on curriculum, science, and climate change education. She has a background in educational resource development, academic writing, and art. Sarah joined the BIOZONE team at the start of 2022.

Lissa Bainbridge-Smith *Author*

Lissa graduated with a Masters in Science (hons) from the University of Waikato. After graduation she worked in industry in a research and development capacity for eight years. Lissa joined BIOZONE in 2006 and is hands-on developing new curricula. Lissa has also taught science theory and practical skills to international and ESL students.

Contents

Theme A: Unity and Diversity

Chapter 1: A1 - Molecules

Chapter 2: A2-Cells

Chapter 3: A3-Organisms

Chapter 4: A4-Ecosystems

Theme B: Form and Function

Chapter 5: B1-Molecules

Chapter 6: B2-Cells

CODING Activity is marked: ☐ to be done ☑ when completed   **AHL content**   ● AOS   ● NOS

CODING: Activity is marked: ☐ to be done ☑ when completed | **AHL content** ● AOS ● NOS

CODING Activity is marked: ☐ to be done   ☑ when completed   **AHL content**   ● AOS   ● NOS

CODING: Activity is marked: ▣ to be done ☑ when completed **AHL content** ● AOS ● NOS

Using This Worktext

The worktext is structured on the two-year International Baccalaureate (IB) Diploma Programme Biology syllabus. The content is divided into four units, each containing four chapters based around the four levels of organization within a theme. Both Standard Level and Additional Higher Level material is covered. The material is clearly marked so you know what level of the syllabus it is written for. Chapter 17 provides support for completing your scientific investigation. The structure of a theme unit and chapter are described below.

Structure of a theme unit

Theme introduction
Identifies the understandings in the theme.

Chapter
Each theme unit consists of four **organizational level chapters** related to the theme. These cover molecules, cells, organisms, or ecosystems.

Summary Assessment
A summative assessment task covering an entire theme.

Chapter introduction
Short statements summarize the **learning outcomes** for the chapter.

Did You Get It?
A formative or summative assessment task covering the content in the chapter.

Activity Pages
The activity pages have been designed to address the **content statements** of the course. **Application of skills** and **nature of science** are also included where applicable.
Most activities have questions for you to answer. This allows you to form a record of work and demonstrate your understanding of the content.

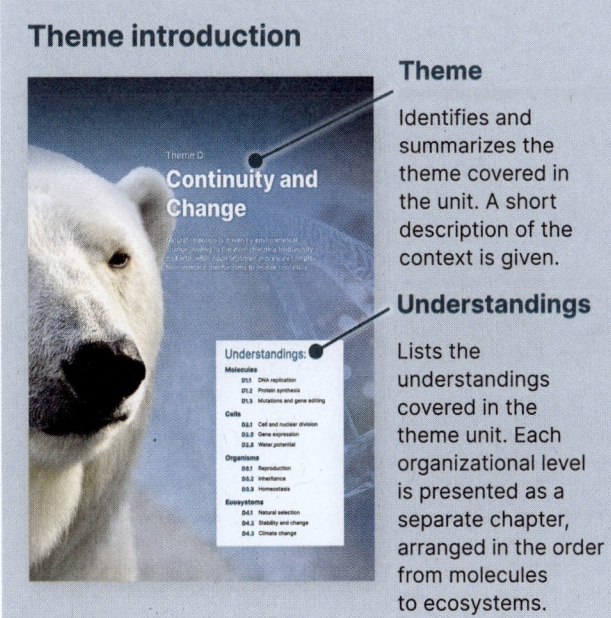

Theme introduction

Theme
Identifies and summarizes the theme covered in the unit. A short description of the context is given.

Understandings
Lists the understandings covered in the theme unit. Each organizational level is presented as a separate chapter, arranged in the order from molecules to ecosystems.

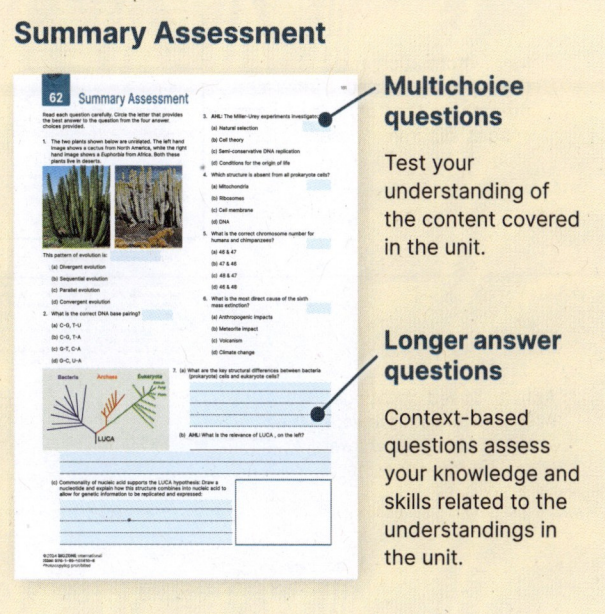

Summary Assessment

Multichoice questions
Test your understanding of the content covered in the unit.

Longer answer questions
Context-based questions assess your knowledge and skills related to the understandings in the unit.

Chapter Introductions

Use the chapter introductions to identify and navigate through the learning outcomes. Useful features of the chapter introductions are provided below.

BIOZONE Resource Hub

Quick link to the BIOZONE Resource Hub using QR code or bit.ly URL

Chapter Title

Identifies the chapter number, theme, and level of organization.

Numbered Understandings

Learning outcome statements indicate what is required to effectively cover each content statement in the chapter.

Use the check boxes to identify activities to be done (•) and tick them off (✓) when you have completed the learning outcome.

Guiding Questions

These help to identify and focus on important areas of study within this chapter.

Orange **AHL** text and a yellow shaded box identify **Advanced Higher Learning** content statements.

The activity in the book related to this learning outcome.

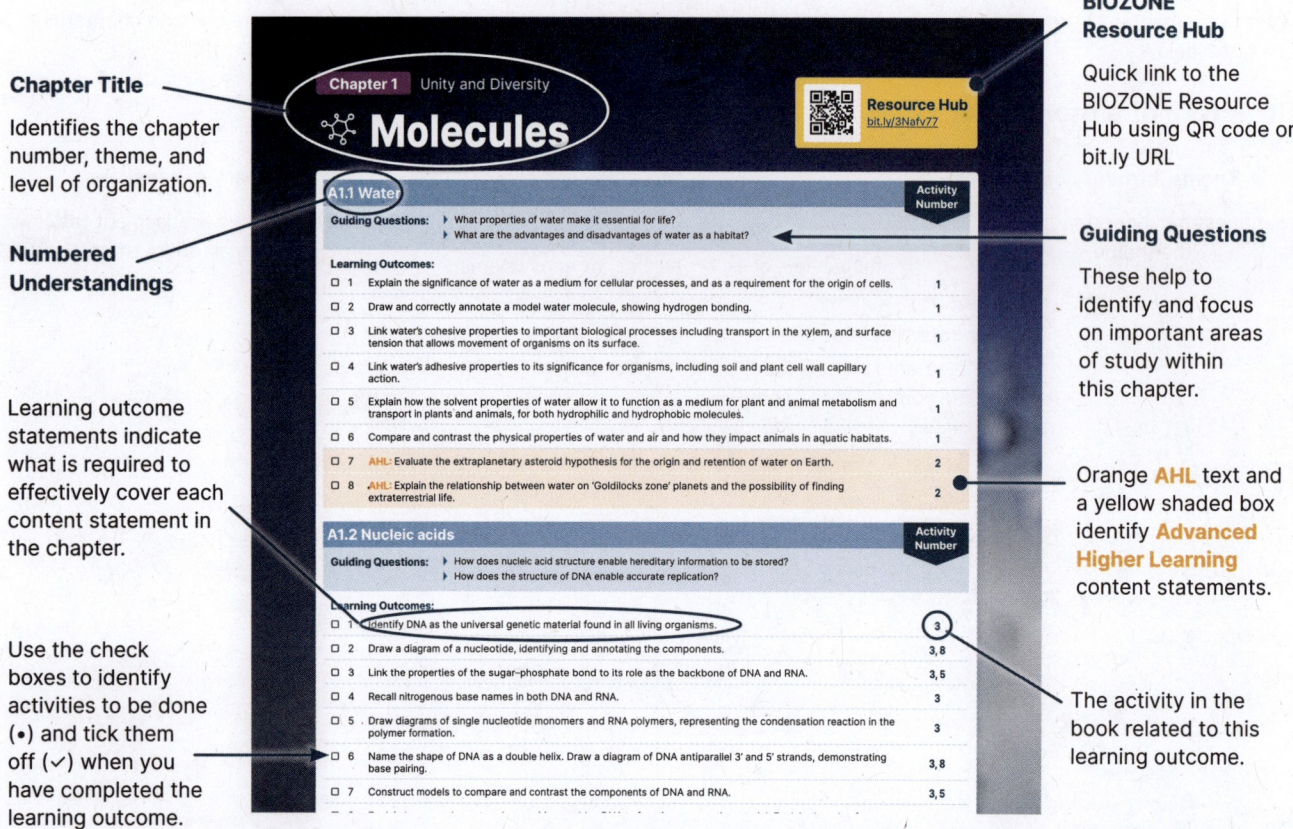

Structure of the activity pages

Activities make up most of the worktext. Be sure to interact with all the elements on the page so you don't miss any valuable information. As you work through the material, answer the questions and complete the tasks provided. Inputting your answers will form a record of work which helps to consolidate your understanding. It can also be used for revision at a later date.

A **Key Idea** provides a focus for the activity.

An orange activity tab indicates an AHL only activity. A blue activity tab identifies standard level content.

An **introductory paragraph** provides background or introductory information to the topic.

More information about the topic is provided through explanatory text, images, diagrams, case studies, and data.

QR codes provide a quick link to engaging interactive 3D models.

Nature of Science (NOS) and **Application of Skills** (AOS) content is embedded within the context of the activity. They are also identified in the tab system.

Activity based questions help you consolidate your learning. Use your answers to review for assessments.

The tabs provide information about the content in the activity and if support material is available on the BIOZONE Resource Hub.

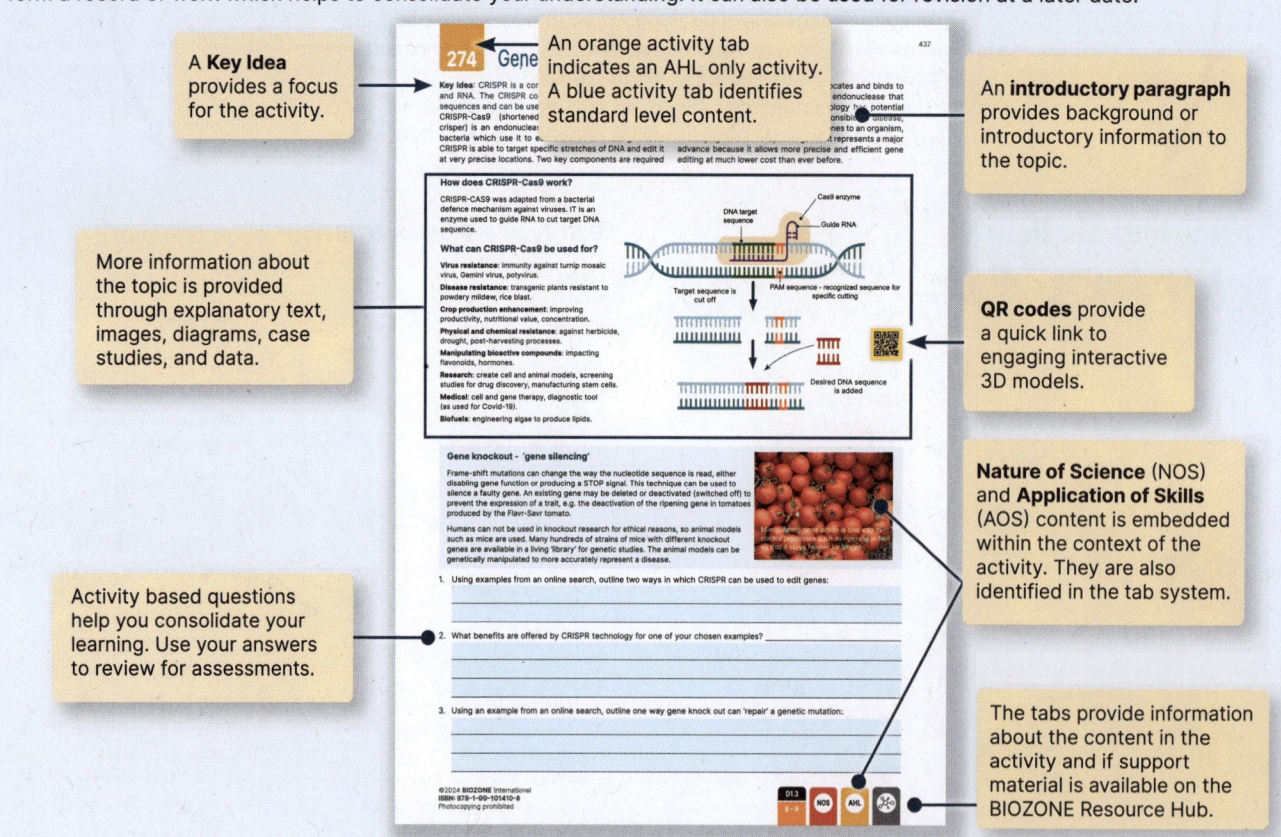

Understanding the Tab System

The tab system at the bottom of the first page of each activity helps you identify the theme and content statement components of the activity. It also indicates if the activity is directed towards AHL content only, and if there are Nature of Science (NOS) and Application of Skills (AOS) content embedded. Additional tabs identify if there are connections to other activities and if support material is provided on BIOZONE's **Resource Hub**. The tab system is explained below.

The **grey hub tab** indicates that the activity has online support via the BIOZONE's **Resource Hub**. This may include videos, animations, articles, 3D models, and interactives.

The **orange AHL** tab indicates that the activity addresses Additional Higher Learning content. This activity will also have an orange activity tab at the top of the first page.

The **black top** of the tab indicates the theme and understanding that the activity addresses. The lower, **coloured section** reflects the theme and the numerical identification of one or more content statements covered in the activity.

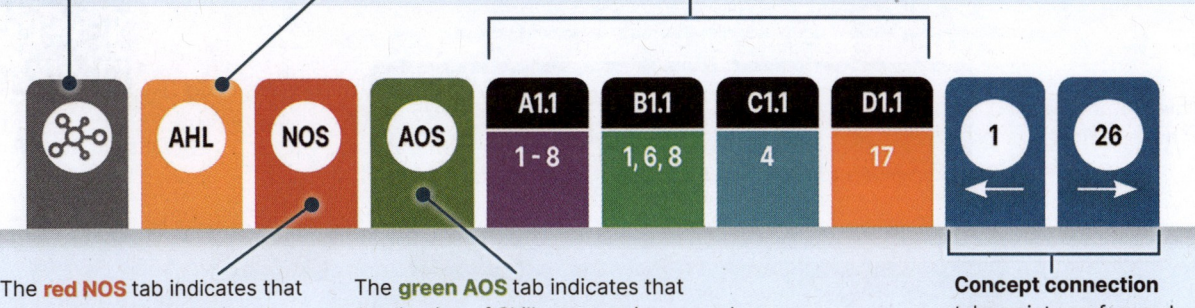

The **red NOS** tab indicates that a Nature of Science theme is covered in the activity.

The **green AOS** tab indicates that Application of Skills content is covered in the activity.

Concept connection tabs point you forward or back to activities with related concepts.

Practical Investigations

Practical investigations form an important component of the IB Biology syllabus. Throughout the worktext you will notice green investigation panels like the one shown on the right. Each investigation has been designed using simple equipment found in most high school laboratories. The investigations provide opportunities for you to carry out hands-on science exploration yourself.

The investigations will help you develop:

▸ Skills in observation
▸ Skills in critical analysis and problem solving
▸ Skills in mathematics and numeracy
▸ Skills in collecting and analysing data and maintaining accurate records
▸ Skills in working both independently and collaboratively as part of a group
▸ Skills in communicating and contributing to group discussions

Detailed equipment lists for each investigation are provided at the back of the worktext.

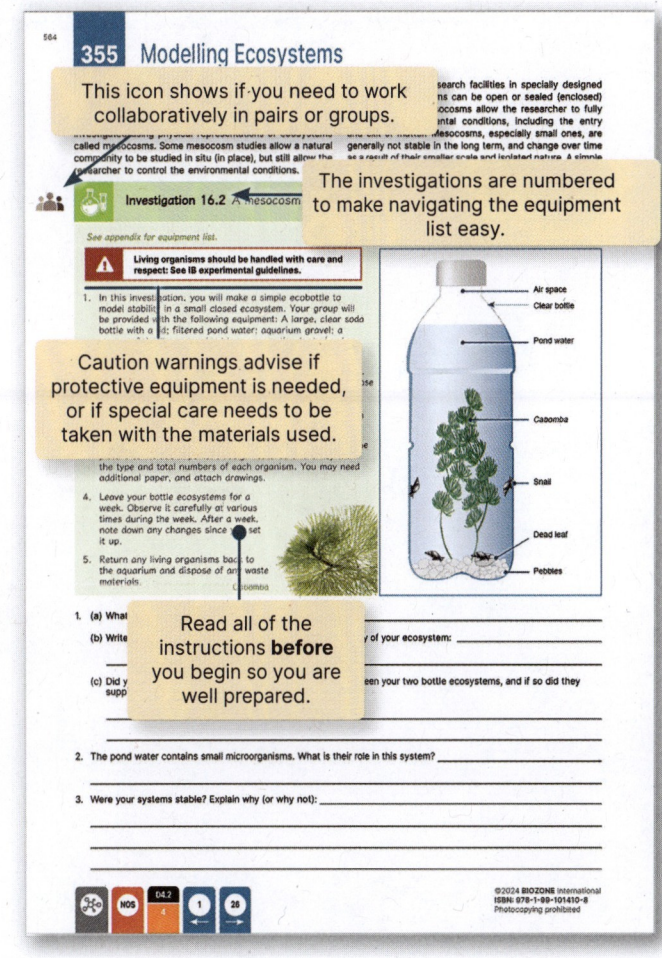

This icon shows if you need to work collaboratively in pairs or groups.

The investigations are numbered to make navigating the equipment list easy.

Caution warnings advise if protective equipment is needed, or if special care needs to be taken with the materials used.

Read all of the instructions **before** you begin so you are well prepared.

Using BIOZONE's Resource Hub

▸ BIOZONE's **Resource Hub** provides links to online content supporting the activities in the worktext. This includes videos, weblinks, 3D models, spreadsheets, animations, and more.

▸ A grey tab (below) on the tab panel indicates there is support on the Resource Hub for the activity. Most activities have resources to support them.

▸ You can also check for any errata or clarifications to the book or answers on the Resource Hub.

www.BIOZONEhub.com

Then enter the code in the text field

IB3-4108

Or scan this QR code

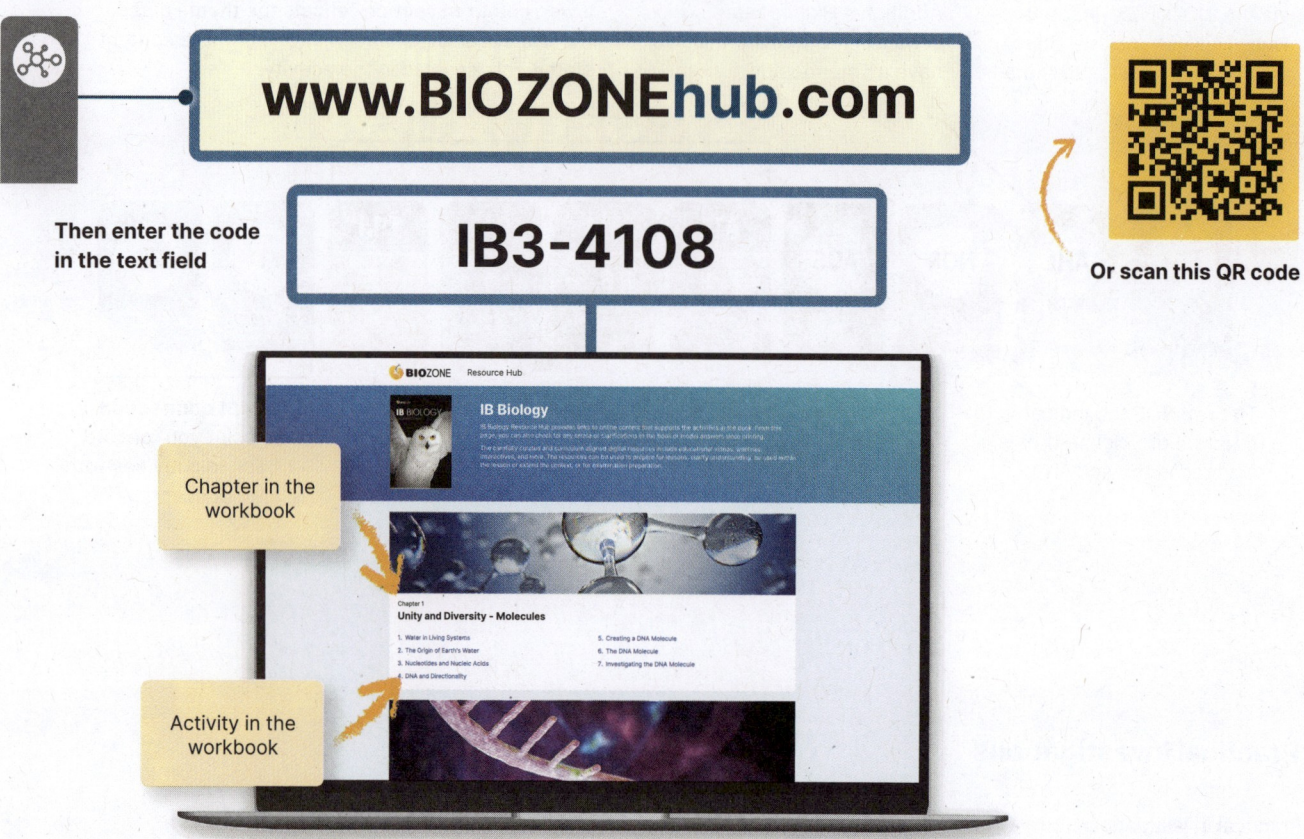

Chapter in the workbook

Activity in the workbook

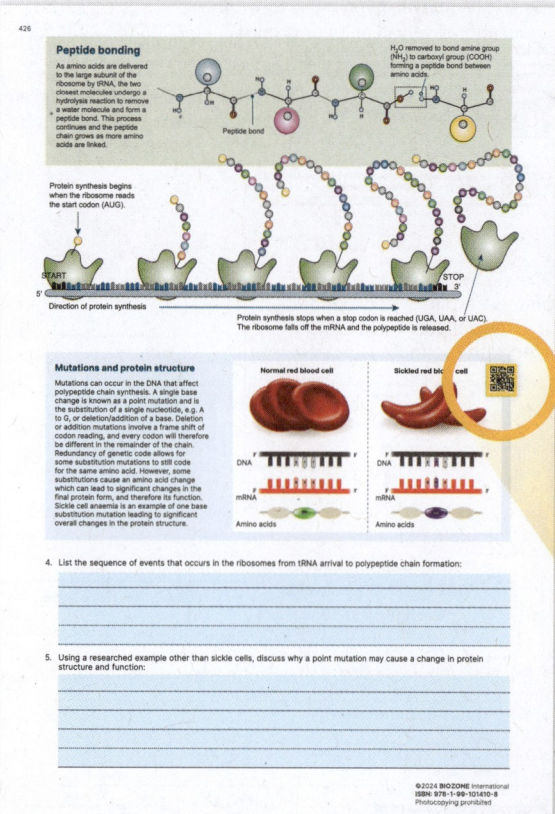

Direct access to 3D models through QR Codes

Some activities have QR codes on the pages (circled, left). These link directly to informative and engaging 3D models. All models can be rotated and zoomed, and some contain informative annotations.

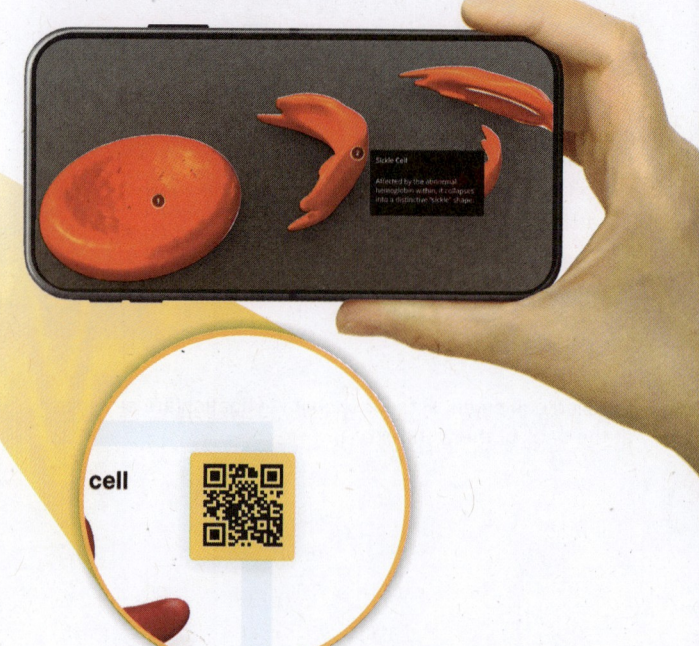

Themes in IB Diploma Biology

▶ The IB syllabus is grouped into four themes (A - D), with each theme developing related concepts. Each theme is further explored at organizational level: 1. molecules, 2. cells, 3. organisms, 4. ecosystems.

▶ The worktext is sequenced into 16 chapters, A1 to D4. Each chapter is further broken down into understandings, for example, A.1.1, A1.2 and so on.

▶ The syllabus is spiralling, and many phenomena are examined in different contexts, in different themes, and at different organizational levels.

▶ The syllabus can be covered in any order to meet differing requirements. Your teacher will let you know the order you will be covering the activities in.

▶ The reference table below indicates key understandings covered in each chapter and where the chapters can be found.

▶ You may wish to record the start date of the chapter and the proposed date of the test for each chapter.

Theme A — Unity and Diversity

1. Molecules — page 1	2. Cells — page 15	3. Organisms — page 46	4. Ecosystems — page 75
A1.1 Water	A2.1 Origins of Cells	A3.1 Diversity of Organisms	A4.1 Evolution and Speciation
A1.2 Nucleic acids	A2.2 Cell Structure	A3.2 Classification and Cladistics	A4.2 Conservation of Biodiversity
	A2.3 Viruses		
Start Date:	Start Date:	Start Date:	Start Date:
Test Date:	Test Date:	Test Date:	Test Date:

Theme B — Form and Function

5. Molecules — page 104	6. Cells — page 122	7. Organisms — page 156	8. Ecosystems — page 208
B1.1 Carbohydrates and Lipids	B2.1 Membranes and Membrane Transport	B3.1 Gas Exchange	B4.1 Adaptation to Environment
B1.2 Proteins	B2.2 Organelles and Compartmentalization	B3.2 Transport	B4.2 Ecological Niches
	B2.3 Cell Specialization	B3.3 Muscle and Motility	
Start Date:	Start Date:	Start Date:	Start Date:
Test Date:	Test Date:	Test Date:	Test Date:

Theme C — Interaction and Interdependence

9. Molecules — page 239	10. Cells — page 283	11. Organisms — page 306	12. Ecosystems — page 356
C1.1 Enzymes and Metabolism	C2.1 Chemical Signalling	C3.1 Integration of Body Systems	C4.1 Populations and Communities
C1.2 Cell Respiration	C2.2 Neural Signalling	C3.2 Defence Against Disease	C4.2 Transfers of Energy and Matter
C1.3 Photosynthesis			
Start Date:	Start Date:	Start Date:	Start Date:
Test Date:	Test Date:	Test Date:	Test Date:

Theme D — Continuity and Change

13. Molecules — page 411	14. Cells — page 441	15. Organisms — page 470	16. Ecosystems — page 535
D1.1 DNA Replication	D2.1 Cell and Nuclear Division	D3.1 Reproduction	D4.1 Natural Selection
D1.2 Protein Synthesis	D2.2 Gene Expression	D3.2 Inheritance	D4.2 Stability and Change
D1.3 Mutations and Gene Editing	D2.3 Water Potential	D3.3 Homeostasis	D4.3 Climate Change
Start Date:	Start Date:	Start Date:	Start Date:
Test Date:	Test Date:	Test Date:	Test Date:

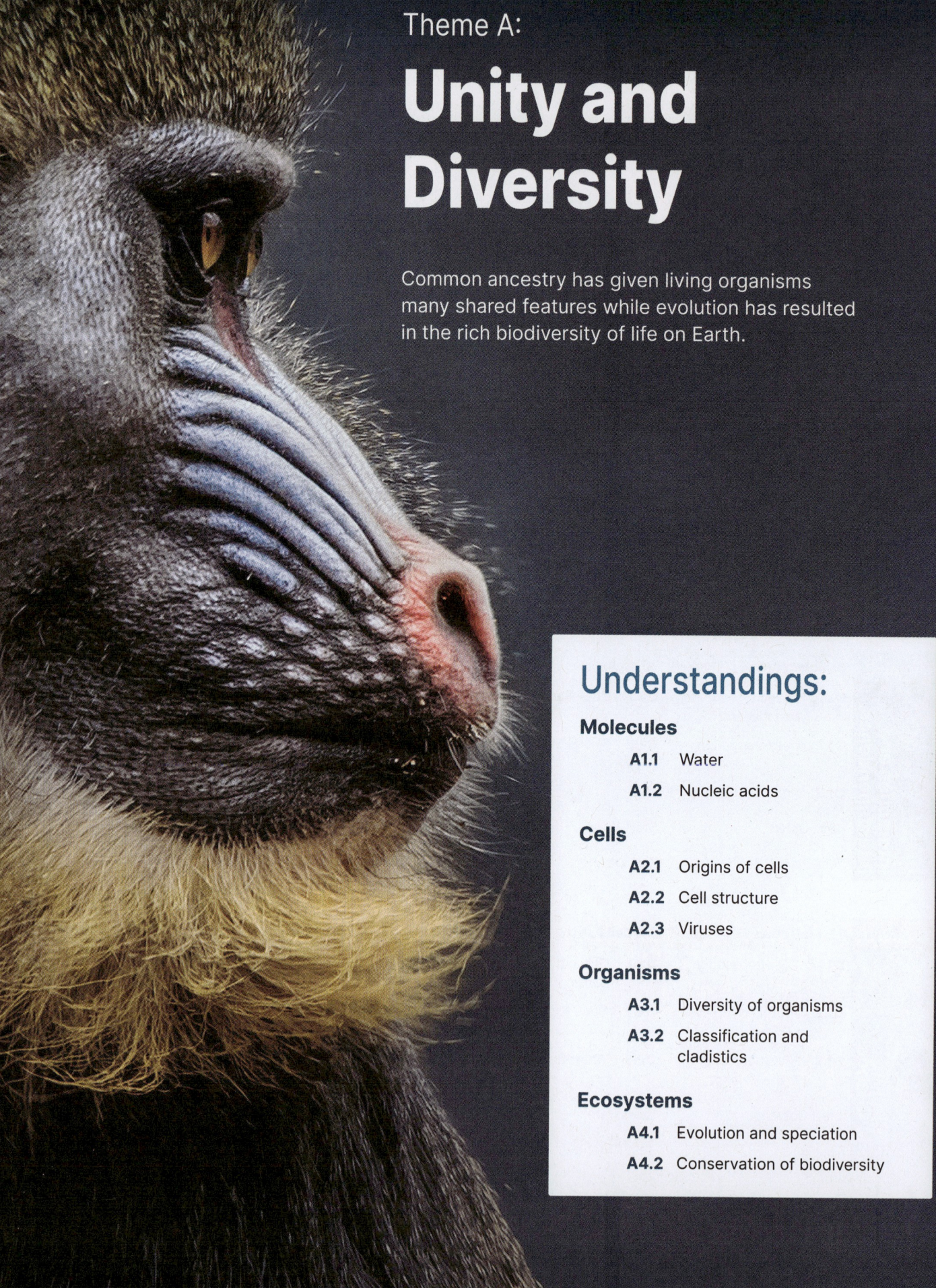

Theme A:

Unity and Diversity

Common ancestry has given living organisms many shared features while evolution has resulted in the rich biodiversity of life on Earth.

Understandings:

Molecules

A1.1   Water

A1.2   Nucleic acids

Cells

A2.1   Origins of cells

A2.2   Cell structure

A2.3   Viruses

Organisms

A3.1   Diversity of organisms

A3.2   Classification and cladistics

Ecosystems

A4.1   Evolution and speciation

A4.2   Conservation of biodiversity

 Molecules

A1.1 Water

Activity Number

Guiding Questions:
▶ What properties of water make it essential for life?
▶ What are the advantages and disadvantages of water as a habitat?

Learning Outcomes:

☐	1	Explain the significance of water as a medium for cellular processes, and as a requirement for the origin of cells.	1
☐	2	Draw and correctly annotate a model water molecule, showing hydrogen bonding.	1
☐	3	Link water's cohesive properties to important biological processes including transport in the xylem, and surface tension that allows movement of organisms on its surface.	1
☐	4	Link water's adhesive properties to its significance for organisms, including soil and plant cell wall capillary action.	1
☐	5	Explain how the solvent properties of water allow it to function as a medium for plant and animal metabolism and transport in plants and animals, for both hydrophilic and hydrophobic molecules.	1
☐	6	Compare and contrast the physical properties of water and air and how they impact animals in aquatic habitats.	1
☐	7	**AHL:** Evaluate the extraplanetary asteroid hypothesis for the origin and retention of water on Earth.	2
☐	8	**AHL:** Explain the relationship between water on 'Goldilocks zone' planets and the possibility of finding extraterrestrial life.	2

A1.2 Nucleic acids

Activity Number

Guiding Questions:
▶ How does nucleic acid structure enable hereditary information to be stored?
▶ How does the structure of DNA enable accurate replication?

Learning Outcomes:

☐	1	Identify DNA as the universal genetic material found in all living organisms.	3
☐	2	Draw a diagram of a nucleotide, identifying and annotating the components.	3, 8
☐	3	Link the properties of the sugar–phosphate bond to its role as the backbone of DNA and RNA.	3, 5
☐	4	Recall nitrogenous base names in both DNA and RNA.	3
☐	5	Draw diagrams of single nucleotide monomers and RNA polymers, representing the condensation reaction in the polymer formation.	3
☐	6	Name the shape of DNA as a double helix. Draw a diagram of DNA antiparallel 3' and 5' strands, demonstrating base pairing.	3, 8
☐	7	Construct models to compare and contrast the components of DNA and RNA.	3, 5
☐	8	Explain how complementary base pairing enables DNA to function as genetic material. Explain the role of hydrogen bonds connecting base pairs, and therefore strands, together.	3
☐	9	Link the structure of DNA to its ability to economically store huge quantities of information with almost limitless sequence combinations.	3
☐	10	Explain how the universality of genetic code in DNA of all living organisms is evidence of common ancestry.	3
☐	11	**AHL:** Connect DNA and RNA 5' to 3' linkage directionality to the processes of replication, transcription, and translation.	4
☐	12	**AHL:** Explain the purpose of purine-to-pyrimidine bonding in enabling DNA helix stability.	4
☐	13	**AHL:** Identify histone proteins as the molecule forming the core of a nucleosome. **AOS:** Use digital molecular visualization to investigate the structure of a nucleosome.	6
☐	14	**AHL:** Provide evidence from the Hershey Chase experiment to support the conclusion that DNA is the genetic material. **NOS:** Explain how technological developments, such as use of radioisotopes, enabled Hershey and Chase to carry out their innovative investigation into DNA.	7
☐	15	**AHL: NOS:** Investigate Chargaff's pyrimidine and purine data and explain how their ratios addressed the 'problem of induction' and falsified the tetranucleotide hypothesis.	7

1 Water in Living Systems

Key Idea: Water's molecular structure accounts for its unique properties and its central role in life's processes.

Water (H_2O) is the main component of living things and typically makes up about 70% of any organism. Water is important in cell chemistry as it takes part in, and is a common product of, many reactions. Its cohesive, adhesive, thermal, and solvent properties come about because of its polarity and its ability to form hydrogen bonds with other polar molecules. Water's physical and chemical properties are essential for sustaining life.

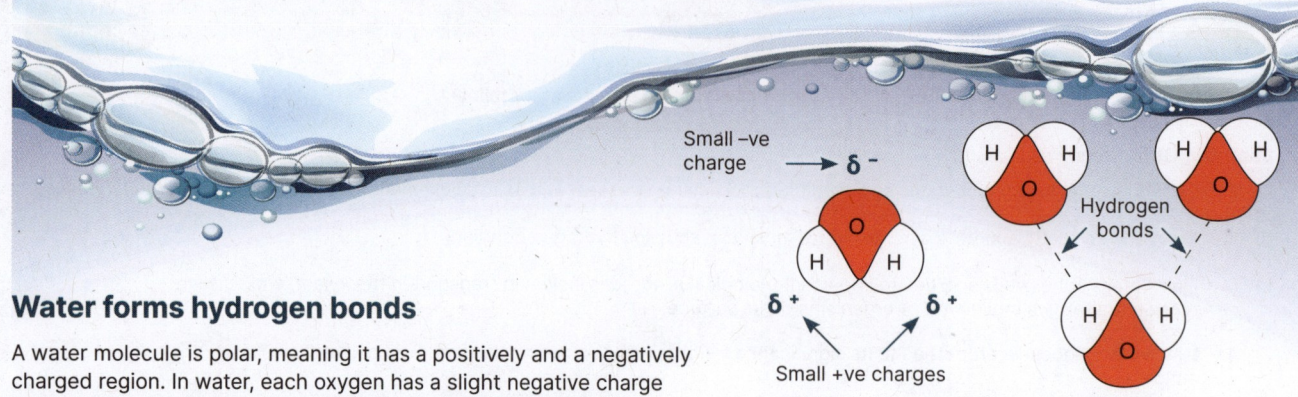

Water forms hydrogen bonds

A water molecule is polar, meaning it has a positively and a negatively charged region. In water, each oxygen has a slight negative charge (δ^-) and each hydrogen has a slight positive charge (δ^+). Water molecules form large numbers of weak hydrogen bonds with other water molecules (right). Individually, hydrogen bonds are weak, but collectively, they are strong enough to account for the unique properties of water, including its cohesion, viscosity, high boiling point, high heat of fusion (energy required to cause a change of state from solid to liquid), and high latent heat of vaporization.

Water in a liquid state has enough energy that hydrogen bonds are continually breaking and reforming. When water cools and loses energy, the hydrogen bonds are strong enough to hold the molecules in place, forming a lattice; this causes water to expand when it freezes. The expansion causes ice to be less dense than liquid water, hence it floats.

Intermolecular bonds between water and other polar molecules or ions are important for biological systems. Inorganic ions may have a positive or negative charge, e.g. positive sodium ion (Na^+) and negative chloride ion (Cl^-). The charged water molecules are attracted to charged ions and surround them. This formation of intermolecular bonds between water and the ions keeps ions dissolved in water. Polar molecules, such as amino acids and carbohydrates, also dissolve readily in water.

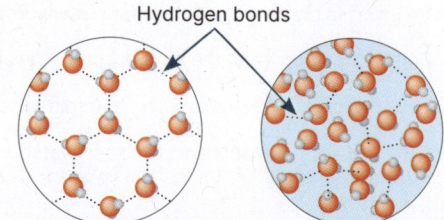

Small –ve charge → δ^-

Small +ve charges → δ^+ δ^+

Hydrogen bonds

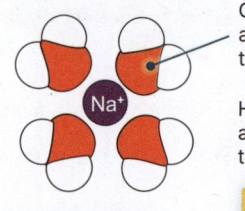

Hydrogen bonds

Ice: H-bonds are fixed in an interconnected framework.

Liquid water: H-bonds constantly break and reform.

Oxygen is attracted to the Na^+

Hydrogen is attracted to the Cl^-

Water surrounding a positive ion (Na^+)

Water surrounding a negative ion (Cl^-)

Cohesive properties

Water molecules are cohesive and stick together because hydrogen bonds form between them. Cohesion allows water to form droplets and is responsible for the surface tension that small organisms rely on to 'walk' on water.

Example: The cohesive and adhesive properties of water allow it to move as an unbroken column through the xylem of plants. This process is essential to water uptake from the soil.

Adhesive properties

Water is attracted to other molecules because of its polar nature. Water will form thin films and 'climb' up surfaces when the molecular forces between them (adhesive forces) are greater than the cohesive forces.

Example: Adhesion enables capillary action, i.e. the ability of a liquid to flow against gravity in a narrow space. This property is also shown by the meniscus of a liquid in a tube.

Solvent properties

Water's polarity allows it to dissociate ions in salts and bond to other polar substances, e.g. alcohols and acids, dissolving them. In contrast, non-polar substances, such as fats and oils, are not water soluble.

Example: Blood plasma in humans and other animals is largely water and transports many water-soluble substances, including ions, glucose, and amino acids, around the body.

Thermal properties

Water has the highest specific heat capacity of all liquids, so it takes a lot of energy before it will change temperature. It also has high latent heat of vaporization, so it takes a lot of energy to transform it from the liquid to the gas phase.

Examples: High specific heat capacity means that large water bodies will maintain a relatively stable temperature. High heat of vaporization makes sweating a very effective cooling mechanism.

A1.1

1-6

©2024 **BIOZONE** International
ISBN: 978-1-99-101410-8

The importance of water in biological systems

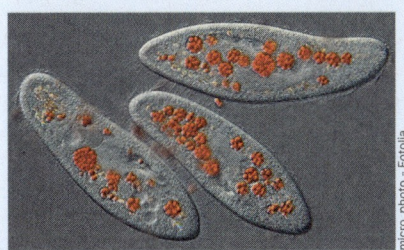

Life originated in water and it still plays a major role in most of life's mechanisms such as metabolic processes, which depend on dissolved reactants (solutes) coming into contact. Water can also act as an acid (donating H^+) or a base (receiving H^+) in chemical reactions.

Water's cohesion is responsible for its high specific heat capacity (a function of its many hydrogen bonds). This means water bodies heat up and cool down only slowly, providing a relatively stable thermal environment. The greater the body of water, the more thermally stable it is.

Water's high latent heat of vaporization means that a change of state from liquid to gas absorbs a lot of energy. When water in sweat evaporates from the skin's surface, it transfers heat from the body to the air, producing cooling. Panting, in animals that do not sweat, operates the same way.

Aquatic animals need to regulate their buoyancy to maintain their position in the water column. Bony fish do this with a swim bladder. Some aquatic birds, such as the black throated loon, are able to compress the air in their lungs to reduce buoyancy and help diving.

Water is known as the universal solvent, because many substances will dissolve in it. In natural waters, dissolved minerals such as calcium are available to aquatic organisms, e.g. shell building organisms such as the hard corals above.

Water's high thermal conductivity means it can quickly remove body heat from animals that are not insulated. Seals, e.g. ringed seal (above) and other marine mammals have insulating layers of blubber to help retain their body heat. Out of water, these animals can run the risk of overheating.

1. Explain how water's molecular structure accounts for each of the following:

   a) Water's cohesion and high heat capacity: _____

   _____

   _____

   b) Water's solvent properties: _____

   _____

   c) Water's high latent heat of vaporization: _____

   _____

2. Use the diagrams opposite to explain why water is less dense in its solid form (as ice) than in its liquid form:

   _____

   _____

3. Summarize the ways in which living systems depend on the properties of water arising from its molecular structure:

   _____

   _____

   _____

   _____

4. Why do marine mammals risk overheating when out of water? _____

   _____

   _____

2 The Origin of Earth's Water

Key Idea: As far as we know, Earth is the only place in the Universe where life exists. The presence of liquid water on the surface is an important factor for the presence of life. There is one thing that life absolutely must have to survive: liquid water. Water is important as a medium for dissolved molecules and ions to carry out the reactions of life. Earth's water is thought to have been acquired through collisions with icy bodies and accretion of minerals and molecules from the original gas cloud from which the solar system formed. Water is also present on many moons in the solar system.

▶ It is thought that much of Earth's water arrived via collisions with icy bodies. Because of its distance from the Sun, liquid water is unlikely to have condensed during or soon after Earth's formation, although hydrated minerals may have been present.

▶ Evidence from the study of meteorites and comets shows that Earth's water is likely to have come from the impact of icy asteroids and planetesimals. The deuterium/hydrogen ratios of water in carbonaceous chondrite asteroids are similar to the ratios found on Earth. Comets, however, have a much higher deuterium/hydrogen ratio and so are unlikely to have delivered much water. It is thought that comets may have contributed only around 10% of Earth's water.

All three phases of water exist on Earth's surface.

Earth's is positioned in the 'Goldilocks zone': not so far from the Sun that water on the surface freezes but not so close that it vaporizes.

Water and extraterrestrial life

Water is essential for life as we know it. Earth's position in the habitable zone of the solar system means it has a surface temperature where water can exist in all three phases. Earth's gravity is important in being able to hold on to its water, stopping it from evaporating into space (unlike Mars for example).

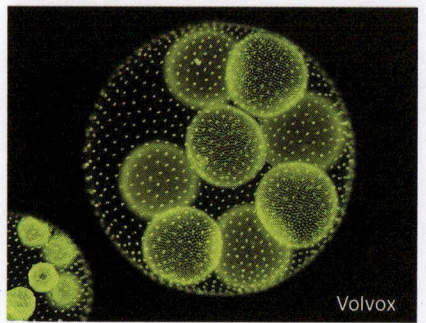

Volvox

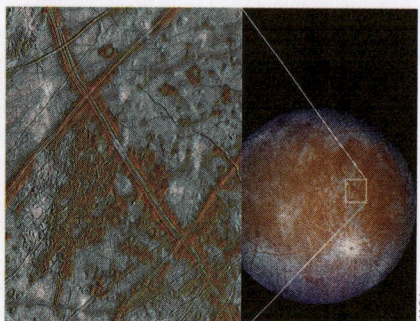

NASA

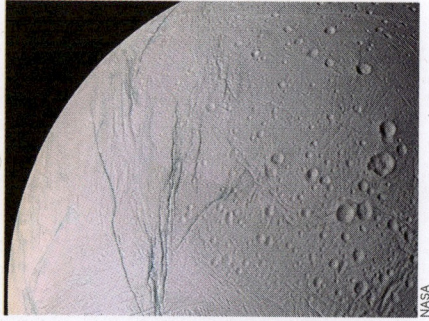

NASA

Water plays a role in many of the metabolic processes of life on Earth. Because of this, it is essential for life as we know it. Earth is not the only place in the solar system with liquid water on its surface. There are 23 moons known or suspected to have large bodies of liquid water on them.

Jupiter's moons Ganymede and Europa (above) both have oceans containing far more liquid water than all of Earth's oceans combined. These are covered by ice sheets many kilometres thick. Evidence shows that they may be heated by hydrothermal vents and tidal stretching from Jupiter's gravity.

Saturn's moon Enceladus (above) also has a vast ocean beneath its icy surface. The presence of hydrothermal vents and mineral rich water lends weight to the idea life could evolve there, as it is thought life may have evolved on Earth near hydrothermal vents.

1. Why is water needed for life? _____

_____

2. What are some factors that help maintain liquid water on Earth's surface? Explain: _____

_____

_____

_____

3. Explain why Earth's water is likely to have come from asteroids rather than comets: _____

_____

_____

_____

4. What is the Goldilocks zone? _____

_____

_____

 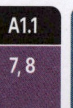 

AHL

A1.1

7, 8

9

©2024 **BIOZONE** International
ISBN: 978-1-99-101410-8
Photocopying prohibited

3 Nucleotides and Nucleic Acids

Key Idea: Nucleotides are the building blocks of DNA and RNA. Nucleic acids are long chains of nucleotides that store and transmit genetic information.

A nucleotide has three components: a base, a sugar, and a phosphate group. They are the building blocks of nucleic acids (DNA and RNA), which are involved in the transmission of inherited information in all living organisms. Some viruses (non living) use RNA to store and transfer genetic

instructions. Nucleic acids have the capacity to store the information that controls cellular activity. The central nucleic acid is called deoxyribonucleic acid (DNA). Ribonucleic acids (RNA) are involved in the 'reading' of the DNA information. All nucleic acids are made up of nucleotides linked together to form chains or strands. The strands vary in the sequence of the bases found on each nucleotide. It is this sequence that provides a cell's 'genetic instructions'.

Chemical structure of a nucleotide

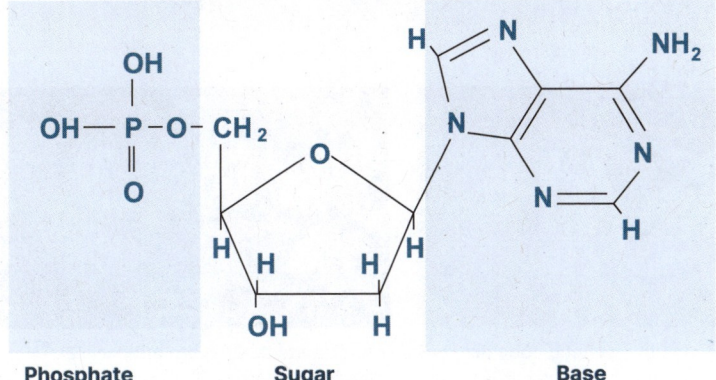

Phosphate   Sugar   Base

Symbolic form of a nucleotide

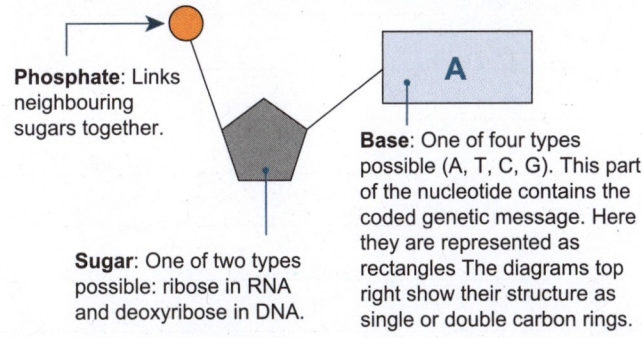

Phosphate: Links neighbouring sugars together.

Base: One of four types possible (A, T, C, G). This part of the nucleotide contains the coded genetic message. Here they are represented as rectangles The diagrams top right show their structure as single or double carbon rings.

Sugar: One of two types possible: ribose in RNA and deoxyribose in DNA.

Bases

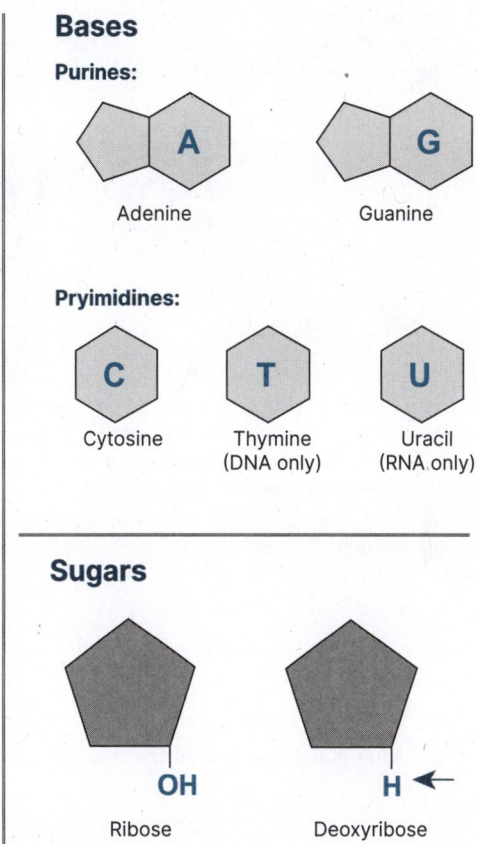

Purines:

A — Adenine   G — Guanine

Pryimidines:

C — Cytosine   T — Thymine (DNA only)   U — Uracil (RNA only)

Sugars

Ribose — OH   Deoxyribose — H

RNA molecule

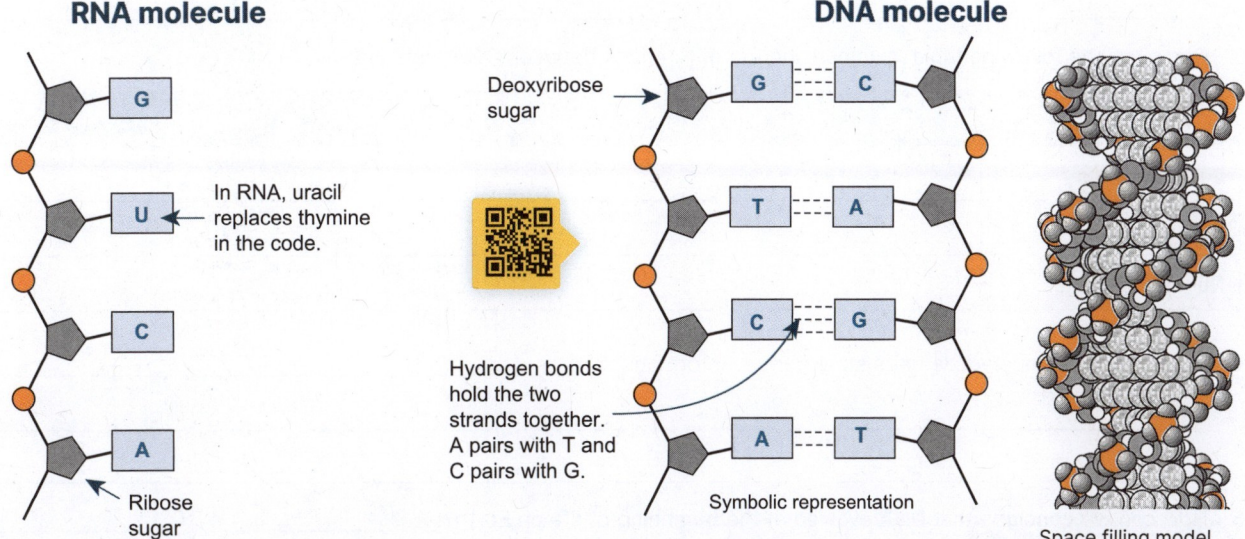

In RNA, uracil replaces thymine in the code.

Ribose sugar

DNA molecule

Deoxyribose sugar

Hydrogen bonds hold the two strands together. A pairs with T and C pairs with G.

Symbolic representation

Space filling model

Ribonucleic acid (RNA) is made up of a single strand of nucleotides linked together. Although it is single stranded, it is often found folded back on itself, with complementary bases joined by hydrogen bonds.

Deoxyribonucleic acid (DNA) is made up of a double strand of nucleotides linked together. It is shown unwound in the symbolic representation (above left). The DNA molecule takes on a double helix shape as shown in the space filling model (above right).

A1.2
1-10

Nucleotides are joined by condensation polymerization

Formation of a nucleotide

Condensation
(water removed)

H_2O

H_2O

A

A

A nucleotide is formed when phosphoric acid and a base are chemically bonded to a sugar molecule. In both cases, water is given off, and they are therefore condensation reactions. In the reverse reaction, a nucleotide is broken apart by the addition of water (hydrolysis).

Formation of a dinucleotide

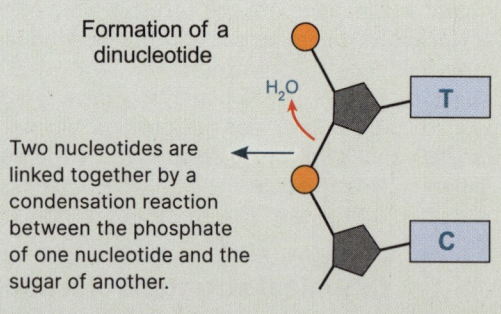

H_2O

T

C

Two nucleotides are linked together by a condensation reaction between the phosphate of one nucleotide and the sugar of another.

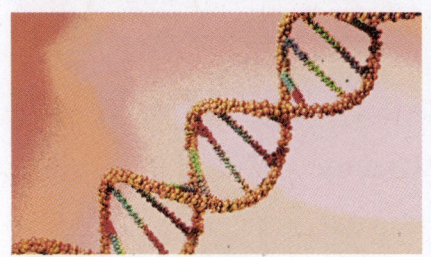

The four bases in DNA (A, T, C, G) can be arranged in any order to make any length of DNA. Thus, an almost endless amount of genetic information can be stored.

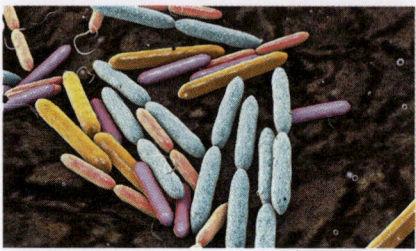

Because DNA and the genetic code are the same in all forms of life, we can conclude that it appeared at the beginning of life on Earth.

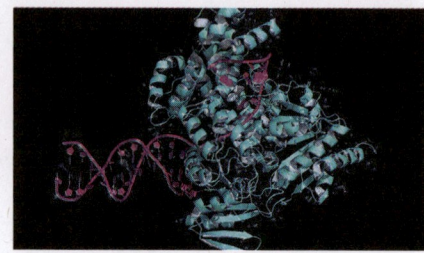

The strands of DNA are antiparallel, i.e. they run in opposite directions to each other. This means they are copied in opposite directions during replication.

1. a) Explain the base-pairing rule that applies in double-stranded DNA: _____

    _____

    b) How is the base-pairing rule for mRNA different? _____

    _____

    c) What is the purpose of the hydrogen bonds in double-stranded DNA? _____

    _____

2. Describe the functional role of nucleotides: _____

    _____

    _____

3. Complete the following table, summarizing the differences between DNA and RNA:

	DNA	RNA
Sugar present		
Bases present		
Number of strands		

4. How can simple nucleotide units store genetic information? _____

    _____

    _____

5. How can we conclude that DNA evolved at the beginning of life on Earth? _____

    _____

    _____

    _____

4 DNA and Directionality

Key Idea: DNA has a specific direction that affects its structure, replication, and transcription.

DNA has directionality. The nucleotides that make up DNA have a specific direction that is based on the position of the carbon atoms. The carbon atoms in the sugar are numbered 1 to 5, starting at the carbon nearest the base. This carbon atom is called 1' (one prime). The carbon atoms to which the phosphate groups attach are the 3' and 5' atoms. The direction of the DNA, and RNA molecules means the molecules that replicate, transcribe, and translate DNA and RNA have to work in the 5' to 3' direction (new nucleotides are added to the 3' end).

Carbon atoms' directionality in DNA

DNA is made up of nucleotides joined together by phosphate groups bonding to sugar molecules, below left. The asymmetric structure gives a DNA strand direction. Each strand runs in the opposite direction to the other, below right. Here the bases are represented using their purine/pyrimidine schematics.

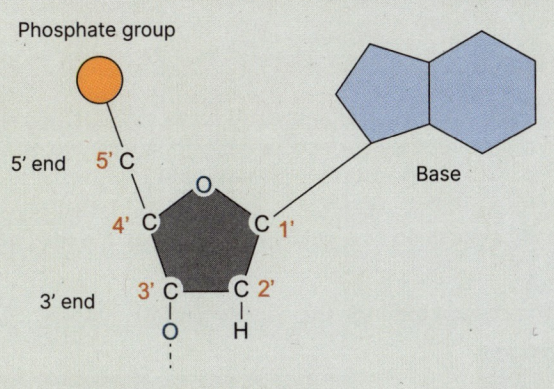

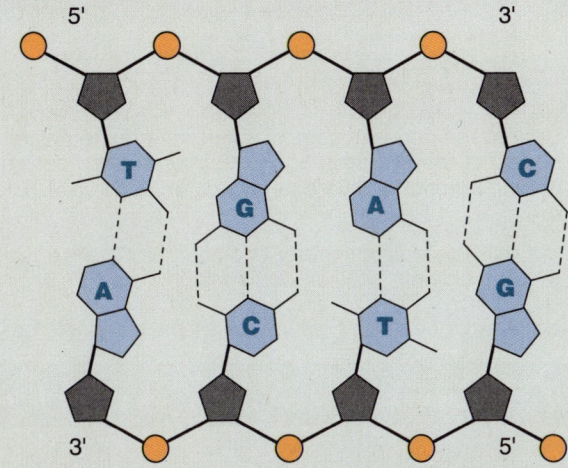

▶ The ends of a DNA strand are labelled the 5' (five prime) and 3' (three prime) ends. The 5' end has a terminal phosphate group (off carbon 5); the 3' end has a terminal hydroxyl group (off carbon 3). During DNA replication, new nucleotides can only be added to the 3' end.

▶ Bases are held together by hydrogen bonds. In the case of A-T pairing, the bases are held together by two hydrogen bonds. C-G base pairs are held together by three hydrogen bonds. Although these hydrogen bonds provide some stability for the DNA molecule, they are not the only or most significant factor.

▶ DNA molecules rich in C-G base pairs have a higher melting point than DNA molecules rich in A-T base pairs. Thus, extremophile bacteria found in hot springs tend to have DNA with a high proportion of C-G base pairs, and the parts of DNA that often separate, e.g. TATA boxes found in promoter regions of a gene, have high proportions of A-T base pairs.

▶ The length of the C-G and A-T base pairs are equal so that the width of the DNA molecule is uniform.

1. a) Why do the DNA strands have an asymmetric structure? _____

   _____

   _____

   b) What are the differences between the 5' and 3' ends of a DNA strand? _____

   _____

   _____

2. What is the significance of the differences in hydrogen bonding between A-T and C-G pairs?

   _____

   _____

   _____

   _____

3. How is the width of the DNA molecule maintained? _____

   _____

   _____

©2024 **BIOZONE** International
ISBN: 978-1-99-101410-8
Photocopying prohibited

5 Creating a DNA Molecule

Key Idea: When base pairing rules are applied, the structure of DNA can be modelled.

Recall that DNA is made up of structures called nucleotides. Two primary factors control the way in which these nucleotide building blocks are linked together: the available space within the DNA double helix and the hydrogen-bonding capability of the bases. These factors cause the nucleotides to join together in a predictable way, referred to as the base pairing rule, which states that %A = %T and %C = %G for both strands of the DNA.

DNA base pairing rule			
Adenine	is always attracted to	**Thymine**	A ←→ T
Thymine	is always attracted to	**Adenine**	T ←→ A
Cytosine	is always attracted to	**Guanine**	C ←→ G
Guanine	is always attracted to	**Cytosine**	G ←→ C

1. a) Cut out the opposite page. Cut out the grey template strand. Dark black lines should be cut. The dashed grey lines represent the hydrogen bonds. Fold on the red dotted lines so that the grey surfaces are facing (a valley fold). Do not cut around the hydrogen bonds on each base. These are just to show you where you will join your bases.

   b) Cut out the complementary strand. The first base (G) is already in position as a guide. Again, fold on the red line so that the blue surfaces are facing each other.

2. Fill in the table below to help you place the remaining bases in the correct order on the complementary strand:

Template strand	Complementary strand
Cytosine (C)	Guanine (G)
Guanine (G)	(a)
Thymine (T)	(b)
Adenine (A)	(c)
Thymine (T)	(d)
Adenine (A)	(e)
Thymine (T)	(f)
Thymine (T)	(g)
Cytosine (C)	(h)
Guanine (G)	(i)

A finished model

3. Cut out the bases and put them into the slots on the complementary strand using the order in the table above. Use tape to fix them in position. Make sure the blue surfaces are facing and the base is in the same orientation as the guide (already in place, G).

4. Line up the first base pairs (C and G) and stick them together with tape. Note that the bases are facing in opposite directions.

5. Continue sticking base pairs together, working your way around the helix, to complete the DNA molecule. Note that the strands are antiparallel.

6. What does antiparallel mean? _____

   _____

7. Explain how the concept of base pairing allowed you to easily model a DNA molecule using the template provided:

   _____

   _____

   _____

A1.2

3 - 10

©2024 **BIOZONE** International
ISBN: 978-1-99-101410-8
Photocopying prohibited

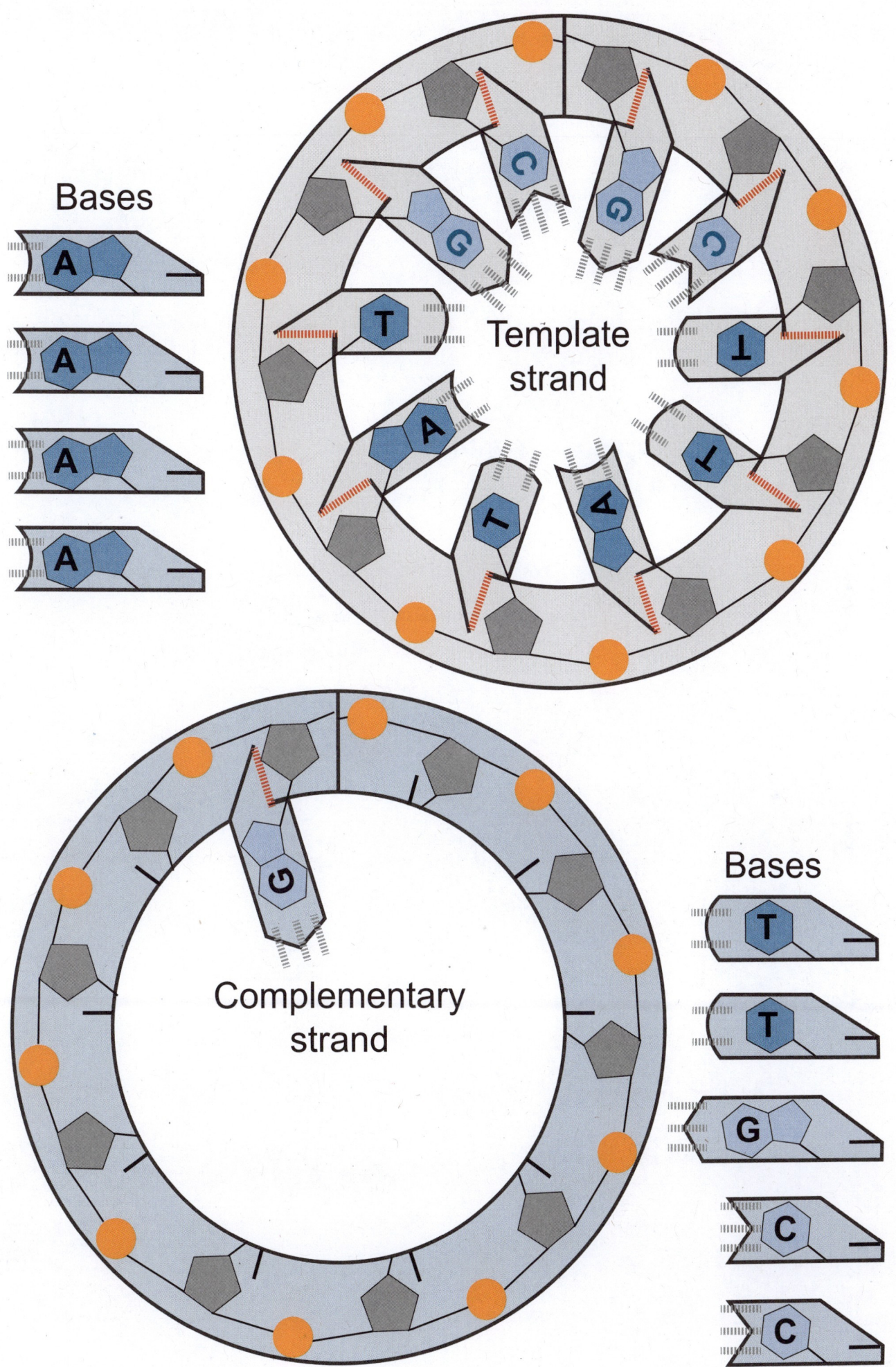

Bases

Template strand

Complementary strand

Bases

The page has been deliberately left blank

6 The DNA Molecule

Key Idea: DNA is packaged around proteins called histones. The DNA in eukaryotes is packaged as discrete linear chromosomes that vary in number from species to species.

The extent of DNA packaging changes during the life cycle of the cell, but classic chromosome structures (below) appear during metaphase of mitosis.

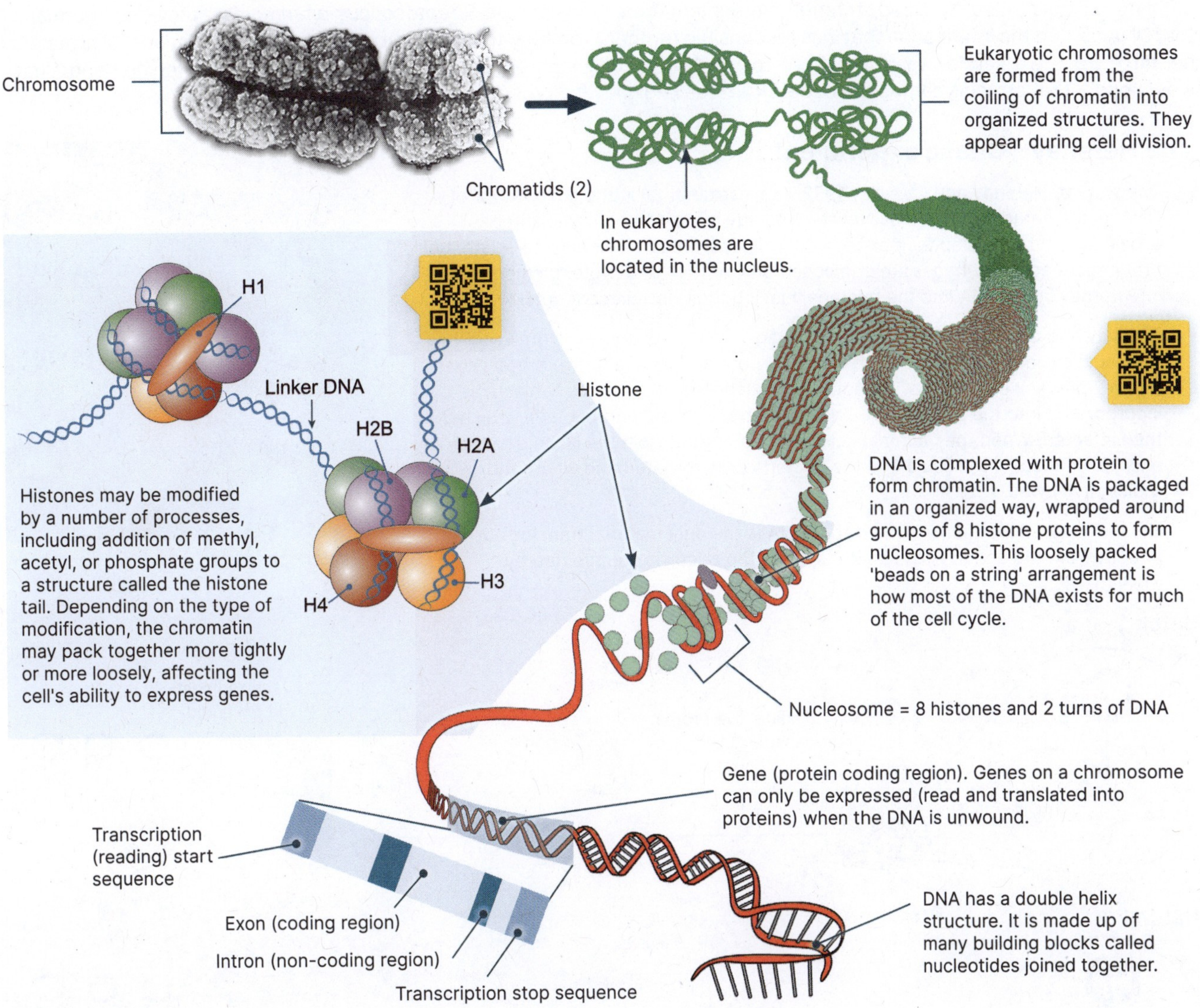

Chromosome

Chromatids (2)

Eukaryotic chromosomes are formed from the coiling of chromatin into organized structures. They appear during cell division.

In eukaryotes, chromosomes are located in the nucleus.

H1

Linker DNA

H2B

H2A

Histone

H4

H3

Histones may be modified by a number of processes, including addition of methyl, acetyl, or phosphate groups to a structure called the histone tail. Depending on the type of modification, the chromatin may pack together more tightly or more loosely, affecting the cell's ability to express genes.

DNA is complexed with protein to form chromatin. The DNA is packaged in an organized way, wrapped around groups of 8 histone proteins to form nucleosomes. This loosely packed 'beads on a string' arrangement is how most of the DNA exists for much of the cell cycle.

Nucleosome = 8 histones and 2 turns of DNA

Gene (protein coding region). Genes on a chromosome can only be expressed (read and translated into proteins) when the DNA is unwound.

Transcription (reading) start sequence

Exon (coding region)

Intron (non-coding region)

Transcription stop sequence

DNA has a double helix structure. It is made up of many building blocks called nucleotides joined together.

1. Explain why eukaryotic DNA needs to be packaged to fit inside a cell nucleus:

_____

_____

2. How do histone proteins help in the coiling up of DNA? _____

_____

_____

3. Suggest why a cell coils up its chromosomes into tight structures when it is going to divide:

_____

_____

4. Explain how the packaging of DNA in an organized way enables closer regulation of gene expression:

_____

_____

_____

©2024 **BIOZONE** International
ISBN: 978-1-99-101410-8
Photocopying prohibited

A1.2
12, 13

7 Investigating the DNA Molecule

Key Idea: Following on from the findings of other scientists, Hershey and Chase performed a classic experiment to demonstrate that DNA was the molecule of heredity.

Many years before Watson and Crick discovered the structure of DNA, biologists had deduced, through experimentation, that DNA carried the information that was responsible for the heritable traits we see in organisms. Prior to the 1940s, it was thought that proteins carried the code. Little was known about nucleic acids. The variety of protein structures and functions suggested they could account for the many traits we see in organisms. Two early experiments, one by Griffith and another by Avery, MacLeod, and McCarty, using the bacterium *Streptococcus pneumoniae,* provided information on how traits could be passed on and what cellular material was responsible. Later experiments by Hershey and Chase helped confirm DNA as the genetic material.

The Hershey - Chase experiment

▸ The work of Hershey and Chase in 1952 was instrumental in the acceptance of DNA as the hereditary material. New technology in the form of radioisotopes allowed them to develop their experiments. They worked on viruses called phages that infect bacteria. Phages are composed only of DNA and protein. When they infect, they inject DNA into the bacteria, leaving their protein coat outside.

▸ Hershey and Chase had two batches of phage. Batch 1 phage were grown with radioactively labelled sulfur, which was incorporated into the phage protein coat. Batch 2 phage were grown with radioactively labelled phosphorus, which was incorporated into the phage DNA. The phage were mixed with bacteria, which they then infected. When the bacterial cells were separated from the liquid, Hershey and Chase studied the liquid and the bacteria cells to determine where the radioactivity was.

▸ Hershey and Chase demonstrated that DNA is the only material transferred directly from the phage to the bacteria when the bacteria are infected by the viruses.

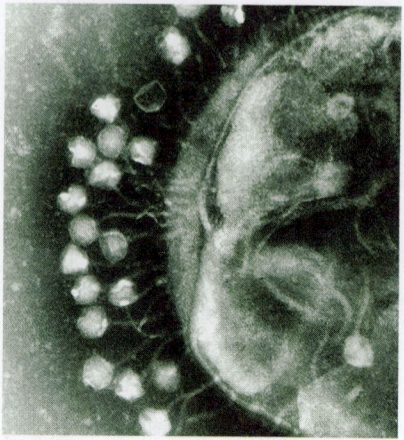

Bacteriophages attached to a bacterium

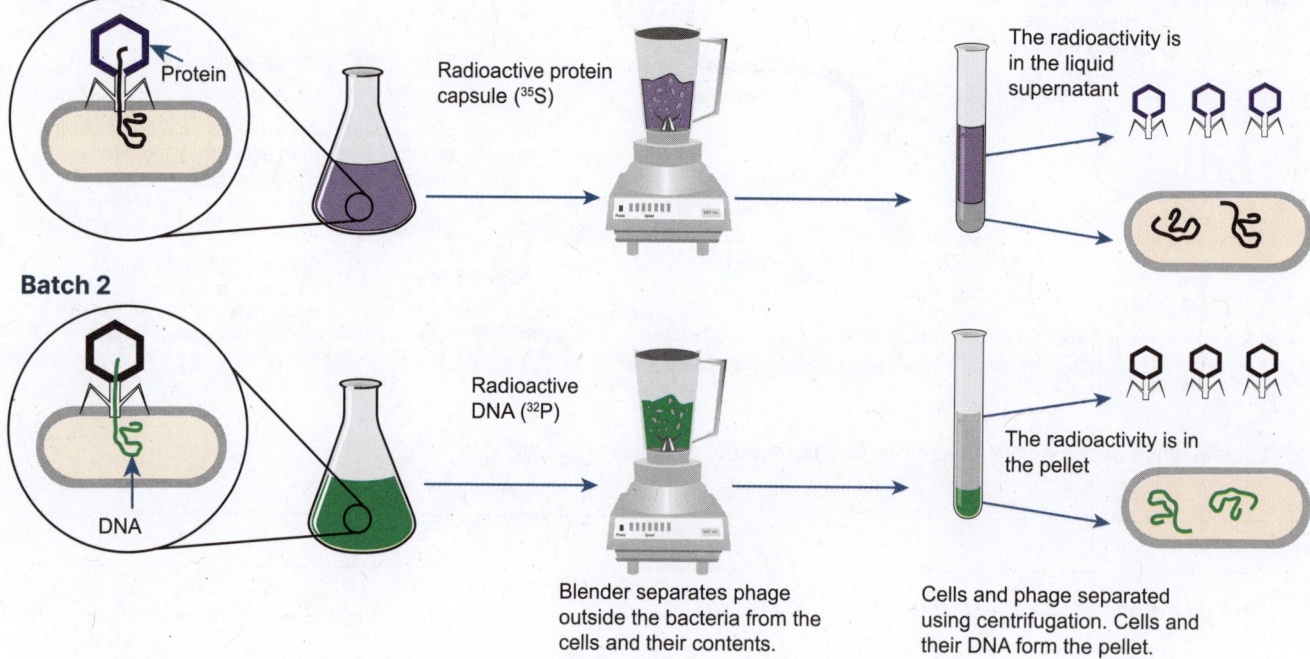

Batch 1

Protein

Radioactive protein capsule (^{35}S)

The radioactivity is in the liquid supernatant

Batch 2

DNA

Radioactive DNA (^{32}P)

The radioactivity is in the pellet

Blender separates phage outside the bacteria from the cells and their contents.

Cells and phage separated using centrifugation. Cells and their DNA form the pellet.

1. a) How did the Hershey-Chase experiment provide evidence that nucleic acids, not protein, are the hereditary material?

_____

_____

_____

   b) What assumptions were made about the role of DNA in viruses and bacteria and the role of DNA in eukaryotes?

_____

   c) How would the results of the experiment have differed if proteins carried the genetic information?

_____

   A1.2 14, 15

©2024 **BIOZONE** International
ISBN: 978-1-99-101410-8

Chargaff's experiment

▸ In 1881, Albrecht Kossel isolated the five nucleotide bases: A, T, C, G and U. He received the Nobel prize for this work in 1910. Soon afterwards, Phoebus Levene proposed the tetranucleotide hypothesis: this stated that DNA was built up of repeating units of A, T, C, and G. This was partially because Levene thought DNA was a single stranded molecule, rather than the double stranded molecule as we know it today.

▸ If the tetranucleotide hypothesis was correct, then the amount of all four bases would be about the same in any DNA, the amount of purines and pyrimidines would be equal and, with the same repeating pattern in all DNA, it wouldn't be able to transfer or store information. This was the dominant hypothesis for nearly three decades. This has been called a 'scientific catastrophe,' as the understanding of DNA was held back decades. It wasn't shown to be incorrect until Erwin Chargaff carried out his experiments in the late 1940s. It is also an example of the 'problem of induction,' in which, from Levene's results, it was reasoned that DNA had no hereditary value, without any further investigation.

▸ Chargaff extracted DNA from cattle and separated the bases from the rest of the DNA using sulfuric acid. The different bases were separated using paper chromatography. Finally, Ultra Violet-visible (UV-vis) spectroscopy was used to determine the relative amounts of each of the bases in the sample (below).

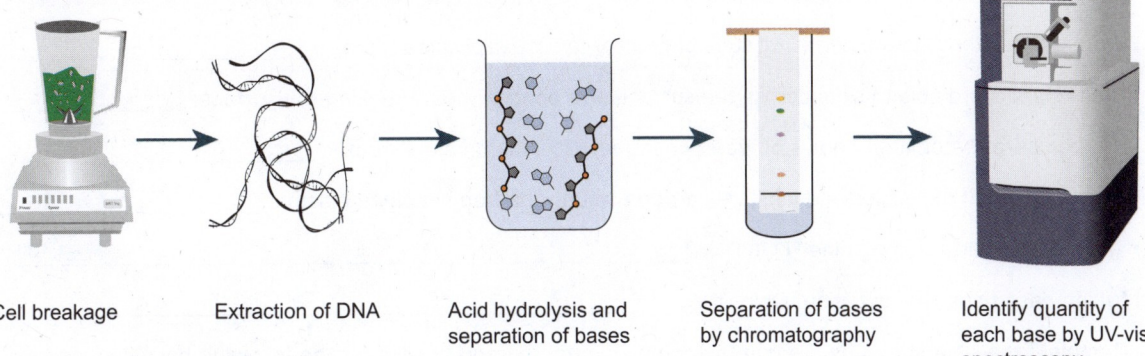

| Cell breakage | Extraction of DNA | Acid hydrolysis and separation of bases | Separation of bases by chromatography | Identify quantity of each base by UV-vis spectroscopy |

▸ Chargaff then examined the DNA from many different organisms. The condensed results of the experiment are shown below.

Species	Ox		Human		Yeast	Bacilli bacteria
Tissue	Thymus	Spleen	Thymus	Liver		
Adenine	0.28*	0.26	0.28	0.27	0.27	0.12
Guanine	0.22	0.20	0.19	0.18	0.16	0.28
Cytosine	0.17	0.16	0.16	0.15	0.14	0.26
Thymine	0.25	0.24	0.28	0.27	0.27	0.11
A/T	*1.1*					
C/G	*0.77*					
A+T/C+G	*1.4*					

Chemical Specificity of Nucleic Acids and Mechanism of their Enzymatic Degradation. E. Chargaff, 1950

*Note the specific units are not given here as these are not necessary for the interpretation of the data.

2. Complete the results table above. The first column is done for you:

3. What can be said about the ratio of A+T:C+G within and between species? _____

_____

4. From the data, write down 2 important rules about the DNA bases: _____

_____

_____

_____

5. How does Chargaff's data falsify the prediction of the tetranucleotide hypothesis that DNA is, 1: made up of repeating ATCG units, 2: not able to act as a store of information?

_____

_____

_____

_____

8 Did You Get It?

1. Using examples, explain how the hydrogen bonds between water molecules affect many of its physical properties:

   _____

   _____

   _____

   _____

   _____

2. For the following examples of the properties of water, match the terms: *adhesive property, cohesive property, thermal property, solvent property* to the correct description:

   (a) Water moves through the narrow tissues of plant xylem by capillary action: _____

   (b) Small insects are able to skim across the surface of a pond without breaking the surface: _____

   (c) The temperature of large bodies of water changes very little on a daily basis: _____

   (d) Many important molecules and ions are carried through the body in blood plasma: _____

3. (a) Which base pairs with thymine (T) in DNA? _____

   (b) Which base pairs with cytosine (C) in DNA? _____

   (c) Complete the DNA strand by filling in the complementary strand of DNA for the following base sequence:

4. Draw and identify the three components of a nucleotide:

   [box]

5. Identify two differences between DNA and RNA: _____

   _____

   _____

6. **AHL**: Why is the presence of liquid water on the surface of a planet important in the search for extraterrestrial life?

   _____

   _____

   _____

7. **AHL**: (a) Bases are held together by what type of bond? _____

   AHL: (b) How many of these bonds hold A and T together? _____

   AHL: (c) How many of these bonds hold C and G together? _____

8. **AHL**: If a molecule of DNA was found to be 29% adenine (A), what % of T, C, and G could be expected to be found?

   _____

 # Cells

A2.1 Origins of Cells

	Activity Number

Guiding Questions:
▸ What hypothesis can offer a viable mechanism to explain the origin of life?
▸ What are the probable stages in between organic substances and the formation of the first cells?

Learning Outcomes:

			Activity Number
☐	1	**AHL:** Evaluate how the unique conditions on early Earth led to the spontaneous formation of pre-biotic carbon compounds.	9
☐	2	**AHL:** Define living and non-living and explain why viruses are considered non-living.	13
☐	3	**AHL: NOS:** Evaluate evidence from theories and hypotheses for cellular complexity and spontaneous origin, considering difficulties in replicating prebiotic Earth conditions and a lack of fossilized specimens.	9
☐	4	**AHL:** Evaluate how the Miller–Urey investigation provided evidence for the origin of carbon compounds.	10
☐	5	**AHL:** Explain how membrane bilayers are formed from fatty acids to allow compartmentalization of the cell.	9
☐	6	**AHL:** Discuss evidence for the RNA world hypothesis.	11
☐	7	**AHL:** Explain how the genetic code is evidence for a last universal common ancestor (LUCA).	12
☐	8	**AHL:** Compare different methods for dating the origin of cells and LUCA and the relative position of these events in the geologic time scale.	12
☐	9	**AHL:** Evaluate evidence for origin and evolution of LUCA in undersea hydrothermal vents.	12

A2.2 Cell Structure

Guiding Questions:
▸ What structures do all cells have in common and what distinguishes one group of cells from another?
▸ How does microscopy enable a view of life at the cellular level?

	Activity Number

Learning Outcomes:

			Activity Number
☐	1	**NOS:** Using cell theory, explain how deductive reasoning can be used to generate predictions on the basic structural unit of any living organism.	17
☐	2	**NOS: AOS:** Demonstrate microscopy skills, including making and viewing cell and tissue slides and correct use of a microscope, to view and calculate actual size and magnification as a form of quantitative observation.	14-15
☐	3	Investigate recent developments and innovations in microscopy.	16
☐	4	Identify cellular structures shared by all organisms.	17
☐	5	Describe key features of prokaryotic cell structure, recognizing variation across different groups.	18
☐	6	Describe key features of eukaryotic cell structure including details of organelles.	18
☐	7	Identify and define the life processes common to unicellular organisms: homeostasis, metabolism, nutrition, movement, excretion, growth, response to stimuli, and reproduction.	19
☐	8	Compare and contrast eukaryotic cellular structure between animals, fungi and plants.	20
☐	9	Investigate examples of atypical eukaryote cell structure in eukaryotes, including aseptate fungal hyphae, skeletal muscle, red blood cells and phloem sieve tube elements.	20
☐	10	Use micrographs to classify cell types as prokaryote, plant, or animal, and identify cellular structures.	18, 21, 22
☐	11	Construct a scientific drawing of a cell's structure, including organelles, from a micrograph, with annotations on function.	22
☐	12	**AHL: NOS:** Evaluate evidence for eukaryote complexity provided by the endosymbiosis theory.	23
☐	13	**AHL:** Investigate how the effect of environmental factors on gene expression can lead to cell differentiation and specialization.	24
☐	14	**AHL:** Link the multiple instances of the evolution of multicellularity to its advantages as a common life form for organisms.	25

A2.3 Viruses

Guiding Questions:
▶ How do a small number of genes still allow viruses to function and reproduce?
▶ How can we classify viruses?

Learning Outcomes:		Activity Number
☐ 1	**AHL:** Identify key structural features common to viruses.	26
☐ 2	**AHL:** Classify viruses based on shape and structure.	26
☐ 3	**AHL:** Describe key stages in the lytic cycle of a bacteriophage virus.	27
☐ 4	**AHL:** Describe key stages in the lysogenic cycle of a bacteriophage virus.	27
☐ 5	**AHL:** Analyze evidence for repeated origin of viruses as an example of convergent evolution.	26
☐ 6 .	**AHL:** Examine reasons for rapid evolution in viruses and implications for treating viral diseases.	28

9 The Origin of Life On Earth

Key Idea: The conditions on primitive Earth provided the precursor molecules needed to produce small organic molecules.

The conditions on a newly formed Earth provided the environment for organic molecules to form and supplied the building blocks for more complex molecules. Early organic molecules served as building blocks for more complex molecules, including proteins and nucleic acids.

How did life on Earth arise?

▶ Most origin of life on Earth theories propose a sequence of events starting with the spontaneous formation of simple organic molecules, formation of more complex molecules, and finally the formation of self-sustaining biological molecules.

▶ In the 1920s, scientists Oparin and Haldane suggested that life on Earth arose through a process of gradual chemical evolution. They proposed that simple inorganic molecules reacted in the reducing atmosphere (oxygen poor) of early Earth to form simple organic molecules. These accumulated in the oceans to produce a 'primordial soup', eventually reacting and combining to form more complex molecules. The energy for the reactions would have been provided from lightning or the Sun.

▶ In the 1950s, Miller and Urey tested Oparin and Haldane's theory by building a closed experimental system to see if organic molecules could be produced under the conditions thought to resemble early Earth. They were successful.

▶ Scientists now think the early atmosphere was not the same as the mix used by Miller and Urey so their theories are not fully supported. However, several recent experiments have shown that organic building blocks can be produced under a wide range of conditions, lending weight to the theory that spontaneous formation of simple organic molecules led to more complex ones.

Scientific hypotheses on the origin of life

It is impossible to know the exact conditions on a pre-biotic Earth that resulted in self-replicating molecules and eventually life. Therefore, we cannot replicate the conditions that produced the first biomolecules or the first cells, only hypothesize as to how and when life appeared. Also, no evidence of this first life has survived to the present: the first protocells did not fossilize.

Conditions on early Earth

Simple organic molecules
Earth formed around 4.6 billion years ago and acquired volatile organic chemicals by collision with comets and meteorites. Earth's early atmosphere was likely made up of methane and high levels of carbon dioxide and lacked free oxygen and ozone.

Methane

Complex organic molecules
High temperatures, along with considerable penetration of UV radiation on early Earth, stimulated the synthesis and accumulation of simple organic molecules. These included amino acids, purines, pyrimidines, sugars, and lipids.

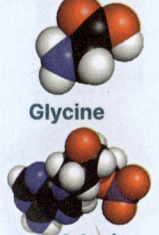

Glycine

Adenine

Nucleic acids
Dehydration synthesis produced polymers of amino acids (proteins) and nucleic acids (RNA or DNA). Evidence suggests RNA was probably the first self-replicating molecule on Earth. It carries genetic information which can be passed on to the next generation.

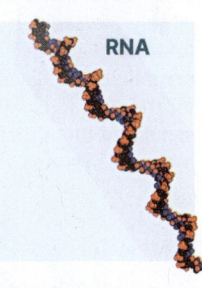

RNA

Lipid-encased probiont
Later, larger lipids were synthesized. These had the ability to self-assemble into double-layered membranes (liposomes). Liposomes could enclose organic, self-replicating and catalytic molecules such as RNA to form a protobiont.

The protobiont is very simple, and is often regarded as a very primitive cell-like structure, likely an immediate precursor to the first living systems.

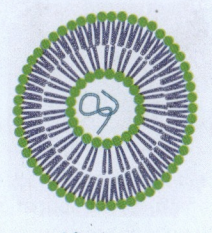

Liposome

---

1. What gases are thought to have been most abundant in the atmosphere of early Earth? _____

_____

2. What conditions on early Earth may have stimulated the synthesis of simple organic molecules? _____

_____

3. What is thought to have been the first self replicating molecule? _____

4. Why are protobionts regarded as primitive cells? _____

_____

_____

©2024 **BIOZONE** International
ISBN: 978-1-99-101410-8
Photocopying prohibited

A2.1
1, 3, 5

NOS

AHL

10 Prebiotic Experiments

Key Idea: When Earth's early atmospheric conditions are simulated, chemical reactions produce biological molecules. In the 1950s, Stanley Miller and Harold Urey attempted to recreate the conditions of primitive Earth. They hoped to produce the biological molecules that preceded the development of the first living organisms. Researchers at the time believed that the Earth's early atmosphere was made up of methane, water vapour, ammonia, and hydrogen gas so these components were used in the experiments. Many variations on this experiment have produced similar results. It seems that the building blocks of life are relatively easy to create and they have been found on comets.

The Miller-Urey experiment

Miller and Urey's classic experiment (right) was run for a week, then samples were taken from the collection trap for analysis. Up to 4% of the carbon (from methane) had been converted to amino acids. In this, and subsequent experiments, all 20 amino acids commonly found in organisms have been formed, along with nucleic acids, sugars, lipids, adenine, and even ATP (if phosphate is added to the flask). Note the absence of free atmospheric oxygen.

It is now thought that the early atmosphere was different from that simulated in the Miller-Urey experiment and instead likely to be similar to the vapours given off by modern volcanoes: carbon monoxide (CO), carbon dioxide (CO_2), and nitrogen (N_2). However, even if the reaction mixture is adjusted to these conditions, the result is largely the same. It has also been noted that the glass walls of the experimental container may have acted as a catalyst for the reactions, although this could have simulated the effect of rocks or minerals in a natural system.

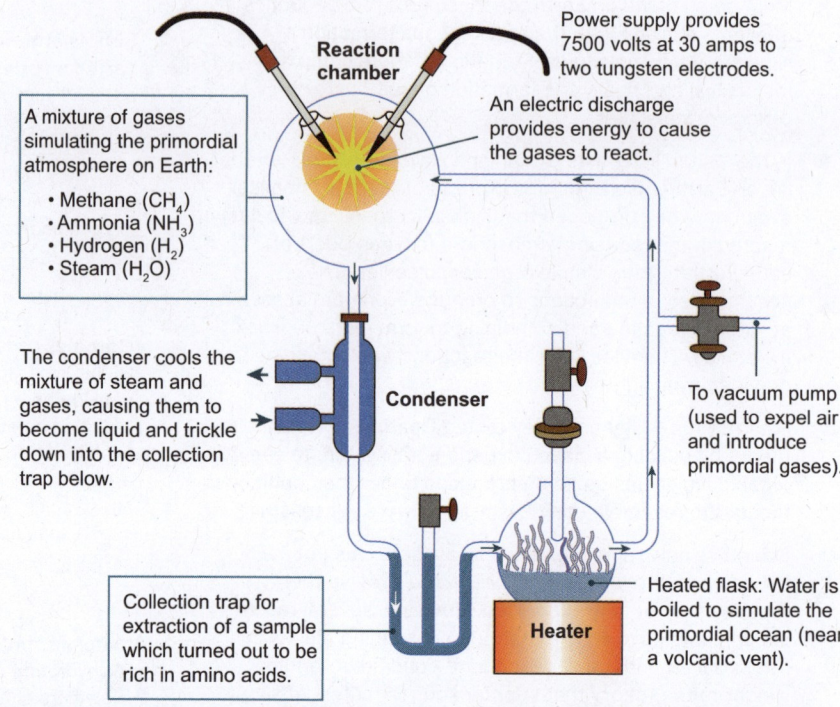

Power supply provides 7500 volts at 30 amps to two tungsten electrodes.

Reaction chamber

An electric discharge provides energy to cause the gases to react.

A mixture of gases simulating the primordial atmosphere on Earth:
- Methane (CH_4)
- Ammonia (NH_3)
- Hydrogen (H_2)
- Steam (H_2O)

The condenser cools the mixture of steam and gases, causing them to become liquid and trickle down into the collection trap below.

Condenser

To vacuum pump (used to expel air and introduce primordial gases).

Collection trap for extraction of a sample which turned out to be rich in amino acids.

Heater

Heated flask: Water is boiled to simulate the primordial ocean (near a volcanic vent).

Some scientists envisage a global winter scenario for the formation of life. Organic compounds are more stable in colder temperatures and could combine in a lattice of ice.

Lightning is a natural phenomenon associated with volcanic activity. It may have supplied energy for the formation of new compounds, e.g. nitrogen oxides, which were incorporated into organic molecules.

Early Earth was universally volcanic. At sites such as deep sea hydrothermal vents and geysers like the one above, gases delivered vital compounds to the surface, where reactions took place.

1. State a hypothesis for the Miller-Urey experiment: _____

_____

2. Discuss whether the Miller-Urey experiment supported the theory that conditions on early Earth could have produced simple organic molecules:

_____

_____

_____

3. Why is it highly unlikely that life could ever begin again on present day Earth? _____

_____

_____

AHL   A2.1   4

©2024 **BIOZONE** International
ISBN: 978-1-99-101410-8
Photocopying prohibited

11 An RNA World

Key idea: Life could begin with self replicating RNA.
An ability to replicate requires the presence of complex molecules, such as proteins, which did not exist on early Earth. The discovery of ribozymes in 1982 helped to solve the problem of how biological information was stored and replicated. Ribozymes are enzymes formed from RNA. They catalyze peptide bond formation during protein synthesis in cells. The ribozymes can catalyze the replication of the original RNA molecule. This mechanism for self replication has led to the theory of an 'RNA world'.

1 Pre RNA world

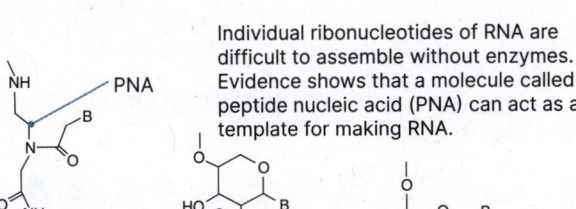

Individual ribonucleotides of RNA are difficult to assemble without enzymes. Evidence shows that a molecule called peptide nucleic acid (PNA) can act as a template for making RNA.

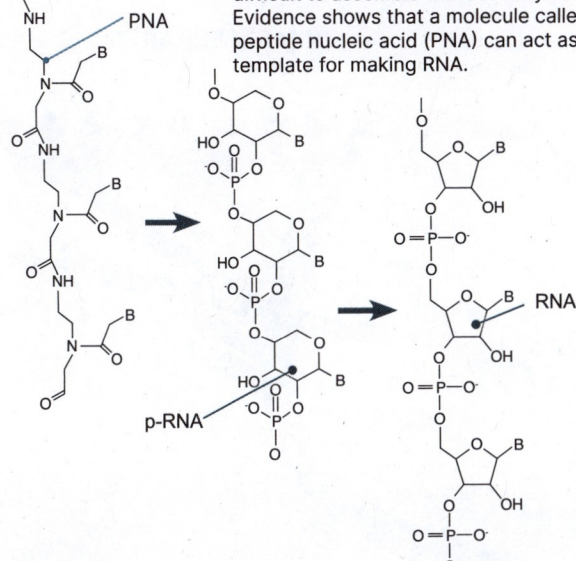

2 RNA world

RNA can be used both for information storage and catalysis. It therefore provides a way around the problem that genes require enzymes to form, and enzymes require genes to form. The first stage of evolution may have been RNA molecules folding to form ribozymes, which could then catalyze the replication of other similar RNAs.

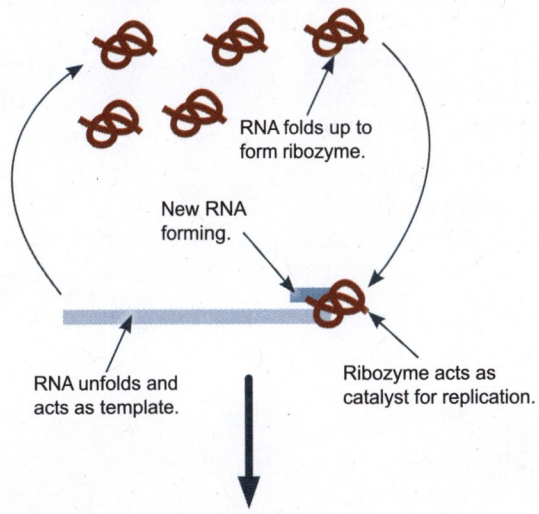

RNA folds up to form ribozyme.

New RNA forming.

RNA unfolds and acts as template.

Ribozyme acts as catalyst for replication.

4 Proto-cells

Certain types of organic molecule, such as fatty acids, spontaneously form micelles (assemblies of lipid molecules) when placed in an aqueous solution.

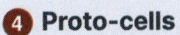
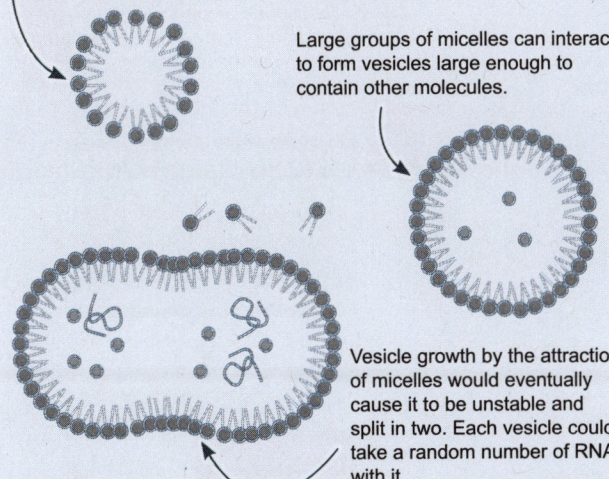

Large groups of micelles can interact to form vesicles large enough to contain other molecules.

Vesicle growth by the attraction of micelles would eventually cause it to be unstable and split in two. Each vesicle could take a random number of RNAs with it.

3 Mutation and competition

From the establishment of self-replicating RNA molecules, there would have been competition of a sort. Incorrect copies of the original RNA produced new varieties of RNA.

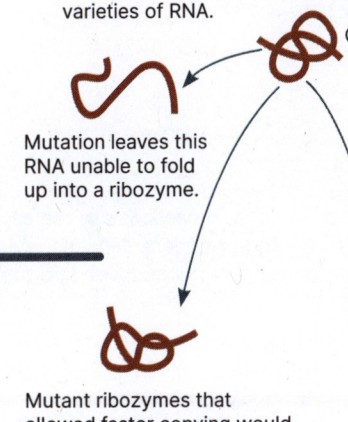

Original RNA

Mutation leaves this RNA unable to fold up into a ribozyme.

Some ribozymes may have added more ribonucleotides to the chain, allowing the RNA to grow in length.

Mutant ribozymes that allowed faster copying would have been able to gather resources faster than the original, becoming more prevalent.

Mutant ribozyme is able to translate RNA into proteins.

1. How did the discovery of ribozymes provide evidence for the RNA world hypothesis? _____

_____

_____

2. Explain how mutations in RNA templates led to the first form of evolution: _____

_____

_____

12 The Common Ancestry of Life

Key Idea: The ancestry of all organisms on Earth today can be traced back to a common ancestor known as LUCA.

It is believed that life first appeared on Earth around 3.8 billion years ago. The fact that the genetic code is universal, and that genes perform the same functions in different organisms, point to the evolution of all life from one Last Universal Common Ancestor (LUCA). It is believed that, over time, different forms of life evolved. Most have become extinct. It is difficult to identify exactly where and when life evolved, but fossil evidence from ancient rocks points to LUCA likely evolving in hydrothermal vents similar to those found in deep ocean trenches. Molecular clock technology that dates specific mutations also gives us pointers as to when life evolved.

The tree of life

Genetic evidence shows three main groups or domains of life: bacteria, archaea, eukarya. Evidence points to multiple gene transfers between the various domains of life. Many life-forms could have emerged before LUCA and then become extinct, so did not contribute further.

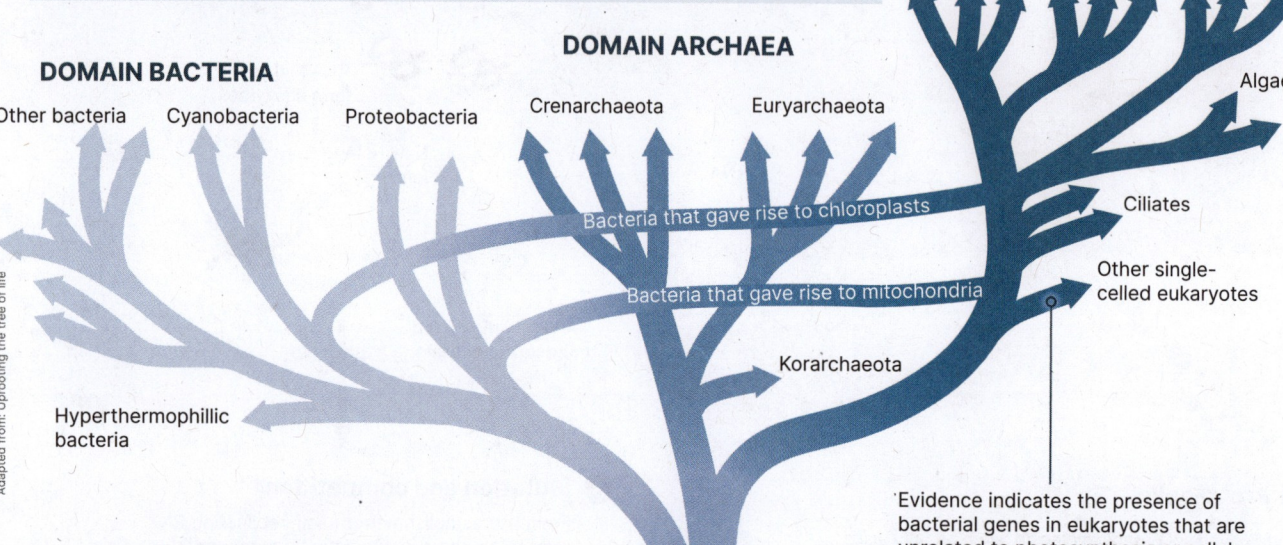

Adapted from: Uprooting the tree of life

DOMAIN EUKARYA
Animals · Fungi · Plants · Algae · Ciliates · Other single-celled eukaryotes

DOMAIN ARCHAEA
Crenarchaeota · Euryarchaeota · Korarchaeota

DOMAIN BACTERIA
Other bacteria · Cyanobacteria · Proteobacteria · Hyperthermophillic bacteria

Bacteria that gave rise to chloroplasts
Bacteria that gave rise to mitochondria

Last Universal Common Ancestor (**LUCA**)

All living organisms on earth today share the same genetic code as LUCA. If there were any other variants in early life, scientists have not yet found any examples of an alternative genetic code.

Evidence indicates the presence of bacterial genes in eukaryotes that are unrelated to photosynthesis or cellular respiration. These could be explained by gene transfers during evolution. A revised tree might include a complex network of connections to indicate single and multiple gene transfers across domains.

Evidence for LUCA

The genetic code is universal, i.e. all organisms use the same four nucleotide bases in their DNA and each base sequence codes for the same amino acid. The mechanisms of transcription and translation from DNA to mRNA to protein are also universal.

In addition, the use of ATP as an energy source to fuel chemical reactions in cells and the role of ribosomes in protein synthesis are also common to all life forms.

Because microbes do not leave fossilized remains, other techniques such as biomolecular analysis must be used to provide further evidence for the existence of a last universal common ancestor. The more genes that organisms have in common, the more closely related they are. By looking for gene sequences common to all organisms, scientists have deduced that 355 genes present in existing organisms were also present in LUCA.

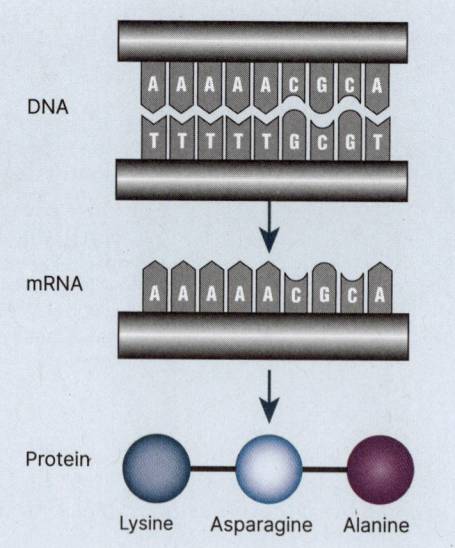

DNA
A A A A A C G C A
T T T T T G C G T

mRNA
A A A A A C G C A

Protein
Lysine · Asparagine · Alanine

1. What does it mean when we say that the genetic code is universal? _____

_____

2. What evidence is there for a Last Universal Common Ancestor? _____

_____

AHL
A2.1
7 - 9

43 →

©2024 **BIOZONE** International
ISBN: 978-1-99-101410-8

Evolution of different life forms

Since the beginning of life on Earth, many life-forms have evolved and developed into the myriad around us today. Over the course of evolutionary time, many organisms would have evolved and then, for various reasons, would have become extinct. Early forms of life were simple, self-replicating cells. More complex organisms evolved as time passed. As can be seen from the diagram below, modern humans have only existed on Earth since relatively recent times.

Timeline of appearance of life-forms on Earth

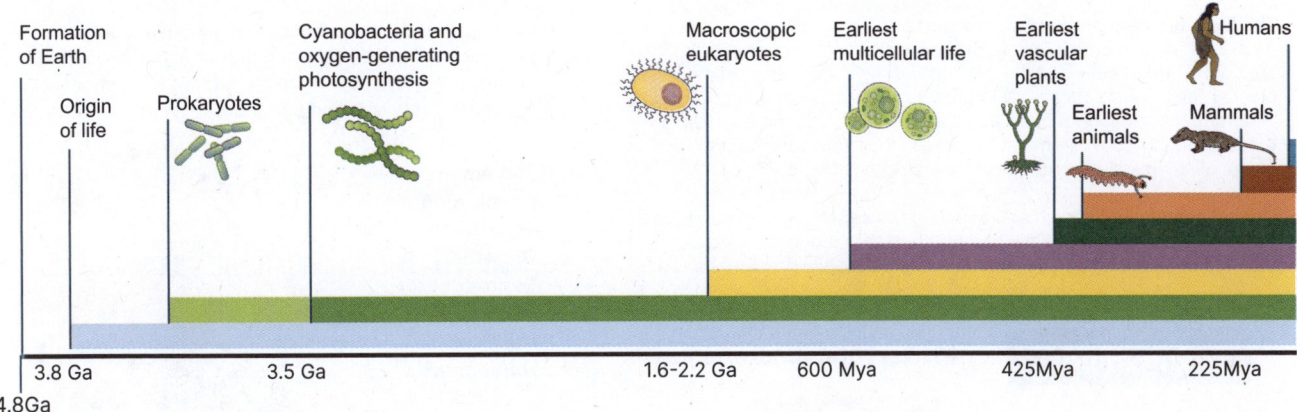

| 3.8 Ga | 3.5 Ga | 1.6-2.2 Ga | 600 Mya | 425Mya | 225Mya |

4.8Ga

Where and when did life evolve?

Evidence of early life on Earth comes from several sources including stromatolite fossils located in western Greenland. Carbon 'signatures' found in these formations point to the existence of early life forms. Other evidence of early life has been found in ancient rocks from Western Australia where some of the oldest existing rocks on Earth remain. Ancient rocks are rare as most are destroyed by geological activity. Many scientists believe that LUCA most likely originated in conditions such as exist in deep sea hydrothermal vents where suitable energy forms existed for early life-forms to use.

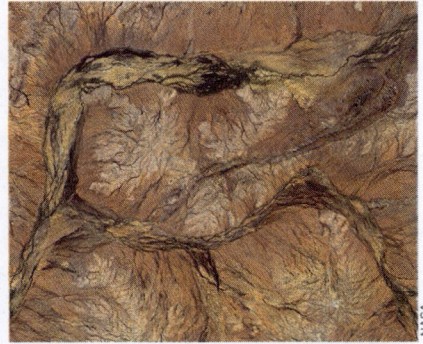

Genetic analysis of protein coding genes from bacteria and archaea indicates that LUCA most likely evolved in the conditions found in deep sea hydrothermal vents. Hydrogen was available as an energy source, and carbon dioxide and nitrogen gases provided further component molecules for life.

Stromatolites are layered sedimentary structures that were formed mainly by cyanobacteria and trapped sediments over geological timescales. They have been present on Earth for billions of years.

Ancient rocks, like those found in Jack Hills, Western Australia have chemical signatures that suggest early forms of life existed. The graphite found in the rocks has a low ratio of heavy to light carbon isotopes which could indicate that living processes took place there.

3. What evidence points to LUCA likely evolving in the vicinity of hydrothermal vents? _____

_____

_____

4. (a) What are stromatolites? _____

_____

(b) What feature of stromatolites can be used as evidence of early life? _____

_____

_____

5. Research an example of an early plant or animal that still exists today. When did this species first appear on Earth?

_____

©2024 **BIOZONE** International
ISBN: 978-1-99-101410-8
Photocopying prohibited

13 The Cell is the Unit of Life

Key Idea: All living organisms are composed of cells. Cells are broadly classified as prokaryotic or eukaryotic.
The cell theory is a fundamental idea in biology. The idea of all living things being composed of cells developed over many years. As viruses are not composed of cells and cannot replicate without a host, they are classed as non-living.

Making predictions with the cell theory

The idea that cells are fundamental units of life is part of the cell theory. The basic principles of the theory are: 1. All living things are composed of cells and cell products. 2. New cells are formed only by the division of pre-existing cells. 3. The cell contains inherited information (genes) that is used as instructions for growth, functioning, and development. The cell is the functioning unit of life; all chemical reactions of life take place within cells.

Using deductive reasoning and applying the cell theory, we can then predict that any newly discovered life form must comprise one or more cells. Theories can be used to make deductive predictions.

All cells show the functions of life

Cells use food, e.g. glucose, to maintain a stable internal environment, grow, reproduce, and produce wastes. The sum total of all the chemical reactions that sustain life is called metabolism.

Movement
Respiration
Sensitivity
Growth
Reproduction
Excretion
Nutrition
Homeostasis

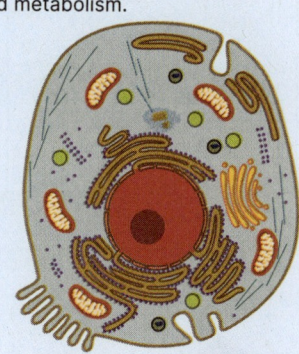

Living things

Cells

Prokaryotic (bacterial) cells

- Autotrophic or heterotrophic.
- Single celled.
- Lack a membrane-bound nucleus and membrane-bound organelles.
- Cells 0.5-10 μm
- DNA a single, circular chromosome. There may be small accessory chromosomes called plasmids.
- Cell walls containing peptidoglycan.

Viruses are non-cellular

- Non-cellular.
- Typical size range: 20-300 nm.
- Contain no cytoplasm or organelles.
- No chromosome, just RNA or DNA strands.
- Enclosed in a protein coat.
- Depend on cells for metabolism and reproduction (replication).

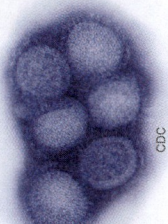

Influenzavirus

Eukaryotic cells

• Cells 30-150 μm • Membrane-bound nucleus and membrane-bound organelles • Linear chromosomes

Plant cells

- Exist as part of multicellular organism with specialization of cells into many types.
- Autotrophic (make their own food): photosynthetic cells with chloroplasts.
- Cell walls of cellulose.

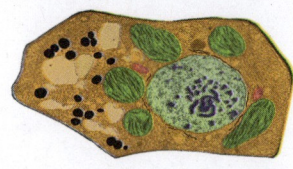

Generalized plant cell

Animal cells

- Exist as part of multicellular organism with specialization of cells into many types.
- Lack cell walls.
- Heterotrophic (rely on other organisms for food).

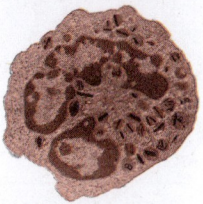

White blood cell

Protist cells

- Mainly single-celled or exist as cell colonies.
- Some are autotrophic and carry out photosynthesis.
- Some are heterotrophic.

Amoeba cell

Fungal cells

- Rarely exist as discrete cells, except for some unicellular forms, e.g. yeasts.
- Plant-like, but lack chlorophyll.
- Rigid cell walls containing chitin.
- Heterotrophic.

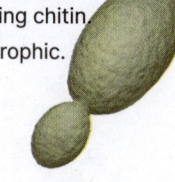

Yeast cell

1. What are the characteristic features of a prokaryotic cell? _____

_____

2. What are the characteristic features of a eukaryotic cell? _____

_____

3. Why are viruses considered to be non-cellular (non-living)? _____

_____

©2024 **BIOZONE** International
ISBN: 978-1-99-101410-8

14 The Light Microscope

Key Idea: We can examine cells using an optical microscope. Microscopes produce an enlarged (magnified) image of an object allowing it to be observed in greater detail than is possible with the naked eye. We can use optical microscopes in the laboratory to observe cell structure and e cell movement. High power compound light microscopes use visible light and a combination of lenses to magnify objects up to several hundred times. The resolution of light microscopes is limited by the wavelength of light and specimens must be thin and mostly transparent so that light can pass through. No detail will be seen in specimens that are thick or opaque.

▶ The light (or optical) microscope (LM) is an important tool in biology. Light microscopy involves illuminating a sample and passing the light that is transmitted or reflected through lenses to give a magnified view of the sample.

▶ Compound light microscopes use visible light and a combination of lenses to magnify objects which are prepared by placing in solution on a slide and covered with a thin coverslip. Both living and dead cells can be examined. Bright field microscopy (below) is the simplest, and involves illuminating the specimen from below and viewing from above.

▶ The wavelength of light limits the resolution of light microscopes to around 0.2 μm (μm = micrometer, 1 millionth of a meter). Objects closer than this will not be distinguished as separate.

Typical compound light microscope

Word list: *In-built light source, arm, coarse focus knob, fine focus knob, condenser, mechanical stage, eyepiece lens, objective lens*

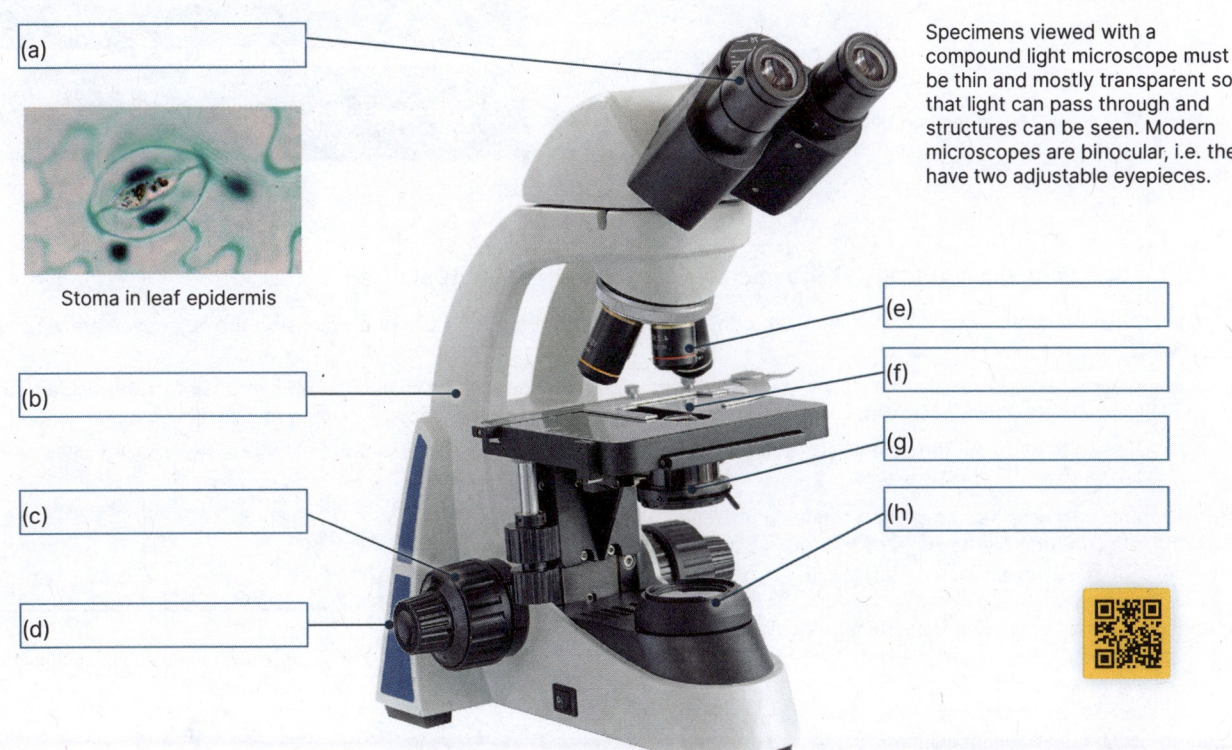

(a)

Stoma in leaf epidermis

(b)

(c)

(d)

Specimens viewed with a compound light microscope must be thin and mostly transparent so that light can pass through and structures can be seen. Modern microscopes are binocular, i.e. they have two adjustable eyepieces.

(e)

(f)

(g)

(h)

What is Magnification?

Magnification refers to the number of times larger an object appears compared to its actual size. Magnification is calculated as follows:

Objective lens power   X   Eyepiece lens power

What is Resolution?

Resolution is the ability to distinguish between close together but separate objects. Examples of high and low resolution for separating two objects viewed under the same magnification are given below.

High resolution   Low resolution

---

1. Label the image above of the compound light microscope (a) to (h). Use words from the list supplied above.

2. Determine the magnification of a microscope using:

    (a) 15 X eyepiece and 40 X objective lens: _____

    (b) 10 X eyepiece and 60 X objective lens: _____

A2.2
2

NOS

▸ Magnification refers to the number of times larger an object appears compared to its actual size.

▸ Linear magnification is calculated by taking a ratio of the image height to the object's actual height. If this ratio is greater than one, the image is enlarged. If it is less than one, it is reduced.

▸ To calculate magnification, all measurements are converted to the same units. Often, you will be asked to calculate an object's actual size, in which case you will be told the size of the object and the magnification.

▸ Calculating magnification is a form of quantitative observation.

Calculating linear magnification: A worked example

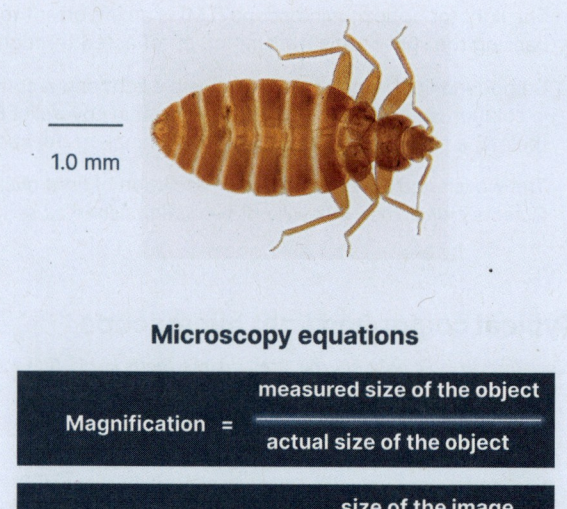

1.0 mm

1 1. Measure the body length of the bed bug image (right). Your measurement should be 40 mm (not including the body hairs and antennae).

2 2. Measure the length of the scale line marked 1.0 mm. You will find it is 10 mm long. The magnification of the scale line can be calculated using equation 1 (below right).

The magnification of the scale line is 10 (10 mm ÷ 1 mm)

*NB: The magnification of the bed bug image will also be 10x because the scale line and image are magnified to the same degree.

3 3. Calculate the actual (real) size of the bed bug using equation 2 (right):

The actual size of the bed bug is 4 mm (40 mm ÷ 10 x magnification).

Microscopy equations

$$\text{Magnification} = \frac{\text{measured size of the object}}{\text{actual size of the object}}$$

$$\text{Actual object size} = \frac{\text{size of the image}}{\text{magnification}}$$

Using an eyepiece graticule and stage micrometer to calculate actual size

▸ A graticule is a measurement scale that fits into the eyepiece. It shows a scale bar of divisions on the object you are viewing down the microscope.

▸ A stage micrometer is a scale that sits on the microscope stage. By comparing the number of divisions of the graticule to the known scale of the stage micrometer, you can calculate the size of the object you are viewing at a particular magnification.

▸ You will need to carry out the below calibration for each magnification of your microscope.

This diagram (right) shows a stage micrometer with 3 lines 0.1 mm (100 µm) apart.
Each 0.1 mm (100 µm) division covers 30 graticule divisions.
Therefore, 30 graticule divisions = 100 µm

1 graticule division = number of micrometres ÷ number of graticule divisions which is the magnification factor.

For this example, at this magnification: 100 µm ÷ 30 divisions = 3.33 µm
Each graticule line at this magnification is therefore equal to 3.33 µm.
If your viewed object spanned 40 graticule lines at this magnification, its actual size is:

40 X 3.33 = 133.20 µm in diameter.

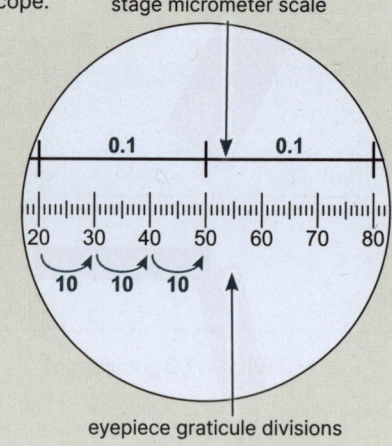

stage micrometer scale

eyepiece graticule divisions

3. Cheek epidermal cells were viewed under brightfield microscopy using the graticule scale above. The measured length of the cells was from 22 to 31 on the graticule divisions. Calculate the size of the cell:

_____

_____

4. The image of an *E.coli* cell is measured as 43 mm. Its actual size is 2 µm. Use this information to calculate the magnification of the image:

_____

_____

©2024 **BIOZONE** International
ISBN: 978-1-99-101410-8

15 Preparing a Slide

Key Idea: Correctly preparing and mounting a specimen on a slide is important if structures are to be seen clearly under a microscope. A wet mount is suitable for most slides. Specimens are usually prepared in some way before viewing in order to highlight features and reveal details. A wet mount is a temporary preparation in which a specimen and a drop of fluid are trapped under a thin coverslip. Wet mounts are used to view thin tissue sections, live microscopic organisms, and suspensions such as blood. A wet mount improves a sample's appearance and enhances visible detail. Sections must be made very thin for two main reasons: a thick section stops light shining through, making it appear dark when viewed; it also has too many layers of cells, making it difficult to see any detail.

 Investigation 2.1 Preparing an onion slide

See appendix for equipment list.

> ⚠   **Caution is required when using scalpels or razors. Iodine stains skin and clothes, and irritates the eyes. You should wear protective eyewear and gloves.**

1. Onions make good subjects for preparing a simple wet mount. Cut a square segment from a thick leaf of the bulb using a razor or scalpel.

2. Bend the segment towards the upper epidermis until the lower epidermis and inner leaf tissue (the parenchyma) snaps so that just the upper epidermis is left attached.

3. Carefully peel off the parenchyma from one side of the snapped leaf and then the other, leaving a peel of just the upper epidermis.

4. Place peel in the centre of a clean glass microscope slide and cover it with a drop of water.

5. Carefully lower a coverslip over the peel. A mounted needle can be used for better precision. This avoids including air in the mount.

6. Use a small piece of tissue or filter paper to remove any excess water.

7. Place the slide on the microscope tray. Locate the specimen or region of interest at the lowest magnification. Focus using the lowest magnification first (remembering to move the lens away from the slide) before switching to the higher magnifications.

8. After viewing the slide under various magnifications, remove the slide and place it on the bench.

9. At the edge of the coverslip place a small drop of iodine stain.

10. On the opposite side of the coverslip use a piece of tissue or filter paper to draw the water out from under the coverslip. The iodine will be drawn under the coverslip.

11. Replace the slide on the microscope and view the stained onion peel.

12. Take a photo of your cell, using your cellphone, if allowed in class. If possible, print your photo on to a sheet of paper and attach it to this page along with your calculations of 13. below.

13. If your microscope is fitted with a stage micrometer and eyepiece graticule, calculate the diameter and length of one cell. Attach your workings to this page, on a separate sheet of paper. Draw a scale bar on your photo.

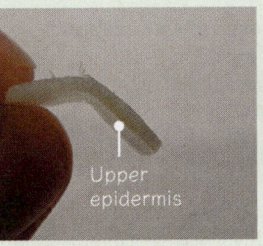
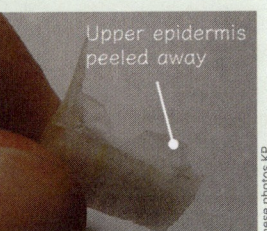
Upper epidermis
Upper epidermis peeled away
These photos KP

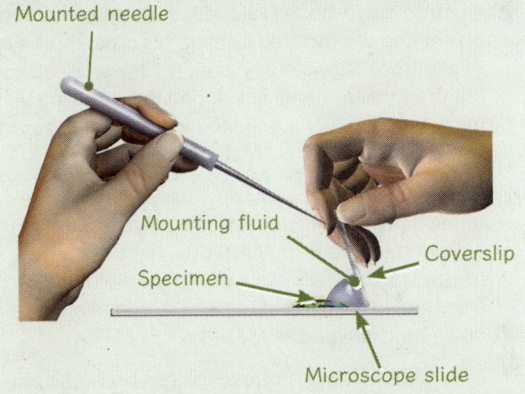
Mounted needle
Mounting fluid
Specimen
Coverslip
Microscope slide

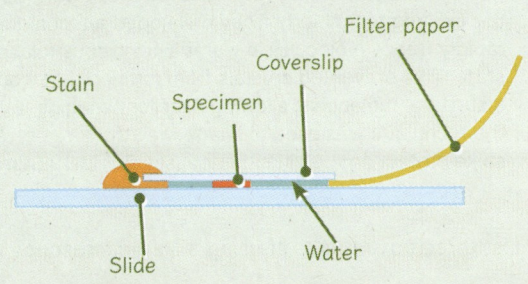
Filter paper
Stain
Coverslip
Specimen
Slide
Water

1. Why must sections viewed under a microscope be very thin? _____

_____

2. Why would no chloroplasts be visible in an onion epidermis cell slide? _____

_____

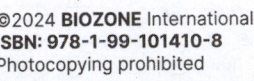

A2.2 / 2 | AOS

16 Developments in Microscopy

Key Idea: Microscopes are used to view objects that cannot be viewed in detail with the naked eye. Microscopes have become increasingly sophisticated over time, with improvements in both magnification and resolution.

Lenses of various descriptions have been used to view objects for around 4000 years, but only in the last few hundred years have techniques developed to build sophisticated devices for viewing in high resolution. Light microscopes are affordable and are suitable for many common laboratory requirements. The development of sophisticated electron microscopes has made it possible to image objects to the atomic level. Stains allow researchers to see specific parts of cells in great detail.

Developments in microscopy

▸ Early microscopes allowed scientists to see beyond the limits of the naked eye. In the 1600s, Anton van Leewenhoek produced over 500 single lens microscopes that could magnify up to 270 times.

▸ Better microscopes provided images that proved the existence of single celled organisms. Scientists could prove that all living organisms were made of cells.

▸ Compound light microscopes are affordable for most laboratories. Both living and dead cells can be observed and specimens can be prepared relatively quickly. However, their powers of magnification are limited.

▸ Electron microscopes (EM) focus beams of electrons using electromagnets to give high resolution images. They are complex and expensive. Specimens must be specially prepared and consequently living specimens cannot be used. In 1938, Ernst Ruska developed the transmission electron microscope (TEM). Electrons pass through an object and are focused by magnets. The short wavelength of electrons allows study of incredibly small objects. Manfred von Ardenne developed the scanning electron microscope (SEM) around the same time, allowing the surface of objects to be imaged.

▸ Developments in microscopy, particularly freeze-fracture techniques, supported the lipid bilayer model of the cellular membrane. In the 1960s, Hans Moor developed a freeze fracture technique that froze a biological specimen, fractured it, and generated a carbon/platinum 'cast' that could be examined by EM.

▸ Cryogenic electron microscopy (Cryo-EM) can be used to determine molecular structures at almost atomic levels. In 1981, Jacques Dubochet and Alasdair McDowall improved the existing technology. They developed a technique to rapidly freeze molecules in water in a vitreous (glass like) state. This prevented crystals from forming that would distort the molecules, allowing excellent imaging. By 2020, the technique was used to image the structure of the spike protein on the virus that causes Covid-19 (right).

A van Leeuwenhoek microscope c. 1673 was only a glorified magnifying glass by today's standards.

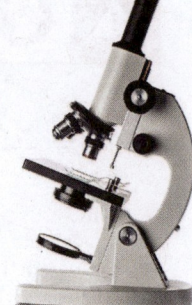

Light microscopes are relatively inexpensive, versatile and portable. They can be used to examine a wide variety of specimens that require limited preparation. Both living and dead specimens can be examined. This example uses a mirror which is a less expensive option than one with a built in light and is easily moved.

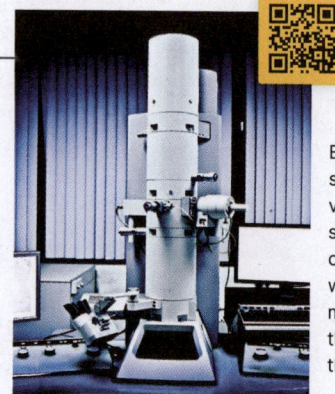

Electron microscopes require samples to be examined under vacuum. Because electrons have a shorter wavelength than light, we can see much higher resolution than with a light microscope. Images must be examined on a computer; the viewer does not look directly at the specimen

Details of cellular structures could be observed that were too small or detailed for even the electron microscope e.g. tight junctions between blood vessels and neurons (right).

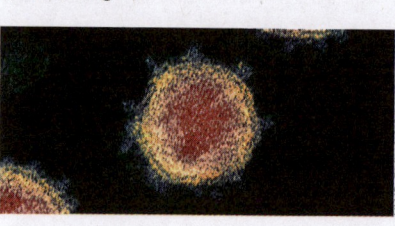

NIAID-RML 2.0 CC

Wikipedia 2.5 CC

1. List three advantages of using a light microscope over a transmission electron microscope (TEM):

_____

_____

2. What are the advantages of a transmission electron microscope (TEM) over a light microscope?

_____

_____

3. What feature of Cryo-EM enables the display of extremely high resolution images?

_____

_____

A2.2

3

©2024 **BIOZONE** International
ISBN: 978-1-99-101410-8
Photocopying prohibited

Stains and their uses

▸ Staining material for viewing under a microscope can make it easier to distinguish particular cell structures.

▸ Stains and dyes can be used to highlight specific components or structures. Stains contain chemicals that interact with molecules in the cell. Some stains bind to a particular molecule, making it easier to see where those molecules are. Others cause a change in a target molecule which changes their colour, making them more visible.

▸ Most stains are non-viable, and are used on dead specimens, but harmless viable stains can be applied to living material.

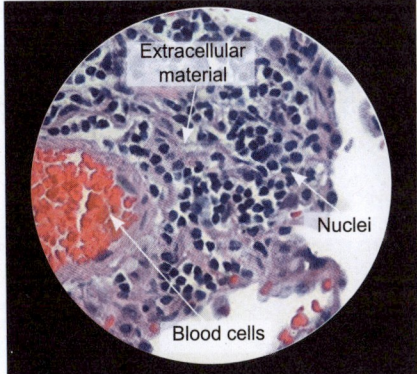

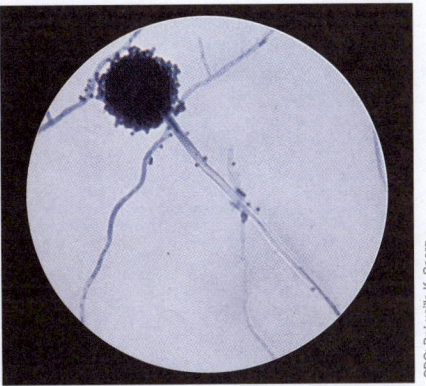

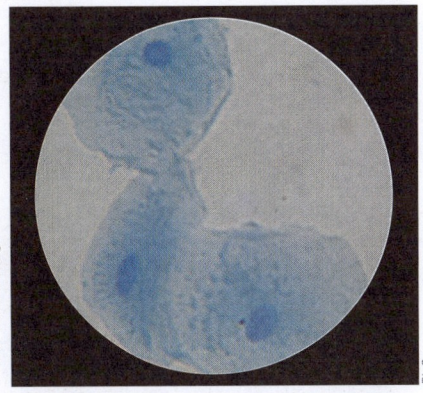

H&E stain is one of the most common stains for animal tissues. Nuclei stain dark blue, whereas proteins, extracellular material, and red blood cells stain pink or red.

Viable stains do not immediately harm living cells. Trypan blue stains dead cells blue but is excluded by live cells. It is also used to study fungal hyphae.

Methylene blue is a common temporary stain for animal cells, such as these cheek cells. It stains DNA and makes the nuclei more visible.

Immunofluorescent staining

▸ Immunofluorescent stains use the principle of specific antigen-antibody recognition to give very bright images of target molecules within cells.

▸ Fluorophores are chemical compounds that emit light when exposed to specific light wavelengths.

▸ Different antibodies can be tagged with different fluorophores to allow different molecules within cells to be detected by immunofluorescence microscopy.

▸ The tagged antibody binds to the target molecule of interest to the researcher and the attached fluorophore is then detected using fluorescent microscopy.

▸ The technique is used to study the location and expression of particular proteins of interest.

▸ It can be used on sections of tissue, cultured cells or cells that have been 'fixed' by a variety of methods.

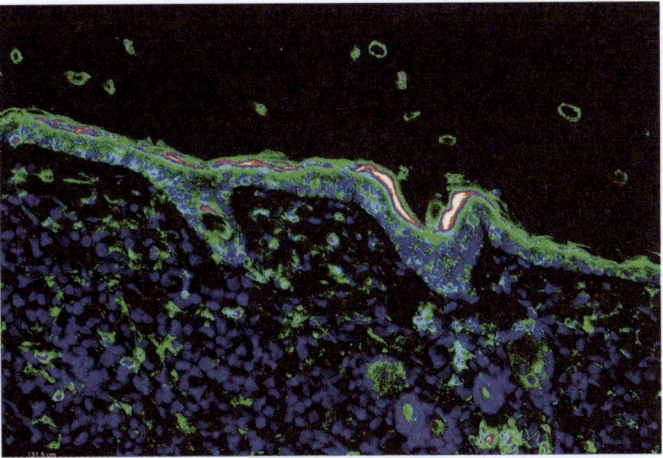

Immunofluorescent image of blue tumour cells being attacked by T-cells of the immune system.

4. Describe the difference the iodine stain makes when viewing onion cells under the microscope, compared to when they are viewed without the stain:

_____

_____

5. What is the main purpose of using a stain? _____

_____

6. What is the difference between a viable and non-viable stain? _____

_____

_____

7. Identify a stain or type of stain that would be appropriate for distinguishing each of the following:

(a) Live vs dead cells: _____ (c) An antigen in a cultured cell preparation: _____

(b) Red blood cells in a tissue preparation: _____ (d) Nuclei in cheek cells: _____

17 Common Features of Cells

Key Idea: All living things are made of cells. All cells have cytoplasm, ribosomes, DNA, and a plasma membrane.

The bulk of the cell is its aqueous cytoplasm, providing an ideal environment for metabolic reactions to take place. DNA provides the cell with an information bank to carry out all reactions. The plasma membrane separates the contents of the cell from the exterior. Ribosomes allow the synthesis of polypeptides through the process of translation. Louis Pasteur proved, by use of his famous swan necked flask experiment, that all life comes from existing life and does not spontaneously appear, as had been previously suggested by other scientists of the time.

Cell theory

Cells which are the basic units of life. Therefore, all new cells must come from existing cells. Although different types of cells have different components, all cells contain four common structures: DNA, cytoplasm, ribosomes, and a plasma membrane.

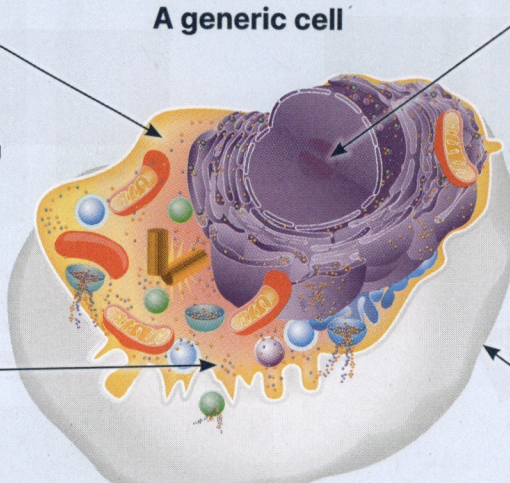

A generic cell

The cytoplasm is a watery solution that contains dissolved substances, including proteins and enzymes. In eukaryotic cells, the nucleus and other cell organelles are kept separate from the cytoplasm by their own surrounding membranes. The cytoplasm is actually highly organised. A scaffold of proteins called the cytoskeleton provides the cell with structural support.

DNA contains the information required by the cell to carry out all processes including growth, repair, and replication. All cells start out with DNA although some, such as red blood cells, will lose it as part of their developmental process.

Ribosomes are responsible for the synthesis of polypeptides by translation. They are made up of 2 subunits and may be found free in the cytoplasm or associated with endoplasmic reticulum if the cell has this.

The plasma membrane separates the cell from its exterior environment. It is composed of a phospholipid bilayer and has proteins moving freely within it. It is semi-permeable and allows some substances to pass through, either passively or via active transport.

Pasteur's swan neck flask experiment

Prior to the observation of Rudolf Virchow that cells divide, scientists did not have a clear understanding of how new organisms arose. For example, when a piece of meat was left on a bench, maggots appeared a day or two later. The spontaneous generation theory suggested maggots had spontaneously generated from the components of the meat. Louis Pasteur disproved this when he carried out his very simple swan neck flask experiments (below).

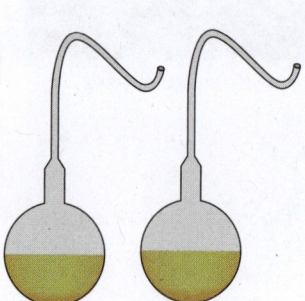
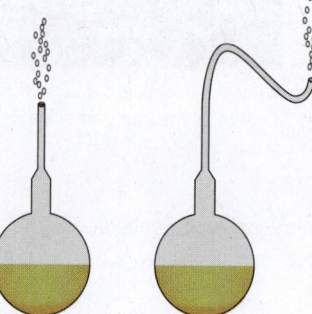
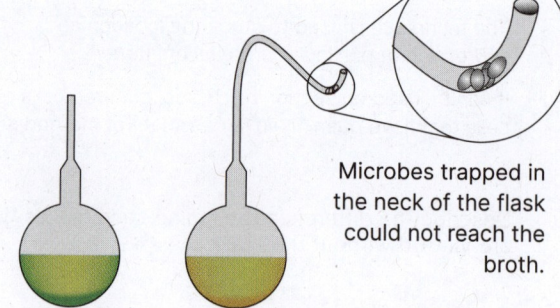

Microbes trapped in the neck of the flask could not reach the broth.

1. Pasteur filled two swan necked flasks with a nutrient broth and boiled it to kill any microbes present.

2. He broke off the neck of one of the flasks, allowing air and dust (on which microbes are carried) to fall straight down onto the broth. The neck of the other flask prevented dust falling in.

3. The broth in the broken flask eventually turned dark, indicating microbial growth. The broth in the unbroken flask remained unchanged. Pasteur concluded that spontaneous generation could not have occurred, or both flasks would have turned dark.

1. Consider the common components of cells above. Predict some substances cells need to remain functioning and the importance of those substance:

_____

_____

_____

_____

A2.2 1, 4

©2024 **BIOZONE** International
ISBN: 978-1-99-101410-8

18 Prokaryote and Eukaryote Cells

Key Idea: Cells are classified as either prokaryotic or eukaryotic and are distinguished on the basis of their size, internal organization, and complexity.

Cells are divided into two broad groups based on their size and organization. Prokaryotic cells (all bacteria and archaea) are small, single cells with a simple internal structure. Eukaryotic cells are larger, more complex cells. All multicellular and some unicellular organisms are eukaryotic.

Prokaryotic cells

▶ Prokaryotic cells are small (~0.5-10 μm) single cells.

▶ They lack any membrane-bound organelles.

▶ They are relatively basic cells and have very little cellular organization. Their DNA, ribosomes, and enzymes are free floating within the cell cytoplasm.

▶ The ribosomes (70S) are smaller than eukaryotic ribosomes.

▶ Photosynthetic bacteria have enzymes and light capturing membranes.

▶ Single, circular chromosome of naked DNA.

▶ Prokaryotes have cell walls, but it is different from the cell walls that some eukaryotes have.

Eukaryotic cells

▶ Eukaryotic cells are large (30-150 μm). They may exist as single cells or as part of a multicellular organism.

▶ Eukaryotes have membrane-bound organelles, e.g. mitochondria and (photosynthetic organisms) chloroplasts.

▶ Eukaryotic cells are complex and have a membrane-bound nucleus.

▶ Ribosomes (80S) are larger than in prokaryotes, except those in mitochondria and chloroplasts, which are 70S.

▶ Photosynthesis occurs only in chloroplast organelles.

▶ Multiple linear chromosomes consisting of DNA and associated proteins.

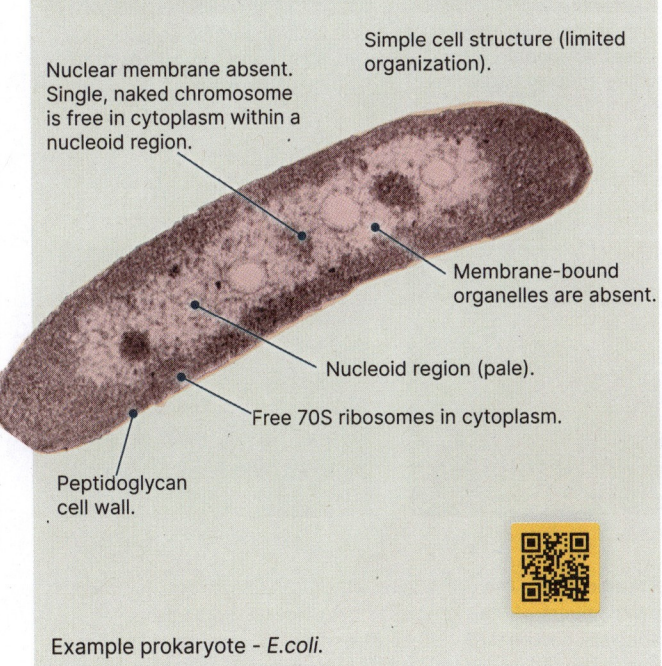

Nuclear membrane absent. Single, naked chromosome is free in cytoplasm within a nucleoid region.

Simple cell structure (limited organization).

Membrane-bound organelles are absent.

Nucleoid region (pale).

Free 70S ribosomes in cytoplasm.

Peptidoglycan cell wall.

Example prokaryote - *E.coli*.

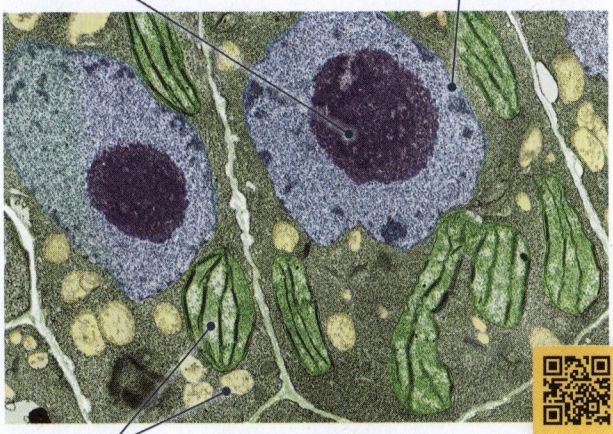

Chromosomes contained in nucleus.

Nuclear membrane present.

Presence of membrane-bound organelles.

Complex cell structure (high degree of organization).

Example eukaryote - plant palisade mesophyll

1. Describe cellular features that enable the identification of a prokaryote cell from a micrograph:

_____

_____

_____

2. Describe cellular features that enable the identification of a eukaryote cell from a micrograph:

_____

_____

3. Draw a scientific diagram of a prokaryote cell based off the micrograph, above left. Include the following labels; nucleoid region, cell wall, and cytoplasm, and annotate the diagram with the function of each structure. Attach the diagram to this page.

©2024 **BIOZONE** International
ISBN: 978-1-99-101410-8
Photocopying prohibited

A2.2

5, 6, 10

19 The Processes of Life in Unicellular Organisms

Key Idea: Unicellular organisms demonstrate the eight core processes of life found in all living organisms.

Living organisms all display the core processes of life: growth, metabolism, movement, response to stimuli, obtaining energy, excretion of wastes, homeostasis, and reproduction. These processes are all demonstrated in both multicellular and unicellular organisms, but not in viruses. All cells carry out these functions at some levels but in unicellular organisms the cell must be able to carry out these functions correctly and consistently by itself.

Growth is the increase in mass and volume of an organism over time.

Growth in unicellular organisms begins after cell division. Growth occurs during G1 and G2 phases of the cell cycle. Yeast is a commonly studied single celled eukaryote. It can double its mass in around 100 minutes in good conditions.

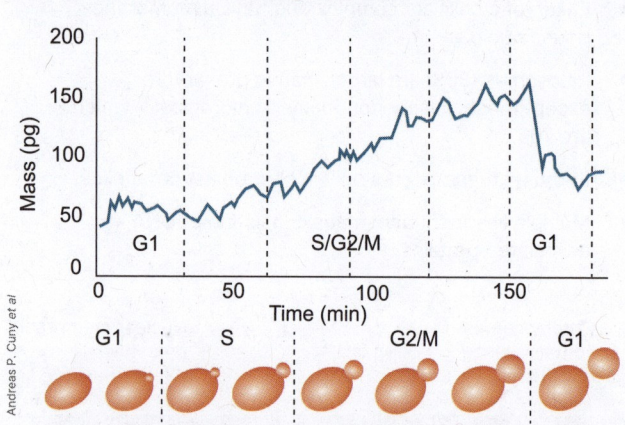

Yeast growth over time

Andreas P. Cuny et al

Metabolism is the sum of all chemical activity taking place within an organism to maintain life.

Organisms can produce energy carrying ATP by aerobic (using oxygen) or anaerobic (not using oxygen) respiration. ATP is used to power all other metabolic activities. Organisms can be obligate (no choice) aerobes or anaerobes or they can be facultative anaerobes (they can choose). A wide variety of anaerobic mechanisms are used by unicellular organisms. They include fermentation, methanogenesis, and reducing sulfur.

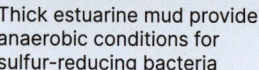

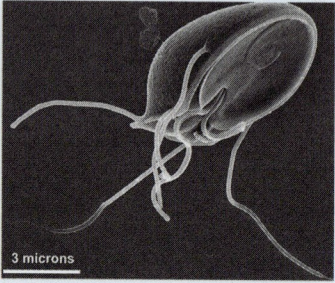

Thick estuarine mud provides anaerobic conditions for sulfur-reducing bacteria

Giardia is anaerobic parasite that can infect human intestines from contaminated water.

Movement allows organisms to find food and shelter, avoid predation, and find mates in order to reproduce. Unicellular organisms have a range of methods to move including pseudopodia, cilia, and flagella.

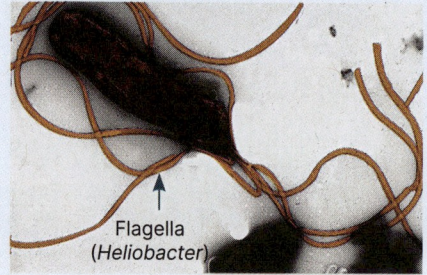

Flagella (*Heliobacter*)

Cilia (*Paramecium*)

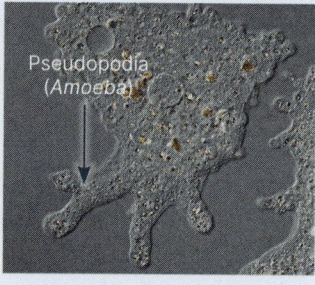

Pseudopodia (*Amoeba*)

Living organisms **respond to stimuli** from their environment. Most unicellular organisms do not have specialized sensory organelles, but will respond to chemical signals in the environment. These may be engulfed and processed or diffuse across the plasma membrane. The rate at which the chemical signal is encountered affects the rate of movement.

The protist *Euglena* is able to detect and respond to light. The eyespot is located near the flagellum. Pigment rods shield a light sensitive area in such a way that only light from the opposite end of the cell reaches the light sensitive area. This area influences flagella movement allowing *Euglena* to move towards the light.

Reproduction is the process by which living organisms produce new members of their species. Most unicellular organisms, especially bacteria, reproduce by asexual reproduction. They produce genetically identical copies of themselves. Sexual reproduction is much rarer in unicellular organisms than in multicellular organisms.

Asexual reproduction may be by budding, such as in yeast, or by binary fission as in bacteria.

In unicellular eukaryotes, sexual reproduction may occur by a cell forming or transforming into a haploid gamete. These then combine with gametes from another cell to produce a new diploid individual.

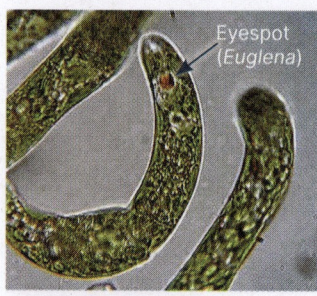

Eyespot (*Euglena*)

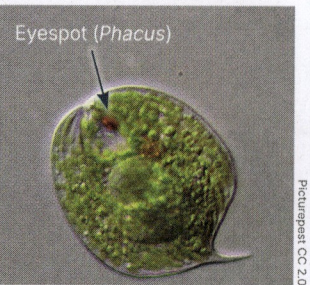

Eyespot (*Phacus*)

Picturepest CC 2.0

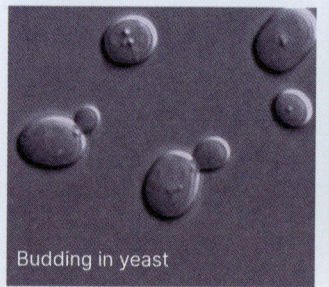

Budding in yeast

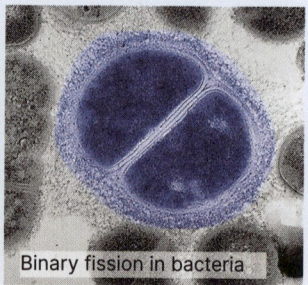

Binary fission in bacteria

All organisms need to obtain **nutrition**, the chemical ingredients needed for growth, reproduction and all other processes. Some unicellular organisms are autotrophic, being able to produce their own food. This can be either by photosynthesis or chemosynthesis.

Others unicellular organisms are heterotrophic and must find food. Food or prey is captured and normally engulfed so that digestion is internal.

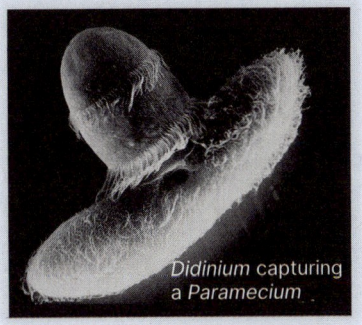

Didinium capturing a *Paramecium*

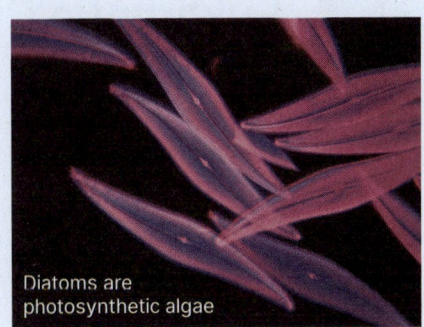

Diatoms are photosynthetic algae

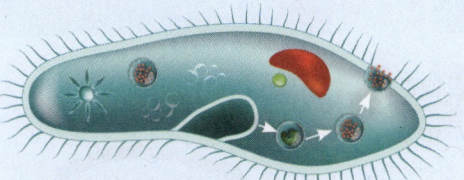

Diagram of phagocytosis of waste in *Paramecium*

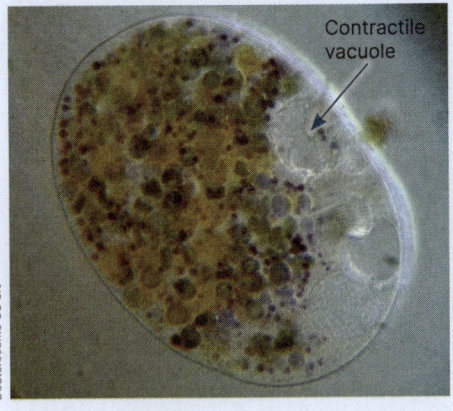

Contractile vacuole

Deuterosome CC 3.0

The build up of the waste products of metabolism can be deadly for a cell. They are removed from the cell via the process of **excretion**.

Unicellular organisms excrete wastes through the cell membrane, as they do not have any specialized excretory organs, with the exception of the contractile vacuole organelle, as seen in *Nassula* (bottom left).

Carbon dioxide, a waste product of cellular respiration, can diffuse freely across the plasma membrane, as can substances like ammonia.

Larger particles tend to be encased in a vacuole, often called a phagosome, and released to the cell surface by phagocytosis (top left). For unicellular organisms that have a contractile vacuole, water soluble wastes are 'collected' by this organelle and released to the external environment.

Homeostasis is an organism's ability to maintain its internal environment despite changes in its external environment.

For unicellular organisms that live in water, there is a need to maintain a stable concentration of substances inside the cell despite the environment having a different solute concentration than the cell. A contractile vacuole (CV) is used to maintain osmolarity (solute concentration) in the cytoplasm. Water moves into the CV by osmosis. It then contracts to expel the water out of the organism via a cytoplasm pore.

Contractile vacuoles are found in many other unicellular organisms such as *Amoeba*, *Paramecium*, and *Chlamydomonas*.

1. Define each of the following life processes:

(a) Response to stimuli: _____

_____

(b) Reproduction: _____

_____

(c) Excretion: _____

_____

(d) Metabolism: _____

_____

2. Approximately by how many times does a yeast cell increase its mass over one cell cycle? _____

3. Why are anaerobic unicellular organisms commonly found in deeper layers of estuary mud?

_____

_____

4. Why are viruses not classed as living organisms? _____

_____

5. Why is a contractile vacuole important in homeostasis of many unicellular organisms? _____

_____

20 Eukaryotic Cell Structures

Key Idea: Eukaryotic cells have many features in common. However, there are several differences between plant, animal and fungal cells. Some specialized cells have atypical structures.

Animal cells, unlike plant cells, do not have a regular shape. In fact, some animal cells, such as phagocytes, are able to alter their shape for various purposes, e.g. engulfing foreign material. Plant and fungal cells have a cell wall, giving rigidity and structure. Some specialized cells, e.g. erythrocytes, have lost many cell components in order to maximize the cell's capacity to carry oxygen. Other specialized cells, such as aseptate fungal hyphae and skeletal muscle fibres have lost their individuality and become fused to become multinucleated cells. Sieve tube elements in plant cells lose their nuclei in development to maximize their capacity for transport of substances through the plant.

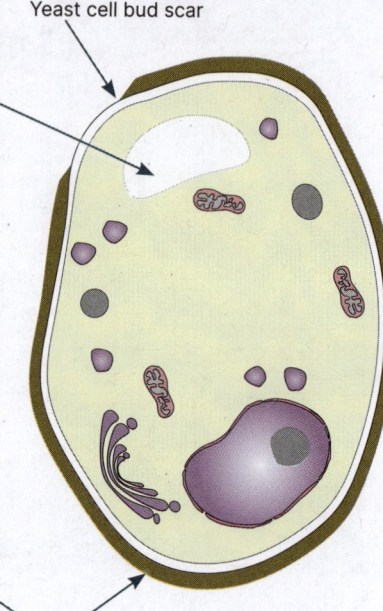

Generic animal cell

Centriole

Generic plant cell

Vacuole

Chloroplasts

Other plastids (here, an amyloplast storing starch granules)

Generic fungal cell

Yeast cell bud scar

Cell wall

Animal cell components

▶ Cell wall - not present.
▶ Vacuole - small and temporary. Used to expel waste products from cells.
▶ Chloroplasts not present.
▶ Other plastids - not present.
▶ Centrioles - composed of a protein called tubulin. Used in spindle fibre formation during cell division.
▶ Cilia/flagella - yes, in some cells but not shown in above image. Found in sperm cells to provide motility and in mucosal membrane cells, to help move mucus that can contain pathogens, out of the body.

Plant cell components

▶ Cell wall - composed of cellulose.
▶ Vacuole - large and permanent. Maintains turgor pressure for cell rigidity.
▶ Chloroplasts - yes, many or few, depending on the type of cell.
▶ Other plastids - double membraned structures for manufacturing/storing food. Amyloplasts store starch. Leucoplasts (in root cells) can synthesize fatty acids and some amino acids.
▶ Centrioles - found in lower plants but absent from conifers/ flowering plants.
▶ Cilia/flagella - found in lower plants but absent from conifers and flowering plants.

Fungi cell components

▶ Cell wall - composed of chitin.
▶ Vacuole- large and permanent. Maintains turgor pressure to assist in cell rigidity.
▶ Chloroplasts - not present.
▶ Other plastids - not present.
▶ Centrioles - not present (except for a few exceptions).
▶ Cilia/flagella - not present in true fungi. An exception is the primitive fungi, *Cryptomycota*, which possesses a flagellum but no cell wall.

1. What is the difference between vacuoles in plant and animal cells? _____

_____

_____

2. Both plant and fungal cells have a cell wall. What would indicate that the structure evolved independently in each?

_____

_____

A2.2

8,9

©2024 **BIOZONE** International
ISBN: 978-1-99-101410-8
Photocopying prohibited

Atypical eukaryotic cells

While eukaryotic cells have much in common, some cells, as they differentiate, change to suit their specialist roles. The resulting changes in structure make them look quite different in appearance from the 'generic' models of eukaryotic cells described previously. The number of nuclei in some specialized cells can be used as examples of their deviation from the generic structure. Some specialized cells with atypical nuclei are: erythrocytes (red blood cells), skeletal muscle cells, phloem sieve tube elements in plants, and aseptate fungal hyphae.

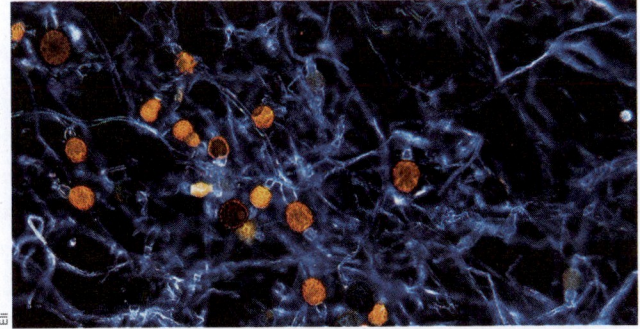

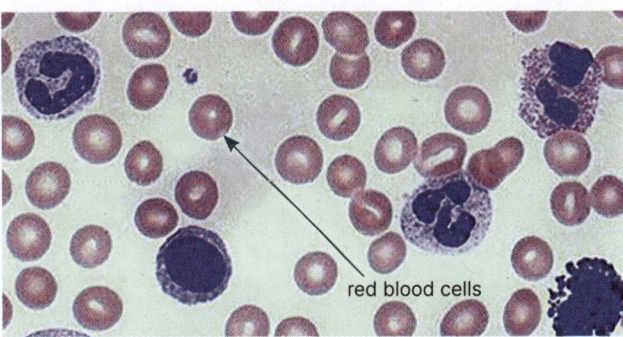

red blood cells

Fungal hyphae form a dense network of filamentous cells that can cover vast areas, especially in rotting wood, or in the soil. These are the blue filaments visible in the above photo. The long cells can be divided by septae, which separate them into individual cells. However, the cells can also be aseptate (without septae) and form a long, continuous cell body with many nuclei present along the length of the filament.

Erythrocytes (red blood cells) are adapted for their specialized role in transporting oxygen around the body. They have no nucleus and the cytoplasm of each erythrocyte typically contains around 270 million molecules of haemoglobin, each made up of four haem groups. This makes them very efficient oxygen transporters. By losing space in the cell normally used up by other organelles, space has been maximized for carrying haemoglobin.

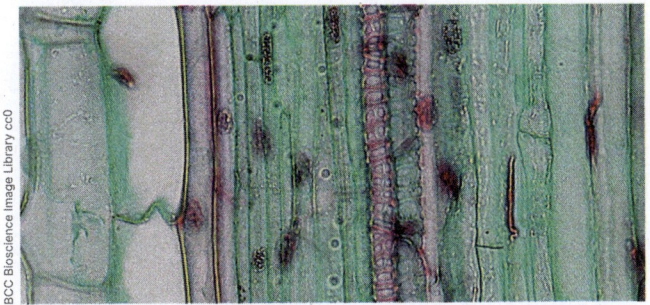

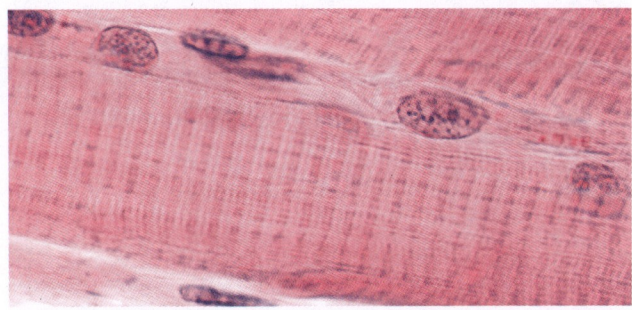

Sieve tubes in plants are made up of sieve elements. These long, continuous cells, divided by sieve plates are responsible for the transport of sugars around the plant. During development, these cells lose their nuclei and other organelles to maximize the space available for their specialized role. Companion cells, which sit alongside the sieve elements, allow for exchange of ions, metabolites, RNA and other proteins.

Skeletal muscle fibres are composed of long, specialized cells. During myogenesis (the formation of muscle cells during embryonic development), the individual cells fuse to form continuous cells with many nuclei distributed along the entire length. A single muscle fibre can have thousands of nuclei. The multiple nuclei provide the cell with all of the enzymes and other proteins required for optimal functioning.

3. Which cell and tissue types from above have lost their nuclei, and what advantage does that adaptation present?

_____

_____

_____

4. Why would large, fused cells (e.g. skeletal muscle fibres and fungal hyphae) have multiple nuclei?

_____

_____

_____

5. Summarize how we know that plant, animal, and fungi are all eukaryotes, rather than prokaryotes: _____

_____

_____

_____

_____

21 Light Microscopy and Cells

Key Idea: Light microscopy allows for the identification and classification of cells into major groups.

Features of cells, such as the presence or absence of a nucleus, large organelles with membranes such as chloroplasts and mitochondria, cell walls, and microvilli, can be observed using a light microscope. These structures can enable the identification of cells as prokaryotes or eukaryotes, and usually as plants, animals, protists, or fungi.

Identifying photosynthetic organisms from light micrographs

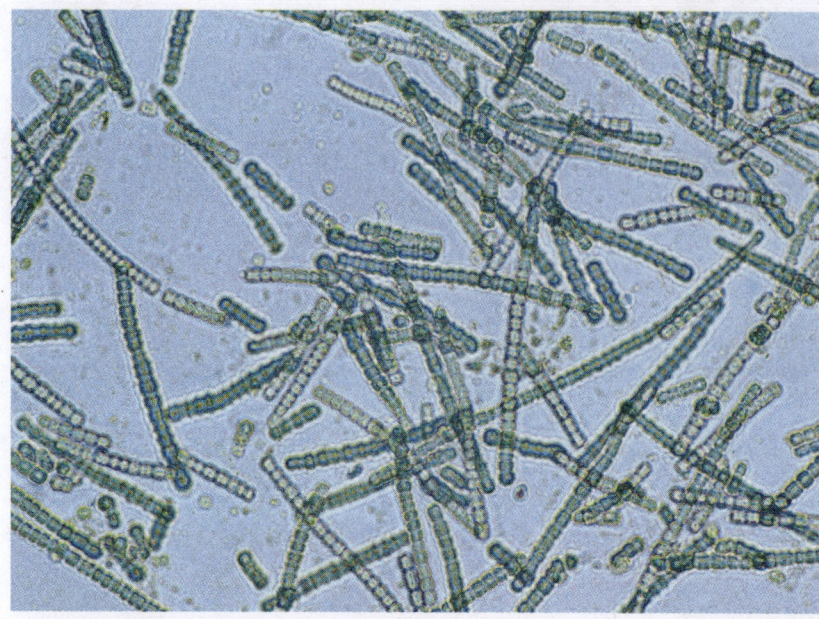

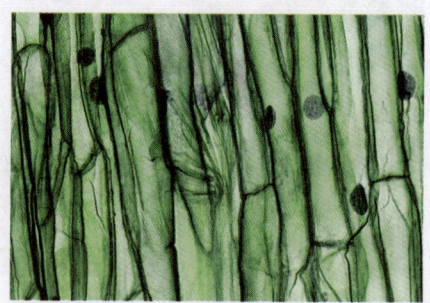

1. The images above are photosynthetic organisms: they can form glucose from carbon dioxide and water via photosynthesis.

(a) The micrograph on the left is of cyanobacteria, a prokaryotic organism. What cellular features would you expect to see or not see to identify them as such:

_____

_____

_____

(b) The light microscopic micrographs on the right are of eukaryotic organisms. What cellular features would you expect to see or not see to distinguish them from prokaryotes?

_____

_____

_____

Medical diagnosing from light microscopy

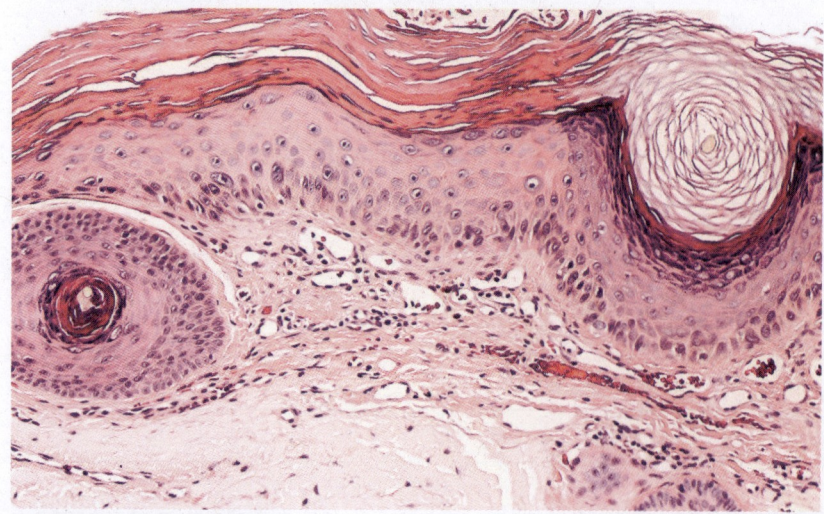

2. The light micrograph on the left shows a cross section of sun-damaged skin. It has a keratosis, a build-up of keratin in a hair follicle that can be a precursor to skin cancer. Suggest which part of the image shows the keratosis and explain your reasoning?

_____

_____

_____

_____

_____

_____

_____

A2.2
10

©2024 **BIOZONE** International
ISBN: 978-1-99-101410-8

22 Electron Microscopy and Cells

Key Idea: Transmission Electron Microscopy gives high resolution images of cells, showing the major organelles.
Transmission electron microscopy (TEM) is the most frequently used technique for viewing cellular organelles.

When viewing TEMs, the cellular organelles may have quite different appearances depending on whether they are in transverse or longitudinal sections. Because images from the microscope are not coloured, colour is added to the images.

1. Identify and label the structures in the animal cell below using the following list of terms: *plasma membrane, rough and smooth endoplasmic reticulum, mitochondrion, nucleus, secretory vesicle, Golgi apparatus, chromosomes*

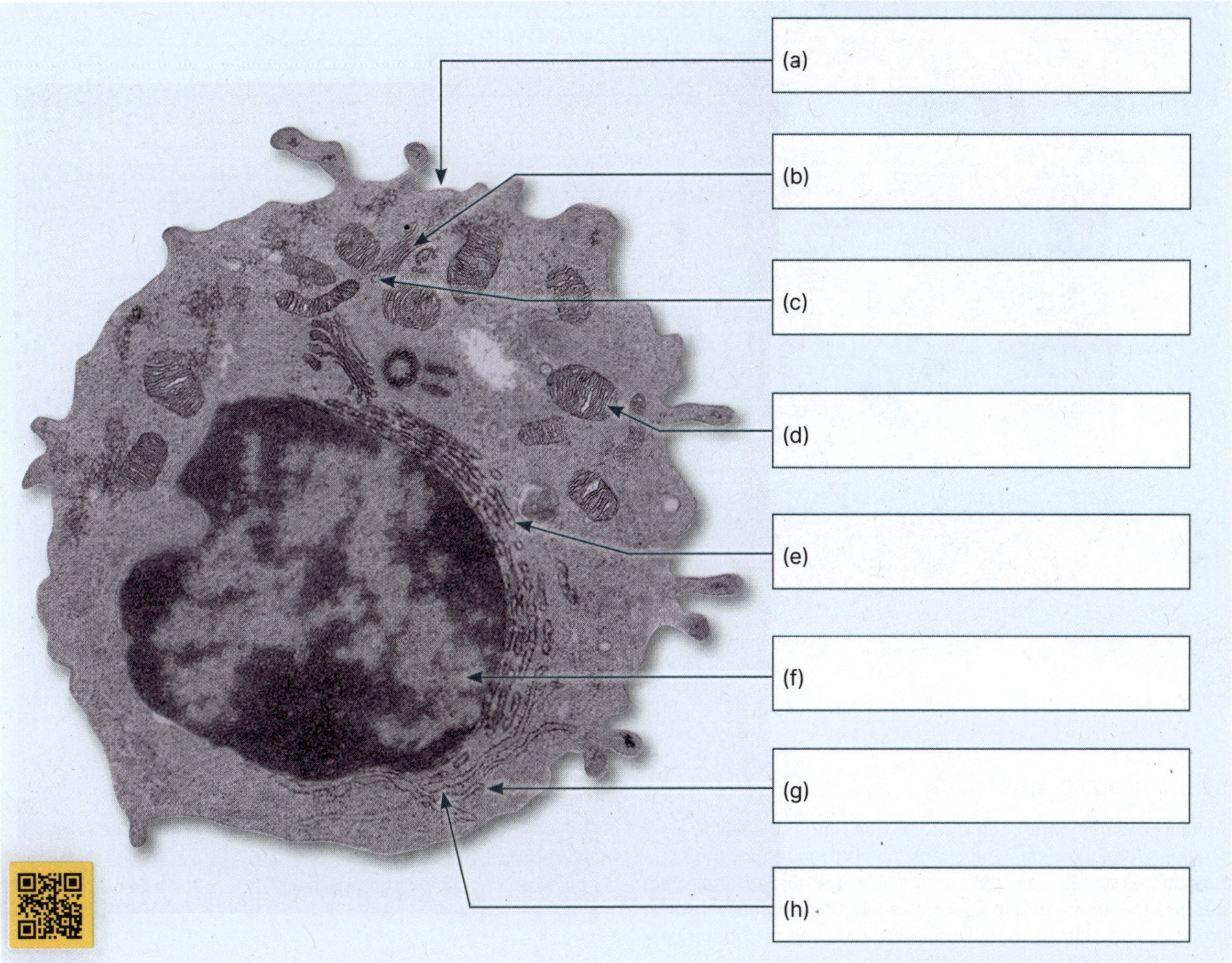

(a)

(b)

(c)

(d)

(e)

(f)

(g)

(h)

2. Which organelles can usually be observed in electron microscope micrographs, but rarely in light micrographs?

_____

3. The cell above is a mobile phagocytic cell, yet still part of a multicellular organism. What features identify it as mobile?

_____

_____

4. What is the structural functional difference between the smooth and rough endoplasmic reticulum? _____

_____

_____

_____

5. What is the relationship between the Golgi apparatus and the secretory vesicle? _____

_____

_____

_____

©2024 **BIOZONE** International
ISBN: 978-1-99-101410-8
Photocopying prohibited

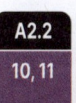 A2.2 10, 11
 AOS

6. Study the diagrams on the other pages in this chapter to familiarise yourself with the structures found in eukaryotic cells. Identify the ten structures in the plant cell below using the following word list: *cytoplasm, endoplasmic reticulum, mitochondrion, starch granule, chromosome, nuclear membrane, sap vacuole, plasma membrane, cell wall, chloroplast:*

(a) _____

(b) _____

(c) _____

(d) _____

(e) _____

(f) _____

(g) _____

(h) _____

(i) _____

(j) _____

Micrographs: microvilli

Animal cells can differentiate into a wide range of cellular shapes, depending on what adaptation is required for their function. Epithelial cells form an outside layer and often project out into the lumen (organ spaces) as villi if the increased surface area is an advantage for adsorption or diffusion of substances across the membrane. Villi may also have even smaller projections, called microvilli, which cover the inside cellular layer of the large intestines (right).

Microvilli resemble a barely observable layer on a light microscope micrograph, and require the use of a transmission electron microscope to view with enough resolution to distinguish them.

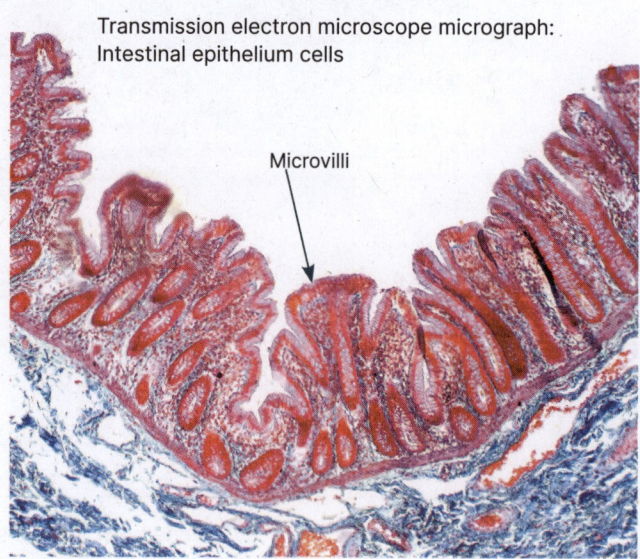

Transmission electron microscope micrograph: Intestinal epithelium cells

Microvilli

7. Describe two structures, pictured in the plant cell above, that are associated with storage: _____

_____

_____

_____

8. Draw a scientific diagram of a eukaryotic cell based on one of the micrographs in this activity. Include appropriate following labels for your cell of choice, selecting from this list: nucleus, mitochondria, chloroplasts, sap vacuole, Golgi apparatus, rough and smooth endoplasmic reticulum, chromosomes, cell wall, plasma membrane, secretory vesicles, microvilli. Annotate the diagram with the function of each structure. Attach the diagram to this page.

©2024 **BIOZONE** International
ISBN: 978-1-99-101410-8
Photocopying prohibited

23 Endosymbiosis Theory

Key Idea: The origin of complexity in eukaryotic cells can be explained by endosymbiosis theory.

Endosymbiosis theory (from endo: internal; symbiosis: relationship) is used to explain the evolution of eukaryotic cells by the engulfment of prokaryotic cells in early common ancestors. It is thought that eukaryotic cells evolved from pre-eukaryotic (bacterial) cells that ingested other free-living bacteria. They formed a symbiotic relationship with the cells they engulfed. The two organelles that evolved in eukaryotic cells as a result of bacterial endosymbiosis were mitochondria for aerobic respiration, and chloroplasts for photosynthesis in aerobic conditions. Primitive eukaryotes probably acquired mitochondria by engulfing purple bacteria. Similarly, chloroplasts may have been acquired by engulfing photosynthetic cyanobacteria. Other organelles may have formed from infolding of the plasma membrane.

Evolution of eukaryotic cells

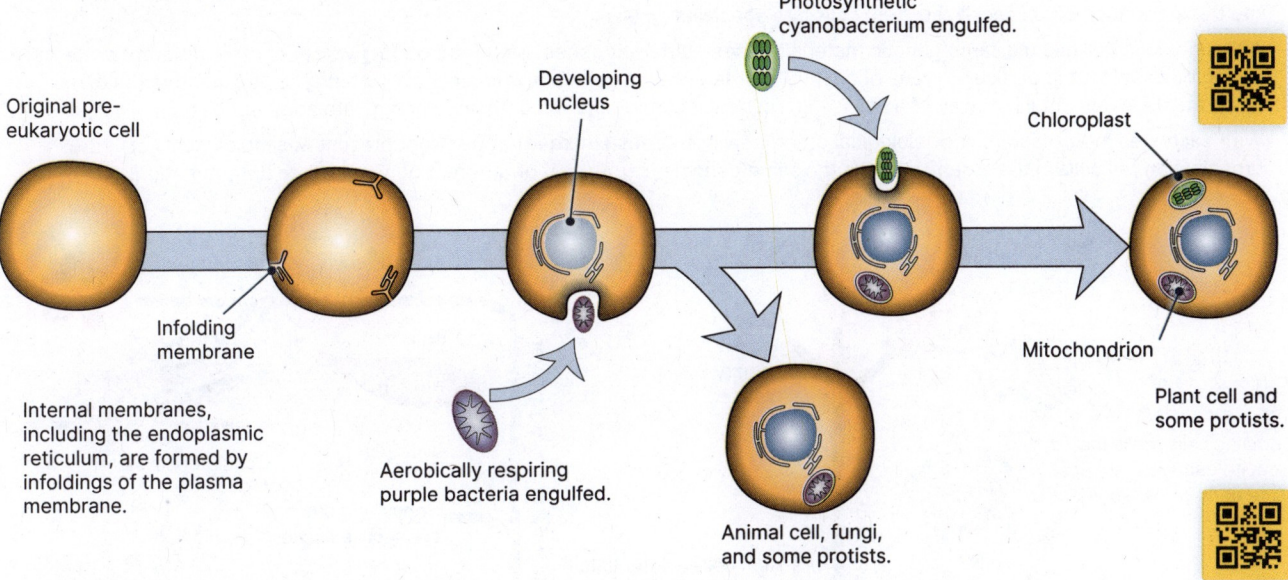

Original pre-eukaryotic cell

Infolding membrane

Internal membranes, including the endoplasmic reticulum, are formed by infoldings of the plasma membrane.

Developing nucleus

Aerobically respiring purple bacteria engulfed.

Photosynthetic cyanobacterium engulfed.

Animal cell, fungi, and some protists.

Chloroplast

Mitochondrion

Plant cell and some protists.

Evidence for the endosymbiosis theory:

Many observations about cells are supported by endosymbiosis theory. Multiple streams of evidence strengthen the theory and include:

▶ Mitochondria and chloroplasts have a similar morphology (structure) to bacteria.

▶ Mitochondria and chloroplasts divide by binary fission, splitting in half to form new organelles just like bacteria. Thus, new mitochondria and chloroplasts arise from pre-existing ones; they are not manufactured by the cell.

▶ Both mitochondria and chloroplasts have a chemically distinct inner membrane. The outer membrane is similar to the plasma membrane (as if a vesicle formed around the engulfed cell) but the inner membrane is similar to the bacterial membrane.

▶ Bacterial DNA is a single circular molecule. Mitochondria and chloroplasts also have their own, single, circular DNA. Like bacterial DNA, this DNA has no intervening, non-protein-coding regions or associated proteins. Also, the organelle DNA mutates at a different rate from the nuclear DNA.

▶ Mitochondria and chloroplasts contain ribosomes that are more similar in size (70S) to bacterial ribosomes than ribosomes in the cytoplasm (80S).

▶ Antibiotics that inhibit protein synthesis in bacteria also inhibit protein synthesis in mitochondria and chloroplasts.

▶ Analysis of chloroplast DNA has shown that they are related to cyanobacteria.

1. How do the 70S ribosomes in mitochondria and chloroplast provide evidence for the endosymbiosis theory?

_____

_____

_____

_____

A2.2
12

NOS

AHL

24 Cellular Differentiation

Key Idea: A zygote divides and produces all the cell types in the body by cellular differentiation. Specific patterns of gene switching on or off determine what cell type develops.

Multicellular organisms consist of many different cell types, each specialized to carry out a particular role. A zygote and its first few divisions are totipotent and can differentiate to form any cell type in the body. During development, these cells divide and follow different developmental pathways.

The process of specialized cells developing from general ones is called cellular differentiation and is achieved through switching genes on and off in particular sequences. Once fully differentiated, the cell cannot turn into another cell type. The environment can also cause changes in gene expression, such as affecting the sex of hatchlings. Other effects can lead to phenotypic changes that may be permanent or temporary.

The pathway of cellular differentiation

▶ When a cell divides by mitosis, it produces genetically identical cells. However, a multicellular organism is made up of many different types of cells, each specialized to carry out a particular role. How can it be that all of an organism's cells have the same genetic material, but the cells have a wide variety of shapes and functions? The answer is through cellular differentiation, the transformation of unspecialized stem cells into specialized cells.

▶ Although each cell has the same genetic material (genes), different genes are turned on (activated) or off in different patterns during development in particular types of cells. The differences in gene activation controls what type of cell forms (below). Once the developmental pathway of a cell is determined, it cannot alter its path and change into another cell type.

▶ With each step in this hierarchy of biological order, new properties emerge that were not present at simpler levels of organization. All cells in the organism share the same genome. Expression of different genes leads to differentiation.

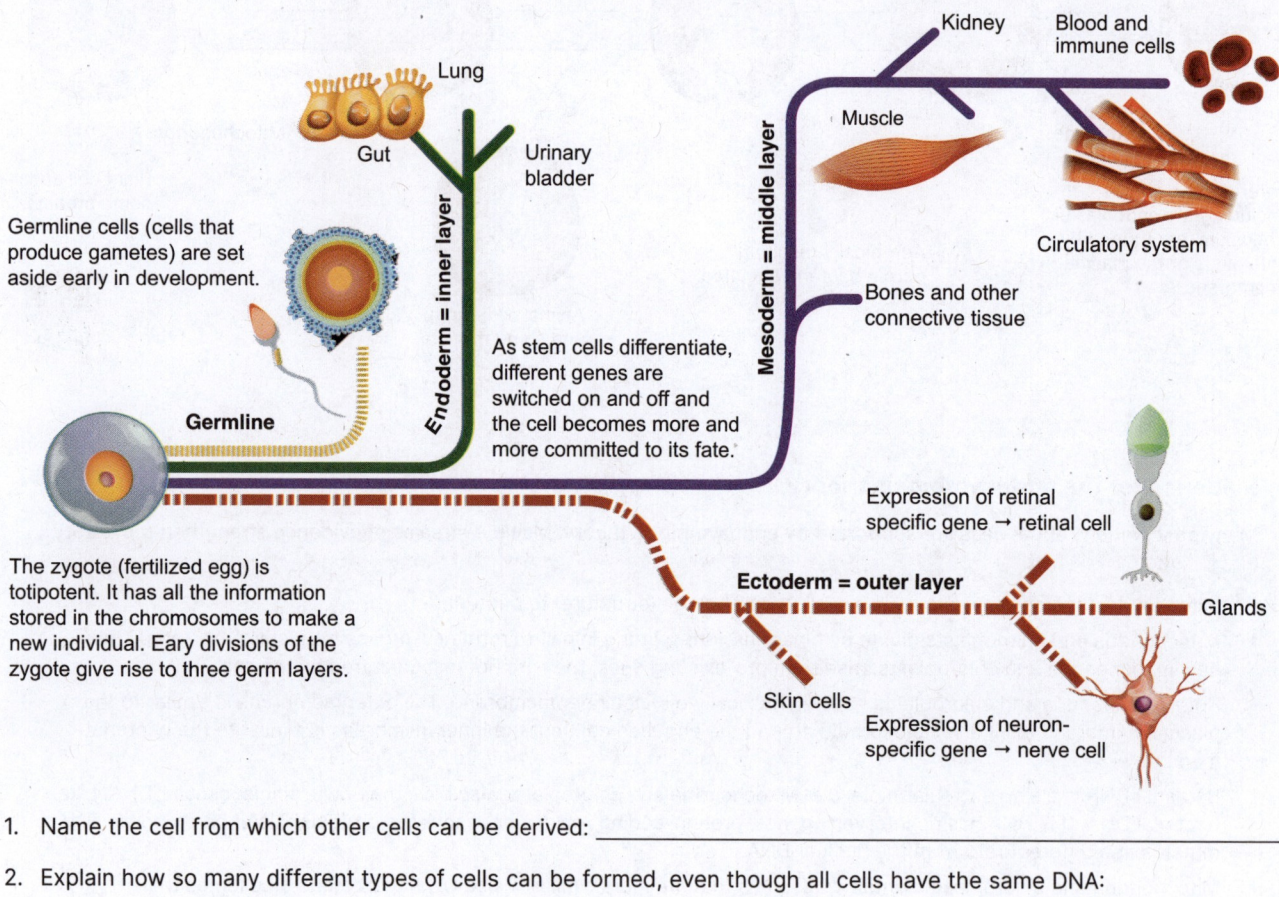

1. Name the cell from which other cells can be derived: _____

2. Explain how so many different types of cells can be formed, even though all cells have the same DNA:

   _____

   _____

   _____

3. (a) What are totipotent cells? _____

   _____

   (b) What are the two defining properties of stem cells?

   i _____

   ii _____

©2024 **BIOZONE** International
ISBN: 978-1-99-101410-8
Photocopying prohibited

Specialized cells in multicellular organisms

Regulation of gene expression, i.e. specific genes in the genome being 'turned on or off' is caused by proteins binding to specific parts of the DNA. This causes cells to differentiate into specialized cells. These can, in turn, form specialized tissues and organs. Cell development and differentiation is encoded by genes. Cellular differentiation is a product of the genes themselves, and their internal and external environment, and the variations in the way those genes are controlled.

Examples of specialized cells in plants

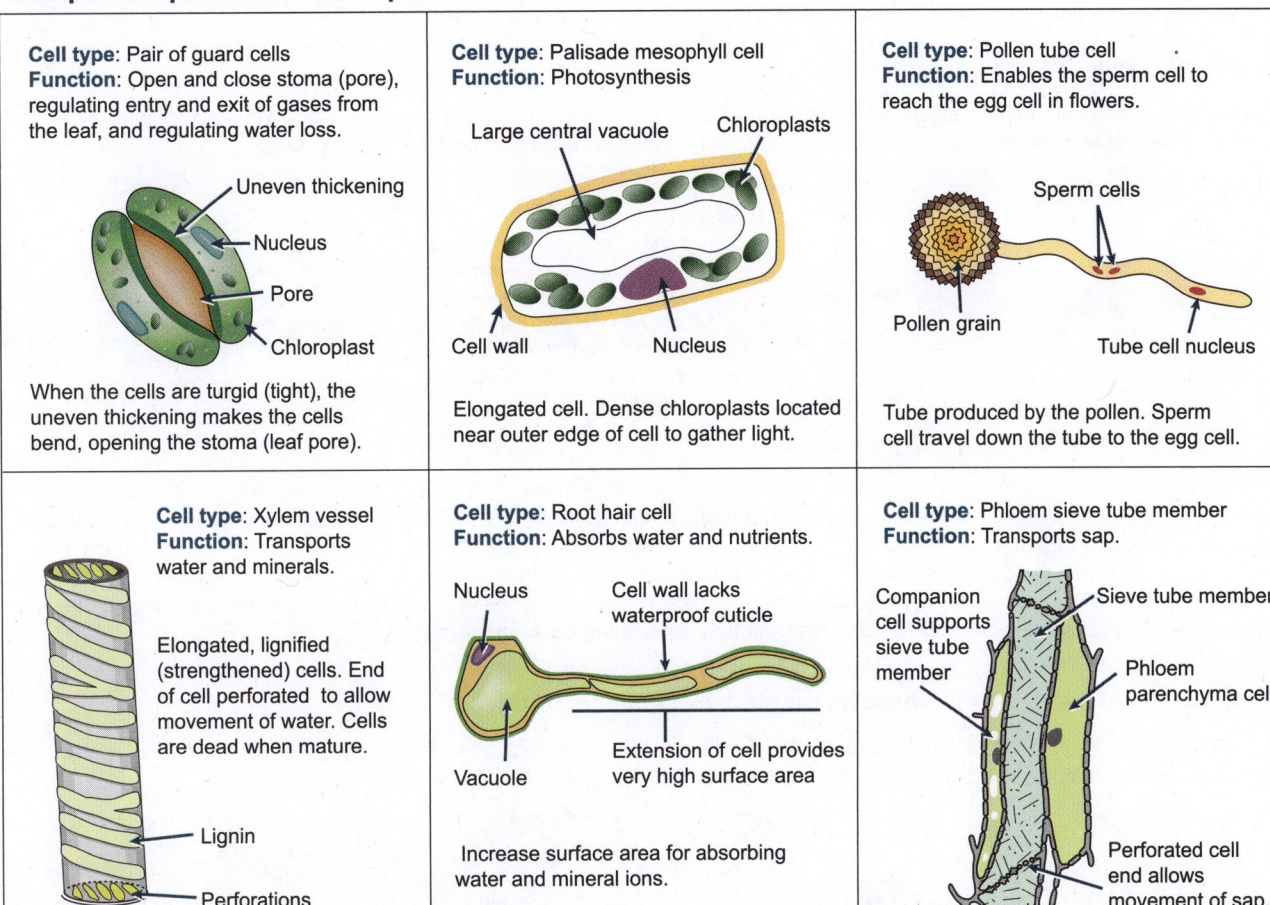

Cell type: Pair of guard cells
Function: Open and close stoma (pore), regulating entry and exit of gases from the leaf, and regulating water loss.

Uneven thickening
Nucleus
Pore
Chloroplast

When the cells are turgid (tight), the uneven thickening makes the cells bend, opening the stoma (leaf pore).

Cell type: Palisade mesophyll cell
Function: Photosynthesis

Large central vacuole
Chloroplasts
Cell wall
Nucleus

Elongated cell. Dense chloroplasts located near outer edge of cell to gather light.

Cell type: Pollen tube cell
Function: Enables the sperm cell to reach the egg cell in flowers.

Sperm cells
Pollen grain
Tube cell nucleus

Tube produced by the pollen. Sperm cell travel down the tube to the egg cell.

Cell type: Xylem vessel
Function: Transports water and minerals.

Elongated, lignified (strengthened) cells. End of cell perforated to allow movement of water. Cells are dead when mature.

Lignin
Perforations

Cell type: Root hair cell
Function: Absorbs water and nutrients.

Nucleus
Cell wall lacks waterproof cuticle
Extension of cell provides very high surface area
Vacuole

Increase surface area for absorbing water and mineral ions.

Cell type: Phloem sieve tube member
Function: Transports sap.

Companion cell supports sieve tube member
Sieve tube member
Phloem parenchyma cell
Perforated cell end allows movement of sap.

Effects of the environment on gene expression.

The phenotype of an organism is its observable, physical characteristics, or traits. Environmental factors can influence gene expression and so modify cellular differentiation and hence the phenotype, without changing the genotype. Environmental factors that affect cellular differentiation include nutrients or diet, temperature, altitude or latitude, and the presence of other organisms. This can occur both during development and later in life.

Temperature and the American alligator

The sex of some animals is determined by the incubation temperature during their embryonic development. American alligator (*Alligator mississippiensis*) eggs incubated above 33 ºC will all develop into males. A difference of only 3 ºC lower (30 ºC)when the eggs are incubating will produce mostly female alligators. Temperature regulated sex determination may provide an advantage by preventing inbreeding, since all siblings will tend to be of the same sex. The alligators have a temperature sensitive protein (TRPV4) that is crucial for the alligators to develop male characteristics in the developing embryo. Temperatures lower than around 34-37 ºC won't activate this protein.

American alligator infants emerging from their eggs

4. Research another species that shows cellular differentiation /sex determination triggered by temperature. Explain the effects of both hot and cold temperatures, as well as the gene or protein affected:

_____

_____

_____

25 Multicellularity

Key Idea: It is likely that multicellularity evolved multiple times over the course of evolution. Being multicellular confers many advantages compared to being unicellular.

Although many single celled organisms exist, being larger and multicellular (with many cells) has advantages. All plants and animals are multicellular organisms, as are many fungi and algae. It is likely that multicellularity evolved multiple times over the course of evolution. Cells can clump together to form a colony and it is a short evolutionary leap for different cells within the colony to become specialized and have different roles. This is the precursor to becoming complex, multicellular organisms.

Prokaryotic cells can work together

▶ Most prokaryotic cells behave as unicellular individuals. However, many form colonies, which may act in a basic multicellular way. Some prokaryotes are able to form groups, with some cells specializing to perform specific tasks.

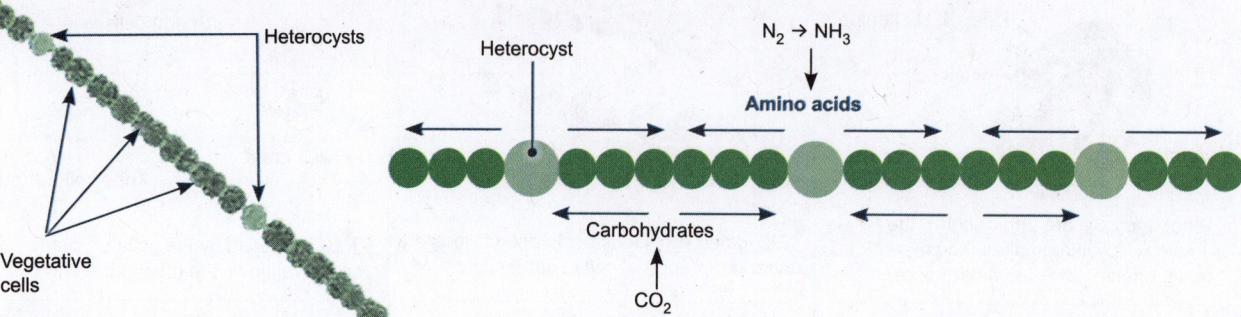

Cyanobacteria, e.g. *Anabaena* (shown above), are a group of bacteria that can photosynthesize. They often form long filaments of individual cells joined together. Under low-nitrogen conditions, some of these cells will specialize to form heterocysts which are able to fix nitrogen from molecular nitrogen (N_2). They show quite different gene expression from neighbouring, unspecialized cells in that they produce the enzyme nitrogenase and cannot photosynthesize. In addition, the heterocysts share the nitrogen they fix with neighbouring cells and receive other nutrients from them, indicating basic cooperation.

Unicellular eukaryotes can behave as a multicelllular single organism

▶ Although most eukaryotes are multicellular, some are permanently unicellular, and some spend some of their life cycle as single, independent cells. In certain conditions, they can come together and behave as a single entity.

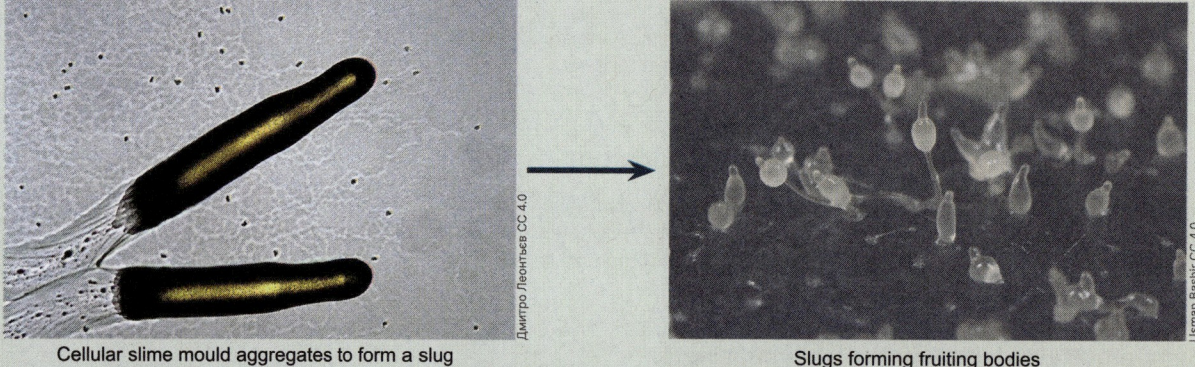

Cellular slime mould aggregates to form a slug

Slugs forming fruiting bodies

Dictyostelium discoideum is a soil living amoeba. It spends part of its life cycle as a free living single cell. When food (bacteria) becomes scarce the individual cells begin to congregate forming a slug composed of about 100,000 cells. This moves to a more favoured area and forms a fruiting body, producing spores that produce new amoeba.

1. What are some similarities in how prokaryotic and eukaryotic unicellular organisms gain an advantage by joining together to form multicellular organisms?

_____

_____

_____

2. Research one of the following examples of multicellularity, and describe the advantages this state provides: Colonization in *Scenedesmus* (green alga); *Volvox* (green alga).

_____

_____

_____

AHL   A2.2   14

©2024 **BIOZONE** International
ISBN: 978-1-99-101410-8
Photocopying prohibited

26 Viruses

Key Idea: Viruses are infectious, highly specialized intracellular parasites. They are acellular and non-living.

Viruses are disease-causing agents (pathogens) that replicate (reproduce themselves) only inside the living cells of other organisms. Viruses are small and acellular, meaning they are not made up of cells. A typical virus contains genetic material (DNA or RNA) encased in a protein

coat (capsid). Some viruses have an additional membrane, called an envelope, surrounding the capsid. Many viruses have glycoprotein receptor spikes on their envelopes that help them to attach to the surface of the host cell they are infecting. Viruses vary greatly in their appearance and the type of host they infect (below). Several theories suggest different origins of viruses.

Viral structure

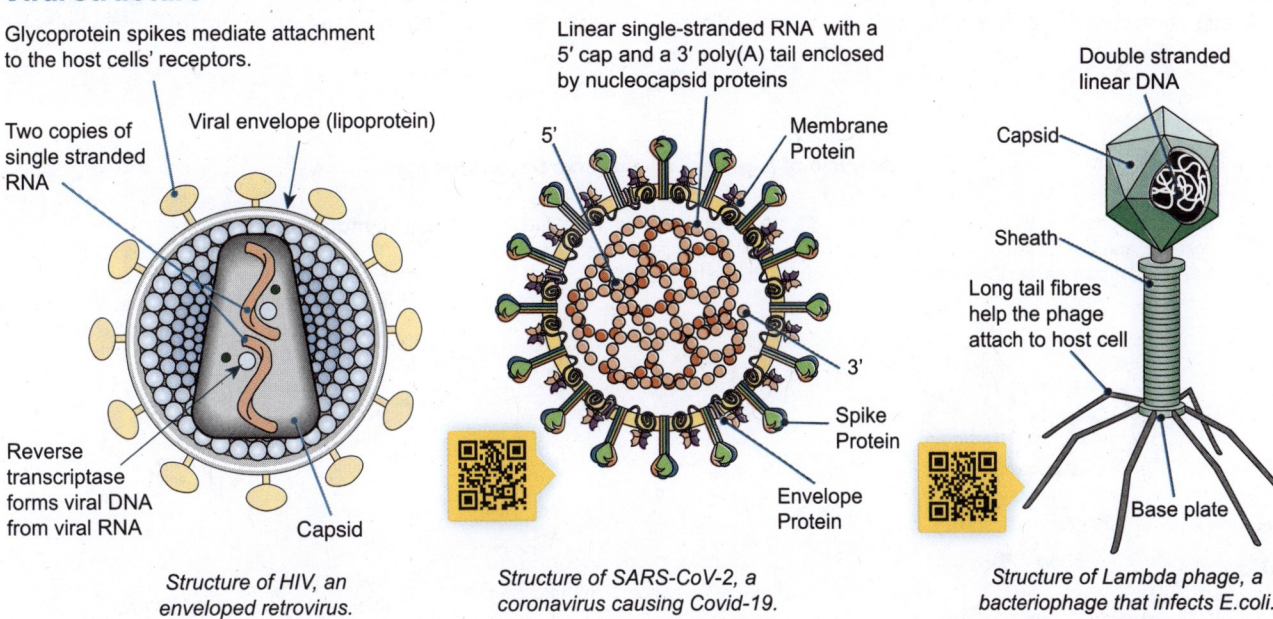

Structure of HIV, an enveloped retrovirus.

Structure of SARS-CoV-2, a coronavirus causing Covid-19.

Structure of Lambda phage, a bacteriophage that infects E.coli.

Theories on virus evolution

Several theories offer suggestions for the origins of viruses. They include:

Virus first hypothesis: could viruses have evolved before cells? This is unlikely, as all viruses require a host cell to reproduce.

Progressive hypothesis: could viruses have arisen as a result of 'escaping' from an existing cell, taking some components with them, such as genetic material, and proteins to form the capsid coat? The genetic structure of retroviruses offers some support to this theory.

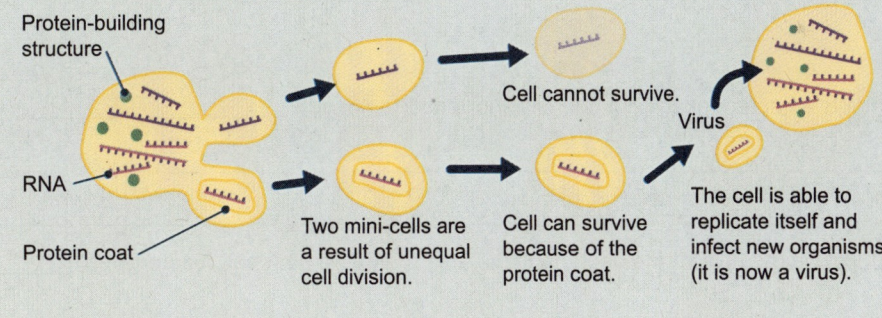

Regressive hypothesis: could viruses have arisen by an existing cell losing most of its components to become a virus? Nucleocytoplasmic large DNA viruses (NCLDVs), including smallpox virus, best support this theory.

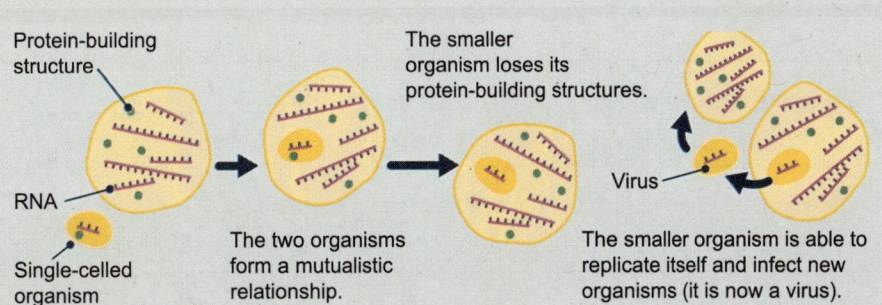

All viruses have several common features. They all require a host to replicate and they use the same genetic code as cells. This is likely the result of convergent evolution, where the different virus types have evolved similar features because of their functions.

1. Write a short report on one example of a virus. Include the viral classification type, a drawn and labelled diagram of the virus, the host it invades, the impacts on the host, and the impacts on humans, if any. Attach the report to the page.

©2024 **BIOZONE** International
ISBN: 978-1-99-101410-8
Photocopying prohibited

A2.3 1, 2, 5 AHL

27 Replication in Viruses

Key Idea: Viruses infect living cells using the metabolic processes of the host cell to produce new viral particles.

Viruses must infect a host cell to reproduce. They do so by using the host's metabolism to produce new virus particles. In bacteriophages (viruses that infect bacterial cells) this process may not immediately follow infection. Instead, the virus may integrate its nucleic acid into the host cell's DNA, forming a provirus or prophage. This type of cycle, called lysogenic, does not kill the host cell outright. Instead, the host cell is occupied by the virus and used to replicate viral genes. During this time, the viral infection is said to be latent. The virus may be transduced into becoming active again, entering the lytic cycle and utilizing the host's cellular mechanisms to produce new virions. The lytic cycle results in death of the host cell through cell lysis. Animal viruses follow a similar pattern to bacteriophage multiplication but there are notable differences. Animal viruses have different mechanisms by which they enter host cells and, once inside the cell, the production of new virions is different. This is partly because of differences in host cell structure and metabolism and because the structure of animal viruses is variable.

Life cycle of λ–phage, a lysogenic bacteriophage

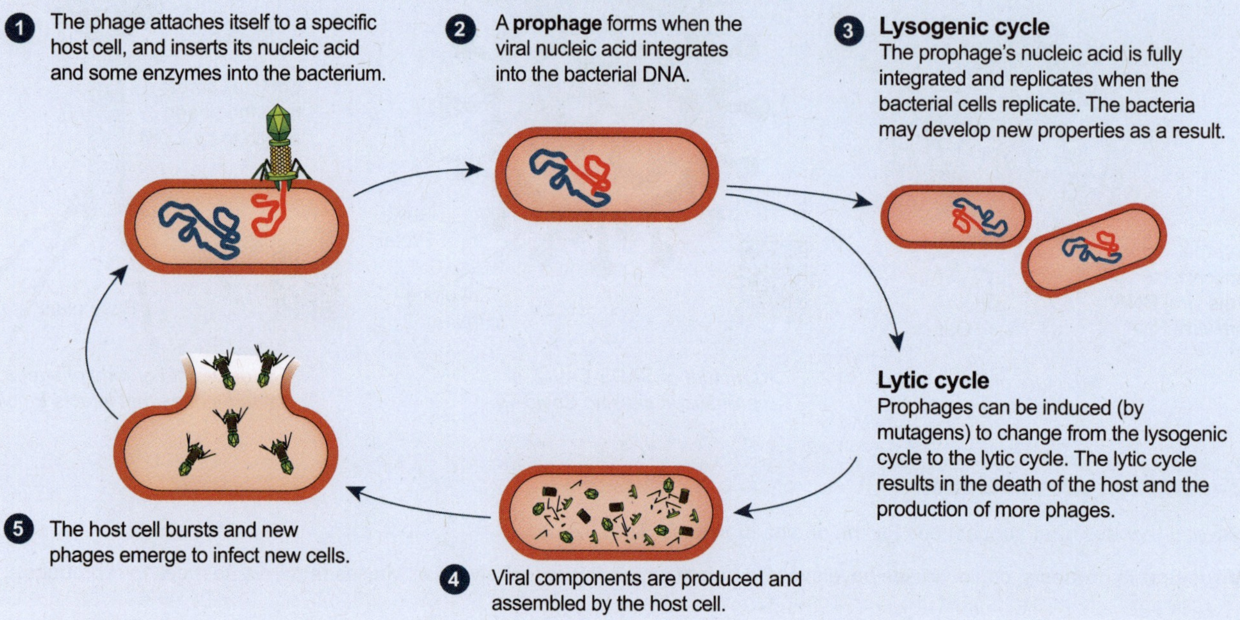

1 The phage attaches itself to a specific host cell, and inserts its nucleic acid and some enzymes into the bacterium.

2 A **prophage** forms when the viral nucleic acid integrates into the bacterial DNA.

3 Lysogenic cycle
The prophage's nucleic acid is fully integrated and replicates when the bacterial cells replicate. The bacteria may develop new properties as a result.

Lytic cycle
Prophages can be induced (by mutagens) to change from the lysogenic cycle to the lytic cycle. The lytic cycle results in the death of the host and the production of more phages.

5 The host cell bursts and new phages emerge to infect new cells.

4 Viral components are produced and assembled by the host cell.

1. What is a bacteriophage (sometimes shortened to phage)? _____

_____

_____

2. (a) How is the RNA of the bacteriophage able to enter a host cell (bacterial cell)? _____

_____

_____

(b) State where a bacteriophage virus replicates its viral DNA: _____

_____

(c) State where an bacteriophage virus synthesizes its proteins: _____

_____

3. What is the significance of viral reproduction to the ability of a virus to cause disease? _____

_____

_____

_____

_____

AHL

A2.3

3, 4

©2024 **BIOZONE** International
ISBN: 978-1-99-101410-8
Photocopying prohibited

28 Rapid Virus Evolution

Key Idea: Rapid evolution in pathogens, especially viruses, helps them evade the human immune system.

The immune system has evolved to 'remember' previous infections and prevent reinfection by the same pathogen. Pathogens, especially viruses, have evolved mechanisms to evade this response by changing their surface features so that a reinfection by the same pathogen is treated as a new infection by the immune system. This gives the pathogen time to replicate before the immune system can respond. Viruses make these changes by having very high mutation rates and being able to recombine genetic material with other strains of the virus to produce new strains, often never encountered by humans. For example, seasonal flu mutates so rapidly that vaccinations only last a year.

Influenzavirus

▸ Influenza (flu) is a disease of the upper respiratory tract caused by the viral genus *Influenzavirus*. Three types of *Influenzavirus* (A, B, and C) affect humans.

▸ The most common and most virulent of these is *Influenzavirus* A, (right). Influenza viruses undergo genetic changes continually, either by antigenic drift (small continual changes) or by antigenic shift (two strains recombining to create a new subtype).

▸ These genetic changes result in changes to the proteins on the viral surface which prevent the human immune system from detecting the virus easily, and allow the virus to reinfect people who may have previously had flu.

▸ Structure of Influenzavirus

▸ Viral strains are identified by the variation in their haemagglutinin (H) and neuraminidase (N) surface antigens. If two different virus subtypes infect a cell they are able to recombine and readily rearrange (reassort) their RNA segments, which alters the protein composition of their H and N glycoprotein spikes. Eight RNA segments (genes) code for the viral proteins.

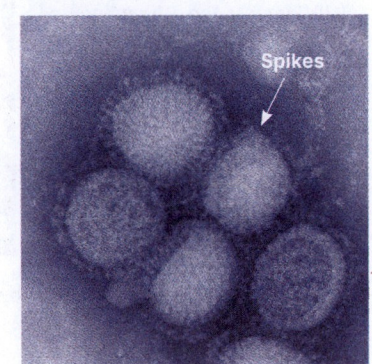

Spikes

Antigenic drifts are small incremental changes (caused by mutations) in the virus that happen continually over time. The changes affect the H and N surface antigens. Accumulated changes result in the immune system not recognizing the virus. As a result, the influenza vaccine, which prepares the immune system for infection, must be adjusted each year to include the most recently circulating influenza viruses.

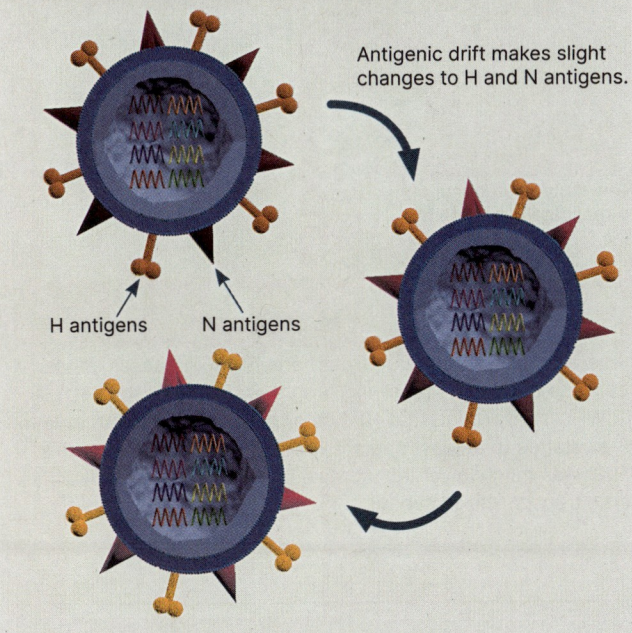

Antigenic drift makes slight changes to H and N antigens.

H antigens   N antigens

Antigenic shift occurs when two or more different viral strains (or different viruses) infect the same cell and recombine to form a new subtype. The changes are large and sudden, and most people lack immunity to the new subtype. New influenza viruses arising from antigenic shift have caused influenza pandemics that have killed millions people over the last century. Influenzavirus A is dangerous to human health because it is capable of antigenic shift.

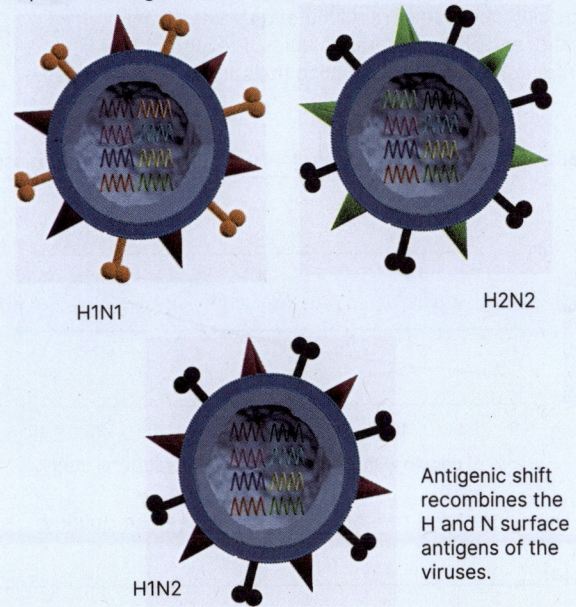

H1N1   H2N2

H1N2

Antigenic shift recombines the H and N surface antigens of the viruses.

1. The Influenzavirus is able to mutate readily and alter the composition of H and N spikes on its surface.

   (a) Why is the virus able to mutate so rapidly? _____

   _____

   _____

   (b) How does this affect the ability of the immune system to recognise and respond to the virus?

   _____

   _____

   _____

A2.3   6   AHL

Human Immunodeficiency Virus (HIV) evolution

▶ HIV mutates rapidly and has a short generation time, allowing it to quickly evolve.

▶ It has one of the highest mutation rates of any known biological system and a generation time of between 1.2 and 2.6 days.

▶ A single, infected cell can produce 1×10^4 new viruses per day. HIV also has a very small genome and because it is a retrovirus there is little error checking during replication by RNA polymerase. These factors contribute to HIV's ability to produce new mutations that avoid the body's immune response and make it difficult to produce an effective vaccine against infection. There is no cure and HIV infection requires a multi-drug treatment to target HIV in several ways.

HIV

Human immunodeficiency virus (HIV) infects the T lymphocytes of the immune system, eventually causing AIDS, a condition that impairs the body's ability to fight disease.

Budding HIV

HIV replicates quickly. Infected cells each produce thousands of copies of HIV per day so, throughout the body, billions of copies are produced daily. HIV shows high genetic variability and mutates frequently. It can also combine its genetic material with other HIV viruses to form new strains. These factors have important consequences for the prevention and treatment of HIV. Any vaccine would quickly become ineffective because the virus changes so rapidly and so many different strains are present. Resistance to drugs used for treatment also arises quickly because of rapid mutation rates and short generation times. Preventing new HIV infections is critical to halting the spread of the disease.

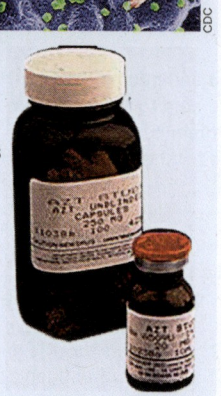

How does HIV evolve?

HIV enters the body.

The body's immune system responds to HIV and produces corresponding antibodies.

HIV uses T lymphocytes to manufacture new viruses.

HIV mutant avoids antibody because its new surface proteins are not recognized by the immune system.

Mutant enters new cell and replicates, producing new mutants.

Anti-retroviral drugs target most HIV particles, but miss some.

New mutants continue to replicate in new cells.

Relationship between NNRTI resistance and anti-retroviral drug use

(graph: y-axis "Prevalence of NNRTI resistance (%)" from 0 to 6; x-axis "% of people with HIV receiving the anti-retroviral drug" from 0 to 40, curve rising from ~0.8 to ~3.2)

In general, resistance to drug treatment in HIV increases as the percentage of patients receiving drug therapy increases (left). A 2012 WHO study found that 10-17% of patients in Western nations had resistance to at least one anti-retroviral drug. In Africa, this number was much lower but increasing. NNRTI = non-nucleoside reverse transcriptase inhibitor

2. Describe three reasons why HIV evolves so quickly:

(a) _____

_____

(b) _____

_____

(c) _____

_____

3. How many generations of HIV per year can occur in an infected person? _____

4. Suggest why a combination drug therapy is used to treat HIV: _____

_____

_____

©2024 **BIOZONE** International
ISBN: 978-1-99-101410-8
Photocopying prohibited

29 Did You Get It?

1. Referring to the cell model on the right, what type of cell is it, and what evidence can you provide to support your claim?

_____

_____

_____

2. What features of the cell would be different if this was a bacterial cell?

_____

_____

3. **AHL:** The endosymbiosis theory is supported by evidence that the mitochondria and chloroplast in eukaryotes resemble bacteria. Elaborate on this evidence and how it relates to the origin of eukaryotic cells:

_____

_____

_____

4. Cell theory was developed in the 19th century but detailed observation of the organelles did not occur until well over a century later. Explain the reason for the delay and how observation at a organelle level was made possible:

_____

_____

_____

5. (a) **AHL:** What does 'A' on the virus model on the right represent?

_____

   (b) **AHL:** This virus is a tobacco plant mosaic virus (TMV). Suggest why this virus does not cause disease in humans?

_____

_____

_____

_____

6. **AHL:** The tobacco plant's ribosomes manufacture viral particles which assemble in the plant. Is this process part of a lysogenic or lytic cycle? Explain the difference in cycles:

_____

_____

_____

7. Reproduction is a life process that occurs in all living organisms, including unicellular organisms such as yeast, *Amoeba*, and *Euglena*. Summarize the steps of typical asexual reproduction in a unicellular organism:

_____

_____

_____

8. **AHL:** Slime moulds incorporate a multicellular stage during reproduction. Explain the advantages of this stage:

_____

_____

 Organisms

A3.1 Diversity of organisms

Guiding Questions:
▶ How do we define a species?
▶ What patterns in genomes are evident within and between species

Activity Number

Learning Outcomes:

☐ 1	Connect classification and taxonomy to the variation seen in all organisms.	30
☐ 2	Use Linnaeus' original definition of a species as a group with shared traits.	30
☐ 3	Use the system of binomial nomenclature to name organisms.	30
☐ 4	Define 'species' using the biological species concept and compare to other competing species definitions.	31
☐ 5	Distinguish between the terms population and species, acknowledging that speciation can result in an arbitrary decision.	31
☐ 6	Examine examples of chromosome number diversity in different species using humans and chimpanzees as an example.	33-35, 38
☐ 7	**AOS:** Identify features seen in a human karyotype to classify chromosomes specifically as evidence for a testable hypothesis on common ancestry with a primate ancestor. **NOS:** Distinguish between testable hypotheses and non-testable statements.	33-35
☐ 8	Explain how variations such as single-nucleotide polymorphisms result in the genetic diversity of organisms.	36, 37
☐ 9	Compare the relative eukaryote genomic diversity between species and within a population of a species.	37, 38
☐ 10	Gather information from a digital database to compare the genomic size of different taxonomic groups with their varying complexity.	37
☐ 11	Investigate a range of current and potential future uses of whole genome sequencing.	39
☐ 12	**AHL:** Identify the constraints of applying the biological species concept to asexually reproducing species.	32
☐ 13	**AHL:** Explain how chromosome number is a shared trait in species and can act as a reproductive isolating mechanism to prevent fertile hybrid offspring from two different species.	38
☐ 14	**AHL: AOS:** Develop a dichotomous key that can be used to identify plant or animal species in a local habitat.	40, 41
☐ 15	**AHL:** Report on how barcodes are used to Identify species from environmental DNA in a habitat.	42

A3.2 Classification and cladistics

Guiding Questions:
▶ How do scientists justify how organisms are classified into related groups?
▶ What are the key differences between cladistics and classical taxonomy for classifying organisms?

Activity Number

Learning Outcomes:

☐ 1	**AHL:** Discuss the importance of taxonomic classification.	43
☐ 2	**AHL:** Describe the constraints of traditional Linnaean taxonomy and therefore why there has been a theoretical paradigm shift to using cladistics for classification.	43
☐ 3	**AHL:** Discuss the advantages of using cladistics to classify organisms based on evolutionary relationships from a common ancestor.	43
☐ 4	**AHL:** Identify the types of evidence used to assign organisms into clades.	43-45
☐ 5	**AHL:** Evaluate the molecular clock theory as a method to estimate clade divergence.	45
☐ 6	**AHL:** Explain how parsimony analysis is used to select a hypothesis to account for cladogram construction due to genetic sequences in genes or amino acids.	44, 45
☐ 7	**AHL:** Use cladograms to deduce evolutionary relationships, identifying the significance of the root, node and terminal branches.	44, 45
☐ 8	**AHL:** Investigate a case study that used cladistics to reclassify a species based on evolutionary relationships. **NOS:** Explain how the emergence of new evidence can falsify current scientific theories and claims.	44
☐ 9	**AHL:** Link the three domain classification system to the use of evidence from rRNA base sequences.	43

30 Variation in Organisms

Key Idea: Organisms can be placed into groups based on their shared characteristics.

All organisms are different from each other. Their individual genetic makeup produces variation even between closely related individuals. However, all organisms also share various characteristics. More closely related organisms will share more characteristics. Taxonomy is the science of grouping organisms based on shared characteristics. Smaller, more precise groups can be grouped together into larger, less precise groups, forming a hierarchy of taxonomic ranks. Organisms can be classified in various ways, e.g. by morphological or genetic traits. Each has its own advantages and disadvantages. In modern classification, distinction of taxa is more often based on evolutionary relationships (phyletics), as found by molecular studies, including genetic and protein analyses. Different ways of classifying organisms can be contradictory and scientists must carefully analyze all information in order to agree on a classification.

Taxonomic ranks

Traditionally, living organisms are classified into a hierarchy of seven main taxonomic ranks (although commonly today the rank of 'domain' is often included above kingdom as the most encompassing group, making eight major taxonomic ranks). The taxonomic ranks in descending order are: domain, kingdom, phylum, class, order, family, genus, species. Other ranks such as subclass, may also be included. There are also some differences between naming conventions depending on the organism, e.g. animal phyla are equivalent to plant divisions.

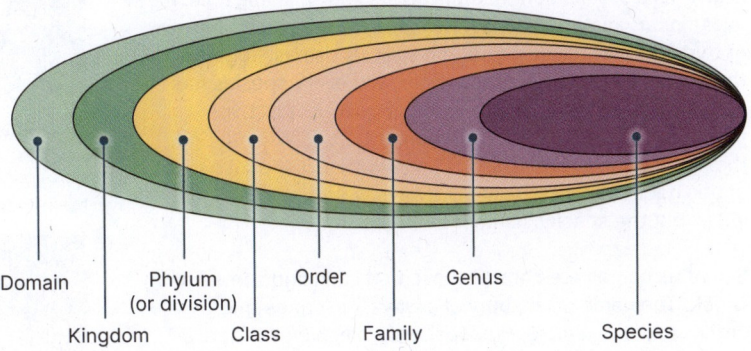

Domain, Kingdom, Phylum (or division), Class, Order, Family, Genus, Species

The example below shows how, as we move through the taxonomic ranks, the organisms we are grouping become more exclusive based on the characteristics of the group. In this case, we are looking at the classification of the grey wolf, *Canis lupus*.

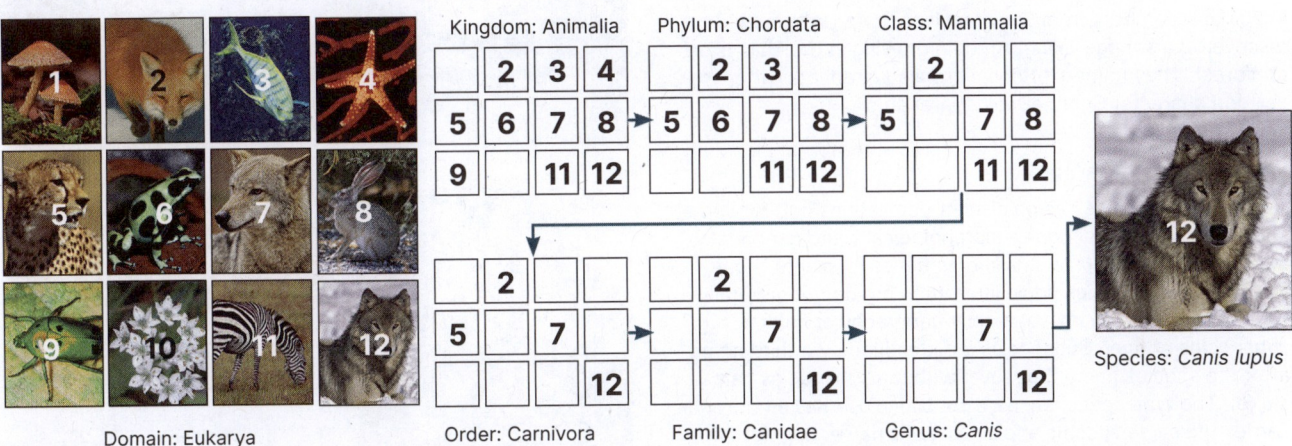

1. The table below shows part of the classification for humans using the eight major levels of classification. For this question, use the example of the classification of the grey wolf above as a guide.

(a) Complete the list of the taxonomic ranks on the left hand side of the table below:

(b) Complete the classification for humans (*Homo sapiens*) on the right hand side of the table below. Use the internet or reference books if you need to:

Taxonomic rank
Human classification

1. _____ _____
2. _____ _____
3. _____ _____
4. _____ _____
5. _____ _____
6. *Family* *Hominidae*
7. _____ _____
8. _____ _____

©2024 **BIOZONE** International
ISBN: 978-1-99-101410-8
Photocopying prohibited

A3.1 1-3

Naming organisms

Many organisms have common names that people use in everyday language. However, common names change from place to place and language to language. Common names may also apply to more than one organism, which causes confusion when people from different parts of the world are referring to particular organisms. For example, the mountain lion, cougar, puma, catamount, and panther are all names given to the same animal, *Puma concolor*. To solve this problem, each species is given a two part (binomial) name, called the scientific name, that is unique to that species. It was pioneered in 1753 by Carl Linnaeus.

Names and meanings

Most species have a common name as well as a scientific name. Common names may change from place to place as people from different areas name species differently based on both language and custom. Scientifically, every species is given a classification that reflects its known lineage, i.e. its evolutionary history. The last two (and most specific) parts of that lineage are the genus and species names. Together, these are called the scientific name, and every species has its own that is specific to it. The term given to this two-part naming system is binomial nomenclature. When typed, the name is always *italicized*. If handwritten, it should be <u>underlined</u>. The genus name is always written with a capital letter, but the species name is not.

Scientific names are normally based on Latin and sometimes Greek. The name often, but not always, describes the organism. For example, *Hippopotamus amphibius* is derived from Greek - hippos meaning horse and potamos meaning river, giving 'river horse'. *Amphibius* meaning two lives - *Hippopotamus amphibius* lives both in the water and out of it.

In 2011, researchers from San Francisco State University discovered a sponge-like mushroom growing in the Malaysian rainforest. They named it *Spongiforma squarepantsii* after the cartoon character SpongeBob SquarePants.

Rangifer tarandus is known as the caribou in North America, but as the reindeer in Europe. The scientific name is unambiguous.

Spongiforma squarepantsii

What are subspecies?

A monotypic species has no distinct populations but, for some species, scientists recognize morphologically and genetically distinct subspecies (in animals) or varieties (in plants, algae, and fungi). Subspecies could interbreed but do not generally do so because they are isolated by geography or habitat, e.g. subspecies of tiger. Subspecies are identified by a third name after the species name (but never with the subspecies name alone). The type specimen, i.e. a specimen used as a reference, carries the same specific and subspecific name. Species with defined subspecies are called polytypic species.

The Bengal tiger
Panthera tigris tigris (India)

The Siberian tiger
Panthera tigris altaica (Russia)

2. (a) What is the two part naming system for classifying organisms called? _____

   (b) What are the two parts of the name? _____

3. Give three reasons why the classification of organisms is important:

   (a) _____

   _____

   (b) _____

   _____

   (c) _____

   _____

4. Which is the type specimen in the tiger examples above? _____

5. Describe what is wrong with the way each of the following scientific names is written:

   (a) Ceratotherium simum: _____

   (b) *Canis Lupus*: _____

31 What is a Species?

Key Idea: A biological species is a group of organisms that can successfully interbreed to produce fertile offspring.

A biological species is defined as a group of individuals capable of interbreeding to produce fertile offspring and reproductively isolated from other such groups. Although simple by definition, species are more difficult to define in reality. For example, some closely related species will interbreed to produce fertile hybrids, e.g. *Canis* species. The concept of a biological species is also difficult to apply to plants, which hybridize easily and can reproduce vegetatively, extinct organisms, and those that reproduce asexually. An alternative to the biological species concept is the phylogenetic species concept, which is based on organisms having a shared evolutionary history.

Distribution of *Canis* species

The global distribution of most species of *Canis* (dogs and wolves) is shown, right. The grey wolf inhabits the forests of North America, northern Europe, and Siberia. The red wolf and Mexican wolf (original distributions shown) were once distributed more widely, but are now extinct in the wild except for reintroductions. In contrast, the coyote has expanded its original range and is now found throughout North and Central America. The range of the three jackal species overlap in the open savannah of eastern Africa. The dingo is distributed throughout Australia.

Distribution of the domestic dog is global as a result of their association with humans. The dog has been able to interbreed with all other members of the genus listed here to form fertile hybrids. Contrast this with members of the horse family, in which hybrids are sterile.

Interbreeding between *Canis* species

The ability of many *Canis* species to interbreed to produce fertile hybrids illustrates one of the problems with the traditional concept of the biological species. Red wolves, grey wolves, Mexican wolves, and coyotes can all form fertile hybrids. Red wolves are very rare, and it is possible that hybridization with coyotes has been a factor in their decline. By contrast, no interbreeding occurs between the three distinct species of jackal, even though their ranges overlap in the Serengeti of eastern Africa. These animals are highly territorial, and simply ignore members of the other jackal species.

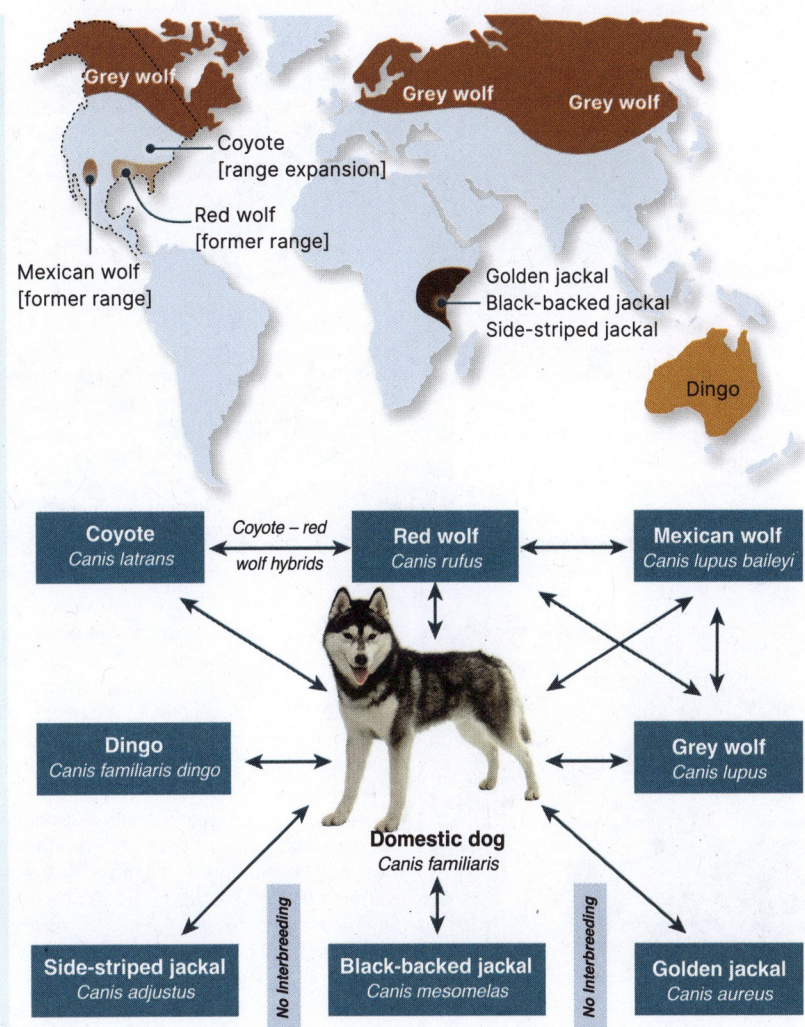

1. (a) Define the term biological species: _____

   _____

   _____

   (b) In what way do the *Canis* species contradict the definition of a biological species?

   _____

   _____

2. What type of barrier prevents the three species of jackal from interbreeding? _____

   _____

3. Describe the factor that has prevented the dingo from interbreeding with other *Canis* species (apart from dogs):

   _____

4. Describe a possible contributing factor to the occurrence of interbreeding between the coyote and red wolf:

   _____

©2024 **BIOZONE** International
ISBN: 978-1-99-101410-8
Photocopying prohibited

A3.1

4, 5

Difficulties in distinguishing species

▸ Greenish warblers (*Phylloscopus trochiloides*) are found in forests across much of northern and central Asia. They inhabit the ring of mountains surrounding the large area of desert which includes the Tibetan Plateau, and Taklamakan and Gobi deserts, and extends into Siberia. In Siberia, two distinct subspecies coexist and do not interbreed but are apparently connected by gene flow around the Himalayas to the south. The greenish warblers may form a rare example of a ring species.

▸ This population of birds may one day split into two separate, genetically isolated groups. In the meantime, they show that species distinctions are somewhat arbitrary and are used to help categorize life, rather than being a true reflection of the world around us.

2 Populations spread both east and west along the Himalayas. Populations developed unique characteristics, but adjacent populations remained able to breed together.

3 East and west populations eventually rejoined in Siberia but, because of morphological, behavioural, and genetic differences, do not interbreed.

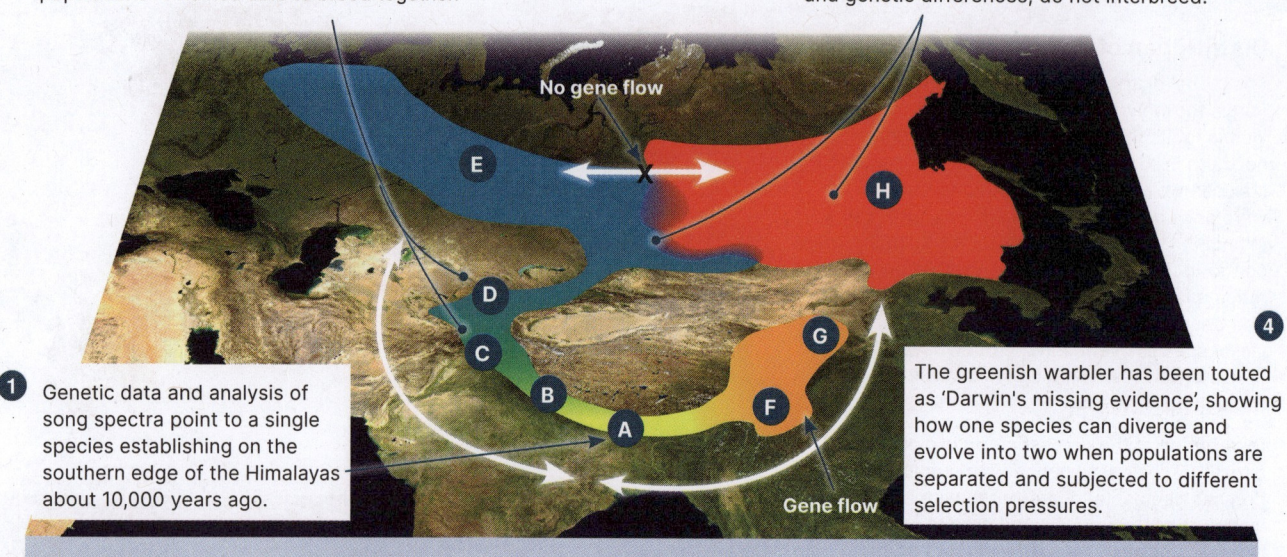

No gene flow

1 Genetic data and analysis of song spectra point to a single species establishing on the southern edge of the Himalayas about 10,000 years ago.

Gene flow

4 The greenish warbler has been touted as 'Darwin's missing evidence', showing how one species can diverge and evolve into two when populations are separated and subjected to different selection pressures.

The two coexisting subspecies of greenish warblers are distinguished by their songs and the number of bars on the wings. The warbler in western Siberia has one light bar across the top of the wing, while the warbler in eastern Siberia has two. Analysis of the songs around the ring show that all songs can be traced to the population labelled A above. Songs become progressively different moving east or west around the ring. The songs of the eastern warblers (H) and western warblers (E) in Siberia are so different that neither recognizes the other. Eastern and western forms have subspecies status.

JM Garg, Wikipedia CC 3.0

5. How do the eastern and western Siberian populations of greenish warblers differ?

_____

_____

6. Explain how these differences might have occurred: _____

_____

_____

_____

7. Greenish warblers are divided into two apparently non interbreeding subspecies. Use the example of the greenish warbler to illustrate the somewhat arbitrary nature of defining a species:

_____

_____

_____

_____

32 Problems in Defining a Species

Key Idea: Applying the biological species concept to extinct, asexually reproducing species can be problematic.

Although the biological species concept works for many sexually reproducing species, it is difficult to use in the case of asexually reproducing species in which there is no interbreeding. It is also difficult to usefully apply to extinct species and species for which breeding data is unknown. In some cases, populations of organisms may look very different, but are, in fact, part of a single gene pool due to specific groups interbreeding. Examples of this are ring species, such as the *Ensatina* salamander in California's Central Valley.

Asexually reproducing species

▶ Microorganisms, including bacteria, make up the majority of asexually reproducing species, but there are also examples of plants and, more rarely, animals. These include the desert grassland whiptail lizard, and the orchid *Corunastylis*.

▶ Many microorganisms, including bacteria and yeast, reproduce by binary fission or budding. In addition, bacteria can engage in horizontal gene transfer, i.e. transferring genes between individuals (rather than from parent to offspring, called vertical gene transfer).

▶ The bacteria *E. coli* has thousands of strains. The model strain called K12MG1655 has about 4400 genes. But a comparison with 20 other strains showed just 2000 of those genes were common to all strains out of a total of about 18,000 observed genes.

Extinct species

▶ Extinct species, as their title suggests, are no longer alive. Therefore, they have no interbreeding populations. Extinct species, such as *Tyrannosaurus rex* or *Mammuthus primigenius* (woolly mammoth) are normally defined by morphology(the shape of teeth and bones etc.) However, even these species are uncertain. Mammoth specimens have been found that appear to be hybrids of what were thought to be separate species.

▶ Ancient human ancestors also pose problems, with much debate around the distinction between species. Modern humans carry small amounts of Neanderthal DNA, showing that the two species were capable of interbreeding.

Species that easily hybridize

▶ In general, most species that we define as a species do not interbreed. They look and behave in different ways and remain as separate populations.

▶ Occasionally, however, they hybridize. Plants are well known to do this. Many important crop plants are hybrids, including wheat, potatoes, and apples.

▶ Many birds, particularly ducks, are known to easily form hybrids. Mallard duck hybrids are common. For example, in New Zealand and Australia, hybrids between the Pacific black duck (grey duck) (*Anas superciliosa*) and the mallard duck (*Anas platyrhynchos*) are common, and have increased *A. superciliosa*'s population decline.

▶ In Europe, the carrion crow (in the west) and hooded crow (in the east) often produce hybrids along their zone of species overlap. These hybrids can easily breed with either species.

M. jeffersonii (above) may be a hybrid between *M. columbi* and *M. primigenius*.

Mallard-Pacific black duck hybrid

Bacteria with gene giving resistance to antibiotic.

Plasmid is transferred via a sex pilus between the bacteria.

Plasmid

Concept	Criterion
Biological species	A group of interbreeding or potentially interbreeding organisms or populations that are reproductively isolated.
Morphological species	A group or populations of organisms that have morphological features statistically similar to each other and different from other groups.
Phylogenetic	A group whose members share an evolutionary history (descended from a common ancestor), diagnosable by having shared derived characteristics.
Ecological	A group having a unique ecological niche that is distinct from all others in the ecosystem.

1. Using examples, explain why using one species concept, e.g. the biological species concept, is often inadequate or inappropriate when defining species:

_____

_____

_____

_____

_____

©2024 **BIOZONE** International
ISBN: 978-1-99-101410-8
Photocopying prohibited

A3.1 12 AHL

33 Karyotypes

Key Idea: The karyotype is the number and appearance of chromosomes in the nucleus of a eukaryotic cell. The karyotype can be pictured in a standard format, called a karyogram, in which the chromosomes are ordered by size. Different species of organisms have different numbers or structures of chromosomes. This is called the karyotype. The karyotype is displayed as a karyogram, a standardized layout of the chromosomes. Karyograms are able to show if a cell is diploid (2n) (having pairs of chromosomes), triploid (3n) or aneuploid (2n + 1) and allow comparison between species. The number of chromosomes a species has can range from two to dozens. In humans, the male karyotype has 44 autosomes (non-sex chromosomes), and an X and Y chromosome (44 + XY). The female karyotype has 44 autosomes and two X chromosomes (44 + XX). Karyograms can be prepared from cells in the metaphase stage of mitosis.

Preparing a karyotype

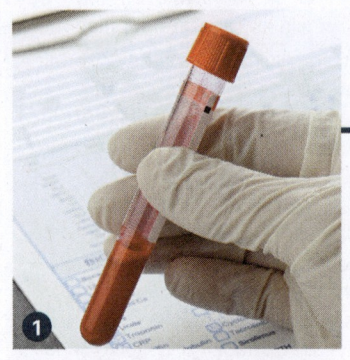

The sample is centrifuged and the lymphocytes (a type of white blood cell) are removed and induced to divide (mitosis).

They are grown for several days in culture and then treated to halt the cycle at the metaphase stage.

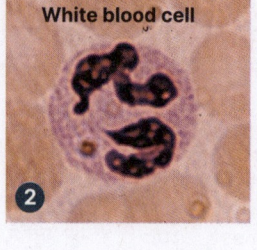

White blood cell

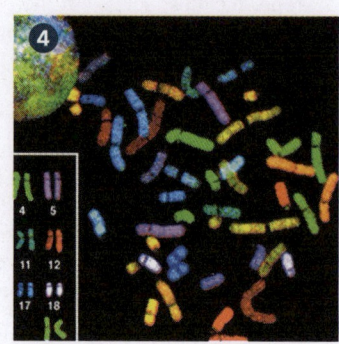

① A sample of cells is taken from the person of interest. This may be from the amniotic fluid surrounding a foetus or from a blood sample from an adult or child.

③ A drop of the cell suspension in preservative is spread on a microscope slide, dried and stained with a dye that causes a banding pattern to appear on each chromosome.

The chromosomes are viewed under a microscope and photographed. Newer techniques use fluorescent probes to colour-code chromosomes and provide a spectral karyogram.

⑤ The photograph is cut up (manually or by computer) so that each chromosome is separate from the others. The chromosomes are then arranged into homologous pairs according to size, shape, and banding pattern.

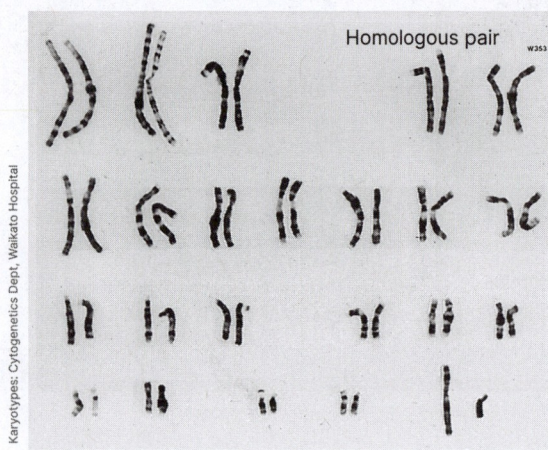

Homologous pair

Karyotypes: Cytogenetics Dept, Waikato Hospital

Conventional karyogram (male): 44 + XY

Close up reveals two chromatids

PLOS cc 2.5

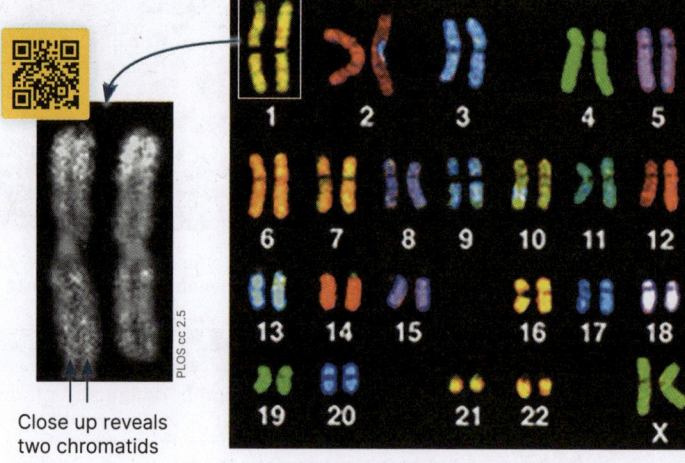

Spectral karyogram (female): 44 + XX

⑥ The karyotype is characteristic of a species. The karyograms below show some of the karyotype diversity of plants and animals. Notice the difference in number and shape of the chromosomes.

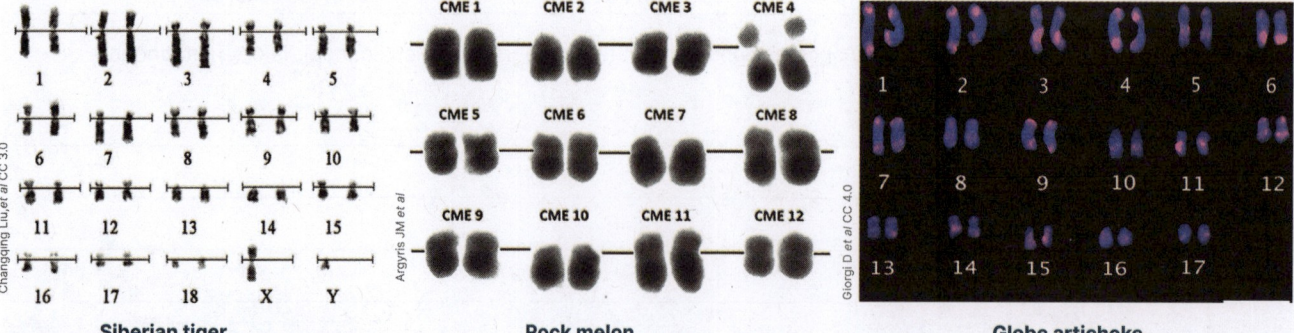

Changqing Liu, et al / CC 3.0

Siberian tiger

Argyris JM et al

Rock melon

Giorgi D et al / CC 4.0

Globe artichoke

A3.1
6, 7

Typical layout of a human karyogram

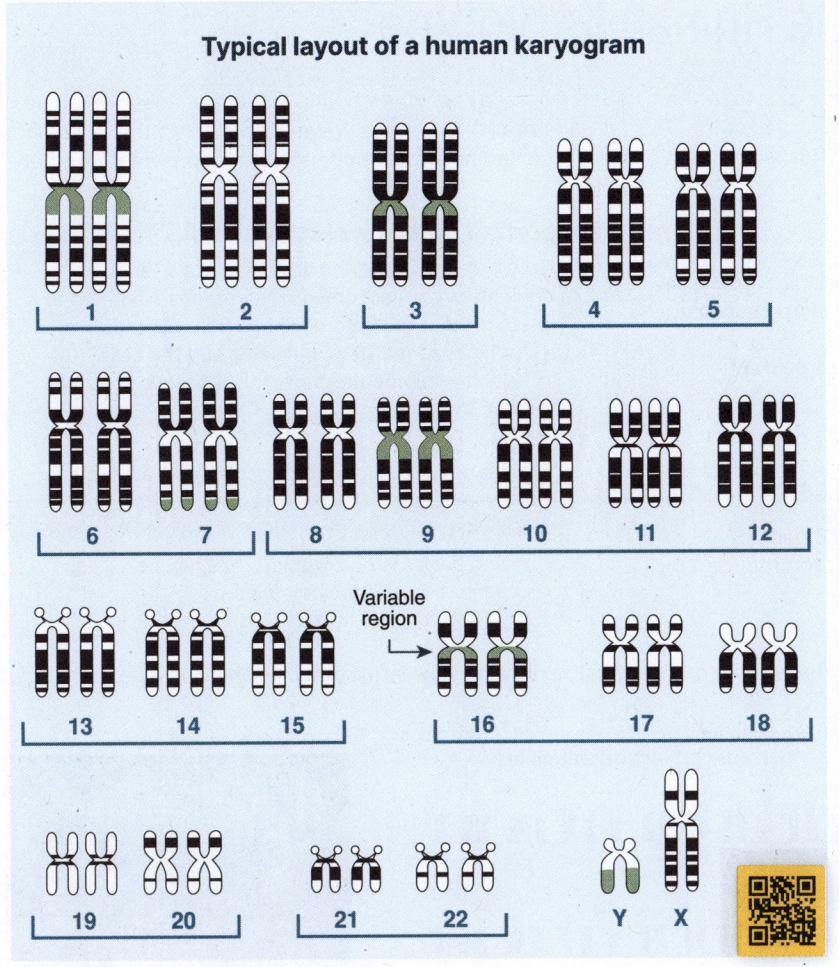

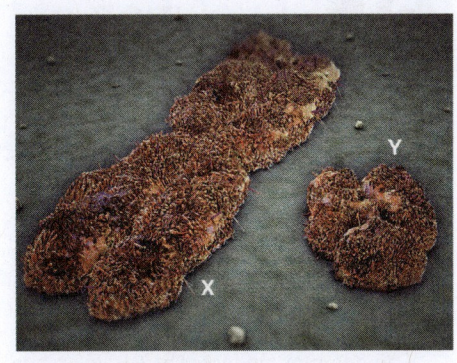

SEM showing human X and Y chromosomes. Although these two are the sex chromosomes, they are not homologous.

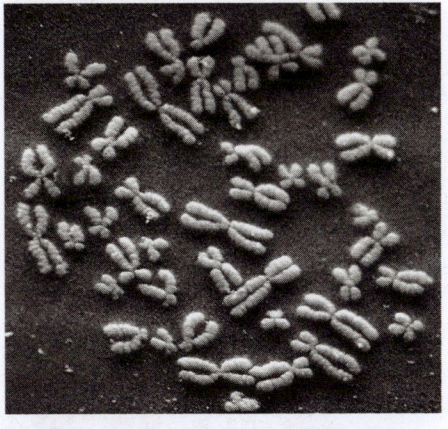

A scanning electron micrograph (SEM) of human chromosomes clearly showing their double chromatids.

1. (a) What is a karyogram? _____

    _____

    (b) What information can it provide? _____

    _____

2. On the male karyogram on the previous page, number each homologous pair of chromosomes. Use the karyogram above as a guide.

3. Circle the sex chromosomes on the female and male karyograms on the previous page.

4. Write down the number of autosomes and the arrangement of sex chromosomes for each sex:

    (a) Female: No. of autosomes: _____   Sex chromosomes: _____

    (b) Male: No. of autosomes: _____   Sex chromosomes: _____

5. State how many chromosomes are found in a:

    (a) Normal human (somatic) body cell: _____   (b) Normal human sperm or egg cell: _____

6. What features of the chromosomes allow them to be paired up for a karyogram? _____

    _____

    _____

7. Why are the X and Y chromosomes not homologous? _____

    _____

8. Write down the diploid (2n) number of chromosomes for:

    (a) Siberian tiger: _____   (b) Rock melon: _____   (c) Artichoke: _____

34 Evolution of the Human Karyotype

Key Idea: Humans have one less chromosome pair than the rest of the extant hominidae.
Humans have 23 pairs of chromosomes. This sets them apart from the rest of the hominidae (great apes) which have 24 pairs of chromosomes. Investigations into this difference have revealed how this happened and approximately when.

The problem of chromosome 2

Humans have 46 chromosomes (23 pairs). Our closest living relatives, the chimpanzees, the gorillas, and the orang-utans, have 48 chromosomes (24 pairs). This could have occurred because:

1. The common ancestor of hominids had 24 pairs of chromosomes and humans lost a pair.

2. The common ancestor of hominids had 24 pairs of chromosomes and two separate chromosomes fused to produce 23 pairs in humans but not in apes.

3. The common ancestor of hominids had 23 pairs of chromosomes and one chromosome split to produce 24 pairs in apes but not in humans.

Important notes on chromosomes:

▸ Telomeres are sections of DNA that are found at the end of chromosomes. Their purpose appears to be to protect the chromosomes from degradation. Over time, as the DNA replicates and the cells divide, the telomeres shorten. It is believed this is related to aging.

▸ Centromeres are sections of DNA where the chromatids are held together after replication but before separation. They are usually, but certainly not always, towards the middle of the chromosome.

▸ Banding patterns on chromosomes can be compared to see if they match up.

To find out, researchers studied the structure in hominid chromosomes. The digram below left shows a comparison of extant (living) hominid chromosomes:

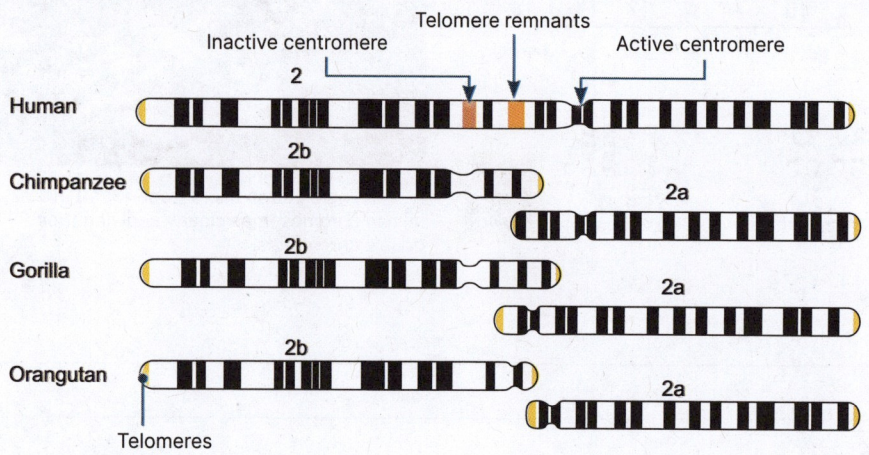

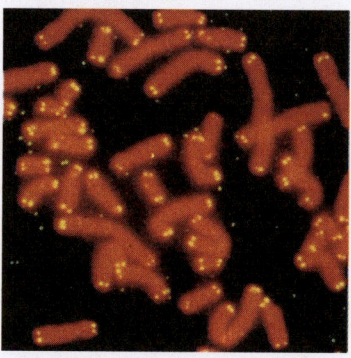

Telomeres, in yellow, at the ends of chromosomes.

▸ Further studies of the chromosome 2 shows it appeared around 1 million years ago, sometime before the split between Neanderthals and modern humans.

1. Which of the three scenarios above is the most likely to have happened to produce chromosome 2?

_____

2. Use the evidence shown to justify your answer to 1 above: _____

_____

_____

_____

_____

_____

3. A student hypothesizes that chromosome 2 split to produce two chromosomes in all hominids other than humans and their direct ancestors. What evidence would need to be seen in the chromosomes to confirm this?

_____

_____

_____

_____

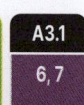

©2024 **BIOZONE** International
ISBN: 978-1-99-101410-8
Photocopying prohibited

35 Making a Karyogram

Key Idea: The distinguishing features of chromosomes can be used to produce a karyogram.

Each chromosome has specific distinguishing features. Chromosomes are stained using a special technique that gives them a banded appearance in which the banding pattern represents regions containing up to many hundreds of genes. Cut out the chromosomes below and arrange them on the record sheet. Use this to determine the sex of the individual. The karyograms presented on the previous pages and the information below on how to recognize chromosome pairs can be used to help you complete this activity.

Distinguishing characteristics of chromosomes

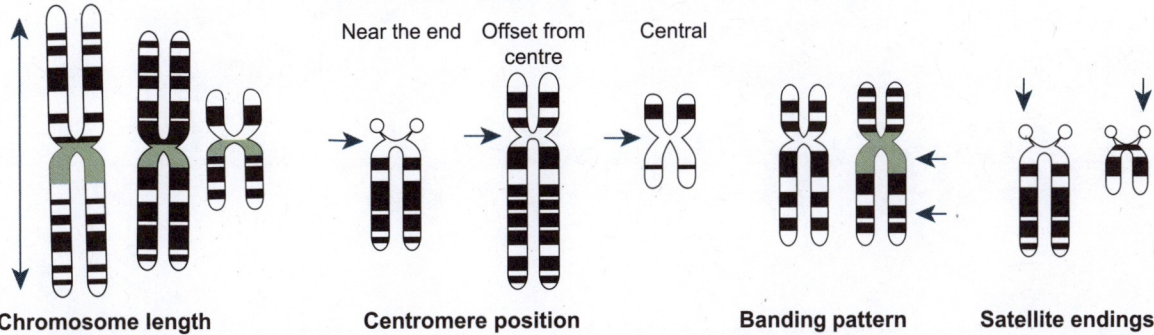

Near the end
Offset from centre
Central
Chromosome length
Centromere position
Banding pattern
Satellite endings

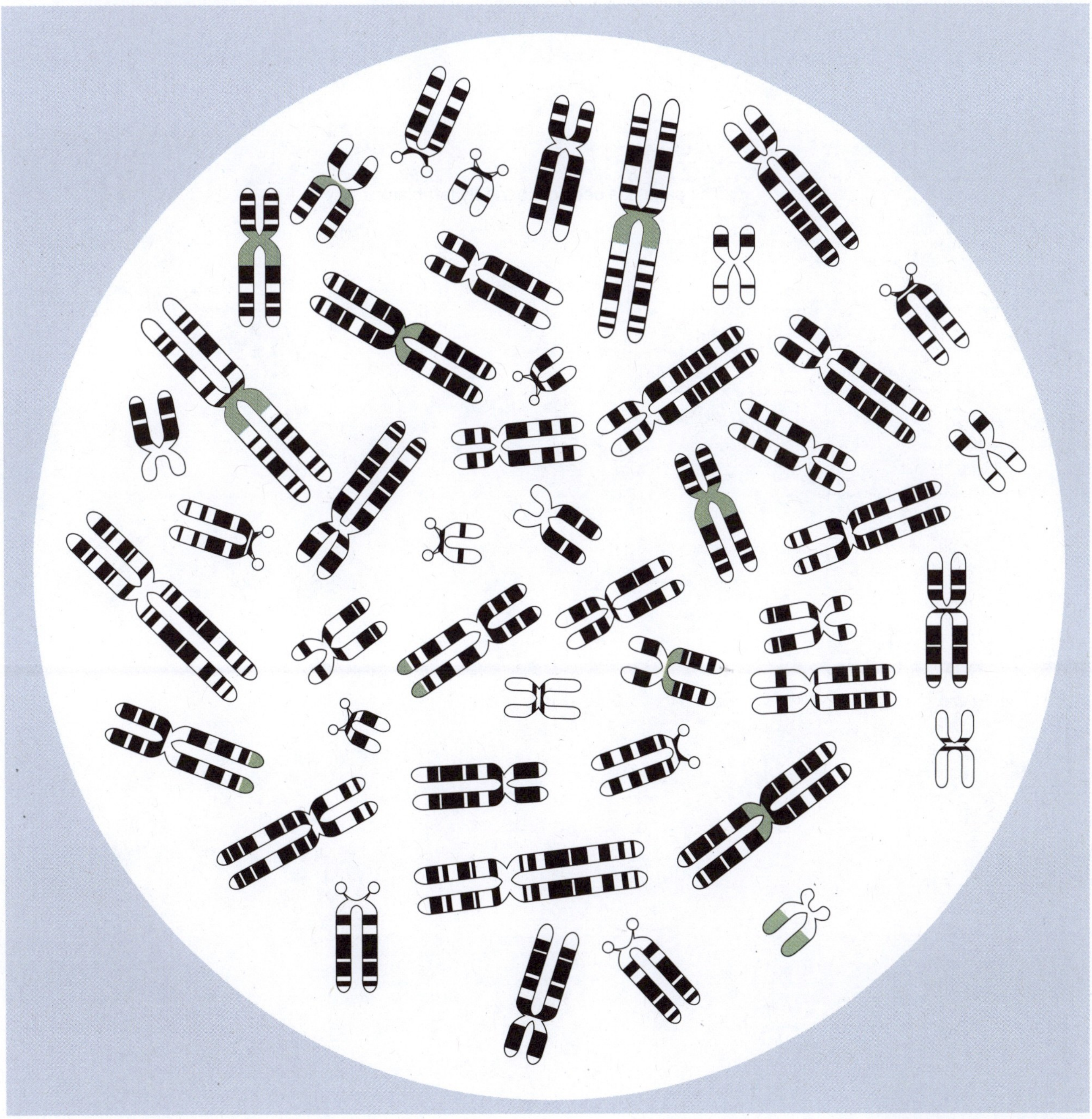

©2024 **BIOZONE** International
ISBN: 978-1-99-101410-8
Photocopying prohibited

The page has been deliberately left blank

1. Cut out the chromosomes on page 55 and arrange them in homologus pairs on the record sheet below:

2. Is the sex of this individual male or female? _____

3. Is the person diploid, triploid or another condition? _____

1	**2**	**3**	**4**	**5**

6	**7**	**8**	**9**	**10**	**11**	**12**

6 **7** **8** **9** **10** **11** **12**

13 **14** **15** **16** **17** **18**

19 **20** **21** **22** **Sex chromosomes**

36 Diversity in Genomes

Key Idea: A genome is the entire haploid amount of genetic material, including all the genes, of a cell or organism. There is variation in genomes both within and between species.

The genome refers to all the genetic material in one haploid set of chromosomes. It contains all of the information the organism needs to function and reproduce. Every cell in an individual has a complete copy of the genome. Within the genome are sections of DNA, called genes, which code for proteins. Variants of these genes are called alleles. This variation accounts for some of the variation between individuals of a species, but some also comes from random changes in the DNA sequence, or variation outside the genes. The extent of genetic variation between species depends on how closely related the species are, with closely related species sharing more DNA sequences and genes than distantly related species.

Variation within species

▸ Genetic variation within species is normally very small. Larger populations tend to have more variation than smaller ones.

▸ The human genome is about 3 billion base pairs long (3 Gbp). Any two random people will share 99.9% of that genome. All the variation seen in the human species comes from just 0.1% of our DNA.

▸ This variation comes from differences in DNA sequences such as having different alleles for genes, e.g. alleles for blue eyes vs brown eyes, or differences in single base pairs called single nucleotide polymorphisms (SNPs). SNPs are a change in a single base pair in the DNA, e.g. from C to T. These changes often have no effect on how the DNA functions. In fact, any two people will have about 3 million (0.001% of DNA) SNP differences between them. Compare this to our closest living relative, the chimpanzee, which shares about 98.8% of our DNA.

SNPs and variation

A single nucleotide polymorphism is a substitution of a single base nucleotide at a specific position in the DNA. For example, the two DNA sequences shown are identical except for one SNP.

GTCGTAATGAA   GTCGTACTGAA

CACCATTACTT   CACCATGACTT

The polymorphism may occur in any place in the genome (but more often outside of genes). If they occur at a specific location at a significant frequency in the population (usually more than 1%), then the SNPs can be considered alleles (different versions of the DNA sequence).

Certain SNPs may occur more often in one population than another, or will only be found in a very specific group. This becomes very useful in researching ancestry or medical issues. For example, it may be possible to link people with a certain set of SNPs with a response to drugs to treat disease, helping doctors to decide on a better course of treatment.

Using SNPs to study species variation

SNPs can be used to study variation within species. There are over 100 million SNPs spread throughout the human genome and there are numerous ways to identify them, such as DNA sequencing. These can be studied to see if they are associated with diseases, groups of people, or to study ancestry.

1 Scientists analyze sections of DNA to find the SNPs in that section.

SNP

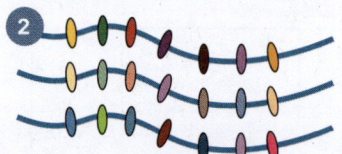

2 The SNPs are found arranged in combinations called haplotypes. Different haplotypes can be associated with different groups of people.

% haplotypes in population

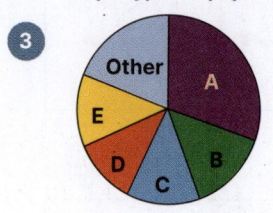

3 Every person has a haplotype profile and these can be placed into groups.

4	
A	Poor
B	Good
C	Fair
D	None
E	Very good

4 The haplotypes can be analyzed based on the researchers' aims. They may be analyzing for a response to certain medicines, or for ancestry. In the example, left, the different haplotypes are analyzed for their response to a specific antibiotic.

1. Define the following terms:

   (a) Genome: _____

   (b) SNP: _____

   (c) Allele: _____

2. Explain how genomic variation can be used to study variation within a species:

   _____

   _____

   _____

   _____

A3.1
8, 9

320

©2024 **BIOZONE** International
ISBN: 978-1-99-101410-8

37 Genome Size

Key Idea: A genome is an organism's complete set of genetic material, including all of its genes. The genomes of many organisms have been sequenced, allowing genes to be compared. These can be searched for on gene databases.

Genome size is often expressed in megabase pairs (Mbp) and can vary in up to 8 orders of magnitude between species. Genome size is also sometimes expressed as a C-value, which is the amount of haploid DNA per cell, in picograms. One picogram is roughly equal to 1 gigabase pair (Gbp). Thousands of different species have now had their genomes sequenced. Many species' genomes are now entered into online databases that can be searched by anyone wishing to find out information about a particular gene. The size of a genome does not reflect the complexity of an organism.

The location and size of the genome varies between organisms

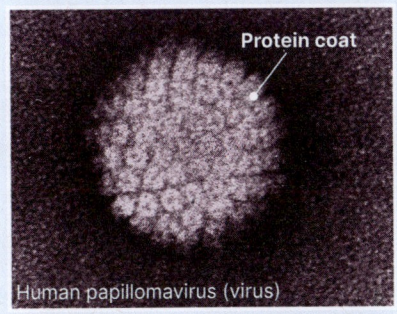
Human papillomavirus (virus)
Protein coat

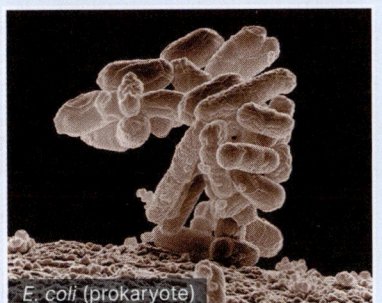
E. coli (prokaryote)

Mouse (eukaryote)

Genome size — Small →→→ Large

Number of genes — Few →→→ Many

- The viral genome is contained within the virus's outer protein coat. Viral genomes are typically small and highly variable. They can consist of single stranded or double stranded DNA or RNA and contain only a small number of genes.

- The human papillomavirus (HPV) genome is a double stranded circular DNA molecule ~8000 bp long.

- In prokaryotes, most of the DNA is located within a single circular chromosome, making them haploid, i.e. one allele, for most genes. Many bacteria also have small accessory chromosomes called plasmids, which carry genes for special functions such as antibiotic resistance.

- The *E. coli* genome is a circular chromosome about 4.6 Mbp long.

- In eukaryotes, most of the DNA is located inside the cell's nucleus, as linear chromosomes. Most eukaryotes are diploid, i.e. two sets of chromosomes (2n). A small amount is found in chloroplasts (in plants) and in mitochondria.

- The mouse genome is ~3 Gbp long in 20 chromosomes (2n = 40). This genome size is similar to humans.

Using databases to find genome sizes

Various internet databases can be used to search for genomes. Three useful sites are found in the **BIOZONE Resource Hub**:

- **Fungal Genome Size Database**: As its name suggests, this focuses on fungal genomes. Although this site has many fields to fill in to refine your search, any of the phylum to species fields can be used. The results are given in picograms and megabase pairs.

- **Kew Gardens Plant DNA C-values Database**: Kew Gardens is a famous botanical garden and research centre in London, England. This site lets you choose if you want to see the search result in picograms or megabase pairs.

- **Animal Genome Size Database**: This database has a simple layout, but holds data on over six thousand genomes. It allows you to manually click down through the taxonomic levels to find your organism. You can also enter the common name of the organism rather than having to use the scientific name. The results are in picograms.

1. Use the databases above to find the genome sizes of the following organisms (in either Mbp or picagrams):

   (a) *Paris japonica* (Japanese canopy plant): _____

   (b) *Oryza sativa* (Rice): _____

   (c) *Saccharomyces cerevisiae* (Yeast): _____

   (d) *Drosophila melanogaster* (Fruit fly): _____

   (e) *Protopterus aethiopicus* (Marbled lungfish): _____

   (f) *Canis lupus* (Grey wolf): _____

A3.1
9, 10

38 Chromosomes and Species

Key Idea: Chromosome number is a shared trait within a species. Every species has its own chromosome number and characteristics.

In most cases, every individual in a species has the same number of chromosomes as every other individual. Also, a haploid (n) set of these chromosomes will be able to pair up with any other haploid set from any other individual from that species during reproduction. This is one of the important traits of a species. The fact that chromosomes must pair up in a cell prevents most species from interbreeding with other species, as mitosis and meiosis (the division of the chromosomes during cell division) will often fail if they do not. Occasionally, species with different chromosome numbers will be able to interbreed and produce offspring, but these offspring are usually sterile and cannot reproduce, e.g. the pairing of a horse and a donkey to produce a sterile mule.

Chromosome numbers for different species

The number of chromosomes between species is extremely variable. Even closely related species can have large differences in the number of chromosomes. For example, in the genus *Bos*, cattle and yaks have 60 chromosomes, whereas gaur have 56.

Organism	Chromosome number (2n)
Vertebrates	
cat	38
rat	42
rabbit	44
human	46
chimpanzee	48
gorilla	48
cattle	60
dog	78
turkey	82
goldfish	94
Invertebrates	
horse roundworm	2
fruit fly *Drosophila*	8
housefly	12
honey bee	32 or 16
Hydra	32
Plants	
broad bean	12
cabbage	18
garden pea	14
rice	24
Ponderosa pine	24
orange	18, 27, or 36
potato	48

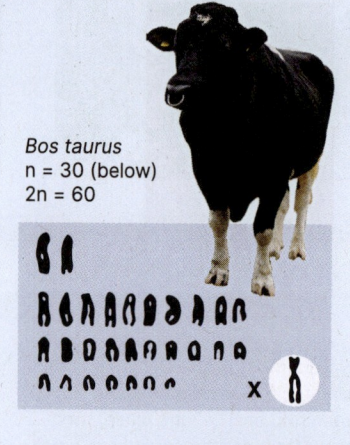

Bos taurus
n = 30 (below)
2n = 60

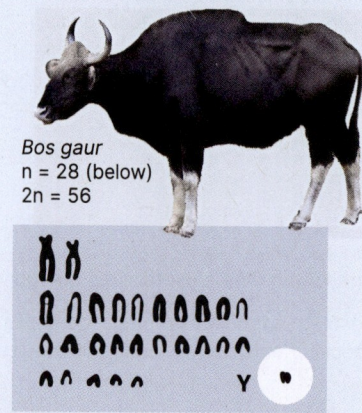

Bos gaur
n = 28 (below)
2n = 56

Hybrids and species

Recall that some species easily form hybrids with other species, e.g. the mallard duck and grey duck, and the carrion and hooded crows. These hybrids are able to interbreed with others of those species. In other cases, species that do interbreed produce sterile hybrids, e.g. mules.

Why is this? Partly, it is because of the inadequate definitions of species. Hooded crows and carrion crows live in different parts of Europe, they behave differently, they only rarely interbreed along an overlapping zone, they are and have been labelled as different species. Yet they are genetically almost identical, with less than 1% difference in their genome.

Donkeys and horses, on the other hand, are genetically very different, each having a different number of chromosomes (64 and 62). Their DNA is also somewhat different, with about 15% of horse genes absent from the donkey genome and about 10% of donkey genes being absent from the horse genome.

Complexity and chromosomes

Every eukaryotic species has at least two chromosomes. The number of chromosomes does not reflect how genetically 'complex' a species is. For example, the largest number of chromosomes known in a eukaryote is found in a primitive plant called the adder's tongue fern, which has 631 pairs. Compare this with a gorilla (a complex organism), which has far fewer chromosomes (24 pairs of chromosomes).

1. (a) What is the difference in the diploid chromosome number between humans and chimpanzees? _____

   (b) Which animal has the minimum number of chromosomes possible? _____

   (c) Which plant has the same number of chromosomes as a gorilla? _____

2. Why is the number of chromosomes in an organism not a reflection of its genetic complexity?

   _____

   _____

   _____

   _____

3. In what way do chromosomes help define a species? _____

   _____

   _____

AHL
A3.1
13

©2024 **BIOZONE** International
ISBN: 978-1-99-101410-8
Photocopying prohibited

39 Using Whole Genome Sequencing

Key Idea: Sequencing whole genomes has uses in medical and evolutionary sciences.

As previously seen in Activity 36, knowing the diversity in a genome can be extremely useful. Many species have now had their entire genomes sequenced and this has provided much new information about how genes have evolved and how they function. In 1990, the Human Genome Project commenced. Its goal was not only to sequence the human genome, but to develop better technology for doing so. In 2003, the first copy of the human genome was produced (about 90% complete). It took 13 years and cost US$2.7 billion. Today, a human genome can be sequenced in less than a day and costs a few thousand dollars. This is likely to become faster and cheaper in the future.

How is genome information used?

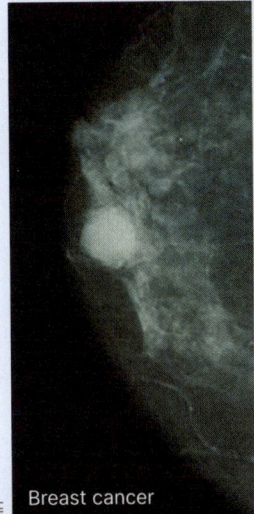
Breast cancer

Identifying and treating disease: By comparing the genomes of healthy individuals to those with a particular disease, it is possible to see if a particular variation is associated with the disease. The information can be used to predict disease risk and to take measures to reduce risk, where possible.

Researchers have developed tests to look for specific diseases and design new medicines to treat disease. For example, 20% of breast cancers are caused by the HER-2 mutation. Herceptin was developed to specifically treat women with the HER-2 mutation and increase patient survival.

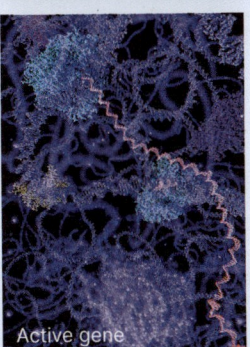
Active gene

Identifying genes and functionality: Before the Human Genome Project, the genome was thought to encode ~100,000 genes. We now know that number is much smaller (~20,000). Projects such as ENCODE identify functional elements of the DNA sequence, helping us to understand how genes (and metabolic processes) are regulated. The knowledge could be used to switch off undesirable genes.

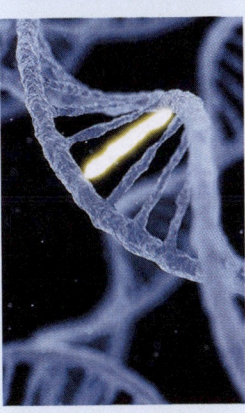

Evolution and species comparisons: The DNA sequences for particular genes between species can be compared to learn about their evolutionary history. Species with fewer differences are more closely related than species with more differences. Researchers can determine the effects of human genes by studying the same gene in other species. For example, studying the genes for DNA repair in rodents has provided information about cancer development in humans.

Cost and speed of genome sequencing over time

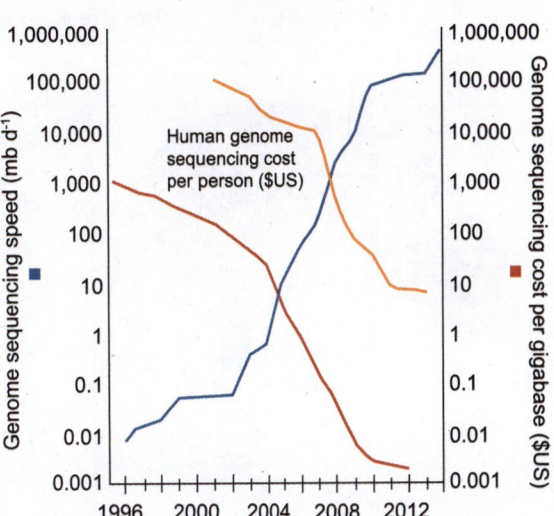

- The speed of genome sequencing has increased as technology has developed. The first DNA sequencing methods could only sequence short lengths of DNA at a time and required bulky equipment and the sequence to be manually read from an electrophoresis gel. Modern sequencers can be very small and can read DNA sequences thousands of bases long automatically.

- The Oxford Nanopore Minion (below) is about the size of a small cellphone and plugs into a computer via USB. The DNA strand is passed through a nanopore in a membrane and the electrical signal produced by each base as it passes through is interpreted by the computer to produce the DNA sequence.

Oxford Nanopore Technologies

1. Compare the speed and cost of genome sequencing over time: _____

_____

_____

2. Use examples to explain the possible uses of genome sequencing: _____

_____

_____

A3.1
11

40 Classification Keys

Key Idea: Classification keys are used to identify an organism based on its distinguishing features and assign it to a species. An organism's classification should include a clear, unambiguous description, an accurate diagram, and its unique name, denoted by the genus and species. Classification keys are used to identify an organism and assign it to the correct species (assuming that the organism has already been formally classified and is included in the key). Typically, keys involve a series of linked steps. At each step, a choice is made between two features (dichotomous key). Each alternative leads to another question until an identification is made. If the organism cannot be identified, it may be a new species or the key may need revision. This activity describes two examples of dichotomous keys. The first describes features for identifying the larvae of genera within the order Trichoptera (caddisflies). The second is a key to the identification of aquatic insect orders.

Caddisfly larvae

Classification key for caddisfly larvae

The key shown here is a simplified version of one commonly used to identify caddisfly larvae. It identifies the organisms to genus level only. To use the key to identify the larvae pictured below, start at the top and branch at each feature until you reach the bottom.

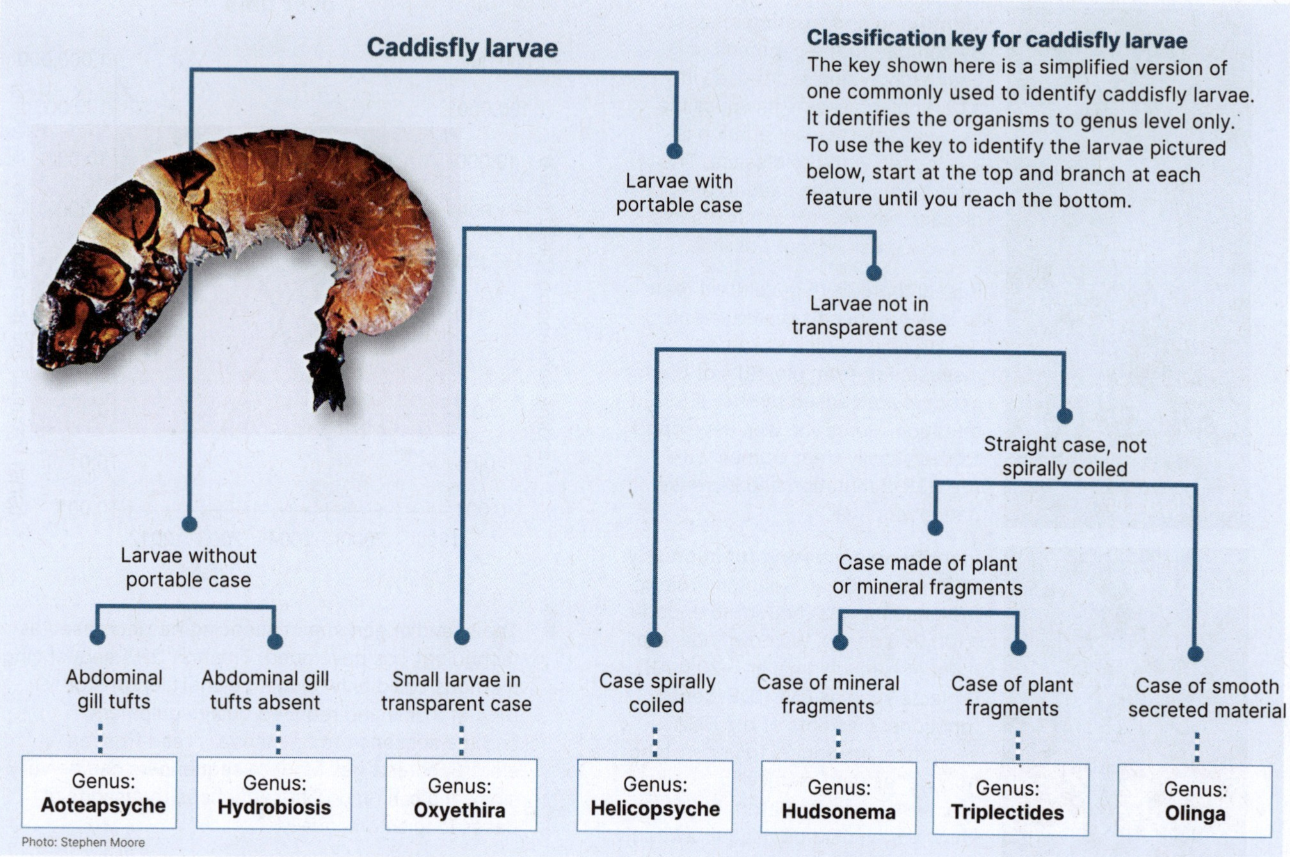

Larvae with portable case

Larvae not in transparent case

Straight case, not spirally coiled

Case made of plant or mineral fragments

Larvae without portable case

Abdominal gill tufts	Abdominal gill tufts absent	Small larvae in transparent case	Case spirally coiled	Case of mineral fragments	Case of plant fragments	Case of smooth secreted material
Genus: **Aoteapsyche**	Genus: **Hydrobiosis**	Genus: **Oxyethira**	Genus: **Helicopsyche**	Genus: **Hudsonema**	Genus: **Triplectides**	Genus: **Olinga**

Photo: Stephen Moore

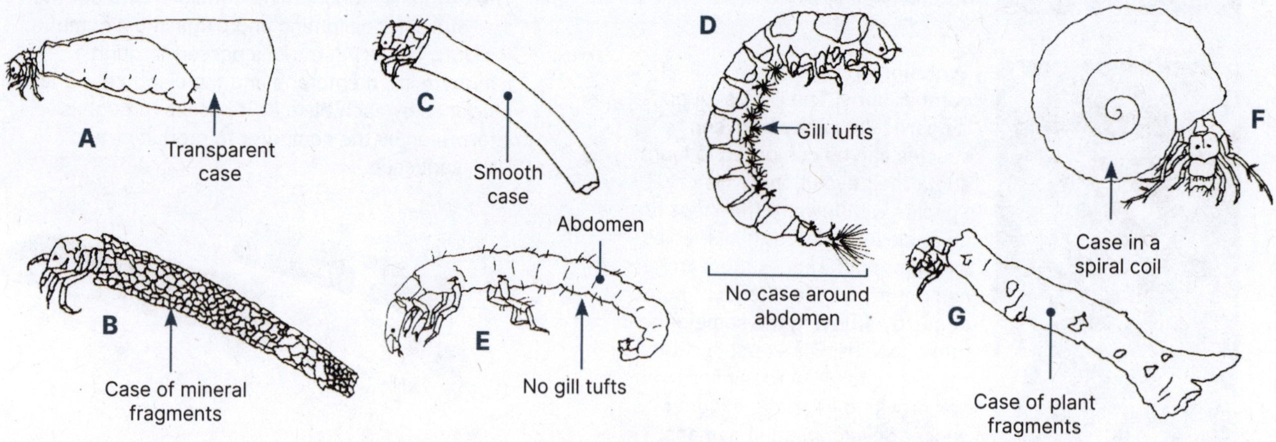

A — Transparent case

B — Case of mineral fragments

C — Smooth case, Abdomen

D — Gill tufts, No case around abdomen

E — No gill tufts

F — Case in a spiral coil

G — Case of plant fragments

1. Describe the main feature used to distinguish the genera in the key above: _____
   _____

2. Use the key above to assign each of the caddisfly larvae (A-G) to its correct genus:

   A: _____ D: _____ G: _____

   B: _____ E: _____

   C: _____ F: _____

©2024 **BIOZONE** International
ISBN: 978-1-99-101410-8
Photocopying prohibited

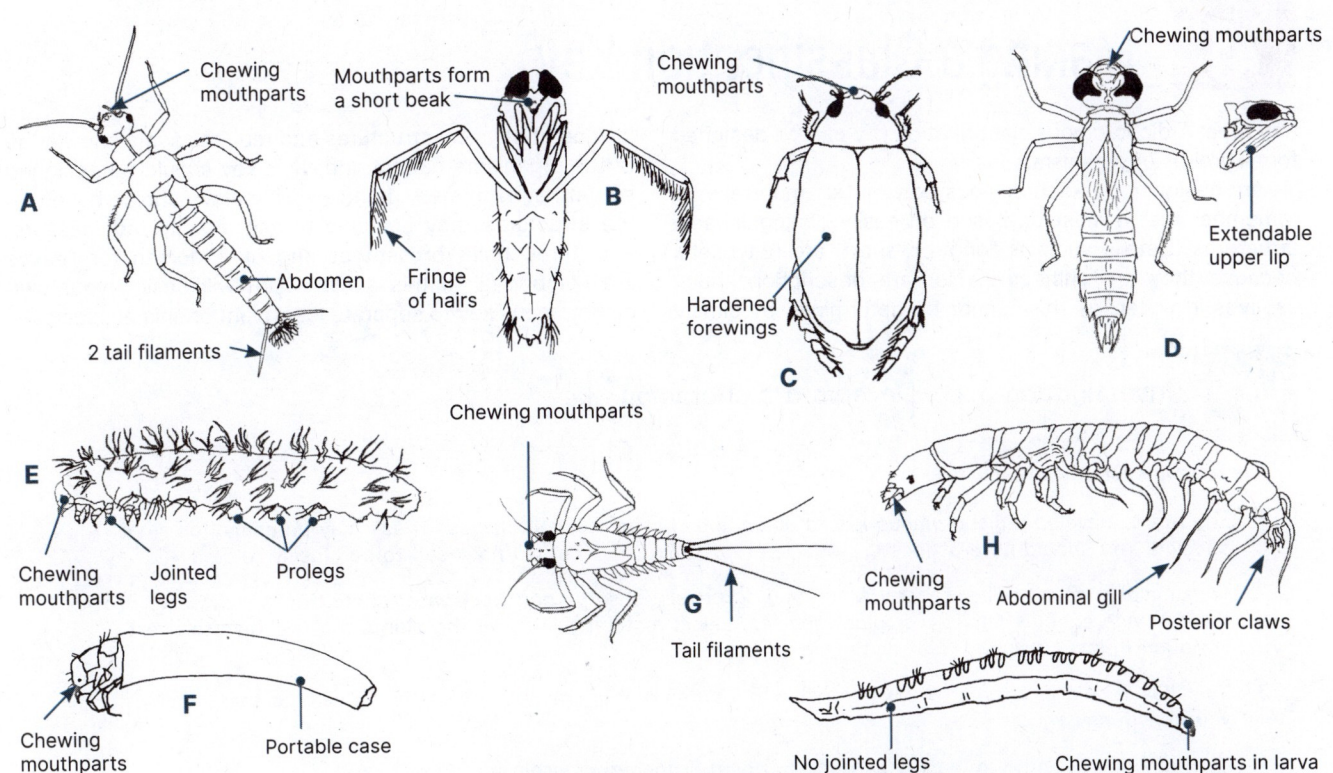

A — Chewing mouthparts, Abdomen, 2 tail filaments

B — Mouthparts form a short beak, Fringe of hairs

C — Chewing mouthparts, Hardened forewings

D — Chewing mouthparts, Extendable upper lip

E — Chewing mouthparts, Jointed legs, Prolegs

F — Chewing mouthparts, Portable case

G — Chewing mouthparts, Tail filaments

H — Chewing mouthparts, Abdominal gill, Posterior claws

No jointed legs, Chewing mouthparts in larva

3. Use the simplified key to identify each of the orders (by order or common name) of aquatic insects (A-I) pictured above:

(a) Order of insect A:

_____

(b) Order of insect B:

_____

(c) Order of insect C:

_____

(d) Order of insect D:

_____

(e) Order of insect E:

_____

(f) Order of insect F:

_____

(g) Order of insect G:

_____

(h) Order of insect H:

_____

(i) Order of insect I:

_____

Key to Orders of Aquatic Insects

#	Description	Order
1	Insects with chewing mouthparts; forewings are hardened and meet along the midline of the body when at rest (they may cover the entire abdomen or be reduced in length).	**Coleoptera** (beetles)
	Mouthparts piercing or sucking and form a pointed cone. With chewing mouthparts, but without hardened forewings.	*Go to 2* *Go to 3*
2	Mouthparts form a short, pointed beak; legs fringed for swimming or long and spaced for suspension on water.	**Hemiptera** (bugs)
	Mouthparts do not form a beak; legs (if present) not fringed or long, or spaced apart.	*Go to 3*
3	Prominent upper lip (labium) extendable, forming a food capturing structure longer than the head.	**Odonata** (dragonflies & damselflies)
	Without a prominent, extendable labium	*Go to 4*
4	Abdomen terminating in three tail filaments which may be long and thin, or with fringes of hairs.	**Ephemeroptera** (mayflies)
	Without three tail filaments	*Go to 5*
5	Abdomen terminating in two tail filaments	**Plecoptera** (stoneflies)
	Without long tail filaments	*Go to 6*
6	With three pairs of jointed legs on thorax	*Go to 7*
	Without jointed, thoracic legs (although non-segmented prolegs or false legs may be present).	**Diptera** (true flies)
7	Abdomen with pairs of non-segmented prolegs bearing rows of fine hooks.	**Lepidoptera** (moths and butterflies)
	Without pairs of abdominal prolegs	*Go to 8*
8	With eight pairs of finger-like abdominal gills; abdomen with two pairs of posterior claws.	**Megaloptera** (dobsonflies)
	Either, without paired, abdominal gills, or, if such gills are present, without posterior claws.	*Go to 9*
9	Abdomen with a pair of posterior prolegs bearing claws with subsidiary hooks; sometimes a portable case.	**Trichoptera** (caddisflies)

41 Making a Classification Key

Key Idea: A dichotomous classification key can be designed for all groups of organisms.

When designing a dichotomous key, it is important to remember that it must be based on easily distinguishable structures. Details such as 'long' or 'short' are not useful because they are subjective. Nor are descriptions such as 'lives in water,' as this cannot be determined by simply looking at physical structures and requires some knowledge of the organism's habitat. Ideally, a key should be designed so that, at each step, the user can make a choice based on the structures they are able to see, count, and measure, e.g. 'three hairs present at end of abdomen', or 'leaves are compound'. In this activity, you will design your own dichotomous key to separate local plant or animal species.

Investigation 3.1 Develop a dichotomous key

See appendix for equipment list.

1. You are to develop a dichotomous key to distinguish between local species. These may be animal or plant species. You should have at least 10 species to distinguish. They do not need to be closely related.

2. For animal species, select several which you can easily observe and work with; photographs may serve provided they provide enough details. For plant species, select matching parts of the plants that will give the best variation in structure, e.g. leaf or flower structure.

3. Your key should be usable by someone else to identify any of the species in the list without them having to ask you questions.

4. Your key can be either a list (as on page 63) or a branched diagram (as on page 62).

5. Use the space below to produce your key:

©2024 **BIOZONE** International
ISBN: 978-1-99-101410-8

42 DNA Barcodes

Key Idea: DNA in the environment can be used to identify the organisms living there, even if they can't be found.

The advent of rapid genome sequencing and the analysis of biological data using computers (bioinformatics) has allowed scientists to identify species based on sequences of DNA. Each species is given a DNA barcode based on short sections of conserved DNA that are unique to each species.

DNA can be shed into the environment by an organism, e.g. in droppings, or from skin. This can be collected from water or soil for some time after the organism is gone. Scientists can identify the presence of the organism via its DNA barcode without ever seeing the organism. For example, this method can be used to monitor diseases such as Covid-19 in a population without having to test every person.

The DNA barcode

DNA barcoding uses short, highly conserved sequences of DNA to produce species-specific information. The sequence of DNA chosen for analysis depends on the type of organism, e.g. plant, animal, fungus, bacterium.

Plant barcoding uses up to three genes in the chloroplasts and one in the nuclear DNA.

The barcode from animals is taken from the cytochrome c oxidase 1 gene (CO1 gene).

Fungal barcoding uses the nuclear ribosomal internal transcribed spacer (ITS).

The goal of DNA barcoding is to enable identification of individual species from short sequences of DNA. This information can then be applied wherever species-specific knowledge is important. Applications include evolutionary biology, conservation, detection of invasive species, dietary analysis (to help describe food webs), and food safety.

Barcoding and environmental sampling

DNA barcoding works on the assumption that each species' DNA is different and can be used to identify a specific species. Closely related species will have similar DNA barcodes.

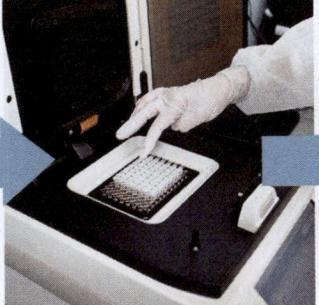
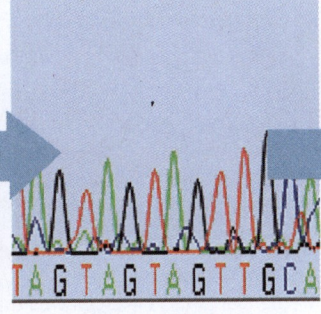
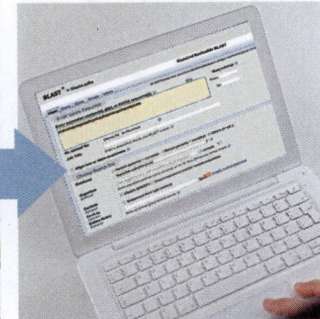

Collect DNA from environment.

Isolate DNA sequences in sample.

Sequence the DNA found

Compare sequence with database to identify species present.

1. What is DNA barcoding? _____

_____

_____

_____

2. Explain how DNA barcoding provides a useful way to:

   (a) Monitor species that are difficult to sample directly: _____

   _____

   (b) Monitor disease in a population: _____

   _____

   (c) Identify the range of species in an ecosystem: _____

   _____

A3.1
15
AHL

43 Classifying Organisms

Key Idea: The classification of biodiversity into groups, or taxa, is constantly being updated in light of new information. Various classification systems exist, each based on different methods and with different advantages and disadvantages. The classification of Earth's biodiversity into formal groups is called taxonomy. As with all science, taxonomic approaches are constantly changing as new information is discovered.

Early classification systems were based on physical appearance. The increasing use of molecular analysis as a taxonomic tool has since led to the reclassification of many taxa, including birds, reptiles, many plants, and primates. Recognizing three domains of life, based on genetic and rRNA analyses, is an example of this. The use of molecular tools has provided new insights into how we group organisms.

Changing classifications

▶ When Linnaeus first published his classification system, he divided life into two kingdoms: plants and animals.

▶ Since then, more groups have been added to reflect our growing understanding of the relationships of species.

▶ Today, people are aware of five or six kingdoms of life. These classifications first appeared around the 1970s and, although still relevant, have been superseded by newer groups and meanings.

▶ In 1990, analysis of rRNA (ribosomal RNA) showed that life actually fitted into three domains: The Bacteria, Archaea, and Eukarya, which sit above the kingdoms in any classification scheme. This was a major shift in classification thinking and is now the dominant theory.

▶ However, these systems continue to be developed and revised.

Whittaker 1969 **Five kingdoms**	Woese *et al.* 1977 **Six kingdoms**	Woese *et al.* 1990 **Three domains**
Monera	Eubacteria	Bacteria
Monera	Archaebacteria	Archaea
Protista	Protista	Eukarya
Fungi	Fungi	Eukarya
Plantae	Plantae	Eukarya
Animalia	Animalia	Eukarya

The three domains

The three domain system is based on differences in rRNA (ribosomal RNA), the cell's lipid membrane structure, and sensitivity to antibiotics. Differences in rRNA are particularly important. Because rRNA plays a critical role in protein synthesis, its structure is highly conserved, thus any changes will be rare and highlight major evolutionary divisions.

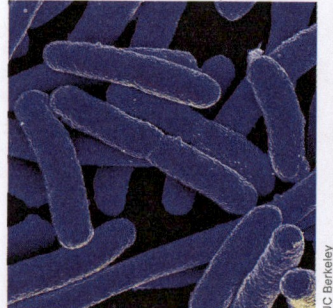
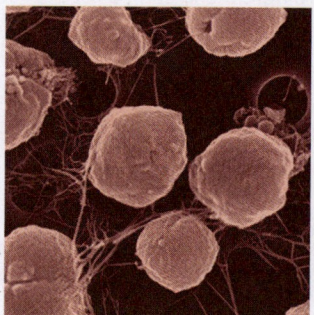

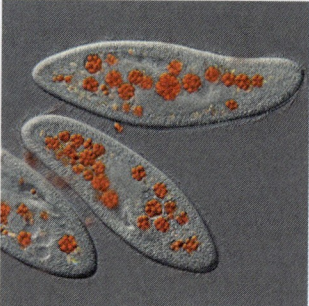

Domain Bacteria

Lack a distinct nucleus and cell organelles. Present in most of Earth's habitats and vital to its ecology. Includes well-known pathogens, many harmless and beneficial species, and the cyanobacteria (photosynthetic bacteria containing the pigments chlorophyll a and phycocyanin).

Domain Archaea

Methanococcus jannaschii was the first archaean genome to be sequenced. The sequencing identified many genes unique to Archaea and provided strong evidence for three evolutionary lineages. Although archaeans may resemble bacteria, they possess several metabolic pathways that are more similar to eukaryotes. Other aspects of their structure and metabolism, such as their membrane lipids and respiratory pathways, are unique. Although once regarded as organisms of extreme environments, such as volcanic springs, archaeans are now known to be widespread, including in the ocean and soil.

Domain Eukarya

Complex cell structure with organelles and nucleus. The three domain classification recognizes the diversity and different evolutionary paths of the unicellular eukaryotes (formerly Protista), which have little in common with each other. The fungi, animals, and plants form the remaining lineages.

1. Using the example of the three domains system, explain how the use of molecular data has led to a more accurate representation of the diversity of life on Earth:

_____

_____

_____

_____

2. What is the purpose of a classification system? _____

_____

©2024 **BIOZONE** International
ISBN: 978-1-99-101410-8
Photocopying prohibited

67

Traditional classification

Based on the Linnaean system (after Linnaeaus), organisms are grouped into taxonomic ranks (levels) on the basis of similarities in physical features (morphology). The scheme is hierarchical. Each taxonomic rank progressively 'sorts' the organisms until the final rank, the species, which includes organisms of just one type. A group within a taxonomic rank is called a taxon (pl. taxa) and is defined by a type, which is often a specimen. Species are named using binomial nomenclature by genus and species (italicized). One difficulty with traditional taxonomy is that ranks are not equivalent for different types of organisms. What's more, unrelated species can be grouped together simply because they look alike. Historically, this resulted in many newly discovered organisms in the New World being misclassified into Old World taxa.

Phylogenetic classification

Phylogenetic classification ties names to clades, so it is often called cladistic. A clade is a taxonomic group that consists of an ancestor and all its descendants (it reflects the evolutionary relationships of the organisms and is monophyletic). The characteristics used for assigning organisms to a clade can be morphological or molecular (DNA or proteins). Molecular data is useful because species that appear similar can be easily distinguished. Phylogenetic classification schemes do not rely on taxonomic rank in the same way as traditional schemes. Using cladistics, many of the taxa with which we are familiar (such as reptiles) do not exist because they do not meet the classification criteria (they are not monophyletic). However, some familiar taxa, e.g. rodents, are clades because the taxon consists of a common ancestor and all its descendants. The rodent clade (below) corresponds to the order Rodentia.

Taxonomic rank	Taxon
Kingdom	Animalia
Phylum	Chordata
Class	Mammalia
Order	Rodentia
Family	Muridae
Genus	*Pseudomys* (Australian mice)
Species	*P. australis* (Plains mouse)

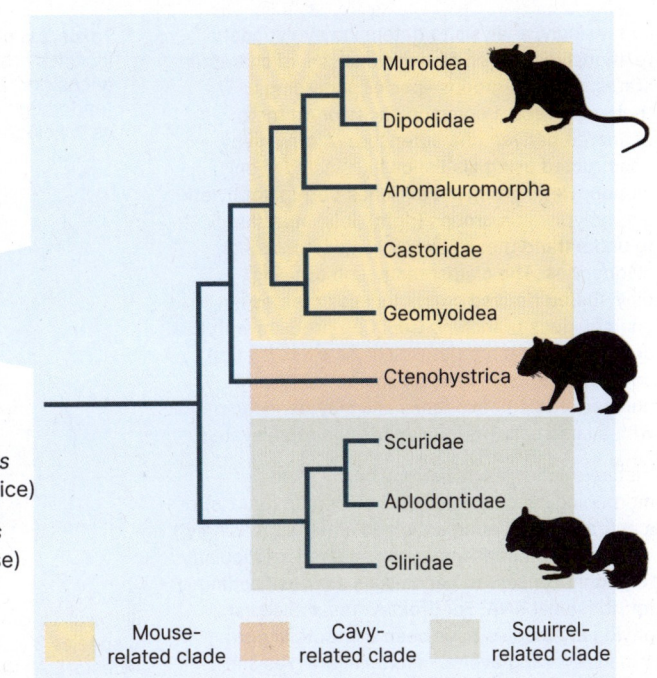

3. Why do classification systems change over time? _____

_____

4. Describe some advantages and disadvantages of:

   (a) Traditional classification systems: _____

   _____

   _____

   _____

   (b) Cladistic classification systems: _____

   _____

   _____

   _____

5. Why do some taxonomic groups from traditional classifications match those of cladistics while others do not?

_____

_____

_____

_____

_____

Determining phylogenetic relationships

Cladistic analysis is one method by which we can construct an evolutionary history for a taxonomic group. Phylogenies constructed using traditional and cladistic methods do not necessarily conflict, but cladistics' emphasis on molecular data has led to reclassification of a number of taxa (including primates and many plants). In natural classifications, all members of the group have descended from a common ancestor (they are monophyletic). Molecular evidence has shown that many traditional groups, e.g. reptiles and figworts, are not descended from a common ancestor, but are paraphyletic or even polyphyletic (see below). Popular classifications will probably continue to reflect similarities and differences in appearance, rather than a strict evolutionary history. In this respect, they are a compromise between phylogeny and the need for a convenient filing system for species diversity.

Crocodiles and birds are traditionally placed in separate classes, but they are in fact more closely related than crocodiles and lizards, both traditionally classified as reptiles.

Increasingly, analysis to determine evolutionary relationships rely on cladistic analysis of character states. Cladism groups species according to their most recent common ancestor on the basis of shared derived characteristics. A phylogeny constructed using cladistics thus includes only monophyletic groups. It excludes both paraphyletic and polyphyletic groups (right). It is important to understand these terms when constructing cladograms. The cladist restriction to using only shared derived characteristics creates an unambiguous branching tree. One problem with this approach is that a strictly cladistic classification could theoretically have an impractically large number of taxonomic ranks and may be incompatible with a Linnaean (traditional classification) system.

Cladistic schemes have traditionally used morphological characteristics, with gain (or loss) of a character indicating a derived state. Increasingly, molecular comparisons are being used, particularly for highly conserved genes such as those coding for ribosomal RNA. For prokaryotes, molecular phylogeny studies have been the most important tool in revealing evolutionary relationships and revolutionizing traditional classification schemes.

Taxon 2 is polyphyletic as it includes organisms with different ancestors. The group 'warm-blooded (endothermic) animals' is polyphyletic as it includes birds and mammals.

Taxon 3 is paraphyletic. It includes species A without including all of A's descendants. The traditional grouping of reptiles is paraphyletic because it does not include birds.

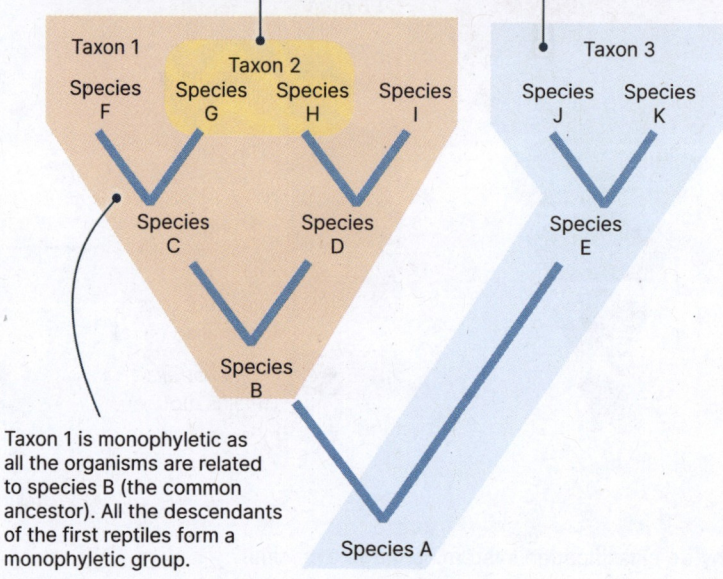

Taxon 1 is monophyletic as all the organisms are related to species B (the common ancestor). All the descendants of the first reptiles form a monophyletic group.

6. What is the meaning of each of the following groups:

(a) Monophyletic: _____

_____

(b) Polyphyletic: _____

_____

(c) Paraphyletic: _____

_____

7. (a) Why are lizards, snakes, turtles, tortoises, and crocodiles all traditionally classed as reptiles?

_____

_____

_____

(b) Explain why this traditional classification of reptiles is incompatible with cladistic classification:

_____

_____

_____

©2024 **BIOZONE** International
ISBN: 978-1-99-101410-8
Photocopying prohibited

44 Cladograms and Phylogenetic Trees

Key Idea: There are many ways to construct phylogenetic trees (evolutionary histories). Cladistics is a method based on shared derived characteristics.

Phylogenetic systematics is a science in which the fields of taxonomy (the study of naming organisms) and phylogenetics (the study of evolutionary history) overlap. Traditional methods for establishing phylogenetic trees have emphasized the physical (morphological) similarities between organisms in order to group species into genera and other higher level taxa. In contrast, cladistics is a method that relies on shared derived characteristics and ignores features that are not the result of shared ancestry. A cladogram is a phylogenetic tree constructed using cladistics. Although cladistics has traditionally relied on morphological data, molecular data, e.g. DNA sequences, are increasingly being used to construct cladograms.

Derived vs ancestral characters

When constructing cladograms, shared derived characters are used to separate the clades (branches on the tree). Using ancestral characters (those that arise in a species that is ancestral to more than one group), would result in distantly related organisms being grouped together and would not help to determine the evolutionary relationships within a clade. Whether or not a character is derived, depends on the taxonomic level being considered. For example, a backbone is an ancestral character for mammals, but a derived character for vertebrates. Production of milk is a derived character shared by all mammals but no other taxa.

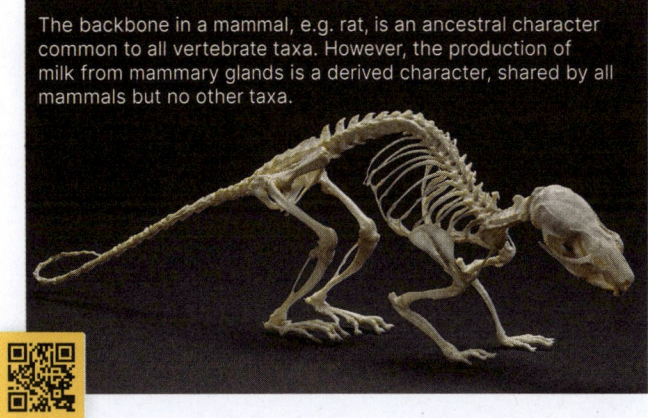
The backbone in a mammal, e.g. rat, is an ancestral character common to all vertebrate taxa. However, the production of milk from mammary glands is a derived character, shared by all mammals but no other taxa.

Constructing a simple cladogram

A table listing the features for comparison allows us to identify where we should make branches in the cladogram. An outgroup (one which is known to have no or little relationship to the other organisms) is used as a basis for comparison.

Comparative features \ Taxa	Jawless fish (outgroup)	Bony fish	Amphibians	Lizards	Birds	Mammals
Vertebral column	✔	✔	✔	✔	✔	✔
Jaws	✘	✔	✔	✔	✔	✔
Four supporting limbs	✘	✘	✔	✔	✔	✔
Amniotic egg	✘	✘	✘	✔	✔	✔
Diapsid skull	✘	✘	✘	✔	✔	✘
Feathers	✘	✘	✘	✘	✔	✘
Hair	✘	✘	✘	✘	✘	✔

▶ The table above lists features shared by selected taxa. The outgroup (jawless fish) shares just one feature (vertebral column), so it gives a reference for comparison and the first branch of the cladogram (tree).

▶ As the number of taxa in the table increases, the number of possible trees that could be drawn increases exponentially. To determine the most likely relationships, the rule of parsimony is used. This assumes that the tree with the least number of evolutionary events is most likely to show the correct evolutionary relationship.

▶ Two possible cladograms are shown on the right. Cladogram 1 requires six events while cladogram 2 requires seven. Applying the rule of parsimony, cladogram 1 must be taken as correct.

▶ Parsimony can lead to some confusion. Some evolutionary events have occurred multiple times. For example, the four chambered heart evolved separately in both birds and mammals. The use of fossil evidence and DNA analysis can help to solve problems like this.

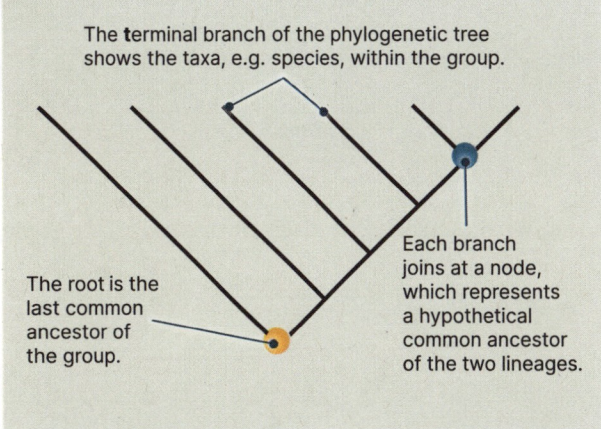
The **t**erminal branch of the phylogenetic tree shows the taxa, e.g. species, within the group.

The root is the last common ancestor of the group.

Each branch joins at a node, which represents a hypothetical common ancestor of the two lineages.

Possible cladograms of vertebrate relationships

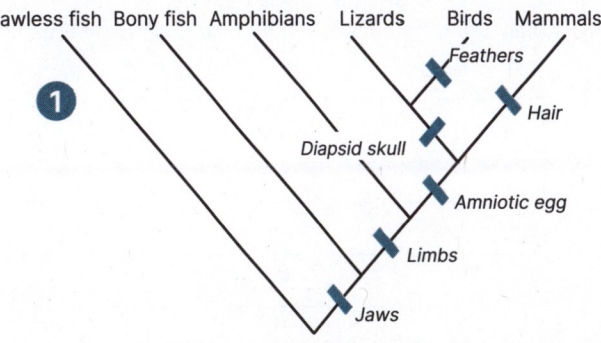

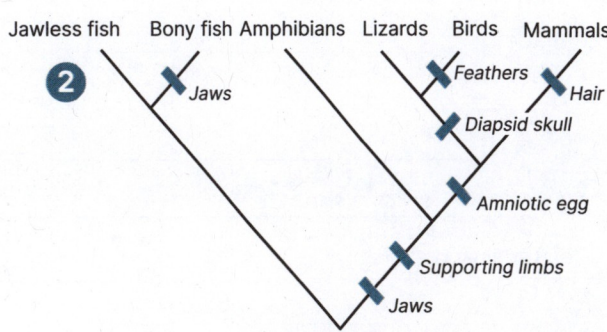

©2024 **BIOZONE** International
ISBN: 978-1-99-101410-8
Photocopying prohibited

Using DNA data

▶ DNA analysis has allowed scientists to confirm many phylogenies and refute or redraw others. In a similar way to morphological differences, DNA sequences can be tabulated and analyzed. The ancestry of whales has been in debate since Darwin. The radically different morphologies of whales and other mammals makes it difficult to work out the correct phylogenetic tree. However, recently discovered fossilized ankle bones, as well as DNA studies, show that whales are more closely related to hippopotamuses than to any other mammal. Coupled with molecular clocks, DNA data can also give the time between each split in the lineage.

▶ The DNA sequences below show part of a nucleotide subset and some of the matching nucleotides used to draw the cladogram. Although whales were once thought most closely related to pigs, based on the DNA analysis the most parsimonious tree disputes this.

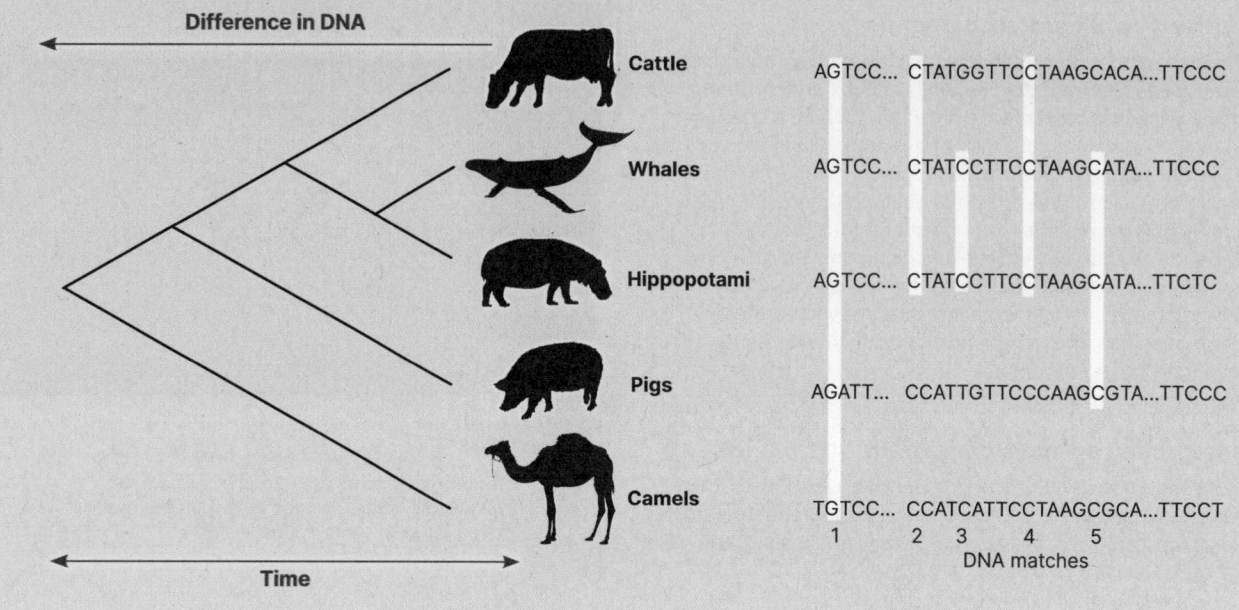

1. (a) What assumption is made when applying the rule of parsimony in constructing a cladogram?

_____

_____

_____

(b) When might this assumption produce an incorrect lineage? _____

_____

2. Explain the difference between a shared characteristic and a shared derived characteristic:

_____

_____

_____

_____

_____

3. Define each of the following terms in relation to phylogenetic trees:

(a) Root: _____

_____

(b) Node: _____

_____

(c) Terminal branch: _____

_____

©2024 **BIOZONE** International
ISBN: 978-1-99-101410-8
Photocopying prohibited

Cladistics and the reclassification of figworts

▶ The angiosperm order Lamiales (which includes lavender, lilac, olive, jasmine, and snapdragons) has around 24,000 members. It contains several families, one of the largest being Scrophulariaceae (figworts). This family once included snapdragons, foxgloves, veronica, and monkeyflowers. While the other families in Lamiales have relatively well defined characteristics, the figworts were mostly assigned to their family based on their lack of characteristic traits. This suggested that the plants in the figwort family were not monophyletic.

▶ Investigations using three genes (*rbcL*, *ndhF*, and *rps2*) from chloroplast DNA falsified the traditional classification of figworts. They revealed that there were at least five distinct monophyletic groups (and probably more) within the figworts, each worthy of family status.

Common figwort, *Scrophularia nodosa*

Green figwort, *Scrophularia umbrosa*

4. In the DNA data for the whale cladogram, identify the DNA match that shows whales and hippopotamuses are more related to each other than to the other groups:

_____

5. What piece of evidence suggested that the original classification of figworts was incorrect?

_____

_____

_____

6. A phylogenetic tree is a hypothesis for an evolutionary history. Use the example of the figworts to show how you could test it:

_____

_____

_____

_____

7. Two possible phylogenetic trees constructed from the same character table are shown below. The numbers next to a blue bar represent an evolutionary event.

(a) Which tree is more likely to be correct?

_____

(b) State your reason:

_____

_____

_____

_____

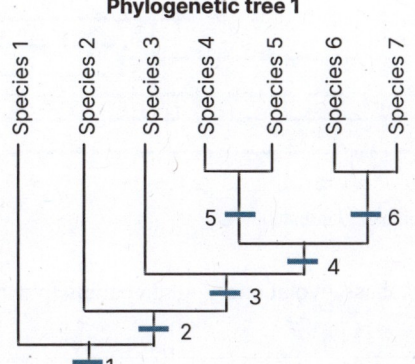

Phylogenetic tree 1

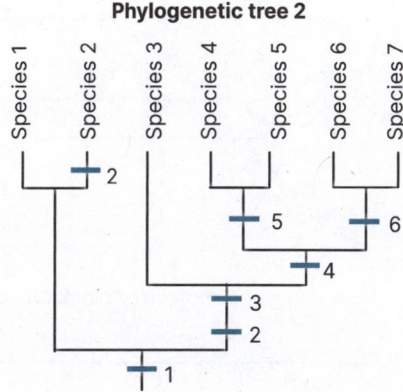

Phylogenetic tree 2

_____

(c) Identify the event that has occurred twice in phylogenetic tree 2: _____

©2024 **BIOZONE** International
ISBN: 978-1-99-101410-8
Photocopying prohibited

45 Molecular Evidence and Cladistics

Key Idea: Molecular evidence is an important way of determining evolutionary relationships and can be used to help determine when species last shared a common ancestor. The molecular clock hypothesis states that mutations occur at a relatively constant rate for any given gene. The genetic difference between any two species can indicate when two species last shared a common ancestor and can be used to construct a phylogenetic tree. The molecular clock for each species and each protein may run at different rates, so molecular clock data is calibrated with other evidence, e.g. morphological, to confirm phylogeny. Molecular clock calculations are carried out on DNA or amino acid sequences.

In a theoretical example, the DNA sequence for a gene in two species (A & B, right) alive today differs by four bases. The mutation rate for the gene is approximately one base per 25 million years. Based on this rate, it can be determined that the common ancestor for these two species lived 50 mya.

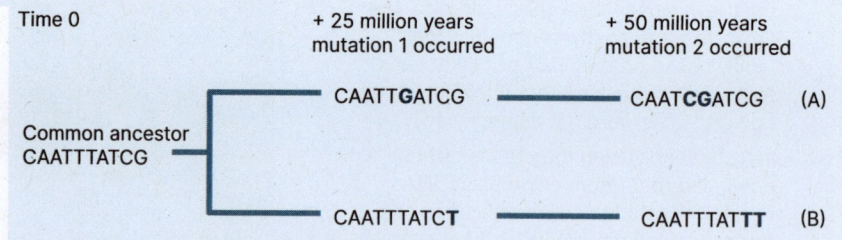

Cytochrome c and the molecular clock theory

		1	2	3	4	5	6	7	8	9	10	11	12	13	14	15	16	17	18	19	20	21	22
Human		Gly	Asp	Val	Glu	Lys	Gly	Lys	Lys	Ile	Phe	Ile	Met	Lys	Cys	Ser	Gln	Cys	His	Thr	Val	Glu	Lys
Pig												Val	Gln			Ala							
Chicken				Ile						Val		Val	Gln			Ala							
Dogfish										Val		Val	Gln			Ala							Asn
Drosophila	<<									Leu		Val	Gln	Arg		Ala							Ala
Wheat	<<		Asn	Pro	Asp	Ala		Ala				Lys	Thr	Arg		Ala						Asp	Ala
Yeast	<<		Ser	Ala	Lys			Ala	Thr	Leu		Lys	Thr	Arg		Glu	Leu						

This table shows the N-terminal 22 amino acid residues of human cytochrome c, with corresponding sequences from other organisms aligned beneath. Sequences are aligned to give the most position matches. A shaded square indicates no change. In every case, the cytochrome's haem group is attached to the Cys-14 and Cys-17. In *Drosophila*, wheat, and yeast, arrows indicate that several amino acids precede the sequence shown.

The sequence homology of cytochrome c (right), a respiratory protein, has been used to construct a phylogenetic tree for some species. Overall, the phylogeny aligns well to other evolutionary data, although the tree indicates that primates branched off before the marsupials diverged from other placental mammals, which is incorrect based on other evidence. Highly conserved proteins, such as cytochrome c, change very little over time and between species because they carry out important roles and if they changed too much they may no longer function properly.

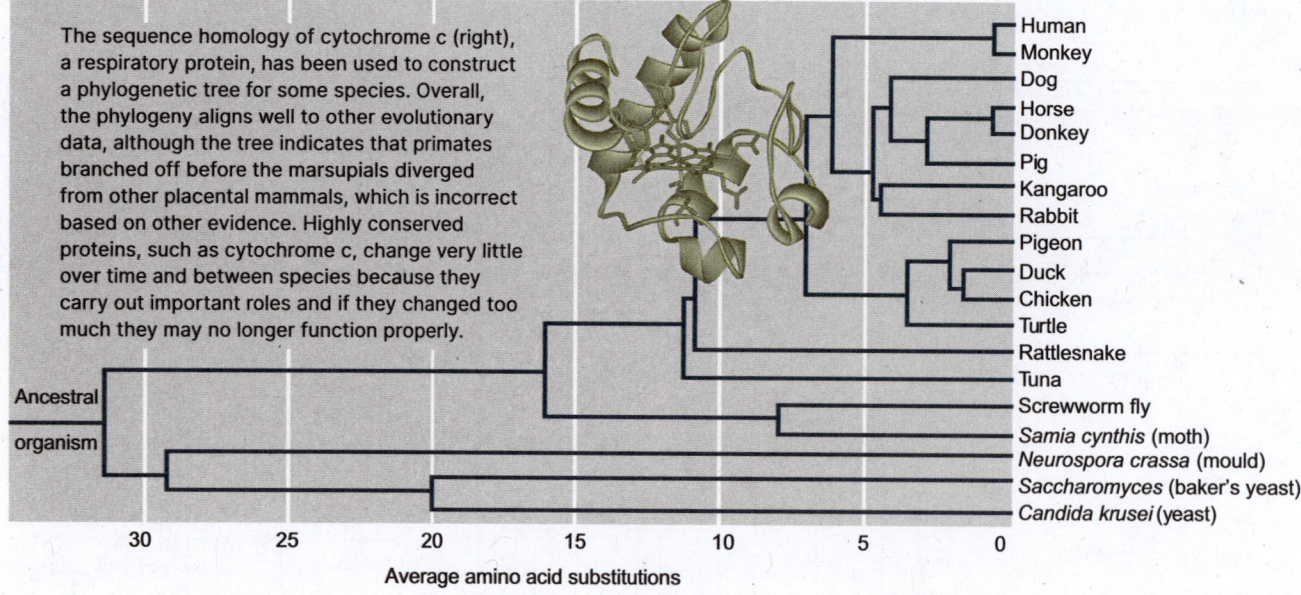

1. How can using molecular clocks help to establish evolutionary relationships (phylogenies) between organisms?

_____

_____

_____

_____

©2024 **BIOZONE** International
ISBN: 978-1-99-101410-8
Photocopying prohibited

Determining relationships between whales and dolphins

▶ On the right are DNA profiles for ten Short INterspersed Elements (SINE) in the whale genome (A-J). The profiles simply show if a SINE is present or not. Study of the presence or absence of SINEs can be used to construct a phylogeny. 15 whale (and dolphin) species and the hippopotamus were profiled at each of the ten sites. The whale species are listed below (1-15).

▶ Note that a present SINE shows as a line towards the top of the profile and an absent SINE shows as a line towards the bottom of the profile.

1 Striped dolphin
2 Risso's dolphin
3 Indo-Pacific bottlenose dolphin
4 Common bottlenose dolphin
5 Long-beaked common dolphin
6 Chinese white dolphin
7 Pantropical spotted dolphin
8 Beluga
9 Finless porpoise
10 Yangtze River dolphin
11 Ginkgo-toothed beaked whale
12 Ganges River dolphin
13 Pygmy sperm whale
14 Omura's whale
15 Common minke whale
16 Hippopotamus

(A) Turt161

(B) Neop28
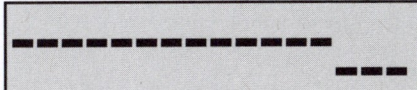

(C) Turt128

(D) Turt94

(E) Turt37
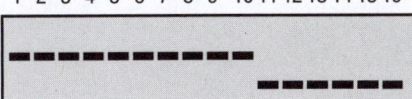

(F) Turt29
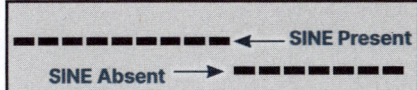

(G) Turt139
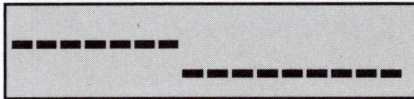

(H) Bala524

(I) Plag113

(J) Turt127
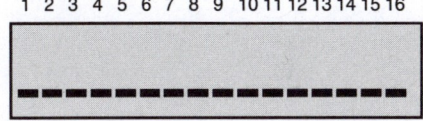

Source: Zhuo Chen et al (2011), see credits for the full reference

2. Working in pairs, use the DNA profiles to complete the table below:

3. Next, use the table to produce a cladogram of whale species:

Taxon	SINE present (1)/absent (0)									
	A	B	C	D	E	F	G	H	I	J
Striped dolphin										
Risso's dolphin										
Indo-Pacific bottlenose dolphin										
Common bottlenose dolphin										
Long-beaked common dolphin										
Chinese white dolphin										
Pantropical spotted dolphin										
Beluga										
Finless porpoise										
Yangtze River dolphin										
Ginkgo-toothed beaked whale										
Ganges River dolphin										
Pygmy sperm whale										
Omura's whale										
Common minke whale										
Hippopotamus										

46 Did You Get it?

1. Define a species according to the biological species concept: _____
_____
_____

2. The preparation of a karyogram involves arranging the chromosomes of an individual into homologous pairs in order.

   (a) Study the karyogram on the right. Circle the sex chromosomes:

   (b) State the sex of this individual: _____

   (c) Determine if the karyotype shown is normal/abnormal: _____

   (d) Explain the reason for the answer you have given in (c):
   _____
   _____
   _____
   _____

NHGRI

3. Use the shapes below to construct the following in the space below (use a separate sheet of paper if you need to and staple it here):
   (a) A dichotomous key to identify each shape:
   (b) **AHL:** A cladogram that shows their phylogenetic relationship (hint: A is the outgroup):

   A B C D E F

4. Put the following taxonomic groups in order from largest to smallest, *Kingdom, Species, Genus, Order, Domain, Class, Family, Phylum*:
_____

5. How many ways can you classify the items in your pencil case? What larger or smaller groups can you think of? Compare your items with a classmate; do they have any groups related to yours in their pencil case? What level of taxon would those items be related at, e.g. species of eraser?

©2024 **BIOZONE** International
ISBN: 978-1-99-101410-8
Photocopying prohibited

 Ecosystems

 **Resource Hub**
bit.ly/3uKaw6J

A4.1 Evolution and Speciation

	Activity Number

Guiding Questions:
▶ What types of evidence are used to support evolution?
▶ Why are both analogous and homologous structures evidence for evolution?

Learning Outcomes:

			Activity Number
☐	1	Distinguish the Darwinian theory of evolution by natural selection from Lamarckism. **NOS:** Justify the use of the term 'theory' as it relates to Darwin's Theory of Evolution by Natural Selection.	47
☐	2	Analyse data from DNA/RNA and amino acid sequences to discuss how similarities and differences can provide evidence for common ancestry.	48
☐	3	Investigate rapid change, occurring due to selective breeding of domestic species, as evidence for evolution.	49
☐	4	Discuss how the pentadactyl limb in different mammals as an example of homologous structures can provide evidence for common ancestry.	50
☐	5	Discuss the phenomenon of convergent evolution using examples of analogous features.	51
☐	6	Explain how the process of speciation leads to new species and contrast with extinction.	52
☐	7	Using specific examples, discuss how reproductive isolation and differential selection can lead to speciation.	52
☐	8	**AHL:** Compare the similarities and differences between sympatric and allopatric speciation, categorizing reproductive isolation as geographic, behavioural, or temporal.	53
☐	9	**AHL:** Explain how adaptive radiation can lead to increased biodiversity in ecosystems.	54
☐	10	**AHL:** Using examples, investigate prezygotic and postzygotic reproductive barriers.	53, 55
☐	11	**AHL:** Analyse examples of hybridization and polyploidy as processes involved in expeditious plant speciation.	56

A4.2 Conservation of Biodiversity

Guiding Questions:
▶ Why are humans the most likely contributor to the sixth mass extinction of species?
▶ What actions can be taken to preserve biodiversity?

	Activity Number

Learning Outcomes:

			Activity Number
☐	1	Define biodiversity at ecosystem, species, and genetic levels.	57
☐	2	**NOS:** Compare different approaches to the taxonomic classification of past and current biodiversity, including splitting or lumping groups.	57
☐	3	Discuss findings from case studies, including the New Zealand North Island giant moa and Caribbean monk seal, investigating the causes of anthropogenic species extinction (the Sixth Mass Extinction).	58
☐	4	Present findings from a case study, preferably local, investigating the causes of anthropogenic ecosystem loss.	59
☐	5	Collate and evaluate evidence for a biodiversity crisis from both verified scientific data and citizen science. **NOS:** Discuss benefits and concerns of using each type of data source.	59
☐	6	Discuss a range of potential causes leading to the current biodiversity crisis.	59
☐	7	Evaluate the appropriateness and effectiveness of different biodiversity conservation methods, including both *in situ* and *ex situ* approaches.	60
☐	8	**NOS:** Debate the ethical, environmental, political, social, cultural, and economic implications for conservation prioritization of evolutionarily distinct and globally endangered species (EDGE).	60

47 The Theory of Evolution

Key Idea: Scientific theories have been developed to explain the diversity of life on Earth, both presently, and in the past. Fossilized remains of organisms have been discovered and puzzled over by humans for thousands of years. Some remains could be identified as being the same as, or very similar to, living organisms. Other fossils seemed to be 'mystery' organisms that no-one could identify. Scientists developed hypotheses and theories to explain the mechanism behind this observed change. Jean-Baptiste Lamarck was one of the first scientists to propose that advantageous physical features from one generation could pass onto another, although he was yet to understand the basic principles of inheritance. Over 50 years later, Charles Darwin published his theory of evolution by natural selection, causing a paradigm shift in the field of evolution that provided a viable explanation supported by evidence.

1. Why did scientists working on the first dinosaur fossils have difficulty interpreting newly discovered examples to determine what type of organism or species they were from, or how they were related to modern species?

   _____
   _____
   _____
   _____
   _____
   _____

Recreated iguanodon dinosaur models (1853)

In the mid 1800s, sculptures that attempted to recreate recently discovered dinosaurs were installed in Crystal Palace, London, England. Scientists at the time had little fossil evidence on which to base their assumptions of dinosaur forms and therefore modelled them on species of living reptiles. Future fossil evidence would lead to completely changing what they knew about dinosaurs.

2. Suggest what scientific evidence, not available in the early 1800s, is available to today's scientists to help them understand how dinosaurs are related to each other, and to current living species?

   _____
   _____
   _____
   _____

▶ Around 1800, French scientist Jean-Baptiste Lamarck proposed a solution to account for how differences between current species, specifically the giraffe, and those found in ancestral fossils developed. Lamarck reasoned that, as food became more scarce for each generation of giraffe ancestor, the animal had to stretch higher. The longer necked giraffe then passed this acquired physical characteristic onto the next generation. This meant that, eventually, the short-necked giraffes ancestors were replaced by the long necked giraffes we have today.

3. Using your knowledge of modern genetics, what might you have explained to Lamarck to help him with his ideas on giraffe neck length 'growing over time'?

   _____
   _____
   _____
   _____
   _____

Lamarck observed giraffe ancestors had shorter necks

4. Although Lamarck's ideas about how change occurred over time in organisms were later proven wrong, why were his ideas still considered to have contributed to understand how organisms change, while sharing common ancestors?

   _____
   _____
   _____
   _____

A4.1
1

©2024 **BIOZONE** International
ISBN: 978-1-99-101410-8

Observations and evidence

▶ In 1831, an aspiring young English naturalist, Charles Darwin, took a job on a British Royal Navy ship, the HMS Beagle. He was to collect and catalogue samples of living and non-living specimens, including any fossils.

▶ Darwin was able to observe a wide range of life-forms, from many diverse and isolated regions. He found that island groups, such as the Galápagos Islands, were particularly rich in biodiversity. After his five year journey, he returned home. He began to hypothesize about how similar, yet different, species could have formed over time, and may have descended from a common ancestor. Darwin did understand that 'traits' were heritable, but the genetic mechanism was still unknown in his time. Unlike Lamarck, Darwin realized the importance of inherited variation already present in a population being the key to differentiated reproduction, where those individuals with the best 'fitness' were more likely to survive and pass on their traits.

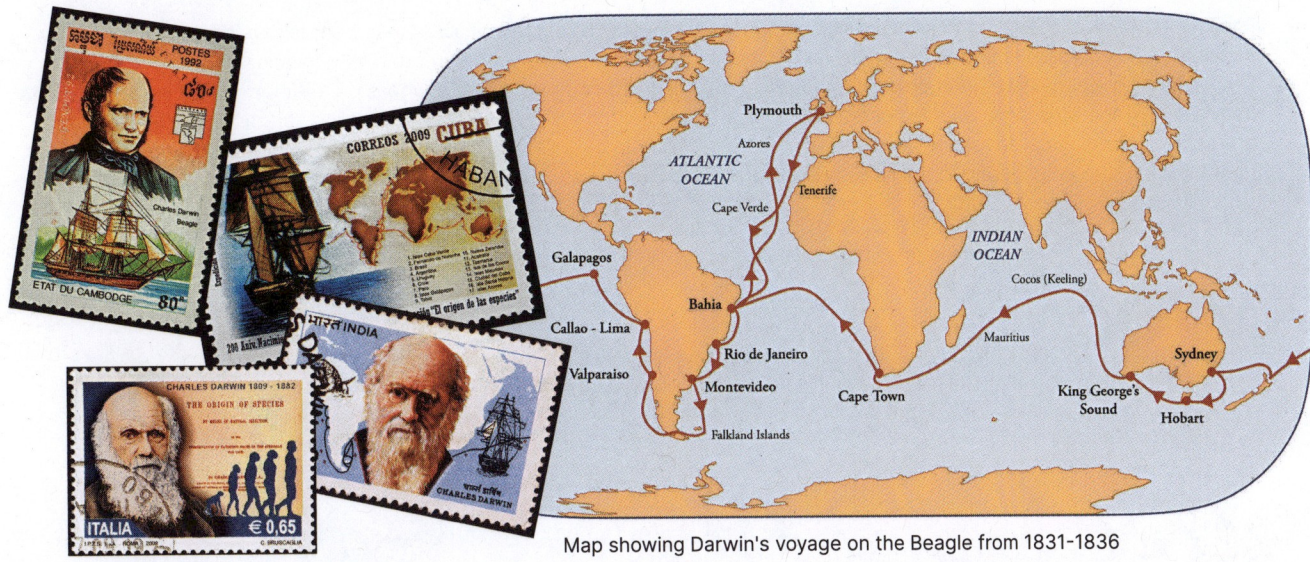

Map showing Darwin's voyage on the Beagle from 1831-1836

5. How might the five year voyage around the world on the Beagle have contributed to Darwin's ideas about evolution?

_____

_____

_____

Developing the theory

▶ In science, a theory is the term used when an explanation of a phenomenon has been developed and supported by significant evidence and observations. It must also be scientifically testable.

▶ Theories are accepted as the current 'truth', but the nature of science dictates that they can be discarded or modified if valid evidence is found to disprove them.

▶ Charles Darwin formulated his theory of evolution privately between 1837-1839. He spent many more years gathering evidence before receiving a letter from Alfred Russel Wallace that covered many of his own ideas. This prompted him to finally publish his theory of evolution by natural selection in 1859, 20 years after he had first thought of it.

▶ With increasing understanding of the role genetics plays in evolution, Neo-Darwinism is a synthesis of both the natural selection theory and Mendelian genetics. Darwin's theory was not refuted, but instead modified to account for the genetic basis of phenotype and the mechanism for passing on traits to the next generation.

Charles Darwin (1809-1882)

6. Why does the term 'theory' have a different meaning in science in comparison to everyday usage, and why was Lamarckism discarded, rather than modified like Darwinism, when new evidence was produced?

_____

_____

_____

_____

_____

_____

©2024 **BIOZONE** International
ISBN: 978-1-99-101410-8

48 Molecular Sequencing as Evidence for Evolution

Key Idea: Relationships between species can be assessed by comparing the sequence difference in DNA and amino acids. Sequencing provides the precise order of nucleotides in a DNA molecule or amino acids in a protein. This information, which can now be analysed using sophisticated computing, allows researchers to compare sequences between species.

Not only can areas of difference be identified, the variation between the nucleotides or amino acids at a certain position can be determined. This information allows researchers to more accurately determine the relatedness between species, even between those with very minor differences, where less difference indicates closer relatedness in individuals.

DNA sequencing

▸ Improved DNA sequencing techniques and powerful computing software have allowed researchers to accurately and quickly sequence and compare entire genomes (all an organism's genetic material) within and between species.

▸ Once DNA sequences have been determined, they are aligned and compared to see where the differences occur (below). DNA sequencing generates large volumes of data and the improvement in computing power has been central to modern sequence analyses. Technological advances have been behind the new field of bioinformatics, which uses computer science, statistics, mathematics, and engineering to analyse and interpret biological data.

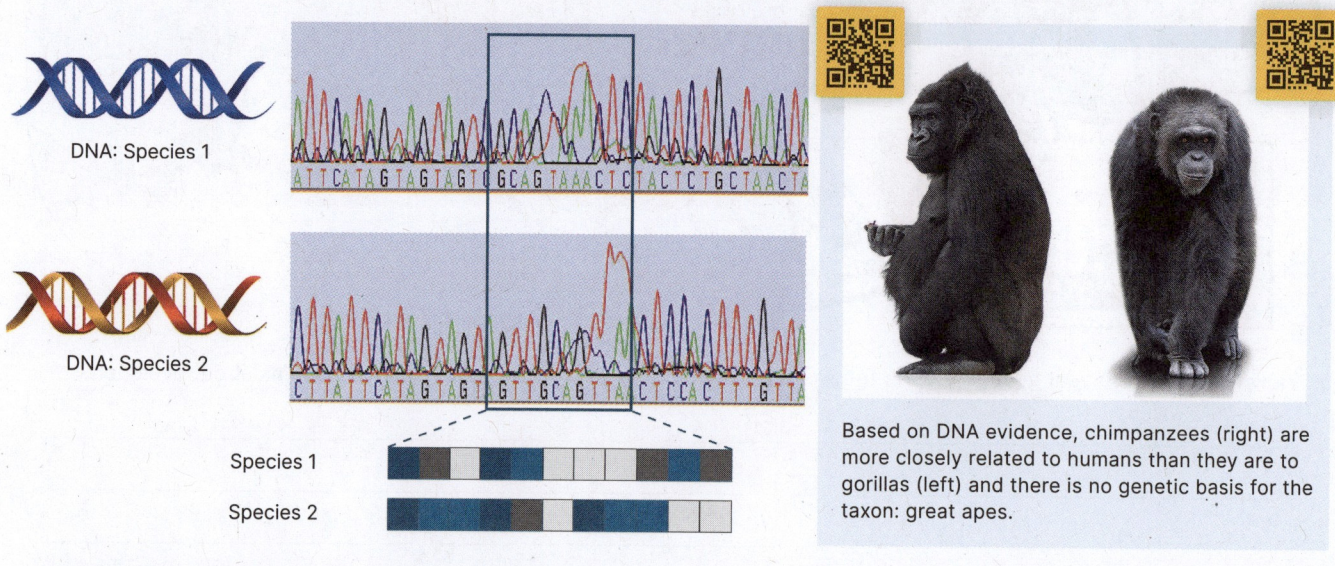

DNA: Species 1

ATTCATAGTAGTAGTGGCAGTAAACTCTACTCTGCTAACTA

DNA: Species 2

CTTATTCATAGTAGTAGTTGCAGTTACTCCACTTTGTTA

Species 1
Species 2

Based on DNA evidence, chimpanzees (right) are more closely related to humans than they are to gorillas (left) and there is no genetic basis for the taxon: great apes.

What type of sequences are compared?

▸ Highly conserved sequences are often used for comparative genomic analysis because they are found in many organisms. The changes (mutations) of the sequences over time can be used to determine evolutionary relationships. As with other forms of molecular analysis, species with fewer nucleotide differences are more closely related than those with many.

▸ Whole genome analysis has been important in classifying the primates. Historical views attributed special status to humans which often confused primate classification schemes. DNA evidence provides impartial quantitative evidence and modern classification schemes have been based on this data.

1. Explain why DNA sequence comparisons are useful in determining evolutionary relationships:

_____

_____

_____

_____

2. Three partial DNA sequences for three different species are presented below.

Species 1	A T G G C C C C C A A C A T T C G A A A A T C G C A C C C C C T G C T C A A A A T T A T C A A C
Species 2	A T G G C A C C T A A C A T C C C C A A C T C C C A C C G T G T A C T C A A A A T C A T C A A G
Species 3	A T G G C A C C C A A T A T C C G C A A A T C A C A C C C C C T G T T A A A A A C A A T C A A C

Based on the number of differences in the DNA sequences:

(a) Identify the two species that are most closely related: _____

(b) Identify the two species that are least closely related: _____

A4.1
2

©2024 **BIOZONE** International
ISBN: 978-1-99-101410-8
Photocopying prohibited

Protein homology

▶ The amino acid sequence of proteins can be used to establish molecular homologies (similarities) between organisms. Any change in the amino acid sequence reflects changes in the DNA sequence. As genetic relatedness increases, the number of amino acid differences due to mutation decreases.

▶ Some proteins are common to many different species. These proteins are often highly conserved, meaning they mutate (change) very little over time. This is because they have critical roles, e.g. in cellular respiration, and mutations are likely to be detrimental to their function.

▶ Evidence indicates that these highly conserved proteins are homologous and have been derived from a common ancestor. Because they are highly conserved, changes in the amino acid sequence are likely to represent major divergences between groups during the course of evolution.

The Hox gene provides evidence for evolution

▶ Hox genes (short for homeobox) determine the position of body 'parts', which develop in the early animal embryo; the order of the genes on the chromosome orders the body plan.

▶ The key protein produced from the gene in the clusters is nearly identical in most animals and plants.

▶ Scientists hypothesize that a single hox gene first arose in the last common bilateral ancestor - and then duplicated into gene clusters in different animals - with different ordering of the genes in different animals radically changing the body plan, however, still using almost identical genes.

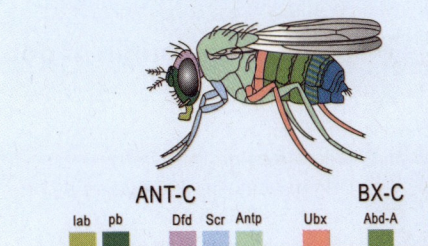

Hox gene cluster in fruit flies, linked to order of body plan. In different animals the order of the genes in clusters is different, resulting in a different ordered body plan.

Haemoglobin sequencing

▶ Haemoglobin is the oxygen-transporting blood protein found in most vertebrates. Haemoglobin DNA sequences from different organisms can be compared to determine evolutionary relationships and common ancestry.

▶ As genetic relatedness decreases, the number of amino acid differences between the haemoglobin chains of different vertebrates increases (below). For example, there are no amino acid differences between humans and chimpanzees, indicating they recently shared a common ancestor. Humans and frogs have 67 amino acid differences, indicating they had a common ancestor a very long time ago.

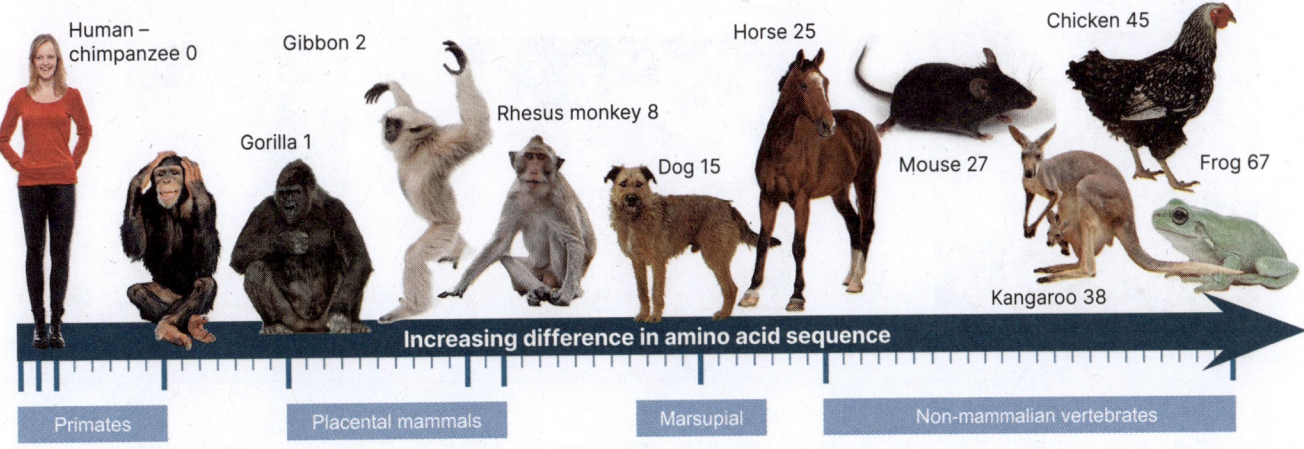

3. Compare the similarities and differences in the haemoglobin sequence of humans, rhesus monkeys, and horses. What do these tell you about the relative relatedness of these organisms to each other, and to other animals?

_____

_____

_____

_____

_____

4. The fossil record shows that amphibians diverged from the lineages that evolved into mammals before the birds did. Why is sequencing data such a valuable tool for providing evidence for evolution in addition to traditional taxonomy?

_____

_____

_____

_____

49 Evolution and Selective Breeding

Key Idea: Selective breeding is the process of breeding organisms with desirable qualities, e.g. high milk yield, so that the trait is reliably passed on to the next generation.

Selective breeding (or artificial selection) is the process by which humans select organisms with desirable traits and breed them so the trait appears in the next generation. The process is repeated over many generations until the characteristic becomes common. Selective breeding often uses reproductive technologies, such as artificial insemination, so that the desirable characteristics of one male can be passed onto many offspring. This increases the rate at which the desirable trait is passed to progeny. Selective breeding is based around the same mechanism that drives natural selection, but occurs at a significantly faster pace. Humans determine which organisms are bred, and therefore which traits are passed on.

The origin of domestic animals

PIG
Wild ancestor: Boar (left)
Origin: Anatolia, 9000 years BP
Now: More than 12 distinct modern breeds, including the Berkshire (meat) and Tamworth (hardiness).

DOMESTIC FOWL
Wild ancestor: Red jungle fowl (right)
Origin: Indus Valley, 4000 BP
Now: More than 60 breeds including Rhode Island red (meat) and leghorn (egg production).

Each domesticated breed has been bred from the wild ancestor. The date indicates the earliest record of the domesticated form (years before present or BP). Different countries have different criteria for selection, based on their local environments and consumer preferences.

Bezoar ibex goat

Mouflon

Zebu: derived from Indian aurochs

GOAT
Wild ancestor: Bezoar goat
Origin: Iraq, 10,000 years BP
Now: approx. 35 breeds including Spanish (meat), Angora (fibre) and Nubian (dairy).

SHEEP
Wild ancestor: Asiatic mouflon
Origin: Iran, Iraq, Levant, 10,000 years BP
Now: More than 200 breeds including Merino (wool), Suffolk (meat), Friesian (milk), and dual purpose (Romney).

CATTLE
Wild ancestor: Auroch (extinct)
Origin: SW Asia, 10,000 years BP
Now: 800 modern breeds including the Aberdeen Angus (meat), Friesian and Jersey (milk), and zebu (draught).

1. (a) How does selective breeding differ from natural selection? _____

   (b) What are the advantages of selective breeding? _____

   _____

   _____

2. How does selective breeding still provide evidence for evolution? _____

   _____

   _____

   _____

   _____

A4.1
3

352

©2024 **BIOZONE** International
ISBN: 978-1-99-101410-8
Photocopying prohibited

Selective breeding in crops

▸ The genetic diversity within crop varieties provides options to develop new crop plants through selective breeding. For thousands of years, farmers have used the variation in wild and cultivated plants to develop crops.

▸ *Brassica oleracea* is a good example of the variety that can be produced by selectively growing plants with desirable traits. Not only are there six varieties of *Brassica oleracea*, but each of those has a number of sub-varieties as well. Although brassicas have been cultivated for several thousand years, cauliflower, broccoli, and Brussels sprouts appeared only in the last 500 years.

Cauliflower (flower)

Broccoli (inflorescence: cluster of flowers on a stem)

Cabbage (terminal buds)

Brussels sprouts (lateral buds)

kale (leaf)

Domestication of *Brassica*

Around 3750 BC in China, the cabbage was probably the first domesticated variety of its wild form to be developed. Selective breeding by humans has produced six separate vegetables from this single species, *Brassica oleracea*. The wild form species is shown in the centre of this diagram. Different parts have been developed by human selection. In spite of the enormous visible differences, if allowed to flower, all six can cross-pollinate. Kale is closer to the wild type than the other related varieties.

Wild form (*Brassica oleracea*)

Kohlrabi (stem)

3. What was an important factor in the populations of wild form *Brassica oleracea* that allowed selective breeding to 'create' many different types of crop?

_____

_____

_____

4. If *Brassica oleracea* was planted in a wide range of different environments, would it be likely that the same types of vegetables evolve through natural selection over a longer time period? Explain your answer:

_____

_____

_____

_____

50 Homologous Structures

Key Idea: Homologous structures (homologies) are structural similarities present as a result of common ancestry. The common structural components have been adapted to different purposes in different taxa.

The bones of the forelimb of air-breathing vertebrates are composed of similar bones arranged in a comparable pattern. This is indicative of common ancestry. The early land vertebrates were amphibians with a pentadactyl limb structure (a limb with five fingers or toes). All vertebrates that descended from these early amphibians have limbs with this same basic pentadactyl pattern. They also illustrate the phenomenon known as adaptive radiation, since the basic limb plan has been adapted to meet the requirements of different niches.

Generalized pentadactyl limb

The forelimbs and hind limbs have the same arrangement of bones but they have different names. In many cases, the basic limb plan has been adapted (e.g. by loss or fusion of bones) to meet the requirements of different niches (e.g. during adaptive radiation of the mammals).

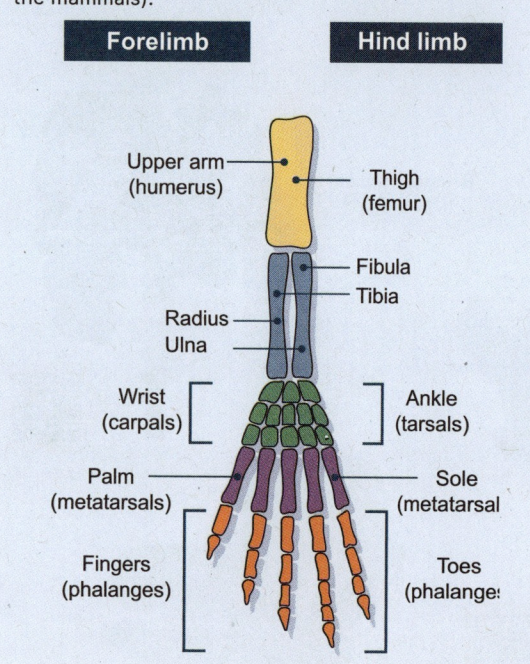

Forelimb	Hind limb

Upper arm (humerus) — Thigh (femur)
Fibula
Tibia
Radius
Ulna
Wrist (carpals) — Ankle (tarsals)
Palm (metatarsals) — Sole (metatarsal
Fingers (phalanges) — Toes (phalange:

Specializations of pentadactyl limbs

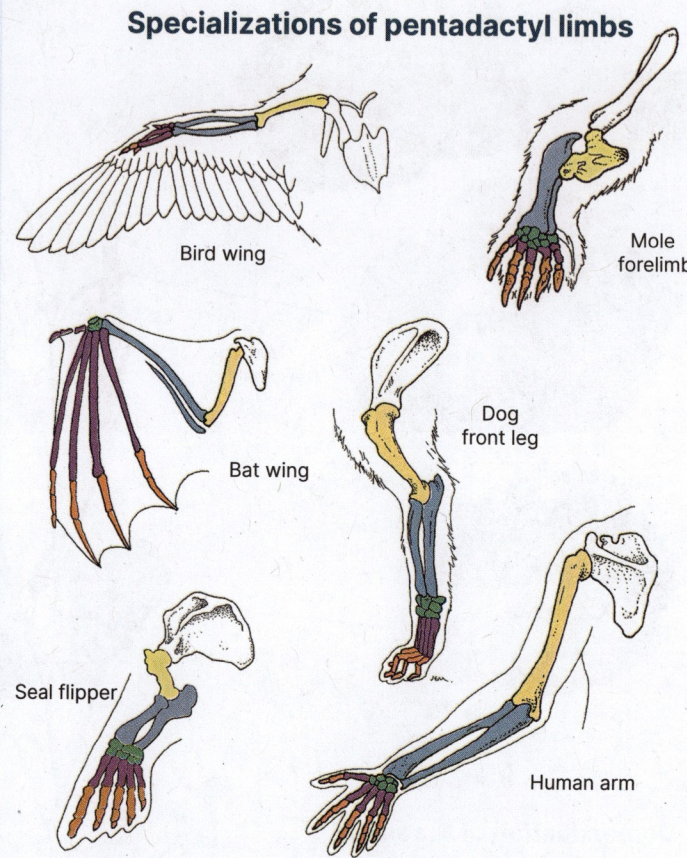

Bird wing
Mole forelimb
Bat wing
Dog front leg
Seal flipper
Human arm

1. Briefly describe the purpose of the major anatomical change that has taken place in each of the limb examples above:

(a) Bird wing: _**Highly modified for flight. Forelimb is shaped for aerodynamic lift and feather attachment.**_

(b) Human arm: _____

(c) Seal flipper: _____

(d) Dog front leg: _____

(e) Mole forelimb: _____

(f) Bat wing: _____

2. Explain how homology in the pentadactyl limb is evidence for evolution: _____
_____
_____
_____

3. Homology is due to divergent evolution. Define this term and explain how discovery of a new species of whale fossil, complete with limb bones, could be classified using evidence from anatomical homology:
_____
_____
_____

©2024 **BIOZONE** International
ISBN: 978-1-99-101410-8
Photocopying prohibited

A4.1

4

51 Convergent Evolution

Key Idea: Some species living in similar environments have evolved similar structural adaptations, called analogous structures, even though they are not closely related. Convergent evolution describes the process by which species from different evolutionary lineages come to resemble each other or evolve analogous structures that serve the same purpose, because they have similar habitats, ecological roles, or selection pressures.

Convergence: same look, different origins

▶ We have seen how artificial selection applies selection pressure to bring about phenotypic change in a population. In natural environments, selection pressures to solve similar problems in particular environments can result in similar phenotypic characteristics in unrelated (or very distantly related) species.

▶ The evolution of succulence in unrelated plant groups (*Euphorbia* and cacti) is an example of convergence in plants. In the example (right), the selection pressures of the aquatic environment have produced a similar streamlined body shape in unrelated vertebrates. Ichthyosaurs, penguins, and dolphins each evolved from terrestrial species that took up an aquatic lifestyle. Their body form has evolved similarities to that of the shark, which has always been aquatic. Note that flipper shape in mammals, birds, and reptiles is a result of convergence, but its origin from the pentadactyl limb is an example of common ancestry.

Fish: Shark

Reptile: Ichthyosaur (extinct)

Mammal: Dolphin

Bird: Penguin

Analogous structures arise through convergence

▶ Analogous structures have the same function and often the same appearance, but different origins.

▶ The example (right) shows the structure of the camera eye in two unrelated taxa (mammals and cephalopod molluscs). The eyes appear similar, but have different embryonic origins and have evolved independently.

▶ The wings of birds and insects are also analogous. The wings have the same function but the two taxa do not share a common ancestor. *Longisquama*, an extinct reptile, also had wing-like structures that probably allowed gliding between trees. These were highly modified, long scales extending from its back and not a modification of the forearm (as in birds).

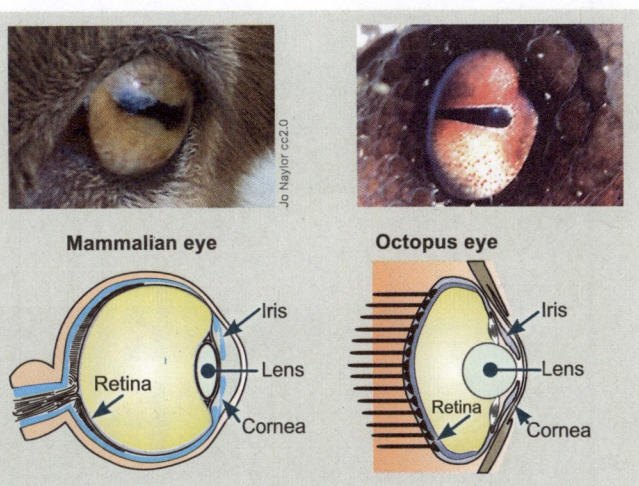

Jo Naylor cc2.0

Mammalian eye   **Octopus eye**

Iris / Lens / Retina / Cornea

1. In the example above, illustrating analogous eye structure, describe two ways in which the structure has evolved in response to the particular selection pressures of its functional requirements.

(a) _____

_____

(b) _____

_____

2. Describe two of the selection pressures that have influenced the body form of the swimming animals above:

(a) _____

(b) _____

3. When early taxonomists encountered new species in the Pacific region and the Americas, they were keen to assign them to existing taxonomic families based on their apparent similarity to European species. In recent times, many of these species have been found to be quite unrelated to the European families they were assigned to. Explain why the traditional approach did not reveal the true evolutionary relationships of the new species:

_____

_____

_____

A4.1

5

52 Speciation

Key Idea: Speciation occurs when a new species evolves from an existing one. This can occur through the mechanisms of reproductive isolation and differential selection.

Species evolve in response to selection pressures from the environment. Gene flow is reduced when populations are separated and the amount of genetic difference between the populations increases over time, with each group becoming increasingly isolated from the other. Continual reduction in gene flow by isolating mechanisms, such as geographical barriers and reproductive isolation may eventually lead to the formation of new species. Speciation is the only means of increasing the total number of species that exist on Earth. Extinction, due to either human activity or natural causes, permanently removes the species.

A species of butterfly lives on a grassland plateau scattered with boulders. During colder weather, some butterflies rest on the sun-heated boulders to absorb the heat, while others move to the lower altitude grassland to avoid the cold.

Continued mountain building raises the altitude of the plateau, separating two sub-populations of butterflies, one in the highlands, the other in the lowlands.

In the highlands, boulder-sitting butterflies (BSBs) do better than grass-sitting butterflies (GSBs). Darker BSBs have greater fitness than light BSBs. In the lowlands, the opposite is true. BSBs only mate on boulders with other BSBs. In the lowlands, light GSBs blend in with the grass and survive better than darker butterflies.

Over time, the boulder-sitting butterflies occur only in the highlands and grass-sitting butterflies in the lowlands. Occasionally, wind brings members of the two populations together, but if they happen to mate (left), the hybrid offspring are not viable or have a much lower fitness

Eventually, gene flow between the separated populations stops, as variation between the populations increases. They fail to recognize each other as members of the same species.

1. The mountain gorilla (*Gorilla gorilla beringei*), found in east-central Africa, has a number of structural features that have been acquired gradually over the last 10, 000 years, through evolution, that distinguishes it from the Eastern lowland gorilla (*Gorilla gorilla graueri*). Why is the mountain gorilla not considered an example of a separate species?

_____

_____

_____

2. If natural events cause the extinction of a species, why is the same species not able to 're-evolve' if identical habitats become available once more?

_____

_____

_____

A4.1
6-7

©2024 **BIOZONE** International
ISBN: 978-1-99-101410-8
Photocopying prohibited

Isolating mechanisms

▶ Isolating mechanisms are barriers to successful interbreeding between species. Reproductive isolation is fundamental to the biological species concept, which defines a species by its inability to breed with other species to produce fertile offspring.

▶ Prezygotic isolating mechanisms act before fertilization occurs, preventing species ever mating, whereas postzygotic barriers take effect after fertilization. Reproductive isolation prevents interbreeding (and therefore gene flow) between species. Any factor that impedes two species from producing viable, fertile hybrids contributes to reproductive isolation.

▶ Single barriers may not completely stop gene flow, so most species commonly have more than one type of barrier. Single barriers to reproduction (including geographical barriers) often precede the development of a suite of reproductive isolating mechanisms (RIMs). Most operate before fertilization (prezygotic RIMs), with postzygotic RIMs being important in preventing offspring between closely related species.

Speciation in bonobo and chimpanzees

▶ Bonobos and chimpanzees last shared a common ancestor around 1 - 1.8 million years ago, prior to one group crossing the Congo River, situated in modern day Democratic Republic of Congo.

▶ Research indicates the crossing took place when the river levels were unusually low. Since then, the Congo River has acted as a geographical barrier to the apes, who are poor swimmers.

▶ Differential selection pressures, created by more abundant food resources on the southern side of the river, favoured individuals that could cooperate together to forage food in groups and show less aggressive traits.

▶ The populations diverged over time, corresponding to shifts in the respective gene pools, to form two separate species: the bonobo on the southern side of the Congo River, and the chimpanzee on the northern side of the River.

Bonobo (Pan paniscus)   **Chimpanzee (Pan troglodytes)**

▶ The two species show some physical differences, with the bonobo being smaller and having darker facial features, but the greatest difference is seen in different behavioural traits.

▶ Scientists have hypothesized that the reduced competition for food led to a 'self-domestication' process in the bonobo, who are organized into peaceful, matriarchal-based social groupings. This is supported with genetic evidence, where genes controlling hormones involved in social cohesion behaviours, including oxytocin, serotonin, and gonadotropin, have been selected for. Additionally, the gene for the enzyme amylase, which breaks down starch, has been selected for.

▶ Chimpanzees form patriarchal social groupings, with dimorphically larger males. They tend to forage individually for scarce resources and are more aggressive, most especially towards other troops when disputing territory.

3. Bonobos and chimpanzees can interbreed and produce fertile offspring when placed together in a zoo but this does not occur in the wild. What does this suggest about multiple isolation mechanisms often being required for speciation?

4. If bonobo and chimpanzee populations continued to be separated, over time, what other types of isolating mechanisms could develop, and what consequences would arise from that?

5. What is the relationship between differential selection, isolating mechanisms, and speciation?

53 Allopatric and Sympatric Speciation

Key Idea: Allopatric speciation requires a geographical barrier to separate populations, while sympatric speciation can occur when two populations occur in the same area.

In all cases of speciation, gene flow between populations is reduced or prevented entirely. A geographic barrier physically prevents gene flow and acts as a reproductive isolating mechanism (RIM) leading to allopatric speciation. A number of RIMs can be involved in sympatric speciation, including behavioural or temporal isolation, all of which stop the flow of genes between populations in the same location.

Geographical isolation

Geographical isolation describes the isolation of a species population (gene pool) by some kind of physical barrier, e.g. mountain range, water body, isthmus, desert, or ice sheet. Geographical isolation is a frequent first step in the subsequent reproductive isolation of a species. (Allopatric)

Example: Geological changes to the lake basins has been instrumental in the proliferation of cichlid fish species in the rift lakes of East Africa (far right). Similarly, many Galápagos Island species, e.g. iguanas, finches, are now quite distinct from the Central and South American mainland species from which they separated.

Lake Victoria
Lake Tanganyika
Lake Malawi

NASA Earth Observatory

Temporal isolation

Temporal isolation means isolated in time, and it prevents species interbreeding because they mate or they are active at different times. For example, individuals from different species do not mate because they are active during different times of the day, e.g. one species is active during the day and the other at night, or in different seasons. (Sympatric)

Example: Closely related animal species may have different breeding seasons or periods of emergence to prevent interbreeding. The periodical cicadas (*Magicicada* genus) are an excellent example of this. Periodical cicadas are found in North America. They spend most of their life underground as juveniles, emerging only to complete their development and mate. To prevent interbreeding, the various species spend either 13 or 17 years underground developing. Emergence of a single species is synchronized so the entire population emerges at the same time to breed.

Periodical cicada

Bruce Marlin

Periodical cicada emerging

Lorax

Behavioural Isolation

In many species, courtship behaviours are a necessary prelude to successful mating. These behaviours may include dances, calls, displays, or the presentation of gifts. The displays are specific and unique to each species. This ensures that males and females of the same species recognize and are attracted to the individual performing the behaviour, but members of other species do not recognize or pay attention to it.

Birds exhibit a wide range of courtship displays. The use of song is widespread but ritualized movements, including nest building, are also common. (Sympatric)

Examples: Galápagos frigatebirds have an elaborate display in which they inflate a bright red throat pouch to attract a mate. Frogs have species-specific calls.

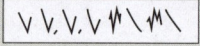

Male frigatebird display

Frog calling

1. Construct a Venn diagram on a sheet of paper, or using a digital programme, comparing and contrasting the features of allopatric and sympatric speciation, including the types of reproductive isolating mechanisms involved in each.

A4.1
8, 10

©2024 **BIOZONE** International
ISBN: 978-1-99-101410-8
Photocopying prohibited

54 Adaptive Radiation in Mammals

Key Idea: Adaptive radiation is diversification among the descendants of a single ancestral group (one lineage) to occupy different niches, leading to increased biodiversity.

Adaptive radiation in mammals has been supported by a large amount of evidence, including fossils and molecular sequencing. Most of the modern mammalian groups became established very early on. The diagram below shows the divergence of the mammals into the major orders that came to occupy the niches left vacant by the extinction of the dinosaurs. Those species that reach the top of the chart have survived to the present day. The width of a shape shows how many species existed at any given time. The dotted lines indicate possible links between the orders.

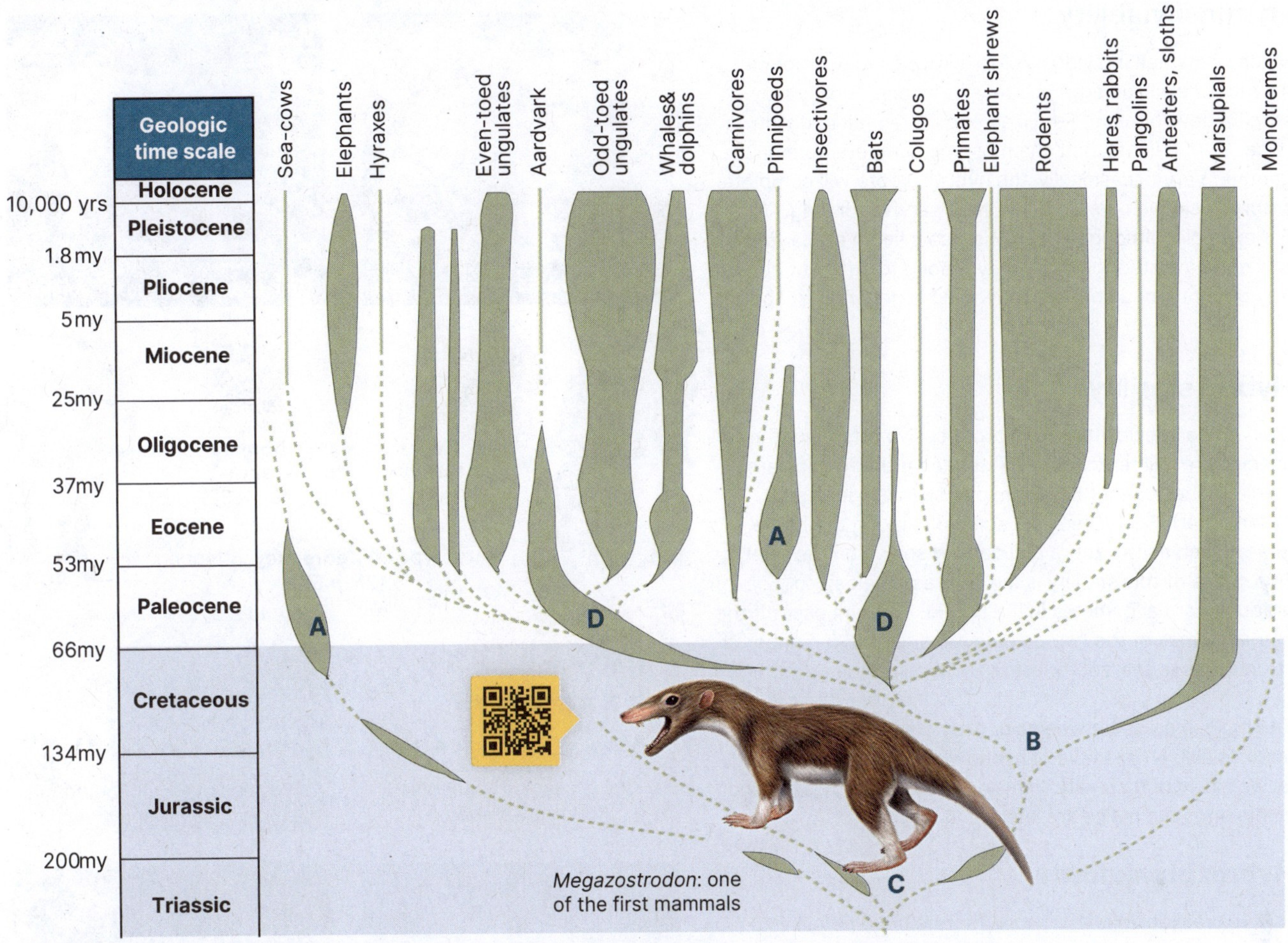

Diagram showing mammalian orders across the geologic time scale: Sea-cows, Elephants, Hyraxes, Even-toed ungulates, Aardvark, Odd-toed ungulates, Whales & dolphins, Carnivores, Pinnipeds, Insectivores, Bats, Colugos, Primates, Elephant shrews, Rodents, Hares, rabbits, Pangolins, Anteaters, sloths, Marsupials, Monotremes.

Geologic time scale: Holocene (10,000 yrs), Pleistocene (1.8 my), Pliocene (5 my), Miocene (25 my), Oligocene (37 my), Eocene (53 my), Paleocene (66 my), Cretaceous (134 my), Jurassic (200 my), Triassic.

Points labelled A, B, C, D on the diagram.

Megazostrodon: one of the first mammals

1. In general terms, explain how adaptive radiation in mammals led to increased diversity:

_____

_____

_____

_____

2. Name the term that you would use to describe the animal groups at point **C** (above): _____

3. Explain what occurred at point **B** (above): _____

_____

4. Describe one thing that the animal orders labelled **D** (above) have in common: _____

_____

5. Identify the two orders that appear to have been most successful in terms of the number of species produced:

_____

6. Explain what has happened to the mammal orders labelled **A** in the diagram above: _____

_____

7. Name the geological time period during which most adaptive radiation occurred: _____

©2024 **BIOZONE** International
ISBN: 978-1-99-101410-8

A4.1
9
AHL

55 Barriers to Hybridization

Key Idea: A range of mechanisms maintain reproductive isolation after fertilization in closely related species. Postzygotic reproductive isolating mechanisms occur after fertilization (formation of the zygote) has occurred. Postzygotic isolating mechanisms are less common than prezygotic mechanisms but are important in maintaining the integrity of closely related species. Several different postzygotic mechanisms operate at different stages. The first prevents development of the zygote. Even if the zygote develops into a viable offspring, further mechanisms prevent long term viability. These include premature death or, more commonly, infertility.

Hybrid inviability

Mating between individuals of two species may produce a zygote (fertilized egg) but genetic incompatibility may stop its development. Fertilized eggs often fail to divide because of mismatched chromosome numbers from each gamete. Very occasionally, the hybrid zygote will complete embryonic development but will not survive for long. For example, although sheep and goats seem similar (right) and can be mated together, they belong to different genera. Any offspring of a sheep-goat pairing is generally stillborn.

Sheep (*Ovis*) 54 chromosomes

Goat (*Capra*) 60 chromosomes

Hybrid sterility

Even if two species mate and produce live offspring, the species are still reproductively isolated if the hybrids are sterile (genes cannot flow from one species' gene pool to the other). Such cases are common among the horse family (such as the zebra and donkey shown on the right). One cause of this sterility is the failure of meiosis to produce normal gametes in the hybrid. This can occur if the chromosomes of the two parents are different in number or structure (see the 'zebronkey' karyotype on the right).

The mule, a cross between a donkey stallion and a horse mare, is also an example of hybrid vigour (they are robust) as well as hybrid sterility. Female mules sometimes produce viable eggs but males are infertile.

  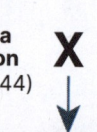 

Zebra stallion (2N = 44) X **Donkey jenny** (2N = 62)

Karyotype of '**Zedronkey**' offspring (2N = 53)

Chromosomes contibuted by zebra stallion | Chromosomes contibuted by donkey jenny

Hybrid breakdown

Hybrid breakdown is a common feature of some plant hybrids. The first generation (F$_1$) may be fertile, but the second generation (F$_2$) is infertile or non-viable. Examples include hybrids between species of cotton (near right), species within the genus *Populus*, and strains of the cultivated rice *Oryza* (far right).

Cotton (*Gossypium*)

Cultivated rice (*Oryza*)

1. Postzygotic isolating mechanisms are said to reinforce prezygotic ones. Explain why this is the case:

2. Briefly explain how each of the postzygotic isolating mechanisms below maintains reproductive isolation:

   (a) Hybrid non-viability:

   (b) Hybrid sterility:

   (c) Hybrid breakdown:

 A4.1 10

©2024 **BIOZONE** International
ISBN: 978-1-99-101410-8
Photocopying prohibited

56 Abrupt Speciation in Plants

Key Idea: An increase in the number of chromosome sets can result in instant speciation and is common in plants.

Polyploidy is a condition in which an organism's cells contain three or more times the haploid number of chromosomes (3n or more). Polyploidy is rare in animals but common in plants. It may result in speciation without geographic separation of populations. Polyploidy occurs when chromosomes fail to separate properly during meiosis and are carried by only one gamete. Union with a normal 'n' gamete produces a triploid (3n). Union with another 2n polyploid gamete produces a tetraploid (4n). An estimated 15% of angiosperm speciation events are accompanied by polyploidy events.

Instant speciation by polyploidy

Polyploidy may result in the formation of a new species without physical isolation from the parent species. This event, a result of faulty meiosis, produces sudden reproductive isolation because the polyploid is unable to interbreed with its diploid parent. Animals are rarely able to achieve new species status this way because the sex-determining mechanism is disturbed (they are effectively sterile, e.g. a tetraploid would have four X chromosomes). Many plants, on the other hand, are able to self-pollinate or reproduce vegetatively. This ability to reproduce on their own enables such polyploid plants to produce a viable population.

Speciation by allopolyploidy

This type of polyploidy results from a hybridization of two species. The resulting hybrid may be sterile to begin with, but a doubling event during meiosis may produce a viable chromosome number. Self fertilization may then produce a fertile hybrid. <u>Examples:</u> Modern wheat. Swedes are a polyploid species formed as a result of hybridization between a type of cabbage and a type of turnip.

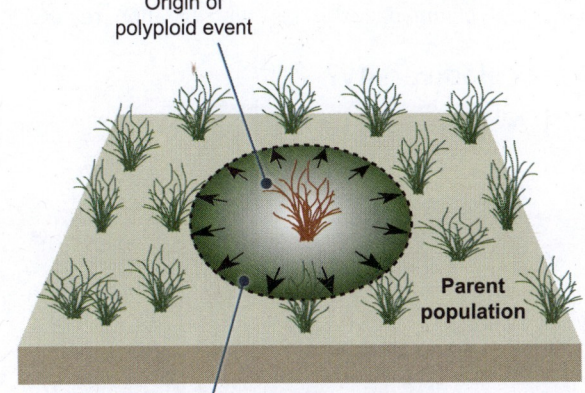

Origin of polyploid event

Parent population

New polyploid plant species spreads out through the existing parent population

Banana 3n = 27   Boysenberry 7n = 49   Strawberry 8n = 56

Many commercial plant species are polyploids.

Polyploidy in *Fallopia*

Fallopia is a genus of plant, commonly called knotweed. There are two species of knotweed that reproduce asexually, the Japanese knotweed, and giant knotweed. Both are gynodioecious, making the plants effectively sterile. A third species, bohemian knotweed, is a hybrid of the other two species, and can breed sexually. This species is an example of allopolyploidy. All three species are invasive pests.

Giant knotweed
4n = 44 chromosomes

Japanese knotweed
8n = 88 chromosomes

Bohemian knotweed
6n = 66 chromosomes (usually)

1. Using the example of *Fallopia*, explain how polyploidy can rapidly result in the formation of a new species:

_____

_____

_____

_____

2. Bohemian knotweed is officially a nothospecies: a hybrid. Its scientific name is still being disputed: it can be known as *Fallopia × bohemica*, *Reynoutria × bohemica*, or *Polygonum × bohemicum*. Why might it be difficult for scientists to name hybrids as a new species?

_____

_____

_____

3. Explain the difference between allopolyploidy and polyploidy: _____

_____

_____

A4.1
11
AHL

57 Earth's Biodiversity

Key Idea: Biodiversity is important in maintaining healthy species populations.

The vast majority of the genetic material in two unrelated individuals in a species is identical. It has to be or they would likely not be the same species. However, a small fraction of the genetic material (about 0.1% in humans) varies between individuals (compared to about 1% difference between humans and chimpanzees). This tiny variation results in the differences between individuals of the same species. These variations allow a species to adapt to changes in the environment, such as changes in food availability or the ability of the immune system to react to a disease. Larger differences in genetic variation are seen between species. Scientists can compare biodiversity of species, living and extinct, between different time periods, and use different methods to classify species, by 'lumping' or 'splitting'.

What is biodiversity?

Genetic diversity is the total number of genetic characteristics in a species. Genetic diversity is an important consideration in studies of biodiversity because species with high genetic diversity (low inbreeding) are generally less susceptible to disease and extinction.
Example: Coyotes have a high level of genetic diversity due to their abundance, wide distribution across North America, and hybridizations with grey wolves.

Species diversity is the number of different species (species richness) that are represented in a given community and their relative abundance (species evenness). High species diversity is associated with stable ecosystems and a large number of biotic interactions.
Example: The Raja Ampat Islands in Indonesia are considered the centre of marine biodiversity. The region is home to 75% of all known species of hard corals.

Habitat diversity describes the number of different habitats provided by a particular region. Habitat diversity is often described as heterogeneity and is associated with species diversity. More heterogeneous ecosystems can support a larger number of species with different habitat needs.
Example: The tropical climate of the Raja Ampat Islands provides an enormous range of marine and terrestrial habitats. It is also relatively undisturbed by humans.

Genetic diversity affects ecosystems

Genetic diversity is important in determining how well an ecosystem functions. Important components include:

- Fitness (survival and reproductive success) of individuals
- Long term viability and adaptability of populations
- Evolution of new species or traits
- Community structure and stability

Genetic diversity affects how members of the same species behave, e.g. some may be more aggressive or less curious. This, in turn, affects how they interact with other organisms and with the ecosystem as a whole. The millions of interactions from genetically diverse organisms ultimately affect how stable and resilient the ecosystem is.

1. What is the difference between genetic diversity and species diversity? _____

_____

_____

_____

_____

2. Why is it advantageous for a population to have high genetic diversity? _____

_____

_____

_____

_____

A4.2

1 - 2

©2024 **BIOZONE** International
ISBN: 978-1-99-101410-8
Photocopying prohibited

Changes in Earth's biodiversity over time

▶ The number of species, and hence the biodiversity of Earth, has changed over time. Evidence for the number of species on Earth in the past has been based on fossil evidence.

▶ Currently, around eight million species exist on Earth. However, this is an estimate as less than 20% have been identified and named, with most of the unknown species likely to be invertebrates.

▶ Scientists have calculated that 99% of total species that have existed have become extinct. Periods when there has been a sudden and large drop in biodiversity are known as mass extinction events. There have been five mass extinction events due to natural causes since the first organisms appeared in the fossil record, around 3.5 billion years ago.

Major events in biodiversity change

▶ An event known as the 'Cambrian explosion' occurred over a relatively short geologic time period (C - on graph, right) of around 20 million years, during the Early Cambrian Period (538 mya). The abrupt appearance in the fossil record of complex multicellular organisms represent most of the major phylums present on Earth today. Almost all life occupied aquatic ecosystems (shown as blue in graph) for the following 200 million years.

▶ Although scientists now hypothesize that simple plants and fungi may have been present on land for 700 million years, the Silurian (S),(420mya) marked the start of the first major terrestrial ecosystems (shown as brown in the graph). This included the evolution of more complex plant species, gradually followed by other terrestrial organisms

▶ A reduction in biodiversity followed the Permian-Triassic extinction (252 mya), brought about by the consequences of large scale environmental changes.

▶ The K-T extinction (66 mya) saw the disappearance of non-avian dinosaurs and many other species. However, this was followed by adaptive radiation of many groups, including the mammals, to fill vacant niches.

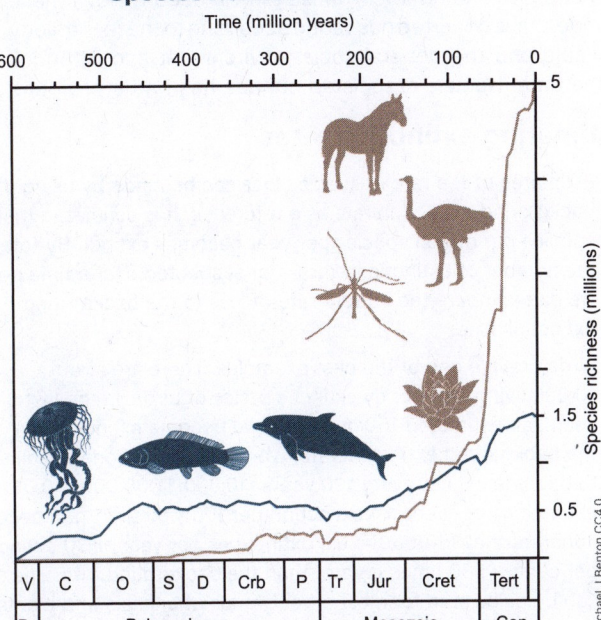

Species richness over time

Species richness in aquatic habitats show in blue, species richness in terrestrial habitats show in brown.

Lumping or splitting?

▶ Many species exhibit wide variation in physical traits between individuals in the same population. These differences can be even more obvious between male and females, and individuals in different life stages. However, scientists have access to tools that allow them to examine genetic differences to confirm the classification of species.

▶ Classification of extinct species is more difficult. Aside from a few frozen remains found in permafrost, usually the only evidence of past species comes from fossils. Therefore, classification is based on morphology alone.

▶ Hominid classification, a family including the great apes and humans, has mainly used the splitting approach, where each new fossil that shows a difference from those already classified is assigned as a new species.

▶ A recent find of five hominid skulls, located in a Republic of Georgia cave, and all slightly different, was initially declared to be evidence for multiple species occupying the same location, at the same time. However, with the principle of niche exclusivity in mind, another group of scientists hypothesized that the five skulls were all from one species and lumped them together. Their rationale was that a similar range of variation could also be seen in modern day ape species and humans. The classification method is still in dispute, and has ramifications for how we configure the tree of life.

3. Considering that the first organisms on Earth were present around 3.5 billion years ago, what are some possible explanations for why biodiversity in terrestrial habitats appeared to gradually increase only from around 400 mya?

_____

_____

_____

4. When classifying new fossil finds, scientists are often divided on a 'lumping" or 'splitting' methodology. Why is ongoing debate on classification advantageous to science?

_____

_____

_____

_____

58 The Sixth Mass Extinction

Key Idea: Evidence suggests that a current, sixth mass extinction event is due to anthropogenic (human) causes.

Human (anthropogenic) activity dominates Earth today. Humans can be found almost everywhere on the globe. As humans have spread across the planet, from Africa into Europe and Asia, and across into the Americas and beyond, and expanded in population size, they have changed the environment around them to suit their needs. How these changes have occurred has varied according to the technology available and the general social environment and attitudes at the time. Human-associated change has had a profound impact on the globe's physical and biological systems. Only in the past century have we begun to fully evaluate the impact of human activity on the Earth. Humans have caused the deliberate or accidental extinction of numerous species and brought many more to the brink. So many species have been lost as a direct or indirect result of human activity that this period in history has been termed the Sixth Extinction. There is debate over when this extinction began, its extent, and the degree of human involvement. However, it is clear that many species are being lost and many that existed before humans appeared in their domains are no longer with us.

Estimating extinction rates

▶ Estimates of the rate of species loss can be made by using the background extinction rate as a reference. It is estimated that one species per million species per year becomes extinct. By totalling the number of extinctions known or suspected over a time period, we can compare the current rate of loss to the background extinction rate.

▶ Birds provide one of the best examples. There are about 10,000 living or recently extinct species of birds. In the last 500 years, an estimated 150 or more have become extinct. From the background extinction rate, we would expect one species to become extinct every 100 years (10,000/1,000,000 = 0.01 extinctions per year = 1 extinction per 100 years). It then becomes apparent that 150/500 = 0.3 extinctions per year or 30 extinctions per century, 30 times greater than the background rate. The same can be calculated for most other groups of animals and plants.

Organism*	Total number of species (approx)*	Known extinction (since ~1500 AD)*
Mammals	5487	87
Birds	9975	150
Reptiles	10,000	22
Amphibians	6700	39
Plants	300,000	114

*These numbers vastly underestimate the true numbers because so many species are undescribed.

North Island giant moa

North Island giant moa (*Dinornis novaezealandiae*) were one of nine species of flightless moa, living exclusively in New Zealand. All moa were herbivores, filling the niche of terrestrial megafauna left vacant due to the absence of mammals. The 2-3m tall moa evolved from an ancestral ratite (flightless bird), present when New Zealand separated from Gondwana, 80 mya.

Humans are estimated to have arrived in New Zealand, as Polynesian settlers, just after 1300 AD, with evidence pointing to an initial group size of no more than 400 people. Research indicates that there was a stable and widespread population of the North Island giant moa prior to human arrival, but within 150 years they became extinct, along with all other moa species. The early age of Māori culture was known as the moa-hunter period.

Human activities such as hunting, egg gathering, the use of fire to destroy habitat, and the introduction of predators, e.g. the Polynesian dog, are believed to be the cause of moa extinction.

North Island giant moa reconstruction
DinosaursRoar CC-BY-SA

The rapid extinction of the moa provides insight to the Quaternary Megafauna extinction occurring between 10,000-50,000 years ago, when over 178 large mammal species, such as mammoth, giant kangaroo, and ground sloths, disappeared. It is thought that the arrival of humans in certain parts of the world correlates to the extinction of species in those areas.

1. Define the Sixth Mass Extinction: _____

_____

_____

2. Use the data in the table to calculate (1) the rate of species extinction per century for each of the groups, and (2) how many times greater this is than the background extinction rate:

(a) Mammals: _____

(b) Reptiles: _____

(c) Amphibians: _____

(d) Plants: _____

A4.2

3

Why is the sixth mass extinction different from previous mass extinctions?

There have been five previous major mass extinctions: short (by geological measurement) periods of time in which many distinct species die out. Evidence for the causes of these point to extreme temperature swings, ocean acidification, extensive volcanism, and asteroid strike, all occurring without any human influence (as humans hadn't yet evolved). The last mass extinction, 66 mya, saw the end of the non-avian dinosaurs and most marine reptiles. The Sixth Mass Extinction is the apparent, human-induced loss of much of Earth's biodiversity. Several human activities contribute to this: 40% of all land is used for food production; agriculture uses 70% of the world's available freshwater; 90% of deforestation occurs because of a need for more agricultural land. In fact, human activities to produce enough food to feed the population is the biggest threat to Earth's biodiversity. According to the World Wildlife Organisation, the rate of species extinction is 1,000 - 10,000 times higher than it would be if humans did not exist.

Caribbean monk seal

The Caribbean monk seal (*Neomonachus tropicalis*) is one of three species of monk seal adapted for warmer water. The seal clade evolved about 23 million years ago. Elephant seals, Antarctic seals, and monk seals separated from the other seal groups around 17 million years ago. Mitochondrial DNA sequencing has shown that the monk seals are the most distantly related to other seal species.

Caribbean monk seal reconstruction

Dinosaurs20 CC-BY-SA

The Caribbean monk seal was originally widely distributed across the Caribbean and the populations were numbered in the hundreds of thousands. Hunting for the seals started to increase once Europeans arrived in the area and huge numbers were slaughtered in the 17th-19th centuries for oil and food. By the 1950s, the Caribbean monk seal was extinct.

Two other species, the closely related Hawaiian monk seal and the Mediterranean monk seal, are both endangered and at risk of also becoming extinct. Monk seals are an important source of information for seal phylogeny. A recent fossil find of a possible fourth species of monk seal, now extinct, has been found on the western coast of New Zealand and sheds light on the origin of the group.

3. Humans are not always directly responsible for the extinction of a species, yet recently extinct species have often become so after humans arrive in their habitat. Discuss the reasons for these extinctions:

_____
_____
_____
_____

4. Why are findings about the North Island giant moa so significant to understanding more about the Quaternary Megafauna extinction, and what are some key features of the moa extinction?

_____
_____
_____
_____

5. How does the extinction of the Caribbean monk seal highlight the loss of valuable biodiversity?

_____
_____
_____
_____

6. Select one species of animal that has become extinct due to human activity, preferably local, and conduct a literature search. Present a case study, similar to the two provided above, and include information on your species origin, an image, details on the species' previous distribution and population number, and causes of extinction linked to humans.

59 Human Activity and Ecosystem Loss

Key Idea: Ecosystem loss around the world is accelerating and, in many cases, this is due to human activity.

Currently, around 50% of natural ecosystems have been lost in comparison to their original states. Some ecosystems are in danger of future collapse due to anthropogenic climate change, including coral reef systems and polar regions. The world is facing a biodiversity crisis that will require committed efforts to reverse. Data to inform scientists and conservationists is generated from many sources, such as the Intergovernmental Science-Policy Platform on Biodiversity and Ecosystem Services, and also citizen science surveys. Each type of data and associated methodologies has its advantages and disadvantages. Ecosystem loss is being driven by human population growth, and this is leading to many environmental issues, as demand on ecosystems is affected by human competition for resources and land.

Is there a current ecosystem loss crisis?

▸ The health of an ecosystem is dependent on complex interactions between biotic factors. The removal of any one element, such as the key species, the photosynthetic producers, or prey can result in eventual ecosystem collapse.

▸ Humans have demonstrably impacted the global environment. Over the past decades, scientists have documented the population decline of a growing number of plant and animal species. Habitat and ecosystem loss is the leading cause of these population declines globally. This comes from logging and burning forests, clearing land for agriculture, damming of rivers, and draining wetlands, destroying the habitat of many resident species. Species with restricted ranges are particularly vulnerable.

▸ It is clear that the world's ecosystems and biodiversity are under threat. The many factors causing this can be summarized as HIPPCO (Habitat destruction, Invasive species, (human) Population growth, Pollution, Climate change, and Over exploitation). Habitat destruction, and its consequent ecological changes, are a major component of HIPPCO.

Loss of mixed dipterocarp forest in Southeast Asia

Forest ecosystems in Southeast Asia are home to some of the most biodiverse habitats in the world. Unfortunately, these areas also have some of the highest rates of deforestation and ecosystem loss.

Mixed dipterocarp forests have a significant Dipterocarpaceae component. This family of plants is made up of around 700 species which provide a wide range of fruits, nectar, pollen, and habitats for countless other species. Some dipterocarp trees are over 80m tall and their valuable wood resource has made them a prime target for forestry.

Some areas of Southeast Asia have already lost over half of their original forestry cover to both deforestation for wood and conversion of land to monoculture plantations. These plantations have arisen due to demand for palm oil, rubber, and Eucalyptus and Acacia wood pulp. Mining and hydroelectric schemes also contribute to forest loss and degradation, and pollution of waterways. Encroaching urbanization from human population growth is spreading into forested regions.

Southeast Asian forest

Aside from habitat loss, the biodiversity in these forests has been impacted by hunting for food and trade. This often involves illegal poaching of protected species to profit from bushmeat sales and supplying the pet-trade.

1. Explain why the current rate of ecosystem loss across the globe is so significant:

_____
_____
_____
_____

2. Mixed dipterocarp forests are economically and socially important to Southeast Asia. How does this complicate efforts to reduce or prevent the loss of these ecosystems?

_____
_____
_____
_____

3. Select one ecosystem that has been lost due to direct or indirect human activity, preferably local. Present a case study report on the location, history, and key aspects of this lost ecosystem. Discuss the range of human activity that has been the main cause of the ecosystem loss and provide details on the consequences. The report can be in any format, including digital and paper. You may choose to work individually or in small groups.

NOS A4.2 4-6

©2024 **BIOZONE** International
ISBN: 978-1-99-101410-8
Photocopying prohibited

Evidence for a biodiversity crisis

▶ Evidence to support the claim that the world is facing a biodiversity crisis has come from many different researchers, government organizations, conservationists, and citizen science surveys.

▶ A wide ranging report recently released by the Intergovernmental Science-Policy Platform on Biodiversity and Ecosystem Services (IPBES) identified around 680 species of vertebrates becoming extinct due to human actions since the 1700s. The current extinction rate is tens to hundreds of times faster than the average rate of the last 10 million years.

▶ The International Union for Conservation of Nature (IUCN) also uses research to inform both governments and private companies about the current biodiversity crisis.

▶ Another source of evidence is from data collected by citizen science, where volunteers participate in projects by contributing surveys and observations in collaboration with scientists.

▶ Evidence, such as data from IUCN on insect biodiversity change (below), can highlight the extent of the biodiversity crisis and be used to promote new conservation policy and projects, which can prevent or reverse the current ecosystem loss.

Insect declines as an indicator for biodiversity loss

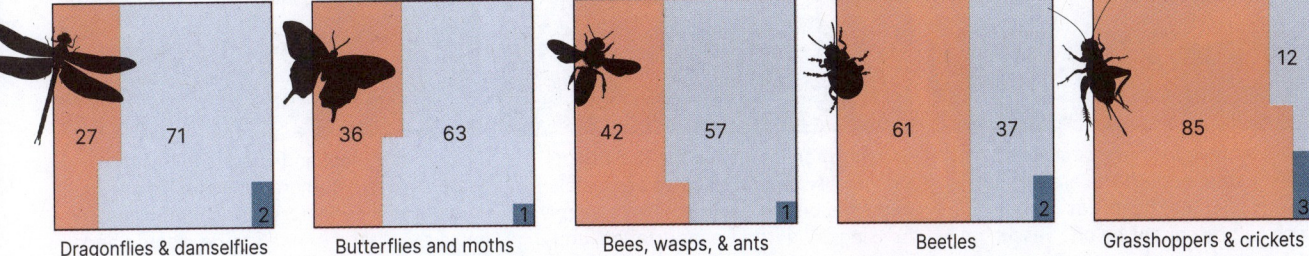

Dragonflies & damselflies	Butterflies and moths	Bees, wasps, & ants	Beetles	Grasshoppers & crickets
27 71 2	36 63 1	42 57 1	61 37 2	85 12 3

- 🟧 % of species decreasing
- ⬜ % of species stable
- 🟦 % of species increasing

Species in the five major insect orders (above) have all declined in recent decades. Of the 2200 species tracked by the IUCN, almost half are declining. These 2200 species represent just a tiny proportion of total insect biodiversity. Even with today's technological advancement, only 20% of insect species are even identified. This means some species will be lost without us knowing they even existed.

Verifying evidence

▶ Publishing scientific papers is a rigorous process involving peer-review. The results must be replicable by other scientists following the same methodology. Evidence from research that does not fulfil these requirements is rejected. This process adds validity to the collected data, but can be expensive, and time-consuming.

▶ Citizen science can involve millions of participants in a project. Many volunteers have a vested interest in protecting their local area and are prepared to spend large amounts of time and energy to help. Common digital devices can make data collection and sharing easier, leading to huge databases of valuable information. However, citizen science can increase bias due to errors in collection methods or species identification, or lack of repeated surveys. This can lower the validity of the data.

Observation recording for a citizen science project

4. Why is scientific evidence a vital factor when considering if a biodiversity crisis is occurring, and what is the importance of including repeated data collection in the methodology?

_____

_____

_____

_____

5. Summarize the advantages and disadvantages of using citizen science to measure ecosystem and biodiversity loss:

_____

_____

_____

_____

6. Become involved in a citizen science project which measures biodiversity, either in your local area or on a more global scale. Digital programmes such as Zooniverse or iNaturalist can be used to search for a wide range of suitable opportunities for participation. More links for citizen science projects are available in **BIOZONE Resource Hub**.

Potential causes leading to the current biodiversity crisis

▶ The human population is still expanding. According to the United Nations, it reached 8.0 billion in late 2022 and could peak at almost 10.4 billion by the mid 2080s. The current population is using up Earth's resources faster as they can be renewed by natural processes. This is a primary contributor to the current biodiversity crisis.

▶ Human demand for resources and the resulting activity is in direct competition with other species on Earth. Land must be developed in order to provide enough food, housing, and resources to sustain growth of the human population. Humans depend on Earth's ecosystems for the services they provide. These ecosystem services include resources such as food and fuel, as well as processes such as purification of the air and water. These directly affect human health.

▶ The biodiversity of an ecosystem affects its ability to provide these services. Biologically diverse and resilient ecosystems that are managed in a sustainable way are better able to provide the ecosystem services on which we depend.

Fishing techniques have become so sophisticated, and efforts are on such a large scale, that hundreds of tonnes of fish can be caught by one vessel, on one fishing cruise. As a result, some fish stocks have plummeted as the human population has increased and fish stocks have been over exploited. Many fish species are on the brink of collapse. The same trends can be seen in terrestrial populations, as wild animals are hunted for food, trophies, and pet supply.

Urbanization increases the density of the human population in an area, making demands on the environment to supply housing, industry, transportation, and waste management. Urban sprawl is the development of large areas of land for housing. As the human population grows, especially around cities and in wealthier countries, more houses are needed. These often use undeveloped or previously agricultural land, reducing habitats for other organisms.

Prior to ten thousand years ago, humans were a hunter-gatherer species. The change to a sedentary, agricultural lifestyle meant that land needed to be cleared and protected from wildlife for the exclusive use of humans and their livestock. One third of the world's land area is now dedicated exclusively to producing food for humans, other parts are mined for resources, and deforestation is driven by demands for timber, land for plantations, and urbanization.

Humans produce a huge amount of waste. Not all is, or can be, recycled and so it is dumped in landfills. Pollution is a major problem in parts of the oceans and on land. Activities causing pollution include deliberate dumping of rubbish, runoff from the land and contaminated discharge, and industry waste. 12.7 million tonnes of plastic finds its way into the sea each year, harming or killing wildlife. Microplastics can also have severe detrimental effects on marine life and are now found in every corner in the world.

Disease spread is facilitated by global transport of plants, animals, and humans; often inadvertently. Food and plants carry diseases across international borders, which is then spread to species that originally were protected by geographical barriers. Some countries, such as New Zealand, have strict biosecurity regulations to prevent spread. Foot and Mouth disease, carried in unpermitted meat, would decimate the meat and dairy industry. Similarly, a wider spread of rabies in rabies-free countries would be harmful.

Kudzu clambers over existing vegetation

Introduced species are those that have evolved at one place in the world and have been transported by humans, either intentionally or inadvertently, to another region. Some of these introductions are beneficial, e.g. introduced agricultural plants and animals. Invasive species are those that have a detrimental effect on the ecosystems into which they have been imported. They can include plants, like the kudzu above; or pests such as insects which damage plants; or animals, which out-compete or kill off existing species.

7. From the information above, summarize how human activities are contributing to the biodiversity crisis. Try to include local examples in your discussion (place longer answers on a sheet of paper and attach to this page):

_____

_____

_____

_____

_____

_____

8. Identify ONE biodiversity issue created by increasing human population. Research in depth and present your findings.

60 Conservation Strategies

Key Idea: *In-situ* (on site) conservation methods manage ecosystems to protect diversity within the natural environment. *Ex-situ* conservation methods operate away from the natural environment and are useful where species are critically endangered.

A variety of strategies are used to protect at-risk species and help the recovery of those that are threatened. Ecological restoration is a long term process and often involves collaboration between scientific institutions and the local communities involved.

Conservation strategies in the ecosystem

▶ *In-situ* conservation means conservation on site and it focuses on ecological restoration, including re-introduction of lost species, and legislation to protect ecosystems of special value.

▶ Nature reserves are areas designated as having wildlife, habitat, or natural formations needing protection, and often contain rare or nationally important species of plants or animals.

▶ Rewilding and reclamation of degraded ecosystems is a key component of *in-situ* conservation, and can include plants and animal species. It can also involve removing introduced species and pests.

A case study in *in-situ* conservation

Sanctuary Mountain Maungatautari is a 3400 hectare conservation 'mainland island', surrounded by 47 km of predator proof fence, situated in the North Island of New Zealand. Endemic New Zealand birds face a unique challenge, having evolved without mammalian predators, and many species are endangered due to introduced pests, such as possums, rats, and mustelids, as well as habitat loss. A project to recreate a pre-human temperate New Zealand ecosystem was established in 2001. The sanctuary has become an 'ark' to preserve birds, reptiles, insects, fungi, and plants.

How the project is proceeding

▶ Predator-proof fence built and maintained.

▶ Initial pest eradication and constant monitoring, both by rangers and a large volunteer group.

▶ Replanting of bare land into native species, vital for food sources and habitat for the New Zealand species.

▶ Reintroduction of endangered species originally extinct from the area, such as kākāpō, tuatara, hihi, kiwi, kākā, New Zealand robin, and takahē.

▶ Educate schools and visitors to promote conservation.

Advantages of *in-situ* conservation

▶ Species left in the protected area have access to their natural resources and breeding sites.

▶ Species will continue to develop and evolve in their natural environment, thus conserving their natural behaviour.

▶ *In-situ* conservation is able to protect many species at once and allow them greater space than those in captivity.

▶ *In-situ* conservation protects larger breeding populations.

▶ *In-situ* conservation is less expensive and requires fewer specialized facilities than captive breeding.

Disadvantages of *in-situ* conservation

▶ Controlling illegal exploitation of *in-situ* populations is difficult.

▶ Habitats that shelter *in-situ* populations may need extensive restoration, including pest eradication, and ongoing control.

▶ Populations may continue to decline during restoration.

Maungatautari mountain surrounded by farmland and lakes

Tuatara have been introduced back into Maungatautari. This unique reptile is the only remnant of its order, present before dinosaurs. It was once widespread through New Zealand, but was eventually reduced to living on a few small islands.

1. Explain why *in-situ* conservation commonly involves collaboration of many people and organizations to protect species:

_____

_____

_____

_____

_____

_____

 A4.2 7-8 NOS

Ex-situ conservation: out of the ecosystem

Ex-situ conservation is the process of protecting an endangered species outside its natural habitat. It is used when a species has become critically low in numbers or *in-situ* methods have been, or are likely to be, unsuccessful. Zoos, aquaria, and botanical gardens are the most conventional facilities for *ex-situ* conservation. They house and protect specimens for captive breeding programmes, with a focus on increasing genetic diversity, and can reintroduce them into the wild to restore natural populations. The maintenance of seed banks by botanic gardens and breeding registers by zoos ensures that efforts to conserve species are not impaired by problems of inbreeding.

The important role of zoos and aquaria

In the past, zoos were created as a form of entertainment for people who were unlikely to view the species in the wild. The animals were often exotic and kept in unsuitable small cages or enclosures.

Modern zoos tend to concentrate on particular species and their key goal has now become conservation and potential reintroduction of bred animals back into their natural habitats. Zoos are part of global programmes that work together to help retain genetic diversity in captive bred animals.

In addition to their role in captive breeding programmes and as custodians of rare species, zoos have a major role in public education. They raise awareness of the threats facing species in their natural environments and engender public empathy for conservation work.

Some species, such as the New Guinea singing dog, Przewalski's wild horse, and the European bison would most likely be extinct without the conservation efforts made by the zoos in which they have been bred.

Above: The okapi is a species of rare forest antelope related to giraffes. Okapi are only found naturally in the Ituri Forest, in the northeastern Congo rainforests. An okapi calf was born to Bristol Zoo Gardens (UK) in 2009, one of only about 100 okapi in captivity.

The role of botanic gardens

Botanic gardens have years of collective expertise and resources and play a critical role in plant conservation. They maintain seed banks, nurture rare species, maintain a living collection of plants, and help to conserve indigenous plant knowledge. They also have an important role in both research and education.

The Royal Botanic Gardens at Kew, London (above) contains an estimated 25,000 species, 2,700 of which are classified by the ICUN as rare, threatened, or endangered. Kew Gardens is involved in both national and international projects associated with the conservation of botanical diversity and is the primary advisor to CITES on threatened plant species. Kew's Millennium Seed Bank partnership is the largest *ex-situ* plant conservation project in the world: Working with a network of over 50 countries, they have banked 10% of the world's wild plant species.

Seed banks and gene banks

Seedbanks and gene banks around the world have a role in preserving the genetic diversity of species.

A seedbank stores seeds as a source for future planting in case seed reserves elsewhere are lost. The seeds may be from rare species whose genetic diversity is at risk, or they may be the seeds of crop plants, in some cases of ancient varieties no longer used in commercial production. The Svalbard Global Seed Vault, Norway (above), has over 1.2 million seeds, with capacity for millions more.

2. Compare and contrast *in-situ* and *ex-situ* methods of conservation, including reference to the advantages and disadvantages of each approach (place longer answers on a sheet of paper and attach to this page):

    _____

    _____

    _____

    _____

    _____

    _____

Evolutionarily distinct and globally endangered species (EDGE)

▸ Some endangered species are phylogenetically unique, and may be one of only a few, or even the only, remaining representatives of their family or order. Examples already discussed are the New Zealand tuatara and the two species of monk seal. Other examples include the pygmy sloth, Chinese giant salamander, and the Ganges River Dolphin.

▸ For some species, extinction has already occurred and therefore their unique genetics has been lost to science. Recently extinct species include the thylacine, commonly known as the Tasmanian tiger, which was the largest known example of carnivorous marsupial that had survived into modern times. The last individual died in 1936 in an Australian zoo.

▸ The EDGE of Existence programme has been developed by The Zoological Society of London and promotes prioritized conservation efforts to save valuable species at high risk of extinction. They engage in research, support conservation efforts and the work of conservationists, and lobby policy makers to protect targeted species.

Saving the kākāpō

▸ The kākāpō is the only flightless, nocturnal parrot species on Earth. It is endemic to New Zealand and is found there on a few highly protected offshore islands. It has recently been introduced to Sanctuary Mountain Maungatautari.

▸ Kākāpō were once widespread across New Zealand but the first Polynesian settlers in the mid 1300s found them easy to catch as food. The birds had adapted to freeze in place when they felt threatened due to the absence of co-evolved predators.

▸ By the time of European arrival, 500 years later, kākāpō population numbers had already dwindled. Introduced pests decimated the remaining birds which were thought to be extinct until an isolated population was discovered.

▸ Intensive conservation measures have helped numbers recover, from 51 individuals in 1995 when the Kākāpō Recovery Plan began, to 248 as of 2023.

▸ Conservation methods include pest free reserves, vigilant monitoring using pest detector dogs (right insert), radio tracking, supplementary feeding, hand rearing, and artificial reproduction.

New Zealand kākāpō

Prioritizing conservation efforts

▸ Saving species and conserving ecosystems can be a huge financial task, often involving input from scientists, organizations, and volunteers. Lack of resources, including funding, means that some species are preferentially conserved over others.

▸ The EDGE of Existence programme is an organization that shines a spotlight on more 'flagship' species such as the kākāpō, above, to lobby for their protection.

▸ Deciding which species to prioritize for conservation efforts raises many ethical, environmental, political, social, cultural, and economic questions. Different groups have vested interests in conserving different species.

3. Kākāpō are a 'flagship species': an ambassador for conservation, much like the giant panda. Their popular status has raised millions of dollars from businesses, government, and the general public to fund their hugely expensive conservation programme. What might be some advantages and disadvantages of this type of funding model?

_____

_____

_____

_____

4. Read more about the goals of The EDGE of Existence programme using the link in the **BIOZONE Resource Hub**. Do you consider the organization's approach to be an effective means of conservation? Explain your rationale.

_____

_____

_____

_____

_____

 5. Using a debate format, identify an endangered species, preferably local, to prioritize for conservation. Divide into two groups, and plan your debate points to cover the implications listed on this page. Argue yay or nay for prioritization.

©2024 **BIOZONE** International
ISBN: 978-1-99-101410-8
Photocopying prohibited

61 Did You Get It?

1. Molecular sequencing evidence, such as DNA similarity (below, right), has increasingly been used to understand common ancestry. Prior to this technology becoming available, only observations of living species and anatomical homology of fossilized remains of extinct species could be utilized.

(a) Why is molecular sequencing so useful as evidence for evolution?

_____

_____

_____

(b) How can anatomical homology be useful to classify fossils, where molecular sequencing cannot?

_____

_____

_____

Similarity of human DNA to that of other primates

DNA similarity (%)

Primate species	DNA similarity
Human	100%
Chimpanzee	97.6%
Gibbon	94.7%
Rhesus monkey	91.1%
Vervet monkey	90.5%
Capuchin monkey	84.2%
Galago	58.0%

(c) **NOS:** The number of species, including all extinct species, of the hominid family (apes and humans) has fluctuated over the past century. Explain the difference in 'lumping' and 'splitting' approaches:

_____

_____

2. Charles Darwin published his theory of evolution by natural selection in 1859. In the years preceding, he was an enthusiast of 'fancy' domesticated pigeons (right), that could be bred to a wide range of types with different traits in shape and colour.

(a) Fancy pigeon breeding is a type of selective breeding. Define this term:

_____

_____

(b) Darwin saw selective breeding as a 'speeded up version' of natural selection. What features are the same in both?

_____

_____

(c) **AHL:** Domesticated pigeons originated from the Rock dove. Populations of the wild-type bird can be found in the Northern United Kingdom. The domestic pigeon and rock dove frequently hybridize but less so in isolated Scottish islands. Explain why speciation between the rock dove and domestic pigeon is more likely to occur on these islands:

_____

_____

_____

3. Caribbean coral reefs are at extreme risk of ecosystem loss, due to climate change, overfishing, and pollution.

(a) Scientists are collecting coral sperm and eggs, and raising embryos in controlled nurseries on land. What type of conservation is this, and why might this be a more successful method than just protecting the immediate reef area?

_____

_____

_____

(b) The coral sperm and eggs come from many sites to ensure genetic diversity. Why is this important?

_____

_____

62 Summary Assessment

Read each question carefully. Circle the letter that provides the best answer to the question from the four answer choices provided.

1. The two plants shown below are unrelated. The left hand image shows a cactus from North America, while the right hand image shows a *Euphorbia* from Africa. Both these plants live in deserts.

Cactus

Euphorbia

This pattern of evolution is:

(a) Divergent evolution

(b) Sequential evolution

(c) Parallel evolution

(d) Convergent evolution

2. What is the correct DNA base pairing?

(a) C-G, T-U

(b) C-G, T-A

(c) G-T, C-A

(d) G-C, U-A

3. **AHL:** The Miller-Urey experiments investigated:

(a) Natural selection

(b) Cell theory

(c) Semi-conservative DNA replication

(d) Conditions for the origin of life

4. Which structure is absent from all prokaryote cells?

(a) Mitochondria

(b) Ribosomes

(c) Cell membrane

(d) DNA

5. What is the correct chromosome number for humans and chimpanzees?

(a) 46 & 47

(b) 47 & 46

(c) 48 & 47

(d) 46 & 48

6. What is the most direct cause of the sixth mass extinction?

(a) Anthropogenic impacts

(b) Meteorite impact

(c) Volcanism

(d) Climate change

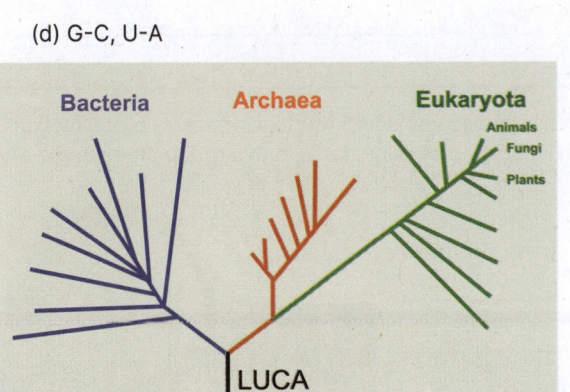

7. (a) What are the key structural differences between bacteria (prokaryote) cells and eukaryote cells?

_____

_____

_____

_____

(b) **AHL:** What is the relevance of LUCA , on the left?

_____

_____

_____

(c) Commonality of nucleic acid supports the LUCA hypothesis: Draw a nucleotide and explain how this structure combines into nucleic acid to allow for genetic information to be replicated and expressed:

_____

_____

_____

_____

©2024 **BIOZONE** International
ISBN: 978-1-99-101410-8

8. (a) There are around 129 species of butterflyfish. The yellow masked butterflyfish (*Chaetodon semilarvatus*) (labelled 24, below, left) is also called the bluecheek butterfly fish. Why is the scientific name important for identification?

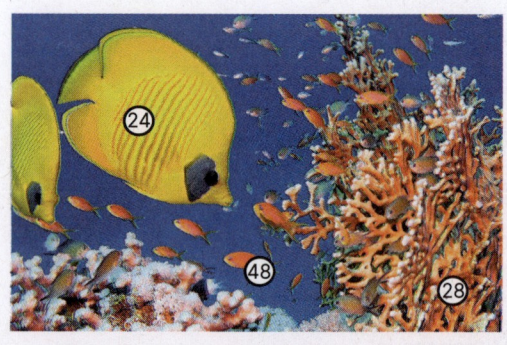

_____

_____

_____

(b) **AHL:** All the butterflyfish form their own clade. However, the red anthias (numbered 48, left) and the butterfly fish could also be in a clade. Explain how this could be:

_____

_____

_____

(c) The butterfish, red anthias, and fire coral (A. millepora) all have the chromosome numbers given in the photograph above. Explain why the chromosome number does not necessarily indicate the relatedness between species:

_____

_____

_____

(d) **AHL:** How could evidence from amino acid sequences be used to develop a cladogram of the three species?

_____

_____

9. Many coral ecosystems, such as the one above, are at risk of anthropogenic extinction. How is this different from mass extinction events in the past?

_____

_____

10. Why is *in situ* conservation, rather than *ex situ*, the more preferred method conservation for coral reef ecosystems?

_____

_____

_____

11. (a) The Rabdophorus group of butterflyfish is made up of around 35 species, including the black-backed butterfly fish (*Chaetodon melannotus*) (right) and the yellow masked butterflyfish (top of page, left). Is this group an example of convergent or divergent evolution? Explain your answer:

_____

_____

_____

(b) The spotfin butterflyfish (*Chaetodon ocellatus*) is found in the Atlantic Ocean (red line), as is the banded butterflyfish (*C. striatus*), but the foureye butterflyfish (*C. capistratus*) is located in the Caribbean (yellow line). Explain how the process of speciation could account for all 3 species of butterflyfish:

_____

_____

_____

_____

_____

The black-backed butterflyfish (top, right) and the Spotfin butterflyfish (bottom, left).

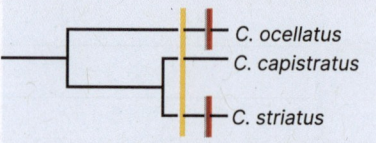

C. ocellatus
C. capistratus
C. striatus

©2024 **BIOZONE** International
ISBN: 978-1-99-101410-8
Photocopying prohibited

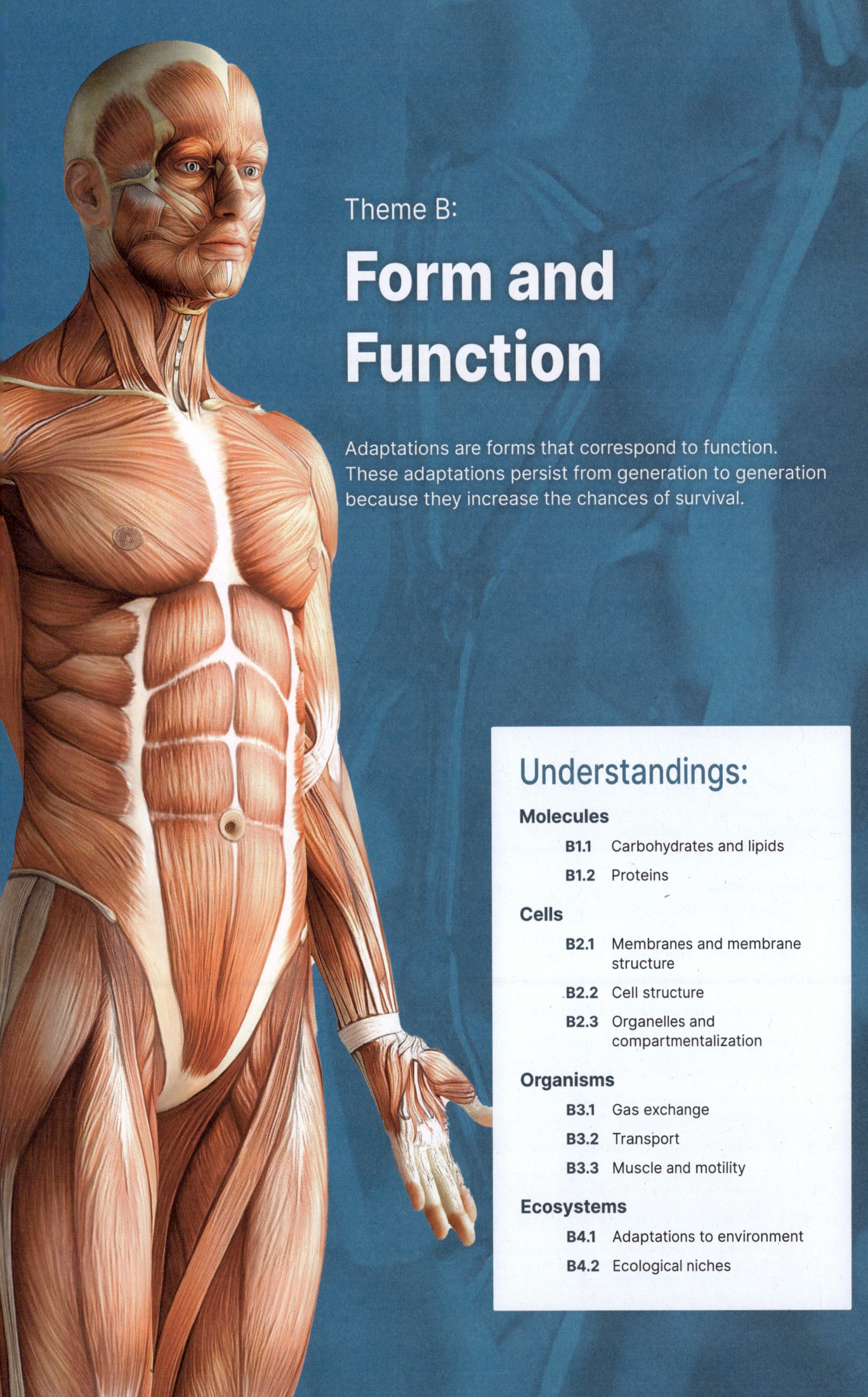

Theme B:

Form and Function

Adaptations are forms that correspond to function. These adaptations persist from generation to generation because they increase the chances of survival.

Understandings:

Molecules

    B1.1 Carbohydrates and lipids

    B1.2 Proteins

Cells

    B2.1 Membranes and membrane structure

    B2.2 Cell structure

    B2.3 Organelles and compartmentalization

Organisms

    B3.1 Gas exchange

    B3.2 Transport

    B3.3 Muscle and motility

Ecosystems

    B4.1 Adaptations to environment

    B4.2 Ecological niches

Molecules

Resource Hub
bit.ly/47G272u

B1.1 Carbohydrates and lipids

	Activity Number

Guiding Questions:
▶ In what ways do variations in form allow diversity of function in carbohydrates and lipids?
▶ How do carbohydrates and lipids compare as energy storage compounds?

Learning Outcomes:

☐ 1	Describe the nature of a covalent bond. Relate the covalent bond to the wide variety of carbon based (organic) molecules.	63
☐ 2	Identify the reactants and products of condensation polymerization reactions: polysaccharides, polypeptides, and nucleic acids.	64, 65
☐ 3	Identify the reactants and products of hydrolysis reactions including the splitting of water molecules.	64, 65
☐ 4	Recognize pentoses and hexoses as monosaccharides and link their structure to their function.	64, 65
☐ 5	Link the physical properties of polysaccharides to their function as energy storage substances.	66, 67
☐ 6	Relate the structure of cellulose to its function as a structural polysaccharide in plants.	66, 67
☐ 7	Explain the role of glycoproteins in cell–cell recognition using the example of ABO antigens.	67
☐ 8	Describe the hydrophobic properties of lipids, including fats, oils, waxes, and steroids.	68
☐ 9	Identify the reactants and products of condensation reactions in forming triglycerides and phospholipids.	68
☐ 10	Compare and contrast the structure and physical properties of saturated, monounsaturated and polyunsaturated fatty acids, relating these to their function as energy storage substances in plants and endotherms.	68
☐ 11	Link the properties of triglycerides in adipose tissues to their role in energy storage and thermal insulation.	68
☐ 12	Use the term amphipathic to describe the nature of phospholipids and relate this to phospholipid bilayers.	69
☐ 13	Identify steroids from molecular diagrams and explain the mechanism for steroids to pass through a phospholipid bilayer.	70

B1.2 Proteins

	Activity Number

Guiding Questions:
▶ What is the relationship between amino acid sequence and diversity in the form and function of proteins?
▶ How are protein molecules affected by their chemical and physical environments?

Learning Outcomes:

☐ 1	Draw a labelled structure of an amino acid showing the main groups.	71
☐ 2	From models, write word equations and draw the reactants and products of condensation reactions forming dipeptides and longer chains of amino acids.	71
☐ 3	Justify the dietary requirements for amino acids, distinguishing between essential and non-essential sources.	72
☐ 4	Link the order and type of amino acids to the formation of infinite possible peptide chains.	72
☐ 5	Investigate the effect of pH and temperature on protein denaturation.	72
☐ 6	**AHL:** Classify amino acids into groups based on their properties, which are determined by different R-groups.	73
☐ 7	**AHL:** Discuss the relationship between the primary protein structure and the complexity of the final 3-dimensional form.	74
☐ 8	**AHL:** Detail the process of secondary structure protein formation.	74
☐ 9	**AHL:** Explain the significance of hydrogen, ionic, and disulfide covalent bonds and hydrophobic interactions for forming the tertiary structure of proteins.	73, 74
☐ 10	**AHL:** Investigate the effect that polar and non-polar amino acids have on the tertiary structure of proteins.	74
☐ 11	**AHL:** Contrast conjugate and non-conjugate proteins using insulin and haemoglobin as examples. **NOS:** Investigate new technologies used to calculate the structure of proteins to the atomic level, such as cryogenic electron microscopy.	74
☐ 12	**AHL:** Contrast globular and fibrous proteins, relating their form to their function, using insulin and collagen as examples.	75

63 Carbon Chemistry

Key Idea: The carbon atom can form four covalent bonds with many other elements. Chains of carbons atoms are the backbone of organic molecules.

Molecular biology is a branch of science that studies the molecular basis of biological activity. All life is based around carbon, which is able to combine with many other elements to form a large number of carbon-based (or organic) molecules. Specific groups of atoms, called functional groups, attach to a C-H core and determine the specific chemical properties of the molecule. The organic macromolecules that make up living things can be grouped into four classes: carbohydrates, lipids, proteins, and nucleic acids.

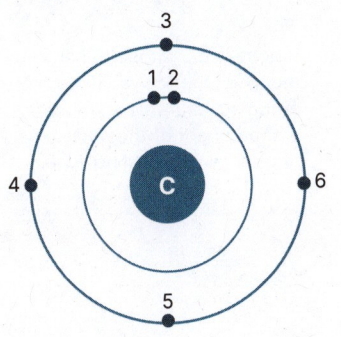

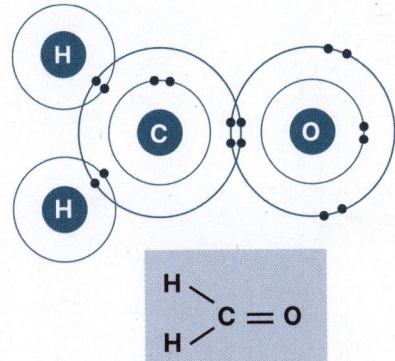

Organic macromolecule	Structural unit	Elements
Carbohydrates	Sugar monomer	C, H, O
Proteins	Amino acid	C, H, O, N, S
Lipids	Not applicable	C, H, O
Nucleic acids	Nucleotide	C, H, O, N, P

A carbon atom (above) has four electrons that are available to form up to four covalent bonds with other atoms. A covalent bond forms when two atoms share a pair of electrons. The number of covalent bonds formed between atoms in a molecule determines the shape and chemical properties of the molecule.

Carbon can form simple organic molecules. In methanal (CH_2O), a carbon (C) atom bonds with two hydrogen (H) atoms and an oxygen (O) atom. In the structural formula (blue box), the bonds between atoms are represented by lines.

The most common elements found in organic molecules are carbon, hydrogen, and oxygen, but organic molecules may also contain other elements, such as nitrogen, phosphorus, and sulfur. Most organic macromolecules are built up of one type of repeating unit or monomer (building block), except lipids, which are quite diverse in structure.

The variety of carbon molecules

Carbon atoms can form covalent bonds with other carbons atoms and nonmetals. In doing so, it can form a virtually endless number of complex molecules including long chains, rings, and sheets. It is this variety of organic molecules that has allowed the complex chemistry of life to evolve. Note that in molecular modelling, by convention, C is shown as black, H as white, O as red, and N as blue.

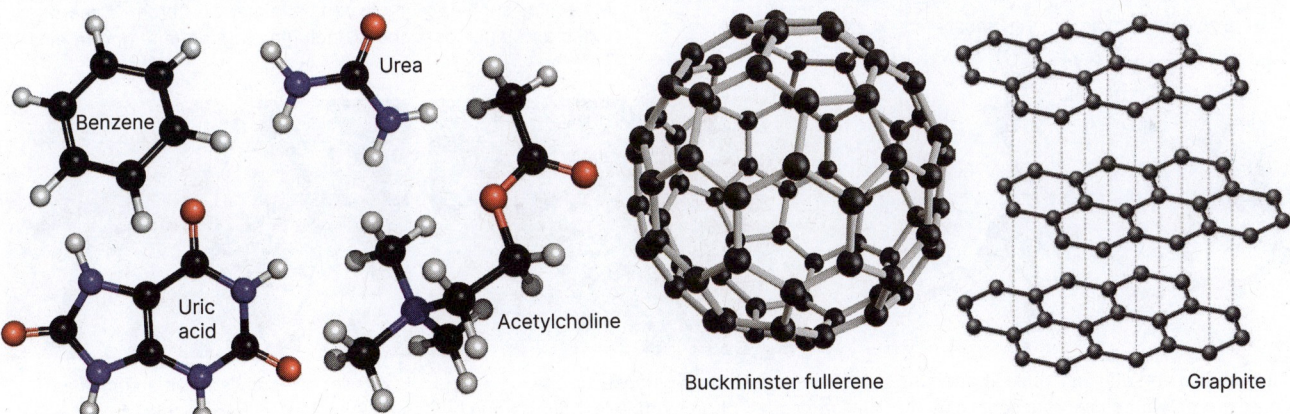

Benzene

Urea

Uric acid

Acetylcholine

Buckminster fullerene

Graphite

1. What feature of the carbon atom allows it to be the basis of so many different kinds of molecules?

_____

2. On the diagram of the carbon atom (top left) identify which of the six electrons (labelled 1-6) are available to form covalent bonds with other atoms?

_____

3. Using resources available on **BIOZONE's Resource Hub**, identify and draw/sketch ways in which carbon molecules can be represented:

B1.1
1

64 Carbohydrate Chemistry

Key Idea: Monosaccharides are the building blocks for larger carbohydrates. They can exist as isomers.

Sugars (monosaccharides and disaccharides) play a central role in cells, providing energy and joining together to form carbohydrate macromolecules, such as starch and glycogen.

Monosaccharide polymers form the major component of most plants (as cellulose). Monosaccharides are important as a primary energy source for cellular metabolism. Carbohydrates have the general formula $C_x(H_2O)_y$, where x and y are variable numbers (often but not always the same).

Monosaccharides

▸ Monosaccharides are single-sugar molecules and include glucose (grape sugar and blood sugar), fructose (honey and fruit juices), and galactose (dairy products). They are used as a primary energy source for fuelling cell metabolism.

▸ They can be joined together to form disaccharides (two monomers) and polysaccharides (many monomers).

▸ Monosaccharides can be classified by the number of carbon atoms they contain. Some important monosaccharides are the hexoses (6 carbons) and the pentoses (5 carbons). The most common arrangements found in sugars are hexose (6 sided) or pentose (5 sided) rings (below).

▸ The commonly occurring monosaccharides contain between three and seven carbon atoms in their carbon chains and, of these, the 6C hexose sugars occur most frequently. All monosaccharides are reducing sugars (they can participate in reduction reactions).

Examples of monosaccharide structures

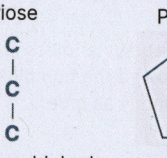
Triose — e.g. glyceraldehyde

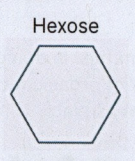
Pentose — e.g. ribose deoxyribose

Hexose — e.g. glucose, fructose, galactose

Ribose: a pentose monosaccharide

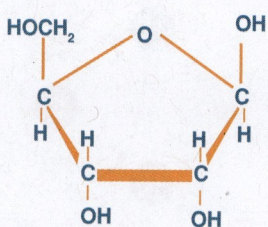

Ribose is a pentose (5 carbon) monosaccharide which can form a ring structure (left). Ribose is a component of the nucleic acid ribonucleic acid (RNA).

Glucose isomers

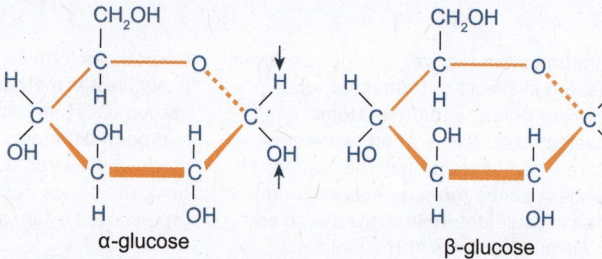

α-glucose   β-glucose

Isomers are compounds with the same chemical formula (same types and numbers of atoms) but different arrangements of atoms. The different arrangement of the atoms means that each isomer has different properties.

Molecules such as glucose can have many different isomers, e.g. α and β glucose, above, including straight and ring forms.

Glucose is a versatile molecule. It provides energy to power cellular reactions, can form energy storage molecules such as glycogen, or it can be used to build structural molecules.

Plants make their glucose via the process of photosynthesis. Animals and other heterotrophic organisms obtain their glucose by consuming plants or other organisms.

Fructose, often called fruit sugar, is a simple monosaccharide. It is often derived from sugar cane (above). Both fructose and glucose can be directly absorbed into the bloodstream.

1. Describe the two major functions of monosaccharides:

   a) _____

   b) _____

2. Describe the structural differences between the ring forms of glucose and ribose: _____

   _____

3. Using glucose as an example, define the term isomer and state its importance: _____

   _____

©2024 **BIOZONE** International
ISBN: 978-1-99-101410-8
Photocopying prohibited

B1.1
2-4

65 Condensation and Hydrolysis of Sugars

Key Idea: Condensation and hydrolysis reactions form and break down disaccharides, respectively.

Monosaccharides can be linked together by condensation reactions to produce disaccharides (and polysaccharides) by dehydration synthesis (building by loss of water). The reverse reaction, hydrolysis, breaks compound sugars down into their constituent monomers. Disaccharides are produced when two monosaccharides are joined together. Different disaccharides are formed by joining together different combinations of monosaccharides.

Condensation and hydrolysis reactions

Monosaccharides can combine to form compound sugars in what is called a condensation reaction. Compound sugars can be broken down by hydrolysis to simple monosaccharides.

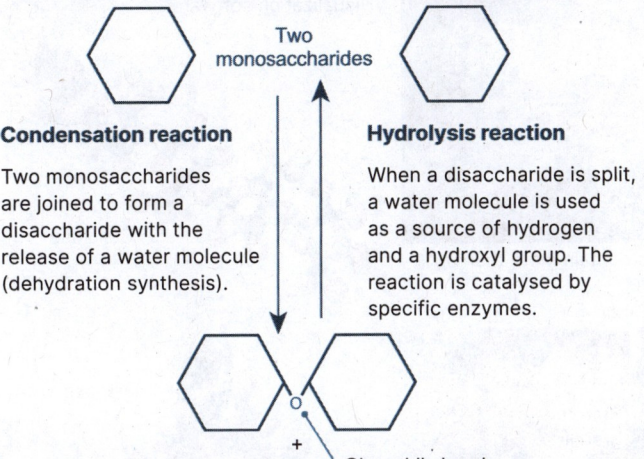

Condensation reaction

Two monosaccharides are joined to form a disaccharide with the release of a water molecule (dehydration synthesis).

Hydrolysis reaction

When a disaccharide is split, a water molecule is used as a source of hydrogen and a hydroxyl group. The reaction is catalysed by specific enzymes.

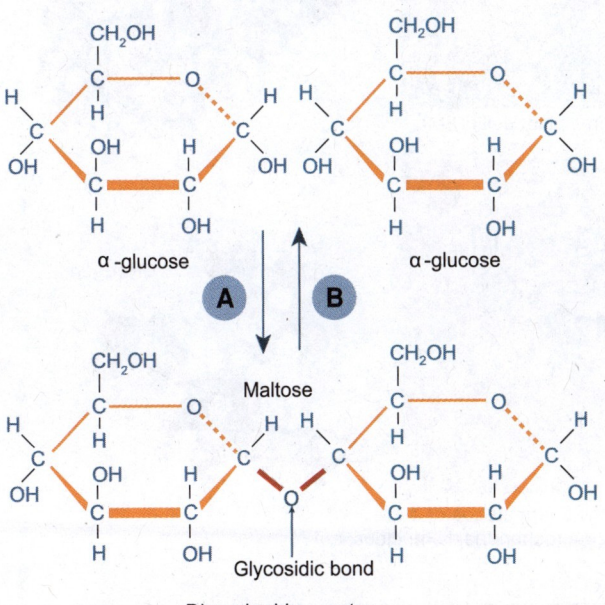

Disaccharides

Disaccharides (below) are double-sugar molecules and are used as energy sources and as building blocks for larger molecules. They are important in human nutrition and are found in milk (lactose), table sugar (sucrose), and malt (maltose).

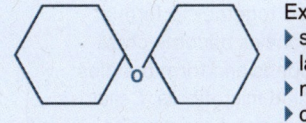

Examples
▶ sucrose
▶ lactose
▶ maltose
▶ cellobiose

The type of disaccharide formed depends on the monomers involved and whether they are in their α- or β- form. Only a few disaccharides, e.g. lactose, are classified as reducing sugars. Some common disaccharides are described below.

Lactose, a milk sugar, is made up of β-glucose + β-galactose. Milk contains 2-8% lactose by weight. It is the primary carbohydrate source for suckling mammalian young.

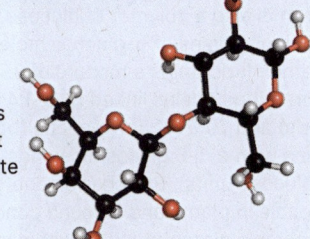

Maltose is composed of two α-glucose molecules. Germinating seeds contain maltose because the plant breaks down its starch stores to mobilize glucose as an energy source for growth.

Sucrose (table sugar) is a simple sugar derived from plants such as sugar cane, sugar beet, or maple sap. It is composed of an α-glucose molecule and a β-fructose molecule.

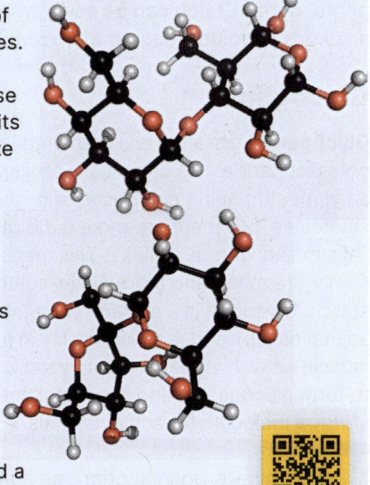

1. Explain briefly how disaccharide sugars are formed and broken down: _____

_____

_____

_____

2. On the diagram above, name the reaction occurring at points **A** and **B** and name the product that is formed:

_____

_____

3. What determines the disaccharide made by condensation? _____

©2024 **BIOZONE** International
ISBN: 978-1-99-101410-8
Photocopying prohibited

B1.1

2-4

66 Polysaccharides

Key Idea: Polysaccharides consist of many monosaccharides joined together by condensation reactions. Their composition and isomerization alter their functional properties.

Polysaccharides (complex carbohydrates) are straight or branched chains of many monosaccharides joined together. They can consist of one or more types of monosaccharide.

The most common polysaccharides are cellulose, starch, and glycogen. All contain only glucose, but their properties are very different. These differences are a function of the glucose isomer involved and the types of linkages joining them. Different polysaccharides can either be a source of readily available glucose or, alternatively, a structural material that resists digestion.

Cellulose

Cellulose is a structural material found in the cell walls of plants. It is made up of unbranched chains of β-glucose molecules held together by β-1,4 glycosidic links. As many as 10,000 glucose molecules may be linked together to form a straight chain. Parallel chains become cross-linked with hydrogen bonds and form bundles of 60-70 molecules called microfibrils. Cellulose microfibrils are very strong and are a major structural component of plants, e.g. as the cell wall. Few organisms can break the β-linkages, so cellulose is an ideal structural material.

Starch

Starch is also a polymer of glucose, but it is made up of long chains of α-glucose molecules linked together. It contains a mixture of 25-30% amylose (unbranched chains linked by α-1,4 glycosidic bonds) and 70-75% amylopectin (branched chains with α-1, 6 glycosidic bonds every 24-30 glucose units). Starch is an energy storage molecule in plants and is found concentrated in insoluble starch granules within specialized plastids called amyloplasts in plant cells (see photo, right). Starch can be easily hydrolyzed by enzymes to soluble sugars when required.

Glycogen

Glycogen, like starch, is a branched polysaccharide. It is chemically similar to amylopectin, being composed of α-glucose molecules, but there are more α-1,6 glycosidic links mixed with α-1,4 links. This makes it more highly branched and more water-soluble than starch. Glycogen is a storage compound in animal tissues and is found mainly in liver and muscle cells. It is readily hydrolyzed by enzymes to form glucose, making it an ideal energy storage molecule for active animals.

Cotton fibres contain more than 90% cellulose fibre.

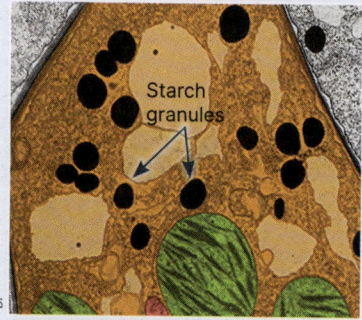

Starch granules in a plant cell (TEM).

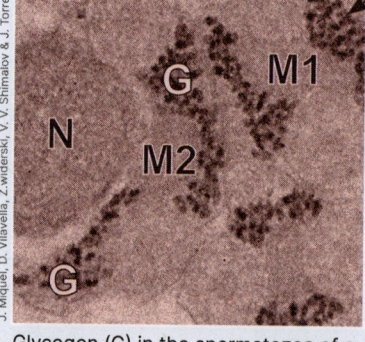

Glycogen (G) in the spermatozoa of a flatworm. M1, M2=mitochondria, N=nucleus.

The structure of polysaccharides can be compared using molecular visualization software

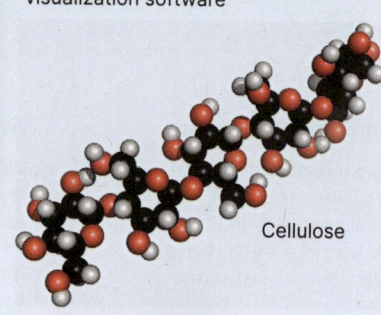

Cellulose

Amylose

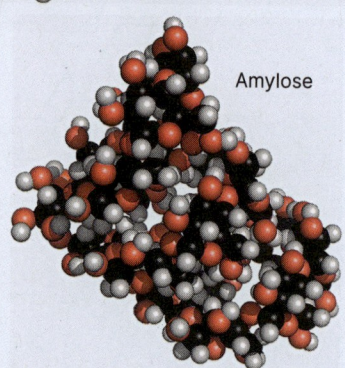

Glycogen

1. a) Why are polysaccharides such a good source of energy? _____

_____

b) How is the energy stored in polysaccharides mobilized? _____

_____

2. Contrast the properties of the polysaccharides starch, cellulose, and glycogen and relate these to their roles in the cell:

_____

_____

_____

_____

_____

B1.1

5, 6

©2024 **BIOZONE** International
ISBN: 978-1-99-101410-8
Photocopying prohibited

67 Functions of Saccharide Polymers

Key Idea: Isomers of glucose play important functional and structural roles as polymers in plants and animals.
Glucose monomers can be linked in condensation reactions to form large structural and energy storage polysaccharides. The glucose isomer involved and the type of glycosidic linkage determines the properties of the molecule. Starch is a storage carbohydrate made up of two α-glucose polymers. Cellulose is a structural β-glucose polymer. In animals, glucose is stored as glycogen. Short glucose polymers called oligosaccharides play important roles as signalling molecules in the cell's plasma membrane. Oligosaccharides attach to proteins to form glycoproteins, or lipids to form glycolipids.

Structure

Glucose and glucose derivatives play important roles in producing structural polymers. Two of the most important are cellulose (plants) and chitin (arthropods). Cellulose is a polymer of glucose. Chitin is a polymer of the glucose derivative N-Acetylglucosamine. In both cases, the glucose molecules are joined with β-1, 4 glycosidic bonds. Cellulose and chitin are the two most abundant polymers in nature.

Storage

In plants, glucose is stored as starch. Starch is manufactured and stored in amyloplasts: non-pigmented storage organelles within plant cells. Starch consists of two types of molecules: the linear and helical amylose and the branched amylopectin. In animals, glucose is stored as glycogen. Glycogen and amylopectin have very similar structures except that glycogen is more highly branched with shorter branches.

Signalling

Derivatives of glucose attached to proteins in the plasma membrane of cells play a role in cell to cell signalling and as binding sites for hormones. Unlike storage and structural polymers of glucose, glycoproteins use short chains of mostly galactose molecules called oligosaccharides. These may only incorporate four or five monomers.

In humans, the glycoproteins are important in the ABO blood group system, forming the A, B, or O antigens on a blood cell's surface. In the diagram below, the differences between the glycoproteins of the A, B, or O blood types can be seen. Note that for the O blood type, the terminal galactose monomer is missing. The differences in these glycoproteins result in the cell's ability to be recognized by the anti-A or anti-B antigens of the immune system.

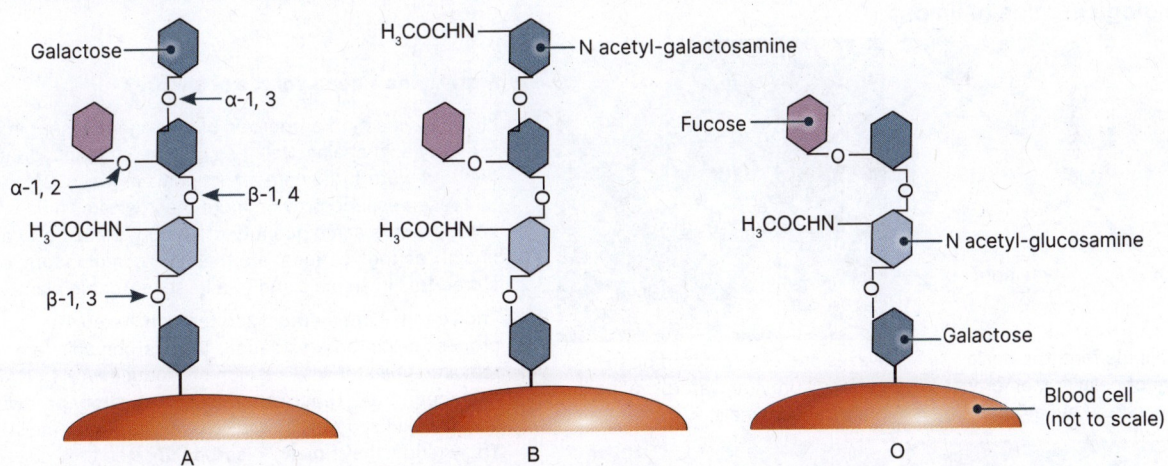

1. What are the differences in structure and bonding between cellulose and amylose?

2. Cellulose is indigestible to humans. Based on structure and bonding, would you expect chitin to be digestible or indigestible. Why?

3. What are the major differences between the structures of the glycoproteins that make up the ABO blood group?

©2024 BIOZONE International
ISBN: 978-1-99-101410-8
Photocopying prohibited

68 Lipids

Key Idea: Lipids are large, non polar, hydrophobic molecules that can provide energy and insulation.

Lipids are organic compounds that are mostly nonpolar (have no overall charge) and hydrophobic, so they do not dissolve in water. Simple lipids (fats and waxes) are distinct from complex lipids, such as phospholipids and fat-soluble cell components such as steroids. Fatty acids are a major component of neutral fats and phospholipids. Most fatty acids consist of an even number of carbon atoms, with hydrogen bound along the length of the chain.

Triglycerides (triacylglycerols)

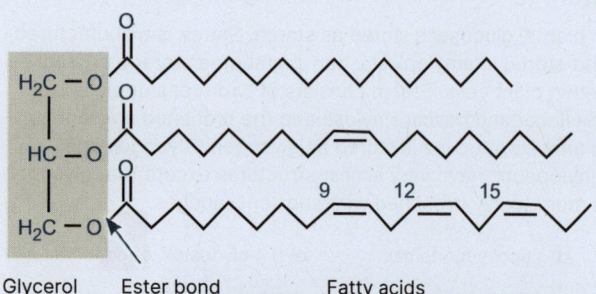

Glycerol   Ester bond   Fatty acids

Triglyceride: an example of a neutral fat

Neutral fats and oils are the most abundant lipids in living things. They make up the fats and oils found in plants and animals. Neutral fats and oils consist of a glycerol attached to one (mono-), two (di-) or three (tri-) fatty acids by ester bonds.

Esterification: A condensation reaction of an alcohol, e.g. glycerol, with an acid, e.g. fatty acid, to produce an ester and water. In the diagram (right), the ester bonds are indicated by thick red lines.

Lipolysis: The breakdown of lipids. It involves hydrolysis of triglycerides into glycerol molecules and free fatty acids.

Triglycerides are formed by condensation

Triglycerides form when glycerol bonds with three fatty acids. Glycerol is an alcohol containing three carbons. Each of these carbons is bonded to a hydroxyl (-OH) group. When glycerol bonds with the fatty acid, an ester bond is formed and water is released. Three separate condensation (or dehydration synthesis) reactions are involved in producing a triglyceride.

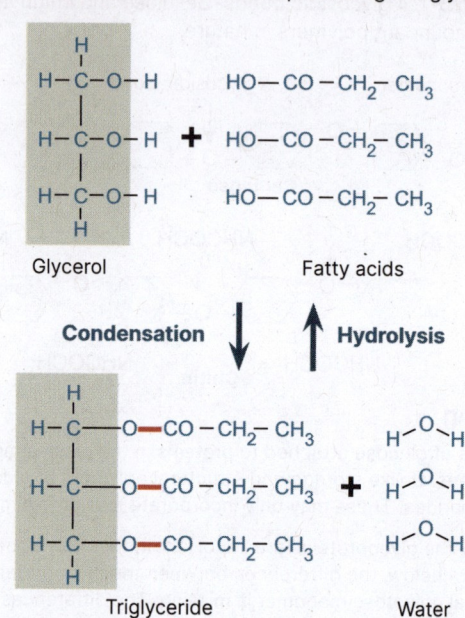

The biological roles of lipids

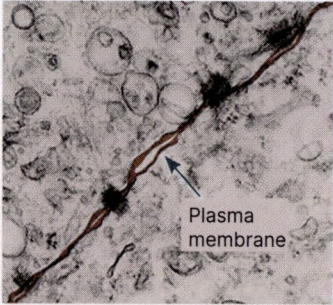

Phospholipids form the basic structure of cellular membranes in prokaryotes and eukaryotes.

Waxes and oils secreted on to surfaces provide waterproofing in plants and animals.

Oxidation of fat provides large amounts of energy and metabolic water. In some desert dwelling animals, e.g. kangaroo rat, this provides all the water they need.

As well as storing energy, fat stores provide insulation, reducing heat losses to the environment. Fat also absorbs shocks, cushioning and protecting internal organs.

Why are lipids a good source of energy?

- Lipids have a high proportion of hydrogen present in the fatty acid chains. Being so reduced and anhydrous (without water), they are an economical way to store fuel reserves, and provide more than twice as much energy as the same quantity of carbohydrate. Fatty acids (mainly as triglycerides) are the most common form of stored fuel in animals and to a lesser extent in plants.
- Triglycerides are metabolized through a stepwise process called beta oxidation. Two-carbon units are removed in a stepwise fashion to produce $FADH_2$, NADH, and acetyl CoA. These enter usual respiratory pathways and are oxidized to CO_2 and H_2O as with other fuels. Triglycerides have many 2-carbon units so they can provide large amounts of energy and water (below).

β-oxidation of fatty acids

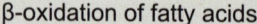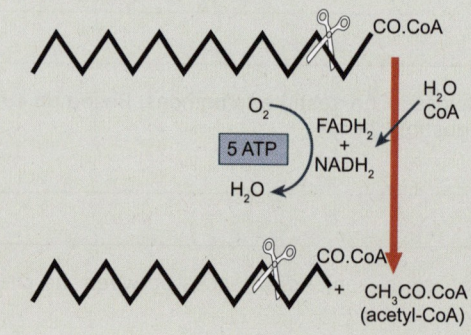

©2024 BIOZONE International
ISBN: 978-1-99-101410-8
Photocopying prohibited

Saturated and unsaturated fatty acids

Fatty acids are carboxylic acids (meaning they have a terminal -COOH) with long hydrocarbon chains. They are classed as either saturated or unsaturated. Saturated fatty acids contain the maximum number of hydrogen atoms. Unsaturated fatty acids contain some double-bonds between carbon atoms and are not fully saturated with hydrogens. A chain with only one double bond is called monounsaturated, whereas a chain with two or more double bonds is called polyunsaturated.

Palmitic acid: a saturated fatty acid

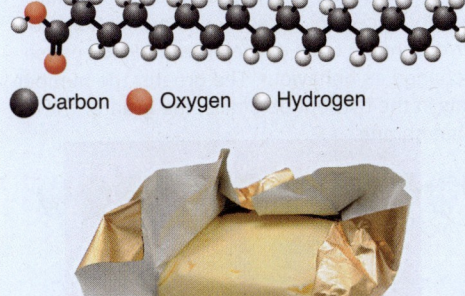

● Carbon ● Oxygen ○ Hydrogen

Butter

Structural formulae (above) and ball and stick models for a saturated fatty acid, palmitic acid (left) and an unsaturated fatty acid, linoleic acid (right). The red arrows (above right) indicate double bonded carbon atoms that are not fully saturated with hydrogens.

Lipids containing a high proportion of saturated fatty acids tend to be solids at room temperature, e.g. butter. Lipids with a high proportion of unsaturated fatty acids are oils and tend to be liquid at room temperature, e.g. olive oil. This is because the unsaturation causes kinks in the straight chains so that the fatty acid chains do not pack closely together.

Linoleic acid: an unsaturated fatty acid

Olive oil

1. Identify the main components (a-c) of the symbolic triglyceride on the right:

   (a) _____   (b)_____   (c)_____

2. Why do lipids have such a high energy content? _____

   _____

   _____

3. (a) Distinguish between saturated and unsaturated fatty acids: _____

   _____

   _____

   (b) Relate the properties of a neutral fat to the type of fatty acid present: _____

   _____

   _____

4. (a) Describe what happens during the esterification (condensation) process to produce a triglyceride:

   _____

   _____

   (b) Describe what happens when a triglyceride is hydrolyzed: _____

   _____

5. Discuss the biological role of lipids: _____

   _____

   _____

   _____

   _____

   _____

69 Phospholipids

Key Idea: Phospholipids have polar and non-polar parts that cause them to form bilayers in aqueous solutions.

A phospholipid is structurally similar to a triglyceride except that a phosphate group and a nitrogen-containing compound replace one of the fatty acids attached to the glycerol. This makes the molecule amphipathic, meaning it has hydrophilic and hydrophobic ends. Phospholipids naturally form bilayers in aqueous solutions and are the main component of cellular membranes. The fatty acid tails can be saturated (straight chains) or unsaturated (kinked chains). The proportion of saturated versus unsaturated fatty acids affects the fluidity of the phospholipid bilayer.

Phospholipids

Phospholipids consist of a glycerol attached to two fatty acid chains and a phosphate (PO_4^{3-}) group. The phosphate end of the molecule is attracted to water (hydrophilic) while the fatty acid end is repelled (hydrophobic). The hydrophobic ends turn inwards to form a phospholipid bilayer.

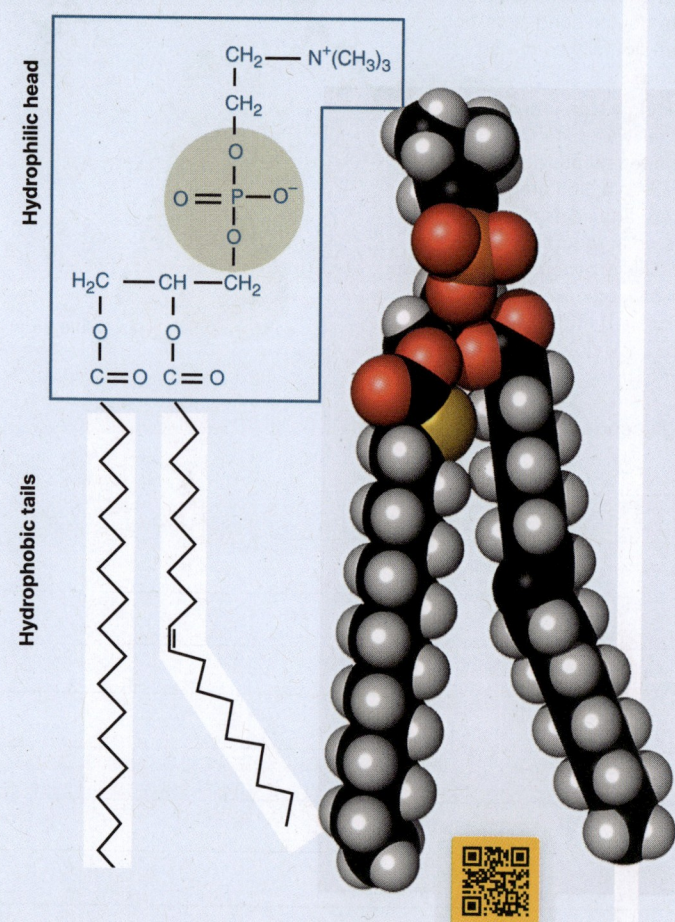

Hydrophilic head

Hydrophobic tails

Phospholipids and membranes

The amphipathic (having hydrophobic and hydrophilic ends) nature of phospholipids means that, when in water, they spontaneously form bilayers. This bilayer structure forms the outer boundary of cells or organelles. Modifications to the different hydrophobic ends of the phospholipids cause the bilayer to change its behaviour. The greater the number of double bonds in the hydrophobic tails, the greater the fluidity of the membrane.

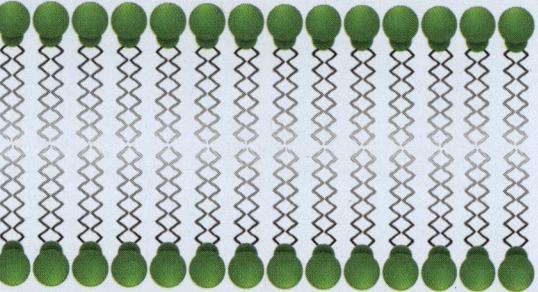

Membrane containing only phospholipids with saturated fatty acid tails.

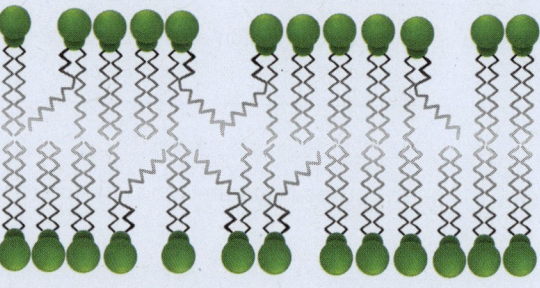

Membrane containing phospholipids with unsaturated fatty acid tails. The fact that the phospholipids do not stack neatly together produces a more fluid membrane.

1. (a) Describe the major structural difference between a triglyceride and a phospholipid:

_____

_____

(b) Relate the structure of phospholipids to their chemical properties and their functional role in cellular membranes:

_____

_____

_____

_____

2. Explain why phospholipid bilayers containing many phospholipids with unsaturated tails are particularly fluid:

_____

_____

_____

B1.1
12

77

©2024 **BIOZONE** International
ISBN: 978-1-99-101410-8
Photocopying prohibited

70 Crossing the Lipid Bilayer

Key Idea: Simple, non-polar molecules can pass directly through the lipid bilayer.

The lipid bilayer that is the basis of a cell's plasma membrane acts as a selective barrier to extracellular molecules. Large molecules and ions are unable to directly cross the plasma membrane but small polar and non-polar molecules can. These include hormones such as testosterone and oestradiol.

What can cross a lipid bilayer?

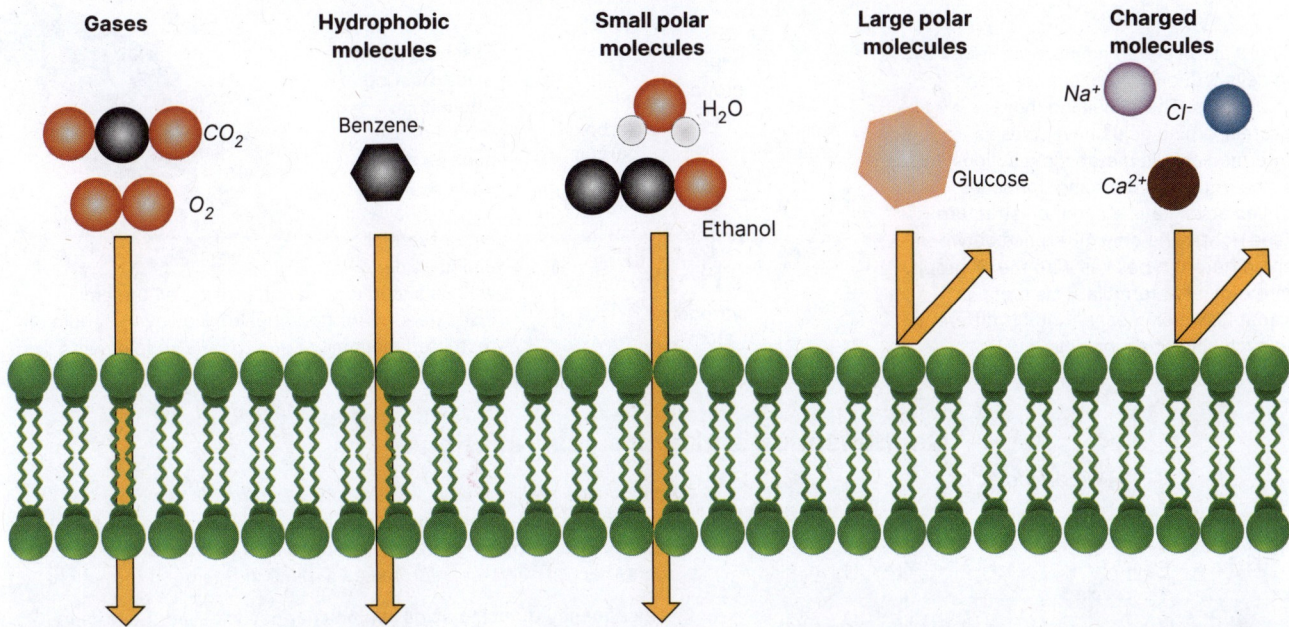

Gases	Hydrophobic molecules	Small polar molecules	Large polar molecules	Charged molecules
CO_2, O_2	Benzene	H_2O, Ethanol	Glucose	Na^+, Cl^-, Ca^{2+}
Small, uncharged molecules can diffuse easily through the membrane.	Lipid soluble molecules diffuse into and out of the membrane unimpeded.	Polar molecules are small enough to diffuse through. Aquaporins increase the rate of water movement.	Cannot directly cross the membrane. Transport (by facilitated diffusion or active transport) involves membrane proteins.	Ions can be transported across the membrane by ion channels (passive) or ion pumps (active).

Steroids

Although steroids are classified as lipids, their structure is quite different from that of other lipids. Steroids have a basic structure of three rings made of 6 carbon atoms each and a fourth ring containing 5 carbon atoms. Examples of steroids include the male and female sex hormones (testosterone and oestradiol).

Because they are non-polar and lipophilic, they can pass through the lipid bilayer even though they are considerably larger than most molecules that move easily through the membrane.

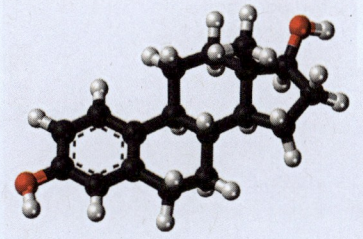
Oestradiol

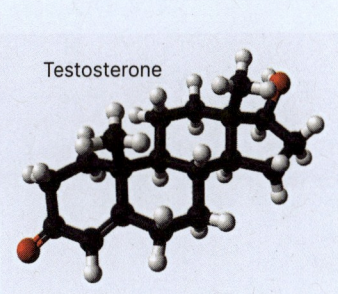
Testosterone

1. Identify the molecule(s) that:

    (a) Can diffuse through the plasma membrane on their own: _____

    _____

    (b) Can diffuse through the membrane via channel proteins: _____

    _____

    (c) Must be transported across the membrane by carrier proteins: _____

    _____

2. Why can non-polar steroids such as testosterone and oestradiol pass through the lipid bilayer?

    _____

    _____

    _____

71 Amino Acids

Key Idea: Amino acids can be joined together by condensation reactions to form polypeptides. Proteins are made up of one or more polypeptide molecules.

Amino acids are the basic units from which proteins are made. Twenty amino acids commonly occur in proteins and they can be linked in many different ways by peptide bonds

to form a huge variety of polypeptides. Peptide bonds are formed by condensation reactions between amino acids. Of the 20 amino acids used by life forms, humans cannot manufacture nine. These are called essential amino acids and must be obtained in food. Not all foods contain these amino acids so a wide variety of foods must be eaten.

There are over 150 amino acids found in cells, but only 20 occur commonly in proteins. The remaining, non-protein, amino acids have roles as intermediates in metabolic reactions, or as neurotransmitters and hormones. All amino acids have a common structure (see right). The only difference between the different types lies with the 'R' group in the general formula. This group is variable, which means that it is different in each kind of amino acid.

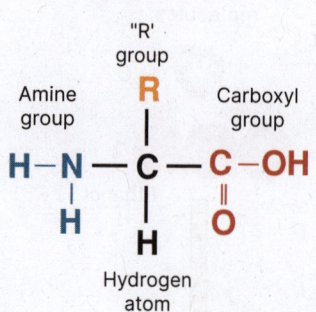

Essential amino acids must be obtained in food. Foods such as meat, eggs, seafood, and soy have all nine essential amino acids. A well balanced diet should provide all the amino acids needed for good health. However, people on restricted diets may need to take supplements to obtain them all.

Condensation and hydrolysis reactions

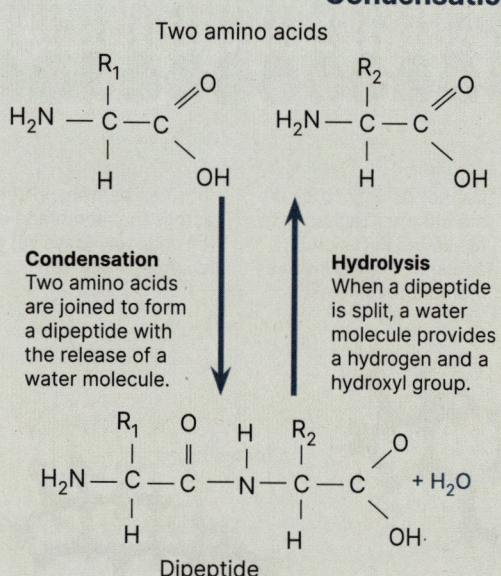

Two amino acids

Condensation
Two amino acids are joined to form a dipeptide with the release of a water molecule.

Hydrolysis
When a dipeptide is split, a water molecule provides a hydrogen and a hydroxyl group.

Dipeptide

Amino acids are linked by peptide bonds to form long polypeptide chains of up to several thousand amino acids. Peptide bonds form between the carboxyl group of one amino acid and the amine group of another (left). Water is formed as a result of this bond formation.

The sequence of amino acids in a polypeptide is called the primary structure and is determined by the order of nucleotides in DNA and mRNA. The linking of amino acids to form a polypeptide occurs on ribosomes in the cytoplasm. Once released from the ribosome, a polypeptide will fold into a structure determined by the composition and position of the amino acids making up the chain.

A polypeptide chain

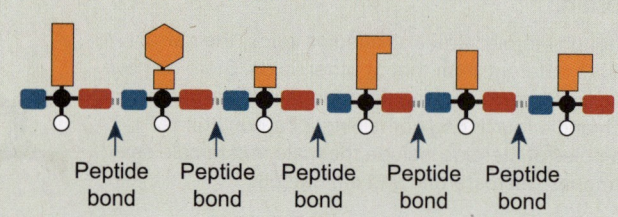

Peptide bond | Peptide bond | Peptide bond | Peptide bond | Peptide bond

1. (a) Describe the general structure of an amino acid: _____

_____

_____

(b) What makes each of the 20 amino acids found in proteins unique? _____

_____

_____

2. Discuss the biological functions of amino acids: _____

_____

_____

_____

_____

_____

©2024 **BIOZONE** International
ISBN: 978-1-99-101410-8
Photocopying prohibited

B1.2
1-3

72 Amino Acids and Proteins

Key Idea: The three dimensional shape of a protein reflects its role. When a protein is denatured, it loses its functionality. A protein may consist of one or several polypeptide chains linked together. Hydrogen bonds and interactions between R-groups cause the polypeptide chain to fold up into a three dimensional shape, held by ionic bonds and disulfide bridges (bonds formed between sulfur containing amino acids). This then carries out a role, such as an enzyme, or a structural element. If bonds are broken (through denaturation), the protein loses its tertiary structure, and its functionality.

The shape of a protein reflects its biological role

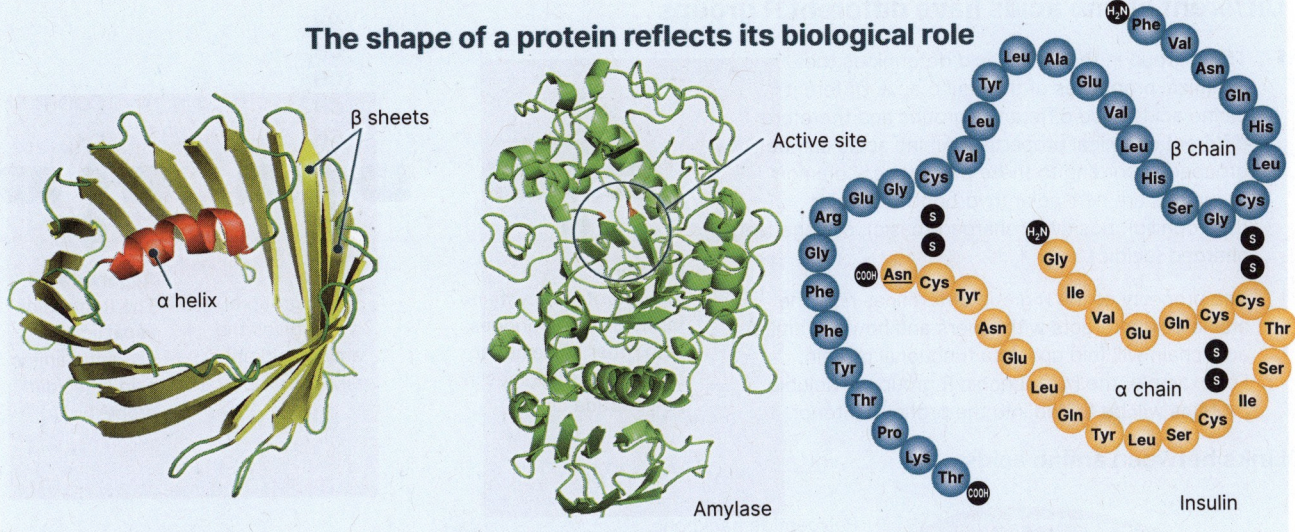

α helix
β sheets
Active site
Amylase
Insulin
β chain
α chain

Channel proteins

Proteins that fold to form channels in the plasma membrane present non-polar R groups to the membrane and polar R groups to the inside of the channel. Hydrophilic molecules and ions are then able to pass through these channels into the interior of the cell. Ion channels are found in nearly all cells and many organelles.

Enzymes

Enzymes are globular proteins that catalyse specific reactions. Enzymes that are folded to present polar R groups at the active site will be specific for polar substances. Non-polar active sites will be specific for non-polar substances. Alteration of the active site by extremes of temperature or pH causes a loss of function.

Sub-unit proteins

Many proteins, e.g. insulin and haemoglobin, consist of two or more sub-units in a complex quaternary structure, often in association with a metal ion. Active insulin is formed by two polypeptide chains stabilized by disulfide bridges between neighbouring cysteines. Insulin stimulates glucose uptake by cells.

Protein denaturation

When the chemical bonds holding a protein together become broken, the protein can no longer hold its three dimensional shape. This process is called denaturation, and the protein usually loses its ability to carry out its biological function.

There are many causes of denaturation, including exposure to heat or pH outside of the protein's optimum range. The main protein in egg white is albumin. It has a clear, thick fluid appearance in a raw egg (right). Heat (cooking) denatures the albumin protein and it becomes insoluble, clumping together to form a thick, white substance (far right).

Raw (native) egg white

Cooked (denatured) egg white

1. Using the example of insulin, explain how interactions between R groups stabilize the protein's functional structure:

   _____

   _____

   _____

2. Why do channel proteins often fold with non-polar R groups to the channel's exterior and polar R groups to its interior?

   _____

   _____

   _____

79

B1.2
4, 5

73 R-Groups

Key Idea: The variable R group of the amino acid determines its properties and, ultimately, the final protein shape.

All amino acids have a common structure, but the R group is different in each kind of amino acid. The property of the R group determines how it will interact with other amino acids and ultimately determines how the amino acid chain folds up into a functional protein. For example, the hydrophobic R groups of soluble proteins are folded into the protein's interior, while hydrophilic groups are arranged on the outside.

Different amino acids have different R groups

▶ The R group in the amino acid determines the chemical properties of the amino acid. Different amino acids have different R groups and therefore different chemical properties. Amino acids can be grouped according to these properties. Common groupings are non-polar (hydrophobic), polar (hydrophilic), positively charged (basic), or negatively charged (acidic).

▶ The property of the R group determines how the amino acid interacts with others and how the amino acid chain will fold up into a functional protein. For example, the hydrophobic R groups of soluble proteins will be folded into the protein's interior.

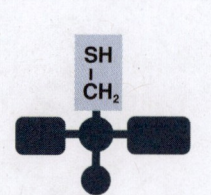

Cysteine
The R group of cysteine forms disulfide bridges with other cysteines to create cross linkages in a polypeptide chain.

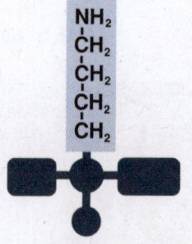

Lysine
The R group of lysine gives the amino acid an alkaline property.

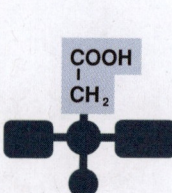

Aspartic acid
The R group of aspartic acid gives the amino acid an acidic property.

Links between amino acids

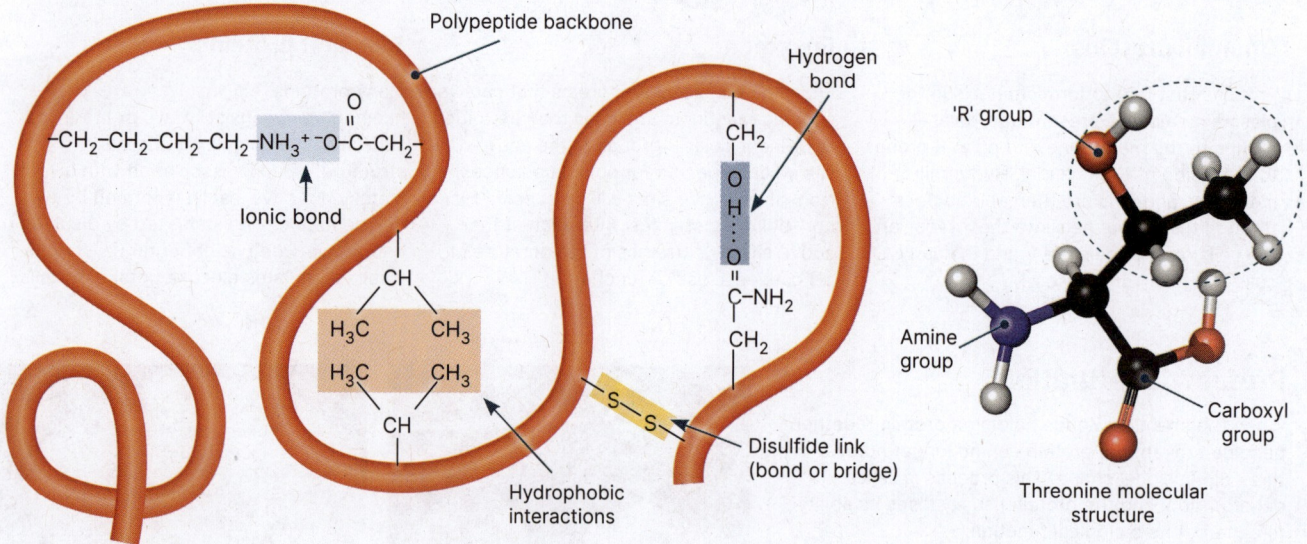

1. (a) Name the different interactions that can shape the polypeptide: _____

_____

   (b) Which of the interactions would be the strongest? _____

2. Do some research to assign each of the 20 amino acids found in proteins to one of the four groups below. Use a standard 3-letter code to identify each amino acid:

   (a) Nonpolar (hydrophobic): _____

   (b) Polar (hydrophilic): _____

   (c) Positively charged (basic): _____

   (d) Negatively charged (acidic): _____

3. Which type(s) of amino acids would you find on the surface of a soluble protein? Which type(s) would you find in the interior? Explain:

_____

_____

_____

©2024 **BIOZONE** International
ISBN: 978-1-99-101410-8
Photocopying prohibited

74 Protein Structure

Key Idea: The sequence and types of amino acids in a protein determine its shape and function.

Proteins are large, complex macromolecules, built up from a linear sequence of amino acids. Proteins account for more than 50% of the dry weight of most cells and are important in virtually every cellular process. The various properties of the amino acids, which are conferred by the different R groups, determine how the polypeptide chain folds up. This three dimensional tertiary structure gives a protein its specific chemical properties and determines its functionality.

1. Describe the main features in the formation of each part of a protein's structure:

a) Primary structure:_____
_____
_____

b) Secondary structure: _____
_____
_____
_____
_____
_____
_____

c) Tertiary structure: _____
_____
_____
_____
_____
_____
_____
_____

d) Quaternary structure: _____
_____
_____
_____
_____

2. How are proteins built up into a functional structure?
_____
_____
_____
_____
_____
_____
_____

Primary (1°) structure (amino acid sequence)

Phe – Glu – Tyr – Ser – Iso – Met – Ala – Ala – Ser – Glu

Peptide bond   Amino acid

Hundreds of amino acids are linked by peptide bonds to form polypeptide chains. The attractive and repulsive charges on the amino acids determine the higher levels of organization in the protein and its biological function.

Secondary (2°) structure (α-helix or β-pleated sheet)

Secondary (2°) structure is maintained by hydrogen bonds between neighbouring CO and NH groups. The hydrogen bonds are individually weak but collectively strong.

α-helix

Hydrogen bonds   Amino acid chain

β-pleated sheet

Polypeptide chains fold into a secondary (2°) structure based on H bonding. The coiled α-helix and β-pleated sheet are common 2° structures. Most globular proteins contain regions of both 2° configurations.

Tertiary (3°) structure (folding of the 2° structure)

Tertiary (3°) structure is maintained by more distant interactions such as disulfide bridges between cysteine amino acids, ionic bonds, and hydrophobic interactions.

α-helix
Aspartic acid
Ionic bond
Lysine
Disulfide bond

A protein's 3° structure is the three-dimensional shape formed when the 2° structure folds up and more distant parts of the polypeptide chains interact.

Quaternary (4°) structure

Some complex proteins are only functional when present as a group of polypeptide chains. Haemoglobin has a 4° structure made up of two alpha and two beta polypeptide chains, each enclosing a complex iron-containing prosthetic (or haem) group.

Alpha chain
Prosthetic (haem) group
Beta chain

A protein's 4° structure describes the arrangement and position of each of the subunits in a multi-unit protein. The shape is maintained by the same sorts of interactions as those involved in 3° structure.

© 2024 **BIOZONE** International
ISBN: 978-1-99-101410-8
Photocopying prohibited

B1.2
7-11

 NOS
 AHL

Conjugated and non-conjugated proteins

▸ Some proteins are functional in their tertiary structure, e.g. lysozyme and glucagon. Others, such as many antibodies or the plant enzyme RuBisCo, are functional only in the quaternary structure. In both cases, there are proteins that are conjugated or non-conjugated (also called simple).

▸ Non-conjugated proteins are proteins that consist of only amino acid residues. Conjugated proteins consist of amino acid residues and other permanently associated chemical groups, called prosthetic groups.

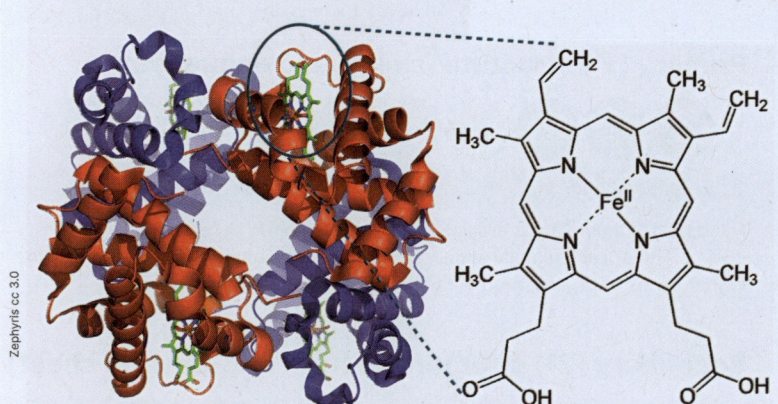

Haemoglobin is a conjugated protein. Each of the four chains that make up the protein have an iron containing haem group (above) at their centre.

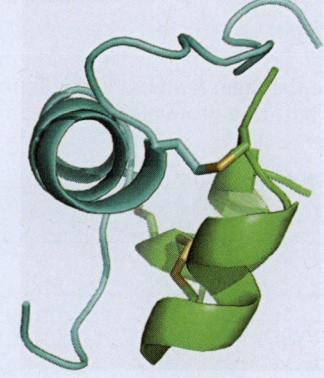

Insulin is a simple non-conjugated protein. It is made of one polypeptide chain that is cleaved to produce two separate chains. These are then joined by two disulphide bridges.

Determining protein structure

▸ How do we know the structure of these proteins? Advances in technology now allow us to work out and visualize the tertiary and quaternary structures of proteins to the atomic level and to predict these structures from the amino acids present in the polypeptide, i.e. from the primary structure.

▸ Traditionally, the structure of a protein is determined by x-ray crystallography. X-rays are aimed at a crystallized sample of a protein. The x-rays produce scattered patterns on film or a detector. By rotating the sample, multiple images at different angles to the x-ray beam can be produced. These images are then analysed by computer to determine the protein's structure.

▸ New methods for determining a protein's structure include cryogenic electron microscopy (right). In this method, a protein is purified and dissolved in very pure water. A thin film is frozen in ethane at cryogenic temperatures. In the frozen film, the protein molecules will be orientated in different directions. An electron microscope is used to take hundreds of images. A computer analyses the images and, from the various orientations of the protein, a three dimensional structure is computed (similar to photogrammetry in which 3D models are computed from multiple photos).

▸ Another developing method in protein analysis is the use of AI (artificial intelligence). Programmes, e.g. AlphaFold, are able to take a sequence of amino acids and predict how the sequence will fold up to produce a three dimensional structure.

Cryogenic electron microscopy

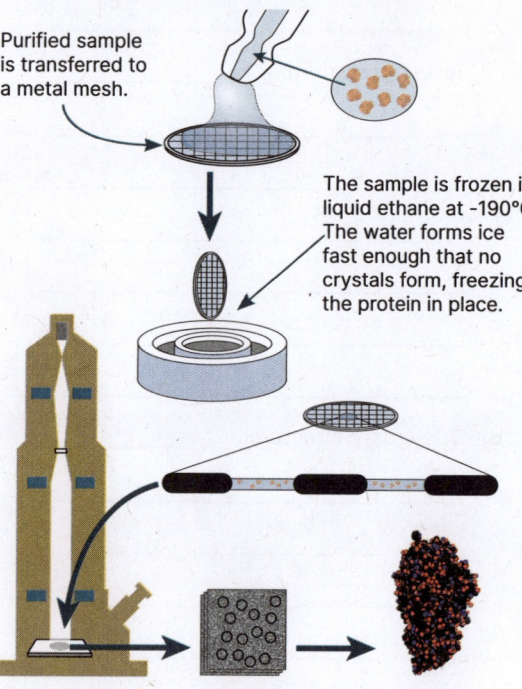

Purified sample is transferred to a metal mesh.

The sample is frozen in liquid ethane at -190°C. The water forms ice fast enough that no crystals form, freezing the protein in place.

Images are taken with an electron microscope. A computer analyses multiple parts of each image and calculates the protein's structure.

3. Distinguish between conjugated and non-conjugated proteins: _____

_____

_____

4. How has the advance of technology helped in determining protein structure? _____

_____

_____

_____

©2024 **BIOZONE** International
ISBN: 978-1-99-101410-8
Photocopying prohibited

75 Comparing Globular and Fibrous Proteins

Key Idea: Protein structure is related to biological function. Proteins can be classified according to their structure or their function. Globular proteins are spherical and soluble in water, e.g. enzymes. Fibrous proteins have an elongated structure and are not water soluble. They are often made up of repeating units and provide stiffness and rigidity to the more fluid components of cells and tissues. They have important structural and contractile roles.

Globular proteins

The shape of globular proteins is a function of their tertiary structure. Some proteins, e.g. actin and tubulin, are globular and soluble as monomers, but polymerize to form long, stiff fibres.

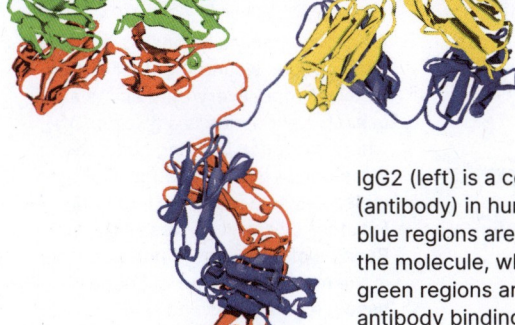

IgG2 (left) is a common immunoglobulin (antibody) in human serum. The red and blue regions are the constant regions of the molecule, whereas the yellow and green regions are variable and determine antibody binding specificity.

Properties of globular proteins

▶ Easily water soluble

▶ Tertiary structure critical to function

▶ Polypeptide chains folded into a spherical shape

Functions of globular proteins

▶ Catalytic, *e.g. enzymes*

▶ Regulatory, *e.g. hormones (insulin)*

▶ Transport, *e.g. haemoglobin*

▶ Protective, *e.g. immunoglobulins (antibodies)*

▶ Structural (rarely), *e.g. actin and tubulin monomers (cytoskeletal elements)*

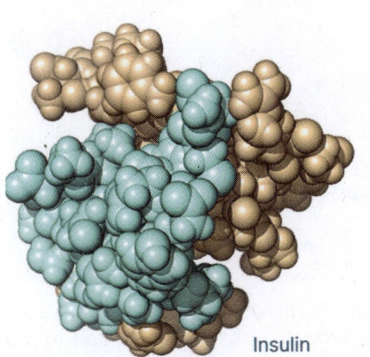

Insulin

RuBisCO

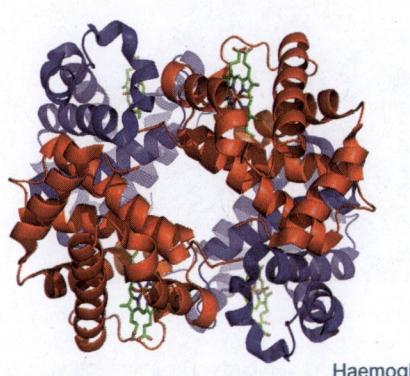

Haemoglobin

Zephyris CC 3.0

Insulin is a peptide hormone involved in the regulation of blood glucose. Insulin is composed of two peptide chains linked together by two disulfide bonds.

RuBisCo is a large multi-unit enzyme. It catalyses the first step of carbon fixation in photosynthesis. It consists of eight large and eight small subunits and is the most abundant protein on Earth.

Haemoglobin is a multi-unit oxygen-transporting protein found in vertebrate red blood cells. One haemoglobin molecule consists of four polypeptide subunits (red and blue). Each subunit holds an oxygen-binding haem group (green).

1. How are globular proteins involved in the functioning of organisms? Use examples to help illustrate your answer:

_____

_____

_____

_____

_____

2. (a) Explain how the shape and properties of a globular protein relate to its functional role: _____

_____

_____

_____

(b) How would its function be affected by a change in tertiary structure? _____

_____

_____

Fibrous proteins

Fibrous proteins are elongated and fibrous in nature or have a sheet like structure. These fibres and sheets are strong and water insoluble. Some, such as keratin, are even insoluble in organic solvents. They have important structural roles.

Properties of fibrous proteins	Functions of fibrous proteins
▶ Water insoluble	▶ Structural role in cells and organisms *e.g. collagen in connective tissues, skin, and blood vessel walls.*
▶ Very tough physically; may be supple or stretchy	
▶ Parallel polypeptide chains in long fibres or sheets	▶ Contractile *e.g. myosin and actin polymers in muscles*

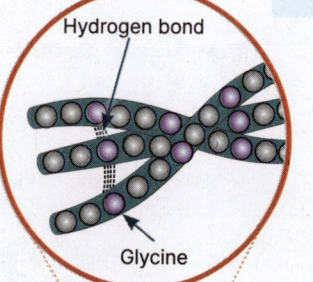

Hydrogen bond

Glycine

Covalent cross links between the collagen molecules

Collagen space filling model

Many collagen molecules form fibrils and the fibrils group together to form larger fibres.

A collagen molecule consists of three polypeptides wound together to form a helical 'rope'. Every third amino acid in each polypeptide is a glycine (Gly), where hydrogen bonding holds the three polypeptides together. Collagen molecules self-assemble into fibrils held together by covalent cross linkages. Bundles of fibrils form fibres. Collagen is the main component of connective tissue, e.g. tendons and skin.

Mammalian hair and claws are α keratin

The scales, beak, and feathers of birds are β keratin

Elastin from an artery

Christian Schmelzer CC4.0

Keratins are found in hair, nails, claws, horn, wool, feathers, and the outer layers of skin. They fall into two classes: α keratins, found in all vertebrates, and the harder β keratins, found in reptiles and birds. The polypeptide chains are arranged in parallel sheets held together by hydrogen bonding. A distinguishing feature of keratins is the high sulfur content, with large numbers of disulfide bridges between cysteine residues. These form permanent, thermally stable, covalent cross linkages and provide additional strength and rigidity.

Elastin is a connective tissue protein with elastic properties that enable tissues to resume their shape after stretching. Elastin has many hydrophobic amino acids, which form mobile hydrophobic regions flanked by covalent cross links between lysine residues.

3. How are fibrous proteins involved in the functioning of organisms? Use examples to help illustrate your answer:

_____

_____

_____

_____

4. Using an example, explain how the shape and properties of a fibrous protein relate to its functional role:

_____

_____

_____

_____

5. What common feature contributes to the strength and stability of collagen, keratin, and elastin?

_____

_____

_____

©2024 **BIOZONE** International
ISBN: 978-1-99-101410-8
Photocopying prohibited

76 Did You Get It?

1. The structure on the right represents a phospholipid bilayer.

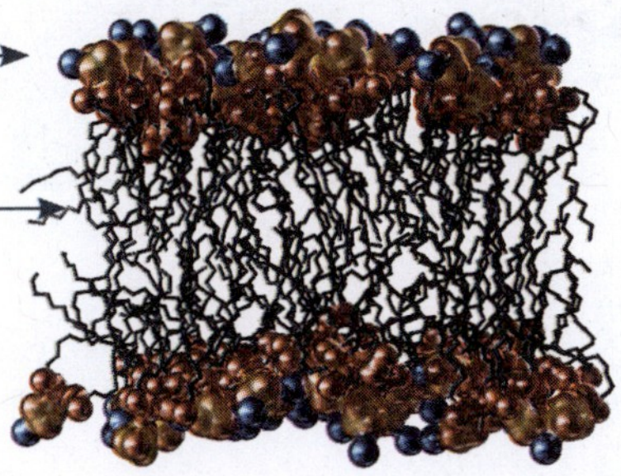

   (a) What does label A represent? _____

   _____

   (b) What does label B represent? _____

   _____

   (c) Explain how the properties of the phospholipid molecule result in the bilayer structure of membranes:

   _____

   _____

   _____

2. The organic molecule on the right is haemoglobin.

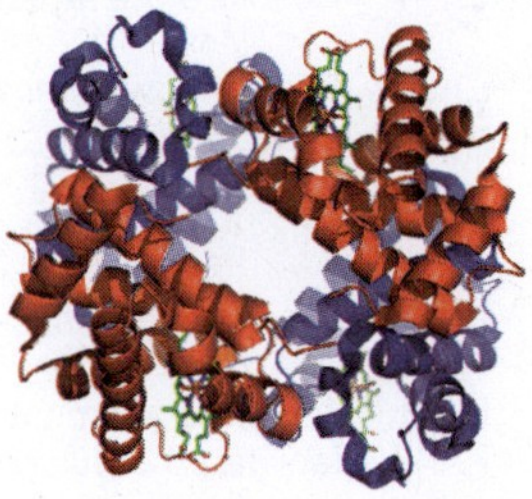

   (a) What class of organic molecules does it belong to? Explain how you decided this:

   _____

   (b) What factors could cause this molecule to lose its shape? _____

   _____

   (c) What would a loss of shape do to the functionality of this molecule? _____

   _____

   (d) **AHL**: Which order of structure does the haemoglobin molecule represent? _____

   _____

3. (a) What general reaction combines two molecules to form a larger molecule? _____

   (b) What general reaction cleaves a larger molecule by the addition of water? _____

   (c) Describe what happens to water in each of the reactions described above: _____

   _____

   _____

   _____

4. In the polypeptide chain below identify (a), (b), and (c):

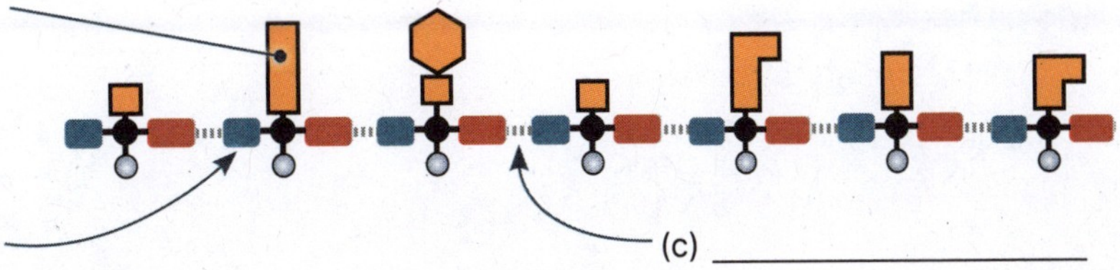

   (a) _____

   (b) _____

   (c) _____

5. **AHL**: Is the protein shown on the right a conjugated or non-conjugated protein? Explain your answer:

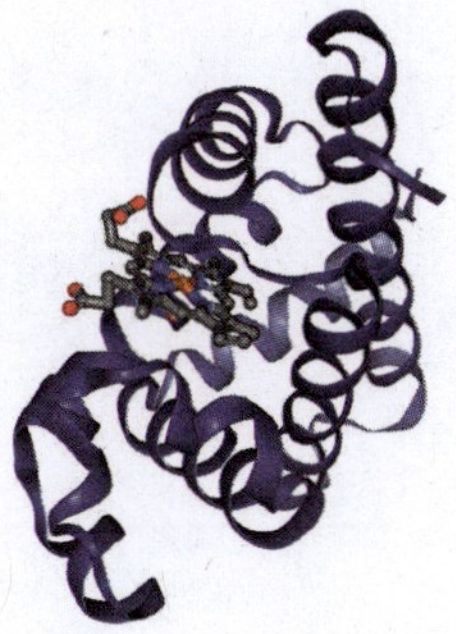

   _____

   _____

   _____

 # Cells

Resource Hub
bit.ly/3tb7K9U

B2.1 Membranes and Membrane Transport

Guiding Questions:
▶ What properties of lipid and protein molecules enable the formation of biological membranes?
▶ What properties of substances enable or inhibit movement across a biological membrane?

Learning Outcomes:

		Activity Number
☐ 1	Identify that lipid bilayers are the basis of cell membranes.	77
☐ 2	Explain how lipid bilayers function as barriers between aqueous solutions.	77
☐ 3	Investigate simple diffusion of molecules, including oxygen and carbon dioxide, across membranes.	77
☐ 4	Describe the structure, location, and function of integral and peripheral proteins in membranes.	77-78
☐ 5	Explain how osmosis facilitates the movement of water molecules across membranes, including the role of aquaporins.	79
☐ 6	Link the structure of channel proteins to the process of facilitated diffusion.	79
☐ 7	Detail the process of active transport, including the role of pump proteins.	80
☐ 8	Compare and contrast selective membrane permeability in simple diffusion, facilitated diffusion, and active transport.	79
☐ 9	Explain the structure and function of glycoproteins and glycolipids.	78
☐ 10	Draw a two-dimensional fluid mosaic model, labelling the key structures of the membrane.	77
☐ 11	**AHL:** Explain the relationship between the fatty acid composition of lipid bilayers and their fluidity.	81
☐ 12	**AHL:** Analyse how cholesterol functions to adjust membrane fluidity in animal cells, allowing them to adapt to different temperatures.	81
☐ 13	**AHL:** Explain how membrane fluidity allows for the fusion and formation of vesicles, through the processes of endocytosis and exocytosis.	82
☐ 14	**AHL:** Describe how gated ion channels function in neurons, including nicotinic acetylcholine receptors and sodium and potassium channels.	83
☐ 15	**AHL:** Explain how exchange transporters generate membrane potentials using the example of a sodium–potassium pump.	84
☐ 16	**AHL:** Explain how indirect active transport using sodium-dependent glucose cotransporters enables glucose absorption in the small intestine and glucose reabsorption in the nephron.	84
☐ 17	**AHL:** Recognize the role cell-adhesion molecules (CAMs) have in the adhesion of cells between cell junctions to form tissues.	85

B2.2 Organelles and Compartmentalization

Guiding Questions:
- ▶ What key adaptations in cell organelles allow them to function effectively?
- ▶ Why is compartmentalization so beneficial to cell function?

Learning Outcomes:

☐	1	**NOS:** Investigate how new technology, including ultracentrifuges and cell fractionation, has allowed for more detailed study into the function of cellular organelles.	86-87
☐	2	Explain the advantages of compartmentalization of the nucleus and cytoplasm as it relates to gene transcription and translation.	86
☐	3	Explain the advantages of cytoplasm compartmentalization for some cell processes.	86
☐	4	**AHL:** Describe how mitochrondrial adaptations enable ATP production during aerobic cell respiration.	88
☐	5	**AHL:** Describe how chloroplast adaptations enable the process of photosynthesis.	88
☐	6	**AHL:** Discuss how the double membrane of the nucleus provides functional benefits during mitosis and meiosis.	89
☐	7	**AHL:** Compare and contrast the structure and function of free ribosomes and those found on the rough endoplasmic reticulum.	89
☐	8	**AHL:** Explain the structure and function of the Golgi apparatus.	90
☐	9	**AHL:** Explain the structure, function, and formation of cell vesicles in cells, including the role of clathrin.	90

B2.3 Cell Specialization

Guiding Questions:
- ▶ What different functions do stem cells have in multicellular organisms?
- ▶ What is the relationship between differentiated cells and specialized cellular functions?

Learning Outcomes:

☐	1	Explain how the process of differentiation, including the impact of gradients on gene expression, results in the development of specialized cells.	91
☐	2	Describe the unique properties of stem cells as they relate to their role.	91
☐	3	Describe the location and function of stem cell niches, such as bone marrow and hair follicles, in adult humans.	91
☐	4	Compare differences between totipotent, pluripotent, and multipotent stem cells.	91
☐	5	Identify cell size as a result of specialization, illustrating with a range of examples.	92
☐	6	**NOS:** Use models to investigate how surface area-to-volume ratios place limitations on cell size.	93 - 94
☐	7	**AHL:** Investigate adaptations in specialized cells that increase surface area-to-volume ratios.	95
☐	8	**AHL:** Explain how the adaptations of type I and type II pneumocytes in alveoli are related to their function.	96
☐	9	**AHL:** Explain how the adaptations of cardiac muscle cells and striated muscle fibres, including branched or unbranched and number of nuclei, are related to their function.	96
☐	10	**AHL:** Explain how the adaptations of human sperm and egg cells are related to their function.	96

77 The Plasma Membrane

Key Idea: The plasma membrane is composed of a lipid bilayer containing freely moving proteins. It is the partially permeable boundary between the internal and external cell environments.

All cells have a plasma membrane forming the outer limit of the cell. Cellular membranes are also found inside eukaryotic cells as part of organelles, such as the endoplasmic reticulum. The accepted model of membrane structure is the fluid-mosaic model (below). The plasma membrane is more than just a passive envelope; it is a dynamic structure, actively involved in cellular activities, e.g. transport. The lipid bilayer functions as a barrier between aqueous solutions due to the internal hydrophobic core of the plasma membrane having limited permeability to both large molecules and polar particles, such as ions.

The fluid mosaic model of membrane structure

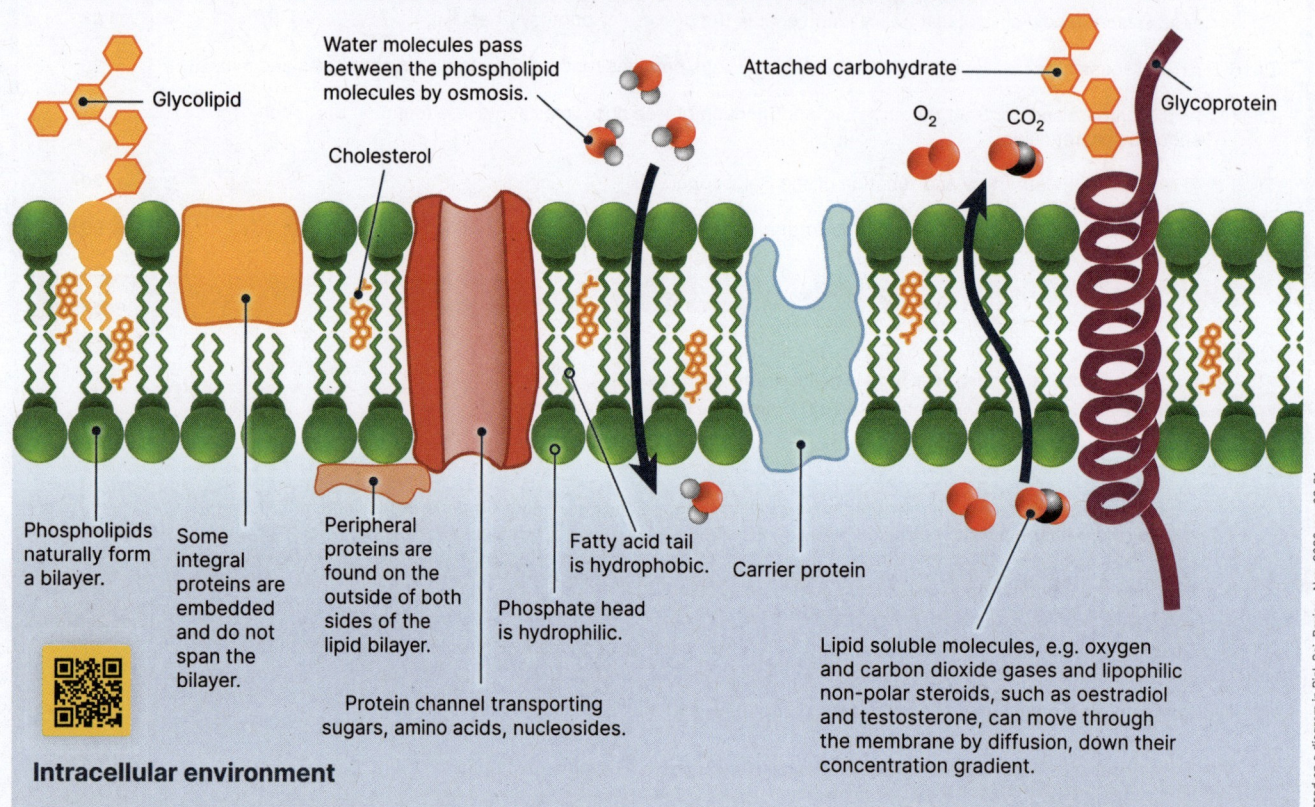

- Glycolipid
- Water molecules pass between the phospholipid molecules by osmosis.
- Cholesterol
- Attached carbohydrate
- O_2
- CO_2
- Glycoprotein
- Phospholipids naturally form a bilayer.
- Some integral proteins are embedded and do not span the bilayer.
- Peripheral proteins are found on the outside of both sides of the lipid bilayer.
- Fatty acid tail is hydrophobic.
- Carrier protein
- Phosphate head is hydrophilic.
- Protein channel transporting sugars, amino acids, nucleosides.
- Lipid soluble molecules, e.g. oxygen and carbon dioxide gases and lipophilic non-polar steroids, such as oestradiol and testosterone, can move through the membrane by diffusion, down their concentration gradient.

Intracellular environment

Based on a diagram in Biol. Sci. Review, Nov. 2009, pp. 20-21

1. List the important components of the plasma membrane:

   _____

2. Identify the kind of molecule on the diagram above that:

   (a) Can move through the plasma membrane by diffusion:

   _____

   (b) Can form a channel through the membrane: _____

   (c) Must be transported across the membrane by carrier proteins:

   _____

3. (a) On the diagram (right) label the hydrophobic and hydrophilic ends of the phospholipid and indicate which end is attracted to water:

   (b) How does the lipid bilayer allow the membrane to act as a barrier?

   _____

   _____

   _____

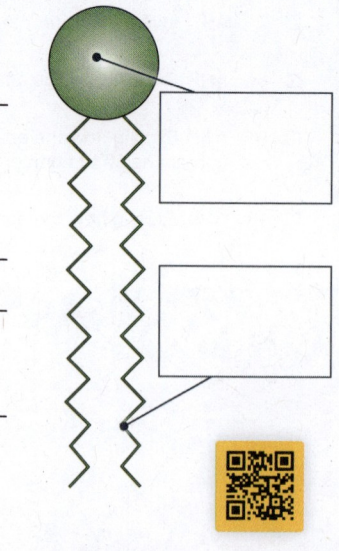

B1.1	B2.1
13	1-4 10

What is diffusion?

Diffusion is the movement of particles down a concentration gradient. It is a passive process, meaning it needs no input of energy to occur. During diffusion, molecules move randomly about, eventually becoming evenly dispersed.

If molecules can move freely, they move from high to low concentration (down a concentration gradient) until evenly dispersed. Each molecule moves down its own concentration gradient, independent of the concentration of other types of molecule (diagram, right).

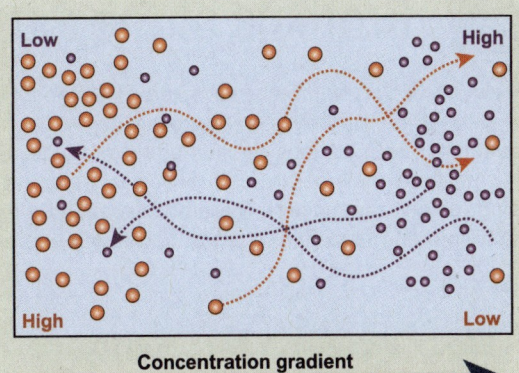

Concentration gradient

4. (a) Why are all molecules not able to diffuse across the membrane freely?

_____

_____

(b) Compare the transport of simple gases / non-polar steroids and water across the plasma membrane:

_____

_____

_____

5. How do models, like the fluid-mosaic model, help us understand structure and processes in systems?

_____

_____

_____

6. Explain how the fluid mosaic model accounts for the observed properties of cellular membranes:

_____

_____

_____

_____

7. Use the symbol for a phospholipid molecule (below) to draw a two-dimensional fluid mosaic model, showing hydrophobic and hydrophilic regions (include features such as include peripheral and integral proteins, glycoproteins, glycoproteins, phospholipids, and cholesterol):

Symbol for phospholipid

78 Proteins of the Plasma Membrane

Key Idea: A cellular membrane is made of a phospholipid bilayer with different proteins embedded in it.

The structure and locations of membrane proteins enable them to perform their particular function in transport, cell signalling, or cell recognition. Proteins associated with the plasma membrane are either integral and found embedded in the membrane, or peripheral and found on the surface of the membrane. Glycolipids and glycoproteins are attached to the external side of the membrane and are involved in cell adhesion and cell recognition. Carrier proteins and channel proteins are transport proteins that allow substances that are unable to utilize diffusion or osmosis to cross the membrane.

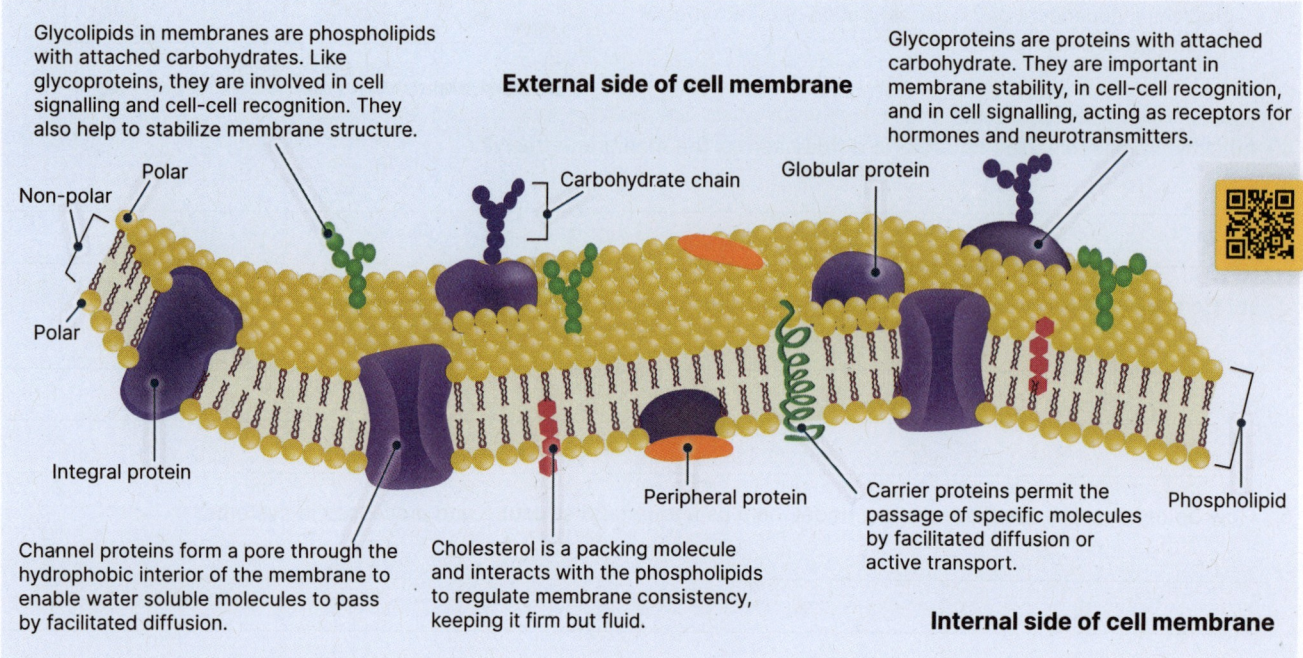

Glycolipids in membranes are phospholipids with attached carbohydrates. Like glycoproteins, they are involved in cell signalling and cell-cell recognition. They also help to stabilize membrane structure.

Glycoproteins are proteins with attached carbohydrate. They are important in membrane stability, in cell-cell recognition, and in cell signalling, acting as receptors for hormones and neurotransmitters.

External side of cell membrane

Carbohydrate chain · Globular protein · Non-polar · Polar · Polar · Integral protein

Channel proteins form a pore through the hydrophobic interior of the membrane to enable water soluble molecules to pass by facilitated diffusion.

Cholesterol is a packing molecule and interacts with the phospholipids to regulate membrane consistency, keeping it firm but fluid.

Peripheral protein

Carrier proteins permit the passage of specific molecules by facilitated diffusion or active transport.

Phospholipid

Internal side of cell membrane

The structure, location, and functions of integral and peripheral proteins

Integral proteins can span the entire width of the cell membrane (transmembrane) or be partially embedded. Carrier and channel proteins are both integral. Examples include the GLUT1 glucose transporter (far right) and the G-protein coupled receptors (right). Channel proteins form a protected pathway and neither bind to transported substances, nor require energy to operate

Peripheral, or extrinsic, proteins are found on either outer surface of the membrane. They can be attached directly to the phospholipids or to other integral proteins and their bonding is only temporary. Peripheral proteins have a range of functions, including communication, molecule transport, membrane support, and anchoring the membrane to the cell's cytoskeleton.

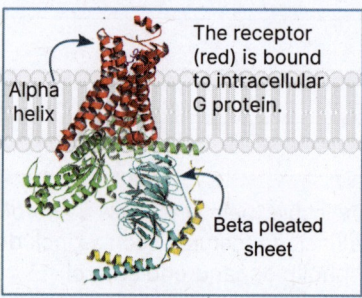

The receptor (red) is bound to intracellular G protein. Alpha helix · Beta pleated sheet

G protein coupled receptors are integral proteins involved in signalling pathways. A signal molecule binds to the receptor protein outside the cell to trigger a reaction involving intracellular G protein. In this example, the receptor binds to adrenaline.

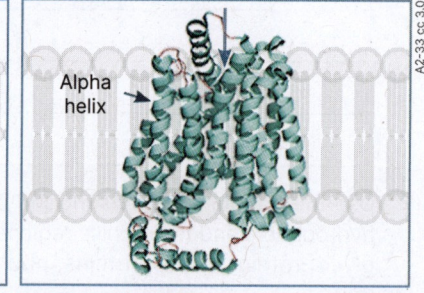

Alpha helix

The GLUT1 glucose transporter is a carrier protein that facilitates the transport of glucose across the plasma membranes of mammalian cells. It increases the rate of glucose transport by 50,000x (high enough to supply the cell's energy needs).

1. (a) Distinguish between the location of integral and peripheral proteins, using examples:

___

(b) Compare and contrast the functions of integral and peripheral proteins, using researched examples:

___

©2024 **BIOZONE** International
ISBN: 978-1-99-101410-8
Photocopying prohibited

The structure and function of the glycoproteins and glycolipids

Glycoproteins and glycolipids are membrane associated molecules composed of a carbohydrate component covalently bonded to either an integral protein (glycoproteins) or a lipid (glycolipids).

Glycoproteins facilitate attachment and adhesion to other cells. They also enable cell to cell communication and act as receptors for chemical signals. Other functions include memory consolidation in neuron cells, cell differentiation leading to changes in cell phenotype, and involvement in cancer growth. The carbohydrate sugar connects to an amino side chain on the protein.

Glycolipids support the structure of the membrane and assist in connecting cells to form tissues. They also have roles in cellular communication and recognition, especially important to the immune system, where they act as recognition sites and antigens. They can identify pathogens, while recognizing the body's own cells.

Membrane-bound glycoproteins and glycolipids

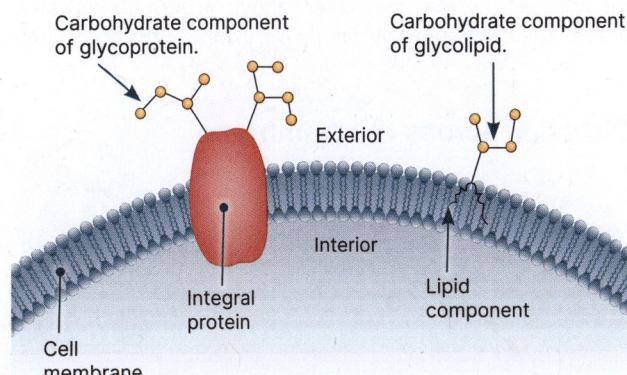

Carbohydrate component of glycoprotein.

Carbohydrate component of glycolipid.

Exterior

Interior

Integral protein

Lipid component

Cell membrane

Glycoproteins, glycolipids, and blood group type

Erythrocytes (red blood cells) can be classified by the ABO blood group type. Glycoproteins and glycolipids are found on the surface of the red blood cell membrane.

The glycoproteins and glycolipids have specific sites that act as antigens, i.e. substances that react with antibodies free floating in the plasma of the body. Opposing antigens and antibodies that mix, such as in a blood transfusion, will cause clumping of the blood, followed by haemolysis and destruction of the blood cells, leading to serious health issues or death.

The carbohydrate component of both glycoproteins and glycolipids is an oligosaccharide (3-10 sugar units). Humans with Type A blood have Type A oligosaccharide, Type B blood have Type B oligosaccharide, Type O blood have a short, nonfunctioning oligosaccharide, and Type AB blood have both Type A and Type B oligosaccharide.

Each blood type has one or two (AB blood) antibodies that react with the antigen on the glycoprotein or glycolipid on the red blood cell membrane, or in the case of Type AB, neither.

Type A cannot mix with Type AB or B, Type B cannot mix with Type AB or A, Type AB can mix with any type of blood, and Type O can only mix with Type O blood.

The consequence of an incompatible blood transfusion

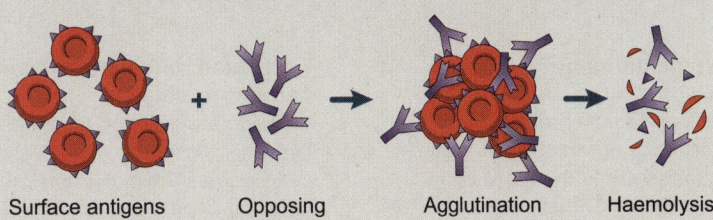

Surface antigens

Opposing antibodies

Agglutination (clumping)

Haemolysis

2. Where are glycoproteins and glycolipids located in the membrane? _____

_____

3. Compare and contrast the structure of glycoproteins and glycolipids, including bonding: _____

_____

_____

_____

_____

_____

4. Compare and contrast the function of glycoproteins and glycolipids: _____

_____

_____

_____

_____

_____

_____

_____

_____

79 Movement Across the Plasma Membrane

Key Idea: Cell membranes facilitate and control diffusion of molecules into and out of the cell.

In biological systems, most diffusion occurs across membranes. Some molecules move freely (unassisted) across the membrane by simple diffusion. Diffusion of other molecules is facilitated by specific carrier and channel proteins in the membrane. Diffusion is important in allowing cells to make exchanges with their extracellular environment, i.e. the blood and fluids that bathe them, and is crucial to the regulation of water content.

Diffusion across the membrane

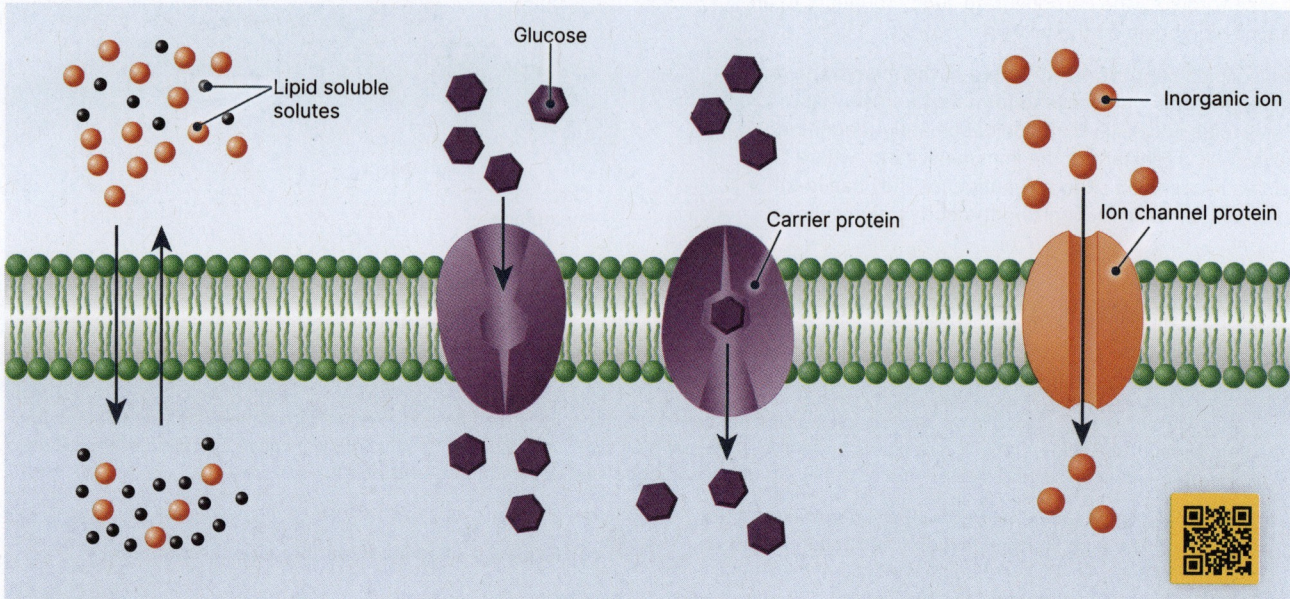

Simple diffusion
Molecules move directly through the plasma membrane without assistance or selectivity. <u>Example:</u> O_2 diffuses into the blood and CO_2 diffuses out. Diffusion gradients are maintained because substances are constantly being imported, made, or used by the cell.

Facilitated diffusion involving carrier proteins
Carrier proteins in the membrane allow large, lipid-insoluble molecules that cannot cross the membrane by simple diffusion to be transported into the cell. <u>Example:</u> The transport of glucose into red blood cells.

Facilitated diffusion involving channel proteins (hydrophilic pores)
Channel proteins (water-filled pores) allow selective permeability of substances such as inorganic ions to pass through. Aquaporins are special channel proteins for rapid diffusion of water. <u>Example:</u> K^+ ions exiting nerve cells to restore resting potential.

1. What do the three types of diffusion described above all have in common? _____

_____

_____

2. How does facilitated diffusion differ from simple diffusion with regard to selective permeability?

_____

_____

_____

3. Some channel proteins are gated and can open and close, such as those for sodium, potassium, and calcium ions. Why would this type of diffusion be necessary for electrical signal transmission in nerve and muscle cells?

_____

_____

_____

4. Why would a thin, flat cell have a greater rate of diffusion to and from its centre than a thick spherical cell?

_____

_____

B2.1

5, 6, 8

©2024 **BIOZONE** International
ISBN: 978-1-99-101410-8
Photocopying prohibited

Osmosis

▸ Osmosis is the diffusion of water molecules from regions of lower solute concentration (higher free water concentration) to regions of higher solute concentration (lower free water concentration) across a partially permeable membrane. A partially permeable membrane allows some molecules, but not others, to pass through.

▸ Water molecules will diffuse across a partially permeable membrane, but in a random movement, until an equilibrium is reached and net movement is zero. The plasma membrane of a cell is an example of a partially permeable membrane. Osmosis is a passive process and does not require any energy input.

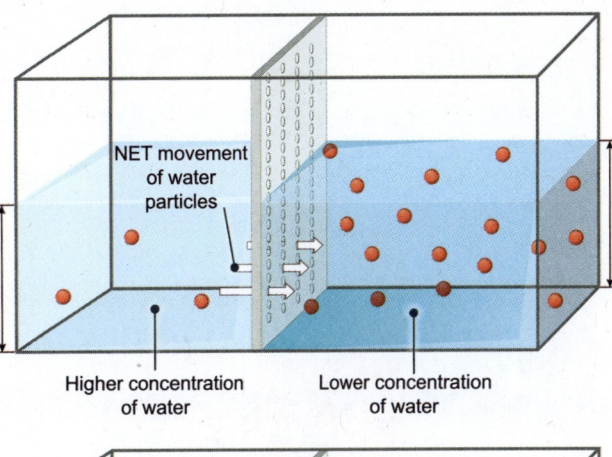

NET movement of water particles

Higher concentration of water   Lower concentration of water

5. What is osmosis? _____

_____

_____

_____

_____

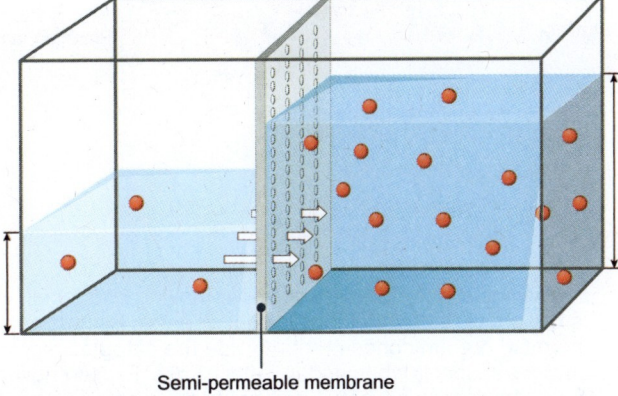

Semi-permeable membrane

6. (a) In which direction (left or right) will be the net movement of water in the diagram (right)?

_____

(b) Why did water move in this direction?

_____

_____

_____

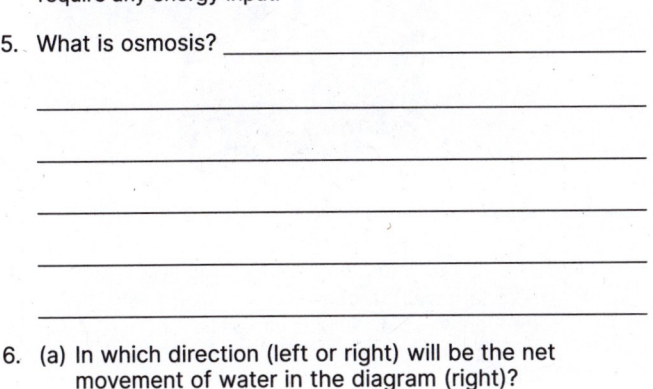

Isotonicity

Aquaporin

Solution

Water molecule

Solute

Membrane

Cytoplasm

Osmolarity and tonicity

Osmotic pressure is created when the solute concentration is different on either side of a semi-permeable membrane. Osmolarity (osmotic concentration) takes into account the total concentration of penetrating and non-penetrating solutes. The greater the solute concentration, the higher the osmolarity. Osmolarity is measured in the number of particles per litre in the solution - $mmol \ L^{-1}$.

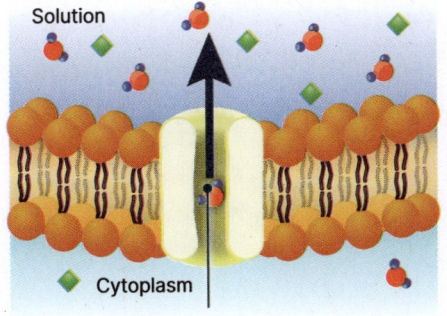

Hypertonicity

Solution

Cytoplasm

Net osmotic movement

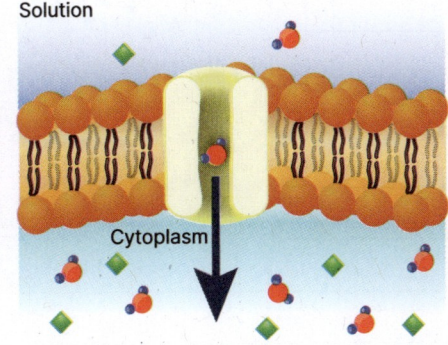

Hypotonicity

Solution

Cytoplasm

Tonicity is the measure of the osmotic pressure gradient between two solutions. Unlike osmolarity, tonicity is only influenced by solutes that cannot cross the semipermeable membrane because these are the only solutes influencing the osmotic pressure gradient and osmosis must occur for the two solutions to reach equilibrium. Solutions are usually categorized as isotonic, hypotonic, or hypertonic depending on whether they have the same, lower, or higher solute concentration relative to another solution across a membrane. Hypertonicity results in cells losing water, and the net osmotic flow is out, while hypotonicity results in cells gaining water.

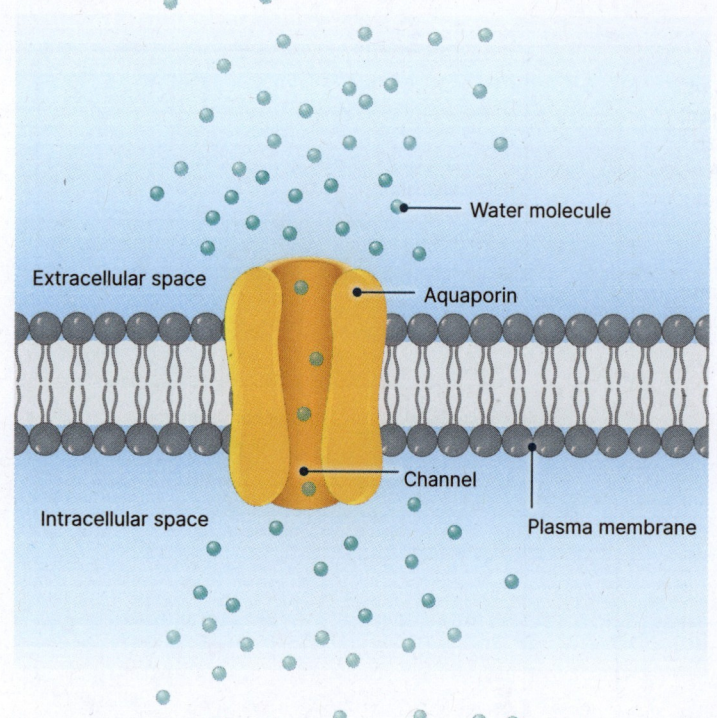

Extracellular space

Water molecule

Aquaporin

Channel

Intracellular space

Plasma membrane

Osmosis and aquaporins

Aquaporins are a special type of channel protein that facilitates the osmotic passage of water molecules across the membrane. They are found in many structures of the body including the kidney, the eye, blood vessels, and secretory glands. Aquaporin channels are always open to water but in some cells, such as those in kidney collecting ducts, the number of aquaporins embedded in the membrane can be altered to increase or decrease osmotic flow.

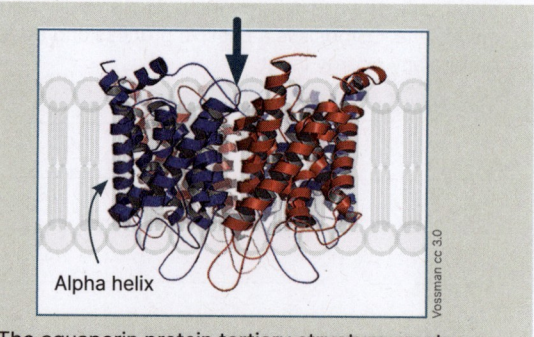

Alpha helix

The aquaporin protein tertiary structure creates a pore (arrowed) through the centre of the protein through which molecules can pass.

7. Endosmosis is the process of water moving into a cell, while exosmosis is the process of water moving out of a cell. Plant root hair cells can increase the ion concentration inside the cytoplasm of the hair cells, by active transport, when under water stress. Explain how this effects the NET osmotic movement and benefits the plant:

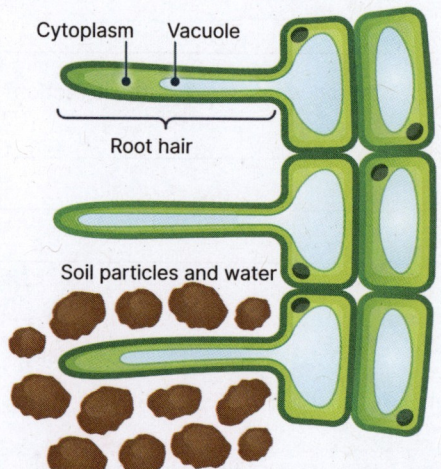

Cytoplasm   Vacuole

Root hair

Soil particles and water

_____

_____

_____

_____

_____

_____

_____

8. Link the function of the aquaporin channel protein to the requirements of some structures in which they are found, e.g. cells in the kidney or blood vessels:

_____

_____

_____

_____

9. Describe the type of movement occurring at A, B, and C across the membrane:

Water

A   B   C   Ions

_____

_____

_____

_____

_____

_____

_____

_____

©2024 **BIOZONE** International
ISBN: 978-1-99-101410-8
Photocopying prohibited

80 Active Transport and Pump Proteins

Key Idea: Active transport uses energy to transport molecules against their concentration gradient across a partially permeable membrane through pump proteins.
Active transport is the movement of molecules (or ions) from regions of low concentration to regions of high concentration across a cellular membrane by a transport protein. Active transport needs energy to proceed because molecules are being moved against their concentration gradient.

▶ The energy for active transport comes from ATP (adenosine triphosphate). Energy is released when ATP is hydrolyzed (water is added) forming ADP (adenosine diphosphate) and inorganic phosphate (Pi).

▶ Transport (carrier) pump proteins in the membrane are used to actively transport molecules from one side of the membrane to the other (below).

▶ Active transport can be used to move molecules into and out of a cell.

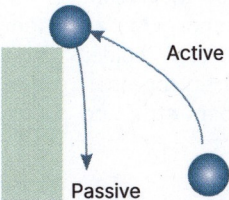

A ball falling is a passive process (it requires no energy input). Replacing the ball requires active energy input.

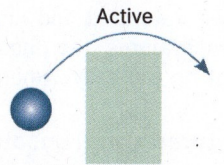

It requires energy to actively move an object across a physical barrier.

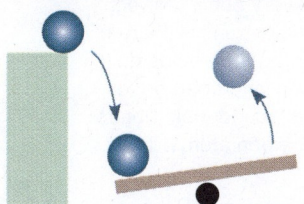

Sometimes the energy of a passively moving object can be used to actively move another. For example, a falling ball can be used to catapult another (left).

Active transport

1 ATP binds to a transport protein.

2 A molecule or ion to be transported binds to the transport protein.

3 ATP is hydrolyzed and the energy released is used to transport the molecule or ion across the membrane.

4 The molecule or ion is released and the transport protein reverts to its previous state.

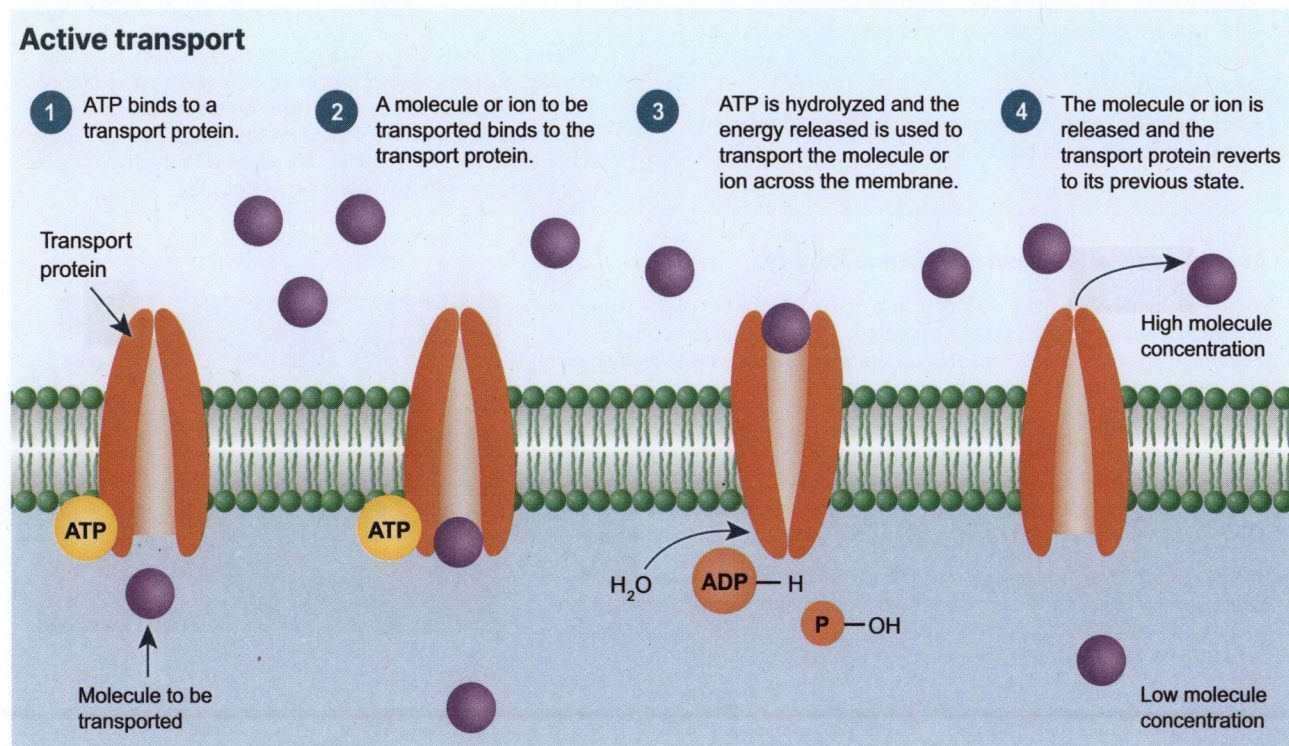

1. (a) What is the essential feature of active transport? _____

(b) How is active transport used in the cell? _____

2. Where does the energy for active transport come from? _____

3. Why is the analogy between how a pump protein functions and a mechanical pump a useful one? _____

©2024 **BIOZONE** International
ISBN: 978-1-99-101410-8
Photocopying prohibited

B2.1
7

81 Membrane Fluidity

Key Idea: The fatty acid component of phospholipids and embedded cholesterol help regulate membrane fluidity as temperature changes.

Membrane fluidity is an important property that allows the cell to alter its shape and allows for continued movement of molecules through the lipid bilayer. Phospholipids can have saturated and unsaturated fatty acid tails. At colder temperatures, the phospholipids are compressed together, making the bilayer more rigid. However, the bent structure of the unsaturated fatty acids help maintain a less-compressed, more fluid membrane. Additionally, cholesterol molecules embedded in the membrane act as a buffer for both higher and lower temperatures, reducing the effect on the membrane and maintaining the fluidity despite temperature fluctuations. Cholesterol has both hydrophobic and hydrophilic regions. It is an integral part of the cellular membrane and provides stability, whilst regulating membrane fluidity during fluctuating temperatures.

Saturated and unsaturated fatty acids

▶ Fatty acid tails are composed of a carbon chain covalently bonded to hydrogen atoms.

▶ Recall that saturated fatty acids have the maximum number of hydrogen atoms bonded, therefore every bond between each carbon atom is single (one pair of electrons is shared). This makes the chain straight. Unsaturated fatty acids do not have the maximum number of carbon-hydrogen bonds, and therefore have one (mono) or more (poly)double bonds (2 pairs of electrons shared) between carbon atoms. This makes the fatty acid tail bend.

▶ Unsaturated fatty acid tails resist compression, which occurs in cooler temperatures. The bent tails push against each other and spread the phospholipids out. The membrane remains more fluid when the bilayer is less compressed.

▶ Phospholipids can have a combination of none, one, or two unsaturated fatty acid tails.

Cholesterol in the cellular membrane

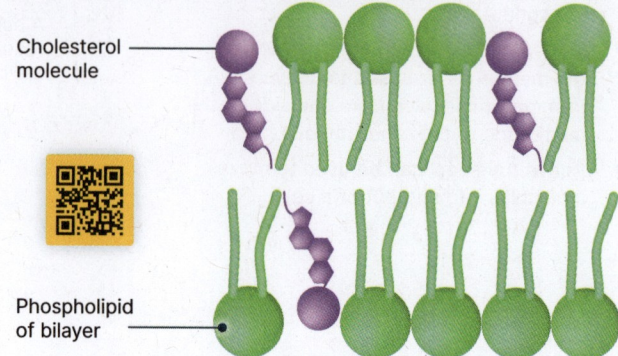

Cholesterol molecule

Phospholipid of bilayer

Cholesterol molecules within the lipid bilayer help modulate the fluidity of the cellular membrane and maintain its stability. The polar head of the cholesterol molecule is attracted to the hydrophilic heads of the phospholipids: it reduces the fluidity in this region. Its hydrophobic tail aligns with the hydrophobic tails of the phospholipids and separates them.

Animal adaptations and membrane fluidity

Some poikilothermic (cold-blooded) animals have adapted to colder temperatures by having a higher proportion of unsaturated fatty acids as part of their membrane phospholipids. Recall that unsaturated fat has a lower melting point than saturated fats. This allows the bilayer to retain fluidity in cold environments.

The Antarctic icefish (family Channichthyidae) has a higher proportion of phospholipids with unsaturated fatty acids in its cellular membranes compared to tropical fish. This enhances membrane fluidity in the cold Antarctic environment. Fish that encounter seasonal changes of temperature can become acclimatized, and increase the proportion of unsaturated fatty acids in a warmer environment.

The high metabolism of homeothermic (warm blooded) species, such as mammals and birds, demands a more fluid membrane for cellular reactions to take place. They have adapted by having a higher proportion of polyunsaturated fatty acids in their cellular membrane.

Antarctic icefish

1. (a) What is the relationship between saturated and unsaturated fatty acids and membrane fluidity?

_____

_____

_____

(b) What is the link between the proportion of unsaturated fatty acids and adaptations for colder environments?

_____

_____

2. How does cholesterol contribute to membrane fluidity stabilization during temperature fluctuations?

_____

_____

AHL | B2.1 11-12 | 68

©2024 **BIOZONE** International
ISBN: 978-1-99-101410-8
Photocopying prohibited

82 Cytosis and Membrane Fluidity

Key Idea: Endocytosis and exocytosis are active transport processes. Endocytosis involves the cell engulfing material. Exocytosis involves the cell expelling material.

Most cells carry out cytosis, a type of active transport in which the plasma membrane folds around a substance to transport it. The ability of cells to do this is a function of the fluidity of the plasma membrane. Cytosis results in bulk transport of substances into or out of the cell and is achieved through the localized activity of the cell cytoskeleton. Endocytosis involves material being engulfed and taken into the cell. It typically occurs in protozoans and some white blood cells of the mammalian defence system (phagocytes). Exocytosis is the reverse of endocytosis and involves expelling material from the cell in vesicles or vacuoles that have fused with the plasma membrane. Exocytosis is common in cells that export material (secretory cells).

Endocytosis

Material (solids or fluids) that are to be brought into the cell are engulfed by an infolding of the plasma membrane.

Plasma membrane

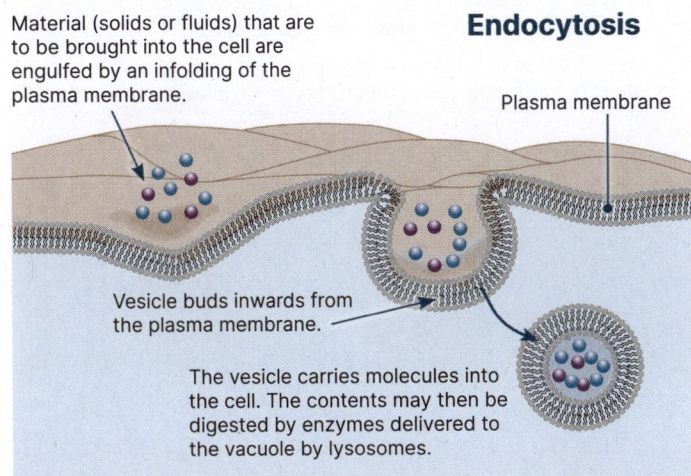

Vesicle buds inwards from the plasma membrane.

The vesicle carries molecules into the cell. The contents may then be digested by enzymes delivered to the vacuole by lysosomes.

Endocytosis occurs by invagination (infolding) of the plasma membrane, which then forms vesicles or vacuoles that become detached and enter the cytoplasm. There are two main types of endocytosis:

Phagocytosis involves the cell engulfing solid material to form large vesicles or vacuoles, e.g. food vacuoles. Examples: Feeding in *Amoeba*, phagocytosis of foreign material and cell debris by neutrophils and macrophages (types of white blood cells). Some endocytosis is receptor mediated and triggered when receptor proteins on the extracellular surface of the plasma membrane bind to specific substances. Examples include the uptake of lipoproteins by mammalian cells.

Pinocytosis involves the non-specific uptake of liquids or fine suspensions into the cell to form small pinocytic vesicles. Pinocytosis is used primarily for absorbing extracellular fluid. Examples: Uptake in many protozoa, some cells of the liver, and some plant cells.

Exocytosis

The contents of the vesicle are expelled into the extracellular space.

Plasma membrane

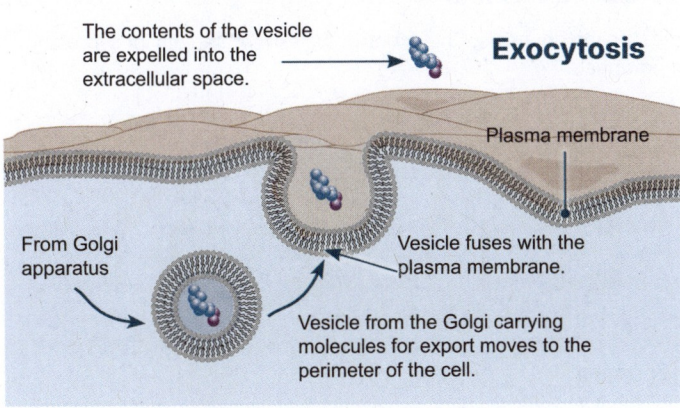

From Golgi apparatus

Vesicle fuses with the plasma membrane.

Vesicle from the Golgi carrying molecules for export moves to the perimeter of the cell.

Exocytosis occurs by fusion of the vesicle membrane and the plasma membrane, followed by release of the vesicle's contents to the outside of the cell.
Example: In multicellular organisms, several types of cells, e.g. lymphocytes, are specialized to manufacture and export products, such as proteins, from the cell to elsewhere in the body or outside it.

1. Distinguish between phagocytosis and pinocytosis: _____

   _____

   _____

2. Describe an example of phagocytosis and identify the cell type involved: _____

   _____

3. Describe an example of exocytosis and identify the cell type involved: _____

   _____

4. How does membrane fluidity allow for the fusion and formation of vesicles? Explain, using scientific terminology:

   _____

   _____

   _____

   _____

©2024 **BIOZONE** International
ISBN: 978-1-99-101410-8
Photocopying prohibited

B2.1
13
AHL

83 Gated Ion Channels

Key Idea: Gated ion channels are transmembrane proteins that allow the movement of ions and molecules through the cellular membrane in a controlled process.

Gated ion channels control the flow of ions through the membrane. The gates to the channels are either open to allow movement of ions and therefore send an action potential signal through a neuron, or closed to allow resting potentials to build across membranes. Signals to control the gates can be either from neurotransmitters or a voltage change. The neurotransmitter-gated ion channels are located in the nervous system and enable rapid information transfer across neuron synapses, initiating ion movement across the membrane. Sodium and potassium voltage gated ion channels have an important role in nerve impulses.

Neurotransmitter-gated ion channels: Nicotinic acetylcholine receptors

▶ Nicotinic acetylcholine receptors are located in the membranes of various cells and tissues. This includes the central nervous system and muscular tissue, including neuromuscular junctions between neurons and muscle cells.

▶ These ligand-gated ion channels are controlled by the neurotransmitter, acetylcholine, but they also respond to drugs such as nicotine. Once the acetylcholine has bound to the nicotinic receptor, the ion channel is opened, allowing the rapid diffusion of Na^+ and K^+ ions. At the neuromuscular junction, this results in muscular contraction of the muscle tissue.

▶ Nicotinic acetylcholine receptors in the brain can be altered by nicotine, which acts as a proxy for acetylcholine. Research has discovered a link between this process and the development of nicotine addiction.

▶ Nicotinic acetylcholine receptors are also involved in the modulation (controlling influence) of other neurotransmitters, including dopamine and serotonin.

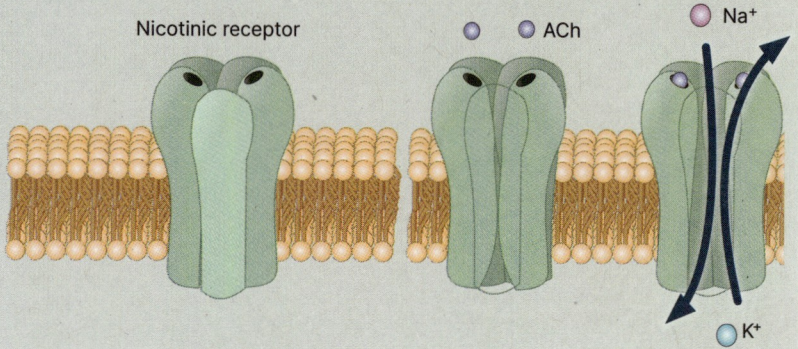

Voltage-gated ion channels: Sodium and potassium channels

▶ Sodium and potassium ion channels are examples of voltage gated ion channels. They control the flow of sodium ions into and potassium ions out of the cell during a nerve impulse (action potential).

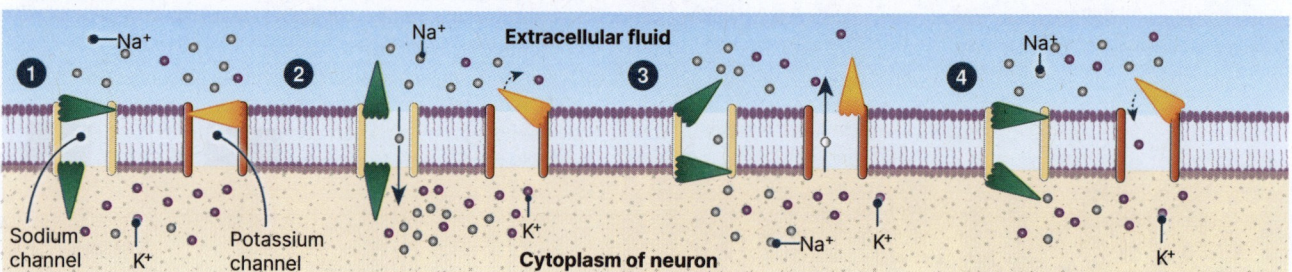

Resting state:
Voltage activated Na^+ and K^+ channels are closed. Negative interior is maintained by the Na^+/K^+ pump.

Depolarization:
Voltage activated Na^+ channels open and there is a rapid influx of Na^+ ions. The interior of the neuron becomes positive relative to the outside.

Repolarization:
Voltage activated Na^+ channels close and the K^+ channels open; K^+ moves out of the cell, restoring the negative charge to the cell interior.

Returning to resting state:
Voltage activated Na^+ and K^+ channels close and the Na^+/K^+ pump restores the original balance of ions, returning the neuron to its resting state (3Na^+ out for 2K^+ in).

1. What is the main function of the gates on the ion channels? _____

_____

_____

2. How do nicotinic acetylcholine receptors work and what would be the effect of introducing nicotine into the body?

_____

_____

_____

_____

©2024 **BIOZONE** International
ISBN: 978-1-99-101410-8
Photocopying prohibited

84 Exchange Transporters and Cotransporters

Key Idea: Exchange transporters, such as the sodium-potassium pump, can function in tandem with cotransporters to transport additional substances across the cell membrane. Exchange transporters move ions across the membrane. Ions are charged, so their movements can create a potential difference (voltage) across membranes. The combination of concentration gradient and voltage that affects an ion's movement is called an electrochemical gradient. Active transport can be either primary or secondary. Primary active transport directly uses ATP for energy to transport molecules, such as ion movement in the sodium-potassium pump. Secondary transport occurs when one molecule is coupled to the movement of another down its concentration gradient. ATP is not directly involved in the transport process.

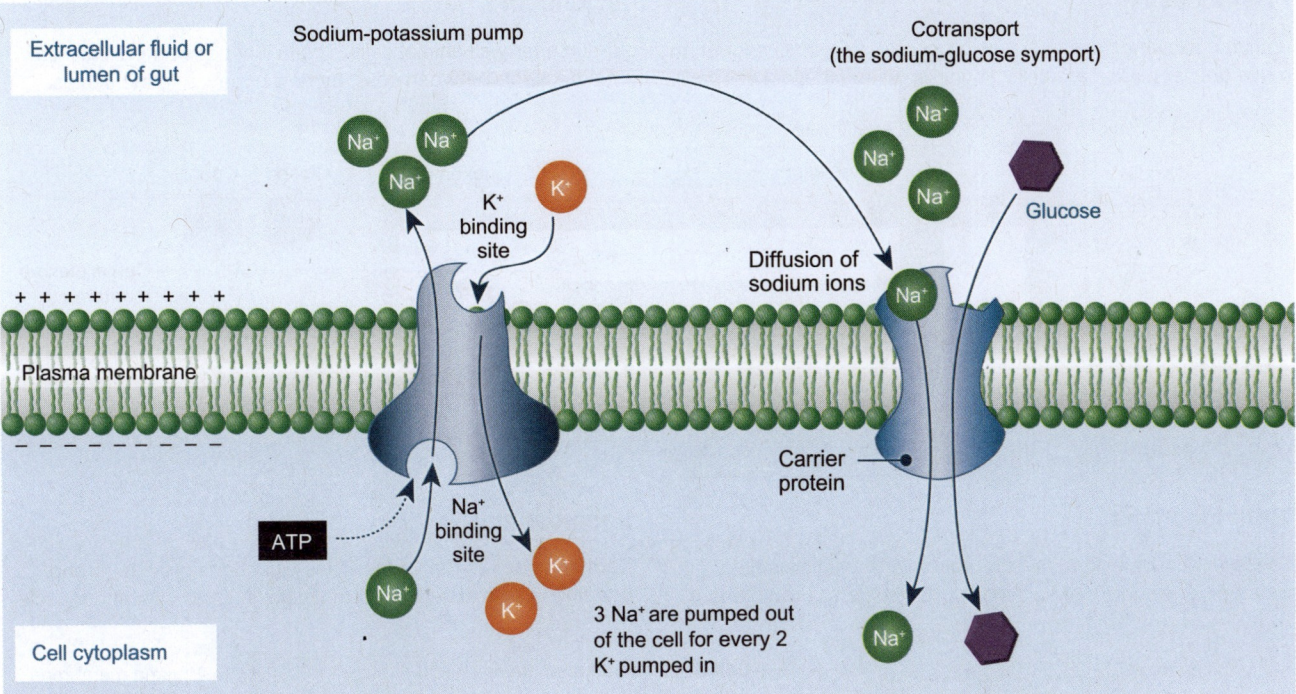

Sodium-potassium (Na⁺/K⁺) pump

The Na⁺/K⁺ pump is a protein in the membrane that uses energy in the form of ATP to exchange sodium ions (Na⁺) for potassium ions (K⁺) across the membrane. The unequal balance of Na⁺ and K⁺ across the membrane creates large concentration gradients that can be used to drive transport of other substances, e.g. cotransport of glucose. The Na⁺/K⁺ pump also helps to maintain the right balance of ions and so helps regulate the cell's water balance.

Cotransport (coupled transport)

A gradient in sodium ions drives the active transport of glucose into intestinal epithelial cells. The specific transport protein couples the return of Na⁺ down its concentration gradient to the transport of glucose into the intestinal epithelial cell across the cell membrane in contact with the gut lumen. Glucose diffuses from the epithelial cells across the opposite surface and is transported away in the blood. A low intracellular concentration of Na⁺ (and therefore the concentration gradient) is maintained by a sodium-potassium pump.

1. What is the purpose of the sodium-potassium pump? _____

_____

2. (a) Explain what is meant by cotransport: _____

_____

_____

(b) How is cotransport used to move glucose into the intestinal epithelial cells?

_____

_____

_____

(c) What happens to the glucose that is transported into the intestinal epithelial cells?

_____

3. (a) The sodium-potassium pump uses primary/secondary (delete one) active transport.

(b) The sodium-glucose symport uses primary/secondary (delete one) active transport.

85 Cell-Adhesion Molecules and Junctions

Key Idea: Cell-adhesion molecules (CAMs) are membrane embedded proteins that function to bind cells together at junctions to form tissues.

Cell Adhesion Molecules have a number of different roles. They are grouped, depending on their function and the type of junction at which they are found. A cellular junction is the portal between two cells, and can be a simple doorway, or a strong and sealed connection. Junctions include plant plasmodesmata, openings that allow the flow of cytoplasm between cells; and animal cell gap junctions that can create open and closed channels between cells: connexon in vertebrates, innexin in invertebrates.

Plasmodesmata

Junction allowing passive transport of particles small enough to move between adjacent cells. Found in plant and algal cell walls.

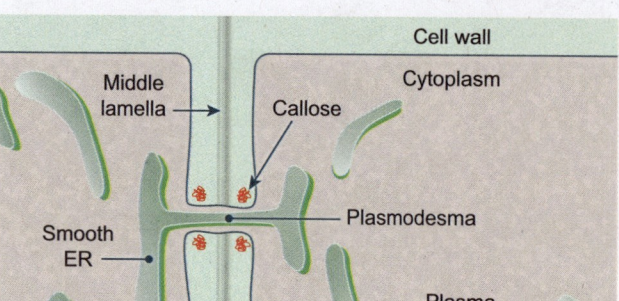

Gap junctions

Found in nearly all animal cells. These junctions and CAMs allow communication between cells, by flow of ions and molecules.

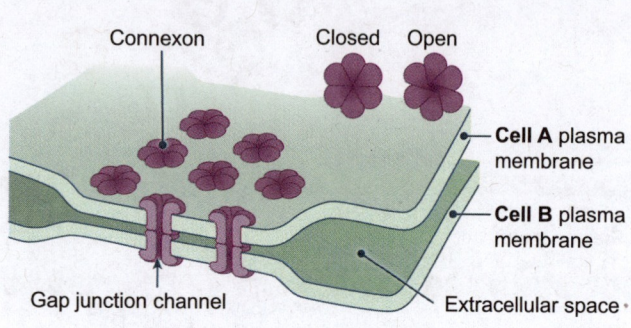

Tight junctions

These form a watertight layer of epithelial animal cells held together by claudin CAMs, seen in organs such as the bladder.

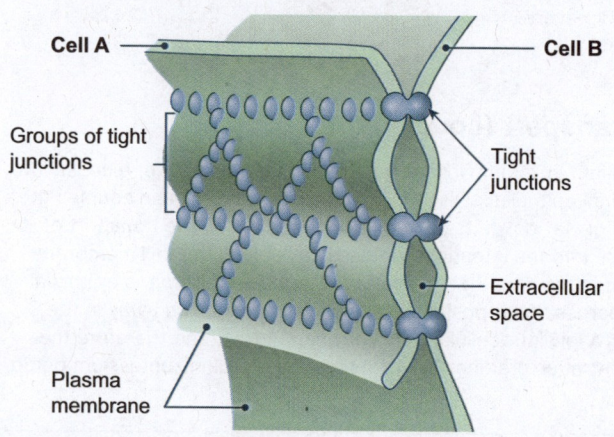

Desmosomes

Cadherins act as the CAMs. These junctions provide strong bonding in animal cells to form tissue, such as skin and muscle.

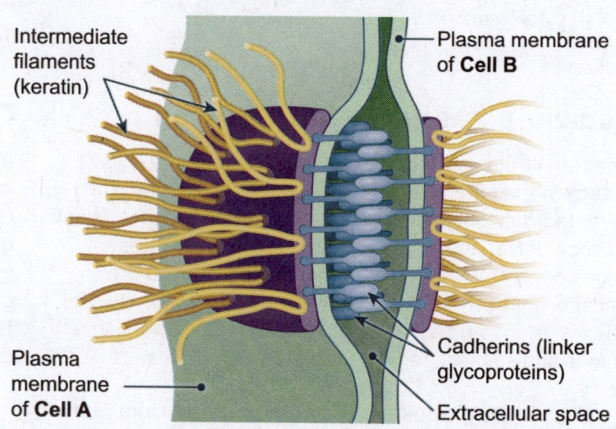

1. What is the difference between a junction and a CAM? _____
_____

2. Explain the purpose for multicellular organisms being able to connect cells together:
_____
_____
_____
_____

3. List some types of cells or tissues where each of the following CAMS / junctions are found:

(a) Plasmodesmata: _____

(b) Tight Junctions: _____

(c) Gap junctions: _____

(d) Desmosomes: _____

AHL   B2.1   17

©2024 **BIOZONE** International
ISBN: 978-1-99-101410-8

86 Compartmentalization in Cells

Key Idea: Eukaryotic cells use organelles to compartmentalize specific functions which increases the cell's efficiency.

Cellular organelles are adapted to perform one or more functions in the cell. In eukaryotic cells, some organelles are bound by a double membrane. These include mitochondria, nuclei and, in plants, chloroplasts. Other organelles are bound by a single membrane. These include the Golgi apparatus, endoplasmic reticulum, and vesicles. Ribosomes have no membrane. Membranes allow compartmentalization into organelles, which contain and regulate certain operations. This increases the cell's overall efficiency because specific areas are focused on specific tasks. Like the plasma membrane, the membranes of organelles control entry and exit of materials to and from their compartments. Membranes also allow attachment of proteins for specific tasks and help create chemical gradients to power the biochemical reactions necessary to sustain life.

Compartments and processes in an animal cell

Cellular respiration

mitochondria
Glucose is broken down, supplying the cell with energy to carry out the many other reactions involved in metabolism.

Protein synthesis

nucleus, rough endoplasmic reticulum, free ribosomes
Genetic information in the nucleus is translated into proteins by attached or free ribosomes.

Transport in and out of the cell

plasma membrane
Diffusion and active transport mechanisms move substances across the plasma membrane.

Containment of damaging oxidative reactions

peroxisomes
Isolate damaging oxidation reactions, such as beta oxidation. Peroxisomes are derived from the endoplasmic reticulum.

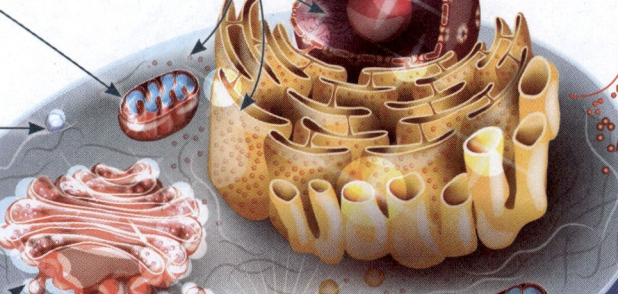

Structural elements of the cell, including the cytoplasm and cytoskeleton are not considered organelles.

Secretion

Golgi apparatus, plasma membrane
The Golgi produces secretory vesicles (small membrane-bound sacs) that are used to modify and move substances around and export them from the cell, e.g. hormones, digestive enzymes.

Cytosis

plasma membrane, vacuoles
Material can be engulfed to bring it into the cell (endocytosis) or the plasma membrane can fuse with secretory vesicles to expel substances from the cell (exocytosis). In animal cells, cytosis may involve vacuoles.

Breakdown

lysosomes
Contain hydrolytic enzymes to destroy unwanted cell organelles and foreign material. Lysosomes are derived from the Golgi.

Cell division

nucleus, centrioles
Centrioles are microtubular structures involved in key stages of cell division. They are part of a larger organelle called the centrosome.

Plant cells carry out photosynthesis

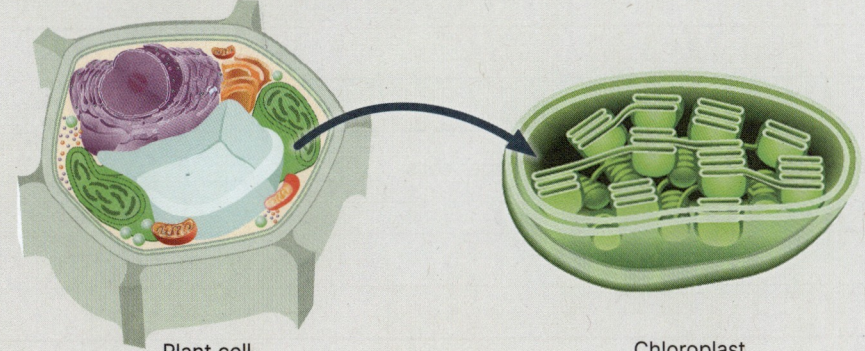

Plant cell

Chloroplast

Chloroplasts capture light energy and convert it into useful chemical energy (as sugars).

B2.2

1 - 3

Membranes allow compartmentalization of reactions and processes

▶ Membranes play an important role in separating regions within the cell (and within organelles) where particular reactions occur. Specific enzymes are therefore often located in particular organelles. The reaction rate is controlled by regulating the rate at which substrates enter the organelle and therefore the availability of the raw materials required for the reactions.

▶ Nucleus and cytoplasm separation: The nucleus acts as a distinct region of biochemical reactions, including transcription of DNA into RNA. While the RNA is still within the nucleus, post-transcriptional modification occurs, removing segments of RNA that are not required as proteins, and stabilizing mRNA prior to translation by the ribosomes. Prokaryotes have no nucleus, therefore any post-transcriptional modification must occur in the cytoplasm.

▶ Process separation: The membrane around the organelles allows for a concentration of metabolites and enzymes required for particular processes, where they are moved across the membrane and contained. The increased surface area of the internal membrane facilitates reactions. Harmful substances, often a by-product of metabolic reactions, can be contained by membranes, and removed from the cells by vesicles.

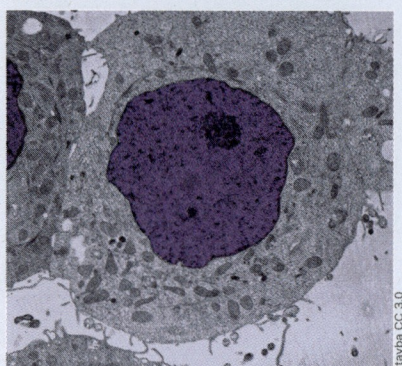

The nucleus is surrounded by a double-membrane structure called the nuclear envelope, which forms a separate compartment containing the cell's genetic material (DNA).

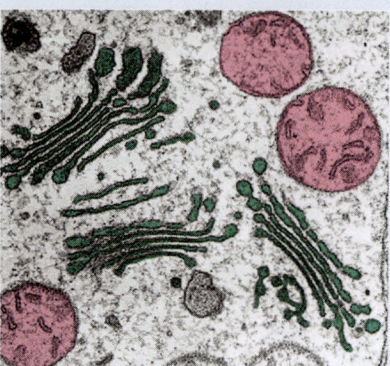

The Golgi apparatus (green) is a specialized membrane-bound organelle that compartmentalizes the modification, packing, and secretion of substances such as proteins and hormones.

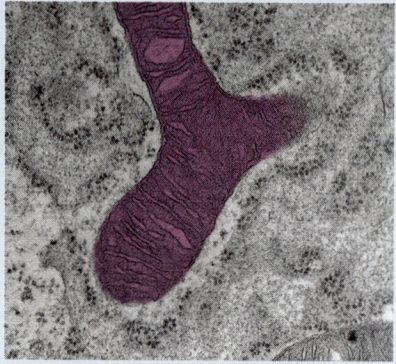

The inner membrane of a mitochondrion provides attachments for enzymes involved in cellular respiration. It allows ion gradients to be produced that can be used in the production of ATP.

1. Select one cellular process from the previous page and summarize how the role of the organelles contribute to it:

   _____

   _____

   _____

2. Identify two examples of intracellular membranes and describe their functions:

   (a) _____

   _____

   (b) _____

   _____

3. Explain how compartmentalization within the cell is achieved and how it contributes to functional efficiency:

   _____

   _____

   _____

   _____

4. Explain how compartmentalization has enabled the evolution of larger cells: _____

   _____

   _____

   _____

   _____

©2024 **BIOZONE** International
ISBN: 978-1-99-101410-8
Photocopying prohibited.

87 Techniques in Cellular Visualization

Key Idea: Technology has enabled cellular components to be studied at a greater depth.

Technology and science are closely linked: new understanding and research in science can lead to new technology being designed. Likewise, the development of new technology and its use by scientists results in a better understanding of processes that were previously uncertain. Technology, including the use of centrifuges and cell fractionation, has allowed researchers to separate the cellular components and organelles from the cell, divide them into specific organelle types and then collect them for study.

Ultracentrifugation and fractionation

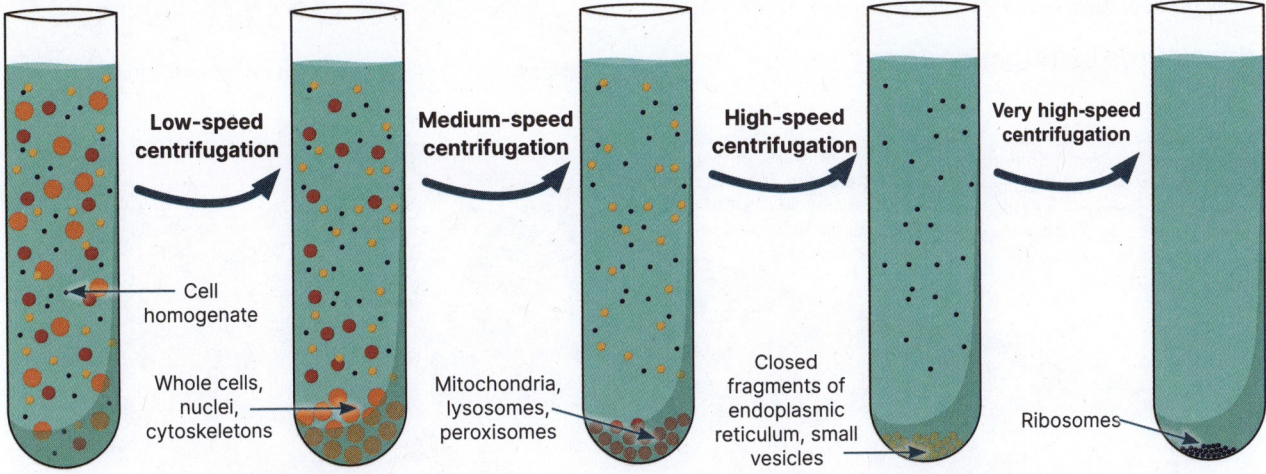

Differential centrifugation involves test tubes containing a homogenate solution of free floating organelles from broken cells being spun at progressively higher speeds to separate cell components on the basis of their size and density.

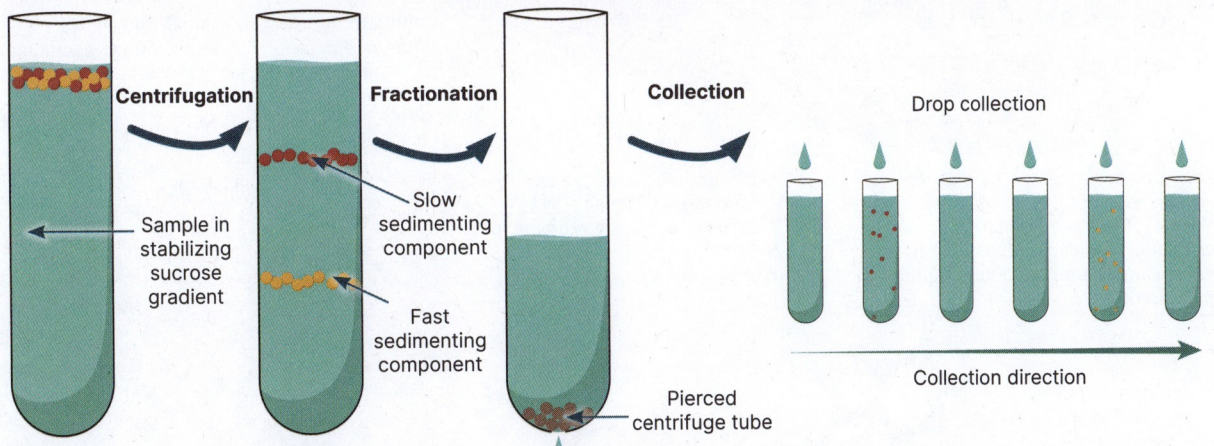

Once cellular components have been centrifuged, the denser components (faster sedimenting) settle lower down the test tube, and the less dense are higher up (slow sedimenting). This fractionates different cellular components based on their density. These are then separated into different tubes by drop collection.

1. Explain the purpose of ultracentrifugation: _____

_____

_____

2. How does the property of density allow for collection of components by fractionation? _____

_____

_____

3. Research another example of new technology leading to further cellular scientific discovery. Summarize your findings:

_____

_____

_____

88 Adaptations in Mitochondria and Chloroplasts

Key Idea: Mitochondria and chloroplasts are cellular organelles that have adapted to carry out the process of cellular respiration and photosynthesis, respectively.

Chloroplasts and mitochondria are organelles involved in the production of energy storage molecules in cells. Both are membranous organelles in which specialized biochemical reactions occur. Chloroplasts are found only in plant cells and some protists, whereas mitochondria are found in all eukaryotic cells. Mitochondria contain proteins (including ATP synthase) involved in the production of ATP, the energy storage molecule of cells. They are compartmentalized, with a double membrane, providing an increased surface area for respiration reactions to occur. Chloroplasts are the organelles responsible for photosynthesis. Chloroplasts have an internal structure characterized by a system of thylakoid membranous structures with bound light-capturing pigments. These absorb light of specific wavelengths, capturing light energy, which is then used to fix carbon into carbohydrates. Enzymes and substrates are compartmentalized within the stroma, where the Calvin cycle occurs.

Adaptations in mitochondrion

The inward foldings are called cristae. Due to invagination (folding in), they form a large surface area for reactions to occur. The electron transport chain and ATP synthesis occur on the inner membrane, catalyzed by the enzyme ATP synthase.

The space enclosed by the inner membrane is called the matrix. It contains enzymes and a pH stabilized solution to allow the Krebs cycle (part of cellular respiration) to occur effectively.

Mitochondria are much smaller than chloroplasts, ranging from about 0.75 to 3 µm.

Intermembrane space is small and allows a difference in proton concentration to accumulate on either side of the inner membrane. This is used to drive ATP synthesis.

Mitochondria are enclosed by a double membrane envelope (inner and outer membrane). The inner membrane is highly folded. Pyruvate moves from the cytoplasm into the mitochondria through the membrane.

Like chloroplasts, mitochondria have their own circular DNA (plasmids), containing their own 70S ribosomes (similar to bacterial ribosomes).

ATP is produced in the mitochondrion. It is an energy carrying molecule used to drive chemical reactions in the body.

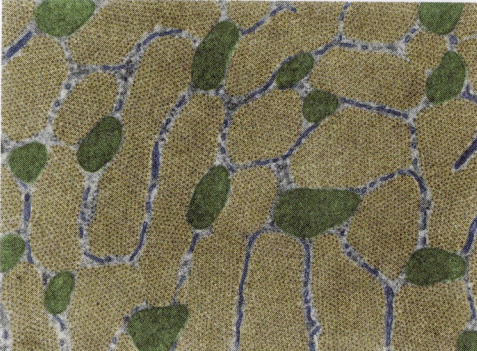

False color TEM showing cross-sectioned muscle myofibrils (yellow) and many mitochondria (green).

1. What adaptations of the cristae make cellular respiration more efficient? _____

_____

_____

_____

2. What is found in the matrix, and how do these substances contribute to cellular respiration? _____

_____

_____

3. How does the intermembrane space enable ATP production during aerobic cell respiration? _____

_____

_____

©2024 **BIOZONE** International
ISBN: 978-1-99-101410-8
Photocopying prohibited

Adaptations in chloroplasts

The internal structure of chloroplasts is characterized by a system of membranous structures called thylakoids arranged into stacks called grana.

Liquid stroma contains the enzymes for the light independent phase (Calvin cycle) as well as the chloroplast's DNA.

Stroma lamellae connect the grana. They account for 20% of the thylakoid membrane.

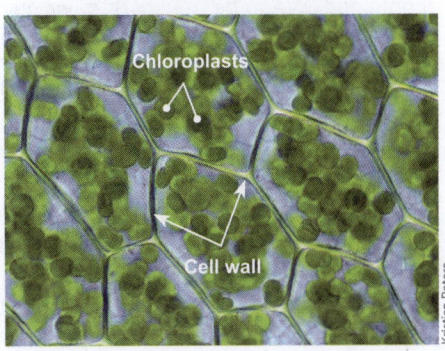

3D model of a chloroplast. Size ranges from 4-6 µm.

Lipid droplet

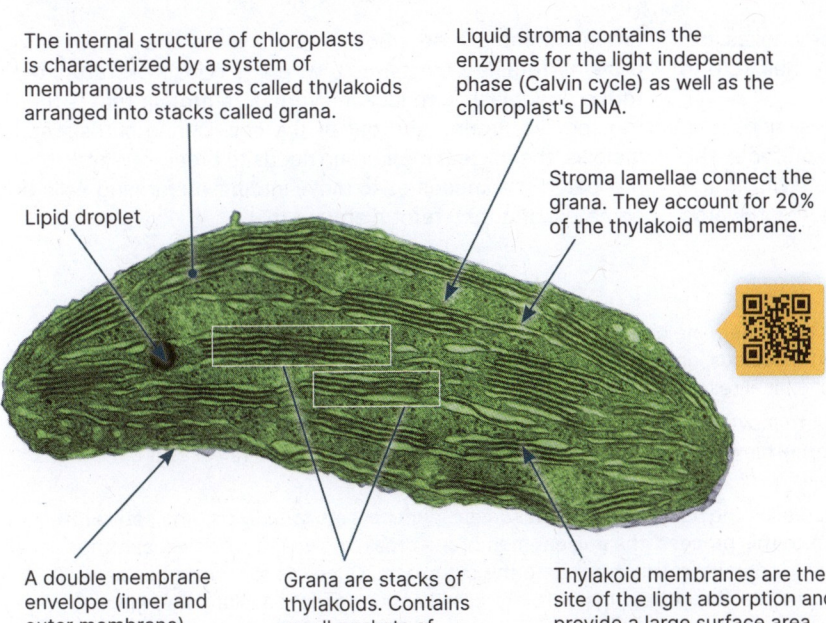

A double membrane envelope (inner and outer membrane) encloses the chloroplast.

Grana are stacks of thylakoids. Contains small pockets of concentrated enzymes.

Thylakoid membranes are the site of the light absorption and provide a large surface area, organized so as not to shade each other.

False colour TEM image of a single chloroplast

Chloroplasts visible in leaf cells. They appear green because they absorb blue and red light, reflecting green light. The chloroplasts are generally aligned so that their broad surface runs parallel to the cell wall to maximize the surface area available for light absorption.

4. Label the transmission electron microscope image of a chloroplast below:

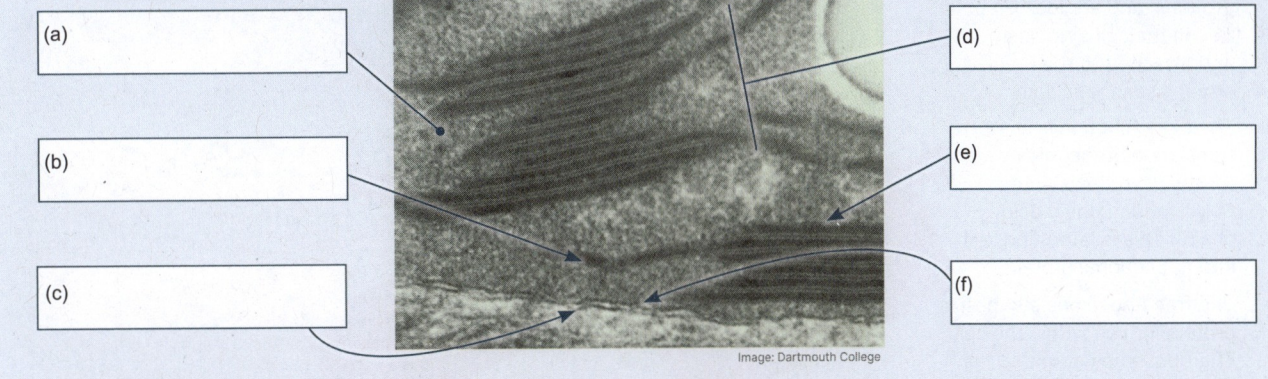

(a)

(b)

(c)

(d)

(e)

(f)

Image: Dartmouth College

5. Explain how the adaptations of the thylakoid membranes help absorb the maximum amount of light:

_____

_____

_____

6. Describe the difference in functions of the stroma and the thylakoid membranes: _____

_____

_____

_____

_____

7. How does the stroma compartmentalization of enzymes and substrates assist the Calvin cycle?

_____

_____

_____

_____

©2024 BIOZONE International
ISBN: 978-1-99-101410-8
Photocopying prohibited

89 The Nucleus and Endoplasmic Reticulum

Key Idea: The nucleus and the ribosomes are organelles and have adaptations that allow them to perform their role in protein synthesis, mitosis, or meiosis.

The nucleus is found in all eukaryotic cells and is a compartmentalized organelle with a double membrane. The nuclear membrane must allow mRNA transcribed from the contained DNA to be transported, via nuclear pores, towards the ribosomes. Free ribosomes translate the mRNA into proteins retained by the cell, whereas ribosomes attached to the endoplasmic reticulum synthesize protein that is then transported around and out of the cell. During mitosis and meiosis, the nuclear membrane needs to break down to allow replicated chromosomes to move into newly forming cells or gametes, and then reform around them.

Double membrane of the nucleus

▸ The nucleus has a double membrane: an inner and outer. The outer membrane also forms the nearest part of the endoplasmic reticulum. The membrane compartmentalizes the nucleus, and prevents cytoplasm, and its contents, from contacting the genetic material. This allows the different biochemical reactions that occur in the nucleus and cytoplasm to remain separate.

▸ The nuclear pores allow the transcribed mRNA to move out of the nucleus and towards the ribosomes. The mRNA transport is controlled by exporter proteins, exportin (exportin-t in vertebrates), which attach to the mRNA and then releases it once it has travelled through the nuclear pore.

▸ During mitosis and meiosis, DNA replication increases the number of chromosomes: identical sets for mitosis, and sets with variation for meiosis. In higher eukaryotes, prior to this process, the nuclear membrane breaks down into vesicles, and the nuclear pore proteins dissociate so all the nucleus contents are released into the cytoplasm. Once the chromosomes have shifted into their new cells or gametes, the nuclear membrane reforms from replicated endoplasmic reticulum around the chromosomes. In some unicellular eukaryotes, such as yeast, the nuclear membrane remains intact throughout (closed) mitosis.

Ribosomes

▸ Ribosomes are organelles without membranes. They are composed of RNA and other proteins, and are located both free in the cell cytoplasm, and attached to the rough endoplasmic reticulum.

▸ Their key function is translation of the mRNA (protein synthesis) into a polypeptide (amino acid chain). This is later modified into a functional protein.

▸ The free ribosomes synthesize protein for use within the cell. The ribosomes bound to the rough endoplasmic reticulum synthesize protein that is either integrated into the cell membrane, or transported out of the cell to be used elsewhere.

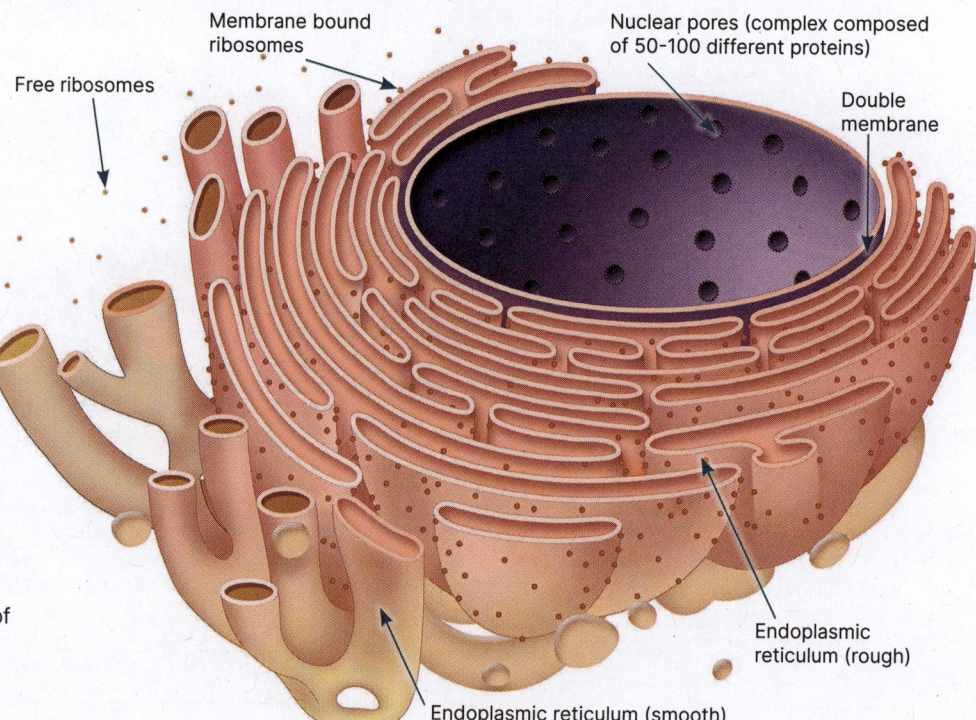

Membrane bound ribosomes

Free ribosomes

Nuclear pores (complex composed of 50-100 different proteins)

Double membrane

Endoplasmic reticulum (rough)

Endoplasmic reticulum (smooth)

1. Why is it important that the nucleus is compartmentalized from the cytosol, the aqueous contents of the cytoplasm?

_____

_____

2. Why does the nuclear membrane break down and then re-form during the prophase stage of mitosis and meiosis?

_____

_____

3. Compare and contrast the location and function of free ribosomes and bound ribosomes:

_____

_____

_____

  AHL B2.2 6-7

©2024 **BIOZONE** International
ISBN: 978-1-99-101410-8
Photocopying prohibited

90 Membranes and the Production of Proteins

Key Idea: The synthesis, packaging and movement of macromolecules inside the cell involves coordination between several membrane-bound organelles.

Many proteins need to be modified in order to become functional. This modification takes place in the rough endoplasmic reticulum (rER). From the rER, proteins are transported to the Golgi where the protein is further modified before being packaged and shipped to its final destination. The protein is encased by a vesicle, an organelle with a lipid bilayer membrane, and transported out of the cell.

Protein secretory pathway

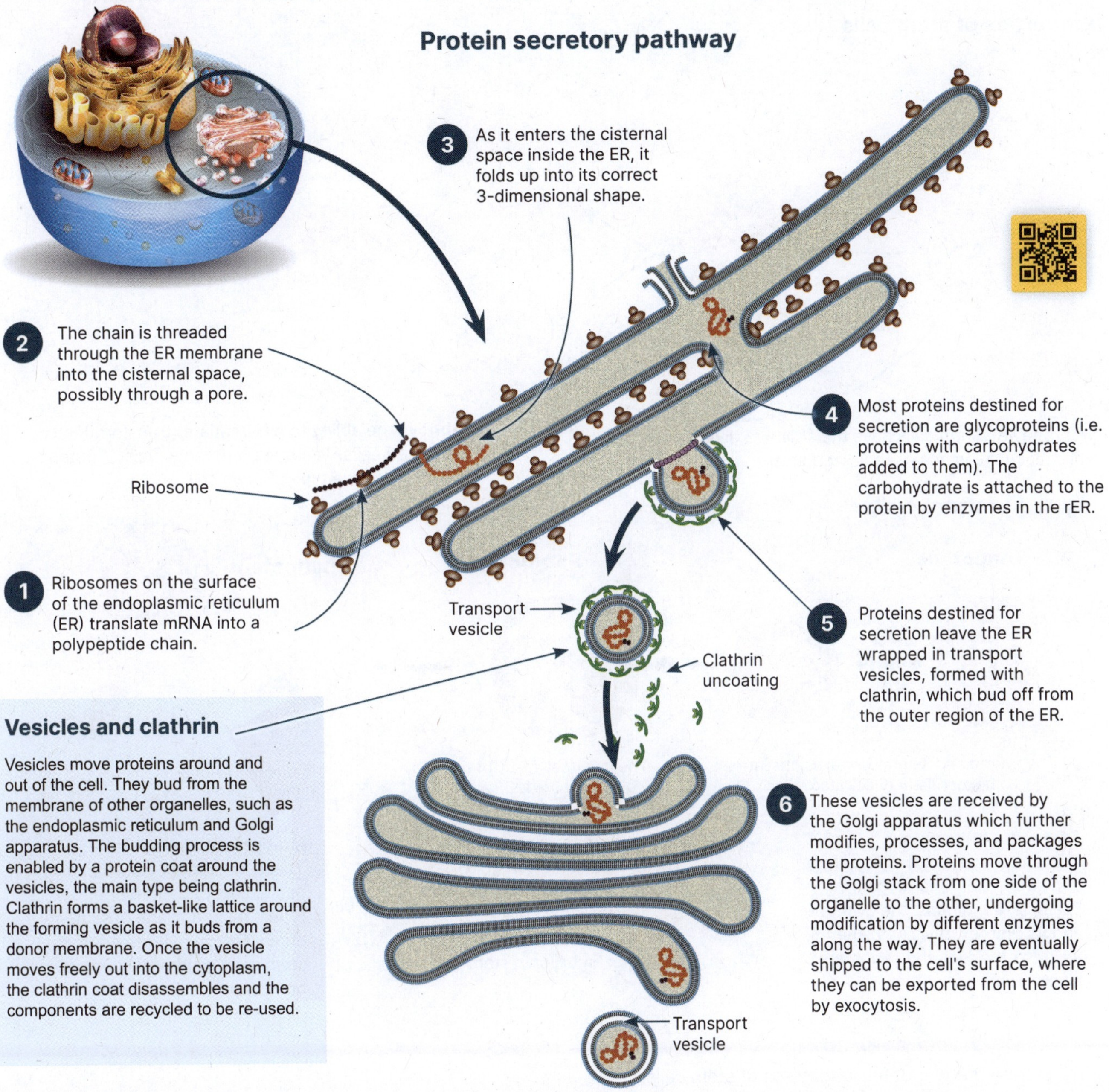

3 As it enters the cisternal space inside the ER, it folds up into its correct 3-dimensional shape.

2 The chain is threaded through the ER membrane into the cisternal space, possibly through a pore.

Ribosome

4 Most proteins destined for secretion are glycoproteins (i.e. proteins with carbohydrates added to them). The carbohydrate is attached to the protein by enzymes in the rER.

1 Ribosomes on the surface of the endoplasmic reticulum (ER) translate mRNA into a polypeptide chain.

Transport vesicle

5 Proteins destined for secretion leave the ER wrapped in transport vesicles, formed with clathrin, which bud off from the outer region of the ER.

Clathrin uncoating

Vesicles and clathrin

Vesicles move proteins around and out of the cell. They bud from the membrane of other organelles, such as the endoplasmic reticulum and Golgi apparatus. The budding process is enabled by a protein coat around the vesicles, the main type being clathrin. Clathrin forms a basket-like lattice around the forming vesicle as it buds from a donor membrane. Once the vesicle moves freely out into the cytoplasm, the clathrin coat disassembles and the components are recycled to be re-used.

6 These vesicles are received by the Golgi apparatus which further modifies, processes, and packages the proteins. Proteins move through the Golgi stack from one side of the organelle to the other, undergoing modification by different enzymes along the way. They are eventually shipped to the cell's surface, where they can be exported from the cell by exocytosis.

Transport vesicle

1. Explain the role of each of the following, related to the production and transport of proteins:

   (a) Endoplasmic reticulum: _____

   _____

   (b) Clathrin: _____

   _____

   (c) Transport vesicles: _____

   _____

   (d) Golgi apparatus: _____

   _____

 B2.2 8-9
 AHL

91 Stem Cells and Cell Specialization

Key Idea: Stem cells are unspecialized cells found in multicellular organisms. They are characterised by the properties of self renewal and potency.

A zygote can differentiate into all the cell types of the body because its early divisions produce stem cells. Stem cells can divide repeatedly while remaining unspecialized. They give rise to the many cell types that make up the tissues of a multicellular organism. Concentration gradients of substances in the embryo influence gene expression, and therefore cell differentiation into multipotent (or adult) stem cells, which are found in most organs, where they replace old or damaged cells and replenish cells throughout life.

Properties of stem cells

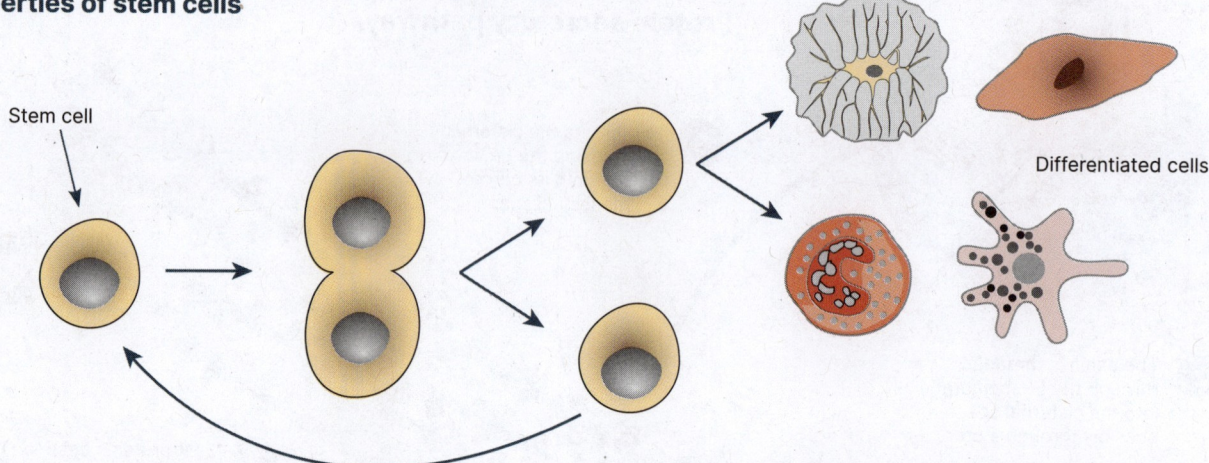

Stem cell

Differentiated cells

Self renewal: Stem cells have the ability to divide many times while maintaining an unspecialized state.

Potency: The ability to differentiate (transform) into specialized cells. There are different levels of potency, depending on the type of stem cell.

Types of stem cells

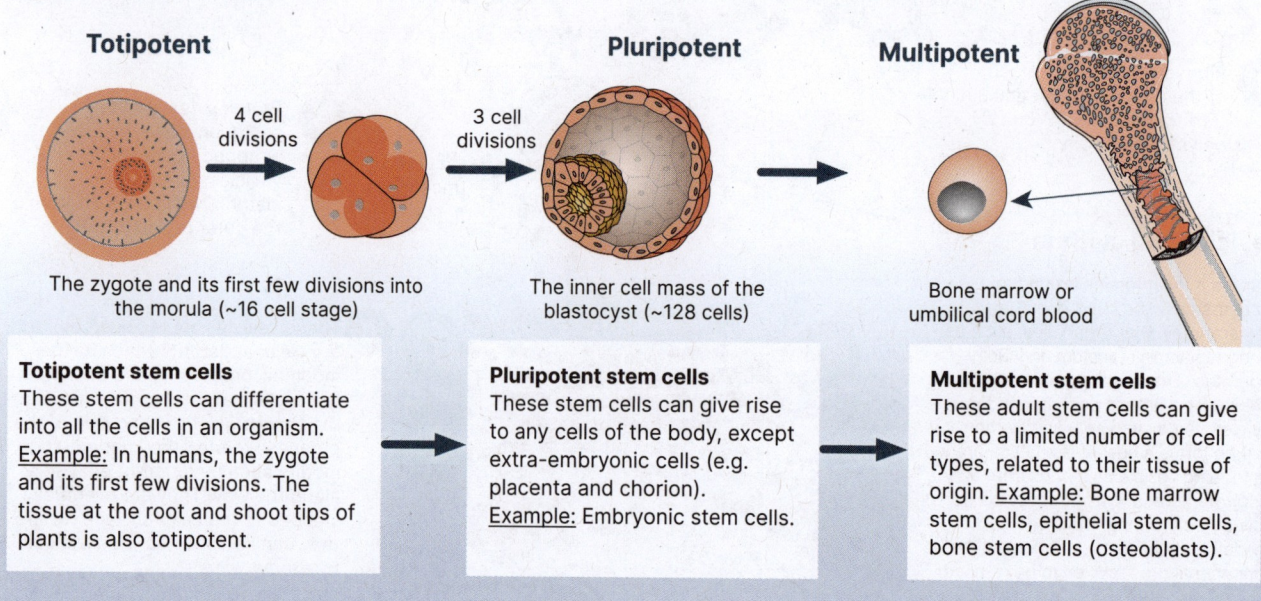

Totipotent

4 cell divisions

3 cell divisions

Pluripotent

Multipotent

The zygote and its first few divisions into the morula (~16 cell stage)

The inner cell mass of the blastocyst (~128 cells)

Bone marrow or umbilical cord blood

Totipotent stem cells
These stem cells can differentiate into all the cells in an organism. <u>Example:</u> In humans, the zygote and its first few divisions. The tissue at the root and shoot tips of plants is also totipotent.

Pluripotent stem cells
These stem cells can give rise to any cells of the body, except extra-embryonic cells (e.g. placenta and chorion). <u>Example:</u> Embryonic stem cells.

Multipotent stem cells
These adult stem cells can give rise to a limited number of cell types, related to their tissue of origin. <u>Example:</u> Bone marrow stem cells, epithelial stem cells, bone stem cells (osteoblasts).

1. Describe the two defining features of stem cells:

   (a) _____

   (b) _____

2. Describe the potency of stem cells and where they are found:

   (a) Totipotency: _____

   _____

   (b) Pluripotency: _____

   _____

   (c) Multipotency: _____

   _____

©2024 **BIOZONE** International
ISBN: 978-1-99-101410-8
Photocopying prohibited

Embryonic stem cells

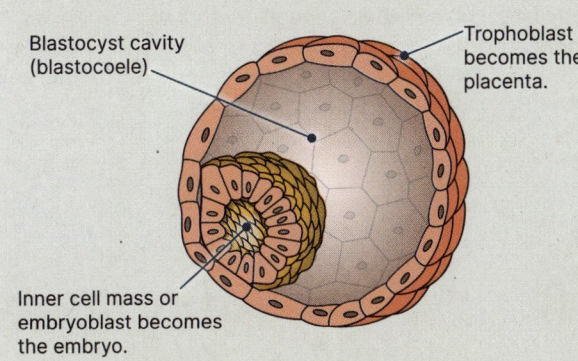

Blastocyst cavity (blastocoele)

Trophoblast becomes the placenta.

Inner cell mass or embryoblast becomes the embryo.

▶ Embryonic stem cells (ESC) are derived from the inner cell mass of blastocysts (above). Blastocysts are 5 day old embryos consisting of a hollow ball of 50-150 cells.

▶ Cells derived from the inner cell mass are pluripotent. They can become any cells of the body, with the exception of placental cells.

▶ When cultured without any stimulation to differentiate, ESC retain their potency through multiple cell divisions. This means they have great potential for therapeutic use in regenerative medicine and tissue replacement.

▶ However, the use of ESC involves the deliberate creation and destruction of embryos and creates ethical issues.

Adult stem cells

▶ Adult stem cells (ASC) are undifferentiated cells found as stem cell niches in several types of tissues (e.g. brain, bone marrow, hair follicles, fat, and liver) in adults, children, and umbilical cord blood.

▶ Unlike ESCs, they are multipotent and can only differentiate into a limited number of cell types, usually related to the tissue of origin.

▶ Fewer ethical issues are associated with using ASCs for therapeutic purposes, because no embryos are destroyed. For this reason, ASCs are already widely used to treat a number of diseases including leukaemia and other blood disorders.

Impact of gradients on gene expression

▶ Different concentration gradients of morphogens, substances that influence differential tissue development, influence gene expression in the stem cells of embryos. The extent of gene expression is controlled by the concentration of the morphogen.

▶ Bilateral eukaryotes (have 2 mirror-image sides) use morphogenic signalling to give each stem cell a positional value, or information. This informs what tissues should develop and where, in the embryo.

▶ The 'influenced' gene expression then enables the formation of structures, such as the nose forming on the face, or the correct length finger in the right order.

▶ The bicoid gene is important in *Drosophila* (fruit fly) development (below). After fertilization, bicoid mRNA from the mother is passed to the egg where it is translated into bicoid protein, a morphogen. Bicoid protein forms a concentration gradient in the developing embryo (see Below), which determines where the anterior (front) and posterior (rear) develop.

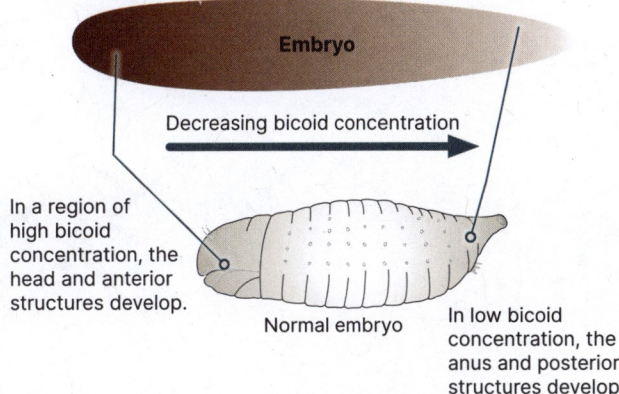

Embryo

Decreasing bicoid concentration

In a region of high bicoid concentration, the head and anterior structures develop.

Normal embryo

In low bicoid concentration, the anus and posterior structures develop.

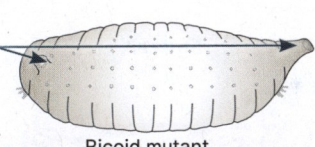

If bicoid is absent from the egg, the embryo develops two posterior ends and no head. Development stops.

Bicoid mutant

3. Distinguish between embryonic stem cells and adult stem cells with respect to their potency:

_____

_____

_____

4. Describe the location and function of stem cell niches in adult humans: _____

_____

_____

_____

_____

5. Explain how the impact of gradients on gene expression results in the development of specialized cells in the embryo:

_____

_____

_____

92 Comparing Human Cell Sizes

Key Idea: Different types of cells vary in size and this is a function of their specialization.

Humans have around 200 different specialized cell types, and each type has adaptations to allow it to perform a particular role. One adaptation is cell size. Different types of cells are different sizes. These vary from the tiny blood cells, which need to be small enough fit through capillaries, to long neurons that need to connect to other cells throughout the body. Size is one factor that is integral to the cell's differentiation, and therefore its function.

Typical sizes of various human cells

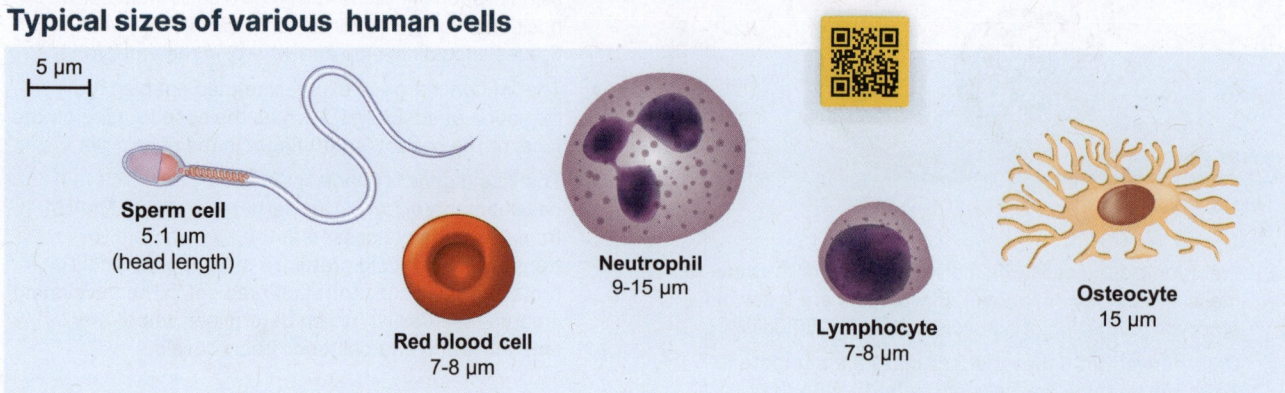

5 μm

Sperm cell
5.1 μm
(head length)

Red blood cell
7-8 μm

Neutrophil
9-15 μm

Lymphocyte
7-8 μm

Osteocyte
15 μm

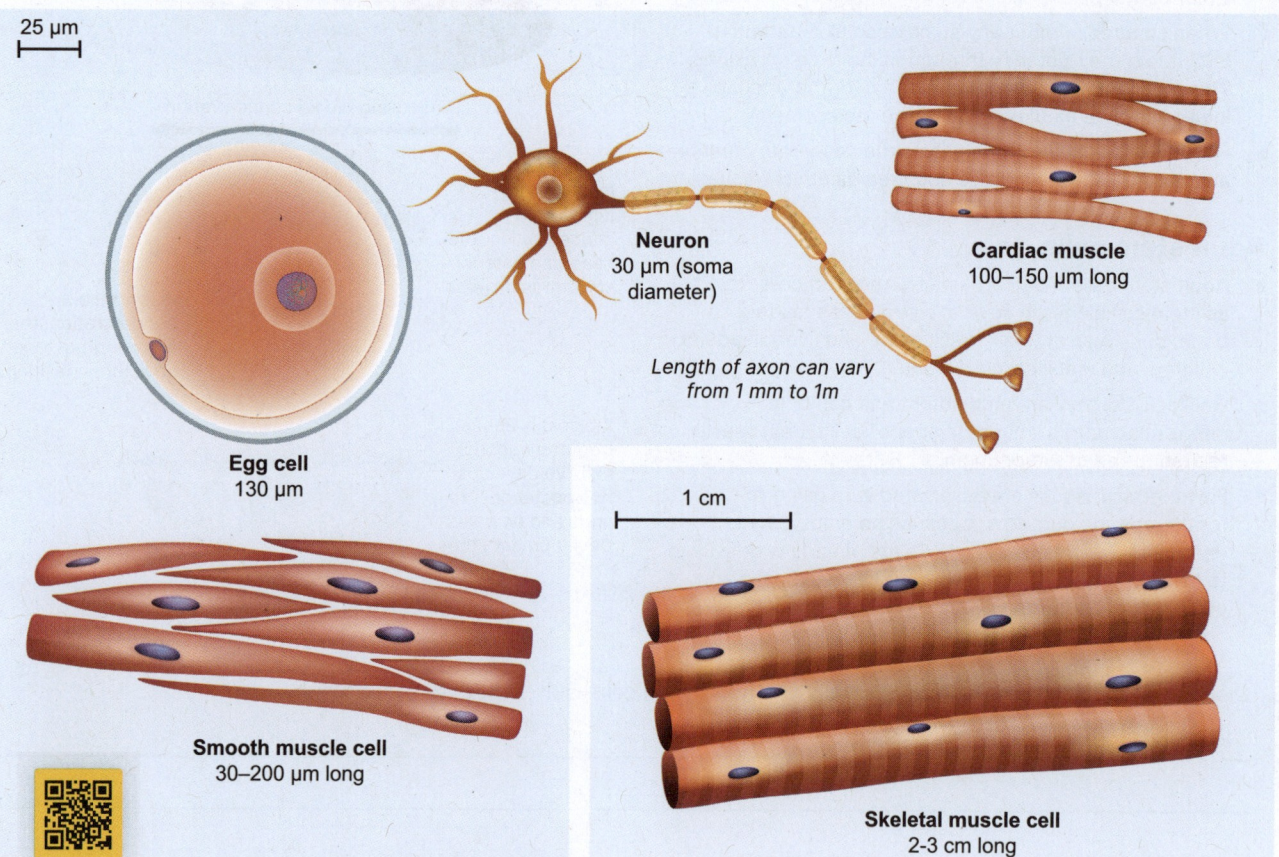

25 μm

Egg cell
130 μm

Neuron
30 μm (soma diameter)

Length of axon can vary from 1 mm to 1m

Cardiac muscle
100–150 μm long

1 cm

Smooth muscle cell
30–200 μm long

Skeletal muscle cell
2-3 cm long

Unit of length (international system)		
Unit	Meters	Equivalent
1 metre (m)	1 m	= 1000 millimetres
1 millimetre (mm)	10^{-3} m	= 1000 micrometres
1 micrometre (μm)	10^{-6} m	= 1000 nanometres
1 nanometre (nm)	10^{-9} m	= 1000 picometres

Micrometres are sometimes referred to as microns. Smaller structures are usually measured in nanometres (nm) e.g. molecules (1 nm) and plasma membrane thickness (10 nm).

1. (a) Compare length of the three muscle cells in μm:

_____

_____

(b) Why do scientific diagrams require the inclusion of a scale as part of them?

_____

_____

_____

_____

©2024 **BIOZONE** International
ISBN: 978-1-99-101410-8

NOS

B2,3
5

93 Constraints to Cell Size

Key Idea: A cell's size and shape directly affects its surface area to volume ratio and its ability to exchange materials with the environment.

In order to function, a cell must obtain the raw materials it needs for metabolism and dispose of the waste products of metabolism. These exchanges must occur across the cellular membrane. In a spherical cell, the cell volume increases faster than the corresponding surface area. As the cell becomes larger, it becomes more and more difficult for it to obtain all the materials it needs to sustain its metabolism. The size of cells can be constrained when material requirements exceed the maximum exchange rate. The surface-area-to-volume relationship of the cells can be estimated using the appropriate formula for their shape.

Calculating surface area to volume ratios

▶ Mathematical formula can be used to calculate the surface area and volume, and consequently the ratio between them.

▶ Models of spheres and cylinders can be used to approximate different types of cells.

	Sphere	Cube	Cylinder
Biological example	*Staphylococcus* bacterial cell	Kidney tubule cell	Axon of neuron
Surface area: The sum of all areas of all shapes that cover the surface of an object.	$4\pi r^2$	$6w^2$	$(2\pi r^2) + (2\pi rh)$
Volume: The amount that a 3-dimensional shape can hold.	$\frac{4}{3}\pi r^3$	w^3	$\pi r^2 h$

Cell A

Sphere
r = 0.78 cm

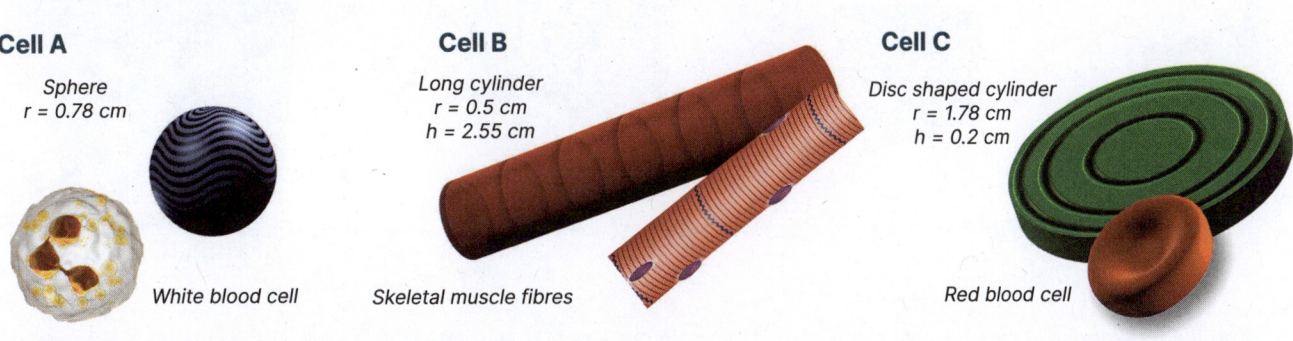

White blood cell

Cell B

Long cylinder
r = 0.5 cm
h = 2.55 cm

Skeletal muscle fibres

Cell C

Disc shaped cylinder
r = 1.78 cm
h = 0.2 cm

Red blood cell

r = radius   l = length   w = width   h = height   π = 3.14

1. Use the formulas for a sphere and a cylinder above to calculate the surface area of cell A, B, and C.

   (a) SA cell A: _____   (c) SA cell C: _____

   (b) SA cell B: _____

2. Use the formulas for a sphere and a cylinder above to calculate the volume of cells A, B, and C.

   (a) Volume cell A: _____   (c) Volume cell C: _____

   (b) Volume cell B: _____

3. Which of the cells above (A, B, C) has the greater surface area to volume ratio? Describe how changing the shape of a cell affects its surface area and its ability to obtain nutrients and dispose of wastes:

   _____
   _____
   _____
   _____
   _____

©2024 **BIOZONE** International
ISBN: 978-1-99-101410-8
Photocopying prohibited

Using models in science

Models are important ways of representing scientific concepts and ideas visually using a simplistic version. They help to visualize trends and patterns in data and can be used to show the complexity of relationships within a system.

Models can describe the relationship between variables, such as the surface area to volume of a cell, or comparing cell sizes to each other.

Scientific information can be represented visually or modelled in many different ways. Representations vary widely, depending on what type of information is being conveyed. The ability to describe and explain visual representations helps you to communicate information about the biological principles, concepts, and processes they involve.

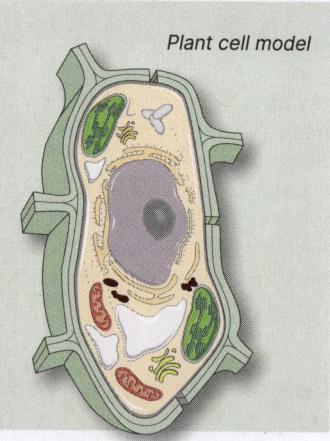
Plant cell model

The effect of increasing size

▸ The size and shape of a cell reflects its function and the need for essential molecules to move in and out. The greater the spherical diameter of a cell, the more material it contains and the further molecules have to move in order to reach the centre. At the same time, its metabolic requirements for raw materials increase. Molecules diffusing into the cell are used up faster than they can be supplied and may not reach the cell's centre, leaving it starved of essential molecules, e.g. oxygen.

▸ The transport of substances across membranes allows cells to exchange matter with their environment. Simple diffusion and transport involving membrane proteins are both affected by cell size and shape because these things affect the amount of surface area available relative to the cell's volume. The larger a cell is the more materials, e.g. oxygen, it needs and the further molecules need to move to reach their destination within the cell.

▸ A cell's surface area is important in determining how many molecules it can obtain. Its volume is important in determining how quickly molecules can reach certain parts of the cell. Surface area to volume ratio is therefore crucial to cell function.

(a) 2 cm cube   (b) 3 cm cube   (c) 4 cm cube   (d) 5 cm cube

4. The diagram above shows four hypothetical cells of different sizes. They range from a small 2 cm cube to a 5 cm cube (not to scale). Explain in general terms how a change in size effects the surface area and the volume of the cell:

_____

_____

_____

_____

5. Use the formulas provided on the previous page to calculate the surface area (SA), volume (V), and the ratio of surface area to volume (SA:V) for each of the four cuboidal cells **(a)-(d)** above. Show your calculations:

(a) SA: _____ V: _____ SA:V _____

(b) SA: _____ V: _____ SA:V _____

(c) SA: _____ V: _____ SA:V _____

(d) SA: _____ V: _____ SA:V _____

6. What are the advantages of using a model to demonstrate the surface-area-to-volume relationship of cells:

_____

_____

_____

_____

©2024 **BIOZONE** International
ISBN: 978-1-99-101410-8
Photocopying prohibited

94 Investigating the Effect of Cell Size

Key Idea: The effect of cell size on the efficiency of diffusion can be investigated using model cells of different sizes.

As described in the previous activity, the efficiency of diffusion decreases as cell size increases. This can be demonstrated easily in a model system. In this activity you will design an experiment to demonstrate the effect of surface area to volume ratios on diffusion in model cells. Think about how you will plan your investigation and analyse your data to obtain meaningful results. This will help you to make valid conclusions about your findings.

Background information

Oxygen, water, cellular waste, and many nutrients are transported into and out of cells by diffusion. However, at a certain surface area to volume ratio, diffusion becomes inefficient. In this activity you will create model cells of varying sizes from agar and use them to test the relationship between cell size and rate or efficiency of diffusion.

▸ The diffusion of molecules into a cell can be modelled by using agar cubes infused with phenolphthalein indicator and soaked in sodium hydroxide (NaOH).

▸ Phenolphthalein is an acid/base indicator and turns pink in the presence of a base.

▸ As the NaOH diffuses into the agar, the phenolphthalein changes to a pink colour and thus indicates how far into the agar block the NaOH has diffused (right).

▸ By cutting an agar block into cubes of various sizes, it is possible to investigate the effect of cell size on diffusion.

A phenolphthalein-infused agar cube after exposure to NaOH.

Equipment list

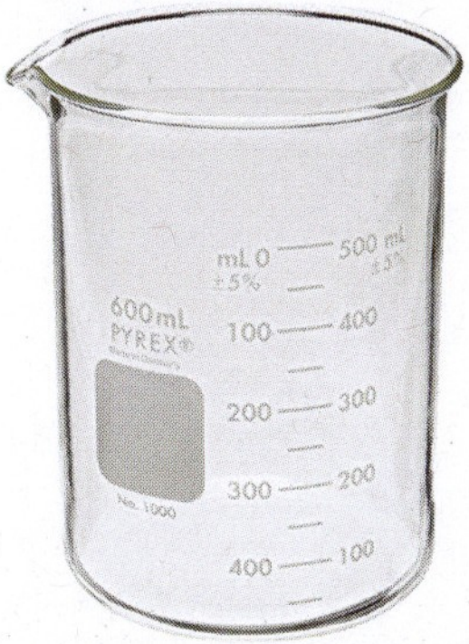

Glass beaker

Paper towel

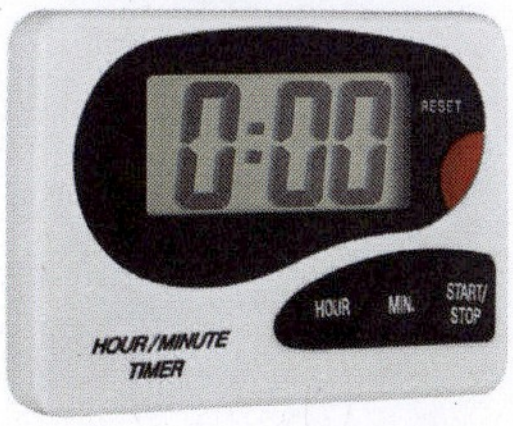

Timer

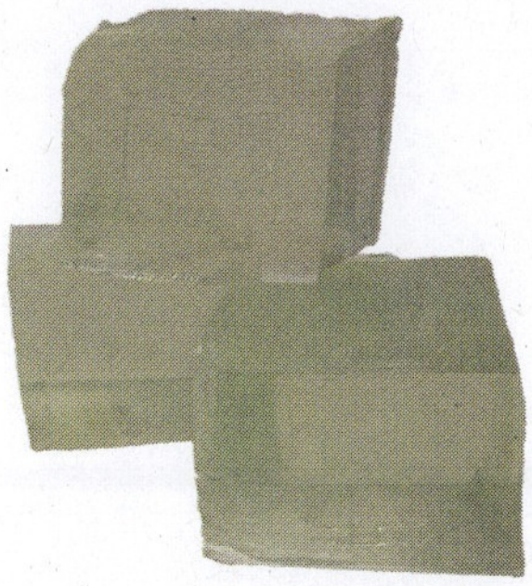

Agar blocks infused with phenolphthalein

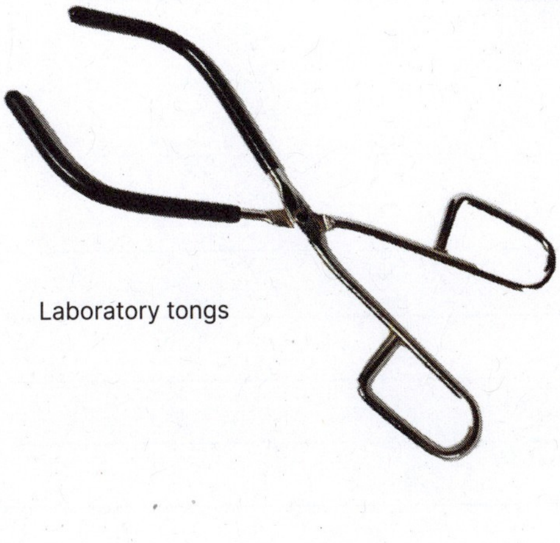

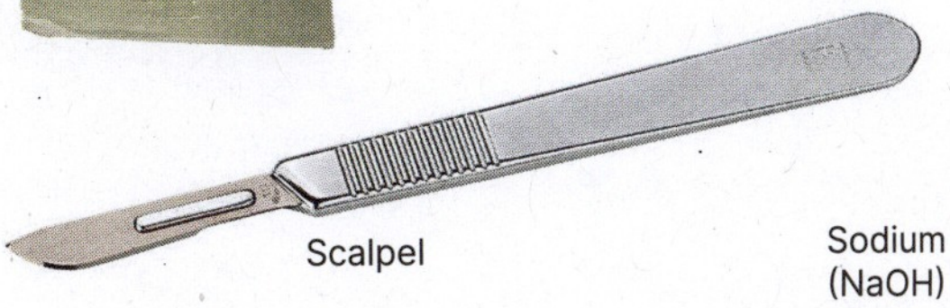

Laboratory tongs

Scalpel

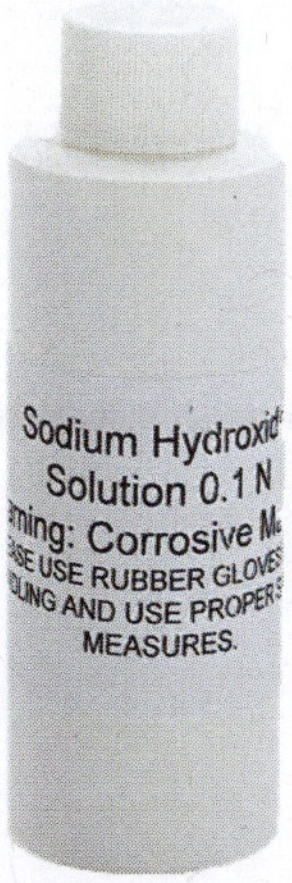

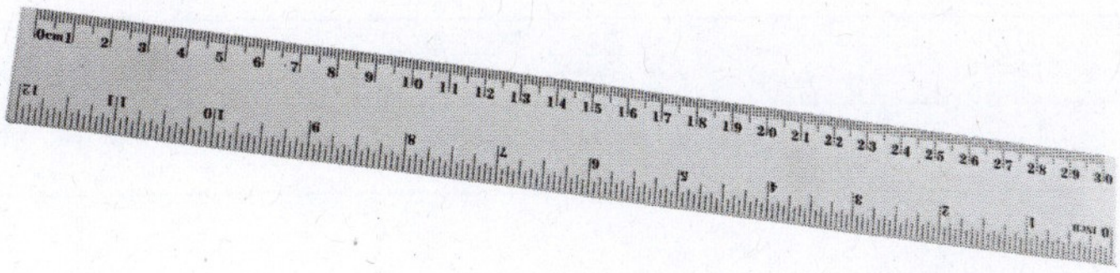

Sodium hydroxide (NaOH) solution

Ruler

Based on the aim of this experiment "To investigate the effect of cell size on diffusion in a model cell", the background information provided, and the equipment list provided, design your own experiment using the questions below to guide you.

1. State a hypothesis for your experiment: _____

_____

_____

2. Write your method as step by step instructions: _____

_____

_____

_____

_____

_____

_____

_____

_____

_____

_____

3. In the space below, draw a table to record your results. Remember to record cell volume and include space to work out how much diffusion has occurred:

4. Write your conclusions here: _____

_____

_____

_____

_____

_____

_____

95 Cellular Adaptations to Increase Surface Area

Key Idea: Some cells have adaptations to their shape that increase the surface area to volume ratio.

The problem of supply and demand can be solved by reducing the diameter of the cell along one axis, and flattening it or elongating it along another. Elongated spheres or cylinders can project to form microvilli, or form a cavity to create invagination and has a greater surface area than a sphere of the same volume. In this way, a cell can grow larger while still gaining the materials it needs. The cells of multicellular organisms are often highly specialized to maximize SA : V.

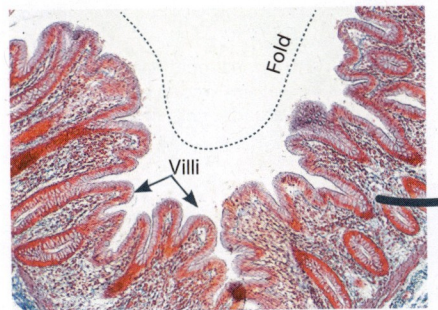

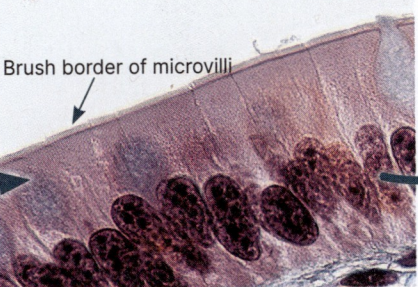

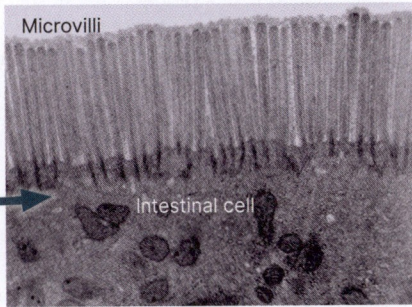

Cells can be organized into tissues and are also organized to increase surface area. Here, the intestinal wall has many folds and on those folds are projections or villi (image above left). Epithelial cells line each villus, each with a brush border of microvilli (image above centre). The tiny microvilli are extensions of the cell's membrane (image above right) and greatly increase the surface area for absorbing nutrients and binding digestive enzymes.

Microvilli and invagination in the mammal nephron

The functional unit of the kidney is the nephron. It is a selective filter element, comprising a renal corpuscle and its associated tubules and ducts. The proximal convoluted tubule is lined with cells that reabsorb around 90% of filtrate received from the blood, including glucose and valuable ions, so they can be recycled for use in the body. The cells are lined on the inside of the tubule lumen with microvilli, to increase surface area for diffusion of substances. The cells also retrieve proteins from the filtrate. These are transported across the cell through endocytosis, where part of the cell membrane wraps around the protein molecule. The invaginations of the membrane increase the surface area for this process to take place.

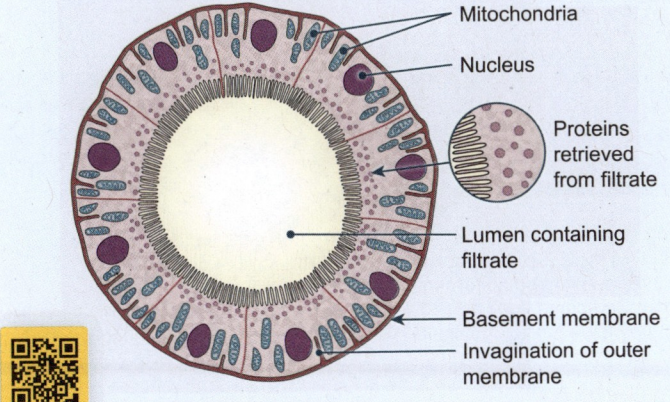

- Mitochondria
- Nucleus
- Proteins retrieved from filtrate
- Lumen containing filtrate
- Basement membrane
- Invagination of outer membrane

Flattening of mammal erythrocytes

Red blood cells (erythrocytes) make up 38-485 of total blood volume. Every mm^3 of blood contains 5-6 million RBCs. These transport oxygen

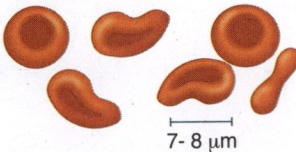

7- 8 μm

(O_2) and a small amount of carbon dioxide (CO_2). The oxygen is carried by haemoglobin (Hb) in the cells. Each Hb molecule can bind four molecules of oxygen. The flattened disk shape of the cells increases the SA:V ratio to maximize the diffusion rate of oxygen across the cellular membrane.

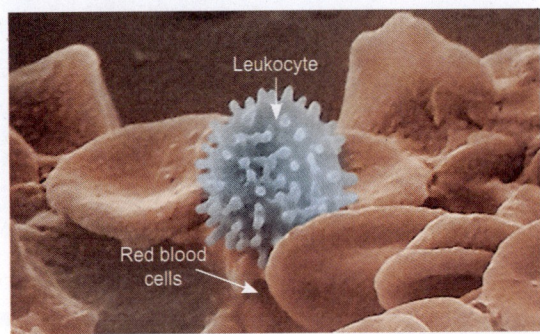

Leukocyte

Red blood cells

SEM of red blood cells and a leukocyte.

1. Explain why the formation of microvilli on cells enables them to function effectively, using an example of a cell type:

_____

_____

_____

2. What function do the adaptations for surface area in the proximal convoluted tubule assist with?

_____

_____

_____

_____

B2.3
7
AHL

152

96 Adaptations of Mammalian Cells

Key Idea: There are many different types of animal cells, each with a specific role in the body. Animal cells are often highly modified for their specific role. There are over 200 different types of cells in the human body. Animal cells lack a cell wall, so they can take on many different shapes. Some, e.g. white blood cells, are even mobile. The shape, size, and the internal structure of a specialized cell reflects its functional role in the body.

▶ Specialized cells often have modifications or exaggerations to a normal cell feature to help them do their job. For example, nerve cells have long, thin extensions to carry nerve impulses over long distances in the body.

▶ Specialization improves efficiency because each cell type is specialized to perform a specific task. They may have more (or fewer) of a particular organelle in order to perform their role most efficiently.

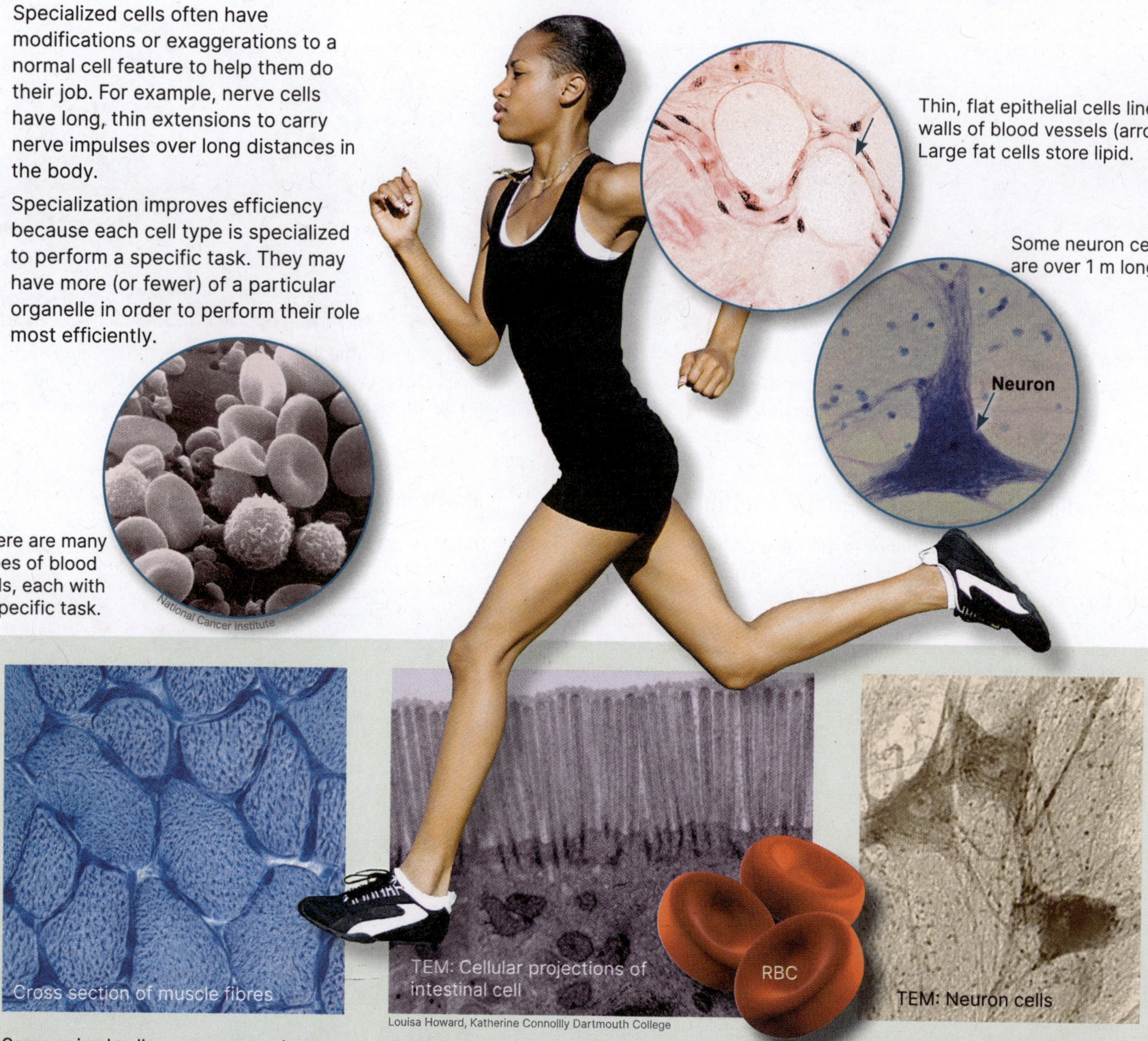

Thin, flat epithelial cells line the walls of blood vessels (arrow). Large fat cells store lipid.

Some neuron cells are over 1 m long.

Neuron

There are many types of blood cells, each with a specific task.

National Cancer Institute

Cross section of muscle fibres

TEM: Cellular projections of intestinal cell

Louisa Howard, Katherine Connollly Dartmouth College

RBC

TEM: Neuron cells

Some animal cells can move or change shape. Muscle cells, called fibres, are able to contract (shorten) as protein fibres within the cell move past each other. This action causes the movement of limbs, and of organs, such as the heart and intestine.

Cells that line the intestine have extended cellular membranes. This increases their surface area so that food (nutrients) can be absorbed quickly and efficiently. Red blood cells (RBCs) have no nucleus so they have more room inside to carry oxygen.

Neuron cells conduct impulses in the form of changes in membrane potential. Impulses are carried from receptors, e.g. eye, to effectors such as muscles, allowing responses to the environment.

1. What is the advantage of cell specialization in a multicellular organism? _____

_____

_____

_____

2. For each of the following specialized animal cells, describe a feature that helps it carry out its function:

(a) Intestinal cell: _____

_____

(b) Nerve cell: _____

_____

AHL · B2.3 · 8 - 10

©2024 **BIOZONE** International
ISBN: 978-1-99-101410-8
Photocopying prohibited

Adaptations of type I and type II pneumocytes

▶ Alveoli are structures in the lungs where gas exchange occurs. Two key cell types, Type I and Type II pneumocytes, integrate to form the alveoli sac. The cells have specialized adaptations to enable the diffusion of oxygen and carbon dioxide gas to diffuse between the capillary and alveoli.

▶ Type I pneumocyte cells are extremely thin, reducing the distance the gases have to diffuse, and they line the alveoli as one single-celled epithelium (layer).

▶ In order to prevent the alveoli sacs collapsing during breath exhalation, a surfactant lines the alveoli lumen (space), reducing surface tension between the air-liquid interface. Type II pneumocytes comprise about 7% of the surface area of the epithelium, and are secretory cells that release the protein and phospholipid surfactant, that lines the alveoli. The hydrophobic properties of the phospholipids prevent water from forming a tightly bonded layer against the epithelium.

▶ Around 25% of the Type II pneumocyte cytoplasm contains secretory vesicles, called lamellar bodies. The lamellar bodies are membrane lined, and one of the largest types of vesicles found in mammals. The surfactant produced is moved into the lumen of the alveoli by exocytosis. There is a controlled and continuous recycling of surfactant, which is predominantly returned back into the Type II pneumocyte cells by endocytosis.

▶ The alveoli epithelium is one example where more than one type of cell is found, with the different cells having specialized adaptations in order to perform a multi-function task, in this case aid diffusion and secrete surfactant.

Cellular structure of an alveolus

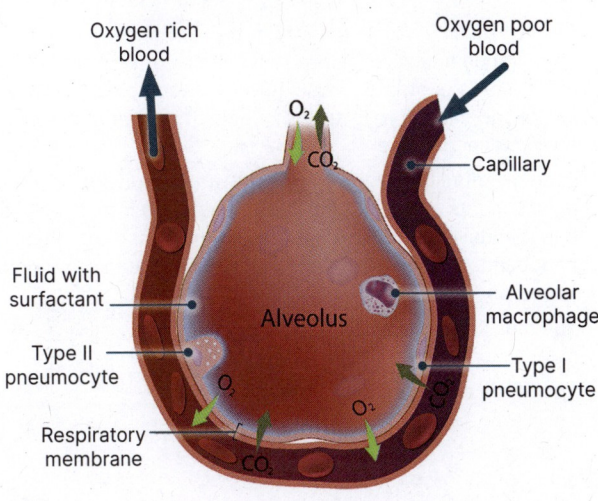

Adaptations of human sperm and egg cells

▶ Gametes (sex cells) are produced for the purposes of sexual reproduction. The gametes of males and females differ greatly in size, shape, and number. These cellular differences reflect their different roles in fertilization and reproduction. Male gametes (sperm) are highly motile and are produced in very large numbers throughout life. Female gametes (eggs or ova) are large, relatively few in number, and are not produced throughout life.

▶ The sperm's cell structure reflects its purpose, which is to swim through the female reproductive tract to the ovum, penetrate the ovum's protective barrier, and donate its genetic material. A sperm cell comprises three regions: a headpiece, containing the nucleus and penetrative enzymes, an energy-producing mid-piece, and a tail for propulsion.

▶ The ovum is a simpler structure than the sperm cell. It has no propulsion mechanism and moves as a result of the wave-like motion of the ciliated cells lining the fallopian tubes of the female reproductive tract. The ovum is required to survive for a much longer time than a sperm, so it contains many more nutrients and metabolites and, as a result, is much larger than a sperm cell (100µm in comparison to the sperm's 5µm).

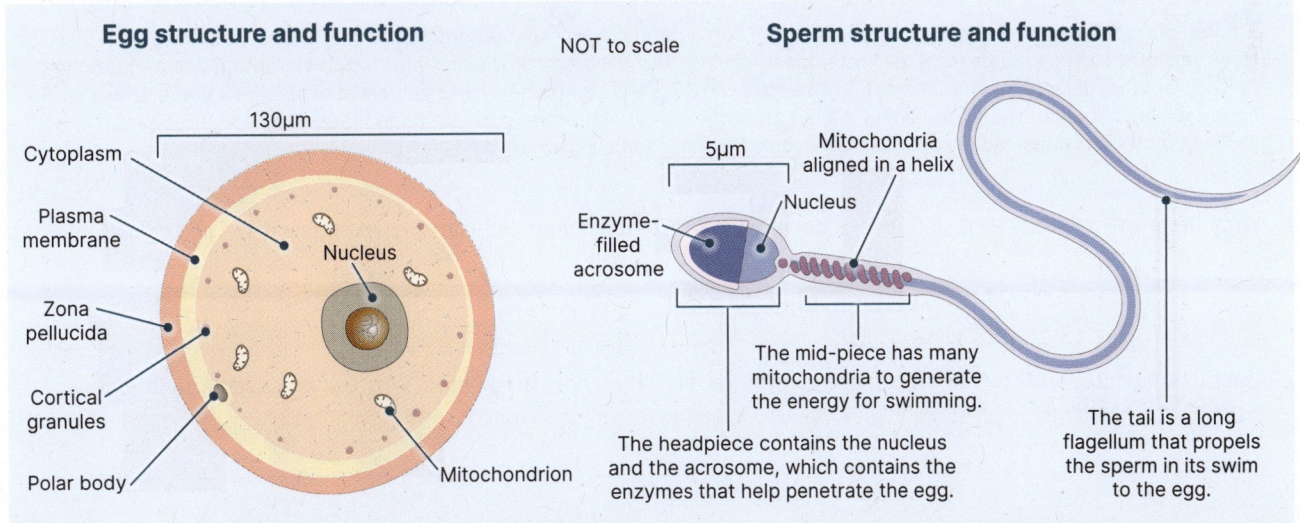

3. Provide another example of different cell types in the same tissue and explain the function of each related to function:

_____

_____

_____

_____

Adaptations of striated muscle fibre and cardiac muscle cells

Striated muscle fibres

▶ Skeletal muscle has a striated (striped or banded) appearance and is organized into bundles of contractile muscle cells or striated muscle fibres. Each fibre has many nuclei and each fibre is itself a bundle of smaller myofibrils arranged lengthwise.

▶ Each myofibril is, in turn, composed of two kinds of myofilaments (thick and thin), that overlap to form light and dark bands. It is the orderly alternation of these light and dark bands that gives skeletal muscle its striated or striped appearance.

▶ The muscle fibres do not have any cell-cell junctions.

▶ The fibres are unbranched.

Cardiac muscle cells

▶ Cardiac muscle cells are also striated and contain myofibrils, but it does not fatigue because the muscle has a built in rest period after each contraction.

▶ The muscle cells have a larger number of mitochondria per cell than other types of muscles, as they have a continuous demand of energy for repeated contraction.

▶ The cells are joined together with desmosomes to tether myofibrils to each other between cells to ensure structural integrity, and with gap junctions for ion flow, so they can synchronize contractions in other borders of the cell.

▶ The muscle fibres are branching and the cells are uninucleate (have only one nucleus).

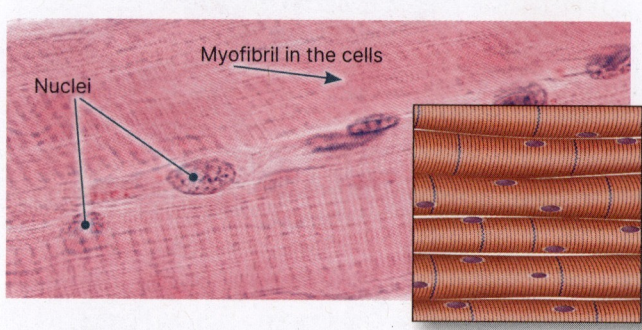

Nuclei — Myofibril in the cells

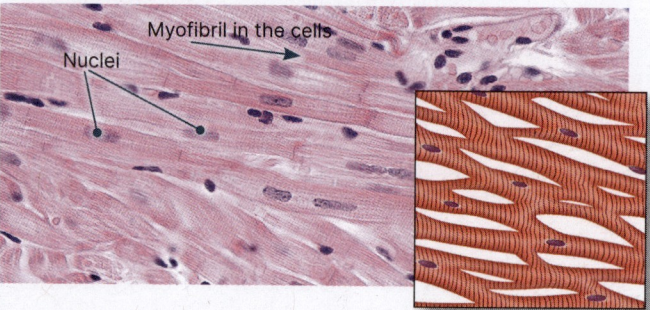

Nuclei — Myofibril in the cells

Comparing cell adaptations in striated muscle fibres and cardiac muscle cells

▶ Research has led to a number of hypotheses being proposed to account for the differences between types of muscle cells.

▶ Striated muscle fibres are not usually considered single cells, unlike cardiac muscle cells. When the muscle fibres undergo mitosis replication, the 'cells' do not separate. This accounts for the multiple nuclei present in each fibre, and the long length of the fibres themselves. Cardiac muscle cells do separate into single cells during replication, so are shorter and have just one nucleus per cell.

▶ Cardiac muscle cells are short and branched, with one cell being connected to three or four other cells. This allows rapid and widespread action potential propagation through cell-cell gap junctions, allowing ion flow, therefore the muscle cells contract in a coordinated pattern. Cardiac muscle cells are controlled by the autonomic nervous system.

▶ Striated muscle fibres are long and unbranched, and can potentially exert more contractile force than if they were branched. However, muscle tissue tends to degenerate due to ageing, and will break down and reform with increasingly more branched fibres over time, decreasing the contraction strength. Striated muscle fibres have no cell-cell gap junctions present, as the contractile myofibrils run unbroken along the length of the fibres. Therefore, contraction of fibres does not spread from one "cell" to another, and instead is controlled by signals from individual motor neurons.

4. What are some key structural differences between the myofibril connections in striated muscle fibres and cardiac muscle?

_____

_____

5. Unbranched muscle fibres can exert more contractile force than branched fibres. Why is it an advantage for cardiac muscle fibres to be branched?

_____

_____

_____

6. Why are striated muscle fibres not considered to be composed of cells but cardiac muscle is, and what is the evidence to support this hypothesis?

_____

_____

_____

©2024 **BIOZONE** International
ISBN: 978-1-99-101410-8
Photocopying prohibited

97 Did You Get It?

1. Label the components of the fluid mosaic model of the membrane below:

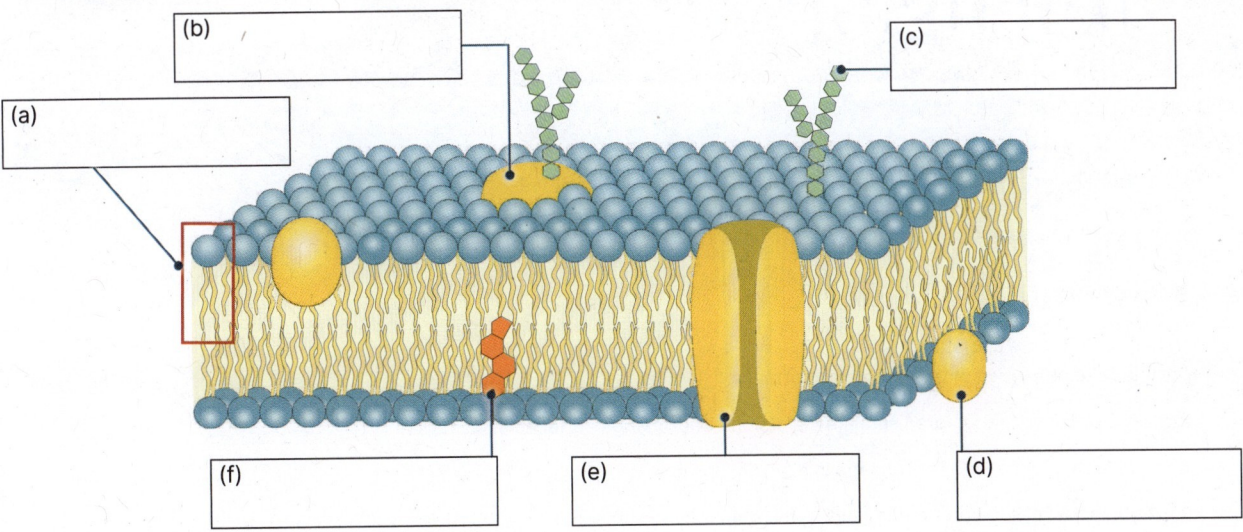

(a)

(b)

(c)

(f)

(e)

(d)

2. What function does structure (e) above have in the cell?

_____

_____

3. **AHL:** Why is membrane fluidity important, and how is it maintained in changing environmental conditions?

_____

_____

_____

4. Compare differences in location and differentiation ability between totipotent, pluripotent, and multipotent stem cells:

_____

_____

_____

5. What is the purpose of compartmentalization in cells?

_____

_____

6. How can eukaryotic cells efficiently obtain the raw materials they need for metabolism, even as they become larger?

_____

_____

_____

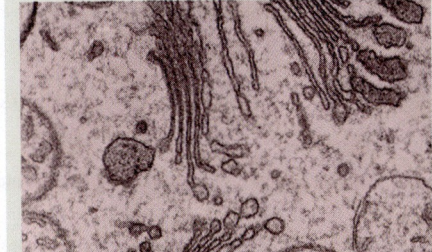

The membrane-bound compartments of the Golgi are responsible for modifying and packaging proteins for secretion. These events are localized for greater efficiency.

7. **AHL:** Describe the process secretion in vesicles from the Golgi apparatus:

_____

_____

_____

8. **AHL:** Why are some tissues comprised of more than one type of specialized cell? Give an example in your answer:

_____

_____

Organisms

B3.1 Gas Exchange

	Activity Number

Guiding Questions:
▶ What adaptations enable gas exchange in multicellular organisms?
▶ What features of gas exchange in plants and mammals are the same, and which are unique to each group?

Learning Outcomes:

			Activity Number
☐	1	Explain the importance of gas exchange in all organisms, factoring in surface area-to-volume ratios	98 - 99
☐	2	Describe the properties of gas-exchange surfaces.	98 - 99
☐	3	Explain how concentration gradients are maintained at exchange surfaces in animals.	100
☐	4	Describe adaptations for gas exchange in mammalian lungs.	100
☐	5	Relate the role of lung structures, including the diaphragm, intercostal muscles, abdominal muscles, and ribs, to the process of ventilation.	101
☐	6	**AOS:** Investigate and measure lung volumes to calculate tidal volume, vital capacity, and inspiratory and expiratory reserves.	102
☐	7	Describe adaptations for gas exchange in leaves	104
☐	8	Draw and label a transverse section of a dicotyledonous leaf, showing the distribution of tissues.	104
☐	9	Link the processes of transpiration and gas exchange in a leaf, including the factors affecting the rate of transpiration.	105
☐	10	**AOS:** Use microscopes to take counts of stomata from micrographs or leaf casts to determine stomatal density. **NOS:** identifying the value of repeated counts in increasing data reliability.	106
☐	11	**AHL:** Describe adaptations of foetal and adult haemoglobin for the process of oxygen transport.	103
☐	12	**AHL:** Explain how the process of Bohr shift in respiring tissues causes an increased dissociation of oxygen due to an increase in carbon dioxide.	103
☐	13	**AHL:** Interpret features of oxygen dissociation curves displaying affinity of haemoglobin for oxygen at different oxygen concentrations.	103

B3.2 Transport

Guiding Questions:
▶ What key plant and animals adaptations enable fluid transport?
▶ What features of transport do animals and plants have in common, and what features differ?

Learning Outcomes:

☐	1	Explain how the adaptations of capillaries allow for material exchange between blood and the internal or external environment.	107
☐	2	**AOS:** Use micrographs to identify structural features in arteries and veins, and therefore, distinguish between the two blood vessels.	108
☐	3	Describe adaptations in arteries that enable the transport of blood away from the heart.	109
☐	4	**AOS:** Measure carotid or radial pulse rates, to determine heart rate	110
☐	5	Describe adaptations in veins that enable the return of blood to the heart.	108
☐	6	**AOS:** Evaluate epidemiological data, **NOS:** including an interpretation of correlation coefficients, relating to factors influencing the incidence of coronary heart disease.	111
☐	7	Investigate the processes and structures involved in water transportation from roots to leaves.	115
☐	8	Describe adaptations of xylem vessels allowing for transport of water.	116
☐	9	**AOS:** Draw and label a diagram of the transverse section of the stem of a dicotyledonous plant, from a micrograph, and annotate with the functions of key structures.	117
☐	10	**AOS:** Draw and label a diagram of the transverse section of the root of a dicotyledonous plant, from a micrograph, and annotate with the functions of key structures.	117

B3.2 Transport

		Activity Number
Guiding Questions:	▶ What key plant and animal adaptations enable fluid transport?	
	▶ What features of transport do animals and plants have in common, and what features differ?	

Learning Outcomes:

☐	11	**AHL:** Explain the processes occurring in capillaries that allow the release and reuptake of tissue fluid.	112
☐	12	**AHL:** Discuss how substances are exchanged between tissue fluid and cells in tissues, including the significance of plasma and tissue fluid composition.	112
☐	13	**AHL:** Describe how excess tissue fluid is drained into lymph ducts, including the structures involved.	112
☐	14	**AHL:** Distinguish between the single circulation of bony fish and the double circulation of mammals, using simple circuit diagrams.	113
☐	15	**AHL:** Identify key heart structures and direction of blood flow from a diagram, and use the information to explain how adaptations of the mammalian heart enable pressurized blood to be delivered to the arteries.	114
☐	16	**AHL: AOS:** Interpret systolic and diastolic blood pressure data to identify stages in the cardiac cycle.	114
☐	17	**AHL:** Explain how the active transport of mineral ions is able to generate root pressure in xylem vessels.	118
☐	18	**AHL:** Discuss how adaptations found in phloem sieve tubes and companion cells enable the translocation of sap.	119 - 120

B3.3 Muscle and Motility

		Activity Number
Guiding Questions:	▶ How is movement enabled by contracting muscles?	
	▶ Why is muscle tissue an advantage in animal bodies?	

Learning Outcomes:

☐	1	**AHL:** Investigate adaptations that enable movement, an essential life process, in both sessile and motile species.	121
☐	2	**AHL:** Explain the process of muscle contraction using the sliding filament model.	126
☐	3	**AHL:** Describe the role of the protein titin and antagonistic muscles in the process of muscle relaxation.	125
☐	4	**AHL:** Link the structure to the function of motor units, and connecting structures, in skeletal muscle.	125
☐	5	**AHL:** Explain how skeletons act as both anchorage for muscles and as levers.	123
☐	6	**AHL:** Describe how structures in a synovial joint enable movement, using the human hip joint as an example.	127
☐	7	**AHL: AOS:** Measure the range of motion in a joint using computer analysis of images or a goniometer.	127
☐	8	**AHL:** Explain how internal body movements are facilitated by antagonistic muscle action, using Internal and external intercostal muscles as an example.	124 - 125
☐	9	**AHL:** Elaborate on reasons for locomotion, providing examples.	121
☐	10	**AHL:** Describe a range of swimming adaptations found in marine mammals.	122

Key Idea: Animal gas exchange systems are suited to the animal's environment, body form, and metabolic needs.

To meet the demands of aerobic metabolism, organisms must exchange gases with the environment. Some organisms can exchange gases directly across their body surface, but most

...organisms have specialized gas exchange systems adapted to function in their specific environment. The type and complexity of the exchange system reflects the demands of metabolism for gas exchange (oxygen delivery and carbon dioxide removal) and the environment (aquatic or terrestrial).

...of every cell in the body. Glucose is broken down to provide energy as ATP. Respiration creates a constant demand for ... (CO_2). These gases are delivered and removed via diffusion across gas exchange surfaces.

Flat organisms, such as this flatworm, use the body surface as the gas exchange surface. Most multicellular organisms have specialized gas exchange systems.

Glucose

Water

Oxygen (O_2)

Carbon dioxide (CO_2)

Energy

Gas exchange is the process by which gases enter and leave the body of animals across gas exchange surfaces. To achieve effective gas exchange ... surface area. They are moist because gases must dissolve before diffusing across. In animals with gas exchange systems, the gas exchange surfaces lie close to capillary networks, and respiratory gases enter and leave the circulatory fluid by diffusion.

Bronchus

Alveoli

...gases cells within a capillary. The blood carries O_2 and CO_2...

In mammalian lungs, the alveoli (microscopic air sacs) provide a large surface area for gas exchange. The walls of the alveoli are only one cell thick (tissue section above) and are covered by capillaries (model, centre). Respiratory gases move across the gas exchange surface by diffusion. Effective gas exchange relies on maintaining a concentration gradient for gas diffusion. Oxygen is transported away from the gas exchange surface by the blood (above right), reducing its concentration relative to the environmental side of the gas exchange surface. CO_2 is transported to the gas exchange surface, increasing its concentration relative to the environmental side of the membrane. It then diffuses out of the blood, across the membrane, and into the external environment.

1. What is the purpose of gas exchange?

2. How are gases exchanged with the environment?

3. How are gradients for diffusion maintained in a simple organism (one without a gas exchange system)?

4. How are gradients for diffusion maintained in an organism with a gas exchange system?

B3.1
1 - 2

©2024 **BIOZONE** International
ISBN: 978-1-99-101410-8

Gas exchange systems and environment

The way an animal exchanges gases with its environment is influenced by the animal's body form and by the environment in which the animal lives. Small or flat organisms, such as sponges and flatworms, living in moist or aquatic environments require no specialized structures for gas exchange. Larger or more complex animals have specialized systems to supply the oxygen to support their metabolic activities. The type of environment presents different gas exchange challenges to animals. In air, gas exchange surfaces will dry out. In water, the oxygen content is much lower than in air.

5. Describe two reasons why most animals require specialized gas exchange structures and systems:

   (a) _____

   _____

   _____

   (b) _____

   _____

   _____

6. Describe three ways the gas exchange surfaces of air breathers are kept moist:

   (a) _____

   _____

   (b) _____

   _____

   (c) _____

   _____

7. Explain why gills would not work in a terrestrial environment:

   _____

   _____

   _____

   _____

   _____

8. Why do animals have to ventilate their gas exchange surfaces:

   _____

   _____

   _____

9. Describe a difficulty associated with gas exchange:

   (a) In air: _____

   _____

   _____

   (b) In water: _____

   _____

   _____

   _____

Simple organisms

The high surface area to volume ratio of very flat or very small organisms, such as this nematode, enables them to use the body surface as the gas exchange surface.

Oxygen   Carbon dioxide

Air breathing vertebrates

The gas exchange surface in mammals and other air breathing vertebrates is located in internal lungs. Their location within the body protects the lungs from the dry environment of the air and keeps the exchange surfaces moist. The many alveoli of the lungs provide a large surface area to maximize gas exchange rates. Exchange rates for the diffusion of gases are maintained by ventilation of the gas exchange surface (breathing in and out).

Carbon dioxide   Oxygen

Mucus and water vapour produced by metabolism both help to keep the gas exchange surface moist.

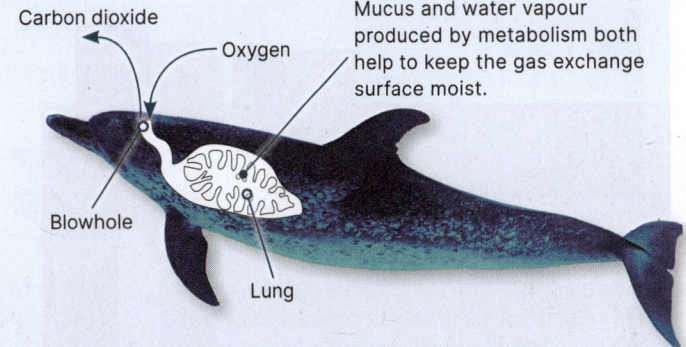

Blowhole

Lung

Bony fish, sharks, and rays

Fish extract oxygen dissolved in water using gills. Gills achieve high extraction rates of oxygen from the water. This is important because water contains only 1% dissolved oxygen by volume, whereas air (at sea level) is 21% oxygen. Bony fish ventilate the gill surfaces by movements of the gill cover. The water supports the gills, and the gill lamellae (the gas exchange surface) can be exposed directly to the environment without drying out.

Oxygen

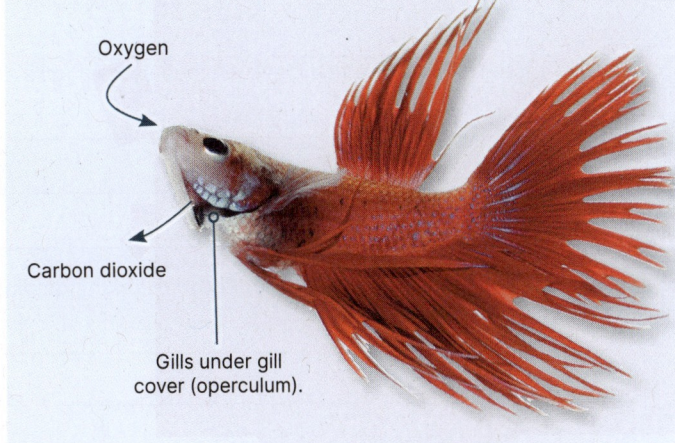

Carbon dioxide

Gills under gill cover (operculum).

99 Gas Exchange in Fish

Key Idea: Fish gills are thin, vascular structures just behind the head. Countercurrent flow enables efficient exchange of gases between the water and the blood in the gill capillaries. Fish obtain the oxygen they need from the water using gills, which are membranous structures supported by cartilaginous or bony struts. As water flows over the gill surface, respiratory gases are exchanged between the blood and the water. In fish, high oxygen extraction rates are achieved using countercurrent exchange and by pumping water across the gill surface (most bony fish) or swimming continuously with the mouth open (called ram ventilation, seen in sharks, rays, and some bony fish, e.g. tuna).

Fish gills

The gills of fish are very thin, filamentous structures, with individual filaments supported and kept apart from each other by the water. This gives them a high surface area for gas exchange. The outer surface of the gill is in contact with the water, and blood flows in vessels inside the gill. Gas exchange occurs by diffusion between the water and blood across the gill membrane and capillaries. The gill cover (operculum) permits exit of water and acts as a pump, drawing water past the gill filaments. The gills of fish are very efficient and achieve an 80% extraction rate of oxygen from water; over three times the rate of human lungs from air.

Gill cover (operculum)

Ventilation of the gills

Most bony fish ventilate the gills by opening and closing the mouth while opening and closing the gill cover. The mouth opens, increasing the volume of the mouth cavity, causing water to enter. The gill cover bulges slightly, moving water into the opercular cavity. The mouth closes and the gill cover opens and water flows out over the gills. These pumping movements keep oxygenated water flowing over the gills, maintaining the concentration gradient for diffusion. Other fish, e.g. sharks and tuna, must swim continuously to achieve the same gill ventilation.

Breathing in bony fish

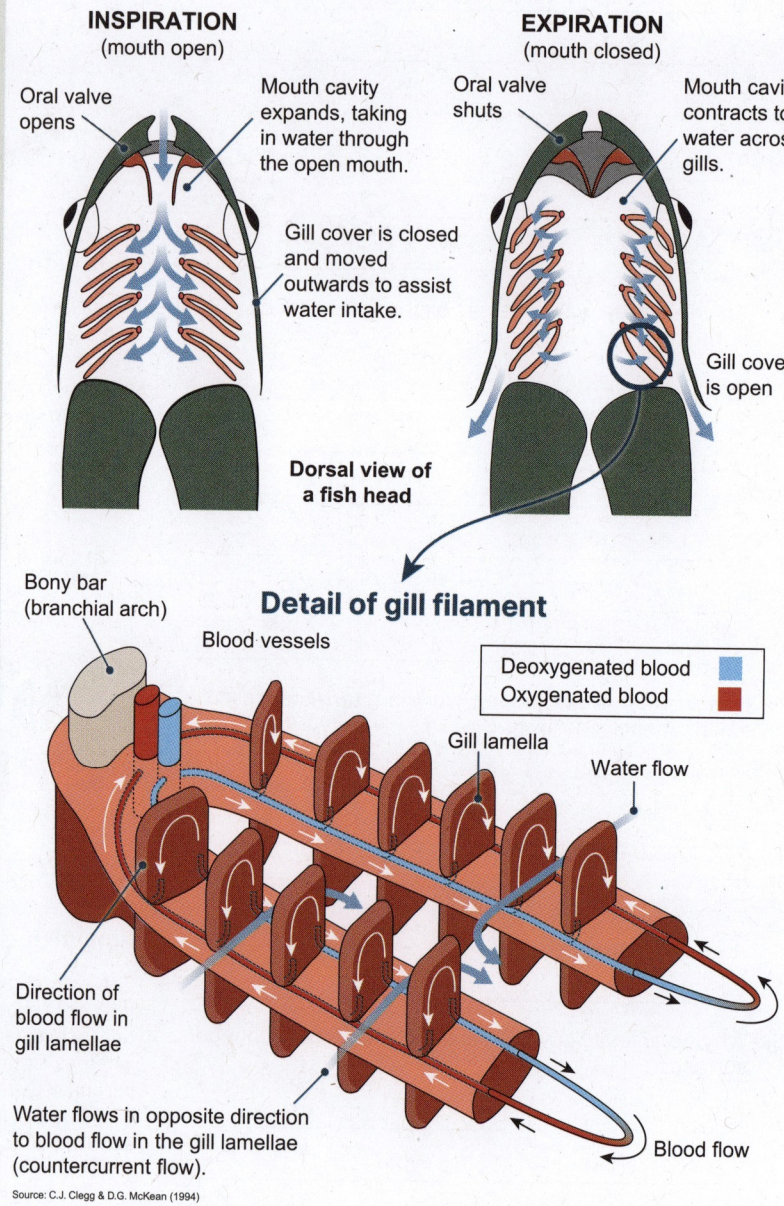

INSPIRATION (mouth open)

Oral valve opens

Mouth cavity expands, taking in water through the open mouth.

Gill cover is closed and moved outwards to assist water intake.

EXPIRATION (mouth closed)

Oral valve shuts

Mouth cavity contracts to force water across the gills.

Gill cover is open

Dorsal view of a fish head

Detail of gill filament

Bony bar (branchial arch)

Blood vessels

Gill lamella

Water flow

| Deoxygenated blood | ☐ |
| Oxygenated blood | ☐ |

Direction of blood flow in gill lamellae

Water flows in opposite direction to blood flow in the gill lamellae (countercurrent flow).

Blood flow

Source: C.J. Clegg & D.G. McKean (1994)

1. Describe three features of a fish gas exchange system (gills and related structures) that facilitate gas exchange:

   (a) _____

   (b) _____

   (c) _____

2. Explain how fish achieve adequate ventilation through pumping their mouth and gill cover, and constant swimming:

   _____

   _____

   _____

B3.1
3

©2024 **BIOZONE** International
ISBN: 978-1-99-101410-8
Photocopying prohibited

Concentration gradient across gill surface

▶ The structure of fish gills and their physical arrangement in relation to the blood flow maximizes gas exchange rates. A constant stream of oxygen-rich water flows over the gill filaments in the opposite direction from the blood flowing through the gill filaments.

▶ This is called countercurrent flow (right and below left) and it is an adaptation for maximizing the amount of oxygen removed from the water. Blood flowing through the gill capillaries encounters water of increasing oxygen content. The concentration gradient (for oxygen uptake) across the gill is maintained across the entire distance of the gill lamella and oxygen continues to diffuse into the blood (CO_2 diffuses out at the same time).

▶ A parallel current flow (below, far right) could not achieve the same oxygen extraction rates because the concentrations across the gill would quickly equalize).

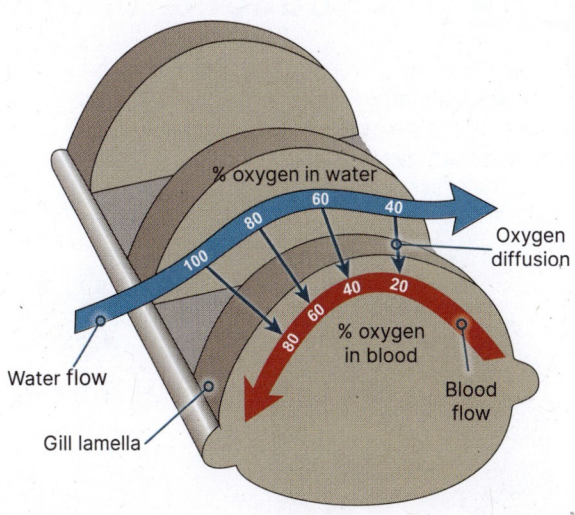

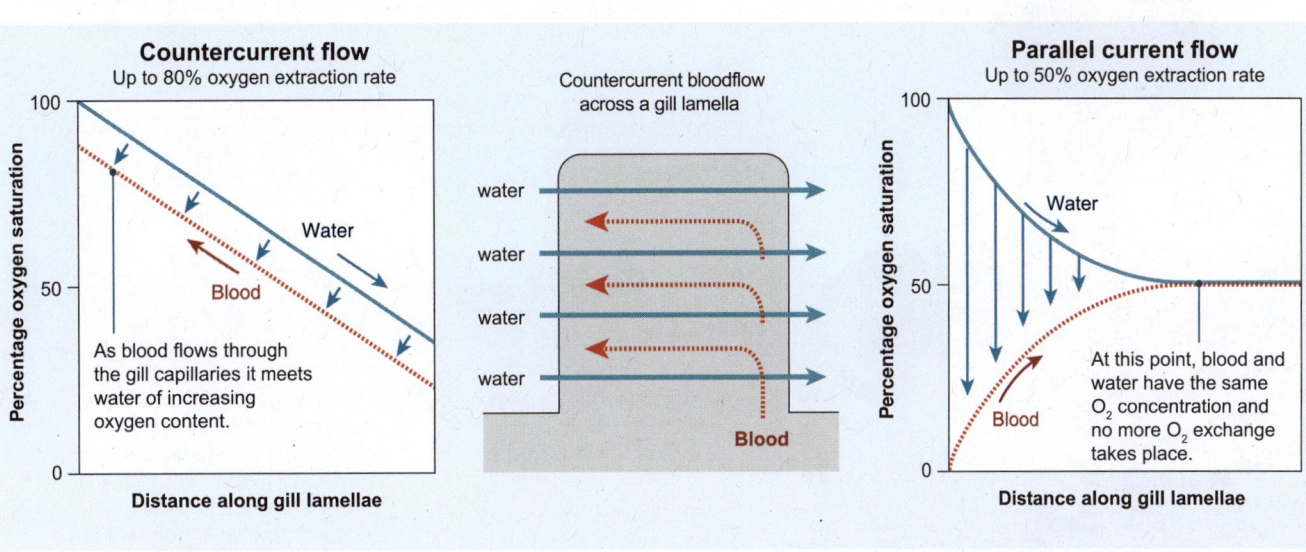

3. Describe countercurrent flow: _____

_____

_____

4. (a) How does the countercurrent system in a fish gill increase the efficiency of oxygen extraction from the water?

_____

_____

_____

   (b) Explain why parallel flow would not achieve the same rates of oxygen extraction: _____

_____

_____

_____

5. Describe the adaptations in gills so they are able to act as exchange surfaces: _____

_____

_____

_____

_____

100 Adaptations for Mammalian Gas Exchange

Key Idea: The tissues and organs of the human gas exchange system work together to enable the exchange of gases between the body's cells and the environment.

The mammalian gas exchange system consists of a range of cells and tissues that are specialized to perform a particular role in the organ system's overall function, which is to exchange respiratory gases (O_2 and CO_2) between the body's cells and the environment. A concentration gradient of these gases is maintained at the exchange surface in the alveoli to facilitate this process.

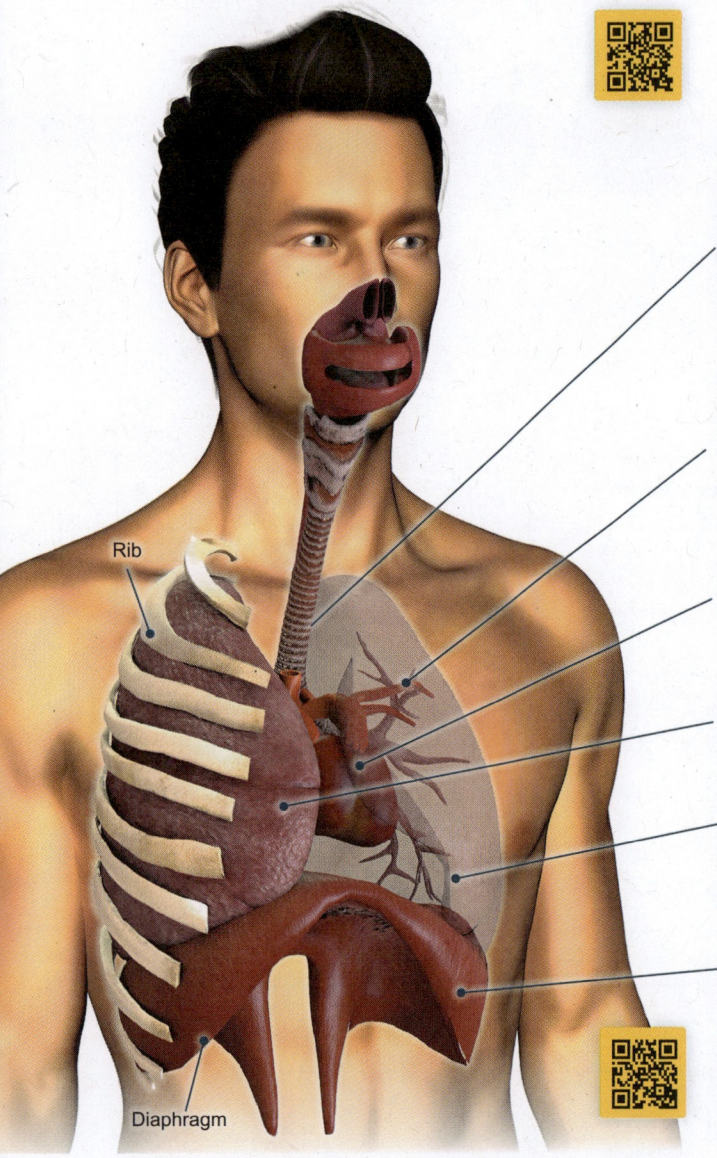

The trachea (windpipe) transfers air to the lungs. It is strengthened with C-shaped bands of stiff cartilage. The trachea divides into two bronchi, also supported by cartilage bands.

Bronchioles branch from the bronchi and divide into progressively smaller branches. Cartilage is gradually lost as the bronchioles decrease in diameter.

The cardiac notch in the left lung makes space for the heart.

The right lung is slightly larger than the left. It takes up 55-60% of the total lung volume.

The smallest respiratory bronchioles subdivide into the alveolar ducts. The alveoli are found at the end of these.

The diaphragm is a dome shaped muscle that works with the intercostal muscles of the ribcage to bring about lung ventilation (breathing). When it contracts, it moves down, reducing pressure in the lung so that air flows in.

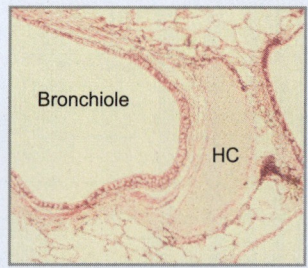

Rings of hyaline cartilage (HC) provide support for the trachea, bronchi, and the larger bronchioles.

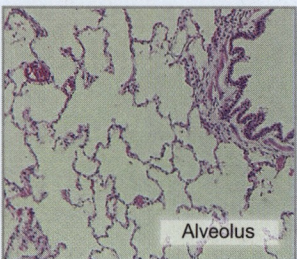

The lungs contain air spaces surrounded by alveolar epithelial cells (pneumocytes), forming alveoli (air sacs), where gas exchange takes place. The alveoli receive air from tubes, called bronchioles.

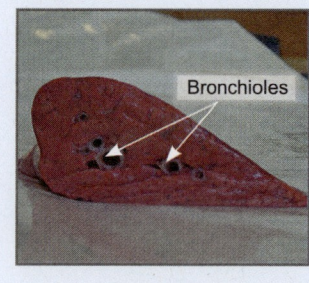

1. Explain how each of these adaptations enables gas exchange:

   (a) High surface area of gas exchange surface: _____

   _____

   (b) Branched network of bronchioles: _____

   _____

   (c) Pervasive capillary bed: _____

   _____

   (d) Surfactant coating of alveoli epithelium: _____

   _____

   _____

B3.1

3 - 4

Cross section through an alveolus

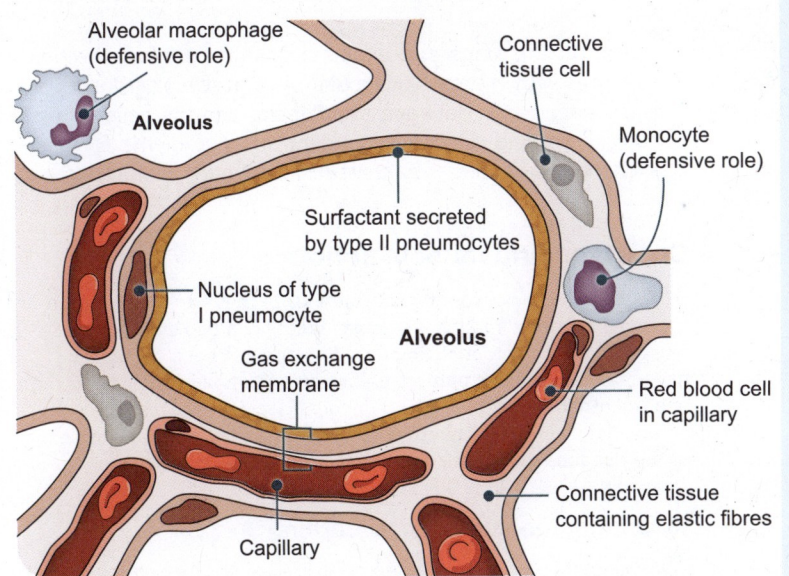

The physical arrangement of the alveoli to the capillaries through which the blood moves. The alveolus is lined with a thin, single-celled layer of Type I pneumocytes (alveolar epithelial cells), across which gases are exchanged. Type II pneumocytes secrete surfactant to reduce surface tension of the alveoli to prevent them collapsing during exhalation. Phagocytes (monocytes and macrophages) are present to protect the lung tissue. Elastic connective tissue gives the alveoli their ability to expand and recoil.

The gas exchange membrane

Surfactant is a phospholipid produced by type II pneumocytes in the alveolar walls.

$0.5 \mu m$

O_2

CO_2

The gas exchange membrane is the layered junction between the alveolar epithelial cells, the endothelial cells of the capillary, and their associated basement membranes (thin connective tissue layers under the epithelia). Gases move freely across this membrane.

2. Describe the structure and purpose of the alveolar-capillary (gas exchange) membrane:

_____

_____

_____

3. The diagram below shows gas exchange across the respiratory membrane in the alveoli. A concentration gradient is maintained to facilitate diffusion of oxygen and carbon dioxide.

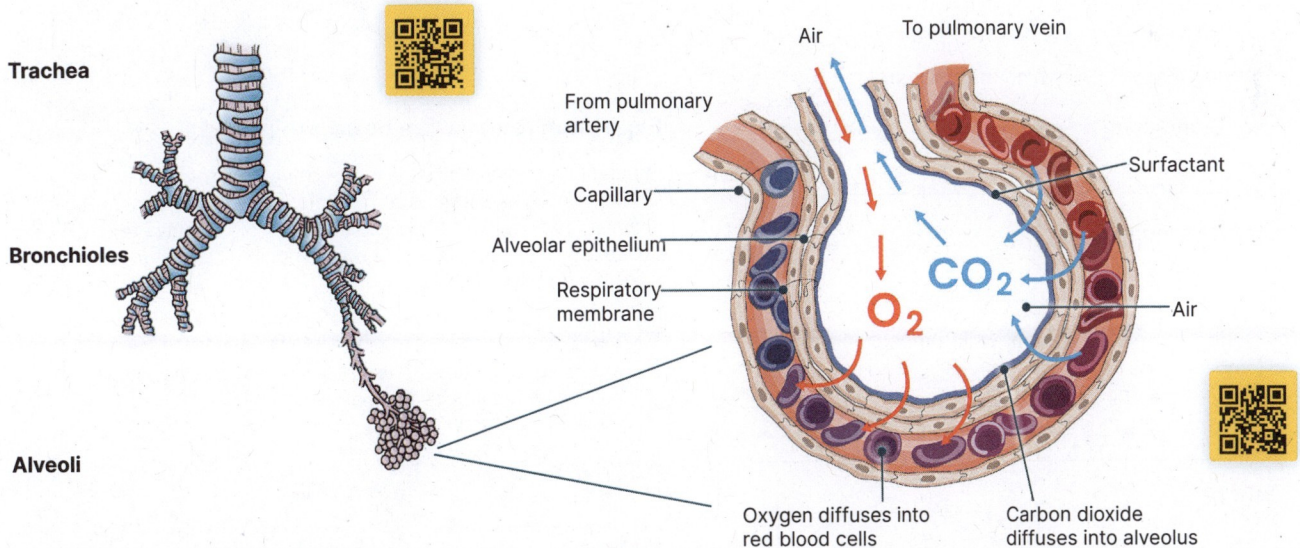

Oxygen diffuses into red blood cells

Carbon dioxide diffuses into alveolus

(a) Explain how a concentration gradient contributes to diffusion of oxygen during gas exchange:

_____

_____

(b) How is the concentration of both gases maintained? _____

_____

_____

101 Lung Ventilation

Key Idea: Breathing provides a continual supply of air to the lungs to maintain the concentration gradients for gas exchange. Different muscles are used in inspiration and expiration to force air in and out of the lungs.

Breathing (ventilation) provides a continual supply of oxygen-rich air to the lungs and expels air high in carbon dioxide. Together with the cardiovascular system, which transports respiratory gases between the alveolar and the cells of the body, breathing maintains concentration gradients for gas exchange. Breathing is achieved by the action of muscles.

1. Explain the purpose of breathing: _____

_____

_____

2. In general terms, how is breathing achieved?

_____

_____

_____

_____

3. (a) Describe the sequence of events involved in quiet breathing:

_____

_____

_____

_____

_____

(b) What is the essential difference between this and the situation during forced breathing?

_____

_____

4. During inspiration, which muscles are:

(a) Contracting: _____

_____

(b) Relaxed: _____

_____

5. During forced expiration, which muscles are:

(a) Contracting: _____

_____

(b) Relaxed: _____

_____

6. Explain the role of antagonistic muscles in breathing:

_____

_____

_____

_____

Breathing and muscle action

Muscles can only do work by contracting, so they can only perform movement in one direction. To achieve motion in two directions, muscles work as antagonistic pairs. Antagonistic pairs of muscles have opposing actions and create movement when one contracts and the other relaxes. Breathing in humans involves two sets of antagonistic muscles. The external and internal intercostal muscles of the ribcage, and the diaphragm and abdominal muscles.

Inspiration (inhalation or breathing in)

During quiet breathing, inspiration is achieved by increasing the thoracic volume (therefore decreasing the pressure inside the lungs). Air flows into the lungs in response to the decreased pressure inside the lung. Inspiration is always an active process involving muscle contraction.

1 External intercostal muscles contract, causing the ribcage to expand and move up. Diaphragm contracts and moves down.

2 Thoracic volume increases, lungs expand, and the pressure inside the lungs decreases.

3 Air flows into the lungs in response to the pressure gradient.

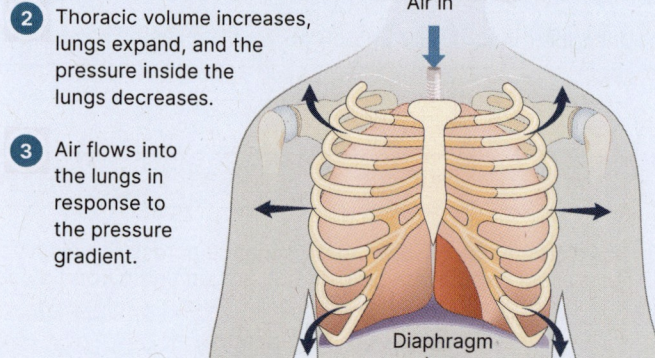

Air in

Diaphragm down

Expiration (exhalation or breathing out)

In quiet breathing, expiration is a passive process, achieved when the external intercostals and diaphragm relax and thoracic volume decreases. Air flows passively out of the lungs to equalize with the air pressure. In active breathing, muscle contraction is involved in bringing about both inspiration and expiration.

1 In quiet breathing, external intercostals and diaphragm relax. The elasticity of the lung tissue causes recoil. In forced breathing, the internal intercostals and abdominal muscles contract to compress the thoracic cavity and increase the force of the expiration.

2 Thoracic volume decreases and the pressure inside the lungs increases.

3 Air flows passively out of the lungs in response to the pressure gradient.

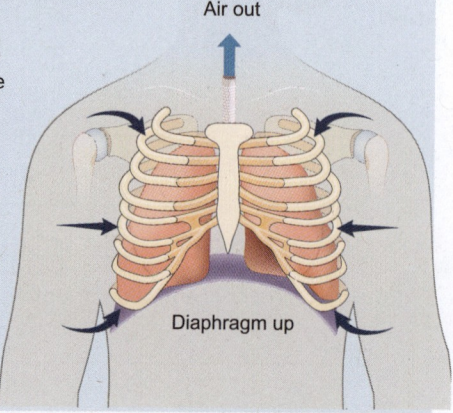

Air out

Diaphragm up

B3.1
5

205

©2024 **BIOZONE** International
ISBN: 978-1-99-101410-8
Photocopying prohibited

102 Measuring Lung Volumes

Key Idea: Vital capacity is the greatest volume of air expelled from the lungs after taking the deepest possible breath. Vital capacity is easily measured using a spirometer or a bell jar system, as described below. In healthy adults, vital capacity ranges between 4-6 L, but is influenced by several factors including gender, age, height, ethnicity, and fitness.

Vital capacity

▸ Vital capacity can be measured using a 6 L calibrated glass bell jar, supported in a sink of water (right). The jar is calibrated by inverting it, pouring in known volumes of water, and marking the level on the bell jar with a marker pen.

▸ To measure vital capacity, a person breathes in as far as possible (maximal inhalation), and then exhales as far as possible (maximal exhalation) into a mouthpiece connected to tubing. The drop in volume within the bell jar is measured; this gives the vital capacity in L.

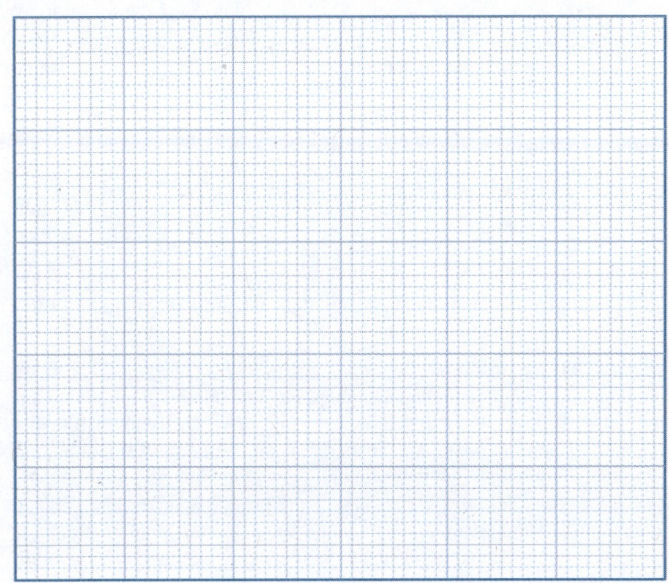

Bell jar

Tubing. Disposable mouthpiece attached for each subject.

Wedge: allows water to circulate.

Investigating vital capacity

▸ A class of high school biology students investigated the vital capacity of the whole class using the bell jar method described above.

▸ The students recorded their heights as well as their vital capacity. The results are presented on the table (right).

Females		Males	
Height cm	Vital capacity L	Height cm	Vital capacity L
156	2.75	181	4.00
145	2.50	163	2.50
155	3.25	167	4.00
170	4.00	174	4.00
162	2.75	177	4.00
164	2.75	177	3.75
163	3.40	176	3.75
158	2.75	177	3.25
167	4.00	178	4.00
165	3.00	178	3.75

1. Calculate the mean vital capacity for:

(a) Females: _____

(b) Males: _____

(c) Explain whether these results are what you would expect:

_____

_____

_____

_____

2. (a) Plot height versus vital capacity as a scatter graph on the grid provided (right). Use different symbols or colours for each set of data (female and male). Include a line of best fit through each set of points. For a line of best fit, the points should fall equally either side of the line.

(b) Describe the relationship between height and vital capacity:

_____

_____

_____

_____

_____

_____

B3.1
6

AOS

Determining changes in lung volume using spirometry

▶ A lung function test, called spirometry, measures changes in lung volume and can be used diagnostically.

▶ The volume of gases exchanged during breathing varies according to the physiological demands placed on the body, e.g. by exercise, and an individual's lung function.

▶ Spirometry measures changes in lung volume by measuring how much air a person can breathe in and out and how fast the air can be expelled. Spirometry can measure changes in ventilation rates during exercise and can be used to assess impairments in lung function, as might occur as a result of disease. In humans, the total adult lung capacity varies between 4 and 6 L and is greater in males.

▶ A simple spirometer consists of a weighted drum, containing oxygen or air, inverted over a chamber of water. A tube connects the air-filled chamber with the subject's mouth, and soda lime in the system absorbs the carbon dioxide breathed out. Breathing results in a trace called a spirogram, from which lung volumes can be measured directly.

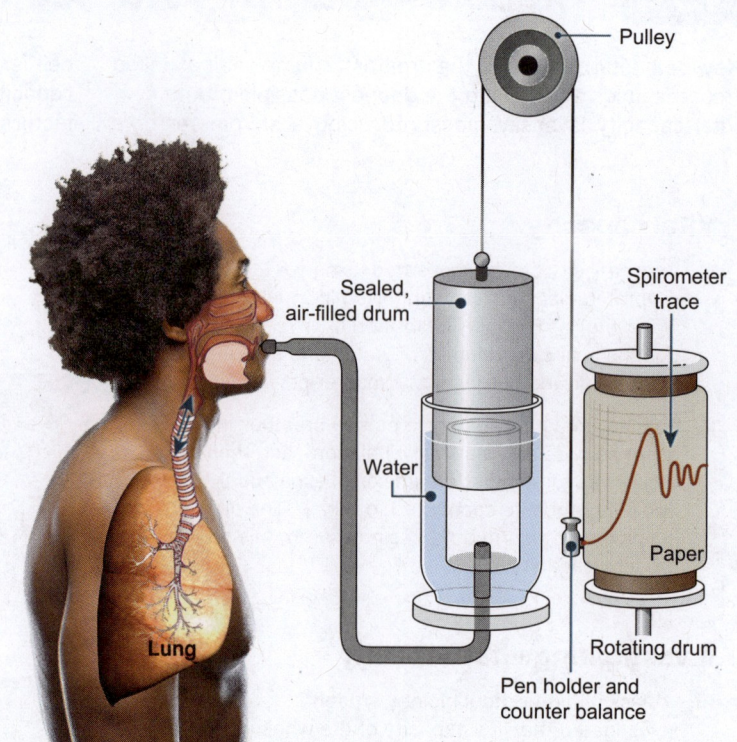

Investigation 7.1   Measuring Lung Volumes

See appendix for equipment list.

1. To measure lung volume metrics in a class laboratory, a simple balloon can be used.

2. Volume (in litres) of the inflated balloon can be calculated by measuring the diameter of the balloon, then v(balloon) $= (4/3)\pi r^3$

3. Alternatively, the inflated balloon can be completely submerged in a large measuring container, with the displacement recorded in L (3 significant figures).

4. In small groups, use a spirometer, or inflated balloon, to make the following measurements (steps 5-7) for each student, and record them in the table below. Repeat measurements 3 times and average results:

5. Tidal volume (TV): Volume of air breathed in and out in a single breath.

6. Inspiratory reserve volume (IRV): Volume breathed in by a maximum inspiration at the end of a normal inspiration.

7. Expiratory reserve volume (ERV): Volume breathed out by a maximum effort at the end of a normal expiration.

Measurement	Individual measurement (in L)			Average (L)
Tidal volume (TV)				
Inspiratory reserve volume (IRV)				
Expiratory reserve volume (ERV)				

3. Calculate inspiratory capacity (IC): Volume breathed in by a maximum inspiration at the end of a normal expiration.

(IC) = TV + IRV   IC = _____

4. If the typical value for residual volume (RV), the volume of air remaining in the lungs at the end of a maximum expiration, is 1.2L, calculate the total lung capacity (TLC), the total volume of the lungs (note: only a fraction of TLC is used in normal breathing).

(TLC) = VC + RV = _____

5. Calculate vital capacity (VC), volume that can be exhaled after a maximum inspiration:

(VC) = IRV + TV + ERV = _____

©2024 **BIOZONE** International
ISBN: 978-1-99-101410-8

103 Oxygen Transport and Haemoglobin

Key Idea: Haemoglobin is an oxygen carrying protein complex that is found in erythrocytes (red blood cells).

Oxygen does not easily dissolve in blood, but is carried in a chemical combination with haemoglobin (Hb) in erythrocytes (red blood cells). The most important factor determining how much oxygen is carried by Hb is the level of oxygen in the blood. The greater the oxygen tension, the more oxygen will combine with Hb. This relationship can be illustrated with an oxygen-haemoglobin dissociation curve. In the lung capillaries, (high O_2), a lot of oxygen is picked up and bound by Hb. In the tissues, (low O_2), oxygen is released. Foetal haemoglobin has an even higher affinity for oxygen. In skeletal muscle, myoglobin picks up oxygen from haemoglobin and therefore serves as an oxygen store when oxygen tensions begin to fall. The release of oxygen is enhanced by the Bohr effect, due to an increased pressure of CO_2 gas and/or increasing pH.

Adult haemoglobin, foetal haemoglobin, and myoglobin

Haemoglobin is found in the blood of nearly all vertebrates, and some invertebrate tissue. The iron present in the complex gives blood its red colour. Haemoglobin is composed of four protein subunits, each with a haem group for binding oxygen; two alpha and two beta subunits in the main form of adult haemoglobin (HbA), and two alpha and two gamma subunits in foetal haemoglobin (HbF). Foetal haemoglobin, which has a higher affinity for oxygen, is present in the embryo from 10 weeks of gestation. It is completely replaced by the adult form at around 6 months of age.

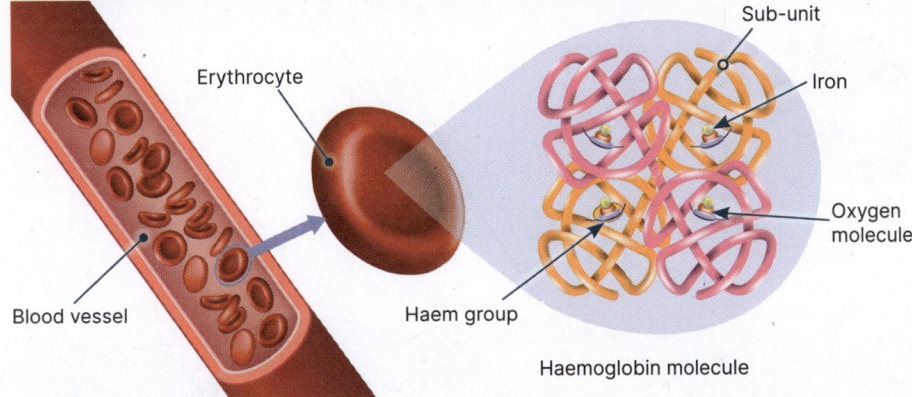

Erythrocyte
Sub-unit
Iron
Oxygen molecule
Blood vessel
Haem group
Haemoglobin molecule

Myoglobin is another oxygen carrying molecule but has one subunit rather than four, with one haem group. Myoglobin has an even higher affinity for oxygen than haemoglobin, and is found in muscle cells. The myoglobin accepts oxygen that has been transported by the haemoglobin, and then stores it until required for respiration by the muscle cells.

Allostery in haemoglobin

The acceptance of one oxygen molecule binding to the haem group of a subunit causes structural changes in the haemoglobin, known as allostery. This increases the affinity between further oxygen molecules and the remaining subunits of the haemoglobin. This process of increasing O_2 affinity is known as cooperative binding. Carbon dioxide bonds to another region of the haemoglobin, and also promotes allosteric binding. When a CO_2 molecule has bound to a subunit, the T-state is stabilized and the molecule has reduced affinity for O_2, which is then released. Hydrogen ions, produced when CO_2 reduces pH, also bind allosterically, and increase O_2 release.

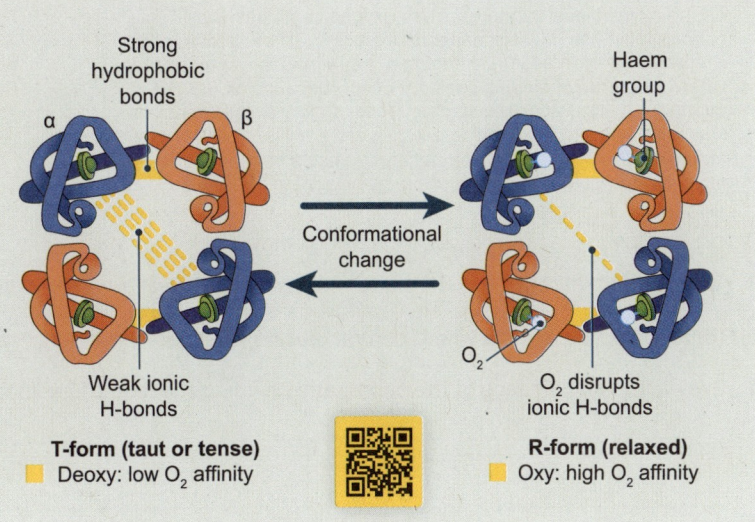

Strong hydrophobic bonds
Haem group
α
β
Conformational change
Weak ionic H-bonds
O_2
O_2 disrupts ionic H-bonds
T-form (taut or tense)
Deoxy: low O_2 affinity
R-form (relaxed)
Oxy: high O_2 affinity

1. Distinguish between foetal haemoglobin (HbF) and adult haemoglobin (HbA) for the following factors:

(a) Location: _____

_____

(b) Structure: _____

_____

(c) Affinity for oxygen: _____

_____

2. Define the term 'allostery' in relation to haemoglobin function: _____

_____

_____

©2024 **BIOZONE** International
ISBN: 978-1-99-101410-8

74
B3.1
11 - 13

AHL

Bohr shift

▶ As carbon dioxide increases in the tissues as a by-product of cellular respiration, the blood pH decreases. This results in a higher concentration of H^+ ions. As the H^+ ions bind to haemoglobin, the dissociation curve shifts to the right (see below, right), and oxygen is released into the tissue. This phenomenon is known as the Bohr shift.

▶ The beneficial consequences of the Bohr shift are that oxygen is transported and released to the tissues that are actively respiring, such as muscle cells during activity. The cells with the most need for oxygen receive the most oxygen.

Dissociation curves

Fig.1: Dissociation curves for haemoglobin and myoglobin at normal body temperature for fetal and adult human blood.

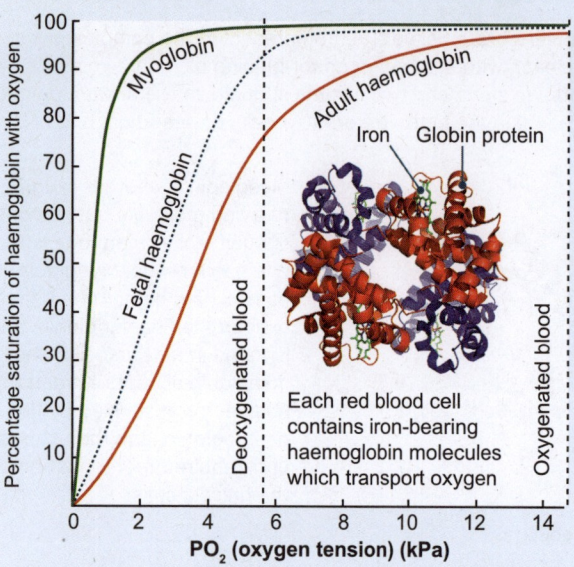

Fig.2: Oxygen-haemoglobin dissociation curves for human blood at normal body temperature at different blood pH.

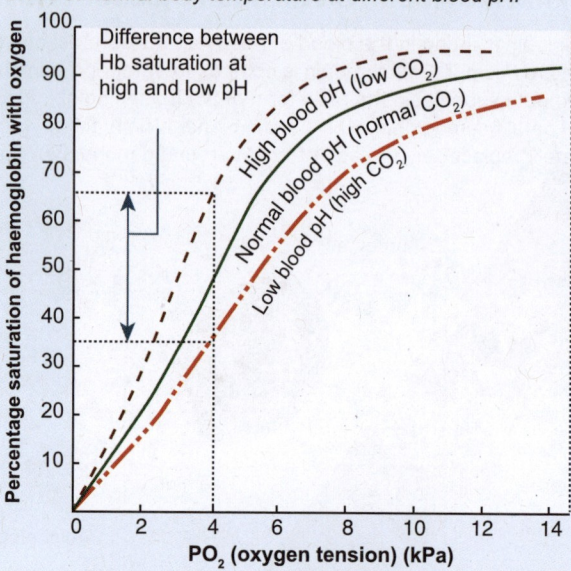

As the oxygen level increases, more oxygen combines with haemoglobin (Hb) due to cooperative binding. This is shown as an S-shaped curve in adult haemoglobin. In comparison, in myoglobin, where cooperative binding does not occur (because of just one subunit), Hb saturation remains high, even at low oxygen tensions. Foetal Hb has a high affinity for oxygen and carries 20-30% more than maternal Hb. Myoglobin in skeletal muscle has a very high affinity for oxygen and will take up oxygen from haemoglobin in the blood.

As pH increases (lower CO_2), more oxygen combines with Hb. As the blood pH decreases (higher CO_2), Hb binds less oxygen and releases more to the tissues (the Bohr effect). The difference between Hb saturation at high and low pH represents the amount of oxygen released to the tissues.

3. (a) Identify two regions in the body where oxygen levels are very high: _____

   (b) Identify two regions where carbon dioxide levels are very high: _____

4. Explain the significance of the cooperative binding reaction of haemoglobin (Hb) to oxygen:

   _____

   _____

5. (a) Explain why the adult haemoglobin shows an S-shaped disassociation curve when oxygen tension increases:

   _____

   _____

   (b) Why does myoglobin not show an S-shaped dissociation curve? _____

   _____

   _____

6. Explain how the process of Bohr shift in respiring tissues causes an increased dissociation of oxygen due to an increase in carbon dioxide, and the benefits of this phenomenon to respiring tissue:

   _____

   _____

   _____

Leaf Adaptations for Gas Exchange

Key Idea: Plants have adaptations to aid gas exchange, including stomata and leaf tissue structure.

Respiring tissues require oxygen, and the photosynthetic tissues of plants also require carbon dioxide in order to produce the sugars for growth and maintenance. The main gas exchange organs in plants are the leaves, and sometimes the stems. In most plants, gases cannot diffuse directly across the leaf surface because of its waxy cuticle, so gases enter and leave the leaf via stomata (pores) in the leaf surface. The stomata have guard cells to control gas exchange, as the plant has to balance its need for CO_2 against its need to reduce water loss, when stomata are open.

Gas exchanges and the function of stomata

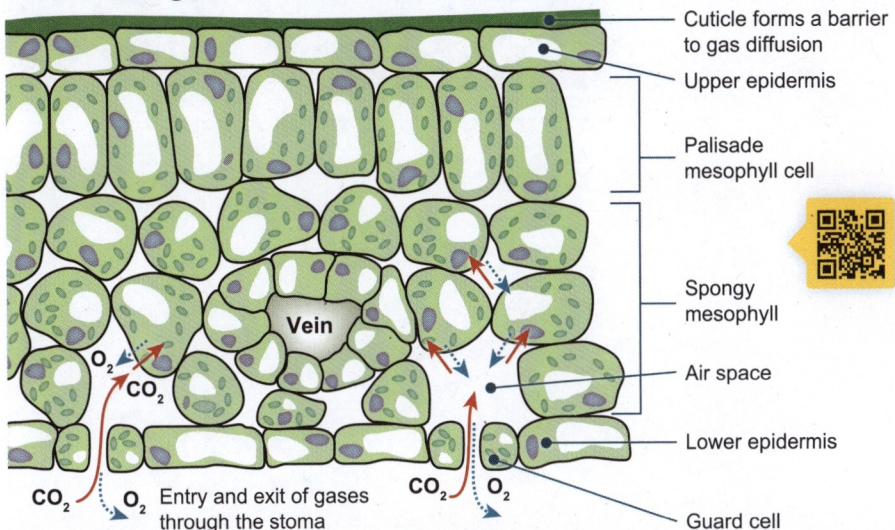

- Cuticle forms a barrier to gas diffusion
- Upper epidermis
- Palisade mesophyll cell
- Spongy mesophyll
- Vein
- Air space
- O_2
- CO_2
- Lower epidermis
- CO_2 O_2 Entry and exit of gases through the stoma
- CO_2 O_2
- Guard cell

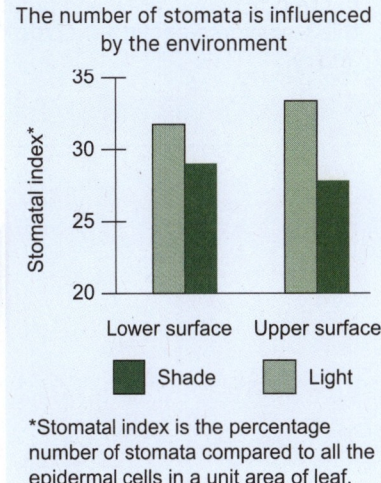

The number of stomata is influenced by the environment

Stomatal index*

| Lower surface | Upper surface |

■ Shade   ■ Light

*Stomatal index is the percentage number of stomata compared to all the epidermal cells in a unit area of leaf.

An increase in light intensity on mature leaves increases the number of stomata developing on young leaves.

Net gas exchanges in a photosynthesizing dicot leaf

▶ Gases enter and leave the leaf through stomata. Inside the leaf (as illustrated for a dicot, above), the large air spaces and loose arrangement of the spongy mesophyll facilitate the diffusion of gases and provide a large surface area for gas exchanges.

▶ Respiring plant cells use oxygen (O_2) and produce carbon dioxide (CO_2). These gases move in and out of the plant and through the air spaces by diffusion.

▶ When the plant is photosynthesising, the situation is more complex. Overall, there is net consumption of CO_2 and net production of oxygen. Fixation of CO_2 maintains a gradient in CO_2 concentration between the atmosphere (high) and the leaf tissue (low).

▶ Oxygen is produced in excess of respiratory needs and diffuses out of the leaf. These net exchanges are indicated by the arrows on the diagram.

Adaptations for gas exchange in plants

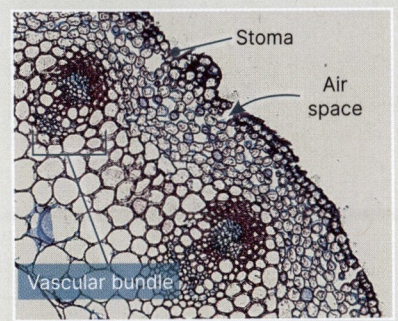

- Stoma
- Air space
- Vascular bundle

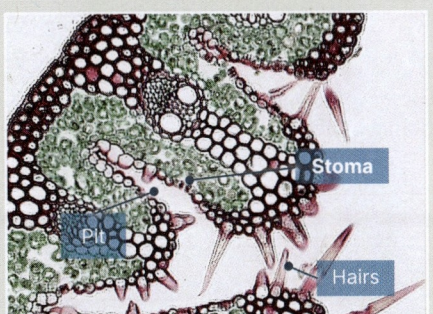

- Stoma
- Pit
- Hairs

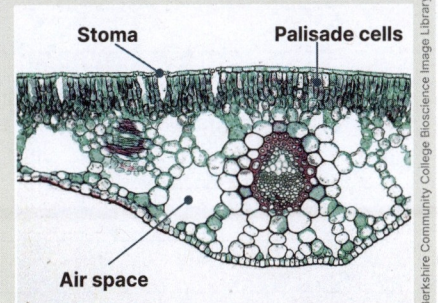

- Stoma
- Palisade cells
- Air space

Berkshire Community College Bioscience Image Library

Some herbaceous plants, e.g. buttercup, above, have photosynthetic stems and CO_2 enters freely into the stem tissue through stomata in the epidermis.

Ammophila (above) is a xerophyte (an arid-adapted plant) and displays many water conserving features. The stomata are found in pits in the leaf's curled surface. The pits restrict water loss to a greater extent than they reduce CO_2 uptake.

Hydrophytes (aquatic plants) have large air spaces in their leaves to help them float. The stomata are on the upper surface where gas exchange can take place. The palisade cells have numerous chloroplasts to increase photosynthesis.

1. What functions do stomata have and why are they necessary? _____

_____

_____

_____

2. Leaves on plants have adaptations to enable gas exchange. To the right is a micrograph taken from a transverse leaf section of a dicotyledon plant. The micrograph shows the distribution of tissue including waxy cuticle, epidermis, spongy mesophyll, stomatal guard cells, and veins, as well as the resulting air spaces.

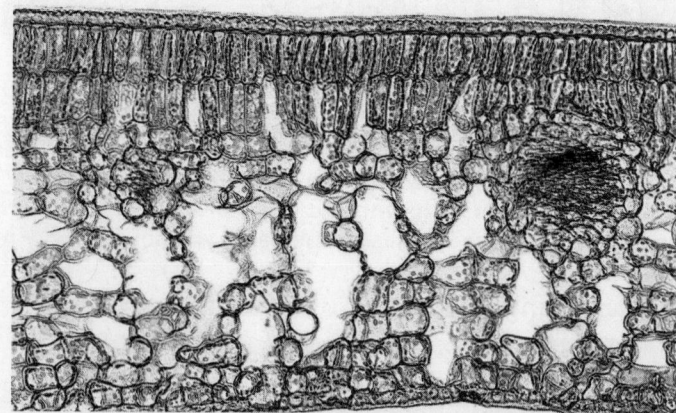

In the box below, draw and label a plan diagram of this transverse section. Include the features listed above.

3. Use the information from the 'adaptations for gas exchange' section on the previous page. Describe a leaf/stomata adaptation in plants for both arid and aquatic environments:

_____

_____

_____

_____

4. The example of a photosynthetic stem on the previous page is from a buttercup, a plant in which the leaves are still the primary organs of photosynthesis.

(a) Identify an example of a plant where the stem is the only photosynthetic organ: _____

(b) Explain how gas exchange can still occur effectively in a plant without leaves? _____

_____

5. Why are palisade and spongy mesophyll arranged differently? _____

_____

_____

©2024 **BIOZONE** International
ISBN: 978-1-99-101410-8
Photocopying prohibited

105 Gas Exchange and Transpiration

Key Idea: Gas exchange through lenticels and stomata is associated with water losses. Guard cells help regulate these water losses.

The leaf epidermis of angiosperms is covered with tiny pores, called stomata, as well as air spaces between the cells of the stems, leaves, and roots. These air spaces are continuous and gases are able to move freely through them and into the plant's cells via the stomata. Each stoma is bounded by two guard cells, which regulate the entry and exit of gases (including water vapour). Although stomata permit gas exchange between the air and the photosynthetic cells inside the leaf, they are also the major routes for water loss through transpiration. Factors that influence the transpiration rate include water supply, wind speed, temperature and humidity. Woody plants have pores called lenticels in their stems and roots that allow gas exchange to take place in these tissues.

Requirements for gas exchange lead to transpiration

▸ Carbon dioxide and oxygen gas, required for photosynthesis and respiration respectively, typically diffuse in and out of a plant through stomata found on leaves. Stomata need to be open for gas exchange to occur. However, evaporative water loss from the stomata, through transpiration, also occurs when stomata are open. Plants must balance their need for gas exchange with their need for water.

▸ The rate of transpiration increases in warmer temperatures, and with lower humidity, increased wind, and increased sunlight.

▸ Many plants can control the opening and closing of the stomata, using active transport to shift K^+ ions across the membrane of surrounding guard cells. This adjusts the turgidity (fullness), the more of which 'bends' the guard cells to open stomata.

In woody plants, the wood prevents gas exchange. A lenticel is a small area in the bark where the loosely arranged cells allow entry and exit of gases into the stem tissue underneath.

Stomatal opening and closing

▸ The opening and closing of stomata depend on both time of day and environmental factors.

▸ Stomata tend to open during daylight in response to light, and close at night.

▸ Environmental factors that influence stomata are light levels, CO_2 concentration in the leaf tissue, and water supply.

▸ Low CO_2 levels promote stomatal opening. Conditions such as high temperatures that induce water stress cause the stomata to close, regardless of light or CO_2 levels.

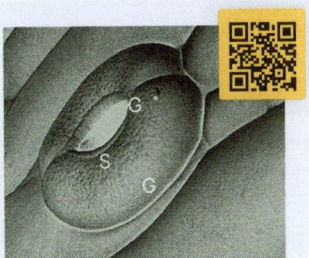

The image above shows a scanning electron micrograph (SEM) of a single stoma from the leaf epidermis of a dicot. The guard cells (G), which are swollen tight and open the pore (S) to allow gas exchange between the leaf tissue and the external environment.

Stomata	Guard cells	Daylight	CO_2	Soil water
Open	Turgid	Light	Low	High
Closed	Flaccid	Dark	High	Low

The effect on stomatal opening and closing depends on guard cell position, light levels, CO_2 concentrations and soil water availability.

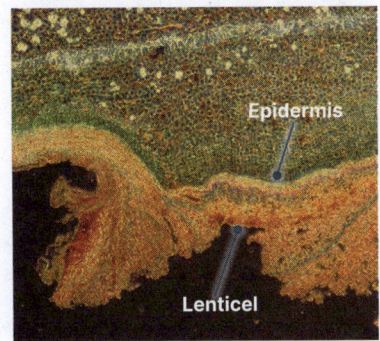

Most gas exchange in plants occurs through the leaves, but some also occurs through the stems and roots. Leaves are very thin with a high surface area which assists gas exchange by diffusion.

1. By which process does oxygen enter the plant tissues? _____

2. Where does most gas exchange occur in plants? _____

3. What is the role of lenticels in plant gas exchange? _____
_____

4. Link the processes of transpiration and gas exchange in a leaf, including the factors affecting the rate of transpiration.
_____
_____
_____
_____
_____
_____

B3.1
9

106 Investigating Stomatal Density

Key Idea: The density and distribution of leaf stomata in different plant species are related to the rate of water loss. Different plant species have different leaf shapes and structures and these can be correlated with the environment in which they are found. Comparing the leaf area and stomatal density of different plant species helps to explain observed differences in transpiration rate but factors in the environment, such as shading and wind, are also important.

Plant species show different leaf shapes and structures associated with their environments

Aloe (agave)
A succulent

Pine
A conifer

Eucalyptus
An Australian gum tree

Sunflower
A perennial dicot with large leaves

Tropical species with thick, fleshy leaves. Its physiology allows it to fix CO_2 during the night and keep stomata closed during the day.

Temperate species with thin, needle-like leaves and a thick waxy leaf cuticle. Stomata are sunken into pits.

Sub-tropical drought tolerant species with a deep root system and waxy leaves that hang downwards.

Widespread cultivated North American dicot with a showy flower head and very large, soft leaves.

Investigation 7.2 Comparing stomatal density

See appendix for equipment list.

1. Your teacher will have up to four leaf types from four dicot plants adapted to different environments, or you may need to obtain samples of your own.

2. The number of stomata per mm^2 on the surface of a leaf can be determined by counting the stomata visible under a microscope*. Use clear nail varnish to paint over the lower surface of a leaf. Leave it to dry. This creates a cast with impressions of the leaf surface.

3. Carefully peel off the layer of nail varnish and place on a clean microscope slide.

4. Calculate the diameter of the area viewable under a microscope using the field of view divided by the magnification of the eyepiece, multiplied by the magnification of the objective lens.
For example, if the eyepiece magnification is 10, the objective lens magnification 40, and the field of view 18, then $18/(10 \times 40) = 0.045$ mm diameter. The area viewable is then πr^2.

5. You could also use a micrometer to measure the diameter of the field of view or use a thin, clear ruler.

6. Place the slide with the layer of nail varnish on it under the microscope and count the number of stomata you see. If there are too many stomata, then count one quarter of the field of view and multiply by four. Do this in several places. Enter your results in the table and calculate a mean.

7. You should also take note of where the stomata are on the leaf. Are they scattered randomly or in specific places?

8. Repeat on the upper surface of the leaf.

9. Repeat for the other leaf types.

10. *A digital microscope can be used to capture images on a computer, which may improve counting.

Magnification — WF10X
Field of view — 18mm

©2024 **BIOZONE** International
ISBN: 978-1-99-101410-8
Photocopying prohibited

| Plant name/type | Number of stomata per mm² lower surface | | | | | Number of stomata per mm² upper surface | | | | |
| | Count number | | | | | Count number | | | | |
	1	2	3	4	Mean	1	2	3	4	Mean

1. (a) Write an aim for the investigation: _____

    _____

    (b) Write a hypothesis for the investigation: _____

    _____

2. Complete the table above:

3. (a) Which plant has the highest stomatal density? _____

    (b) Which plant has the lowest stomatal density? _____

4. (a) Is there a relationship between the number of stomata per mm² and the type of leaf or plant?

    _____

    (b) Explain your answer: _____

    _____

    _____

5. (a) Where are the majority of stomata located In a typical dicot leaf? _____

    (b) Suggest why this might be the case: _____

    _____

    _____

6. Explain your results in terms of the environment the plants are adapted for and the need to regulate water loss:

    _____

    _____

    _____

    _____

    _____

    _____

7. Explain how you increased the reliability of the quantitative data you collected: _____

    _____

    _____

    _____

107 Arteries

Key Idea: Arteries are thick-walled blood vessels that carry blood away from the heart to the capillaries.

Large arteries leave the heart and divide into medium-sized (distributing) arteries. Within the tissues and organs, these distributing arteries branch to form arterioles, which deliver blood to capillaries. Arterioles lack the thick layers of arteries and consist only of an endothelial layer wrapped by a few smooth muscle fibres at intervals along their length. Blood flow to the tissues is altered by contraction (vasoconstriction) or relaxation (vasodilation) of the blood vessel walls. Vasoconstriction increases blood pressure, whereas vasodilation has the opposite effect.

Arteries

Arteries, regardless of size, can be recognized by their well-defined, rounded lumen (internal space) and the muscularity of the vessel wall. Arteries have an elastic, stretchy structure that gives them the ability to withstand the high pressure of blood being pumped from the heart. At the same time, they help to maintain pressure by having some contractile ability themselves (a feature of the central muscle layer). Arteries nearer the heart have more elastic tissue, giving greater resistance to the higher blood pressures of the blood leaving the left ventricle. Arteries further from the heart have more muscle to help them maintain blood pressure. Between heartbeats, the arteries undergo elastic recoil and contract. This tends to smooth out the flow of blood through the vessel.

Arteries comprise three main regions (right):

1. A thin, inner layer of epithelial cells called the tunica intima (endothelium) lines the artery.

2. A thick central layer (the tunica media) of elastic tissue and smooth muscle that can both stretch and contract.

3. An outer connective tissue layer (the tunica externa) has a lot of elastic tissue.

Structure of an artery

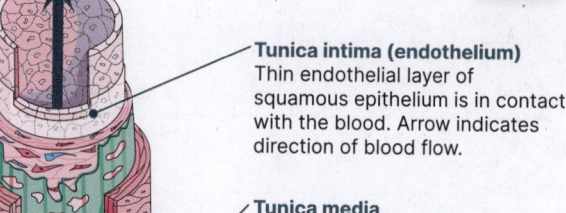

Tunica intima (endothelium)
Thin endothelial layer of squamous epithelium is in contact with the blood. Arrow indicates direction of blood flow.

Tunica media
Thick layer of elastic tissue and smooth muscle tissue allows for both stretch and contraction, maintaining blood flow without loss of pressure.

Tunica externa
Layer of elastic connective tissue (collagen and elastin) anchor the artery to other tissues and allow it to resist overexpansion. Relatively thinner in larger, elastic arteries and thicker in muscular, distributing arteries.

Cross section through a large artery

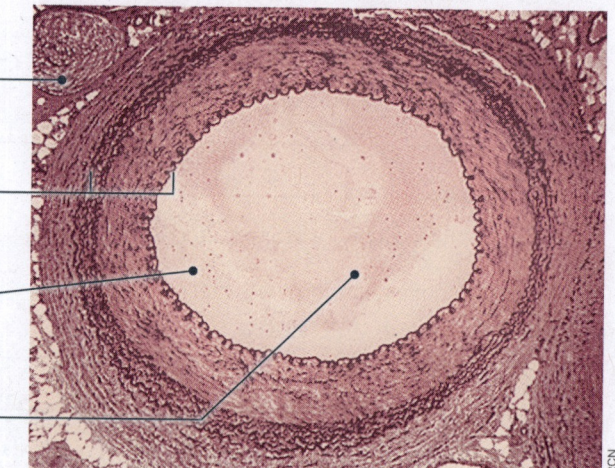

(a)

(b)

(c)

(d)

1. Using the information above to help you, label the micrograph (a)-(d) of the cross section through an artery (above):

2. Why do the walls of arteries need to be thick, with a lot of elastic tissue? _____

_____

3. What is the purpose of the smooth muscle in the artery walls? _____

_____

4. How do arteries contribute to the regulation of blood pressure? _____

_____

_____

_____

B3.2
2 - 3

©2024 **BIOZONE** International
ISBN: 978-1-99-101410-8
Photocopying prohibited

108 Veins

Key Idea: Veins are blood vessels that return the blood from the tissues to the heart.

The smallest veins (venules) return blood from the capillaries to the veins. Veins and their branches contain about 59% of the blood in the body. The structural differences between veins and arteries are mainly in the relative thickness of the vessel layers, and the diameter of the lumen (veins have a large lumen). Both are related to the vessel's functional role.

Veins

When several capillaries unite, they form small veins called venules. The venules collect the blood from capillaries and drain it into veins. Veins are made up of the same three layers as arteries but they have less elastic and muscle tissue, a relatively thicker tunica externa, and a larger, less defined lumen. The venules closest to the capillaries consist of an endothelium and a tunica externa of connective tissue. As the venules approach the veins, they also contain the tunica media characteristic of veins (right). Although veins are less elastic than arteries, they can still expand enough to adapt to changes in the pressure due to muscular action regulating blood flow. Blood flowing in the veins has lost a lot of pressure because it has passed through the narrow capillary vessels. The low pressure in veins means that many veins, especially those in the limbs, need to have valves to prevent back-flow of the blood as it returns to the heart.

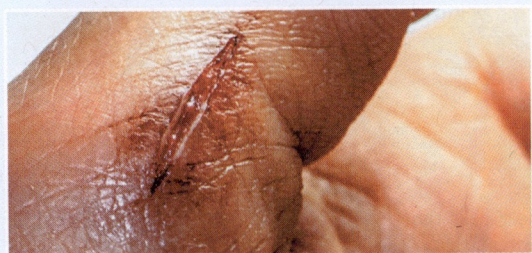

If a vein is cut, as is shown in this finger wound, the blood oozes out slowly in an even flow, and usually clots quickly as it leaves. In contrast, arterial blood spurts rapidly and requires pressure to slow the flow.

Structure of a vein

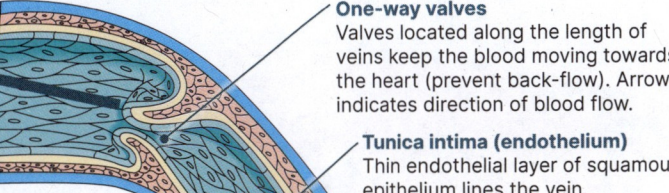

One-way valves
Valves located along the length of veins keep the blood moving towards the heart (prevent back-flow). Arrow indicates direction of blood flow.

Tunica intima (endothelium)
Thin endothelial layer of squamous epithelium lines the vein.

Tunica media
Layer of smooth muscle tissue with collagen fibres (connective tissue). The tunica media is much thinner relative to that of an artery and the smaller venules may lack this layer.

Direction of blood flow

Tunica externa
Layer of connective tissue (mostly collagen). Relatively thicker than in arteries and thicker than the tunica media.

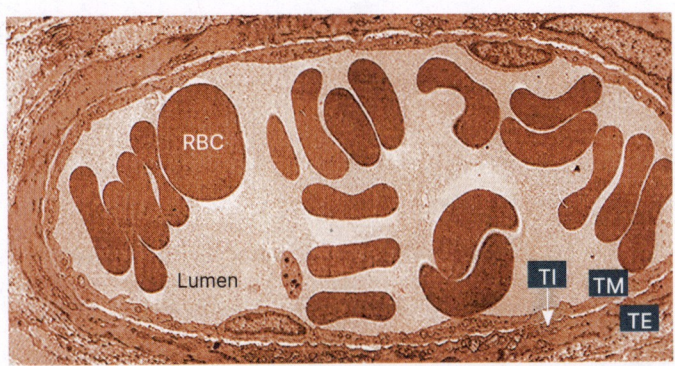

TEM of a vein showing red blood cells (RBC) in the lumen, and the tunica intima (TI), tunica media (TM), and tunica externa (TE).

1. Contrast the structure of veins and arteries for each of the following properties:

   (a) Thickness of muscle and elastic tissue: _____

   (b) Size of the lumen (inside of the vessel): _____

2. With respect to their functional roles, explain the differences you have described above: _____

   _____

   _____

   _____

3. What is the role of the valves in assisting the veins to return blood back to the heart? _____

   _____

4. Why does blood ooze from a venous wound rather than spurting, as it does from an arterial wound?

   _____

   _____

   _____

   _____

B3.2
2, 5

AOS

109 Capillaries

Key Idea: Capillaries are small, thin-walled vessels that allow the exchange of material between the blood and the tissues. Capillaries are very small vessels that connect arterial and venous circulation and allow efficient exchange of nutrients and wastes between the blood and tissues. Capillaries form networks or beds and are abundant where metabolic rates are high. Fluid that leaks out of the capillaries has an essential role in bathing the tissues.

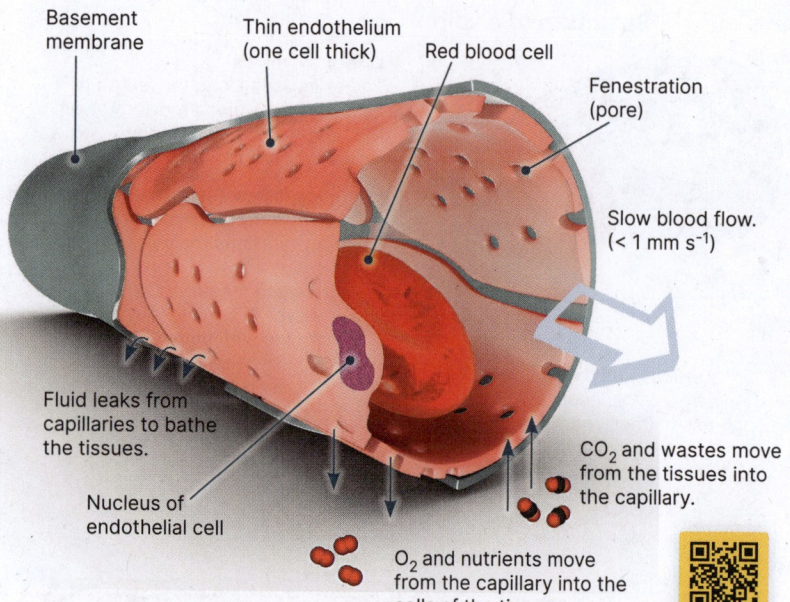

Basement membrane

Thin endothelium (one cell thick)

Red blood cell

Fenestration (pore)

Slow blood flow. (< 1 mm s^{-1})

Fluid leaks from capillaries to bathe the tissues.

Nucleus of endothelial cell

CO_2 and wastes move from the tissues into the capillary.

O_2 and nutrients move from the capillary into the cells of the tissues.

Exchanges in capillaries

▶ Blood passes from the arterioles into the capillaries where the exchange of materials between the body cells and the blood takes place. Capillaries are small blood vessels with a diameter of just 4-10 μm. The only tissue present is an endothelium of squamous epithelial cells. Capillaries are so numerous that no cell is more than 25 μm from any capillary.

▶ Blood pressure causes fluid to leak from capillaries through small gaps where the endothelial cells join. This fluid bathes the tissues, supplying nutrients and oxygen, and removing wastes.

▶ The density of capillaries in a tissue is an indication of that tissue's metabolic activity. For example, cardiac muscle relies heavily on oxidative metabolism. It has a high demand for blood flow and is well supplied with capillaries. Smooth muscle is far less active than cardiac muscle, relies more on anaerobic metabolism, and does not require such an extensive blood supply.

Fenestration in capillaries

▶ Some types of capillaries are fenestrated: they have small holes in their walls that allow larger molecules to pass through. The pore is covered by a membrane that can open or close to allow molecules to pass through.

▶ This is important for function in organs like kidneys, where filtrate containing waste passes from the capillaries to the nephron. The small intestine capillaries have fenestrations to allow broken down food to be absorbed.

▶ Especially wide capillaries in the liver, called sinusoids (right, SEM), have gaps between epithelial cells, and fenestrations around 100 nm in diameter. These vessels are particularly porous and allow rapid transport of molecules and nutrients.

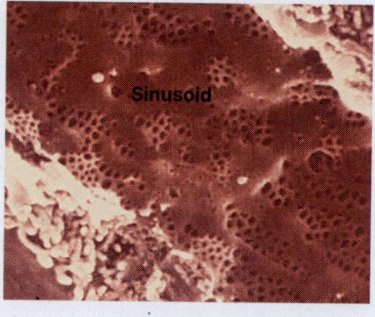

Sinusoid

1. What is the role of capillaries? _____

_____

2. Explain how the capillaries are adapted to enable transport of substances across their walls, to and from the surrounding environment:

_____

_____

_____

_____

_____

3. Why are fenestrations in the epithelial walls of the capillaries important for the function of some tissues and organs?

_____

_____

_____

B3.2

1

©2024 **BIOZONE** International
ISBN: 978-1-99-101410-8

110 Measuring Changes in Pulse Rate

Key Idea: The heart rate can be measured by counting the pulse rate in the carotid or radial artery.

Heart rate in beats per minute (bpm) can be obtained from measuring the pulse for 30 seconds and doubling the number of pulses recorded. The pulse can be easily felt by placing your fingertips on the wrist just under the thumb, on top of the radial artery, or on the carotid artery on the neck just beneath the jaw.

Investigation 7.3 Investigating effect of exercise on heart rate.

See appendix for equipment list.

In this practical, you will work in groups of three to see how exercise affects heart rates. The body's response to exercise can be measured by monitoring changes in heart rate before and after a controlled physical effort. Choose one person to carry out the exercise and one person each to record heart rate.

Heart rate (beats per minute) is obtained by measuring the pulse (right) for 15 seconds and multiplying by four.

CAUTION: The person exercising should have no known pre-existing heart or respiratory conditions.

Gently press your index and middle fingers, not your thumb, against the carotid artery in the neck (just under the jaw) or the radial artery (on the wrist just under the thumb) until you feel a pulse.

Measuring the carotid pulse

Measuring the radial pulse

1. Resting measurements: The person carrying out the exercise should sit on a chair for 5 minutes and try not to move. After 5 minutes of sitting, measure their heart rate. Record the resting data in the table (below).

2. Exercising measurements: Choose an exercise to perform. Some examples include: step ups onto a chair, skipping rope, jumping jacks, or running in place.

3. Begin the exercise and take measurements after 1, 2, 3, and 4 minutes. The person exercising should stop just long enough for the measurements to be taken. Record the results in the table.

4. Post-exercise measurements: After the exercise period has finished, the exerciser should sit down on a chair. Take pulse measurements 1 and 5 minutes after finishing the exercise. Record the results in the table, below.

Activity	Resting before exercise	At 1 minute during exercise	At 2 minutes during exercise	At 3 minutes during exercise	At 4 minutes during exercise	1 minute after exercise	5 minutes after exercise
Pulse Rate							

1. Graph your results on a separate piece of paper and attach to this page. You will use the vertical axis to plot heart rate. In your small groups, develop a conclusion from the results then, as a whole class, compare variation across groups.

Digital tools for heart rate monitoring

Many athletes use heart rate (HR) data to monitor their training and racing to ensure they are pacing themselves at the correct intensity and volume over time. HR zones can allow cyclists to train, for example, at anaerobic threshold, base, or recovery to achieve desired training targets. Unusual HR measurements can also be used to indicate fatigue or illness. A typical athlete system can involve a digital HR monitor chest strap which is paired with a wrist monitor (right). The logged HR data can be uploaded to a computer app in real time, and analysed. Additionally, a digital ring, worn on the finger, can collect continuous HR data.

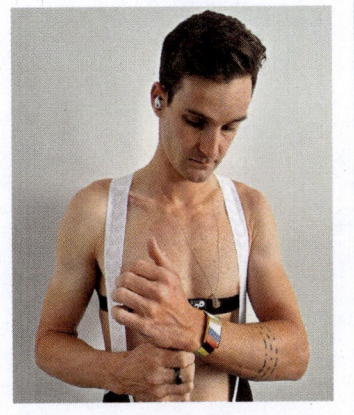

2. What are the advantages and disadvantages of using traditional methods compared to digital tools to monitor heart rate?

_____

_____

_____

_____

_____

_____

_____

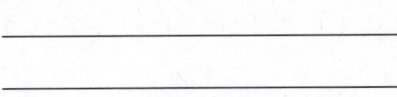

111 Coronary Occlusions

Key Idea: A coronary occlusion is the partial or complete obstruction of blood flow in a coronary artery which often leads to coronary heart disease.

Coronary occlusions are blockages of the coronary arteries that carry blood to the heart muscle. They may be partial or complete, and may occur suddenly or develop over time, often leading to heart attacks. Occlusions restrict blood flow to the heart and, without adequate oxygen and nutrients, the heart tissue is damaged and may even die. Epidemiological data can be analysed for correlation to contributing factors.

Arteries are lined with endothelium, a thin layer of cells which makes the artery smooth and allows the blood to flow through it easily. If fatty deposits (plaques) form on the artery, the blood flow through the artery is disrupted. The flow of blood to the heart can become limited as the plaque increases in size. An inadequate supply of blood to the heart may result in angina. People suffering angina often feel breathless and have chest pain, as the heart beats harder and faster to meet its oxygen demands.

Risk factors for coronary artery disease

▶ High blood pressure damages arterial walls.

▶ High levels of LDL cholesterol contribute to plaque formation because less cholesterol is transported to the liver and more is deposited in the artery walls.

▶ Smoking damages blood vessels, raises blood pressure, and reduces oxygen availability.

▶ High blood sugar levels damage blood vessels.

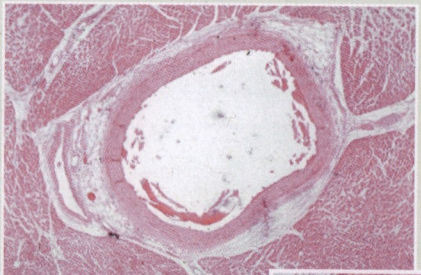

PEIR Digital Library

Normal, unobstructed coronary artery (left), and a coronary artery with moderately severe blockage (below). Note the formation of the plaque in the arterial wall.

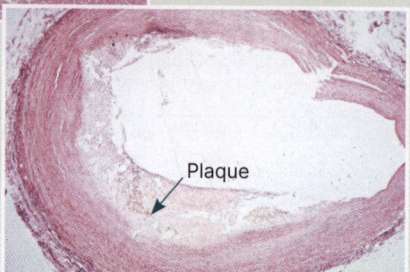

Plaque

Plaques may break off causing a blood clot to form on the plaque's surface. If it is large enough, the blood clot can completely block the artery and cause a heart attack. If blood flow is not restored to the heart within 20-40 minutes, the heart muscle begins to die.

Plaque Development and Coronary Occlusions

Initial lesion
Atherosclerosis is triggered by damage to an artery wall caused by blood borne chemicals or persistent hypertension.

Fatty streak
Low density lipoproteins (LDLs) accumulate beneath the endothelial cells. Macrophages follow and absorb them, forming foam cells.

Intermediate lesion
Foam cells accumulate, forming greasy, yellow lesions called atherosclerotic plaques.

Atheroma
A core of extracellular lipids under a cap of fibrous tissue forms.

Fibroatheroma
Lipid core and fibrous layers. Accumulated smooth muscle cells die. Fibres deteriorate and are replaced with scar tissue.

Complicated plaque
Calcification of plaque. Arterial wall may ulcerate. Hypertension may worsen. Plaque may break away and blood cells may then collect at the damaged site, forming a clot.

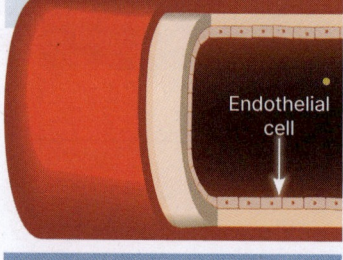

Endothelial cell

Earliest onset

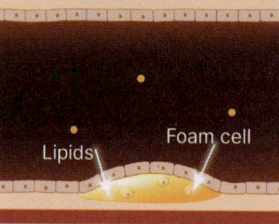

Lipids   Foam cell

From first decade

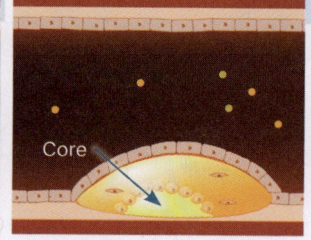

Core

From third decade

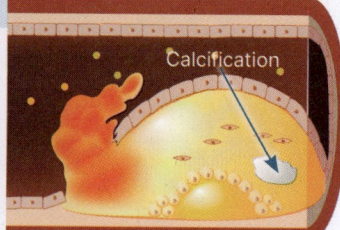

Calcification

From fourth decade

1. (a) What is a coronary occlusion? _____

    (b) How can a coronary occlusion lead to heart damage and failure? _____

    _____

    _____

    _____

2. List some factors that increase the chances of a coronary occlusion: _____

    _____

    _____

NOS   AOS   B3.2   6

©2024 **BIOZONE** International
ISBN: 978-1-99-101410-8
Photocopying prohibited

Interpreting epidemiological data on coronary heart disease

Extensive research has been undertaken in populations around the world to collect data on the incidence and prevalence of coronary heart disease. Possible contributing factors including age, sex, lifestyle habits, environment, and underlying conditions have been examined to understand how they have influenced the development of this disorder.

Self-reported coronary heart disease among persons aged 18 and over, by age and sex, 2017-2018

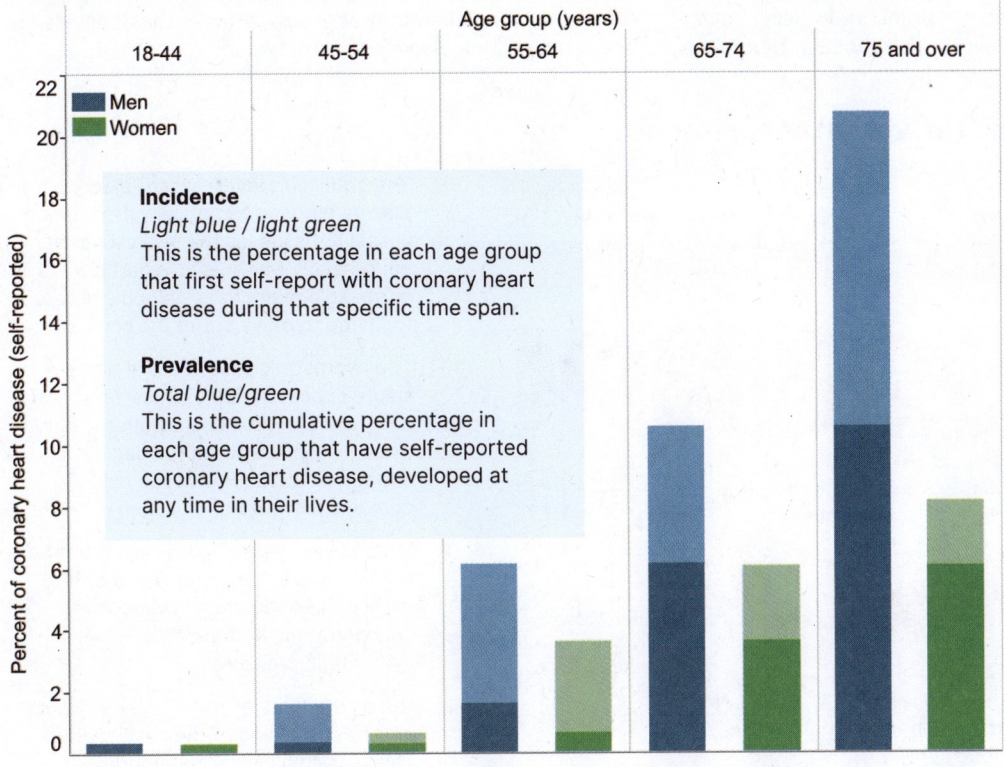

Incidence
Light blue / light green
This is the percentage in each age group that first self-report with coronary heart disease during that specific time span.

Prevalence
Total blue/green
This is the cumulative percentage in each age group that have self-reported coronary heart disease, developed at any time in their lives.

Causes of fatal coronary heart disease
Recent studies indicate that most heart attacks are caused by the body's inflammatory response to a plaque. The inflammatory process causes young, soft, cholesterol-rich plaques to rupture and break into pieces. If these block blood vessels, they can cause fatal heart attacks, even in previously healthy people.

Aorta opened lengthwise showing build up of plaque.

3. (a) Compare trends in male and female coronary heart disease prevalence over time: _____

_____

_____

(b) Compare trends in male and female coronary heart disease incidence over time: _____

_____

_____

(c) Evaluate any difference between incidence and prevalence trends: _____

_____

_____

_____

Correlation coefficients

Correlation coefficients are a numerical statistical measure of relationship strength between two variables. Values are calculated between 1 and -1, where 1 is a very strong positive correlation, 0 shows no correlation, and -1 shows a very strong negative relationship. However, correlation does not imply causation, where one event is directly the result of another.

4. Research on factors related to the incidence of coronary heart disease included cholesterol intake and cigarette smoking. Researchers used the data to estimate likely incidence in particular age groups of men. Compared to actual observed incidence, the following correlation coefficients were calculated: Men (39-49) = 0.974 Men (50-59) = 0.826. Explain how this data supported the research findings and why it does not necessarily prove a causal link:

_____

_____

_____

112 The Formation of Tissue Fluid

Key Idea: Tissue fluid, also known as interstitial fluid, originates from the blood in capillaries and surrounds cells, before circulating back into the capillaries, and lymph.

Blood transports nutrients, wastes, and respiratory gases to and from the tissues. Tissue fluid helps transport these between the blood and the tissues. As with all cells, substances can move into and out of the endothelial cells of the capillary walls in several ways: by diffusion, by cytosis,

and through gaps where the membranes are not joined by tight junctions. Capillaries are leaky, so fluid flows across their plasma membranes an into surrounding cellular tissue. Whether fluid moves into or out of a capillary depends on the balance between the blood (hydrostatic) pressure and the concentration of solutes at each end of a capillary bed. The remaining tissue fluid that does not re-enter the capillaries is drained by lymph vessels to form lymph.

Movement of tissue fluid in and out of capillaries

At the arteriolar end of a capillary bed, hydrostatic (blood) pressure (HP) forces fluid out of the capillaries and into the tissue fluid.

At the venous end of a capillary bed, hydrostatic pressure drops and most (90%) of the leaked fluid moves back into the capillaries.

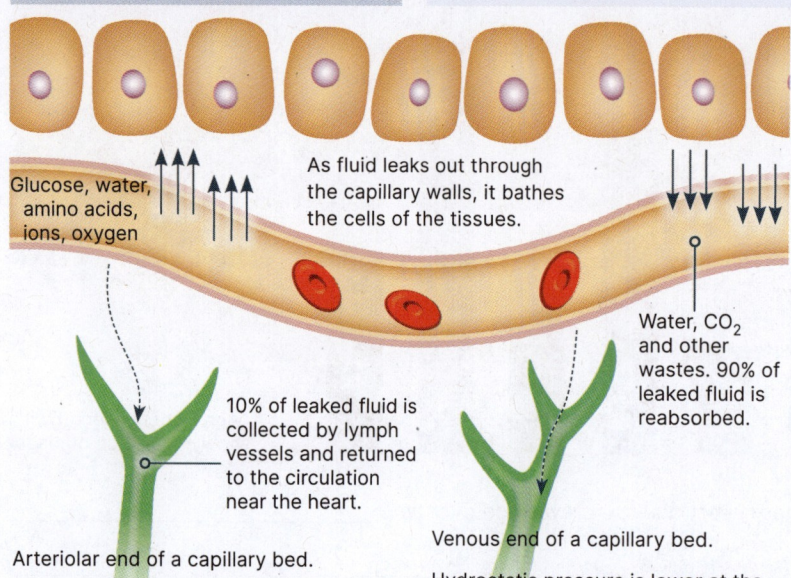

Glucose, water, amino acids, ions, oxygen

As fluid leaks out through the capillary walls, it bathes the cells of the tissues.

Water, CO_2 and other wastes. 90% of leaked fluid is reabsorbed.

10% of leaked fluid is collected by lymph vessels and returned to the circulation near the heart.

Arteriolar end of a capillary bed.

Hydrostatic pressure and solute concentration is higher at the arteriolar end.

RESULT: Net outward pressure. Water and solutes leave the capillary.

Venous end of a capillary bed.

Hydrostatic pressure is lower at the venous end. Solute concentration is slightly higher in the capillary too because of retained proteins.

RESULT: Net inward pressure. Water and solutes re-enter the capillary.

▶ The aqueous (water-based) blood plasma diffuses osmotically, from blood vessel to tissue, as the 'leaky' capillary endothelial layer acts as a partially permeable membrane - blood cells are too large to move out of the capillary.

▶ The hydrostatic pressure and higher solute concentration at the arterial end of a capillary forces fluid from the blood through gaps between the capillary endothelial cells. This tissue fluid contains nutrients and oxygen.

▶ Most of the tissue fluid is then returned to the blood at the venous end of the capillary bed, due to a lower solute concentration, and therefore lower hydrostatic pressure.

▶ Not all the fluid re-enters the capillaries at the venous end of the capillary bed. This extra fluid is collected by the lymphatic vessels, a network of vessels alongside the blood vessels. Once the fluid enters the lymphatic vessels it is called lymph.

1. What is the purpose of the tissue fluid? _____

_____

_____

2. Explain how hydrostatic (blood) pressure and solute concentration operate to cause fluid movement at:

(a) The arteriolar end of a capillary bed: _____

_____

_____

_____

(b) The venous end of a capillary bed: _____

_____

_____

_____

AHL

B3.2
11 - 13

©2024 **BIOZONE** International
ISBN: 978-1-99-101410-8
Photocopying prohibited

Substance movement in tissue fluid

▶ All cells need a supply of O_2 and nutrients and removal of CO_2 and waste products. Capillaries form a network around cells but tissue fluid must circulate between the vessels and cells to transport these substances directly to and from the cell membranes.

▶ Most blood cells cannot travel through the capillary cell walls. Tissue fluid is a similar consistency to plasma.

▶ Substances such as glucose, amino acids, and oxygen that are found in higher concentrations in plasma, are transported in the tissue fluid and captured by cells, so they are in low concentration in the returning fluid.

▶ Conversely, carbon dioxide is in low concentrations in the arterial end of capillaries, but is dissolved in high concentration in returning tissue fluid.

▶ Lymph is similar to tissue fluid but has more lymphocytes and has a role in the immune system.

Comparing blood, tissue fluid, and lymph composition

	Blood	Tissue fluid	Lymph
Cells	Erythrocytes, leukocytes, platelets	Some leukocytes	Lymphocytes
Proteins	Hormones and plasma proteins	Some hormones and proteins	Few
Glucose	High	Used by body cells	Low
Amino acids	High	Used by body cells	Low
Oxygen	High	Used by body cells	Low
Carbon dioxide	Low	Produced by body cells	High

Lymph drainage

▶ The lymphatic system is a network of tissues and organs that collects the tissue fluid leaked from the blood vessels and transports it to the heart.

▶ The lymph vessels are held in place between cells with collagen filaments joined to connective tissue between cells.

▶ The excess tissue fluid moves into the initial lymph capillaries through gaps between the thin cells, where it becomes known as lymph fluid.

▶ The lymph capillaries merge into pre-collector and then collector lymphatic vessels, which contain one-way valves to direct the lymph in one direction.

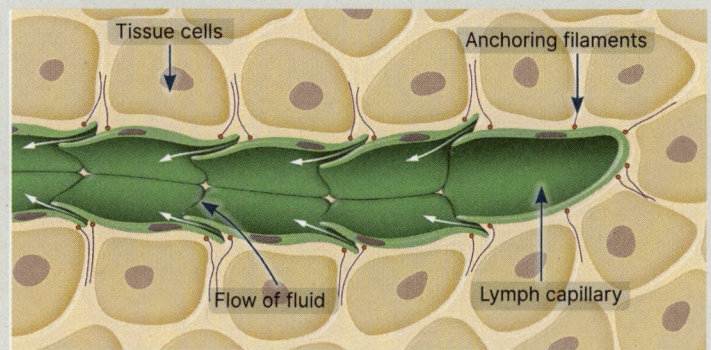

Tissue cells · Anchoring filaments · Flow of fluid · Lymph capillary

▶ Lymph passes through lymph nodes as it collects together from all areas of the body, and the lymphatic vessels eventually drain into the subclavian vein near the heart.

▶ The lymphatic system has an important role in immunity because the fluid (lymph) transported by the lymphatic system is rich in infection-fighting white blood cells called lymphocytes.

3. Describe the two ways in which tissue fluid is returned to the general circulation:

(a) _____

(b) _____

4. Discuss how substances are exchanged between tissue fluid and cells in tissues, including the significance of plasma and tissue fluid composition:

_____

_____

_____

_____

5. Describe how excess tissue fluid is drained into lymph ducts, including the structures involved:

_____

_____

_____

_____

113 Single and Double Circulatory Systems

Key Idea: Closed circulatory systems occur as single or double circuit systems. The single circuit system of fish operates at lower pressure than the double circuit system of mammals.

All vertebrates have closed circulatory systems in which the body's blood flows entirely within blood vessels. Exchanges between the blood and tissues occur by gas diffusion across thin capillary walls. In fish, blood moves in a single circuit, leaving the gills at low pressure to flow around the body before returning to the heart. In all other vertebrates (amphibians, reptiles, birds, and mammals), there is a double circuit system and blood passes from the heart to the lungs (the pulmonary circuit), returning to the heart before being pumped to the body's tissues (the systemic circuit). This double pump system produces much higher pressure in the systemic circuit than in the pulmonary circuit, preventing fluid accumulation in the lungs and providing the pressure to supply the brain and maintain kidney filtration rates.

Closed, single circulatory systems

▶ In the single circulatory system of fish, blood is pumped from the heart to the gills, then directly to the body.

▶ The blood loses pressure at the gills and flows at low pressure around the body. The low pressure reduces blood flow through the body and thus reduces the rate of oxygen delivery to the body's cells. However, most fish have relatively low metabolic rates and this system adequately meets their needs.

▶ Water has a much lower oxygen content than air (about 12 parts per million maximum compared with 210,000 ppm in air). This limits the amount of oxygen fish can extract, even with efficient gills. The low oxygen content in the water would not support a higher metabolic rate but because most fish do not use metabolism to maintain body temperature (a large energy cost), a lower metabolic rate still allows for a relatively active lifestyle.

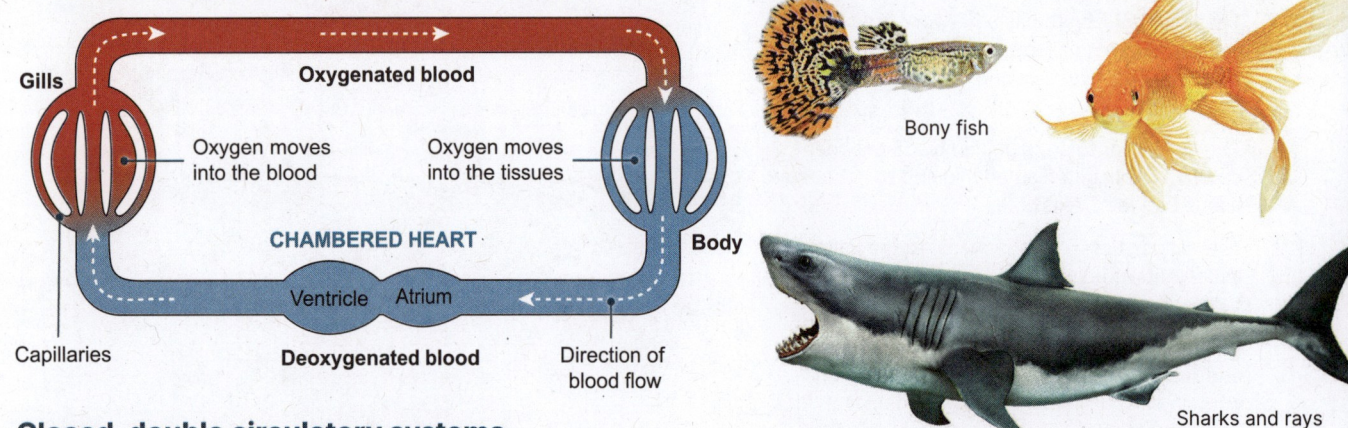

Bony fish

Sharks and rays

Closed, double circulatory systems

▶ Because oxygen is relatively abundant in the air, metabolic rates in air-breathing animals can be relatively high (although not necessarily so). Double circulatory systems develop higher pressure than single circuit systems, delivering oxygenated blood to the body at a rate sufficient to meet higher metabolic demands.

▶ Double circulatory systems occur in all vertebrates other than fish. They are most efficient in mammals and birds, where the heart is fully divided into two halves and the two circuits are completely separate. These animals rely on metabolism to maintain body temperature, so their metabolic demands are necessarily high.

▶ Double circulatory systems have two distinct circuits: the pulmonary circuit, which circulates blood between the lungs and the heart, and the systemic circuit, which pumps oxygenated blood to the rest of the body. The return of oxygenated blood from the lungs to the heart means that the blood can be pumped to the rest of the body at the higher pressures needed to supply organs and maintain kidney filtration rates. Meanwhile, the blood in the lungs (the pulmonary circuit) remains at a low pressure, suitable for facilitating gas exchange.

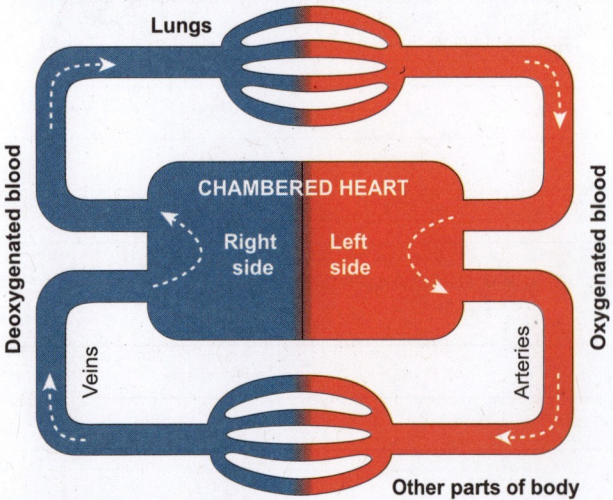

Birds

Mammals

Amphibians

Reptiles

Double systems are also found in birds, amphibians, and reptiles. Birds, like mammals, use metabolism to maintain body temperature (a high energy cost) and maintain high metabolic rates.

©2024 **BIOZONE** International
ISBN: 978-1-99-101410-8
Photocopying prohibited

Fish heart

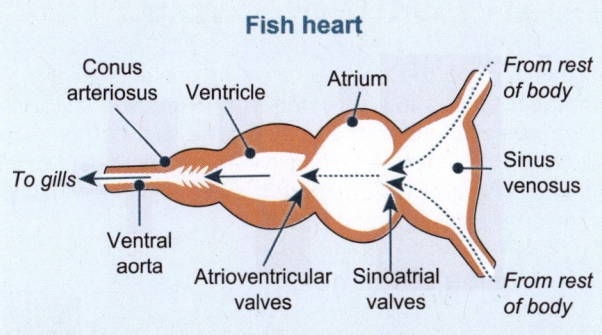

Conus arteriosus · Ventricle · Atrium · From rest of body · Sinus venosus · To gills · Ventral aorta · Atrioventricular valves · Sinoatrial valves · From rest of body

The fish heart is linear, with a sequence of chambers in series. There are two main chambers (atrium and ventricle) as well as an entry (the sinus venosus) and sometimes a smaller exit chamber (the conus). Blood from the body first enters the heart through the sinus venosus, then passes into the atrium and the ventricle. A series of one-way valves between the chambers prevents reverse blood flow. Blood leaving the heart travels to the gills.

Mammalian heart

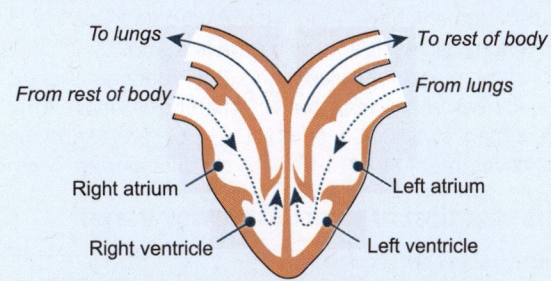

To lungs · To rest of body · From rest of body · From lungs · Right atrium · Left atrium · Right ventricle · Left ventricle

In birds and mammals, the heart is fully partitioned into two halves, resulting in four chambers. Blood circulates through two circuits, with no mixing of the two. Oxygenated blood from the lungs is kept separated from the de-oxygenated blood returning from the rest of the body.

1. What is the main difference between single and double closed systems of circulation? _____

_____

_____

_____

2. (a) Where does the blood flow to immediately after it has passed through the gills in a fish?

_____

_____

(b) How does the single system impact the pressure at which the blood flows in the systemic circulation?

_____

_____

3. (a) Where does the blood flow to immediately after it has passed through the lungs in a mammal?

_____

_____

(b) How does the double system impact the pressure at which the blood flows in the systemic circulation?

_____

_____

4. Explain the higher functional efficiency of a double circuit system relative to a single circuit system:

_____

_____

_____

5. How does a closed circulatory system give an animal finer control over the distribution of blood to tissues and organs?

_____

_____

_____

_____

©2024 **BIOZONE** International
ISBN: 978-1-99-101410-8
Photocopying prohibited

114 The Mammalian Heart and the Cardiac Cycle

Key Idea: Humans have a four chambered heart divided into left and right halves. It acts as a double pump.

It is a hollow, muscular organ made up of four chambers (two atria and two ventricles) that alternately fill and empty of blood, acting as a double pump. The tricuspid and bicuspid heart valves prevent back-flow of the blood from ventricle to atrium, while the semi-lunar valve prevents back-flow from the arteries into the ventricles. The left side (systemic circuit) pumps blood to the body tissues and the right side (pulmonary circuit) pumps blood to the lungs. The adaptations of the heart enable the structure to send pressurized blood, via the arteries, around the body.

Cross-section of heart (anterior view)

Sinoatrial node (SAN) is the heart's pacemaker. It is a small mass of specialized muscle cells on the wall of the right atrium, near the entry point of the superior vena cava. It starts the cardiac cycle, spontaneously generating action potentials that cause the atria to contract. The SAN sets the basic heart rate, but this rate is influenced by hormones and impulses from the autonomic nervous system.

- - - ▶ Spread of impulses across atria
- - ▶ Spread of impulses to ventricles

Atrioventricular node (AVN) at the base of the atrium briefly delays the impulse to allow time for the atrial contraction to finish before the ventricles contract.

Bundle of His (atrioventricular bundle) A tract of conducting (Purkyne) fibres that distribute the action potentials over the ventricles causing ventricular contraction.

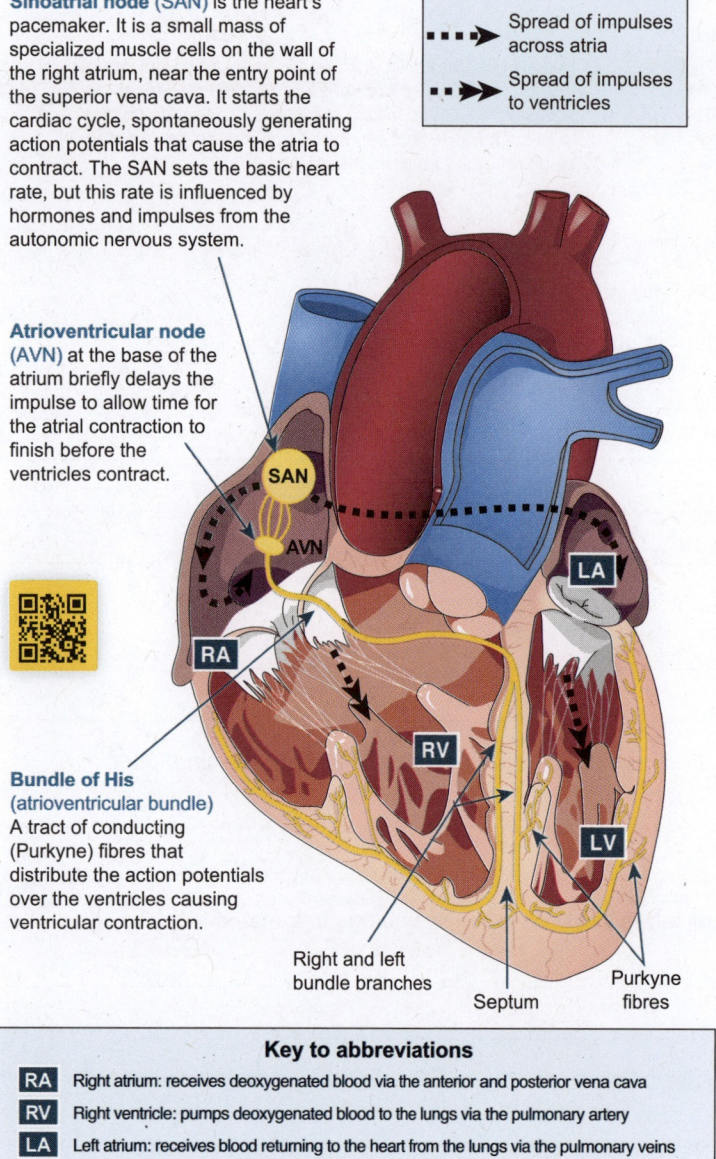

Right and left bundle branches
Septum
Purkyne fibres

Key to abbreviations

RA	Right atrium: receives deoxygenated blood via the anterior and posterior vena cava
RV	Right ventricle: pumps deoxygenated blood to the lungs via the pulmonary artery
LA	Left atrium: receives blood returning to the heart from the lungs via the pulmonary veins
LV	Left ventricle: pumps oxygenated blood to the head and body via the aorta

Valves of the heart

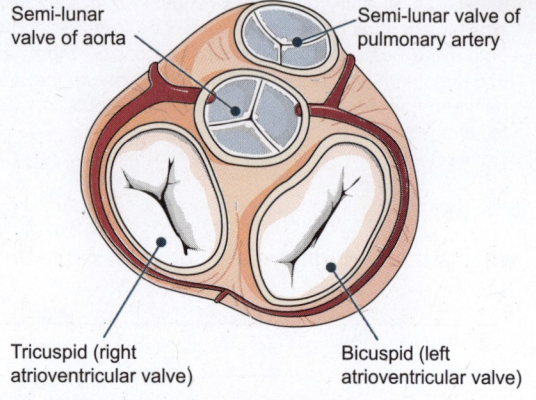

Semi-lunar valve of aorta
Semi-lunar valve of pulmonary artery
Tricuspid (right atrioventricular valve)
Bicuspid (left atrioventricular valve)

External view of heart

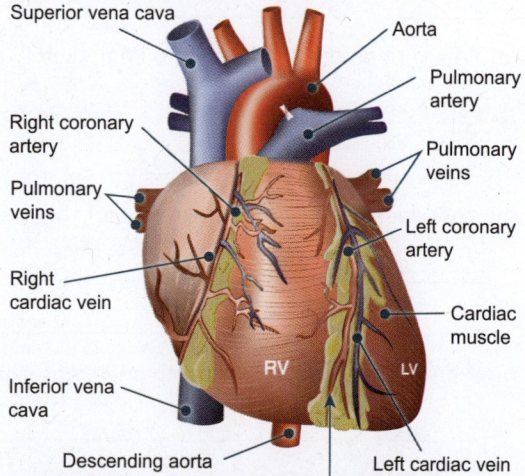

Superior vena cava
Aorta
Pulmonary artery
Right coronary artery
Pulmonary veins
Pulmonary veins
Left coronary artery
Right cardiac vein
Cardiac muscle
Inferior vena cava
RV
LV
Descending aorta
Left cardiac vein

Coronary arteries: The high oxygen demands of the heart muscles are met by a dense capillary network. Coronary arteries arise from the aorta and spread over the surface of the heart supplying the cardiac muscle with oxygenated blood. Deoxygenated blood is collected by cardiac veins and returned to the right atrium via a large coronary sinus.

1. In the schematic diagram of the heart below, label the four chambers and the main vessels entering and leaving them. The arrows indicate the direction of blood flow.

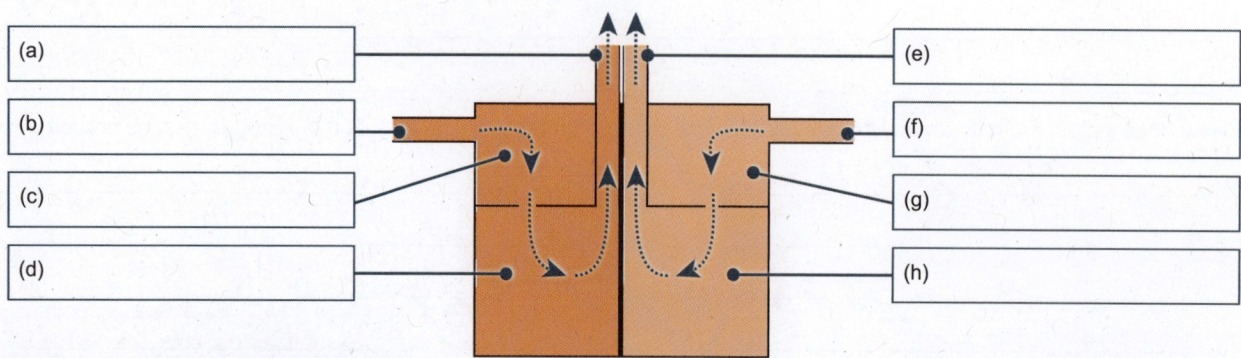

(a)
(b)
(c)
(d)
(e)
(f)
(g)
(h)

   **B3.2** 15 - 16  204

©2024 **BIOZONE** International
ISBN: 978-1-99-101410-8
Photocopying prohibited

The cardiac cycle

The cardiac cycle refers to the sequence of events of a heartbeat and involves three main stages. The heart pumps with alternate contractions (systole) and relaxations (diastole). Pressure changes within the heart's chambers generated by the cycle of contraction and relaxation are responsible for blood movement and cause the heart valves to open and close, preventing the back-flow of blood. The noise of the blood when the valves open and close produces the heartbeat sound (lubb-dupp). The heart beat occurs in response to electrical impulses, which can be recorded as a trace called an electrocardiogram or ECG. The ECG pattern is the result of the different impulses produced at each phase of the cardiac cycle, and each part is identified with a letter code. An ECG provides a useful method of monitoring changes in heart rate and activity and detection of heart disorders. The electrical trace is accompanied by volume and pressure changes (below).

The pulse results from the rhythmic expansion of the arteries as the blood spurts from the left ventricle. Pulse rate therefore corresponds to heart rate.

Stage 1: Atrial contraction and ventricular filling
The ventricles relax and blood flows into them from the atria. Note that 70% of the blood from the atria flows passively into the ventricles. It is during the last third of ventricular filling that the atria contract.

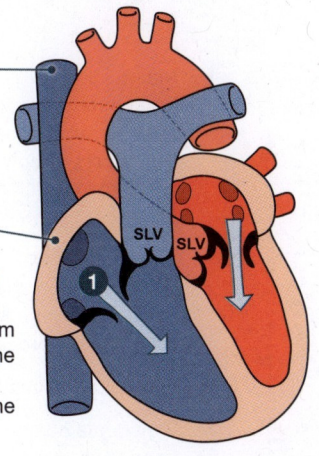

Heart during ventricular filling

Stage 2: Ventricular contraction
The atria relax, the ventricles contract, and blood is pumped from the ventricles into the aorta and the pulmonary artery. The start of ventricular contraction coincides with the first heart sound.

Stage 3: (not shown) There is a short period of atrial and ventricular relaxation. Semilunar valves (SLV) close to prevent backflow into the ventricles (see diagram, left). The cycle begins again. For a heart beating at 75 beats per minute, one cardiac cycle lasts about 0.8 seconds.

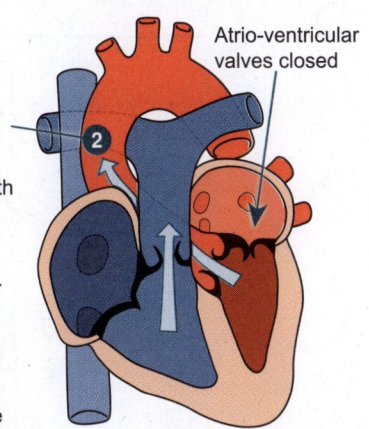

Atrio-ventricular valves closed

Heart during ventricular contraction

Cardiac cycle events in the left ventricle

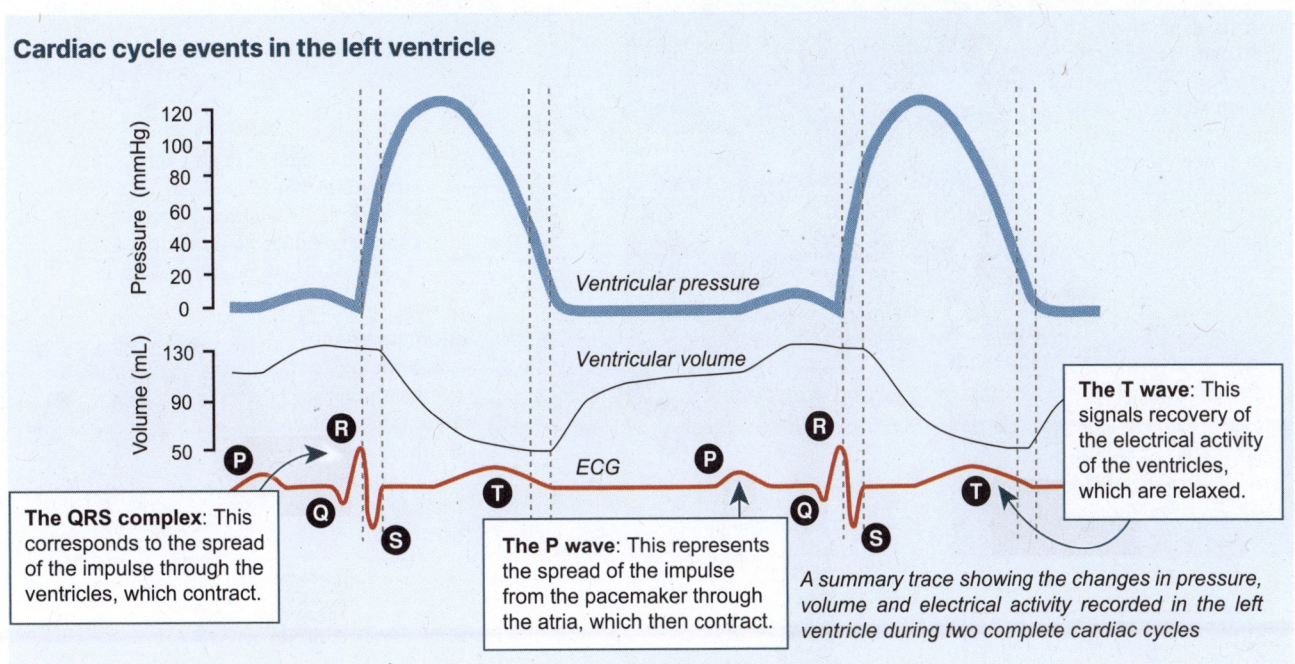

The QRS complex: This corresponds to the spread of the impulse through the ventricles, which contract.

The P wave: This represents the spread of the impulse from the pacemaker through the atria, which then contract.

The T wave: This signals recovery of the electrical activity of the ventricles, which are relaxed.

A summary trace showing the changes in pressure, volume and electrical activity recorded in the left ventricle during two complete cardiac cycles

2. Identify each of the following phases of an ECG by its international code:

    (a) Excitation of the ventricles and ventricular systole: _____

    (b) Electrical recovery of the ventricles and ventricular diastole: _____

    (c) Excitation of the atria and atrial systole: _____

3. Suggest the physiological reason for the period of electrical recovery experienced each cycle: _____

4. Identify the points on the trace above corresponding to each of the following, indicating which phase code(s) it is near:

    (a) Ejection of blood from the ventricle _____

    (b) Closing of the atrioventricular valve _____

    (c) Filling of the ventricle _____

    (d) Opening of the atrioventricular valve _____

115 Transport of Water Through a Plant

Key Idea: Water moves through the xylem primarily as a result of evaporation from the leaves and the cohesive and adhesive properties of water molecules.

Plants lose water all the time. Approximately 99% of the water a plant absorbs from the soil is lost by evaporation from the leaves and stem. This loss, mostly through stomata, is called transpiration and the flow of water through the plant is called the transpiration stream. Plants rely on a gradient in solute concentration that increases from the roots to the air to move water through their cells. Water flows passively from soil to air along this gradient of increasing solute concentration. The gradient is the driving force for the movement of water up a plant. Transpiration has benefits to the plant because evaporative water loss cools the plant and the transpiration stream helps the plant to take up minerals. Water movement factors are described below.

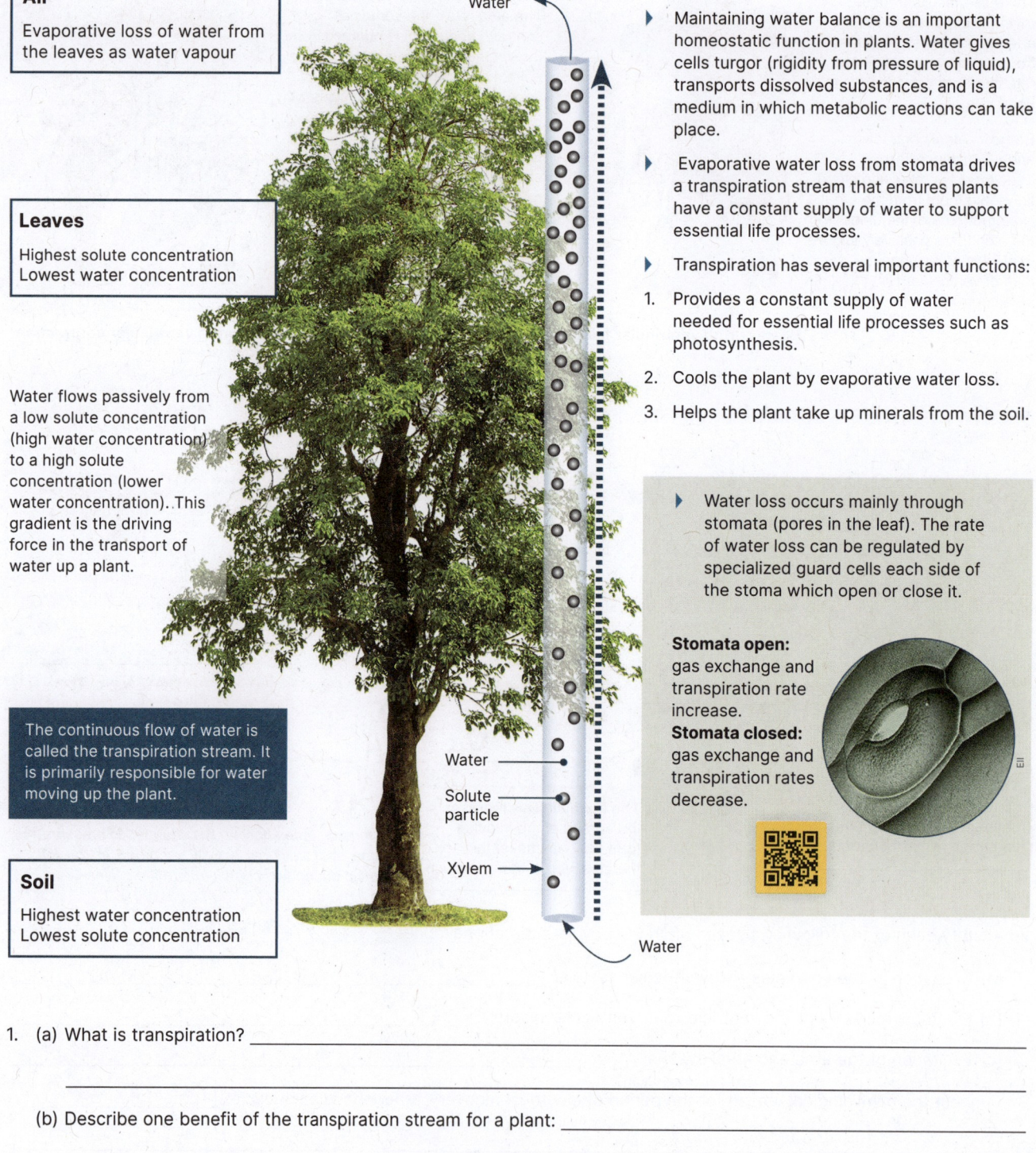

Air
Evaporative loss of water from the leaves as water vapour

Leaves
Highest solute concentration
Lowest water concentration

Water flows passively from a low solute concentration (high water concentration) to a high solute concentration (lower water concentration). This gradient is the driving force in the transport of water up a plant.

The continuous flow of water is called the transpiration stream. It is primarily responsible for water moving up the plant.

Soil
Highest water concentration
Lowest solute concentration

Water · Solute particle · Xylem

▸ Maintaining water balance is an important homeostatic function in plants. Water gives cells turgor (rigidity from pressure of liquid), transports dissolved substances, and is a medium in which metabolic reactions can take place.

▸ Evaporative water loss from stomata drives a transpiration stream that ensures plants have a constant supply of water to support essential life processes.

▸ Transpiration has several important functions:
1. Provides a constant supply of water needed for essential life processes such as photosynthesis.
2. Cools the plant by evaporative water loss.
3. Helps the plant take up minerals from the soil.

▸ Water loss occurs mainly through stomata (pores in the leaf). The rate of water loss can be regulated by specialized guard cells each side of the stoma which open or close it.

Stomata open: gas exchange and transpiration rate increase.
Stomata closed: gas exchange and transpiration rates decrease.

1. (a) What is transpiration? _____

(b) Describe one benefit of the transpiration stream for a plant: _____

2. How does the plant regulate the amount of water lost from the leaves? _____

©2024 **BIOZONE** International
ISBN: 978-1-99-101410-8
Photocopying prohibited

B3.2 7

Moving water through the xylem

1 Transpiration pull
Water is lost from the air spaces by evaporation through stomata and is replaced by water from the mesophyll cells. The constant loss of water to the air (and production of sugars) creates a solute concentration in the leaves that is higher than elsewhere in the plant. Water is pulled through the plant along a gradient of increasing solute concentration.

2 Cohesion-tension
The transpiration pull is assisted by the special cohesive properties of water. Water molecules cling together as they are pulled through the plant. They also adhere to the walls of the xylem (adhesion). This creates one unbroken column of water through the plant. The upward pull on the cohesive sap creates a tension (a negative pressure). This helps water uptake and movement up the plant.

3 Root pressure
Water entering the stele from the soil creates a root pressure; a weak 'push' effect for the water's upward movement through the plant. Root pressure can force water droplets from some small plants under certain conditions (guttation), but generally it plays a minor part in the ascent of water.

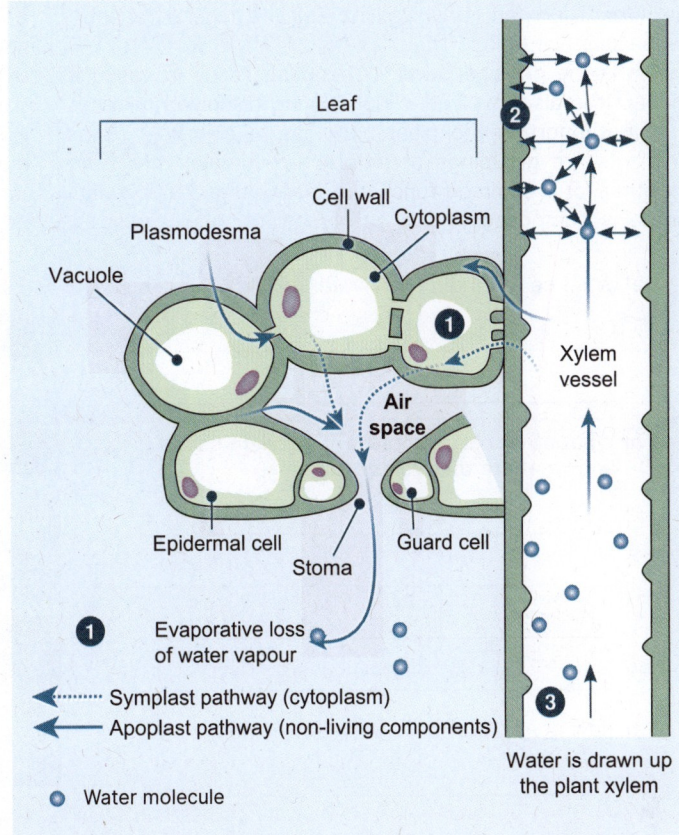

3. (a) What would happen if too much water was lost from the leaves? _____

_____

(b) When might this happen? _____

_____

4. Describe the three processes that assist the transport of water from the roots of the plant upward:

(a) _____

(b) _____

(c) _____

5. The maximum height water can move up the xylem by cohesion-tension alone is about 10 m. How then does water move up the height of a 40 m tall tree?

_____

©2024 **BIOZONE** International
ISBN: 978-1-99-101410-8
Photocopying prohibited

116 Xylem Tissue and Water Transport

Key Idea: The xylem is involved in water and mineral transport in vascular plants.

Xylem is the principal water conducting tissue in vascular plants. It is also involved in conducting dissolved minerals and in supporting the plant body. As in animals, tissues in plants are groupings of different cell types that work together for a common function. In flowering plants, xylem tissue is composed of five cell types: tracheids, vessels, xylem parenchyma, sclereids (short sclerenchyma cells), and fibres. The tracheids and vessel elements form the bulk of the tissue. They are heavily strengthened and are the conducting cells of the xylem. Parenchyma cells are involved in storage, while fibres and sclereids provide support. When mature, xylem is dead. Water can enter and exit the xylem through pits which are covered in water permeable cell walls, but lack the lignified layer found in the rest of the xylem.

1. (a) What cells conduct the water in xylem?

   _____

   (b) What other cells are present in xylem tissue and what are their roles?

   _____

   _____

   _____

2. (a) How does water pass between vessels?

   _____

   _____

   _____

   (b) How does water pass between tracheids?

   _____

   _____

   _____

   (c) Which cell type do you think provides the most rapid transport of water and why?

   _____

   _____

   _____

   _____

   (d) Why do you think the tracheids and vessel elements have/need secondary thickening?

   _____

   _____

   _____

3. How can xylem vessels and tracheids be dead when mature, yet still functional?

   _____

   _____

   _____

   _____

Photos: McKDandy cc 2.5

Water moves through the continuous tubes made by the cell vessel elements of the xylem, which have no end wall.

Smaller tracheids are connected by pits in the walls but do not have end wall perforations.

Vessels

Xylem is dead when mature. Note how the cells have lost their cytoplasm.

As shown in these SEM and light micrographs of xylem, the tracheids and vessel elements form the bulk of the xylem tissue. They are heavily strengthened and are involved in moving water through the plant. The transporting elements are supported by parenchyma (packing and storage cells) and sclerenchyma cells (fibres and sclereids), which provide mechanical support to the xylem.

BCC Bioscience Image Library cc0

400X

The xylem cells form continuous tubes through which water is conducted.

Spiral thickening of lignin around the walls of the vessel elements give extra strength and rigidity.

Vessel element
Diameter ~ 500 µm
Secondary walls of cellulose are laid down after the cell has elongated or enlarged and lignin is deposited to add strength. This thickening is a feature of tracheids and vessels.

Vessels connect end to end. The end walls of the vessels are perforated to allow rapid water transport.

Tip of tracheid
Diameter ~80 µm

Pits and bordered pits allow transfer of water between cells but there are no end wall perforations.

No cytoplasm or nucleus in mature cell.

Tracheids are longer and thinner than vessels.

Vessel elements and tracheids are the two water conducting cell types in the xylem of flowering plants. Tracheids are long, tapering hollow cells. Water passes from one tracheid to another through thin regions in the wall called pits. Vessel elements are much larger cells with secondary thickening in different patterns, e.g. spirals. Vessel end walls are perforated to allow efficient conduction of water.

©2024 **BIOZONE** International
ISBN: 978-1-99-101410-8

117 Distribution of Plant Tissue

Key Idea: The vascular tissue in dicots can be identified by its appearance in sections viewed with a light microscope. The structure of the vascular tissue in dicotyledons (dicots) has a very regular arrangement with the xylem and phloem found close together. In the stem, the vascular tissue is distributed in a regular fashion near the outer edge of the stem. In the roots, the vascular tissue is found near the centre of the root.

The structure of the vascular tissue in dicotyledons (dicots) has a very regular arrangement, with xylem and phloem found close together. In the stem, the vascular tissue is distributed in a regular fashion near the outer edge of the stem. In the roots, the vascular tissue is found near the centre of the root.

Dicot stem structure

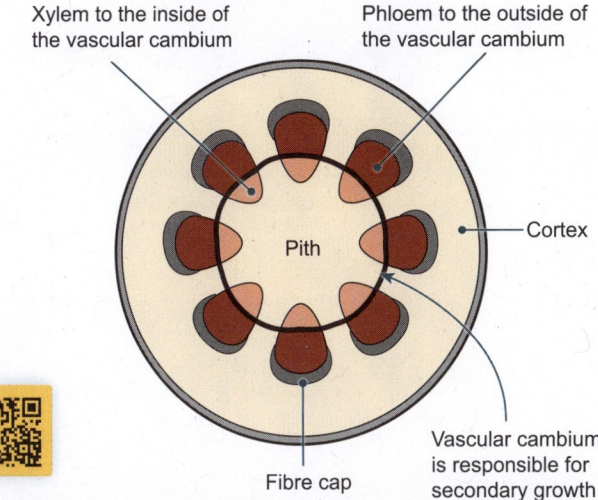

Xylem to the inside of the vascular cambium

Phloem to the outside of the vascular cambium

Pith

Cortex

Fibre cap

Vascular cambium is responsible for secondary growth

In dicots, the vascular bundles (xylem and phloem) are arranged in an orderly fashion around the stem. Each vascular bundle contains xylem (to the inside) and phloem (to the outside). Between the phloem and the xylem is the vascular cambium. This is a layer of cells that divide to produce the thickening of the stem.

Dicot root structure

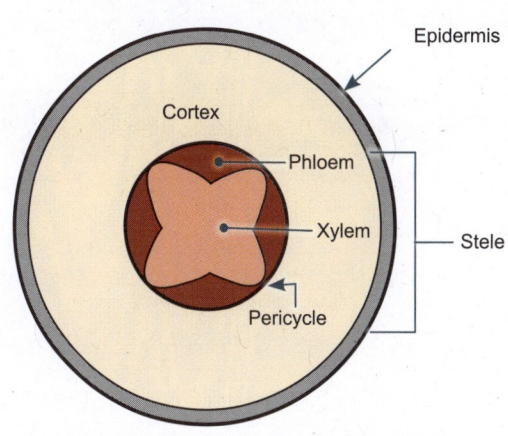

Epidermis

Cortex

Phloem

Xylem

Stele

Pericycle

In a dicot root, the vascular tissue, (xylem and phloem) forms a central cylinder through the root called the stele. The large cortex is made up of parenchyma (packing) cells, which store starch and other substances. Air spaces between the cells are essential for aeration of the root tissue, which is non-photosynthetic.

1. In the stem micrograph below, identify the structures labelled A, B, and C:

    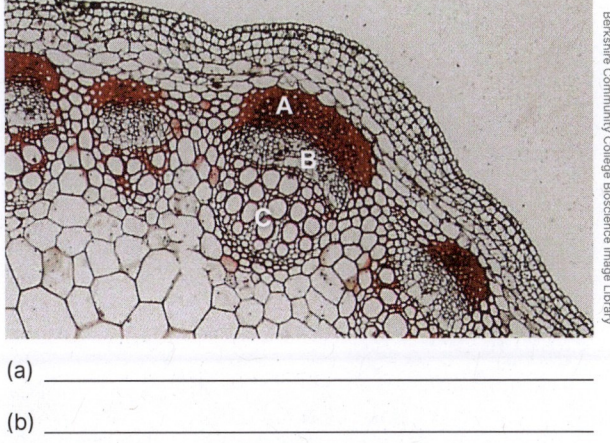

    Berkshire Community College Bioscience Image Library

    (a) _____

    (b) _____

    (c) _____

2. In the root micrograph below, identify the structures labelled A and B:

    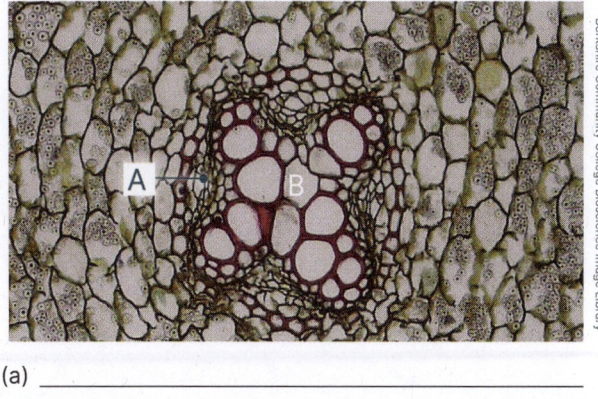

    Berkshire Community College Bioscience Image Library

    (a) _____

    (b) _____

3. Describe the differences in the structure of the vascular tissue in stems and roots: _____

    _____

    _____

    _____

    _____

4. What is the role of the vascular cambium? _____

    _____

B3.2
9 - 10

AOS

Distribution of vascular tissue

The angiosperms (flowering plants) are commonly divided into two groups: the monocots (plants that produce seeds with one embryonic leaf), and the dicots (plants that produce seeds with two embryonic leaves). The arrangement of the vascular tissue is quite different in these two groups. In the stem, this difference can be clearly seen. The vascular bundles in dicots are arranged around the periphery of the stem, while in monocots they appear more scattered. These arrangements can best be seen by making slides of herbaceous plants for viewing under a microscope.

Cherry tree: dicot

Tulip: monocot

Investigation 7.4   Investigating vascular tissue

See appendix for equipment list.

1. Your teacher may provide you with prepared slides or you may need to make them yourself. Refer to investigation 1.2 if you need to make your own slides. You will also be provided with two unknown, prepared slides. One will be a monocot and one will be a dicot. They will be slides showing vascular tissue in either stems or roots.

2. You will need to prepare slides with transverse sections (cut across the stem or root) of stems and roots or use the slides provided by your teacher. Useful dicot plants for this are buttercups and sunflowers and useful monocot plants are corn/maize.

3. Go to the BIOZONE Resource Hub and look at the 4 'Interacting Systems in Plants' images:
   Angiosperm morphology: monocotyledonous roots,
   Angiosperm morphology: monocotyledonous stems,
   Angiosperm morphology: herbaceous dicotyledonous roots,
   Angiosperm morphology: herbaceous dicotyledonous stems.
   These links have many high quality images. They will help you identify the vascular tissue in your slides.

bit.ly/3L13n7Z

4. Place a slide on the microscope stage and focus on low power, and then on high power. In the space below record whether the specimen is a dicot or monocot, stem or root, and write a brief description of or draw the arrangement of the vascular tissue. Use extra paper if needed. Include a scale for your diagrams.

5. Repeat for all the plant roots and stems available.

6. Identify the unknown slides provided by your teacher. For each slide, state whether it is a monocot or dicot, a root or stem slide, and the reasons for your decision.

_____

_____

_____

_____

_____

_____

©2024 **BIOZONE** International
ISBN: 978-1-99-101410-8
Photocopying prohibited

Uptake at the Root

Key Idea: Water uptake by the root is a passive process. Mineral uptake can be passive or active.

Plants need to take up water and minerals constantly. They must compensate for the continuous loss of water from the leaves and provide the materials the plant needs to make food. The uptake of water and minerals is mostly restricted to the younger, most recently formed cells of the roots and the root hairs. Water uptake occurs by osmosis, whereas mineral ions enter the root by diffusion and active transport. The active transport of ions generates root pressure, so water moves into the plant when transpiration rates are low due to high humidity, or in deciduous trees during winter.

Water and mineral uptake by roots

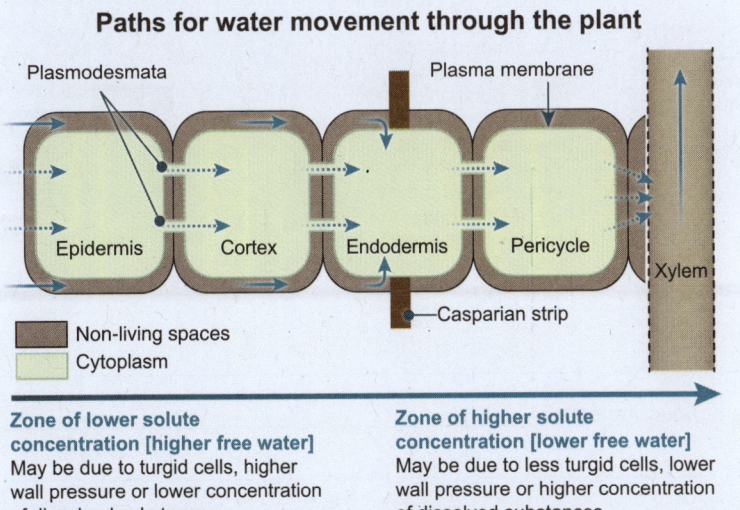

Root hairs have a thin cuticle, so water enters the root easily

Cortex cells of root

Epidermal cell

Stele (vascular cylinder). The outer layer of the stele, the pericycle, is next to the endodermis.

Root hair

Water moves by osmosis

Schematic cross-section through a dicot root

The endodermis is the central, innermost layer of the cortex. It is a single layer of cells with a waterproof band of suberin, called the **Casparian strip**, which encircles each cell.

Root hairs are extensions of the root epidermal cells and provide a large surface area for absorbing water and nutrients.

Root hairs provide a large surface area for absorption. They lack the waxy cuticle found on leaves so there is no barrier to water uptake.

1. What two mechanisms do plants use to absorb nutrients?

_____

_____

_____

_____

2. Transpiration pull of water molecules is the main method of drawing water and dissolved minerals through up the plant's xylem. Explain why plants may also need to 'supplement' the water transportation process by increasing root pressure:

_____

_____

_____

_____

_____

_____

3. Explain how the active transport of mineral ions is able to generate root pressure in xylem vessels:

_____

_____

_____

_____

_____

Paths for water movement through the plant

Plasmodesmata

Plasma membrane

Epidermis | Cortex | Endodermis | Pericycle

Xylem

Casparian strip

■ Non-living spaces
□ Cytoplasm

Zone of lower solute concentration [higher free water]
May be due to turgid cells, higher wall pressure or lower concentration of dissolved substances

Zone of higher solute concentration [lower free water]
May be due to less turgid cells, lower wall pressure or higher concentration of dissolved substances

▶ The uptake of water and minerals is mostly restricted to the younger, most recently formed cells of the roots and the root hairs.

▶ Water moves into the roots because the solute concentration is higher in the root tissue than in the soil. When transpiration rates are too low to 'pull in' water, due to lost leaves in winter or high environmental humidity, active transport through specific transport proteins in the root hair cell membranes can increase the root pressure.

▶ Some water moves through the plant tissues via cytoplasmic connections between cells (the plasmodesmata), but most passes through the free spaces outside the plasma membranes of the cells.

119 Phloem Tissue

Key Idea: Phloem is the principal food (sugar) conducting tissue in vascular plants, transporting dissolved sugars around the plant.

Like xylem, phloem is also a complex tissue, made up of a variable number of cell types. The bulk of phloem tissue is made up of the sieve tubes (sieve tube elements and sieve cells) and their companion cells. The sieve tubes are the main conducting cells in phloem and are closely associated with the companion cells which support them. Parenchyma cells, concerned with storage, occur in phloem, and strengthening fibres and sclereids (short sclerenchyma cells) may also be present. Unlike xylem, functional, mature phloem is alive. All these adaptations in the phloem allow the plant to load the sap at the source and then transport and unload at the sink.

Longitudinal section through a sieve tube end plate

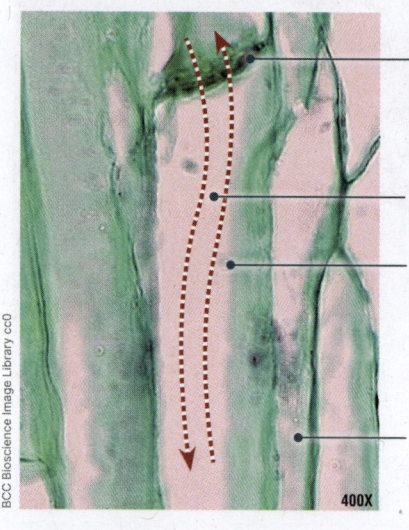

Tiny holes (arrowed in the photograph below) perforate the sieve tube elements allowing the sugar solution to pass through.

Sugar solution flows in both directions.

The sieve tube elements (also called sieve tube members) lose most of their organelles but are still alive when mature.

Companion cell
A cell next to the sieve tube member, responsible for keeping it alive.

BCC Bioscience Image Library cc0

400X

Transverse section through a sieve tube end plate

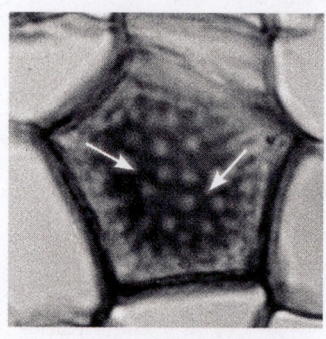

RCN

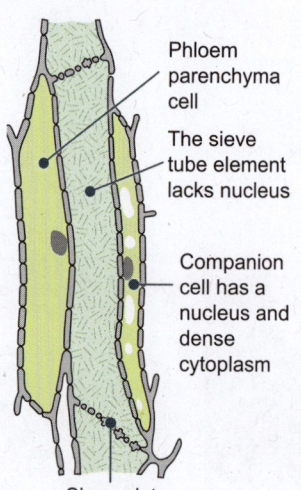

Phloem parenchyma cell

The sieve tube element lacks nucleus

Companion cell has a nucleus and dense cytoplasm

Sieve plate

▶ Adjacent sieve tube elements are connected through sieve plates through which phloem sap flows.

▶ Adjacent companion cells are connected with plasmodesmata.

The structure of phloem tissue

Phloem is alive at maturity and functions in the transport of sugars and minerals around the plant. Like xylem, it forms part of the structural vascular tissue of plants.

Fibres are associated with phloem as they are in xylem. Here, they are seen in cross section where you can see the extremely thick cell walls and the way the fibres are clustered in groups.

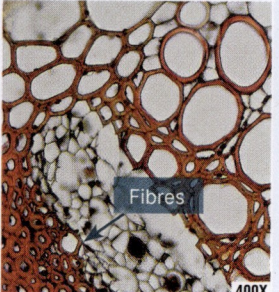

Fibres

400X

In this cross section through the vascular bundle of a corn stem, the smaller companion cells can be seen lying alongside the sieve tube members. It is the sieve tube elements that, end on end, produce the sieve tubes. They are the conducting tissue of phloem. They have reduced cytoplasm and organelles.

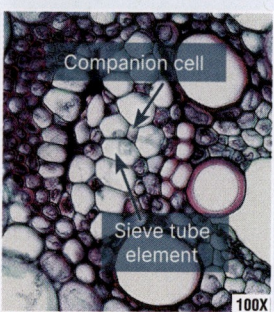

Companion cell

Sieve tube element

100X

In this longitudinal section of a corn stem, each sieve tube element has a thin companion cell associated with it. Companion cells retain their nucleus and have many mitochondria. The cells control the metabolism of the sieve tube member next to them, and also have a role in the loading and unloading of sugar into the phloem.

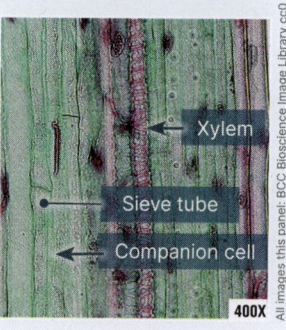

Xylem

Sieve tube

Companion cell

400X

All images this panel: BCC Bioscience Image Library cc0

1. (a) What is the conducting cell type in phloem? _____

   (b) What other cell type is associated with these conducting cells? _____

   (c) Describe two roles of these associated cells: _____

   _____

2. Mature phloem is a live tissue, whereas xylem (the water transporting tissue) is dead when mature. Why is it necessary for phloem to be alive to be functional, whereas xylem can function as a dead tissue?

   _____

   _____

3. What is the role of fibres and sclereids in phloem? _____

   _____

AHL

B3.2
18

120 Translocation

Key Idea: Phloem transports the organic products of photosynthesis (sugars) through the plant by translocation. In vascular plants, the products of photosynthesis move as phloem sap. Apart from water, phloem sap contains mainly sucrose (up to 30%). It may also contain minerals, hormones, and amino acids in transit around the plant. Movement of sap in the phloem is from a source (an organ where sugar is made or mobilized) to a sink (an organ where sugar is stored or used). The sap moves through the phloem sieve-tube members, which are arranged end-to-end and perforated with sieve plates. Loading sucrose into the phloem at a source (leaf) involves energy expenditure. We know this because it is slowed or stopped by high temperatures or respiratory inhibitors. In some plants, unloading the sucrose at the sinks also requires energy, although in others unloading into the cells of the sink organ, e.g. root, occurs by diffusion alone.

Phloem transport

Phloem sap moves from source to sink at rates as great as 100 m/h, which is too fast to be accounted for by cytoplasmic streaming. The most acceptable model for phloem movement is the mass flow hypothesis (also known as the pressure flow hypothesis). Phloem sap moves by bulk flow, which creates a pressure (hence the term 'pressure-flow'). The key elements in this model are outlined below and right. For simplicity, the cells that lie between the source (and sink) cells and the phloem sieve-tube have been omitted.

1. Loading sugar into the phloem increases the solute concentration inside the sieve-tube cells. This causes the sieve-tubes to take up water by osmosis.

2. The water uptake creates a hydrostatic pressure that forces the sap to move along the tube, just as pressure pushes water through a hose.

3. The pressure gradient in the sieve tube is reinforced by the active unloading of sugar and consequent loss of water by osmosis at the sink, e.g. root cell.

4. Xylem recycles the water from sink to source.

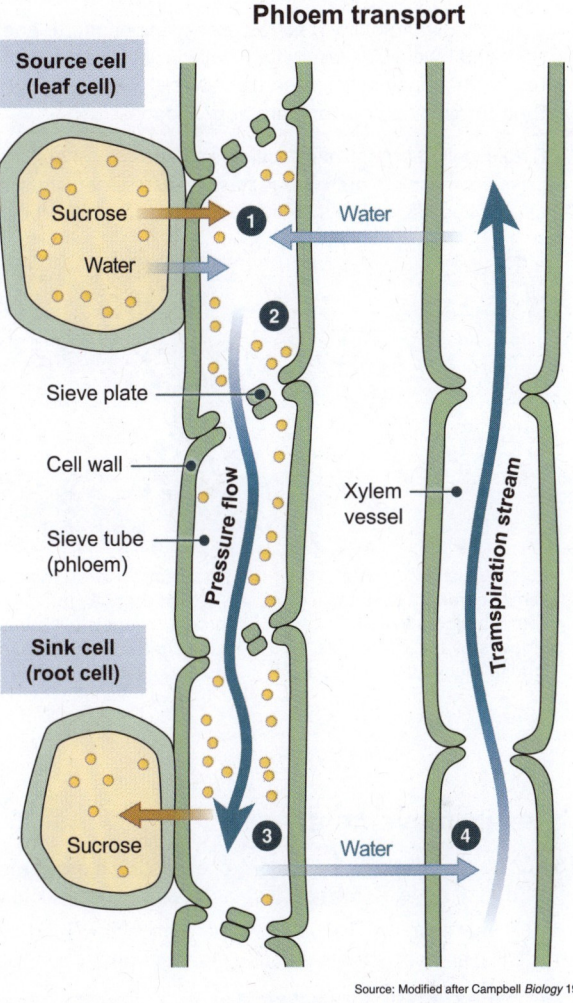

Phloem transport

Measuring phloem flow
Aphids can act as natural phloem probes to measure phloem flow. The sucking mouthparts (stylet) of the insect penetrates the phloem sieve-tube cell. While the aphid feeds, it can be severed from its stylet, which remains in place and continues to exude sap. Using different aphids, the rate of flow of this sap can be measured at different locations on the plant.

Source: Modified after Campbell *Biology* 1993

1. (a) From what you know about osmosis, explain why water follows the sugar as it moves through the phloem:

(b) What is meant by 'source to sink' flow in phloem transport?

2. Discuss how adaptations found in phloem sieve tubes and companion cells enable the translocation of sap:

121 Movement

Key Idea: Movement is a key life function in organisms and they exhibit a range of adaptations to enable it.

All organisms move in their environment so they can obtain resources, such as food or mates, or to avoid harm, such as predation. Motile organisms have adaptations that allow movement from one location to another. This group includes most multicellular animals, but also some single-celled organisms, and indeed some specialized cell types within animals, such as phagocytes. Sessile organisms are fixed in one location, and include some simple animals, plants, and fungi. However, they have adaptations that allow them movement to utilize resources in their direct environment, such as moving tentacles in anemone which direct food into their mouths, and plant leaves that rotate towards light.

Motile movement

▸ Self-propelled motility requires energy expenditure. Energy is provided by cellular respiration, which allows muscular tissue to contract against the fixed framework to which it is attached, therefore moving one or more appendages.

▸ Single-celled organisms, such as bacteria or protists, do not possess muscles, but instead have cellular extensions, such as cilia or flagella, which they move in a coordinated process.

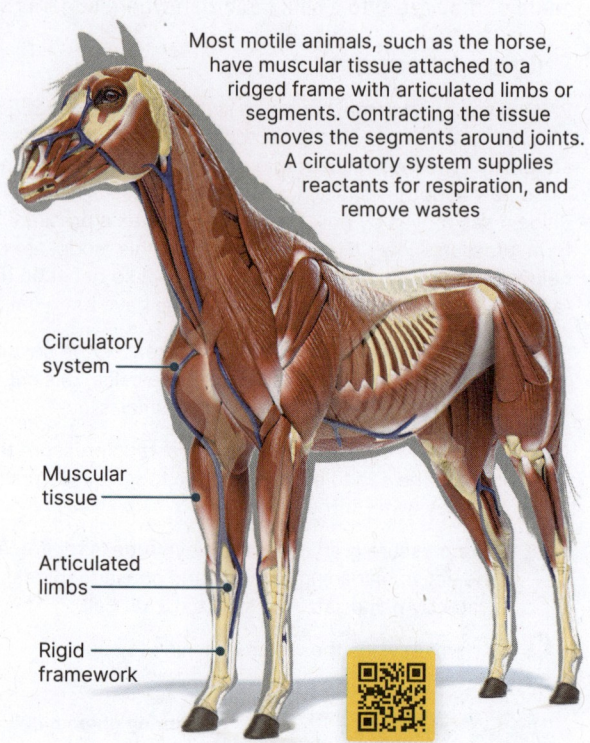

Most motile animals, such as the horse, have muscular tissue attached to a ridged frame with articulated limbs or segments. Contracting the tissue moves the segments around joints. A circulatory system supplies reactants for respiration, and remove wastes

Circulatory system

Muscular tissue

Articulated limbs

Rigid framework

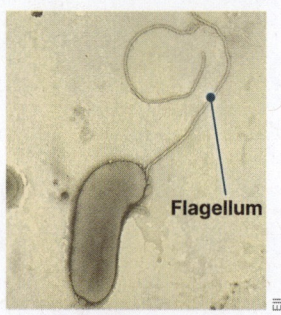

Flagellum

Helicobacter pylori, is a vibrio bacterium that causes stomach ulcers in humans. It moves by means of polar flagella (flagellum = singular).

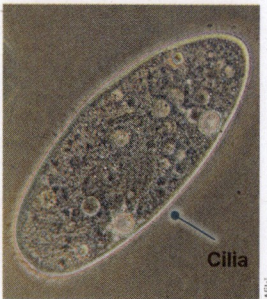

Cilia

Paramecium aurelia, is a protist that can move through water by beating the thousands of cilia surround the outer layer.

Sessile movement

▸ Sessile organisms are fixed in one location but can move parts of their structures, either rapidly or slowly, in order to obtain food or mate. Some species have a motile planktonic larval stage in order to assist with dispersal.

▸ Animal examples include sponges, which filter feed by drawing particles in with coordinated movement of flagellated collar cells, and sea squirts that use muscular propulsion to suck in water containing food particles.

Barnacles can be found on coasts throughout the world. They are sessile, and permanently attached to seashore rocks, so use protruding feather-like cirri to waft in food (above).

Sea anemones fix themselves to the rock by their adhesive foot. Unlike barnacles and sponges, they can move slowly over the rock surface. Feeding is done with a ring of tentacles around the mouth.

Sponges are completely immobile. Special cells called choanocytes have flagella that produce a water current within the sponge that draws water and nutrients into and out of the sponge.

1. Name one example of a motile organism and one example of a sessile organism. Prepare a report, including text and diagrams, to explain the movement adaptations that each species has that allow them to obtain food or for defence.

AHL

B3.3
1, 9

19

Reasons for locomotion

▸ Species have evolved specialized locomotion adaptations to enhance their ability to survive. Locomotion requires energy expenditure, so the advantages of its use need to outweigh the expense of obtaining the energy required. Advantages include seeking prey or avoiding predators, finding mates, and migration to areas with better resources.

Migrating slime moulds

Cellular slime moulds spend much of their life cycle as individual, amoeboid-like cells, often living in the soil and feeding on bacteria. They move by extending areas of their cell membrane with cytoplasm. When food becomes scarce, the single cells group together to form a motile slug, which is capable of movement. The slug may move some distance before finding a suitable area to form a sessile fruiting body. The fruiting body is held a few millimetres off the ground by a stalk. The cells in the stalk die, whereas the cells in the fruiting body form spores, which will be dispersed to new areas.

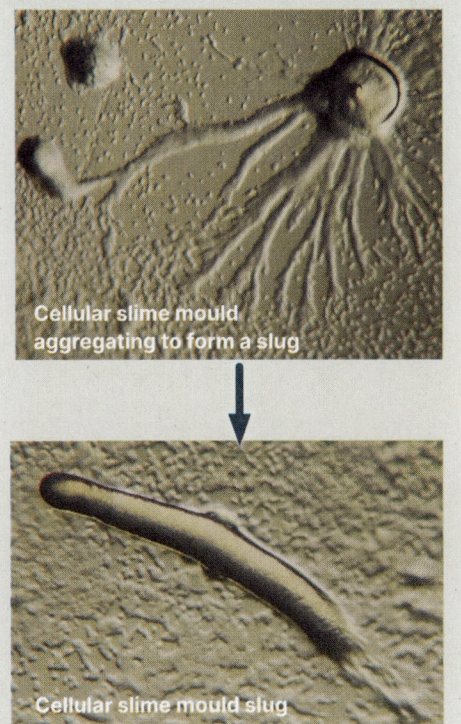

Cellular slime mould aggregating to form a slug

Cellular slime mould slug

Speed begets speed

▸ Cheetahs are the fastest terrestrial animal and can reach top speeds of up to 120 km per hour while chasing prey. They have the longest legs to body proportion of all cats, a lengthened, flexible spine, and a long tail to add balance when turning sharply. These features enable extremely fast locomotion.

▸ Prey such as antelope also have adaptations for speed to avoid being caught. Long slender legs, and a small streamlined body, similar to features seen in the cheetah, allow them to be fast and manoeuvrable.

Endurance to find mates

▸ Tiny, ruby-throated hummingbirds migrate over 3000 km from Central America to summer breeding grounds in mid-North America in just two weeks. To fuel this flight, the hummingbird almost doubles its body weight before travelling.

▸ The birds have a tiny wingspan of 10 cm but beat their wings up to 80 times per second, using their strong, short humerus bone to stabilize them. Their shoulder joint is extremely flexible, allowing the wings to rotate nearly 180 degrees, producing an efficient, figure-of-eight movement.

2. Suggest how both a predator such as a cheetah, and prey such as antelope have co-evolved to move at high speeds:

_____

_____

_____

_____

3. Compare advantages and disadvantages of locomotion in animals, using some examples to illustrate your answer:

_____

_____

_____

_____

122 Swimming Adaptations in Marine Mammals

Key Idea: Marine mammals have adaptations that allow them to swim efficiently.

All marine mammals have adaptations that allow them to maintain an oxygen supply to the tissues while submerged. The need for efficient swimming has created selection pressures for a streamlined body plan, where the limbs originally used on land in ancestral groups are modified into propelling flippers and tail flukes. The same selection pressure on marine mammals often results in convergence of external appearance, despite origin from different groups.

Sperm whales (3000 m)

Weddell seals (1000 m)

Pantropical spotted dolphins (300 m)

Dugong (10 m)

Adaptations for diving

▶ Diving mammals have physiological adaptations that enable them to stay underwater. Dolphins, whales, seals and, to a lesser extent, dugongs (Australia) and manatees (Northern Hemisphere), are among the most well adapted diving animals (diving depth in captions above). They exhale before diving, expelling most of the air from their lungs. In deep divers, the flexible rib cage allows the lungs to be compressed at depth so that only the trachea contains air. This stops nitrogen entering the blood and prevents decompression sickness ('the bends') when surfacing.

▶ During dives, heart rate slows and blood flow is redistributed to critical organs (plot, below left). Most diving mammals have high levels of myoglobin, an oxygen-binding protein found in skeletal muscle. Sperm whales are the deepest divers (3000 m) and Weddell seals dive to 1000 m for 40 minutes or more (plot below, right). During these dives, heart rate drops to 4 or 5 beats per minute (4% of the rate at the surface).

▶ Dugongs and manatees, which graze on the ocean floor, are also well adapted for diving, but their dives are generally shallow feeding dives (~3 m) and their muscles do not contain the high concentrations of myoglobin, typical of deep divers.

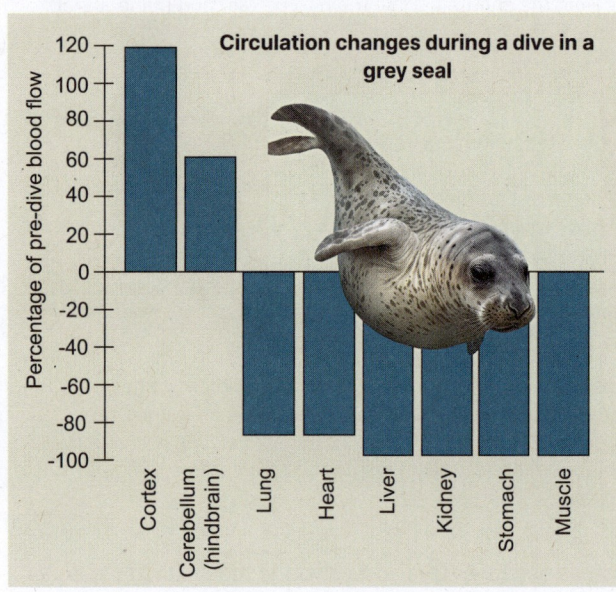
Circulation changes during a dive in a grey seal
(y-axis: Percentage of pre-dive blood flow; x-axis categories: Cortex, Cerebellum (hindbrain), Lung, Heart, Liver, Kidney, Stomach, Muscle)

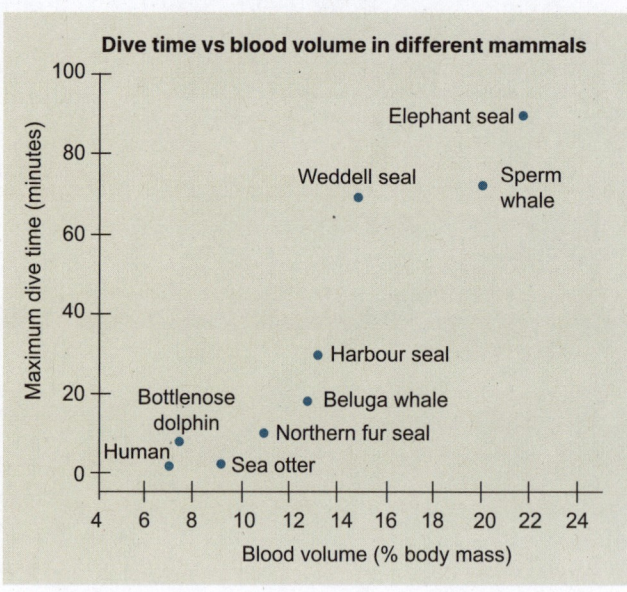
Dive time vs blood volume in different mammals
(y-axis: Maximum dive time (minutes); x-axis: Blood volume (% body mass))
Elephant seal, Weddell seal, Sperm whale, Harbour seal, Beluga whale, Northern fur seal, Bottlenose dolphin, Human, Sea otter

1. Mammals that have evolved adaptations for diving exhale before they dive (humans don't do this, we inhale). How might this be an advantage to diving mammals?

_____

2. What is the relationship between blood volume as a percentage of body mass and the maximum dive time in mammals with adaptations for diving?

_____

3. Explain the changes in blood flow in a grey seal during a dive: _____

_____

_____

B3.3
10

©2024 **BIOZONE** International
ISBN: 978-1-99-101410-8
Photocopying prohibited

Adaptations for swimming

▸ Mammals that have evolved for a marine niche can have similar adaptations that enable them to swim efficiently.

▸ The extent of similarity between adaptations can depend upon the length of time spent submerged in water or the length of time since the last common ancestor with a land mammal. Time submerged can range from from almost 100% of the time for whales and dolphins to 5 min bursts in sea otters,

▸ Pectoral fins have been adapted from forelimbs. Cetaceans (whales, dolphins) and sirenians (sea cows, manatees, and dugongs) have fully lost their hind limbs, and are fully aquatic. Pinnipeds (seals, sea lions, and walruses) have retained their hind limbs, albeit reduced in size, and spend time on land. Marine fissipeds (sea otters and polar bears) are non-related species, but both are excellent swimmers, yet spend significant time on land and have retained fully functional front and hind limbs with no flippers.

Adaptations in humpback whale (*Megaptera novaeangliae*)

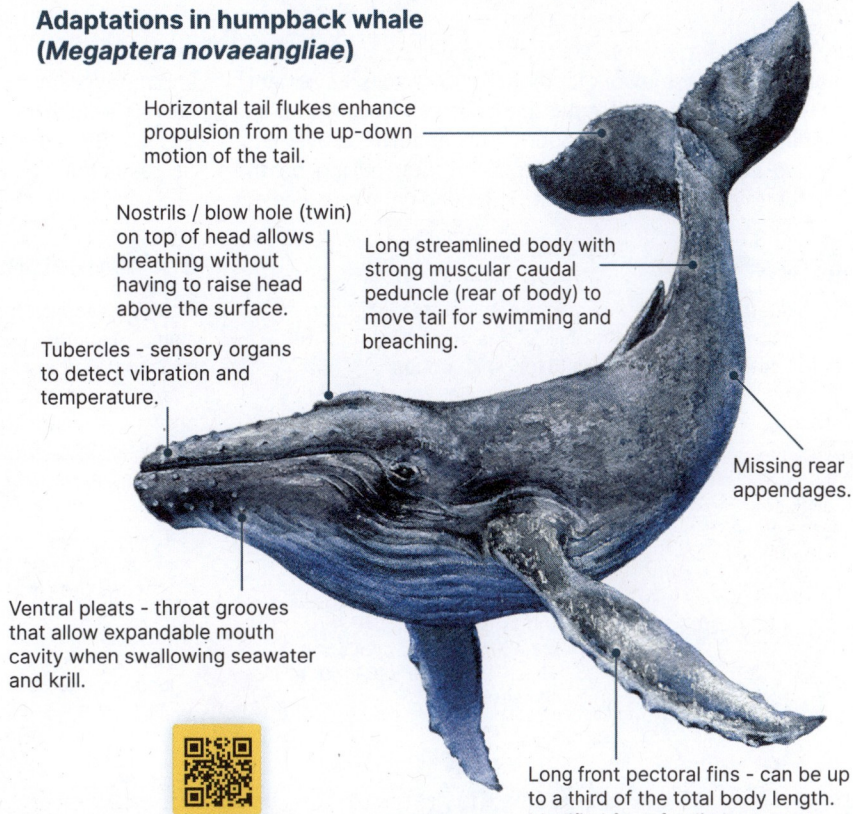

Horizontal tail flukes enhance propulsion from the up-down motion of the tail.

Nostrils / blow hole (twin) on top of head allows breathing without having to raise head above the surface.

Long streamlined body with strong muscular caudal peduncle (rear of body) to move tail for swimming and breaching.

Tubercles - sensory organs to detect vibration and temperature.

Missing rear appendages.

Ventral pleats - throat grooves that allow expandable mouth cavity when swallowing seawater and krill.

Long front pectoral fins - can be up to a third of the total body length. Modified from forelimbs.

Tail fluke movement of mammals

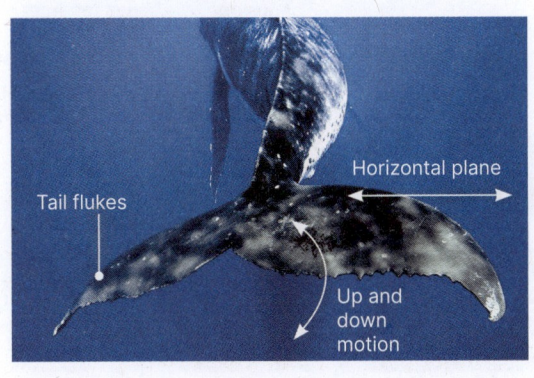

Tail flukes

Horizontal plane

Up and down motion

▸ All marine mammals with tail flukes propel themselves with an up and down motion. This is retained from their terrestrial ancestors that flexed their spine up and down when they walked, with their four limbs underneath their bodies. The tail fluke lies along the horizontal plane and provides propulsion from the movement of the tail.

▸ This motion is in contrast to many fish, who have tail fins with a vertical plane, and lateral undulation (wave like motion). These fins propel from side-to-side and are not adapted from a terrestrial ancestor with limbs.

▸ Swimming reptiles such as crocodiles, sea-snakes, and marine iguanas, swim with lateral undulations. Like all reptiles they have retained the horizontal flexing of the spine (as seen in fish) to advance the limbs and provide propulsion.

4. What key adaptations are required for a marine habitat in mammals? _____

_____

_____

5. Why have pinnipeds retained their external hind limbs, while cetaceans and sirenians have lost theirs?

_____

_____

6. Explain the reason for difference in motion of horizontal-plane mammalian flukes and vertical-plane fish and reptile fins:

_____

_____

_____

_____

123 Exoskeletons and Endoskeletons

Key Idea: Skeletons give support to organisms and provide an attachment framework for muscles to enable movement. The skeleton is a rigid structure and has many functions including structure and support, and enabling movement. Muscles attached to the skeleton contract, pulling on the skeleton to generate movement. The bones and muscles act as a system of levers, with joints acting as a fixed point of leverage (or fulcrum), and the muscles applying the effort to move the load or resistance, e.g. the bone and associated tissue. Skeletons may be external to the body wall (exoskeleton) as in arthropods, or internal (endoskeleton) and lying inside the body wall, as in vertebrates.

Exoskeletons

Exoskeletons are found in many invertebrates including arthropods, corals, and molluscs. The composition of the exoskeleton varies between taxonomic groups.
In animals with exoskeletons, the muscles for movement are attached to the inner surface of the exoskeleton.

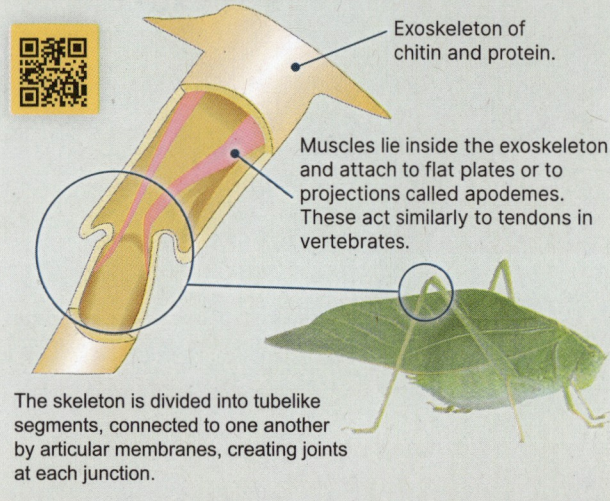

Exoskeleton of chitin and protein.

Muscles lie inside the exoskeleton and attach to flat plates or to projections called apodemes. These act similarly to tendons in vertebrates.

The skeleton is divided into tubelike segments, connected to one another by articular membranes, creating joints at each junction.

Endoskeletons

A bony endoskeleton is the internal support structure of some animals including vertebrates, sponges, and echinoderms. In vertebrates, it is composed mostly of calcium phosphate. The endoskeleton functions as an attachment site for muscles and provides a means to transmit muscular forces. Muscles pull on the skeleton to create movement about joints. Muscles work in opposing pairs to create opposing movement.

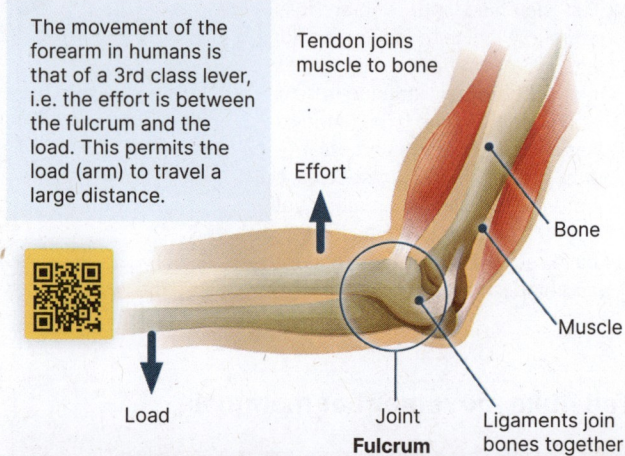

The movement of the forearm in humans is that of a 3rd class lever, i.e. the effort is between the fulcrum and the load. This permits the load (arm) to travel a large distance.

Tendon joins muscle to bone

Effort

Bone

Muscle

Load

Joint
Fulcrum

Ligaments join bones together

Insect flight

Insect moulting

Snake skeleton

Sea urchin

Arthropods are capable of many different types of movement, e.g. flight, walking, or burrowing through soil. All modes of locomotion are achieved through the action of muscles contracting and moving the jointed exoskeleton. The exoskeleton is rigid, so must be shed periodically to allow for growth. The new, larger skeleton is initially soft and hardens after the moult.

The rhythmic contraction of muscles in a snake act on the bones of its skeleton, allowing it to move across the ground. Although a sea urchin (an echinoderm) may look as though it has an exoskeleton, the spines are actually projections of the endoskeleton, which lies just below a layer of skin and muscles. Endoskeletons, being internal, can grow with the organism.

1. Compare and contrast the functional structure of endoskeletons and exoskeletons: _____

_____

_____

_____

2. Explain how skeletons act as both anchorage for muscles and as levers: _____

_____

_____

_____

_____

AHL
B3.3
5

©2024 **BIOZONE** International
ISBN: 978-1-99-101410-8
Photocopying prohibited

124 The Sliding Filament Theory

Key Idea: Muscles contract when thick and thin filaments slide past each other.

Muscle contraction is achieved by the thick and thin muscle filaments sliding past one another, made possible by their structure and arrangement. The ends of the thick myosin filaments are studded with heads or cross bridges that can link to the thin filaments next to them. The thin filaments contain the protein actin, but also a regulatory protein complex. When the cross bridges of the thick filaments connect to the thin filaments, a shape change moves one filament past the other. Two things are necessary for cross bridge formation: calcium ions, which are released from the

sarcoplasmic reticulum when the muscle receives an action potential, and ATP, which is hydrolyzed by ATPase enzymes on the myosin. When cross bridges attach and detach in sarcomeres throughout the muscle cell, the cell shortens. Although a muscle fibre responds to an action potential by contracting maximally, skeletal muscles can produce varying levels of contractile force. These graded responses are achieved by changing the frequency of stimulation and by changing the number and size of motor units recruited. Maximal contractions of a muscle are achieved when nerve impulses arrive at the muscle at a rapid rate and a large number of motor units are active at once.

The sliding filament theory

Muscle contraction requires calcium ions (Ca^{2+}) and energy (in the form of ATP) in order for the thick and thin filaments to slide past each other. The steps are:

1. The binding sites on the actin molecule, to which myosin 'heads' will locate, are blocked by a complex of two protein molecules: tropomyosin and troponin.

2. Prior to muscle contraction, ATP binds to the heads of the myosin molecules, priming them in an erect, high energy state. Arrival of an action potential causes a release of Ca^{2+} from the sarcoplasmic reticulum. The Ca^{2+} binds to the troponin and causes the blocking complex to move so that the myosin binding sites on the actin filament become exposed.

3. The heads of the cross-bridging myosin molecules attach to the binding sites on the actin filament. Release of energy from the hydrolysis of ATP accompanies the cross bridge formation.

4. The energy released from ATP hydrolysis causes a change in shape of the myosin cross bridge, resulting in a bending action (the power stroke). This causes the actin filaments to slide past the myosin filaments towards the centre of the sarcomere.

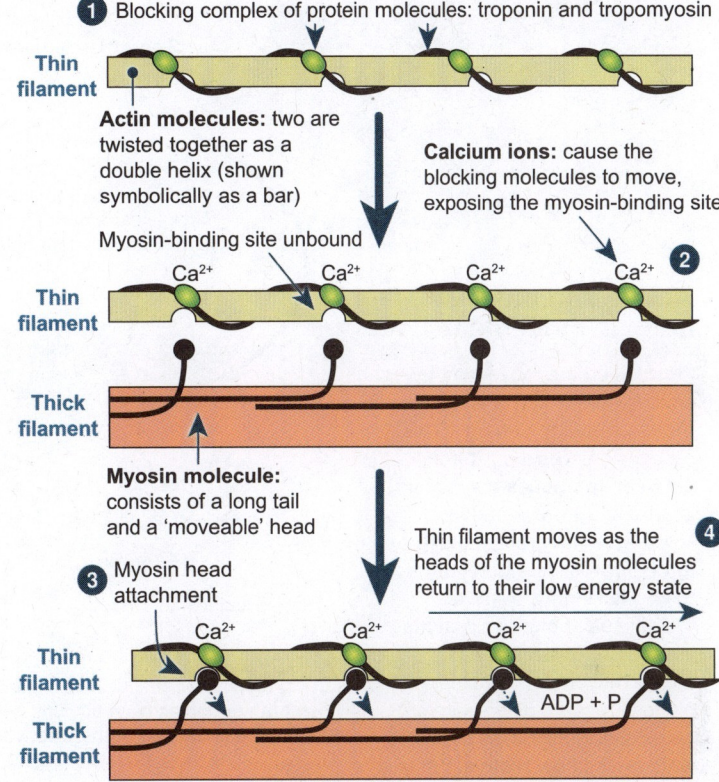

5. (Not illustrated). Fresh ATP attaches to the myosin molecules, releasing them from the binding sites and repriming them for a repeat movement. They become attached further along the actin chain as long as ATP and Ca^{2+} are available.

1. Test your vocabulary by matching each term to its correct definition, as identified by writing the letter in the correct box.

(i) Myosin ☐ A Bind to the actin molecule in a way that prevents myosin head from forming a cross bridge

(ii) Actin ☐ B Supplies energy for the flexing of the myosin 'head' (power stroke)

(iii) Calcium ions ☐ C Has a moveable head that provides a power stroke when activated

(iv) Troponin-tropomyosin ☐ D Two protein molecules twisted in a helix shape that form the thin filament of a myofibril

(v) ATP ☐ E Cause binding site to become 'unblocked'

2. Describe the two ways in which a muscle, as a whole, can produce contractions of varying force:

(a) _____

(b) _____

_____

125 Skeletal Muscle Structure and Function

Key Idea: Skeletal muscle is organized into bundles of muscle cells or fibres. The muscle fibres are made up of repeating contractile units called sarcomeres.

Skeletal muscle is organized into bundles of muscle cells or fibres. Each fibre is a single cell with many nuclei and each fibre is itself a bundle of smaller myofibrils arranged lengthwise. Each myofibril is, in turn, composed of two kinds of myofilaments (thick and thin), which overlap to form light and dark bands. It is the orderly alternation of these light and dark bands which gives skeletal muscle its striated or striped appearance. The sarcomere, bounded by dark Z lines, forms one complete contractile unit.

Structure of muscle

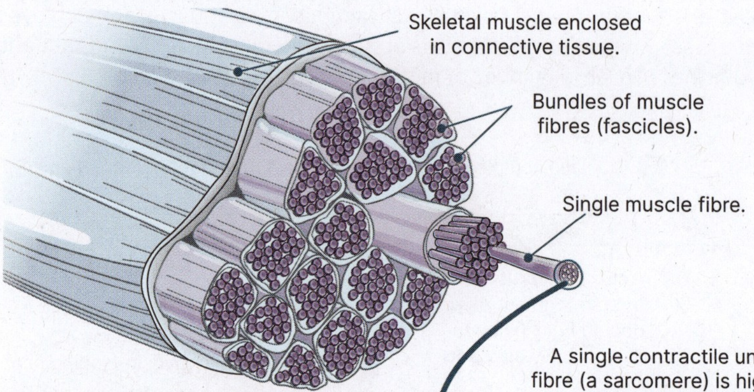

Skeletal muscle enclosed in connective tissue.

Bundles of muscle fibres (fascicles).

Single muscle fibre.

Skeletal muscle

Skeletal muscle is controlled by the peripheral central nervous systems, and is under voluntary control for contraction.

Skeletal muscle is involved with purposeful movement, maintenance of posture, stability of joints, and homoeostatic body temperature control (i.e. shivering when cold).

Structure of a muscle fibre (cell)

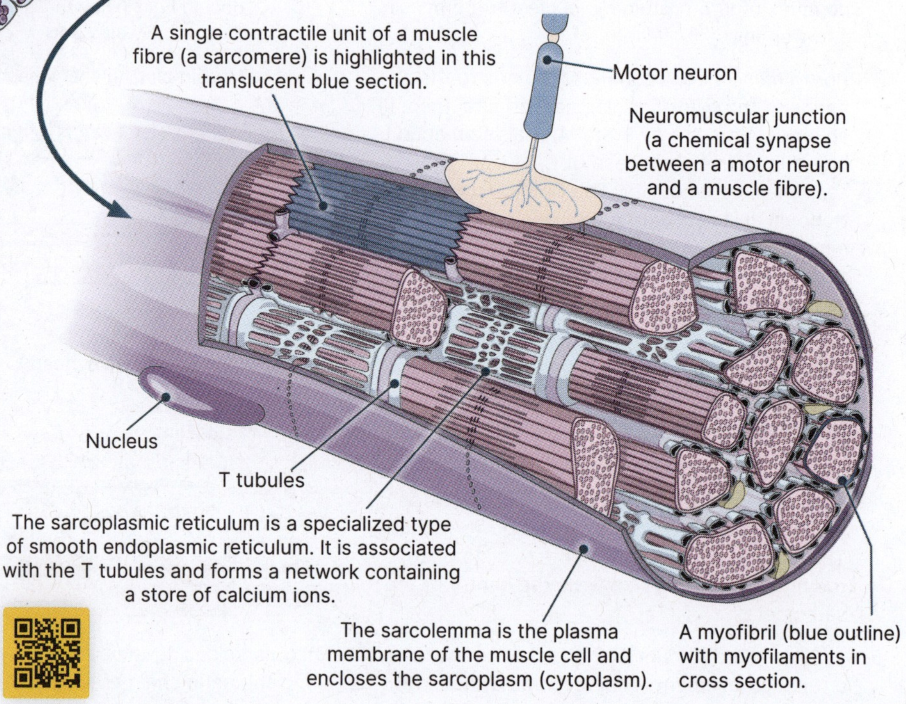

A single contractile unit of a muscle fibre (a sarcomere) is highlighted in this translucent blue section.

Motor neuron

Neuromuscular junction (a chemical synapse between a motor neuron and a muscle fibre).

Nucleus

T tubules

The sarcoplasmic reticulum is a specialized type of smooth endoplasmic reticulum. It is associated with the T tubules and forms a network containing a store of calcium ions.

The sarcolemma is the plasma membrane of the muscle cell and encloses the sarcoplasm (cytoplasm).

A myofibril (blue outline) with myofilaments in cross section.

Active and passive muscle movement

Muscles create movement by contraction, which is an active process requiring energy. This process creates a pull force between the bones the muscle is attached to, via tendons. Muscle cannot create a push force but will instead relax to original length passively. Multiple pairs of muscles acting in different directions can create movement in multiple planes. These muscle pairs are called antagonistic, as they work against each other. While one muscle contracts, the other relaxes, and vice-versa.

The function of titin

Titin is a large protein molecule attached to the end of each thick myosin filament and connected to the Z line. It provides stability to the sarcomere.

Titin has 'elastic' properties: when put under force, it can expand in length, storing potential energy. It can return to its original length once the force is removed.

This property of titin is important when the sarcomere is stretched during active contraction of an antagonist muscle or through gravitational forces creating stretch. The titin molecule 'unfolds' to lengthen and prevents overstretching of the sarcomere, and then refolds to provide a passive method of sarcomere recoil.

The titin folding / unfolding mechanism works in synchrony with the actin / myosin system to add power output.

The complete process is still yet to be fully understood. Ongoing research will assist with understanding how all the components work together to create movement.

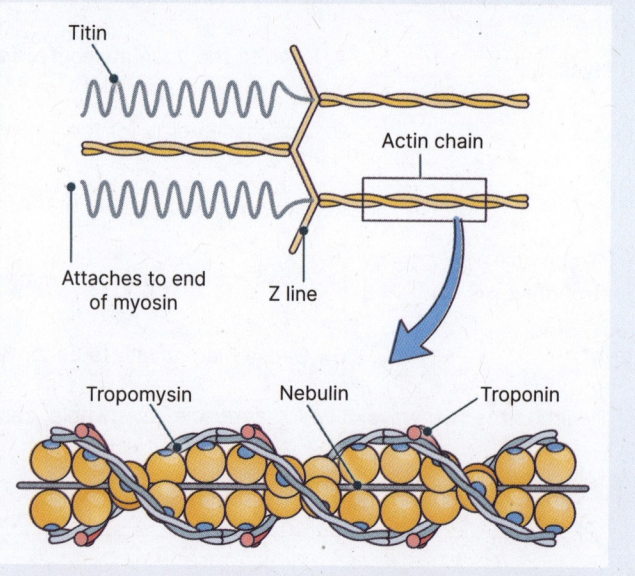

Titin

Actin chain

Attaches to end of myosin

Z line

Tropomysin

Nebulin

Troponin

B3.3

3 - 4

©2024 **BIOZONE** International
ISBN: 978-1-99-101410-8
Photocopying prohibited

The banding pattern of myofibrils

Within a myofibril, the thin filaments, held together by the Z lines, project in both directions. The arrival of an action potential sets in motion a series of events that cause the thick and thin filaments to slide past each other. This is called contraction and it results in shortening of the muscle fibre. It is accompanied by a visible change in the appearance of the myofibril: the I band and the sarcomere shorten and H zone shortens or disappears (below).

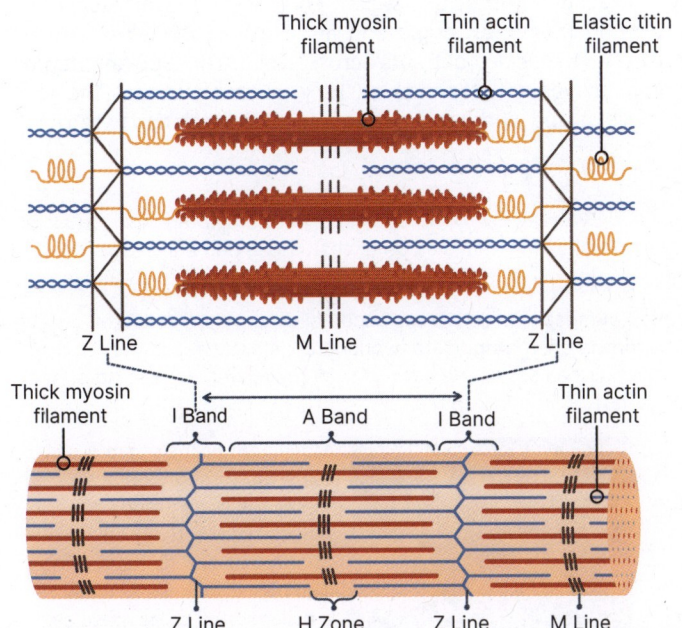

Thick myosin filament · Thin actin filament · Elastic titin filament

Z Line · M Line · Z Line

Thick myosin filament · I Band · A Band · I Band · Thin actin filament

Z Line · H Zone · Z Line · M Line

Longitudinal section of a sarcomere

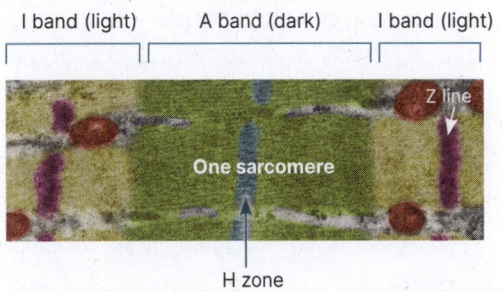

I band (light) · A band (dark) · I band (light)

One sarcomere · Z line

H zone

The photograph of a sarcomere (above) shows the banding pattern arising as a result of the highly organized arrangement of thin and thick filaments. It is represented schematically in longitudinal section and cross section.

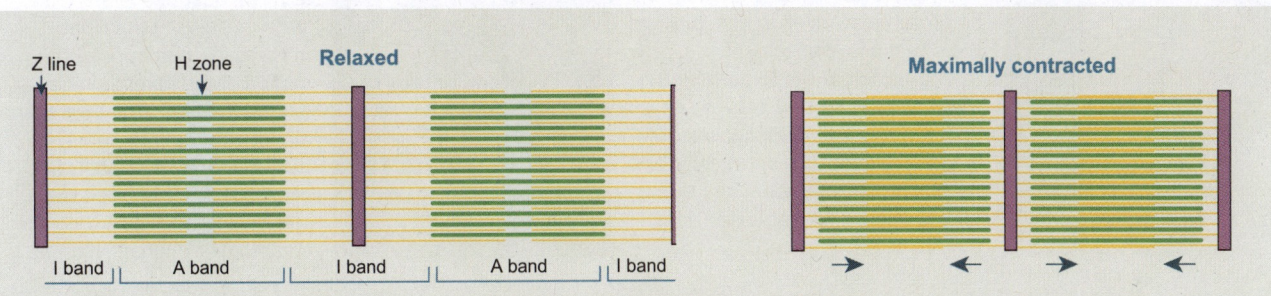

Relaxed

Z line · H zone

I band · A band · I band · A band · I band

Maximally contracted

1. (a) Explain the cause of the banding pattern visible in striated muscle: _____

    _____

    (b) Explain the change in appearance of a myofibril during contraction with reference to the following:

    The I band _____

    The H zone _____

    The sarcomere _____

2. Study the electron micrograph of the sarcomere (top, right).

    (a) Is it in a contracted or relaxed state (use the diagram, in the box, to help you decide)? _____

    (b) Explain your answer: _____

3. Describe the role of the protein titin in the process of muscle relaxation: _____

    _____

    _____

    _____

    _____

    _____

126 Antagonistic Muscles

Key Idea: Antagonistic muscles are muscle pairs that have opposite actions to each other. Together, their opposing actions bring about movement of body parts.

In both vertebrates and invertebrates, muscles provide the contractile force to move body parts. Muscles create movement of body parts when they contract across joints. Because muscles can only pull and not push, most body movements are achieved through the action of opposing sets of muscles called antagonistic muscles. Antagonistic

muscles function by producing opposite movements: as one muscle contracts (shortens), the other relaxes (lengthens). Skeletal muscles are attached to the skeleton by tough connective tissue structures (tendons in vertebrates or attachment fibres in insects). They always have at least two attachments: an origin and an insertion. Body parts move when a muscle contracts across a joint. The type and degree of movement depends on how much movement the joint allows and where the muscle is located in relation to the joint.

Opposing movements require opposing muscles

▶ The flexion (bending) and extension (unbending) of limbs is caused by the action of antagonistic muscles. Antagonistic muscles work in pairs and their actions oppose each other. During movement of a limb, muscles other than those primarily responsible for the movement may be involved to fine tune the movement.

▶ Every coordinated movement in the body requires the application of muscle force. This is accomplished by the action of agonists, antagonists, and synergists. The opposing action of agonists and antagonists (working constantly at a low level) also produces muscle tone. Note that either muscle in an antagonistic pair can act as the agonist or prime mover, depending on the particular movement (for example, flexion, or extension).

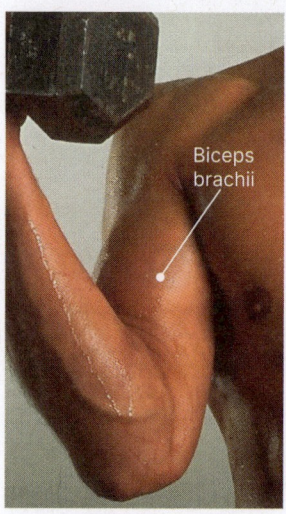

Biceps brachii

Agonists or prime movers are primarily responsible for the movement and produce most of the force required.
Antagonists oppose the prime mover. They may also play a protective role by preventing over-stretching of the prime mover.
Synergists assist the prime movers and may be involved in fine-tuning the direction of movement. During flexion of the forearm (left) the brachialis muscle acts as the prime mover and the biceps brachii is the synergist. The antagonist, the triceps brachii at the back of the arm, is relaxed. During extension, their roles are reversed.

Quadriceps

Hamstrings

Movement of the upper leg is achieved through the action of several large groups of muscles, collectively called the quadriceps and the hamstrings. The hamstrings are actually a collection of three muscles, which act together to flex the leg. The quadriceps are four large muscles at the front of the thigh that oppose the motion of the hamstrings and extend the leg. When the prime mover contracts forcefully, the antagonist also contracts very slightly. This stops over-stretching and allows greater control over thigh movement.

Internal and external intercostal muscles

▶ Ribs and costal cartilage comprise the thoracic cage, protecting the internal organs but also to enable mechanical breathing.

▶ The muscles between the ribs are antagonistic pairs. When we actively breathe - during exercise or higher demand for oxygen, inhalation occurs when the external intercostal muscle contracts to pull the ribs up and outwards to draw air into the lungs.

▶ The internal intercostal muscles contract to 'depress' the ribs and by reducing the thoracic space they cause exhalation. The innermost intercostal muscles also help with exhalation.

▶ The layers of muscle fibre for external and internal intercostal muscles are in different planes of orientation - therefore pulling ribs in different directions when contracting.

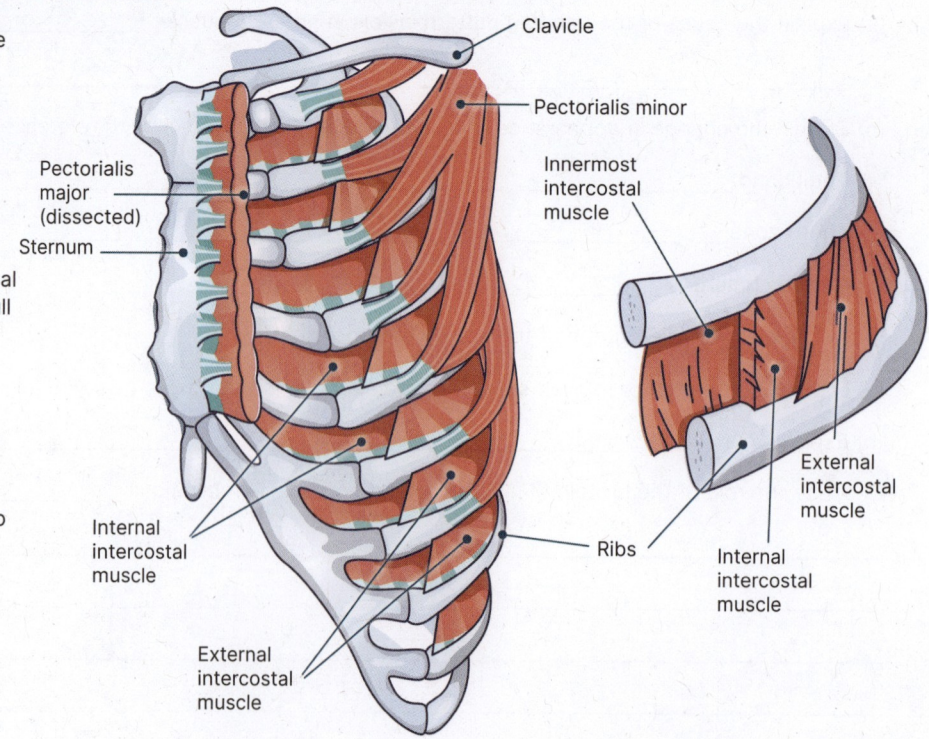

Clavicle

Pectorialis minor

Pectorialis major (dissected)

Sternum

Innermost intercostal muscle

External intercostal muscle

Internal intercostal muscle

Ribs

Internal intercostal muscle

External intercostal muscle

  B3.3 8

©2024 **BIOZONE** International
ISBN: 978-1-99-101410-8
Photocopying prohibited

Mechanism of antagonistic movement of the ribcage

▶ The internal and external intercostal muscles lie in different orientations to each other between the rib bones.

▶ The contraction of the intercostal muscles attached to the ribs generates torque. The direction of the torque determines whether the ribcage is lifted (external intercostal) or lowered (internal intercostal).

▶ Therefore, contraction of one set of intercostal muscles causes stretching in the other. As detailed in the previous activity, the protein titin, attached to the end of each myosin filament in the sarcomere, can store up potential energy when stretched and release it once more to assist the next cycle of contraction. This passive mechanism provides a 'passive' power boost to antagonistic muscle contraction without the use of active energy.

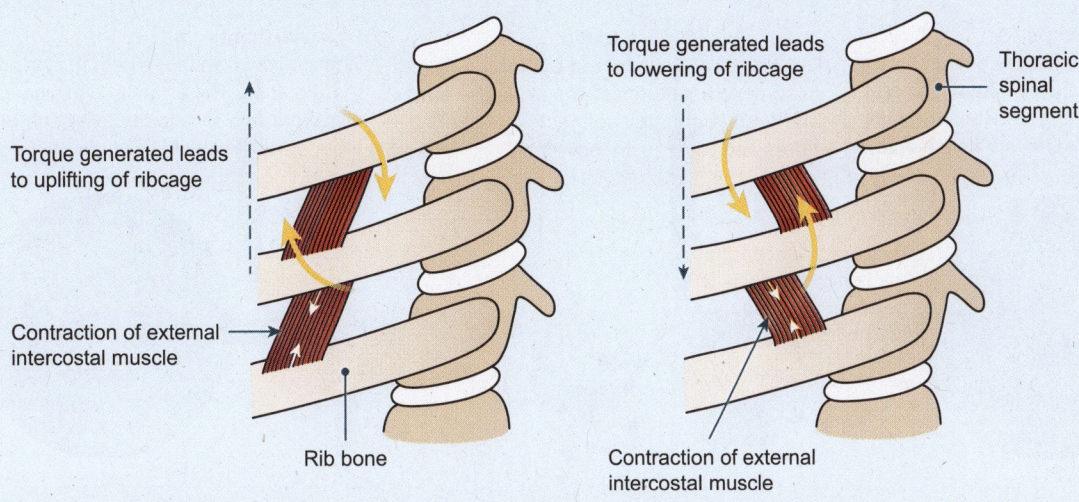

1. Name the muscle and describe the role of each of the following muscles in moving the ribcage up and down in humans:

   (a) Prime mover: _____

   _____

   (b) Antagonist: _____

   _____

   (c) Synergist: _____

   _____

2. Explain why the muscles that cause movement of body parts tend to operate as antagonistic pairs:

   _____

   _____

3. Summarize the action of the antagonistic muscle pairs in enabling inhalation and exhalation:

   _____

   _____

   _____

   _____

4. Why is it important that internal and external intercostal muscles are in different orientations?

   _____

   _____

5. How does titin contribute to the effectiveness of skeletal muscle contraction in antagonistic muscles?

   _____

   _____

127 Movement About Joints

Key Idea: A joint is the junction where two or more bones meet. All movements of the skeleton occur at joints.

Bones are too rigid to bend. To allow movement, the human skeletal system consists of bones held together at joints by flexible connective tissues called ligaments. Joints are points of contact between bones or between cartilage and bones. Joints may be classified structurally as fibrous, cartilaginous, or synovial (below). Each of these joint types allows a certain degree of movement. Bones move about a joint by the force of muscles acting upon them. Synovial joints are the most common and most movable joints in the body and have a fluid-filled joint capsule surrounding the articulating surfaces of the bones. The most freely movable synovial joints are also the least stable and the most prone to injury. Restricting the amount of movement gives less freedom, but also makes the joint more stable. Range of movement in a joint can be measured with a goniometer, a simple instrument that measures angles, or from images.

Cartilaginous Joints

Here, the bone ends are connected by cartilage. Most allow limited movement, although some, e.g. between the first ribs and the sternum, are immovable.

Immovable Fibrous Joints

The bones are connected by fibrous tissue. In some, e.g. sutures of the skull, the bones are tightly bound by connective tissue fibres and there is no movement.

Synovial Joints

These allow free movement in one or more planes. The articulating bone ends are separated by a joint cavity containing lubricating synovial fluid (see next page).

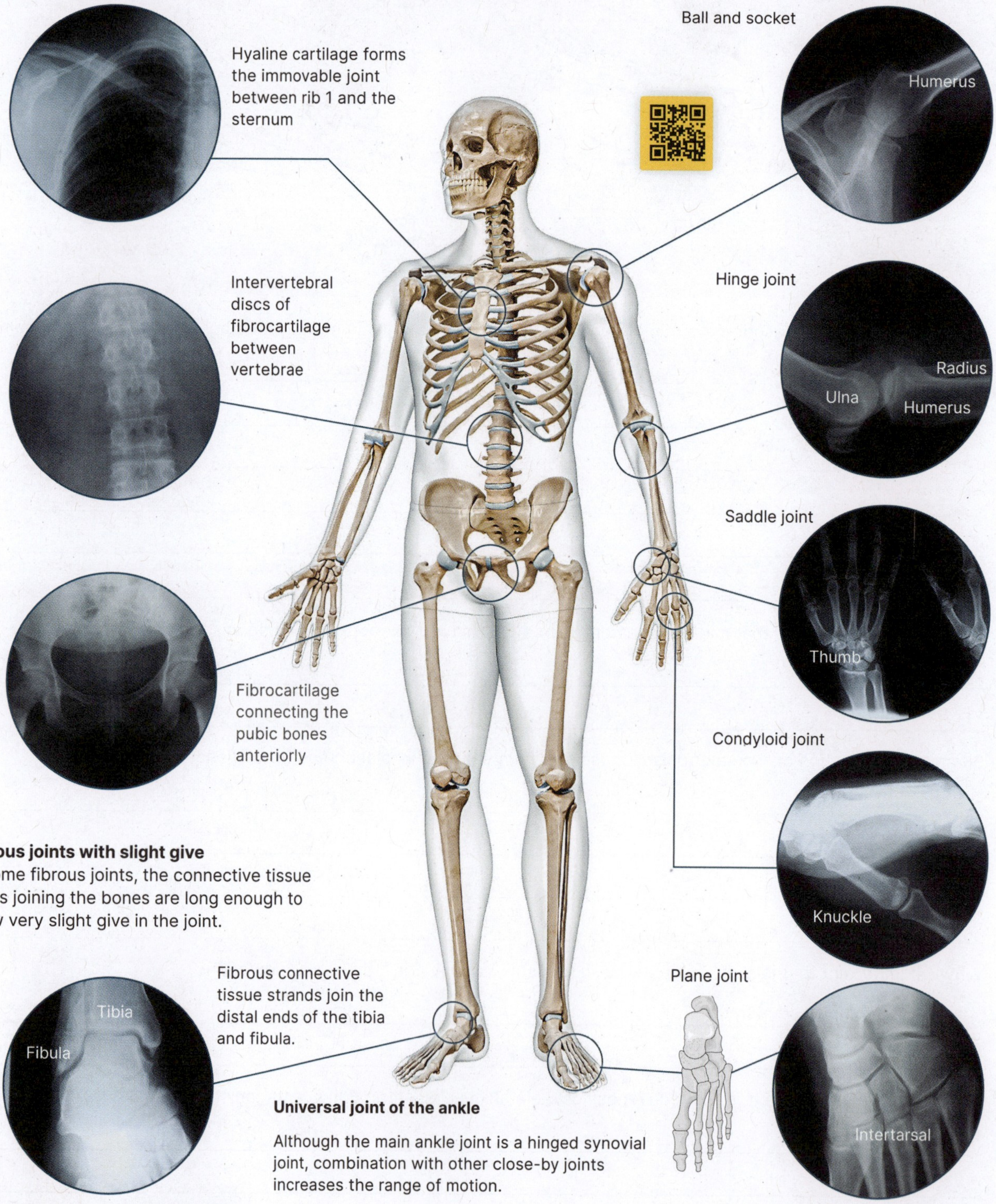

Hyaline cartilage forms the immovable joint between rib 1 and the sternum

Intervertebral discs of fibrocartilage between vertebrae

Fibrocartilage connecting the pubic bones anteriorly

Ball and socket

Humerus

Hinge joint

Radius

Ulna   Humerus

Saddle joint

Thumb

Condyloid joint

Knuckle

Plane joint

Intertarsal

Fibrous joints with slight give

In some fibrous joints, the connective tissue fibres joining the bones are long enough to allow very slight give in the joint.

Tibia

Fibula

Fibrous connective tissue strands join the distal ends of the tibia and fibula.

Universal joint of the ankle

Although the main ankle joint is a hinged synovial joint, combination with other close-by joints increases the range of motion.

AHL

AOS

B3.3
6 - 7

©2024 **BIOZONE** International
ISBN: 978-1-99-101410-8
Photocopying prohibited

Synovial joints at the hip

▸ The joint between the femur and pelvis is an example of a synovial joint, sometimes called a ball and socket joint. The synovial joints of the skeleton allow free movement in one or more planes. The articulating bone ends are separated by a joint cavity containing lubricating synovial fluid.

▸ Cartilage is a flexible connective tissue. It protects a joint surface against wear, but has no blood supply (it is avascular). It covers the femoral head and lines the hip socket. The hip joint also has a number of bursa, which are fluid filled cavities lined with synovial membrane. It acts as a cushion, e.g. between tendon and bone, or between bones.

▸ Muscles have an origin on one (less moveable) bone and an insertion on another (more moveable) bone. Tendons connect the muscle to different points of the bone. There are numerous muscles involved with the hip joint (see insert). Some muscles are present to move the femur, through extension, rotation, abduction (away from body), adduction (towards body), while others act to stabilize the hip joint.

▸ The articular capsule is composed of ligaments, a strong connective tissue, and attach from the neck of the femur to the lower region of the pelvis in a spiral fashion. The ligaments secure the femur to the pelvis, and prevent over extension of the hip.

▸ The fusion of pelvic bones (ischium, pubis, ilium) provides rigidity for the femur to lever against. Weight-bearing is the most important function of the pelvic girdle, so the bones are large and thick.

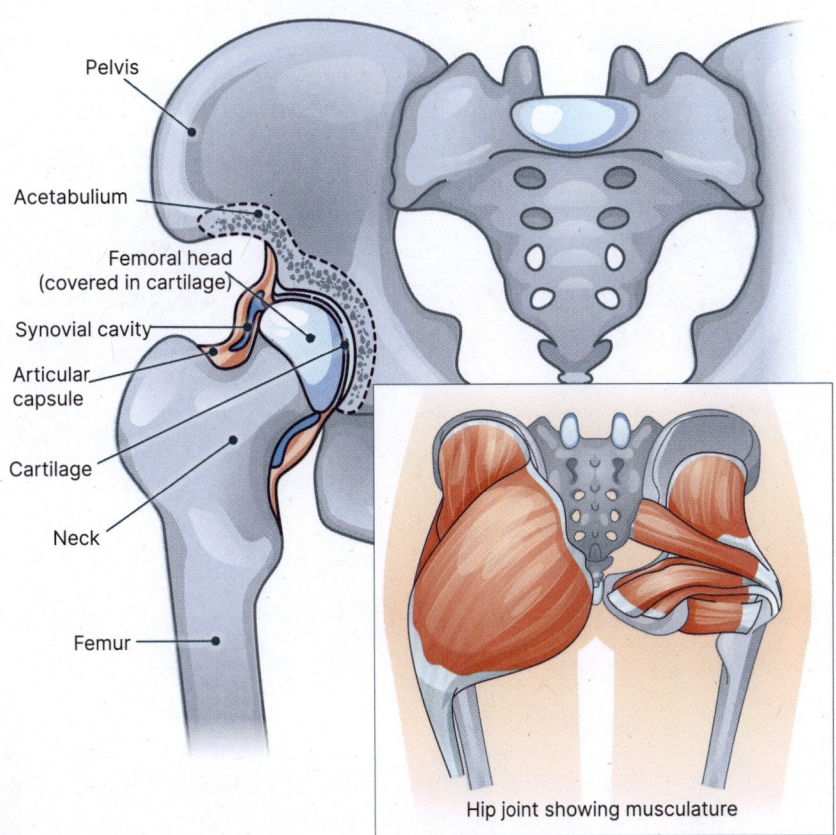

Pelvis
Acetabulium
Femoral head (covered in cartilage)
Synovial cavity
Articular capsule
Cartilage
Neck
Femur

Hip joint showing musculature

1. Classify each of the synovial joint models (A-E) below, according to the descriptors below:

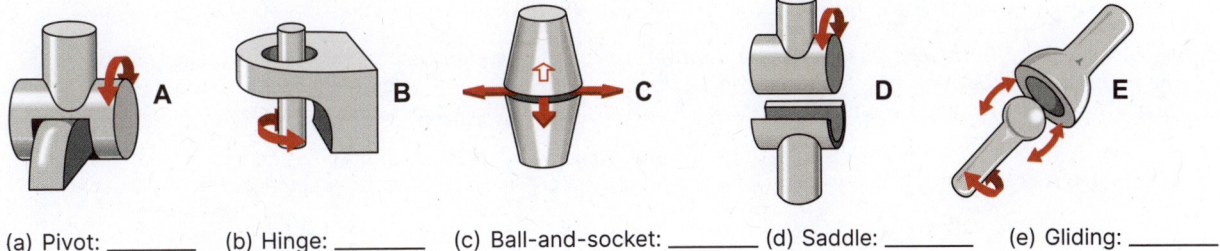

A   B   C   D   E

(a) Pivot: _____   (b) Hinge: _____   (c) Ball-and-socket: _____   (d) Saddle: _____   (e) Gliding: _____

2. Explain the role that synovial fluid and cartilage play in the structure and function of the hip joint:

_____

_____

_____

_____

_____

3. The femur needs to move through a wide range of motion without moving out of position in the pelvis. Explain how the muscles, tendons, and ligaments enable both movement and stability:

_____

_____

_____

_____

Range of motion in a joint

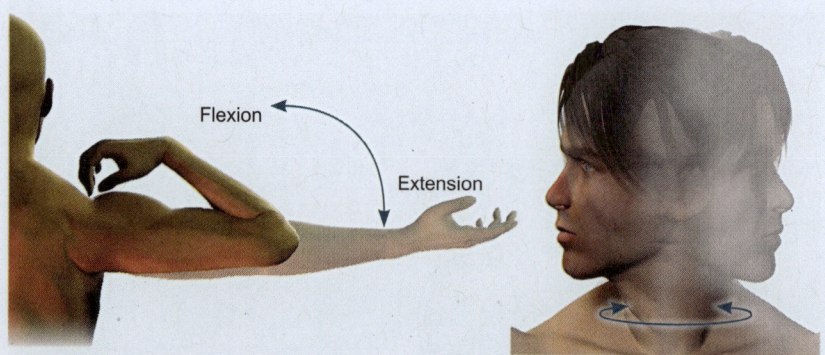
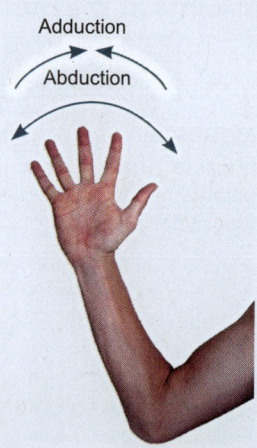

Flexion decreases the angle of the joint and brings two bones closer together. Extension is its opposite. Extension more than 180° is called hyperextension.

Rotation is movement of a bone around its longitudinal axis. It is a common movement of ball and socket joints and the movement of the atlas around the axis.

Abduction is a movement away from the mid line, whereas adduction describes movement towards the mid line. The terms also apply to opening and closing the fingers.

Investigation 7.5 Investigating range of motion in a shoulder joint

See appendix for equipment list.

In this practical, you will work in groups of three to measure the range of motion in a shoulder joint using a goniometer.

One person volunteers to be measured, one person will measure using a goniometer, and the third person will record.

CAUTION: The person exercising should have no known pre-existing medical conditions or injuries involving shoulders.

When measuring, keep the fulcrum of the goniometer on the fulcrum of the shoulder.

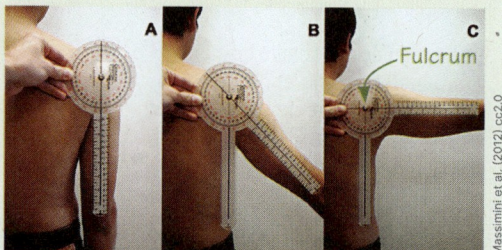

Measuring the motion of shoulder joints using a goniometer.

1. Shoulder abduction: Start with the arm in the downward position (A), with palm touching thigh, then continue to raise the arm sideways (B-C). Some students may be able to raise their arm above point C without difficulty. Take a measurement of the angle when fully flexed and record. Repeat three times and average measurement.

2. Shoulder flexion: Start with the arm in the downward position, rotated so that the palm is facing front of body, then continue to raise the arm forward. This position should result in a greater angle than step 1. Take a measurement of the angle when fully flexed and record. Repeat three times and calculate the mean.

3. Backwards flexion: Place arm by side (A). Flex the arm backwards, behind the body - keeping the arm straight. Take a measurement of the angle when fully flexed and record. Repeat three times and average measurement.

4. Given time, the group may wish to swap around roles and repeat measurements for each member.

Group member	1.1	1.2	1.3	Ave	2.1	2.2	2.3	Ave	3.1	3.2	3.3	Ave

4. Explain how the range of motion in the shoulder differs depending on the plane or position it is moved to, including the role of muscles, ligaments, and bones:

_____

_____

_____

©2024 **BIOZONE** International
ISBN: 978-1-99-101410-8
Photocopying prohibited

128 Did You Get It?

1. Cellular respiration requires gas exchange of oxygen gas from the external environment to move into the circulatory system. Compare and contrast the processes in fish and mammals, including the structures and concentration gradients at exchange surfaces.

Gas exchange in fish   Gas exchange in mammals

_____

_____

_____

_____

_____

_____

_____

2. What adaptations do the alveoli have for gas exchange? _____

_____

3. What type of blood vessels are found in the structures above, and how are they adapted to transport materials?

_____

_____

_____

4. **AHL:** At low blood pH, less oxygen is bound by haemoglobin and more is released to the tissues. Name this effect and comment on its significance for oxygen delivery to respiring tissue:

_____

_____

_____

5. Plants also respire. What role do stomata have in the control of gas exchange, and what are the consequences of this control relating to transpiration?

_____

_____

_____

6. **AHL:** What type of joint is shown to the right, and what are its defining features?

_____

_____

_____

_____

Clavicle

Subacromial bursa

Articular capsule

Tendon sheath

Synovial fluid

Head of humerus

Articular cartilage

Tendon

Synovial membrane

Biceps brachii muscle

7. **AHL:** Explain how the antagonistic muscles function to create movement in the shoulder:

_____

_____

_____

_____

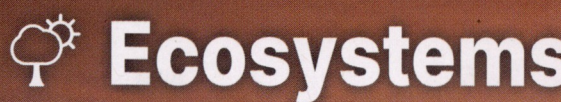

Ecosystems

B4.1 Adaptation to environment

		Activity Number

Guiding Questions:
▶ How are the adaptations and habitats of species related?
▶ What causes the similarities between ecosystems within a terrestrial biome?

Learning Outcomes:

		Activity Number
☐ 1	Define the term 'habitat' and provide a range of examples.	129
☐ 2	Investigate adaptations in organisms that have evolved in response to the abiotic environment, including a grass species adapted to sand dunes and a tree species adapted to mangrove swamps.	130
☐ 3	Analyse how abiotic variables can affect species distribution, determining a range of tolerance.	131
☐ 4	**AOS:** Collect data from a transect sample to correlate the distribution of plant or animal species with an abiotic variable. Use relevant sensors to collect abiotic data.	132-134
☐ 5	Describe the conditions required for the formation of coral reefs, including the water depth, temperature, pH, salinity, and clarity.	137
☐ 6	Link terrestrial biome distribution to the abiotic factors present in the environment using graphed climatic variable data.	135
☐ 7	Define the features of Earth's main biomes, including tropical forest, temperate forest, taiga, grassland, tundra, and hot desert biomes.	136
☐ 8	Investigate examples of plant and animal adaptations to hot deserts and tropical rainforest biomes.	138, 139

B4.2 Ecological niches

		Activity Number

Guiding Questions:
▶ What are the advantages of specialized modes of nutrition to living organisms?
▶ How are the adaptations of a species related to its niche in an ecosystem?

Learning Outcomes:

		Activity Number
☐ 1	Define an ecological niche, including the biotic and abiotic factors that influence it.	140
☐ 2	Classify organisms as obligate anaerobes, facultative anaerobes, or obligate aerobes, based on their tolerance of oxygen in their environment.	141
☐ 3	Identify the key groups of organisms, including plants, algae, and several groups of photosynthetic prokaryotes, that use photosynthesis as their mode of nutrition	142
☐ 4	Distinguish between holozoic and heterotrophic nutrition in animals.	142
☐ 5	Define mixotrophic nutrition, as used by some protists, and classify mixotrophic organisms as obligate or facultative.	142
☐ 6	Define saprotrophic nutrition, as used by some fungi and bacteria.	142
☐ 7	Investigate the diversity of nutrition in archaea, one of the three branches (domains) of life.	142
☐ 8	**AOS:** Examine models of hominidae skulls to investigate the relationship between dentition and diet. **NOS:** Extrapolate data from living mammals to extinct species.	143
☐ 9	Investigate a range of plant adaptations to resist herbivory, and a range of herbivore adaptations to overcome plant defences.	144
☐ 10	Investigate predator and prey adaptations, including chemical, physical, and behavioural adaptations.	145
☐ 11	Describe a range of plant adaptations for harvesting light, using rainforest plants as examples.	146
☐ 12	Distinguish between fundamental and realized niches.	140, 147
☐ 13	Explain the relationship between competitive exclusion and the uniqueness of ecological niches.	140, 148

129 Habitat

Key Idea: The environment in which an organism lives is its habitat. The habitat may not be homogeneous.

The environment in which an organism (or species) lives (including all the physical and biotic factors) is its habitat. The term habitat can be considered species specific. Each species has its own specific habitat which contains all it needs for its survival and is determined by factors such as temperature, moisture, light, and the availability of food and shelter. This is different from an ecosystem, which includes the interactions between the organisms and the environment. Habitats can vary widely in scale but they are always smaller than an ecosystem, as an ecosystem may include many habitats.

▶ A habitat may be a specific description of the area in which a species or organism lives or is found, such as the sandy edges of a lake. This includes, the range in temperature and light, and the possible community.

▶ A habitat may also be described as a type, e.g desert, boreal forest, or rocky shore. The more specific the description, the smaller the range of organisms that may live in that habitat type. For example, a forest habitat is a very general description. More specifically, a habitat type could be a tropical rainforest. However, within the tropical rainforest are again more specific habitats, such as the canopy, or the forest floor. Within those again, smaller more specific habitats can be described, such as the leaf litter and the various physical factors those include.

A forest (right) may contain several smaller habitats. The lake in the foreground forms one habitat, e.g. for fish, the clearing beyond it forms another, e.g. for grasses and flowering herbs. However, the forest as a whole is also a habitat, e.g. for deer.

The scale of available habitats

A habitat may be vast and relatively homogeneous for the most part, as is the open ocean. Barracuda (above) occur around reefs and in the open ocean where they are aggressive predators.

For non-motile organisms, such as the fungus pictured above, a suitable habitat may be defined by the particular environment in a relatively small area, such as on this decaying log.

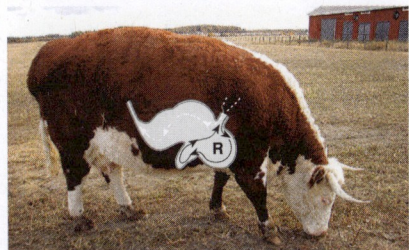

For microbial organisms, such as the bacteria and protozoans of the ruminant gut, the habitat is defined by the chemical environment within the rumen (R) of the host animal, in this case, a cow.

1. Distinguish clearly between a habitat and an ecosystem: _____

   _____

   _____

   _____

2. Explain how habitats can vary in scale: _____

   _____

   _____

   _____

3. A city can be described as both a habitat and an ecosystem. Explain how: _____

   _____

   _____

   _____

B4.1
1

130 Plant Adaptations

Key Idea: Habitats with extreme physical conditions require special adaptations to overcome.

Extreme environments place special physiological stresses on the organisms living in them, such as the requirement to deal with extreme heat, lack of water, high salt content, or extreme cold. Plants adapted to survive in these environments have special adaptations lacking in plants from more moderate environments. These adaptations can help the plant overcome the stress from the environment to such a degree, they often thrive in and dominant their environment.

Adaptations to conserve water

Plants adapted to dry conditions are called xerophytes. Xerophytes are found in a number of environments, but all show adaptations to conserve water. These adaptations include small, hard leaves, an epidermis with a thick cuticle, sunken stomata, succulence, and permanent or temporary absence of leaves.

▶ Most xerophytes are found in deserts, but they may be found in humid environments, provided that their roots are in dry micro-environments, e.g. the roots of epiphytic plants that grow on tree trunks or branches.

▶ Many xerophytes have a succulent morphology. Their stems are often thickened and retain a large amount of water in the tissues, e.g. Aloe.

▶ Many xerophytes have a low surface area to volume ratio, reducing the amount of water lost through transpiration.

▶ Salt tolerant plants and many alpine species may show xeromorphic features in response to the lack of free water and high transpirational losses in these often windy, exposed environments.

Adaptations to sand dunes

Sand dunes produce a number of challenges to overcome, including shifting sands, lack of water, exposure to winds and heat and, if coastal, high levels of salt. Adaptations vary, but as well as having many of the adaptations typical of xerophytes, dune plants may have a deep taproot or extensive shallow roots to anchor them in place (far right), grow from rhizomes, and grow low to the ground.

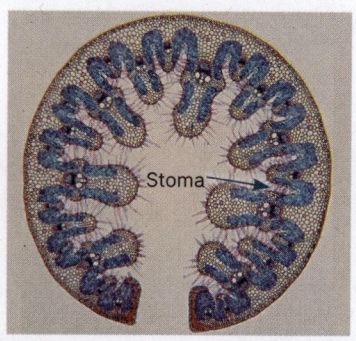

Grasses on coastal sand dunes, e.g. marram grass, above, curl their leaves. Stomata are sunken in pits, creating a moist microclimate around the pore, which reduces transpiration rate.

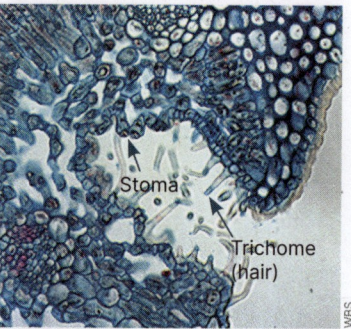

Oleander has a thick, multi-layered epidermis and the stomata are sunken in trichome-filled pits on the leaf underside which restrict water loss. Trichomes (leaf 'hairs') maintain a layer of still air at the leaf surface.

An outer surface coated in fine hairs traps air close to the surface and reduces the transpiration rate.

Dune grasses, e.g. marram grass, above, often have extensive rhizome systems. These help to hold their position and also stabilize the dune.

1. What is a xeromorphic adaptation? _____

_____

2. Describe three xeromorphic adaptations of plants that reduce water loss:

_____

_____

_____

3. (a) How does creating a moist microclimate around areas of water loss reduce the transpiration rate?

_____

_____

(b) How do trichomes contribute to reducing transpiration rate? _____

_____

_____

4. How does a low surface area to volume ratio in a plant such as a cactus reduce water loss?

_____

_____

B4.1
2

©2024 **BIOZONE** International
ISBN: 978-1-99-101410-8
Photocopying prohibited

Adaptations to high salt environments

Mangroves are halophytes, a group of plants with adaptations for growth in seawater or salty, water-logged soil. They grow in the upper part of the intertidal zone, but also extend further inland to form salt marshes and other coastal wetland communities.

Mangroves grow from the upper part of the intertidal zone to the high water mark, forming some of the most complex and productive ecosystems on Earth. The high salt environment would kill most other kinds of plants as high salt levels cause water to flow out of the cells. Mangroves overcome this by storing salt in their cell vacuoles and maintaining a high concentration of solutes in the cytoplasm of their cells. This reverses the osmotic gradient and maintains the transpiration stream.

Salt crystals

Salt may be secreted through salt glands in the surface layer of the leaves or stored in older leaves before they fall.

Pneumatophores are specialized 'breathing' roots that grow 25-30 cm above the mud surface. They allow the mangrove to obtain oxygen. They are composed of spongy tissue with numerous air spaces. Oxygen enters the pneumatophores through lenticels (pits) in the waterproof bark. It diffuses through the spongy tissue to the rest of the plant.

Oxygen

Lenticels

Water level at high tide

A waxy coating of suberin on the root cells excludes 97% of salt from the water taken up by the roots.

Only the top few centimetres of the mud contains oxygen. Beneath, the mud is anaerobic (lacking oxygen), black, and foul-smelling. A deep root system is of no use here.

Cable roots radiate from the trunk, about 20-30 cm below the surface. Growing off these radial roots are fine feeding-roots (not shown), which create a stable platform.

Prop roots that descend from the trunk act like buttresses, providing additional support for the tree in the soft mud and supplement the oxygen uptake from the pneumatophores.

The mangrove propagule is a partially developed seedling adapted for dispersal in water. It is able to quickly take root once it reaches a suitable site.

1. What two physical adaptations of mangroves provide support for the plant in soft mud?

    (a) _____

    (b) _____

2. What is the purpose of pneumatophores? _____

    _____

    _____

3. Describe a physiological problem associated with living in a high-salt substrate: _____

    _____

    _____

4. Describe three methods by which various mangrove species solve the problem of a high salt environment:

    _____

    _____

    _____

131 Tolerance and Population Distribution

Key Idea: Organisms will be able to tolerate some degree of variation in the abiotic factors of a habitat, e.g. temperature range.

Within any habitat, each species has a range of tolerance to variations in its environment. Within the population, individuals will have slightly different tolerance ranges based on small differences in genetic make-up, age, and health. The wider an organism's tolerance range for any one factor, e.g. temperature, the more likely it is that the organism will survive variations in that factor. For the same reasons, species with a wider tolerance range are likely to be more widely distributed. Organisms have a narrower optimum range within which they function best. This may vary seasonally or during development. Organisms are usually most abundant where the abiotic factors are closest to the optimum range.

Habitat occupation and tolerance range

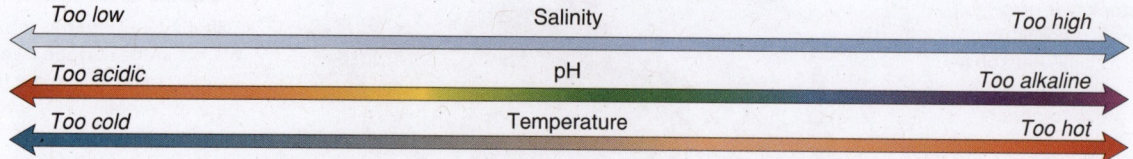

The law of tolerances states that "*for each abiotic factor, a species population (or organism) has a tolerance range within which it can survive. Toward the extremes of this range, that abiotic factor tends to limit the organism's ability to survive*".

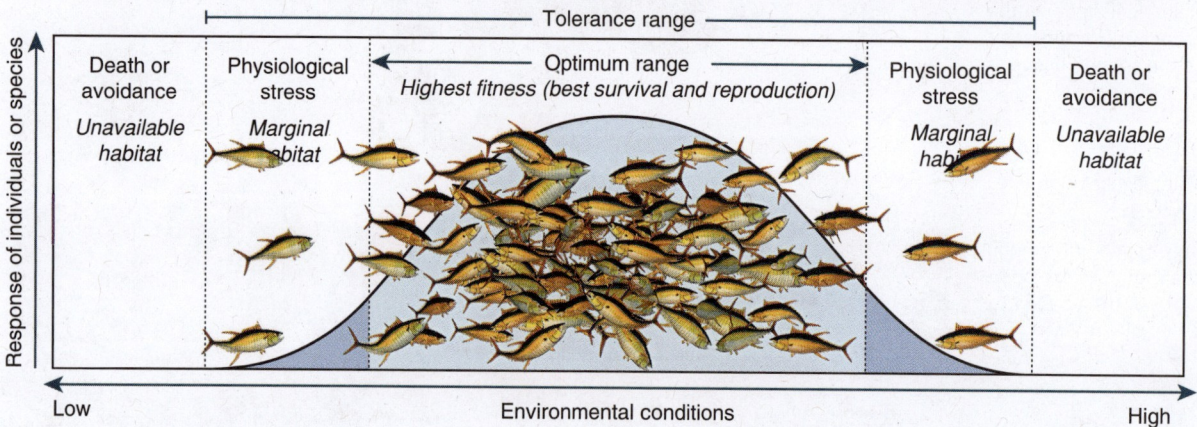

Seasonal changes in vertical distribution of zooplankton in Lake Johnson (1970)

Lake Johnson, in New Zealand's South Island, is a nutrient-rich lake which develops a strong, thermal stratification with deoxygenation of bottom waters during summer to autumn (December - May). A study of the lake during 1969-1972 showed a strong seasonal pattern of zooplankton distribution (below).

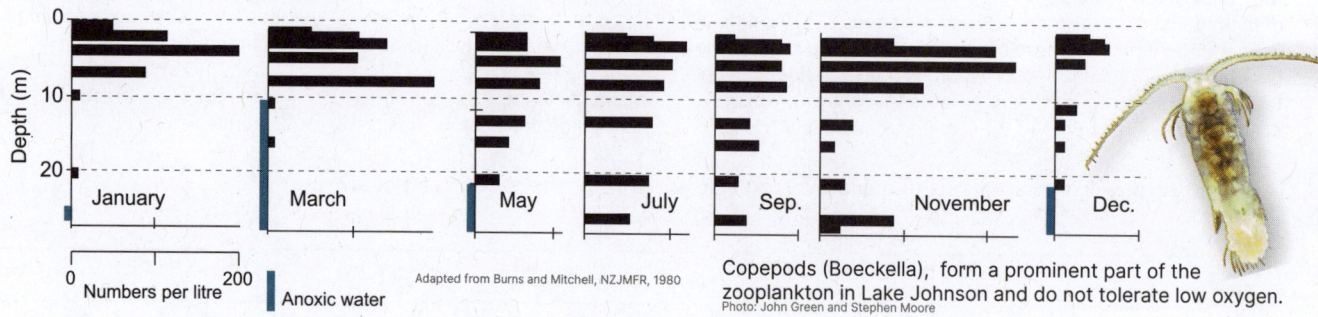

Adapted from Burns and Mitchell, NZJMFR, 1980

Copepods (Boeckella), form a prominent part of the zooplankton in Lake Johnson and do not tolerate low oxygen.
Photo: John Green and Stephen Moore

1. What is the relationship between an organism's tolerance range and the habitat it occupies?

_____

_____

_____

2. (a) What is the limiting factor in the distribution of zooplankton in Lake Johnson? _____

   (b) How does the change in this factor affect the distribution of zooplankton? _____

   _____

   _____

3. A rainforest tree would not survive in the desert. What abiotic variable(s) prevent the tree growing in a desert?

   _____

B4.1
3

©2024 **BIOZONE** International
ISBN: 978-1-99-101410-8
Photocopying prohibited

132 Sampling and Sensors

Key Idea: Sampling provides information about both biotic and abiotic factors in an environment.

When studying plant and animal populations not only does information need to be gathered about the population numbers and distribution, but also about the environment those populations are found in. This might include measurements of the temperature, light levels, wind speed, rainfall, or dissolved oxygen and solids in water. Most of these abiotic factors can be measured with sensors, which can provide either single measurements at the time of sampling, or continuous measurements for longer term analysis.

Using dataloggers in field studies

Usually, when we collect information about populations in the field, we also collect information about the physical environment. This provides important information about the local habitat and can be useful in assessing habitat preference. With the advent of dataloggers, collecting this information is straightforward.

Dataloggers are electronic instruments that record measurements over time. They are equipped with a microprocessor, data storage facility, and sensor. Different sensors are used to measure a range of variables in water or air. The limits of the datalogger's operation can be set before use, (e.g. the sampling interval). It can then be used remotely to record and store data. Dataloggers in remote environments can be linked to computers at a base of operations via cellphone networks or satellite making data collection quick, accurate, and simple. They also enable data collection from multiple sites without the need for researchers to continually travel to them and allow prompt data analysis from those sites.

Stream

Dataloggers fitted with sensors are portable and easy to use in a wide range of aquatic (left) and terrestrial (right) environments. Different variables can be measured by changing the sensor attached to the logger.

Light meter: Measures light intensity levels but not light quality (wavelength). Light levels can change dramatically from a forest floor to its canopy. A light meter provides a quantitative measure of these changes, many of which are not detectable with our own visual systems.

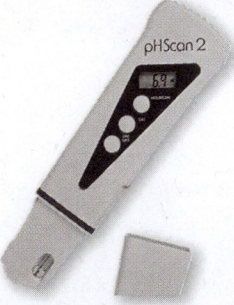

pH meter: Measuring pH of water gives important information about water quality and the tolerance of organisms in it. Pure water has a pH of 7 but bodies of water generally have a pH range from 6 to 8. Modern digital meters can give very precise readings to measure small changes in pH.

Digital thermometers: Digital thermometers fitted with probes give very precise measurements of temperature for water, air, soil etc. Simple meters and probes are relatively cheap and easy to use in multiple environments, providing useful information of changes over time or between locations.

1. Explain how dataloggers make sampling abiotic factors a much simpler process than it might otherwise be:

2. Explain the need to sample abiotic factors when sampling animal or plant populations:

3. Imagine sampling a rocky shore habitat, from the water's edge to one metre above the high water mark. List the types of meters you might want to use to build up a picture of the environment:

B4.1
4

133 Transect Sampling

Key Idea: Transect sampling is useful for providing information about species distribution along an environmental gradient.

A transect is a line placed across a community of organisms. Transects provide information on the distribution of species in the community. They are particularly valuable when the transect records community composition along an environmental gradient, e.g. up a mountain or across a seashore. The usual practice for small transects is to stretch a string between two markers. The string is marked off in measured distance intervals and the species at each marked point are noted. The sampling points along the transect may also be used for the siting of quadrats, so that changes in density and community composition can be recorded. Belt transects are essentially a form of continuous quadrat sampling. They provide more information on community composition but can be difficult to carry out. Some transects provide information on the vertical, as well as horizontal, distribution of species, e.g. tree canopies in a forest.

Continuous belt transect

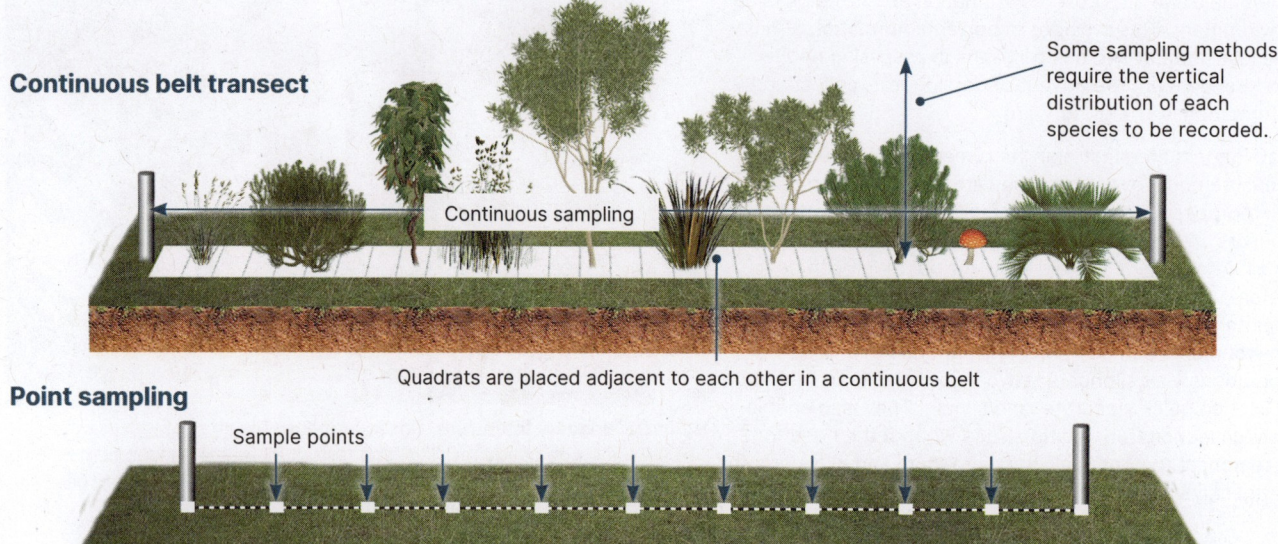

Some sampling methods require the vertical distribution of each species to be recorded.

Continuous sampling

Quadrats are placed adjacent to each other in a continuous belt

Point sampling

Sample points

Interrupted belt transect

4 quadrats across each sample point

Line of transect

1. Belt transect sampling uses quadrats placed along a line at marked intervals. In contrast, point sampling transects record only the species that are touched or covered by the line at the marked points.

   (a) Describe one disadvantage of belt transects: _____

   _____

   (b) Why might line transects give an unrealistic sample of the community in question? _____

   _____

   _____

   (c) How do belt transects overcome this problem? _____

   _____

   (d) When would it not be appropriate to use transects to sample a community? _____

   _____

2. How could you test whether or not a transect sampling interval was sufficient to accurately sample a community?

   _____

   _____

©2024 **BIOZONE** International
ISBN: **978-1-99-101410-8**
Photocopying prohibited

A kite graph is a good way to show the distribution of organisms sampled using a belt transect. Data may be expressed as abundance or percentage cover along an environmental gradient. Several species can be shown together on the same plot so that the distributions can be easily compared.

3. The data on the right were collected from a rocky shore field trip. Four common species of barnacle were sampled in a continuous belt transect from the low water mark to a height of 10 m above that level. The number of each of the four species in a 1 m^2 quadrat was recorded.

Plot a kite graph of the data for all four species below. Be sure to choose a scale that takes account of the maximum number found at any one point and allows you to include all the species on the one plot. Include a scale on the diagram so that the number at each point on the kite can be calculated.

An example of a kite graph

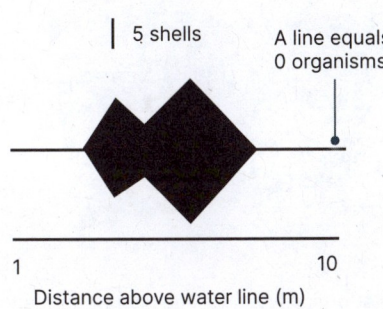

| 5 shells

A line equals 0 organisms

1   10

Distance above water line (m)

Distribution of 4 common barnacle species on a rocky shore

Height above low water (m)	Barnacle species			
	Plicate barnacle	Columnar barnacle	Brown barnacle	Sheet barnacle
0	0	0	0	65
1	10	0	0	12
2	32	0	0	0
3	55	0	0	0
4	100	18	0	0
5	50	124	0	0
6	30	69	2	0
7	0	40	11	0
8	0	0	47	0
9	0	0	59	0
10	0	0	65	0

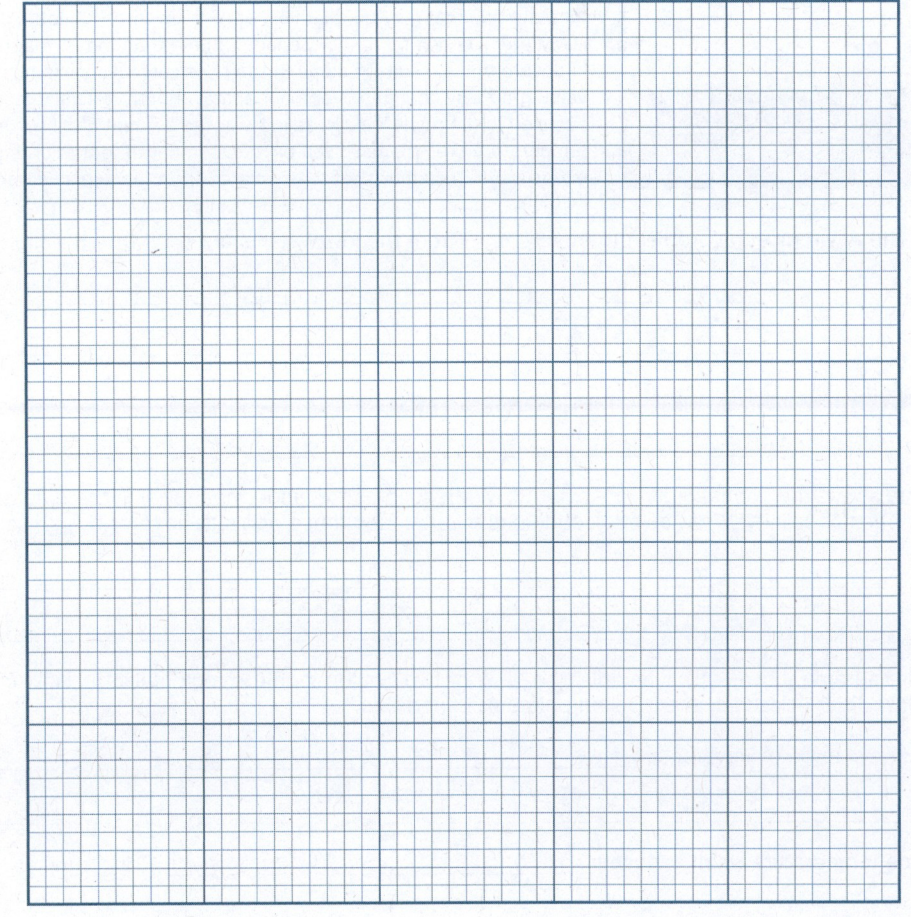

134 Abiotic Factors and Population Distribution

Key Idea: Measuring abiotic factors along a transect can help in understanding how population distribution is affected by changes in abiotic factors.

Gradients in abiotic factors are found in almost every environment. They influence habitats and are important in determining community patterns. By measuring population numbers and abiotic factors, it is possible to correlate population distribution with changes in abiotic factors.

Abiotic changes on a rocky shore

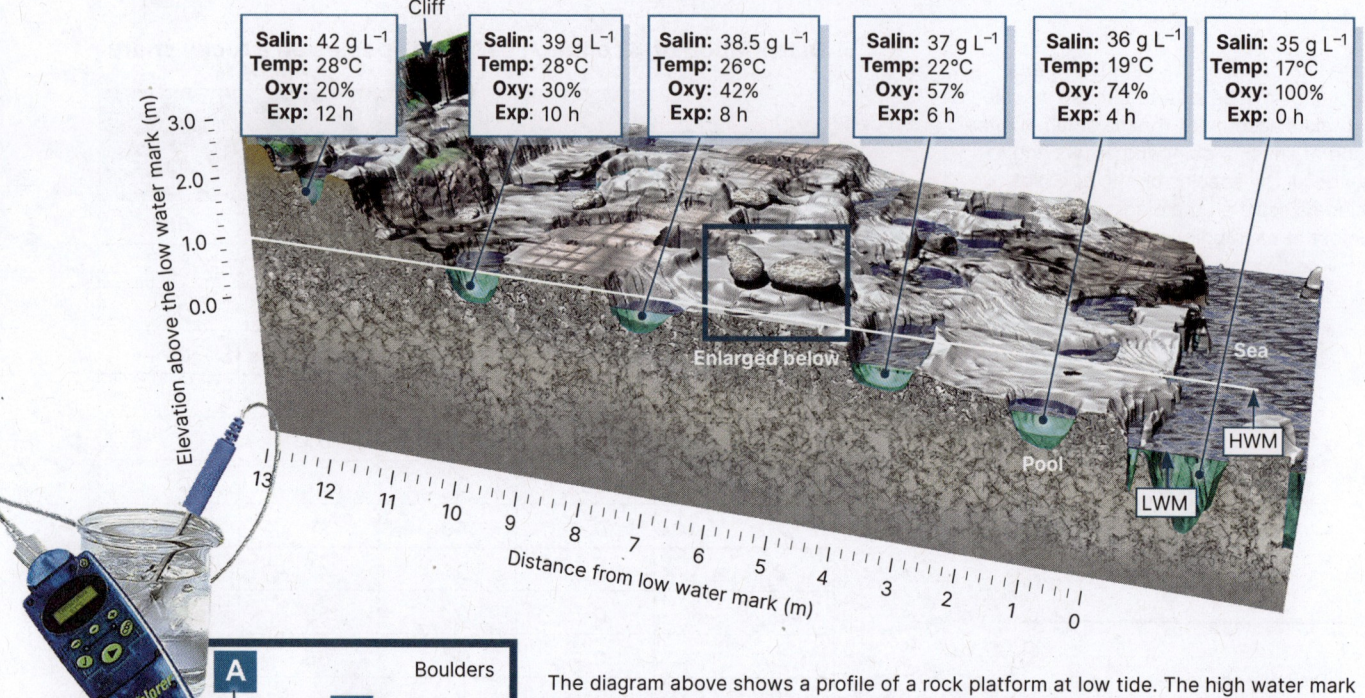

| Salin: 42 g L⁻¹ | Salin: 39 g L⁻¹ | Salin: 38.5 g L⁻¹ | Salin: 37 g L⁻¹ | Salin: 36 g L⁻¹ | Salin: 35 g L⁻¹ |

Salin: 42 g L^{-1}
Temp: 28°C
Oxy: 20%
Exp: 12 h

Salin: 39 g L^{-1}
Temp: 28°C
Oxy: 30%
Exp: 10 h

Salin: 38.5 g L^{-1}
Temp: 26°C
Oxy: 42%
Exp: 8 h

Salin: 37 g L^{-1}
Temp: 22°C
Oxy: 57%
Exp: 6 h

Salin: 36 g L^{-1}
Temp: 19°C
Oxy: 74%
Exp: 4 h

Salin: 35 g L^{-1}
Temp: 17°C
Oxy: 100%
Exp: 0 h

The diagram above shows a profile of a rock platform at low tide. The high water mark (HWM) shown is the average height of the spring tide. In reality, the high tide level varies with the phases of the moon. The low water mark (LWM) is an average level subject to the same variations. The rock pools vary in size, depth, and position on the platform. They are isolated at different elevations, trapping water from the ocean for time periods that may be relatively brief or up to 10-12 hours duration. Pools near the HWM are exposed for longer periods of time than those near the LWM. The difference in exposure times results in some of the physical factors exhibiting a gradient; the factor's value gradually changes over a horizontal and/or vertical distance. Physical factors sampled in the pools include salinity, or the amount of dissolved salts (g) per liter (Salin), temperature (Temp), dissolved oxygen compared to that of open ocean water (Oxy), and exposure, or the amount of time isolated from the ocean water (Exp).

1. Describe the environmental gradient (general trend) from the low water mark (LWM) to the high water mark (HWM) for:

   (a) Salinity: _____

   (b) Temperature: _____

   (c) Dissolved oxygen: _____

   (d) Exposure: _____

2. Rock pools above the normal high water mark (HWM), such as the uppermost pool in the diagram above, can have wide extremes of salinity. What abiotic conditions might cause these pools to have very high salinity?

   _____

   _____

3. The inset diagram (above, left) is an enlarged view of two boulders on the rock platform. How might the abiotic factors listed below differ at each of the labelled points A, B, and C?

   (a) Mechanical force of wave action: _____

   _____

   (b) Surface temperature when exposed: _____

   _____

©2024 **BIOZONE** International
ISBN: 978-1-99-101410-8
Photocopying prohibited

Investigation 8.1 Correlating abiotic factors and population distribution

See appendix for equipment list.

1. Select a natural or semi-natural environment to carry out this study. This could be a rocky or sandy shore, river or stream edge, forest edge or bush reserve, or mountainside.

2. Decide what organism(s) you will sample. You could, for example, sample all the tree species present at specific altitudes on a mountain, the number and species of limpet or barnacle at specific distances between the low water and high water mark on a rocky shore, or the number and species of invertebrates in the leaf litter at specific distances from a rotting log in a forest. An environment with a gradient of change (e.g. in temperature or light) will give a better result than one in which abiotic factors remain the same in all places.

3. Whatever population(s) you choose to sample, you will need to record as many abiotic factors as possible. Include the intensity of light, the temperature of the water, air, or soil, wind speed, soil moisture, altitude, height about the ground, salinity etc. The more data you gather, the more likely you may find a correlation between one or more factors and the distribution of the population.

4. In a logbook, draw up a table to record your data. Remember to record the location of each sample site (e.g. distance from the forest edge, or altitude up the mountain), any relevant abiotic factors, and the number of organisms or species.

5. Be systematic with your sampling (see Activities 228 - 230 for more information on sampling procedures). Use the same sampling technique at each location (e.g. same sized quadrat or same distance between sample points along a transect line, same method of counting.

6. Analyse your data. You could do this using a scatter plot, e.g. number of organisms per quadrat vs altitude, or spreadsheet. Calculating a correlation coefficient (r) is a useful and simple way to identify any correlation between abiotic values and population distribution (see below). More complex analysis could include chi-squared testing (Activities 244 and 245).

Calculating a correlation coefficient 'r'

▸ Pearson's linear correlation (r) relates the value of one variable to another for a simple linear relationship. For a perfect correlation in which the change in one variable is perfectly matched or caused by a corresponding change in another variable, r =1 (positive correlation) or -1 (negative correlation). If there is no correlation between variables, then r = 0.

▸ Remember that correlation is not causation. Just because two variables (e.g. temperature and population number) are correlated it does not mean one causes the other.

▸ The simplest way to calculate a correlation coefficient is to use a spreadsheet such as Microsoft Excel:

  1. Enter the data into two columns.

  2. Type the formula =CORREL(Range Column1, Range Column 2).

Plotting a graph and displaying R^2

▸ The R^2 value, the coefficient of determination, is the square of the correlation coefficient. It quantifies the proportion of variance in the dependent variable that is explained by the independent variable. Unlike r, R^2 has no direction and determines how a model or equation fits the data.

▸ R^2 can be shown on a graph in Excel

1. Enter the data into two columns.

2. Select the columns and insert a scatter plot.

3. Select any data point on the plot and right click the mouse. Select 'Add trendline'. Check the box 'Display R squared value on chart'. For the data shown above right, the line will cross the y axis at a value of about 5. R^2 will equal the square of r calculated for the data shown (middle right).

4. In the case of this data, we know that this species of crab (*Chiromantes dehaani*) does not live in fresh water (salinity 0). Therefore the intercept can be set to 0. This will shift the trendline and recalculate the R^2 accordingly (bottom right). If you calculate the square root of this new value, a much larger correlation coefficient is obtained.

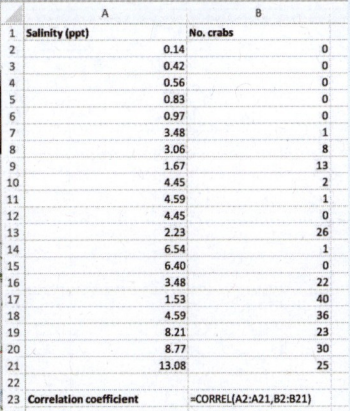

The number of mudflat crabs
***Chiromantes dehaani* vs water salinity**

	A	B
1	Salinity (ppt)	No. crabs
2	0.14	0
3	0.42	0
4	0.56	0
5	0.83	0
6	0.97	0
7	3.48	1
8	3.06	8
9	1.67	13
10	4.45	2
11	4.59	1
12	4.45	0
13	2.23	26
14	6.54	1
15	6.40	0
16	3.48	22
17	1.53	40
18	4.59	36
19	8.21	23
20	8.77	30
21	13.08	25
22		
23	Correlation coefficient	=CORREL(A2:A21,B2:B21)

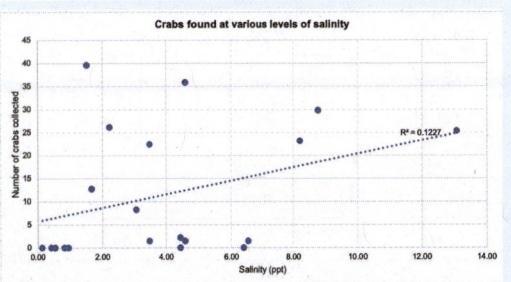

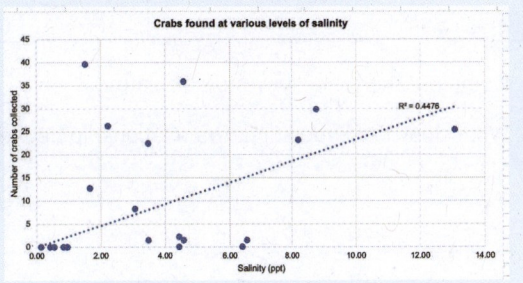

Data adapted from Effects of reduced salinity caused by reclamation on population and physiological characteristics of the sesarmid crab Chiromantes dehaani, Weiwei Lv et al, Nature.com

A correlation example

Pearson's coefficient correlation can be calculated using the formula:

$$r = \frac{\Sigma xy - n\bar{x}\bar{y}}{ns_x s_y}$$

To find s_x and s_y use the population standard deviation:

$$s = \sqrt{\frac{\Sigma(x - \bar{x})^2}{n}}$$

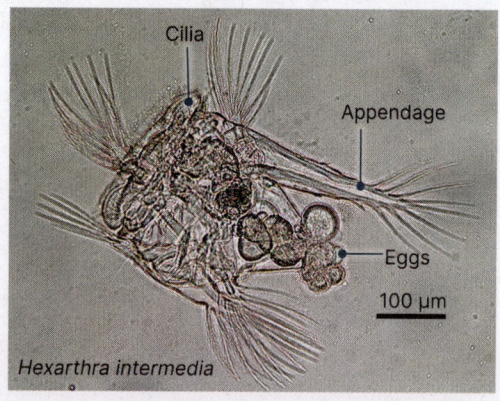

Hexarthra intermedia (right) is a species of rotifer. Rotifers are small, ciliated animals found in fresh water ponds. Most feed on small algae. A study was carried out in order to understand how changes in their abundance might be related to seasonal changes in environmental factors. The data below records abundance per litre of pond water against pond temperature at the time that sample was taken. Note that the *Hexarthra* counts are not in whole numbers because number per litre was calculated from a larger, filtered sample volume. n = 25.

Hexarthra intermedia

Cilia · Appendage · Eggs · 100 µm

Data kindly supplied by Dr Ian Duggan, University of Waikato

Hexarthra no. (x) / per L	Temperature (y) / °C	$(x - \bar{x})^2$	$(y - \bar{y})^2$	xy
36.21	19.75			
33.76	17.53			
10.83	15.05			
1.88	14.40			
0.33	11.73			
2.40	11.05			
0.35	9.23			
0.08	8.75			
0.00	12.35			
0.04	13.13			
0.00	14.15			
0.21	14.63			
0.29	15.98			
5.72	19.63			
4.39	18.00			
7.42	19.80			
72.87	23.33			
443.38	23.30			
34.38	22.30			
147.58	25.88			
947.64	24.58			
573.47	22.90			
444.63	20.95			
338.25	21.10			
34.33	18.90			
$\bar{x}=$	$\bar{y}=$	$\Sigma(x - \bar{x})^2 =$	$\Sigma(y - \bar{y})^2 =$	$\Sigma xy =$
Standard deviation x =		Standard deviation y =	$r =$	

4. Complete the table above to calculate *r*:

5. What does *r* tell you about the relationship between *Hexarthra* numbers and temperature? _____

_____

_____

_____

©2024 **BIOZONE** International
ISBN: 978-1-99-101410-8
Photocopying prohibited

135 Abiotic Factors and Biome Distribution

Key Idea: Temperature and rainfall play an important role in determining the geographical location of terrestrial biomes. Temperature and precipitation are excellent predictors of biome distribution. Temperature decreases from the equator to the poles. Temperature and precipitation act together as limiting factors to determine the type of desert, grassland, or forest biome in a region. Latitude directly affects solar input and temperature.

Within a single latitudinal region, the level of precipitation (rainfall) governs the type of plant community found. Note that the effect of altitude is similar to that of latitude (ice will occur at high altitudes even at low-latitudes).

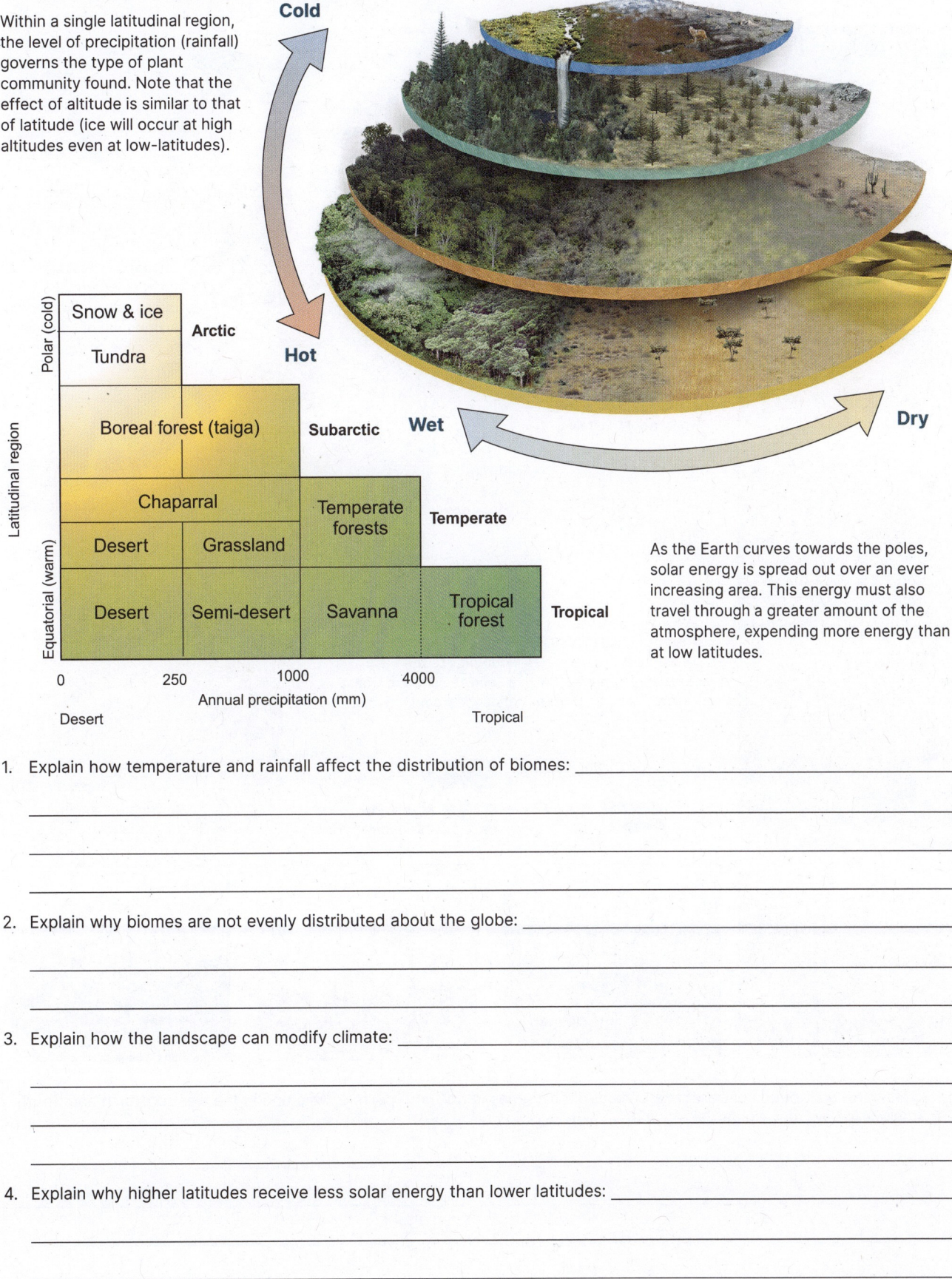

As the Earth curves towards the poles, solar energy is spread out over an ever increasing area. This energy must also travel through a greater amount of the atmosphere, expending more energy than at low latitudes.

1. Explain how temperature and rainfall affect the distribution of biomes: _____
_____
_____
_____

2. Explain why biomes are not evenly distributed about the globe: _____
_____
_____

3. Explain how the landscape can modify climate: _____
_____
_____
_____

4. Explain why higher latitudes receive less solar energy than lower latitudes: _____
_____
_____
_____

B4.1

6

136 The World's Terrestrial Biomes

Key Idea: The climate plays an important role in determining the location of Earth's biomes.

Global patterns of vegetation distribution are closely related to climate. Although they are complex, major vegetation biomes can be recognized. These are large areas where a distinctive vegetation type has formed in response to a particular physical environment. Biomes have characteristic features, but the boundaries between them are not distinct. The same biome may occur in widely separated regions wherever the climatic and soil conditions are similar.

Arctic

Tundra, Alaska

Taiga (boreal forest)

Temperate forest, USA

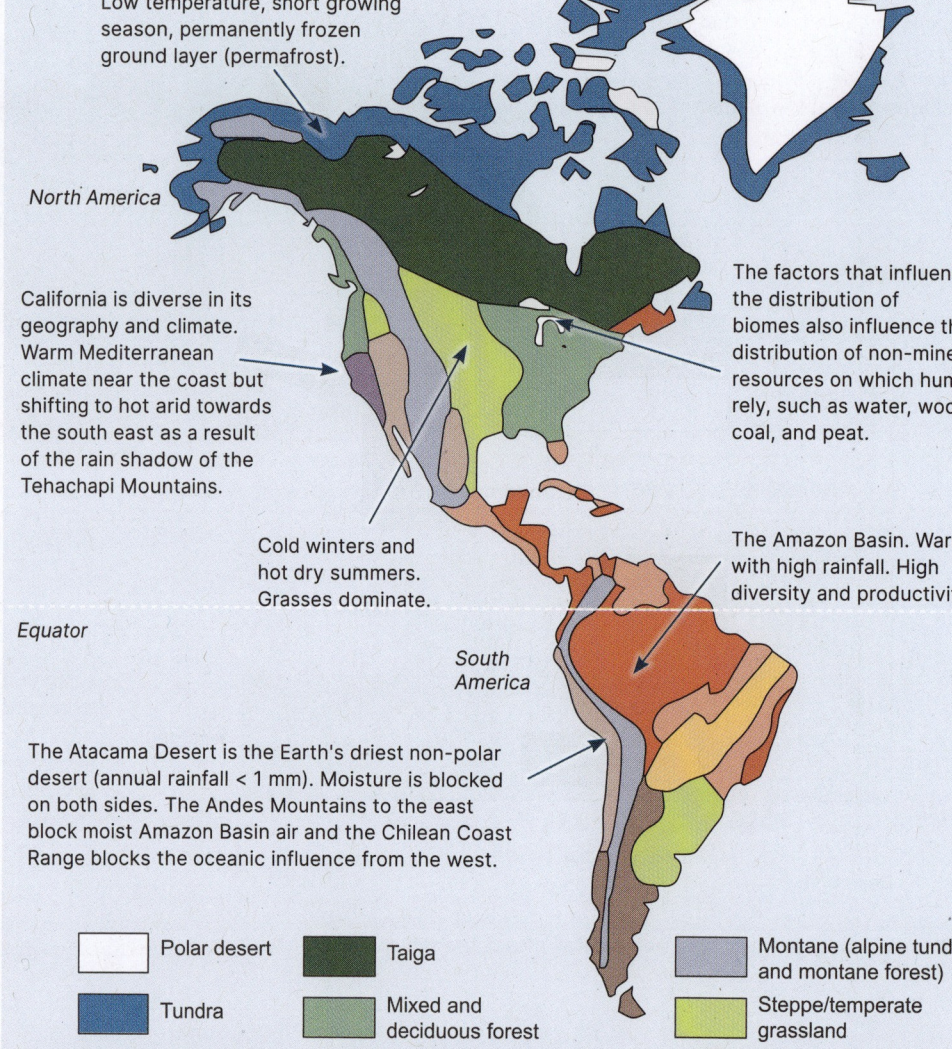

Low temperature, short growing season, permanently frozen ground layer (permafrost).

North America

California is diverse in its geography and climate. Warm Mediterranean climate near the coast but shifting to hot arid towards the south east as a result of the rain shadow of the Tehachapi Mountains.

Cold winters and hot dry summers. Grasses dominate.

Equator

South America

The factors that influence the distribution of biomes also influence the distribution of non-mineral resources on which humans rely, such as water, wood, coal, and peat.

The Amazon Basin. Warm with high rainfall. High diversity and productivity.

The Atacama Desert is the Earth's driest non-polar desert (annual rainfall < 1 mm). Moisture is blocked on both sides. The Andes Mountains to the east block moist Amazon Basin air and the Chilean Coast Range blocks the oceanic influence from the west.

Legend:
- Polar desert
- Tundra
- Taiga
- Mixed and deciduous forest
- Montane (alpine tundra and montane forest)
- Steppe/temperate grassland

Alpine tundra, Colombia

Prairie, USA

Savanna, East Africa

Rainforest, Western Ghats, India

1. Explain the distribution of deserts and semi-desert areas in northern parts of Asia and in the west of North and South America (away from equatorial regions):

_____

_____

_____

_____

_____

B4.1
7
©2024 **BIOZONE** International
ISBN: 978-1-99-101410-8

Vegetation patterns are determined largely by climate, which in turn is heavily influenced by topography and proximity to the ocean. Where there are large mountain ranges, wind is deflected upwards causing rain on the windward side and a drier 'rain shadow' on the leeward side. Rain shadowing governs the occurrence of many deserts globally, and some of the world's driest regions, including the Atacama Desert in Chile and Death Valley in California, are in rain shadows. Wherever they occur, montane regions are associated with their own altitude adapted vegetation. Biome classification may vary considerably and is not necessarily static as environments shift under patterns of changing climate and human influence. However, most classifications recognize desert, tundra, grassland and forest types and distinguish them on the basis of latitude.

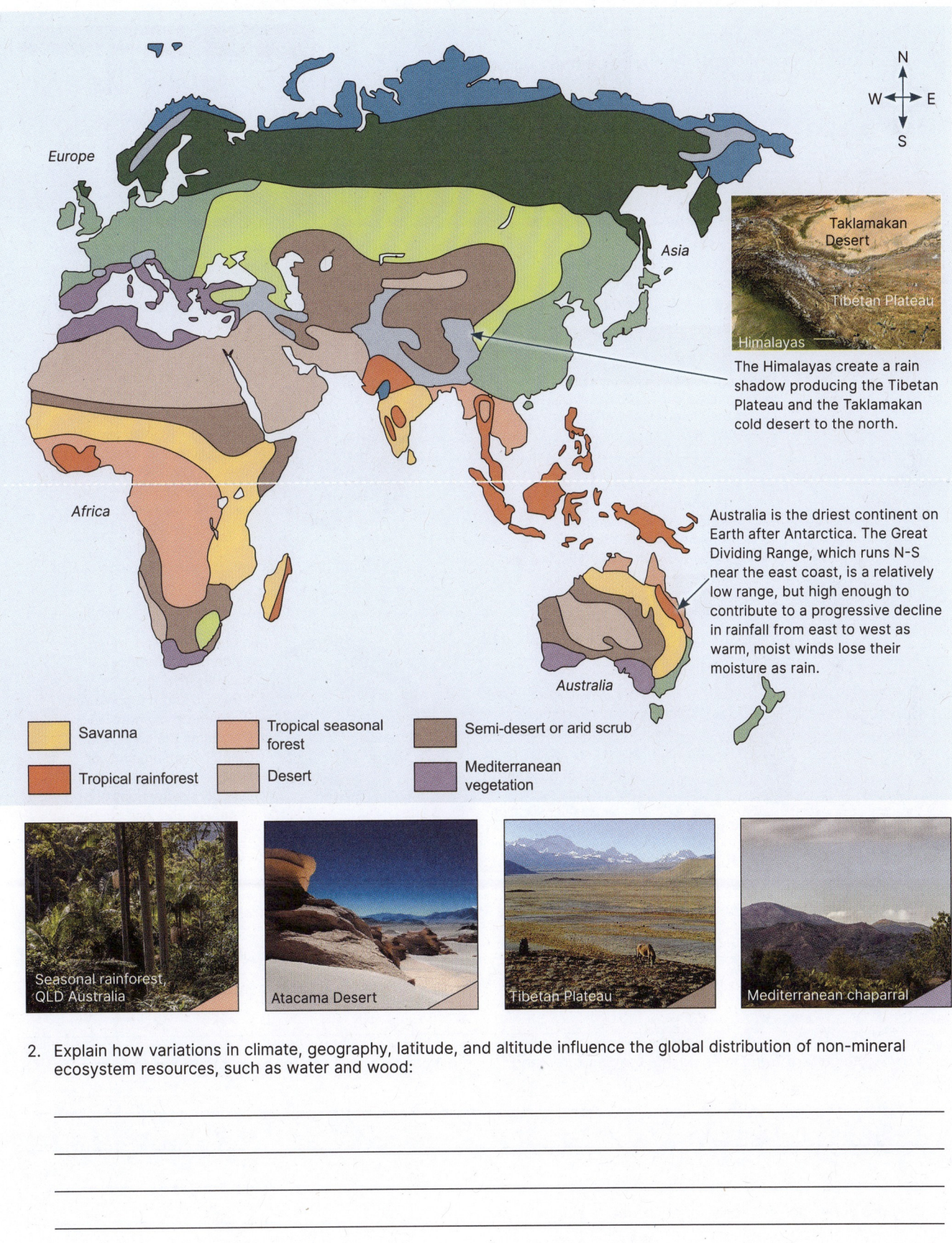

The Himalayas create a rain shadow producing the Tibetan Plateau and the Taklamakan cold desert to the north.

Australia is the driest continent on Earth after Antarctica. The Great Dividing Range, which runs N-S near the east coast, is a relatively low range, but high enough to contribute to a progressive decline in rainfall from east to west as warm, moist winds lose their moisture as rain.

Legend:
- Savanna
- Tropical rainforest
- Tropical seasonal forest
- Desert
- Semi-desert or arid scrub
- Mediterranean vegetation

Seasonal rainforest, QLD Australia

Atacama Desert

Tibetan Plateau

Mediterranean chaparral

2. Explain how variations in climate, geography, latitude, and altitude influence the global distribution of non-mineral ecosystem resources, such as water and wood:

_____

_____

_____

_____

_____

137 Marine Biomes

Key Idea: Coral reefs are an important marine ecosystem. They need specific conditions to form and flourish.

Water covers ~70% of Earth's surface, so aquatic biomes are a major component of the global environment. Aquatic biomes include all those environments that are dominated by water. These environments include deep oceans, shallow seas and reefs, swamps and estuaries, and rivers and lakes. Coral reefs are found in areas with high sunlight and warm shallow water. They cover just 0.1% of the ocean's area but around 25% of all marine species are found in them. Coral reefs are under threat as the ocean conditions are changing beyond their tolerance range.

The open oceans are characterized by saline (salty) waters, waves, and currents. Five oceans are recognized but they are all interconnected as one global ocean.

Coral reefs occur in tropical and subtropical regions and are of biological (not geological) origin. Their skeletons of calcium carbonate are made by living organisms.

Estuaries are regions where fresh water of rivers meets tidal flows from the ocean. They are extremely productive areas, gaining from both the land and sea.

Coral reefs

Light: Corals need sunlight so that zooxanthellae can photosynthesize. This means they need clear waters, normally to a depth of 60 m.

Temperature: Corals prefer water with a temperature range of between 16-34°C (depending on the coral).

Salinity: Corals can tolerate salinity between 23 and 42 ppt.

pH: Corals prefer seawater with a pH of about 8.1. Below this, calcium carbonate becomes difficult to obtain so corals can not build their calcium carbonate skeletons.

Corals are colonial organisms. A colony is composed of individual animals called polyps. Each polyp secretes and is surround by a casing of calcium carbonate, producing the colony's hard skeleton. The polyp also harbours unicellular photosynthetic organisms called zooxanthellae, which produce oxygen and nutrients the polyp uses.

When physical factors range outside the preference of corals, they become stressed and the polyp will expel the zooxanthellae. This turns the coral white, called bleaching, and can cause the polyp to die. As the ocean's pH slowly drops and surface temperatures rise, greater areas of coral bleaching are occurring every year.

1. Given that coral reefs need warm water and good quality light, where in the world are coral reefs most likely to form?

_____

2. Why would coral reefs not be able to form in turbid waters (water with high particle concentrations)?

_____

3. Why is the current trend in rising sea temperatures and falling pH concerning, both for coral and the wider ocean?

_____

_____

_____

4. Why are corals normally only found in waters above 60 m deep? _____

_____

©2024 **BIOZONE** International
ISBN: 978-1-99-101410-8
Photocopying prohibited

138 Adaptations to Tropical Environments

Key Idea: Tropical rainforests have the greatest biodiversity on Earth, with organisms showing a vast array of adaptations. Tropical environments have a large amount of light, warmth, and moisture, ideal for plant growth. This combination of factors has produced tropical rainforests that have the highest biodiversity of any terrestrial environment. A single hectare may have over 42,000 different species of plants and animals. With such large numbers of organisms all competing for space and nutrients, it is unsurprising that the inhabitants of a tropical rainforest have evolved a vast array of adaptations including camouflage, mimicry, and specialized diets.

Plant adaptations

Plants in tropical rainforests have adaptations to deal with excessive rain, low soil nutrients, low light levels, and other competing plants.

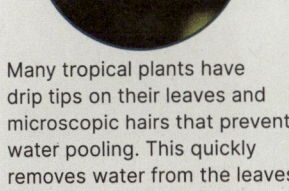

Lianas and epiphytes are adapted to live high on branches or climb up tree trunks in order to reach the light.

Bark helps reduce water loss. This isn't a problem in tropical rainforests so many tropical trees have much thinner, smooth bark than temperate trees. This also helps in stopping vines getting a grip.

Many tropical plants have drip tips on their leaves and microscopic hairs that prevent water pooling. This quickly removes water from the leaves and stops organisms such as fungi growing on them.

Tropical soils are nutrient poor so most trees have shallow roots. Large trees like the kapok have massive buttresses to spread their weight and provide support.

Animal adaptations

Tropical rainforest animals have adaptations to take advantage of the variety of habitats, including mimicry, camouflage and developing poisons.

Many animals have specialized in foraging for specific foods. Toucans have specialized in eating fruit which is available throughout the year.

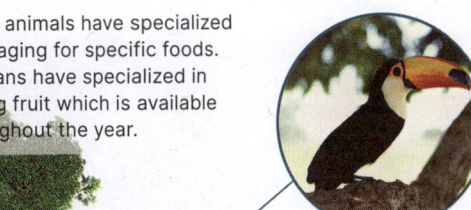

Many insects mimic other types of animal either for defence or for predation, such as the ant mimicking spider (left).

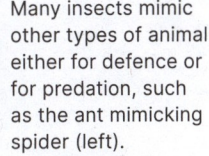

Many animals (and plants) have developed poisons for defence, e.g. poison arrow frog above, or for predation.

Many animals in tropical rainforests show an extraordinary degree of adaptation for camouflage. The dead leaf butterfly (left) looks exactly as its name suggests.

Atudu CC 4.0

1. In a group of four, research plant and animal adaptations in tropical rainforests. The resources on **BIOZONE's Resource Hub** may be useful. Each person should identify one adaptation in a named plant and one in a named animal. Report back to your group with your findings and record all four plant and four animal adaptations below:

_____

_____

_____

_____

_____

B4.1
8

139 Adaptations to Desert Environments

Key Idea: Hot deserts are some of the most extreme environments on Earth. The plants and animals that live there have evolved special adaptations to deal with this extreme environment.

Hot deserts experience extreme lack of water, scorching heat during the day, and often near freezing temperatures during the night. Solar radiation is also very high due to the lack of clouds and the Sun being almost directly overhead. The plants and animals that live in deserts have adaptations for conserving water and keeping their bodies cool. These include physiological and behavioural adaptations which are just as important as any structural adaptations.

Animal adaptations

Adaptations in animals that live in hot deserts include becoming nocturnal, having extremely concentrated urine, and having body coverings, e.g. fur, that both shields the body from the sun but allows heat to be lost for cooling.

Plant adaptations

In a hot desert, plants need to conserve water and reduce the amount of direct solar radiation received by their leaves. Adaptations include reduced leaves, ribbed stems and trucks, and closing stomata during the day.

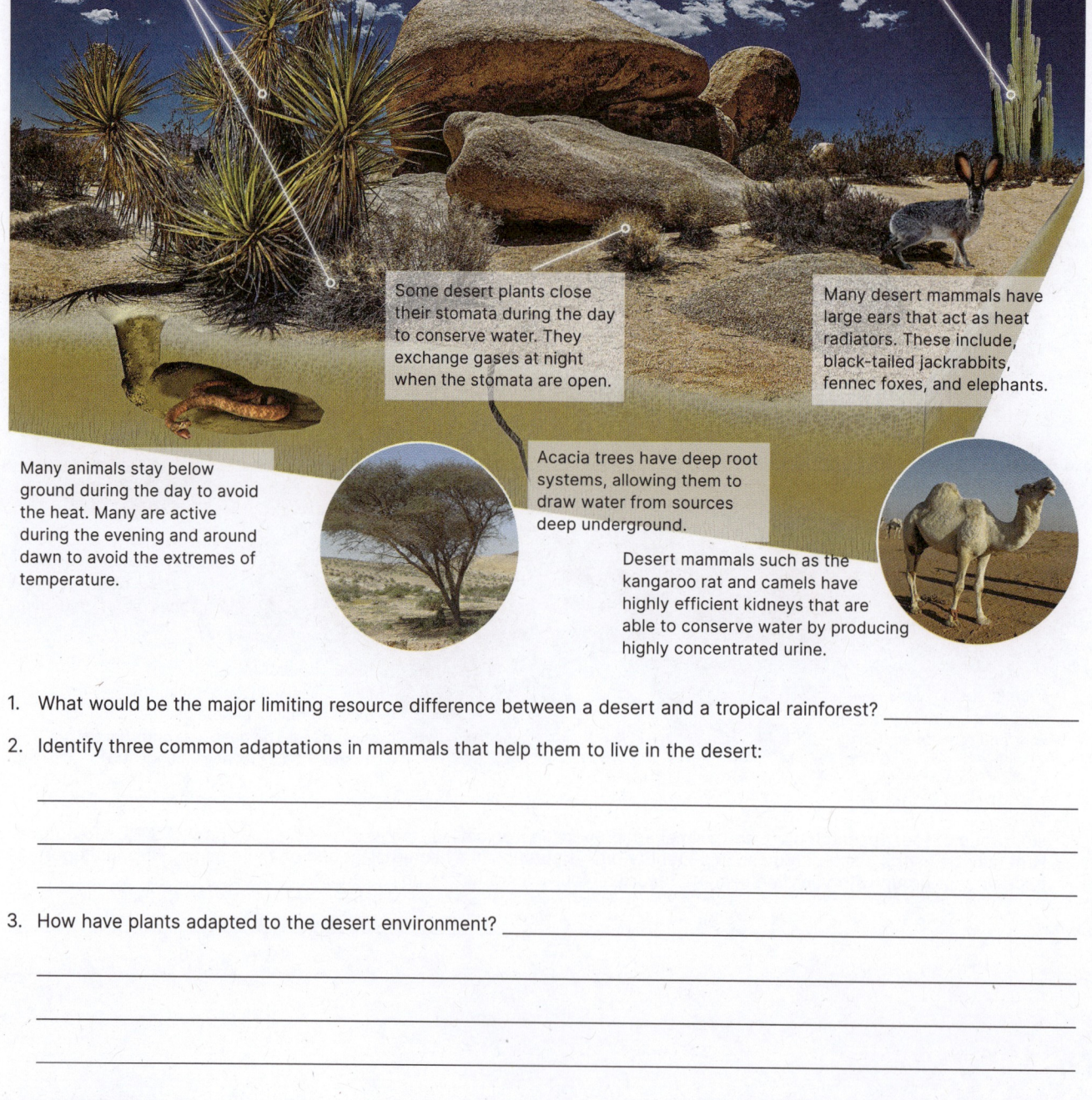

Many desert plants have fine hairs over the leaves or protect their leaves inside a lattice of branches that reduce wind and raise humidity.

Plants such as cacti are able to store large volumes of water after rain. Their pleated trucks expand as they soak up water.

Some desert plants close their stomata during the day to conserve water. They exchange gases at night when the stomata are open.

Many desert mammals have large ears that act as heat radiators. These include, black-tailed jackrabbits, fennec foxes, and elephants.

Many animals stay below ground during the day to avoid the heat. Many are active during the evening and around dawn to avoid the extremes of temperature.

Acacia trees have deep root systems, allowing them to draw water from sources deep underground.

Desert mammals such as the kangaroo rat and camels have highly efficient kidneys that are able to conserve water by producing highly concentrated urine.

1. What would be the major limiting resource difference between a desert and a tropical rainforest? _____

2. Identify three common adaptations in mammals that help them to live in the desert:

   _____

   _____

   _____

3. How have plants adapted to the desert environment? _____

   _____

   _____

   _____

   _____

B4.1

8

©2024 **BIOZONE** International
ISBN: 978-1-99-101410-8
Photocopying prohibited

140 The Ecological Niche

Key Idea: An organism's niche describes its functional role within its environment.

The ecological niche describes the functional role of an organism in an ecosystem, including its habitat and all its interactions with the environment. It includes how the species responds to the distribution of resources and how it alters those resources for other species. The full range of environmental conditions under which an organism can exist describes its fundamental niche. As a result of interactions with other organisms, species usually occupy a realized niche that is narrower than this. Central to the niche concept is the idea that two species with exactly the same niche cannot coexist, because they would compete for the same resources and one would exclude the other. This is Gause's competitive exclusion principle. More often, species compete for only some of the same resources.

The physical conditions influence the habitat. A factor may be well suited to the organism, or present it with problems to be overcome.

Adaptations enable the organism to exploit the resources of the habitat. The adaptations take the form of structural, physiological and behavioural features of the organism.

Physical conditions
- Substrate
- Humidity
- Sunlight
- Altitude
- Aspect
- Salinity
- pH
- Exposure
- Temperature
- Depth

Resources offered by the habitat
- Food sources • Shelter • Mating sites
- Nesting sites • Predator avoidance

Adaptations for:
- Locomotion
- Activity (day/night)
- Tolerance to physical conditions
- Predator avoidance
- Self defence
- Defence of range
- Reproduction
- Feeding

Resource availability is affected by the presence of other organisms and interactions with them: competition, predation, parasitism, and disease.

The habitat provides opportunities and resources for the organism. The organism may or may not have the adaptations to exploit them fully.

The realized niche

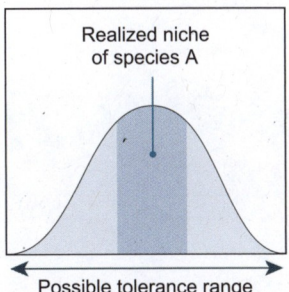

The tolerance range represents the fundamental niche of a species. The realized niche of a species is narrower than this because of competition with other species.

Intraspecific competition

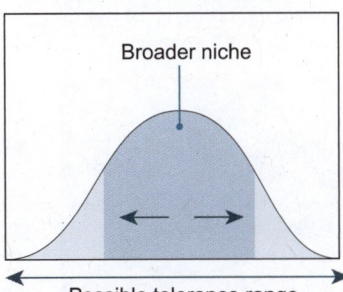

Individuals of the same species exploit the same resources so competition is intense. Individuals must use resources at the extremes of their tolerance range and the realized niche expands.

Interspecific competition

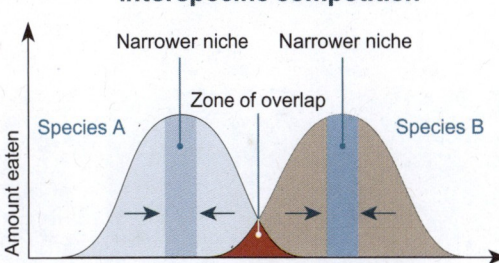

When two (or more) species compete for some of the same resources, their resource use curves will overlap and competition will be intense in this zone. Selection will favour niche specialization so that one or both species occupy a narrower niche.

1. (a) In what way could the realized niche be regarded as flexible? _____

_____

_____

   (b) What factors might further constrain the extent of the realized niche? _____

_____

_____

2. Contrast the effects of interspecific competition and intraspecific competition on niche breadth:

_____

_____

141 Dealing With Different Levels of Oxygen

Key Idea: The presence or absence of oxygen can define where an organism can or cannot live.

Earth has not always had free oxygen in its atmosphere. During the first billion or so years of life on Earth, oxygen was not freely available. It was not until 2.5 billion years ago that oxygen, produced by microbes splitting water for photosynthesis, first began to appear in the atmosphere. Since then, obligate aerobes have come to dominate life on Earth. However, parts of the environment, such as the deep mud of estuaries and parts of the deep ocean, and the inner guts and tissues of some animals, lack any oxygen. In these environments, obligate anaerobes can be found. Instead of using oxygen as a final electron acceptor in their respiration pathway, obligate anaerobes may use sulfates, nitrates, and even carbon monoxide. Placing them in an oxygen rich environment will kill them.

Determining the oxygen requirements of microorganisms

Thioglycolate is a nutrient medium that consumes oxygen and produces anaerobic conditions. When placed in a test tube, oxygen from the air can diffuse into the top layer producing an oxygen gradient from top (oxygen present) to bottom (no oxygen). This can then be used to determine the oxygen conditions required for bacterial growth.

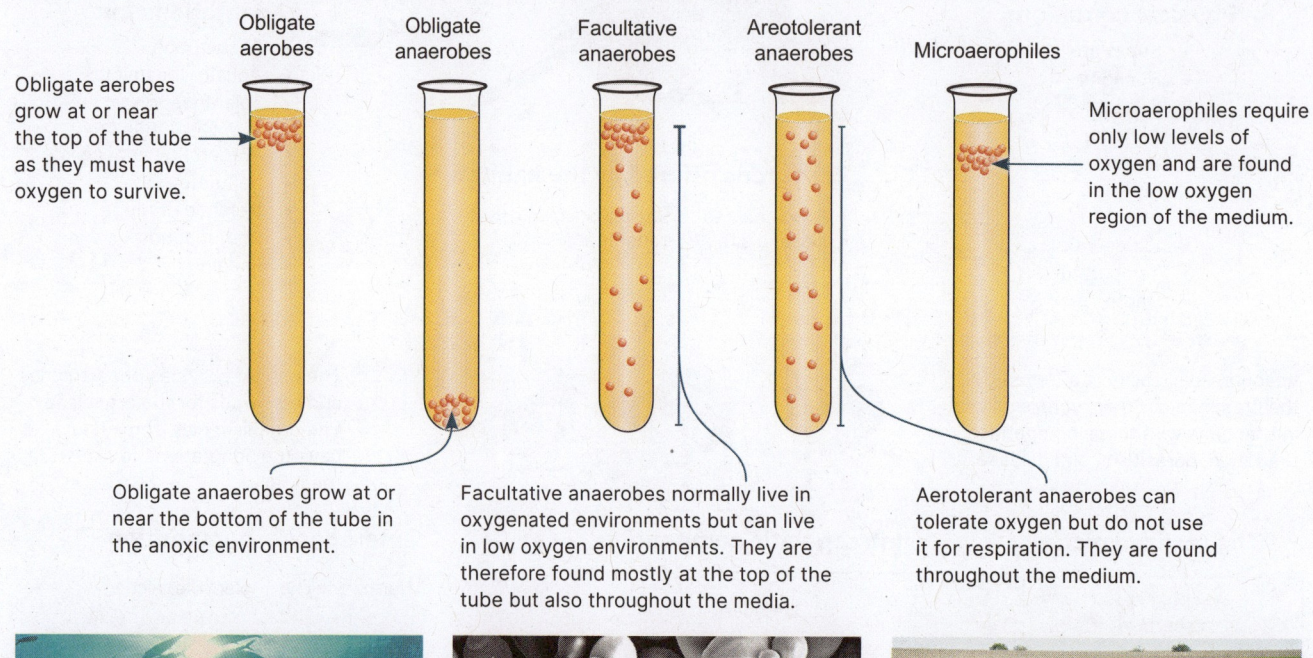

Obligate aerobes | Obligate anaerobes | Facultative anaerobes | Areotolerant anaerobes | Microaerophiles

Obligate aerobes grow at or near the top of the tube as they must have oxygen to survive.

Microaerophiles require only low levels of oxygen and are found in the low oxygen region of the medium.

Obligate anaerobes grow at or near the bottom of the tube in the anoxic environment.

Facultative anaerobes normally live in oxygenated environments but can live in low oxygen environments. They are therefore found mostly at the top of the tube but also throughout the media.

Aerotolerant anaerobes can tolerate oxygen but do not use it for respiration. They are found throughout the medium.

Most large organisms are obligate aerobes because of their energetic lifestyle. Some smaller eukaryotes, such as some molluscs, are facultative anaerobes.

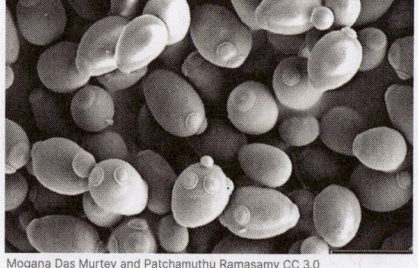

Mogana Das Murtey and Patchamuthu Ramasamy CC 3.0

Facultative anaerobes can switch to anaerobic respiration or fermentation in low oxygen environments, e.g. *Saccharomyces* yeast used in baking and alcohol production.

For obligate anaerobes oxygen is poisonous. Obligate anaerobes can be found in many environments in the gut of animals. Obligate anaerobic microbes play an important role in digesting cellulose in ruminants.

1. The photograph, right, shows three test tubes of thioglycolate inoculated with microbes. Identify the microbes in the tubes A, B, and C as either an obligate aerobe, an obligate anaerobe, or a facultative anaerobe.

   (a) Tube A: _____

   (b) Tube B: _____

   (c) Tube C: _____

2. Explain the advantage of being a facultative anaerobe: _____
_____
_____
_____

Eunice Laurent CC 4.0

A   B   C

B4.2
2

©2024 **BIOZONE** International
ISBN: 978-1-99-101410-8
Photocopying prohibited

142 Modes of Nutrition

Key Idea: Autotrophs (self feeders) use light or chemical energy to make their own food. Heterotrophs feed on other things to gain energy and carbon.

The nutritional mode of an organism describes how it obtains its energy and carbon. Autotrophs make their food from simple inorganic substances using the free energy in sunlight or chemical energy. Heterotrophs feed on other organisms to obtain their energy. They depend either directly on other organisms (dead or alive) or their by-products, e.g. faeces, cell walls, or food stores.

Autotrophic nutrition (autotrophs)

Green plant

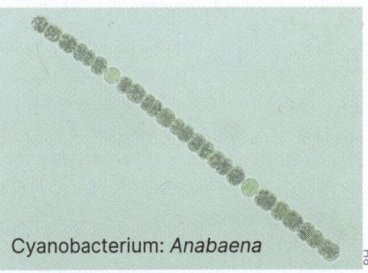
Cyanobacterium: *Anabaena*

Photoautotrophs (photosynthetic organisms) use light as their energy source, and carbon dioxide as a source of carbon to make their own food. They include bacteria, cyanobacteria, algae, and green plants, and some archaea.

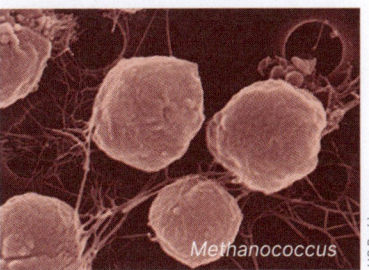
Methanococcus

Chemoautotrophs (chemosynthetic organisms) use inorganic compounds (e.g. elemental hydrogen) as a source of energy, and CO_2 as a source of carbon. Most are bacteria or archaea that live in hostile environments, such as geothermal and deep sea vents, e.g. *Methanococcus* (above) uses hydrogen to reduce CO_2 to methane.

Heterotrophic nutrition (chemoheterotrophs)

Chemoheterotrophs need an organic source of carbon and energy. This is usually glucose (from plants, dead material, or other animals). They include all animals, and many bacteria and protists. The ingestion of solid or liquid organic material from other organisms (holozoic nutrition) is the main feeding mode of animals. All fungi (above right) are chemoheterotrophs and most are decomposers (saprotrophs), obtaining nutrition from the extracellular digestion of dead organic material.

Mixotrophic nutrition

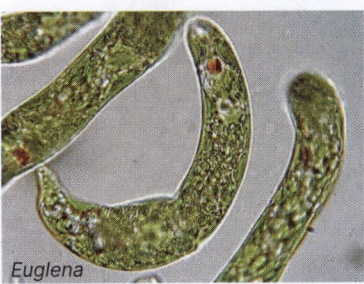
Euglena

Coral can be said to mixotrophic due to their symbiotic relationship with zooxanthellae.

Mixotrophic organisms can use both autotrophic and heterotrophic modes of nutrition. The protist *Euglena* is a well known example and is a facultative mixotroph rather than obligate. Mixotrophs may comprise half of all plankton.

1. Distinguish between photoautotrophs, chemoautotrophs, and chemoheterotrophs in terms of their sources of energy and carbon:

   (a) Photoautotroph

      i. Source of energy: _____

      _____

      ii. Source of carbon: _____

      _____

   (b) Chemoautotrophs

      i. Source of energy: _____

      _____

      ii. Source of carbon: _____

      _____

   (c) Chemoheterotroph

      i. Source of energy: _____

      _____

      ii. Source of carbon: _____

      _____

2. What would be the advantage of being mixotrophic?

   _____
   _____
   _____
   _____
   _____
   _____
   _____

3. What is the main difference between holozoic feeders and saprotrophs in the way that they obtain their nutrition?

   _____
   _____
   _____
   _____
   _____
   _____

B4.2

3-7

143 Dentition and Diet in Hominidae

Key Idea: Hominids show dental adaptations for both herbivorous and omnivorous diets.

The taxonomic family Hominidae, also known as the great apes, includes humans, chimpanzees, gorillas, and orangutans. Humans and chimpanzees (including bonobos) are omnivores. Orangutans are herbivores expect for occasionally eating insects and birds eggs. Gorillas are entirely herbivorous. The dentition of the groups within hominidae in part reflect their dietary choices, with either smaller teeth adapted for a diet with generally softer food, or larger teeth adapted for coarser food and a greater amount of chewing.

Dentition in Hominidae

Teeth are very hard structures, specialized for chewing food (mastication). Their shape and size is indicative of diet, as is the size of the upper and lower jaw (mandible). In humans (below), teeth are relatively small and adapted for a generalized diet of mostly soft food.

Adult (permanent) teeth per quadrant

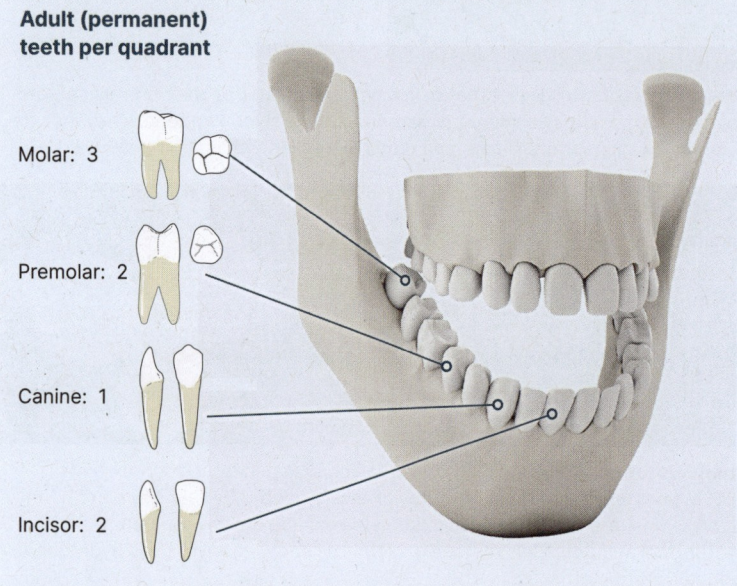

Molar: 3
Premolar: 2
Canine: 1
Incisor: 2

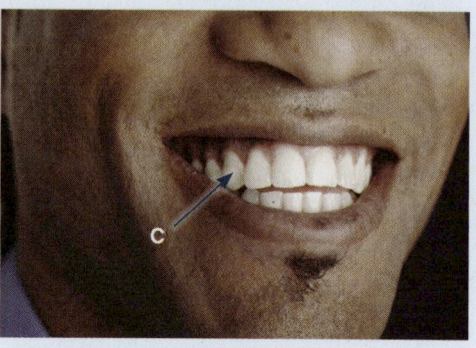

Incisors are adapted for biting, clipping, or nipping food. They are generally large in herbivores and small in carnivores. In humans, the canine (C) teeth are greatly reduced. However in other hominids, such as chimpanzees, the canine teeth are much larger. This is in part because they are still used in threat displays (below)

The oral cavity is divided into quadrants and the number and type of teeth is described by a **dental formula**. In a hominids the formula is written as:

$$\frac{2\ 1\ 2\ 3}{2\ 1\ 2\ 3}$$

indicating the number of teeth in the top and bottom jaw (one side of the mouth). Numbers indicate incisors, canines, premolars, molars. Other mammal groups have different formulae.

The molars sit at the back of the mouth. They are adapted for crushing and grinding food. The tougher the food, the bigger the molars.

Dentition of human relatives

The diagrams below show the mandible of ancient human ancestors and relatives. It can be seen that the jaw structure has changed over time. As primates, they all have a dental formula of 2123, but the size of the jaw and teeth vary. Use the QR codes or go to the **BIOZONE Resource Hub** to examine the skulls in 3D.

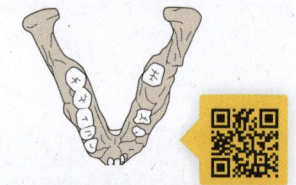
Australopithecus afarensis
3.9-2.9 mya

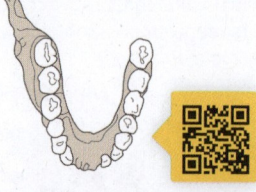
Homo habilis
2.31-1.65 mya

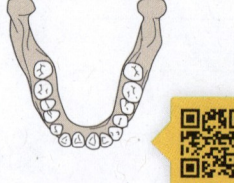
Homo erectus
2-0.1 mya

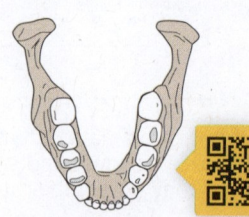
Paranthropus robutus
2.3-0.8 mya

1. Describe the purpose of the:

   (a) Incisors: _____

   (b) Canine teeth: _____

   (c) Molars: _____

B4.2
8

©2024 **BIOZONE** International
ISBN: 978-1-99-101410-8

Comparing dentition and diets

‣ The type of food eaten greatly affects the evolution of the teeth and jaw structure. Mammals that eat meat e.g. cats and dogs, have teeth adapted to grip their prey, slice flesh, and crush or cut through bone. Mammals eating a more varied diet, e.g. pigs and rodents, tend to have much less specialized teeth adapted to dealing with a wide range of foods. Herbivorous mammals, especially those adapted to grazing, have teeth evolved to clip vegetation and grind up coarse fibres.

‣ Primates, including hominids, are mostly herbivorous. Of the group, humans and chimpanzees have a much more omnivorous diet, humans especially so. This is reflected in the shape of the jaw and teeth.

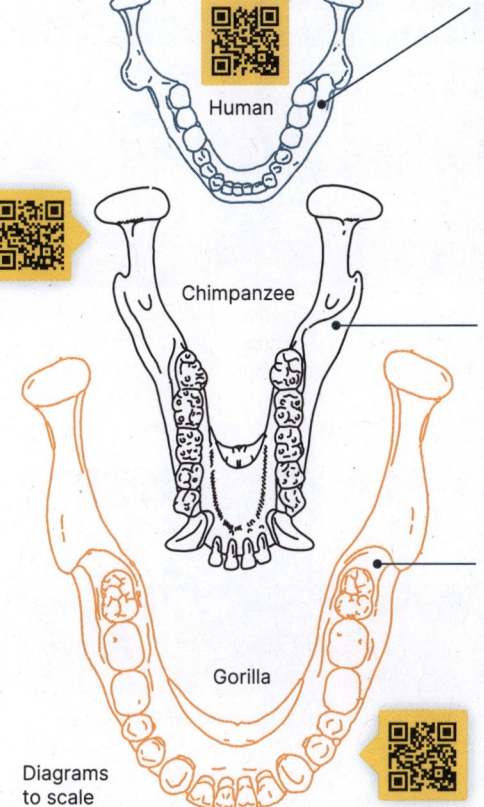

Human

Chimpanzee

Gorilla

Diagrams to scale

Chimpanzee picking fruit

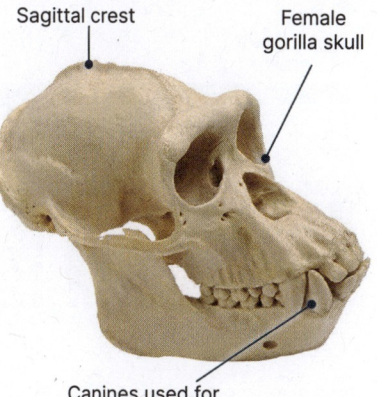

Sagittal crest

Female gorilla skull

Canines used for display (in males)

Humans have a relatively small mandible and teeth. Humans began to use fire to cook food about 1 million years ago. Cooking softens food and so the jaw needs to produce much less force to chew food than if the food was raw. A modern human can produce a bite force of about 700 Newtons (bite force is a measure of the force produced by the jaw muscles rather than the pressure exerted by the teeth).

Chimpanzees eat a variety of foods but fruit makes up the majority of their diet. They also eat termites and, on occasion, they will obtain meat by hunting monkeys. Their food is generally much softer than that of gorillas but chimpanzee jaws and teeth are still quite robust.

Gorillas have a massive mandible and teeth compared to chimpanzees and humans. These are adapted to chew coarse vegetation, including shoots, leaves, bark, and fruit. The muscles for the jaw bone are also massive, connecting from the jaw to the sagittal crest running along the middle of the skull (right). Based on analysis of their teeth, gorillas can produce a bite force of 2,865 Newtons.

From tooth to tiger

Teeth are the hardest structures in the body. As such, they are often all that is left in fossils. From the shape of the tooth, it is possible to deduce the diet, and therefore habit, of an extinct animal. This is based on comparing fossil teeth with the teeth of living animals and the foods they eat.

From the teeth of this animal, we can deduce what it eats and its habit.

- The carnassials (specialized molars) indicate the shearing of meat. This animal is therefore a carnivore.
- The large, sharp canines indicate the need to grip and throttle. The animal is an active hunter.
- The large teeth will need a large skull and body. Long legs are needed for running after prey and claws for grip.
- And so from a tooth, we get a tiger.

2. How can the diet of an animal be deduced from its teeth? Use the hominids as examples:

_____

_____

_____

_____

3. Use the QR codes or go to the **BIOZONE Resource Hub** to examine the 3D skulls of the human ancestors and relatives. Which has the most coarse diet? Which has a more general diet. Explain your answer:

_____

_____

_____

144 Herbivores and Resisting Herbivory

Key Idea: Animals possess a wide range of adaptations for feeding on plants. In response, plants have developed many different ways to dissuade herbivores.

Plants produce fruit and seeds, have sugary sap, and nutritious leaves and roots, all of which provide important food sources for animals with the adaptations to harvest them. However, many plants also have adaptations to deter herbivores from feeding on them, including producing poisons and growing structures such as spines.

Adaptations for herbivory

Piercing mouthparts
Modified mouthparts pierce the phloem to allow feeding on sugary sap. Low nutrients means large volumes of sap are required.

Chewing mouthparts
Many insects feed on leaves by eating them directly. The mouthparts are modified to efficiently bite and chew.

Inhibiting toxins
Many herbivores produce enzymes that break down plant toxins. These include mixed function oxidases.

Grazing
Many large mammals eat grasses. Their incisor teeth are modified for clipping, and the molars for chewing.

Ruminants
Ruminants, e.g. cattle, house microbes in a special part of the stomach (the rumen) that break down cellulose, helping further digestion.

Browsing
Browsers often have large guts for storing and fermenting leaves. Leathery lips prevent damage from thorns. Mobile tongues reach between leaves.

Preventing herbivory

Spines and thorns
Thorns and spines protect leaves by causing herbivores pain and physical damage.

Chemical defences
Plants produce a range of chemicals to deter herbivores, including caffeine, strychnine, tannins, and resins.

Hairs and waxes
Many plants cover their leaves and stems with hairs and waxes that affect texture and irritate herbivore mouthparts.

Using animal defenders
Plants, such as *Vachellia*, use food rewards to recruit animals e.g. ants, to defend them against other herbivores.

Mimicry
Some plants are able to mimic objects to deter herbivores. The passion flowers mimic butterfly eggs to prevent them laying on their leaves.

Timing bud burst
Waiting for as long as possible before bud burst, or timing it to coincide with others, can starve then overwhelm herbivores.

1. Identify two modifications of insect mouthparts for herbivory: _____

2. Describe how ruminants are able to digest cellulose from plant material: _____
_____
_____

3. Using two examples from those above, describe two sets of 'arms races' between herbivores and plants:
_____
_____
_____

B4.2
9

©2024 **BIOZONE** International
ISBN: 978-1-99-101410-8
Photocopying prohibited

145 Predator-Prey Strategies

Key Idea: For every predator adaptation for capturing prey, there is a prey adaptation to prevent being caught. Predators have numerous adaptations for locating, identifying, and subduing prey. Prey can avoid being eaten by using passive defences, such as hiding, or active ones, such as escaping or defending themselves against predators.

Predator avoidance strategies

Wasp beetle

Monarch butterfly

Mimicry
Harmless prey gain immunity from attack by mimicking harmful animals. This is called Batesian mimicry.

Poisonous
Poisonous animals often advertise their unpalatability by using brightly coloured and gaudy markings.

Meerkat

Skunk

Standing guard
Many animals in family groups, e.g. meerkats, post lookouts to spot predators and warn others of the danger.

Chemical defence
Some animals can produce offensive smelling chemicals. American skunks squirt a nauseous fluid at attackers.

Deer

Leaf insect

Offensive weapons
Offensive weapons such as antlers, claws, and horns are essential if prey are to fend off a predator's attack.

Camouflage
Cryptic shape and colouration allow some animals to blend into their background, like this insect above.

Prey capturing strategies

Praying mantis

Basking shark

Concealment
Some animals camouflage themselves in their habitat, and strike when the prey comes within reach.

Filter feeding
Many marine species, e.g. barnacles, baleen whales, manta rays, sponges, filter the water to extract tiny plankton.

Chimpanzee

Rattlesnake
Infrared pit

Tool use
Some animals are gifted tool users. Chimpanzees use prepared twigs to extract termites from mounds.

Stealth
The night hunting ability of some poisonous snakes is greatly helped by the presence of infrared sensors.

Frog fish

Spider with prey

Lures
This angler fish, glow worms, and a type of spider all use lures to attract prey within striking range.

Traps
Spiders have developed a unique method of trapping their prey. Strong, sticky silk threads trap flying insects.

1. Identify a predator avoidance behaviour: _____

2. Describe the behaviour of a named predator that facilitates prey capture: _____

_____

3. How does Batesian mimicry benefit the mimic (the one that is actually edible)? _____

_____

4. Identify two structures prey might use to ward off predators: _____

5. Describe a chemical defence use by a prey species: _____

_____

6. How is heat detection, used by some snakes, an advantage when hunting? _____

_____

_____

B4.2
10

146 Plant Adaptations for Gathering Light

Key Idea: There is great competition between plants for light. This has caused plants to evolve many different strategies to gather light.

In plants, the leaf is the primary photosynthetic organ and is adapted to maximize light capture. The structure of the leaf reflects the environment of the leaf (sun or shade, terrestrial or aquatic), as well as its resistance to water loss and the importance of the leaf relative to other parts of the plant that may be photosynthetic, such as the stem. To allow the leaf to gather as much light as possible, plants have a vast array of adaptations. These depend on the environment, but generally, the higher the plant can get its leaves, the better for it. This is especially so in places of high competition for light and space, such as tropical rainforests.

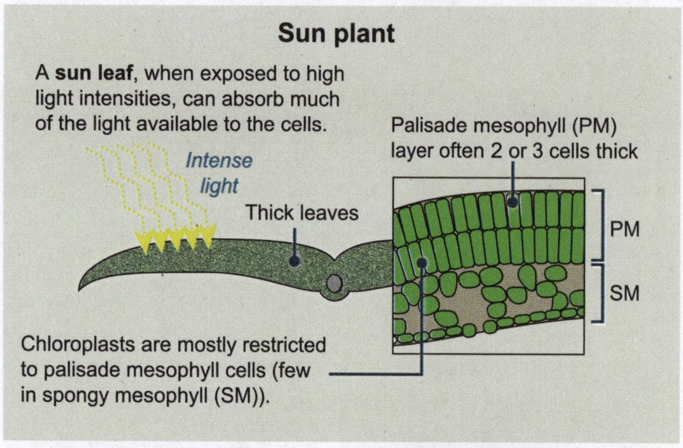

Sun plant

A **sun leaf**, when exposed to high light intensities, can absorb much of the light available to the cells.

Intense light

Thick leaves

Palisade mesophyll (PM) layer often 2 or 3 cells thick

PM

SM

Chloroplasts are mostly restricted to palisade mesophyll cells (few in spongy mesophyll (SM)).

Epiphytes grow on branches high in the canopy. Their roots may hang free in the moist air or anchor the plant in the crevices between the branches and the truck.

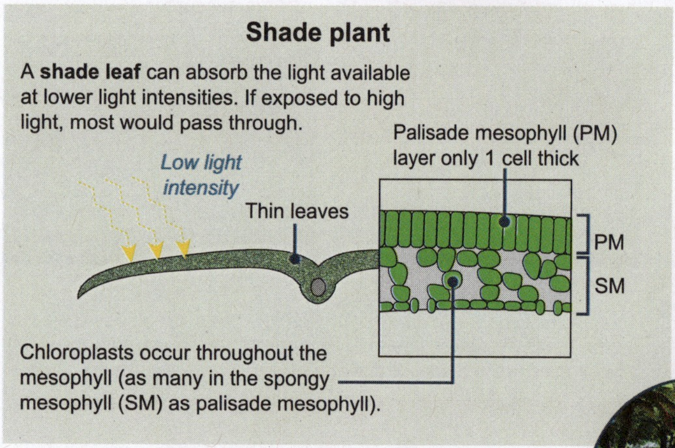

Shade plant

A **shade leaf** can absorb the light available at lower light intensities. If exposed to high light, most would pass through.

Low light intensity

Thin leaves

Palisade mesophyll (PM) layer only 1 cell thick

PM

SM

Chloroplasts occur throughout the mesophyll (as many in the spongy mesophyll (SM) as palisade mesophyll).

Shade tolerant plants live the lower layers of the forest. Their leaves can be very large and often relatively thin. The leaves are specially adapted to the low light of the forest floor.

One strategy to gather light is to simply outgrow other plants by growing as tall as possible. These emergents are the tallest in the rainforest.

Lianas are long stemmed, woody vines that root in the soil and climb into the canopy using trees and shrubs as support. There, they produce flowers and leaves.

Vinayaraj CC 4.0

Hemiepiphytes start life as an epiphyte but their roots eventually grow down to the ground and envelop the host plant, eventually killing it. The strangler fig is a famous example.

1. Explain how the strangler fig can effectively bypass the effort of building a trunk as a sapling, yet still form extremely large trees:

_____

_____

_____

2. How do plants on the forest floor gather light? _____

_____

_____

B4.2

11

©2024 **BIOZONE** International
ISBN: 978-1-99-101410-8
Photocopying prohibited

147 Fundamental and Realized Niches

Key Idea: Evidence from studies of natural populations shows that competition can reduce niche breadth.
Seashores provide a wide range of habitats and opportunities. This results in a high diversity and abundance of organisms competing for the benefits of living there. Competition in species of barnacle, a common crustacean on rocky shores, was studied by J.H. Connell in Scotland. By removing one barnacle species and observing the effect on another, it was possible to determine the extent of the fundamental niches and compare them to the realized niches.

Competitive exclusion in barnacles

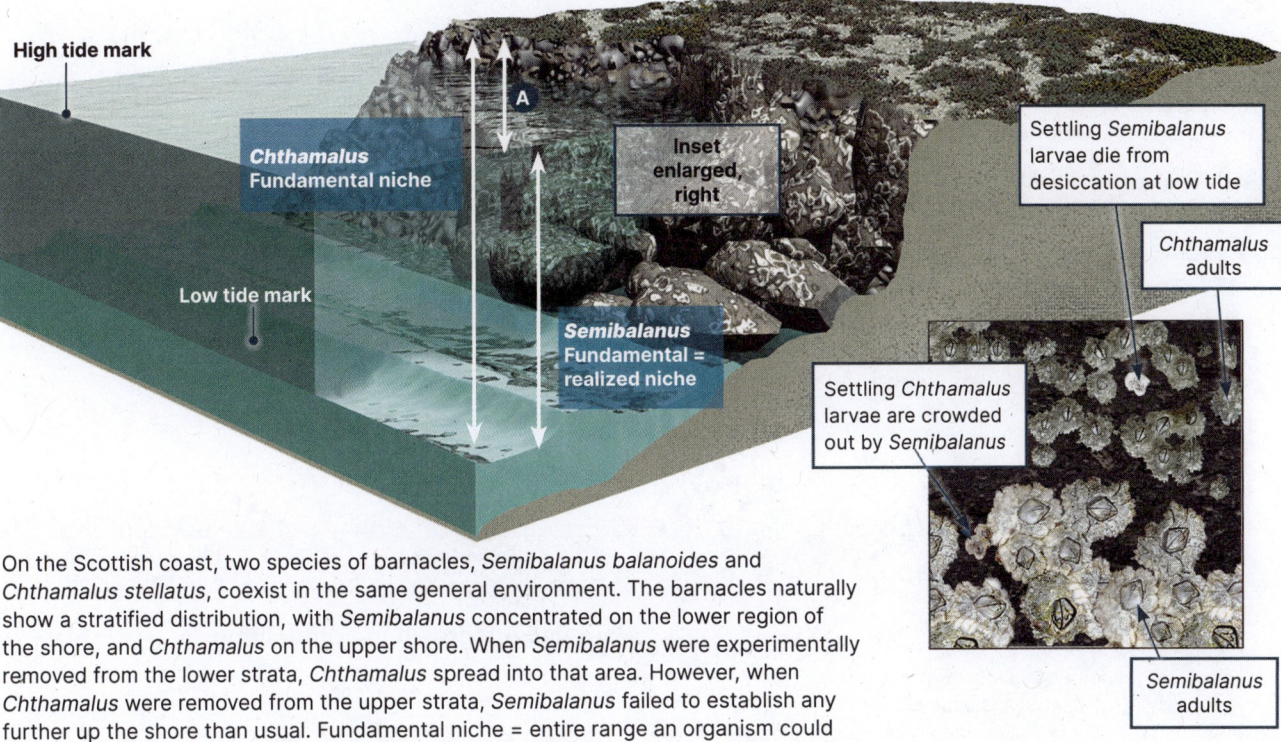

High tide mark

A

Chthamalus Fundamental niche

Inset enlarged, right

Low tide mark

Semibalanus Fundamental = realized niche

Settling *Semibalanus* larvae die from desiccation at low tide

Chthamalus adults

Settling *Chthamalus* larvae are crowded out by *Semibalanus*

Semibalanus adults

On the Scottish coast, two species of barnacles, *Semibalanus balanoides* and *Chthamalus stellatus*, coexist in the same general environment. The barnacles naturally show a stratified distribution, with *Semibalanus* concentrated on the lower region of the shore, and *Chthamalus* on the upper shore. When *Semibalanus* were experimentally removed from the lower strata, *Chthamalus* spread into that area. However, when *Chthamalus* were removed from the upper strata, *Semibalanus* failed to establish any further up the shore than usual. Fundamental niche = entire range an organism could occupy. Realized niche = range the organism actually occupies.

1. (a) In the example of the barnacles above, describe what is represented by the zone labelled with the arrow A:

    _____

    _____

    _____

    (b) Outline the evidence for the barnacle distribution being the result of competitive exclusion:

    _____

    _____

    _____

    _____

2. (a) What keeps *Semibalanus* larvae from establishing at higher shore levels? _____

    _____

    (b) What is the consequence of this to the realized niche compared to the fundamental niche of *Semibalanus*?

    _____

    _____

3. There are many studies underway about the effect global warming and rising sea levels might have on marine communities. What effect might a rise in sea level have on the *Chthamalus/Semibalanus* community?

    _____

    _____

    _____

B4.2

12

148 Competition and Competitive Exclusion

Key Idea: Two species with the same niche cannot inhabit the same area as they will compete for resources and one will eventually exclude the other.

Interspecific competition (competition between different species) is usually less intense than intraspecific (same species) competition because coexisting species have evolved slight differences in their realized niches. However, when two species with very similar niche requirements are

brought into direct competition, e.g. through the introduction of a foreign species, one usually benefits at the expense of the other, which is excluded (the competitive exclusion principle). The introduction of alien species is implicated in the competitive displacement and decline of many native species. Displacement of native species by introduced ones is more likely if the introduced competitor is adaptable and hardy, with high fertility.

Competition in *Paramecium*

In 1934, Georgii Gause, a Russian biologist, carried out a series of experiments on *Paramecium*. The results led him to propose the competitive exclusion principle, a fundamental idea in ecology.

In the first stage of the experiments, he grew three species of *Paramecium* in isolation in a nutritive medium containing their essential resource (bacterial food). Their growth curves are shown below:

Paramecium grown in isolation

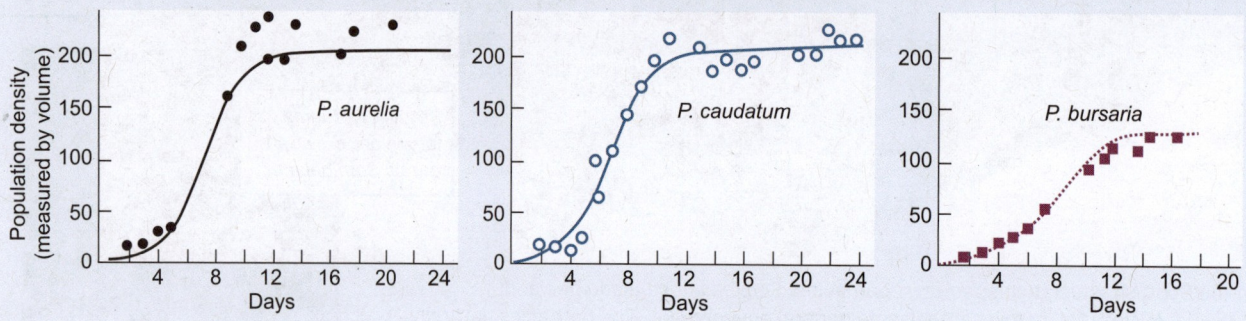

In the second stage of the experiment Gause grew *P. aurelia* and *P. caudatum* together. He found that *P. caudatum* was always out-competed and became extinct from the culture. Gause then grew *P. caudatum* with *P. bursaria*. He found they were able to exist together (but at lower numbers). Investigation found that *P. caudatum* occupied the oxygen rich top half of the culture tube, whereas *P. bursaria* retreated to the lower, poorly oxygenated region. *P. bursaria* contains symbiotic algae, which release oxygen in photosynthesis. This allows *P. bursaria* to remain in the anoxic zone.

Paramecium grown in competition

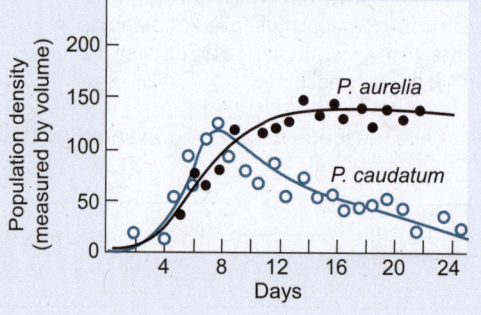
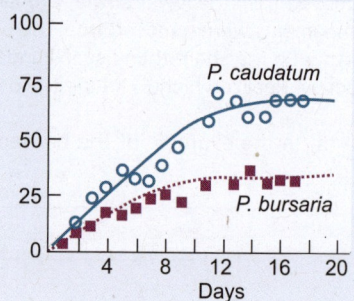

1. What is the competitive exclusion principle? _____

    _____

    _____

2. (a) What was the fundamental niche of all the *Paramecium* species? _____

    _____

    (b) What were the realized niches of *P. caudatum* and *P. bursaria*? _____

    _____

    _____

    (c) How do the experiments support Gause's competitive exclusion principle? In what way do they not?

    _____

    _____

    _____

©2024 BIOZONE International
ISBN: 978-1-99-101410-8

B4.2
13

149 Did You Get It?

1. The figure below shows changes in vegetation cover along a 2 m vertical transect up the trunk of an oak tree. Changes in the physical factors: light, humidity, and temperature along the same transect were also recorded.

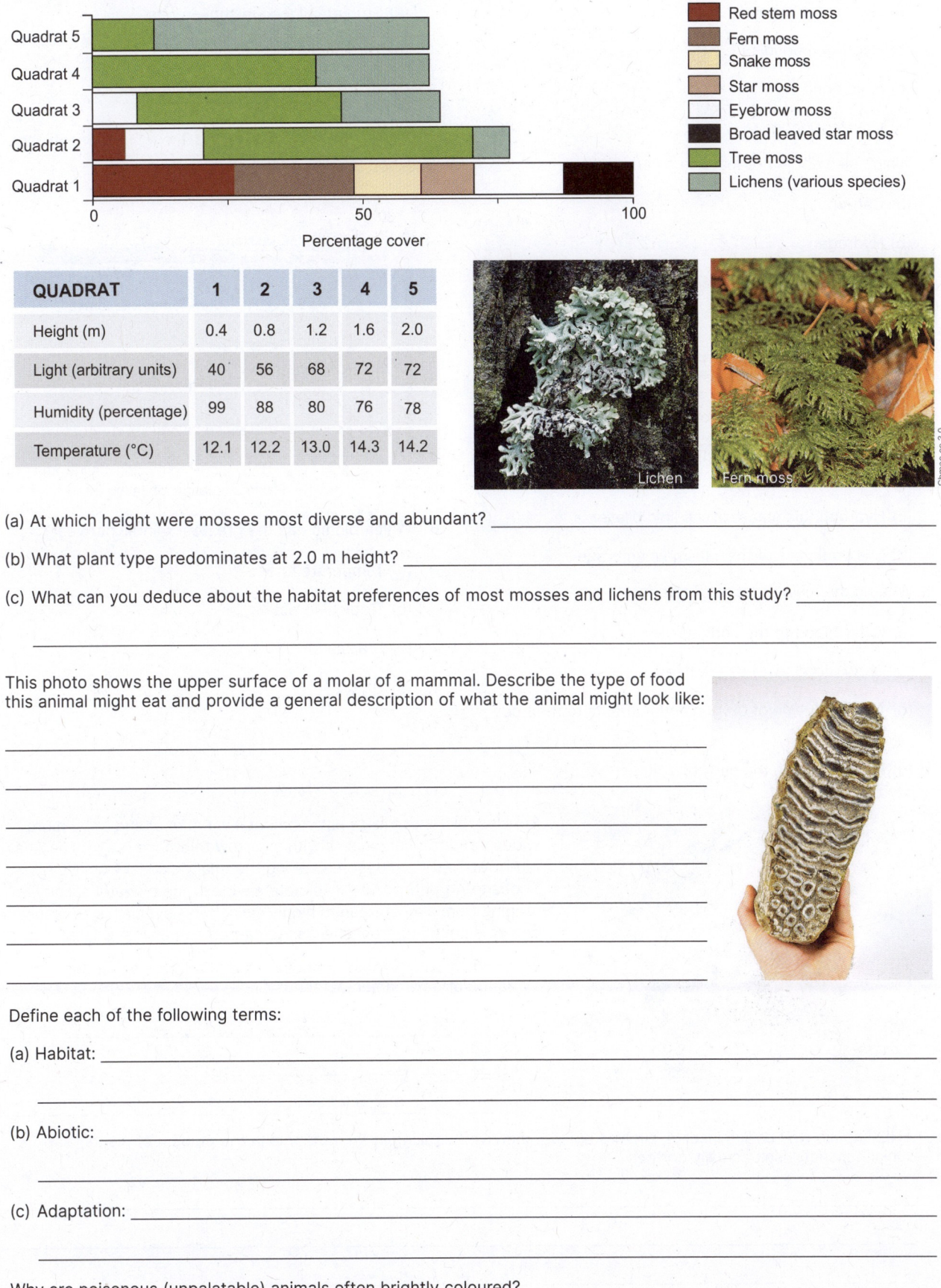

Legend:
- Red stem moss
- Fern moss
- Snake moss
- Star moss
- Eyebrow moss
- Broad leaved star moss
- Tree moss
- Lichens (various species)

Percentage cover

QUADRAT	1	2	3	4	5
Height (m)	0.4	0.8	1.2	1.6	2.0
Light (arbitrary units)	40	56	68	72	72
Humidity (percentage)	99	88	80	76	78
Temperature (°C)	12.1	12.2	13.0	14.3	14.2

Lichen   Fern moss

Chmee cc 3.0

(a) At which height were mosses most diverse and abundant? _____

(b) What plant type predominates at 2.0 m height? _____

(c) What can you deduce about the habitat preferences of most mosses and lichens from this study? _____

_____

2. This photo shows the upper surface of a molar of a mammal. Describe the type of food this animal might eat and provide a general description of what the animal might look like:

_____

_____

_____

_____

_____

_____

_____

3. Define each of the following terms:

(a) Habitat: _____

_____

(b) Abiotic: _____

_____

(c) Adaptation: _____

_____

4. Why are poisonous (unpalatable) animals often brightly coloured? _____

_____

_____

150 Summary Assessment

1. Which type of bond involves sharing of electron pairs between atoms:

    (a) Hydrophobic bond

    (b) Hydrogen bond

    (c) Ionic bond

    (d) Covalent bond

2. Which element in not found in a carbohydrate?

    (a) Carbon

    (b) Hydrogen

    (c) Nitrogen

    (d) Oxygen

3. The primary barrier to the passage of ions and polar molecules through the plasma membrane is the:

    (a) Boundary layer of carbohydrates

    (b) Hydrophobic nature of the proteins in the plasma membrane

    (c) Hydrophobic nature of the lipid bilayer

    (d) The thickness of the plasma membrane

4. What is the main purpose of arteries?

    (a) Carry blood to the body tissues

    (b) Carry blood towards the heart

    (c) Allow diffusion of oxygen and nutrients to the body tissues

    (d) Pump blood to the body

5. Which of the following adaptations is likely to be found in a xerophtyic plant?

    (a) Stomata placed in sunken pits

    (b) Trichomes

    (c) Reduced leaf size

    (d) All of the above

Question 6 refers to the graph below

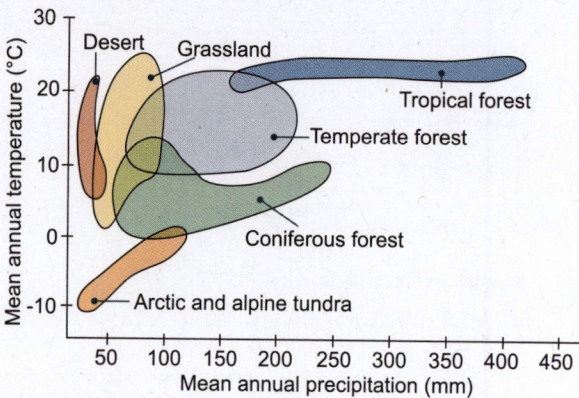

6. Which biome has the greatest rainfall per year?

    (a) Temperate forest

    (b) Tropical forest

    (c) Grassland

    (d) Tundra

Species tolerant of large environmental variations tend to be more widespread than organisms with a narrow tolerance range. The Atlantic blue crab (left) is widespread along the Atlantic coast from Nova Scotia to Argentina. Adults tolerate a wide range of water salinity, ranging from almost fresh to highly saline. This species is an omnivore and eats anything, from shellfish to carrion and animal waste.

7. Suggest an advantage to being able to tolerate variations in a wide range of environmental factors?

    _____

    _____

    _____

8. **AHL**: Explain why calling insects, such as aphids, that live by taking in the nutrients in the phloem of plants, 'sap sucking insects' isn't actually correct:

    _____

    _____

    _____

    _____

    _____

©2024 **BIOZONE** International
ISBN: 978-1-99-101410-8
Photocopying prohibited

9. **AHL**: People with iron-deficient anaemia lack haemoglobin in the blood. The graph right shows the oxygen-haemoglobin dissociation curves for a person with iron deficient anaemia compared to a person with normal haemoglobin levels.

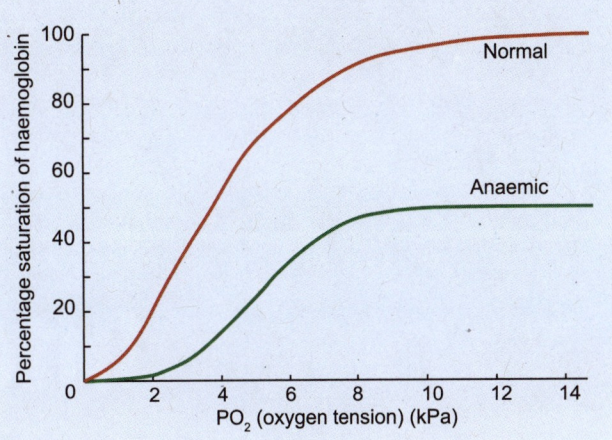

(a) What does an oxygen-haemoglobin dissociation curve show?

_____

_____

_____

_____

_____

(b) What has happened to the oxygen-haemoglobin dissociation curve in the anaemic person and why?

_____

_____

_____

10. Transpiration in a hydrangea shoot was investigated using a potometer (measures water loss). The experiment was set up and the plant left to stabilize (environmental conditions: still air, light shade, 20°C). The plant was then placed in different environmental conditions and the water loss was measured each hour. Finally, the plant was returned to original conditions, allowed to stabilize and transpiration rate measured again. The results are presented below:

Experimental conditions	Temperature (°C)	Humidity (%)	Transpiration rate (g/h)
(a) Still air, light shade, room temperature	20	70	1.20
(b) Moving air, light shade	20	70	1.60
(c) Still air, bright sunlight	23	70	3.75
(d) Still air and dark, moist chamber	19.5	100	0.05

(a) Describe the three processes in which water is transported from the roots to the leaves:

_____

_____

_____

_____

(b) What conditions acted as the control in this experiment? _____

(c) Which factors increased transpiration rate and why? _____

_____

_____

(d) Why did the plant have such a low transpiration rate in humid, dark conditions? _____

_____

_____

_____

11. How can an organisms be both saprotrophic and heterotrophic at the same time? _____

_____

_____

©2024 **BIOZONE** International
ISBN: 978-1-99-101410-8
Photocopying prohibited

Interaction and Interdependence

Effective systems at all biological organizational levels require the coordinated interaction and interdependence of components, leading to the emergence of new properties.

Understandings:

Molecules

C1.1 Enzymes and metabolism

C1.2 Cell respiration

C1.3 Photosynthesis

Cells

C2.1 Chemical signalling

C2.2 Neural signalling

Organisms

C3.1 Integration of body systems

C3.2 Defence against disease

Ecosystems

C4.1 Populations and communities

C4.2 Transfers of energy and matters

⚛ Molecules

Resource Hub
bit.ly/3sTtZRV

C1.1 Enzymes and Metabolism

		Activity Number
Guiding Questions:	▶ How can we classify the different enzyme-molecule reactions?	
	▶ How do key components of metabolism interact together?	

Learning Outcomes:

☐ 1	Explain the importance of enzymes to human metabolism.	**151**
☐ 2	Define metabolism and explain the requirement for many different enzymes.	**151**
☐ 3	Distinguish between anabolic and catabolic reactions providing examples of each.	**151**
☐ 4	Classify enzymes as globular proteins and explain the mechanism for catalysis at the active site.	**151**
☐ 5	Describe the induced fit model of enzyme activity.	**151**
☐ 6	Explain the role of molecular motion and substrate-active site collisions during enzyme catalysis.	**151**
☐ 7	Explore the relationships between the active site structure, enzyme–substrate specificity, and denaturation.	**151**
☐ 8	**AOS:** Interpret graphed data on the effects of temperature, pH, and substrate concentration on the rate of enzyme activity. **NOS:** Recognize that generalized sketches of graphs are models and can be verified by experimental results.	**152**
☐ 9	**AOS:** Investigate reaction rates in enzyme-catalysed reactions including the use of secondary data.	**153**
☐ 10	**AOS:** Interpret data from graphs to determine the effect of enzymes on activation energy.	**152**
☐ 11	**AHL:** Elaborate on intracellular and extracellular enzyme-catalysed reactions including glycolysis, the Krebs cycle, and chemical digestion as examples.	**154**
☐ 12	**AHL:** Explain why heat energy is generated by the reactions of metabolism and how this heat is used by some animals.	**155**
☐ 13	**AHL:** Use glycolysis, the Krebs cycle, and the Calvin cycle to compare different cyclical and linear metabolic pathways.	**156**
☐ 14	**AHL:** Explain how allosteric sites function resulting in non-competitive inhibition.	**157**
☐ 15	**AHL:** Distinguish between non-competitive and competitive inhibition using statins as an example.	**157**
☐ 16	**AHL:** Explain how feedback inhibition is used to regulate metabolic pathways using the production of isoleucine as an example.	**158**
☐ 17	**AHL:** Describe the process of mechanism-based inhibition using penicillin and its subsequent resistance as examples.	**158**

C1.2 Cell Respiration

		Activity Number
Guiding Questions:	▶ Why does cellular energy release rely on hydrogen and oxygen?	
	▶ What are the main energy transformations in a cell?	

Learning Outcomes:

☐ 1	Link the properties of adenosine triphosphate (ATP) to its function as a cellular energy currency.	**159**
☐ 2	Elaborate on life processes at cellular level that require a supply of energy from ATP.	**159**
☐ 3	Describe the energy transfers involved in ATP and ADP interconversions.	**160**
☐ 4	Distinguish between gas exchange and cell respiration and elaborate on the reactants and products of respiration involving the production of ATP.	**161**
☐ 5	Compare and contrast anaerobic and aerobic cell respiration in humans including use of word equations.	**161**
☐ 6	**AOS:** Investigate variables affecting the rate of cell respiration including calculating the rate from raw data or secondary data.	**162**

C1.2 Cell Respiration

Guiding Questions:
▶ Why does cellular energy release rely on hydrogen and oxygen?
▶ What are the main energy transformations in a cell?

Learning Outcomes:

☐	7	**AHL:** Explain the role of NAD during cell respiration including products of reactions.	163
☐	8	**AHL:** Use annotated models to explain key steps in the glycolysis pathway.	163
☐	9	**AHL:** Describe how pyruvate is converted to lactate to generate NAD in anaerobic cell respiration.	165 - 166
☐	10	**AHL:** Explain key steps in the anaerobic respiration pathway including the significance of NAD regeneration.	165 - 166
☐	11	**AHL:** Elaborate on the respiration link reaction where pyruvate oxidation and decarboxylation occurs.	163
☐	12	**AHL:** Elaborate on acetyl oxidation and decarboxylation in the Krebs cycle.	163
☐	13	**AHL:** Explain how transfer of energy is facilitated in the electron transport chain process.	163
☐	14	**AHL:** Elaborate on the formation of a proton gradient during the electron transport chain process.	163
☐	15	**AHL:** Describe the processes of chemiosmosis and ATP synthesis in the mitochondria.	164
☐	16	**AHL:** Explain the role of oxygen acting as a terminal electron acceptor in aerobic cell respiration.	164
☐	17	**AHL:** Contrast the structure and function of lipids and carbohydrates when acting as respiratory substrates.	164

C1.3 Photosynthesis

Guiding Questions:
▶ How does radiant energy enter into a photosynthetic reaction, and how is this energy used?
▶ How is photosynthesis influenced by abiotic factors?

Learning Outcomes:

☐	1	Summarize how light energy is transformed to chemical energy during the process of photosynthesis including its importance to life processes in ecosystems.	167
☐	2	Write a word equation for photosynthesis showing all reactants and products.	167
☐	3	Elaborate on the production of oxygen in photosynthesis.	167
☐	4	**AOS:** Use collected data from chromatographic separation of photosynthetic pigments to calculate R_f values.	168
☐	5	Analyse data to explain how photosynthetic pigments absorb specific light wavelengths.	169
☐	6	**AOS:** Analyse data to determine photosynthetic rates against different light wavelengths and plot an action spectrum.	169
☐	7	**AOS:** Develop a hypothesis to investigate the effect of limiting factors, including varying concentrations of carbon dioxide, light intensity, or temperature on the rate of photosynthesis. **NOS:** Identify the dependent and independent variables.	170
☐	8	Analyse how free-air carbon dioxide enrichment (FACE) and enclosed greenhouse experiments can be used to investigate rates of photosynthesis and plant growth. **NOS:** Identify the challenges to controlling variables in some investigations.	170
☐	9	**AHL:** Explain how photosystems are able to generate and emit excited electrons.	171 - 172
☐	10	**AHL:** Discuss the advantages a photosystem consisting of different types of pigment molecules has over a single molecule of chlorophyll or any other pigment.	172
☐	11	**AHL:** Elaborate on the process of photosystem II where oxygen is generated by the photolysis of water.	172
☐	12	**AHL:** Elaborate on the process of chemiosmosis where ATP is produced, including the source of electrons.	172
☐	13	**AHL:** Elaborate on the process of photosystem I where NADP is reduced.	172
☐	14	**AHL:** Explain how thylakoids act as systems for light-dependent reactions of photosynthesis.	172
☐	15	**AHL:** Explain how enzyme Rubisco enables carbon fixation including the substrates and products involved.	173
☐	16	**AHL:** Explain how the reduced forms NADP and ATP are able to synthesize triose phosphate.	173
☐	17	**AHL:** Elaborate on how ATP is involved in the regeneration of RuBP as part of the Calvin cycle.	173
☐	18	**AHL:** Describe how the products of the Calvin cycle and mineral nutrients are involved in the synthesis of carbohydrates, amino acids and other carbon compounds.	173
☐	19	**AHL:** Discuss the interdependence of the light-dependent and light-independent reactions and the conditions that regulate them.	171

151 The Properties of Enzymes

Key Idea: Enzymes are biological catalysts. The active site is critical to this functional role.

Enzymes are globular proteins that act as biological catalysts, speeding up biochemical reactions in living organisms. The enzyme itself remains unchanged in the process. It is not used up and can help carry out the same reaction many times. Enzymes have a specific shape that is determined by the protein's tertiary structure. Enzyme specificity ensures that enzymes usually only act on one specific chemical or substrate (reactant). In some instances, an enzyme will act on a small group of similar substrates. An enzyme's specificity is a result of the precise configuration of the catalytic region, or active site, where the substrate binds and the chemical reaction occurs to form the product.

The role of enzymes

▶ The sum of cellular reactions is known as metabolism. The rates of chemical reactions increase as temperature increases. The activation rate of a chemical reaction is the minimum amount of energy required to energize reactants to take part in the reaction. Enzymes lower the activation rate of reactions to oocur at body temperature.

▶ A multitude of complex interacting reactions and metabolic pathways occur in living organisms. Therefore, a wide range of specific enzymes, each only functioning for one or a few types of reactants, are present in the organism.

▶ Where different metabolic pathways share a similar reaction step or reactant, the same enzymes are often used for each, i.e Calvin cycle and photo-respiration use RuBisCo.

▶ Enzymes control the rate at which metabolic reactions occur, with a range of inhibitory substances or feedback loops increasing or decreasing the rate of reaction products being formed.

▶ Metabolism can broadly be grouped into catabolic reactions which break apart macromolecules, and anabolic reactions which join substances together.

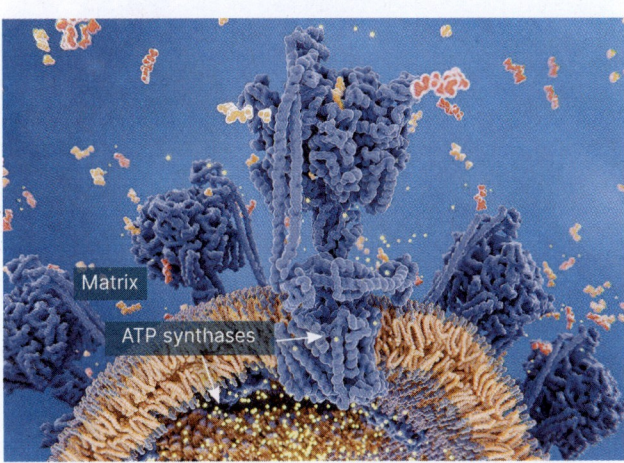

Matrix

ATP synthases

ATP synthase is a transmembrane enzyme that catalyses the synthesis of ATP from ADP and inorganic phosphate, driven by a proton gradient generated by electron transfer. The image above represents ATP synthase in the membrane of a mitochondrion, but it is also found in the membranes of chloroplasts, where ATP is generated in the light dependent reactions of photosynthesis. ATP is the source of cellular energy and 'fuels' the metabolism.

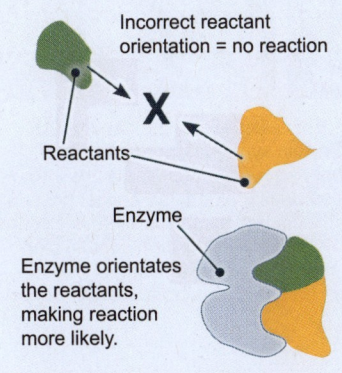

Incorrect reactant orientation = no reaction

X

Reactants

Enzyme

Enzyme orientates the reactants, making reaction more likely.

Collisions can immobilize reactants. Enzymes can also be fixed and embedded within cell membranes.

Enzymes and collisions

▶ When two or more substances join (in an anabolic reaction), the reaction requires reactants to collide with the correct orientation (diagram left). Enzymes enhance reaction rates by providing a site for reactants to come together in such a way that a reaction will occur. They do this by orienting the reactants so that the reactive regions are brought together. In some reactions, one of the reactants is immobilized in the cell and the enzymes move towards them; in others, the enzymes are immobilized, often within the cell membrane, and the reactants move to collide with them.

▶ Collisions must also occur with enough energy in order to be successful and result in a reaction. The higher the temperature, the faster the particles move, and the more energy they have on impact.

▶ Additionally, enzymes may also destabilize the bonds within the reactants, making it easier for a reaction to occur.

1. Explain the link between enzymes and metabolism: _____

_____

_____

2. What is meant by enzyme specificity and why does that result in so many different enzymes in organisms?

_____

_____

_____

_____

_____

Catabolic and anabolic reactions

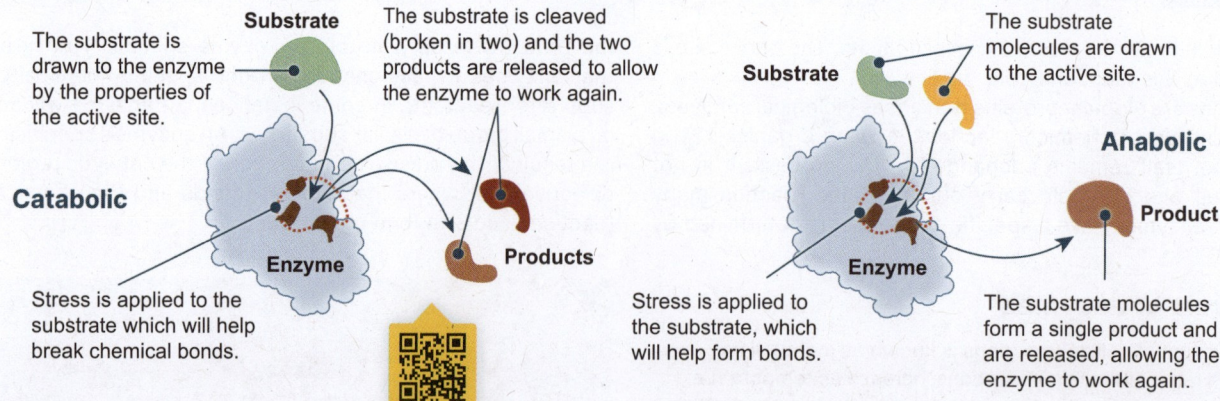

Catabolic

The substrate is drawn to the enzyme by the properties of the active site.

Substrate

The substrate is cleaved (broken in two) and the two products are released to allow the enzyme to work again.

Enzyme

Products

Stress is applied to the substrate which will help break chemical bonds.

Anabolic

The substrate molecules are drawn to the active site.

Substrate

Enzyme

Product

Stress is applied to the substrate, which will help form bonds.

The substrate molecules form a single product and are released, allowing the enzyme to work again.

Enzymes can catalyse the breakdown of molecules

The properties of an enzyme's active site can draw in a single substrate molecule. Chemical bonds are broken, cleaving the substrate to form two separate molecules. Reactions that break down complex molecules into simpler ones are called catabolic reactions and involve a net release of energy (they are exergonic). Examples: Hydrolysis reactions to break macromolecules into monomers, as seen in cellular respiration and digestion.

Enzymes can catalyse the building of molecules

The properties of an enzyme's active site can draw in two substrate molecules. The two substrate molecules form bonds and become a single molecule. Reactions that build more complex molecules and structures from simpler ones are called anabolic reactions and involve a net use of energy (they are endergonic). Examples: Condensation reactions that form macromolecules from monomers, as seen in photosynthesis, protein synthesis, and glycogen formation.

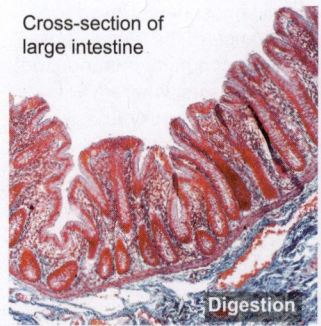

Cross-section of large intestine

Digestion

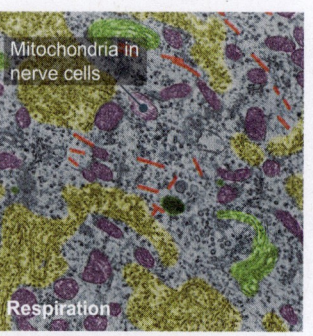

Mitochondria in nerve cells

Respiration

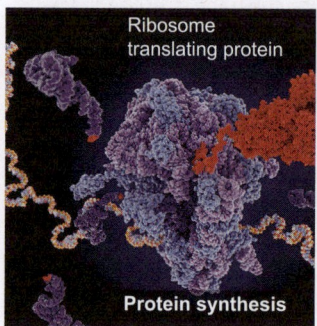

Ribosome translating protein

Protein synthesis

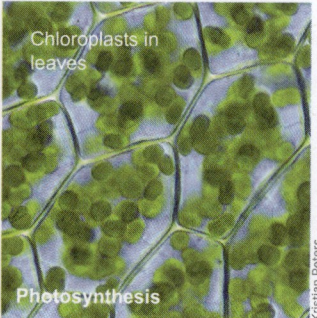

Chloroplasts in leaves

Photosynthesis

Organisms obtain energy through the catabolic reaction of respiration, which breaks glucose macromolecules apart in catabolic reactions. This yields ATP, the universal energy source that ultimately provides energy to fuel cellular reactions. Photosynthetic organisms can manufacture their own glucose sources, but heterotrophs need to consume other organisms to obtain the reactants for respiration. These substances need to be broken down, through the process of digestion, into small enough molecules that can be transported to all cells.

Many macromolecules inside the cells need to be manufactured through anabolic reactions. Protein synthesis, controlled by genetic material, joins together sequences of amino acids into long polypeptide chains that are eventually folded into an inexhaustible variety of proteins. Photosynthetic organisms combine small carbon dioxide and water molecules into glucose, through a multi-step enzyme-controlled metabolic pathway. Although some steps in photosynthesis are catabolic, the final products are made from the original reactants.

3. Distinguish between anabolic and catabolic reactions, with examples: _____

_____

_____

_____

_____

_____

4. What is the link between condensation reactions and anabolism, and hydrolysis reactions and catabolism?

_____

_____

_____

_____

The induced fit model of enzyme action

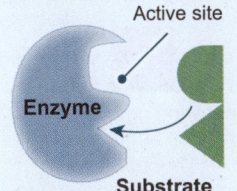

1 A substrate molecule is drawn into the enzyme's active site by its particular properties (resulting from its amino acid side chains). The active site is like a cleft into a three-dimensional protein which the substrate molecule(s) fit.

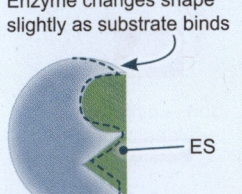

Enzyme changes shape slightly as substrate binds

2 The enzyme changes shape as the substrate binds, forming an enzyme-substrate (ES) complex. Chemical and electrostatic interactions are important in forming the ES complex. The shape change makes a change in the substrate more likely. In this way, the enzyme's interaction with its substrate is an induced fit.

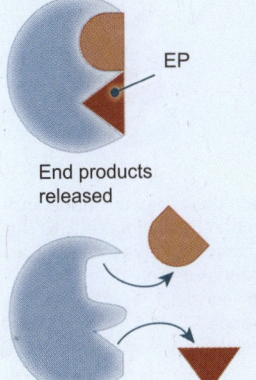

3 The ES interaction results in an intermediate enzyme-product (EP) complex. The substrate becomes bound to the enzyme by weak chemical bonds, straining bonds in the substrate and allowing the reaction to proceed more readily.

End products released

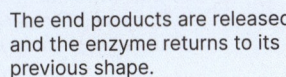

4 The end products are released and the enzyme returns to its previous shape.

Once the substrate enters the active site, the shape of the active site changes to form an active complex. The formation of an ES complex strains substrate bonds and lowers the energy required to reach the transition state, which allows the reaction to proceed. The induced-fit model is supported by X-ray crystallography, chemical analysis, and studies of enzyme inhibitors, which show that enzymes are flexible and change shape when interacting with the substrate.

The active site

An enzyme acts on a specific chemical called the **substrate**. The substrate binds to a specific part of the enzyme called the **active site**, a cluster of a few amino acids that interact together. This site is specific to an enzyme and is a function of its globular, 3-dimensional tertiary structure. The active site accounts for an enzyme's specificity for its substrate (the substrate has a shape and charge to interact correctly with the active site of the enzyme).

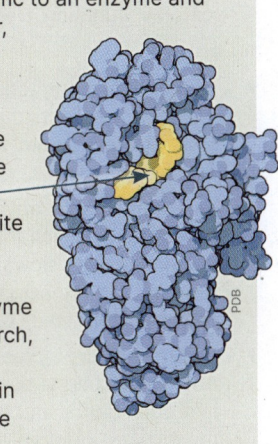

During digestion, the enzyme amylase (blue) breaks starch, a polymer, (yellow) into smaller 2-3 glucose units in a hydrolysis reaction at the enzyme's active site.

Short protein chain (blue and purple) - there are 8 in a RuBisCo enzyme molecule

Large protein chain (yellow and orange) - there are 8 in a RuBisCo enzyme molecule

RuBisCo is a spherical (globular) enzyme composed of 16 protein subunits. It is involved in the first main step of carbon fixation in plants and other photosynthetic organisms. It catalyses the attachment of CO_2 to a 5-C sugar derivative called RuBP. The three-dimensional shape enables just a few amino acids to interact and form specific active sites.

5. Explain how an enzyme interacts with its substrate in an 'induced' fit model: _____

_____

_____

_____

_____

6. (a) What is meant by the active site of an enzyme? Relate this to the enzyme's tertiary structure: _____

_____

_____

_____

(b) Explain how RuBisCo can combine RuBP with both CO_2 in photosynthesis and with O_2 in photorespiration:

_____

_____

Denaturation of enzymes

▶ When an enzyme (or any protein) loses its specific tertiary structure, it also loses its ability to carry out its functional role. The change in shape with associated loss of function is called denaturation and it can occur when environmental factors, e.g. extremes of pH or temperature, disrupt the chemical bonds between amino acids.

▶ In enzymes, the physical and chemical nature of the active site are also altered. Recall that the active site will only interact with a specific substrate. When the shape of the active site is altered, the substrate can no longer bind and the enzyme loses its catalytic function. Denaturation can be reversible, depending on the enzyme and the nature of the denaturing agent.

▶ Hydrophobic interactions and chemical bonds between amino acids hold a protein in its three dimensional functional shape. Bonds (red lines on the diagram, right) include hydrogen bonds, ionic bonds, and disulfide bonds.

▶ All enzymes have optimal conditions, i.e. the environmental conditions in which they work best. Outside the optimal conditions, enzyme activity slows down and will stop completely if the enzyme is denatured (stabilizing bonds are broken). Denaturation can be caused by extremes of temperature or pH.

▶ As temperature increases, the enzyme molecule gains kinetic energy. Above a certain temperature, the energy is so great that stabilizing hydrogen bonds and hydrophobic interactions are shaken apart and the enzyme unfolds.

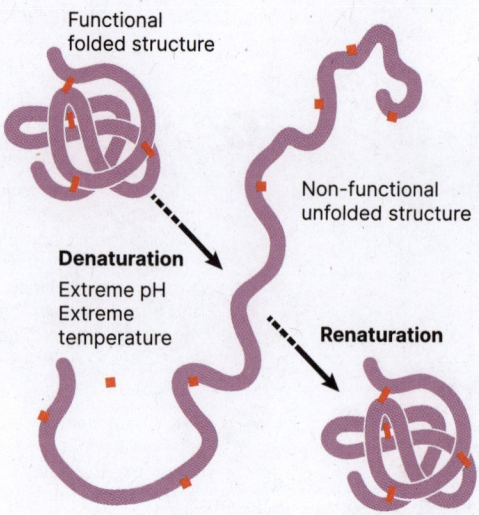

Functional folded structure

Denaturation
Extreme pH
Extreme temperature

Non-functional unfolded structure

Renaturation

Refolded structure

pH is a measure of the number of hydrogen ions in solution.

$$pH = -\log[H^+]$$

Changes in the concentrations of H^+ ions (pH) will affect those bonds that contain a charge, i.e. hydrogen bonds and ionic bonds. These bonds help maintain the tertiary structure and the active site so they are very vulnerable to changes in hydrogen ion concentration. The reaction does not proceed because the substrate does not enter the active site or does not align properly.

Sometimes enzymes can regain their original shape and activity. This is called renaturation (refolding back to the original shape). Renaturation is more likely to occur if a moderate pH change causes denaturation and the environment has returned to a suitable pH. Temperature induced denaturation is usually irreversible.

7. Pepsin is a protein digesting enzyme in the stomach. It is most active between pH 1.5-2.5 (the pH of the stomach). Pepsin is inactive at pH 6.5 or above, but does not become fully denatured until pH 8.0.

(a) When the stomach empties, the partially digested food and enzyme mixture is released into the small intestine. The pH of the small intestine ranges between 6.0-7.4. Would pepsin be active in the small intestine? Explain your answer:

_____

_____

(b) Predict what would happen to pepsin activity under the following conditions. Explain your answer:

(i) A pepsin solution at pH 7.0 was adjusted to pH 1.5: _____

_____

(ii) A pepsin solution at pH 8.5 was adjusted to pH 1.5: _____

_____

8. The graph (right) shows enzyme reaction rate versus temperature for hypothetical enzyme X.

(a) In terms of enzyme activity, what is happening at point A? _____

_____

(b) Enzyme X catalyses a reaction in a critical metabolic pathway. Predict what would happen if the temperature remained high (close to point A):

_____

_____

_____

Rate of reaction

Temperature (°C)

A

©2024 **BIOZONE** International
ISBN: 978-1-99-101410-8
Photocopying prohibited

152 Factors Affecting Enzyme Reaction Rate

Key Idea: Enzymes operate most effectively within a narrow range of conditions and increase the rate of biological reactions by lowering the reaction's activation energy. The rate of enzyme-catalysed reactions is influenced by both enzyme and substrate concentration.

Chemical reactions in cells are accompanied by energy changes. The amount of energy released or taken up is directly related to the tendency of a reaction to run to completion (for all the reactants to form products). Any reaction needs to raise the energy of the substrate to an unstable transition state before the reaction will proceed (below). The amount of energy needed to do this is the activation energy (Ea). Enzymes lower the Ea by destabilizing bonds in the substrate so that it is more reactive. Enzyme reactions can break down a single substrate molecule into simpler substances (catabolic reactions), or join two or more substrate molecules together (anabolic reactions). Enzymes usually have an optimum set of conditions, e.g. of pH and temperature, under which their activity is greatest. Many plant and animal enzymes show little activity at low temperatures.

Enzymes lower the activation energy for biochemical reactions

▸ The presence of an enzyme simply makes it easier for a reaction to take place. All catalysts speed up reactions by influencing the stability of bonds in the reactants. They may also provide an alternative reaction pathway, thus lowering the activation energy (E_a) needed for a reaction to take place (below). These reactions are accompanied by energy changes.

▸ When the reactants have lower energy than the product, the reaction requires energy input and is endergonic. When the reactants have higher energy than the product, energy is released in product formation and the reaction is exergonic.

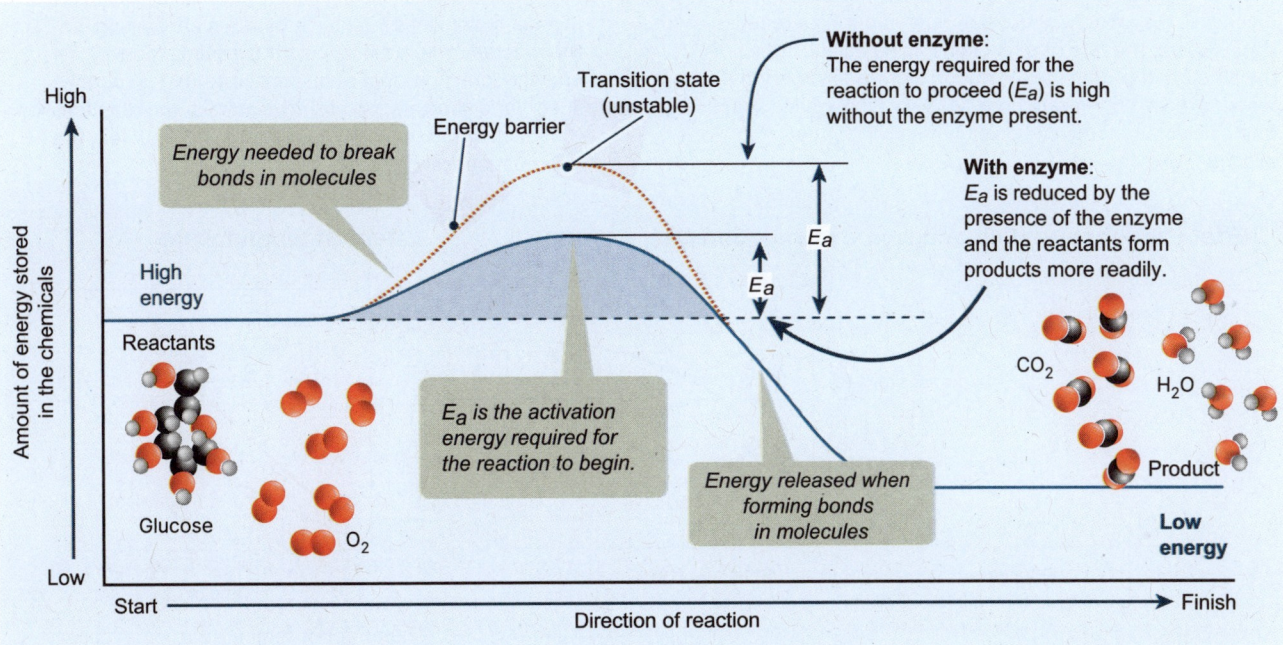

1. Why do reactants need energy added to them in order for them to react? _____

_____

_____

2. Explain how enzymes lower the activation energy for a reaction: _____

_____

_____

_____

3. Respiration is considered an exothermic reaction (NET energy release). Explain this in terms of bonds broken and formed, using information from the graph above:

_____

_____

_____

_____

Interpreting an enzyme catalysed reaction

▶ Enzyme activity increases with increasing temperature, but falls off after the optimum temperature is exceeded and the enzyme is denatured, and the enzyme no longer functions.

▶ Extremes in pH can also cause denaturation. Within their normal operating conditions, enzyme reaction rates are influenced by enzyme and substrate concentration in a predictable way.

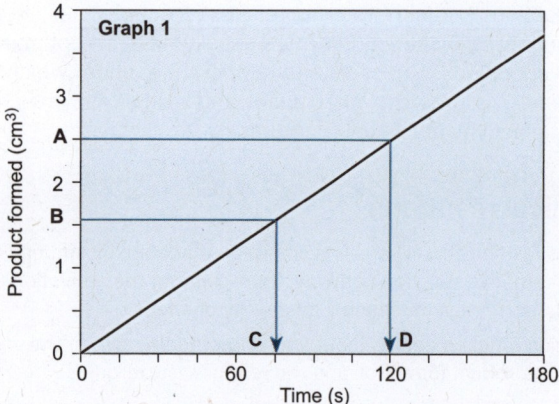

If you plot the amount of product formed during a reaction against time, the rate of a reaction can be calculated from the amount of product made during a given time period. For a reaction in which the rate does not vary (above), the reaction rate calculated at any one point in time will be the same. For example: B ÷ C = A ÷ D.

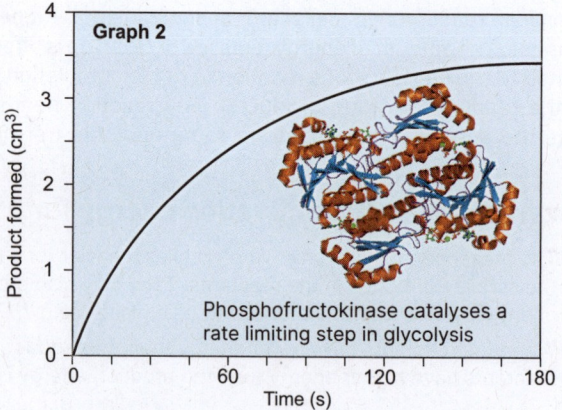

Phosphofructokinase catalyses a rate limiting step in glycolysis

In most biological systems, substrates are limiting, so the reaction rate often levels off over time (above). The enzyme forms product at an initial rate that is roughly linear for a short period after the start of the reaction. As the reaction proceeds and substrate is consumed, the rate continuously slows.

Effect of concentration of enzyme on reaction rate

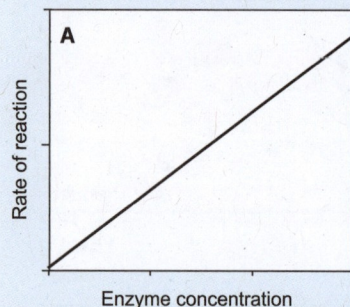

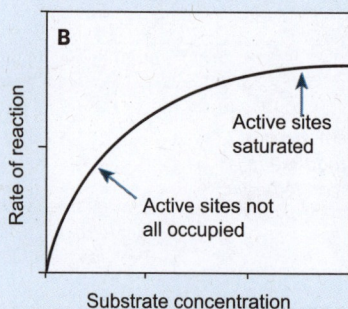

Given an unlimited amount of substrate, the rate of reaction will continue to increase as enzyme concentration increases. More enzyme means more reactions between substrates can be catalysed in any given time (graph A).

If there is unlimited substrate but the enzyme is limited, the reaction rate will increase until the enzyme is saturated, at which point the rate will remain static (graph B).

Effect of temperature

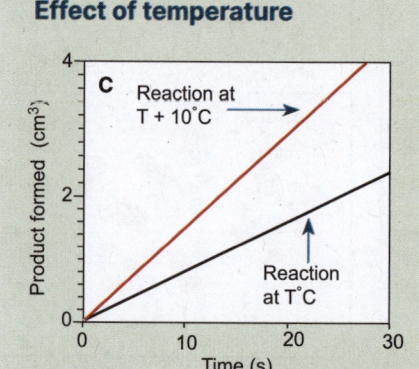

The effect of temperature on a reaction rate is expressed as the temperature coefficient, usually given as the Q_{10}, which expresses the increase in reaction rate for every rise of 10°C. It is a useful way to express the temperature dependence of a process. For most biological systems, its value is ~2 to 3.

4. Use A ÷ D to calculate the reaction rate in graph 1 (top left): _____

5. (a) What must be happening to the reaction mix in graph 1 to produce the straight line (constant reaction rate)?

_____

_____

(b) Explain why the reaction rate in graph 2 changes over time: _____

_____

6. Explain why a reaction rate might drop off as the enzyme-catalysed reaction proceeds over time:

_____

_____

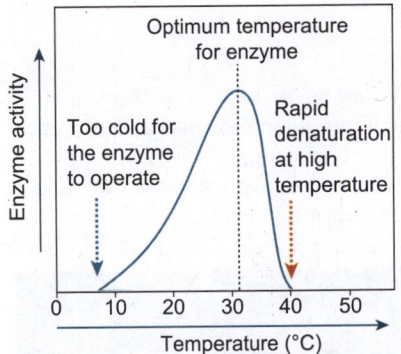

Professor Dr. habil. Uwe Kils CC3.0

Antarctic icefish

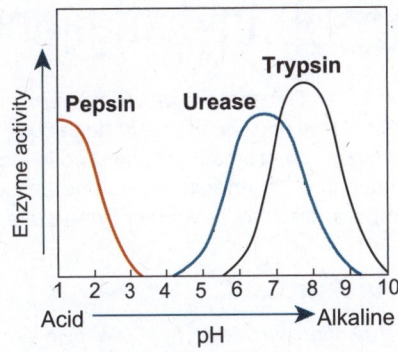

Higher temperatures increase the speed with which molecules in a solution move. This increases the frequency of collisions between a substrate and the active site, increasing reaction rate. However, few enzymes can tolerate temperatures higher than 50–60°C, and enzymes become denatured. The temperature at which an enzyme works at its maximum rate is the optimum temperature.

Enzymes performing the same function in species in different environments are very slightly different in order to maintain optimum performance. For example, the enzyme acetylcholinesterase has an optimum temperature of -2°C in the nervous system of an Antarctic icefish but an optimum temperature of 25°C in grey mullet found in the Mediterranean.

Like all proteins, enzymes are denatured by extremes of pH (very acid or alkaline). Within these extremes, most enzymes have a specific pH range for optimum activity. For example, digestive enzymes are specific to the region of the gut where they act: pepsin in the acid of the stomach and trypsin in the alkaline small intestine. Urease catalyses the hydrolysis of urea at a near neutral pH.

7. (a) Describe the change in reaction rate when the enzyme concentration is increased and the substrate is not limiting:

_____

_____

(b) Suggest how a cell may vary the amount of enzyme present:

_____

_____

8. Describe the change in reaction rate when the substrate concentration is increased (with a fixed amount of enzyme):

_____

_____

9. (a) Describe what is meant by an optimum temperature for enzyme activity:

_____

(b) Predict the effect on enzyme activity if an enzyme with a temperature optimum of 37°C was added to its substrate and incubated at 25°C. What would happen if the temperature was slowly increased to the optimum?

_____

_____

_____

(c) For graph C on the previous page, calculate the Q_{10} for the reaction: _____

_____

_____

10. (a) State the optimum pH for each of the enzymes:

Pepsin: _____ Trypsin: _____ Urease: _____

(b) Explain how the pH optimum of each of these enzymes is suited to its working environment: _____

_____

_____

_____

_____

_____

153 Investigating Peroxidase Activity

Key Idea: The factors affecting peroxidase activity can be measured using the indicator guaiacol.

Enzymes control all the metabolic activities required to sustain life. Changes to environmental conditions, e.g. pH or temperature, may alter an enzyme's shape and functionality.

This may result in a reduction or loss of activity. In this exercise you will use the information provided and your own understanding of enzymes to investigate the effect of pH on enzyme activity and then design an experiment to investigate the effect of inhibitors on enzyme function.

Background

Hydrogen peroxide (H_2O_2) is a toxic by-product of respiration and must be broken down in order to avoid cellular damage. **Peroxidase** acts in the presence of naturally occurring organic reducing agents (electron donors) to catalyse the breakdown of H_2O_2 into water and oxidized organic molecules.

$$2H_2O_2 + 2AH_2 \xrightarrow{\text{Peroxidase}} 4H_2O + A_2$$

Like all enzymes, the activity of peroxidase is highest within specific ranges of pH and temperature, and activity drops off or is halted altogether when the conditions fall outside the optimal range. The conversion of H_2O_2 is also influenced by other factors such as the levels of substrate and enzyme.

The effect of peroxidase on H_2O_2 breakdown can be studied using a common reducing agent called guaiacol. Oxidation of guaiacol (as in the equation above) forms tetraguaiacol, which is a dark orange colour. The rate of the reaction can be followed by measuring the intensity of the orange colour as a function of time.

Increasing levels of oxygen production over time (minutes)

A time-colour palette is shown above. You can use it as a reference against which to compare your own results from the investigation below. The palette was produced by adding a set amount of peroxidase to a solution containing hydrogen peroxide and water. The colour change was recorded at set time points (0-6 minutes).

Investigation 9.1   Investigating peroxidase activity

See appendix for equipment list.

1. Prepare six substrate tubes by adding to a boiling tube 7 mL of distilled water, 0.3 mL of 0.1% H_2O_2 solution, and 0.2 mL of prepared guaiacol solution. Cover the tubes with parafilm and mix.

2. Prepare six enzyme tubes by adding 6.0 mL of prepared buffered pH solution (one of pH 3, 5, 6, 7, 8, and 10) and 1.5 mL of prepared turnip peroxidase solution. Cover the tubes with parafilm and mix.

3. Combine the contents of substrate and enzyme tubes and cover with parafilm. Mix and place back on the rack.

4. Begin timing immediately. Record the colour change every minute (1-6 based on the colour palette above).

5. You can take photos with your phones or keep a written record of the colour changes.

	Colour reference number						
	0 min	1 min	2 min	3 min	4 min	5 min	6 min
pH 3							
pH 5							
pH 6							
pH 7							
pH 8							
pH 10							

1. The colour palette (above) shows the relative amounts of tetraguaiacol formed when guaiacol is oxidized. How can this be used to determine enzyme activity?

_____

_____

©2024 **BIOZONE** International
ISBN: 978-1-99-101410-8
Photocopying prohibited

2. Graph your results on the grid (right).

3. (a) Describe the effect of pH on peroxidase activity:

_____

_____

_____

_____

(b) Was there a colour change at pH 10? Explain the result at this pH and relate it to the enzyme's structure and the way it interacts with its substrate:

_____

_____

_____

_____

_____

4. In your experiment, the rate of enzyme activity is measured by comparing against a ranked colour palette. How could you have measured the results more quantitatively?

_____

5. How might the results be affected if you did not begin timing immediately after mixing the contents of the enzyme and substrate tubes together?

_____

_____

6. Why is peroxidase written above the arrow in the equation for enzymatic breakdown of H_2O_2? _____

_____

_____

7. Based on the information provided and your answer to question 4, design an experiment to investigate the effect of lead nitrate (an enzyme inhibitor) on the activity of turnip peroxidase. Summarize your method as step by step instructions below. Note how you will record and display the data and calculate the reaction rate. Include reference to any limitations or sources of potential error in your design:

_____

_____

_____

_____

_____

_____

_____

_____

_____

_____

_____

_____

154 Intracellular and Extracellular Enzyme Reactions

Key Idea: Enzyme-catalysed reactions can be classified as intracellular or extracellular.

Enzymes can be defined based on where they are produced relative to where they are active. An Intracellular enzyme performs its functions within the cell that produces it. Most enzymes are intracellular enzymes, such as respiratory enzymes involved in glycolysis and in the Krebs cycle. An extracellular enzyme is an enzyme that functions outside the cell from which it originates, i.e. it is produced in one location but active in another, such as those involved in digestion.

Comparing intracellular and extracellular enzymes

▶ Both types of enzymes are globular proteins, as are all enzymes and are also involved with digestion, breaking larger molecules into smaller molecules. Intracellular enzymes break large polymer chains into many small chains, while extracellular enzymes break off individual monomers, one-by-one, from the end of polymer chains.

▶ Intracellular enzymes function in cellular lysosomes to break down the contents, and are bound within the membrane. They are activated by the higher pH inside the lysosomes. Extracellular enzymes travel to the site of digestion in the body, such as the small intestine and stomach, and are activated by different enzymes or low pH, depending on the enzyme.

▶ Aside from lysosome digestion, intracellular enzymes have a much broader range of functions than extracellular enzymes, including facilitating cellular metabolic pathways. The majority of enzymes are intracellular.

Intracellular enzymes

Glycolysis and Krebs cycle enzymes

These enzyme-catalysed reactions are part of the respiration metabolic pathway. Ten enzymes are involved in the reaction steps of glycolysis, and eight enzymes in the Krebs cycle (also known as the Citric acid cycle).

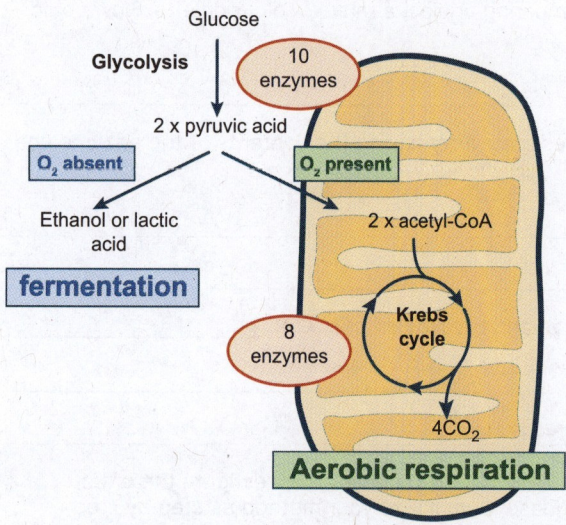

Extracellular enzymes

Trypsin

Trypsin is a protein-digesting enzyme and hydrolyses the peptide bond immediately after a basic residue (e.g. arginine). It is produced in an inactive form (called trypsinogen) and secreted into the small intestine by the pancreas. It is activated in the intestine by the enzyme enteropeptidase to form trypsin. Active trypsin can convert more trypsinogen to trypsin.

Trypsinogen with amino acids (green) making the enzyme inactive by blocking active site.

Trypsin with inhibitor (red) that remains in place until the enzyme is ready to be used.

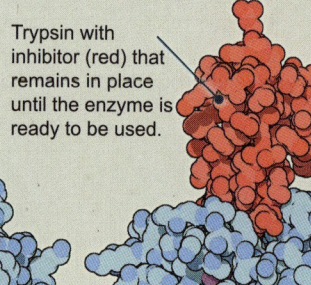

1. (a) Explain the key difference between an intracellular and extracellular enzyme: _____

   _____

   _____

   _____

   (b) Suggest why the majority of enzymes are intracellular: _____

   _____

   _____

2. What type of enzymes are required for the reactions in the respiration metabolic pathway, and why are so many different enzymes required?

   _____

   _____

   _____

   _____

©2024 **BIOZONE** International
ISBN: 978-1-99-101410-8
Photocopying prohibited

155 Generating Heat

Key Idea: Metabolic reactions generate heat and this is used by some animals to maintain homeostatic body temperature. The chemical reactions of metabolism involve breaking and reforming bonds between atoms and molecules, and therefore energy transfer between reactants and products.

Not all of the energy will be transferred at each reaction, and some potential energy held in chemical bonds will be transformed into heat energy and lost to surroundings. This heat energy can then be utilized by the organism to maintain a body temperature suitable for metabolism to operate.

Metabolism and heat energy loss

▶ All organisms require energy in order to perform the metabolic processes required for them to function and reproduce. Energy input must exceed energy loss in order to power cellular processes. This energy is obtained by cellular respiration: a set of metabolic reactions that convert biochemical energy from 'food' into the nucleotide ATP (adenosine triphosphate).

▶ ATP is considered to be a universal energy carrier, transferring chemical energy within the cell for use in metabolic processes.

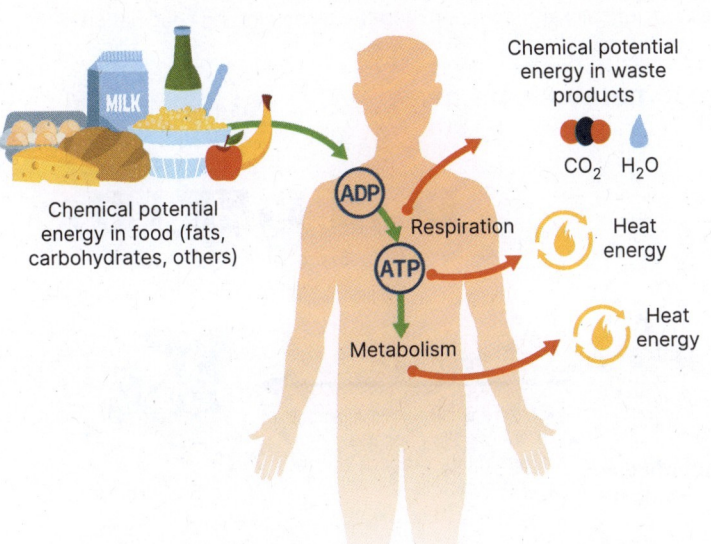

Chemical potential energy in food (fats, carbohydrates, others)

Chemical potential energy in waste products

CO_2 H_2O

Heat energy

Heat energy

ADP / ATP / Respiration / Metabolism

Heat energy is lost during the series of reactions to transfer potential energy from food to potential energy in ATP, via glycolysis, the Krebs cycle, and the electron transport reactions of respiration. Potential energy from the food is used to convert ADP to ATP, as well as transferred into the chemical bonds of respiratory waste products, CO_2 and H_2O, and some energy is 'lost' as heat energy.

When ATP is converted back to ADP, the potential energy difference is used to 'power' the multitude of cellular reactions that comprise the organism's metabolism. Energy transfer is not 100% efficient and a percentage is lost as heat energy.

Why is body heat important?

The essential processes of life are regulated by enzymes and require a certain temperature in order to operate most efficiently. This optimal temperature varies, depending on the organism. A bacterium living in a geothermal hot pool will have a very different optimal temperature from a soil bacterium in a woodland! Below the optimal temperature, metabolic reactions proceed very slowly. Above the optimal temperature, the enzymes may be denatured and the reaction cannot proceed.

The average body temperature of mammals is ~38°C. For birds, it is ~40°C. Most snakes and lizards operate best in the range 24-35°C, although some operate in the mammalian range.

All birds are endothermic and generate most of their body heat from metabolic processes. Even in cold Antarctic temperatures, the metabolic activity of these emperor penguins provides the warmth to sustain life processes. Endothermy requires a lot of energy so endotherms cannot go without food for long.

1. Why is metabolic heat energy generation important for organisms? _____

_____

_____

_____

2. Explain the following statement: metabolic reactions are not 100% efficient in energy transfer:

_____

_____

_____

3. What is the relationship between demand for heat generation, high metabolism, and high energy demand in animals?

_____

_____

_____

C1.1
12
AHL

156 Metabolic Pathways

Key Idea: A metabolic pathway is a series of linked and enzyme enabled biochemical reactions.

Metabolic pathways are linked biochemical reactions that occur within living organisms to maintain life. Enzymes activate (catalyse) each step of a metabolic pathway and, in turn, each enzyme is encoded by specific genes. Metabolic pathways are controlled by regulating the amount of enzyme present (by switching the genes encoding that enzyme on or off) or by controlling enzyme activity. Each step in a metabolic pathway is part of a sequence, where the product from one step becomes the substrate for the next. Metabolic pathways can be linear, such as that seen in glycolysis, when glucose is converted to pyruvic acid, or pathways can be cyclical, such as the Krebs cycle that occurs in the mitochondria in respiration, and the Calvin cycle that occurs in the chloroplasts as part of the photosynthetic pathway.

Linear pathways in metabolism

▸ Linear metabolic pathways have a number of steps which vary depending on the reaction. Sometimes, these linear pathways are also known as chains, where the product of each step acts as the reactant for the next.

▸ The pathway begins with a substrate that acts as the initial reactant, and ends when the final product has been produced.

▸ Examples of linear metabolic pathways include glycolysis, as part of cellular respiration, and the blood clotting coagulation cascade that occurs in blood vessels.

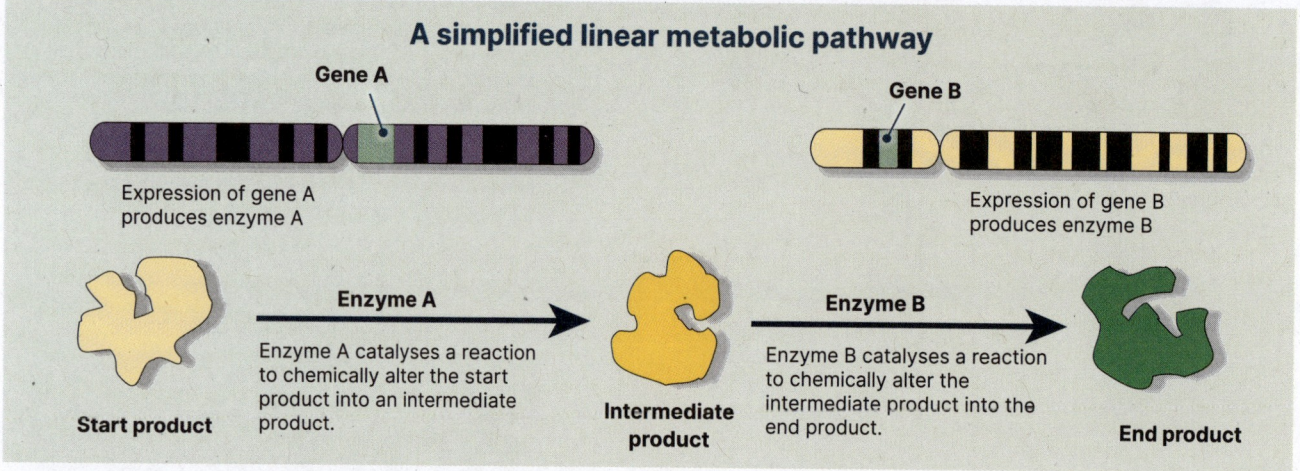

A simplified linear metabolic pathway

Gene A — Expression of gene A produces enzyme A

Gene B — Expression of gene B produces enzyme B

Start product — **Enzyme A** Enzyme A catalyses a reaction to chemically alter the start product into an intermediate product. → **Intermediate product** — **Enzyme B** Enzyme B catalyses a reaction to chemically alter the intermediate product into the end product. → **End product**

Glycolysis has a linear metabolic pathway

▸ Glycolysis converts glucose to two 3-carbon pyruvate molecules. It is considered a linear metabolic pathway as there is a final product. There are ten different enzymes involved in the glycolysis metabolic pathway.

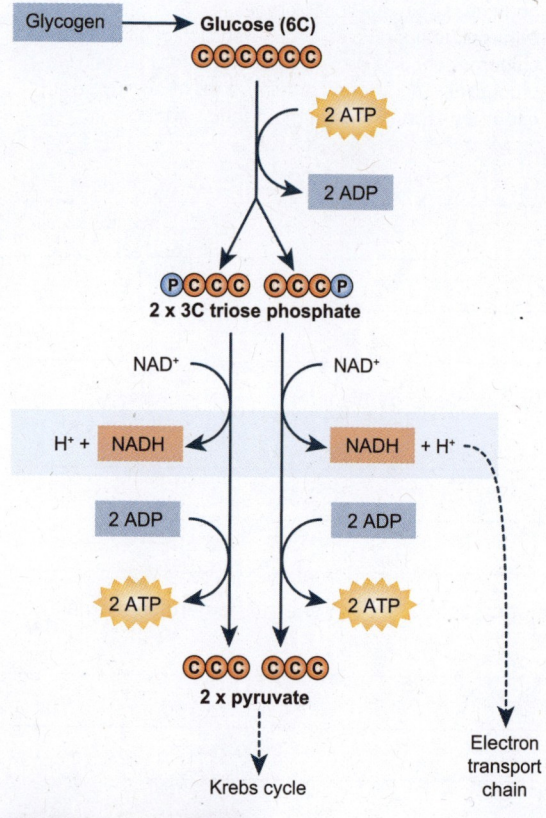

1. Define the term, 'metabolic pathway':

_____

_____

_____

_____

2. What is the role of enzymes in metabolic pathways?

_____

_____

3. Why is glycolysis considered a linear metabolic pathway even though it is branched?

_____

_____

_____

_____

_____

AHL

C1.1

13

©2024 **BIOZONE** International
ISBN: 978-1-99-101410-8
Photocopying prohibited

Cyclic pathways in metabolism

▸ Cyclic metabolic pathways progress in a way so that the end product becomes the start product to continue the cycle. Like linear pathways, cyclic pathways are often controlled by feedback mechanisms.

The Krebs cycle (respiration)

The Krebs cycle occurs in the mitochondrial matrix. It begins by combining the 2 carbons from acetyl-coA, derived from pyruvate via the link reaction, with its own end product, oxaloacetate, to produce a 6 carbon molecule to restart the cycle. The Krebs cycle requires a series of eight enzymes in the cyclic pathway.

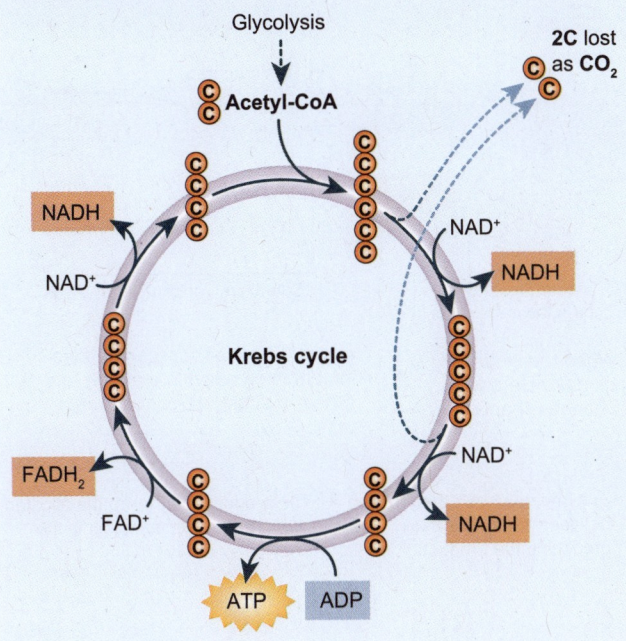

The Calvin cycle (photosynthesis)

The Calvin cycle converts carbon dioxide to glucose in a series of reactions. The cyclic pathway involves 11 different enzymes that enable 13 different reactions. This reaction occurs within the chloroplasts of plants and photosynthetic organisms. The Calvin cycle is also known as the light-independent reactions, because they occur without the direct input of light.

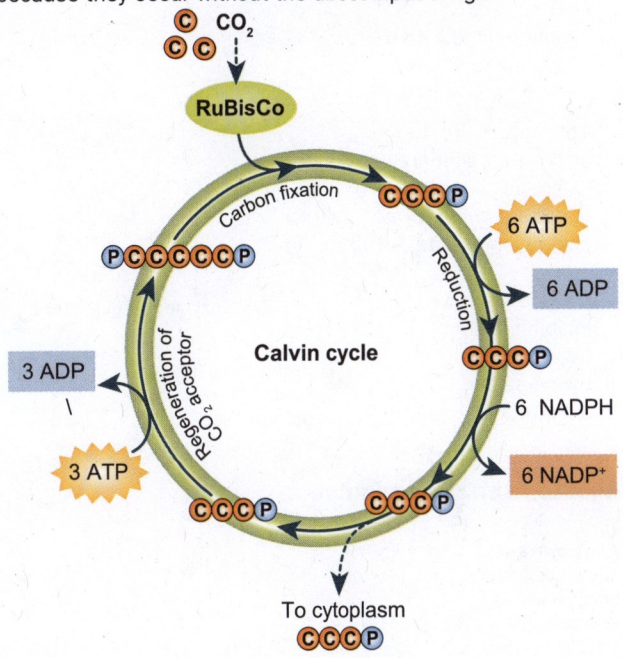

4. How does ATP act as a supplier of energy to power metabolic reactions? _____

_____

_____

_____

5. In what way are the processes pictured above (photosynthesis and cellular respiration) connected?

_____

_____

_____

6. Summarize the difference between metabolic cyclical and linear pathways, by completing the table:

	Linear pathways	Cyclic pathways
Named examples		
Relationship of beginning reactant and ultimate product		
Complexity		

157 Enzyme Inhibitors

Key Idea: Enzyme activity is controlled by different types of enzyme inhibition.

Enzyme activity can be stopped, temporarily or permanently, by chemicals called enzyme inhibitors. Irreversible inhibitors bind tightly to the enzyme and are not easily displaced. Reversible inhibitors can be displaced from the enzyme and have a role as enzyme regulators in metabolic pathways.

Competitive inhibitors compete directly with the substrate for the active site and their effect can be overcome by increasing the concentration of available substrate. A non-competitive inhibitor does not occupy the active site, but distorts it so that the substrate and enzyme can no longer interact. Some inhibitors may be irreversible, in which case the inhibitors act as poisons, including fungal antibiotics.

Non-competitive inhibition (reversible)

Non-competitive inhibitors bind with the enzyme at a site other than the active site.

They inactivate the enzyme by altering its shape so that the substrate and enzyme can no longer interact.

Non-competitive inhibition cannot be overcome by increasing the substrate concentration.

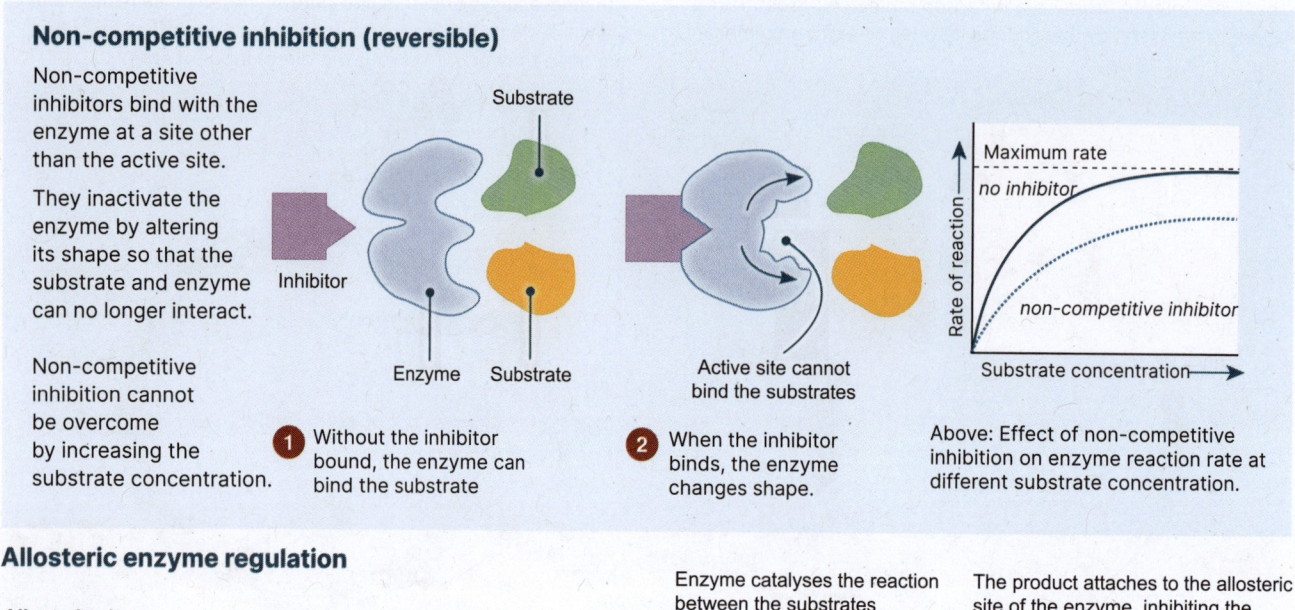

1 Without the inhibitor bound, the enzyme can bind the substrate

2 When the inhibitor binds, the enzyme changes shape.

Above: Effect of non-competitive inhibition on enzyme reaction rate at different substrate concentration.

Allosteric enzyme regulation

Allosteric site: The place on an enzyme where a molecule that is not a substrate may bind. The allosteric binding site is never the active site, but is specific for only certain substances.

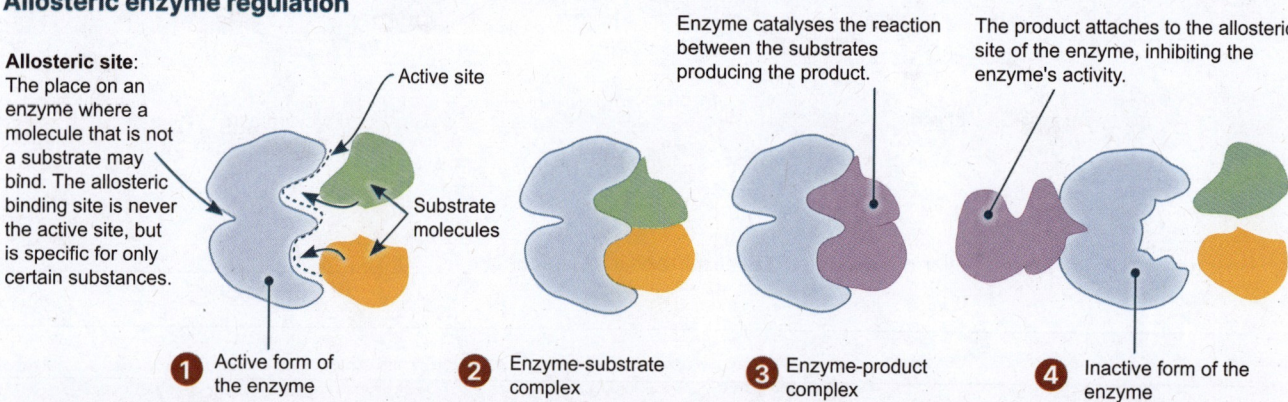

Enzyme catalyses the reaction between the substrates producing the product.

The product attaches to the allosteric site of the enzyme, inhibiting the enzyme's activity.

1 Active form of the enzyme

2 Enzyme-substrate complex

3 Enzyme-product complex

4 Inactive form of the enzyme

Allosteric regulation is a form of non-competitive regulation where the regulatory molecule (an activator or inhibitor) binds to an enzyme somewhere other than the active site. Most non-competitive inhibition is a form of allosteric regulation. Allosteric regulation controls the enzymes in many metabolic pathways, and enzyme activity can be regulated by the products produced in the pathway. The action is usually by feedback inhibition (negative feedback). When the concentration of the end product is high, the end product will bind to the allosteric site of the first enzyme in the pathway, inhibiting the enzyme and shutting down the pathway. When the concentration of the end product is reduced, the allosteric site is vacated and the pathway is activated again.

Competitive inhibition (reversible)

Competitive inhibitors compete with the normal substrate for the enzyme's active site.

Competitive inhibitors compete directly with the substrate for the active site, and their effect can be overcome by increasing the substrate concentration.

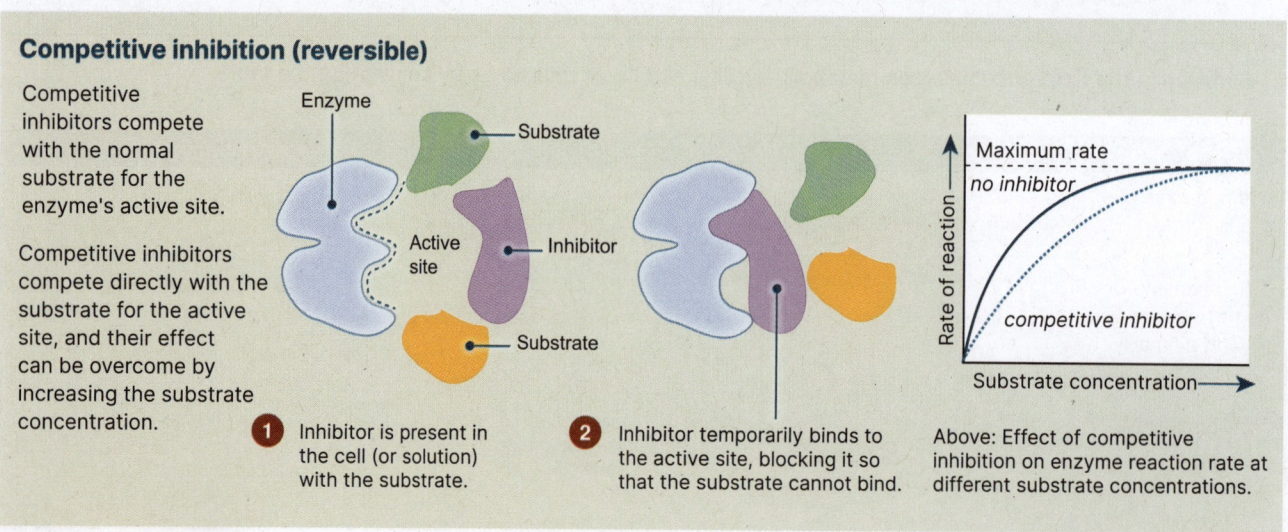

1 Inhibitor is present in the cell (or solution) with the substrate.

2 Inhibitor temporarily binds to the active site, blocking it so that the substrate cannot bind.

Above: Effect of competitive inhibition on enzyme reaction rate at different substrate concentrations.

AHL

C1.1
14 - 15, 17

©2024 **BIOZONE** International
ISBN: 978-1-99-101410-8
Photocopying prohibited

Poisons are irreversible mechanism-based inhibitors

Some enzyme inhibitors are poisons because the binding of enzyme and inhibitor is irreversible. Irreversible inhibitors form strong covalent bonds with an enzyme. These inhibitors may act at near, or remotely from the active site and modify the enzyme's structure to such an extent that it ceases to work. For example, the poison cyanide is an irreversible enzyme inhibitor that combines with the copper and iron in the active site of cytochrome c oxidase and blocks cellular respiration.

Since many enzymes contain sulfhydryl (-SH), alcohol, or acidic groups as part of their active sites, any chemical that can react with them may act as an irreversible inhibitor. Heavy metals, Ag^+, Hg^{2+}, or Pb^{2+}, have strong affinities for -SH groups and destroy catalytic activity. Most heavy metals are non-competitive inhibitors.

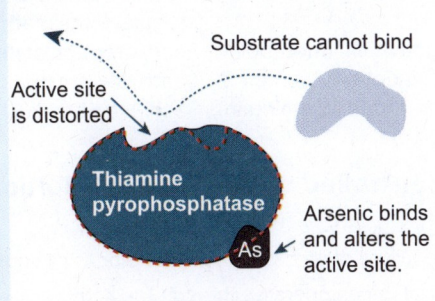

Substrate cannot bind

Active site is distorted

Thiamine pyrophosphatase

Arsenic binds and alters the active site.

Arsenic and phosphorus share some structural similarities so arsenic will often substitute for phosphorus in biological systems. It targets a wide variety of enzyme reactions. Arsenic can act as either a competitive or a non-competitive inhibitor (as above) depending on the enzyme.

Penicillin

Many drugs work by irreversible inhibition of a pathogen's enzymes. Penicillin (below) and related antibiotics inhibit transpeptidase, a bacterial enzyme that forms some of the linkages in the bacterial cell wall. Susceptible bacteria cannot complete cell wall synthesis and cannot divide. Human cells are unaffected by the drug.

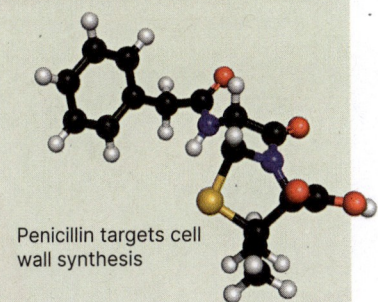

Penicillin targets cell wall synthesis

1. Distinguish between competitive and non-competitive inhibition: _____

_____

_____

_____

2. Hypercholesterolaemia, or excessive levels of low-density lipoprotein (LDL) cholesterol in the blood, can lead to fatty deposits blocking blood vessels, increasing the risk of dangerous bloodclots or strokes. Patients can be given statins, medication that acts as a competitive inhibitor, blocking activity of HMG-CoA reductase, an enzyme utilized in the production of cholesterol in the liver. Considering the properties of a competitive inhibitor, discuss why statins are considered a reasonably safe medication, even at high doses:

_____

_____

_____

3. Describe how an allosteric regulator can regulate enzyme activity: _____

_____

_____

4. Explain why heavy metals, such as lead and arsenic, are poisonous: _____

_____

_____

5. (a) In the context of enzymes, explain how penicillin is exploited to control human diseases: _____

_____

_____

   (b) Explain why the drug is poisonous to the target organism, but not to humans: _____

_____

_____

158 Control of Metabolic Pathways

Key Idea: The end product of a metabolic pathway can regulate the pathway itself.

Metabolism refers to all the chemical activities (metabolic reactions) of life. They form a tremendously complex network of reactions that is necessary in order to maintain the organism. Often, the products of a metabolic pathway regulate the pathway itself. This might be achieved by the end product of the pathway inhibiting the reactions in the pathway so that no more product is produced. This can be achieved by allosteric enzyme regulation (below).

Metabolic pathways can be controlled by feedback inhibition

▸ Metabolic pathways consist of a series of sequential steps. The product of one step in the pathway is generally the reactant (substrate) for the following step. The failure of one step therefore blocks all subsequent steps.

▸ You have already seen how feedback inhibition operates through the activity of allosteric inhibitors. Metabolic pathways are commonly regulated by feedback inhibition in this way, as illustrated by the end-product inhibition of the pathway that converts the amino acid threonine (one of 9 essential amino acids) to isoleucine.

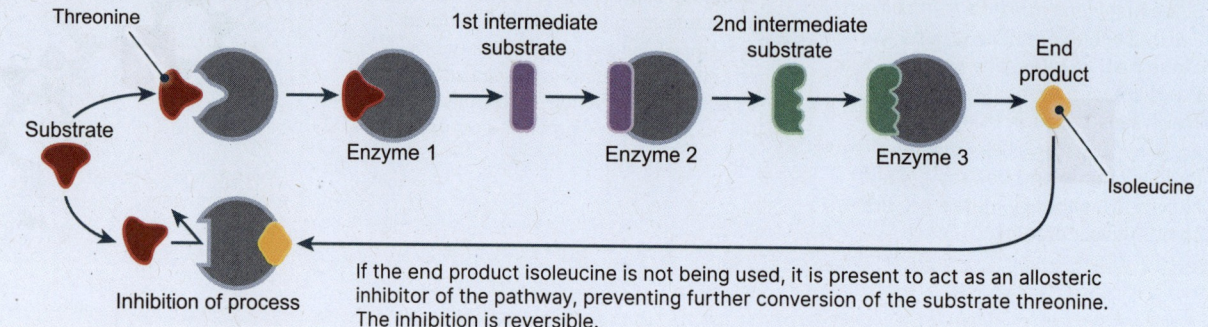

If the end product isoleucine is not being used, it is present to act as an allosteric inhibitor of the pathway, preventing further conversion of the substrate threonine. The inhibition is reversible.

Isoleucine Synthesis from Threonine

Threonine

↓ threonine deaminase

Intermediate A

↓ Enzyme 2

Intermediate B

↓ Enzyme 3

Intermediate C

↓ Enzyme 4

Intermediate D

↓ Enzyme 5

Isoleucine

The pathway for the biosynthesis of isoleucine from the amino acid threonine is controlled by end product inhibition (negative feedback).

Threonine is converted to a series of different intermediate products by different enzymes.

Enzymes are shown in blue text.

Threonine deaminase is inhibited by isoleucine. When there is a high concentration of isoleucine the pathway is inhibited but as the concentration of isoleucine decreases, the threonine deaminase is no longer inhibited and the pathway begins again.

Threonine

Isoleucine

1. Define feedback inhibition in enzymes: _____

_____

_____

_____

2. Explain how feedback inhibition regulates the threonine-isoleucine pathway: _____

_____

_____

_____

AHL

C1.1

16

©2024 **BIOZONE** International
ISBN: 978-1-99-101410-8

159   The Role of ATP in Cells

Key Idea: ATP transports chemical energy within the cell for use in metabolic processes.

All organisms require energy to perform the metabolic processes required to function and reproduce. This energy is obtained by cellular respiration, a set of metabolic reactions which ultimately convert biochemical energy from 'food' into the nucleotide adenosine triphosphate (ATP). ATP is the energy carrier found in all organisms, transferring chemical energy within the cell for use in metabolic processes such as biosynthesis, cell division, cell signalling, thermoregulation, cell mobility, and active transport of substances across cell membranes.

Adenosine triphosphate (ATP)

▸ The ATP molecule consists of three components: a purine base (adenine), a pentose sugar (ribose), and three phosphate groups that attach to the 5' carbon of the pentose sugar. Adenine + ribose form adenosine (the 'A' in ATP). The structure of ATP is shown right.

▸ The bonds between the phosphate groups contain electrons in a high energy state which store a large amount of energy. This is released during ATP hydrolysis. Typically, hydrolysis is coupled to another cellular reaction to which the energy is transferred. The end products of the reaction are adenosine diphosphate (ADP) and an inorganic phosphate (Pi).

▸ Note that energy is released during the formation of bonds during the hydrolysis reaction, not the breaking of bonds between the phosphates (which requires energy input).

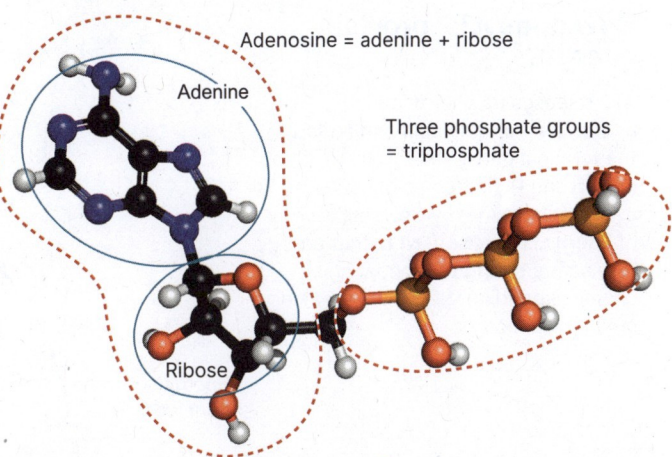

Adenosine = adenine + ribose

Adenine

Three phosphate groups = triphosphate

Ribose

ATP powers life processes in the cell

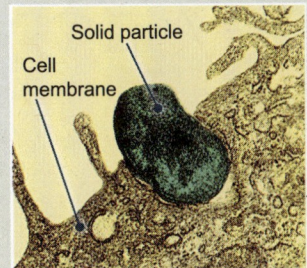

Cell membrane

Solid particle

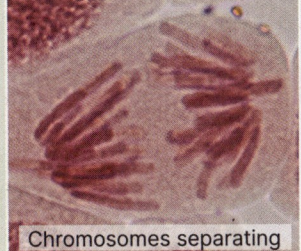

Chromosomes separating

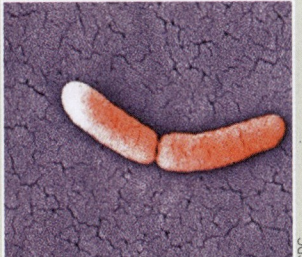

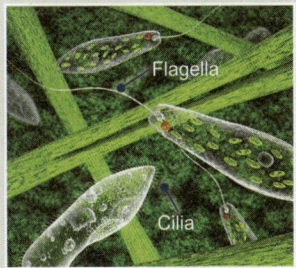

Flagella

Cilia

The energy released from the removal of a phosphate group of ATP is used for active transport of molecules and substances across the plasma membrane e.g. phagocytosis (above) and other active transport processes.

Mitosis, as seen in the stained onion cell above, requires ATP to proceed. Formation of the mitotic spindle and chromosome separation both require the energy provided by ATP hydrolysis to occur.

ATP is required when bacteria divide by binary fission (above). Synthesis of new macromolecules formed during DNA replication and components of the cell wall and other cellular components requires ATP.

Single-celled organisms use ATP as an energy source to move external flagella or cilia for movement. Filaments inside these motile appendages use ATP powered motor proteins called dyneins to slide past each other, causing bending.

1. What process produces ATP in a cell? _____

2. Identify the three distinct elements of the space-filling model of ATP, labelled (a)-(c) below right:

   (a) _____   (b) _____   (c) _____

3. Which two of the elements you labelled in question 2 make up adenosine? _____

4. Elaborate on life processes at the cellular level to which ATP supplies energy, using named examples:

   _____

   _____

   _____

   _____

   _____

   _____

(a)

(b)

(c)

©2024 **BIOZONE** International
ISBN: 978-1-99-101410-8
Photocopying prohibited

C1.2

1 - 2

160 | ATP and Energy

Key Idea: ATP is the universal energy carrier in cells. Energy is stored in the covalent bonds between phosphate groups. The nucleotide ATP (adenosine triphosphate) is the universal energy carrier for the cell. ATP can release its energy quickly by hydrolysis of the terminal phosphate. This reaction is catalysed by the enzyme ATPase. Once ATP has released its energy, it becomes ADP (adenosine diphosphate), a low energy molecule that can be recharged by adding a phosphate. The energy to do this is supplied by the controlled breakdown of glucose in cellular respiration.

How does ATP provide energy?

ATP releases its energy during hydrolysis. Water is split and added to the terminal phosphate group resulting in ADP and Pi (inorganic phosphate). For every mole of ATP hydrolysed, 30.7 kJ of energy is released. Note that energy is released during the formation of chemical bonds not from the breaking of chemical bonds.

Mitochondrion

The reaction to convert ADP to ATP occurs inside the mitochindria.

ATP is reformed during the reactions of cellular respiration (i.e. glycolysis, Krebs cycle, and the electron transport chain).

ATP

Adenine

Triphosphate

Ribose

Phosphate

Energy absorbed from food

ATP-ADP cycle

Phosphate

Energy released for cell

Hydrolysis is the addition of water. ATP hydrolysis gives ADP + Pi (HPO_4^{2-}) + H^+.

Adenine

Ribose

Diphosphate

Note: The phosphate bonds in ATP are often referred to as high energy bonds. This can be misleading. The bonds contain *electrons in a high energy state* (making the bonds themselves relatively weak). A small amount of energy is required to break the bonds but when the intermediates recombine and form new chemical bonds, a large amount of energy is released. The final product is less reactive than the original reactants.

In many textbooks, the reaction series above is simplified and the intermediates are left out:

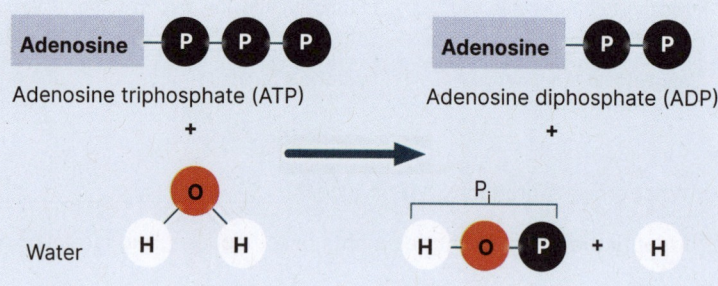

Adenosine — P P P
Adenosine triphosphate (ATP)
+
Water   H   H

Adenosine — P P
Adenosine diphosphate (ADP)
+
P_i
H — O — P   +   H

1. (a) How does ATP supply energy to power metabolism? _____

_____

_____

   (b) In what way is the ADP/ATP system like a rechargeable battery? _____

_____

_____

2. What respiratory substrate provides the energy for reforming ATP? _____

3. During the many metabolic reactions occurring in the body, most of the energy in the initial respiratory substrate is lost as heat. What is the purpose of this heat?

_____

C1.2
3

©2024 **BIOZONE** International
ISBN: 978-1-99-101410-8
Photocopying prohibited

161 Cellular Respiration Overview

Key Idea: Cellular respiration is a metabolic process that produces ATP.

Aerobic respiration releases the energy in glucose via a series of connected reactions. In eukaryotes, the first stage (glycolysis) occurs in the cytoplasm and subsequent stages occur within the mitochondrion. In prokaryotes, all stages except the last occur in the cytoplasm. The final stage, electron transport, is associated with proteins attached to the plasma membrane. In both prokaryotes and eukaryotes, the processes are essentially identical.

An overview of cellular respiration

▶ Cellular respiration involves three metabolic stages, plus a link reaction, summarized below.

▶ The first two stages are catabolic pathways that decompose glucose and other organic molecules. These two sets of reactions are connected by pyruvate decarboxylation (the link reaction).

▶ In the third stage, the electron transport chain accepts electrons from the first two stages and passes these from one electron acceptor to another. The energy released at each stepwise transfer is used to make ATP. The final electron acceptor in this process is molecular oxygen.

1 Glycolysis. In the cytoplasm, glucose is broken down into two molecules of pyruvate.

L The link reaction. In the mitochondrial matrix, pyruvate is split and added to coenzyme A.

2 Krebs cycle. In the mitochondrial matrix, a derivative of pyruvate is decomposed to CO_2.

3 Electron transport and oxidative phosphorylation. This occurs in the inner membranes of the mitochondrion and accounts for almost 90% of the ATP generated by respiration.

Respiratory substrates

Glucose is the main source of energy that enters the respiratory reaction chain. However, other organic molecules, such as fatty acids from lipids and other carbon based substances can be utilized as well, when glucose is in short supply.

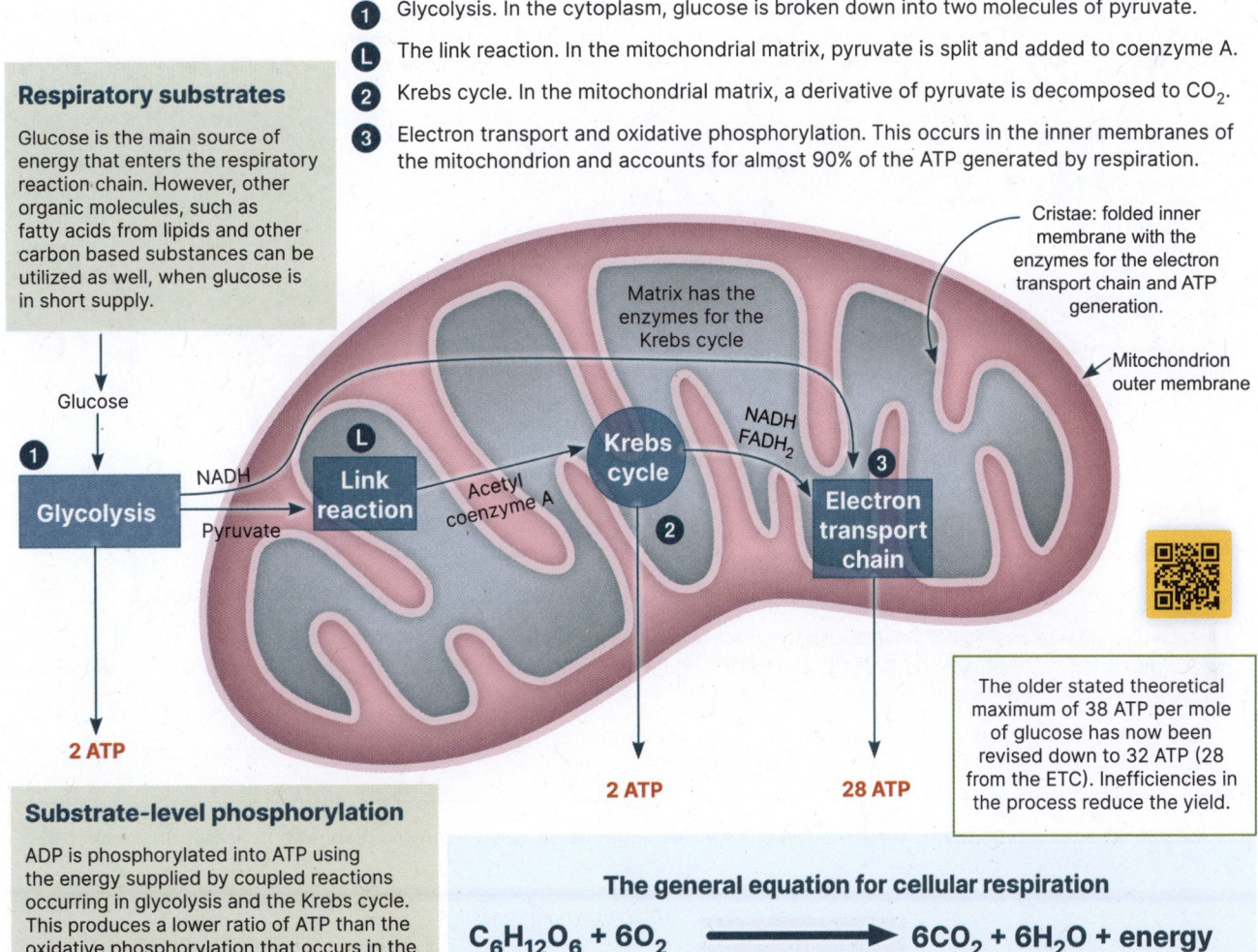

The older stated theoretical maximum of 38 ATP per mole of glucose has now been revised down to 32 ATP (28 from the ETC). Inefficiencies in the process reduce the yield.

Substrate-level phosphorylation

ADP is phosphorylated into ATP using the energy supplied by coupled reactions occurring in glycolysis and the Krebs cycle. This produces a lower ratio of ATP than the oxidative phosphorylation that occurs in the electron transport chain.

The general equation for cellular respiration

$$C_6H_{12}O_6 + 6O_2 \longrightarrow 6CO_2 + 6H_2O + energy$$

1. Describe precisely in which part of a eukaryotic cell the following take place:

   (a) Glycolysis: _____

   (b) The link reaction: _____

   (c) Krebs cycle reactions: _____

   (d) Electron transport chain: _____

2. What is the link between cellular respiration, glucose, and ATP? _____

   _____

   _____

C1.2

4 - 5

Aerobic and anaerobic pathways for ATP production

A Aerobic respiration

B Lactic acid fermentation

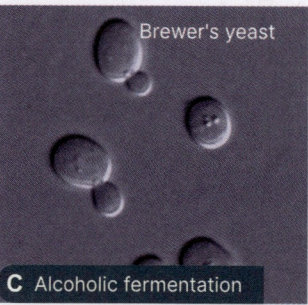

Brewer's yeast

C Alcoholic fermentation

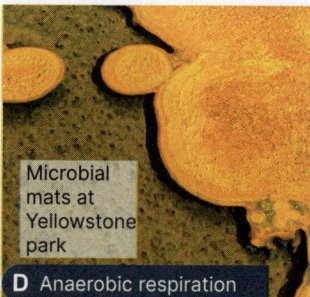

Microbial mats at Yellowstone park

D Anaerobic respiration

Aerobic respiration produces the energy (as ATP) needed for metabolism. The rate of aerobic respiration is limited by the amount of oxygen available. In animals and plants, most of the time the oxygen supply is sufficient to maintain aerobic metabolism. Aerobic respiration produces a high yield of ATP per molecule of glucose (path A).

During maximum physical activity, when oxygen is limited, anaerobic metabolism provides ATP for working muscle. In mammalian muscle, metabolism of a respiratory intermediate produces lactate, which provides fuel for working muscle and produces a low yield of ATP. This process is called lactic acid fermentation (path B).

The process of brewing utilises the anaerobic metabolism of yeasts. Brewer's yeasts preferentially use anaerobic metabolism in the presence of excess sugars. This process, called alcoholic fermentation, produces ethanol and CO_2 from the respiratory intermediate pyruvate. It is carried out in vats that prevent entry of O_2 (path C).

Many bacteria and archaea are anaerobic, using molecules other than oxygen, e.g. nitrate or sulfate, as a terminal electron acceptor of their electron transport chain. These electron acceptors are not as efficient as oxygen (less energy is released per oxidized molecule) so the energy (ATP) yield from anaerobic respiration is generally quite low (path D).

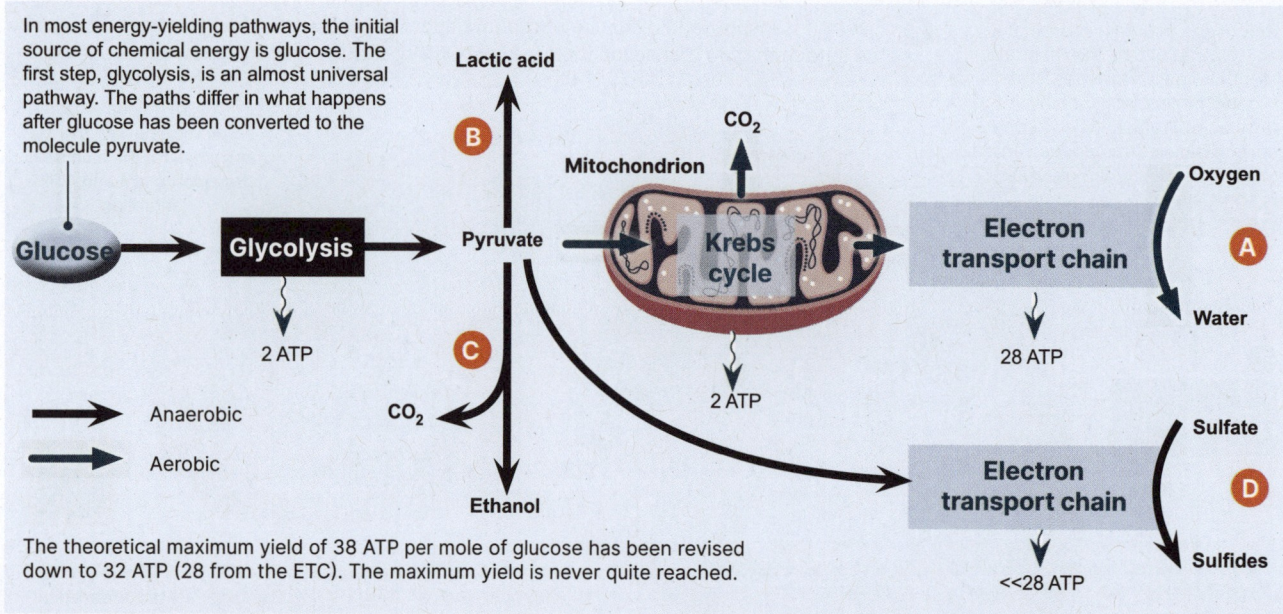

In most energy-yielding pathways, the initial source of chemical energy is glucose. The first step, glycolysis, is an almost universal pathway. The paths differ in what happens after glucose has been converted to the molecule pyruvate.

The theoretical maximum yield of 38 ATP per mole of glucose has been revised down to 32 ATP (28 from the ETC). The maximum yield is never quite reached.

3. What do all the ATP yielding pathways above have in common? _____

_____

4. Distinguish between anaerobic pathways in eukaryotes, e.g. yeasts, and anaerobic respiration in anaerobic microbes:

_____

_____

_____

_____

5. When brewing alcohol, why is it important to prevent entry of oxygen to the fermentation vats? _____

_____

_____

6. Write the word equation for anaerobic respiration and compare with aerobic respiration:

_____

_____

_____

©2024 **BIOZONE** International
ISBN: 978-1-99-101410-8
Photocopying prohibited

162 Measuring Respiration

Key Idea: Respiration is the process by which cells convert energy in glucose to usable energy, which is stored in the molecule ATP. The process uses oxygen, which can be quantified using a simple respirometer.

A respirometer can be used to measure the amount of oxygen consumed by an organism during cellular respiration and so can be used to measure respiration rate. A simple respirometer is shown in the diagram below. The carbon dioxide produced during respiration is absorbed by the potassium hydroxide. As the oxygen is used up, the coloured bubble in the glass tube moves. Measuring the movement of the bubble allows an estimation of the change in volume of gas and therefore the rate of cellular respiration. Various versions of these simple respirometers exist. You may use a set-up that uses micro respirometers, which are easy to construct and work quickly because of their small volumes

Investigation 9.2   Measuring respiration in germinating seeds

See appendix for equipment list.

⚠ 👁 🧤 Caution is required when handling potassium hydroxide as it is caustic and can cause chemical burns. You should wear protective eyewear and gloves.

1. Work in groups of four to set up three respirometers using the set-up shown right as a guide.

2. Collect three boiling tubes and place two cotton balls in the bottom of each. Label the vials 1, 2, and 3.

3. Use a dropper to add 15% potassium hydroxide (KOH) solution on the cotton balls until they are saturated (there should be no liquid in the boiling tube). Add the same amount of KOH to the cotton balls in each vial.

4. Place gauze on top of the cotton balls in each vial. This prevents the KOH coming into contact with the seeds and killing them.

5. Quarter fill vial 1 with glass beads made up to the same volume as in the other two vials. Determine the volumes using a water displacement method.

6. Quarter fill vial 2 with 25 germinated bean seeds. These seeds will be damp as they have been germinated under damp paper towels for four days.

7. Quarter fill vial 3 with the same number of non-germinating (dormant) seeds into vial 3 (making up the volume with glass beads).

8. Once the vials have been prepared, cap each vial with a stopper fitted with a 1 mL pipette, tip pointing outward. Add a weight (two metal washers) to each vial.

9. Use a dropper or fine pipette to place a drop of food colouring into the tip of each pipette to seal off the system.

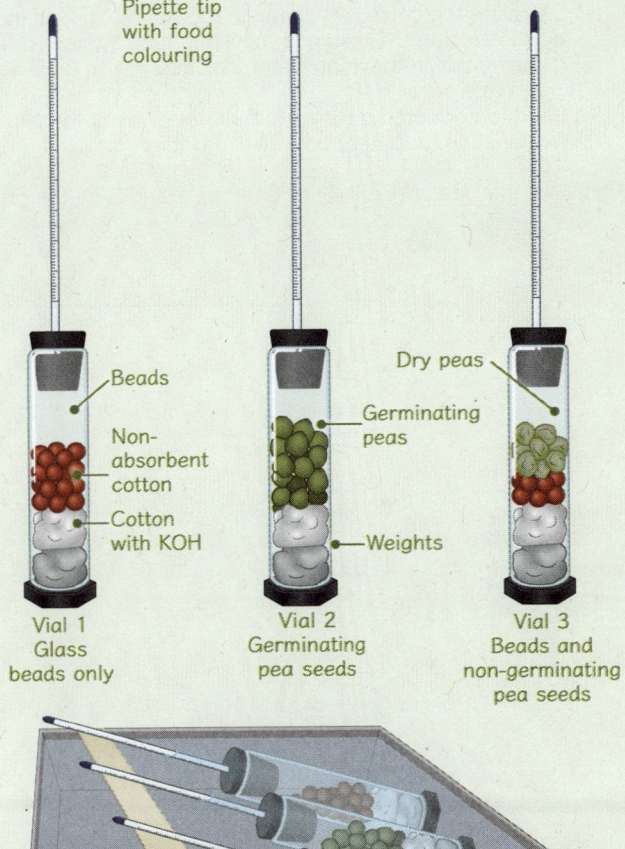

Pipette tip with food colouring

Beads
Non-absorbent cotton
Cotton with KOH

Vial 1
Glass beads only

Dry peas
Germinating peas
Weights

Vial 2
Germinating pea seeds

Vial 3
Beads and non-germinating pea seeds

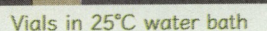

Vials in 25°C water bath

10. Place all three respirometers in a 25°C water bath supported so that the pipette tips stay out of the water (see diagram right). Leave for 10 minutes to equilibrate.

11. After 10 minutes of equilibration, record the position of the food colouring bubble and start a timer.

12. The recorded movement of the coloured bubble in tube 1 (glass beads) account for any pressure fluctuations. Subtract the changes recorded in tube 1 in each of tube 2 and 3 to obtain the corrected volume changes caused by the use of oxygen.

13. Record your results on the table at the top of the next page.

©2024 **BIOZONE** International
ISBN: 978-1-99-101410-8
Photocopying prohibited

C1.2
6

Results

Time X	Tube 1: Beads alone		Tube 2: Germinating peas				Tube 3: Dry peas and beads			
	Reading at time X (mL)	Difference (correction)	Reading at time X	Difference	Corrected volume (mL)	Respiration rate (mL/min)	Reading at time X	Difference	Corrected volume (mL)	Respiration rate (mL/min)
0										
5										
10										
15										
20										
25										

1. (a) Calculate the corrected distance the bubble moved in tubes 2 (germinating seeds) and 3 (non germinating seeds) by subtracting the distance moved in tube 1 (glass beads) from each value (pale green column above). Record these values in the appropriate columns in the table above.

    (b) Use the corrected distance the bubble moved to calculate the rate of respiration for germinating and non-germinating (dormant) peas. Record these values in the appropriate columns in the table above.

    (c) Plot the rate of respiration for tubes 2 and 3 on the grid (below). Include an appropriate title and axes labels.

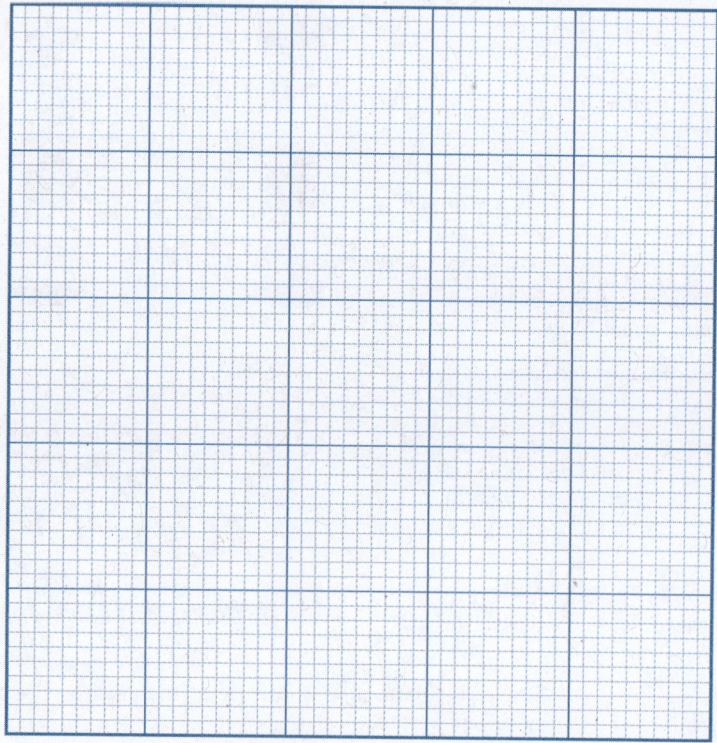

(d) What does your plot show? _____

_____

_____

_____

©2024 **BIOZONE** International
ISBN: 978-1-99-101410-8

2. Explain the purpose of the following in the experiment:

(a) KOH: _____

_____

(b) Equilibration period: _____

_____

(c) The vial with the ungerminated seeds: _____

_____

(d) The vial with the glass beads: _____

_____

3. Why does the bubble in the capillary tube move? _____

_____

_____

4. Where are the sources of error in measuring respiration rate with respirometers? How could you minimize them?

_____

_____

_____

5. What conclusion can you make about cellular respiration in germinated and ungerminated (dormant) seeds?

_____

_____

_____

_____

6. How would you have to modify the experiment if you were measuring respiration in a plant instead of seeds?

_____

_____

_____

7. A student repeated the respirometer experiment using maggots instead of seeds. Their results are tabulated below:

Time (minutes)	Distance bubble moved (mm)	Rate (mm/min)
0	0	
5	25	
10	65	
15	95	
20	130	
25	160	

(a) Calculate the rates and record them in the table above:

(b) Graph the rates on the grid above right.

(c) Describe the results: _____

_____

163 The Biochemistry of Aerobic Respiration

Key Idea: During cell respiration, the energy in glucose is transferred to ATP in a series of enzyme controlled steps.
The oxidation of glucose is a catabolic, energy yielding pathway. The breakdown of glucose and other organic fuels (such as fats and proteins) to simpler molecules releases energy for ATP synthesis. Glycolysis and the Krebs cycle supply electrons to the electron transport chain, which drives oxidative phosphorylation. Glycolysis nets two ATP. The conversion of pyruvate (the end product of glycolysis) to acetyl CoA links glycolysis to the Krebs cycle. One 'turn' of the cycle releases carbon dioxide, forms one ATP, and passes electrons to three NAD+ and one FAD. Most of the ATP generated in cellular respiration is produced by oxidative phosphorylation when NADH + H+ and FADH2 donate electrons to the series of electron carriers in the electron transport chain. At the end of the chain, electrons are passed to molecular oxygen, reducing it to water. Electron transport is coupled to ATP synthesis.

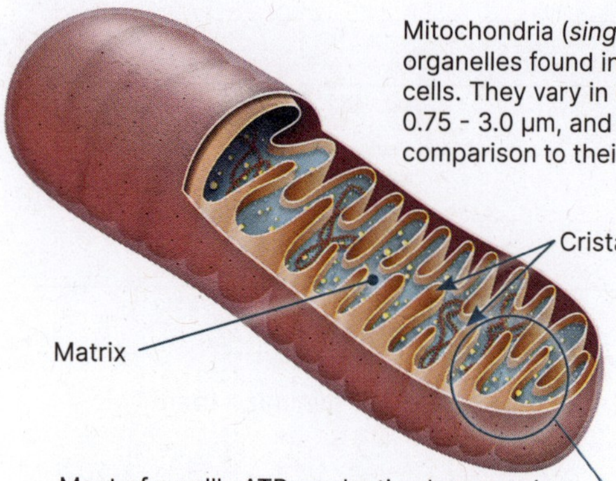

Mitochondria (*sing.* mitochondrion) are organelles found in most eukaryotic cells. They vary in diameter from 0.75 - 3.0 μm, and can be quite long in comparison to their diameter.

Cristae

Matrix

Most of a cell's ATP production happens in the mitochondria. The Krebs cycle and the electron transport chain occur here.

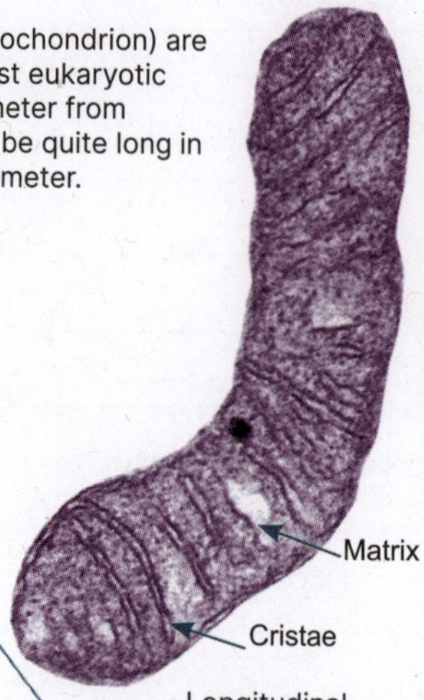

Matrix

Cristae

Longitudinal section of mitochondrion.

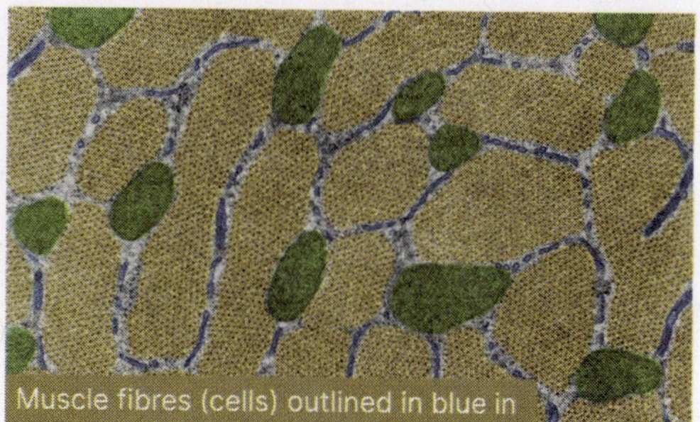

Muscle fibres (cells) outlined in blue in cross section, with mitochondria green

Cells that require a lot of ATP for cellular processes have a lot of mitochondria. Sperm cells contain a large number of mitochondria near the base of the tail. In skeletal muscle fibres (above) mitochondria (green) can occupy up to 20% of the cytoplasmic space. In liver cells, the figure is about 25% and in heart muscle cells, it can be as high as 40%.

Respiratory substrates

Glucose is derived from carbohydrates and is the main source of energy that enters the respiratory reaction chain. Anaerobic respiration requires carbohydrate substrates. However, when they are in short supply, other organic molecules, such as lipids, can provide alternative aerobic respiratory substrates. Lipids are hydrocarbon based, and have more oxidisable substances to yield more energy per gram. Lipids hydrolyse into triglycerides and then glycerol and fatty acids. Fatty acids break down into 2-carbon acetyl groups, entering into the link reaction.

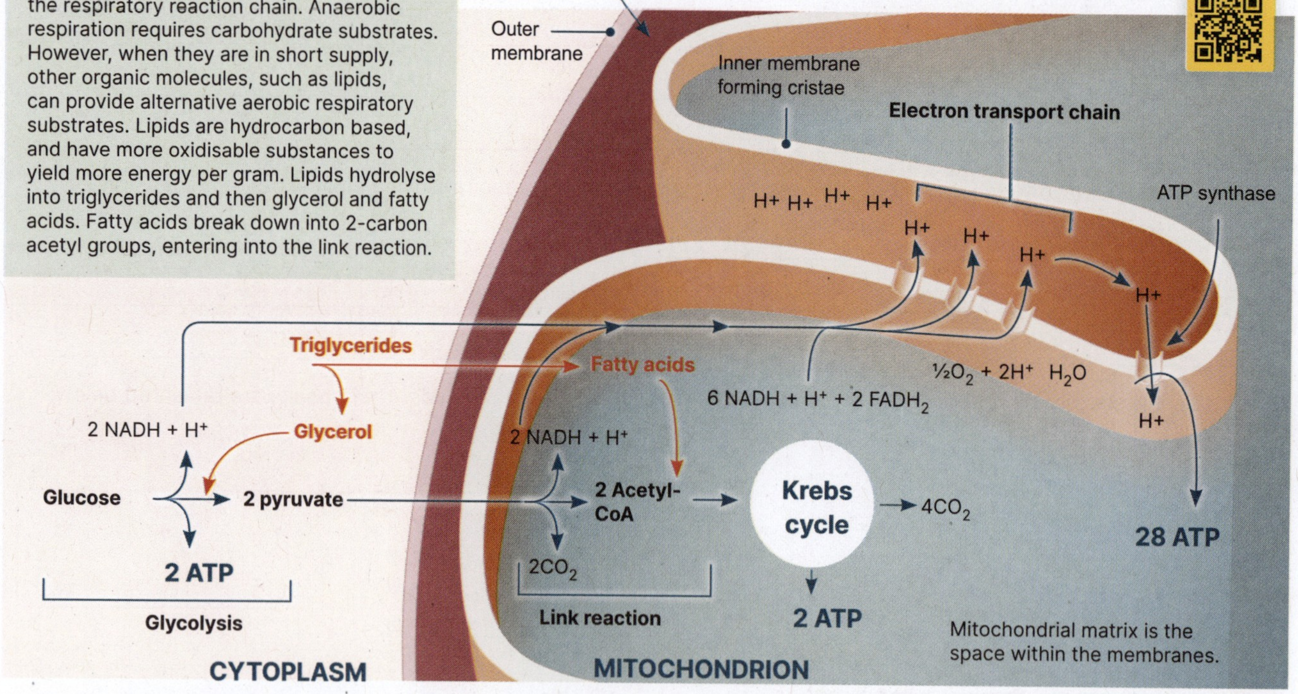

Outer membrane

Inner membrane forming cristae

Electron transport chain

H^+ H^+ H^+ H^+

H^+ H^+ H^+

ATP synthase

H^+

$½O_2 + 2H^+$ H_2O

H^+

Triglycerides

Fatty acids

6 NADH + H+ + 2 FADH2

2 NADH + H+

Glycerol

2 NADH + H+

2 Acetyl-CoA

Krebs cycle

→ 4CO2

28 ATP

Glucose → **2 pyruvate**

2CO2

2 ATP

Glycolysis

Link reaction

2 ATP

Mitochondrial matrix is the space within the membranes.

CYTOPLASM

MITOCHONDRION

The role of NAD in respiration

NAD (nicotinamide adenine dinucleotide) is a coenzyme and a hydrogen carrier. The reduced NAD+ (and FAD) are hydrogen acceptors, and have a critical role in cellular respiration.
When hydrogen becomes available from glycolysis, the link reaction, and the Krebs cycle, the NAD molecule will accept them and then transport hydrogen (hydrogen ions and electrons) to the electron transport chain. NAD+ is reduced to NADH + H+ (often just written as NADH).
Redox reactions are paired reactions of both oxidation and reduction. Once the hydrogen is delivered to the electron transport chain, the NADH undergoes an oxidation reaction and is recycled once more as NAD.

Reduction

$NAD^+ + 2H$ → $NADH + H^+$

Oxidation

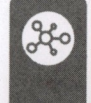

©2024 BIOZONE International
ISBN: 978-1-99-101410-8
Photocopying prohibited

Glycolysis (cytoplasm)

The first part of respiration involves the breakdown of glucose in the cytoplasm. Glucose (a 6-carbon sugar) is broken into two molecules of pyruvate (also called pyruvic acid), a 3-carbon acid. A total of 2 ATP and 2NADH + 2H$^+$ are generated from this stage. NADH is produced for use in the electron transport chain. Glycolysis initially uses two ATP but produces four ATP. It therefore produces a net of two ATP molecules. It also produces two pyruvate molecules each of which then enters the Krebs cycle. No oxygen is required (the process is anaerobic). Ten different enzymes are involved in the glycolysis reaction pathway.

Link reaction and Krebs cycle (mitochondria)

The link reaction takes pyruvate from glycolysis and converts it into acetyl-CoA which enters the Krebs cycle. Pyruvate loses a carbon atom (decarboxylation) which forms carbon dioxide. The remaining 2C intermediate is oxidized, losing hydrogen and electrons to NAD+, which is reduced and forms NADH + H. Coenzyme A is added to the 2C intermediate to form acetyl CoA. NADH + H is used in the electron transport chain.

In the Krebs cycle, acetyl coenzyme A is attached to the 4C molecule oxaloacetate and coenzyme A is released. Oxaloacetate is eventually remade in a cyclic series of redox reactions that produce more NADH and FADH$_2$ for the electron transport chain (four oxidations total per cycle). Two ATP are also made by substrate level phosphorylation.

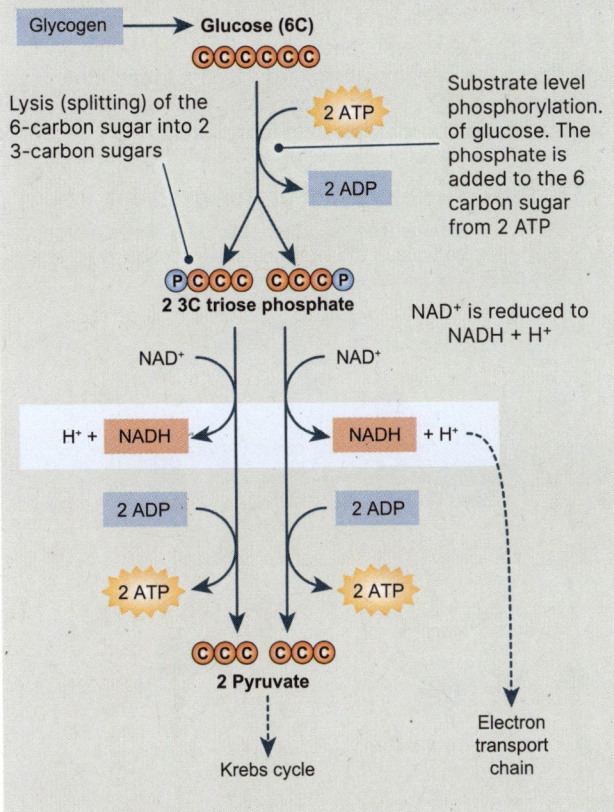

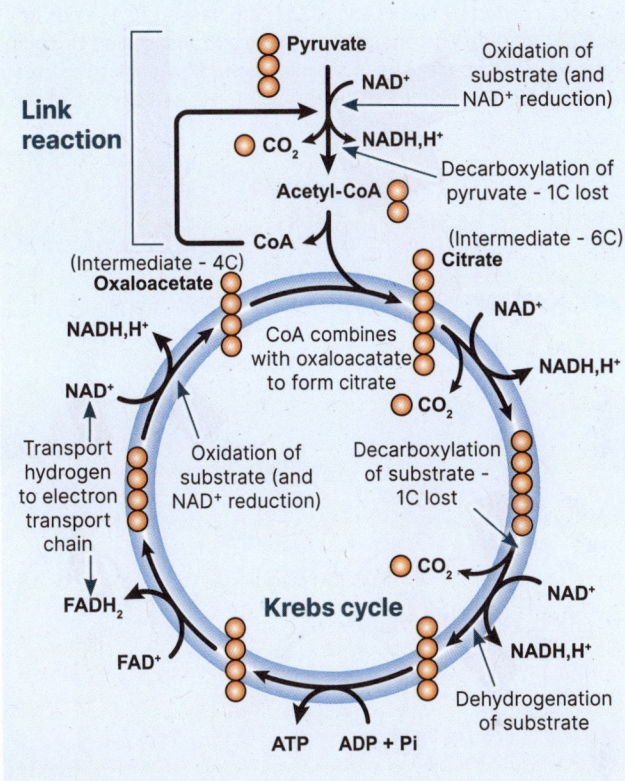

1. Define the following terms, explaining where they occur during cell respiration:

Term	Definition
(a) NAD$^+$ reduction	
(b) Substrate oxidation	
(c) Dehydrogenation	
(d) Phosphorylation	
(e) Decarboxylation	
(f) Intermediates	

164 The Electron Transport Chain and Chemiosmosis

Key Idea: Chemiosmosis is the process in which electron transport is coupled to ATP synthesis.

Chemiosmosis occurs in the membranes of mitochondria and involves the establishment of a proton (hydrogen) gradient across a membrane. The concentration gradient is used to drive ATP synthesis. Chemiosmosis has two key components: an electron transport chain (ETC) sets up a proton gradient as electrons pass along it to a final electron acceptor, and an enzyme called ATP synthase uses the proton gradient to catalyse ATP synthesis. In cellular respiration, electron carriers on the inner membrane of the mitochondria oxidize NADH + H^+ and $FADH_2$, derived from glycolysis, the link reaction, and the Krebs cycle, where NAD^+ and FAD^+ were reduced. The energy released from this process is used to move protons against their concentration gradient, from the mitochondrial matrix into the space between the two membranes. The return of protons to the matrix via ATP synthase is coupled to ATP synthesis.

Electron transport chain

Electrons carried by NADH and $FADH_2$ are passed to a series of electron carrier enzymes embedded in the inner membrane of the mitochondria. The energy from the electrons is used to pump H^+ ions across the inner membrane from the matrix into the intermembrane space. These are allowed to flow back to the matrix via the enzyme ATP synthase which uses their energy to produce ATP. The electrons are coupled to H^+ and oxygen at the end of the electron transport chain to form water.

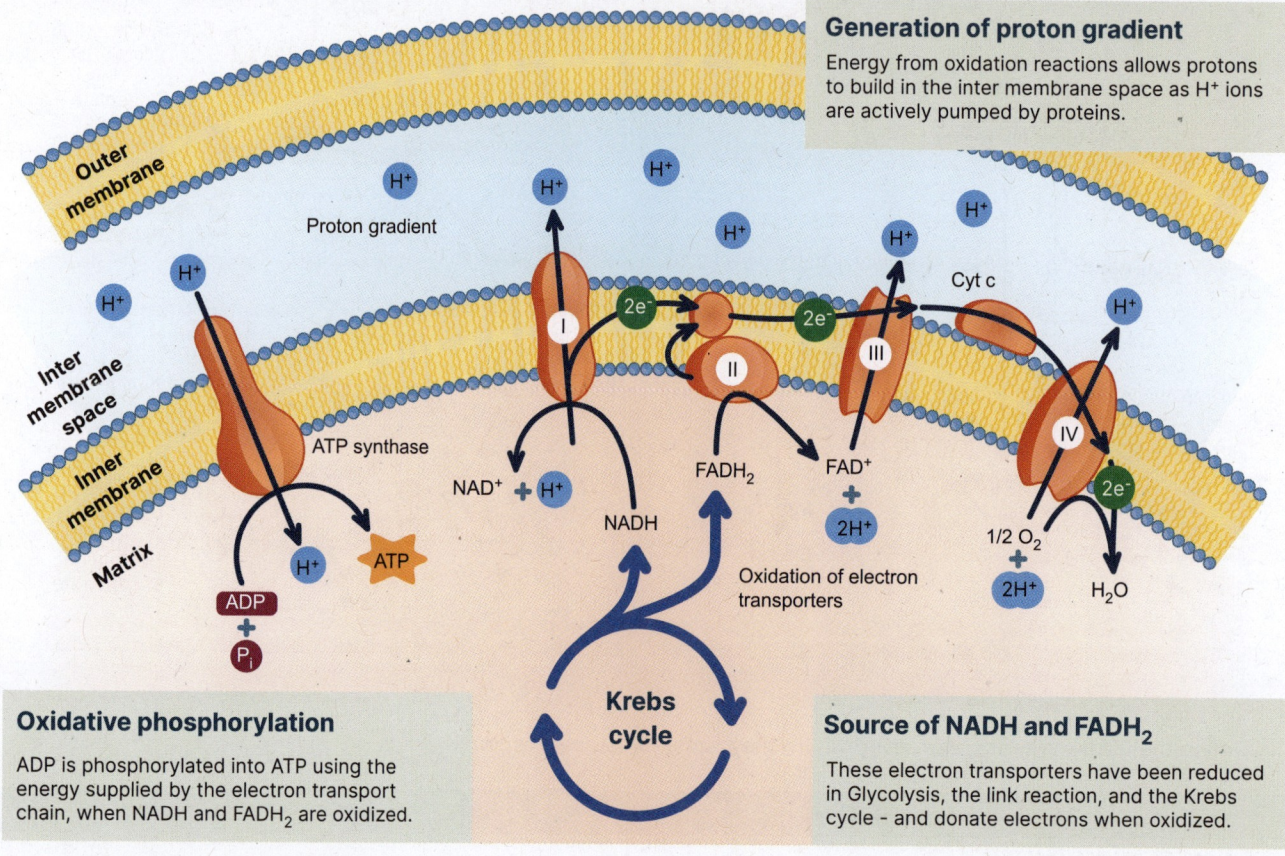

Generation of proton gradient

Energy from oxidation reactions allows protons to build in the inter membrane space as H^+ ions are actively pumped by proteins.

Oxidative phosphorylation

ADP is phosphorylated into ATP using the energy supplied by the electron transport chain, when NADH and $FADH_2$ are oxidized.

Source of NADH and $FADH_2$

These electron transporters have been reduced in Glycolysis, the link reaction, and the Krebs cycle - and donate electrons when oxidized.

1. (a) What is substrate level phosphorylation? _____

_____

(b) How many ATP are produced this way during cellular respiration (per molecule of glucose)? _____

2. (a) What is oxidative phosphorylation? _____

_____

(b) How many ATP are produced this way during cellular respiration (per molecule of glucose)? _____

3. A pair of electrons is donated from each electron transporter when they are oxidized. Write the balanced reaction for the oxidation of NADH and $FADH_2$ and link this to the generation of a proton gradient:

_____

_____

_____

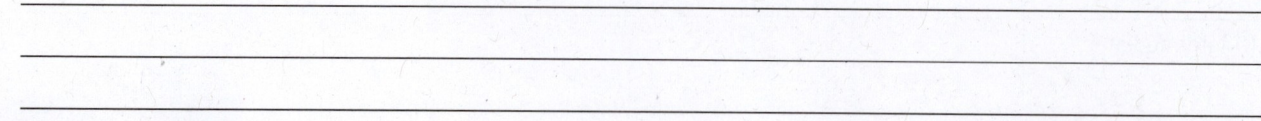

©2024 **BIOZONE** International
ISBN: 978-1-99-101410-8
Photocopying prohibited

AHL

C1.2

13 - 16

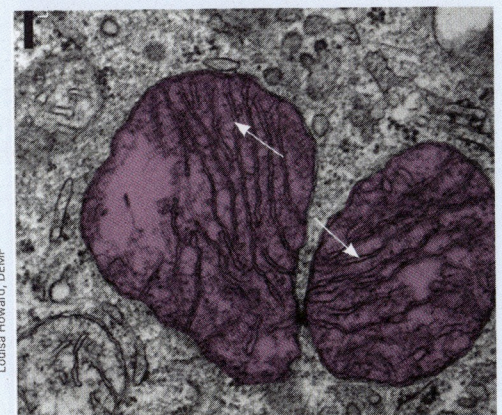

The intermembrane spaces can be seen (arrows) in this transverse section of mitochondria.

Louisa Howard, DEMF

The Evidence for Chemiosmosis

British biochemist, Peter Mitchell, proposed the chemiosmotic hypothesis in 1961. He proposed that, because living cells have membrane potential, electrochemical gradients could be used to do work, i.e. provide the energy for ATP synthesis. Scientists at the time were sceptical, but the evidence for chemiosmosis was extensive and came from studies of isolated mitochondria and chloroplasts. Evidence included:

▶ The outer membranes of mitochondria were removed, leaving the inner membranes intact. Adding protons to the treated mitochondria increased ATP synthesis.

▶ When isolated chloroplasts were illuminated, the medium in which they were suspended became alkaline.

▶ Isolated chloroplasts were kept in the dark and transferred first to a low pH medium (to acidify the thylakoid interior) and then to an alkaline medium. They then spontaneously synthesized ATP (no light was needed).

Chemiosmosis

Chemiosmosis is a process where there is build up in the gradient of substances across a membrane, in this case protons (H⁺ ions). The high concentration of protons in the inner membrane space then flows passively through an ATP synthase protein channel to the lower concentration in the matrix of the mitochondria. The flow of H⁺ ions causes the cylindrical portion of the ATP synthase to spin. This movement provides energy for oxidative phosphorylation, which is energy-coupled to this proton flow and allows ADP to be phosphorylated (adding phosphate) into ATP.

Structure of ATP synthase

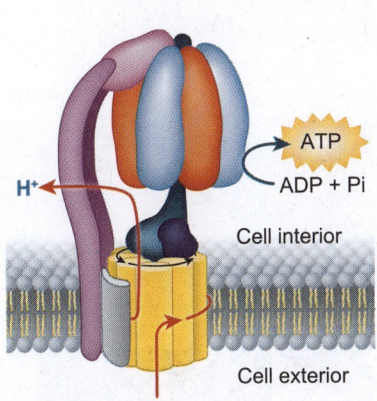

Role of oxygen

Electrons released by NADH and FADH$_2$ are accepted by oxygen, the terminal electron acceptor, and in combination with the H⁺ ions released into the matrix, the final product is water.

Removing the electrons enables the proton (positive charge) gradient to form - which would otherwise be neutralized. The process of producing water also removes excess H⁺ ions from the matrix - to maintain the lower concentration required for the passive flow of ions from inter membrane space to the matrix.

4. Summarize the pathway of hydrogen from the oxidation in glycolysis to combining with oxygen to form water in the matrix at the end of the electron transport chain:

_____

_____

_____

_____

_____

5. How does chemiosmosis enable oxidative phosphorylation to occur in the electron transport chain?

_____

_____

_____

_____

6. Explain the role of oxygen acting as a terminal electron acceptor in aerobic cell respiration:

_____

_____

_____

165 Anaerobic Pathways

Key Idea: Glucose can be metabolized aerobically and anaerobically to produce ATP. The ATP yield from aerobic processes is higher than from anaerobic processes.
Aerobic respiration occurs in the presence of oxygen. Organisms can also generate ATP when oxygen is absent by using a molecule other than oxygen as the terminal electron acceptor for the pathway. In alcoholic fermentation in yeasts, the electron acceptor is ethanal. In lactic acid fermentation, which occurs in mammalian muscle even when oxygen is present, the electron acceptor is pyruvate itself.

Alcoholic fermentation

In alcoholic fermentation, the H^+ acceptor is ethanal which is reduced to ethanol with the release of carbon dioxide (CO_2). Yeasts respire aerobically when oxygen is available but can use alcoholic fermentation when it is not. At ethanol levels above 12-15%, the ethanol produced by alcoholic fermentation is toxic and this limits their ability to use this pathway indefinitely. The root cells of plants also use fermentation as a pathway when oxygen is unavailable but the ethanol must be converted back to respiratory intermediates and respired aerobically.

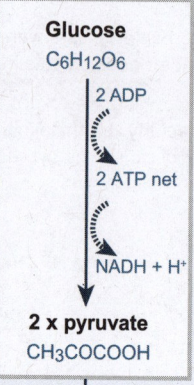

Glucose
$C_6H_{12}O_6$

2 ADP

2 ATP net

NADH + H^+

2 x pyruvate
$CH_3COCOOH$

Lactic acid fermentation

Skeletal muscles produce ATP in the absence of oxygen using lactic acid fermentation. In this pathway, pyruvate is reduced to lactic acid, which dissociates to form lactate and H^+. The conversion of pyruvate to lactate is reversible and this pathway operates alongside the aerobic system all the time to enable greater intensity and duration of activity. The regeneration of NAD^+ produced during lactate fermentation can then be recycled and used in the glycolysis reactions, which ultimately yields 2 ATP per glucose molecule. This regeneration is the key difference in otherwise similar anaerobic respiration in all living organisms.

Glucose
$C_6H_{12}O_6$

2 ADP

2 ATP net

NADH + H^+

2 x pyruvate
$CH_3COCOOH$

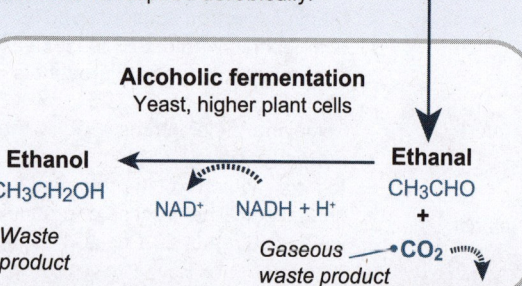

Alcoholic fermentation
Yeast, higher plant cells

Ethanol
CH_3CH_2OH

NAD^+ NADH + H^+

Waste product

Ethanal
CH_3CHO
+

Gaseous → CO_2
waste product

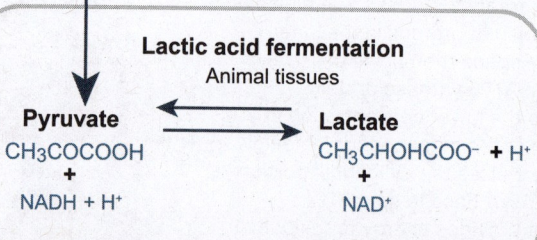

Lactic acid fermentation
Animal tissues

Pyruvate
$CH_3COCOOH$
+
NADH + H^+

Lactate
$CH_3CHOHCOO^- + H^+$
+
NAD^+

The alcohol and CO_2 produced from alcoholic fermentation form the basis of the brewing and baking industries. In baking, the dough is left to ferment and the yeast metabolizes sugars to produce ethanol and CO_2. The CO_2 causes the dough to rise.

Yeasts are used to produce almost all alcoholic beverages, e.g. wine and beers. The yeast used in the process breaks down the sugars into ethanol (alcohol) and CO_2. The alcohol produced is a metabolic by-product of fermentation by the yeast.

Andrea Braakhius, Wintec

The lactate shuttle in vertebrate skeletal muscle works alongside the aerobic system to enable maximal muscle activity. Lactate moves from its site of production to regions within and outside the muscle, e.g. liver, where it can be respired aerobically.

1. Describe the key difference between aerobic respiration and fermentation (anaerobic respiration):

_____

_____

2. What is the importance of NAD^+ regeneration in lactic acid fermentation?

_____

_____

_____

_____

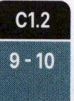

C1.2
9 - 10

©2024 **BIOZONE** International
ISBN: 978-1-99-101410-8
Photocopying prohibited

166 Investigating Yeast Fermentation

Key Idea: Brewer's yeast preferentially uses alcoholic fermentation when there is excess sugar. The CO_2 released can be collected as a measure of fermentation rate.

Brewer's yeast is a facultative anaerobe (meaning it can respire aerobically or use fermentation). One would expect glucose to be the preferred substrate, as it is the starting molecule in cellular respiration, but brewer's yeast can use a variety of sugars, including disaccharides (two unit sugars), which can be broken down into single units. The rate at which yeast (*S cerevisiae*) metabolizes carbohydrate substrates is influenced by temperature, solution pH, and type of carbohydrate available. High levels of sugars suppress aerobic respiration in yeast, so yeast will preferentially use fermentation in the presence of excess substrate.

Investigation 9.3 Investigating fermentation in yeast

See appendix for equipment list.

Work in pairs for this activity. Your teacher will assign you a substrate to investigate.

1. Make a yeast culture by adding 10 g of active yeast to 50 mL of water at 24°C.

2. In a conical flask, boil 225 mL of tap water then cool to room temperature (24°C). This removes any dissolved oxygen from the water.

3. Add 25 g of substrate (glucose, maltose, sucrose, lactose, or none). Stir carefully to dissolve (stirring too vigorously will cause oxygen to dissolve back into the water).

4. Add 25 mL of the source yeast culture to the conical flask solution.

5. Add a thin layer of paraffin oil over the solution in the conical flask to create an anaerobic environment.

6. Stopper the conical flask and set up a measuring cylinder to capture any gas as in the diagram (right).

7. Start timing and record the change in gas volume every five minutes for 1 hour. Record the results for your substrate in the table. Pool data as a class and use it to complete the table below.

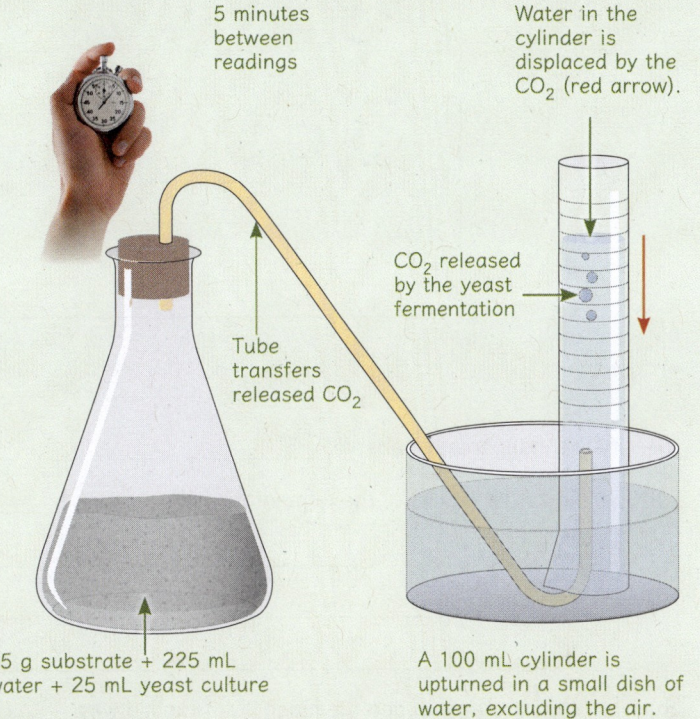

5 minutes between readings

Water in the cylinder is displaced by the CO_2 (red arrow).

CO_2 released by the yeast fermentation

Tube transfers released CO_2

25 g substrate + 225 mL water + 25 mL yeast culture

A 100 mL cylinder is upturned in a small dish of water, excluding the air.

Time (min)	Cumulative volume of carbon dioxide collected (mL)				
	None	Glucose	Maltose	Sucrose	Lactose
0					
5					
10					
15					
20					
25					
30					
35					
40					
45					
50					
55					
60					

1. Write the equation for the fermentation of glucose by yeast:

_____

_____

2. Using the final values (60 minutes) collected from the class, calculate the rate of CO_2 production per minute for each substrate:

(a) None: _____

(b) Glucose: _____

(c) Maltose: _____

(d) Sucrose: _____

(e) Lactose: _____

C1.2
10
AHL

270

3. Use the tabulated data to plot an appropriate graph of the results on the grid provided:

4. Identify the independent variable: _____

5. (a) Summarize the results of the fermentation experiment: _____

    _____

    _____

    _____

    (b) Which substrate produced the most CO_2? Explain why: _____

    _____

    _____

    _____

    (c) Were fermentation rates lower on maltose and sucrose than on glucose? Was this what you expected? Suggest an explanation (you may have to do some research on these molecules to find out the answer):

    _____

    _____

    _____

    (d) Did any substrate produce no CO_2? Can you suggest why?

    _____

    _____

6. Predict what would happen to CO_2 production rates if the yeast cells were respiring aerobically:

    _____

    _____

    _____

©2024 **BIOZONE** International
ISBN: 978-1-99-101410-8
Photocopying prohibited

167 Photosynthesis Overview

Key Idea: Photosynthesis is the process of converting sunlight, carbon dioxide, and water into glucose and oxygen. Photosynthetic organisms use pigments called chlorophyll to absorb sunlight of specific wavelengths and capture its energy. During photosynthesis, this light energy from the sun is used to split water molecules and react them with carbon dioxide. Glucose, a source of food for plants, is produced along with oxygen gas as a waste product.

Photosynthesis general equation	$6CO_2 + 6H_2O + energy \xrightarrow[\text{Chlorophyll}]{\text{Light}} C_6H_{12}O_6 + 6O_2$

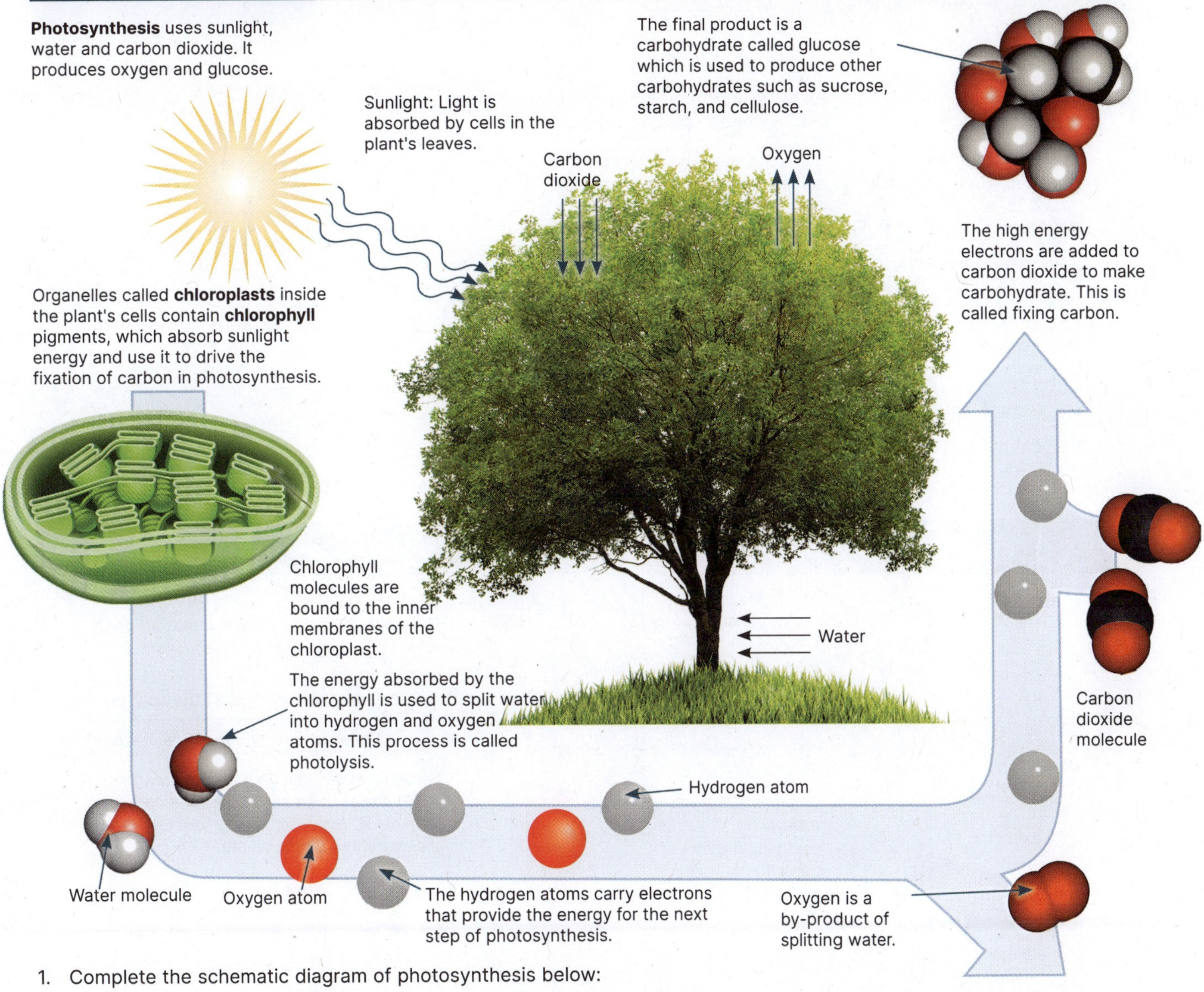

Photosynthesis uses sunlight, water and carbon dioxide. It produces oxygen and glucose.

Sunlight: Light is absorbed by cells in the plant's leaves.

Carbon dioxide

Oxygen

The final product is a carbohydrate called glucose which is used to produce other carbohydrates such as sucrose, starch, and cellulose.

The high energy electrons are added to carbon dioxide to make carbohydrate. This is called fixing carbon.

Organelles called **chloroplasts** inside the plant's cells contain **chlorophyll** pigments, which absorb sunlight energy and use it to drive the fixation of carbon in photosynthesis.

Chlorophyll molecules are bound to the inner membranes of the chloroplast.

The energy absorbed by the chlorophyll is used to split water into hydrogen and oxygen atoms. This process is called photolysis.

Water

Carbon dioxide molecule

Hydrogen atom

Water molecule   Oxygen atom

The hydrogen atoms carry electrons that provide the energy for the next step of photosynthesis.

Oxygen is a by-product of splitting water.

1. Complete the schematic diagram of photosynthesis below:

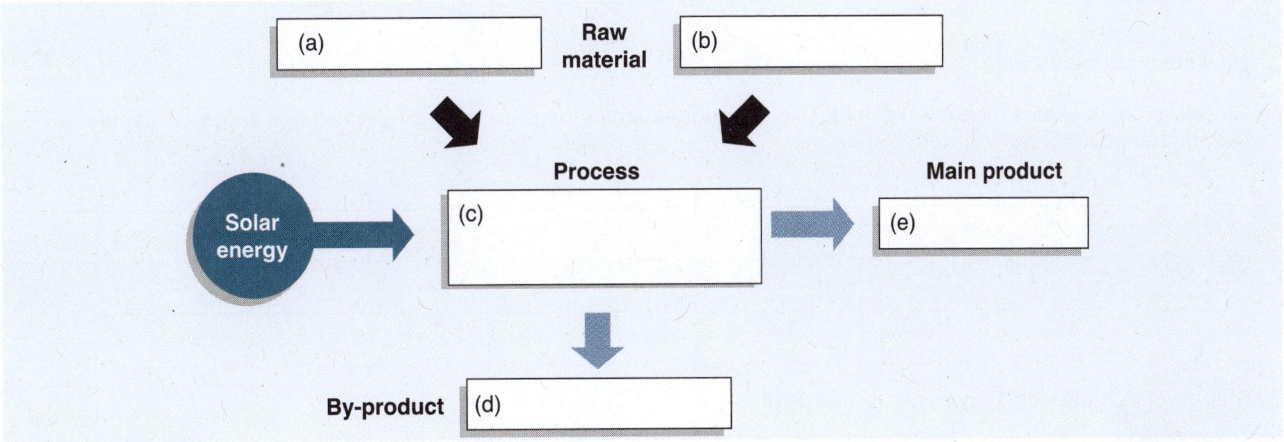

Raw material

(a) _____   (b) _____

Solar energy →

Process

(c) _____

Main product

(e) _____

By-product (d) _____

C1.3
1 - 3

168 Separation of Pigments by Chromatography

Key Idea: Photosynthetic pigments can be separated from a mixture using chromatography.

Chromatography involves passing a mixture dissolved in a mobile phase (a solvent) through a stationary phase, which separates the molecules according to their specific characteristics, e.g. size or charge. In thin layer chromatography, the stationary phase is a thin layer of adsorbent material, e.g. silica gel or cellulose, attached to a solid plate. A sample is placed near the bottom of the plate which is placed in an appropriate solvent (the mobile phase).

Investigation 9.4   Separating photosynthetic pigments

See appendix for equipment list.

1. Tear leaves, e.g. spinach or silverbeet/chard, into small sections and place in a pestle. Add a pinch of sand and 10 mL of ethanol. Grind up the leaves to form a dark green mixture.

2. Pour the mixture into a beaker or boiling tube, cover with plastic-wrap and leave for 5-10 minutes. This gives time for the chlorophyll pigments to better dissolve into the ethanol.

3. Cut a piece of filter paper or chromatography paper into a strip 1-2 cm wide. It should be long enough to reach from the top of a beaker or boiling tube to the bottom.

4. Use a pencil to draw a line across the width of the paper 1 cm from the bottom to mark the start position.

5. Use a micropipette to place a drop of the ground leaf mixture onto the middle of the line. You may need to do this a few times and air dry between each application to concentrate pigments on the spot.

6. Pour ethanol into a beaker or boiling tube to a depth of just over 1 cm. Set up the chromatography paper as in the diagram (below left).

7. Leave for long enough that the solvent front (ethanol) travels nearly to the top of the paper, or the pigments are well spread out. This may take up to 20 minutes.

8. Remove the paper and air dry. Calculate the R_f value for each pigment (below right).

Chromatography set up

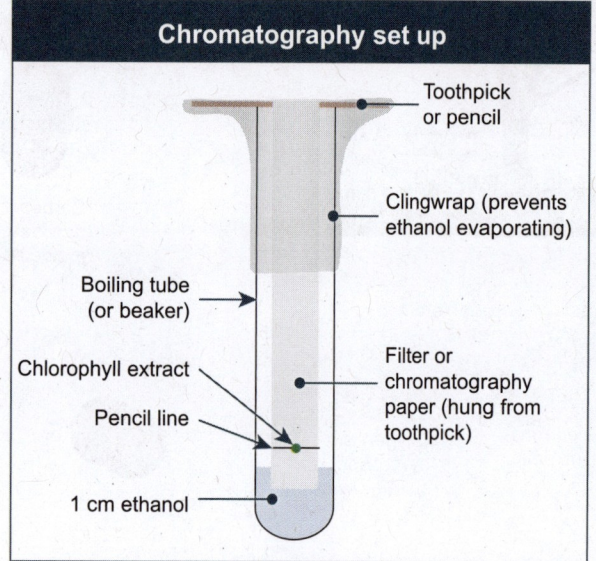

Labels: Toothpick or pencil; Clingwrap (prevents ethanol evaporating); Boiling tube (or beaker); Chlorophyll extract; Filter or chromatography paper (hung from toothpick); Pencil line; 1 cm ethanol

Determining Rf values

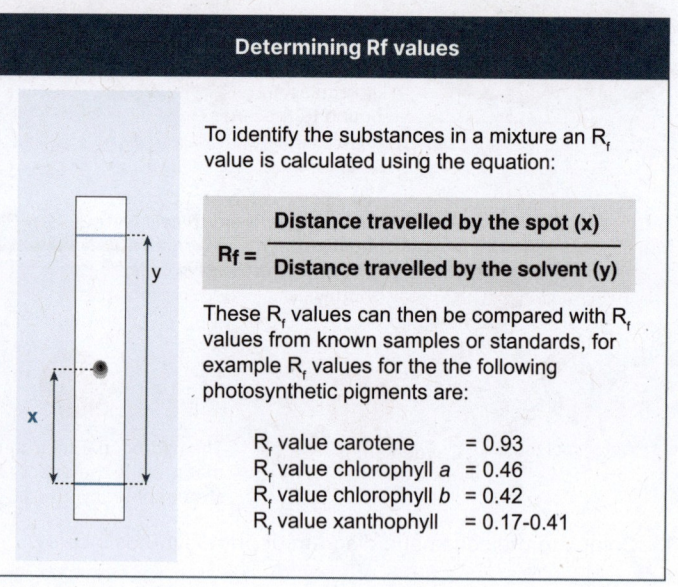

To identify the substances in a mixture an R_f value is calculated using the equation:

$$R_f = \frac{\text{Distance travelled by the spot (x)}}{\text{Distance travelled by the solvent (y)}}$$

These R_f values can then be compared with R_f values from known samples or standards, for example R_f values for the the following photosynthetic pigments are:

R_f value carotene = 0.93
R_f value chlorophyll *a* = 0.46
R_f value chlorophyll *b* = 0.42
R_f value xanthophyll = 0.17-0.41

---

1. How many pigments were you able to separate? _____

2. Calculate the R_f values for each pigment. Use the information in the *Determining R_f values* box above to identify each pigment on your chromatography paper:

_____

_____

_____

_____

3. Staple your chromatography paper to this page.

AOS   C1.3 / 4   169

©2024 **BIOZONE** International
ISBN: 978-1-99-101410-8

169 Pigments and Light Absorption

Key Idea: Chlorophyll pigments absorb light and capture light energy for photosynthesis.

The ability of phototrophic organisms to capture light energy is a function of the membrane-bound pigments they possess. Pigments are substances that absorb visible light, and different pigments absorb light of different wavelengths. The amount of light absorbed vs the wavelength of light is called the absorption spectrum of that pigment. The absorption spectrum of different photosynthetic pigments provides clues to their role in photosynthesis, since light can only perform work if it is absorbed. An action spectrum profiles the effectiveness of different wavelengths of light in fuelling photosynthesis. It is obtained by plotting wavelength against a measure of photosynthetic rate, e.g. O_2 production.

The electromagnetic spectrum

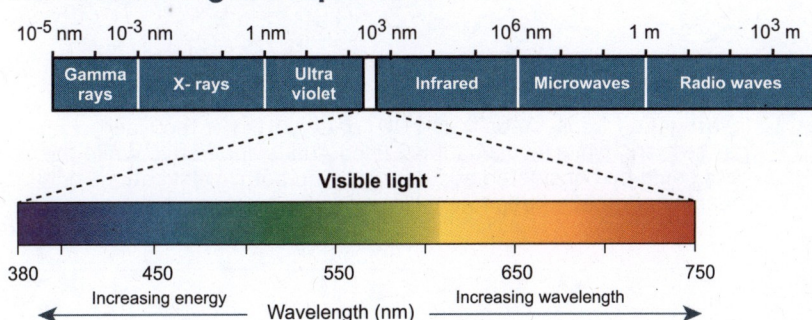

Light is a form of energy known as electromagnetic radiation (EMR). The segment of the electromagnetic spectrum most important to life is the narrow band between about 380 nm and 750 nm. This radiation is known as visible light because it is detected as colours by the human eye. It is visible light that drives photosynthesis. EMR travels in waves, where wavelength provides a guide to the energy of the photons. The greater the wavelength of EMR, the lower the energy of the photons in that radiation.

Absorption spectra of photosynthetic pigments
(Relative amounts of light absorbed at different wavelengths)

Chlorophyll b
Carotenoids
Chlorophyll a

Action spectrum for photosynthesis
(Effectiveness of different wavelengths in fuelling photosynthesis)

The action spectrum and the absorption spectrum for the photosynthetic pigments (combined) match closely.

The photosynthetic pigments of plants

The photosynthetic pigments of plants fall into two categories: chlorophylls (which absorb red and blue-violet light) and carotenoids (which absorb strongly in the blue-violet and appear orange, yellow, or red). The pigments are located on the chloroplast membranes (the thylakoids) and are associated with membrane transport systems.

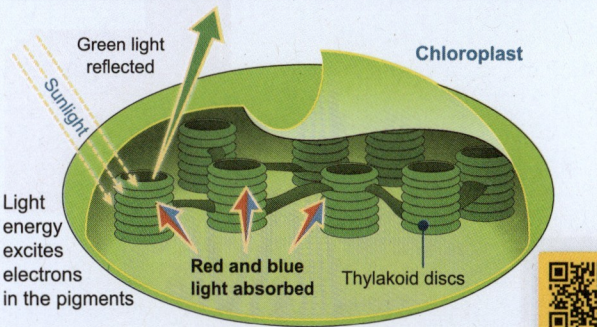

The pigments of chloroplasts in higher plants (above) absorb blue and red light, and the leaves therefore appear green (which is reflected). Each photosynthetic pigment has its own characteristic absorption spectrum (top left). Only chlorophyll a participates directly in the light reactions of photosynthesis. The accessory pigments (chlorophyll b and carotenoids) can absorb wavelengths that chlorophyll a cannot and pass the energy (photons) to chlorophyll a, broadening the spectrum that can drive photosynthesis.

Left: Graphs comparing absorption spectra of photosynthetic pigments compared with the action spectrum for photosynthesis.

1. What is meant by the absorption spectrum of a pigment? _____

_____

2. Use the graphs above to explain why the action spectrum for photosynthesis doesn't exactly match the absorption spectrum of chlorophyll a:

_____

_____

_____

C1.3
5 - 6
AOS

Investigating the effect of wavelength on photosynthetic rate

▶ The rate of photosynthesis can be investigated by tracking the substances involved in photosynthesis. These include measuring the uptake of carbon dioxide, the production of oxygen, or the change in biomass over time.

▶ Measuring the rate of oxygen production provides a good approximation of the photosynthetic rate and is relatively easy to carry out. *Cabomba* pondweed can be submerged with a syringe attached to a capillary tube (Audus apparatus) exposed to light - movement of an air bubble can indicate volume of oxygen produced over time. Alternatively, the *Cabomba* can be inverted in a $NaHCO_3$ solution, with oxygen production relating to the number of oxygen bubbles per unit time.

▶ Measuring the rate of CO_2 uptake can take place in a laboratory using a Infra-Red Gas Analyser (IRGA) which can detect the concentration of CO_2 passing into a sealed chamber around a plant or leaf, as well as the concentration leaving the leaf or plant. Alternatively, a CO_2 monitor can be placed inside a plastic bag around a plant and results recorded over time.

Analysing data

Two separate investigations on the effect of light wavelength on the rate of photosynthesis yielded the following data:

Investigation 1

Light colour	O_2 production (ppt/min)	CO_2 depletion (ppt/min)
white	0.245	-0.189
blue	0.115	-0.103
green	0.004	-0.002
red	0.065	-0.032
cell phone light	0.152	-0.140
none	0.001	+0.001

3. Investigation 1 collected O_2 and CO_2 data (parts per thousand) with gas sensors, while investigation 2 used Audus apparatus. While the data could be considered accurate, why do both investigations only measure photosynthetic rate indirectly?

_____

_____

_____

_____

Investigation 2

Colour (nm)	400	420	440	460	480	500	520	540	560	580	600	620	640	660	680	700
O_2 (mL/h)	22.2	30.4	44.3	53.4	28.6	23.4	20.0	18.3	18.3	20.0	23.4	23.4	35.4	42.6	38.8	23.5

4. Plot the data from investigation 2 to make an action spectra on the graph below

5. What does the action spectra indicate about the likely composition of photosynthetic pigments in the plants that were tested and why? (hint: refer to previous, page):

_____

_____

_____

_____

_____

_____

_____

_____

_____

6. (a) Construct a separate graph showing both O_2 and CO_2 data from investigation 1 for blue, green, and red light.

(b) What colour light is the cell phone likely to produce, and what evidence do you have? _____

_____

(c) What could explain the (small) readings with no light? _____

_____

_____

©2024 **BIOZONE** International
ISBN: 978-1-99-101410-8
Photocopying prohibited

170 Factors Affecting Photosynthesis

Key Idea: Environmental factors, such as light availability and carbon dioxide level, affect photosynthetic rate.

The rate at which plants can photosynthesise is dependent on environmental factors, particularly light and carbon dioxide (CO_2), but also temperature. In the plant's natural environment, fluctuations in these factors (and others) influence photosynthetic rate, so that the rate varies daily and seasonally. The effect of each factor can be tested experimentally by altering one while holding the others constant. Usually, either light or CO_2 levels are limiting. Humans can overcome the limitations of low light or CO_2 by growing plants in a controlled environment.

▶ The rate of photosynthesis is affected by abiotic (non-living) factors in the environment, particularly light intensity, temperature, and CO_2 level. These factors are termed limiting factors, because they influence achievable photosynthetic rate.

▶ The two graphs (right) illustrate the effect of these variables on the rate of photosynthesis in cucumber plants. Figure A shows the effect of increasing light intensity at constant temperature and CO_2 level. Figure B illustrates how this response is influenced by CO_2 concentration and temperature. 30°C is at the upper range of tolerance for many plants.

Gaps in forest canopy allow rapid growth in plants receiving light.

Figure A: Effect of light intensity on photosynthetic rate

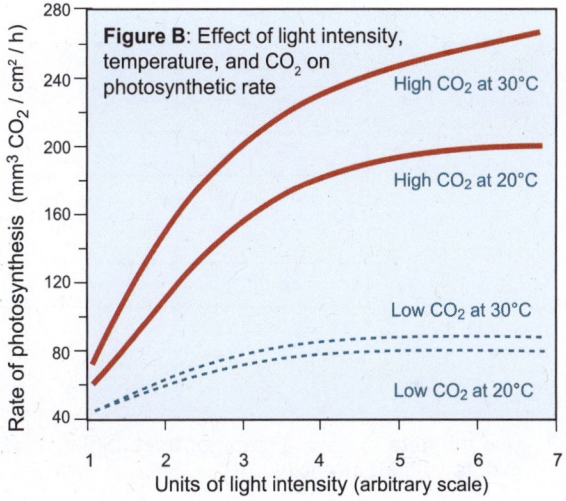
Figure B: Effect of light intensity, temperature, and CO_2 on photosynthetic rate

High CO_2 at 30°C
High CO_2 at 20°C
Low CO_2 at 30°C
Low CO_2 at 20°C

Units of light intensity (arbitrary scale)

Developing hypotheses

A hypothesis is a testable statement and can be stated as an alternative hypothesis: claiming there will be a measurable effect on the dependent variable if the independent variable is changed. Conversely, a null hypothesis states that there will be no measurable effect of the dependent hypothesis if the independent variable is changed. Valid hypotheses need to be based on prior knowledge, either from published research or developed from investigation results.

1. Based on the data above, develop a hypothesis to investigate the effect of limiting factors, on the following:

   (a) CO_2 concentration: _____

   _____

   (b) Light intensity: _____

   _____

   (c) Temperature: _____

   _____

2. What limiting factor from the three above appears to have the most effect on photosynthetic rate? Justify your answer:

   _____

   _____

   _____

   _____

   _____

©2024 **BIOZONE** International
ISBN: 978-1-99-101410-8
Photocopying prohibited

C1.3
7 - 8

Investigation 9.5 Investigating the effect of light intensity on photosynthetic rate

See appendix for equipment list.

1. Weigh 0.8-1.0 grams of *Cabomba aquatica* stem on a balance. Cut the stem underwater and invert to ensure a free flow of oxygen bubbles.

2. Place into a beaker filled (at approximately 20°C) with a solution containing 0.2 mol/L sodium hydrogen carbonate (to supply carbon dioxide).

3. Invert a funnel over the *Cabomba* and then invert a test tube filled with the sodium hydrogen carbonate solution on top to collect any gas produced.

4. Place the beaker at distances 20, 25, 30, 35, 40, 45, and 50 cm from a 60W light source. Measure the light intensity with a lux meter at each interval.

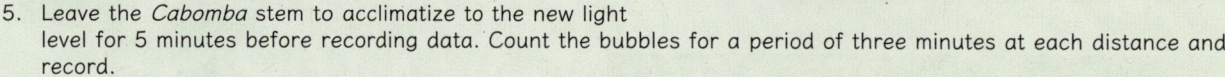

- Oxygen bubbles
- Test tube with NaHCO₃ solution
- Inverted funnel
- Beaker with NaHCO₃ solution at 20°C
- *Cabomba* stem

5. Leave the *Cabomba* stem to acclimatize to the new light level for 5 minutes before recording data. Count the bubbles for a period of three minutes at each distance and record.

6. Alternatively, the volume of gas captured in the test tube can be measured and recorded.

Distance (cm)	Light intensity (lx)	Bubbles counted in three minutes	Bubbles per minute	Volume (mL)
50 cm				
45 cm				
40 cm				
35 cm				
30 cm				
25 cm				
20 cm				

3. Use the data to draw a graph of the bubbles produced per minute vs light intensity:

4. Why is measuring light intensity directly in lux better than inferring light intensity from the measured distance?

_____

_____

_____

5. The sample of gas collected during the experiment was tested with a glowing splint. The splint reignited when placed in the gas. What does this confirm about the gas produced?

_____

6. Why is measuring gas collected, rather than bubbles produced, a more accurate why of recording data?

_____

_____

7. Was your hypothesis from 1b rejected or supported? _____

_____

_____

©2024 **BIOZONE** International
ISBN: 978-1-99-101410-8
Photocopying prohibited

Carbon dioxide enrichment experiments

▶ Carbon dioxide is a limiting factor for photosynthesis. Generally, the rate of photosynthesis can be increased if the concentration of carbon dioxide around the immediate vicinity of plants is increased, up to a certain point.

▶ Understanding the CO_2 levels to maximize the growing potential of the plants can be beneficial to commercial crop growers.

▶ Scientists use two main methods to investigate future photosynthesis and growth rates: experimenting inside contained greenhouses and experimenting in the field with free-air CO_2 enrichments (FACE).

CO_2 enrichment in greenhouses

▶ Experiments in greenhouses have found that increasing CO_2 can benefit plant growth and photosynthesis. The optimal CO_2 concentration for plants ranges between 700 - 1200 ppm, much higher than ambient air.

▶ Plants can capture CO_2 through the process of photosynthesis. Multi-disciplinary teams of scientists are investigating how to use technology to extract CO_2 from the ambient air (carbon capture) and deliver it directly to plants inside enclosed greenhouses.

Glasshouse environments can artificially boost CO_2 levels

Plants produce CO_2 through respiration during darkness, which builds to levels higher than ambient air inside contained greenhouses. However, photosynthesis quickly utilizes this during daylight hours, bringing the CO_2 levels well below the optimal level required for plant growth.

Controlling variables

CO_2 concentration can be a limiting factor for photosynthesis. As an independent variable, it can be altered in various enrichment experiments to measure its effect on photosynthetic rate. A number of other limiting factors, including water and nitrogen availability, also affect photosynthesis rate /plant growth, and can therefore be experimentally controlled. This is straightforward in a greenhouse setting, but less so in field studies. Rain or drought in deserts is unpredictable, and natural levels of nitrogen in soils can fluctuate. Each variable needs to be considered when interpreting CO_2 enrichment data when more than one limiting factor is involved.

Free-air CO_2 enrichments (FACE)

▶ For a number of decades, scientists have considered how plants can contribute to reducing CO_2 from the ambient air (carbon capture). Various long term FACE experiments have taken place to investigate how plant growth is affected by introducing additional CO_2 into the direct environment around plants.

▶ In the Nevada Desert, the FACE experiment was carried out in a natural environment, with no other variables artificially changed.

▶ The rate of photosynthesis did increase, but only in combination with a sufficient wet season. In drought years, lack of water became the significant limiting factor and extra CO_2 pumped in had no effect on plant growth.

Nevada Desert FACE test site

United States Department of Energy

▶ Another FACE experiment, at Oak Ridge National Laboratory, also found that elevated CO_2 levels increased the rate of photosynthesis, and therefore plant growth. However, in this experiment, nitrogen was quickly used by the faster growing plants and without the addition of nitrogen fertilizer, lack of this substance became the limiting factor.

8. What advantages can be gained by using CO_2 enrichment for crop growth? _____

_____

_____

9. How could variables be more effectively controlled in the Nevada Desert FACE experiment?

_____

_____

_____

171 Details of Photosynthesis

Key Idea: Photosynthesis is the process by which light energy is used to convert CO_2 and water into glucose and oxygen. Photosynthesis is of fundamental importance to living things because it transforms sunlight energy into chemical energy stored in molecules, releases free oxygen gas, and absorbs carbon dioxide (a waste product of cellular metabolism). Photosynthesis has two sets of reactions, the light dependent phase and the light independent phase. In the light dependent

phase, light energy is converted to chemical energy (ATP and NADPH). This phase occurs in the thylakoid membranes of chloroplasts. In the light independent phase, the ATP and NADPH are used to synthesize carbohydrates. This phase occurs in the stroma of chloroplasts. In photosynthesis, water is split and electrons are transferred together with hydrogen ions from water to CO_2, reducing it to triose phosphates (then converted to sugars).

Light dependent phase (LDP): Thylakoid

In the first phase of photosynthesis, chlorophyll captures light energy which is used to split water, producing O_2 gas (waste) and H^+ ions that are transferred to the molecule NADPH. ATP is also produced.

Light independent phase (LIP): Stroma

The second phase of photosynthesis occurs in the stroma and uses NADPH and ATP to drive a series of enzyme controlled reactions (the Calvin cycle) that fix carbon dioxide to produce triose phosphate. This phase does not need light to proceed.

Interdependence of phases

Although each phase of photosynthesis occurs in different areas of the chloroplast, both are reliant on the continued cycling of the ADP/ATP, and $NADP^+$/NADPH species, to transfer energy from light to ultimately chemical energy.

CO_2 is required as a reactant for the Calvin cycle, but also as a terminal electron acceptor in the light dependent phase.

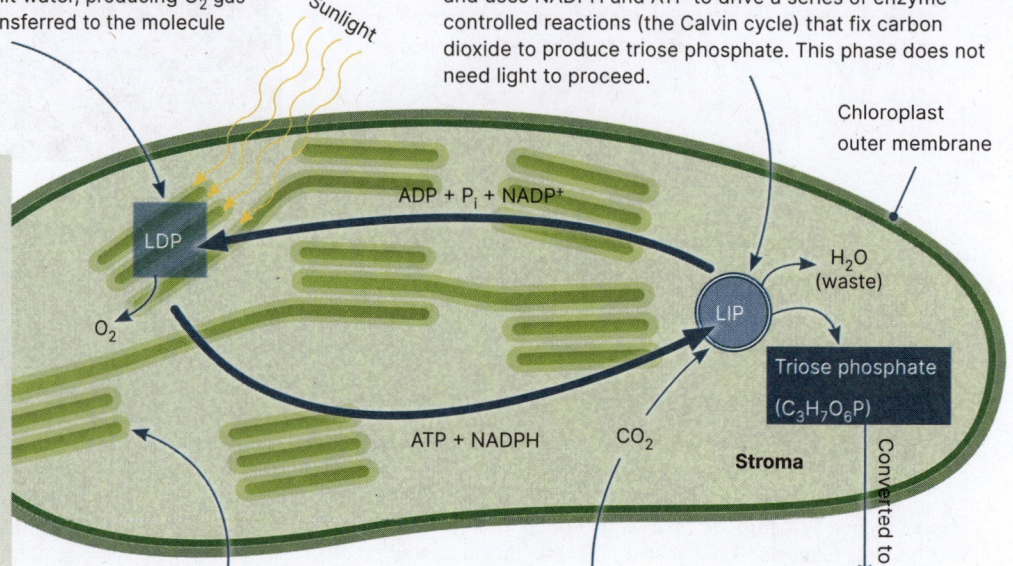

Sunlight

ADP + P_i + $NADP^+$

LDP

O_2

ATP + NADPH

CO_2

LIP

H_2O (waste)

Triose phosphate ($C_3H_7O_6P$)

Chloroplast outer membrane

Stroma

Converted to

The light dependent phase occurs in the thylakoid membranes of the grana.

CO_2 from the air provides raw materials for glucose production.

Monosaccharides, e.g. glucose, and other carbohydrates, lipids, and amino acids.

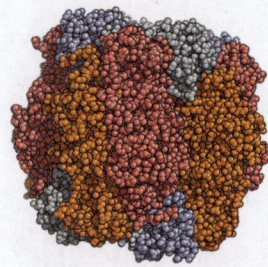

Rubisco is the central enzyme in the LIP of photosynthesis (carbon fixation) catalysing the first step in the Calvin cycle. However, it is remarkably inefficient, processing just three reactions a second. To compensate, rubisco makes up almost half the protein content of chloroplasts.

Photosynthesis can be summarised in the equation:

$$6CO_2 + 12H_2O \xrightarrow[\text{Chlorophyll}]{\text{Light}} C_6H_{12}O_6 + 6O_2 + 6H_2O$$

The apparently extra 6 H_2Os are to show that O_2 comes from the H_2O, not the CO_2.

1. Identify the two phases of photosynthesis and their location in the cell:

(a) _____

(b) _____

2. (a) What is the role of the enzyme RuBisCo? _____

(b) RuBisCo is the most abundant protein on Earth. Suggest a reason for this: _____

_____

_____

3. State the origin and fate of the following molecules involved in photosynthesis:

(a) Carbon dioxide: _____

(b) Oxygen: _____

(c) Hydrogen: _____

©2024 **BIOZONE** International
ISBN: 978-1-99-101410-8

172 Light Dependent Reactions

Key Idea: Light energy is used to drive the reduction of $NADP^+$ and the production of ATP.

Photosynthesis is a redox process where water is split, and electrons and hydrogen ions are transferred from water to CO_2. The electrons increase in potential energy as they move from water to sugar. The energy to do this is provided by light. Photosynthesis has two phases. In the light dependent reactions, light energy is converted to chemical energy (ATP and NADPH). In the light independent reactions, the chemical energy is used to synthesize carbohydrates. The light dependent reactions most commonly involve non-cyclic phosphorylation, which produces ATP and NADPH in roughly equal quantities. The electrons lost are replaced from water. In cyclic phosphorylation, the electrons lost from photosystem II are replaced by those from photosystem I. ATP is generated, but not NADPH.

Photosystems

Photosystems are embedded in the thylakoid membranes of cyanobacteria and photosynthetic eukaryotes and are the site of light dependent reactions. Different pigments are arranged in arrays along with proteins into a complex. Pigments can absorb light photons and then pass this energy on from pigment to pigment. When the energy is collected at the site called the special pair, an excited electron is emitted and, via the primary electron acceptor, it will then move into the electron transport chain. The chlorophyll a special pair absorbs light of different wavelengths, depending on whether they are in photosystem I or photosystem II.

As you will recall, different pigments capture light of different wavelengths. To utilize the full spectrum of visible wavelengths in white light, the photosystem arrays have a variety of pigments. The array allows for sufficient energy to be captured and collected at the special pair. Additional pigments also function as coenzymes in some of the photosynthesis reactions.

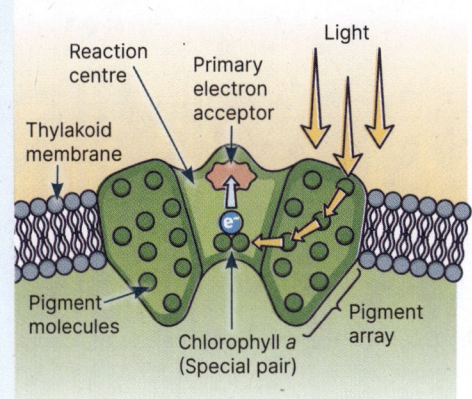

Non-cyclic phosphorylation in photosystems I and II

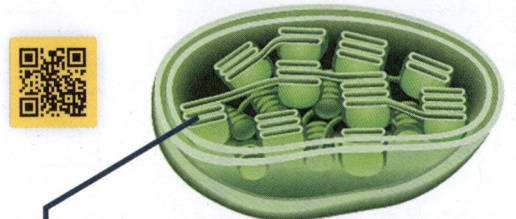

Part of a thylakoid disc is shown below. The chlorophyll molecules are part of the photosystem complexes (I and II) in the thylakoid membrane.

Reducing power (NADPH) and energy (ATP) for the light independent reactions.

Each electron is passed from one electron carrier to another, losing energy as it goes. This energy is used to pump H^+ across the thylakoid membrane to build a proton gradient.

Light strikes the chlorophyll pigment molecules in the thylakoid membrane. Each photosystem is made of many pigment molecules.

Electrons are used to reduce $NADP^+$ to NADPH.

3 $NADP^+$ is the final electron acceptor

$NADP^+$ reductase → NADPH ← $NADP^+$

5 ATP / ADP
NADPH
ATP
Flow of H^+ back across the membrane from high concentration to low concentration is coupled to the synthesis of ATP (steps 4 and 5), a process called chemiosmosis.

ATP synthase

1 e⁻
2 H^+
Chlorophyll

e⁻
Chlorophyll

H^+ **4**
Chemiosmosis

Importance of photolysis **6**

Photolysis of water releases oxygen gas and hydrogen ions. The H^+ ions are used in chemiosmosis and the oxygen becomes a 'waste' product.

The first photosynthetic organisms to emit oxygen appeared around 2.7 billion years ago. They emitted oxygen oxidized iron ions (mostly Fe^{2+}) in the ocean, producing banded iron formations. The eventual build-up of free oxygen in the Earth's oceans and atmosphere allowed the evolution of oxygen respiring organisms. Free atmospheric oxygen eventually produced the ozone layer around 600 million years ago, shielding terrestrial life from harmful UV light.

H_2O → $\frac{1}{2}O_2 + 2H^+$ → H^+
PHOTOSYSTEM II

PHOTOSYSTEM I

Thylakoid space: Hydrogen reservoir, low pH

Photosystem II absorbs shorter wavelength light energy (photons) less than 680 nm. Electrons are extracted from the photolysis of water, and elevated to a moderate energy level. Chlorophyll b is more common than chlorophyll a.

Photosystem I absorbs longer wavelength light energy to elevate electrons to an even higher level. Its electrons are replaced by electrons from photosystem II. It uses both cyclic and non-cyclic photophosphorylation, and does not involve photolysis of water. Chlorophyll a is more common than chlorophyll b.

©2024 **BIOZONE** International
ISBN: 978-1-99-101410-8

C1.3
9 - 14

Cyclic phosphorylation in photosystem I

Cyclic phosphorylation involves only photosystem I and no NADPH is generated. Electrons from photosystem I are shunted back to the electron carriers in the membrane, so this pathway produces ATP only. The Calvin cycle uses more ATP than NADPH, so cyclic phosphorylation makes up the difference. It is activated when NADPH levels build up, and remains active until enough ATP is made to meet demand.

Electrons are cycled through a pathway that takes them away from $NADP^+$ reductase.

ATP

ADP

ATP is produced while NADPH production ceases.

ATP synthase

H^+

e^-

Thylakoid membrane

Chlorophyll

e^-

Chlorophyll

H^+

H^+

Thylakoid membrane: bound pigment molecules and ATP synthase

PHOTOSYSTEM II is not active. Photolysis of water stops. O_2 is not released.

PHOTOSYSTEM I

1. Describe the role of the carrier molecule $NADP^+$ in photosynthesis: _____

_____

2. Explain the role of chlorophyll molecules in photosynthesis: _____

_____

_____

_____

3. (a) Where do the light dependent reactions occur? _____

   (b) Summarize the events of the light dependent reactions: _____

_____

_____

_____

4. Explain how ATP generation is linked to the light dependent reactions of photosynthesis: _____

_____

_____

_____

_____

_____

_____

5. (a) Explain what you understand by the term non-cyclic phosphorylation: _____

_____

_____

_____

   (b) Suggest why this process is also known as non-cyclic photophosphorylation: _____

_____

_____

©2024 **BIOZONE** International
ISBN: 978-1-99-101410-8
Photocopying prohibited

173 Light Independent Reactions

Key Idea: The Calvin cycle uses ATP and NADPH from the light dependent (dark) reactions to produce organic carbon molecules.

In the light independent reactions (commonly called the Calvin cycle), hydrogen (H⁺) is added to CO_2 and a 5C intermediate to make carbohydrates. The H⁺ and ATP are supplied by the light dependent reactions. The Calvin cycle uses more ATP than NADPH, but the cell uses cyclic phosphorylation (which does not produce NADPH) when it runs low on ATP to make up the difference.

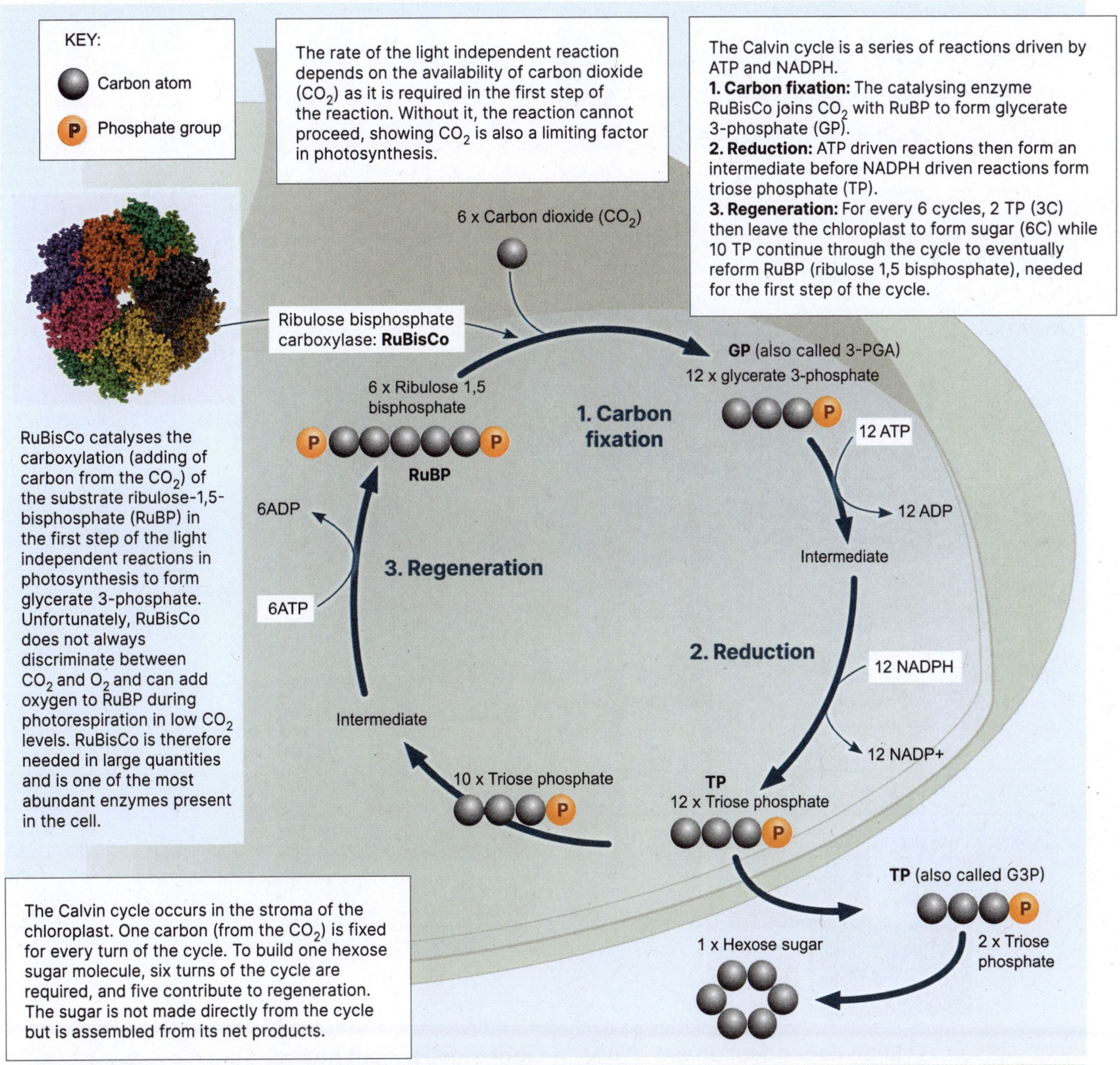

KEY:
- Carbon atom
- P Phosphate group

The rate of the light independent reaction depends on the availability of carbon dioxide (CO_2) as it is required in the first step of the reaction. Without it, the reaction cannot proceed, showing CO_2 is also a limiting factor in photosynthesis.

The Calvin cycle is a series of reactions driven by ATP and NADPH.
1. Carbon fixation: The catalysing enzyme RuBisCo joins CO_2 with RuBP to form glycerate 3-phosphate (GP).
2. Reduction: ATP driven reactions then form an intermediate before NADPH driven reactions form triose phosphate (TP).
3. Regeneration: For every 6 cycles, 2 TP (3C) then leave the chloroplast to form sugar (6C) while 10 TP continue through the cycle to eventually reform RuBP (ribulose 1,5 bisphosphate), needed for the first step of the cycle.

RuBisCo catalyses the carboxylation (adding of carbon from the CO_2) of the substrate ribulose-1,5-bisphosphate (RuBP) in the first step of the light independent reactions in photosynthesis to form glycerate 3-phosphate. Unfortunately, RuBisCo does not always discriminate between CO_2 and O_2 and can add oxygen to RuBP during photorespiration in low CO_2 levels. RuBisCo is therefore needed in large quantities and is one of the most abundant enzymes present in the cell.

6 x Carbon dioxide (CO_2)

Ribulose bisphosphate carboxylase: **RuBisCo**

6 x Ribulose 1,5 bisphosphate — **RuBP**

1. Carbon fixation

GP (also called 3-PGA)
12 x glycerate 3-phosphate

12 ATP → 12 ADP

Intermediate

3. Regeneration
6ADP
6ATP

2. Reduction
12 NADPH → 12 NADP+

Intermediate

10 x Triose phosphate

TP 12 x Triose phosphate

TP (also called G3P)

1 x Hexose sugar

2 x Triose phosphate

The Calvin cycle occurs in the stroma of the chloroplast. One carbon (from the CO_2) is fixed for every turn of the cycle. To build one hexose sugar molecule, six turns of the cycle are required, and five contribute to regeneration. The sugar is not made directly from the cycle but is assembled from its net products.

Synthesis of substances from the Calvin cycle and minerals

The Calvin cycle produces 3-carbon sugars from the continual addition of carbon from atmospheric CO_2. These sugars can then undergo further reactions to form a wide range of carbohydrate macromolecules, including glucose, which is further metabolically processed to form starch and cellulose (plants). Starch biosynthesis normally occurs during the day within the chloroplast stroma. You will recall that carbohydrates are used to produce ATP through the respiration pathway, but they are also used for cellular functioning and structure.

Intermediates formed in the Calvin cycle undergo separate metabolic pathways to form precursors for amino acids and therefore proteins via protein synthesis, as well as lipids, used in membrane formation.

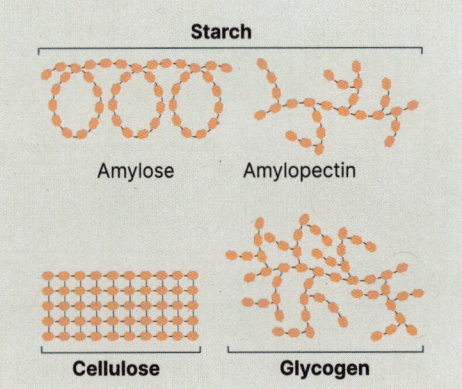
Starch — Amylose, Amylopectin, Cellulose, Glycogen

1. On a separate blank sheet, and either individually or in small groups, construct a schematic model showing the main pathway of the light dependent and light independent pathways, adding key reactants and products.

174 Did You Get It?

1. What is the link between the enzyme's active site and the induced fit model? _____

_____

2. Suggest why digestion is largely performed by extracellular enzymes: _____

_____

_____

3. **AHL:** Explain the difference between linear and cyclic metabolic pathway, giving examples: _____

_____

_____

_____

4. Complete the diagram of aerobic cellular respiration below by filling in the boxes:

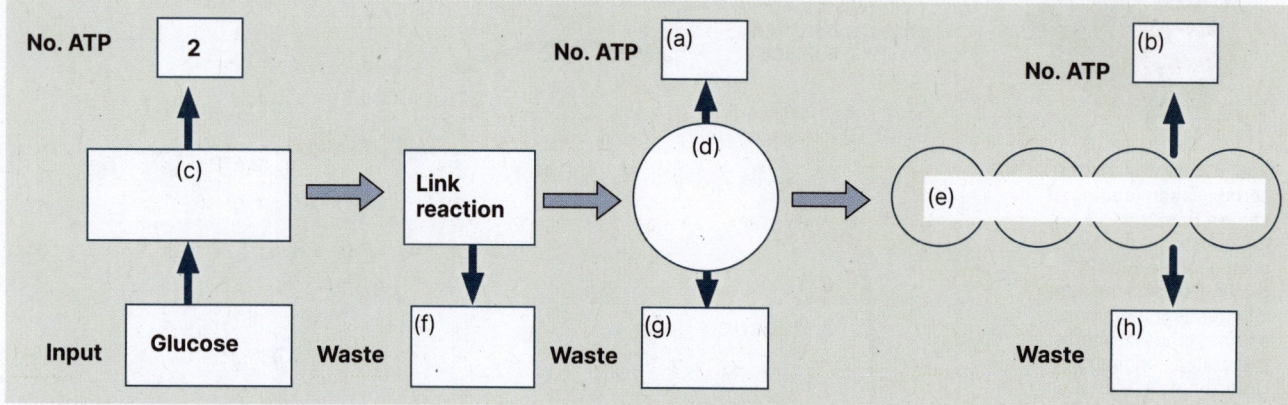

5. **AHL:** Explain how chemiosmosis allows ATP synthase to produce ATP: _____

_____

_____

_____

6. Complete the diagram below, including the raw material (inputs), products (outputs), and processes.

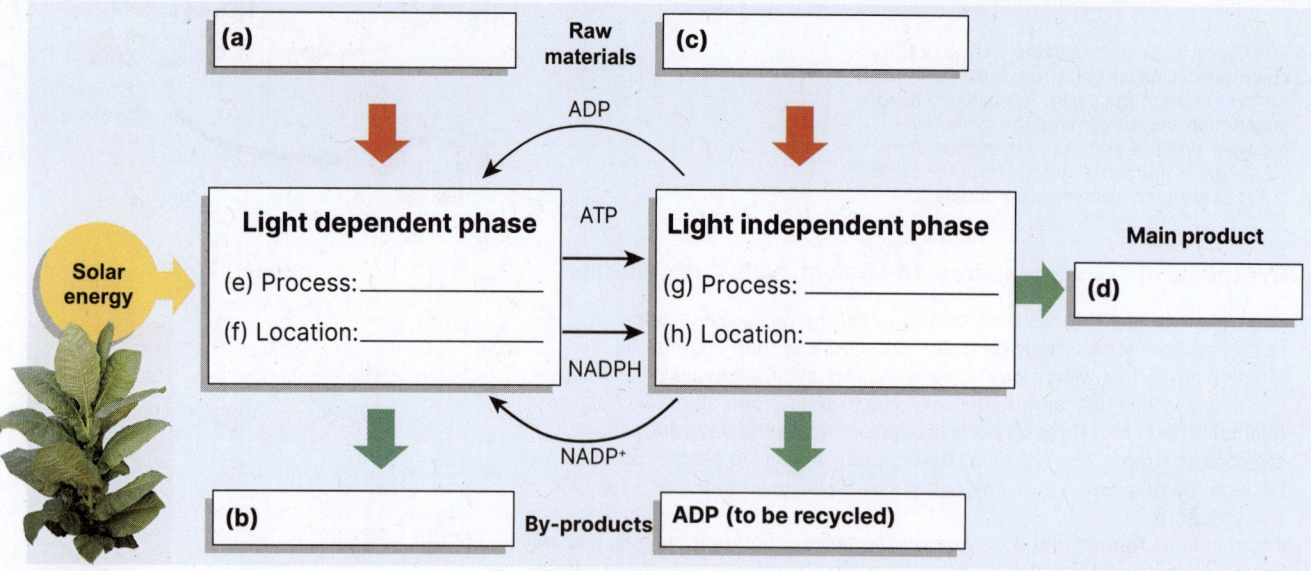

7. **AHL:** Explain the importance of RuBisCo in the Calvin cycle: _____

_____

_____

©2024 **BIOZONE** International
ISBN: 978-1-99-101410-8
Photocopying prohibited

Cells

C2.1 Cell signalling

	Activity Number

Guiding Questions:
▸ How does cell signalling reach and stimulate the correct cells?
▸ What effect do chemical signals have on cellular responses?

Learning Outcomes:

			Activity Number
☐	1	**AHL:** Explain how proteins act as receptors with binding sites for specific ligands (signalling chemicals).	175
☐	2	**AHL:** Investigate the process cell signalling in bacterial quorum sensing including bioluminescence in *Vibrio fischeri*.	175
☐	3	**AHL:** Distinguish between different categories of signalling chemicals in animals including hormones, neurotransmitters, cytokines, and calcium ions.	176
☐	4	**AHL:** Provide explanations for the chemical diversity of hormones and neurotransmitters including named examples.	176
☐	5	**AHL:** Contrast between localized and distant effects of signalling molecules.	176, 177
☐	6	**AHL:** Contrast between processes occurring in transmembrane receptors in a plasma membrane and intracellular receptors in the cytoplasm or nucleus.	178
☐	7	**AHL:** Explain how signal transduction pathways are initiated by receptors.	175, 178
☐	8	**AHL:** Explain how neurotransmitters binding to transmembrane receptors results in changes to membrane potential.	180
☐	9	**AHL:** Explain how transmembrane receptors activate G-proteins when conveying a signal into cells.	178, 180
☐	10	**AHL:** Describe the mechanism of action of epinephrine (adrenaline) receptors **NOS:** Demonstrate understanding of the etymology of the terms epinephrine and adrenaline.	178, 180
☐	11	**AHL:** Describe the mechanism of action of transmembrane receptors with tyrosine kinase activity using the protein hormone insulin as an example.	181
☐	12	**AHL:** Explain how intracellular receptors can affect gene expression using oestradiol, progesterone, and testosterone as signalling chemical examples.	182
☐	13	**AHL:** Describe the effects that oestradiol and progesterone have on target cells.	183
☐	14	**AHL:** Compare and contrast the regulation of cell signalling pathways by positive and negative feedback.	183

C2.2 Neural signalling

	Activity Number

Guiding Questions:
▸ How do neurons generate electrical signals and then facilitate the movement of those signals?
▸ What processes allow interaction between neuron cells?

Learning Outcomes:

			Activity Number
☐	1	Discuss the structure and function of neurons.	184
☐	2	Explain how concentration gradients of sodium and potassium ions result in the generation of the resting potential.	185
☐	3	Explain how nerve impulses are propagated along nerve fibres.	185
☐	4	Investigate the variation in speed of a nerve impulse along an axon. **AOS:** Apply the coefficient of determination (R^2) to evaluate the degree to which nerve impulse speed is affected by nerve fibre diameter, and determine the effect of mylenated and non-mylenated nerves on nerve impulse speed using cat and squid neurons as examples respectively.	186
☐	5	Describe how junctions between neurons and between neurons and effector cells can act as synapses.	187
☐	6	Elaborate on the release of neurotransmitters from a presynaptic membrane.	187
☐	7	Elaborate on the generation of an excitatory postsynaptic potential using acetylcholine as an example.	187

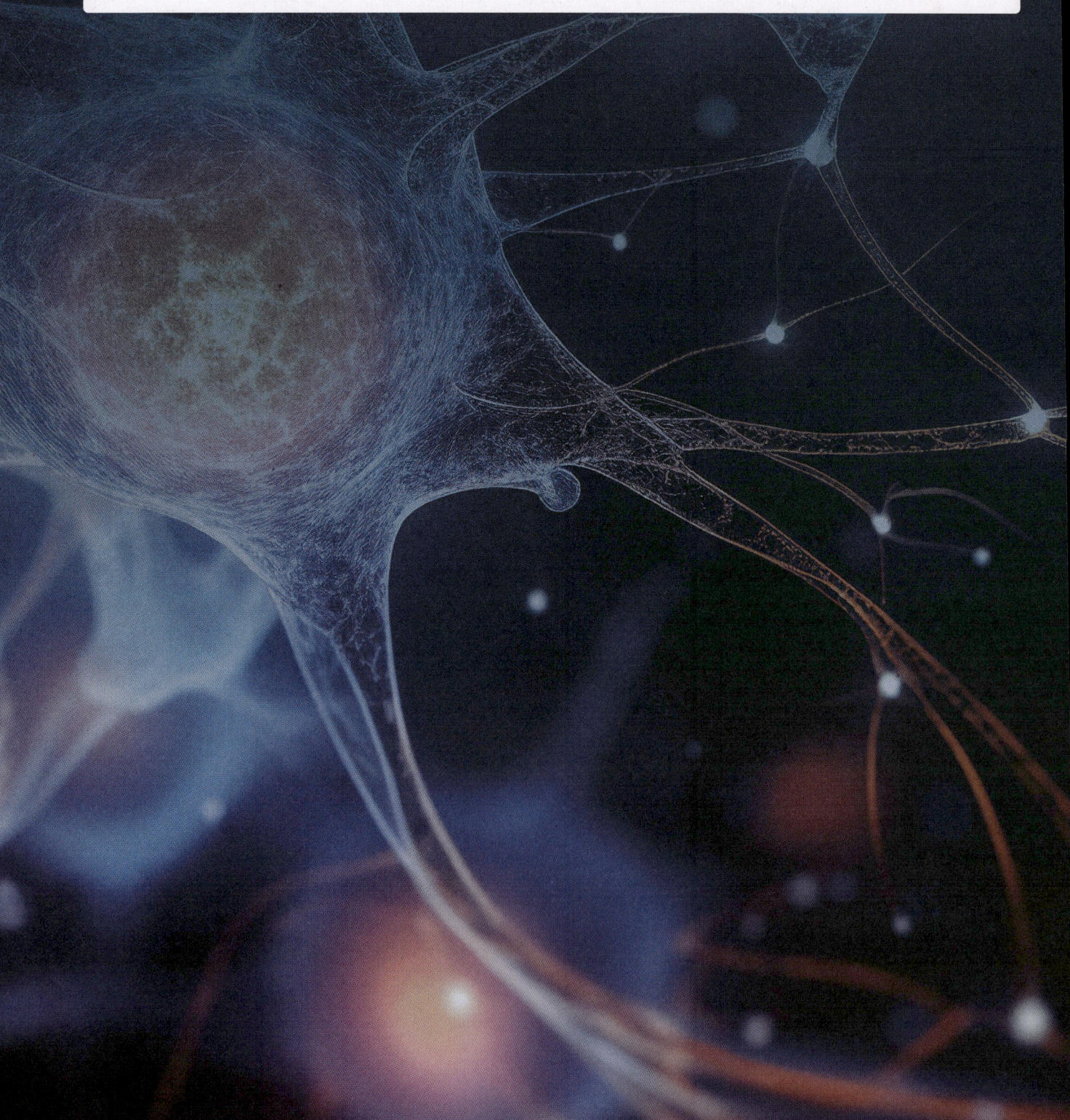

175 Signals and Signal Transduction

Key Idea: Signal transduction converts external signals into functional change within the cell.

Cells can send messages to (signal) other cells by means of signalling molecules, called ligands. Ligands are any molecule that binds to a receptor protein in the plasma membrane and initiates signal transduction, e.g. a hormone.

Signal transduction is the process by which molecular signals are converted from one form to another so that they can be transmitted from outside the cell to inside and bring about a response. The transduction involves an external signal molecule binding to a receptor and triggering a series of biochemical reactions.

An overview of signal transduction

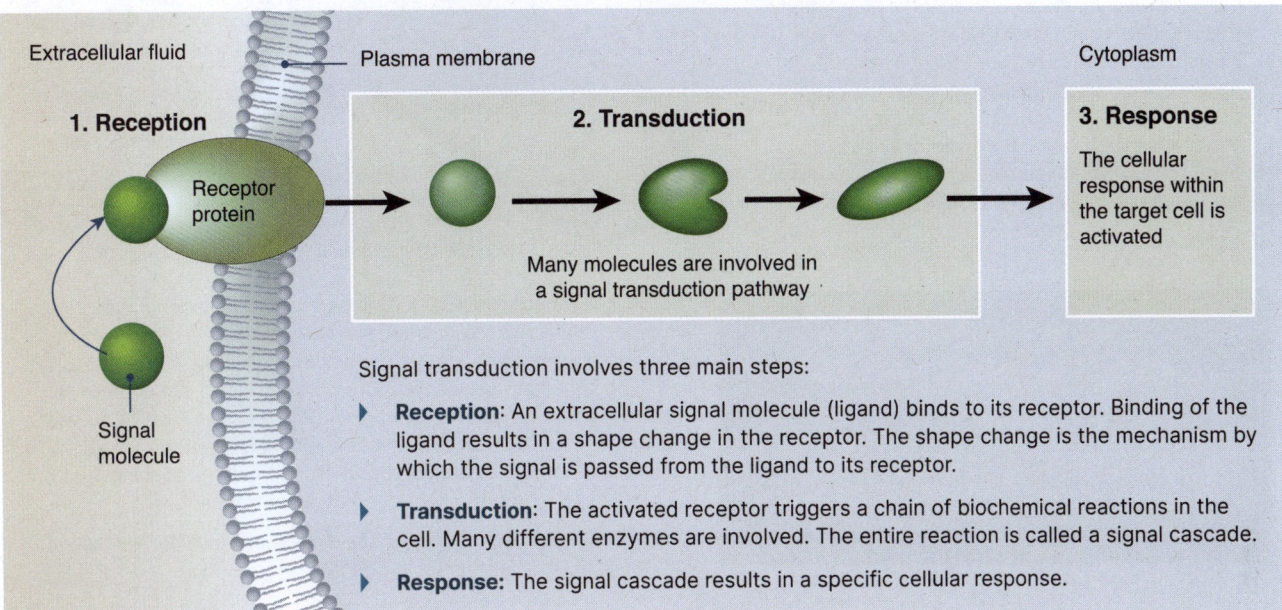

Signal transduction involves three main steps:

▶ **Reception**: An extracellular signal molecule (ligand) binds to its receptor. Binding of the ligand results in a shape change in the receptor. The shape change is the mechanism by which the signal is passed from the ligand to its receptor.

▶ **Transduction**: The activated receptor triggers a chain of biochemical reactions in the cell. Many different enzymes are involved. The entire reaction is called a signal cascade.

▶ **Response**: The signal cascade results in a specific cellular response.

What is a receptor?

Chemical signals must be received by a cell in order to exert their effect. Reception of signal molecules (ligands) is the job of proteins called receptors, which bind ligands and link them to a biochemical pathway to cause a cellular response. The specificity of receptors to their particular signal molecule increases the efficiency of cellular responses and saves energy. The binding sites of cell receptors are specific only to certain ligands. This stops them reacting to every signal the cell encounters. Receptors generally fall into two main categories:

Intracellular receptors
Intracellular (cytoplasmic) receptors are located within the cell's cytoplasm and bind ligands that can cross the plasma membrane.

Extracellular receptors
Extracellular (transmembrane) receptors span the plasma membrane with regions both inside and outside the cell. They bind ligands that cannot cross the olasma membrane (right).

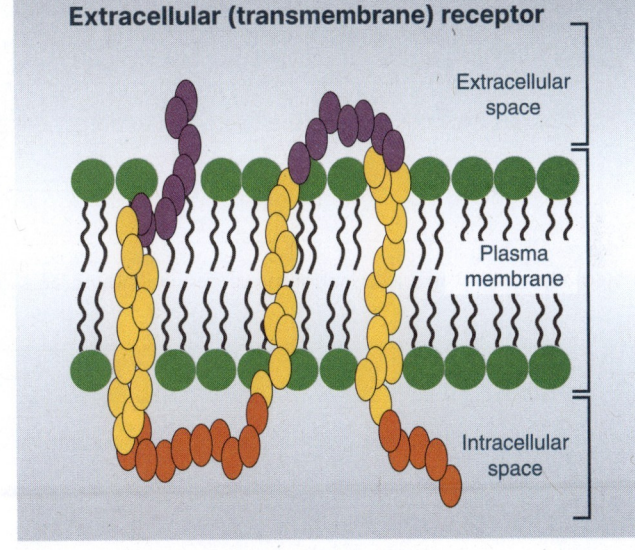

1. Describe the three stages of signal transduction:

   (a) _____

   _____

   (b) _____

   _____

   (c) _____

   _____

2. Why doesn't every cell respond to a particular signal molecule? _____

_____

C2.1

1, 2

Quorum sensing and bioluminescence in bacteria

▶ Quorum sensing is a process of cell to cell communication between bacterial cells. It involves extracellular signalling molecules called autoinducers. Bacteria share information about cell density in their environment and then alter their cellular response through changes in gene expression.

▶ Quorum sensing allows the bacterial population to act together in a coordinated way. A critical number (quorum) is required for the action to be beneficial. Quorum sensing helps to ensure that energy and resources are not expended unnecessarily. It regulates many activities in bacteria, including symbiosis, virulence, motility, antibiotic production, and biofilm formation.

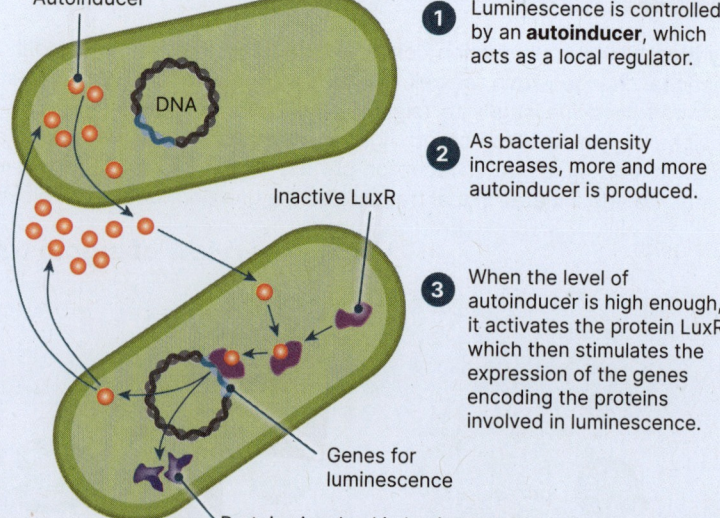

Autoinducer

DNA

Inactive LuxR

Genes for luminescence

Proteins involved in luminescence

1 Luminescence is controlled by an **autoinducer**, which acts as a local regulator.

2 As bacterial density increases, more and more autoinducer is produced.

3 When the level of autoinducer is high enough, it activates the protein LuxR, which then stimulates the expression of the genes encoding the proteins involved in luminescence.

1 cm

Margaret McFall-Ngai cc 4.0

The Hawaiian bobtail squid (left) lives in a mutually beneficial symbiotic relationship with the bacterium *Aliivibrio fischeri*, which luminesces when the bacterial population reaches a certain density. Hatchling squid capture bacteria from the environment and house them at high densities in a light organ. The bacteria receive sugars and amino acids and the squid uses the luminescent bacteria to provide camouflage through counter-illumination (producing light on their bellies to match the light coming down from above).

Luminescent bacteria produce light as the result of a chemical reaction that converts chemical energy to light energy.

3. (a) Explain how the autoinducer molecule in luminescent bacteria signals when to luminesce:

_____

_____

(b) Explain how this enables the bacterium to detect the population density:

_____

_____

(c) How might this information be of survival advantage to the bacterium? _____

_____

_____

4. How does the bobtail squid make use of the quorum sensing abilities of *Aliivibrio fischeri*?_____

_____

_____

5. The images below show a signalling pathway, but the steps are out of order. Assign each image a number (1-4) to indicate its correct position in the sequence:

(a) _____ (b) _____ (c) _____ (d) _____

ISBN: 978-1-99-101410-8

176 Signal Molecules

Key Idea: There is a wide variety of cell signalling molecules (ligands). They may have long or short term effects.

Signalling molecules bind to specific receptors to cause a response in a target cell. There is a huge number of different signal molecules, and each has a specific effect. Some (such as hormones) tend to be slow acting and long lasting, while others (such as neurotransmitters) take effect rapidly, but the effect is short lived as the chemical is quickly broken down. Chemical signals can affect cells within an organism or, in the case of microbes, can be used to communicate with nearby cells. Pheromones are chemical signals secreted into the environment. They may travel over long distances to influence members of the same species.

Hormones help animals prepare for and adjust to seasonal changes

In mammals, hormones are secreted by endocrine glands, e.g. the pituitary, and carried in the blood to target cells. Hormones are very potent, and effective at low concentrations. Hormonal responses tend to be slow because it takes time for the signal to reach its target, and generally long lasting because they induce metabolic changes.

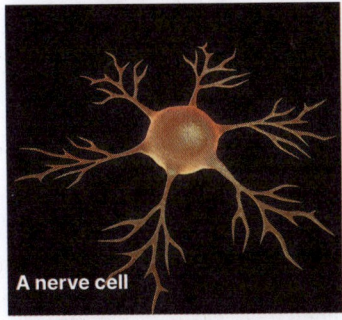
A nerve cell

Neurotransmitters are chemicals that carry signals between nerve cells or between a nerve cell and another type of cell such as a muscle or gland. Neurotransmitters act on the cell immediately next to it. They are released into a synapse (gap between the cells) and bind to receptors on the receiving cell. The response is rapid and short lived.

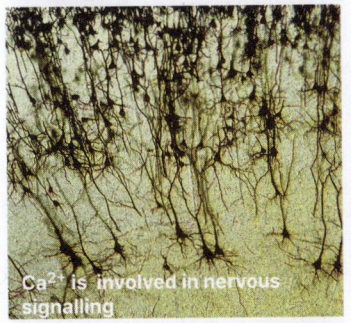
Ca^{2+} is involved in nervous signalling

Calcium ions (Ca^{2+}) act as second messengers in all eukaryotes. Ca^{2+} has a wide range of effects in a cell including signalling muscle contraction, neuronal transmission, and cell growth. Calcium ions bind to the regulatory protein calmodulin, which may then activate other protein complexes. Ca^{2+} can signal between cells via gap junctions: channels that permit cell to cell communication.

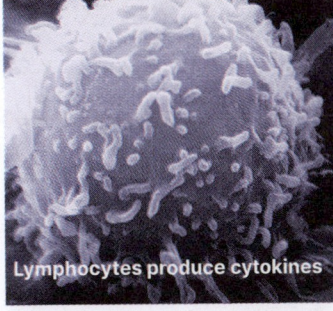
Lymphocytes produce cytokines

Cytokines are a large group of peptides and small proteins involved in coordinating the response of cells in the immune system both within their immediate vicinity or over large distances. Cytokines are produced by a wide range of cells, including immune cells and endothelial cells. They include interferons, interleukins, and tumour necrosis factors.

Pheromones are chemical signals released into the external environment and are common in social insects and mammals. Pheromones have different roles, e.g. aggression, aggregation, reproduction, territoriality, but all act to generate a specific response in members of the same species. In mammals, pheromones are used to signal sexual receptivity and attract mates.

1. What is the difference between hormones and neurotransmitters?

2. (a) Identify a type of long lasting chemical signal:

   (b) Identify a type of short lasting chemical signal:

3. What is the main function of cytokines?

4. Cytokines and hormones are both involved in endocrine signalling. How do they differ?

5. Identify the smallest signalling molecule:

6. How is Ca^{2+} able to quickly move between cells?

7. How do pheromones differ from the other types of chemical signals?

C2.1 3-5 AHL

The diversity of signal molecules: hormones and neurotransmitters

A wide variety of chemicals act as ligands. Hormones and neurotransmitters are two important groups. Each has many different types of ligand. Some are able to cross the plasma membrane, while others cannot, giving the cells ways of communicating with different and specific parts of other cells.

Hormones

Hormones can be steroids, peptides (proteins), or amines.

▶ Steroid hormones are lipophilic ('fat loving'). They are able to diffuse across the plasma membrane and so are able to bind to receptors in the cytoplasm or nucleus. The activated receptor can then bind to DNA and affect transcription. Examples include testosterone and oestrogen.

Testosterone

▶ Peptide hormones are hydrophillic ('water loving') and lipophobic ('fat hating'). They cannot cross the plasma membrane and therefore must bind to a receptor on the surface of the cell. This will then relay the signal into the cell via second messengers, such as Ca^{2+} or cyclic AMP (cAMP).

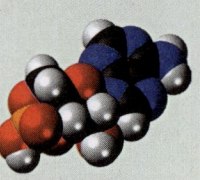
cAMP

▶ Amine hormones are derived from the amino acid tyrosine. They include epinephrine (adrenaline) and thyroxine. Amine hormones have properties of both steroid and peptide hormones.

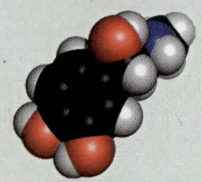
Epinephrine

Neurotransmitters

Neurotransmitters can be amino acids, peptides, amines, or nitric oxide.

▶ Amino acid neurotransmitters are amino acids that can transmit a nerve message across a synapse. They can be excitatory or inhibitory. Examples include glutamate (glutamic acid), an excitatory neurotransmitter in the brain, and gamma-aminobutyric acid (GABA), the main inhibitory neurotransmitter in the brain.

Glutamic acid

▶ Peptide neurotransmitters (neuropeptides) are peptides 3 to 36 amino acids long. They include oxytocin (produced in the hypothalamus) and both α and β endorphins.

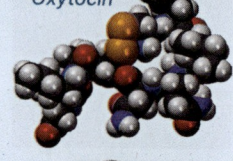
Oxytocin

▶ Many amines are important neurotransmitters, including dopamine, serotonin, and norepinephrine (noradrenaline). These play a part in sleep, arousal, and regulation of heart rate and blood pressure.

Dopamine

▶ Nitric oxide is a vasodilator produced by the endothelium of the blood vessels, nitric oxide tells blood vessels to contract or dilate, regulating blood pressure. Note: nitric oxide (NO) is different from nitrous oxide (NOS) which is used as an anaesthetic. Unlike other neurotransmiters, NO is produced on demand, rather than stored in vesicles.

Nitric oxide

8. Suggest why nitric oxide is produced when needed rather than being stored in vesicles:

_____

_____

_____

9. Suggest a need for excitatory and inhibitory neurotransmitters: _____

_____

_____

_____

10. Identify a functional difference between how steroid hormones and peptide hormones produce cell signals:

_____

_____

_____

11. Suggest why there are so many different kinds of cell signalling molecules: _____

_____

_____

_____

©2024 **BIOZONE** International
ISBN: 978-1-99-101410-8
Photocopying prohibited

177 Communication Over Short and Long Distances

Key Idea: How do cells use signal molecules to communicate over short and long distances?

Cells in close proximity can communicate using local regulators, i.e. signal molecules that travel only a short distance from the secreting cell to the target cell before being inactivated. A number of cells in the immediate vicinity of the emitting cell may be stimulated, causing a specific cellular response, such as a nerve impulse. If distantly separated cells within an organism are to communicate, signal molecules (hormones) must be transported in a medium, usually blood. Target cells with receptors that can bind the hormone will be able to respond to these endocrine signals. Target cells may be concentrated within a specific tissue, but they may be widespread throughout the body.

Local regulation between animal cells

▶ Local regulation across synapses occurs in the nervous system and takes place between a neuron (nerve cell) and another cell, e.g. another neuron or a muscle cell.

▶ In this 'synaptic signalling', an electrical impulse travels along a long conducting process of the neuron called an axon. It eventually reaches the synapse (the gap between the two cells) where it triggers the release of chemical signal molecules called neurotransmitters from the axon terminals.

▶ The neurotransmitter travels across the synapse to affect the target cell. The effect can increase a response (excitatory) or decrease a response (inhibitory) depending on the cell and neurotransmitter involved. Chemical changes in the cell bring about a range of specific cellular responses, e.g. opening ion channels, causing muscle contraction, and regulation of heart rate and blood pressure.

Axon of neuron (stimulating cell)

Axon terminal

Neurotransmitter released from vesicle by exocytosis

Synaptic cleft

Receptor

Membrane of target cell

The neurotransmitter is broken down (inactivated) or reabsorbed after diffusing across the synapse. This prevents continued response by the target cell.

Endocrine signalling: communicating over distance

▶ Recall how hormones are chemical messengers in endocrine signalling pathways. Hormones are secreted from endocrine glands and distributed throughout the body in the blood. This allows hormones to reach the same targets at much the same time, even when those targets are at different places in the body.

▶ Endocrine control acts more slowly than signalling through local regulators (as in nerve pathways) because the distances involved are greater. The effects of hormones are often wide ranging, affecting many different tissues. Their effects may be long-lasting, affecting development, e.g. the oestrogens, or more immediate, e.g. insulin promoting cellular uptake of glucose.

INSULIN

▶ Produced by beta endocrine cells in the pancreas. Insulin signalling allows the body's cells to take up glucose and promotes glucose storage (as glycogen) in the liver. Its effect is to lower blood glucose level.

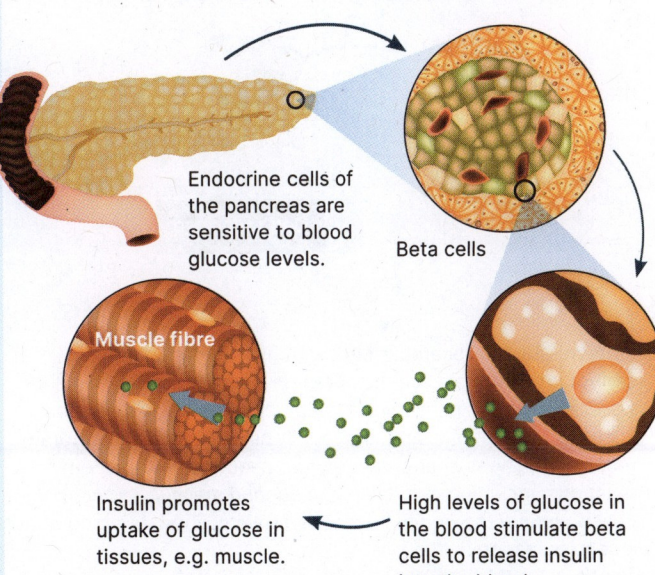

Endocrine cells of the pancreas are sensitive to blood glucose levels.

Beta cells

Muscle fibre

Insulin promotes uptake of glucose in tissues, e.g. muscle.

High levels of glucose in the blood stimulate beta cells to release insulin into the blood.

1. Explain how local regulation is used in cell to cell communication in animals: _____

_____

_____

_____

2. How do cells communicate over longer distances? _____

_____

_____

©2024 **BIOZONE** International
ISBN: 978-1-99-101410-8
Photocopying prohibited

C2.1
5
AHL

178 Transmembrane and Intracellular Receptors

Key Idea: Cell signal receptors may be either embedded in the plasma membrane or free floating in the cytoplasm.

Cell receptors fall into two broad classes. Transmembrane (extracellular) receptors bind hydrophilic ligands (signal molecules) outside of the cell. The ligand does not pass across the plasma membrane to cause the cellular response. Most cell receptors are extracellular receptors. Intracellular receptors bind hydrophobic ligands that pass into the cell directly across the plasma membrane. Intracellular receptors may be located in the cytoplasm or on the nuclear membrane.

Transmembrane receptors

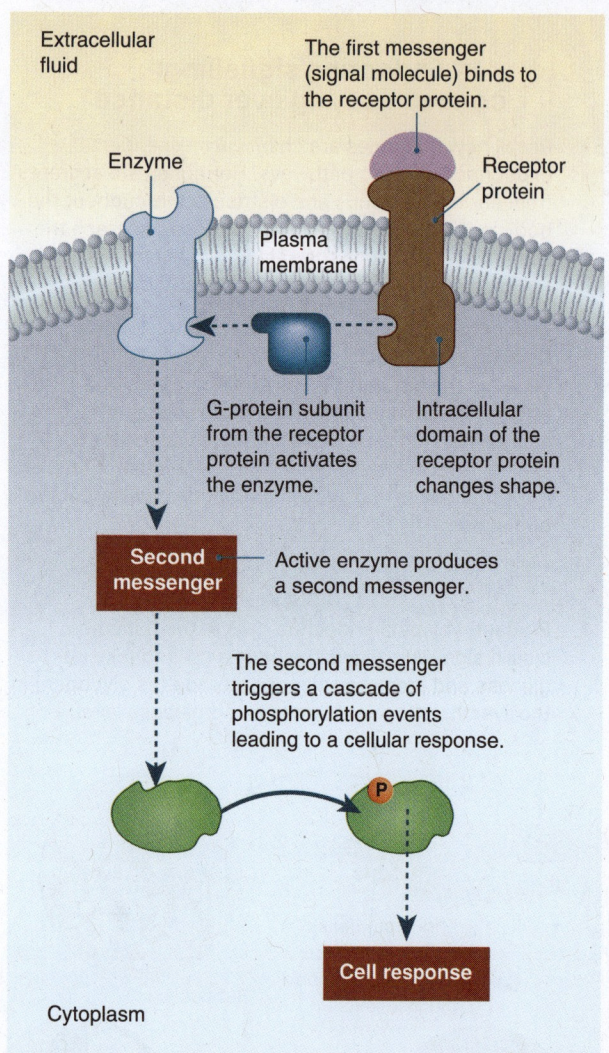

Intracellular receptors

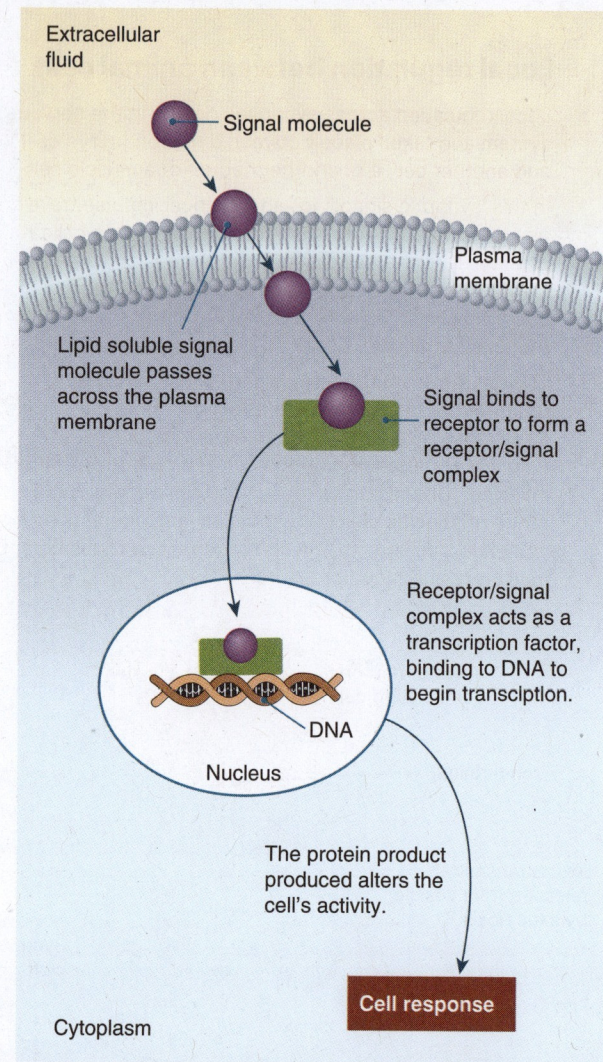

Transmembrane receptors bind with hydrophilic ligands which cannot cross the plasma membrane. Hydrophilic signals include water soluble hormones such as epinephrine and insulin. The ligand is the first messenger. When it binds, the transmembrane receptor changes shape, triggering a sequence of biochemical reactions, including activation of a second messenger. As a consequence, the original signal can be amplified, bringing about a cellular response.

Intracellular receptors bind with hydrophobic ligands, normally hormones, such as steroids, e.g. oestrogen, that diffuse freely across the plasma membrane and into the cytoplasm of target cells. The intracellular receptor and ligand form a receptor-signal complex. The complex moves to the cell nucleus where it binds directly to the DNA and acts as a transcription factor, resulting in the transcription of a one or more specific genes.

1. Describe the differences between an intracellular receptor and a transmembrane receptor:

_____

_____

_____

_____

2. What is the important difference between the ligands that bind to intracellular and transmembrane receptors?

_____

_____

©2024 **BIOZONE** International
ISBN: 978-1-99-101410-8
Photocopying prohibited

C2.1
6,7,9,10

179 Transmembrane Receptors for Neurotransmitters

Key Idea: Ligand-gated ion channels play an important role in the transmission of nerve impulses.

Neurotransmitter receptors can be divided into two main groups, the ionotropic receptors and the metabotropic receptors. When activated, ionotropic receptors allow ions to pass through the receptor, whereas metabotropic receptors require a second messenger to relay the signal into the cell. Ionotropic receptors are also called ligand-gated ion channels, as once a ligand is bound they open a channel that allows ions through the plasma membrane (below).

Acetylcholine receptor

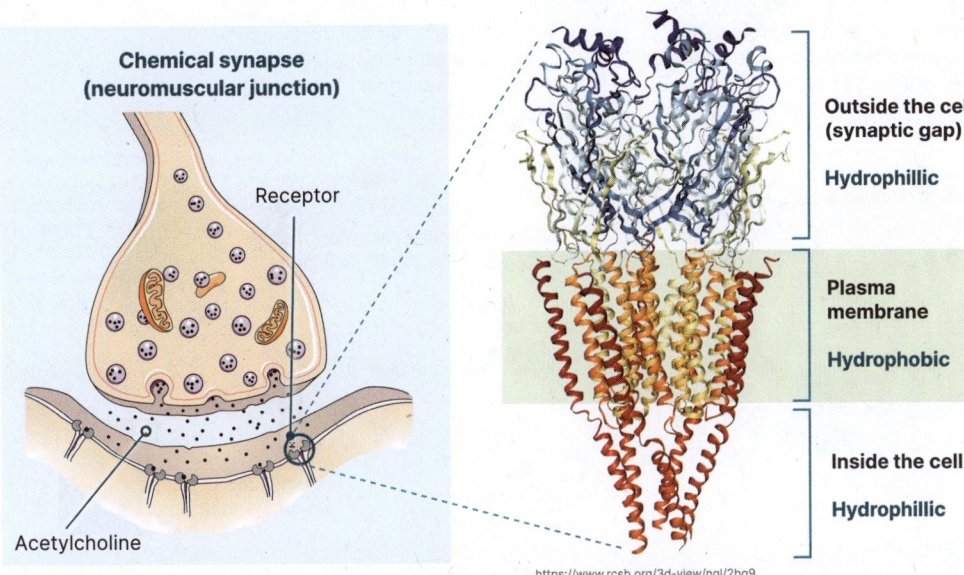

The nicotinic acetylcholine receptor (left and below) is a ligand-gated ion channel that is located on the surface of the post synaptic cell. It is activated by the neurotransmitter acetylcholine, secreted from the presynaptic cell.

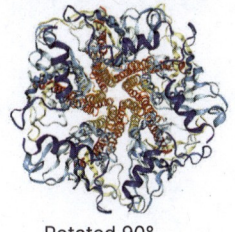

Rotated 90°. (View from top)

https://www.rcsb.org/3d-view/ngl/2bg9

Propagating the neurotransmitter signal

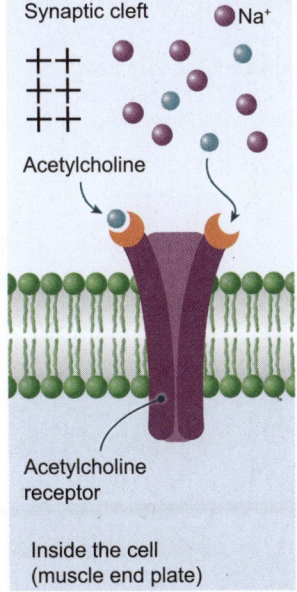 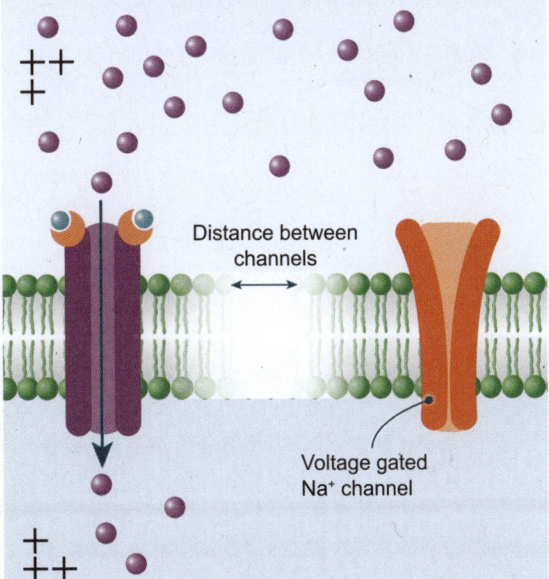 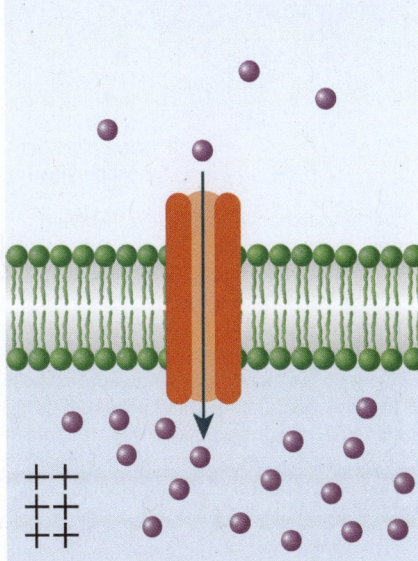

Acetylcholine diffuses from the axon terminal into the synaptic cleft. It binds to nicotinic acetylcholine receptors on the postsynaptic cell.

The nicotinic acetylcholine receptors are ligand-gated ion channels. Once acetylcholine is bound, the channel opens and lets Na⁺ through. The influx of Na⁺ changes the membrane potential, making it less negative. This is called an end plate potential.

If the end plate potential is strong enough, voltage gated ion channels open, letting more Na⁺ through. This larger influx of Na⁺ depolarizes then reverses the membrane potential, causing an action potential which propagates the nerve signal.

1. How does the action of the neurotransmitter acetylcholine cause the production of an action potential?

180 G-Coupled Protein Receptors

Key Idea: Large proteins require second messengers to pass their message through the plasma membrane.

G-coupled proteins transfer signals into a cell from a first messenger signal molecule. The first messenger, e.g. adrenaline (epinephrine), binds to the receptor and activates it, releasing bound subunits that bring about an intracellular response through a signal transduction pathway. This occurs via a second messenger system involving cAMP as the second messenger. The signal molecule binds to the specific receptor of a target cell and starts a cascade of reactions that amplifies the original signal and brings about a response, e.g. enzyme activation.

Signalling via adrenaline

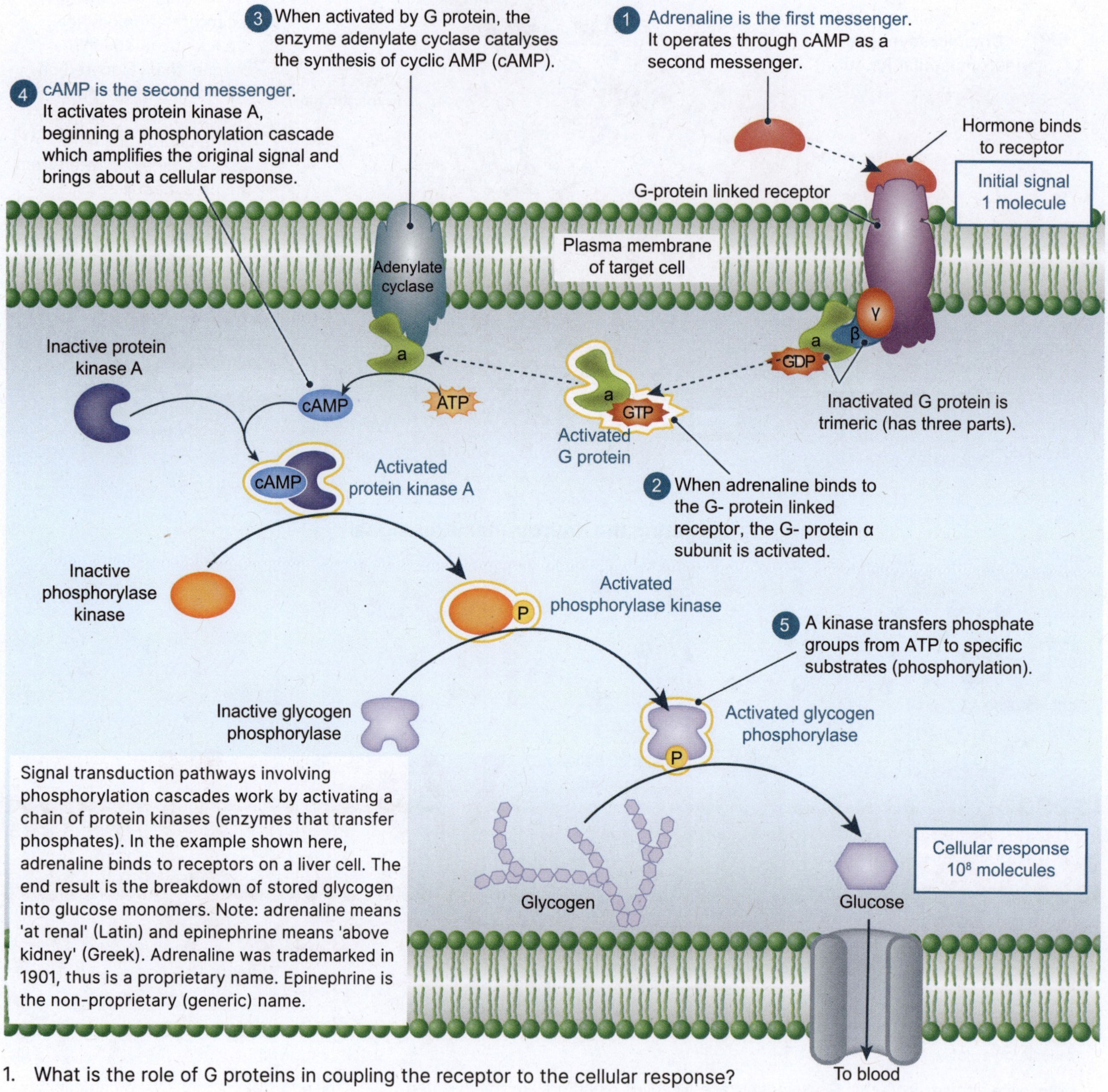

3 When activated by G protein, the enzyme adenylate cyclase catalyses the synthesis of cyclic AMP (cAMP).

1 Adrenaline is the first messenger. It operates through cAMP as a second messenger.

4 cAMP is the second messenger. It activates protein kinase A, beginning a phosphorylation cascade which amplifies the original signal and brings about a cellular response.

Hormone binds to receptor

Initial signal 1 molecule

G-protein linked receptor

Plasma membrane of target cell

Adenylate cyclase

Inactive protein kinase A

cAMP

ATP

Inactivated G protein is trimeric (has three parts).

Activated G protein

cAMP

Activated protein kinase A

2 When adrenaline binds to the G- protein linked receptor, the G- protein α subunit is activated.

Inactive phosphorylase kinase

P

Activated phosphorylase kinase

5 A kinase transfers phosphate groups from ATP to specific substrates (phosphorylation).

Inactive glycogen phosphorylase

Activated glycogen phosphorylase

P

Signal transduction pathways involving phosphorylation cascades work by activating a chain of protein kinases (enzymes that transfer phosphates). In the example shown here, adrenaline binds to receptors on a liver cell. The end result is the breakdown of stored glycogen into glucose monomers. Note: adrenaline means 'at renal' (Latin) and epinephrine means 'above kidney' (Greek). Adrenaline was trademarked in 1901, thus is a proprietary name. Epinephrine is the non-proprietary (generic) name.

Glycogen

Glucose

Cellular response 10⁸ molecules

To blood

1. What is the role of G proteins in coupling the receptor to the cellular response?

_____

_____

_____

_____

2. Explain why a second messenger is needed to convey a signal inside a cell from a water soluble first messenger:

_____

_____

_____

   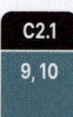

AHL   NOS   C2.1 9, 10

©2024 **BIOZONE** International
ISBN: 978-1-99-101410-8

181 Tyrosine Kinase Activity in Receptors

Key Idea: Activation of the insulin receptor by insulin causes a signal cascade that results in cellular glucose uptake.

Insulin is a peptide hormone secreted by the pancreas. It is involved in regulating blood glucose levels by promoting the uptake of glucose by cells. Insulin circulates in the blood where it binds to transmembrane insulin receptors on the surface of cells and triggers a signal cascade that results in activation of the membrane transporters that bring glucose into the cell. The insulin receptor is made up of four subunits (two extracellular α subunits, and two β subunits spanning the plasma membrane). The β subunit is bound to the protein tyrosine kinase. Recall that kinases catalyse the transfer of phosphates between molecules, giving the target molecule energy to carry out biological activity.

Glucose uptake pathway

A signal transduction pathway often involves protein modification and phosphorylation cascades, as shown for insulin signalling, leading to glucose uptake by the cell.

Insulin circulating in the blood

1 Two molecules of insulin must bind to the extracellular domain of the insulin receptor to activate it.

Bound insulin

Unbound insulin

Glucose in the blood

Glut4 glucose transporter

α   α

Insulin receptor

Extracellular environment

β   β

Intracellular environment

Tyrosine kinase

2 Once the insulin is bound, phosphate groups are added to the tyrosine kinases by a process called autophosphorylation.

Signalling transduction pathways result in the activation of many proteins, enabling the cell to produce a large response to the signalling from just a few ligands.

Inactive molecules

Active molecules

Glut4 secretory vesicle

5 Glut4 glucose transporters insert into the membrane allowing the uptake of glucose.

3 Once phosphorylated the tyrosine kinase then phosphorylates other molecules to begin a signal cascade.

4 The cascade sequence results in the activation of many Glut4 secretory vesicles, which produce the Glut4 glucose transporters.

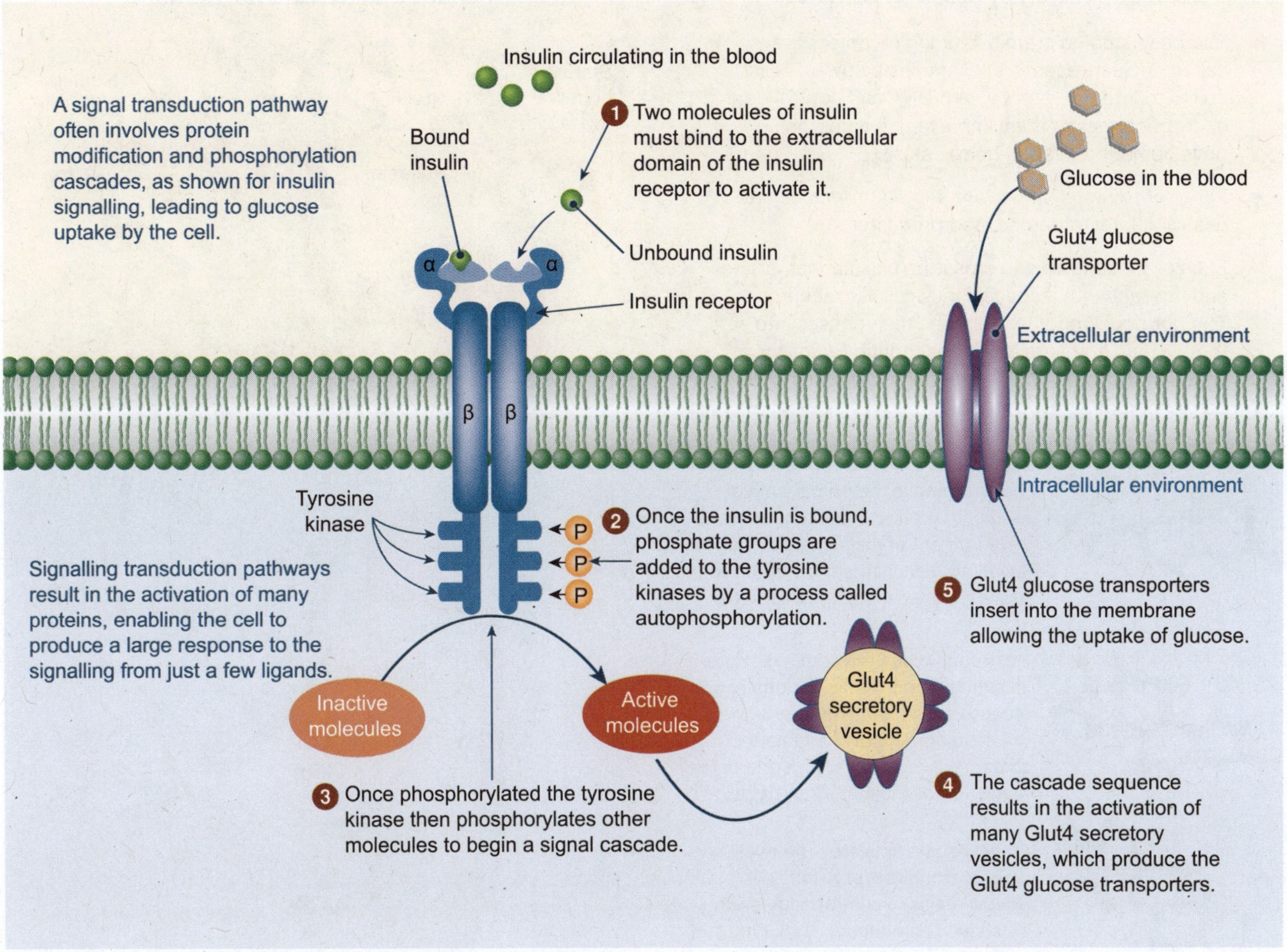

1. (a) What type of signalling does this example represent: autocrine, local regulation, or endocrine? _____

   (b) Explain why you chose this answer: _____

   _____

   _____

2. What is the role of tyrosine kinase in the glucose uptake pathway?

   _____

   _____

   _____

3. How does the signal cascade increase the response of the insulin receptor? _____

   _____

   _____

   _____

   _____

©2024 **BIOZONE** International
ISBN: 978-1-99-101410-8
Photocopying prohibited

C2.1
11

AHL

182 Intracellular Receptors

Key Idea: Steroid hormones are able to pass through the plasma membrane and bind to receptors inside the cell. Steroid hormones are steroids (fatty molecules) that act as hormones. They have wide-ranging metabolic effects and are involved in regulating metabolic activity, inflammation, immune function, salt and water balance, development of sexual characteristics, and response to stress and injury.

Steroid hormones are lipid soluble and diffuse freely from the blood through the plasma membrane and into the cytoplasm of target cells. In the cytoplasm, the steroid binds to a specific steroid receptor (transcription factor) to form a functional hormone-receptor complex. The complex is able to enter the nucleus and bind to specific DNA sequences to induce transcription of its target genes.

Gene activation by steroid hormones

▶ The best-studied steroid-binding receptors are called nuclear receptors (examples below). Their ability to interact directly with DNA and control gene expression makes them important in embryonic development and adult homeostasis.

▶ Three of the most important steroid hormones are oestradiol, testosterone, and progesterone.

▶ Each of these is able to cross the plasma membrane and interact with a specific intracellular receptor. The hormone-receptor complex then passes into the nucleus and acts as a transcription factor for a specific part of the DNA, initiating or regulating the production of a new protein.

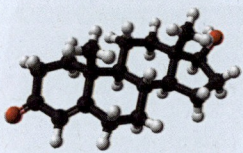 Testosterone activates the androgen receptor and is the primary male hormone. It plays a key role in the development of male characteristics, including development of reproductive tissues, bone, and muscle growth.

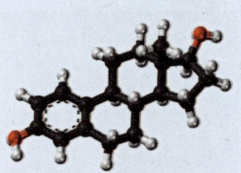 Oestradiol binds to various receptors including, importantly, the oestrogen receptors ERα and ERβ. It promotes the development of the female characteristics and plays a role in the regulation of the menstrual cycle.

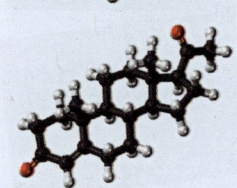 Progesterone activates the nuclear progesterone receptor (nPR). It plays a major role in the regulation of female reproduction, including the menstrual cycle.

Action of steroid hormone

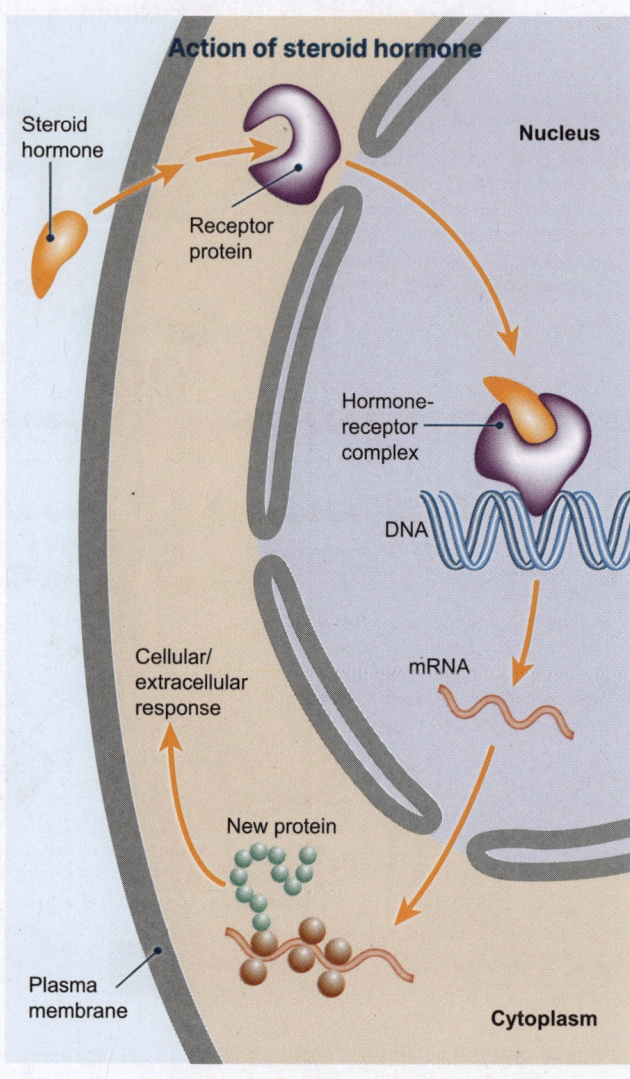

1. How does the action of steroid hormones differ from that of water soluble signalling molecules?

   _____

   _____

   _____

   _____

2. Suggest why steroid hormones are important in developmental processes:

   _____

   _____

   _____

3. Identify the two hormones that play major roles in the development of male and female characteristics:

   _____

©2024 **BIOZONE** International
ISBN: 978-1-99-101410-8
Photocopying prohibited

AHL   C2.1   12

183 Feedback and Hormones

Key Idea: Hormones from the hypothalamus and anterior pituitary regulate the menstrual cycle. The cycles of hormonal fluctuations are regulated by feedback loops.

Feedback loops are important for regulating mechanisms and processes in the body. Hormone levels during the menstrual cycle are regulated through negative feedback mechanisms, (except for the mid cycle LH surge). The main control centres

for this regulation are the hypothalamus and the anterior pituitary gland. The hypothalamus secretes gonadotropin releasing hormone (GnRH), which is transported in capillaries to the anterior pituitary. Here, it induces the release of two hormones: follicle stimulating hormone (FSH) and luteinising hormone (LH). These two hormones bring about the cyclical changes in the ovaries and uterus.

Positive and negative feedback

Many hormones are regulated by positive and negative feedback loops, which either amplify or reduce their effects:

Positive feedback loops amplify (increase) a response in order to achieve a particular result. Examples include fever, blood clotting, childbirth (labour), and lactation (production of milk).

The mechanisms within a positive feedback loop will cease to function once the end result is achieved, e.g. the baby is born (right) or suckling stops. Positive feedback loops are less common than negative feedback loops in biological systems because the escalation in response is unstable. Unresolved positive feedback responses, e.g. high fevers, can be fatal.

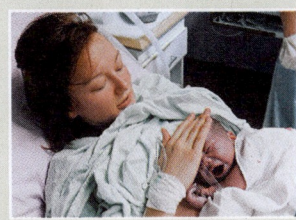

Negative feedback loops are control systems that maintain the body's internal environment at a relatively steady state.

When variations from the norm are detected by the body's receptors, a response or output from the effectors that opposes the stimulus is classified as negative feedback. Negative feedback discourages variations from a set point and returns internal conditions to a steady state. Most physiological systems achieve homeostasis through negative feedback loops, e.g. blood glucose regulation (above).

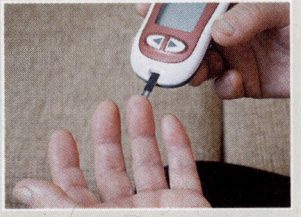

Control of the menstrual cycle

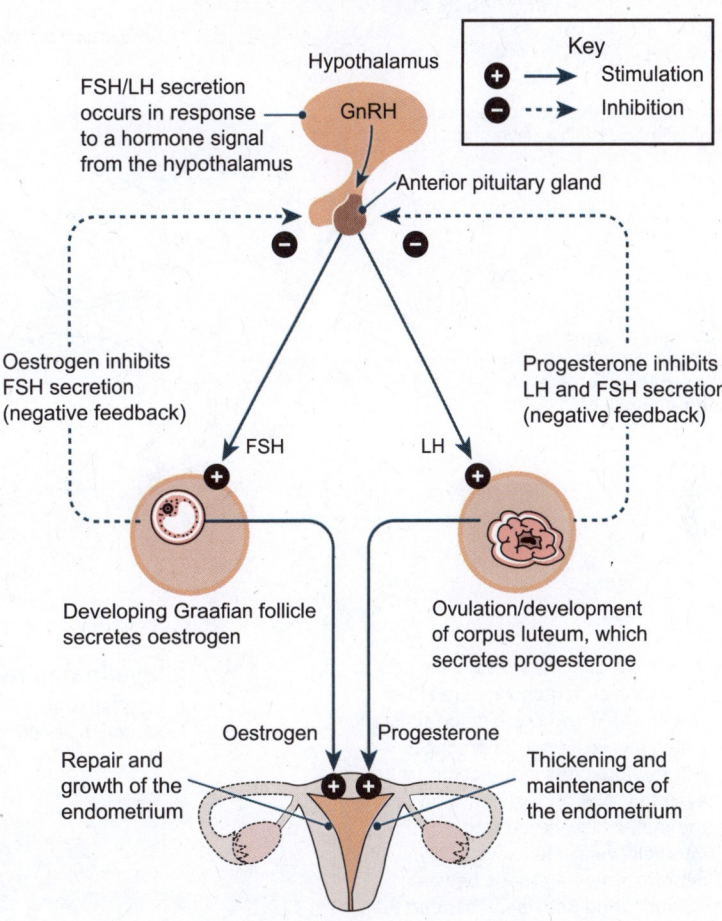

▶ In the first half of the cycle, FSH stimulates follicle development in the ovary. The developing follicle secretes oestrogen which acts on the uterus and, in the anterior pituitary, inhibits FSH secretion.

▶ In the second half of the cycle, LH induces ovulation and development of the corpus luteum. The corpus luteum secretes progesterone which acts on the uterus and also inhibits further secretion of LH (and also FSH).

1. Describe the roles of FSH and LH in the control of the menstrual cycle:

_____

_____

_____

2. How does negative feedback regulate the secretion of FSH and LH in the menstrual cycle?

_____

_____

_____

_____

©2024 **BIOZONE** International
ISBN: 978-1-99-101410-8
Photocopying prohibited

184 Neuron Structure and Function

Key Idea: Neurons are electrically excitable cells that are specialized to process and transmit information via electrical and chemical signals. Increased axon diameter and myelination both increase conduction speed along a neuron. Neurons transmit information in the form of electrochemical signals from receptors (in the central nervous system) to effectors. Neurons consist of a cell body (soma) and long processes (dendrites and axons). Conduction speed increases with axon diameter and with myelination. Faster conduction speeds enable more rapid responses to stimuli.

Motor (efferent) neuron
Transmits impulses from the CNS to effectors (muscles or glands).

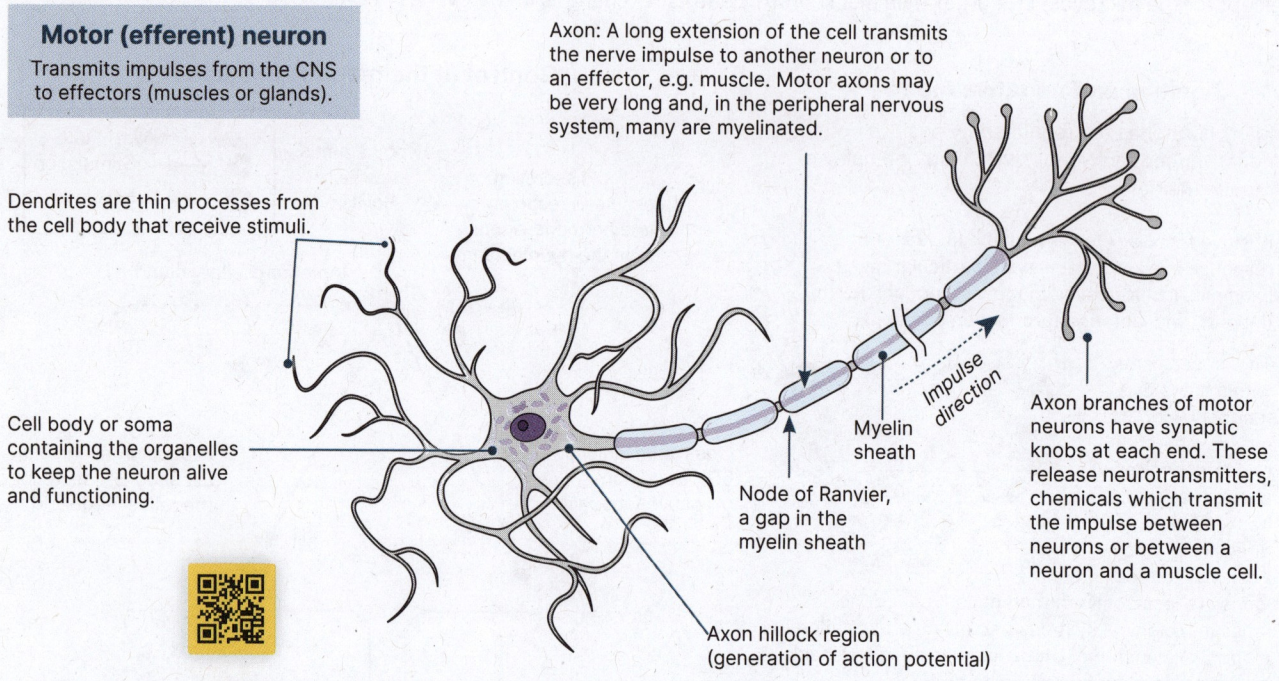

Axon: A long extension of the cell transmits the nerve impulse to another neuron or to an effector, e.g. muscle. Motor axons may be very long and, in the peripheral nervous system, many are myelinated.

Dendrites are thin processes from the cell body that receive stimuli.

Cell body or soma containing the organelles to keep the neuron alive and functioning.

Impulse direction

Myelin sheath

Axon branches of motor neurons have synaptic knobs at each end. These release neurotransmitters, chemicals which transmit the impulse between neurons or between a neuron and a muscle cell.

Node of Ranvier, a gap in the myelin sheath

Axon hillock region (generation of action potential)

Where conduction speed is important, the axons of neurons are sheathed within a lipid and protein rich substance called myelin. Myelin is produced by oligodendrocytes in the central nervous system (CNS) and by Schwann cells in the peripheral nervous system (PNS). At intervals along the axons of myelinated neurons, there are gaps between neighbouring Schwann cells and their sheaths. These are called nodes of Ranvier. Myelin acts as an insulator, increasing the speed at which nerve impulses travel because it prevents ion flow across the neuron membrane and forces the current to 'jump' along the axon from node to node.

Myelinated Neurons
Diameter: 1-25 µm
Conduction speed: 6-120 ms^{-1}

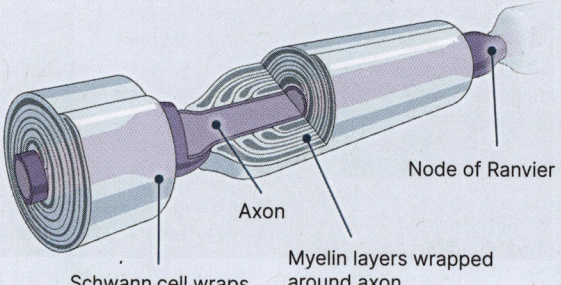

Schwann cell wraps only one axon and produces myelin

Axon

Myelin layers wrapped around axon

Node of Ranvier

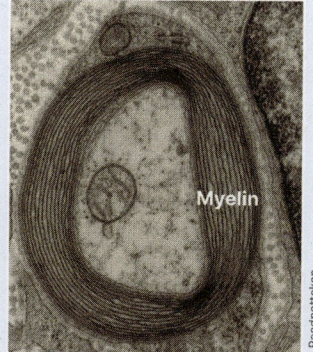

Myelin

TEM cross section through a myelinated axon

Roadnottaken

1. What is the function of a neuron? _____

_____

2. What factors increase the speed of conduction along a neuron? _____

_____

3. How does myelination increase the speed of nerve impulse conduction? _____

_____

_____

4. What is the advantage of faster conduction of nerve impulses? _____

_____

 C2.2 1
 200

©2024 **BIOZONE** International
ISBN: 978-1-99-101410-8

185 The Nerve Impulse

Key Idea: The nerve impulse is an electrical signal that propagates along the axon of a neuron.
Nerve cells (neurons) maintain a polarized membrane when not transmitting a signal. When induced to transmit a signal the membrane quickly depolarizes by allowing positively charged sodium ions to flood into the cell via voltage gated ion channels. To restore and maintain the polarity of the cell sodium-potassium pumps move sodium ions out of the cell.

Membrane potential

The resting neuron

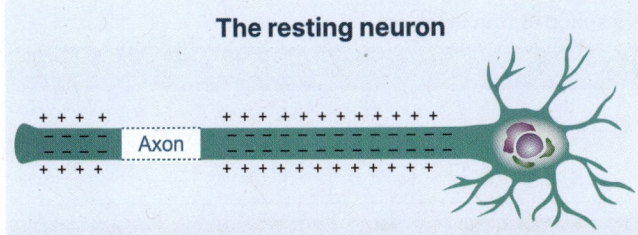

When a neuron is not transmitting an impulse, the inside of the cell is negatively charged relative to the outside and the cell is said to be electrically polarized. The potential difference (voltage) across the membrane is called the resting potential. For most nerve cells, this is about -70 mV. Nerve transmission is possible because this membrane potential exists.

The nerve impulse

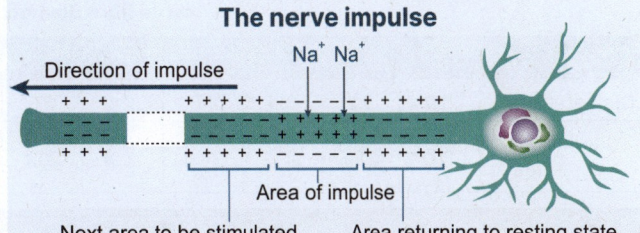

When a neuron is stimulated, the distribution of charges on each side of the membrane briefly reverses. This process of depolarization causes a burst of electrical activity to pass along the axon of the neuron as an action potential. As the charge reversal reaches one region, local currents depolarize the next region and the impulse spreads along the axon.

Maintaining membrane potential

▶ Membrane potential is produced by the use of a sodium/potassium pump. Sodium from inside the cell is pumped out of the cell and potassium from outside the cell is pumped into the cell. However, because three sodium ions are pumped out of the cell for every two potassium ions that are pumped in, a positive charge builds up on the outside of the cell relative to the inside of the cell.

▶ This produces a potential difference (voltage) across the cell membrane, called the resting potential. Because membrane potential is defined by the interior of the cell relative to the exterior of the cell, the membrane potential will be negative.

▶ Note that the sodium-potassium pump produces a sodium ion concentration gradient that would drive sodium into the cell, while also producing a potassium ion concentration that would drive potassium from inside the cell to outside the cell.

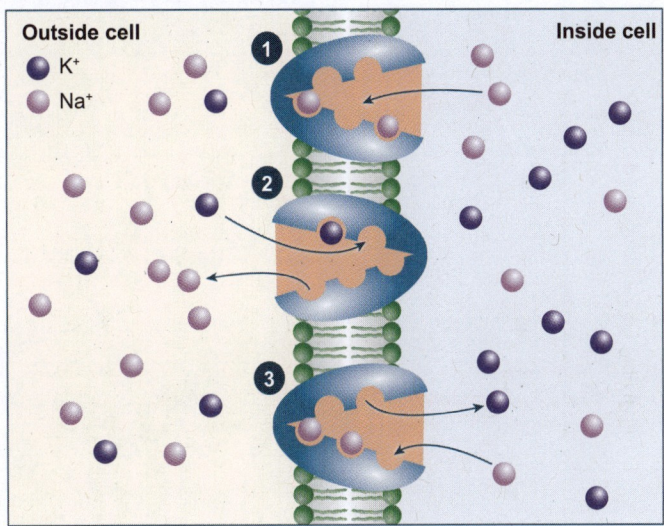

1. (a) Define the term 'resting potential': _____

_____

(b) Define the term 'action potential': _____

_____

2. How is the resting potential produced and maintained? _____

_____

_____

_____

3. How is an action potential propagated along a nerve fibre (axon)? _____

_____

_____

_____

C2.2

2, 3

186 The Speed of Nerve Impulses

Key Idea: The diameter and myelination of neurons affects the speed of the nerve impulse.

Nerve impulses move along a nerve fibre at a variety of speeds, from less than 1 ms⁻¹ to nearly 200 ms⁻¹ in the shrimp giant nerve fibre. The first attempts to measure these speeds were in 1849 by Hermann von Helmholtz, who showed that the speed of transmission in the nerve fibres of a frog was around 30 ms⁻¹. Classic experiments carried out in the 1930s by J.B. Hursh on the myelinated nerve fibres of cats and R. J. Pumphrey *et al* on the unmyelinated nerve fibres of squid showed the effect of myelination and the diameter of the nerve fibre on impulse speed.

Adapted from J.B. Hursh and R. J. Pumphrey et al

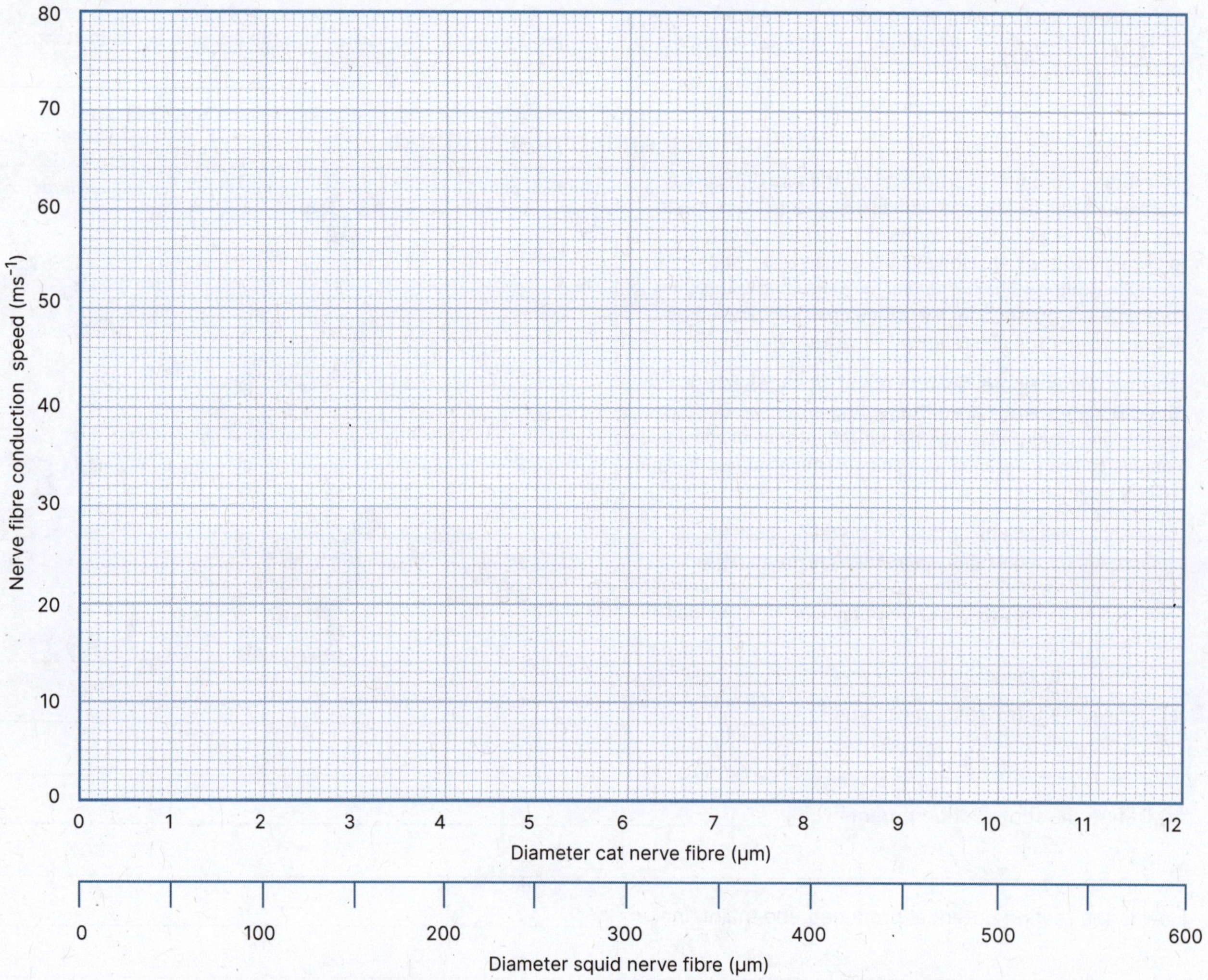

Nerve fibre diameter vs speed of transmission

Cat													
Diameter (µm)	3.0	3.2	3.4	4.0	4.4	5.0	6.2	6.2	6.4	7.6	8.0	8.8	9.0	10.0
Conduction velocity (ms⁻¹)	17.0	21.9	24.3	36.4	30.4	26.7	42.5	46.1	43.7	60.7	46.2	55.4	54.5	80.0

Squid													
Diameter (µm)	30	40	43	50	108	130	130	148	280	294	357	413	415	567
Conduction velocity (ms⁻¹)	2.2	4.7	6.4	5.6	5.9	7.8	5.9	8.2	15.0	11.0	13.8	17.4	18.6	20.0

Graph with y-axis "Nerve fibre conduction speed (ms⁻¹)" scaled 0–80, x-axis "Diameter cat nerve fibre (µm)" scaled 0–12, and secondary axis "Diameter squid nerve fibre (µm)" scaled 0–600.

1. Plot the diameter and speed of transmission of both cat (myelinated) and squid (nonmyelinated) nerves. Include a key:

2. Calculate the coefficient of determination (R^2) for each set of data (see Activity 134 for details):

   (a) Cat _____

   (b) Squid: _____

3. Which appears more important for speed of transmission: myelination, or nerve fibre diameter?

   _____

©2024 **BIOZONE** International
ISBN: 978-1-99-101410-8
Photocopying prohibited

C2.2

4

AOS

187 Chemical Synapses

Key Idea: Synapses are junctions between neurons, or between neurons and receptor or effector cells. Nerve impulses are transmitted across synapses.

Action potentials are transmitted across junctions called synapses. Almost all synapses in vertebrates are chemical synapses, which involve the diffusion of a signal molecule or neurotransmitter from one cell to another. Chemical synapses can occur between two neurons, between a receptor cell and a neuron, or between a neuron and an effector, e.g. muscle fibre or gland cell. The synapse consists of the axon terminal (synaptic knob), a gap called the synaptic cleft, and

the membrane of the post-synaptic (receiving) cell. Arrival of an action potential at the axon terminal causes release of the neurotransmitter, which diffuses across the cleft and produces an electrical response in the post-synaptic cell (an example of signal transduction). Cholinergic synapses are named for the neurotransmitter they release, acetylcholine (ACh). In the example pictured below, ACh results in depolarization (excitation) of the post-synaptic neuron. Unlike electrical synapses, in which transmission can occur in either direction, transmission at chemical synapses is always in one direction (unidirectional).

The structure of a cholinergic synapse

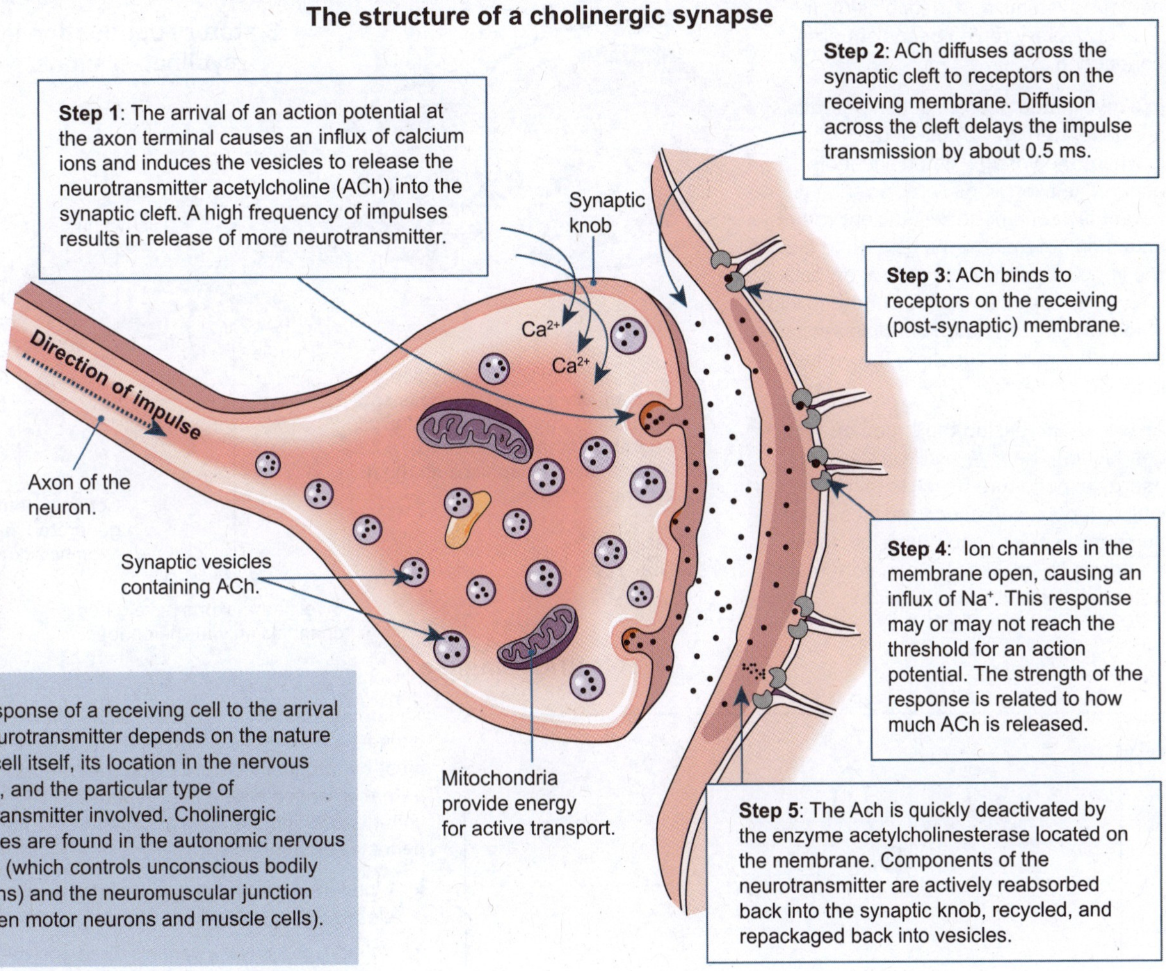

Step 1: The arrival of an action potential at the axon terminal causes an influx of calcium ions and induces the vesicles to release the neurotransmitter acetylcholine (ACh) into the synaptic cleft. A high frequency of impulses results in release of more neurotransmitter.

Step 2: ACh diffuses across the synaptic cleft to receptors on the receiving membrane. Diffusion across the cleft delays the impulse transmission by about 0.5 ms.

Step 3: ACh binds to receptors on the receiving (post-synaptic) membrane.

Step 4: Ion channels in the membrane open, causing an influx of Na^+. This response may or may not reach the threshold for an action potential. The strength of the response is related to how much ACh is released.

Step 5: The Ach is quickly deactivated by the enzyme acetylcholinesterase located on the membrane. Components of the neurotransmitter are actively reabsorbed back into the synaptic knob, recycled, and repackaged back into vesicles.

Synaptic knob

Direction of impulse

Ca^{2+}

Ca^{2+}

Axon of the neuron.

Synaptic vesicles containing ACh.

Mitochondria provide energy for active transport.

The response of a receiving cell to the arrival of a neurotransmitter depends on the nature of the cell itself, its location in the nervous system, and the particular type of neurotransmitter involved. Cholinergic synapses are found in the autonomic nervous system (which controls unconscious bodily functions) and the neuromuscular junction (between motor neurons and muscle cells).

1. (a) What is a synapse? _____

    _____

    (b) What defines a cholinergic synapse? _____

2. What causes the release of neurotransmitter into the synaptic cleft? _____

    _____

3. Why is there a brief delay in impulse transmission across the synapse? _____

    _____

4. What determines the strength of the response in the receiving cell? _____

    _____

    _____

©2024 **BIOZONE** International
ISBN: 978-1-99-101410-8
Photocopying prohibited

C2.2

5-7

188 Electrical Impulses in the Nerve

Key Idea: A nerve impulse occurs in response to a stimulus and involves the transmission of a membrane depolarization along the axon of a neuron.

The plasma membrane of cells, including neurons, contain sodium-potassium ion pumps that actively pump sodium ions (Na$^+$) out of the cell and potassium ions (K$^+$) into the cell. The action of these ion pumps in neurons creates a separation of charge (a potential difference or voltage) either side of the membrane and makes the cells electrically excitable. It

is this property that enables neurons to transmit electrical impulses. When a nerve is stimulated, a brief increase in membrane permeability to Na$^+$ temporarily reverses the membrane polarity (a depolarization). After the nerve impulse passes, the sodium-potassium pump restores the resting potential. In neurons that are unmyelinated, the nerve impulse is propagated along the entire length of the nerve. In myelinated neurons, the nerve impulse jumps between the nodes of Ranvier, speeding up conduction (below).

Axon myelination is a feature of vertebrate nervous systems and it enables them to achieve very rapid speeds of nerve conduction.

In a myelinated neuron, action potentials are generated only at the nodes, which is where the voltage-gated channels occur. The axon is insulated so the action potential at one node is sufficient to trigger an action potential in the next node and the impulse 'jumps' along the axon (called saltatory conduction). This contrasts with a non-myelinated neuron in which voltage-gated channels occur along the entire length of the axon.

As well as increasing the speed of conduction, the myelin sheath reduces energy expenditure because the area over which depolarization occurs is less. The number of sodium and potassium ions that need to be pumped to restore resting potential is, therefore, also less.

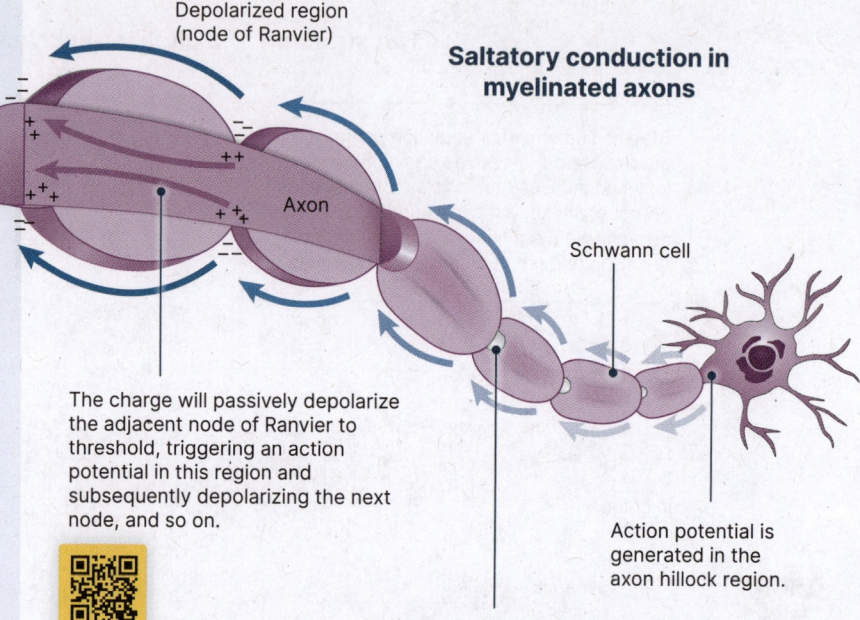

Saltatory conduction in myelinated axons

Depolarized region (node of Ranvier)

Axon

Schwann cell

The charge will passively depolarize the adjacent node of Ranvier to threshold, triggering an action potential in this region and subsequently depolarizing the next node, and so on.

Action potential is generated in the axon hillock region.

Myelinated axons have gated channels only at their nodes.

The electrical impulse

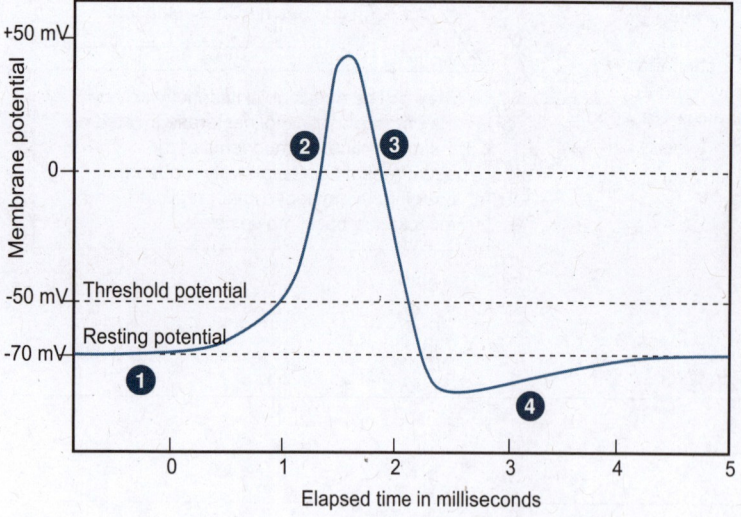

The depolarization in an axon can be shown as a change in membrane potential (in millivolts). A stimulus must be strong enough to reach the threshold potential before an action potential is generated. This is the voltage at which the depolarization of the membrane becomes unstoppable.

▶ When at rest, voltage-gated Na$^+$ channels are closed (1).

▶ Voltage-gated Na$^+$ channels open and the membrane depolarizes as Na$^+$ floods into the cell (2).

▶ Voltage gated Na$^+$ channels close and voltage-gated K$^+$ channels open, allowing K$^+$ to move out of the cell (3).

▶ A delay in closing voltage gated K$^+$ causes hyperpolarization. Na$^+$/K$^+$ pumps restore the membrane potential (4). During this refractory period, the nerve cannot respond, so nerve impulses are discrete.

1. What happens to a nervous signal if the threshold potential is not reached? _____

2. Contrast the placement of voltage-gated ion channels in myelinated and nonmyelinated neurons:

_____

_____

3. Based on the oscilloscope trace above, how many times can a nerve cell respond per second?

_____

AHL

C2.2
8-11

©2024 **BIOZONE** International
ISBN: 978-1-99-101410-8
Photocopying prohibited

189 Chemicals at Synapses

Key Idea: Many exogenous chemicals may increase or decrease the effect of neurotransmitters at synapses. Exogenous chemicals, e.g. drugs, may act at synapses, either mimicking or blocking the usual effect of a neurotransmitter, whether it be excitatory or inhibitory. They may also block the reuptake of neurotransmitters by acting on transporter proteins that remove a neurotransmitter from the synaptic cleft. Many recreational and therapeutic drugs work through their action at synapses, controlling the response of the receiving cell to incoming action potentials.

Blocking transmission

Neonicotinoids are a group of neuro-active insecticides. They mimic the action of acetylcholine on cholinergic receptors, but unlike acetylcholine they are not broken down by the enzyme acetylcholinesterase. This means they can bind to the cholinergic receptors permanently and block the transmission of nerve signals.

This has two effects. Initially, it leads to over-stimulation as the nerve continues to fire until ion exchange ceases, and eventually to paralysis as the nerve can not 'reset'.

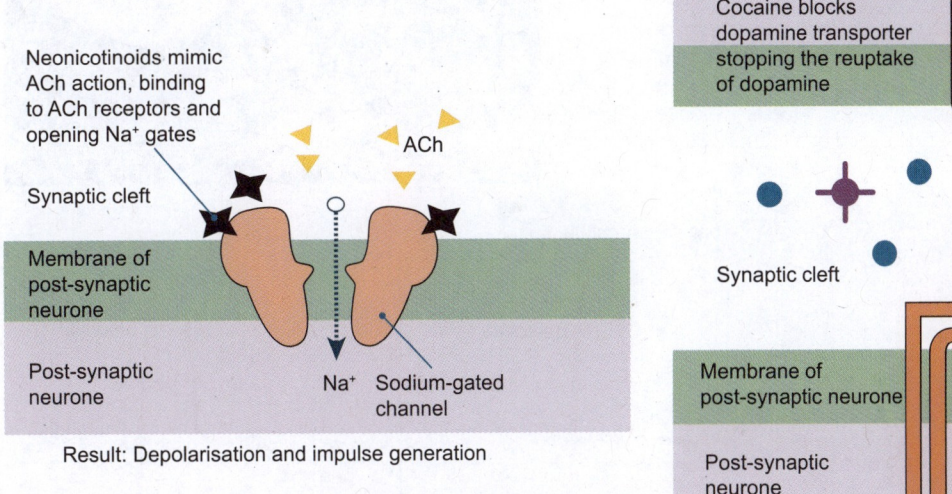

Result: Depolarisation and impulse generation

Neonicotinoids are commonly used as insecticides because of their far greater specificity to insects than other animals. However, they are a broad spectrum insecticide and so can kill both target and non target insects. The use of neonicotinoids is now restricted in many countries due to their (unintended) effects on honey bees.

Blocking reuptake

Once signal transmission is complete, the neurotransmitter is reabsorbed by the presynaptic cell, or broken down by enzymes. Certain drugs can block the reuptake of the neurotransmitter, allowing it to continue to stimulate the receptor. The drug cocaine is an example of this.

Cocaine blocks the reuptake of the neurotransmitter dopamine by binding to the dopamine transporter. This allows dopamine to accumulate in the synaptic cleft, stimulating the dopamine receptors on the postsynaptic cell.

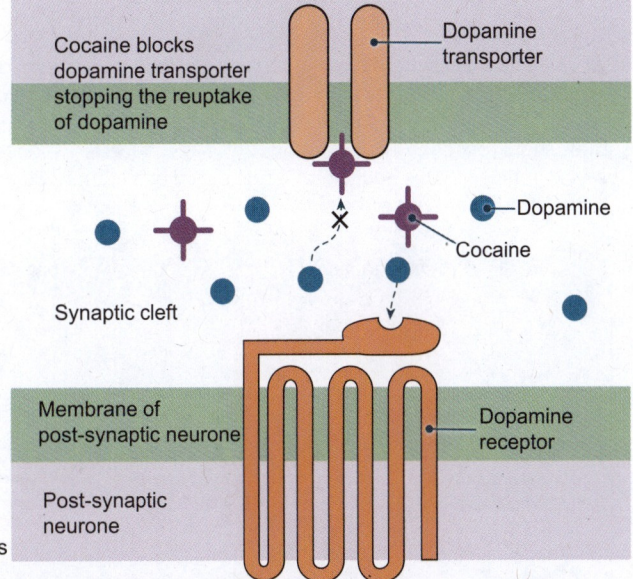

Result: Continued dopamine stimulus

Dopamine is part of the reward-motivation pathways in the brain, so cocaine use produces a euphoric effect and repeated use leads to addiction. However, the body quickly builds a tolerance to cocaine so for each use more is required for the same effect.

---

1. Describe three ways in which exogenous chemicals can affect signalling at synapses:

_____

_____

2. (a) Explain how neonicotinoids work: _____

_____

_____

(b) Explain why neonicotinoid based pesticides have been regulated in many countries: _____

_____

_____

3. Explain how cocaine produces its effects: _____

_____

_____

C2.2
12, 13
AHL

190 Summation at Synapses

Key Idea: The totality of signals from presynaptic cells determines if an action potential is reached in the post synaptic cell.

The nature of synaptic transmission in the nervous system allows the integration (interpretation and coordination) of inputs from many sources. These inputs can be excitatory (causing depolarization) or inhibitory (making an action potential less likely). Excitatory neurotransmitters, e.g. acetylcholine, reduce the polarization of the membrane (decrease membrane potential) by opening Na^+ channels. Inhibitory neurotransmitters e.g. GABA, increase the polarization of the membrane (hyperpolarization, increase membrane potential) by opening Cl^- channels. It is the sum of all excitatory and inhibitory inputs that leads to the final response in a post-synaptic cell. Synaptic integration is behind all the various responses we have to stimuli. It is also the most probable mechanism by which learning and memory are achieved.

Summation at synapses

Graded postsynaptic responses may sum to produce an action potential. Impulse transmission across chemical synapses has several advantages, despite the delay caused by neurotransmitter diffusion. Chemical synapses transmit impulses in one direction to a precise location and, because they rely on a limited supply of neurotransmitter, they are subject to fatigue (inability to respond to repeated stimulation). This protects the system against overstimulation.

Synapses allow inputs from many sources to be integrated. The response of a post-synaptic cell is often not strong enough on its own to generate an action potential. However, because the strength of the response is related to the amount of neurotransmitter released, subthreshold responses can sum together to produce a response in the post-synaptic cell. This additive effect is called summation. Summation can be temporal or spatial (below).

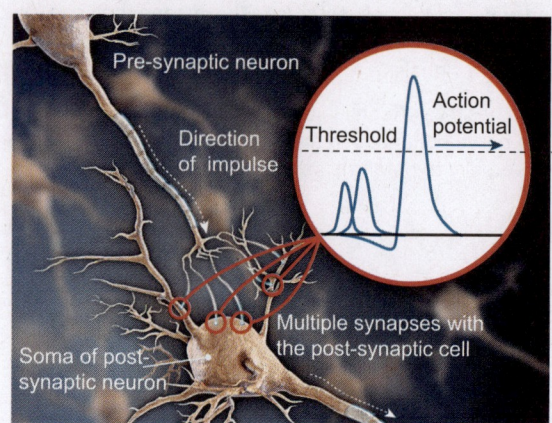

EPSP -IPSP cancellation

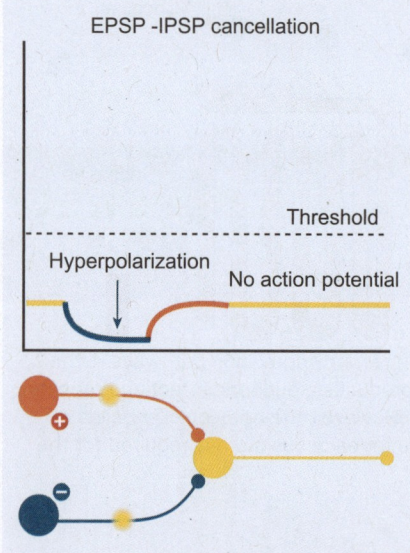

If an inhibitory signal reaches a synapse at the same time as an excitatory signal, the changes to membrane potential cancel out, producing no response.

Spatial summation

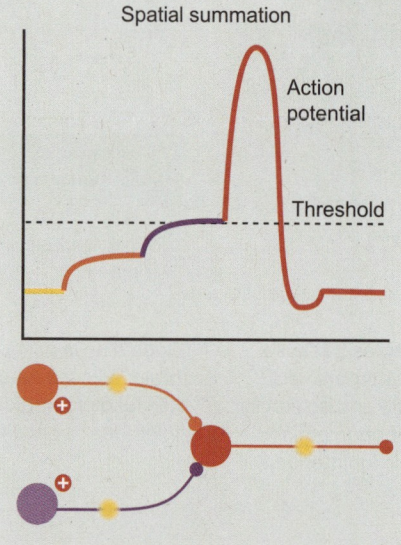

Impulses from spatially separated axon terminals may arrive simultaneously at different regions of the same post-synaptic neuron. The responses from the different places sum to produce an action potential.

Temporal summation

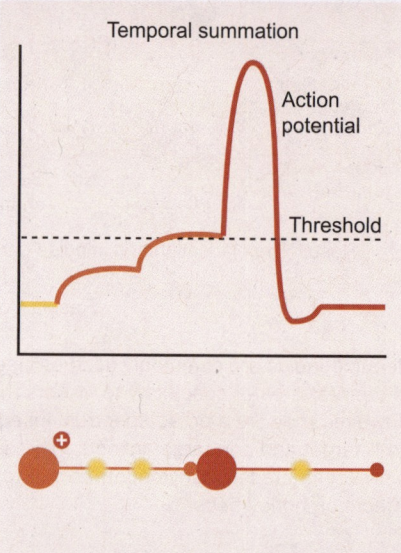

Several impulses may arrive at the **synapse** in quick succession from a single **axon**. The individual responses sum to reach threshold and produce an action potential in the post-synaptic neuron.

1. (a) Explain what is meant by summation: _____

    _____

    (b) In simple terms, distinguish between temporal and spatial summation: _____

    _____

    _____

2. How does hyperpolarization reduce the chance of an action potential occurring? _____

    _____

    _____

C2.2
14

©2024 **BIOZONE** International
ISBN: 978-1-99-101410-8

191 The Perception of Pain

Key Idea: Nociceptors are free nerve endings in the skin that send pain signals to the brain.

Pain is an important stimulus. It tells us when damage has occurred and causes us to either deal with or move away from the object causing pain. Stimuli that cause pain are responded to by free nerve endings called nociceptors. These are found in the skin, muscles, joint, and anywhere where harmful stimuli are likely to be felt. Nociceptors respond to a wide range of stimuli including mechanical, thermal, and chemical stimuli.

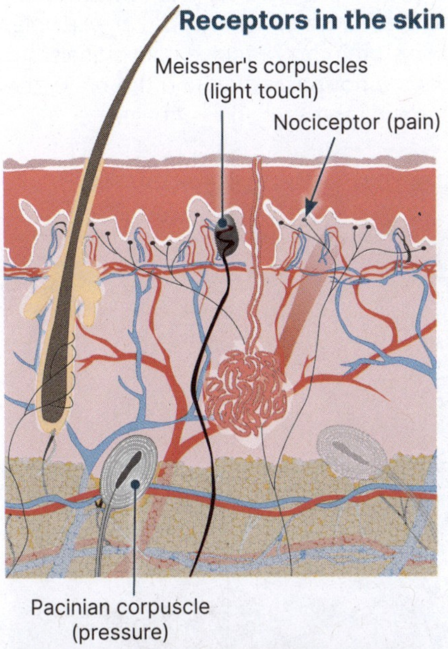

Receptors in the skin

Meissner's corpuscles (light touch)

Nociceptor (pain)

Pacinian corpuscle (pressure)

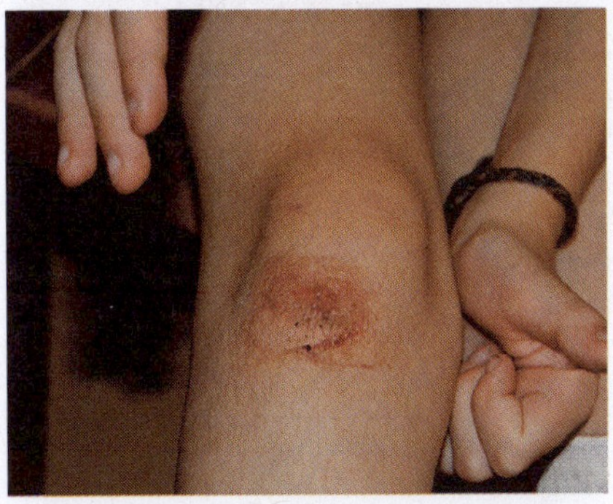

Nociceptors are free nerve endings found in the dermis. They respond to damaging stimuli by sending pain signals to the brain. A graze (above) or burn is painful because they trigger many of these receptors.

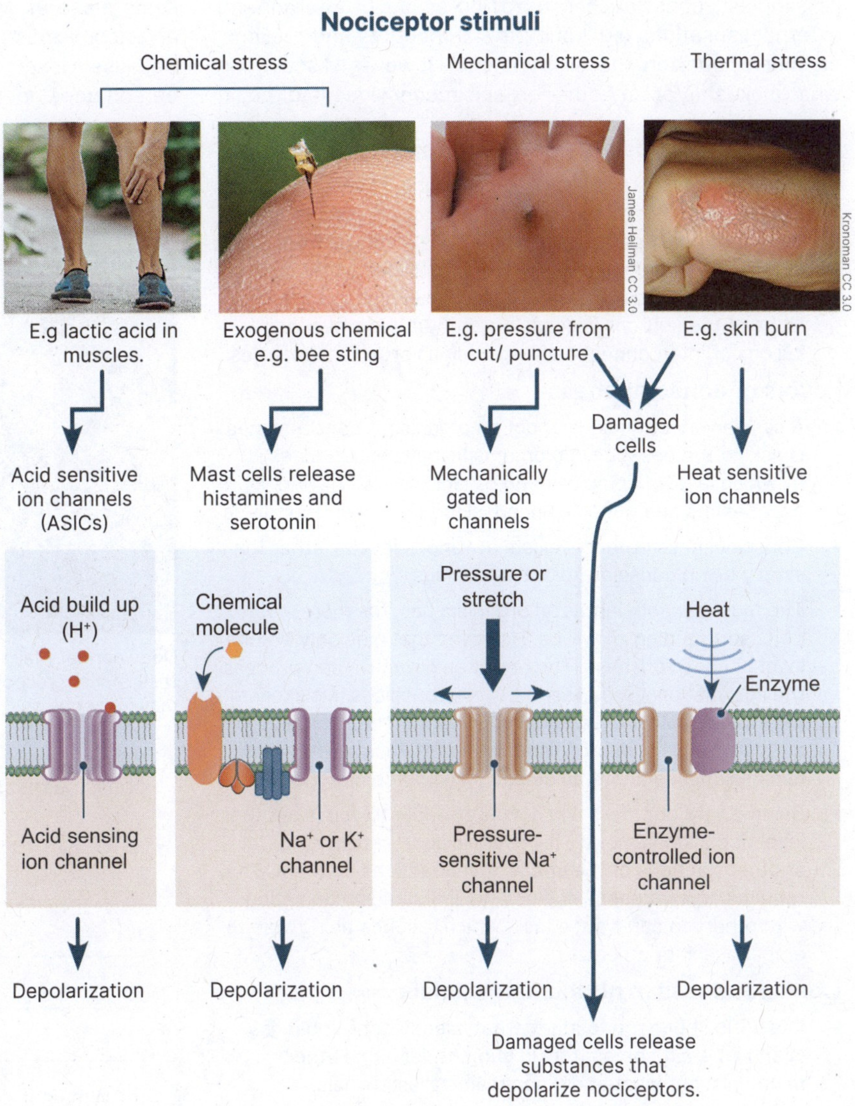

Nociceptor stimuli

Chemical stress		Mechanical stress	Thermal stress
E.g lactic acid in muscles.	Exogenous chemical e.g. bee sting	E.g. pressure from cut/ puncture	E.g. skin burn

James Heilman CC 3.0

Kronoman CC 3.0

Acid sensitive ion channels (ASICs)

Mast cells release histamines and serotonin

Mechanically gated ion channels

Damaged cells

Heat sensitive ion channels

Acid build up (H^+)

Chemical molecule

Pressure or stretch

Heat

Enzyme

Acid sensing ion channel

Na^+ or K^+ channel

Pressure-sensitive Na^+ channel

Enzyme-controlled ion channel

Depolarization

Depolarization

Depolarization

Depolarization

Damaged cells release substances that depolarize nociceptors.

1. What is the purpose of nociceptors? _____

_____

2. What kind of stimuli do nociceptors respond to? _____

_____

3. For each of the following, identify which type of receptor nociceptors responds:

   (a) Placing finger into too hot water: _____

   (b) Banged shin: _____

   (c) Standing on a sharp stone in bare feet: _____

4. Explain why nociceptors are also called polymodal receptors: _____

_____

5. Capsaicin, found in chilli peppers, activates heat activated ions channels in nociceptors. What is the result of this?

_____

©2024 **BIOZONE** International
ISBN: 978-1-99-101410-8

C2.2

15

192 What is Consciousness?

Key Idea: Consciousness is an emergent property of billions of neurons working together.

What exactly consciousness is and what causes it is a matter of intense scientific research and debate. Consciousness is an emergent property of billions of neurons in the brain working together to produce the state of being aware. At its simplest, consciousness is being aware of internal and external sensations e.g. your internal thoughts and feelings, or your effects on the environment. However, discovering what makes us or any other organism conscious is difficult.

Consciousness as an emergent property.

▶ Emergence is the property of small building blocks together forming bigger units that have different properties than just the sum of their parts. It is complexity arising from simplicity. In the case of consciousness, this is a property that emerges from many millions or billions of neurons working together. None of the neurons is conscious but put together with billions of interconnections, a conscious organism emerges.

Levels of consciousness

▶ A simple way of thinking about an organism's consciousness is asking the question, 'do organisms have experiences?'. At which level of complexity do organisms stop having experiences and become unconscious of their surroundings?

▶ Consciousness probably exists at various levels, rather than simply being conscious or not conscious.

▶ The more sensory inputs an organism has, the more aware of its surroundings it will be. However, that may only be one level of consciousness. The better an organism can process the information it receives, the more conscious it is likely to be. For example, instead of simply reacting to a stimulus in a repetitive way, an organism might first decide what that stimulus means and then decide on a course of action.

▶ Being aware of others and how they relate to you is another level of consciousness. This could include anticipating another's actions or the effect of your actions on them. This requires an awareness of time and self, or communicating with others to carry out a task, which requires being able to empathise with others.

Consciousness in nonhuman animals

▶ It is difficult enough to study what exactly consciousness is and where it comes from in humans, let alone other mammals, but what about non mammalian animals, or invertebrates?

▶ We know that many birds display consciousness on similar levels as some mammals, e.g. squirrels and scrub jays show similar behaviours when storing or burying food. They will re-hide food they buried if they think they were being watched, showing that they are conscious of others' possible intentions. Birds such as the Eurasian magpie have also passed the 'mirror test', showing that they are self-aware.

▶ We also know that an octopus can plan ahead. Octopuses have been observed carrying coconut shells or other objects for hiding in at a later time; they can remember people, and can solve puzzles. A study in 2022 seemed to show that they experience pain on an emotional level. However, invertebrate brains are not like mammalian brains. Indeed, although an octopus has a central brain, it is unlike that of a vertebrate's. Each of its arms also contains a dense set of neurons that acts as a 'mini-brain.' These neurons allow the arms to act independently, although the octopus can control them individually if it wants to. This makes the question of where consciousness resides even more difficult.

Studies tend to show that consciousness is not limited to any one part of the brain, although some parts of the brain appear to be more important in this regard than others. How the different parts of the brain communicate is also important. In a conscious person, signals are passed around the brain but in an unconscious person, signals tend to remain in the same place as where they were experienced. For example, an unconscious person's brain may receive pain signals but because those signals are not passed around the brain, the person does not experience the sensation of pain.

Georgia Pinaud PD

Dogs normally fail the mirror test because they don't use sight as their primary recognition sense. Tests using scent show that dogs may in fact be self-aware.

1. What is emergence and how does it relate to consciousness?

   _____

   _____

   _____

   _____

   _____

2. (a) Why is it difficult to determine where consciousness comes from in an organism?

   _____

   _____

   _____

   _____

   _____

   (b) Why is it difficult to determine that consciousness even exists in other animals?

   _____

   _____

   _____

   _____

   _____

  C2.2
16

©2024 **BIOZONE** International
ISBN: 978-1-99-101410-8
Photocopying prohibited

193 Did You Get It?

1. (a) **AHL**: The molecules labelled A-C are signalling molecules. Identify the signal molecule that will bind to the receptor shown:

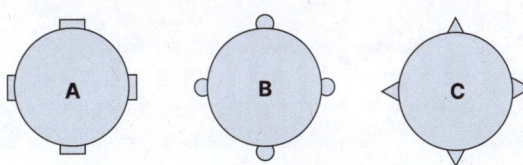

_____

(b) **AHL**: What prevents the other two signal molecules from binding to this receptor?

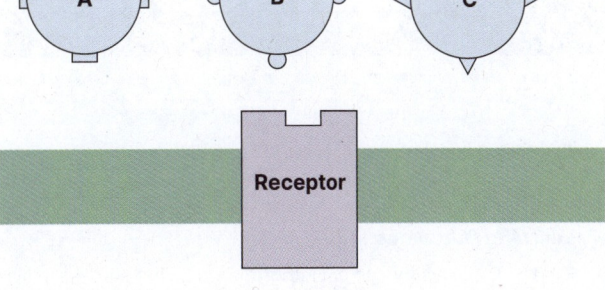

_____

_____

_____

(c) **AHL**: Why is it important that not all cells react to every signal molecule? _____

_____

_____

_____

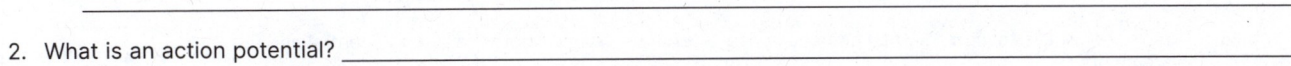

2. What is an action potential? _____

_____

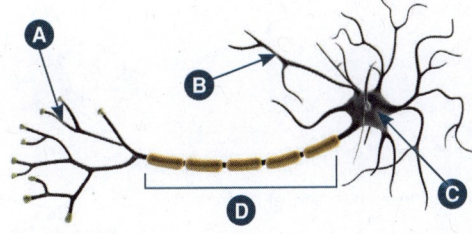

3. Identify structures A-D in the image of the neuron on the left:

A: _____

B: _____

C: _____

D: _____

4. (a) **AHL**: What occurs during saltatory conduction? _____

_____

(b) **AHL**: What influence does this have on conduction speed? _____

5. **AHL**: The graph below shows a recording of the changes in membrane potential in an axon during transmission of an action potential. Match each stage (A-E) to the correct summary provided below.

☐ Membrane depolarization (due to rapid Na⁺ entry across the axon membrane.

☐ Hyperpolarization (an overshoot caused by the delay in closing of the K⁺ channels).

☐ Return to resting potential after the stimulus has passed.

☐ Repolarization as the Na⁺ channels close and slower K⁺ channels begin to open.

☐ The membrane's resting potential.

Voltmeter records change in potential difference across membrane

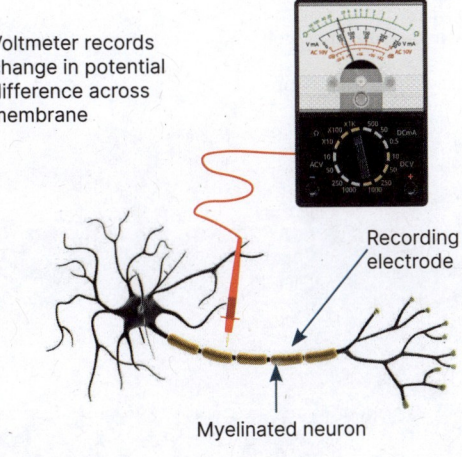

Myelinated neuron

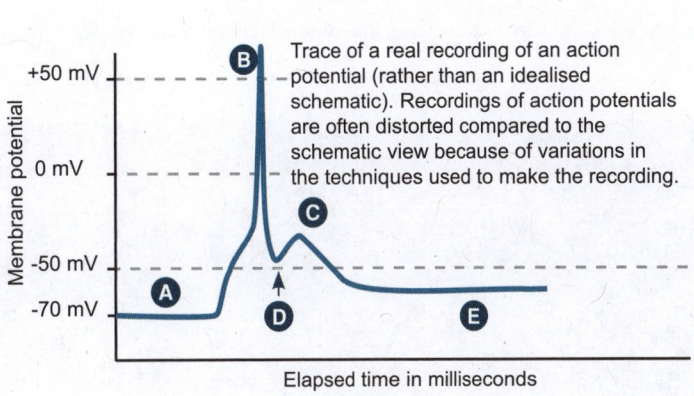

Trace of a real recording of an action potential (rather than an idealised schematic). Recordings of action potentials are often distorted compared to the schematic view because of variations in the techniques used to make the recording.

©2024 **BIOZONE** International
ISBN: 978-1-99-101410-8
Photocopying prohibited

C2.2 10 AOS

Organisms

C3.1 Integration of Body Systems

	Activity Number

Guiding Questions:
▶ How do nerves and hormones work together to enable body system integration?
▶ How are body systems regulated by feedback mechanisms?

Learning Outcomes:

			Activity Number
☐	1	Explain why system integration is a vital process in living systems.	194
☐	2	Use models to investigate the hierarchy of subsystems that are integrated in a multicellular living organism.	194
☐	3	Distinguish between both hormonal and nervous signalling to enable the integration of organs, and the transport of materials and energy.	195 - 196
☐	4	Describe how the brain acts as a central information integration organ.	197
☐	5	Describe how the spinal cord acts as an integrating centre for unconscious processes.	198
☐	6	Explain the process occurring when neurons convey messages from receptor cells to the central nervous system (input).	199
☐	7	Explain the process occurring when muscles are stimulated to contract via motor neurons (output).	199
☐	8	Compare transverse sections of both sensory and motor neurons.	200
☐	9	Explain the mechanism of involuntary responses using pain reflex arcs as an example.	201
☐	10	Discuss the general role of the cerebellum in coordinating skeletal muscle contraction and balance.	199
☐	11	Explain how melatonin secretion is able to modulate sleep patterns as part of circadian rhythms.	202
☐	12	Explain how the body can be prepared for vigorous activity due to epinephrine (adrenaline) secretion from the adrenal glands.	203
☐	13	Explain how the hypothalamus and pituitary gland control the endocrine system.	203
☐	14	Describe the role that baroreceptors and chemoreceptors have in the feedback control of heart rate.	204
☐	15	Describe the role that chemoreceptors have in the feedback control of ventilation rate.	205
☐	16	Describe the role that the central nervous system and the enteric nervous system have in the feedback control of peristalsis in the digestive system.	206
☐	17	**AHL: AOS:** Investigate tropic responses in seedlings, collecting both quantitative and qualitative data. **NOS:** Distinguish between these data types, identifying factors that effect precision, accuracy, and reliability.	207
☐	18	**AHL:** Describe how positive phototropism enables a directional growth response to lateral light in plant shoots.	207
☐	19	**AHL:** Explain how growth, development, and response to stimuli are controlled by phytohormones acting as signalling chemicals.	208
☐	20	**AHL:** Explain how phytohormone concentration gradients are maintained using auxin efflux carriers as an example.	209
☐	21	**AHL:** Elaborate on the auxin promotion of cell growth.	209
☐	22	**AHL:** Explain how root and shoot growth are regulated due to interactions between auxin and cytokinin.	210
☐	23	**AHL:** Discuss the positive feedback mechanism involved in fruit ripening and ethylene production.	211

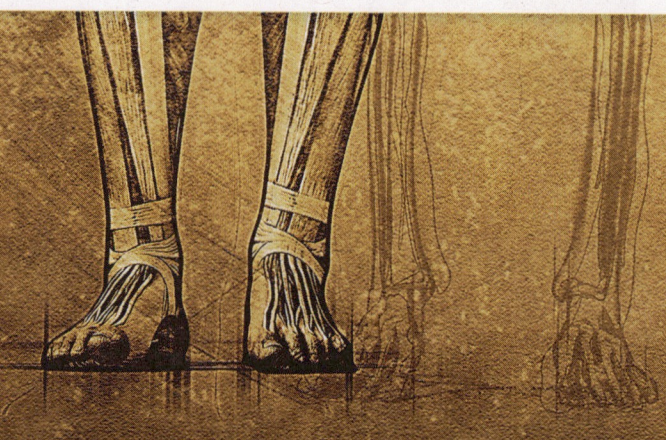

C3.2 Defence Against Disease

Guiding Questions:
- How does the immune systems interact to defend the body from pathogens and infection?
- What environmental factors influence disease transmission, spread, and incidence in a population?

Learning Outcomes:

		Activity Number
☐ 1	Identify and categorize pathogens. **NOS:** Report on important scientific breakthroughs in the recognition of pathogens as the cause of infectious diseases in humans. These will be a result of observations and subsequent development of control methods.	212
☐ 2	Explain how skin and mucous membranes act as a primary defence.	213
☐ 3	Describe the mechanism of blood clotting to seal cuts in the skin.	213
☐ 4	Contrast the responses of the innate and adaptive immune systems.	214
☐ 5	Explain how phagocytes can control infection, including amoeboid movement.	215
☐ 6	Elaborate on the general antibody production function of lymphocyte cells in the adaptive immune system.	216, 218
☐ 7	Explain how antigens act as recognition molecules to trigger antibody production.	217 - 218
☐ 8	Describe how helper T-lymphocytes are involved in the activation of B-lymphocytes and describe the result of this interaction.	218
☐ 9	Explain the importance of multiplication of activated B-lymphocytes to form clones of antibody-secreting plasma cells.	219
☐ 10	Explain how retaining memory cells leads to immunity.	219
☐ 11	Discuss the mechanism of HIV transmission.	220
☐ 12	Explain how HIV infection of lymphocytes eventually leads to AIDS.	220
☐ 13	Describe how antibiotics are effective against bacterial infection but not viral infection.	221
☐ 14	Explain how some strains of pathogenic bacteria have evolved resistance to several antibiotics. **NOS:** Explore the developments of new treatments to counter this resistance.	222
☐ 15	Use a range of examples to demonstrate how zoonoses can transfer from other species to humans, including tuberculosis, rabies, Japanese encephalitis, and COVID-19.	223
☐ 16	Describe how immunization is obtained by use of vaccines.	225
☐ 17	Discuss the mechanism of herd immunity and the prevention of epidemics. **NOS:** Discuss the value of scientific research to inform the public about the risks of vaccines.	225
☐ 18	Evaluate a range of data related to the COVID-19 pandemic. **AOS:** Calculate both percentage difference and percentage change.	224

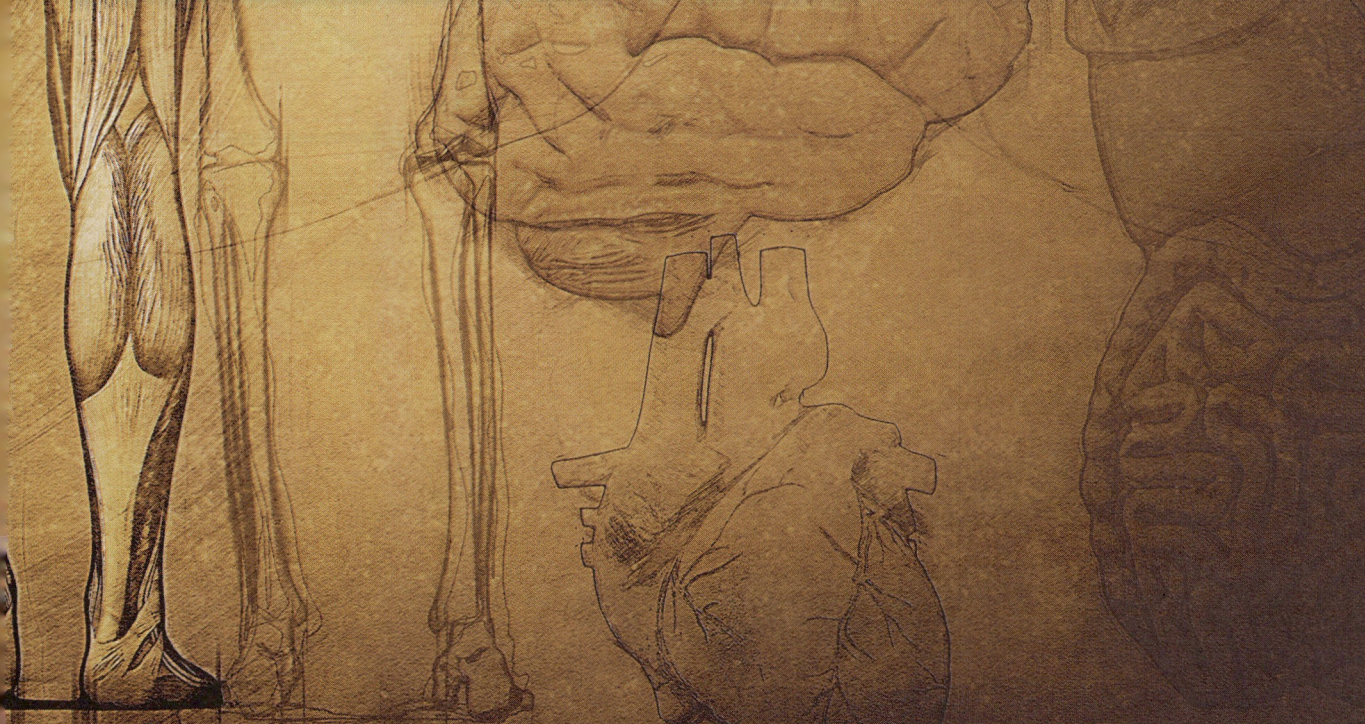

194 Biological Organization

Key Idea: Structural organization in animals, as in all multicellular organisms, is hierarchical.

Organization and the emergence of novel properties in complex systems are two of the defining features of living organisms. Emergent properties, such as circulation, result when components interact together, but are not inherent to the individual parts. Multicellular organisms are organized according to a hierarchy of structural levels. At each level, new emergent properties arise that were absent at simpler levels. Hierarchical organization allows specialized cells to group together into tissues and organs to perform a specific function. This improves efficiency in the organism.

The chemical level

1

All the chemicals essential for maintaining life, e.g. water, ions, fats, carbohydrates, amino acids, proteins, and nucleic acids.

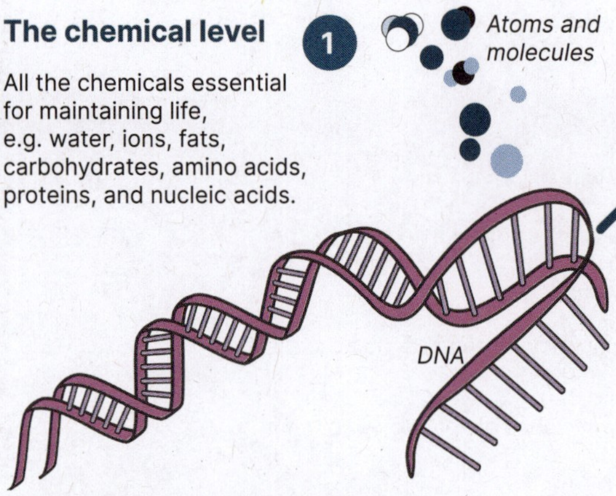

Atoms and molecules

DNA

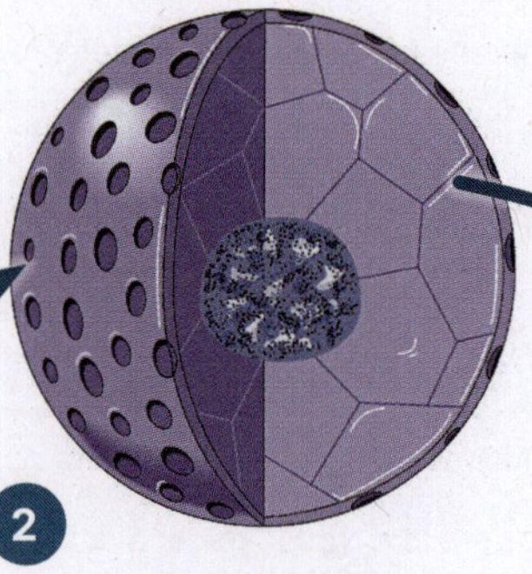

The organelle level

2

Molecules associate together to form the organelles and structural components of cells, e.g. the nucleus (above).

The cellular level

3

Cells are the basic structural and functional units of an organism. Cells are specialized to carry out specific functions, e.g. cardiac (heart) muscle cells (below).

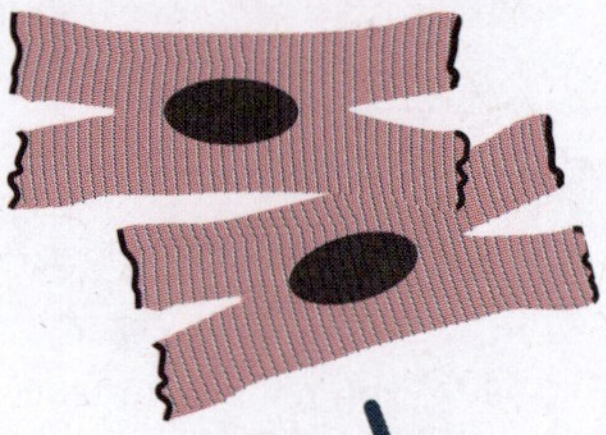

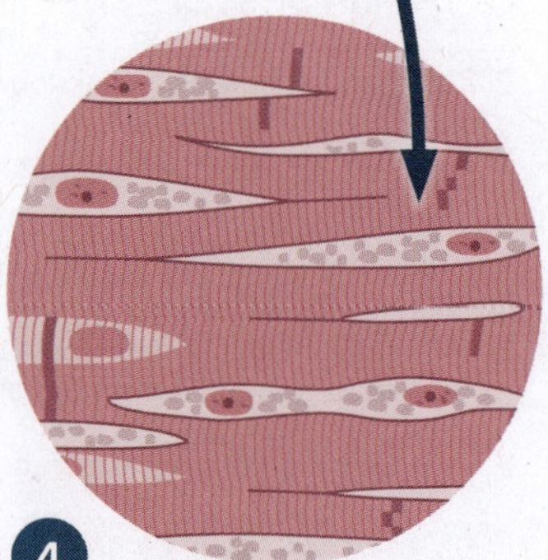

The tissue level

4

Groups of cells with related functions form tissues, e.g. cardiac (heart) muscle (above). The cells of a tissue often have a similar origin.

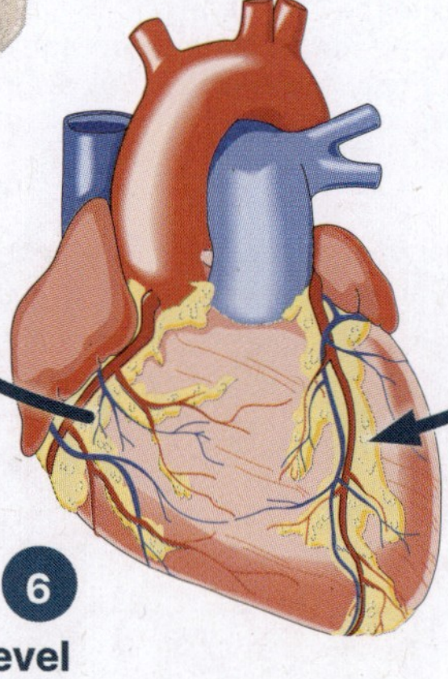

The organ level

5

An organ is made up of two or more types of tissues to carry out a particular function. Organs have a definite form and structure, e.g. heart (left).

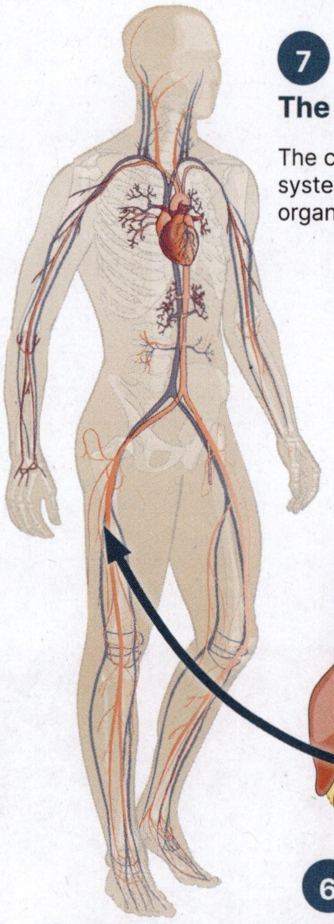

The body system level

6

Groups of organs with a common function form an organ system, e.g. cardiovascular system (left).

The organism

7

The cooperating organ systems make up the organism, e.g. a human.

Systems integration

In order to perform a specific function, the body must co-ordinate processes at many different levels.

For example, the circulatory system is one of many body systems that need to integrate together with each other to enable all life processes to occur.

At a finer level, the organs, including the heart and blood vessels, need to integrate in function so substances are transported around the body effectively, and at the correct pressure. The tissue level must also integrate to allow muscles in the heart to contract in synchrony, and at the cellular level sufficient energy must be supplied to the mitochondria in the cells so ATP is supplied to the muscles fibres when required. A multitude of processes at different levels need to integrate to ensure correct functioning.

1. What is an emergent property? _____

_____

2. Assign each of the following emergent properties to the level at which it first appears:

(a) Metabolism: _____

(b) Behaviour: _____

(c) Replication: _____

C3.1

1-2

©2024 **BIOZONE** International
ISBN: 978-1-99-101410-8
Photocopying prohibited

195 Integration in Nervous and Hormonal Signalling

Key Idea: The nervous and endocrine systems work together to maintain homeostasis.

In mammals, the nervous system and endocrine (hormonal) systems act independently and together to maintain homeostasis. The two systems are quite different in their modes of action, the responses they elicit, and the duration of action. The nervous system stimulates rapid, short-lived responses through electrical signals transmitted directly between adjacent cells. The endocrine system produces a slower, more long-lasting response through blood-borne chemicals called hormones. Hormones control many life processes such as reproduction, growth, and development.

Signalling by neurons (nerve cells)

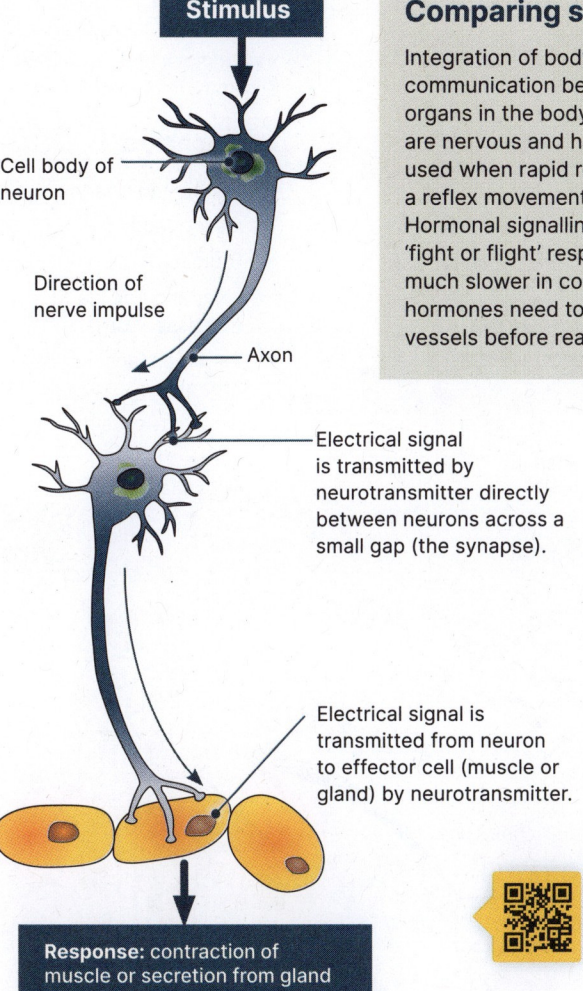

Stimulus

Cell body of neuron

Direction of nerve impulse

Axon

Electrical signal is transmitted by neurotransmitter directly between neurons across a small gap (the synapse).

Electrical signal is transmitted from neuron to effector cell (muscle or gland) by neurotransmitter.

Response: contraction of muscle or secretion from gland

Comparing signalling systems

Integration of body systems requires signalling communication between different cells, tissues, and organs in the body. The two key signalling systems are nervous and hormonal. Nervous signalling is used when rapid response is important, such as a reflex movement in response to heat or pain. Hormonal signalling can still be fast, such as the 'fight or flight' response due to adrenaline, but much slower in comparison to nervous signalling, as hormones need to be transported through the blood vessels before reaching the target cells.

Signalling by hormones

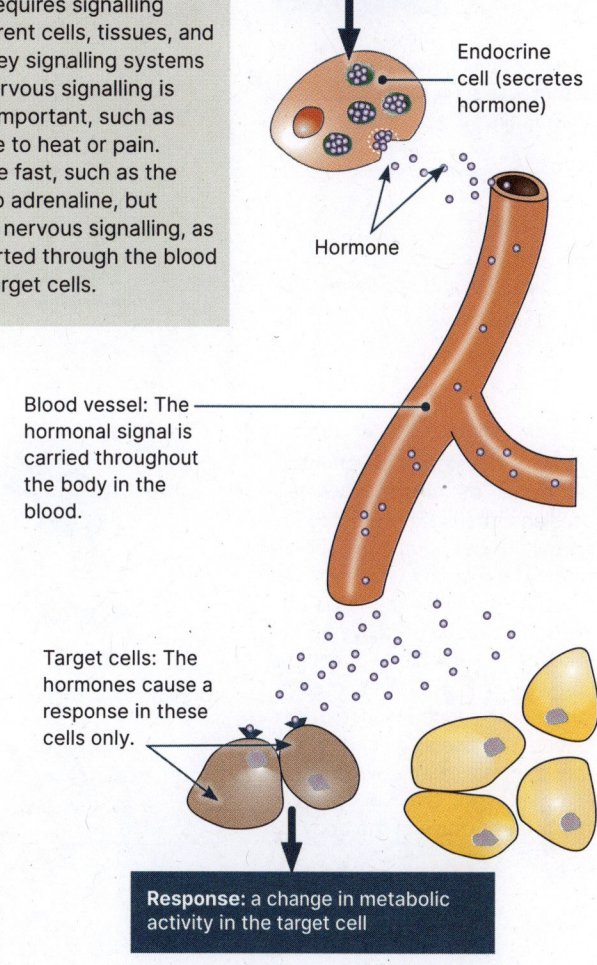

Stimulus

Endocrine cell (secretes hormone)

Hormone

Blood vessel: The hormonal signal is carried throughout the body in the blood.

Target cells: The hormones cause a response in these cells only.

Response: a change in metabolic activity in the target cell

The nervous system transmits electrical impulses directly between cells through electrical junctions or via chemicals called neurotransmitters, which can diffuse across the small gap (synapse) between cells. The response of a cell to nervous stimulation is rapid (milliseconds), short lived, and localized.

Hormones secreted from endocrine cells are carried in the blood throughout the body, where they interact only with target cells carrying the correct receptor to bring about a response. The speed of hormonal signalling is relatively slow, and it exerts its effects over minutes, hours, or days.

1. Complete the table below to show the comparison between nervous and hormonal signalling:

	Nervous control	Hormonal control
Communication	*Impulses directly between cells across cell to cell junctions.*	
Speed		
Duration		
Target pathway		*Carried in blood throughout body to target cells.*
Action		

196 Integration in Materials and Energy Transport

Key Idea: Materials and energy, in the form of glucose, are transported around the body through an integrated system of organs, connected by the blood system.

As heterotrophs, humans must consume substances that supply energy and nutrients to all cells in the body. At the interface of digestive organs, the digested substances must cross the membranes and be absorbed into the blood system, via a network of capillaries, and then transported to the liver. Once processed, materials and glucose can then be distributed to all organs around the body via the blood system once more. Oxygen and carbon dioxide are also transported to and from all organs in the blood system.

Circulatory system

Delivers oxygen (O_2) and nutrients to all cells and tissues. Removes carbon dioxide (CO_2) and other waste products of metabolism.

Components

▶ Heart
▶ Blood vessels: veins, arteries, and capillaries
▶ Blood

In mammals, the digestive system and cardiovascular system interact to transport materials and energy around the body

Digestive system

Digest food and absorb useful molecules from it, and eliminate undigested material.

Components

▶ Mouth and pharynx
▶ Oesophagus
▶ Stomach
▶ Liver and gall bladder (accessory organs)
▶ Pancreas (accessory organ)
▶ Small intestine
▶ Large intestine

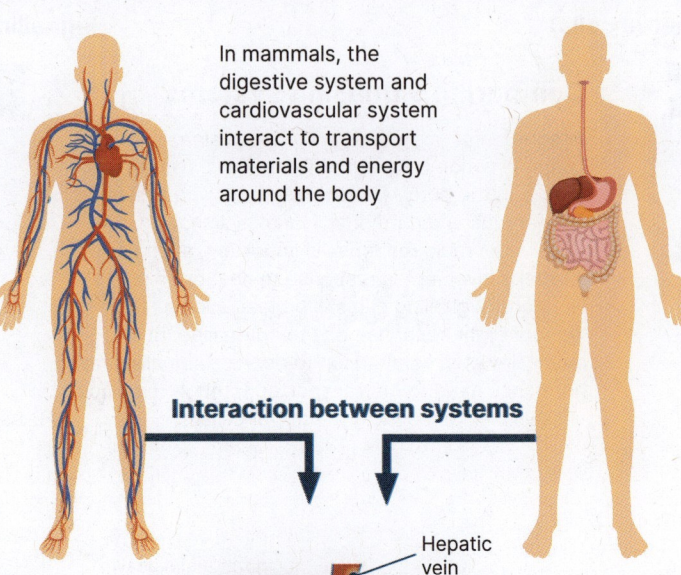

Interaction between systems

Food is digested in the stomach and small intestine from where it is absorbed and passed to the circulatory system. The capillaries around the stomach and intestines collect nutrients and then drain into the hepatic portal vein. This vein carries the blood directly to the liver. The liver then processes this nutrient-rich blood, e.g. it stores glucose as glycogen. The hepatic vein then transports nutrients from the liver to supply the other tissues of the body.

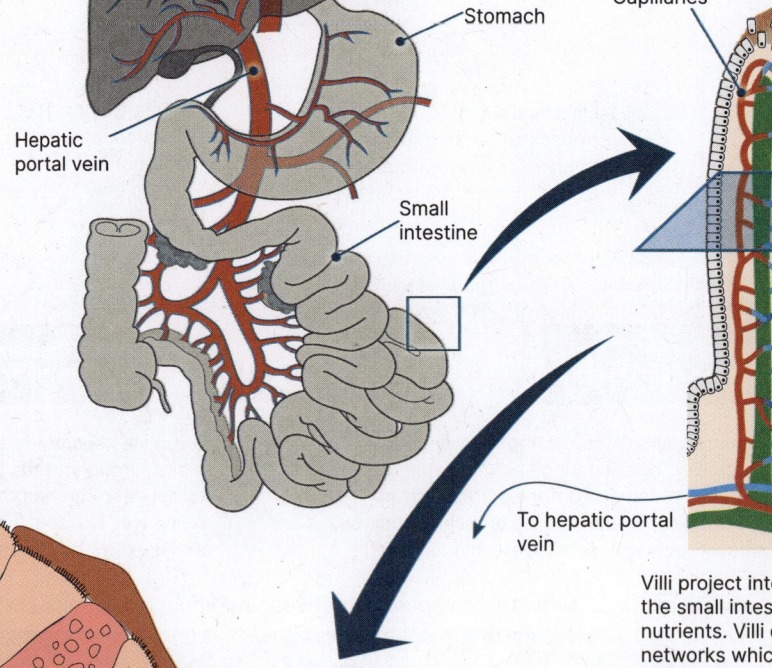

Hepatic vein
Liver
Stomach
Hepatic portal vein
Small intestine
Capillaries
Villus
To hepatic portal vein

Villi project into the lumen of the small intestine and absorb nutrients. Villi contain capillary networks which receive the nutrients and transport them to the hepatic portal system.

Intestinal epithelial cell

CO_2

O_2

Capillary

Glucose

Glucose (which supplies energy to all cells), and other nutrient materials are passed to the blood and transported to other parts of the body. Oxygen, which enters the blood during gas exchange, passes to the intestinal cells, while carbon dioxide passes into the blood.

1. Create a model flowchart on paper representing the transport of materials, including O_2 and CO_2, and energy in the form of glucose, as they are transported around the body. Include the name of the organs involved:

C3.1
3

197 Information Integration in the Brain

Key Idea: The brain is a key structure in the central nervous system and is involved with processing and integrating information, memory formation and storage, and learning. The human brain has a large, well developed cerebrum divided into two hemispheres. It has prominent folds (gyri) and grooves (sulci). Each cerebral hemisphere has an outer region of grey matter and an inner region of white matter, and is divided into four lobes by deep sulci or fissures. These lobes are the temporal, frontal, occipital, and parietal lobes. The cerebrum provides us with the ability to receive and process information which allows us to write, speak, calculate, plan, memorize, and produce new ideas.

Functional regions in the cerebrum

Primary motor area: controls voluntary muscle movement. Stimulation of a point one side of the motor area results in muscular contraction on the opposite side of the body.

Frontal lobe: includes the primary motor cortex but is also responsible for emotion, abstract thought, problem solving, and memory.

Language areas: the motor speech area (Broca's area) is concerned with speech production. The sensory speech area (Wernicke's area) is concerned with speech recognition and coherence.

Olfactory: (smell) area

Auditory areas: interpret the basic characteristics and meaning of sounds.

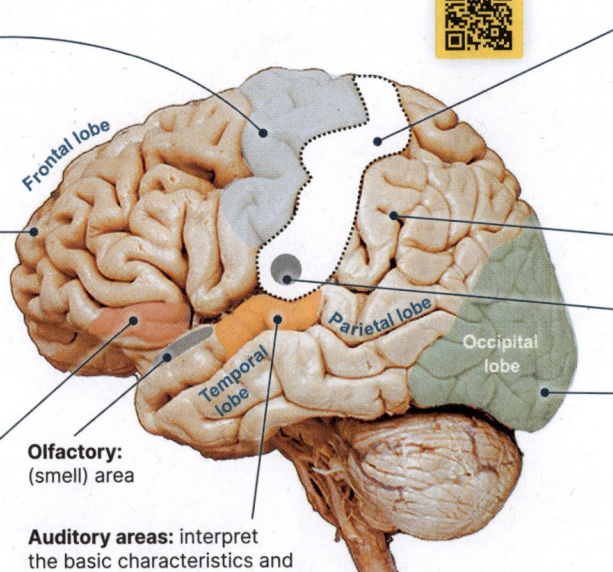

Primary somatosensory area: receives sensations from receptors in the skin, muscles, and viscera, allowing recognition of pain, temperature, or touch. The size of the sensory region for different body parts depends on the number of receptors in that body part.

Sensory association area: gives meaning to the sensations.

Primary gustatory area: interprets sensations related to taste.

Visual areas: within the occipital lobe, these receive, interpret, and evaluate visual stimuli. In vision, each eye views both sides of the visual field but the brain receives impulses from left and right visual fields separately. The visual cortex combines the images into a single impression or perception of the image.

Processing information

▸ Inputs into the brain are received from many different sensory organs both from the external environment and internal signals from other regions of the body, and are transported to the brain via neural signalling.

▸ The information is decoded and processed in different regions of the brain and may elicit a response, or the formation of a memory, leading to learning. Information processing is rapid and requires the coordinated integration of brain structures that have different roles.

▸ A 'sensory memory' may be triggered by a smell, or sound. Unique or significant information may then move through into the short-term memory region of the brain.

1. Why is the brain called a central information integration organ? _____

_____

_____

2. What is the function of the primary somatosensory area? _____

_____

3. What is the function of the primary motor area? _____

_____

_____

4. For each of the following body functions, identify the region(s) of the brain involved in its control:

   (a) Visual processing: _____

   (b) Taste: _____

   (c) Language: _____

   (d) Memory: _____

C3.1

4

Where are memories stored?

▶ Recent research indicates that different types of learning and memory occur in specific regions of the brain.

▶ Short-term working memory is stored mainly in the prefrontal cortex. These memories are brief 'snapshots' of information received from a number of different inputs, such as visual and auditory, and made available for instant recall. However, the memories can be forgotten quickly if not 'embedded' into other regions of the brain by repetition and sleep.

▶ Explicit (conscious) long-term memory, based on real-world 'lived' events (episodic), and 'learnt' information and facts (semantic), are stored in the hippocampus, the amygdala, and the neocortex. The brain can 'flavour' and strengthen the memories if the events occur with associated emotions, and 'fill-in' gaps with self-created memories if events are missing in the sequence.

▶ Implicit (unconscious) long-term memory, automatic routines that can occur automatically without conscious thought, such as experienced car driving, are stored in the basal ganglia and cerebellum.

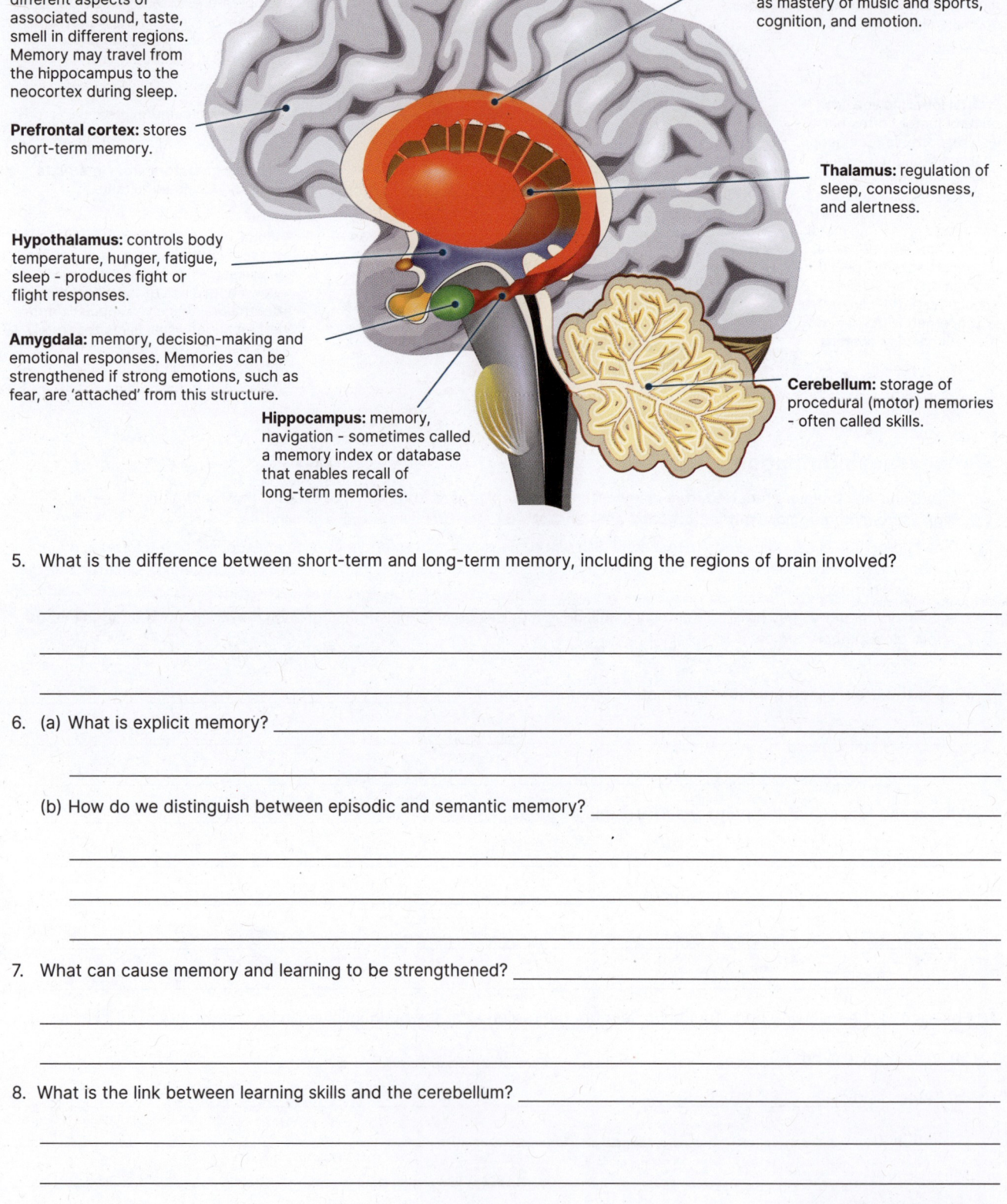

Neocortex: Outside layer of brain, consolidates long-term memory, integrating different aspects of associated sound, taste, smell in different regions. Memory may travel from the hippocampus to the neocortex during sleep.

Prefrontal cortex: stores short-term memory.

Hypothalamus: controls body temperature, hunger, fatigue, sleep - produces fight or flight responses.

Amygdala: memory, decision-making and emotional responses. Memories can be strengthened if strong emotions, such as fear, are 'attached' from this structure.

Hippocampus: memory, navigation - sometimes called a memory index or database that enables recall of long-term memories.

Basal ganglia: control of movements, learning, habit - such as mastery of music and sports, cognition, and emotion.

Thalamus: regulation of sleep, consciousness, and alertness.

Cerebellum: storage of procedural (motor) memories - often called skills.

5. What is the difference between short-term and long-term memory, including the regions of brain involved?

_____

_____

_____

6. (a) What is explicit memory? _____

_____

(b) How do we distinguish between episodic and semantic memory? _____

_____

_____

_____

7. What can cause memory and learning to be strengthened? _____

_____

_____

8. What is the link between learning skills and the cerebellum? _____

_____

_____

©2024 **BIOZONE** International
ISBN: 978-1-99-101410-8
Photocopying prohibited

198 The Autonomic Nervous System

Key Idea: The autonomic system is an involuntary reflex system, and integration of processes occur in the spinal cord. The autonomic nervous system (ANS) regulates unconscious (involuntary) visceral functions through reflexes, which are processes that occur without awareness. Although most autonomic nervous system activity is beyond our conscious control, voluntary control over some basic reflexes, such as bladder emptying and breathing, can be learned. Some people have even learnt to control heart rate. Two branches, the parasympathetic and sympathetic divisions, have broadly opposing actions on the organs they control (excitatory or inhibitory). Nerves in the parasympathetic division release acetylcholine. This neurotransmitter is rapidly deactivated at the synapse and its effects are short lived and localized. Most sympathetic postganglionic nerves release norepinephrine (noradrenaline), which enters the bloodstream and is deactivated slowly. Hence, sympathetic stimulation tends to have more widespread and long lasting effects than parasympathetic stimulation. Aspects of autonomic nervous system structure and function are illustrated below. The lines indicate nerves to organs or ganglia, which are concentrations of nerve cells.

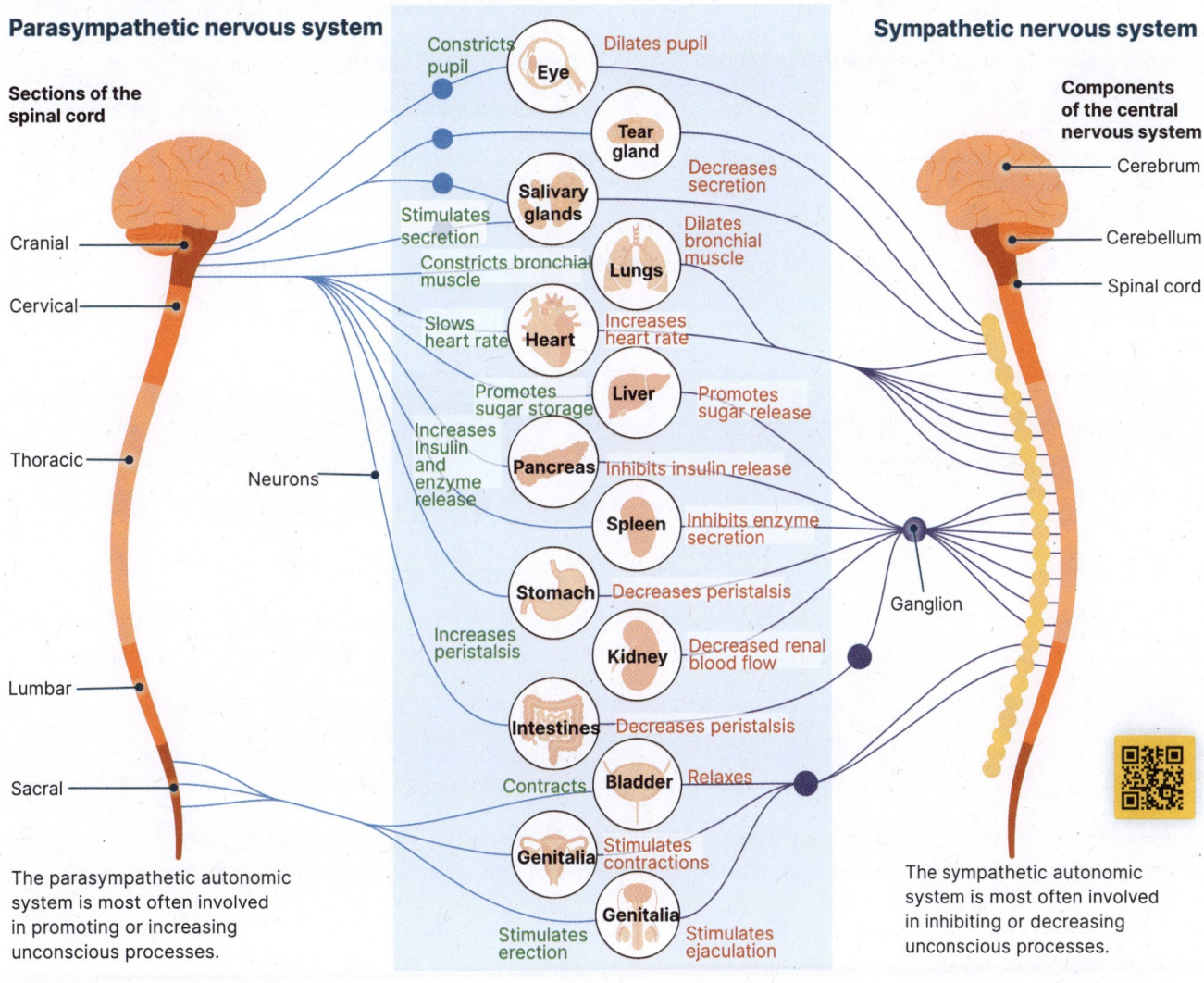

Parasympathetic nervous system

Sections of the spinal cord

Cranial
Cervical
Thoracic
Neurons
Lumbar
Sacral

Constricts pupil — **Eye** — Dilates pupil
Tear gland — Decreases secretion
Stimulates secretion — **Salivary glands** — Dilates bronchial muscle
Constricts bronchial muscle — **Lungs**
Slows heart rate — **Heart** — Increases heart rate
Promotes sugar storage — **Liver** — Promotes sugar release
Increases Insulin and enzyme release — **Pancreas** — Inhibits insulin release
Spleen — Inhibits enzyme secretion
Stomach — Decreases peristalsis
Increases peristalsis — **Kidney** — Decreased renal blood flow
Intestines — Decreases peristalsis
Contracts — **Bladder** — Relaxes
Genitalia — Stimulates contractions
Stimulates erection — **Genitalia** — Stimulates ejaculation

Ganglion

Sympathetic nervous system

Components of the central nervous system

Cerebrum
Cerebellum
Spinal cord

The parasympathetic autonomic system is most often involved in promoting or increasing unconscious processes.

The sympathetic autonomic system is most often involved in inhibiting or decreasing unconscious processes.

1   (a) Why are processes of the autonomic nervous system also considered unconscious processes?

_____

_____

(b) What is the benefit of processes, such as breathing rate, being unconscious processes?

_____

_____

2. Use information from the diagram above to explain the integration of unconscious control of the digestive system?

_____

_____

_____

The effects of the autonomic nervous system on the body

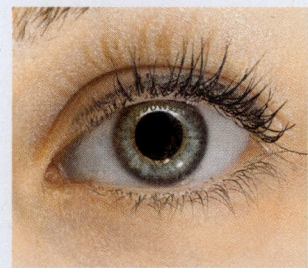   

When a person is fearful, their pupils enlarge as a consequence of sympathetic nervous system activity, i.e. the fight or flight response. The same response occurs during sexual arousal. Sympathetic stimulation also decreases secretion of the lacrimal (tear) glands.

Parasympathetic stimulation is responsible for bladder emptying through contraction of the bladder wall and relaxation of the urethral sphincter. This reflex activity can be inhibited by conscious control, but this ability does not develop until 2-4 years of age.

As part of the fight or flight response, the sympathetic nervous system dilates arteries and increases the rate and force at which the heart contracts. Heart rate is increased when increased blood flow to the heart causes reflex stimulation of the accelerator centre.

The enteric nervous system (ENS) is an interdependent part of the ANS. The ENS regulates itself but is influenced by sympathetic and parasympathetic nerves which are connected to it. The ENS innervates the gut, pancreas, and gallbladder.

3. Explain the structure and role of each of the following divisions of the autonomic nervous system:

   (a) The sympathetic nervous system: _____

   _____

   _____

   (b) The parasympathetic nervous system: _____

   _____

   _____

   _____

   _____

4. Explain why sympathetic stimulation tends to have more widespread and longer lasting effects than parasympathetic stimulation:

   _____

   _____

   _____

5. Using the example of the control of heart rate, describe the role of reflexes in autonomic nervous system function:

   _____

   _____

   _____

   _____

   _____

6. With reference to the emptying of the bladder, explain how conscious control can modify a reflex activity:

   _____

   _____

   _____

   _____

   _____

   _____

©2024 **BIOZONE** International
ISBN: 978-1-99-101410-8

199 Neural Inputs and Outputs

Key Idea: Information is received and sent from the central nervous system through neurons.

The brain is constantly receiving, processing, and prioritizing information and coordinating appropriate responses to stimuli. A range of peripheral sensory neurons collect information from the external environment, such as light, sound, touch, and temperature. Visceral sensory neurons provide information about homeostatic status. Neurons transmit the information via electrical impulses and chemical signals across synapses. After processing information in the central nervous system, if a response is needed, then motor neurons can carry information towards effector organs. The cerebellum has a specific role in coordinating balance. Receptors in the muscles, joints, and skin (proprioceptive system) and the inner-ear (vestibular) send information to the cerebellum about the state of the body's balance. If adjustments are required, then information is sent via motor neurons to skeletal muscles to make required contractions. The cerebellum is able to coordinate body movements and enables learning and fine-tuning of motor skills over time.

Input to central nervous system from the sensory neurons

The body has a range of sensory neurons to register information. The traditional senses of sight, sound, touch, smell, and taste require adapted receptors on peripheral sensory neurons. Internal stimuli, such as hunger, inflammation, blood pressure, and pain are detected by visceral sensory neurons. Temperature change can be detected by both peripheral and visceral receptors.

Sensory neurons in the head input information directly to different parts of the brain, depending on the type of stimuli. Sensory neurons outside the brain enter the spinal cord first, then information travels via the spinal nerves to input into the brain.

Stimuli that could indicate danger (temperature, pain) travel through sensory nerves that cross over in the spinal cord (purple line). Other information, such as proprioception, travels through sensory nerves that cross over in the brain stem (green line). Signals pass through the thalamus for processing, and then input to the cortex for interpretation.

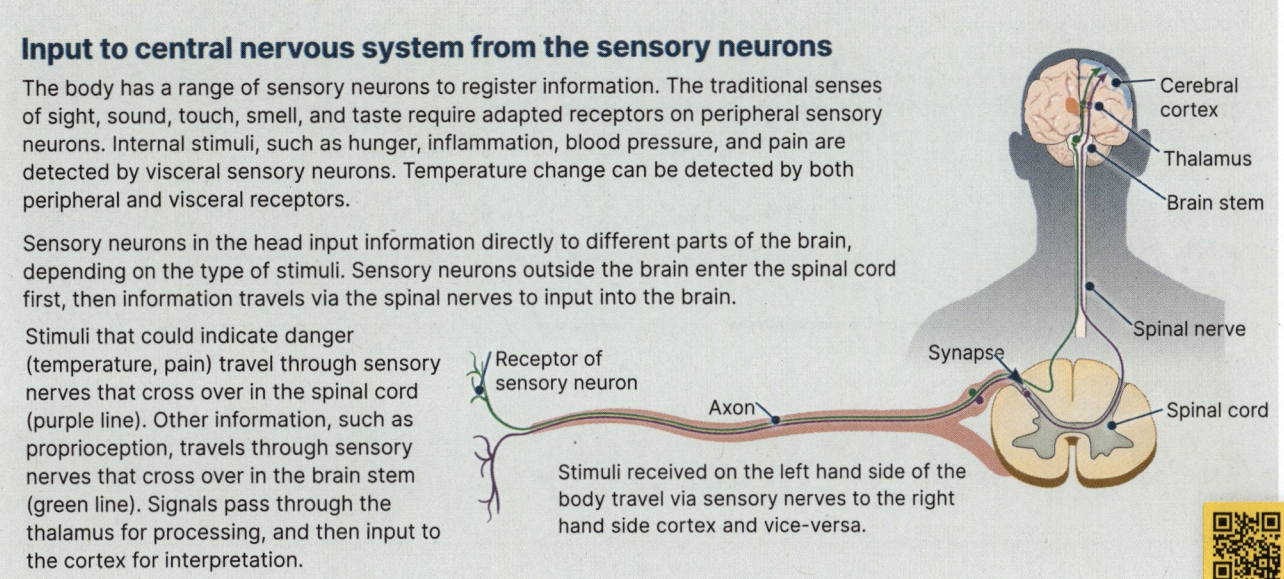

Stimuli received on the left hand side of the body travel via sensory nerves to the right hand side cortex and vice-versa.

Output from central nervous system to muscles via motor neurons

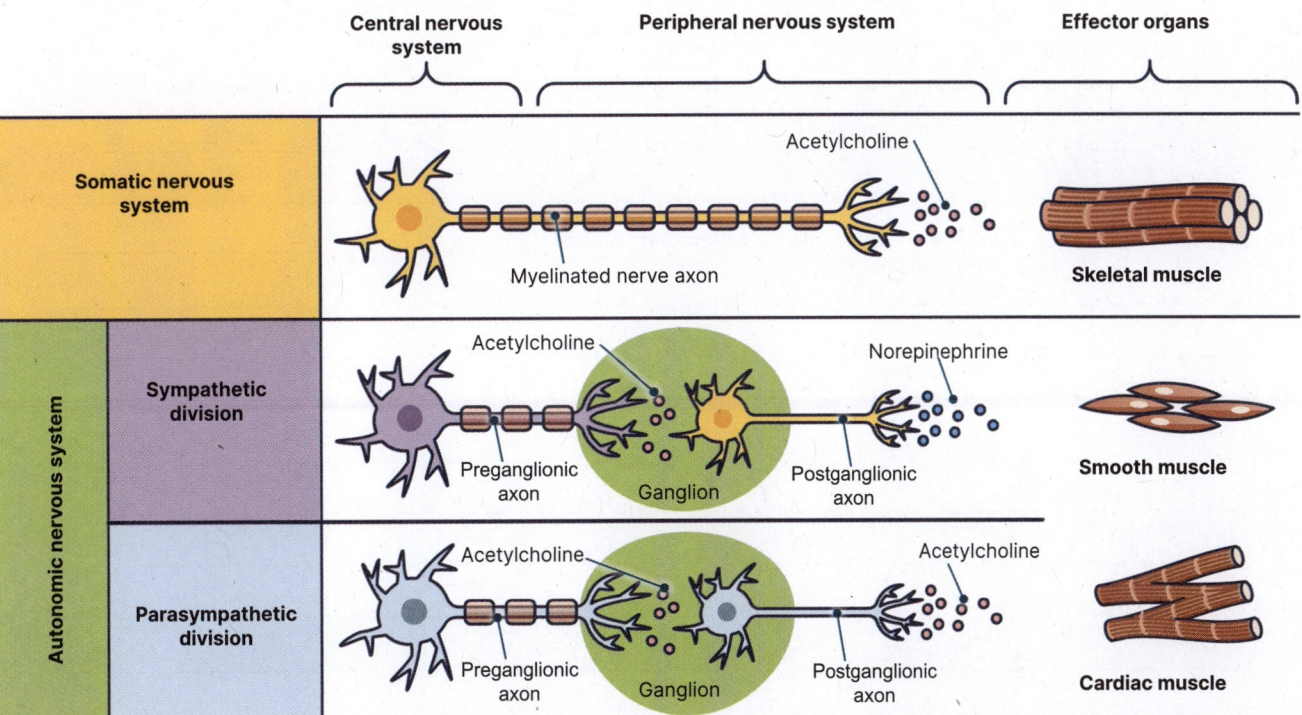

1. Compare between sensory neurons and motor neurons, including location and function:

_____

_____

_____

C3.1

6 - 7, 10

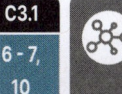

Pathways for the somatosensory system

The somatosensory (somatic-sensory) system is a system of sensory neurons and pathways that monitor and respond to changes within or at the surface of the body. Messages are relayed to the cerebral cortex, where specific areas are responsible for specific senses.

The brain is divided into left and right hemispheres. Any sensory system going to the cerebral cortex must cross over at some point because the cerebral cortex operates on a contralateral (opposite side) basis. Nerve fibres in the left or right halves of the spinal cord therefore cross over in the medulla (lower brain stem) so that signals from sensory neurons on the body's left hand side are processed by the right hand side of the brain and vice versa (right).

Note that voluntary nerves are controlled in the cerebral cortex and pass through the cerebellum. The cerebellum receives information from higher brain centres about what muscles should be doing and from the peripheral nervous system about what the muscles are doing. It has a critical role in providing corrective feedback to minimize any discrepancies and ensure smooth motor activity.

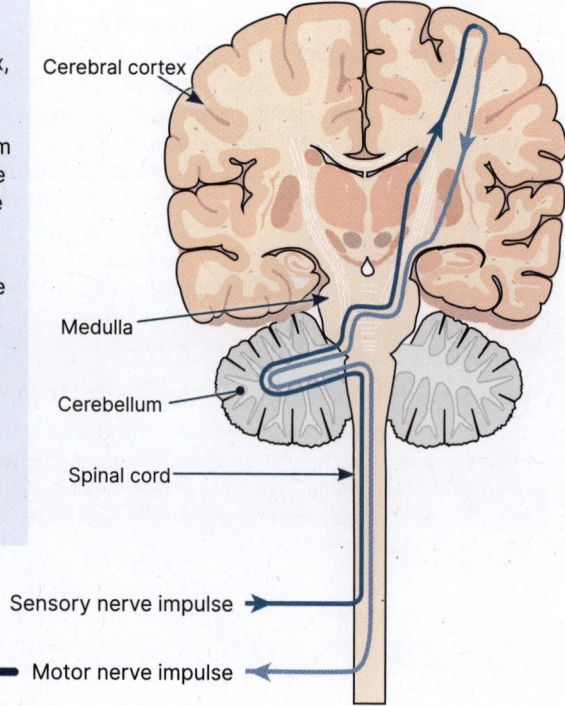

Cerebral cortex

Medulla

Cerebellum

Spinal cord

Uneven surface

Lift leg higher to step up

Stimulus → Receptor → Sensory nerve impulse →

Effector ← Motor nerve impulse ←

Response

2. What does contralateral control mean? _____

_____

_____

3. Why is 'learning' muscular movement a useful role of the cerebellum? _____

_____

4. What part of the brain is involved in conscious thought that is applied to movement, and why is this necessary?

_____

_____

5. What types of input are required by the cerebellum to maintain balance? _____

_____

_____

6. (a) Describe the general stimulus-response pathway in the somatosensory system: _____

_____

_____

_____

   (b) Explain why nerve impulses for muscle control travel through the cerebellum: _____

_____

_____

_____

7. Suggest which region of the brain is important in the perfection of motor skills, and why? ___

_____

_____

©2024 **BIOZONE** International
ISBN: 978-1-99-101410-8
Photocopying prohibited

200 Structure of Nerves

Key Idea: Nerves are bundled together as nerve fibres.
Nervous tissue is made up of two main cell types: neurons (nerve cells), which are specialized to transmit nerve impulses, and supporting cells, which are collectively called neuroglia. Neurons have a recognizable structure with a cell body (soma) and long processes (dendrites and axons). Most long neurons in the peripheral nervous system (PNS) are also supported by a fatty insulating sheath of myelin. Information, in the form of electrochemical impulses, is transmitted along neurons from receptors to effectors. The speed of impulse conduction depends primarily on the axon diameter and whether or not the axon is myelinated or unmyelinated.

Myelinated neurons

Where conduction speed is important, the axons of neurons are sheathed within a lipid-rich substance called myelin. Oligodendrocytes are the cells responsible for synthesizing myelin in the central nervous system (CNS). Outside the CNS, in the PNS, myelin is produced by specialized cells called Schwann cells. At intervals along myelinated axons, there are gaps between neighbouring Schwann cells and their sheaths, called nodes of Ranvier.

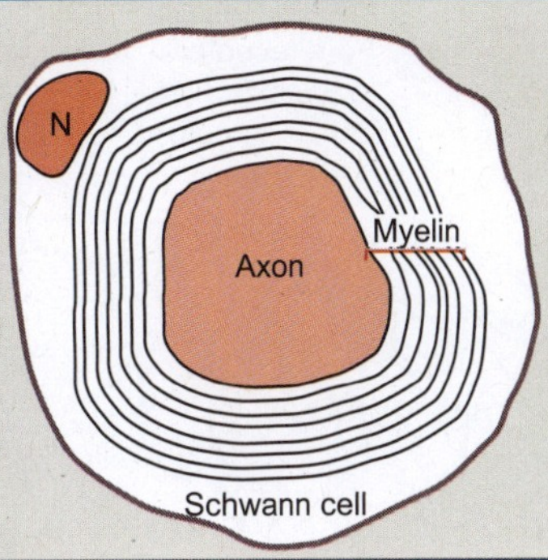

Anatomy of a nerve

Spinal nerve

Epineurium

Perineurium

Unmyelinated nerve fibre

Node of Ranvier

Myelinated nerve fibre

Blood vessels

Fascicle

Nerve fibres

Endoneurium

Transverse section

Transverse section drawing through a myelinated axon.
N = nucleus of Schwann cell.

N

Axon

Myelin

Schwann cell

Myelin acts as an insulator, increasing the speed at which nerve impulses travel because it forces the impulse to 'jump' from one uninsulated region to the next.

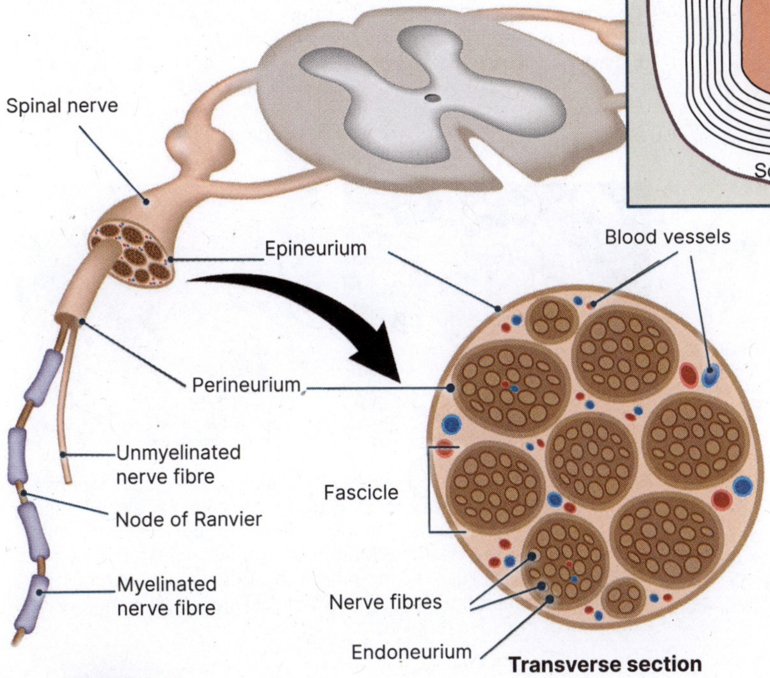

Unmyelinated neurons

Unmyelinated axons are more common in the CNS where the distances travelled are less than in the peripheral nervous system. Here, the axons are protected by Schwann cells, but there is usually no myelin produced. Impulses travel more slowly because the nerve impulse is propagated along the entire axon membrane rather than jumping from node to node as in myelinated neurons.

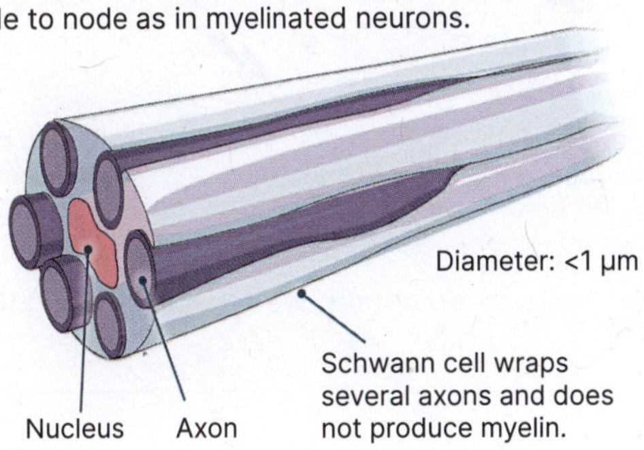

Diameter: <1 µm

Schwann cell wraps several axons and does not produce myelin.

Nucleus   Axon

1. (a) What is the function of myelination in neurons? _____

    _____

    (b) What cell type is responsible for myelination in the CNS? _____

    (c) What cell type is responsible for myelination in the PNS? _____

    (d) Explain why many of the neurons in the peripheral nervous system are myelinated, whereas those in the central nervous system are often not:

    _____

    _____

    _____

2. Identify two types of effectors innervated by motor neurons: _____

    _____

©2024 **BIOZONE** International
ISBN: 978-1-99-101410-8

184

C3.1
8

201 Reflexes

Key Idea: A reflex is an involuntary response to a stimulus. A reflex is an automatic response to a stimulus involving a small number of neurons and a central nervous system (CNS) processing point; usually the spinal cord, but sometimes the brain stem. This type of circuit is called a reflex arc. Reflexes permit rapid responses to stimuli. They are classified according to the number of CNS synapses involved. Monosynaptic reflexes involve only one CNS synapse, e.g. the knee jerk reflex, whereas polysynaptic reflexes involve two or more, e.g. the pain withdrawal reflex. Both are spinal reflexes. The pupil reflex (opening and closure of the pupil) is an example of a cranial reflex.

Pain withdrawal: A polysynaptic reflex arc

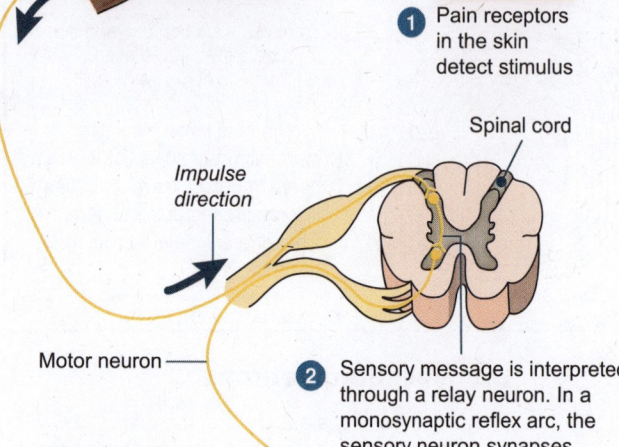

Sensory neuron

Stimulus = pin prick

1 Pain receptors in the skin detect stimulus

Spinal cord

Impulse direction

Motor neuron

2 Sensory message is interpreted through a relay neuron. In a monosynaptic reflex arc, the sensory neuron synapses directly with the motor neuron.

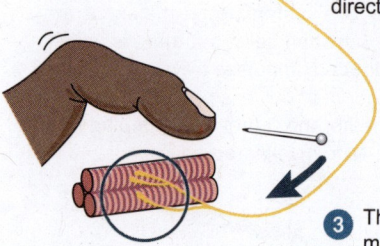

Response = withdraw finger

3 The impulse reaches the motor end plate and causes muscle contraction.

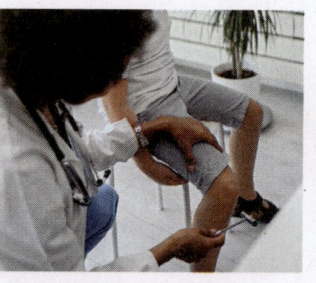

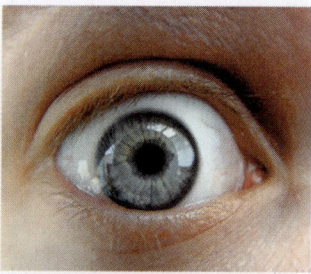

The patella (knee jerk) reflex is a simple deep tendon reflex used to test the function of the femoral nerve and spinal cord segments L2-L4. It helps to maintain posture and balance when walking.

The pupillary light reflex refers to the rapid expansion or contraction of the pupils in response to the intensity of light falling on the retina. It is a polysynaptic cranial reflex and can be used to test for brain death.

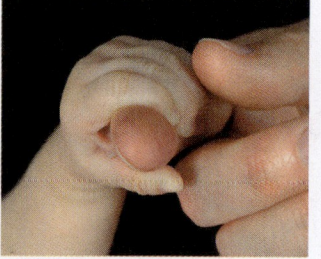

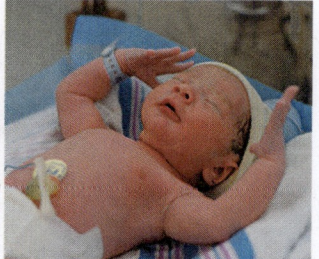

Normal newborns exhibit a number of primitive reflexes in response to particular stimuli. These reflexes disappear within a few months of birth as the child develops. Primitive reflexes include the grasp reflex (above left) and the startle or Moro reflex (above right) in which a sudden noise will cause the infant to throw out its arms, extend the legs and head, and cry. The rooting and sucking reflexes are further examples of primitive reflexes.

1. Reflexes do not require conscious thought to occur. How does this provide a survival advantage? _____
_____
_____

2. (a) Describe the difference between a monosynaptic and a polysynaptic reflex arc: _____
_____
_____
_____

(b) Which would produce the most rapid response, given similar length sensory and motor pathways? Explain:
_____

3. What might be the survival advantage of primitive reflexes in newborns? _____
_____
_____
_____
_____

©2024 **BIOZONE** International
ISBN: 978-1-99-101410-8
Photocopying prohibited

C3.1
9

202 Sleeping and Waking

Key Idea: Sleep patterns are modulated by the release of melatonin to establish a circadian rhythm.

A biological clock is an endogenous (internal) timing system that releases or suppresses hormones, or changes body temperature to control the physiological responses and activities of an organism. Circadian (Latin: around a day) rhythms are the physical and behavioural changes established by the biological clock. The biological clock will continue even in the absence of environmental cues, although the period (duration) of the rhythm may be slightly different from the environmental rhythm. Circadian rhythms use environmental cues, such as daylight, to 'reset' the clock into synchrony with the actual time period. Circadian rhythms have an adaptive function, such as helping anticipate environmental changes and preparing the body for the activities that will predictably follow. Melatonin is a hormone released from the pineal gland to induce sleepiness. High levels of melatonin circulate in the blood stream at night and are low during the day. The rhythm is controlled by the suprachiasmatic nucleus in the pineal gland, which is stimulated by light.

Melatonin regulation

▶ The location of the biological clock varies between organisms. In birds, reptiles, and amphibians it is located in the pineal gland. In insects, each cell has its own biological clock. In mammals, the biological clock is located in the hypothalamus.

▶ For most humans, the biological clock runs at about a 24 hour, 11 minute day. To keep it synchronized with the 24 hour-day cycle, it needs to be reset each day, reacting to outside stimuli such as light and dark, and meal times.

▶ The clock is made up of a collection of cells in the hypothalamus, called the suprachiasmatic nucleus (SCN), just behind the eyes. Light from the eyes stimulates the nerve pathways to the SCN and regulates its activity.

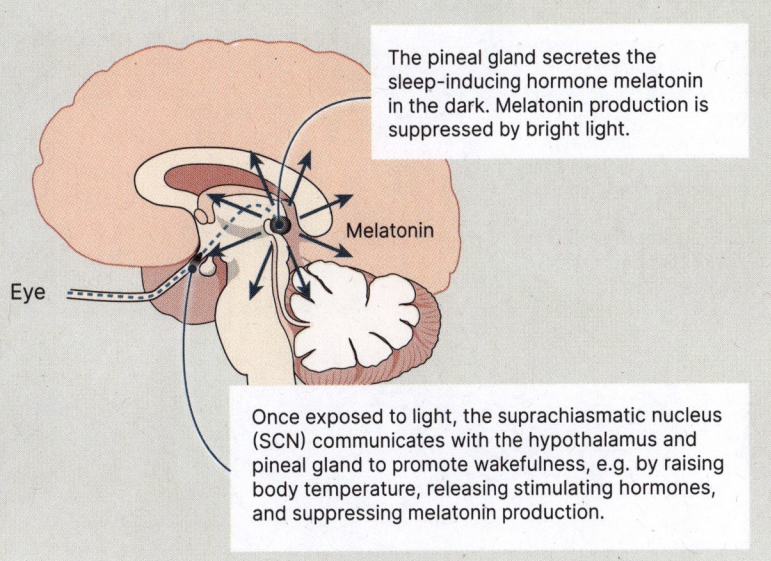

The pineal gland secretes the sleep-inducing hormone melatonin in the dark. Melatonin production is suppressed by bright light.

Eye

Melatonin

Once exposed to light, the suprachiasmatic nucleus (SCN) communicates with the hypothalamus and pineal gland to promote wakefulness, e.g. by raising body temperature, releasing stimulating hormones, and suppressing melatonin production.

Body temperature and sleepiness

▶ An increase in melatonin levels in the body results in heat loss, where an increased skin temperature radiates more heat away to lower core body temperature. The heat is lost primarily from distal (outer) body surface areas, such as the fingertips. A lower body temperature induces sleepiness.

▶ Melatonin helps to regulate the body's 'set-point' baseline core temperature through control of vasoconstriction mechanisms. The extent of relaxation of vascular beds close to the skin, that allow for more heat loss, are stabilized to maintain an effective homeostatic body temperature in rhythm with the 24hr day.

Graph: Level vs time, showing Melatonin and Body temperature curves, labelled Midnight.

1. (a) Where is the biological clock located in humans? _____

   (b) What is the relationship between a biological clock and a circadian rhythm? _____

   _____

   _____

2. Explain how melatonin secretion is able to modulate sleep patterns as part of circadian rhythms:

   _____

   _____

   _____

   _____

   _____

C3.1

11

203 Control of the Endocrine System

Key Idea: The endocrine system is made up of ductless glands that secrete hormones into the blood. These participate in feedback loops and regulate internal functions.

Endocrine glands are scattered widely throughout the body and their positioning does not necessarily reflect the location of their influence. Unlike exocrine glands, e.g. salivary glands, endocrine glands lack ducts and secrete hormones directly into the blood. Hormones are chemical messengers that are produced at one endocrine site and carried in the blood to influence target cells that may be quite distant. The key glands that exert control over the endocrine system are the pituitary gland and hypothalamus, both located in the centre of the cerebrum. The adrenal gland secretes epinephrine (adrenaline), which is a stress related hormone involved in the fight or flight response. The hormone release is initially signalled via neurons connected from the hypothalamus.

Hypothalamus
A small area of the brain that links the nervous and endocrine systems via the pituitary. Information comes to the hypothalamus through sensory pathways from sensory receptors allowing the hypothalamus to control and integrate many basic physiological activities, such as temperature regulation, food and fluid intake, and sleep, including the reflex activity of the autonomic nervous system.

Pituitary gland
A small 'master gland' that produces or releases hormones that control the activity of many other endocrine glands.

Pineal gland
Produces melatonin, called the sleep hormone.

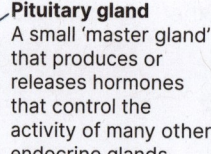

Thyroid

Adrenal glands

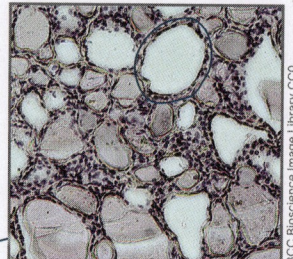

The functional unit of the thyroid is the spherical thyroid follicle (circled left, circular in cross section). These are lined with follicular cells. The thyroid secretes three hormones that influence metabolic rate and protein synthesis.

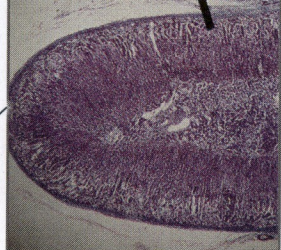

The adrenal glands are pyramid-shaped glands found on top of the kidneys. They secrete many hormones, including epinephrine (adrenaline), which plays a role in the fight or flight response. The stimulus from stress is signalled via neurons to the adrenal glands (see below).

Flight or fight response
short term stress response

1. Increased heart rate through increased muscle contraction
2. Increased blood pressure
3. Liver converts glycogen to glucose; blood glucose levels increase
4. Dilation of bronchioles
5. Blood flow to gut and kidney reduced
   Blood flow to muscles and brain increased
6. Increased metabolic rate

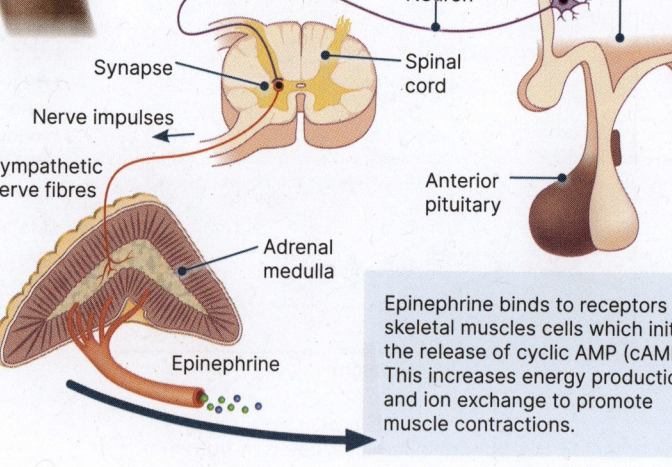

Epinephrine binds to receptors on skeletal muscles cells which initiate the release of cyclic AMP (cAMP). This increases energy production and ion exchange to promote muscle contractions.

1. Explain how the hypothalamus and pituitary gland control the endocrine system: _____

_____

_____

_____

2. Briefly outline what occurs during the flight or fight response, including the role of epinephrine: _____

_____

_____

_____

_____

©2024 **BIOZONE** International
ISBN: 978-1-99-101410-8
Photocopying prohibited

C3.1
12 - 13

204 Control of Heart Rate

Key Idea: The pacemaker sets the basic rhythm of the heart. The heart rate is influenced by the cardiovascular control centre, located in the medulla oblongata that sits above the top of the spinal cord, primarily in response to sensory information from baroreceptors (pressure) receptors in the walls of the blood vessels entering and leaving the heart. Secondary sensory input is received from arterial chemoreceptors, sensory organs which are located in carotid and aortic bodies, regions close to the respective arteries. These organs detect changes in arterial O_2 and CO_2 levels, and send signals back to the cardiovascular centre via the vagus nerve (aortic) and glossopharyngeal nerve (carotid).

Cardiovascular control ┄┄┄┄┄┄

Increase in rate **+**

Decrease in rate **−**

Medulla oblongata

Cardiovascular centre responds directly to norepinephrine (noradrenaline) and to low pH (high CO_2). It sends output to the sinoatrial node (SAN) to increase heart rate. Changing the rate and force of heart contraction is the main mechanism for controlling cardiac output in order to meet changing demands.

Higher brain centres influence the cardiovascular centre, e.g. excitement or anticipation of an event.

Baroreceptors in aorta, carotid arteries, and vena cava give feedback to cardiovascular centre on blood pressure. Blood pressure is directly related to the heart rate.

+ or **−**

Sympathetic output to heart via cardiac nerve increases heart rate. Sympathetic output predominates during exercise or stress. **+**

Parasympathetic output to heart via vagus nerve decreases heart rate. Parasympathetic (vagal) output predominates during rest. **−**

Aortic baroreceptors

Aortic bodies

Extrinsic input to SAN

Opposing actions keep blood pressure within narrow limits.

The intrinsic rhythm of the heart is influenced by the cardiovascular centre, which receives input from sensory neurons and hormones.

Influences on heart rate

Increase	Decrease
Increased physical activity	Decreased physical activity
Decrease in blood pressure	Increase in blood pressure
Secretion of epinephrine or norepinephrine	Re-uptake and metabolism of epinephrine or norepinephrine
Increase in H⁺ or CO_2 concentrations in blood	Decrease in H⁺ or CO_2 concentrations in blood

Reflex responses to changes in blood pressure

Reflex	Receptor	Stimulus	Response
Bainbridge reflex	Baroreceptors in vena cava and atrium	Stretch caused by increased venous return	Increase heart rate
Carotid reflex	Baroreceptors in the carotid arteries	Stretch caused by increased arterial flow	Decrease heart rate
Aortic reflex	Baroreceptors in the aorta	Stretch caused by increased arterial flow	Decrease heart rate

1. Explain how each of the following extrinsic factors influences the basic intrinsic rhythm of the heart:

    (a) Increased venous return: _____

    (b) Release of epinephrine in anticipation of an event: _____

    (c) Increase in blood CO_2: _____

2. How do these extrinsic factors bring about their effects? _____

    _____

3. (a) Identify the nerve that brings about increased heart rate: _____

    (b) Identify the nerve that brings about decreased heart rate: _____

4. Account for the different responses to stretch in the vena cava and the aorta: _____

    _____

    _____

C3.1

14

205 Control of Ventilation rate

Key Idea: The basic rhythm of ventilation is controlled by the respiratory centre, a cluster of neurons located in the medulla oblongata, situated in the brain stem.

This rhythm is adjusted in response to the physical and chemical changes that occur when we carry out different activities. Although control of ventilation is involuntary, we can exert some degree of conscious control over it. The diagram below illustrates these controls. Control of ventilation rate is complex and has some similarities to control of heart rate. This is not surprising, given the importance of both systems in supplying oxygen to the tissues. However, the main trigger for changing the ventilation rate is a change in the blood CO_2, whereas the main trigger for changing the basic rhythm of heart rate is blood pressure change.

The respiratory centre and the control of breathing

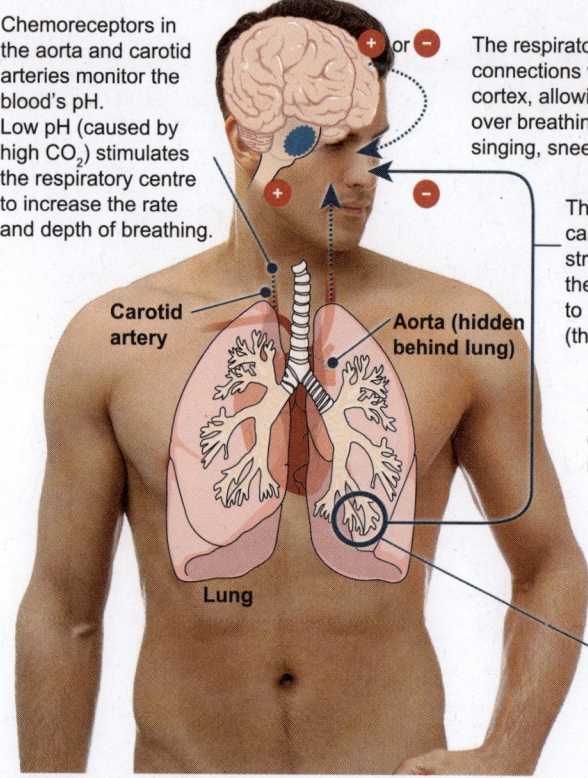

Chemoreceptors in the aorta and carotid arteries monitor the blood's pH.
Low pH (caused by high CO_2) stimulates the respiratory centre to increase the rate and depth of breathing.

The respiratory centre has connections with the cerebral cortex, allowing voluntary control over breathing e.g. when talking, singing, sneezing, and coughing.

The vagus nerve carries impulses from stretch receptors to the respiratory centre to inhibit inspiration (the inflation reflex).

Intercostal nerves from the respiratory centre stimulate inspiration.

Carotid artery

Aorta (hidden behind lung)

Lung

Stretch receptors in the bronchioles and bronchi monitor the amount of lung inflation.

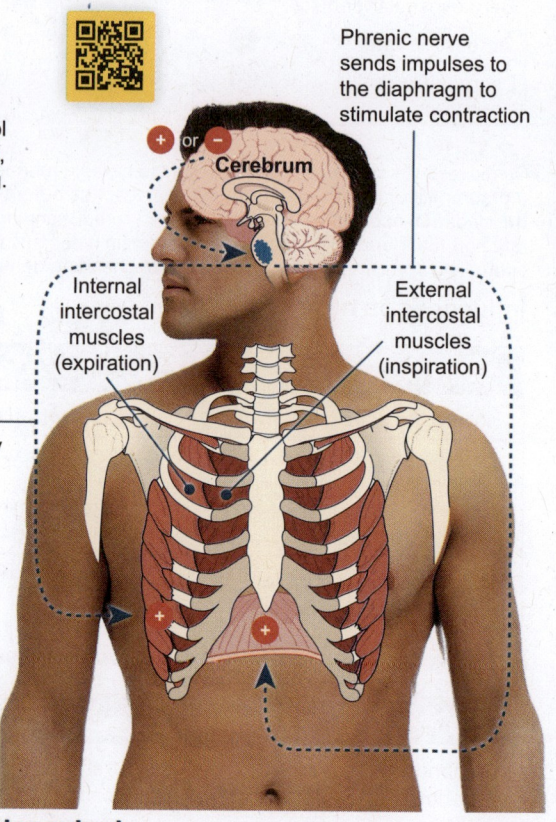

Phrenic nerve sends impulses to the diaphragm to stimulate contraction

Cerebrum

Internal intercostal muscles (expiration)

External intercostal muscles (inspiration)

Sensory input

Stretch receptors in the bronchioles monitor lung inflation and send impulses to inhibit the respiratory centre. Input from sensory receptors and the higher brain centres influence the basic rhythm.

Motor output

The respiratory centre sends rhythmic impulses to the intercostal muscles and the diaphragm to bring about controlled ventilation.

1. Explain how the basic rhythm of ventilation is controlled: _____

_____

2. Describe the role of each of the following in the regulation of ventilation:

   (a) Phrenic nerve: _____

   (b) Intercostal nerves: _____

   (c) Vagus nerve: _____

   _____

   (d) Inflation reflex: _____

3. (a) Describe the effect of low blood pH on the rate and depth of breathing: _____

   _____

   (b) Explain how this effect is mediated: _____

   _____

   (c) Suggest why blood pH is a good mechanism by which to regulate ventilation rate: _____

   _____

C3.1
15

101

©2024 **BIOZONE** International
ISBN: 978-1-99-101410-8
Photocopying prohibited

206 Control of Peristalsis

Key Idea: Solid food is ingested and chewed into a small mass called a bolus. It is then moved through the gut by waves of muscular contraction called peristalsis.

Ingested food is chewed and mixed with saliva to form a small mass called a bolus. Wave-like muscular contractions called peristalsis move the food, first as a bolus and then as semi-fluid chyme, through the digestive tract. Control from the central nervous system (CNS) allows for voluntary control of ingestion through the mouth and egestion through the anus. Between these points in the digestive system, peristalsis is controlled by the autonomous enteric nervous system (ENS), allowing for coordinated (and involuntary) peristalsis movement. The smooth muscle contractions need to occur in a particular sequence so food can move forward.

Peristalsis

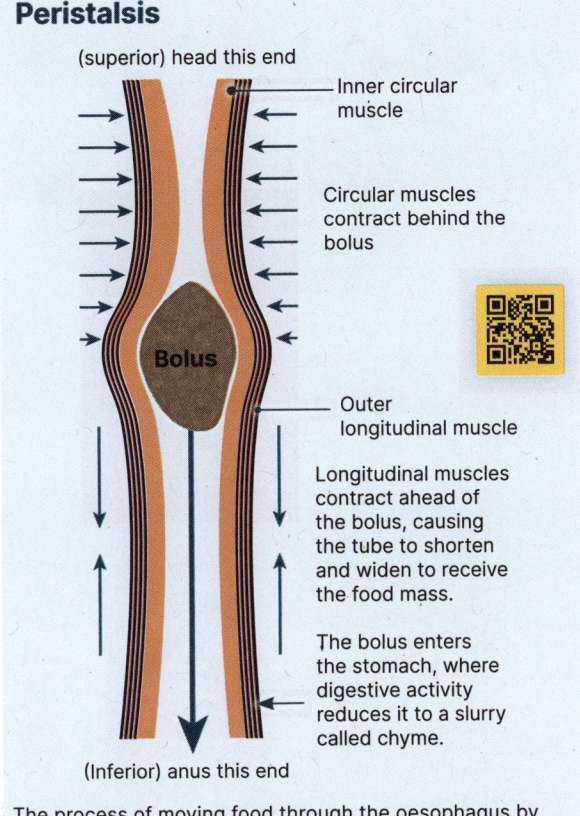

(superior) head this end

- Inner circular muscle
- Circular muscles contract behind the bolus

Bolus

- Outer longitudinal muscle

Longitudinal muscles contract ahead of the bolus, causing the tube to shorten and widen to receive the food mass.

The bolus enters the stomach, where digestive activity reduces it to a slurry called chyme.

(Inferior) anus this end

The process of moving food through the oesophagus by waves of muscular contractions is called peristalsis. Motility requires a coordinated contraction of smooth muscle tissue.

Enteric nervous system

- Mesentary (membrane that holds the intestine in place)
- Enteric neuron
- Myenteric plexus
- Submucosal plexus
- Outer longitudinal muscle
- Inner circular muscle
- Submucosa

Nervous system control of peristalsis

The enteric nervous system (ENS) is a semi-independent section of the autonomic nervous system. The associated neurons control motor functions of peristalsis, but also secretions, localized blood flow, and endocrine functions.

The enteric plexus supplies motor neurons to control gut motility to both the outer longitudinal layer and inner circular layer of intestinal muscles in the form of a mesh of neurons (myenteric plexus) that extend between them. Parasympathetic influence increases peristalsis, while sympathetic influence decreases peristalsis.

The submucosal plexus contacts the submucosa layer and is involved in sensory, absorption, and secretion functions, it also has a role in localized muscle movements, by innervating (stimulating) the smooth muscle and coordinating peristalsis motility.

1. Compare and contrast the nervous control between ingestion/egestion and peristalsis in the digestive system:

2. Why is it necessary that peristalsis is under involuntary autonomic control by the ENS?

3. What is the advantage of swallowing food/ingestion and egestion being under voluntary nervous control?

207 Investigating Phototropism

Key Idea: Tropisms are directional growth responses to external stimuli.

The stimulus direction determines the direction of the growth response. Tropisms are identified according to the stimulus involved, e.g. photo- (light), gravi- (gravity), hydro- (water), and are identified as positive (towards the stimulus) or negative (away from the stimulus). Tropisms act to position the plant in the most favourable growth environment. Phototropism in plants was linked to a growth promoting substance as early as the 1920s. Early experiments investigating phototropism in severed coleoptiles provided evidence for the hypothesis that the plant hormone auxin was responsible for tropic responses in stems. Auxins (a group of plant hormones) promote cell elongation and are inactivated by light. Thus, when a stem is exposed to directional light, auxin becomes unequally distributed on either side of the stem. The stem responds to the unequal auxin concentration by differential growth, i.e. it bends. Qualitative and quantitative data can be collected in phototropism investigations.

▶ Light is an important growth requirement for all plants. Most plants show an adaptive response of growing towards the light. This growth response is called phototropism.

▶ The bending of the plants shown on the right is a phototropism in response to light shining from the left and is caused by the plant hormone auxin. Auxin causes the elongation of cells on the shaded side of the stem, causing it to bend.

Investigation 11.1 Investigating phototropism response to light in seedlings

See appendix for equipment list.

1. Work in groups of four to set up four small petri dishes or pots of seeds (labelled A - D). Mustard or cress seeds are suitable as they are fast growing. Dishes should all be 3/4 filled with damp cotton wool. If using pots, fill with potting mix and dampen.

2. Add the same number of seeds to each - around 20 evenly spaced - and apply more water so all are damp.

3. Set up areas inside a cardboard box or cylinder where the only source of light will come from the 'windows' in the box placed next to a light source. A lamp may be used if there is no access to direct sunlight.

4. Seedlings A will have a slot cut into the box side so the light source will be angled at 90 degrees (light will be from the side).

5. Seedlings B will have a slot cut into the box side, but at the top, so the light source will be at a diagonal to the punnet (Measure angle from light source to punnet).

6. Seedlings C will have a slot cut into the box top so the light source will be directly above the punnet.

7. Seedlings D will be the control, and placed in a section of the box with no light.

8. The seedlings and the box then can be placed near the light source. Check every 2-3 days to ensure the growth medium (cotton wool or potting mix) is damp.

9. After 10-14 days, remove the seedlings. Qualitative data: draw an observational diagram of each of the four seedling dishes. Quantitative data: measure the angle of the stem bend, from the base to the stem top.

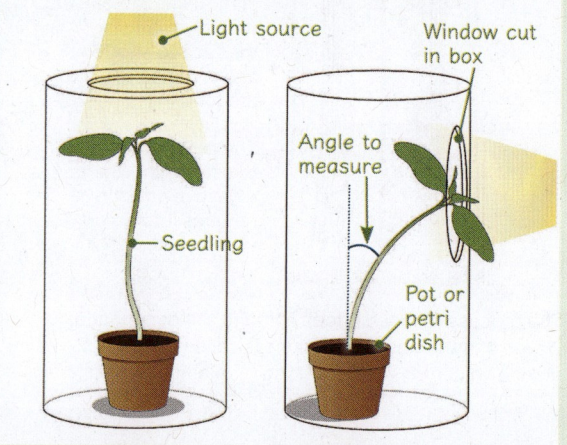

1. What was the purpose of the control in the investigation? _____

_____

2. Distinguish between quantitative and qualitative data, providing an example from this investigation:

_____

_____

3. Prepare a laboratory investigation report. Include: hypothesis, detailed method identifying variable and a description on how your group enabled the precision, accuracy, and reliability of results/measurements. Present your results, including qualitative diagrams and quantitative stem angle measurements, and a conclusion. Attach report to the page.

©2024 **BIOZONE** International
ISBN: 978-1-99-101410-8
Photocopying prohibited

208 Phytohormones as Signalling Chemicals

Key Idea: Phytohormones, such as auxin, gibberellin, and ABA, play important roles in plant responses to stimuli. Phytohormones are signalling chemicals that have a wide range of roles in the growth and developmental responses of vascular plants. Auxin (indole-acetic acid or IAA) is a naturally occurring phytohormone with a role in suppressing the growth of lateral buds. This inhibitory influence of a shoot tip or apical bud on the lateral buds is called apical dominance. Gibberellins are involved in stem and leaf elongation, as well as breaking dormancy in seeds. Specifically, they stimulate cell division and cell elongation. Abscisic acid or (ABA) was originally thought to be involved in accelerating abscission in leaves and fruit. However, it is now thought that ABA is a growth inhibitor with a variety of roles.

Plant responses and the role of auxins, gibberellins, and abscisic acid

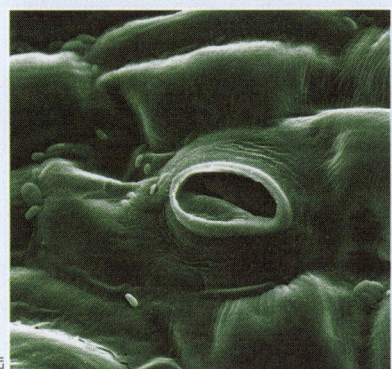

ABA stimulates the closing of stomata in most plant species. Its synthesis is stimulated by water deficiency (water stress). ABA also promotes seed dormancy. It is concentrated in senescent leaves but it is unlikely to be involved in leaf abscission except in a few species.

Gibberellins are responsible for breaking dormancy in seeds and promote the growth of the embryo and emergence of the seedling. Gibberellins are used to hasten seed germination and ensure germination uniformity in the production of barley malt in brewing.

Differential auxin transport, and therefore differential growth, is responsible for phototropism. Auxins also produce apical dominance in shoots. Auxin is produced in the shoot tip and diffuses down the stem to inhibit the development of the lateral buds.

ABA is produced in ripe fruit and induces fruit fall. The effects of ABA are generally opposite to those of cytokinins.

Gibberellins cause stem and leaf elongation by stimulating cell division and cell elongation. They are responsible for bolting in brassicas.

Auxin is required for fruit growth. The maturing seed releases auxin, inducing the surrounding flower parts to develop into fruit.

1. What is the role of gibberellins in stem elongation and in the germination of plants such as barley?

_____

_____

_____

2. What is the role of abscisic acid in closure of stomata? _____

_____

_____

3. What role does abscisic acid have, other than closing stomata? _____

_____

_____

C3.1 19 AHL

209 Auxin and Phototrophism

Key Idea: Concentration gradients of phytohormones in plant tissue lead to phototropic responses to light.

Auxins are plant hormones with a central role in a range of growth and developmental responses in plants. Auxins are responsible for apical dominance in shoots and are produced in the shoot tip. Indole-acetic acid (IAA) is the most potent native auxin in intact plants. The response of a plant tissue to IAA depends on the tissue itself, the hormone concentration, the timing of its release, and the presence of other hormones. Auxins can only diffuse, one way, between cells through cell membrane proteins called auxin efflux carriers, which are asymmetrically distributed. If there is coordinated movement of auxin, in response to a stimulus such as light and gravity, the gradients in auxin concentration during growth prompt differential responses in specific tissues and contribute to directional growth.

Auxin efflux carriers

▸ Auxin efflux (flowing outwards) carriers are cell membrane channel proteins that can be localized in specific regions of a cell. Auxins are able to diffuse freely one way into plant cells through these efflux carriers.

▸ Under dark conditions, auxin moves evenly down the stem. It is transported from cell to cell by diffusion and via auxin efflux carriers (right).

▸ Outside the cell, auxin is a non-ionized molecule (AH) which can diffuse into the cell. Inside the cell, the pH of the cytoplasm causes auxin to ionize, becoming A− and H+. Auxin efflux carriers at the basal end of the cell then transport A− out of the cell where it regains an H+ ion and reforms AH. In this way auxin is transported in one direction through the plant.

▸ When plant cells are illuminated by light from one direction, auxin efflux carriers in the plasma membrane on the shaded side of the cell are activated and auxin is transported to the shaded side of the plant.

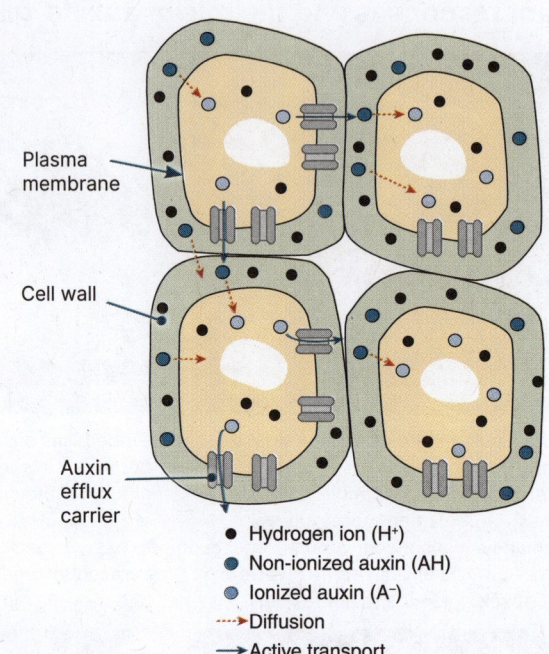

Plasma membrane

Cell wall

Auxin efflux carrier

- ● Hydrogen ion (H⁺)
- ● Non-ionized auxin (AH)
- ● Ionized auxin (A⁻)
- ---▸ Diffusion
- ⟶ Active transport

Maintaining concentration gradients of phytohormones

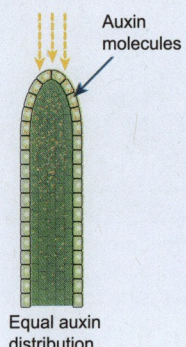

Auxin molecules

Equal auxin distribution

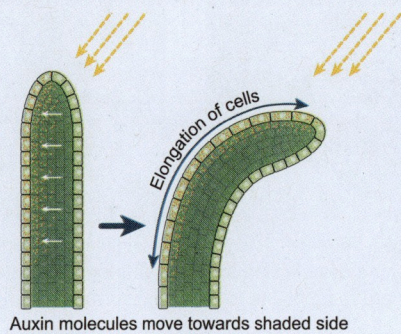

Elongation of cells

Auxin molecules move towards shaded side of the plant and stimulate cell elongation

▸ Coordinated movement of auxins in cells due to a stimulus such as light, enables a concentration gradient to form in plant tissue.

▸ Phototropism responses increase the concentration of auxin on the shaded part of the stem. Cells with increased concentrations of auxins will increase in size, resulting in a bending of the stem, and therefore leaves, towards the light source.

▸ The advantage of this response is that the leaves are able to utilize the most direct rays of light to increase the rate of photosynthesis.

1. Define phototropism: _____

_____

2. Explain how phytohormone concentration gradients are maintained, using auxin efflux carriers as an example:

_____

_____

_____

3. Describe the effect of auxin on cell growth: _____

_____

_____

AHL C3.1 20 - 21

©2024 **BIOZONE** International
ISBN: 978-1-99-101410-8
Photocopying prohibited

Promotion of cell growth by auxin.

▶ The presence of an auxin concentration in a plant cell increases the movement of hydrogen ions from the cytoplasm into the apoplast. The apoplast is the extracellular space formed outside the plasma membrane and between the cell wall, and is filled with water and gas.

▶ The hydrogen ions lower the pH and activate expansins, which are non-enzymatic cell wall proteins that facilitate cell elongation in processes such as fruit development, pollen tube formation during fertilization, and root hair growth.

▶ The activated expansins function by releasing some of the cross links that form between cellulose molecules, allowing the cell to elongate.

▶ Once the cell wall elongates, the centre fills with additional water and cytoplasm to increase the overall size of the plant cell.

Mechanism of auxin action in plant shoots

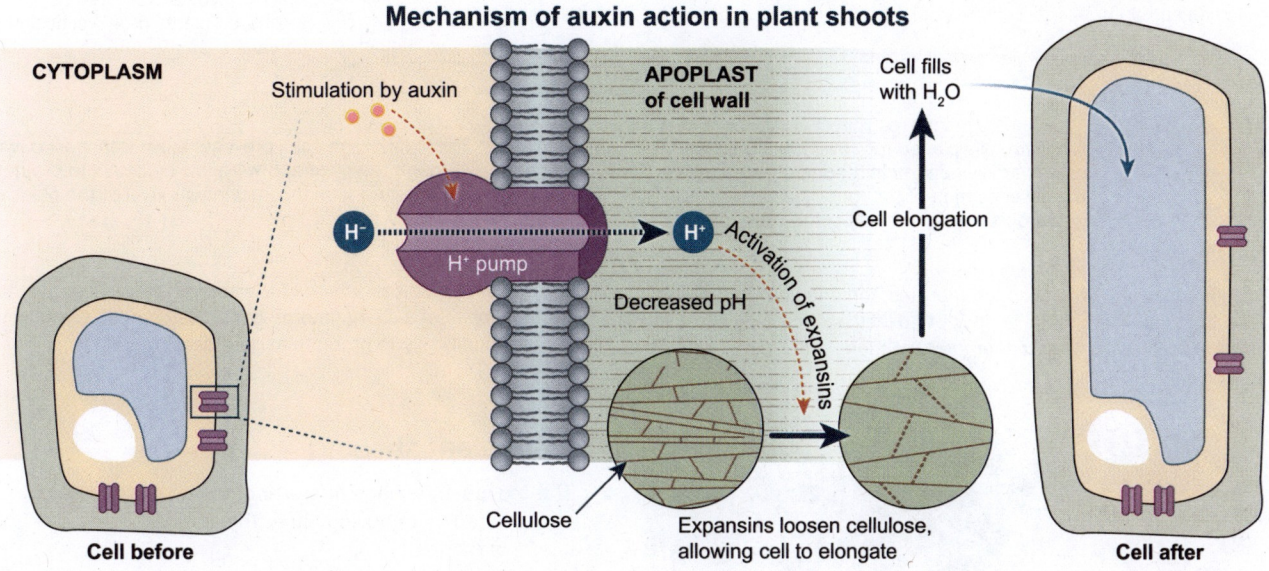

4. Directional light: A pot plant is exposed to direct sunlight near a window and as it grows, the shoot tip turns in the direction of the Sun. When the plant was rotated, it adjusted by growing towards the Sun in the new direction.

   (a) How do the cells behave to bring about this change in shoot direction at:

   Point A? _____

   Point B? _____

   (b) Which side (A or B) would have the highest hormone concentration, and why?

   _____

   _____

   _____

   _____

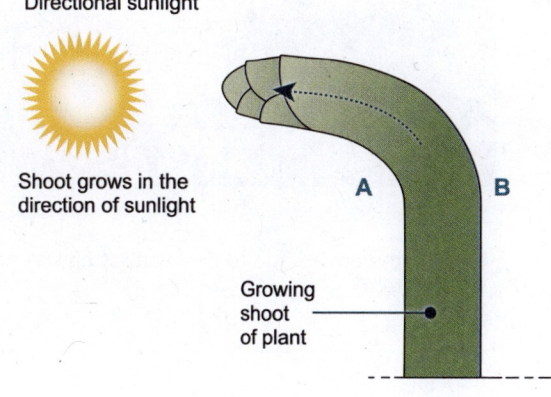

5. Light excluded from shoot tip: When a tin-foil cap is placed over the top of the shoot tip, light is prevented from reaching the shoot tip. When growing under these conditions, the direction of growth does not change towards the light source, but grows straight up. State what conclusion you can come to about the source and activity of the hormone that controls the growth response:

   _____

   _____

   _____

   _____

   _____

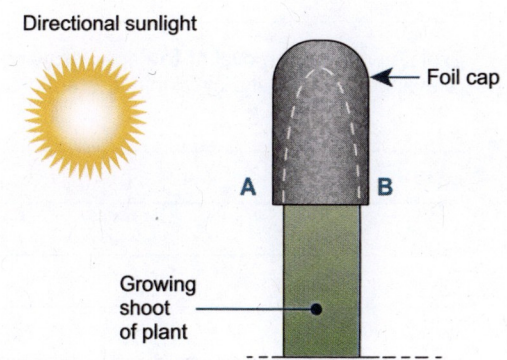

210 Interaction of Plant Hormones

Key Idea: Auxin and cytokinin interact together as growth regulators in plant shoots and roots.

The importance of auxin as a plant growth regulator, as well as its widespread occurrence in plants, led to it being proposed as the primary regulator in the gravitropic response. Auxin is synthesized in undifferentiated meristem cells in the stem tip. Interacting with auxin is another hormone, cytokinin, produced in root tips. The two hormones move in opposite directions through the plant, and interact with each other to regulate shoot and root growth.

Interaction of auxin and cytokinin at root tips and stem

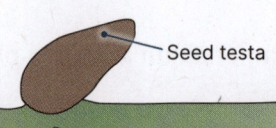

Cytokinin acts as an antagonist to auxin, and prevents bending of the root tip, so lateral roots can spread horizontally outwards.

Seed testa

Apex growth is promoted by auxins, so plants grow vertically upward.

A horizontally placed root (radicle) tip grows downwards when **auxin** is concentrated on the lower section of the root. This is positive gravitropism which occurs in the main root.

Removal of the apex prevents auxin transport down through the plant and will allow cytokinin to promote axillary bud growth, ensuring survival of the plant.

Root tip

Auxin induces cell division in the root tips, forming undifferentiated cells. Cytokinin inhibits the actions of auxin to switch the cells into a differentiated state, adapted for their function.

Shoot tip

Auxins promote apical growth of the shoot (the vertical growth) at the expense of axillary (lateral) bud development. Cytokinins act antagonistically to promote axillary development.

Tomato plant shaping

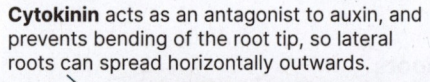

▶ The growth and shape of growing tomato plants can be manipulated by removing either the apex or lateral branches.

▶ By removing the apex, the cytokinins will induce the axillary buds to move out of dormancy and form side shoots, making the tomato plant more 'bushy' and wider.

▶ By removing the side branches, and associated axillary buds, the auxins will induce the tomato plant to continue growing vertically, making a tall, thin plant.

▶ Many horticulturists initially remove the tomato axillary buds so they extend vertically up support ropes in greenhouses. Once the plants are tall enough, the apical tip is removed to encourage lateral shoots to grow, and produce leaves and fruit.

1. What is the advantage to the plant of having both cytokinin and auxin interacting at root tips to develop a functional root system?

_____

_____

_____

2. What is the advantage to the formation of a plant by having cytokinin and auxin interacting at shoot tips?

_____

_____

3. Explain why the removal of the apex growing tip encourages a more 'bushy' plant. Include, in your answer, the relevance of auxin and cytokinin interactions as antagonist hormones:

_____

_____

_____

_____

_____

©2024 **BIOZONE** International
ISBN: 978-1-99-101410-8
Photocopying prohibited

AHL

C3.1

22

211 Positive Feedback Loops and Fruit Ripening

Key Idea: Fruit ripening is controlled by the production of ethylene and its release acts as a positive feedback loop to ensure a synchronous response from nearby fruit.

Fruits are an adaptation of plants to entice a mutualistic relationship with species to disperse seeds away from the plant. Ripened fruits often have a colour that makes them more visible to the species that eat them, and indicate higher sugar levels. Synchronized ripening will increase the chance that the fruit will be 'discovered' and identified by the seed dispersing species (typically birds and mammals), and therefore that all seeds will have an opportunity to be carried away from the parent plant. Ethylene is biosynthesized by plants in response to environmental cues or auxin levels. Higher levels of ethylene promote fruit ripening. The ripened fruit increases ethylene production which, in turn, induces ripening in neighbouring fruit, and so on. Fruit ripening is an example of a positive feedback loop, that ends when fruit are removed from the plant.

Positive feedback loop in fruit ripening

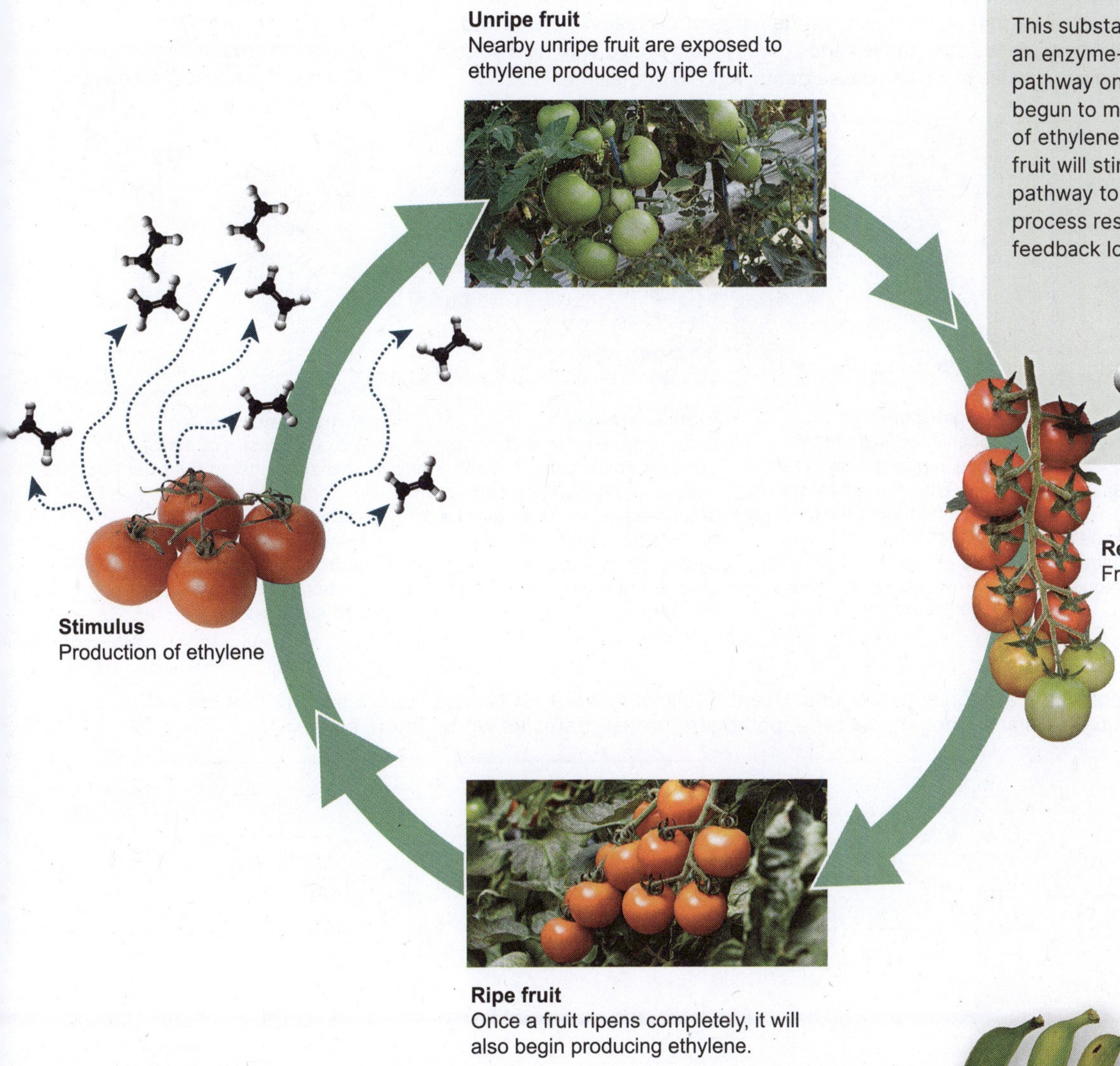

Unripe fruit
Nearby unripe fruit are exposed to ethylene produced by ripe fruit.

Ethylene (ethene)
This substance is produced by an enzyme-facilitated metabolic pathway once the fruit has begun to mature. The presence of ethylene in an unripened fruit will stimulate the metabolic pathway to produce more. This process results in a positive feedback loop for ripening.

Response
Fruit ripens

Stimulus
Production of ethylene

Ripe fruit
Once a fruit ripens completely, it will also begin producing ethylene.

1. Ethylene is used commercially to speed up fruit ripening or promote fruit fall. For example, bananas (right) are often picked unripe and then ripened with ethylene once at their destination. This prevents them from over-ripening during transport. Discuss how the commercial use of plant hormones can be used to the grower's advantage:

   _____

   _____

   _____

   _____

   _____

   _____

C3.1
23
AHL

212 Pathogens and Disease

Key Idea: Pathogens are infectious agents that spread between organisms and cause infectious disease.

Pathogens are disease-causing agents and cause infectious disease. An infectious disease is a disease which can be spread between individuals. Pathogens can be classified in a number of different ways. Those which can be seen with the naked eye are called macroorgansims, while those which are too small to be seen with the naked eye are microorganisms. Pathogens can also be categorized depending on whether they are living organisms (cellular pathogens) or non-living organisms (non-cellular pathogens). Some common groups of pathogens are described below.

Cellular pathogens

Cellular pathogens are living organisms. They possess all the cellular machinery they need to carry out their life process and reproduce. Bacteria and eukaryotic pathogens (fungi, protists, and parasitic worms) are cellular pathogens. However, no pathogenic archaea has ever been identified. Cellular pathogens cause a wide range of diseases in plants and animals. The severity of the diseases they cause varies greatly: some have very mild effects, e.g. athlete's foot or ringworm, while some can cause death, e.g. TB or malaria.

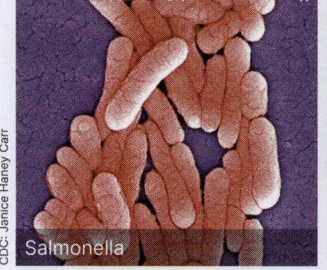
CDC: Janice Haney Carr
Salmonella

'Ringworm' caused by *Tinea*

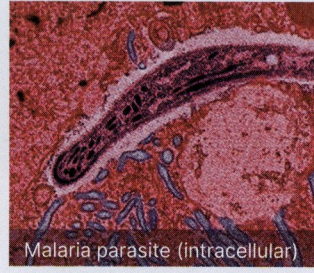
Malaria parasite (intracellular)

Bacterial pathogens
Pathogenic bacteria can be transmitted through food, water, air, or by direct contact. Bacteria have caused widespread, devastating diseases, but the discovery and use of antibiotics and aseptic (sterile) techniques has reduced deaths.

Fungal pathogens
Pathogenic fungi are more common in plants than in animals. They spread by spores and the infections they cause are generally chronic (long-lasting, low grade) infections because fungi grow relatively slowly. However, some can be fatal.

Protist pathogens
Protists are a large and diverse group of eukaryotes. A number of species are significant pathogens of animals or plants. Pathogenic protists have very complex life cycles, often involving a number of different hosts and several life stages.

Non-cellular pathogens

Non-cellular pathogens are non-living entities. Viruses and prions are both non-cellular pathogens. They do not have their own cellular machinery and must use a host's cellular machinery to reproduce.

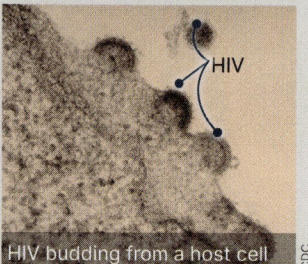
HIV
HIV budding from a host cell
CDC

Viruses and prions
Viruses consist of a protein coat surrounding their genetic material. They cause a wide variety of diseases.
Prions are infectious, abnormal proteins that can damage other proteins. They cause a range of serious diseases.

1. Use the template below to categorize pathogens. Use the following word list to help you: Cellular pathogen, virus, prokaryote, fungi, non-cellular pathogen, bacteria, pathogen, protist, parasitic worm, prion, eukaryote.

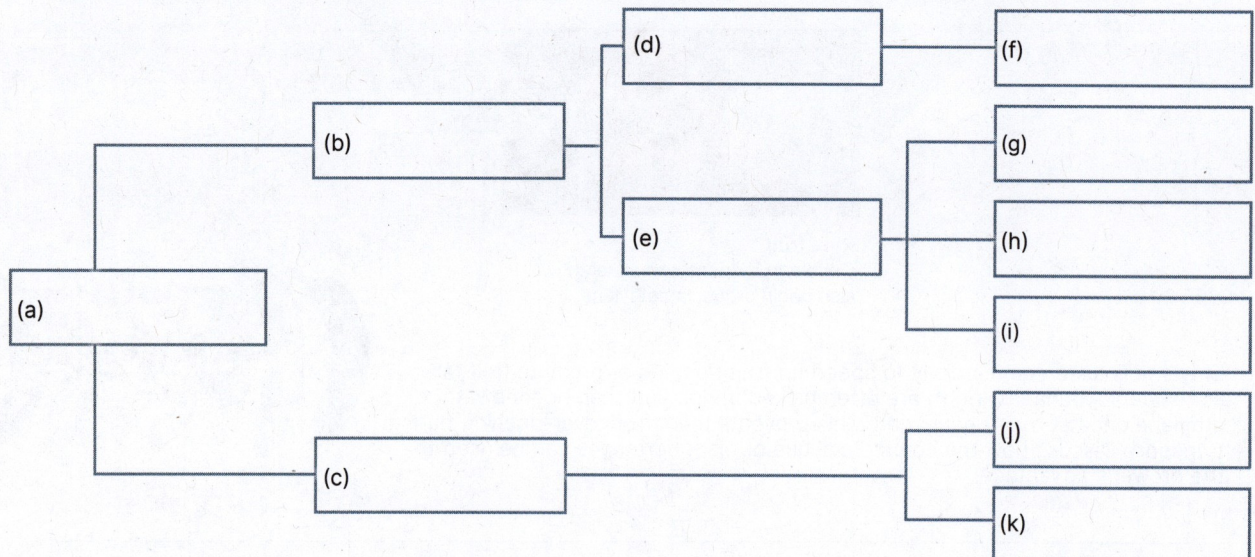

2. Explain the difference between cellular pathogens and non-cellular pathogens: _____

_____

_____

_____

  NOS C3.2
1

©2024 **BIOZONE** International
ISBN: 978-1-99-101410-8
Photocopying prohibited

The control of infectious disease

Infectious disease has always been a part of human civilization. It wasn't until people began to understand that disease could be transferred by microscopic, infectious pathogens that infectious diseases began to be truly controlled. However, before the current medical knowledge that humans have developed, scientists had to rely on careful observation of disease spread and the characteristics of the disease in order to stop the spread. Controlling a disease requires everyone affected to follow rules of hygiene and isolation, or removing the source of the disease. If those rules aren't followed, the disease spreads. In most cases, this means that diseases are controlled at a level where society operates without a disease causing any escalating effect, but the disease never goes away, e.g. flu. When those controls are not enough to suppress the disease, an epidemic, or pandemic can occur.

John Snow, The first epidemiologist

▶ In 1854, the area around Broad Street in London experienced an outbreak of cholera.

▶ The physician John Snow investigated the outbreak and he is often credited with ending it. Snow produced the map shown on the right. The black bars represent the number of cases of cholera in the area. The blue dots indicate the placement of water wells with pumps from which the population obtained water.

▶ He also studied the statistics of cholera deaths. Table 1 shows the number of cholera deaths for the areas of London and compares the number of cases to the elevation above the Thames (which was highly polluted at the time). Higher elevations sourced water upstream from London.

▶ By studying the data, Snow was able to locate the source of the cholera outbreak and convince authorities to take action to reduce its effects.

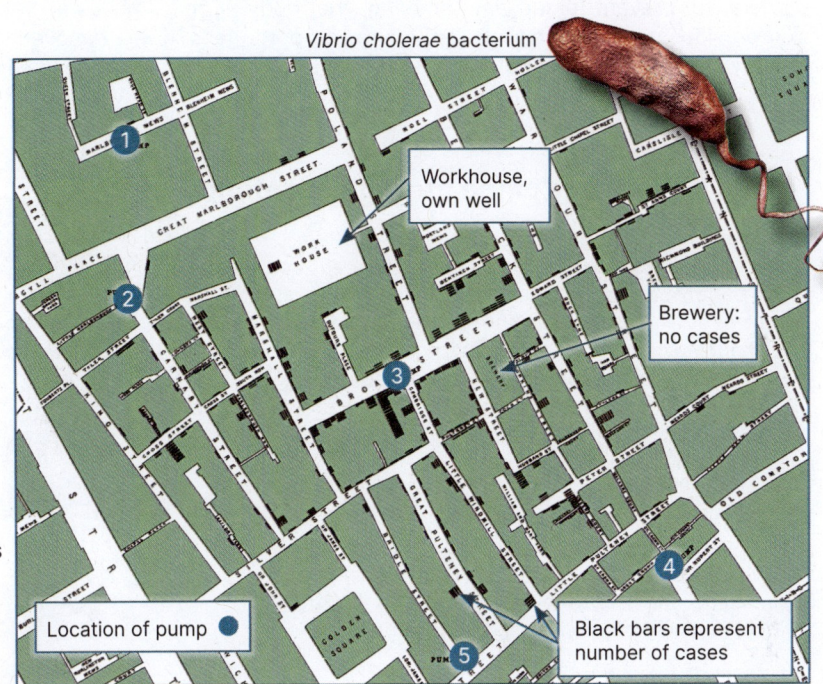

Vibrio cholerae bacterium

Workhouse, own well

Brewery: no cases

Location of pump ●

Black bars represent number of cases

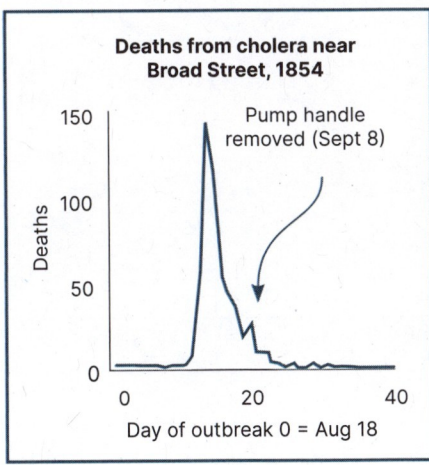

Deaths from cholera near Broad Street, 1854

Pump handle removed (Sept 8)

Deaths

Day of outbreak 0 = Aug 18

Table 1

Location in London	Elevation above Trinity high water mark (feet)	Population (1851)	Deaths from cholera	
			In five weeks ending Aug 12	In week ending Aug 19
West Districts	28	376,427	102	184
North Districts	135	490,396	62	38
Central Districts	49	393,256	58	32
East Districts	26	485, 522	168	105
South Districts	6	616,635	817	370
Total		2,362,236	1207	729

3. Use the data provided to locate the source of the cholera outbreak. write a statement arguing why it is the source, and what should be done to prevent further cases of cholera occurring:

_____

_____

_____

4. The pump handle from the offending well was removed (see graph above). Explain how this made a difference in ending the outbreak?

_____

_____

5. Carry out a small research project of a historical case of observations during the 19th-century enabling the control of a disease; for example, childbed fever in Vienna. Include details of the pathogen and disease, relevant data and images of the disease spread, and an explanation of how observations led to disease control.

213 The Skin as a First Line of Defence

Key Idea: In the body the first line of defence against infection and pathogens is the skin and mucous membrane layers.

The human body has a range of physical, chemical, and biological defences that provide resistance against pathogens. The primary defence consists of external barriers to prevent pathogens entering the body. The skin provides a physical barrier to the entry of pathogens. Healthy skin is rarely penetrated by microorganisms. Its low

pH is unfavourable to the growth of many bacteria and its chemical secretions, e.g. sebum, and antimicrobial peptides, inhibit growth of bacteria and fungi. Mucous membranes (muscosa) line all body cavities that are open to the exterior. They are a thin, epithelial layer that is moist and bathed in mucus secretions. Mucus traps foreign bodies so they can be removed from the body, and it provides a chemically unsuitable environment for pathogens to multiply.

The importance of the first line of defence

The skin is the largest organ of the body. It forms an important physical barrier against the entry of pathogens into the body. A natural population of harmless microbes lives on the skin, but most other microbes find the skin inhospitable. The continual shedding of old skin cells (arrow, right) physically removes bacteria from the surface of the skin. Sebaceous glands in the skin (top right) produce sebum, which has antimicrobial properties, and the slightly acidic secretions of sweat inhibit microbial growth.

Cilia line the epithelium of the respiratory tract (right) and regions of the urinary and reproductive tracts. Their wave-like movement sweeps out mucus containing trapped foreign material and keeps the passage free of microorganisms, preventing them from colonizing the body.

Antimicrobial chemicals are present in many bodily secretions. Tears, saliva, nasal secretions, and human breast milk all contain lysozymes and phospholipases. Lysozymes kill bacterial cells by catalyzing the hydrolysis of cell wall linkages, whereas phospholipases hydrolyze the phospholipids in bacterial cell membranes, causing bacterial death. Low pH gastric secretions also inhibit microbial growth, and reduce the number of pathogens establishing colonies in the gastrointestinal tract.

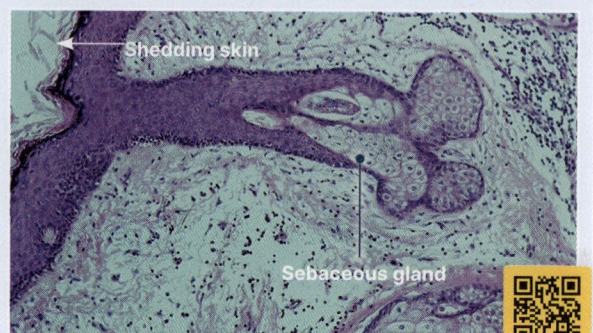

Shedding skin

Sebaceous gland

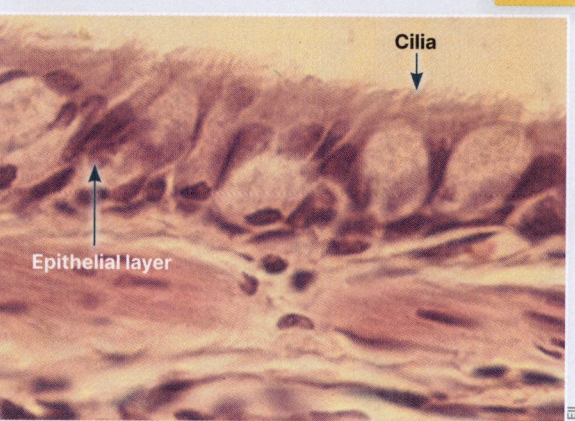

Cilia

Epithelial layer

1. (a) Why are the skin and mucous membranes called the first line of defence? _____

_____

_____

   (b) Why could these defences also be called non-specific? _____

_____

_____

2. How does the skin act as a barrier to prevent pathogens entering the body? _____

_____

_____

_____

3. Describe the role of cilia as a non-specific defence: _____

_____

_____

_____

C3.2
2-3

©2024 **BIOZONE** International
ISBN: 978-1-99-101410-8
Photocopying prohibited

Blood clotting as a form of infection defence

In addition to its transport role, blood also helps in the body's defence against infection and haemostasis (the prevention of bleeding and maintenance of blood volume). The tearing of a blood vessel initiates clotting. Clotting is normally a rapid process that seals off the tear, preventing blood loss and the invasion of bacteria into the site. Clot formation is triggered by the release of clotting factors from the damaged cells at the site of the tear or puncture. A hardened clot forms a scab, which acts to prevent further blood loss and acts as a mechanical barrier to the entry of pathogens.

Blood clotting

1 Injury to the lining of a blood vessels exposes collagen fibers to the blood. Platelets stick to the collagen fibers.

3 As the platelets clump together, more chemicals are released, accelerating the clot formation (positive feedback).

When tissue is wounded, the blood quickly coagulates to prevent further blood loss and maintain the integrity of the circulatory system. For external wounds, clotting also prevents the entry of pathogens. Blood clotting involves a cascade of reactions involving many clotting factors in the blood.

4 A fibrin clot reinforces the seal. The clot traps blood cells and the positive feedback loop ends. The clot eventually dries to form a scab.

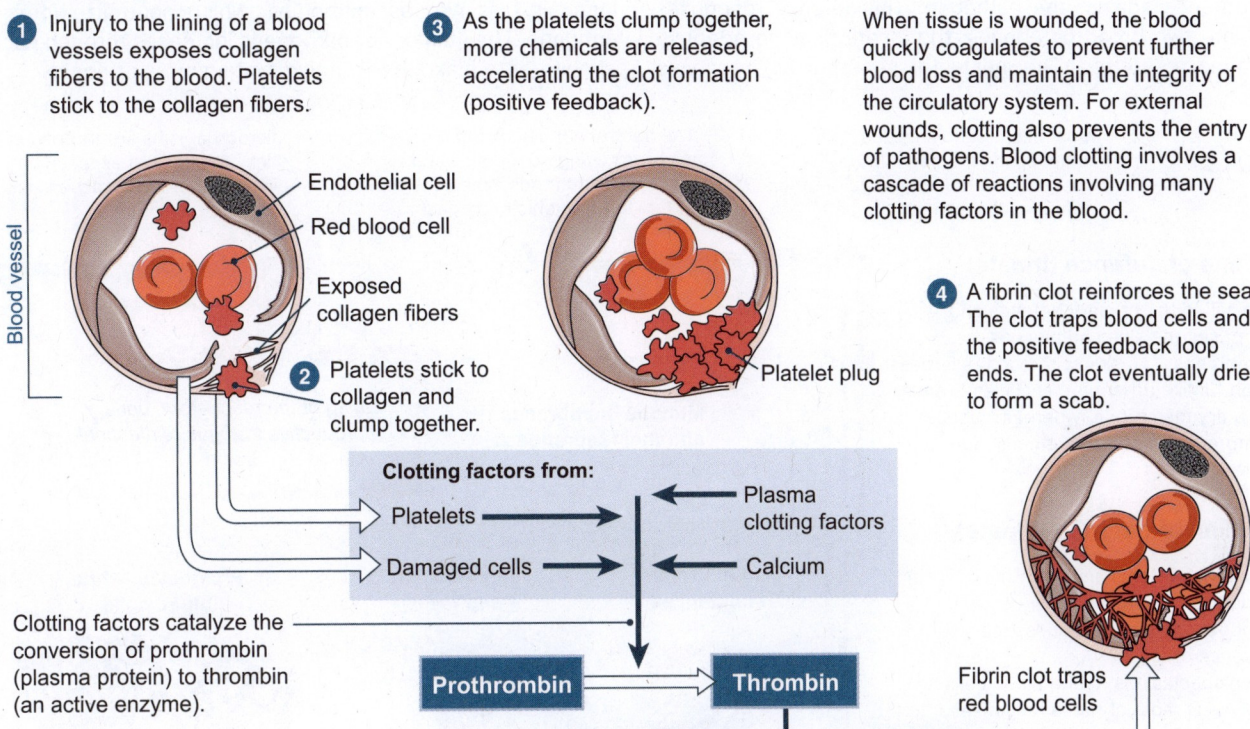

4. Explain two roles of the blood clotting system in internal defence and haemostasis:

   (a) _____

   (b) _____

5. Explain the role of each of the following in the sequence of events leading to a blood clot:

   (a) Injury: _____

   _____

   (b) Release of chemicals from platelets: _____

   _____

   (c) Clumping of platelets at the wound site: _____

   _____

   (d) Formation of a fibrin clot: _____

   _____

6. (a) Explain the role of clotting factors in the blood in formation of the clot: _____

   _____

   (b) Explain why these clotting factors are not normally present in the plasma: _____

   _____

214 Innate and Adaptive Immune Systems

Key Idea: The human body has innate and adaptive defences that, together, provide resistance against disease.

The human body has a range of physical, chemical, and biological defences that provide resistance against pathogens. The primary defence consists of external barriers to prevent pathogens entering the body. If this fails, a second line of defence targets any foreign bodies that enter. Lastly, the specific immune response provides a targeted third line of defence against the pathogen. The defence responses fall into two broad categories: the innate and the adaptive immune responses. The innate (or non-specific) response

makes up the first and second lines of defence. It protects against a broad range of non-specific pathogens. This response is unchanging and present in all animals throughout their lifetimes. It involves blood proteins (e.g. complement), inflammation, and phagocytic white blood cells. The adaptive (or specific) immune response is the third line of defence. It is specific to identified pathogens and is present only in vertebrates. It involves defence by specific T cells (cellular immunity) as well as antibodies, which neutralize foreign antigens. The memory of pathogens increases through each encounter to make the adaptive system more effective.

1st line of defence (Innate)

Also called the primary defence, and comprises the skin and mucous membranes. The skin is the body's largest organ. Unlike other epithelial membranes, it is a dry membrane, whereas mucous membranes (mucosa) are moist due to secretions.

2nd line of defence (Innate)

A range of defence mechanisms operate inside the body to inhibit or destroy pathogens. These responses react to the presence of any pathogen, regardless of which species it is. White blood cells are involved in most of these responses.

The 2nd line of defence includes the complement system, whereby blood plasma proteins work together to bind pathogens and induce inflammation to help fight infection.

3rd line of defence (Adaptive)

Once the pathogen has been identified by the immune system, lymphocytes (specialized white blood cells) launch a range of specific responses to the pathogen, including the production of defensive proteins called antibodies. Each type of antibody is produced by a B cell lymphocyte clone and is specific against a particular antigen.

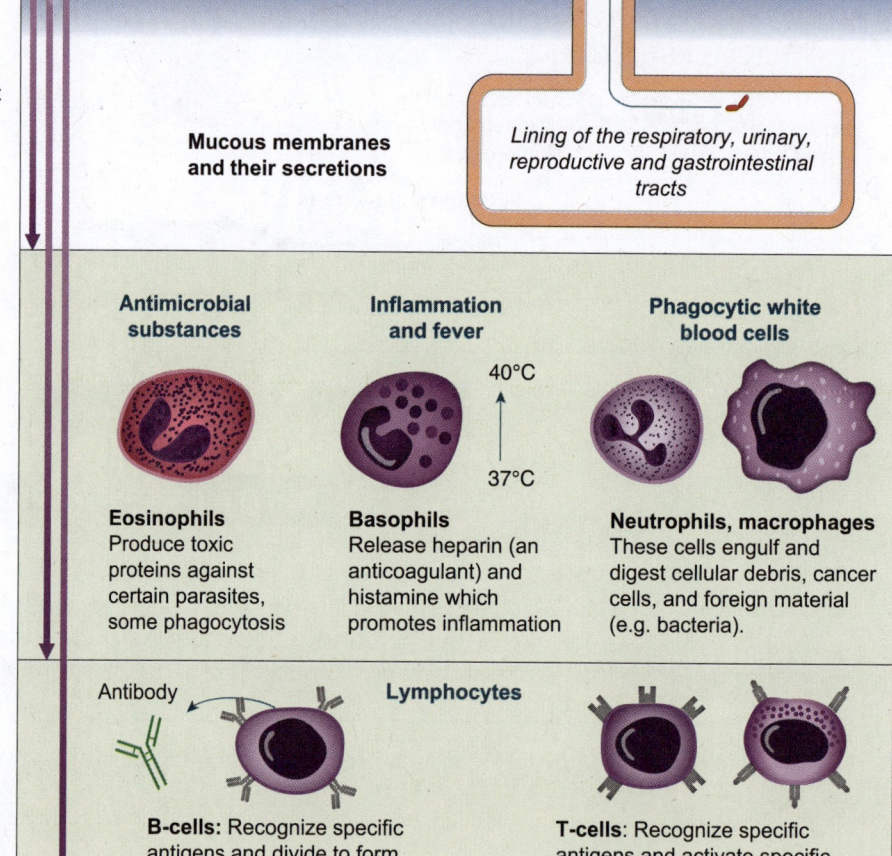

Most microorganisms find it difficult to get inside the body. If they succeed, they face a range of other defences that protect the body.

The natural populations of harmless microbes living on the skin and mucous membranes inhibit the growth of most pathogenic microbes.

Intact skin

Microorganisms are trapped in sticky mucus and expelled by cilia (tiny hairs that move in a wavelike fashion).

Mucous membranes and their secretions

Lining of the respiratory, urinary, reproductive and gastrointestinal tracts

Antimicrobial substances

Eosinophils
Produce toxic proteins against certain parasites, some phagocytosis

Inflammation and fever

40°C
↑
37°C

Basophils
Release heparin (an anticoagulant) and histamine which promotes inflammation

Phagocytic white blood cells

Neutrophils, macrophages
These cells engulf and digest cellular debris, cancer cells, and foreign material (e.g. bacteria).

Antibody

Lymphocytes

B-cells: Recognize specific antigens and divide to form antibody-producing clones.

T-cells: Recognize specific antigens and activate specific defensive cells.

---

1. (a) Define the terms innate and adaptive, as they relate to immunity: _____

   _____

   _____

   (b) Why is it important to have an innate defence system? _____

   _____

   _____

   (c) What is the advantage of adaptive immunity, as distinct from the innate system? _____

   _____

   _____

C3.2

4

©2024 **BIOZONE** International
ISBN: 978-1-99-101410-8
Photocopying prohibited

215 Phagocytes and Phagocytosis

Key Idea: Phagocytes are mobile, white blood cells that ingest and destroy extracellular foreign material and dead or dying cells to control infection by pathogens.

Phagocytosis is the process by which a cell engulfs another cell or particle. Cells that do this are called phagocytes. All types of phagocytes, e.g. neutrophils, dendritic cells, and macrophages, are white blood cells (leukocytes). These specialized cells have receptors on their surfaces that can detect antigenic material, such as microbes. The phagocytes move towards pathogens with independent, amoeboid movement. They then ingest the microbes by endocytosis at the cell membrane and digest them with lysosome produced enzymes, rendering them harmless. As well as destroying microbes, phagocytes also release substances called cytokines which help to coordinate the overall response to an infection. Macrophages and dendritic cells also play an important role in processing and presenting antigens from ingested microbes to other cells of the immune system.

Stages in phagocytosis and destruction of a pathogen

1 Detection and interaction

Microbe coated in chemical markers is detected by the phagocyte, which attaches to it. Chemical markers coating the foreign material, e.g. a bacterial cell, mark it as a target for phagocytosis.

2 Engulfment

The markers trigger engulfment of the microbe by the phagocyte. The microbe is taken in by endocytosis.

3 Phagosome forms

A phagosome forms, enclosing the microbe in a membrane.

4 Fusion with lysosome

Phagosome fuses with a lysosome containing digestive enzymes. The fusion forms a phagolysosome.

5 Digestion

The microbe is broken down into its chemical constituents.

6 Discharge

Indigestible material is discharged from the phagocyte.

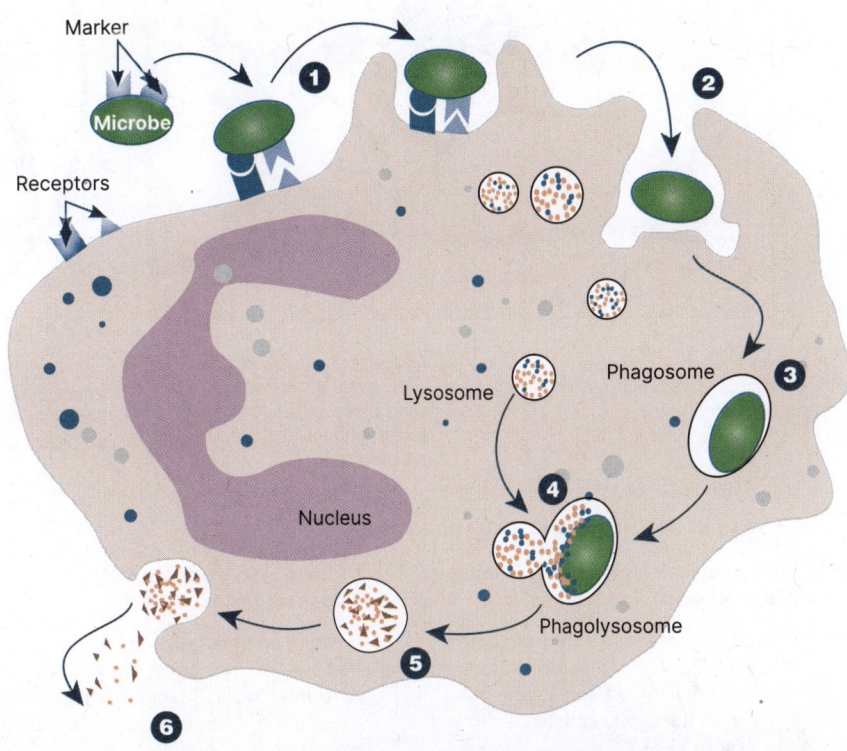

Amoeboid movement in phagocytes

In order to move independently, the phagocytes force their cytoplasm out to form a pseudopodium (false foot), which moves them forward in space and wraps around the pathogen to engulf them (right). Phagocytes can alter their shape to fit between the thin walls of the capillaries and out into interstitial fluid to trap pathogens. Neutrophils mostly remain inside blood vessels, while macrophages are only found at very low levels in the blood as most remain in body tissues, where they mature.

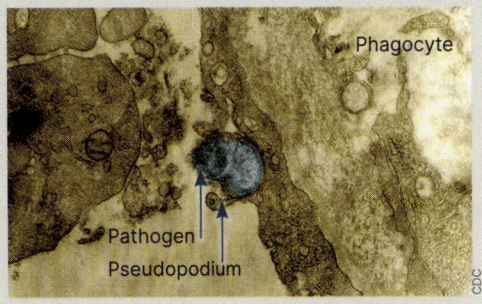

1. Explain the role of chemical markers and phagocyte receptors in enhancing phagocytosis: _____

_____

_____

2. What is the purpose of phagocytosis and how is it involved in internal defence? _____

_____

_____

_____

_____

The inflammatory response

Inflammation is a defensive response to damage. The inflammation process involves pain, redness, heat, and swelling. Damage to the body's tissues can be caused by physical agents, e.g. sharp objects, heat, radiant energy, or electricity; microbial infection, or chemical agents, e.g. gases, acids and bases. The damage triggers a defensive response called inflammation. The inflammatory response is beneficial and the process of inflammation can be divided into three distinct stages. These are described below.

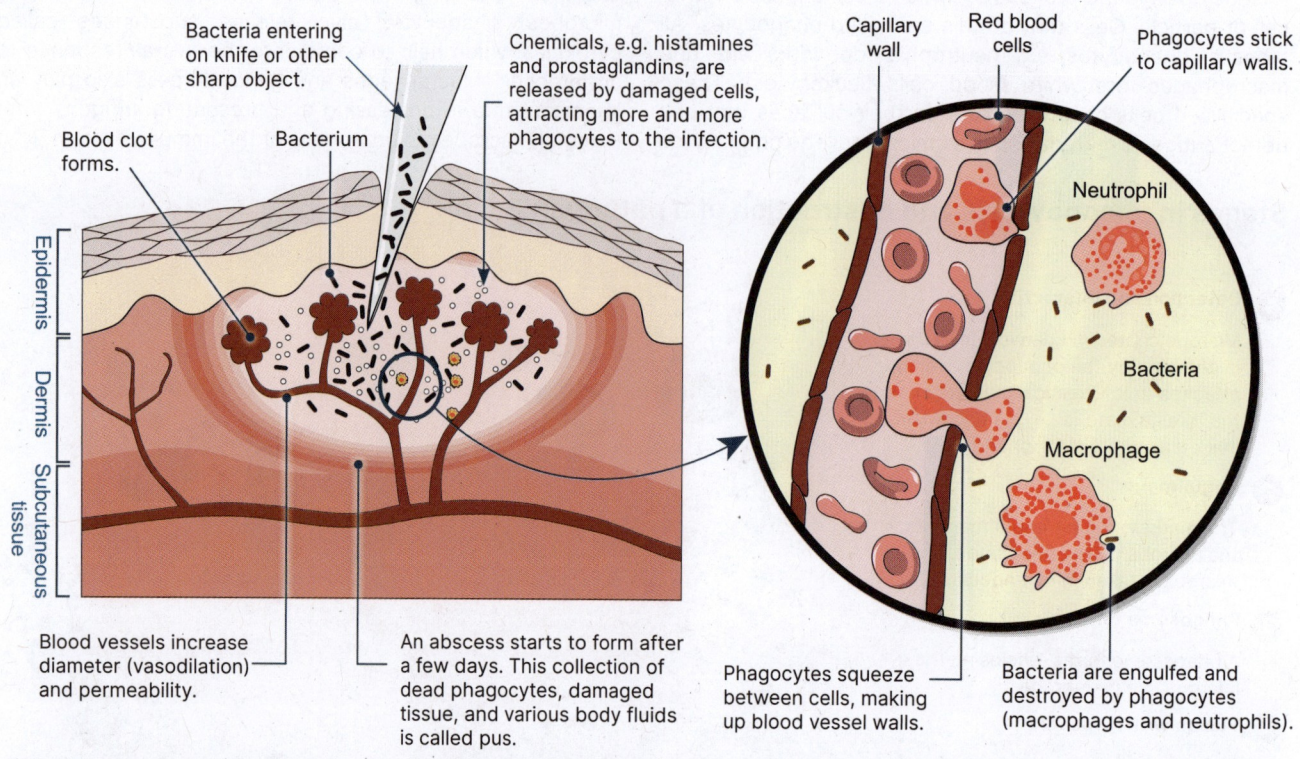

Bacteria entering on knife or other sharp object.

Chemicals, e.g. histamines and prostaglandins, are released by damaged cells, attracting more and more phagocytes to the infection.

Blood clot forms.

Bacterium

Capillary wall

Red blood cells

Phagocytes stick to capillary walls.

Neutrophil

Epidermis
Dermis
Subcutaneous tissue

Bacteria

Macrophage

Blood vessels increase diameter (vasodilation) and permeability.

An abscess starts to form after a few days. This collection of dead phagocytes, damaged tissue, and various body fluids is called pus.

Phagocytes squeeze between cells, making up blood vessel walls.

Bacteria are engulfed and destroyed by phagocytes (macrophages and neutrophils).

Stages in the inflammatory response

Increased diameter and permeability of blood vessels

Blood vessels increase their diameter and permeability in the area of damage. This increases blood flow to the area and allows defensive substances to leak into tissue spaces.

Phagocyte migration and phagocytosis

Within one hour of injury, phagocytes appear on the scene. They squeeze between cells of blood vessel walls to reach the damaged area, where they destroy invading microbes.

Tissue repair

Functioning cells or supporting connective cells create new tissue to replace dead or damaged cells. Some tissue regenerates easily (skin) while others do not at all (cardiac muscle).

3. What is amoeboid movement in phagocytes and how does it assist with the inflammatory response? _____
_____
_____
_____

4. Outline the three stages of inflammation and identify the beneficial role of each stage:

(a) _____
_____

(b) _____
_____

(c) _____
_____

5. Why does pus form at the site of infection? _____
_____

©2024 **BIOZONE** International
ISBN: 978-1-99-101410-8
Photocopying prohibited

216 Antibodies

Key Idea: Antibodies are large, Y-shaped proteins made by B-lymphocyte cells which destroy specific antigens.

Antibodies and antigens play key roles in the response of the immune system. Antigens are foreign molecules that promote a specific immune response. Antibodies (or immunoglobulins) are proteins made in response to antigens. They are secreted from B-lymphocyte cells into the plasma where they can recognise, bind to, and help destroy antigens. They are also found in the lymph nodes. There are five classes of antibodies and each plays a different role in the immune response. Each type of antibody is specific to only one particular antigen. One specific B-lymphocyte makes just one specific antibody; therefore, there are many B-lymphocytes circulating in the blood.

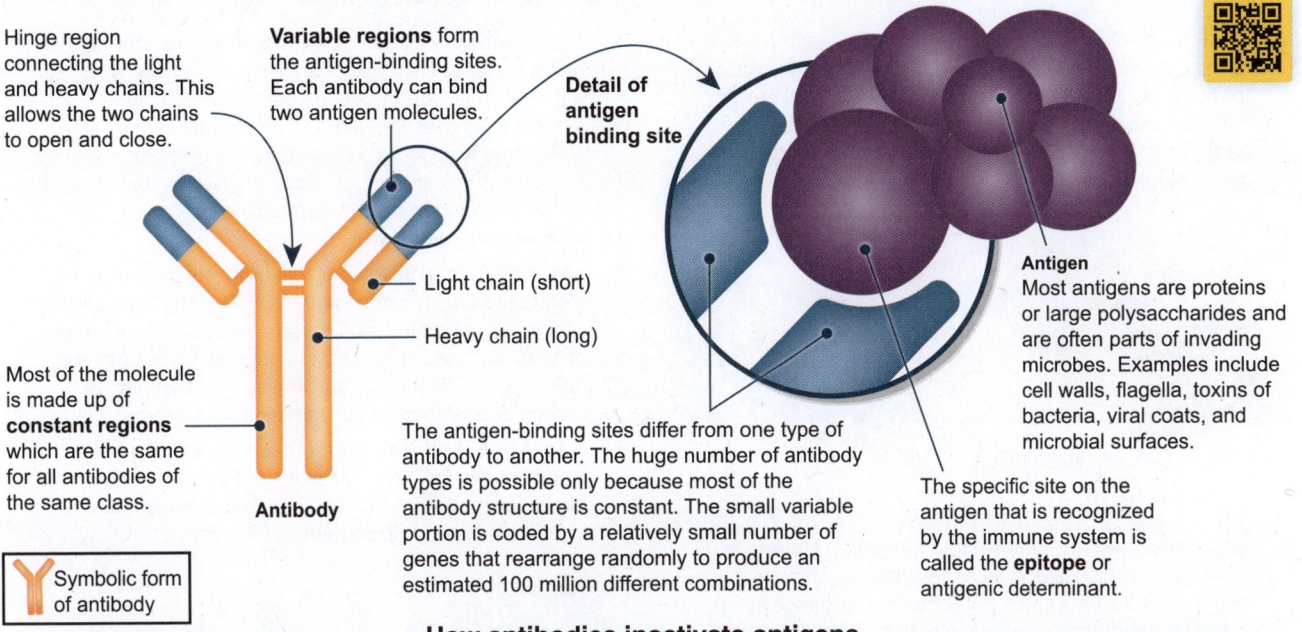

Hinge region connecting the light and heavy chains. This allows the two chains to open and close.

Variable regions form the antigen-binding sites. Each antibody can bind two antigen molecules.

Detail of antigen binding site

Light chain (short)

Heavy chain (long)

Most of the molecule is made up of **constant regions** which are the same for all antibodies of the same class.

Antibody

Symbolic form of antibody

The antigen-binding sites differ from one type of antibody to another. The huge number of antibody types is possible only because most of the antibody structure is constant. The small variable portion is coded by a relatively small number of genes that rearrange randomly to produce an estimated 100 million different combinations.

Antigen
Most antigens are proteins or large polysaccharides and are often parts of invading microbes. Examples include cell walls, flagella, toxins of bacteria, viral coats, and microbial surfaces.

The specific site on the antigen that is recognized by the immune system is called the **epitope** or antigenic determinant.

How antibodies inactivate antigens

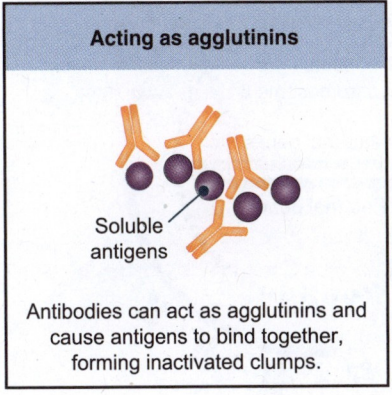

Acting as agglutinins

Soluble antigens

Antibodies can act as agglutinins and cause antigens to bind together, forming inactivated clumps.

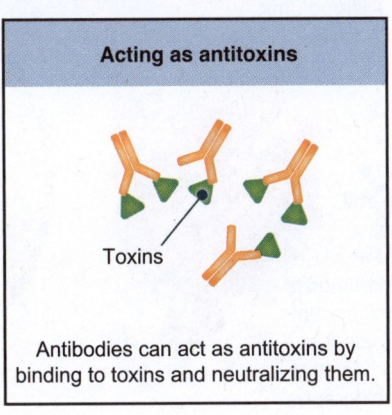

Acting as antitoxins

Toxins

Antibodies can act as antitoxins by binding to toxins and neutralizing them.

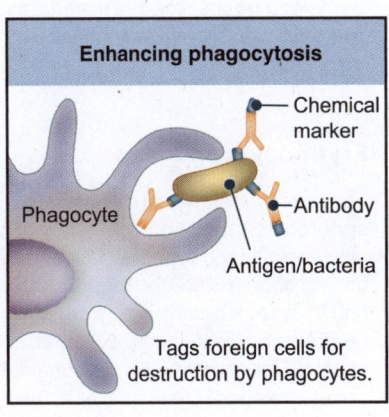

Enhancing phagocytosis

Chemical marker

Phagocyte

Antibody

Antigen/bacteria

Tags foreign cells for destruction by phagocytes.

1. Explain the relationship between B-lymphocytes and antibodies with reference to their proliferation: _____

_____

_____

_____

2. Explain how the following actions by antibodies enhance the immune system's ability to stop infections:

(a) Acting as agglutinins: _____

_____

(b) Acting as antitoxins: _____

_____

(c) Tagging foreign cells with chemical markers: _____

_____

217 Antigens

Key Idea: Antigens are substances capable of producing an immune response leading to antibody production.

An antigen is any substance that causes an immune response in an organism. Most antigens are foreign material and originate from outside the organism, e.g. a microbe, or other disease causing pathogen. Many antigens are glycoproteins embedded in a pathogen membrane or outer wall. They, or their toxins, generate an immune response in the host, which destroys them. However, an antigen can also originate inside the body, with each cell type having its own unique antigen. The immune system can recognize the difference between non-self and self to prevent damage to their own bodies.

Non-self antigens

Any foreign material provoking an immune response is termed a non-self antigen. Disease-causing organisms (pathogens) such as bacteria, viruses, and fungi are non-self antigens. The body recognizes them as foreign and will attack and destroy them before they cause harm.

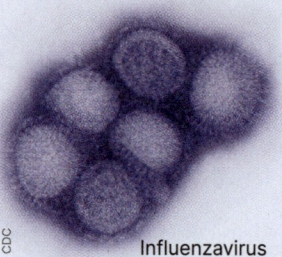
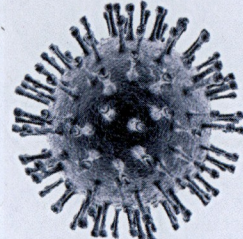

CDC

Influenzavirus   HIV virus

Pathogens have ways of avoiding detection. Mutations result in new surface antigens, delaying the immune response and allowing the pathogen to reproduce in its host undetected for a time, e.g. the flu virus, above. Some pathogens, e.g. the malaria-causing *Plasmodium*, switches off its surface antigens in order to enter cells undetected.

Erythrocyte antigens

The type of antigens present on the surface of a erythrocyte (red blood cell - RBC) determines an individual's blood type. ABO blood group antigens and Rh antigens are the most important in the blood typing system because they are strongly immunogenic, i.e. cause a strong immune response. Blood must be checked for compatibility before a patient can receive donated blood. Transfusion of incompatible blood may cause a fatal transfusion reaction in which RBCs from the donated blood clump together (agglutinate), block capillaries, and rupture (haemolysis). This is because an incompatible antigen induces the formation of antibodies.

Distinguishing self from non-self

▶ Every type of cell has unique protein markers (antigens) on its surface. The type of antigen varies greatly between cells and between species. The immune system uses these markers to identify its own cells (self) from foreign cells (non-self). If the immune system recognizes the antigen markers, it will not attack the cell. If the antigen markers are unknown, the cell is attacked and destroyed.

▶ The system responsible for this property is the major histocompatibility complex (MHC). The MHC is a cluster of tightly linked genes on chromosome 6. The main role of MHC antigens is to bind to antigenic fragments and display them on the cell surface so that they can be recognized by the cells of the immune system.

▶ Class I MHC antigens are found on the surfaces of almost all human cells. Class II MHC antigens occur only on macrophages and B-lymphocytes of the immune system, like the stylized neutrophil below.

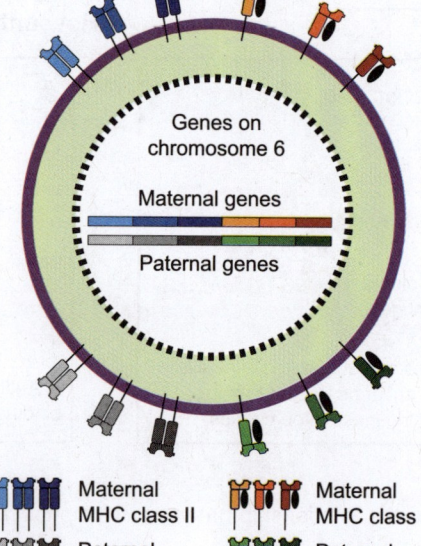

Genes on chromosome 6

Maternal genes

Paternal genes

Maternal MHC class II		Maternal MHC class I
Paternal MHC class II		Paternal MHC class I

1. What is an antigen? _____
_____

2. Distinguish between self and non-self antigens: _____
_____
_____

3. Why is it important that the body detects foreign antigens? _____
_____
_____

©2024 **BIOZONE** International
ISBN: 978-1-99-101410-8
Photocopying prohibited

218 Lymphocytes and Adaptive Immune Response

Key Idea: Antigens activate the immune system's B- and T-lymphocytes against specific pathogens when processed by antigen-presenting cells.

There are two main components of the adaptive immune system: the humoral, and the cell-mediated responses. They work separately and together to protect against disease. The humoral immune response is associated with the serum (the non-cellular part of the blood) and involves the action of antibodies secreted by B-lymphocytes (B cells).

Antibodies are found in extracellular fluids including lymph, plasma, and mucus secretions and protect against viruses, and bacteria and their toxins. The cell-mediated immune response is associated with the production of specialized T-lymphocytes. Antigens are recognised by T-lymphocytes only after antigen processing. The antigen is first engulfed by an antigen-presenting cell, which processes the antigen and presents it on its surface. T helper cells can then recognize the antigen and activate other cells of the immune system.

Lymphocyles and their functions

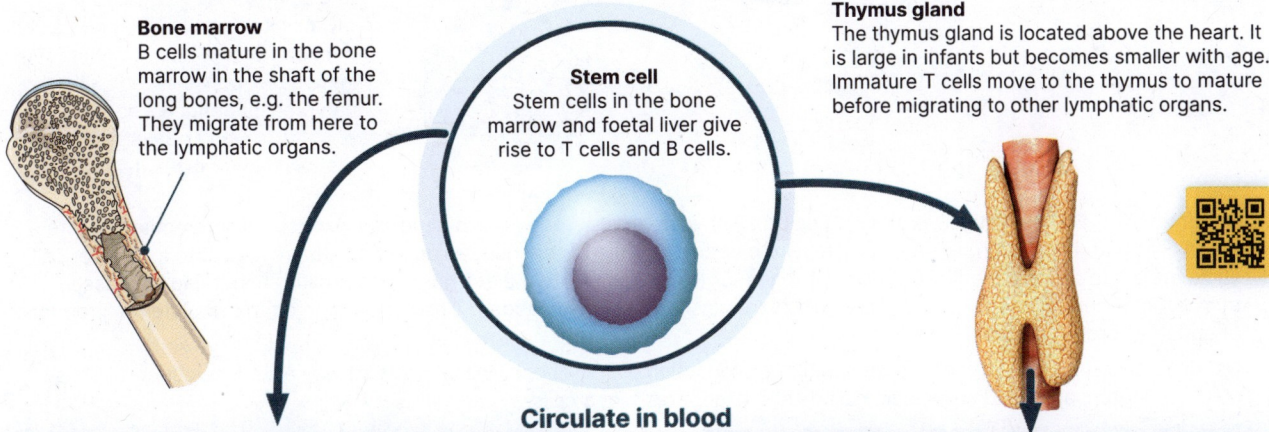

Bone marrow
B cells mature in the bone marrow in the shaft of the long bones, e.g. the femur. They migrate from here to the lymphatic organs.

Stem cell
Stem cells in the bone marrow and foetal liver give rise to T cells and B cells.

Thymus gland
The thymus gland is located above the heart. It is large in infants but becomes smaller with age. Immature T cells move to the thymus to mature before migrating to other lymphatic organs.

Circulate in blood

B-lymphocytes

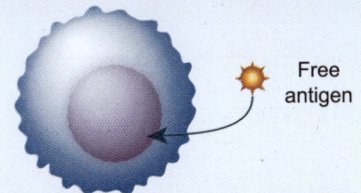

Free antigen

B cells recognize and bind antigens. Each B cell recognizes one specific antigen. Helper T cells recognize specific antigens on B cell surfaces and induce their maturation and proliferation. A mature B cell may carry as many as 100,000 antigenic receptors embedded in its surface membrane. B cells defend against bacteria and viruses outside the cell and toxins produced by bacteria (free antigens).

T-lymphocytes

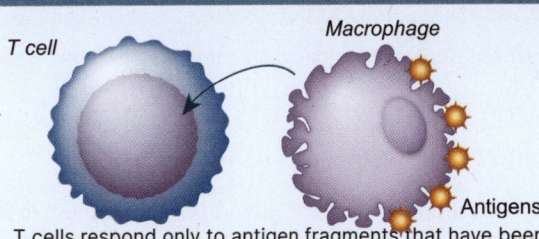

T cell

Macrophage

Antigens

T cells respond only to antigen fragments that have been processed and presented by antigen-presenting cells. (Most cells can present antigens, but only specialized APCs lead to antibodies). They defend against:

▸ Intracellular bacteria and viruses

▸ Protozoa, fungi, flatworms, and roundworms

▸ Cancerous cells and transplanted foreign tissue

Differentiate into two kinds of cells

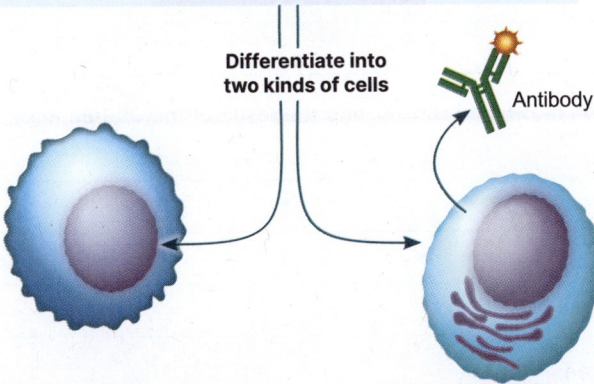

Antibody

Differentiate into various kinds of cells:

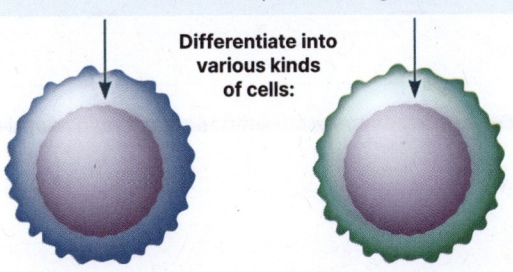

T helper cell activates T killer cells and other helper T cells. They are needed for B cell activation.

T killer cell destroys target cells on contact. Recognises tumour or virus-infected cells by their surface markers. They are also called cytotoxic T cells.

Memory cells
Some B cells differentiate into long-lived memory cells (see Clonal Selection). When these cells encounter the same antigen again (even years or decades later), they rapidly differentiate into antibody-producing plasma cells.

Plasma cells
When stimulated by an antigen (see Clonal Selection), some B cells differentiate into plasma cells, which secrete antibodies into the bloodstream. The antibodies then inactivate the circulating antigens.

There are also other types of T cells:
T memory cells have encountered specific antigens before and can respond quickly and strongly when the same antigen is encountered again.
T regulator cells control the immune response by turning it off when no more antigen is present. They are important in the development of self tolerance.

C3.2
6 - 8

Dendritic cells stimulate the activation and proliferation of lymphocytes

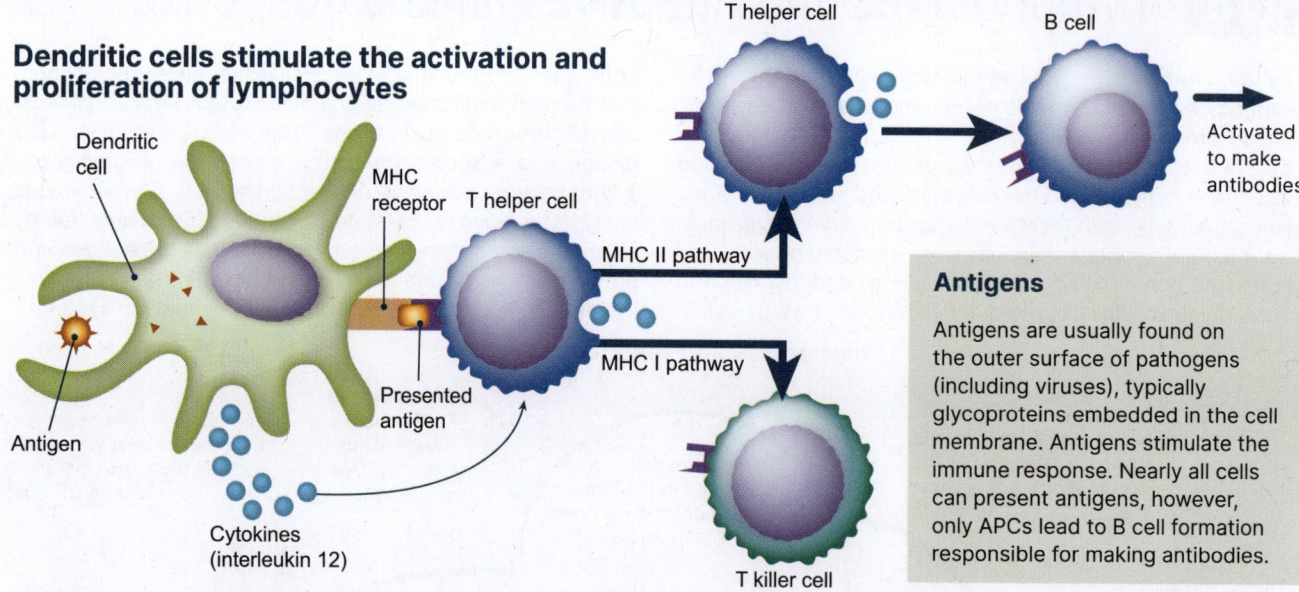

Antigens

Antigens are usually found on the outer surface of pathogens (including viruses), typically glycoproteins embedded in the cell membrane. Antigens stimulate the immune response. Nearly all cells can present antigens, however, only APCs lead to B cell formation responsible for making antibodies.

▶ Dendritic cells (DC) are antigen-presenting cells (APC). Immature DC originate in bone marrow from undifferentiated white blood cells and migrate through the body to lymph nodes. When a DC encounters an antigen, it presents it to a T helper cell, stimulating it to secrete chemicals called cytokines. Cytokines stimulate the activation and proliferation (rapid increase in number) of T cells, activating the immune system against that specific antigen. T helper cells go on to stimulate the production of antibody-producing B cells.

▶ Dendritic cells with MHC I receptors stimulate the production of T killer cells via a MHC pathway from T helper cells. These MHC I receptors are found on all mammalian cells, except red blood cells.

▶ Dendritic cells with MHC II receptors stimulate the production of T helper cells. These MHC II receptors are found on dendritic, macrophage, and B-lymphoctytes - all of which are antigen-presenting cells.

1. Where do B cells and T cells originate (before maturing)? _____

2. (a) Where do B-lymphocytes mature? _____

   (b) Where do T-lymphocytes mature? _____

3. Describe the nature and general action of the two major divisions in the immune system:

   (a) Humoral immune system: _____

   _____

   _____

   (b) Cell-mediated immune system: _____

   _____

   _____

4. Describe how helper T-lymphocytes are involved in the activation of B-lymphocytes, and the result of this interaction.

   _____

   _____

   _____

   _____

   _____

5. Explain the chain of events required for antibodies to be produced: _____

   _____

   _____

6. Describe the function of each of the following cells in the immune system response:

   (a) T helper cells: _____

   (b) T killer cells: _____

©2024 **BIOZONE** International
ISBN: 978-1-99-101410-8
Photocopying prohibited

219 Clonal Selection

Key Idea: Clonal selection theory explains how lymphocytes can respond to a large and unpredictable range of antigens. The clonal selection theory explains how the immune system can respond to the large and unpredictable range of potential antigens in the environment. The diagram below describes clonal selection after antigen exposure for B-lymphocytes. In the same way, a T-lymphocyte stimulated by a specific antigen will multiply and develop into different types of T cells. Clonal selection and differentiation of lymphocytes provide the basis for immunological memory.

Five (a-e) of the many B cells generated during development. Each one can recognize only one specific antigen.

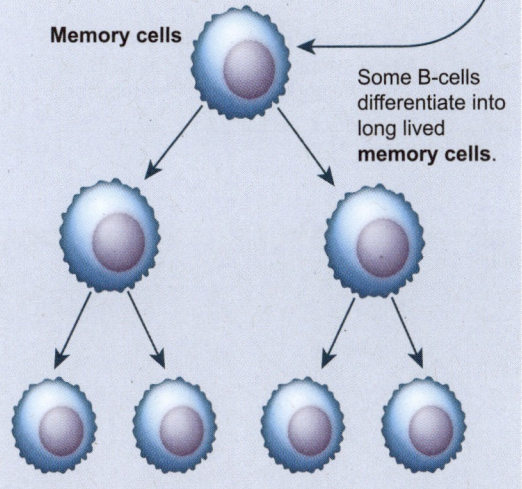

This B-cell encounters and binds an antigen. It is then stimulated to proliferate.

Clonal selection theory

Millions of B-lymphocytes form during development. Antigen recognition is randomly generated, so collectively they can recognize many antigens, including those that have never been encountered. Each B-lymphocyte has receptors on its surface for specific antigens and produces antibodies that correspond to these receptors. When a B-lymphocyte encounters its antigen, it responds by proliferating and producing many clones that produce the same kind of antibody. This is called clonal selection because the antigen selects the B-lymphocytes that will proliferate.

Memory cells

Some B-cells differentiate into long lived **memory cells**.

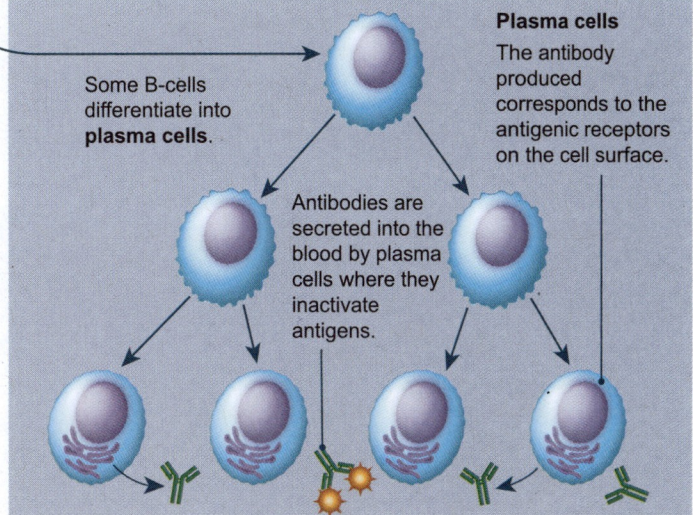

Some B-cells differentiate into **plasma cells**.

Antibodies are secreted into the blood by plasma cells where they inactivate antigens.

Plasma cells

The antibody produced corresponds to the antigenic receptors on the cell surface.

Some B cells differentiate into long lived memory cells. These are retained in the lymph nodes to provide future immunity (immunological memory). If the antigen returns a second time, memory B cells react more quickly and vigorously than the first time the antigen appeared.

Plasma cells secrete antibodies specific to the antigen that stimulated their development. Each plasma cell lives for only a few days, but can produce about 2000 antibody molecules per second. During development, any B cells that react to the body's own antigens are destroyed in a process that leads to self tolerance (acceptance of the body's own tissues).

1. Describe how clonal selection results in the proliferation of one particular B cell clone: _____

_____

_____

2. (a) What is the function of the plasma cells in the immune system response? _____

_____

(b) What is the significance of B cells producing antibodies that correspond to (match) their antigenic receptors?

_____

_____

3. (a) Explain the basis of immunological memory: _____

_____

(b) Why are B memory cells able to respond so rapidly to an encounter with an antigen long after an initial infection?

_____

_____

C3.2
9 - 10

220 HIV/AIDS and the Immune System

Key Idea: The human immunodeficiency virus (HIV) infects lymphocyte cells, eventually causing AIDS, a fatal disease, which acts by impairing immune system function.

HIV (human immunodeficiency virus) is a retrovirus (a type of viral pathogen) which binds to the CD4 receptor on the surface of T helper cells; these are central to cellular immunity and coordinate the immune response. HIV causes immune deficiency by replicating inside T helper cells and destroying them. Over time, a disease called AIDS (acquired immunodeficiency syndrome) develops and the immune system progressively loses its ability to fight infection. HIV is transmitted from person to person in body fluids such as blood, vaginal secretions, semen, breast milk, and across the placenta. Unless there are skin cuts, the risk of HIV transmission and infection through close contact between people remains very low.

HIV infects T helper cells

HIV infects T helper cells through fusing its lipid bilayer, originally sourced from a previously infected cell, with that of the host cell. It uses the cells to replicate itself in great numbers, then the newly formed viral particles exit the cell to infect more T helper cells. Many T helper cells are destroyed by viral replication. Because of their role in cellular immunity, T helper cell destruction recruits more T-lymphocytes, accelerating the infection of new cells. Once the T helper cell population becomes depleted, the immune system's ability to fight infection is severely compromised.

Glycoprotein spikes mediate attachment to the host cells' receptors.

Two copies of single stranded RNA

Viral envelope (lipoprotein)

Reverse transcriptase forms viral DNA from viral RNA

Capsid

The graph below shows the relationship between the level of HIV infection and the number of T helper cells. AIDS is only the end stage of an HIV infection. Shortly after the initial infection, HIV antibodies appear in the blood. There are three clinical categories during progression of the disease. The progressive reduction of T helper cells results in almost no immunity against any other pathogens. Mild illness, such as a respiratory cold, can be fatal for HIV/AIDS patients in the final stages.

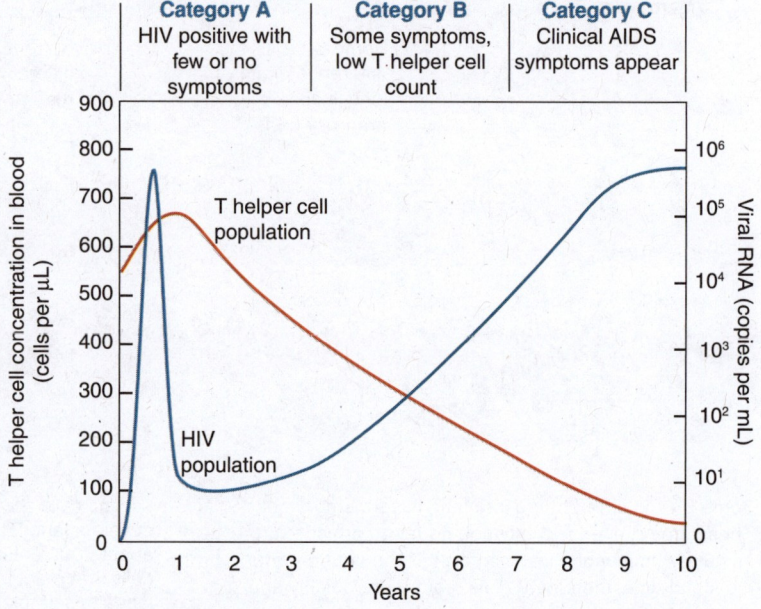

Category A HIV positive with few or no symptoms

Category B Some symptoms, low T helper cell count

Category C Clinical AIDS symptoms appear

T helper cell population

HIV population

HIV uses the cellular machinery of T helper cells to replicate

The genetic material of HIV is a single strand of RNA.

HIV hijacks the T helper cells' machinery to replicate itself. Reverse transcriptase produces double stranded DNA (dsDNA) from the viral RNA. The host cell transcribes the viral genes to produce new viruses.

The new HIV particles bud from the T helper cell. Between 1000 and 3000 new HIV particles can be released from a single infected cell.

The HIV particles mature and infect more T helper cells. As more T helper cells become infected, the body's immune response weakens.

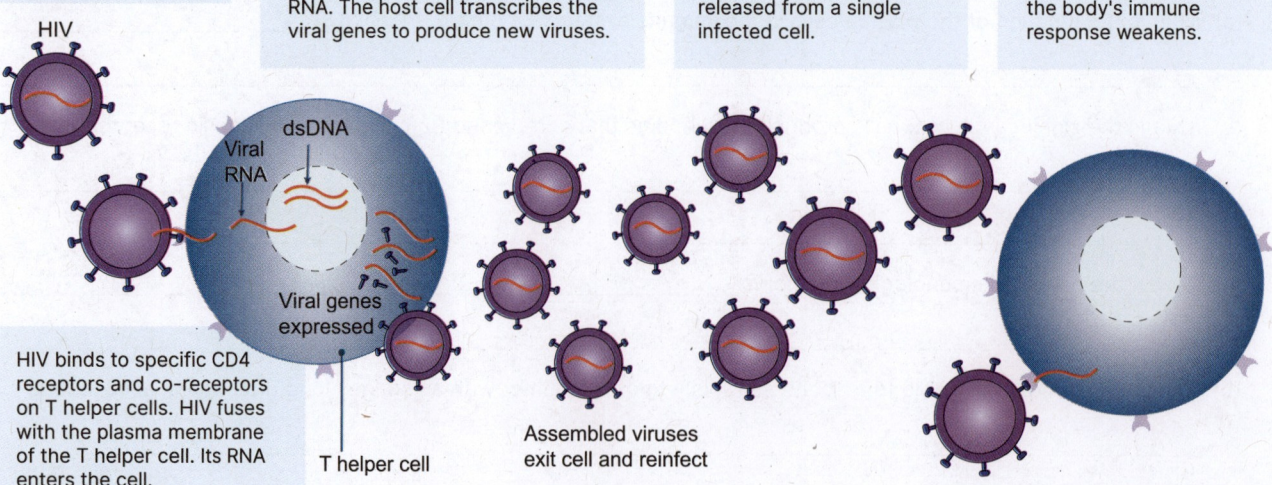

HIV

dsDNA

Viral RNA

Viral genes expressed

HIV binds to specific CD4 receptors and co-receptors on T helper cells. HIV fuses with the plasma membrane of the T helper cell. Its RNA enters the cell.

T helper cell

Assembled viruses exit cell and reinfect

C3.2

11 - 12

©2024 **BIOZONE** International
ISBN: 978-1-99-101410-8
Photocopying prohibited

AIDS: The end stage of an HIV infection

HIV/AIDS is a spectrum of disorders (right) arising as a consequence of impaired immune function, which prevents the body detecting and destroying pathogens or damaged cells. People with healthy immune systems can fight off the challenges of pathogens and are able to detect and destroy damaged (pre-cancerous) cells. However, people with HIV are susceptible to all pathogens because their resistance to disease is so low. What's more, loss of the T cell population compromises the ability of HIV-infected people to detect and destroy pre-cancerous cells. Rare cancers are a common symptom of HIV/AIDS.

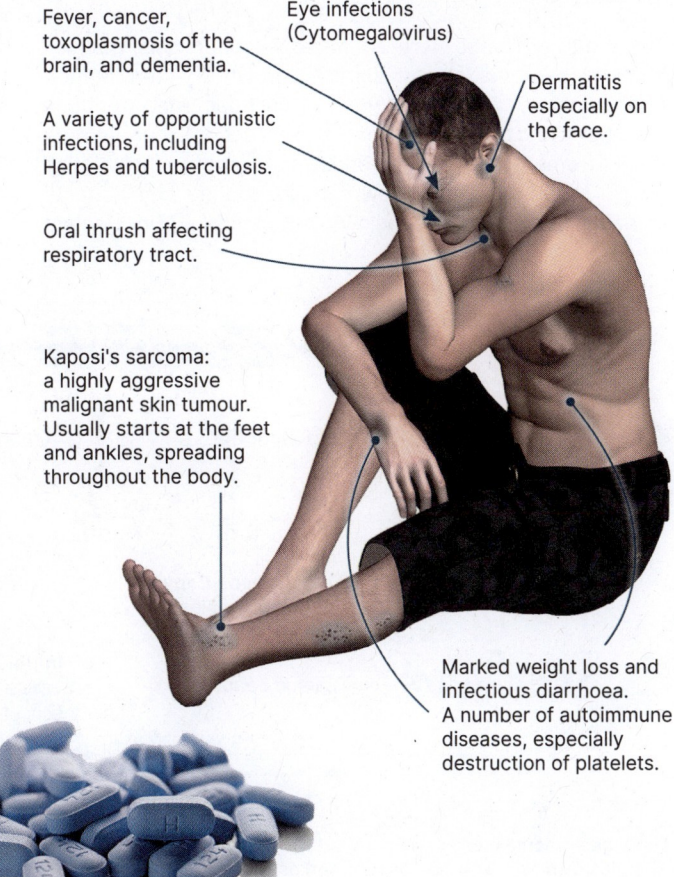

Fever, cancer, toxoplasmosis of the brain, and dementia.

Eye infections (Cytomegalovirus)

Dermatitis especially on the face.

A variety of opportunistic infections, including Herpes and tuberculosis.

Oral thrush affecting respiratory tract.

Kaposi's sarcoma: a highly aggressive malignant skin tumour. Usually starts at the feet and ankles, spreading throughout the body.

Marked weight loss and infectious diarrhoea. A number of autoimmune diseases, especially destruction of platelets.

Medications

Antibiotics can be used to treat some of the infections contracted due to the reduced immune system, e.g. tuberculosis, but they cannot be used to treat the HIV infection itself because antibiotics are ineffective against viruses.

Although there is currently no cure for HIV/AIDS, some antiretroviral drugs can slow the progress of the disease by interfering with the replication of HIV and slowing the advance of the disease.

Pre-exposure prophylaxis (PrEP) medications (right) can be prescribed to people who do not have HIV, but are at risk of contracting it. Correct use of PrEP, along with other precautions, can reduce the risk of contracting HIV by up to 99%.

1. (a) What type of cells does HIV infect? _____

    (b) How does HIV recognize this type of cell? _____

    _____

    (c) What is the role of reverse transcriptase in HIV replication? _____

    _____

    _____

2. Study the graph on the previous page showing how HIV affects the number of T helper cells. Describe how the viral population changes with the progression of the disease:

    _____

    _____

    _____

    _____

3. (a) What effect does HIV have on the cells of the immune system? _____

    _____

    _____

    (b) Describe the effect of this change on the long-term health of a person with HIV: _____

    _____

    _____

    _____

221 Antibiotics

Key Idea: Antibiotics are antimicrobial chemicals that kill bacteria (bactericidal) or inhibit their growth (bacteriostatic). Antibiotics are chemicals that act against bacterial infections by either killing the bacteria (bactericidal action) or preventing them from growing (bacteriostatic action). Antibiotics interfere with bacterial growth by disrupting key aspects of bacterial metabolism (below). Antibiotics are ineffective against viruses because viruses lack the peptidoglycan walls or ribosomes that antibiotics target. Antibiotics do not affect eukaryote cells either, as they are unable to bind to their specific ribosomes. Antibiotics are produced naturally by bacteria and fungi to kill or inhibit competitors or pathogens; however, most modern antibiotics are semi-synthetic modifications of these natural compounds.

How antibiotics work

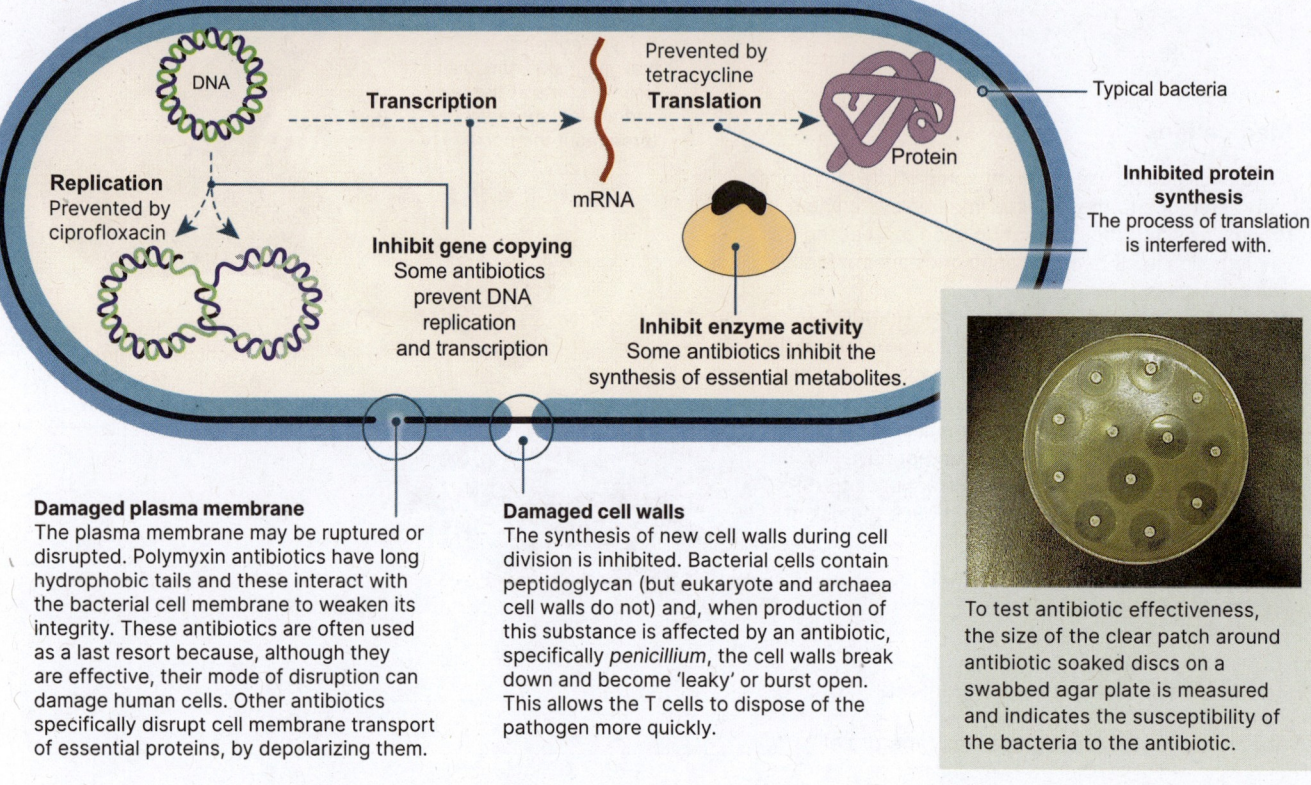

Replication
Prevented by ciprofloxacin

Transcription

Inhibit gene copying
Some antibiotics prevent DNA replication and transcription

Inhibit enzyme activity
Some antibiotics inhibit the synthesis of essential metabolites.

Prevented by tetracycline
Translation

Typical bacteria

Inhibited protein synthesis
The process of translation is interfered with.

Damaged plasma membrane
The plasma membrane may be ruptured or disrupted. Polymyxin antibiotics have long hydrophobic tails and these interact with the bacterial cell membrane to weaken its integrity. These antibiotics are often used as a last resort because, although they are effective, their mode of disruption can damage human cells. Other antibiotics specifically disrupt cell membrane transport of essential proteins, by depolarizing them.

Damaged cell walls
The synthesis of new cell walls during cell division is inhibited. Bacterial cells contain peptidoglycan (but eukaryote and archaea cell walls do not) and, when production of this substance is affected by an antibiotic, specifically *penicillium*, the cell walls break down and become 'leaky' or burst open. This allows the T cells to dispose of the pathogen more quickly.

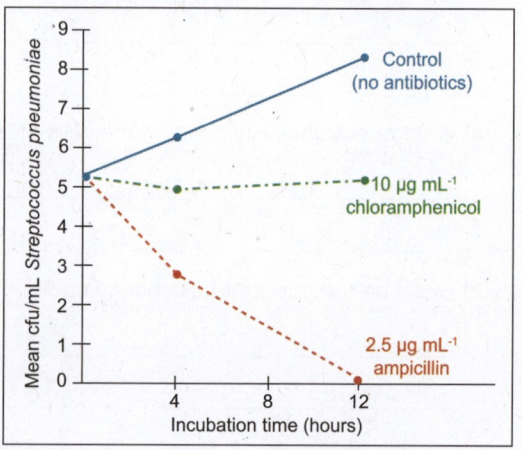

To test antibiotic effectiveness, the size of the clear patch around antibiotic soaked discs on a swabbed agar plate is measured and indicates the susceptibility of the bacteria to the antibiotic.

1. Describe how antibiotics are effective against bacterial infection but not viral infection: _____

_____

_____

_____

2. Explain why antibiotics are reasonably safe for use by humans to fight bacterial infection: _____

_____

_____

3. The graph (right) shows the effects of two antibiotics. Identify the antibiotic with a bacteriostatic action and the antibiotic with a bactericidal action. Explain your choice:

Bacteriostatic: _____

_____

_____

Bactericidal: _____

_____

_____

_____

Effects of bactericidal / bacteriostatic antibiotics

Control (no antibiotics)

$10 \ \mu g \ mL^{-1}$ chloramphenicol

$2.5 \ \mu g \ mL^{-1}$ ampicillin

Mean cfu/mL *Streptococcus pneumoniae*

Incubation time (hours)

©2024 **BIOZONE** International
ISBN: 978-1-99-101410-8
Photocopying prohibited

222 Evolution of Antibiotic Resistance

Key Idea: Widespread use of antibiotics and pesticides has created a selective environment for the proliferation of chemical resistance in microbial populations.

Resistance to antibiotics is becoming a more common and concerning occurrence in the modern world. It arises and spreads when chemical control agents do not remove all the targeted organisms. Those that survive because of their suite of specific inherited characteristics are able to pass on these genes and so resistance becomes more common in subsequent generations, i.e. natural selection. Antibiotic resistance in bacteria, particularly to multiple antibiotics, poses serious threats to human health.

The evolution of antibiotic resistance in bacteria

Antibiotic resistance arises when genetic changes allow bacteria to tolerate levels of antibiotic that would normally inhibit growth. Resistance may arise spontaneously through mutation or by transfer of DNA between microbes (horizontal gene transfer). Genomic analyses from 30,000 year old permafrost sediments show that the genes for antibiotic resistance predate modern antibiotic use. In the current selective environment of widespread antibiotic use, these genes have proliferated and antibiotic resistance has spread. For example, methicillin resistant strains of *Staphylococcus aureus* (MRSA) have acquired genes for resistance to all penicillins. Such strains are called superbugs.

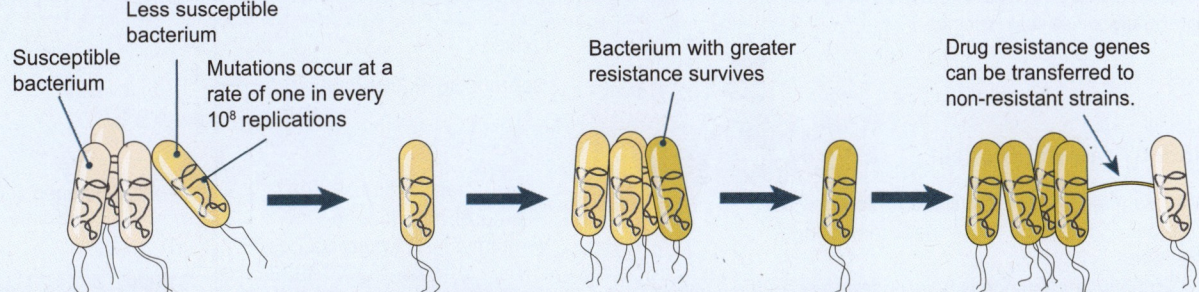

Susceptible bacterium

Less susceptible bacterium

Mutations occur at a rate of one in every 10^8 replications

Bacterium with greater resistance survives

Drug resistance genes can be transferred to non-resistant strains.

Any population, including bacterial populations, includes variants with unusual traits; in this case reduced sensitivity to an antibiotic. These variants arise as a result of typical mutations in the bacterial chromosome.

When a person takes an antibiotic, only the most susceptible bacteria will die. The more resistant cells remain alive and continue dividing. Note that the antibiotic does not create the resistance; it provides the environment in which selection for resistance can take place.

If the amount of antibiotic delivered is too low, or the course of antibiotics is not completed, a population of resistant bacteria develops. Within this population too, there will be variation in susceptibility. Some will survive higher antibiotic levels than others.

A highly resistant population has evolved. The resistant cells can exchange genetic material with other bacteria (via horizontal gene transmission), passing on the genes for resistance. The antibiotic initially used against this bacterial strain will now be ineffective.

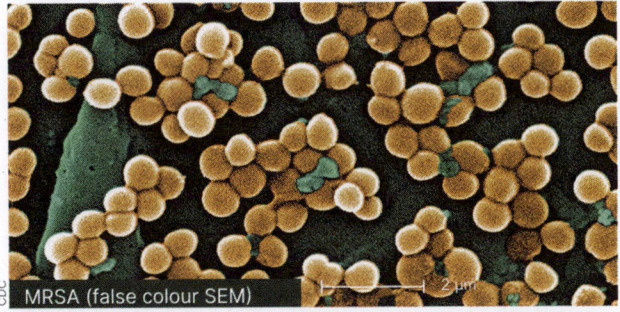

MRSA (false colour SEM)

Staphylococcus aureus is a common bacterium responsible for several minor skin infections in humans. MRSA is a strain that has evolved resistance to penicillin and related antibiotics. MRSA is troublesome in hospital-associated infections because patients with open wounds, invasive devices, e.g. catheters, or poor immunity are at greater risk of infection than the general public.

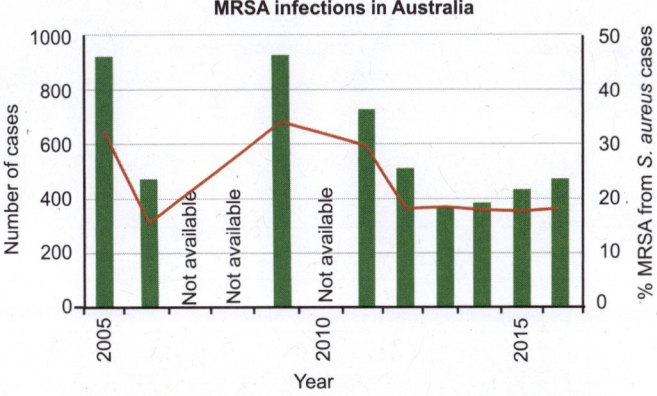

MRSA infections in Australia

In Australia, MRSA cases have remained relatively steady since 2012. Greater reporting measures and strict hygiene standards have reduced the number of cases since the early 2000s.

1. Describe how resistance develops in a population: _____

_____

_____

_____

2. How can antibiotic resistance be transferred between strains of bacteria? _____

_____

_____

C3.2
14
NOS

The development of antibiotics and the evolution of antibiotic resistance

▶ The modern era of antibiotics began in 1928 with Alexander Fleming's discovery of penicillin. By the mid 1940s, penicillin was being produced in vast quantities, mainly to treat World War II soldiers.

▶ However, even then, Alexander Fleming warned about overuse causing antibiotic resistance. Indeed, by the mid 1950s, penicillin resistance in bacteria was already becoming a problem. There are now various levels of resistance in bacteria to all commonly used antibiotics.

▶ Incorrectly prescribed antibiotics contribute to resistance. Studies have shown incorrect antibiotic therapies in up to 50% of cases.

▶ Extensive use of antibiotics in agriculture have also contributed to resistance. Farmers use antibiotics to prevent infection in large, high density, livestock and poultry operations, where crowded conditions are potentially 'disease-inducing'.

Finding new antibiotics

New technology can be utilized to source new antibiotics to replace those that are no longer effective:

▶ Artificial intelligence is being used to scan the database of all human proteins (the proteome) in search of possible peptides that may have antibiotic properties.

▶ High-throughput screening (HTS) of chemical databases, such as the synthetic molecule library or natural product reservoir, allows for rapid identification of any possible substances with antibiotic properties. The testing can be automated and alert scientists to results. Research indicates that, from 10,000 possible compounds in a library, 250 of them could indicate opposable antibiotics properties after screening, and 3-6 compounds will then go onto clinical development, with 1 novel drug marketed.

Historical timeline of antibiotic use

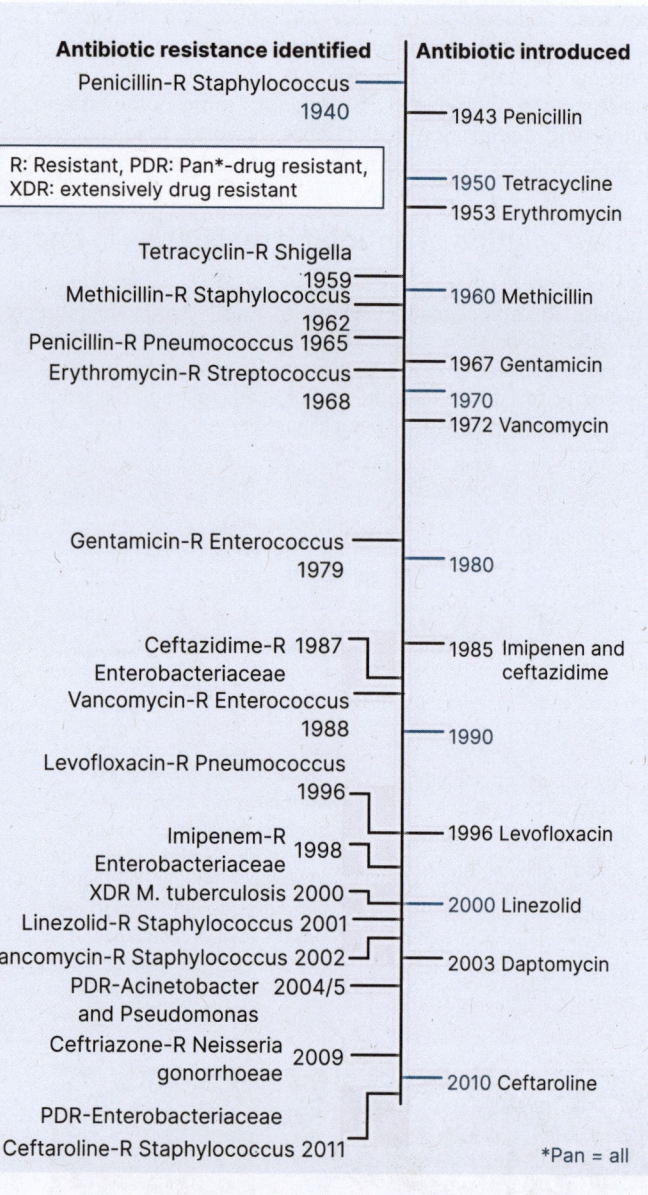

Antibiotic resistance identified	Antibiotic introduced
Penicillin-R Staphylococcus 1940	1943 Penicillin
	1950 Tetracycline
	1953 Erythromycin
Tetracyclin-R Shigella 1959	
Methicillin-R Staphylococcus 1962	1960 Methicillin
Penicillin-R Pneumococcus 1965	
Erythromycin-R Streptococcus 1968	1967 Gentamicin
	1970
	1972 Vancomycin
Gentamicin-R Enterococcus 1979	1980
Ceftazidime-R Enterobacteriaceae 1987	1985 Imipenen and ceftazidime
Vancomycin-R Enterococcus 1988	
	1990
Levofloxacin-R Pneumococcus 1996	
Imipenem-R Enterobacteriaceae 1998	1996 Levofloxacin
XDR M. tuberculosis 2000	2000 Linezolid
Linezolid-R Staphylococcus 2001	
Vancomycin-R Staphylococcus 2002	2003 Daptomycin
PDR-Acinetobacter and Pseudomonas 2004/5	
Ceftriazone-R Neisseria gonorrhoeae 2009	
	2010 Ceftaroline
PDR-Enterobacteriaceae	
Ceftaroline-R Staphylococcus 2011	

R: Resistant, PDR: Pan*-drug resistant, XDR: extensively drug resistant

*Pan = all

3. With reference to MRSA, describe the implications to humans of widespread antibiotic resistance:

_____

_____

_____

_____

4. Comment on any relationship between the year of an antibiotic's production and the year resistance first appears:

_____

_____

_____

5. Novel antibiotics can be found in yet unknown compounds. Why is new technology so important to help find them?

_____

Newly discovered plant species can contain potential antibiotic compounds.

_____

_____

_____

©2024 **BIOZONE** International
ISBN: 978-1-99-101410-8
Photocopying prohibited

223 Zoonoses

Key Idea: Diseases that transmit from animals to humans are known as zoonoses and are caused by a number of pathogens, including viruses and bacteria.

The presence of infectious disease never goes away. Some diseases are very difficult to eradicate and are always present, and some diseases are new (emerging). Diseases that are always present at low levels in a population or region are known as endemic diseases. Occasionally, there may be a sudden increase in the prevalence of a particular disease. On a local level this is known as an outbreak. When an infectious disease spreads rapidly through a nation and affects large numbers of people it is called an epidemic. Some viruses that infect other species can also be spread and infect humans. The diseases they cause in humans are called zoonotic diseases and can be particularly dangerous as humans have no natural immunity to them.

Zoonotic diseases

▶ Zoonotic diseases originate from a natural animal host carrying a pathogen. The pathogen can be viral, bacterial, or parasitic, and usually causes limited damage to the natural host they have co-evolved with.

▶ Close contact, or consumption, between one or more animal species can result in transfer of the pathogen to a transmission host. The pathogen may circulate between one or more different transmission hosts for some time, amplifying the pathogen population.

▶ Many animal specific pathogens pass through humans but they are unable to replicate and are rapidly destroyed. On rare occasions, when a pathogen, particularly bacterial or viral, begins to replicate inside a human, it increases the likelihood of being transmitted to another human. This is because it can mutate and evolve within an internal human environment and adapt to reproduce efficiently. If the pathogen multiplies inside a human cell, there is a heavier pathogen load, which increases the chance of spread.

▶ Once human-to-human transmission has been established with an adapted pathogen, this can lead to an epidemic, and possibly pandemic, because there is no immunity for the novel disease.

Tuberculosis TB is an infectious bacterial disease caused by *Mycobacterium tuberculosis* (MTB). The pathogen is spread through the air when infectious people cough, sneeze, talk, or spit, and another person inhales the particles. A number of species can act as the natural host, depending on the country. For example, it is suggested that badgers transmits the disease to cattle when wandering around farmland. Humans can then contract the disease from consuming infected dairy or meat products.

Rabies is a virus that infects bat populations at a low level (~ 1%). Bats act as reservoir species. Humans can be bitten by some species of bat and contract rabies directly but in many countries various species of animals act as transmission hosts. These include wild animals such as raccoons and skunks in the USA, but the most common transmission host is the domestic dog. Bites from the animals break the skin of humans and allow transmission of the virus through bodily fluids.

Japenese encephalitis disease is caused by a virus which typically cycles between infected birds and pigs by *Culex* mosquitoes that feed off the animals' blood. The virus load is amplified by these animal reservoirs. People become infected and develop Japanese encephalitis when they are bitten by an infected mosquito. Most people show very mild symptoms but 1% of infected people develop encephalitis, of which 20-30% of the cases are fatal.

Covid-19 The SARS-CoV-2 virus led to Covid-19 disease in humans. A wild population of horseshoe bats are hypothesized as the natural host, which then passed the virus on to an intermediate host. One transmission host could be the ant-eating pangolin, but some scientists hypothesize that the raccoon dog or a cat-like civet may have also been involved. Humans became the spillover host in late 2019. A Covid-19 pandemic was declared in March 2020.

1. What is the typical pathway for a zoonotic disease to originate and transmit to humans?

_____

_____

C3.2
15

Zika virus: An example of global zoonotic disease spread and its containment

‣ Zika virus was first isolated from the Zika Forest in Uganda in 1947. The spread of the virus into non-human primates was facilitated by mosquitoes that pierced the skin and mixed pathogen borne blood between individuals. Human zika infections were discovered shortly after. Humans have since acted as a disease reservoir, rather than a spillover host with person-to-person transmission.

‣ Since then, zika has spread slowly across the globe, with outbreaks in the Americas in 2015 and 2016. Due to the disease spreading only when *Aedes* mosquitoes are present, zika has been limited to the same regions, albeit widely spread.

‣ Zika causes a mild fever and rash that is not usually serious in adults. However, in the last few years, infection of pregnant women by Zika has been linked to microcephaly (small head and brain) in newborn babies.

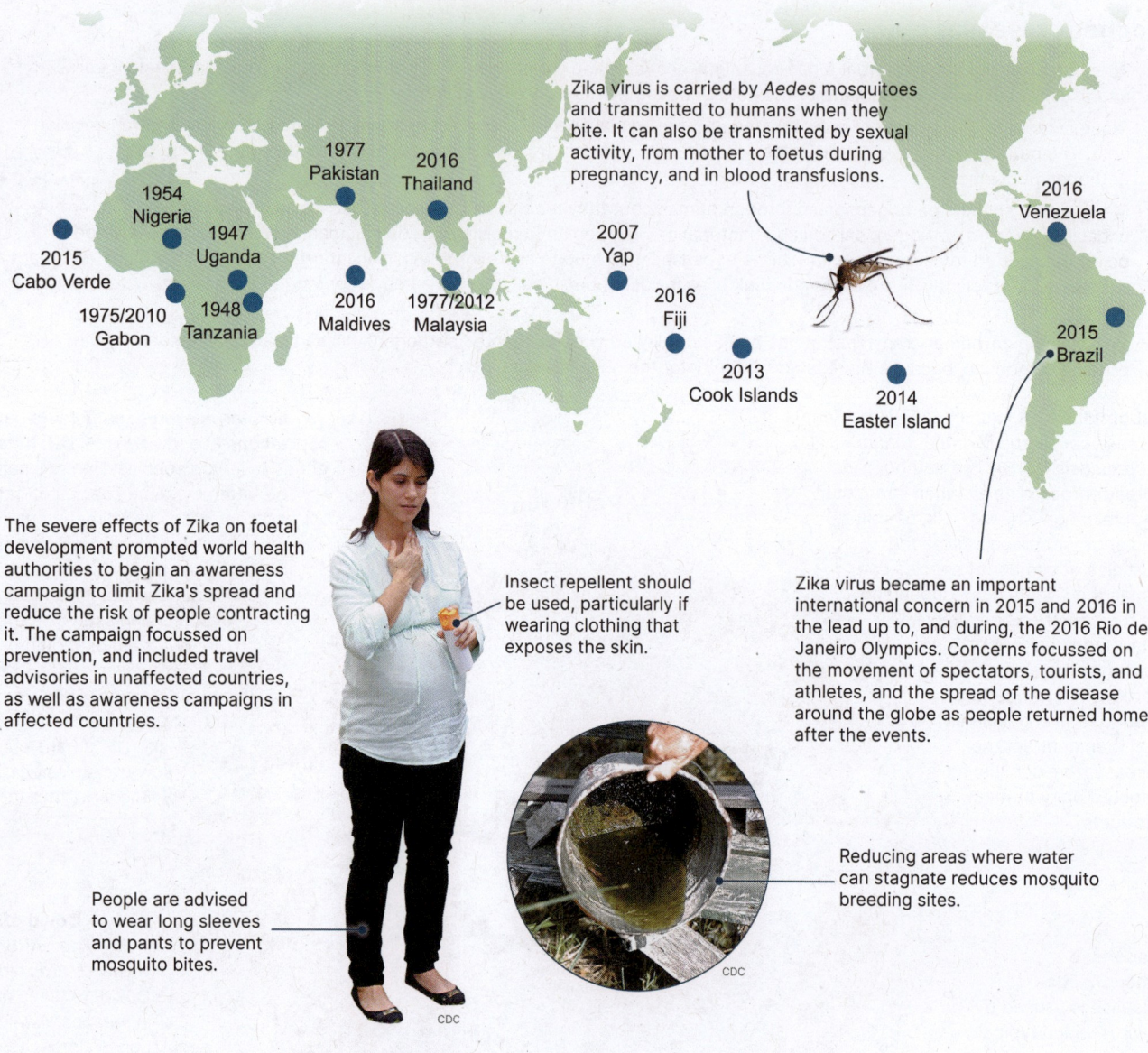

Zika virus is carried by *Aedes* mosquitoes and transmitted to humans when they bite. It can also be transmitted by sexual activity, from mother to foetus during pregnancy, and in blood transfusions.

2015 Cabo Verde
1954 Nigeria
1947 Uganda
1975/2010 Gabon
1948 Tanzania
1977 Pakistan
2016 Thailand
2016 Maldives
1977/2012 Malaysia
2007 Yap
2016 Fiji
2013 Cook Islands
2014 Easter Island
2016 Venezuela
2015 Brazil

The severe effects of Zika on foetal development prompted world health authorities to begin an awareness campaign to limit Zika's spread and reduce the risk of people contracting it. The campaign focussed on prevention, and included travel advisories in unaffected countries, as well as awareness campaigns in affected countries.

Insect repellent should be used, particularly if wearing clothing that exposes the skin.

Zika virus became an important international concern in 2015 and 2016 in the lead up to, and during, the 2016 Rio de Janeiro Olympics. Concerns focussed on the movement of spectators, tourists, and athletes, and the spread of the disease around the globe as people returned home after the events.

People are advised to wear long sleeves and pants to prevent mosquito bites.

Reducing areas where water can stagnate reduces mosquito breeding sites.

2. (a) In which general direction has Zika virus spread across the globe? _____

(b) Describe the area that Zika virus appears to be generally confined to and explain this: _____
_____
_____

3. How is Zika virus transmitted? _____
_____
_____

4. Provide a short case study report on a zoonotic disease that has affected humans. It can be an elaboration on one of the cases already described (excluding zika) or a different disease. Describe the type of pathogen responsible for the disease. Describe the natural host and any transmission host, where the disease originated and where it spread, and the impact of the disease on humans.

©2024 **BIOZONE** International
ISBN: **978-1-99-101410-8**
Photocopying prohibited

224 The Covid-19 Pandemic

Key Idea: Covid-19 is a zoonotic disease caused by infection with the SARS-CoV-2 virus.

In December 2019, a new strain of coronavirus was detected in Wuhan, China. The new virus was named Severe Acute Respiratory Syndrome Coronavirus 2 (SARS-CoV-2). Infection with the virus causes a disease called Covid-19. The WHO declared a pandemic in March 2020 as the virus spread around the world. The Covid-19 pandemic disrupted the world travel and global economies for several years, from 2020-2022. Around 7 million people died from the virus or virus complications in the first three years of the pandemic. The first strain of disease resulted in around 20% of infected people developing severe breathing problems and high level hospital care for the elderly and people with underlying medical problems was required. Vaccines for the disease were developed in an extremely quick time frame, late in 2020. Combined with extensive vaccination programmes and less virulent variants of the virus, the death rate per case fell.

Where did SARS-CoV-2 come from?

▸ There are many questions over the SARS-CoV-2 virus that remain unanswered (below). Extensive investigations are ongoing.

Was the virus circulating earlier than the first reported case in December 2019?
This seems likely but researchers have not found antibodies in blood samples stored in blood banks before the first reported date of the virus. Very early circulation is therefore unlikely.

Did the virus come from bats?
While bats are a likely reservoir, no species has been found that carries viruses that completely match SARS-CoV-2. Also, coronaviruses similar to SARS-CoV-2 have been identified in pangolins (below).

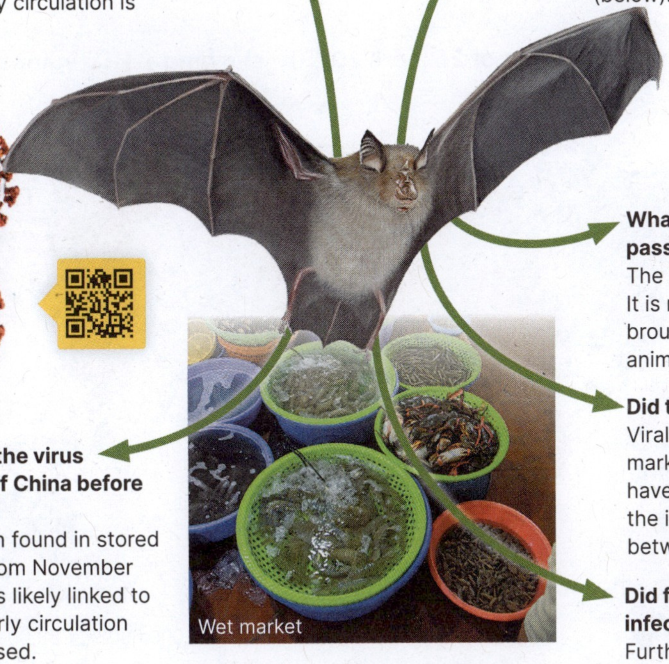

What was the intermediate species that passed the virus to humans?
The intermediate animal is still unknown. It is most likely that a farmed animal, brought to a wet market where live animals are sold, is the direct source.

Did the virus originate in a wet market?
Viral material has been found around wet markets in Wuhan. No firm conclusions have been drawn but wet markets provide the ideal conditions for viruses to jump between species.

To what extent was the virus circulating outside of China before November 2019?
Antibodies have been found in stored blood from Europe from November 2019 onwards. This is likely linked to travel from China. Early circulation could have been missed.

Wet market

Did frozen meat play a part in early infections?
Further analysis is needed but it has been established that the virus jumped to people from live animals.

1. (a) What is the likely source (reservoir) of SARS-CoV-2? _____

   (b) How is the virus spread? _____

   _____

   _____

   (c) Why did investigations focus on wet markets in Wuhan, China? _____

   _____

   _____

2. The virus has since mutated, many times, into different variants that have become more contagious than the original virus. Define virulent and contagious, and relate them to the progress of the Covid-19 pandemic since 2020:

   _____

   _____

   _____

C3.2
18

Different countries, different outcomes

Reports of viral pneumonia (a lung infection) in Wuhan, China were reported on the 31st December 2019. Early in January 2020, a new coronavirus was identified as the cause of the infections.

Despite strict restrictions, including travel bans, being placed on the residents of Wuhan and the surrounding region, the virus began to spread through China. On 13th January 2020, the first case outside of China was recorded in Thailand. Within 10 days, the virus had spread to a number of countries, including the US, as infected travellers flew around the world. During the early stages of the pandemic, some countries were very successful in slowing or containing the spread of the virus, while in other countries the virus spread widely, causing high numbers of infections and deaths. The graph (right) shows the number of confirmed cases (July, 2020) for three countries: China, New Zealand, and the US in the initial stages of the pandemic. The way their governments, health departments, and populations responded to the disease was important in the pattern of Covid-19 spread.

The diagram below shows the number of confirmed cases of Covid-19 by country as of 26 April 2023. The darker shades of blue indicate higher numbers of confirmed cases.

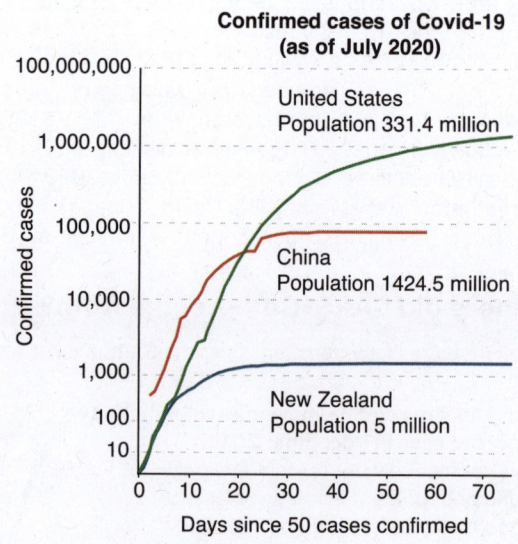

Confirmed cases of Covid-19 (as of July 2020)

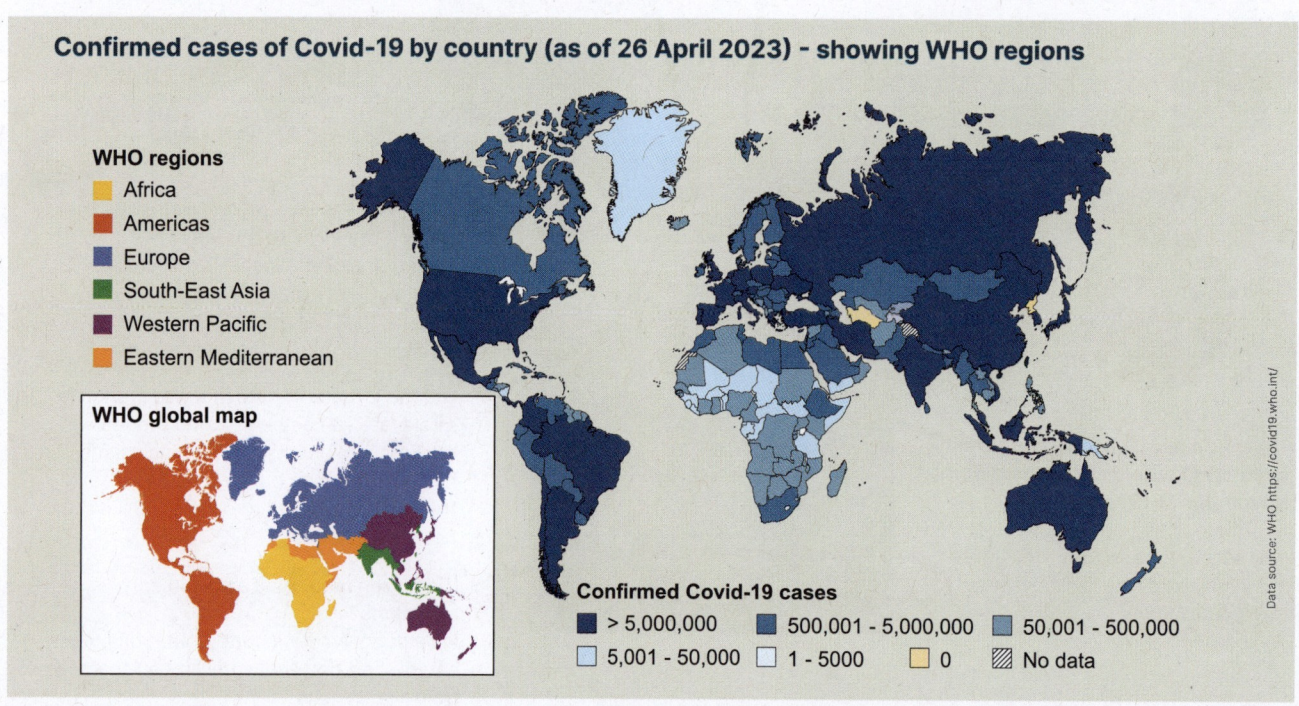

3. (a) Explain the general trends in Covid-19 cases over 60 days between the three countries: United States, China, and New Zealand (see top right graph):

_____

_____

(b) What differences may have occurred in the demographics, including population, and pandemic response, to account for these results in the three countries (you may need to locate further information online):

_____

_____

_____

_____

4. What link might there be between the cumulative cases in the first 50 days between different countries?

_____

_____

©2024 **BIOZONE** International
ISBN: 978-1-99-101410-8
Photocopying prohibited

Evaluating Covid-19 data

▸ Scientists, health organizations, and governments use statistical data to inform them about the spread of Covid-19 over time. Analysis of infectious disease data, such as the change in numbers of cases per day or week in any given area, the percentage of deaths or serious illness, or response to vaccination, can provide a clearer understanding of disease patterns and risk factors, to make treatment and control more effective.

▸ Disease data can be from both primary and secondary sources. Percentage difference and percentage change are two statistics that allow a comparison of data from two time periods. These can be used to evaluate change in the Covid-19 pandemic cases and deaths in different regions.

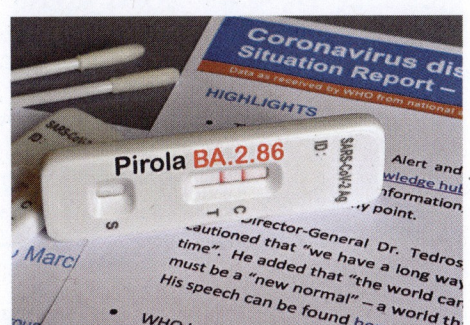

Percentage difference and percentage change

Percentage difference

1. Calculate absolute difference between two values (this will always be a positive number).
2. Calculate the average of the two starting values.
3. Divide the absolute difference by the average (from step 3).
4. Convert the value to a percentage : multiply by 100 (3 significant figures is sufficient).

The two data sets being compared must represent the same metric. The value calculated can be used to compare the difference between two average values.

Percentage change

1. Identify if values have increased over time (percentage increase) or decreased over time (percentage decrease.
2. Percentage increase= final value - initial value   or Percentage decrease = initial value - final value.
3. Divide difference by the initial value.
4. Convert this value to a percentage.

This calculation is used if dealing with old and new values of the same metric. The value calculated measures the change over a time period.

5. (a) Use the World Health Organization data from confirmed cases and deaths from Covid-19 for the week of April 27th 2020 and April 26th 2021 to calculate percentage difference and percentage change in WHO regions (see map on previous page to locate regions).

Region	Week of April 27th 2020		Week of April 26th 2021		Percentage difference	Percentage change	Percentage difference	Percentage change
	Cases	Deaths	Cases	Deaths	Cases	Cases	Deaths	Deaths
Europe	170, 607	23, 260	1, 171, 751	38, 796				
Western Pacific	8, 804	15, 778	134, 558	22, 902				
Americas	297, 359	616	1, 307, 063	25, 276				
South-East Asia	20, 493	288	2, 713, 418	1, 586				
Eastern Mediterranean	39, 237	966	324, 863	6,483				
Africa	9, 138	222	42, 387	1, 033				

Data from covid19.who.int

(b) What trends do you see in percentage difference between the two dates for cases? _____

_____

_____

(c) Do the trends for percentage difference in deaths appear different than that of cases? _____

_____

(d) What information does the percentage change from the two dates provide in addition to percentage difference?

_____

6. What other statistics could be calculated from the above data to provide an indication of the changing situation in different regions during the Covid-19 pandemic over a year? Complete the calculations and discuss:

_____

_____

_____

_____

225 Vaccines and Immunization

Key Idea: A vaccine is a suspension of antigens that is deliberately introduced into the body to protect against disease. If enough of the population is vaccinated, herd immunity provides protection to unvaccinated individuals.

A vaccine is a preparation of a harmless foreign antigen or nucleic acids that is deliberately introduced into the body to protect against a specific disease. The antigen in the vaccine is usually some part of the pathogen. It triggers the immune system to produce antibodies against the antigen but it does not cause the disease. The immune system remembers its response and will produce the same antibodies if it encounters the antigen again. If enough of the population is vaccinated, herd immunity (indirect protection) provides unvaccinated individuals in the population with a measure of protection against the disease.

Types of vaccine

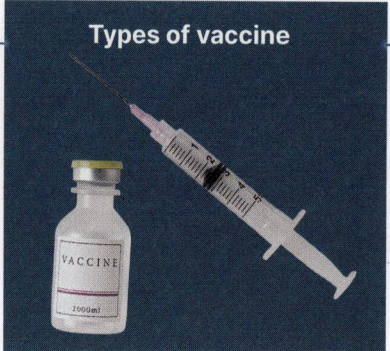

Whole-agent vaccine

Contains whole, non-virulent microorganisms

Inactivated (killed)

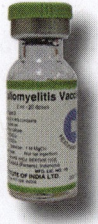

Viruses for vaccines may be inactivated with formalin or other chemicals. They present almost no risk of infection, e.g. most influenza vaccines, Salk polio vaccine.

Attenuated (weakened)

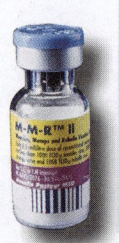

Attenuated viruses are usually strains in which mutations have accumulated during culture. These live viruses can back-mutate to a virulent form, e.g. MMR vaccine.

Subunit vaccine

Contains some part or product of microbes that can produce an immune response. Includes vaccines made using genetic engineering, inactivated toxins, and conjugated and acellular vaccines, e.g. the diphtheria-tetanus-pertussis vaccine and the vaccine against bacterial meningitis.

Why are vaccinations given?

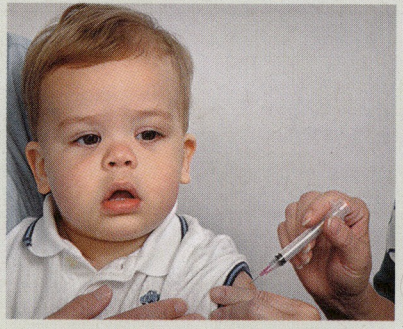

Vaccines against common diseases are given at various stages during childhood according to an immunization schedule. Vaccination has been behind the decline of some once-common childhood diseases, such as mumps and measles.

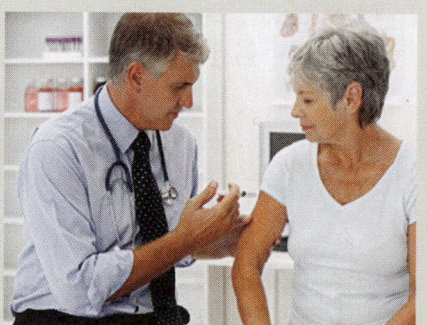

Most vaccinations are given in childhood, but adults may be vaccinated against a disease, e.g. TB, tetanus, if they are in a high risk group, e.g. the elderly or farmers, or to provide protection against seasonal diseases such as influenza.

Tourists may need specific vaccines if the country they are visiting has a high incidence of a certain disease. For example, travellers to South America should be immunized against yellow fever, a disease that does not occur in many other countries.

1. What is a vaccine? _____

_____

2. Describe how immunization is obtained from vaccines, including the role of antigens: _____

_____

_____

_____

_____

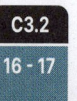

C3.2
16 - 17

Vaccination can provide herd immunity

▸ Herd immunity occurs when the vaccination of a significant portion of a population provides some protection for individuals who have not developed immunity, e.g. have not been vaccinated and are not immunized. In order to be effective for any particular disease, a high percentage of the population needs to be vaccinated against that disease. High vaccination rates make it difficult for the disease to spread because there are very few susceptible people in the population.

▸ Herd immunity is important for people who cannot be vaccinated, e.g. the very young; people with immune system disorders; or people who are very sick, such as cancer patients.

High herd immunity: Most of the population is immunized. The spread of the disease is limited. Only a few people are susceptible and become infected.

Low herd immunity: Only a small proportion of the population is immunized. The disease spreads more readily through the population, infecting many more people.

 Immunized and healthy

 Not immunized and healthy

Not immunized, sick and contagious

Disease variance

The level of vaccination coverage to obtain herd immunity differs for each disease and population density. Highly contagious diseases, such as measles, need a much higher vaccine uptake (95%) than a less contagious disease such as polio (80-85%).

3. Attenuated viruses provide long term immunity to their recipients and generally do not require booster shots. Why might attenuated viruses provide such effective long-term immunity when inactivated viruses do not?

_____

_____

_____

_____

4. (a) What is herd immunity? _____

_____

_____

(b) Why are health authorities concerned when the vaccination rates for an infectious disease fall? _____

_____

_____

5. Some members of the population are unable to be vaccinated. Give an example and explain why herd immunity is very important to them:

_____

_____

_____

Developing a Covid-19 vaccine

▶ The development of a Covid-19 vaccine became a global priority when the seriousness of the virus was recognised. In order to produce the vaccine fast enough for it to be useful, vaccine development was fast tracked.

▶ A number of Covid-19 vaccines have undergone the same rigorous development and testing process as other vaccines, but over a shorter time period. A number have been approved as safe to use.

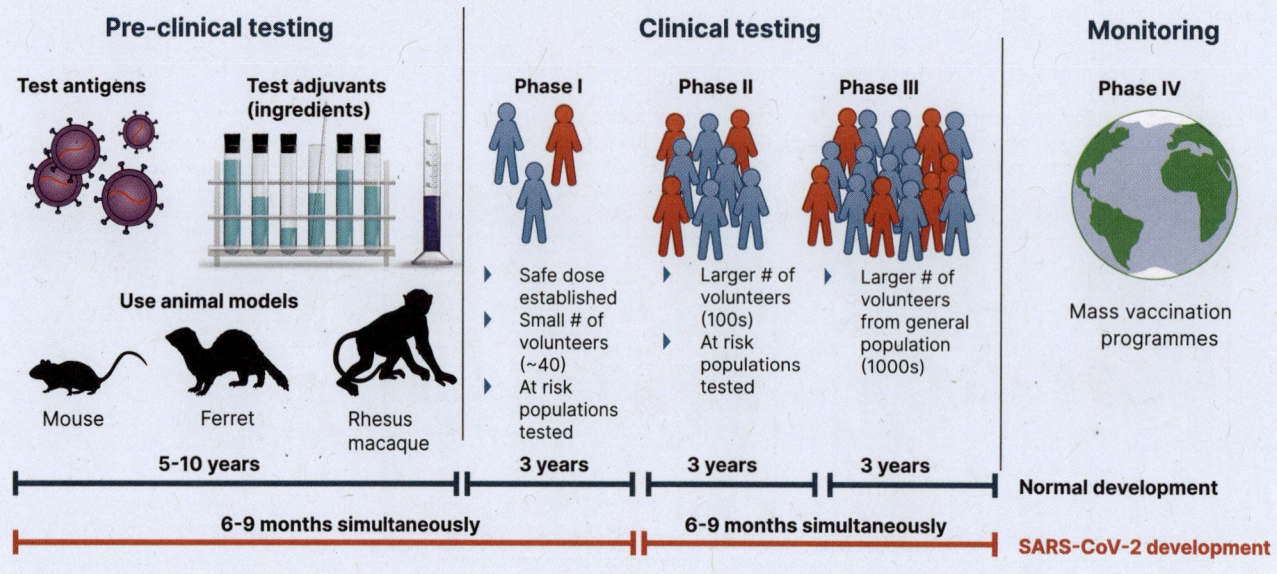

6. Explain how the Covid-19 vaccines were developed in a much shorter time than most other vaccines:

_____

_____

_____

_____

7. Transparency and peer review is an important part of research. Early research results were released from various Covid-19 vaccination development teams so that other scientists could replicate the experiments.

(a) Why is peer review an important part of scientific methodology (process)? _____

_____

_____

(b) What are some potential issues of media publishing early results on Covid-19 vaccination development before the research has been properly evaluated?

_____

_____

_____

8. The vaccines for Covid-19 have a low risk of side effects. Some side effects were more common and potentially less harmful, while more serious, but rarer, side-effects were publicized. Why did health organizations and governments still encourage the general public to be vaccinated for Covid-19 despite the risks of the vaccination?

Covid-19 vaccine side effects

Manufacturers are required to release a list of potential side-effects that may occur, regardless how rare, for the Covid-19 vaccination. Common side effects include a sore arm at the site of injection, a low grade fever, headache, lethargy, nausea, and shortness of breath. Rare side effects include potential myocarditis and allergic reactions.

_____

_____

_____

_____

_____

226 Did You Get It?

1. (a) The nervous system is divided into what two parts? _____
   _____

   (b) Explain the difference between the automatic and voluntary nervous system with reference to some examples of their functions:
   _____
   _____

   (c) Compare the components and function of a typical sensory neural input and a motor neural output from the brain:
   _____
   _____
   _____

2. What system does the nervous system integrate with to enable the fight or flight response? Explain the general pathway of response, including structures involved:
   _____
   _____
   _____
   _____

3. Why is it necessary for peristalsis to be under autonomic control? _____
   _____
   _____

4. **AHL:** Discuss how gradients of auxin result in a phototrophic response in plants. Draw diagrams to explain the process:

5. What type of disease does the Hendra virus (right) cause? Explain your reasoning:
   _____
   _____

6. Structures on the Hendra virus act as antigens in infected humans. Explain the sequence of events that occurs in the immune system when encountering a pathogen:
   _____
   _____
   _____
   _____

Hendra virus, which is carried by fruit bats and infects horses and humans, has emerged in Queensland, Australia.

CSIRO cc 3.0

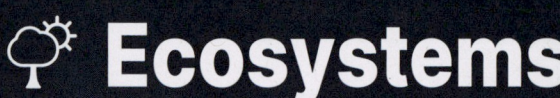

☼ Ecosystems

C4.1 Populations and communities

Activity Number

Guiding Questions:
▶ How are population sizes regulated by species interactions in a community?
▶ How do species interactions result in interdependent populations within a community?

Learning Outcomes:

☐ 1	Define the term 'population', and explain how different populations can be distinguished from each other.	227
☐ 2	Explain valid reasons for using random sampling to estimate population size. **NOS:** Quantify and state the sampling error that can occur due to random sampling in a range of activities.	228, 230-232
☐ 3	Use random quadrat sampling to estimate population size for sessile organisms. **AOS:** Calculate standard deviation to measure the variation in the population estimate and the spread of the population.	228-232
☐ 4	**AOS:** Collect data from capture–mark–release–recapture sampling and estimate the population size using the Lincoln index.	233
☐ 5	Define carrying capacity and include a description of limiting resources.	234
☐ 6	Explain how density-dependent factors act as negative feedback control of population size, including competition, predation, and disease.	235, 236
☐ 7	Use population data to study population growth curves. **NOS:** Explain how idealized models of exponential and sigmoid (logistic) growth curves can be used to help predict population growth. **AOS:** Plot data for population growth on a logarithmic scale and compare against model growth curves to identify the type of population growth.	236, 237
☐ 8	**AOS:** Collect population growth data and use it to plot the population growth curve. Compare against the sigmoid growth curve.	238
☐ 9	Compare and contrast competition and cooperation in intraspecific relationships, illustrating with examples.	240
☐ 10	Define the term 'community' in the context of an ecosystem.	239
☐ 11	Classify interspecific relationships using examples including herbivory, predation, interspecific competition, mutualism, parasitism, and pathogenicity.	241, 242
☐ 12	Elaborate on the interspecific interaction of mutualism using root nodules in legumes, mycorrhizae in orchids, and zooxanthellae in hard corals.	241
☐ 13	Investigate resource competition between both an endemic and invasive species, using a local case as an example.	242
☐ 14	Evaluate a range of different research approaches (laboratory and fieldwork) for testing interspecific competition. **NOS:** Explain how hypotheses can be tested experimentally and by making observations in the field; distinguish between the two.	243
☐ 15	**AOS:** Apply the chi-squared test for association between the presence or absence of two species.	244, 245
☐ 16	Investigate, using a case study and data, how predator–prey relationships act as a density-dependent control of animal populations.	246
☐ 17	Compare both top-down and bottom-up control of populations in communities.	247
☐ 18	Distinguish between allelopathy and secretion of antibiotics using local examples.	248

Guiding Questions:
▶ How are the adaptations and habitats of species related?
▶ What causes the similarities between ecosystems within a terrestrial biome?

Learning Outcomes:

☐ 1	Contrast between open, closed, and isolated systems. Identify examples for each, including ecosystems as open systems.	249
☐ 2	Identify ecosystem exceptions where sunlight is not the principle source of energy, including deep caves and the deep ocean. **NOS:** Contrast scientific laws and theories and understand their usefulness in making generalizations in science.	249 - 251
☐ 3	Explain how chemical energy flows through food chains.	249, 251-253
☐ 4	Construct food chains and food webs using arrows to indicate the flow of energy from organism to organism.	251-253
☐ 5	Discuss the supply of energy to decomposers and position of that trophic group in food chains.	250-253
☐ 6	Explain how external energy sources are used by autotrophs to synthesize carbon compounds from simple inorganic substances.	250
☐ 7	Compare between the use of light energy in photoautotrophs and use of chemical energy from oxidation reactions in chemoautotrophs.	250
☐ 8	Define 'heterotrophs' and describe how they obtain and assimilate carbon compounds.	250
☐ 9	Explain how cellular respiration in autotrophs and heterotrophs is the main process by which energy is released from carbon compounds.	250
☐ 10	Classify organisms into trophic levels using correct terminology. Analyse the different trophic levels organisms can occupy in different food chains.	254
☐ 11	**AOS:** Use research data to construct energy pyramids showing the energy at each level of an ecosystem	254
☐ 12	Use the 10% rule to help explain the loss of energy from each successive trophic level. Explain the role of decomposers in energy transformations in an ecosystem.	250, 251, 254
☐ 13	Explain that energy transfers are not 100% efficient and energy is lost to the environment during cellular respiration as heat in both autotrophs and heterotrophs.	250, 251
☐ 14	Relate the 10% rule to the restrictions on the number of trophic levels in ecosystems.	251, 254
☐ 15	Compare the relative primary production as mass of carbon per unit area per unit time in different biomes.	255
☐ 16	Investigate secondary production using caterpillars as an example of heterotrophs. Calculate the efficiency of energy transfer from producers to consumers.	256
☐ 17	Use and develop models to demonstrate the importance of functioning carbon cycles in an ecosystem.	257
☐ 18	Discuss how ecosystems can act as carbon sinks and carbon sources.	257
☐ 19	Explain how the processes of combustion, biomass, peat, coal, and natural gas release carbon dioxide in the atmosphere and discuss the significance of human influence on these processes.	257
☐ 20	Analyse annual fluctuations and the long-term data from the Keeling Curve, accounting for photosynthesis, respiration, and combustion.	258
☐ 21	Investigate the effect of aerobic respiration and photosynthesis on atmospheric oxygen fluxes.	257
☐ 22	Connect the importance of functioning nutrient cycles in providing chemical elements to organisms in ecosystems.	257

227 Features of Populations

Key Idea: What are the attributes of populations?
Populations are groups of interbreeding individuals within the same area. Individuals may migrate to and from other populations. The attributes of a population can be measured or calculated to provide information about the population. Such attributes include aspects of population composition, distribution and abundance, and dynamics (below). Populations may grow, decline, or remain stable based on the interplay between environmental resources and conditions, migration, births and deaths, and the intrinsic ability of a population to reproduce.

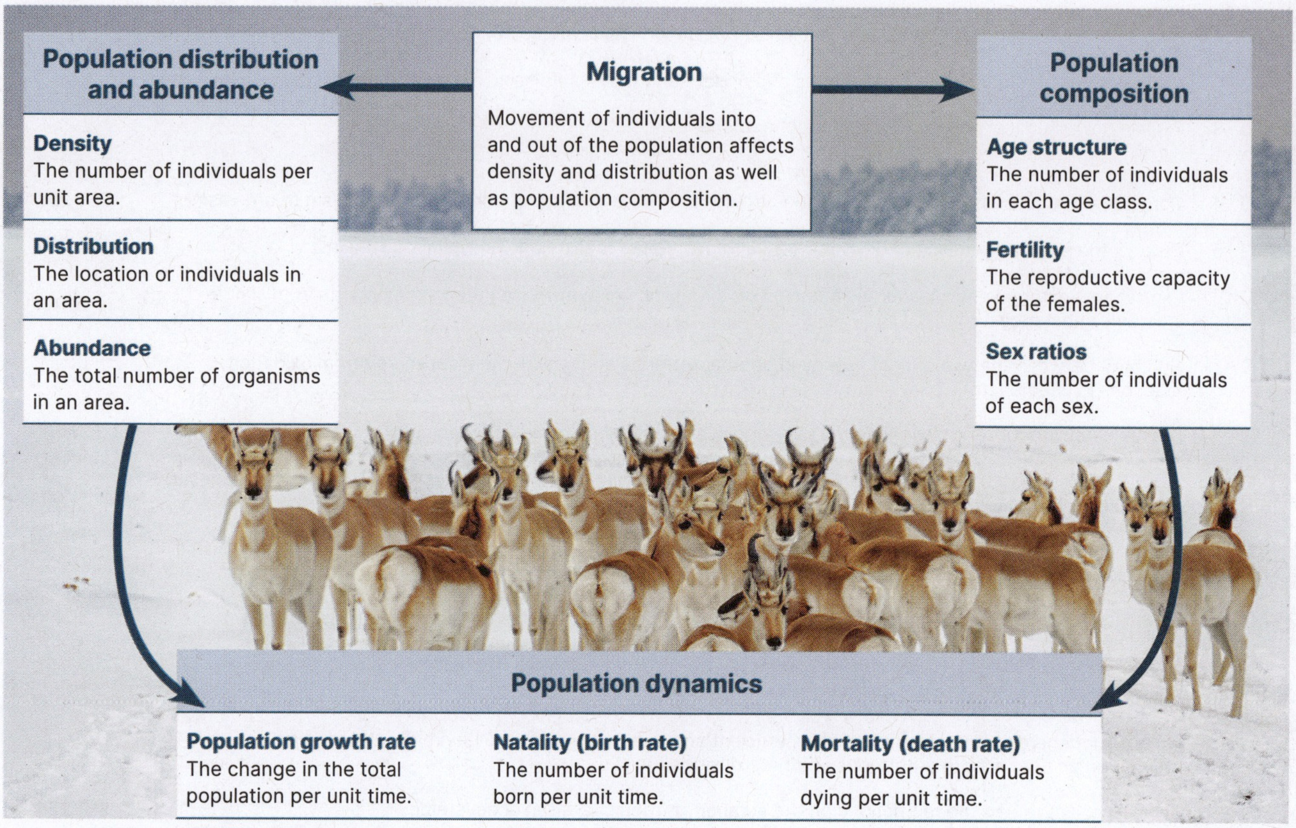

Population distribution and abundance

Density
The number of individuals per unit area.

Distribution
The location or individuals in an area.

Abundance
The total number of organisms in an area.

Migration
Movement of individuals into and out of the population affects density and distribution as well as population composition.

Population composition

Age structure
The number of individuals in each age class.

Fertility
The reproductive capacity of the females.

Sex ratios
The number of individuals of each sex.

Population dynamics

Population growth rate
The change in the total population per unit time.

Natality (birth rate)
The number of individuals born per unit time.

Mortality (death rate)
The number of individuals dying per unit time.

1. Describe one example of a population attribute that would be a good indicator of each of the following:

   (a) Whether the population is increasing or decreasing: _____

   _____

   (b) The ability of the environment to support the population: _____

   _____

2. (a) Identify the population attributes that can be measured directly from the population: _____

   _____

   (b) Identify the population attributes that must be calculated from the data collected: _____

   _____

3. Explain the value of population sampling for each of the following situations:

   (a) Conservation of a population of an endangered species: _____

   _____

   _____

   _____

   (b) Management of a fish stock: _____

   _____

   _____

C4.1
1

©2024 **BIOZONE** International
ISBN: 978-1-99-101410-8
Photocopying prohibited

228 Sampling Populations

Key Idea: Sampling a population provides information about its size, density, distribution, age structure, and more.

Imagine counting or measuring every individual pine tree in a pine forest. Is this feasible? Most likely not, because there are too many individuals and not enough time or resources to count them all. To get around these problems, researchers sample the population. Sampling involves choosing a smaller area that represents the population and counting or measuring the organisms in that area. The information gathered from the sample is used to draw conclusions about that population. But how well does the sample represent the actual population? Sampling a population always comes with a measure of error. Knowing the significance of that error is important when calculating or using sample statistics.

What can sampling tell us?

Population size
How many individuals are in a population of organisms living in a certain area, such as a forest, or a pond? Counting individuals in a measured area and then applying specific equations to that sample allows us to estimate the size of the population over a larger area.

Species distribution
How is a particular species distributed in the ecosystem and does this change over time, e.g. seasonally? Sample data can tell us about the geographical range of the species and how this might be affected by environmental change.

Age structure
What is the distribution of ages in a population? How many individuals are of breeding age, or are juveniles? Information like this is important when determining harvest quotas or breeding plans for endangered animals.

Sampling and sampling error

Estimating population size can be difficult. Populations may be very spread out, making it difficult to find individuals, or they may be mobile, or occupy a vast area. In any sample areas there may be more or fewer individuals than any other area simply by chance. This needs to be taken into account when calculating population statistics. Any statistic, e.g. the estimated population, or the average height of students, will have an error based simply on the fact that some of the population has been sampled and some has not. As a result, our sample statistic will never exactly match the population parameter. The difference between these is called the sampling error. The larger the sample, the smaller the sample error will be.

The sampling error

The sampling error is the difference between the population parameter and the sample statistic, e.g. the actual population number and the estimated population number. Because we almost never know the population parameter (the reason why we have to sample), we can estimate the sampling error using a confidence interval (see following page).

4	3	2	3	2	0
2	3	4	1	2	1
4	6	4	4	3	3
3	2	5	3	6	5
0	4	3	4	3	4
2	3	1	1	0	2

1. (a) The box above right represents a small field divided into 36 squares. The number in each square represents the number of dandelions found in that area. Role a six sided dice to randomly select six of the squares above (first roll for row, second roll for column). Record the numbers in these squares below (not repeating any squares):

   _____

   (b) Calculate the mean number of dandelions in the six squares and then estimate the total population of dandelions.

   Mean dandelions per square: _____ Population estimate: _____ Sampling error: _____

2. Repeat 1. above two more times:

   (a) Mean dandelions per square: _____ Population estimate: _____ Sampling error: _____

   (b) Mean dandelions per square: _____ Population estimate: _____ Sampling error: _____

3. The entire population of dandelions is 102. Use this to calculate the sampling error for each sample above:

4. What can you say about the estimated population of dandelions based on your sampling?

   _____

   _____

C4.1
2, 3

NOS

Expressing confidence in your data

▶ When we take measurements, e.g. fish length, from samples of a larger population, we are using the samples as indicators of what the whole population looks like. Therefore, when we calculate a sample mean for a variable, it is useful to know how close that value is to the true population mean for that same variable, i.e. its accuracy. If you are confident that your data set fairly represents the entire population, you are justified in making inferences about the population from your sample.

▶ You can start by calculating a simple measure of dispersion called standard deviation. Standard deviation is a measure of the amount of variation in a set of values. Are the individual data values all close to the mean, or are the data values highly variable? Standard deviation provides a way to evaluate the distribution of your data which can help you decide the step in your analysis.

Standard deviation

▶ Sample standard deviation (s) is presented as $\bar{x} \pm s$.

▶ In normally distributed data, 68% of all data values will lie within one standard deviation (1s) of the mean. 95% of all values will lie within two standard deviations (2s) of the mean (see the distribution plotted right).

▶ The lower the standard deviation, the more closely the data values cluster around the mean.

▶ The formula for calculating standard deviation is shown in the green box (below).

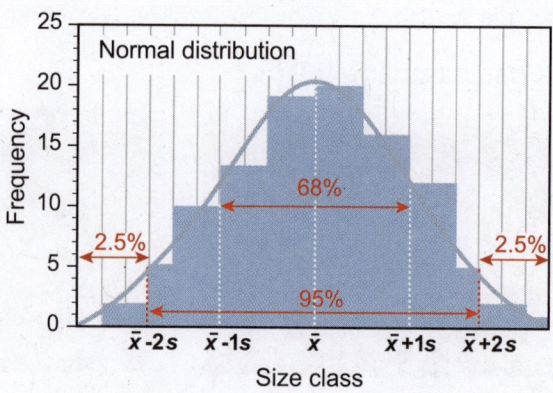

Calculating standard deviation

$$s = \sqrt{\frac{\sum(x - \bar{x})^2}{n - 1}}$$

$\sum(x - \bar{x})^2$ = sum of squared deviations from the mean

n = sample size.

$n - 1$ provides an unbiased s for small sample sizes (large samples can use n).

Both of the histograms below show a normal distribution of data with the values spread symmetrically about the mean. However, their standard deviations are different. In histogram A, the data values are widely spread around the mean. In histogram B, most of the data values are close to the mean. Sample B has a smaller standard deviation than sample A.

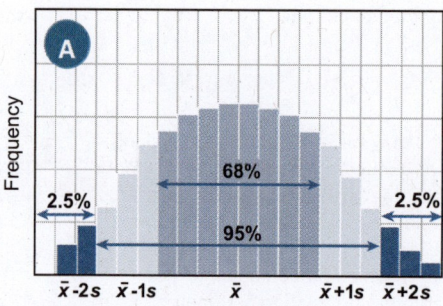

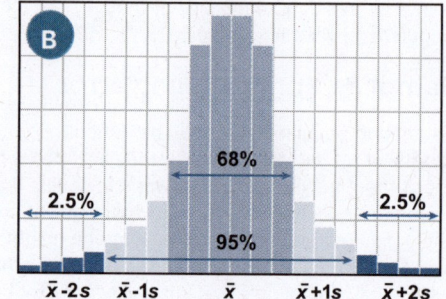

95% confidence intervals

▶ So how can you tell if your sample is giving a fair representation of the entire population? We can do this using the 95% confidence interval. This statistic allows you to make a claim about the reliability your sample data. The mean ± the 95% confidence interval (95% CI) gives the 95% confidence limits (95% CL). This tells us that, on average, 95 times out of 100, the true population mean will lie within the confidence limits. Having a large sample size greatly increases the reliability of your data.

▶ If the 95% confidence limits from two different sampled populations do not overlap, e.g. from fertilized and non fertilized fields, then the populations are statistically different for the feature sampled.

Step 1 is to calculate the standard error of the mean (SE). It is simple to calculate and is usually a small value.

Step 2 is to calculate the 95% confidence interval (95% CI). It is calculated by multiplying SE by the value of t (from t tables) at P = 0.05 for the appropriate degrees of freedom (df) (n-1) for your sample.

$$SE = \frac{s}{\sqrt{n}}$$

$$95\% \ CI = SE \times t_{(P=0.05)}$$

5. Use the data from 1 (a) to calculate the standard deviation of the mean number of dandelions per quadrat.

x	$\bar{x}$	$(x - \bar{x})^2$
$\sum(x - \bar{x})^2$		

$$\frac{\sum(x - \bar{x})^2}{n - 1} = \boxed{}$$

standard deviation
s = $\boxed{}$

6. Calculate the standard error of the mean number of dandelions per quadrat:

_____

7. Calculate the 95% confidence interval of the mean number of dandelions per quadrat (t = 2.571):

_____

8. Multiply the 95% CI by 36 (total possible quadrats) to get the 95% CI for your population estimate:

_____

229 How Do We Sample Ecosystems?

Key Idea: Sampling should provide data that are unbiased and accurate. Choice of sampling method and design should be based on suitability to the populations being sampled, the environment, and the time and resources available.

Most practical exercises in ecology involve collecting data about the distribution and abundance of one or more species in a community. Most studies also measure the physical factors in the environment as these may help to explain the patterns of distribution and abundance observed. There are many sampling options (below), each appropriate to different environments or organisms and with advantages and drawbacks. You must take several factors into account when sampling to make sure the data you collect accurately and impartially represents the ecosystem being investigated.

Sampling designs and techniques

Point sampling
Individual points are chosen using a grid reference or random numbers applied to a map grid. The organisms at each point are recorded. Point sampling is often used to collect data about vegetation distribution.

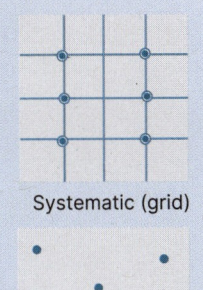
Systematic (grid)

Pros: Point sampling is efficient if time is limited. It is a good method for determining species abundance and community composition.
Cons: May miss organisms in low abundance.

Random

Area sampling using quadrats
A quadrat provides a known unit area of sample, e.g. 1 m². Quadrats are placed randomly or in a grid pattern on the sample area. The presence and abundance of organisms in each square is noted. Quadrat sampling is appropriate for plants and slow moving animals and can be used to evaluate community composition.

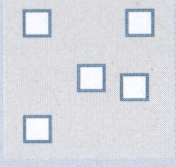

Line and belt transects
In a line transect, a tape or rope marks the line. The species occurring on the line are recorded (all along the line or at regular points). Lines can be chosen randomly (right) or may follow an environmental gradient.
Pros: Low environmental impact and good for assessing the presence/absence of plant species.
Cons: Rare species may be missed.

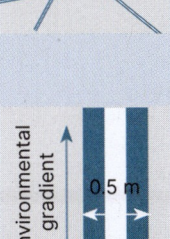

In a belt transect, quadrats are used to sample organisms at regular intervals along a measured strip.
Pros: Provide a lot of information on abundance and distribution as well as presence/absence.
Cons: Time consuming to carry out properly.

Environmental gradient
0.5 m

Mark and recapture sampling
1. Animals are captured, marked, and then released back into the population (right).

2. After a suitable time to allow the marked animals to remix with the population, the population is resampled. The number of marked animals recaptured in a second sample is recorded as a proportion of the total.

Pros: Useful for highly mobile species which are otherwise difficult to record.
Cons: Time consuming to do well.

1: All marked.

2: Proportion recaptured

Sampling considerations

▶ Random sampling methods should be used to avoid bias in the data. In random sampling, every possible sample of a given size has the same chance of selection.

▶ The methods used to sample communities and their populations must be appropriate to the ecosystem being investigated. Communities in which the populations are at low density and have a random or clumped distribution will require a different sampling strategy from those where the populations are uniformly distributed and at higher density.

▶ The sample size, e.g. the number of quadrats, must be large enough to provide data to enable us to make inferences about aspects of the whole population.

1. Name a sampling technique that would be appropriate for determining:

   (a) Percentage cover of a plant species in pasture:

   _____

   _____

   (b) Change in community composition from low to high altitude on a mountain:

   _____

   _____

   (c) Association of plant species with particular soil types in a nature reserve:

   _____

   _____

   (d) The population size of a fish in a lake:

   _____

   _____

2. What are the benefits of collecting information about the physical environment when sampling populations?

   _____

   _____

   _____

   _____

   _____

   _____

   _____

   _____

C4.1
3

Sampling strategies

In most ecological studies, it is not possible to measure or count all the members of a population. Instead, information is obtained through sampling in a manner that provides a fair (unbiased) representation of the organisms present and their distribution. This is usually achieved through random sampling. Sometimes researchers collect information by non-random sampling, a process that does not give all the individuals in the population an equal chance of being selected. While faster and cheaper to carry out than random sampling, non-random sampling may not give a true representation of the population.

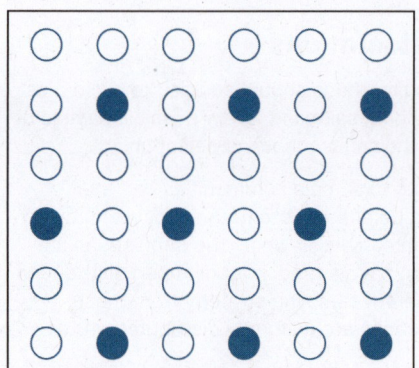 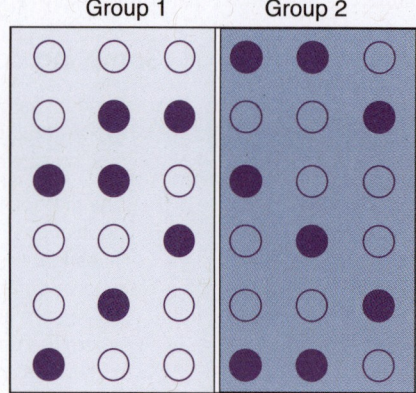 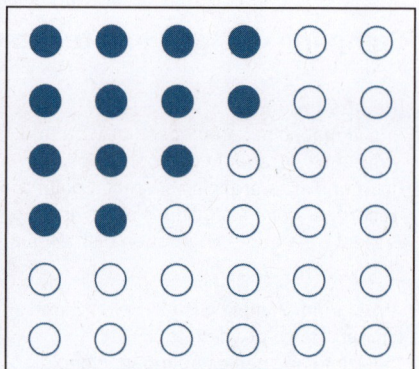

Systematic sampling

Samples from a larger population are selected according to a random starting point and a fixed, periodic sampling interval. For the example above, the sampling period is every fourth individual. Systematic sampling is a random sampling method, provided the periodic interval is determined beforehand and the starting point is random.

Example: Selecting individuals from a patient list.

Stratified sampling

In stratified sampling, the population is divided into subgroups (strata) before sampling. Samples are then taken from a stratum in proportion to its representation in the total population. The strata should be mutually exclusive, and individuals must be assigned to only one stratum. Random or systematic sampling is then applied within each stratum.

Example: Dividing the population into males and females.

Opportunistic sampling

A non-random sampling technique in which subjects are selected because they are easily accessible to the researcher. Opportunistic sampling excludes a large proportion of the population and is usually not representative of the population. It is sometimes used in pilot studies to gather data quickly and with little cost.

Example: Selecting 13 people at a cafe where you are having lunch.

Stratified sampling in ecology

▶ Many study areas are not uniform. Instead, they include a variety of distinct habitats, especially if the study site is large. In stratified sampling, the various habitats are sampled separately in proportion to their representation in the total area. This ensures that the sampling fairly represents the entire habitat.

▶ The sample area is usually divided into groups (strata) based on biophysical features, e.g. landform, soil type, elevation etc., and then by vegetative structure, e.g. forest, woodland, grassland etc.

▶ Proportional sampling is an essential feature of stratified sampling. For example, the ecosystem on the right contained 30% eucalypt woodland and 70% grass. The researcher decided to place 20 random quadrat samples in total. To ensure proportional sampling, they placed six quadrats in the eucalypt woodland and 14 in the grass.

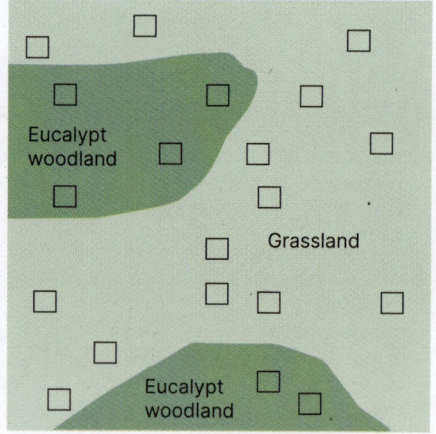

3. A student wants to investigate the incidence of asthma in their school. Describe how they might select samples from the school population using:

(a) Systematic sampling: _____

_____

(b) Stratified sampling: _____

_____

(c) Opportunistic sampling: _____

_____

230 Quadrat Sampling

Key Idea: Quadrat sampling involves a series of random placements of a frame of known size over an area of habitat to assess the abundance or diversity of organisms.

Quadrat sampling is a method by which organisms in a certain proportion (sample) of the habitat are counted directly. It is used when the organisms are too numerous to count in total. It can be used to estimate population abundance (number), density, frequency of occurrence, and distribution. Quadrats may be used without a transect when studying a relatively uniform habitat. In this case, the quadrat positions are chosen randomly using a random number table. The general procedure is to count all the individuals (or estimate their percentage cover) in a number of quadrats of known size and to use this information to work out the abundance or percentage cover value for the whole area.

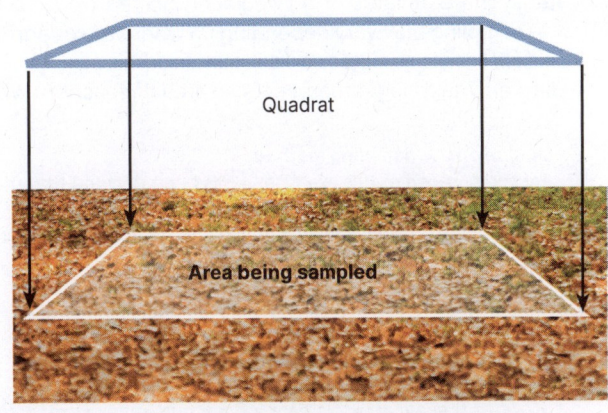

Quadrat

Area being sampled

$$\text{Estimated average density} = \frac{\text{Total number of individuals counted}}{\text{Number of quadrats} \times \text{area of each quadrat}}$$

Guidelines for quadrat use:

1. The area of each quadrat must be known. Quadrats should be the same shape, but not necessarily square.

2. Enough quadrat samples must be taken to provide results that are representative of the total population.

3. The population of each quadrat must be known. Species must be distinguishable from each other, even if they have to be identified at a later date. It has to be decided beforehand what the count procedure will be and how organisms over the quadrat boundary will be counted.

4. The size of the quadrat should be appropriate to the organisms and habitat, e.g. a large size quadrat for trees.

5. The quadrats must be representative of the whole area. This is usually achieved by random sampling (right).

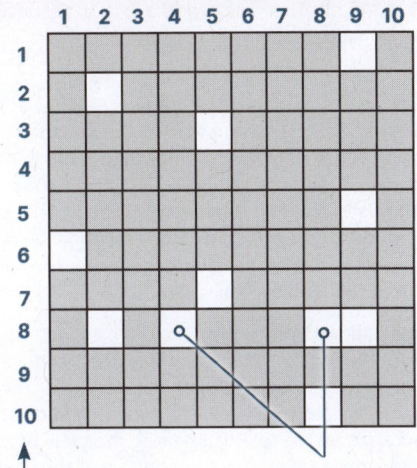

The area to be sampled is divided up into a grid pattern with indexed coordinates.

Quadrats are applied to the predetermined grid on a random basis. This can be achieved by using a random number table.

Sampling a centipede population

A researcher by the name of Lloyd (1967) sampled centipedes in Wytham Woods near Oxford in England. A total of 37 hexagon–shaped quadrats were used, each with a diameter of 30 cm (see diagram on right). These were arranged in a pattern so that they were all touching each other. Use the data in the diagram to answer the following questions.

1. Determine the average number of centipedes captured per quadrat:

_____

2. Calculate the estimated average density of centipedes per square metre (remember that each quadrat is 0.08 square metres in area):

_____

3. Looking at the data for individual quadrats, describe in general terms the distribution of the centipedes in the sample area:

_____
_____
_____

4. Describe one factor that might account for the distribution pattern:

_____
_____

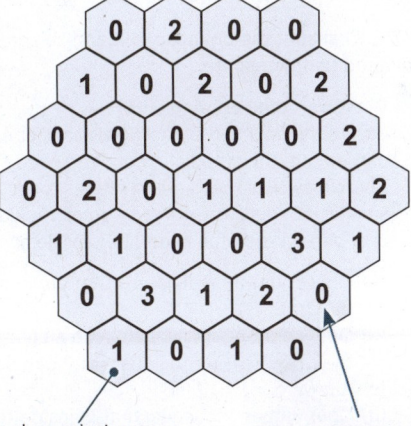

Each quadrat was a hexagon with a diameter of 30 cm and an area of 0.08 square meters.

The number in each hexagon indicates how many centipedes were caught in that quadrat.

Centipede

C4.1
3

231 Sampling a Rocky Shore Community

Key Idea: The estimates of a population gained from using quadrat sampling may vary depending on where the quadrats are placed. Larger samples can account for variation. The diagram (next page) represents an area of seashore with its resident organisms. The distribution of coralline algae and three animal species are shown. This exercise is designed to prepare you for planning and carrying out a similar procedure to practically investigate a natural community.

1. **Decide on the sampling method**
   For the purpose of this exercise, it has been decided that the populations to be investigated are too large to be counted directly. A quadrat sampling method is to be used to estimate the average density of the four animal species as well as that of the algae.

2. **Mark out a grid pattern**
   Use a ruler to mark out 3 cm intervals along each side of the sampling area (area of quadrat = 0.03 × 0.03 m). Draw lines between these marks to create a 6 × 6 grid pattern (total area = 0.18 × 0.18 m). This will provide a total of 36 quadrats that can be investigated.

3. **Number the axes of the grid**
   Only a small proportion of the possible quadrat positions will be sampled. It is necessary to select the quadrats in a random manner. It is not sufficient to simply guess or choose your own on a 'gut feeling'. The best way to choose the quadrats randomly is to create a numbering system for the grid pattern and then select the quadrats from a random number table. Starting at the top left hand corner, number the columns and rows from 1 to 6 on each axis.

4. **Choose quadrats randomly**
   To select the required number of quadrats randomly, use random numbers from a random number table. The random numbers are used as an index to the grid coordinates. Choose 6 quadrats from the total of 36 using the table of random numbers provided for you at the bottom of the next page. Make a note of which column of random numbers you choose. Each member of your group should choose a different set of random numbers (i.e. different column: A–D) so that you can compare the effectiveness of the sampling method.

   Column of random numbers chosen: _____

   NOTE: Highlight the boundary of each selected quadrat with coloured pen/highlighter.

5. **Decide on the counting criteria**
   Before the counting of the individuals for each species is carried out, the criteria for counting need to be established. There may be some problems here. You must decide before sampling begins as to what to do about individuals that are only partly inside the quadrat. Possible answers include:

   (a) Only counting individuals that are completely inside the quadrat.
   (b) Only counting individuals with a clearly defined part of their body inside the quadrat (such as the head).
   (c) Allowing for 'half individuals' (e.g. 3.5 barnacles).
   (d) Counting an individual that is inside the quadrat by half or more as one complete individual.

   Discuss the merits and problems of the suggestions above with other members of the class (or group). You may even have counting criteria of your own. Think about other factors that could cause problems with your counting.

6. **Carry out the sampling**
   Carefully examine each selected quadrat and count the number of individuals of each species present. Record your data in the spaces provided on the next page.

7. **Calculate the population density**
   Use the combined data TOTALS for the sampled quadrats to estimate the average density for each species by using the formula:

$$\text{Density} = \frac{\text{Total number in all quadrats sampled}}{\text{Number of quadrats sampled} \times \text{area of a quadrat}}$$

Remember that a total of 6 quadrats are sampled and each has an area of 0.0009 m^2. The density should be expressed as the number of individuals per square metre (no. m^{-2}).

Plicate barnacle:		Snakeskin chiton:	
Oyster borer		Coralline algae:	

8. Calculate the mean no. of organisms per quadrat:

Plicate barnacle:		Snakeskin chiton:	
Oyster borer		Coralline algae:	

9. Estimate the total population for each species for the area sampled:

$$\text{Population} = \text{Mean number of individuals per quadrat} \times (\text{area of habitat} \div \text{area of quadrat})$$

Plicate barnacle:		Snakeskin chiton:	
Oyster borer		Coralline algae:	

1. (a) Compare your population estimate with others in your class. Discuss any differences in their totals:

   _____

   _____

   (b) Carry out a direct count of all 3 animal species and the algae for the whole sample area (all 36 quadrats). Did your population estimate match the actual population in the area? Discuss:

   _____

   _____

©2024 **BIOZONE** International
ISBN: 978-1-99-101410-8
Photocopying prohibited

Coordinates for each quadrat	Plicate barnacle	Oyster borer	Snakeskin chiton	Coralline algae
1:				
2:				
3:				
4:				
5:				
6:				
TOTAL				

Table of random numbers

A	B	C	D
2 2	3 1	6 2	2 2
3 2	1 5	6 3	4 3
3 1	5 6	3 6	6 4
4 6	3 6	1 3	4 5
4 3	4 2	4 5	3 5
5 6	1 4	3 1	1 4

The table above has been adapted from a table of random numbers from a statistics book. Use this table to select quadrats randomly from the grid above. Choose one of the columns (A to D) and use the numbers in that column as an index to the grid. The first digit refers to the row number and the second digit refers to the column number. To locate each of the 6 quadrats, find where the row and column intersect, as shown below:

Example: 5 2   refers to the 5th row and the 2nd column

232 Estimating Population Size: Quadrats

Key Idea: Sampling using quadrants to count organisms can help provide an estimate of the total population.
In this investigation, you will use the knowledge and skills practised in Activities 228 to 231 and use quadrats to carry out a survey of a specific plant or sessile (non moving) animal in a specific area to calculate its population.

Investigation 12.1 Carry out a sampling survey using quadrats to estimate the population of a plant or sessile animal.

See appendix for equipment list.

> ⚠ **Care should always be taken when working outdoors. Beware of slippery surfaces or steep drop-offs. Care should also be taken not to harm plants and animals when sampling.**

1. Decide on the animal or plant you want to determine the population of, e.g. the population of barnacles on a section of rocky shore or the population of lichen on an exposed rock face. You may wish to find the populations of multiple organisms living next to each other, e.g. different barnacle species on the same rocky shore.

2. Determine the area for sampling. This could be a local park, field, forest remnant, or seashore. Knowing the total area that could potentially be surveyed is needed for a population estimate. This could be calculated from a map that is to scale e.g. an ordinance or topographic map.

   Investigation area: _____

3. Within your survey area, determine the placement of quadrats using a random number generator to produce a set of coordinates or distances along a transect. The number of quadrats you use should be enough to be able to accurately represent the population number, but not so many as to take an unreasonable amount of time. Larger and more diverse survey areas will need more quadrats than smaller, more homogeneous areas.

4. Calculate the number of quadrats that will fit into your study area: (No. quadrats = study area / quadrat area).

   _____

5. Carry out your survey. Use a log book to produce a table that records the quadrat coordinates or number, and the number of your chosen animals or plants in each quadrat.

6. Once you have collected your data, you need to analyse it. Calculate the following statistics. Using a spreadsheet will save time and increase accuracy (*see the* BIOZONE Resource Hub).

   The mean number of organisms per quadrat:

   _____

   The standard deviation for the mean number of organisms per quadrat:

   _____

   The 95% confidence interval for the mean number of organism per quadrat:

   _____

   The mean number of organisms per square metre:

   _____

   The standard deviation for the mean number of organisms per square metre:

   _____

   The 95% confidence interval for the mean number of organism per square metre:

   _____

   The total population of organisms based on your mean number of organisms per quadrat:

   _____

   The 95% confidence interval for the total population (total possible quadrats x confidence interval):

   _____

1. What was the population of organisms in the area you investigated? What was its possible range? Discuss your results and how you could make your estimate of the population more accurate:

   _____

   _____

©2024 **BIOZONE** International
ISBN: 978-1-99-101410-8
Photocopying prohibited

233 Estimating Population Size: Mark and Recapture

Key Idea: Mark and recapture sampling enables estimates of the population size of highly mobile organisms.

The mark and recapture method of estimating population size is used in the study of animal populations in which the individuals are highly mobile. It is of no value where animals do not move or move very little. The number of animals caught in each sample must be large enough to be valid. The technique is outlined in the diagram below.

First capture	**Release back into the natural population**	**Second capture**
In the first capture, a random sample of animals from the population is selected. Each selected animal is marked in a distinctive way.	The marked animals from the first capture are released back into the natural population and left for a period of time to mix with the unmarked individuals.	Only a proportion of the second capture sample will have animals that were marked in the previous capture.

The Lincoln Index

$$\text{Total population (N)} = \frac{\text{Number in 1st sample (all marked) (M)} \times \text{Number in 2nd sample (n)}}{\text{Number of marked individuals recaptured in 2nd sample (R)}}$$

Steps in the mark and recapture technique:

1. Sample the population by capturing as many individuals as is possible and practical. Capture technique will depend on the animal.

2. Mark the captured animals to distinguish them from unmarked animals.

3. Return the marked animals to their habitat and leave them for an extended period to allow them to redistribute themselves in the population.

4. Sample the population again (the sample must be large enough to provide valid data but the sample size can be different to the first).

5. Determine the numbers of marked to unmarked animals in the second sample. Use the equation above to estimate the population size.

Animals may be marked or tagged

USFWS

1. For this exercise, you will need several boxes of matches and a pen. Work in a group of 2-3 students to 'sample' the population of matches in the full box by using the mark and recapture method. Each match will represent one animal.

(a) Take out 10 matches from the box and mark them on 4 sides with a pen so that you will be able to recognise them from the other unmarked matches later.

(b) Return the marked matches to the box and shake the box to mix the matches.

(c) Take a sample of 20 matches from the same box and record the number of marked matches and unmarked matches.

(d) Determine the total population size by using the equation above.

(e) Repeat the sampling 4 more times (steps b–d above) and record your results:

	Sample 1	Sample 2	Sample 3	Sample 4	sample 5
Estimated population					

(f) Count the actual number of matches in the matchbox: _____

(g) Compare the actual number to your estimates and state by how much it differs: _____

C1.1
4
AOS

Researchers at New Zealand's University of Waikato used mark and recapture to obtain estimates of population number and biomass of four fish species and one hybrid in one of the campus lakes. Fish were sampled using electrofishing (right), which temporarily stuns the fish so that they can be netted. Biomass estimates for each species were calculated from the mean mass of all fish sampled in the recapture.

Fish were marked with a fin clip (tagging carries a higher risk of infection at the tagging site). Resampling period was 7 weeks (22 January 2014-14 March 2014). The purpose of recapture was to remove pest species (all species except shortfin eels).

Each of the campus lakes is isolated with no inflow or outflow.

Results are presented in the table below:

Electrofishing, campus lakes, University of Waikato NZ

Data provided by Prof. Brendan Hicks, University of Waikato

Species	Number originally marked (M)	Number caught in recapture (n)	Number of marked recaptures (m)	Population estimate (N = M x n / m)	Mean fish mass (g)	Biomass (kg)	Lake area (ha)	Biomass by area (kg per ha)
Goldfish	32	104	14		365		0.69	
Koi carp-goldfish hybrids	6	9	3		1020		0.69	
Koi carp	9	35	2		114		0.69	
Catfish	7	33	2		303		0.69	
Shortfin eels	45	12	1		189		0.69	

2. (a) Complete the columns in the table above for population estimate, biomass, and biomass by area (kg per hectare).

   (b) What is the significance of the lake being isolated in terms of the reliability of the mark-recapture estimates?

   _____

   _____

   (c) Why is it useful to make biomass estimates for the fish in this lake, especially when the pest fish are being removed:

   _____

   _____

3. Describe some of the problems with the mark and recapture method if the second sampling is:

   (a) Left too long a time before being repeated: _____

   _____

   _____

   (b) Too soon after the first sampling: _____

   _____

   _____

   _____

4. Describe two assumptions in this method of sampling that would cause the method to fail if they were not true:

   (a) _____

   _____

   (b) _____

   _____

5. Some types of animal would be unsuitable for this method of population estimation, i.e. would not work.

   (a) Name an animal for which this method of sampling would not be effective: _____

   (b) Explain your answer: _____

   _____

234 Carrying Capacity

Key Idea: Carrying capacity (K) is the size of a population that can be indefinitely sustained by the environment.

An ecosystem's carrying capacity (K), i.e. the size of population that the available resources can sustain indefinitely, is limited by the ecosystem's resources. Factors affecting carrying capacity (population limiting factors) can be biotic, e.g. food supply, or abiotic, e.g. water, climate, and available space. The carrying capacity is determined by the most limiting factor and can change over time, e.g. as a result of environmental changes. A population at below carrying capacity will increase because resources are not limiting. As the population approaches its carrying capacity (or exceeds it), resources become limiting and environmental resistance increases, thereby decreasing population growth. An understanding of carrying capacity is important to explaining population fluctuations and patterns of population growth.

Limiting factors and carrying capacity

Limiting factors are factors that limit the growth, abundance, or distribution of an organism or a population. Which factor is limiting and its effect may change over time. Ultimately, changes in limiting factors (such as water or food) operate by changing a habitat's carrying capacity. The graph (right) shows how the carrying capacity of a steppe environment varies based on changes to biotic and abiotic limiting factors.

1. A small number of steppe voles move into an area of steppe. The population increases quickly.

2. The population overshoots the carrying capacity.

3. Large numbers damage the environment and food becomes limiting. The carrying capacity falls.

4. The population stabilizes at the new carrying capacity.

5. The steppe experiences a drought and carrying capacity is reduced.

6. The drought breaks and carrying capacity rises to a stable but lower level because of drought damage.

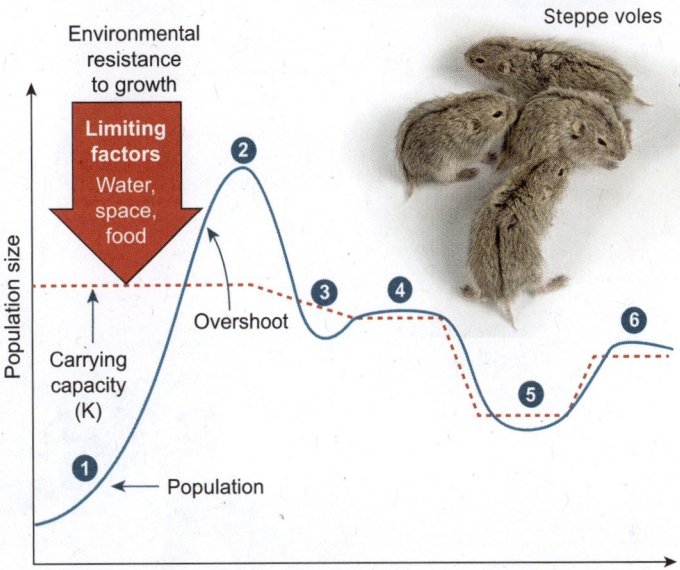

Steppe voles

Large wildfires, especially when they occur with high frequency, can severely reduce the carrying capacity of ecosystems, especially if recovery is slowed by climatic factors such as drought.

Temporary increases in carrying capacity (increase in food supplies) can result in rapid population increases, especially in fast breeding species. When carrying capacity is exceeded, many will die.

Water is a critical limiting factor in many ecosystems. Droughts, especially over several years, can reduce the ecosystem's carrying capacity and result in large losses of crops, wildlife, and livestock.

1. Explain how changes in limiting factors alter carrying capacity? _____

_____

_____

2. What limiting factors have changed at points 3, 5, and 6 in the graph above, and how have they changed?

   (a) 3: _____

   (b) 5: _____

   (c) 6: _____

3. Explain what the scenario above illustrates about carrying capacity: _____

_____

_____

©2024 **BIOZONE** International
ISBN: 978-1-99-101410-8
Photocopying prohibited

C4.1

5

235 Population Regulation

Key Idea: Population size and growth is regulated by factors that are both dependent and independent of the population density.

Very few species show continued exponential growth. Population size is regulated by factors that limit population growth. The diagram below illustrates how population size can be regulated by environmental factors. Density independent factors are those that act independently of population density and affect all individuals more or less equally. Density dependent factors have a greater effect when the population density is higher. They become less important when the population density is low.

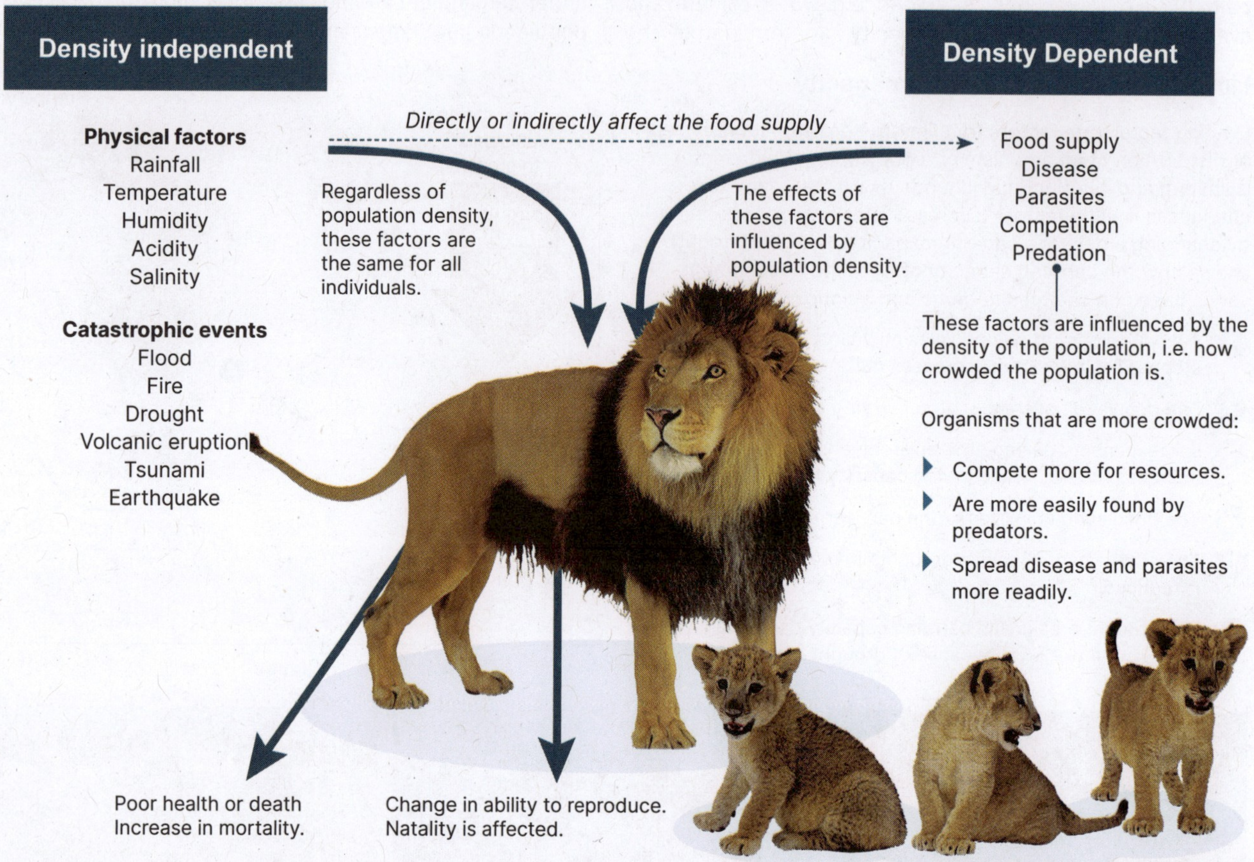

Density independent

Density Dependent

Directly or indirectly affect the food supply

Physical factors
Rainfall
Temperature
Humidity
Acidity
Salinity

Regardless of population density, these factors are the same for all individuals.

The effects of these factors are influenced by population density.

Food supply
Disease
Parasites
Competition
Predation

These factors are influenced by the density of the population, i.e. how crowded the population is.

Catastrophic events
Flood
Fire
Drought
Volcanic eruption
Tsunami
Earthquake

Organisms that are more crowded:

▶ Compete more for resources.

▶ Are more easily found by predators.

▶ Spread disease and parasites more readily.

Poor health or death
Increase in mortality.

Change in ability to reproduce.
Natality is affected.

1. Discuss the role of density dependent factors and density independent factors in population regulation. In your discussion, make it clear that you understand the meaning of each of these terms:

_____
_____
_____
_____
_____
_____

2. Explain how an increase in population density allows disease to have a greater influence in regulating population size:

_____
_____
_____
_____

3. Explain how density dependent factors can form negative feedback controls on populations:

_____
_____
_____

C4.1
6

©2024 **BIOZONE** International
ISBN: 978-1-99-101410-8
Photocopying prohibited

236 Population Growth

Key Idea: The population number is a result of births, death, and migration.

The number of individuals in a population (N) is calculated by knowing the gains and losses to the population. Births, deaths, immigrations (movements into the population) and emigrations (movements out of the population) are events that together determine the number of individuals in a population. Scientists usually measure the rate of these events, which are influenced by the resources available and the biotic potential, which varies among species.

Births, deaths, immigration (movements into the population) and emigration (movements out of the population) are events that determine the population size. Population growth depends on the number of individuals added to the population from births and immigration, minus the number lost through deaths and emigration. This is expressed as:

> **Population growth =**
> **(Births + Immigration) – (Deaths + Emigration)**
> **(B + I)   –   (D + E)**

The difference between immigration and emigration gives net migration. Ecologists usually measure the rate of these events. These rates are influenced by limiting factors in the environment (such as availability of food, water, or habitat) and by the characteristics of the organisms themselves (their biotic potential or natural capacity to increase, *r*).

Rates in population studies are commonly expressed in one of two ways:

▶ **Numbers per unit time**, e.g. 20,150 live births per year. The birth rate is termed the natality, whereas the death rate is the mortality.

▶ **Per capita rate** (number per head of population), e.g. 122 live births per 1000 individuals per year (12.2%).

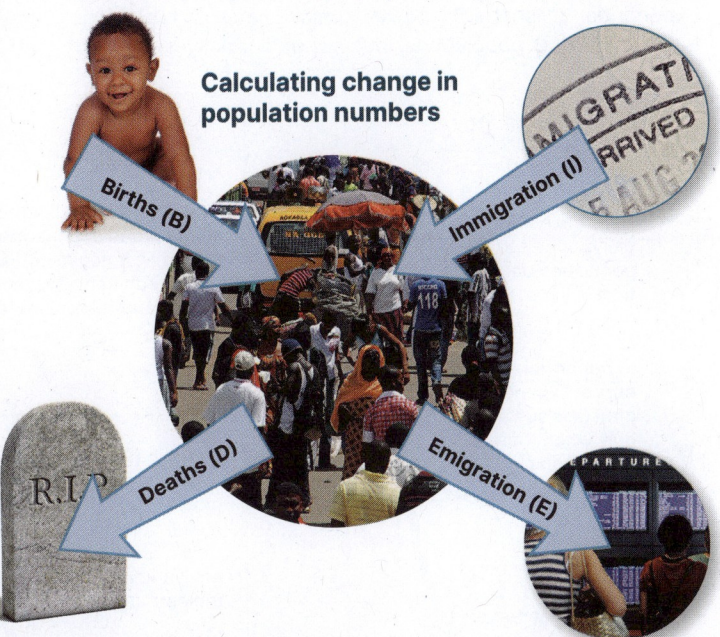

Calculating change in population numbers

Births (B) · Immigration (I) · Deaths (D) · Emigration (E)

The human population is estimated to peak at around 9.7 billion by 2050 as a result of multiple factors, including falling birth rates. Humans have the technology and production efficiency to solve many resource problems and so might appear exempt from the direct influence of limiting factors. However, declining availability of water and land for food production are limiting factors likely to constrain population growth, at least regionally.

1. Define the following terms used to describe changes in population numbers:

   (a) Death rate (mortality): _____

   (b) Birth rate (natality): _____

   (c) Net migration rate: _____

2. Explain how the concept of limiting factors applies to population biology: _____

   _____

   _____

3. Using the terms, B, D, I, and E (above), construct equations to express the following:

   (a) A population in equilibrium: _____

   (b) A declining population: _____

   (c) An increasing population: _____

4. A population started with a total number of 100 individuals. Population data was collected over the following year. Calculate birth rates, death rates, net migration rate, and rate of population change for the data below (as percentages):

   (a) Births = 14: Birth rate = _____ (b) Net migration = +2: Net migration rate = _____

   (c) Deaths = 20: Death rate = _____ (d) Rate of population change = _____

   (e) State whether the population is increasing or declining: _____

237 Population Growth Curves

Key Idea: Colonizing populations tend to show exponential growth, while established populations show logistic growth.

Populations grow by increasing in numbers. Patterns of population growth fall into two main types: exponential or logistic (sigmoidal). These patterns are a reflection of the biological characteristics of the species and the environment. Exponential (J-shaped) growth (below, left) is typical of early colonizing populations, but is not normally sustained indefinitely because the resources of the environment are limited. The populations of established communities tend to show logistic (S-shaped or sigmoid) growth (below, right), and stabilize at a level that can be supported by the environment. Logistic growth is determined by two things: how fast a species reproduces and the carrying capacity (saturation density) of the environment (represented by the letter K).

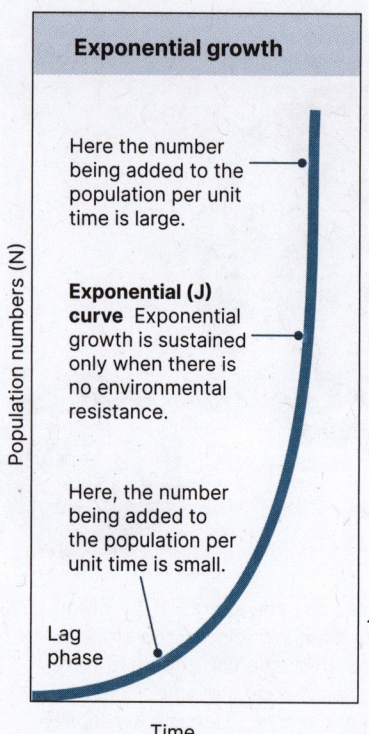

Exponential growth

Here the number being added to the population per unit time is large.

Exponential (J) curve Exponential growth is sustained only when there is no environmental resistance.

Here, the number being added to the population per unit time is small.

Lag phase

Population numbers (N)

Time

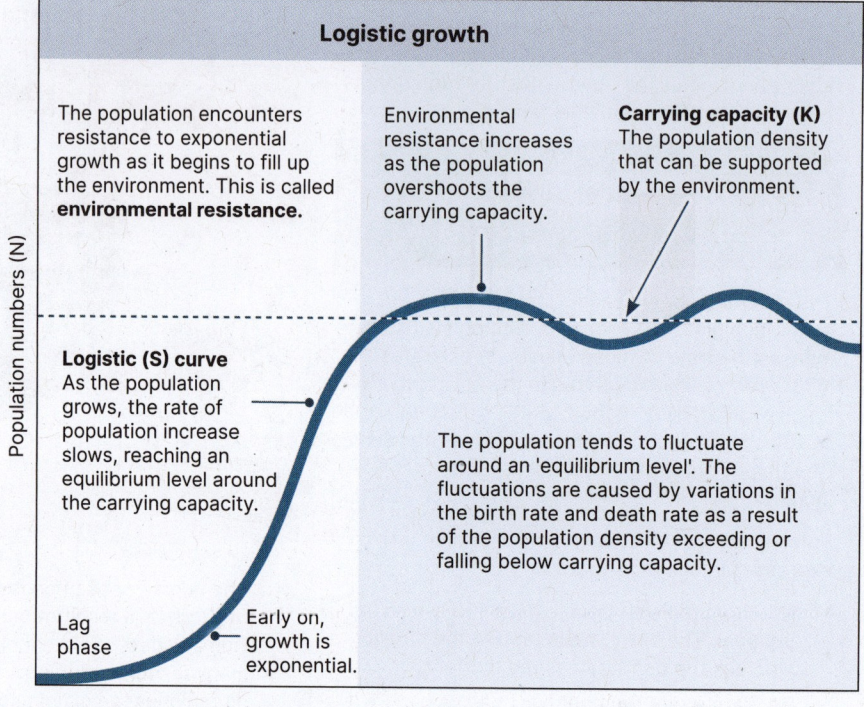

Logistic growth

The population encounters resistance to exponential growth as it begins to fill up the environment. This is called **environmental resistance.**

Environmental resistance increases as the population overshoots the carrying capacity.

Carrying capacity (K) The population density that can be supported by the environment.

Logistic (S) curve As the population grows, the rate of population increase slows, reaching an equilibrium level around the carrying capacity.

The population tends to fluctuate around an 'equilibrium level'. The fluctuations are caused by variations in the birth rate and death rate as a result of the population density exceeding or falling below carrying capacity.

Lag phase

Early on, growth is exponential.

Population numbers (N)

Time

1. Produce a line graph of the grey wolf population on the grid below:

2. What type of growth does the wolf population show? _____

3. What would explain the lag in population growth before 1999? _____

4. Explain why wolves showed this pattern of population growth: _____

_____

_____

Grey wolf population in Montana, USA			
Year	Population	Year	Population
1979	5	1997	52
1981	5	1999	75
1983	9	2001	125
1985	20	2003	190
1987	15	2005	252
1989	16	2007	420
1991	38	2009	525
1993	52	2011	651
1995	60	2013	620
		2014	552

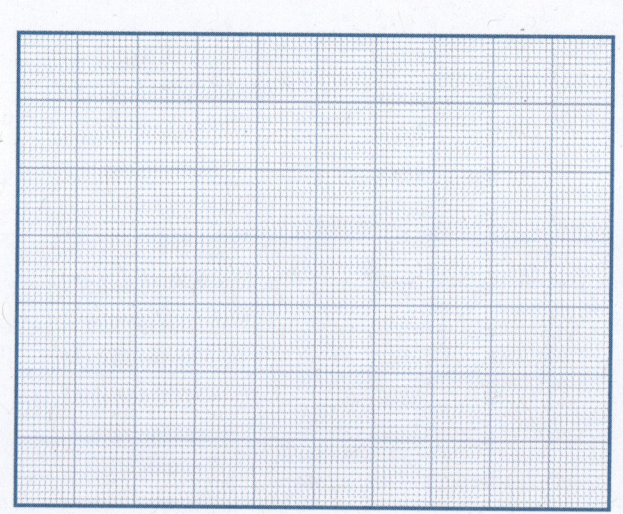

C4.1

7

©2024 **BIOZONE** International
ISBN: 978-1-99-101410-8
Photocopying prohibited

Comparing exponential growth on linear and logarithmic plots

Because numbers from counts change so rapidly for exponential growth, plotting the data on a linear graph normally results in the numbers from earlier counts being very close to the bottom of the graph and difficult to plot and read, while the large numbers from later counts are spread out and take up a much greater part of the graph. This can be remedied by plotting the graph on a logarithmic scale instead of a linear scale (below):

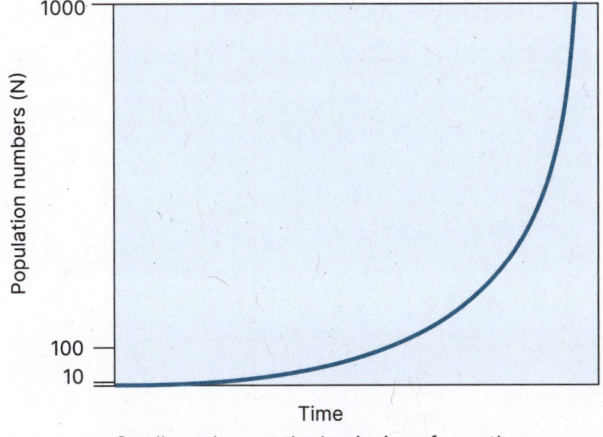

Small numbers at the beginning of growth are difficult to see using a linear scale.

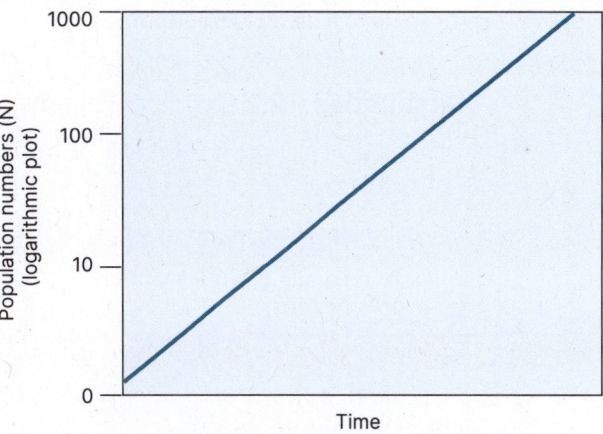

Small numbers at the beginning of growth are much easier to see using a logarithmic scale.

Population growth, West Virginia, USA			
Year	Population	Year	Population
1800	78,592	1920	1,463,701
1810	105,469	1930	1,729,205
1820	136,808	1940	1,901,974
1830	176,924	1950	2,005,552
1840	224,537	1960	1,860,421
1850	308,313	1970	1,744,237
1860	376,688	1980	1,949,644
1870	442,014	1990	1,793,477
1880	618,457	2000	1,807,021
1890	762,794	2010	1,854,265
1900	958,800	2020	1,791,420
1910	1,221,119		

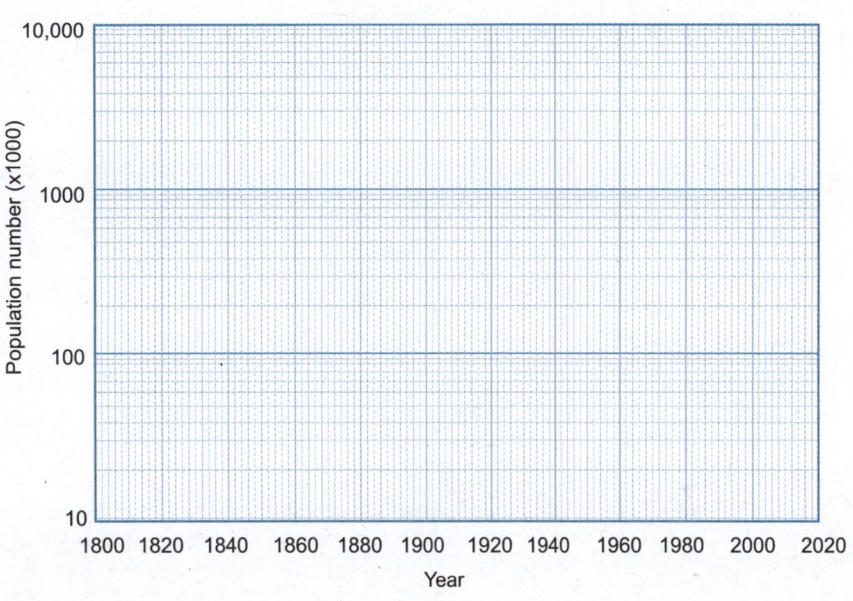

5. Census information provides a good test of the predictions of population growth models. The table of data above records census data for the human population of West Virginia 1800-1990.

(a) Based on what you know so far, what sort of curve would you predict for a new settlement of humans?

(b) Why has this curve occurred?

(c) If the data were plotted on a graph with a linear scale, what shape would the graph be?

(d) If the data were plotted on graph with a logarithmic scale, what shape would the graph be?

(e) Plot the data on the grid provided. The y axis is logarithmic. Both axes have been labelled to help you.
Note: the y axis starts at 10,000.

(f) Did the plotted curve agree with your predictions from 5 (a)?

238 Modelling Population Growth

Key Idea: Population growth can be studied using an organism that reproduces quickly in laboratory conditions. Yeasts are unicellular fungi commonly used in baking and brewing. Yeast can reproduce rapidly in warm environments by asexual reproduction. New cells bud off old cells. A small population of cells can quickly lead to a large population. During this lab, you will make a yeast solution and count and record the population growth over time.

Investigation 12.2 Modelling yeast growth

See appendix for equipment list.

1. You will model the population growth of yeast cells over the course of 2-3 days (depending on lab time).

2. In a 500 mL beaker, add a tablespoon (about 10g) of sugar to 300 mL of tap water at room temperature. Stir until the sugar is dissolved.

3. Add 2.5g (about half a teaspoon) of baker's yeast and stir gently.

4. Using a pipette, remove 1 drop of the solution from the middle of the beaker.

5. Place the drop on a microscope slide and cover with a coverslip.

6. Focus the microscope on low power (10x) and then on high power (40x).

7. Count the number of yeast cells you can see and record. Carefully move the view to another section of the slide and again count and record the number of cells you can see. Do this two more times. Calculate and record the average number of cells. In a logbook create a table for recording the time and average number of yeast cells counted. Make sure you record the amount of time (in hours) elapsed since the experiment started every time you count the number of yeast cells.

8. Cover the beaker with a petri dish or similar to prevent anything further entering the dish. It does not need to be air tight. Place the beaker in a dark place where the temperature is relatively stable (the temperature will not rise above about 40°C or fall below 15°C).

9. After 1 hour (earlier if the lesson is shorter) repeat steps 4-7. Stir the yeast solution gently to ensure the cells are evenly dispersed before taking a sample.

10. Repeat sampling and counting often as possible over the course of three days. Every 2-3 hours is best.

11. If there are too many cells to count, you can dilute your sample in the following way. Remove one drop of yeast solution and place in a test tube. Add one drop of distilled water and mix. Remove one drop and place on the microscope for counting. This has diluted the solution by half, so make sure you multiply your counts by two (if you dilute by three times, multiply your counts by three, etc).

12. Once you have gathered your data, use a spreadsheet, e.g. Excel, to graph the average cell count vs time. Produce a linear and a logarithmic plot of the data. Print them out and attached to this page.

1. How did the yeast population develop over the time of the experiment? Was there exponential or logistic growth?

_____

_____

_____

2. Explain the population growth you saw: _____

_____

_____

_____

3. How might the population growth have differed if the sugar in the original solution was increased?

_____

_____

4. How might the population growth have differed if the sugar in the solution was replenished every day of the experiment?

_____

_____

C4.1
8

©2024 **BIOZONE** International
ISBN: 978-1-99-101410-8
Photocopying prohibited

239 Ecosystems and Communities

Key Idea: A community is all the organisms in an ecosystem. An ecosystem includes all the interactions between the organisms in the community and the abiotic environment.

An ecosystem is a community of living organisms and the physical (non-living) components of their environment. The community (the living components of the ecosystem) is itself made up of a number of populations, these being organisms of the same species living in the same geographical area. The type and availability of resources, such as water, in the environment determine species distribution and survival and are an important influence on how different species interact.

BIOTIC FACTORS	ABIOTIC FACTORS		
The living organisms in the environment, including their interactions, e.g. as competitors, predators, or symbionts. • Plants • Animals • Microorganisms • Fungi • Protists, e.g. algae, protozoans	**Hydrosphere (water)** • Dissolved nutrients • pH • Salinity • Dissolved oxygen • Precipitation • Temperature	**Atmosphere (air)** • Wind speed • Wind direction • Humidity • Light intensity/quality • Precipitation • Temperature	**Geosphere (rock/soil)** • Nutrient availability • Soil moisture • pH • Composition • Temperature • Depth

▶ Ecosystems are natural units made up of a community of living organisms (biotic factors) and the physical conditions (abiotic factors) in an area. Abiotic factors include non-living factors associated with the geosphere, hydrosphere, and atmosphere (above). The living organisms and their activities, e.g. as predators, competitors etc. make up the biotic factors of an ecosystem.

▶ The interactions of living organisms with each other and with the physical environment help determine an ecosystem's features. The components of an ecosystem are linked to each other (and to other ecosystems) through nutrient cycles and energy flows.

1. Distinguish clearly between a community and an ecosystem:

_____

_____

_____

2. The image above depicts buffalo in Yellowstone National Park. From the following list, assign the appropriate term to each of the features described below. Terms: *population, community, ecosystem, physical factor*.

(a) All the buffalo present: _____ (c) All the organisms present: _____

(b) The entire National Park: _____ (d) The river: _____

3. An ecosystem provides resources to its community of living organisms, including food, water, and habitat. In addition, an ecosystem provides essential services such as nutrient recycling and climate regulation. How do you think the availability of resources might influence the distribution and abundance of species present, and affect how species might interact?

_____

_____

_____

_____

_____

240 Interactions Within Species

Key Idea: Interactions within species includes both competition for resources and cooperation to gain those resources.

Intraspecific competition describes competition among individuals of the same species. These individuals share the same resource requirements so competition for resources (food, habitat, mates) is more intense than competition between different species. Resource limitation affects individual and population growth and often determines the distribution of individuals within the environment. However, many species also cooperate, especially those living in social groups, for example, in hunting and defensive actions.

Scramble competition

Direct competition between individuals for a finite resource is called scramble competition. These birch sawfly larvae feed along the birch leaf edges. Larvae compete for the same food and those hatching too late in the season are unlikely to survive.

Contest competition

In contest competition, there is a winner and a loser and resources are obtained completely or not at all. For example, male elephant seals fight for territory and mates. Unsuccessful males may not mate at all.

Competition in social species

In many social species, dominance hierarchies ensure that dominant individuals have priority access to resources. Lower ranked individuals must contest what remains. If food is scarce, only dominant individuals may receive enough to survive.

Resource competition limits population growth

Most frogs hatch from eggs as larvae (pollywogs/tadpoles) and then undergo metamorphosis into the adult form. The larvae must reach a minimum mass to successfully complete metamorphosis, so they depend on getting enough food. Individuals that take too long to reach the minimum required mass decrease their chances of successful metamorphosis and often die before becoming adults.

▸ Researchers Dash and Hota (1980) investigated the growth rate of frog larvae (*Rana*) reared experimentally at different densities (5, 40, 80, and 160) in the same 2 L volume. When density increases, more larvae compete for the same food resources.

▸ The results are plotted right. The increase in mean body mass over time indicates growth rate. The flattening of the curve indicates size at metamorphosis.

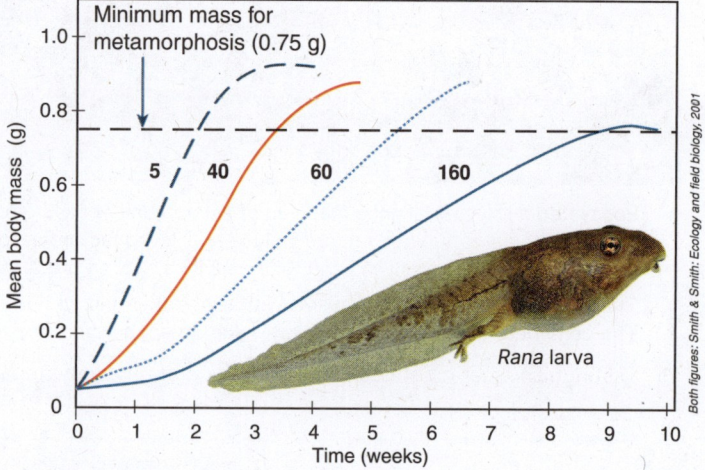

Minimum mass for metamorphosis (0.75 g)

Rana larva

Both figures: Smith & Smith: Ecology and field biology, 2001

1. Describe the difference between scramble competition and contest competition: _____

_____

_____

_____

2. Answer the following questions, based on evidence from the *Rana* growth rate data:

(a) What is the effect of increasing density on individual growth rate of *Rana* larvae? _____

_____

_____

(b) Explain the likely effect of strong resource competition on population size of *Rana*: _____

_____

_____

_____

C4.1
9

©2024 **BIOZONE** International
ISBN: 978-1-99-101410-8
Photocopying prohibited

Cooperative behaviour

▶ **Cooperative behaviour** describes two or more individuals working together to achieve a common goal, such as defence, food acquisition, or rearing young. It increases the probability of survival for all individuals involved. Examples include hunting as a team, e.g. wolf packs, chimpanzee hunts; responding to the actions of others with a similar goal, e.g. migrating mammals; or acting to benefit others, e.g. mobbing in small birds. Cooperation occurs most often between members of the same species.

▶ **Altruism** is an extreme form of cooperative behaviour in which one individual disadvantages itself for the benefit of another. Altruism is often seen in highly social animal groups. Most often, the individual who is disadvantaged receives benefit in some non-material form, e.g. increased probability of passing genes onto the next generation.

Coordinated behaviour is used by many social animals for the purpose of both attack (group hunting) and defence. Cooperation improves the likelihood of a successful outcome, e.g. a successful kill.

Animals may move en masse in a coordinated way and with a common goal, as in the mass migrations of large herbivores. Risks to the individual are reduced by group behaviour.

Kin selection is altruistic behaviour towards relatives. Individual meerkats from earlier litters remain to care for new pups instead of breeding themselves. They help more often when more closely related.

Evidence of cooperation between species

▶ Many small bird species will cooperate to attack a larger predatory species, such as a hawk, and drive it off. This behaviour is called mobbing. It is accompanied by mobbing calls, which can communicate the presence of a predator to other vulnerable species, which benefit from and will become involved in the mobbing.

▶ One example is the black-capped chickadee, a species that often forms mixed flocks with other species. When its mobbing calls in response to a screech owl were played back, at least ten other species of small bird displayed various degrees of mobbing behaviour. The interspecific communication helps to coordinate the community.

3. (a) What is altruism? _____

_____

(b) Why would altruism be more common when individuals are related? _____

_____

_____

_____

_____

4. How do cooperative interactions enhance the survival of both individuals and the group they are part of?

_____

_____

_____

5. What evidence is there that unrelated species can act cooperatively? Why would they do this?

_____

_____

_____

_____

©2024 **BIOZONE** International
ISBN: 978-1-99-101410-8
Photocopying prohibited

241 Interactions Between Species

Key Idea: Species interactions may be beneficial for both, harmful for both, or beneficial for just one species.

Organisms do not live in isolation. The interactions within and between species are an important component of the biotic factors that structure every ecosystem. Many of these relationships involve exploitation: a predator eats its prey and herbivores eat plants. Other relationships involve two or more species being entirely or partly reliant on their very close ecological relationship. Such a relationship is called a symbiosis. Symbioses can be beneficial for both parties (a mutualism) or exploitative (as in parasitism). Resource availability often influences the extent to which species interact. While this is most obvious for species competing for limited resources, it is important in other interactions also.

Type of interaction between species				
Mutualism	**Competition**	**Parasitism**	**Predation/Herbivory**	**Pathogenicity**
A symbiosis in which both species benefit. If both species depend on the symbiosis for survival, the mutualism is obligate. Mutualism can involve more than two species.	Individuals of the same or different species compete for the same limited resources. Both parties are detrimentally affected.	A symbiotic relationship in which the parasite lives in or on the host, taking all its nutrition from it. The host is harmed but not usually killed, at least not directly. Parasites may have multiple hosts and their transmission is often linked to food webs.	Predators eat other organisms. Carnivores eat animals, killing their prey in the process. Herbivory is plant predation. Herbivores eat plants but do not normally end up killing the plant. Predation is a consumer-resource interaction and a type of exploitation.	Pathogenicity is the property of causing disease. This is limited to microorganisms living on or in plants and animals. They include bacteria, protists, and fungi and usually cause an immune response in their host. Pathogens cause disease by living in or on tissues and releasing toxins that harm the organism.
Examples: Flowering plants and their insect pollinators. The flowers are pollinated and the insect gains food. Ruminants and their rumen protozoa and bacteria. The microbes digest the cellulose in plant material and produce short-chain fatty acids, which the ruminant uses as an energy source.	**Examples**: Neighbouring plants of the same and different species compete for light and soil nutrients. Vultures compete for the remains of a carcass. Insectivorous birds compete for suitable food in a forest. Tree-nesting birds with similar requirements compete for nest sites.	**Example**: Parasitic tongue-replacing isopods cut the blood supply to the tongue of the host fish, causing it to fall off. The parasite attaches to what is left of the tongue, feeding on blood or mucus.	**Examples**: Praying mantis consuming insect prey. Canada lynx eating snowshoe hare. Deer browsing on leaves. Cattle grazing on grass.	**Examples**: Ringworm on the skin of a person. Botulism in ducks. Tuberculosis in cattle.

Honeybee and flower — Ringworm

A ⇄ B

Benefits   Benefits

1. In the spaces above, draw a simple model to show whether each species/individual in the interaction described is harmed or benefits. The first one has been completed for you.

2. Although hyenas attack and kill large animals such as wildebeest, they will also scavenge carrion or drive other animals off their kills.

   (a) Identify this type of interaction: _____

   (b) Describe how each species in this interaction is affected by the presence of the other (benefits/harmed/no effect):

   _____

3. Identify the interactions in the table that are exploitative (one species taking advantage of another):

   _____

   _____

C4.1

11, 12

©2024 **BIOZONE** International
ISBN: 978-1-99-101410-8
Photocopying prohibited

Mutualism as a species interaction

Mutualistic relationships can provide both parties with access to a resource that would otherwise be unavailable (usually food) and create new pathways for energy flow in ecosystems. Stable ecosystems depend on the presence of both species in sufficient numbers to maintain healthy populations.

Staghorn coral

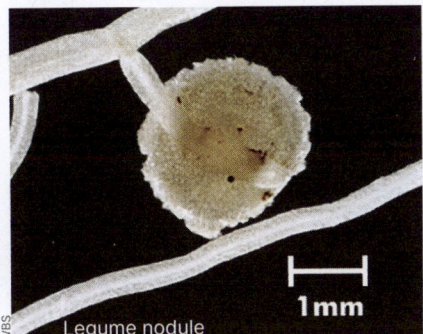
1mm
Legume nodule

Germinating orchid

Reef building corals rely on mutualism with algae in their tissues. The algae supply the coral with energy (glucose and glycerol) and obtain a habitat and the compounds for photosynthesis and growth (CO_2 and nitrogen). This symbiosis is crucial to recycling nutrients within the reef.

Legumes include soybeans, clovers, and peanuts. They are plants that develop root nodules that house the nitrogen-fixing bacteria e.g. *Rhizobia*. The bacteria provides the plant with nitrogen in the form of ammonium. The bacteria receive nutrients and shelter.

Orchids develop critical mutualistic relationships with endomycorrhizal fungi. Orchid seeds cannot germinate without the fungi which provides the seeds' initial nutrients. As the orchid grows, the fungi supply nitrogen and phosphorus, and receives carbohydrates.

4. The honeyeaters are a diverse family of small to medium-sized nectar-feeding birds common in Australia. Many Australian plant species, including proteas and myrtles, are pollinated by honeyeaters.

   (a) Identify this type of interaction: _____

   (b) Describe how each species is affected (benefits/harmed/no effect):

   _____

   _____

5. The squat anemone shrimp, also known as the sexy shrimp, lives among the tentacles of sea anemones, where it gains protection and scavenges scraps of food from the anemone. The anemone is apparently neither harmed nor benefited by the shrimp's presence.

   (a) Identify this type of interaction: _____

   (b) Describe how each species is affected (benefits/harmed/no effect):

   _____

   _____

   _____

6. Dingoes will kill and scavenge a range of species. In groups of two or more, they can attack and kill large animals, such as kangaroos, but will also scavenge carrion, such as this dingo with a fish on Fraser Island.

   (a) Identify this type of interaction: _____

   (b) Describe how each species is affected (benefits/harmed/no effect):

   _____

   _____

7. Detail the vital role in resource creation that each pair of species has between staghorn coral and algae:

   _____

   _____

8. Explain the similarities and differences between a predator and a parasite:

   _____

   _____

242 Interspecific Competition

Key Idea: Different species compete for resources including food and space.

Competition between different species (or interspecific competition) is usually less intense than competition within a species because different species usually do not depend on exactly the same resources. Laboratory studies of interspecific competition in enclosures with limited resources provide evidence that competition can drive one species to extinction. However, in natural populations, competing species usually exist in a more or less stable equilibrium.

Interspecific competition is an important biotic factor in natural communities

Competition between different species takes one of two forms. As with all competition, there is a negative effect on all species involved, because fewer resources are available to each.

Exploitative competition occurs when two or more species use the same resource. This often results in neither species having enough of the resource to meet their needs.

Interference competition involves a direct interaction between competitors. In animals, this is usually aggressive behaviour. In plants, the interaction is passive.

▶ Interspecific competition is usually less intense than competition between members of the same species because competing species have different requirements for at least some resources, e.g. different habitat or food preferences.

▶ Interspecific competition can be an important influence on the species present in an ecosystem or their distribution. However, in naturally occurring populations, it is generally less effective at limiting population size than intraspecific competition, especially in animals. This is because different species usually exploit a different spectrum of resources to avoid direct competition most of the time.

▶ Interspecific competition in natural plant communities is very dependent on nutrient availability and will be greater when soil nutrients are low. Fast growing plants with large, dense root systems can absorb large amounts of nitrogen, depleting soil nitrogen so that other plants cannot grow close to them. Similarly, fast growing plants may quickly grow tall enough to intercept the available light and prevent the germination of plants nearby. Pest plants often have this strategy and become very difficult to control.

▶ Sometimes, humans may introduce a species with the same resource requirements as a native species. The resulting competition can lead to the decline of the native species. For example, the American grey squirrel was introduced to England in the 1800s where it has displaced the smaller, native red squirrel from much of its former range.

In some communities, many different species may be competing for the same resource. This type of competition is called interference competition because the individuals interact directly over a scarce resource. In the example above, three species compete for what remains of a carcass.

The plant species in the forest community above compete for light, space, water, and nutrients. A tree that can grow taller than those around it will be able to absorb more sunlight, and grow more rapidly to a larger size, than the plants in the shade below.

1. (a) What is interspecific competition? Describe an example: _____

_____

_____

(b) Why is interspecific competition usually less intense than intraspecific competition? _____

_____

_____

_____

(c) Why is interspecific competition generally less effective at limiting population size than intraspecific competition?

_____

_____

_____

C4.1

11, 12

©2024 **BIOZONE** International
ISBN: 978-1-99-101410-8
Photocopying prohibited

Competition between red squirrels and grey squirrels in the UK

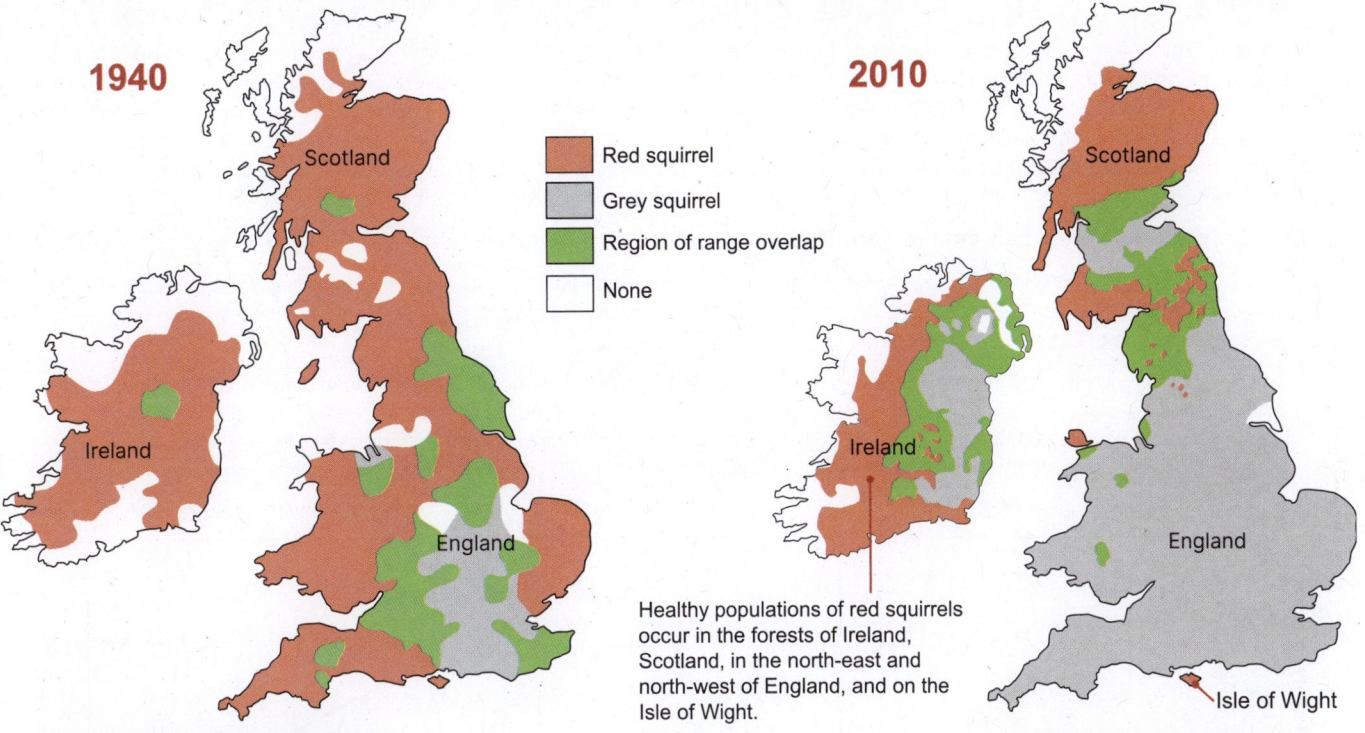

1940

Scotland

Ireland

England

- Red squirrel
- Grey squirrel
- Region of range overlap
- None

2010

Scotland

Ireland

England

Isle of Wight

Healthy populations of red squirrels occur in the forests of Ireland, Scotland, in the north-east and north-west of England, and on the Isle of Wight.

Paul Whipsey cc 3.0

Red squirrel

The European red squirrel was the only squirrel species in Britain until the introduction of the American grey squirrel in 1876. Regular distribution surveys (above) have recorded the reducing range of the reds, with the larger, more aggressive grey squirrel displacing populations of reds over much of England. Grey squirrels can exploit tannin-rich foods, which are unpalatable to reds. In mixed woodland and in competition with greys, reds may not gain enough food to survive the winter and breed. Reds are also very susceptible to several viral diseases, including squirrelpox, which is transmitted by greys.

Whereas red squirrels once occupied a range of forest types, they are now almost solely restricted to coniferous forest. The data suggest that the grey squirrel is probably responsible for the red squirrel decline, but other factors, such as habitat loss, are also likely to be important.

BirdPhotos.com cc 3.0

Grey squirrel

2. (a) What evidence is there that competition with grey squirrels is responsible for the decline in red squirrels in the UK?

_____

_____

_____

(b) Is the evidence conclusive? If not, why not? _____

_____

_____

_____

_____

3. Research a local case of an endemic species in competition with an invasive, introduced species. How has the endemic species been affected by the introduced species? What advantages does the introduced species have that have given it a competitive edge over the endemic species?

_____

_____

_____

243 Testing for Interspecific Competition

Key Idea: The effect of competition can be tested in laboratory conditions or through observation in the field. The competitive exclusion principle states that two competing species cannot occupy the same niche. Recall that in the case of competition between barnacles, *Semibalanus balanoides* out-competed *Chthalamus stellatus*, and that *Chthalamus* stellatus could only thrive in a specific area when *Semibalanus balanoides* was removed. That principle can also be shown in the laboratory; recall Gause's *Paramecium* experiments. The examples below are further evidence that, when competing species try to use the same resource, one will eventually be out-competed and its population will decline.

One competing species can exclude another in a manipulated environment

▶ Experiments in which two competing species are grown in controlled and manipulated conditions typically result in one species out-competing the other. These findings support the competitive exclusion principle, which states that two species with the same resource requirements cannot coexist.

▶ In one laboratory study, two diatom species were grown both alone and together in a nutrient medium that supplied continual silica. Diatoms are algae that require silica to form their cell walls. Grown alone, each species reduced the silica to a low level but the populations were stable. However, *Synedra* depleted the silica to a lower level than did *Asterionella*. When grown together, *Synedra* depleted the silica to a level that caused the *Asterionella* population to die out (below right).

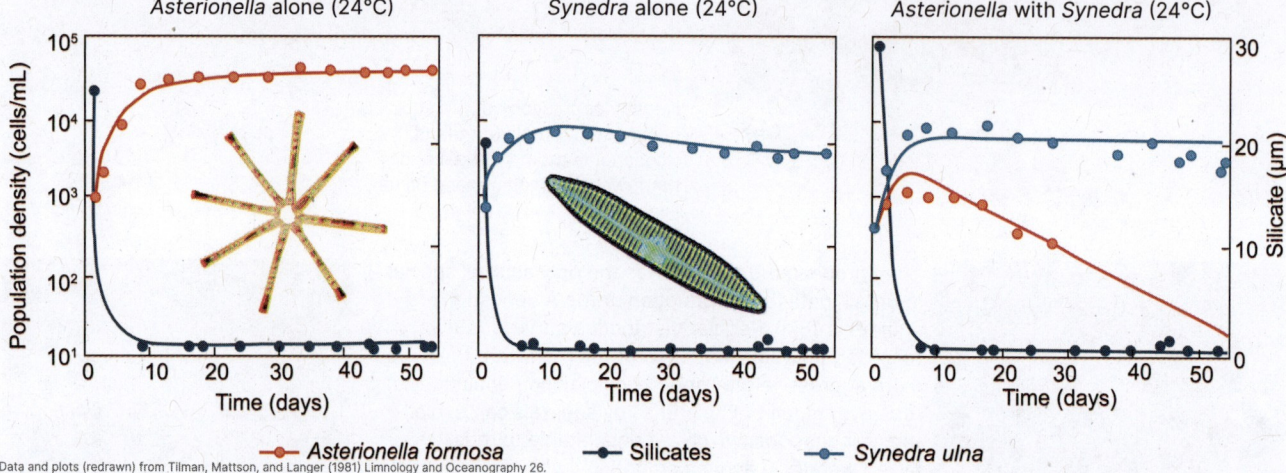

Data and plots (redrawn) from Tilman, Mattson, and Langer (1981) Limnology and Oceanography 26.

1. (a) Using the example of the diatoms above, describe the experimental evidence for the role of resource competition in the exclusion of one species by another:

_____

_____

_____

_____

(b) Suggest why these manipulated environments are unlikely to represent the usual situation in a natural ecosystem:

_____

_____

_____

_____

2. The graph on the right shows the same experiment between *Synedra* and *Asterionella* as that in the top right graph, except in this case the experiment was conducted at 8°C. Describe the effect of this change on the result of the experiment and explain why this might have happened:

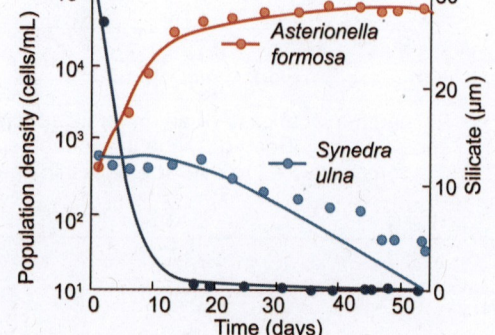

_____

_____

_____

_____

©2024 **BIOZONE** International
ISBN: 978-1-99-101410-8
Photocopying prohibited

Investigating niche overlap and coexistence in gliders

The effect of competition between species can also been shown by observation and gathering data in the field. In large areas of eastern Australia, the distribution of sugar gliders and squirrel gliders overlap. Researchers looked at historical and recent data (below) to see how the two species were ecologically separated where their distribution overlapped.

Gliders in Australia

Gliders are small, nocturnal possum-like marsupials that live most of their life in tree canopies. Seven species of glider are found in Australia, and six of these are found in Queensland. In Australia, gliders of the genus *Petaurus* occupy very similar niches. All are nocturnal, require tree hollows for nesting, and feed on insects, nectar, pollen, honeydew, and plant sap.

Squirrel glider: The squirrel glider is distributed from Victoria to northern Queensland, but is not found on the Cape York Peninsula. Squirrel gliders have a limited habitat range, and are restricted to dry eucalypt forests and woodlands. Squirrel gliders live in family groups of 2-10 individuals and weigh 200-260 g.

Sugar glider: The distribution of the sugar glider is broader than the squirrel glider. It inhabits the eastern and northern coasts of Australia, New Guinea, and the surrounding islands. The sugar glider is found in a wide range of habitats including drier coastal eucalypt forests and woodlands, to wetter rainforest habitats. Sugar gliders live in family groups of 2-10 individuals and weigh 95-160 g.

Squirrel glider

Sugar glider

Table 1. Occurrence of glider species in rainforest and other forest.

	Forest type	
	Rainforest	Other
Historical data		
Sugar glider	77%	23%
Squirrel glider	17%	83%
Recent data		
Sugar glider	64%	36%
Squirrel glider	7%	93%

Table 2. Frequency of glider species at different elevations.

	Records in elevation class			% of records that were rainforest at:	
	< 80 m	80-300 m	> 300 m	< 80 m	> 80 m
Historical data					
Sugar glider	77%	0%	23%	70%	100%
Squirrel glider	85%	12%	3%	14%	33%
Recent data					
Sugar glider	71%	0%	29%	50%	75%
Squirrel glider	85%	13%	2%	6%	13%

Data: Rowston, C & Catterall, C.P. (2004) Habitat segregation, competition and selective deforestation: effects on the conservation status of two similar *Petaurus* gliders. Conservation of Australia's forest fauna http://hdl.handle.net/10072/416

3. Study table 1. What do you notice about the type of forest in which each species is found? _____

_____

_____

_____

4. The majority of both species are found below 80 m (Table 2). How do you think they avoid competition with each other?

_____

_____

5. Suggest why the niche of the sugar glider is more restricted when both species inhabit the same area: _____

_____

_____

_____

6. Using examples in this activity, explain the difference between testing a hypothesis by experiment or by observation:

_____

_____

_____

_____

244 Using the Chi-Squared Test in Ecology

Key Idea: The chi-squared test is used to compare sets of categorical data and evaluate whether differences between them are statistically significant or due to chance.

The chi-squared test (χ^2) is used to determine differences between categorical data sets when working with frequencies (counts). For the test to be valid, the data recorded for each categorical variable, e.g. species, must be raw counts (not measurements or derived data). The chi-squared test is used for two types of comparison: test for goodness of fit and tests of independence (association). A test for goodness of fit is used to compare an experimental result with an expected theoretical outcome. You will perform this test later to compare the outcome of genetic crosses to an expected theoretical ratio. A test for independence evaluates whether two variables are associated. The chi-squared test is not valid when sample sizes are small (<20). Like all statistical tests, it aims to test the null hypothesis; the hypothesis of no difference (or no association) between groups of data. The worked example below uses the chi-squared test for association in a study of habitat preference in mudfish.

Using the chi-squared test for independence

Black mudfish (*Neochanna diversus*) is a small fish species native to New Zealand found in wetlands and swampy streams. Researchers were interested in finding environmental indicators of favourable mudfish habitat. They sampled 80 wetland sites for the presence or absence of mudfish and recorded whether there was emergent vegetation present or absent. Emergent vegetation, defined as vegetation rooted in water but emerging above the water surface, is an indicator of a relatively undisturbed environment. A chi-squared for association was used to test whether mudfish were found more often at sites with emergent vegetation than by chance alone. The null hypothesis was that there is no association. The worked example is below. The table of observed values records the number of sites with or without mudfish and with or without emergent vegetation.

Photo and data: Rhys Barrier, University of Waikato

Black mudfish are able to air-breathe and can survive through seasonal drying of their wetland habitat.

Step 1: Enter the observed values (O) in a contingency table
A χ^2 test for association requires that the data (counts or frequencies) are entered in a contingency table (a matrix format to analyse and record the relationship between two or more categorical variables). Marginal totals are calculated for each row and column and a grand total is recorded in the bottom right hand corner (right).

	Mudfish absent (0)	Mudfish present (1)	Total
Emergent vegetation absent (0)	15	0	15
Emergent vegetation present (1)	26	39	65
Total	41	39	80

Step 2: Calculate the expected values (E)
Calculating the expected values for a contingency table is simple. For each category, divide the row total by the grand total and multiply by the column total. You can enter these in a separate table or as separate columns next to the observed values (right).

	Mudfish absent (0)	Mudfish present (1)	Total
Emergent vegetation absent (0)	7.69	7.31	15
Emergent vegetation present (1)	33.31	31.69	65
Total	41	39	80

Step 3: Calculate the value of chi-squared (χ^2) of $(O - E)^2 \div (E)$
The difference between the observed (O) and expected (E) values is calculated as a measure of the deviation from a predicted result. Since some deviations are negative, they are all squared to give positive values. This step is best done as a tabulation to obtain a value for $(O - E)^2 \div (E)$ for each category. The sum of all these values is the value of chi squared (blue table right).

$$\chi^2 = \sum \frac{(O - E)^2}{E}$$

Where: O = the observed result
E = the expected result
Σ = sum of

Category	O	E	O–E	$(O-E)^2$	$\dfrac{(O-E)^2}{E}$
Mudfish 0/EmVeg 0	15	7.69	7.31	53.44	6.95
Mudfish 1/EmVeg 0	0	7.31	-7.31	53.44	7.31
Mudfish 0/EmVeg 1	26	33.31	-7.31	53.44	1.60
Mudfish 1/EmVeg 1	39	31.69	7.31	53.44	1.69

Total = 80 $\chi^2 \longrightarrow \Sigma = 17.55$

Step 4: Calculate the degrees of freedom (df)
The degrees of freedom for a contingency table is given by the formula: (rows-1) x (columns-1). For this example, degrees of freedom (df) is therefore (2-1) x (2-1) = 1.

Critical values of χ^2 at different levels of probability. By convention, the critical probability for rejecting the null hypothesis (H_0) is 5%. If the test statistic is greater than the tabulated value for P = 0.05 we reject H_0 in favour of the alternative hypothesis.

Step 5: Using the chi squared table
On the χ^2 table (relevant part reproduced in the table right) with 1 degree of freedom, the calculated value for χ^2 of 17.55 corresponds to a probability of less than 0.001 (see arrow). This means that by chance alone a χ^2 value of 17.55 could be expected less than 0.1% of the time. This probability is much lower than the 0.05 value which is generally regarded as significant. The null hypothesis can be rejected and we have reason to believe that black mudfish are associated with sites with emergent vegetation more than expected by chance alone.

	Level of Probability (P)				
df	0.05	0.025	0.01	0.005	0.001
1	3.84	5.02	6.63	7.88	10.83
2	5.99	7.38	9.21	10.60	13.82
3	7.81	9.35	11.34	12.84	16.27

C4.1
14

©2024 **BIOZONE** International
ISBN: 978-1-99-101410-8
Photocopying prohibited

245 Chi-Squared Exercise in Ecology

Key Idea: Chi-squared can be used to determine if an association between two species is statistically significant.

In ecological studies, it is often found that two or more species are found in association. This is usually because of similar environmental requirements or because one species depends on the other. The following example is based on a study in which the presence or absence of two plant species was recorded in a marked area. The two species are sometimes, but not always, found together. The chi-squared test is used to test the significance of the association.

Using chi-squared to test species associations in a successful marsh-meadow community

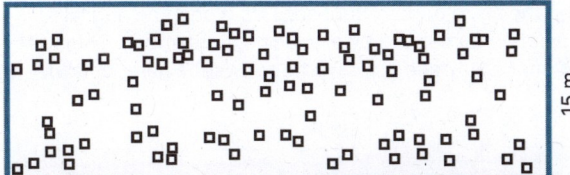

15 m

50 m

Lesser pond sedge (*Carex acutiformis*) is a swamp plant

Marsh bedstraw (*Galium palustre*) grows in ditches and wet meadows

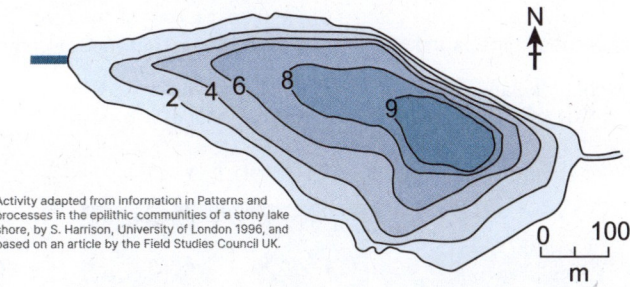

Activity adapted from information in Patterns and processes in the epilithic communities of a stony lake shore, by S. Harrison, University of London 1996, and based on an article by the Field Studies Council UK.

Lake Crosemere (above) is a one of a series of kettle hole lakes in England, formed by glacial retreat at the end of the last glacial period. In a natural process of succession, the lake is gradually infilling from its western edge, and wet meadow and marsh species are replacing the species of the open water. Students investigated the association between two plants previously recorded in studies of the area: the lesser pond sedge (LPS) and marsh bedstraw (MBS). They recorded species presence or absence in 100 quadrats (0.5 m²) placed in an area 15 X 50 m using coordinates generated using a random number function on a spreadsheet. The results are summarized in table 1 below. Follow the steps to complete the analysis.

1. State the null hypothesis (H_0) for this investigation:

   _____

   _____

2. In words, summarize the observed results in table 1:

   _____

   _____

   _____

   _____

	LPS present (1)	LPS absent (0)	Total
MBS present (1)	11	3	14
MBS absent (0)	31	55	86
Total	42	58	100

Table 1: Observed results for presence/absence of lesser pond-sedge (LPS) and marsh bedstraw (MBS).

3. Calculate the expected values for presence/absence of LPS and MBS. Enter the figures in table 2:

4. Complete the table to calculate the χ^2 value: χ^2 = _____

5. Calculate the degrees of freedom: _____

6. Using the χ^2, state the P value corresponding to your calculated χ^2 value (use the χ^2 table opposite):

   _____

7. State whether or not you reject your null hypothesis:

   _____

8. What could you conclude about this plant community:

   _____

   _____

   _____

	LPS present (1)	LPS absent (0)	Total
MBS present (1)			
MBS absent (0)			
Total			

Table 2: Expected results for presence/absence of lesser pond-sedge (LPS) and marsh bedstraw (MBS).

Category	O	E	O–E	(O–E)²	$\frac{(O-E)^2}{E}$
LPS 1/MBS 1					
LPS 0/MBS 1					
LPS 1/MBS 0					
LPS 0/MBS 0					
Total = 100					χ^2 =

C4.1
15

246 Predator-Prey Relationships

Key Idea: Resource availability influences the interactions between populations of predators and their prey.

A predator eating its prey is one of the most obvious ecological interactions we see between species. Predators are well adapted to locate and subdue their prey, and prey species are equally well adapted to avoid being eaten and to maintain their populations despite predation. Vertebrate predators generally do not control their prey populations, which tend to fluctuate seasonally depending on available food supply. However, predator populations are heavily influenced by the availability of prey, especially when there is little opportunity for switching to alternative prey species.

How do predators respond to increases in prey?

Time lagged numerical responses: In many herbivore-carnivore systems, prey populations fluctuate seasonally with changes in vegetation growth. Predators may respond to increases in prey by increasing their rate of reproduction. This type of numerical response shows a time lag because of the time it takes for the predator population to respond by producing more young. The most famous of these time-lagged predator-prey cycles is that recorded for the Canada lynx and its prey, the snowshoe hare (below).

Figure right: Population cycles in Canada lynx and snowshoe hare
Image (below left): Canada lynx, a specialist predator of snowshoe hares, taking almost no other prey.
Image (below right): Snowshoe hare, the primary prey of the lynx.

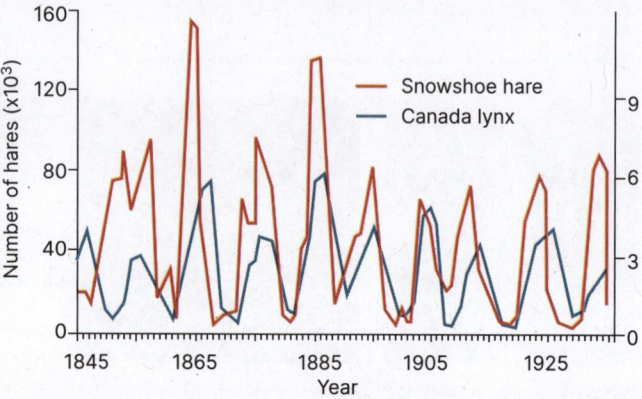

No-lag numerical responses: Predators can show an immediate numerical response to increases in prey availability. In one study, numbers of kestrels fluctuated with the density of voles, their prey (below). The birds could track the vole populations without a time lag because of their high mobility. An increase in vole numbers was accompanied by a rapid immigration of birds into the area.

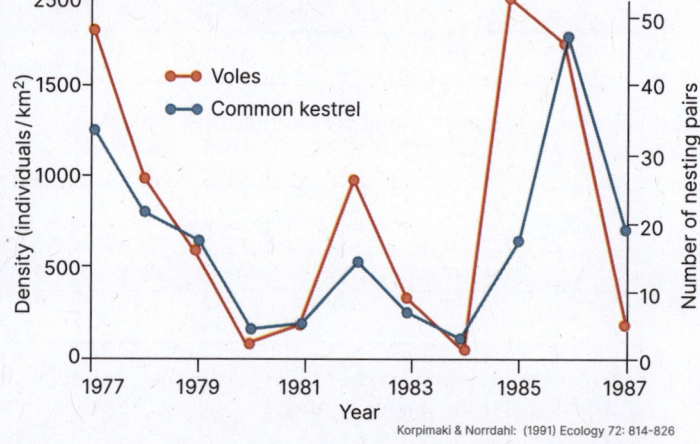

Images: Common or European kestrel (left) and its prey, a vole. Kestrels feed almost exclusively on small rodents like voles.
Graph: Numerical response of kestrels to changes in vole numbers. Study area was 47 km² of farmland in western Finland.

Korpimaki & Norrdahl: (1991) Ecology 72: 814-826

1. (a) Describe how seasonal fluctuations in resources relate to cycles of abundance of Canada lynx and snowshoe hare:

_____

_____

_____

(b) Explain why the fluctuations in lynx numbers lag slightly behind those of the hare: _____

_____

_____

(c) Why do you think the cycle of lynx abundance is so closely linked to the hare abundance? _____

_____

_____

_____

Predator density depends on prey density

▶ Lemmings are small rodents found in the Arctic tundra. They have been subject to many misconceptions about the remarkable ability of their populations to suddenly and rapidly increase to huge densities. Misconceptions range from lemming populations growing because they literally fall from the sky, to jumping off cliffs in deliberate mass suicides.

▶ Studies of lemmings before the 1990s found that the lemming population fluctuated on an approximately four year cycle from extremely low to extremely high densities. Exactly why lemming populations did this is not entirely known although research points to the major influence being predation.

▶ In 2003 researchers published data gathered over 16 years showing the interactions between lemmings and their various predators. The major predators of lemmings are the Arctic fox, the snowy owl, the stoat, and the long tailed skua. Of these predators, the stoat is the most dependent on lemmings for food.

Argus fin PD

Cyclic Dynamics in a Simple Vertebrate Predator–Prey Community Olivier Gilg et al, Science 2003

Lemming predator densities

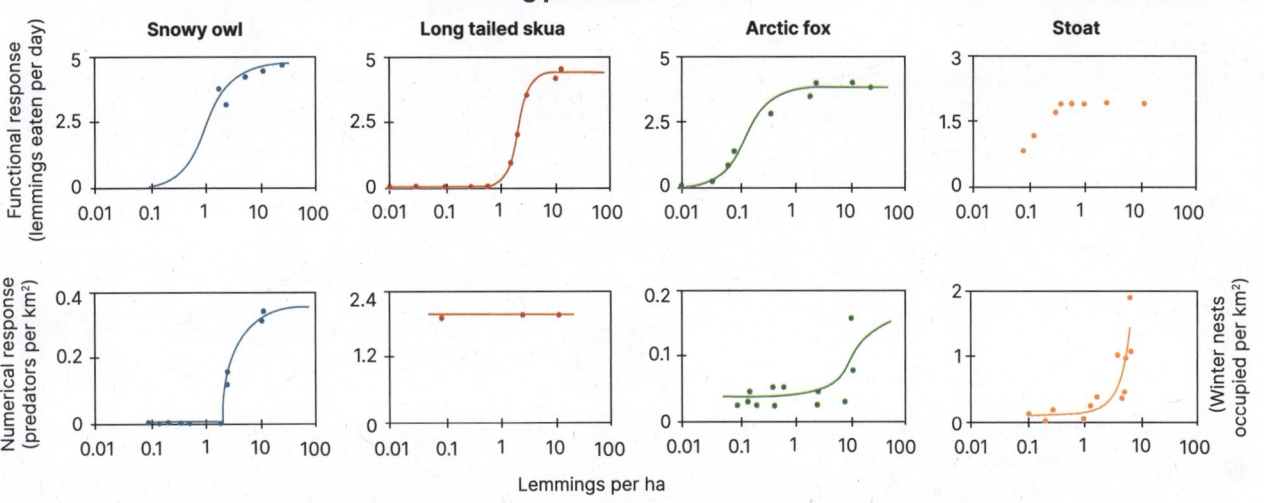

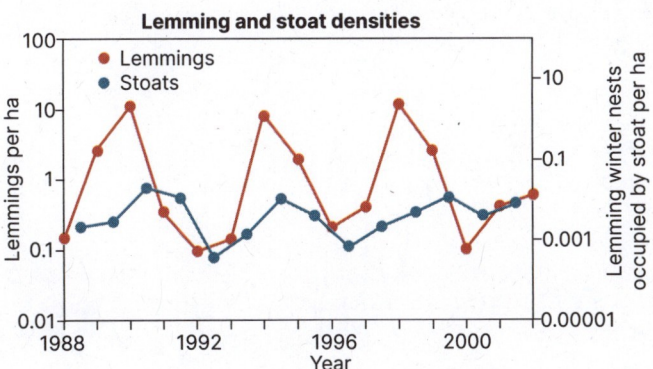

Stoats depend on lemmings

Of the four predators above, the stoat is the most dependent on lemmings. The other three predators are nomadic, migratory, or feed on a much wider variety of food. However, they all show larger increases in lemming predation when lemming densities rise. Stoats overwinter in lemming nests (graph centre right). Continued observation of the lemming population shows that since the 1990s the four year lemming cycle has failed to develop. Researchers think this is possibly due to lower quality winter snow, which helps predators find lemmings.

2. Why do birds of prey, e.g. kestrels, often show no lag in increasing numbers as prey increases?

_____

_____

_____

3. (a) Describe the relationship between predator feeding and lemming populations: _____

_____

_____

(b) Which predator population appears to be unaffected by lemming populations? _____

(c) Why might this be? _____

_____

(d) What might the effect of the failure of the lemming cycle on the predators? _____

_____

_____

247 Control of Populations

Key Idea: Population control can come from the top e.g. predators, or from the bottom, e.g. the availability of food. Populations are regulated by pressures from above their trophic level, e.g. predators, and below their trophic levels, e.g. food sources. These are called top-down and bottom-up controls. Although each control may have more or less influence over the population at any one time, together they regulate how fast a population can grow and how large it can become. Humans exert both bottom-up and top-down controls either directly or indirectly on many populations.

Top-down control

▶ Top-down pressures are applied from a higher trophic level, e.g. predators.

▶ Predators suppress prey abundance or limit the population rate of growth by catching and eating prey.

▶ Top-down controls tend to cause a rise in the population of the higher trophic level, until the controlled population becomes scarce.

▶ Human activity exerts top-down control via hunting and fishing.

In the example above, a decreasing or low secondary predator population leads to an increasing or high primary predator population. This results in a decreasing or low herbivore population, which allows plant growth to increase.

Bottom-up control

▶ Bottom-up pressures result from limited resources to lower trophic levels, e.g. producers or lower order consumers.

▶ Fewer resources at lower trophic levels limits the food and resources to higher levels and suppresses their population growth. All higher trophic levels will be affected.

▶ Humans exert bottom-up controls via competing for resources, e.g. fencing off areas for farms, or logging forests.

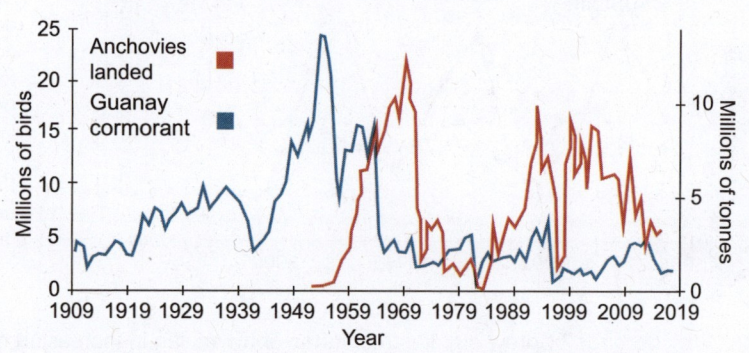

Reduced plant growth, due to low soil nutrients or low sunlight etc, causes a decrease in herbivore populations which in turn cause decreases in predator populations at all levels.

1. The graph on the right shows the population of Peruvian guanay cormorant (*Leucocarbo bougainvilliorum*) and the tonnage of anchovies landed each year from the waters around Peru. Guanay cormorants feed on anchovies.

   (a) Describe the changes that can be seen on the graph:

   _____

   _____

   _____

   _____

Governance analysis of two historical MPAs in northern Peru, Daniela Laínez del Pozoetc 2020

   (b) Has the guanay cormorant experienced top-down or bottom-up control of its population? Explain why:

   _____

   _____

2. Sea otters live along the Northern Pacific coast of North America and have been hunted for hundreds of years for their fur. On the island of Shemya, there are no longer any sea otters, while they have survived on the Amchitka Islands. Around Shemya, there are large areas of sea floor with very little kelp growing and large numbers of sea urchins. Around the Amchitka Islands there are large kelp forests and relatively few sea urchins. Sea otters eat sea urchins.

   (a) What kind of population control(s) are being exerted on these islands? _____

   _____

   _____

   (b) What would happen to the area of kelp around Shemya island if sea otters were reintroduced?

   _____

©2024 **BIOZONE** International
ISBN: 978-1-99-101410-8
Photocopying prohibited

C4.1
17

248 Chemical Defences

Key Idea: Plants and fungi can deter or kill potential competitors by releasing toxic chemicals into their environment.

Some organisms, including plants and fungi, are able to secrete chemicals into the environment that affect the growth of nearby organisms. Fungi such as *Penicillium* are well known for releasing antibiotics that can kill bacteria. Plants such as the black walnut release chemicals into the soil to stop plants growing nearby and so reduce competition. In plants the release of these chemicals is called allelopathy.

Allelopathy

Allelopathy is the phenomenon of certain plants producing chemicals that affect the growth or germination of other plants. These chemicals are called allelochemicals. Examples of allelopathy are often difficult to prove as some effects often overlap with straight competition, e.g. wilting may be caused by competition for water.

Black walnut is a well know example of a plant with allelopathic effects. Black walnuts release the chemical hydrojuglone from roots, leaves, and nut husks. Once in the soil, this produces juglone, affecting the growth of nearby plants, possibly by interfering with the mitochondria. Juglone can be used as herbicide, but is also used as a dye and food colouring.

Sorghum produces an allelochemical called sorgoleone from its root hairs. Sorogleone interferes with photosynthesis in germinating plants. It also inhibits the process of nitrification (converting NH_3 to NO_3^-) by bacteria in legume root nodules. This may help the plant by reducing nitrogen leaching near its own roots, but affects other plants' access to nitrogen.

Coastal she-oak, (*Casuarina equisetifolia*) is native to Australia and many other countries around the India and Pacific region. In nature, the she-oaks' leaf litter releases allelochemicals into the soil. Extract from its leaves show allelopathic effects on seed germination and seedling growth, and have been evaluated for use as a herbicide.

Antibiotics

The powdery mildew infecting this plant is a fungus

Penicillium

Slash pine

Antibiotics are chemicals that act against bacteria. They are produced by plants, fungi, and bacteria to reduce competition or to fight infection. The action of antibiotics is highly variable and can work by inactivating proteins and enzymes, interrupting cell signalling, or disrupting the plasma membrane.

The *Penicillium* fungus is a famous example of an organism producing antibiotics. In nature, the chemical penicillin, named after the fungus, is released by the fungus to kill or reduce the growth of bacteria. This reduces competition or any chance of infection. Cephalosporin, a broad spectrum antibiotic is produced by the fungi and pathogen of wheat, *Cephalosporium*.

Plants produce a wide range of chemical compound that they use against predators and microbes. New chemicals are constantly being studied. The chemical dehydroabietic acid, found in resin from the tree *Pinus elliottii* (slash pine), has been found to have antibiotic effects against MRSA, making it a possibly useful new antibiotic agent.

1. (a) What are allelochemicals and what is their purpose? _____

   _____

   _____

   (b) Why might allelochemicals be useful as herbicides? _____

   _____

2. For what purpose would plants and fungi produce antibiotics? _____

   _____

   _____

C4.1
18

249 Systems and Energy

Key Idea: Ecosystems are open systems. They exchange matter and energy with the wider environment.

Thermodynamics lays out the fundamental laws of energy that govern the universe. Firstly, the energy in an isolated system is constant. Secondly, disorder (entropy) increases over time. In other words, to maintain order within a system, there must be an input of energy. Living organisms follow these laws by using energy in the form of light (in plants) or food (in animals) to drive the chemical reactions that maintain order in their bodies.

What is a system?

Systems are groups of inter-related components working together by way of a driving force. A simple example of a system could be the water in a lake or pond being maintained by the flow of water in and out. A much more complex example would be that of an ecosystem.

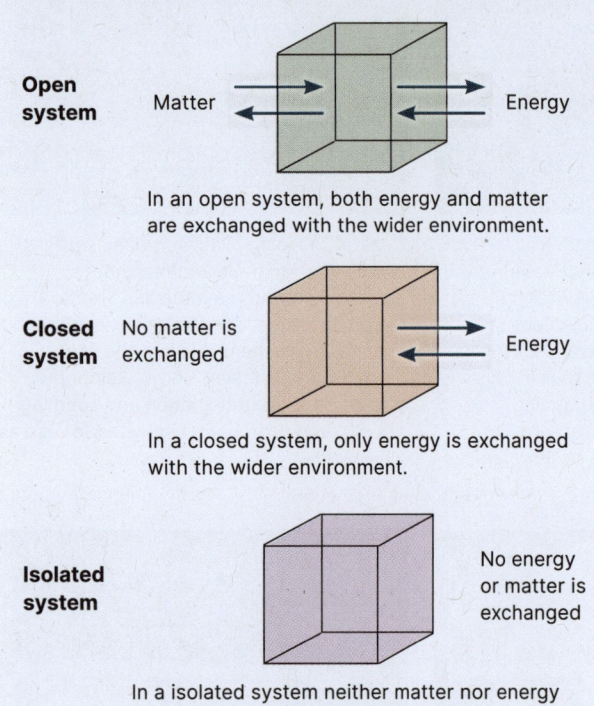

Open system — Matter / Energy

In an open system, both energy and matter are exchanged with the wider environment.

Closed system — No matter is exchanged / Energy

In a closed system, only energy is exchanged with the wider environment.

Isolated system — No energy or matter is exchanged

In a isolated system neither matter nor energy are exchange with the wider environment.

Thermodynamics and living things

▶ Entropy is a measure of the disorder in a system, or the amount of energy not available to do work, and increases with time. The greater the entropy, the greater the disorder in the system.

▶ Living things maintain order by using energy. Producers gain their energy directly from the sun. Consumers, e.g. animals, gain their energy from food. In both cases, glucose is the major energy generating molecule that is either made or obtained. The energy stored in the glucose molecule is released during its stepwise breakdown and is used to drive biochemical reactions. Without this energy, those reactions would stop and life would cease.

▶ However, not all the energy in glucose is used; in fact the majority of it is lost as heat. The ten percent law (or better, the ten percent rule) states that only about 10% of energy from one trophic level, e.g. producers, is available to the next trophic level, e.g. consumers (below). Thus, to maintain life, energy must constantly flow into an ecosystem from the sun to plants as energy is lost as heat to the environment.

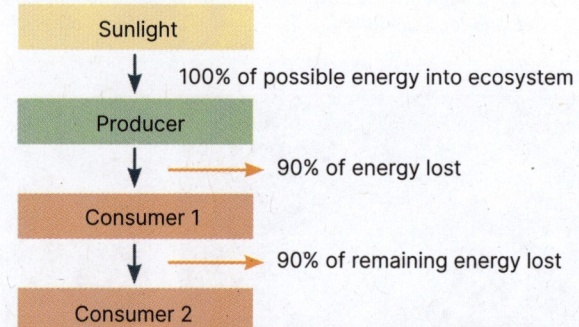

Sunlight

100% of possible energy into ecosystem

Producer

90% of energy lost

Consumer 1

90% of remaining energy lost

Consumer 2

Laws or theories?

▶ The laws of thermodynamics are fundamental laws that govern energy in the universe. In science a law is a description or equation that will predict what will happen given certain conditions. For example $F = ma$ is the mathematical description of Newton's second law of motion.

▶ A theory is an explanation of how or why a phenomenon occurs. Darwin's theory of evolution is a classic example. It lays out a complete explanation of how organisms evolve over time.

▶ Both laws and theories can be used to make predictions. We can use Newton's laws to predict the motion of stars and planets. We can use Darwin's theory of evolution to predict what kinds of fossils we might expect to find in rocks of various ages or to predict the effect of a selective breeding program.

1. Identify each of the following as either an open, closed, or isolated system:

    (a) A large lake: _____   (c) Covered beaker of water: _____

    (b) Thermos of water: _____   (d) Digestive system: _____

2. What law limits the number of consumer levels that can exist in an ecosystem? _____

3. What is the difference between a law and a theory? _____

    _____

    _____

 NOS
 C4.2 1 - 3

©2024 **BIOZONE** International
ISBN: 978-1-99-101410-8

250 Flow of Energy Through Ecosystems

Key Idea: Autotrophs use sunlight or chemical energy to make their own food whereas heterotrophs obtain their food and energy from other living things.

Living things obtain their energy for metabolism in two main ways. Autotrophs (producers) use the energy in sunlight or inorganic molecules to make their own food. Heterotrophs (consumers) rely on other organisms as a source of energy and carbon. All other organisms depend on producers, even if they do not consume them directly. The energy flow into and out of each trophic (feeding) level can be represented on a diagram using arrows of different sizes to represent relative amounts of energy lost from different trophic levels.

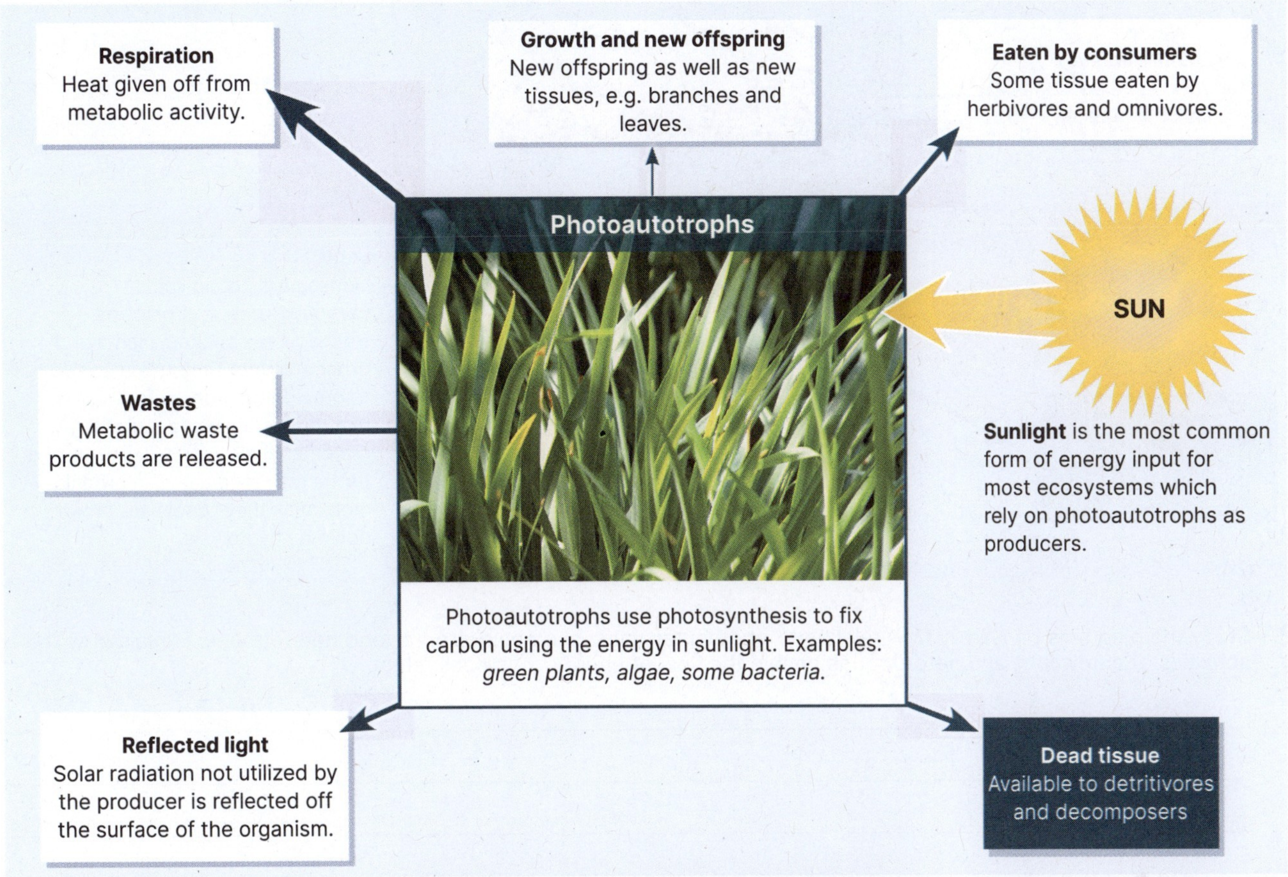

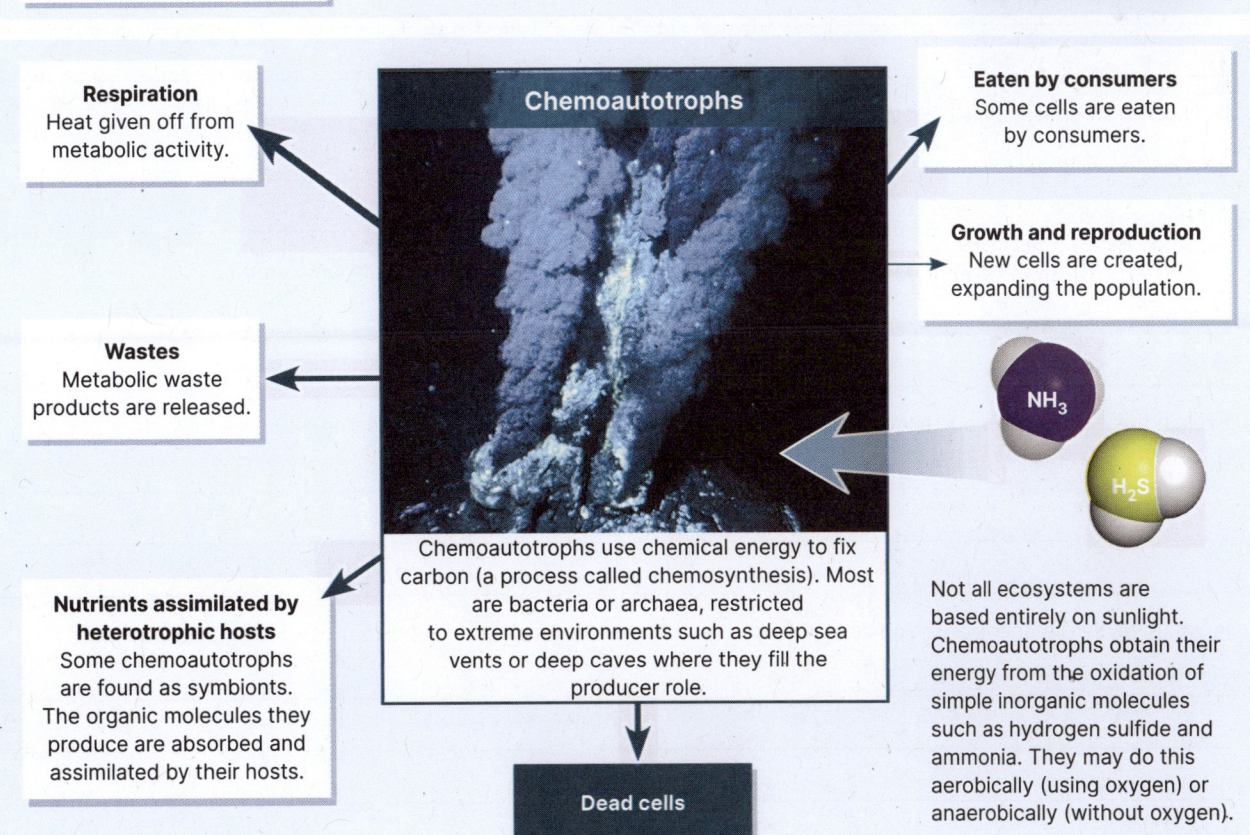

Respiration
Heat given off from metabolic activity.

Growth and new offspring
New offspring as well as new tissues, e.g. branches and leaves.

Eaten by consumers
Some tissue eaten by herbivores and omnivores.

Photoautotrophs

SUN

Sunlight is the most common form of energy input for most ecosystems which rely on photoautotrophs as producers.

Wastes
Metabolic waste products are released.

Photoautotrophs use photosynthesis to fix carbon using the energy in sunlight. Examples: *green plants, algae, some bacteria*.

Reflected light
Solar radiation not utilized by the producer is reflected off the surface of the organism.

Dead tissue
Available to detritivores and decomposers

Respiration
Heat given off from metabolic activity.

Chemoautotrophs

Eaten by consumers
Some cells are eaten by consumers.

Growth and reproduction
New cells are created, expanding the population.

Wastes
Metabolic waste products are released.

NH_3

H_2S

Nutrients assimilated by heterotrophic hosts
Some chemoautotrophs are found as symbionts. The organic molecules they produce are absorbed and assimilated by their hosts.

Chemoautotrophs use chemical energy to fix carbon (a process called chemosynthesis). Most are bacteria or archaea, restricted to extreme environments such as deep sea vents or deep caves where they fill the producer role.

Not all ecosystems are based entirely on sunlight. Chemoautotrophs obtain their energy from the oxidation of simple inorganic molecules such as hydrogen sulfide and ammonia. They may do this aerobically (using oxygen) or anaerobically (without oxygen).

Dead cells

©2024 **BIOZONE** International
ISBN: 978-1-99-101410-8
Photocopying prohibited

C4.2
2, 5-9
12, 13

Respiration
Heat given off from metabolic activity.

Growth and new offspring
New offspring as well as growth and weight gain.

Eaten by carnivores
Some tissue eaten by carnivores and omnivores.

Heterotrophs

Wastes
Metabolic waste products are released (e.g. as urine, faeces, carbon dioxide).

Heterotrophs rely on other living organisms or organic particulate matter for their energy. Heterotrophs include herbivores, carnivores, detritivores, and decomposers.
Examples: *animals, some protists, some bacteria*

Dead tissue
Available to detritivores and decomposers

Food
Consumers obtain lipids, carbohydrates, and proteins from sources such as plants and other animals. They use these to provide the energy for their metabolic processes and the raw materials they need for growth and tissue maintenance. The organic molecules they obtain are broken down by hydrolysis.

1. Study the diagrams on energy flow relating to photoautotrophs, chemoautotrophs, and heterotrophs. Explain how the activities of autotrophs and heterotrophs enable the flow of energy in an ecosystem:

_____

_____

_____

_____

_____

2. Describe how energy may be lost from organisms in the form of:

(a) Wastes: _____

(b) Respiration: _____

3. Explain why so little energy is available for growth and reproduction, regardless of trophic group:

_____

_____

4. Explain the ecological importance of chemoautotrophic organisms in deep sea environments:

_____

_____

_____

5. In what way is the chemoautotrophic system of the deep sea thermal vent linked to other ecological systems?

_____

_____

_____

_____

251 Quantifying Energy Flow in an Ecosystem

Key Idea: Chemical energy in the bonds of molecules flows through an ecosystem between trophic levels. Energy is lost as heat at each trophic level.

Energy cannot be created or destroyed. It can only be transformed from one form, e.g. light energy, into another, e.g. chemical energy. This means that the flow of energy through an ecosystem can be measured. Each time energy is transferred from one trophic level to the next, e.g. by eating, some energy is given out as heat to the environment due to cellular respiration. Living organisms cannot convert heat to other forms of energy, so this heat is effectively lost from the system and so the amount of energy available to one trophic level is always less than at the previous level. Potentially, we can account for the transfer of energy from its input (as solar radiation) to its release as heat from organisms, because energy is conserved. The percentage of energy transferred from one trophic level to the next is the trophic efficiency. It varies between 5% and 20% and measures the efficiency of energy transfer. An average figure of 10% trophic efficiency is often used. This is called the ten percent rule.

Energy flow through an ecosystem

NOTE

Numbers represent **kilojoules** of energy per square metre per year (kJ m^{-2} yr^{-1})

Sunlight falling on plant surfaces
7,000,000

Light absorbed by plants
1,700,000

Energy absorbed from the previous trophic level
100

Energy lost as heat **65** → **Trophic level** → **15** Energy lost to detritus

20

Energy passed on to the next trophic level

The energy available to each trophic level will always equal the amount entering that trophic level, minus total losses to that level (due to metabolic activity, death, excretion etc). Energy lost as heat will be lost from the ecosystem. Other losses become part of the detritus and may be utilized by other organisms in the ecosystem

A

Producers
87,400

50,450

(a)

7,800

22,950

Primary consumers

B

1600

G

4,600

1,330

(b)

2,000

Secondary consumers

Detritus

10,465

D

90

F

Heat loss in metabolic activity

(c)

19,300

(d)

55

Tertiary consumers

C

Decomposers and detritivores

19,200

E

1. Study the diagram above, illustrating energy flow through a hypothetical ecosystem. Use the example at the top of the page as a guide to calculate the missing values (a)–(d) in the diagram. Note that the sum of the energy inputs always equals the sum of the energy outputs. Place your answers in the spaces provided on the diagram.

C4.2
2 - 5
12 - 14

2. What is the original source of energy for this ecosystem? _____

3. Identify the processes occurring at the points labelled A – G on the diagram:

   A. _____   E. _____

   B. _____   F. _____

   C. _____   G. _____

   D. _____

4. (a) Calculate the percentage of light energy falling on the plants that is absorbed at point **A**:

   Light absorbed by plants ÷ sunlight falling on plant surfaces x 100 = _____

   (b) What happens to the light energy that is not absorbed? _____

   _____

5. (a) Calculate the percentage of light energy absorbed that is actually converted (fixed) into producer energy:

   Producers ÷ light absorbed by plants x 100   = _____

   (b) How much light energy is absorbed but not fixed: _____

   (c) Account for the difference between the amount of energy absorbed and the amount actually fixed by producers:

   _____

   _____

6. Of the total amount of energy fixed by producers in this ecosystem (at point **A**), calculate:

   (a) The total amount that ended up as metabolic waste heat (in kJ): _____

   (b) The percentage of the energy fixed that ended up as waste heat: _____

7. (a) State the groups for which detritus is an energy source: _____

   (b) How could detritus be removed or added to an ecosystem? _____

   _____

8. Under certain conditions, decomposition rates can be very low or even zero, allowing detritus to accumulate:

   (a) From your knowledge of biological processes, what conditions might slow decomposition rates?

   _____

   (b) What are the consequences of this lack of decomposer activity to the energy flow? _____

   _____

   (c) Add an additional arrow to the diagram on the previous page to illustrate your answer.

   (d) Describe three examples of materials that have resulted from a lack of decomposer activity on detrital material:

   _____

   _____

   _____

9. The ten percent rule states that the total energy content of a trophic level in an ecosystem is only about one-tenth (or 10%) that of the preceding level. For each of the trophic levels in the diagram on the preceding page, determine the energy passed on to the next trophic level as a percentage:

   (a) Producer to primary consumer: _____

   (b) Primary consumer to secondary consumer: _____

   (c) Secondary consumer to tertiary consumer: _____

   (d) Which of these transfers is the most efficient? _____

©2024 **BIOZONE** International
ISBN: 978-1-99-101410-8
Photocopying prohibited

252 Food Chains

Key Idea: A food chain is a model to illustrate the feeding relationships between organisms.

Organisms in ecosystems interact by way of their feeding (trophic) relationships. These interactions can be shown in a food chain, which is a simple model to illustrate how energy, in the form of food, passes from one organism to the next. Each organism in the chain is a food source for the next. The levels of a food chain are called trophic levels. An organism is assigned to a trophic level based on its position in the food chain. Organisms may occupy different trophic levels in different food chains or during different stages of their life. Arrows link the organisms in a food chain. The direction of the arrow shows the flow of energy through the trophic levels. Most food chains begin with a producer, which is eaten by a primary consumer (herbivore). Higher level consumers (carnivores and omnivores) eat other consumers.

Producers (autotrophs) e.g. plants, algae, and autotrophic bacteria, make their own food from simple inorganic substances, often by photosynthesis using energy from the sun. Inorganic nutrients are obtained from the abiotic environment, such as the soil and atmosphere.

Consumers (heterotrophs) e.g. animals, get their energy from other organisms. Consumers are ranked according to the trophic level they occupy, i.e. 1st order, 2nd order, and classified according to diet, e.g. carnivores eat animal tissue, omnivores eat plant and animal tissue.

Millipede

Detritivores and saprotrophs are both consumers that gain nutrients from digesting dead organic matter (DOM). Detritivores consume DOM, e.g. earthworms, millipedes, whereas saprotrophs secrete enzymes to digest the DOM extracellularly and absorb the products of digestion, e.g. fungi, soil bacteria.

1. Describe how the following obtain their energy:

   (a) Producers: _____

   (b) Consumers: _____

   (c) Detritivores: _____

   (d) Decomposers: _____

2. (a) Draw arrows on the diagram below to show how the energy flows through the organisms in the food chain. Label each arrow with the process involved in the energy transfer. Draw arrows to show how energy is lost by respiration

   (b) What is the original energy source for this food chain? _____

Respiration

Producers
Trophic level: 1

Herbivores
Trophic level: 2

Carnivores
Trophic level: 3

Carnivores
Trophic level: 4

Detritivores and decomposers

C4.2
3 - 5

253 Food Webs

Key Idea: A food web depicts all the interconnected food chains in an ecosystem. Sunlight is the energy source for most ecosystems. Some energy is lost at each trophic level. The different food chains in an ecosystem are interconnected to form a complex web of feeding interactions called a food web. Sunlight is the initial energy source for almost all ecosystems. Sunlight provides a continuous, but variable, energy supply, which is fixed in carbon compounds by photosynthesis. Energy flows through ecosystems in the chemical bonds within organic matter (food) and, in accordance with the second law of thermodynamics, is dissipated as heat as it is transferred through trophic levels. This loss of energy from the system limits how many links can be made in each food chain, as living organisms cannot convert heat to other forms of energy. Two simplified food webs showing the transfer of energy are depicted below.

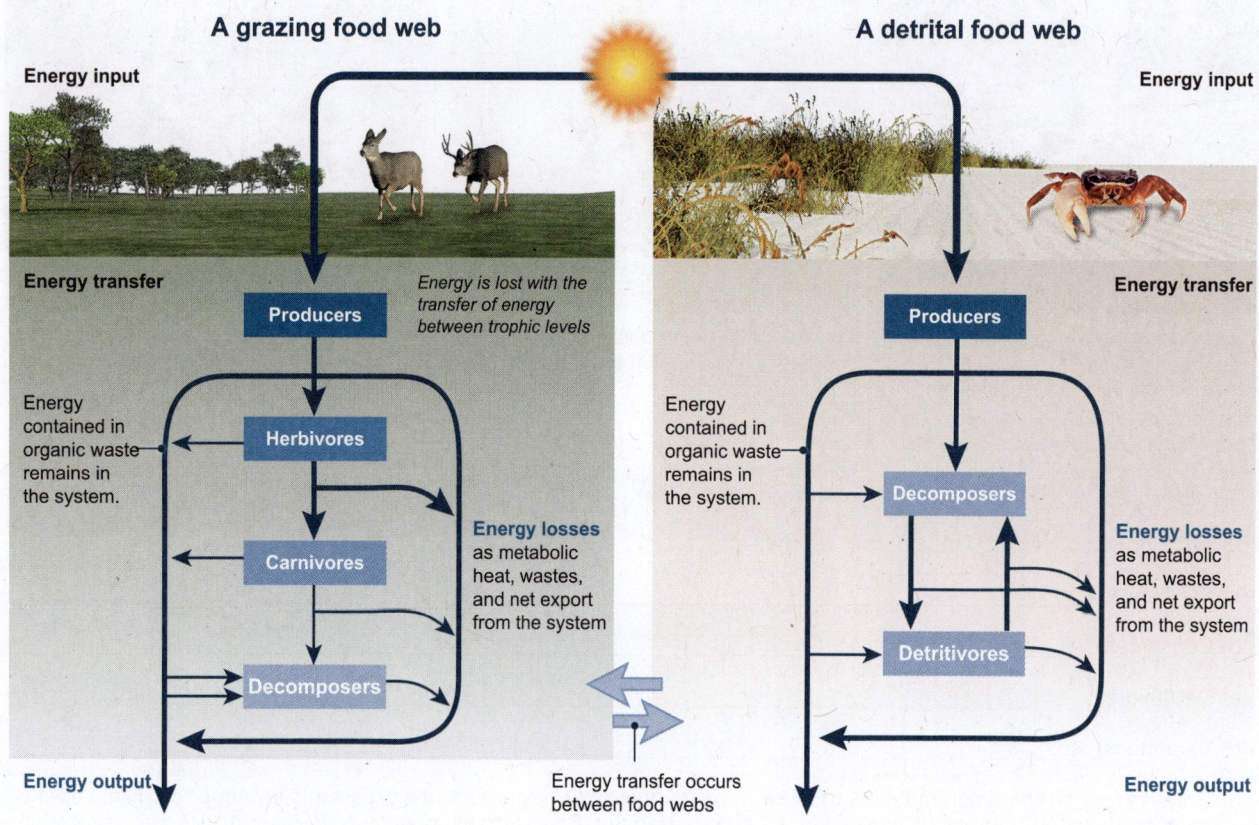

1. Describe how energy is transferred through ecosystems: _____

_____

_____

2. (a) Describe what happens to the amount of energy available to each successive trophic level in a food chain:

_____

   (b) Explain why this is the case: _____

_____

_____

_____

3. With respect to energy flow, describe a major difference between a detrital and a grazing food web: _____

_____

_____

_____

_____

©2024 **BIOZONE** International
ISBN: 978-1-99-101410-8
Photocopying prohibited

Creating a food web

Knowing what the inhabitants of an ecosystem feed on allows food chains to be constructed. Food chains can be used to construct a food web. The organisms below are typical of those found in many lakes. For simplicity, only a few organisms are represented here. Real lake communities have hundreds of different species interacting together. Your task is to assemble the organisms below into a food web in a way that shows how they are interconnected by their feeding relationships.

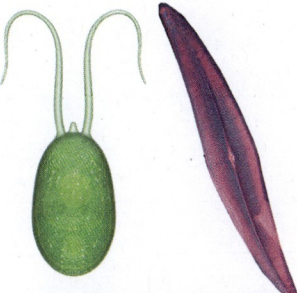

Autotrophic protists (algae)
Chlamydomonas (above left), and some diatoms (above right) are photosynthetic.

Macrophytes
Aquatic green plants are photosynthetic.

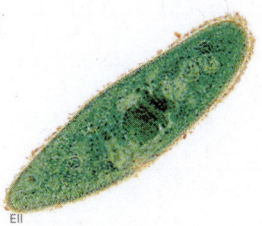

Protozoan , e.g. Paramecium
Diet: Mainly bacteria and microscopic green algae such as *Chlamydomonas*.

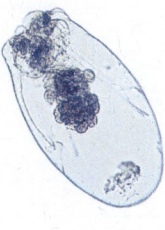

Asplanchna (zooplankton)
A large, carnivorous rotifer.
Diet: Protozoa and young zooplankton, e.g. *Daphnia*.

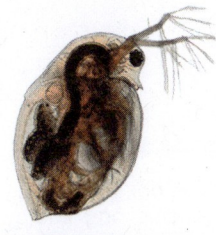

Daphnia (zooplankton)
Small freshwater crustacean.
Diet: Planktonic algae.

Common carp
Diet: Mainly feeds on bottom-living insect larvae and snails, but will also eat some plant material (not algae).

Three-spined stickleback
Common in freshwater ponds and lakes. **Diet**: Small invertebrates such as *Daphnia* and insect larvae.

Diving beetle (adults and larvae)
Diet: Aquatic insect larvae and adult insects. They will also scavenge from detritus. Adults will also take fish fry.

Herbivorous water beetle
Diet: Adults feed on macrophytes. Young beetle larvae are carnivorous, feeding primarily on pond snails.

Dragonfly larva
Large aquatic insect larvae.
Diet: Small invertebrates including *Hydra*, *Daphnia*, insect larvae, and leeches.

Great pond snail
Diet: Omnivorous. Main diet is macrophytes but will eat decaying plant and animal material also.

Leech
Fluid feeding predators.
Diet: Small invertebrates, including rotifers, small pond snails, and worms.

Pike
Diet: Smaller fish and amphibians. They are also opportunistic predators of rodents and small birds.

Mosquito larva
Diet: Planktonic algae.

Hydra
A small, carnivorous cnidarian.
Diet: small *Daphnia* and insect larvae.

Detritus
Decaying organic matter (includes bacterial decomposers).

4. From the information provided for the lake food web components on the previous page, construct twelve different food chains to show the feeding relationships between the organisms. Some food chains may be shorter than others and most species will appear in more than one food chain. An example has been completed for you.

Example 1: Macrophyte ⟶ Herbivorous water beetle ⟶ Carp ⟶ Pike

(a) _____

(b) _____

(c) _____

(d) _____

(e) _____

(f) _____

(g) _____

(h) _____

(i) _____

(j) _____

(k) _____

(l) _____

5. Use the food chains you created above to help you to draw up a food web for this community in the box below. Use the information supplied on the previous page to draw arrows showing the flow of energy between species; only energy from (not to) the detritus is required.

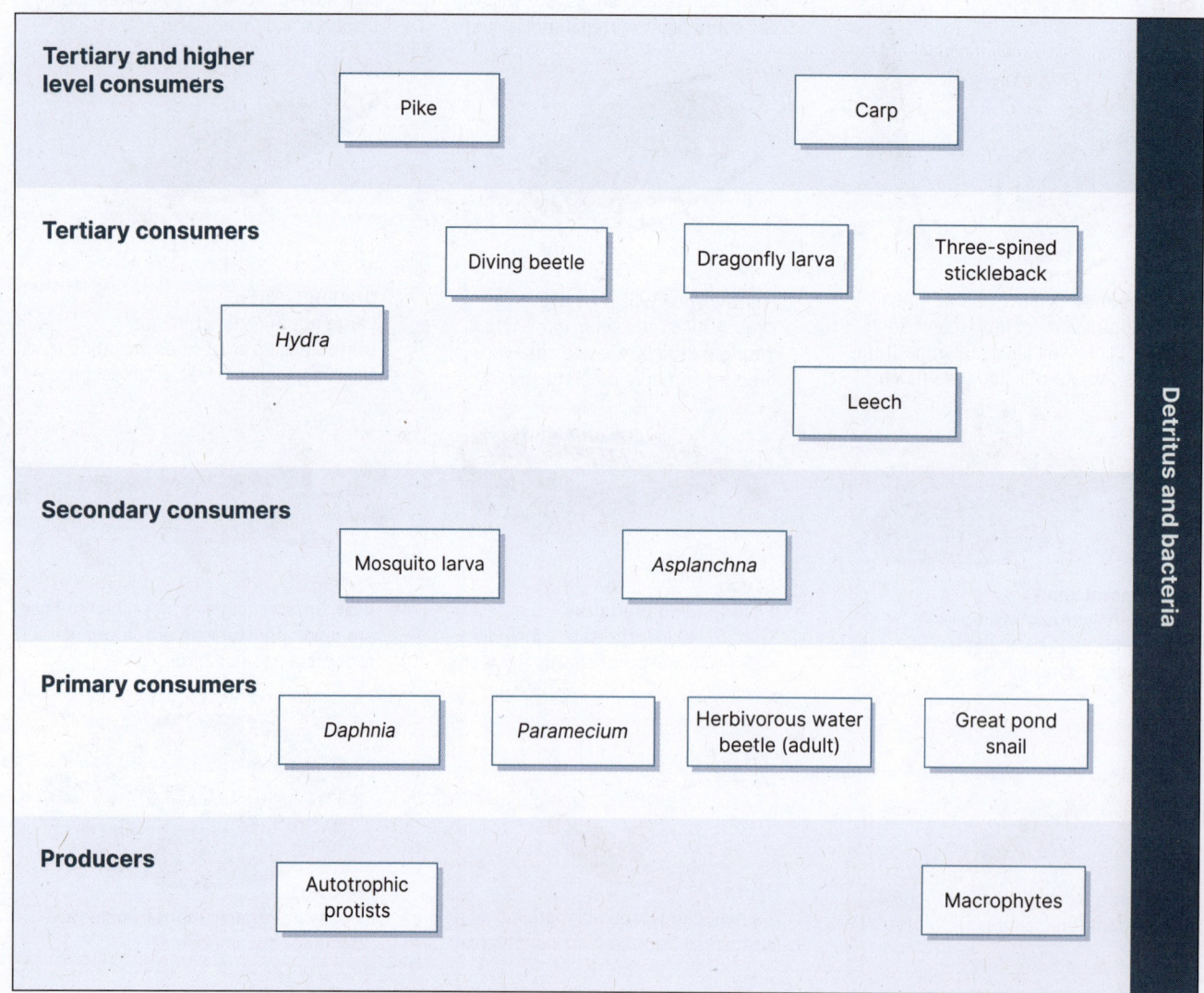

Tertiary and higher level consumers

Pike   Carp

Tertiary consumers

Diving beetle   Dragonfly larva   Three-spined stickleback

Hydra

Leech

Secondary consumers

Mosquito larva   Asplanchna

Primary consumers

Daphnia   Paramecium   Herbivorous water beetle (adult)   Great pond snail

Producers

Autotrophic protists   Macrophytes

Detritus and bacteria

254 Ecological Pyramids

Key Idea: Ecological pyramids can be used to illustrate the amount of energy at each trophic level in an ecosystem. The energy, biomass, or numbers of organisms at each trophic level in any ecosystem can be represented by an ecological pyramid. The first trophic level is placed at the bottom of the pyramid and subsequent trophic levels are stacked on top in their 'feeding sequence'. Ecological pyramids provide a convenient model to illustrate the relationship between different trophic levels in an ecosystem. Pyramids of energy show the energy contained within each trophic level.

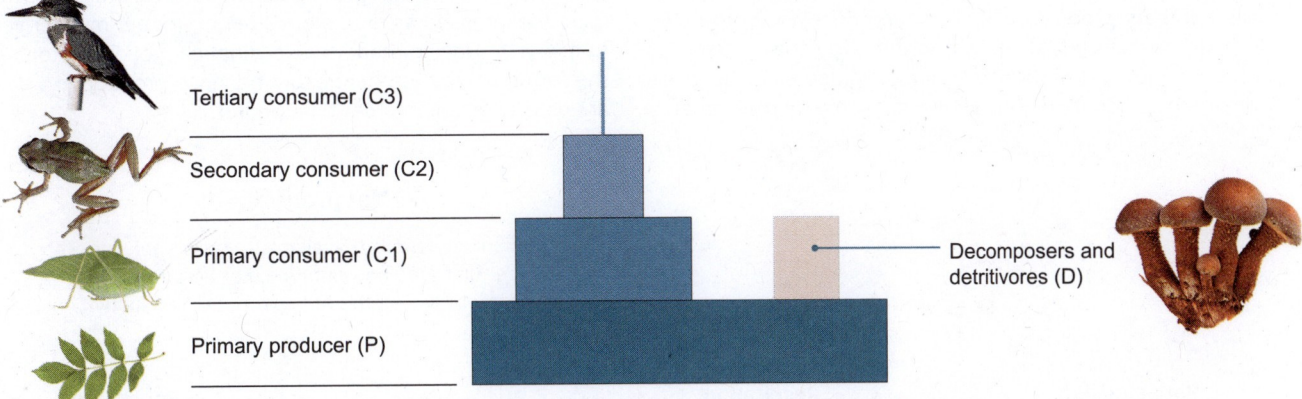

Tertiary consumer (C3)

Secondary consumer (C2)

Primary consumer (C1)

Primary producer (P)

Decomposers and detritivores (D)

▶ The generalized ecological pyramid pictured above shows a conventional pyramid shape, with a large base at the primary producer level and increasingly smaller blocks at subsequent levels. Not all pyramids have this appearance. Decomposers are placed at the level of the primary consumers and off to the side because they may obtain energy from many different trophic levels and so do not fit into the conventional pyramid structure. Pyramids of biomass measure the mass of the biological material at each trophic level. They are usually similar in appearance to pyramids of energy (biomass diminishes along food chains as the energy retained in the food chain diminishes).

Pyramids for a plankton community

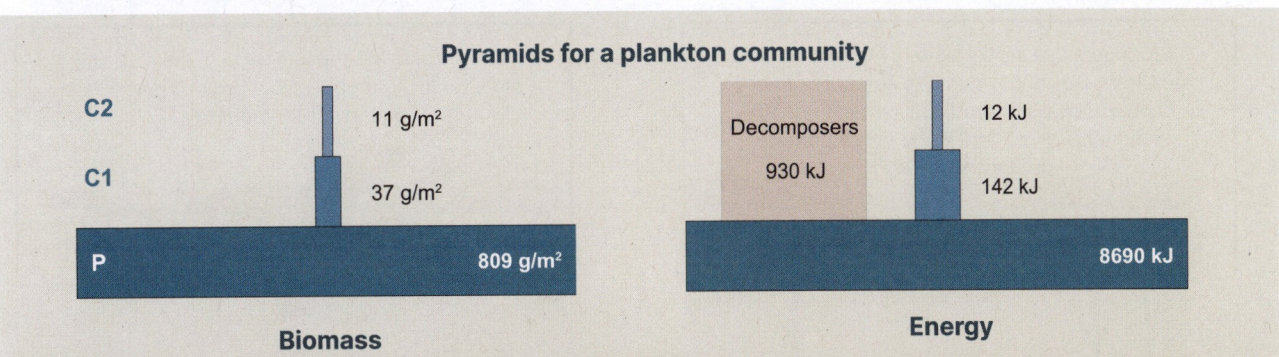

C2 — 11 g/m² | C1 — 37 g/m² | P — 809 g/m²

Decomposers 930 kJ | 12 kJ | 142 kJ | 8690 kJ

Biomass   **Energy**

▶ The two pyramids above relate to the same plankton community. The pyramids of biomass and energy are virtually identical.

▶ A large biomass of producers supports a smaller biomass of consumers. The energy at each trophic level is reduced at each progressive stage in the food chain. As a general rule, a maximum of 10% of the energy is passed on to the next level in the food chain. The remaining energy is lost due to respiration, waste, and heat.

1. The partially completed table below is from an investigation of the ecosystem around Silver Springs, Florida. It shows the energy available in each trophic level.

	Energy in organisms kJ m⁻² yr⁻¹	Energy lost (respiration/ waste)	Energy passed to next level	% available to next level
Producers (P)	31,897	27,279	*4618*	
Herbivores (C1)	*4618*	4154		
Carnivores (C2)		444		
Top carnivores (C3)		20		

(a) Complete the table above:

(b) Draw a energy pyramid from the Silver Springs ecosystem in the space on the right:

(c) Why can energy pyramids never be inverted? _____

_____

Based on data from Odum, 1957, Ecological Monographs 27(1), 55-112

C4.2 10-12 14

255 Primary Productivity

Key Idea: Primary production is the biomass or energy accumulated by producers in an ecosystem. Primary productivity is the rate of biomass or energy accumulation. What do we mean when we say that an ecosystem has a high (or low) productivity? The energy accumulated by plants or other producers in an ecosystem (or measured area) is called primary production. It is the first energy storage step in an ecosystem. All of the sunlight energy that is fixed as chemical energy in carbon compounds (glucose) is the gross primary production (GPP). However, some of this energy is required by the producers themselves for respiration. Subtracting respiration from GPP gives the net primary production (NPP). This represents the energy or biomass available to the primary consumers in the ecosystem. Note that 'production' refers to a quantity of material. Productivity, which is more meaningful in biological systems is a rate, usually expressed as grams or kJ per m^2 (or per m^3) per year ($g\ m^{-2}\ yr^{-1}$). Having made that distinction, you will often see the terms used interchangeably because production values are usually given for a set time period.

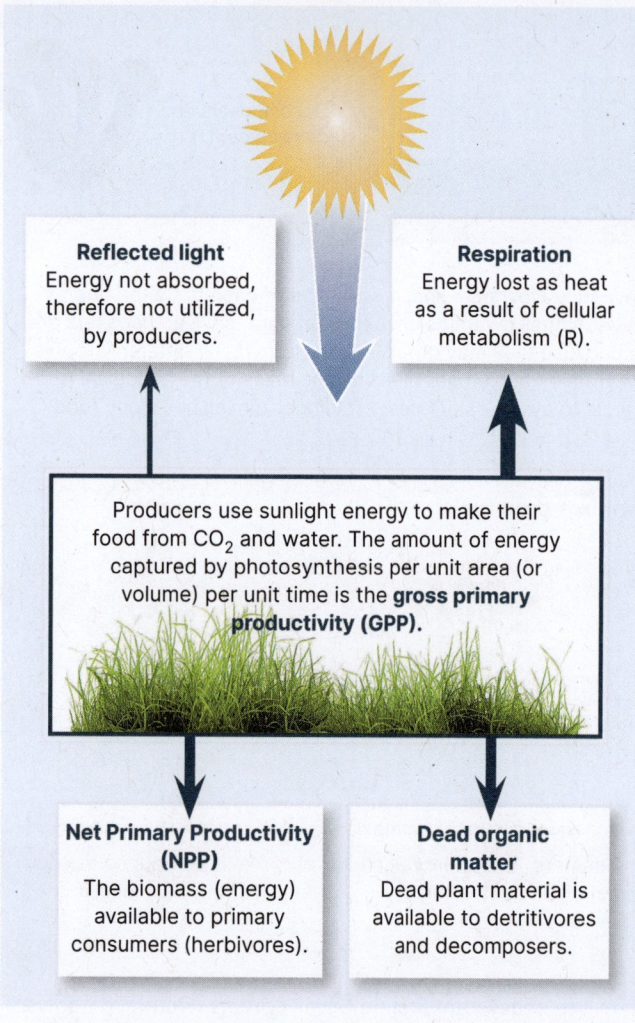

Reflected light
Energy not absorbed, therefore not utilized, by producers.

Respiration
Energy lost as heat as a result of cellular metabolism (R).

Producers use sunlight energy to make their food from CO_2 and water. The amount of energy captured by photosynthesis per unit area (or volume) per unit time is the **gross primary productivity (GPP)**.

Net Primary Productivity (NPP)
The biomass (energy) available to primary consumers (herbivores).

Dead organic matter
Dead plant material is available to detritivores and decomposers.

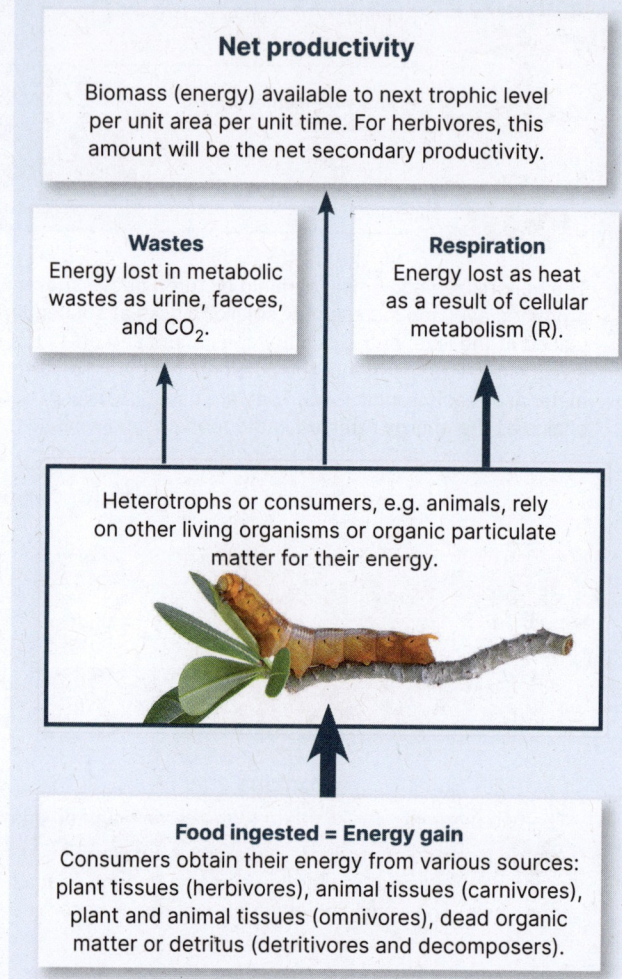

Net productivity
Biomass (energy) available to next trophic level per unit area per unit time. For herbivores, this amount will be the net secondary productivity.

Wastes
Energy lost in metabolic wastes as urine, faeces, and CO_2.

Respiration
Energy lost as heat as a result of cellular metabolism (R).

Heterotrophs or consumers, e.g. animals, rely on other living organisms or organic particulate matter for their energy.

Food ingested = Energy gain
Consumers obtain their energy from various sources: plant tissues (herbivores), animal tissues (carnivores), plant and animal tissues (omnivores), dead organic matter or detritus (detritivores and decomposers).

1. (a) Explain the difference between gross primary productivity (GPP) and net primary productivity (NPP):

_____

_____

_____

(b) What factors do you think could influence the GPP of ecosystems? _____

_____

_____

_____

(c) Why do you think it is important to distinguish between GPP and NPP when studying the productivity of ecosystems?

_____

_____

©2024 **BIOZONE** International
ISBN: 978-1-99-101410-8
Photocopying prohibited

The productivity of ecosystems varies

▶ The energy entering ecosystems is fixed by producers at a rate that depends on limiting factors such as temperature and the availability of light, water, and nutrients such as nitrogen and phosphorus. This means that the gross primary productivity of ecosystems varies across the globe according to the local conditions.

▶ Producers convert the energy they capture into biomass. The biomass produced per area per unit time after respiratory needs are met is the net primary production. This will be the amount of energy (as biomass) available to the next trophic level.

▶ Globally, the least productive terrestrial ecosystems are those that are limited by heat energy and water. The most productive are those with high light and temperature, plenty of water, and non-limiting supplies of soil nitrogen. The primary productivity of oceans is lower overall than that of terrestrial ecosystems because the water reflects (or absorbs) much of the light energy before it reaches and is utilized by producers. Many regions of the open ocean are also low in nutrients.

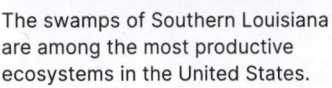

The swamps of Southern Louisiana are among the most productive ecosystems in the United States.

The productivity of arid scrubland ecosystems, like this in eastern Ethiopia, is limited by lack of water.

Chart — Average net primary productivity / x 1000 kJ m^{-2}y^{-1} (x-axis 5 to 50):

- Estuaries
- Swamps and marshes
- Tropical rainforest
- Temperate forest
- Boreal forest
- Savanna
- Agricultural land
- Woodland and shrubland
- Temperate grassland
- Lakes and streams
- Continental shelf
- Tundra
- Open ocean
- Desert scrub
- Extreme desert

2. Studying the ecosystem NPP graph above:

   (a) Identify the three most productive ecosystems: _____

   (b) What factors are likely to contribute to this high productivity? _____

   _____

   _____

   (c) Why do deserts have low NPP? _____

   _____

   _____

   (d) Why is primary productivity of the open ocean lower than that of most terrestrial ecosystems? _____

   _____

   _____

3. Estuaries and wetlands (including coastal wetlands) are among the most productive on Earth. Recall the characteristics of the aquatic biomes you looked at earlier and describe factors that contribute to the high productivity of these systems:

   _____

   _____

   _____

   _____

256 Trophic Efficiencies and Secondary Production

Key Idea: Secondary production is the generation of primary consumer (heterotrophic) biomass in a system.

Consumers eat producers and use the energy and material to generate their own biomass. The efficiency of this process can be investigated by measuring the biomass of the producer and consumer. In the experiment below, students determined the net secondary production and respiratory losses using 12 day old cabbage white butterfly larvae feeding on Brussels sprouts. Of the NPP from the Brussels sprouts that is consumed by the larvae, some will be used in cellular respiration, some will be available to secondary consumers (the net secondary production) and some will be lost as egested waste products (frass).

The method

▶ The wet mass of ten, 12 day old larvae, and approximately 30 g Brussels sprouts was accurately measured and recorded.

▶ The larvae and Brussels sprouts were placed into an aerated container. After three days, the container was disassembled and the wet mass of the Brussels sprouts, larvae, and frass was individually measured and recorded.

▶ The Brussels sprouts, larvae and frass were placed in separate containers and placed in a drying oven and their dry mass was recorded.

Cabbage white caterpillar (larva)

Note: We assume the proportion of biomass of Brussels sprouts and caterpillars on day 1 is the same as the calculated value from day 3.

Table 1: Brussels sprouts

	Day 1	Day 3	
Wet mass of Brussels sprouts	30 g	11 g	g consumed =
Dry mass of Brussels sprouts	–	2.2 g	
Plant proportion biomass (dry/wet)			
Plant energy consumed (wet mass x proportion biomass x 18.2 kJ)			kJ consumed per 10 larvae =
Plant energy consumed ÷ no. of larvae			kJ consumed per larva (E) =

Table 3: Frass

	Day 3
Dry mass frass from 10 larvae	0.5 g
Frass energy (waste) = frass dry mass x 19.87 kJ	
Energy from frass from 1 larva (W)	

Table 2: Caterpillars (larvae)

	Day 1	Day 3	
Wet mass of 10 larvae	0.3 g	1.8 g	g gained =
Wet mass per larva			g gained per larva =
Dry mass of 10 larvae	–	0.27 g	
Larva proportion biomass (dry/wet)			
Energy production per larva (wet mass x proportion biomass x 23.0 kJ)			kJ gained per larva (S) =

Table 4: Respiration

kJ consumed per larva (E) =	
kJ gained per larva (S) =	
Energy (in kJ) from frass from 1 larva (W) =	
Respiratory losses (in kJ) per larva =	

1. Complete the calculations in tables 1-4 above.

2. (a) Write the net secondary production per larva value here: _____

   (b) Write the equation to calculate the percentage efficiency of energy transfer from producers to consumers (use the notation provided) and calculate the value here:

   _____

   (c) Is this value roughly what you would expect? Explain: _____

   _____

3. (a) Calculate the approximate gain in dry biomass of all 10 larvae over 3 days (in grams): _____

   (b) Calculate the approximate rate of secondary production for the larvae in grams per day:

   _____

C4.2
16

©2024 **BIOZONE** International
ISBN: 978-1-99-101410-8

257 The Carbon Cycle

Key Idea: The cycling of carbon through the abiotic and biotic components of ecosystems makes carbon continually available to organisms.

Carbon is an essential element of life and is incorporated into the organic molecules that make up living organisms. Large quantities of carbon are stored in sinks, which include the atmosphere as carbon dioxide gas (CO_2), the ocean as carbonate and bicarbonate, and rocks such as coal and

limestone. Carbon cycles between the biotic and abiotic environment. Carbon dioxide is converted by autotrophs into carbohydrates via photosynthesis and returned to the atmosphere as CO_2 through respiration (fluxes). These fluxes can be measured. Some of the sinks and processes involved in the carbon cycle, together with the carbon fluxes, are shown below. Humans intervene in the carbon cycle through activities such as combustion and deforestation.

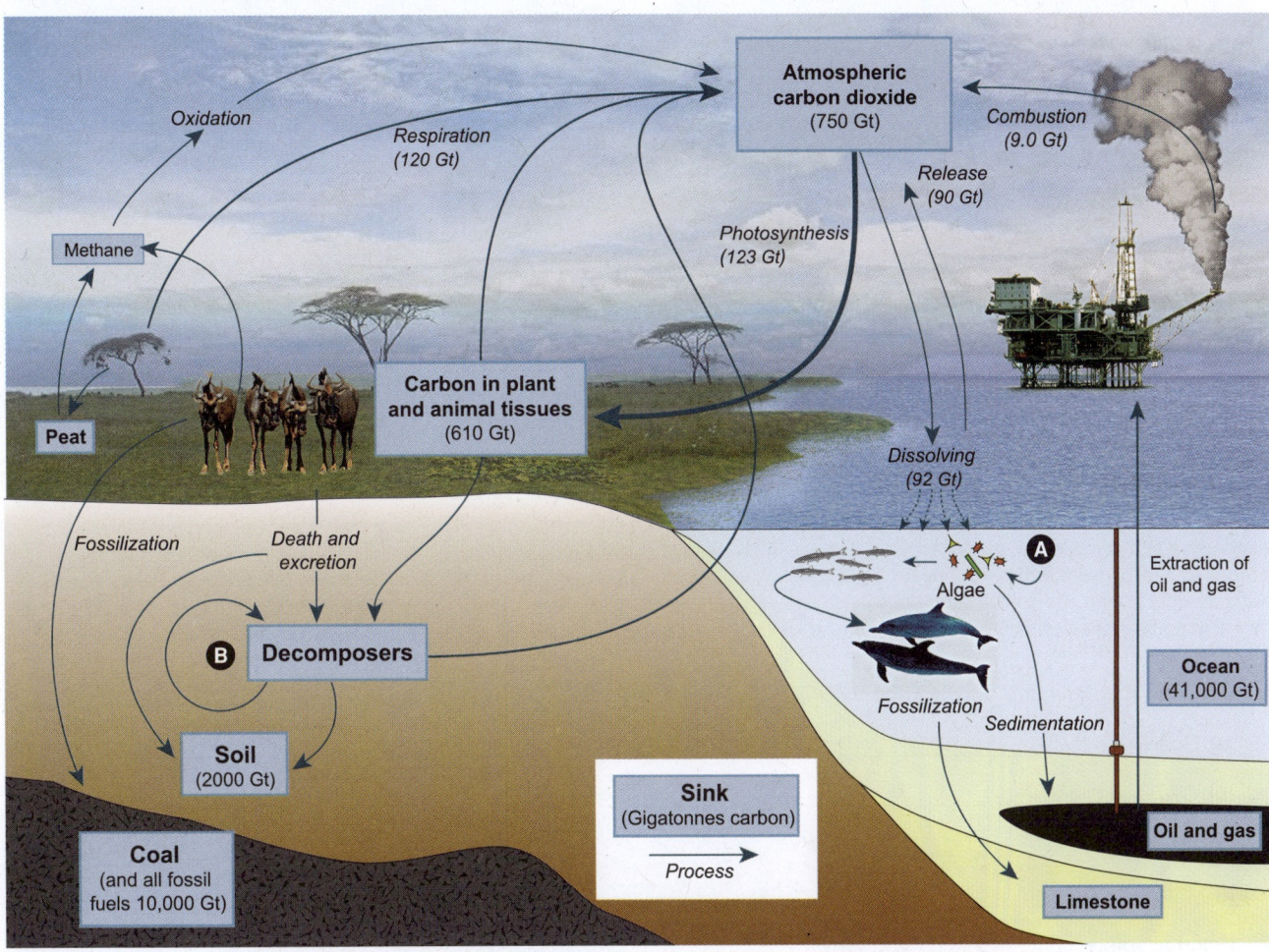

1. Add arrows and labels to the diagram above to show:

   (a) Dissolving of limestone by acid rain
   (b) Release of carbon from the marine food chain
   (c) Mining and burning of coal
   (d) Burning of plant material.

2. (a) Name the processes that release carbon into the atmosphere: _____

   _____

   (b) In what form is the carbon released? _____

3. Name the four geological reservoirs (sinks) in the diagram above that can act as a source of carbon:

   (a) _____ (c) _____

   (b) _____ (d) _____

4. (a) Identify the process carried out by algae at point **A**: _____

   (b) Identify the process carried out by decomposers at **B**: _____

5. What would be the effect on carbon cycling if there were no decomposers present in an ecosystem? _____

   _____

   _____

©2024 **BIOZONE** International
ISBN: 978-1-99-101410-8
Photocopying prohibited

C4.2

17 - 19

21, 22

Bracket fungus on tree trunk

Coal mine in Wyoming

Carbon may be locked up in biotic or abiotic systems for long periods of time, e.g. in the wood of trees or in fossil fuels such as coal or oil. Human activity, e.g. extraction and large scale combustion of fossil fuels, has disturbed the balance of the carbon cycle.

Organisms break down organic material to release carbon. Fungi and decomposing bacteria break down dead plant matter in the leaf litter of forests. Termites, with the aid of symbiotic protozoans and bacteria in their guts, digest the cellulose of woody tissue.

Coal is formed from the remains of terrestrial plant material buried in shallow swamps and subsequently compacted under sediments to form a hard black material. Coal is composed primarily of carbon and is a widely used fuel source.

Oil and **natural gas** formed in the past when dead algae and zooplankton settled to the bottom of shallow seas and lakes. These remains were buried and compressed under layers of non-porous sediment.

Limestone is a type of sedimentary rock composed mostly of calcium carbonate. It forms when the shells of molluscs and other marine organisms with calcium carbonate ($CaCO_3$) skeletons become fossilized.

Peat (partly decayed organic material) forms when plant material is not fully decomposed due to acidic or anaerobic conditions. Peaty wetlands are an efficient carbon sink but are lost through oxidation when land is drained.

6. Describe the biological origin of the following geological deposits:

   (a) Coal: _____

   (b) Oil: _____

   (c) Limestone: _____

   (d) Peat: _____

7. Using examples, compare and contrast the amount of time carbon spends in its various reservoirs:

   _____

   _____

   _____

   _____

8. Explain the role of living organisms in the carbon cycle: _____

   _____

   _____

   _____

9. Accumulated reserves of carbon such as peat, coal, and oil represent a sink or natural diversion from the cycle. In natural circumstances, the carbon in these sinks eventually returns to the cycle through geological processes which return deposits to the surface for oxidation. Explain the effect of human activity on the amount of carbon stored in sinks:

   _____

   _____

   _____

©2024 **BIOZONE** International
ISBN: 978-1-99-101410-8
Photocopying prohibited

Fluxes in the biotic environment affect the carbon cycle

The balance of photosynthesizing and respiring organisms can affect the amount of CO_2 in the atmosphere. If the biomass of photosynthesizing organisms vastly outweighs that of respiring organisms, CO_2 will be removed from the atmosphere and the carbon will be stored as biomass. Respiration returns carbon to the atmosphere.

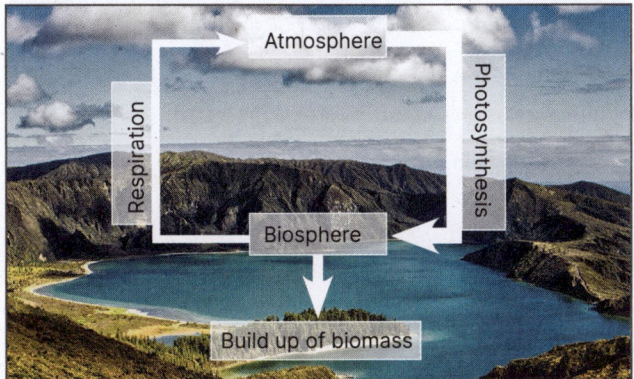

Photosynthesis and carbon

▶ Photosynthesis removes carbon from the atmosphere by fixing the carbon in CO_2 into carbohydrate molecules, e.g. glucose. Plants use the glucose to build structures such as wood.

▶ Respiration in living organisms returns some carbon to the atmosphere. If the amount or rate of carbon fixation exceeds that released in respiration then carbon will build up in the biosphere and be depleted in the atmosphere.

Respiration and carbon

▶ Cellular respiration breaks down glucose and releases carbon into the atmosphere as carbon dioxide.

▶ If the rate of carbon release is greater than that fixed by photosynthesis then carbon may accumulate in the atmosphere over time. Deforestation and the burning of fossil fuels have increased the amount of carbon in the atmosphere and depleted what was stored in the biosphere.

Nutrients cycle through the environment

Like carbon, many other elements cycle through the biotic and abiotic parts of the environment. Nitrogen, sulfur, and phosphorus are some important examples. Most elements cycle in similar patterns to the carbon cycle. All have their own specific pathways and chemical reactions along the way, but most will at some stage be part of the biosphere, geosphere, or atmosphere.

Root nodules on soy bean roots

Guano has a high phosphorus content

Nitrogen is an essential part of nucleic acids and proteins. Atmospheric nitrogen is not accessible to plants and animals. It enters the biosphere via nitrogen fixing bacteria.

Sulfur plays an important role in shaping proteins. Sulfur dioxide is produced by industry and volcanic activity (above). Chemical reactions convert it to sulphates which are taken up by living organisms.

Phosphorus is important in nucleic acids. It cycles as various compounds, e.g. phosphate (PO_4^{3-}). As with other elements, bacteria play an important role in making phosphorus available to plants.

10. Study the diagram at the top of the page. Describe the effect of each of the following:

    (a) Increasing the rate of photosynthesis on atmospheric and biospheric carbon:

    _____

    _____

    (b) Increasing the rate of cellular respiration on atmospheric and biospheric carbon.

    _____

    _____

    (c) Identify human activity with the same effects as increased respiration: _____

11. Why is nutrient cycling important for life on Earth? _____

    _____

    _____

    _____

258 Analyzing Changes in Atmospheric Carbon Dioxide

Key Idea: The Keeling Curve is the graph of the accumulated data on atmospheric carbon dioxide.

Charles Keeling was the first person to systematically measure atmospheric carbon dioxide concentrations. The curve on the graph resulting from these measurements now bears his name: the Keeling Curve. Monitoring began at Mauna Loa (Hawaii) in 1958. The Keeling curve shows not only the rise in atmospheric carbon dioxide over decades due to anthropogenic causes, but also seasonal and daily changes in carbon dioxide due to differences in respiration and photosynthesis. Carbon dioxide concentrations are now monitored in many locations around the world.

▶ The graphs below based on data from Mauna Loa Observatory show the change in carbon dioxide concentration since 1958, and over a two year and one week span.

Carbon dioxide concentration at Mauna Loa Observatory

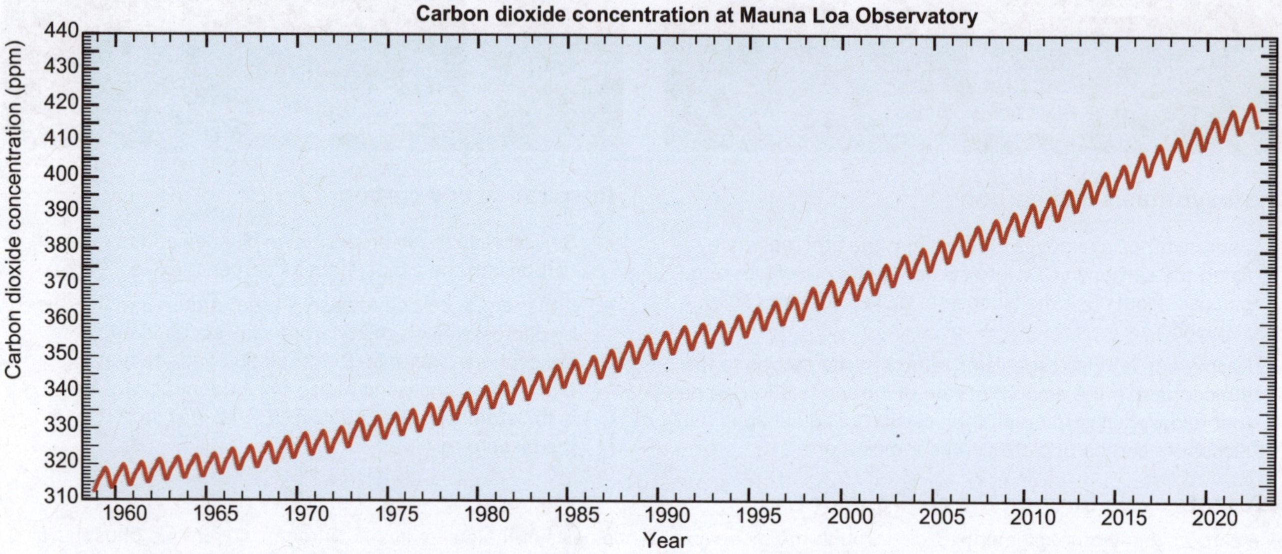

(Graph: Carbon dioxide concentration (ppm) vs Year, showing the Keeling Curve rising from ~315 ppm in 1960 to ~420 ppm in 2020, with seasonal oscillations. Y-axis: 310–440 ppm; X-axis: 1960–2020.)

Two year carbon dioxide concentration data

(Graph: Carbon dioxide concentration (ppm) vs time from Nov 2021 to Sep 2023. Y-axis: 410–425 ppm.)

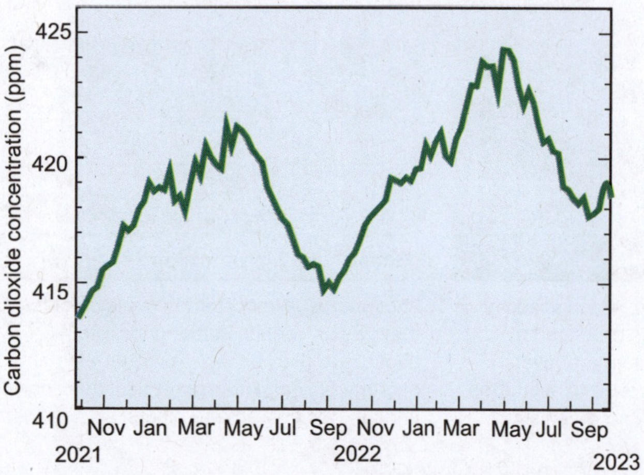

One week carbon dioxide concentration data

(Graph: Carbon dioxide concentration (ppm) vs date 18–24 Oct 2023. Y-axis: 405–440 ppm, with labels 12 PM and 12 AM marked around Oct 19.)

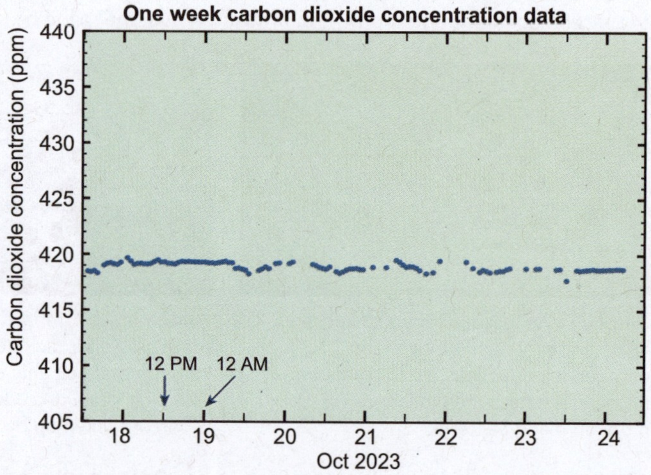

1. By how many parts per million has the CO_2 concentration changed between 1958 and 2022? _____

2. By how many parts per million has the CO_2 concentration changed between October 2021 and October 2023?

   _____

3. Why does the CO_2 concentration reach a peak in May and a minimum in September? (Hint: The islands of Hawaii are in the Northern Hemisphere).

   _____

   _____

4. Why does the CO_2 concentration reach a peak just after midnight every night?

   _____

   _____

5. CO_2 is known to trap heat in the atmosphere. What might the effect of increasing atmospheric CO_2 be?

   _____

C4.2
20

©2024 **BIOZONE** International
ISBN: 978-1-99-101410-8
Photocopying prohibited

259 Did You Get It?

1. Ticks are obligate blood feeders and must obtain blood to pass from one life stage to the next. Ticks attach to the outside of hosts, in this case a cat, where they suck blood and fluids and cause irritation.

   (a) Identify this type of interaction: _____

   (b) Describe how each species is affected (benefits/harmed/no effect):

   _____

   (c) How would the tick population be affected if the host became rare?

   _____

   _____

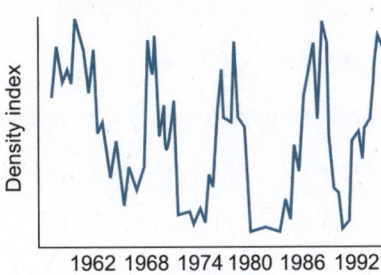

Great gerbil, Bakanas, Kazakstan

Density index

1962 1968 1974 1980 1986 1992
Year

2. The graph shows the population changes in the Great gerbil in Kazakstan. What factors might be causing the population to fluctuate as they do?

   _____

   _____

   _____

   _____

   _____

   _____

3. The diagram below represents a food chain.

$$A \longrightarrow B \longrightarrow C \longrightarrow D$$

   (a) Which letter represents the producer? _____

   (b) Which letter represents the tertiary consumer? _____

   (c) If there is a total of 1 MJ of energy stored within group B how much energy might be expected to be stored within group D?

   _____

4. Where do decomposers fit in food chains? Give reasons: _____

   _____

   _____

   _____

5. (a) The graph (right) shows primary productivity in the oceans. Describe and explain the shapes of the curves:

   _____

   _____

   _____

   (b) About 90% of all marine life lives in the photic zone (the depth to which light penetrates). Suggest why this is so:

   _____

   _____

   _____

   _____

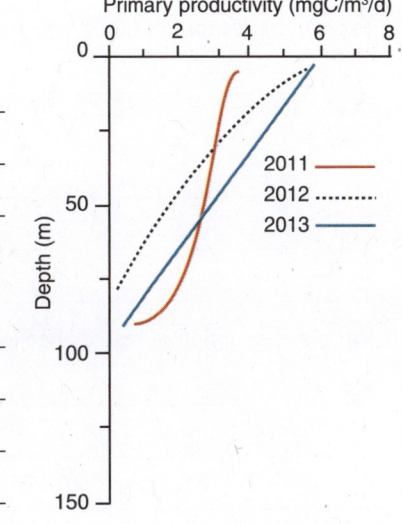

Primary productivity (mgC/m³/d)

2011
2012
2013

Depth (m)

260 Summary Assessment

1. The active site of an enzyme is:

   (a) A specific part of the enzyme that remains active after denaturation.

   (b) Part of the enzyme that allows it to be actively transported into and out of a cell.

   (c) A specific part of the enzyme where the substrate binds.

   (d) None of the above.

2. Enzymes speed up reactions by:

   (a) Reducing the activation energy needed.

   (b) Increasing the activation energy.

   (c) Adding energy to the reaction.

   (d) Taking part in the reaction and forming part of the products.

3. **AHL**: Which of the following is true of the light dependent reactions?

   (a) Cyclic phosphorylation involves electrons cycling within photosystem I.

   (b) Non-cyclic phosphorylation involves electrons being passed from photosystem II to photosystem I.

   (c) The light dependent reactions involve pigments bound on the thylakoid membranes.

   (d) All of the above.

4. Which process belows also occurs in fermentation?

   (a) Glycolysis.

   (b) Link reaction.

   (c) Electron transport chain.

   (d) Krebs cycle.

5. Cytokines can act as:

   (a) Local regulators.

   (b) Endocrine regulators.

   (c) Both (a) and (b).

   (d) None of the above.

6. Signal transduction may involve:

   (a) Phosphorylation cascades.

   (b) Activation of transcription factors.

   (c) Second messengers.

   (d) All of the above.

7. Myelinated nerves conduct nerves signals:

   (a) Faster than unmyelinated nerves.

   (b) by saltatory conduction.

   (c) When an action potential is triggered.

   (d) All of the above.

8. The type of population growth labelled X in the graph below is:

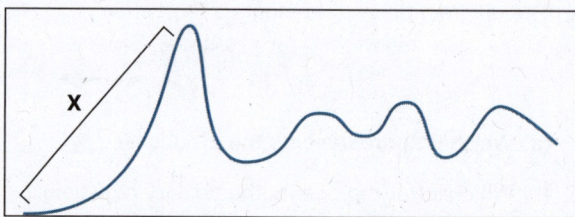

   (a) Logistic growth.

   (b) Exponential growth.

   (c) Stable.

   (d) Density dependent.

9. Light was shone from above onto a maize seedling with a horizontal shoot and root for 12 hours. Measurements of the angle of curvature were taken every two hours. The results are shown in the table:

   Explain the reason for these results:

   _____

   _____

   _____

   _____

   _____

Length of time exposed to light	Angle of curvature (°)	
	Shoot	Root
0	0	0
2	7	-3
4	14	-4
6	18	-8
8	23	-11

10. Write a description of how a chemical synapse allows transmission of nerve signals between cells:

   _____

   _____

   _____

11. The diagram below shows a model of part of the adaptive immune system reaction when an antigen is encountered.

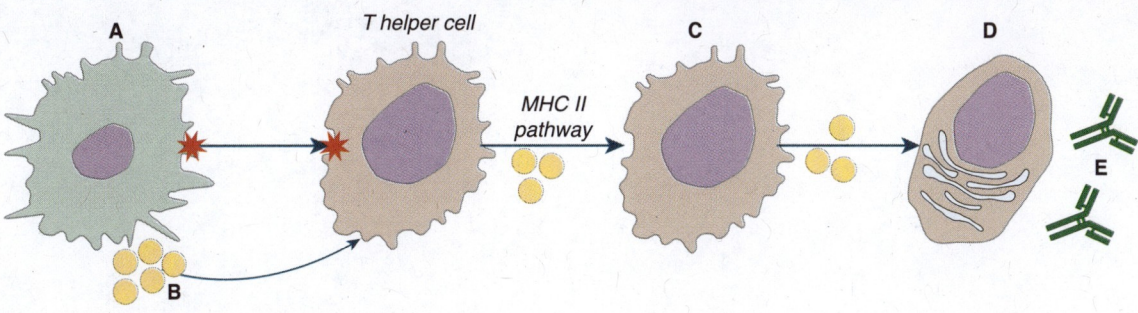

(a) What type of cell is A and what is its role? _____

_____

_____

(b) Identify the structures labelled B and state their role in this part of the process: _____

_____

_____

(c) What type of cell is C? _____

(d) What type of cell is D? _____

(e) Identify the structures labelled E and state their role: _____

_____

12. Clearly explain the differences between aerobic and anaerobic respiration:

_____

_____

_____

13. (a) Name the interaction shown in the image right:

_____

(b) Is an inter or Intra specific interaction shown?

_____

(c) Do the individuals benefit or are they harmed in the interaction?

_____

(d) Identify two examples in which this interaction occurs between and within species:

_____

_____

_____

14. Explain the relationship between photosynthesis and respiration in an ecosystem:

_____

_____

_____

15. Explain the difference between temporal and spatial summation at synapses:

_____

_____

Theme D:

Continuity and Change

Natural selection is driven by environmental change leading to the ever-changing biodiversity on Earth, while each organism processes complex homoeostatic mechanisms to enable continuity.

Understandings:

Molecules

D1.1 DNA replication

D1.2 Protein synthesis

D1.3 Mutations and gene editing

Cells

D2.1 Cell and nuclear division

D2.2 Gene expression

D2.3 Water potential

Organisms

D3.1 Reproduction

D3.2 Inheritance

D3.3 Homeostasis

Ecosystems

D4.1 Natural selection

D4.2 Stability and change

D4.3 Climate change

Molecules

D1.1 DNA Replication

		Activity Number

Guiding Questions:
▶ How is new DNA made?
▶ How have biotechnology applications been advanced by understanding of DNA replication?

Learning Outcomes:

☐ 1	Construct a flow chart of the steps in DNA replication and link the importance of DNA replication to the processes of reproduction and growth.	**261**
☐ 2	Explain how accuracy in copying base sequences is enabled by semi-conservative DNA replication and complementary base pairing.	**261**
☐ 3	Summarize the role of enzymes involved in DNA replication including helicase and DNA polymerase.	**261**
☐ 4	Explain how the polymerase chain reaction (PCR) and gel electrophoresis amplify and separate DNA fragments.	**263 - 264**
☐ 5	Provide examples of polymerase chain reaction and gel electrophoresis applications. **NOS:** Explain methods to enhance reliability of results.	**265**
☐ 6	**AHL:** Explain the process of 3' and 5' directionality of DNA polymerases when adding nucleotides.	**262**
☐ 7	**AHL:** Contrast the replication processes on the leading strand and the lagging strand using relevant terminology.	**262**
☐ 8	**AHL:** Describe the functions of the enzymes DNA primase, DNA polymerase I, DNA polymerase III, and DNA ligase in replication in prokaryotes.	**262**
☐ 9	**AHL:** Explain how the proofreading of DNA is carried out.	**262**

D1.2 Protein Synthesis

		Activity Number

Guiding Questions:
▶ What series of events occur in order for protein synthesis to take place?
▶ How is the mechanism of protein synthesis safeguarded against errors?

Learning Outcomes:

☐ 1	Provide an overview of the transcription process during RNA synthesis from a DNA template.	**268**
☐ 2	Elaborate on the role of hydrogen bonding and complementary base pairing in transcription.	**268**
☐ 3	Explain how DNA templates are able to maintain stability for the lifespan of a cell.	**268**
☐ 4	Link the process of gene expression to the process of transcription.	**268**
☐ 5	Provide an overview of the translation process including polypeptide synthesis from mRNA.	**269**
☐ 6	Describe the specific roles of mRNA, ribosomes, and tRNA during the translation process.	**269**
☐ 7	Explain the process and function of tRNA and mRNA complementary base pairing; define codon and anticodon.	**269**
☐ 8	Provide an overview of features of the genetic code including definitions of degeneracy and universality.	**267**
☐ 9	Decipher an amino acid sequence using an mRNA codon table.	**267**
☐ 10	Provide details on the sequence of events from mRNA to forming a polypeptide chain, occurring in ribosomes.	**269**
☐ 11	Explain how protein structure can change due to mutations using an example of a point mutation.	**269**
☐ 12	**AHL:** Compare the directionality of 5' to 3' transcription to 5' to 3' translation.	**270**
☐ 13	**AHL:** Explain how transcription is initiated at the promoter.	**270**
☐ 14	**AHL:** Discuss the possible function and types of DNA non-coding sequences.	**270**
☐ 15	**AHL:** Explain the process of post-transcriptional modification in eukaryotic cells.	**270**
☐ 16	**AHL:** Describe how alternative splicing of exons in the same gene can lead to protein variants.	**270**
☐ 17	**AHL:** Elaborate on the translation initiation process.	**271**
☐ 18	**AHL:** Explain how polypeptides are modified to form their functional state, including the two-stage modification of pre-proinsulin to insulin.	**271**
☐ 19	**AHL:** Explain how proteasomes enable amino acid recycling.	**271**

D1.3 Mutation and Gene Editing

Guiding Questions:
▶ How is DNA replicated?
▶ How have biotechnology applications been advanced by DNA replication?

Learning Outcomes:

☐ 1	Classify and distinguish between types of gene mutations, including substitutions, insertions, and deletions.	272
☐ 2	Link the formation of single-nucleotide polymorphisms (SNPs) to base substitutions and describe possible consequences of the mutations.	272
☐ 3	Describe some consequences of insertion and deletion mutations including functionality of the polypeptides.	272
☐ 4	Summarize the main causes of gene mutations including the role of radiation and chemical mutagens.	273
☐ 5	Explain why mutation is considered a random event.	273
☐ 6	Compare the consequences of mutation in both germ cells, such as inherited genetic disorders, and somatic cells, such as cancer.	273
☐ 7	Explain how mutation acts as a source of genetic variation and the only means of forming new alleles which is essential for evolution. **NOS:** Evaluate the advantages and disadvantages of commercial genetic testing to potential carriers of genetic mutations.	273
☐ 8	**AHL:** Explain how gene function research can be assisted by gene knockout.	274
☐ 9	**AHL: NOS:** Debate the ethics for gene editing in humans when using CRISPR Cas9 technology.	274
☐ 10	**AHL:** Analyse several hypotheses that account for conserved or highly conserved gene sequences.	275

261 DNA Replication

Key Idea: DNA can produce identical copies through a semi-conservative replication process with the use of enzymes.

Cell division has three purposes: growth from a single fertilized cell into a mature organism, repair or replacement of damaged and old cells, and reproduction. Before a cell can divide, it must double its DNA. It does this by a process called DNA replication. This process ensures that each resulting cell receives a complete set of genetic instructions from the parent cell. After the DNA has replicated, each chromosome is made up of two chromatids that will become separated during cell division to form separate chromosomes. During DNA replication, nucleotides are added at the replication fork. Enzymes are responsible for all of the key events. DNA replication is semi-conservative, meaning each new DNA molecule is made of one strand from the parent DNA and one strand from the daughter DNA.

Step 1
Unwinding the DNA molecule

A normal chromosome consists of an unreplicated DNA molecule. Before cell division, this long molecule of double stranded DNA must be replicated.

For this to happen, it is first untwisted and separated (unzipped) at high speed at its replication fork by an enzyme called helicase. This enzyme breaks the hydrogen bonds between the complementary bases on either strand. Another enzyme relieves the strain that this generates by cutting, winding, and rejoining the DNA strands.

Step 2
Making new DNA strands

The formation of new DNA is carried out mostly by an enzyme complex called DNA polymerase.

DNA polymerase catalyzes the condensation reaction that joins adjacent nucleotides. The strand is synthesized in a 5' to 3' direction, with the polymerase moving 3' to 5' along the strand it is reading. Thus, the nucleotides are assembled in a continuous fashion on one strand but in short fragments on the other strand. These fragments are later joined by an enzyme to form one continuous length. Nucleotides are added through complementary base-pairing. The base pairing rule ensures that nucleotide A is always paired with nucleotide T and nucleotide C is always paired with nucleotide G.

Step 3
Rewinding the DNA molecule

Each of the two new double-helix DNA molecules has one strand of the original DNA (green/white) and one strand that is newly synthesized (blue). The two DNA molecules rewind into their double-helix shape again.

DNA replication is semi-conservative, with each new double helix containing one old (parent) strand and one newly synthesized (daughter) strand. The new chromosome has twice as much DNA as a non-replicated chromosome. The two chromatids will become separated in the cell division process to form two separate chromosomes.

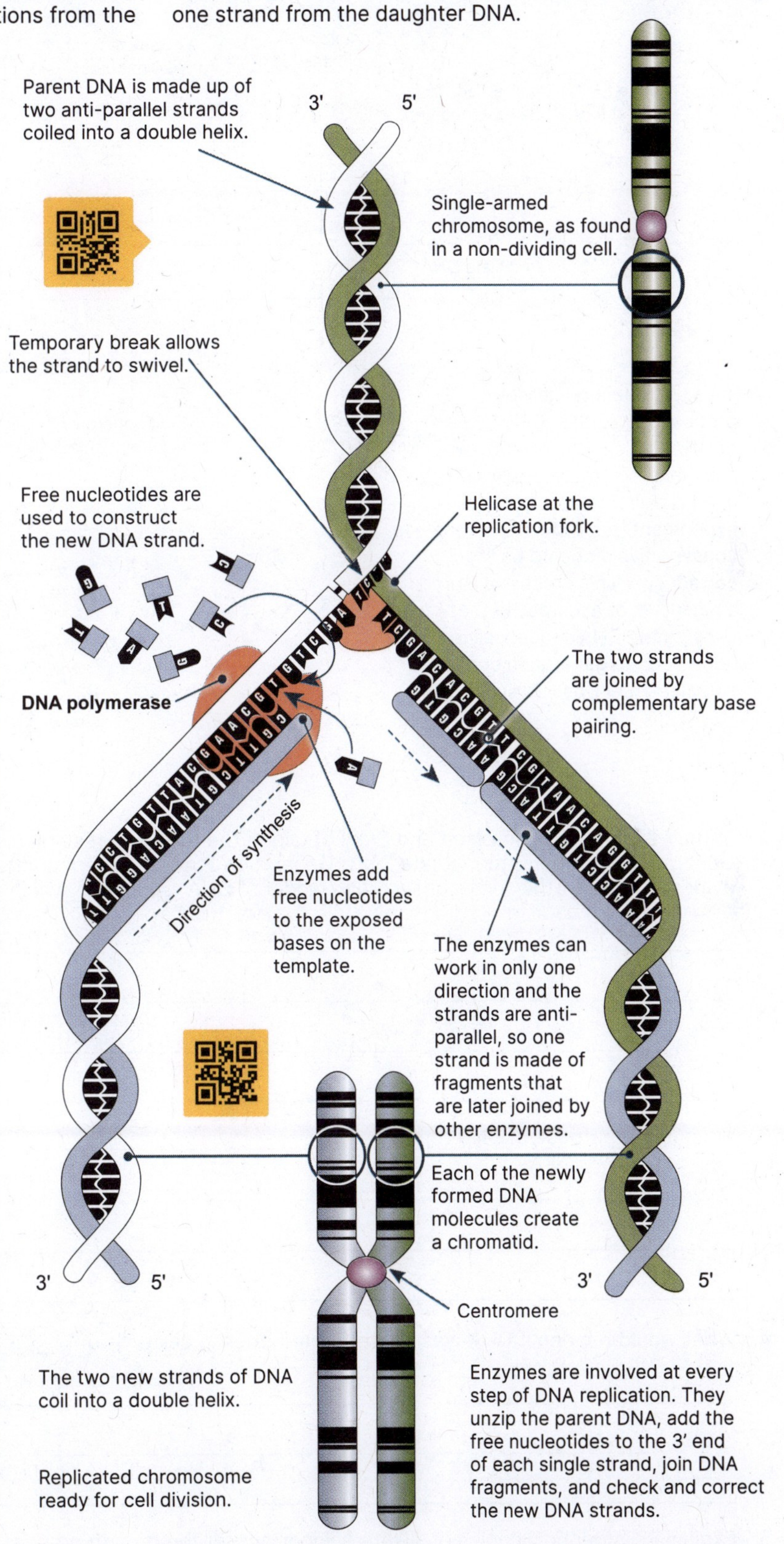

Parent DNA is made up of two anti-parallel strands coiled into a double helix.

Single-armed chromosome, as found in a non-dividing cell.

Temporary break allows the strand to swivel.

Free nucleotides are used to construct the new DNA strand.

DNA polymerase

Helicase at the replication fork.

The two strands are joined by complementary base pairing.

Direction of synthesis

Enzymes add free nucleotides to the exposed bases on the template.

The enzymes can work in only one direction and the strands are anti-parallel, so one strand is made of fragments that are later joined by other enzymes.

Each of the newly formed DNA molecules create a chromatid.

Centromere

3' 5'

3' 5'

The two new strands of DNA coil into a double helix.

Replicated chromosome ready for cell division.

Enzymes are involved at every step of DNA replication. They unzip the parent DNA, add the free nucleotides to the 3' end of each single strand, join DNA fragments, and check and correct the new DNA strands.

1. Create a flow chart on a sheet of paper showing the steps of the DNA replication process. Include the structures and enzymes involved. Staple your flowchart to this page.

Meselson and Stahl's semi-conservative replication experiment

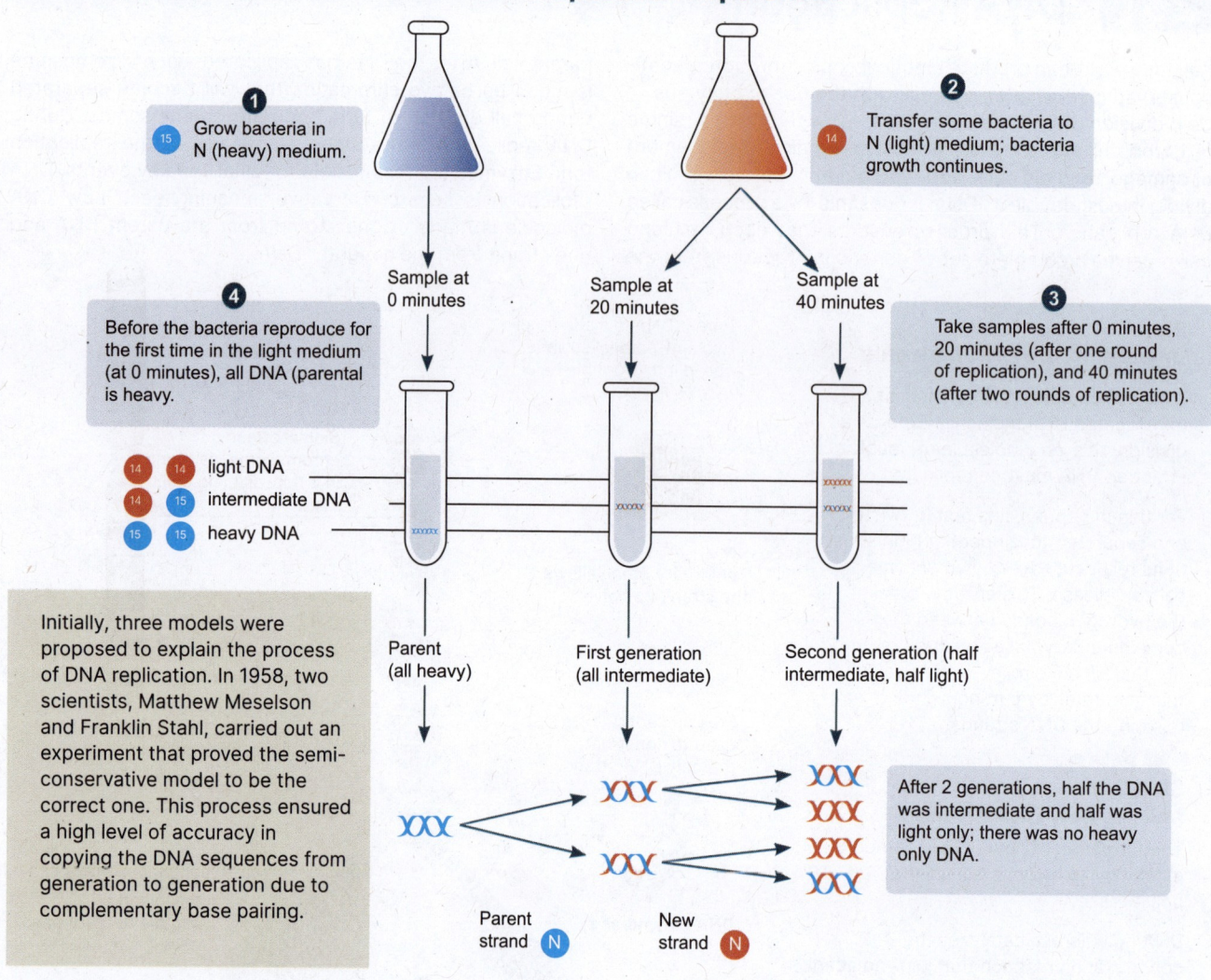

1 Grow bacteria in N (heavy) medium.

2 Transfer some bacteria to N (light) medium; bacteria growth continues.

4 Before the bacteria reproduce for the first time in the light medium (at 0 minutes), all DNA (parental is heavy.

3 Take samples after 0 minutes, 20 minutes (after one round of replication), and 40 minutes (after two rounds of replication).

Sample at 0 minutes

Sample at 20 minutes

Sample at 40 minutes

light DNA
intermediate DNA
heavy DNA

Parent (all heavy)

First generation (all intermediate)

Second generation (half intermediate, half light)

Initially, three models were proposed to explain the process of DNA replication. In 1958, two scientists, Matthew Meselson and Franklin Stahl, carried out an experiment that proved the semi-conservative model to be the correct one. This process ensured a high level of accuracy in copying the DNA sequences from generation to generation due to complementary base pairing.

After 2 generations, half the DNA was intermediate and half was light only; there was no heavy only DNA.

Parent strand N   New strand N

2. In their experiment, Meselson and Stahl obtained the following results: Generation 0 = 100% 'heavy DNA'; Generation 1= 100% 'intermediate DNA'; Generation 2 = 50% 'intermediate DNA', 50% 'light DNA'. How does this evidence support the semi-conservative model of DNA replication which matches the result of Meselson and Stahl?

_____

_____

_____

_____

3. What is the purpose of DNA replication? _____

_____

_____

_____

4. What would happen if DNA was not replicated prior to cell division? _____

_____

_____

_____

5. Explain why identical copies of DNA are important for the growth process to occur correctly: _____

_____

_____

©2024 **BIOZONE** International
ISBN: 978-1-99-101410-8
Photocopying prohibited

262 Details of DNA Replication (Prokaryote)

Key Idea: The process of DNA replication is controlled by many different enzymes.

DNA replication involves many enzyme-controlled steps. They are shown below as separate, but many of the enzymes are clustered together as enzyme complexes. As the DNA is replicated, enzymes 'proof-read' it and correct mistakes. The polymerase enzyme can only work in one direction. This means that one new strand (leading) is constructed as a continuous length and the other new strand (lagging) is made in short segments to be joined together later.

DNA replication occurs at a rate of ~1000 nucleotides per second in *E. coli* and ~50 nucleotides per second in humans.

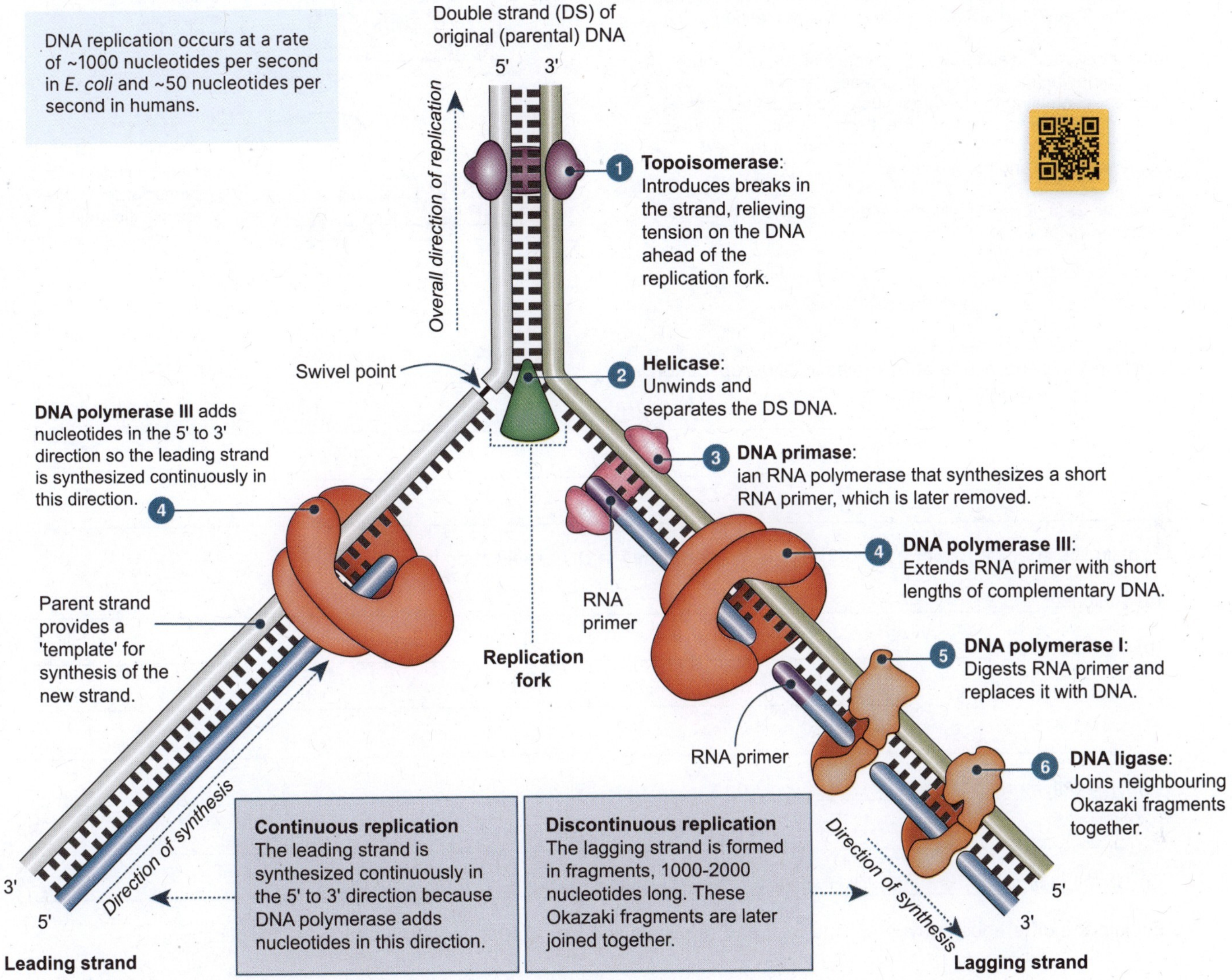

Double strand (DS) of original (parental) DNA

5'   3'

Overall direction of replication

1 Topoisomerase: Introduces breaks in the strand, relieving tension on the DNA ahead of the replication fork.

Swivel point

2 Helicase: Unwinds and separates the DS DNA.

DNA polymerase III adds nucleotides in the 5' to 3' direction so the leading strand is synthesized continuously in this direction.

3 DNA primase: ian RNA polymerase that synthesizes a short RNA primer, which is later removed.

4 DNA polymerase III: Extends RNA primer with short lengths of complementary DNA.

RNA primer

Replication fork

5 DNA polymerase I: Digests RNA primer and replaces it with DNA.

Parent strand provides a 'template' for synthesis of the new strand.

RNA primer

6 DNA ligase: Joins neighbouring Okazaki fragments together.

Direction of synthesis

3'
5'

Leading strand

Continuous replication The leading strand is synthesized continuously in the 5' to 3' direction because DNA polymerase adds nucleotides in this direction.

Discontinuous replication The lagging strand is formed in fragments, 1000-2000 nucleotides long. These Okazaki fragments are later joined together.

Direction of synthesis

5'
3'

Lagging strand

Directionality of DNA and the impact on DNA replication

The deoxyribose sugar that makes the backbone of the DNA molecule is asymmetric and the carbon atoms are numbered clockwise from the oxygen atom. The 'end' at which the phosphate group is joined to the 3rd carbon is labelled the 3' (3 prime) end, and the other side at which the phosphate group is joined to the 5th carbon is labelled the 5' (5 prime) end. This gives the strand directionality. One end runs 3' to 5' and the other runs 5' to 3'.

The DNA polymerases add the 5' end of (new) nucleotides to the 3' end of the nucleotide strands.

The leading DNA strand only requires replication to be initiated once by DNA primase, adding to the 3' end of the strand towards the replication fork. This enzyme catalyzes a short RNA molecule called a primer - and this becomes the point of DNA synthesis. This process results in continuous DNA synthesis.

The lagging DNA strand requires replication to be initiated multiple times, starting from the 3' closest to the replication fork (Note: this is the opposite direction from the leading strand). Only short strands can be made, called Okazaki fragments, and then the DNA primase constructs a new primer. This process results in discontinuous DNA synthesis.

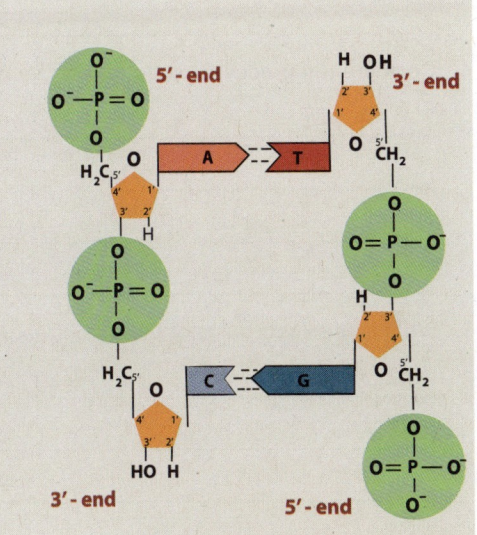

DNA Proofreading

You will recall that DNA polymerase III adds nucleotides to the 3′ end of the forming DNA. Occasionally, the base that is added is incorrect and therefore cannot form the hydrogen bonds with the nucleotide on the template strand.

DNA polymerase has a proofreading function, and can detect the mismatched nucleotide.

A 3′–5′ exonuclease subunit will break the incorrect nucleotide from the 3′ end of the DNA, and allow a new nucleotide to join.

If the incorrect nucleotide is not detected initially, a second line of defence, mismatch repair, can be used. An entire segment containing the incorrect nucleotide is cut out and DNA polymerase adds in the correct sequence of nucleotides. A second enzyme, DNA ligase, seals the ends of the new sequence to the original sequence.

Note: the CRISPR cas-9 system makes use of this process to insert new sequences into RNA.

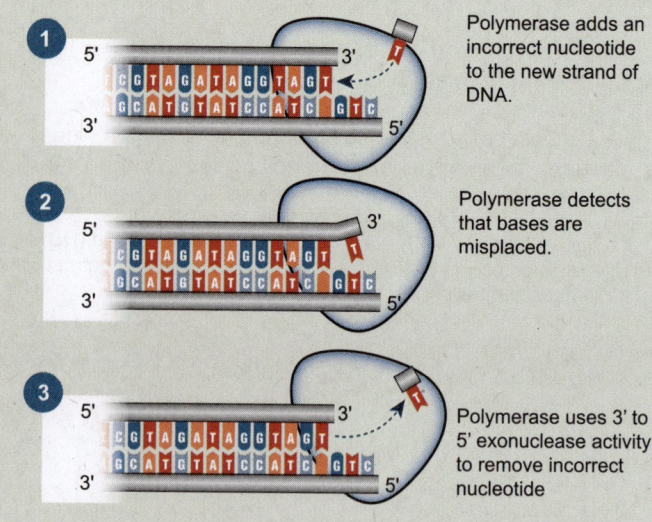

1. Polymerase adds an incorrect nucleotide to the new strand of DNA.

2. Polymerase detects that bases are misplaced.

3. Polymerase uses 3′ to 5′ exonuclease activity to remove incorrect nucleotide

1. Explain the general role of enzymes in DNA replication: _____

_____

_____

_____

2. Explain the specific role of each of the following enzymes in DNA replication in prokaryotes:

    (a) Helicase: _____

    (b) DNA primase: _____

    (c) DNA polymerase I: _____

    _____

    (d) DNA polymerase III: _____

    _____

    (e) DNA ligase: _____

3. Explain the difference between 3′ and 5′ strand terminals and how this impacts the direction of replication:

_____

_____

_____

4. Compare the processes occurring at the leading and lagging strands including the presence of Okazaki fragments:

_____

_____

_____

_____

5. What is the importance of DNA proofreading and how does that process occur? _____

_____

_____

_____

_____

263 Polymerase Chain Reaction

Key Idea: The polymerase chain reaction (PCR) is a process that can make billions of copies of a target DNA sequence of interest so that it can be analyzed.

The polymerase chain reaction technique is carried out *in vitro* rather than in a living organism. An overview of PCR is given below. PCR's ability to amplify small quantities of DNA means it can be used to identify the presence of organisms in an environment even if they are in very low numbers. Examples include identifying COVID-19 infections (even in people showing no or few symptoms), DNA in ancient bone fragments, material from crime scenes, and as part of a genetic engineering process.

A single cycle of PCR

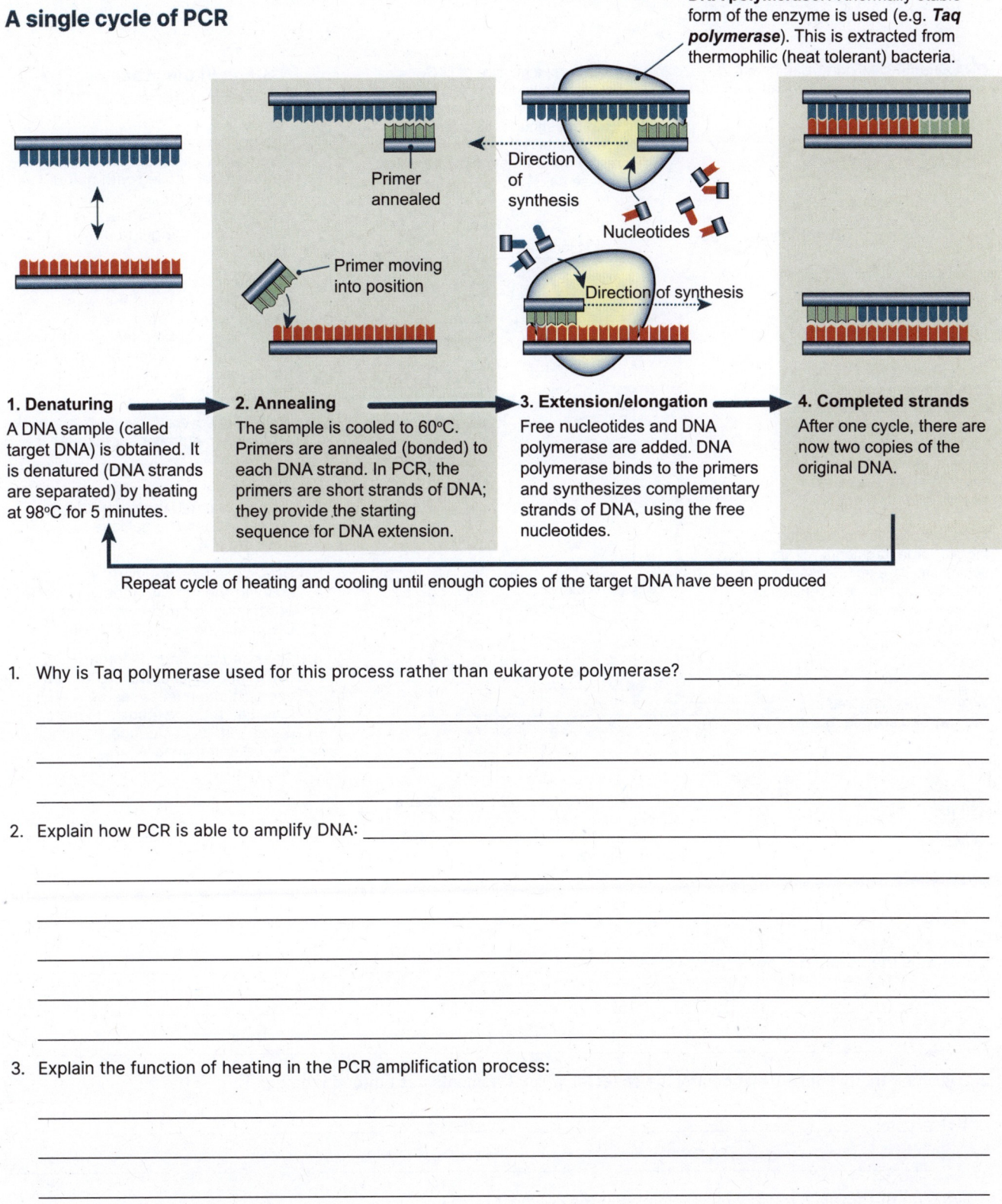

DNA polymerase: A thermally stable form of the enzyme is used (e.g. *Taq polymerase*). This is extracted from thermophilic (heat tolerant) bacteria.

Primer annealed

Direction of synthesis

Primer moving into position

Nucleotides

Direction of synthesis

1. Denaturing
A DNA sample (called target DNA) is obtained. It is denatured (DNA strands are separated) by heating at 98°C for 5 minutes.

2. Annealing
The sample is cooled to 60°C. Primers are annealed (bonded) to each DNA strand. In PCR, the primers are short strands of DNA; they provide the starting sequence for DNA extension.

3. Extension/elongation
Free nucleotides and DNA polymerase are added. DNA polymerase binds to the primers and synthesizes complementary strands of DNA, using the free nucleotides.

4. Completed strands
After one cycle, there are now two copies of the original DNA.

Repeat cycle of heating and cooling until enough copies of the target DNA have been produced

1. Why is Taq polymerase used for this process rather than eukaryote polymerase? _____

_____

_____

_____

2. Explain how PCR is able to amplify DNA: _____

_____

_____

_____

_____

_____

_____

3. Explain the function of heating in the PCR amplification process: _____

_____

_____

_____

4. After only two cycles of replication, four copies of the double-stranded DNA exist. Calculate how much a DNA sample will have increased after:

(a) 10 cycles: _____   (b) 25 cycles: _____

264 Gel Electrophoresis

Key Idea: Gel electrophoresis is a tool used to isolate fragments of DNA for further study.

Gel electrophoresis can be used for DNA profiling (comparing individuals based on their unique DNA banding profiles). DNA has an overall negative charge, so when an electrical current is run through a gel, the DNA moves towards the positive electrode. The rate at which the DNA molecules move through the gel depends primarily on their size and the strength of the electric field. The gel they move through has many pores. Smaller DNA molecules move through the pores more quickly than larger ones. At the end of the process, the DNA molecules can be stained and visualized as a series of bands. Each band contains DNA molecules of a particular size. The bands furthest from the start of the gel contain the smallest DNA fragments.

DNA solutions: Mixtures of different sizes of DNA fragments are loaded in each well in the gel.

DNA markers, a mixture of DNA molecules with known molecular weights (size) are often run in one lane. They are used to estimate the sizes of the DNA fragments in the sample lanes. The figures below are hypothetical markers (bp = base pairs).

Negative electrode (–)

4 lanes

Wells: Holes are made in the gel with a comb, acting as a reservoir for the DNA solution.

DNA fragments move: The gel matrix acts as a sieve for the negatively charged DNA molecules as they move towards the positive terminal. Small fragments move easily through the matrix, whereas large fragments don't.

As DNA molecules migrate through the gel, large fragments will lag behind small fragments. As the process continues, the separation between larger and smaller fragments increases.

Tray: The gel is poured into this tray and allowed to set.

Large fragments

Small fragments

- 50,000 bp
- 20,000 bp
- 10,000 bp
- 5000 bp
- 2500 bp
- 1000 bp
- 500 bp

Positive electrode (+)

Gel: A gel is prepared, which will act as a support for separation of the fragments of DNA. The gel is a jelly-like material, called agarose.

Steps in the process of gel electrophoresis of DNA

1. The gel is placed in an electrophoresis chamber and the chamber is filled with buffer, covering the gel. This allows the electric current from electrodes at either end of the gel to flow through it.

2. DNA samples are mixed with a 'loading dye' to make the DNA sample visible. The dye also contains glycerol or sucrose to make the DNA sample heavy so that it will sink to the bottom of the well.

3. The gel is covered, electrodes are attached to a power supply and turned on.

4. When the dye marker has moved through the gel, the current is turned off and the gel is removed from the tray.

5. DNA molecules are made visible by staining the gel with methylene blue or ethidium bromide which binds to DNA and fluoresces in UV light.

6. The band or bands of interest are cut from the gel and dissolved in chemicals to release the DNA. This DNA can then be studied in more detail, e.g. its nucleotide sequence can be determined.

1. What is the purpose of gel electrophoresis? _____

_____

_____

2. Describe the two forces that control the speed at which fragments pass through the gel:

(a) _____

(b) _____

3. Why do the smallest fragments travel through the gel most rapidly? _____

_____

_____

D1.1
4

©2024 **BIOZONE** International
ISBN: 978-1-99-101410-8
Photocopying prohibited

265 Applications of DNA Tools

Key Idea: DNA tools can be used for a wide range of applications by humans.

Often, it is very hard to obtain enough DNA to analyze. Researchers need to increase the quantity of DNA they have to work with and this is done using polymerase chain reaction (PCR). Gel electrophoresis is used to separate out samples. Repeating the analysis of DNA can increase reliability,

and therefore confidence, of the results. This reliability is particularly important when using DNA as a tool for solving crimes, such as homicide. DNA tools also have several other applications. DNA evidence has been used to identify body parts, solve cases of industrial sabotage and contamination, for paternity testing, and even in identifying animal products illegally made from endangered species.

Using DNA to solve crimes

▸ Although it does not make a complete case on its own, DNA profiling in conjunction with other evidence is one of the most powerful tools in identifying offenders.

▸ A lot of DNA is found at crime scenes and the information collected can be used to help identify the criminal. However, not all of the DNA collected will be from the criminal. Other DNA could belong to the victim, people who came to their aid, e.g. paramedics, or the police investigators if they have not taken correct precautions.

▸ In the example below, the criminal who broke into this home has left behind several samples of their DNA. Samples of material that may contain DNA are taken for analysis. At a crime scene, this may include blood and body fluids as well as samples of clothing or objects that the offender might have touched. Samples from the victim and the investigator are also taken to eliminate them as a possible source of contamination (below right). In this example, the DNA of the people who live in the house and the investigator will also be collected so their profiles can be eliminated. A calibration or standard is run so that the technician knows the profile has run correctly.

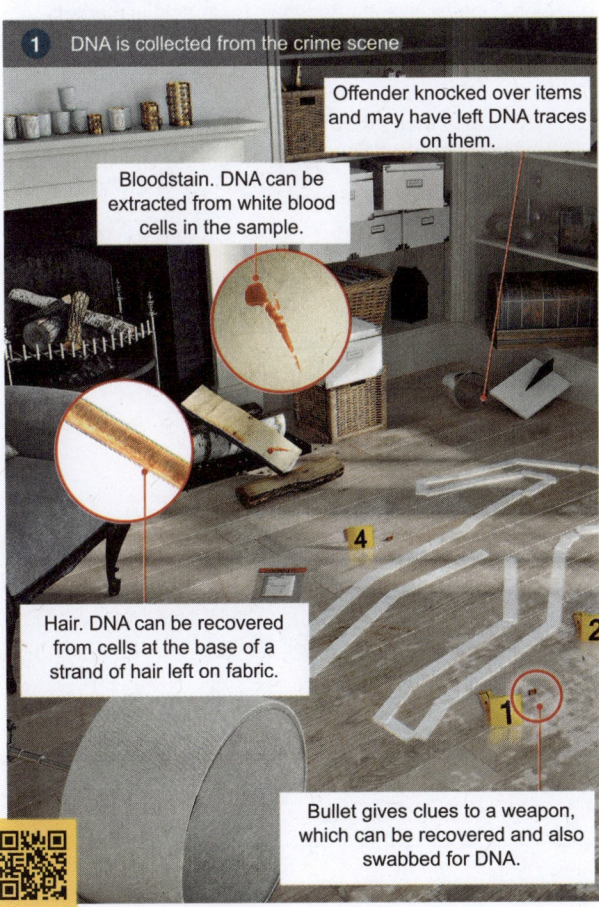

① DNA is collected from the crime scene

Offender knocked over items and may have left DNA traces on them.

Bloodstain. DNA can be extracted from white blood cells in the sample.

Hair. DNA can be recovered from cells at the base of a strand of hair left on fabric.

Bullet gives clues to a weapon, which can be recovered and also swabbed for DNA.

② DNA is isolated and profiles are made from all samples and compared to known DNA profiles such as that of the victim.

A B C D — Calibration, Profiles of collected DNA — Investigator (C) — Victim (D)

③ Unknown DNA samples are compared to DNA databases of convicted offenders and to the DNA of the alleged offender.

Alleged offender — Calibration — A E F G — Profiles from DNA database

④ Although it does not make a complete case, DNA profiling, in conjunction with other evidence, is one of the most powerful tools in identifying offenders.

1. Why are DNA profiles obtained for both the victim and investigator and how can the results be made more reliable?

2. Use the evidence to decide whether the alleged offender is innocent or guilty and explain your decision:

D1.1 5 NOS

Paternity testing

DNA profiling can be used to determine paternity (and maternity) by looking for matches in alleles between parents and children. It is used in situations of child support or inheritance. DNA profiling can establish the certainty of paternity (and maternity) to a 99.99% probability of parentage.

Every STR (short tandem repeat) allele is given the number of its repeats as its name, e.g. 8 or 9. In a paternity case, the mother may be 11, 12 and the father may be 8, 13 for a particular STR. The child will have a combination of these. See table below:

DNA marker	Mother's alleles	Child's alleles	Father's alleles
CSF1PO	7, 8	8, 9	9, 12
D10S1248	14, 15	11, 14	10, 11
D12S391	16, 17	17, 17	17, 18
D13S317	10, 11	9, 10	8, 9

The frequency of each allele occurring in the population is important when determining paternity (or maternity). For example, DNA marker CSF1PO allele 9 has a frequency of 0.0294 making the match between father and child very significant, whereas allele 12 has a frequency of 0.3446, making a match less significant. For each allele, a paternity index (PI) is calculated. These indicate the significance of the match. The PIs are combined to produce a probability of parentage. 10-13 different STRs are used to identify paternity. Mismatches of two STRs between the male and child is enough to exclude the male as the biological father.

Whale DNA: tracking illegal slaughter

Humpback whale

Under International Whaling Commission regulations, some species of whales can be captured for scientific research and their meat can be sold legally. Most whales, including humpback and blue whales, are fully protected and to capture or kill them is illegal.

Between 1999 and 2003, researchers used DNA profiling to investigate whale meat sold in markets in Japan and South Korea. They found 10% of the samples tested were from fully protected whales including western grey whales and humpbacks. They also found that many more whales were being killed than were being officially reported.

3. For the STR D10S1248 in the example above, what possible allele combinations could the child have?

_____

4. A paternity test was carried out and the abbreviated results are shown below:

DNA marker	Mother's alleles	Child's alleles	Man's alleles
CSF1PO	7, 8	8, 9	9, 12
D10S1248	14, 15	11, 14	10, 11
D19S433	9, 10	10,15	14, 16
D13S317	10, 11	9, 10	8, 9
D2S441	7, 15	7, 9	14, 17

(a) Could the man be the biological father? _____

(b) Explain your answer: _____

_____

_____

5. (a) How could DNA profiling be used to refute official claims of the type of whales captured and sold in fish markets?

_____

_____

(b) How could DNA profiling be used to refute official claims of the number of whales captured and sold in fish markets?

_____

_____

_____

©2024 BIOZONE International
ISBN: 978-1-99-101410-8
Photocopying prohibited

266 What is Gene Expression?

Key Idea: Gene expression is the process of transcribing and translating the information stored in DNA into protein products.

Gene expression is the process by which the information in genes is used to synthesize a protein or polypeptide. It involves transcription of the DNA into mRNA and translation of the mRNA into a polypeptide. Eukaryotic genes include non-protein coding regions called introns. Intronic DNA must be edited out before the mRNA can be translated by the ribosomes. Transcription and editing occur in the nucleus. Genes can be switched on or off for expression at this stage. Translation of the mRNA by the ribosomes occurs in the cytoplasm. This flow of information from DNA to RNA to protein is known in biology as the central dogma.

A summary of eukaryotic gene expression

The genome includes all the genetic material in the haploid set of chromosomes of an organism, including all its genes and DNA sequences.

TRANSLATION · Nuclear pore · Nucleus · EDITING · TRANSCRIPTION · mRNA · Ribosome · mRNA · Primary transcript · DNA

Amino acids are linked together at the ribosome to form the polypeptide encoded by mRNA.

Cytoplasm

The primary transcript is edited. The non-protein coding introns are removed and modifications are made to help the mRNA exit the nucleus.

In the nucleus, the gene is rewritten into a single stranded primary RNA transcript, using one strand of DNA as a template. RNA polymerase catalyzes this process.

1. What is the link between genes and proteins? _____

2. (a) What does the term 'gene expression' mean? _____

(b) What are the three stages in gene expression in eukaryotes and what happens in each stage?

(i) _____

(ii) _____

(iii) _____

3. The photograph below right shows an SEM of a giant polytene chromosome. These chromosomes are common in the larval stages of flies, which must grow rapidly before changing to the adult form. They form as a result of repeated cycles of DNA replication without cell division. This creates many copies of genes. Within these chromosomes, visible 'puffs' indicate regions where there is active transcription of the genes.

(a) What is the consequence of active transcription in a polytene chromosome?

_____

(b) Explain why this might be useful in a larval insect: _____

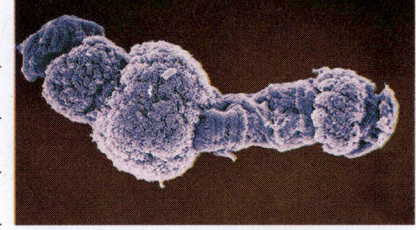

D1.2 · 4

267 The Genetic Code

Key Idea: The genetic code is the set of rules by which the genetic information in DNA or mRNA is translated into proteins. The genetic information for the assembly of amino acids is stored as three-base sequences. These three letter codes on mRNA are called codons. Each codon represents one of 20 amino acids used to make proteins. The code is effectively universal, being the same in organisms (with a few minor exceptions). This universality implies that all life on earth can be linked back to the Last Universal Common Ancestor (LUCA). The genetic code is summarized in a mRNA-amino acid table which identifies the amino acid encoded by each mRNA codon. The code is degenerate, meaning there may be more than one codon for each amino acid. Most of this degeneracy is in the third nucleotide of a codon.

The mRNA - amino acid table

The table on the right is used to 'decode' the genetic code. It shows which amino acid each mRNA codon codes for. There are 64 different codons possible: 61 code for amino acids, and three are stop codons.

Amino acid names are written as three letter abbreviations, e.g. Ser = serine. To work out which amino acid a codon codes for, carry out the following steps:

i Find the first letter of the codon in the row on the left hand side of the table. AUG is the start codon.

ii Find the column that intersects that row from the top, second letter, row.

iii Locate the third base in the codon by looking along the row on the right hand side that matches your codon.

e.g. GAU codes for Asp (aspartic acid)

Read second letter here / Read first letter here / Read third letter here

		Second letter			
First letter		U	C	A	G	Third letter
U	UUU Phe UUC Phe UUA Leu UUG Leu	UCU Ser UCC Ser UCA Ser UCG Ser	UAU Tyr UAC Tyr UAA STOP UAG STOP	UGU Cys UGC Cys UGA STOP UGG Trp	U C A G
C	CUU Leu CUC Leu CUA Leu CUG Leu	CCU Pro CCC Pro CCA Pro CCG Pro	CAU His CAC His CAA Gln CAG Gln	CGU Arg CGC Arg CGA Arg CGG Arg	U C A G
A	AUU Ile AUC Ile AUA Ile AUG Met	ACU Thr ACC Thr ACA Thr ACG Thr	AAU Asn AAC Asn AAA Lys AAG Lys	AGU Ser AGC Ser AGA Arg AGG Arg	U C A G
G	GUU Val GUC Val GUA Val GUG Val	GCU Ala GCC Ala GCA Ala GCG Ala	GAU Asp GAC Asp GAA Glu GAG Glu	GGU Gly GGC Gly GGA Gly GGG Gly	U C A G

1. (a) Use the base-pairing rule to create the complementary strand for the DNA template strand shown below.

(b) For the same DNA template strand, determine the mRNA sequence and (c) use the mRNA - amino acid table to determine the amino acid sequence. Note that in mRNA, uracil (U) replaces thymine (T) and pairs with adenine.

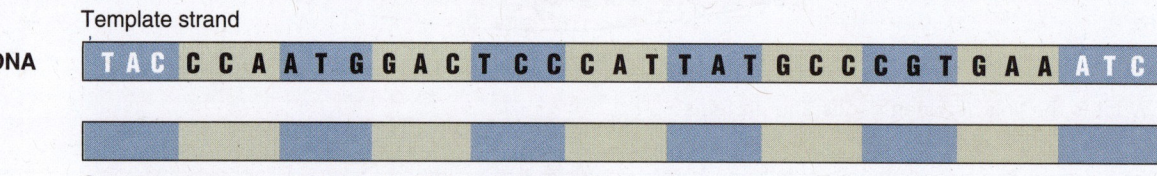

Template strand

DNA T A C C C A A T G G A C T C C C A T T A T G C C C G T G A A A T C

Complementary strand (this is the DNA coding strand)

Gene expression

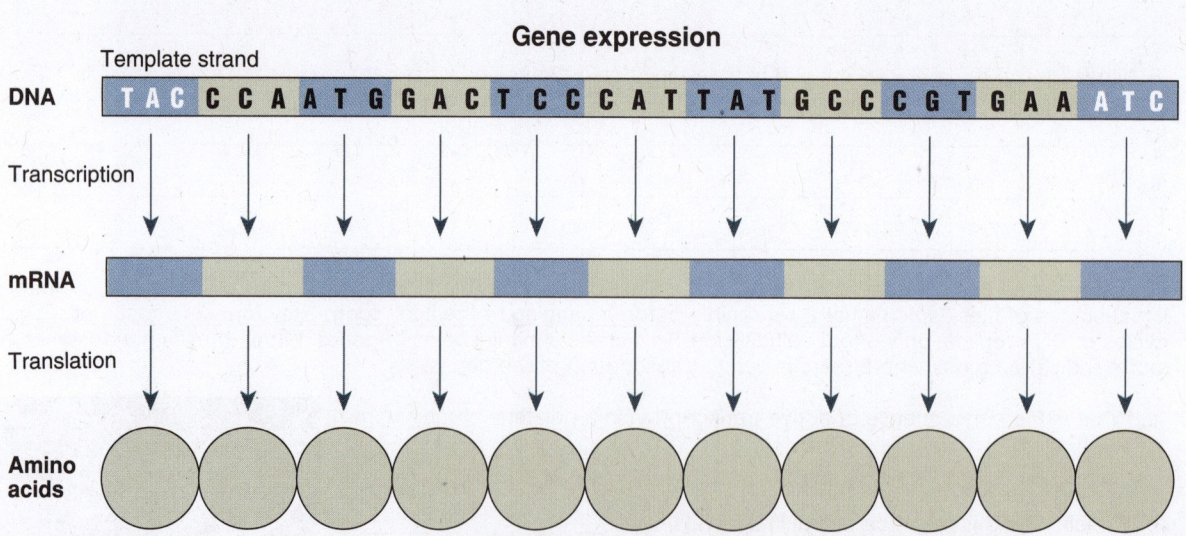

Template strand

DNA T A C C C A A T G G A C T C C C A T T A T G C C C G T G A A A T C

Transcription

mRNA

Translation

Amino acids

2. What do you notice about the sequence on the DNA coding strand and the mRNA strand? _____

_____

 D1.2 8-9
 12 ←

©2024 **BIOZONE** International
ISBN: 978-1-99-101410-8
Photocopying prohibited

Redundancy and degeneracy

Redundancy and degeneracy are important concepts in understanding the genetic code.

▶ Redundancy is when several situations code for, or control, the actions of one specific thing.

▶ Degeneracy is when a particular output can be produced by several different pathways.

Examples of redundancy and degeneracy are illustrated below. In modern aircraft, redundant features add safety by making sure that if one system fails others will ensure a smooth, safe flight. Degeneracy can be seen in proteins when different proteins have the same function.

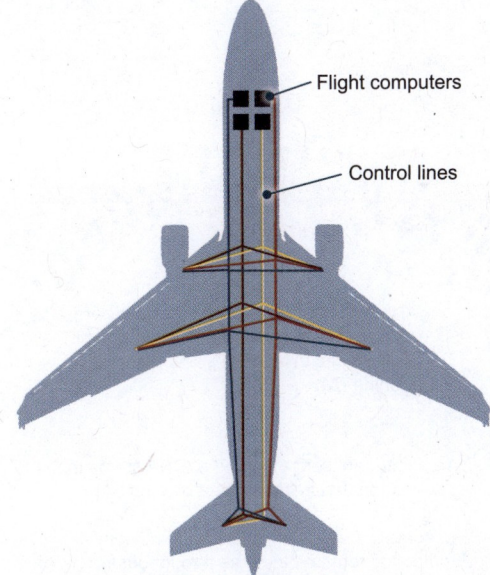

Flight computers

Control lines

Modern aircraft (left) have multiple redundant features for safety. Often there are three or four flight computers linked independently to the flight surfaces and other input/output devices. If one computer or control line fails, the others can continue to fly the plane normally.

Degeneracy is seen in the production of the enzymes salivary and pancreatic amylase. Salivary amylase breaks down carbohydrates in the mouth, whereas pancreatic amylase does so in the small intestine. The enzymes are encoded by different genes (AMY1A and AMY2A) but have the same functional role (right).

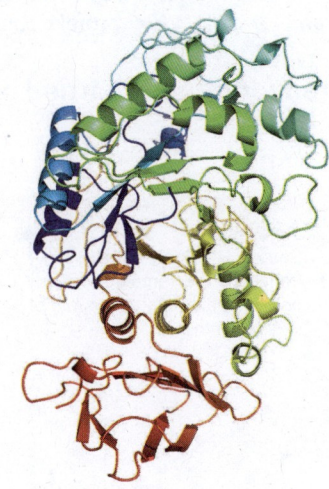

Salivary amylase (above) is structurally different from pancreatic amylase, but has the same function.

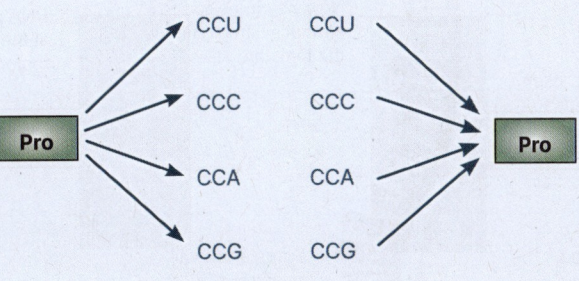

The genetic code shows degeneracy. This means that a number of 3 base combinations specify one amino acid. The codons for the same amino acid often differ by only a single letter (often the second or third). For example, proline is encoded by four different codons.

The degeneracy of the genetic code creates redundancy, so that several codons code for the same amino acid, e.g. CCU, CCC, CCA, and CCG code for proline. Note that, although there is redundancy, there is no ambiguity - none of the codons encodes any other amino acid.

3. Explain how degeneracy adds 'safety' to the coding of protein chains: _____

_____

_____

_____

4. The genetic code shows redundancy but no ambiguity. What does this mean and why is it important? _____

_____

_____

_____

5. Identify the following:

   (a) The codons that encode valine (Val): _____

   (b) The codons that encode aspartic acid (Asp): _____

6. (a) Arginine (Arg) is encoded in how many ways? _____

   (b) Glycine (Gly) is encoded in how many ways? _____

   (c) Which amino acid(s) are encoded in only one way? _____

©2024 **BIOZONE** International
ISBN: 978-1-99-101410-8
Photocopying prohibited

268 Transcription in Eukaryotes

Key Idea: Transcription is the first step of gene expression. It involves the enzyme RNA polymerase rewriting the information into a primary RNA transcript. In eukaryotes, transcription takes place in the nucleus.

When a gene is required to be expressed, then that specific region of the DNA is 'copied'. RNA polymerase rewrites the DNA into a primary RNA transcript using a single template strand of DNA. The protein-coding portion of a gene is bounded by a start (promoter) region and a terminator region. These regulatory regions control transcription by telling RNA polymerase where to start and stop transcription. In eukaryotes, non protein-coding sections called introns must first be removed and the remaining exons spliced together to form the mature mRNA before the gene can be translated into a protein. This editing process also occurs in the nucleus.

Transcription is carried out by RNA polymerase (RNAP)

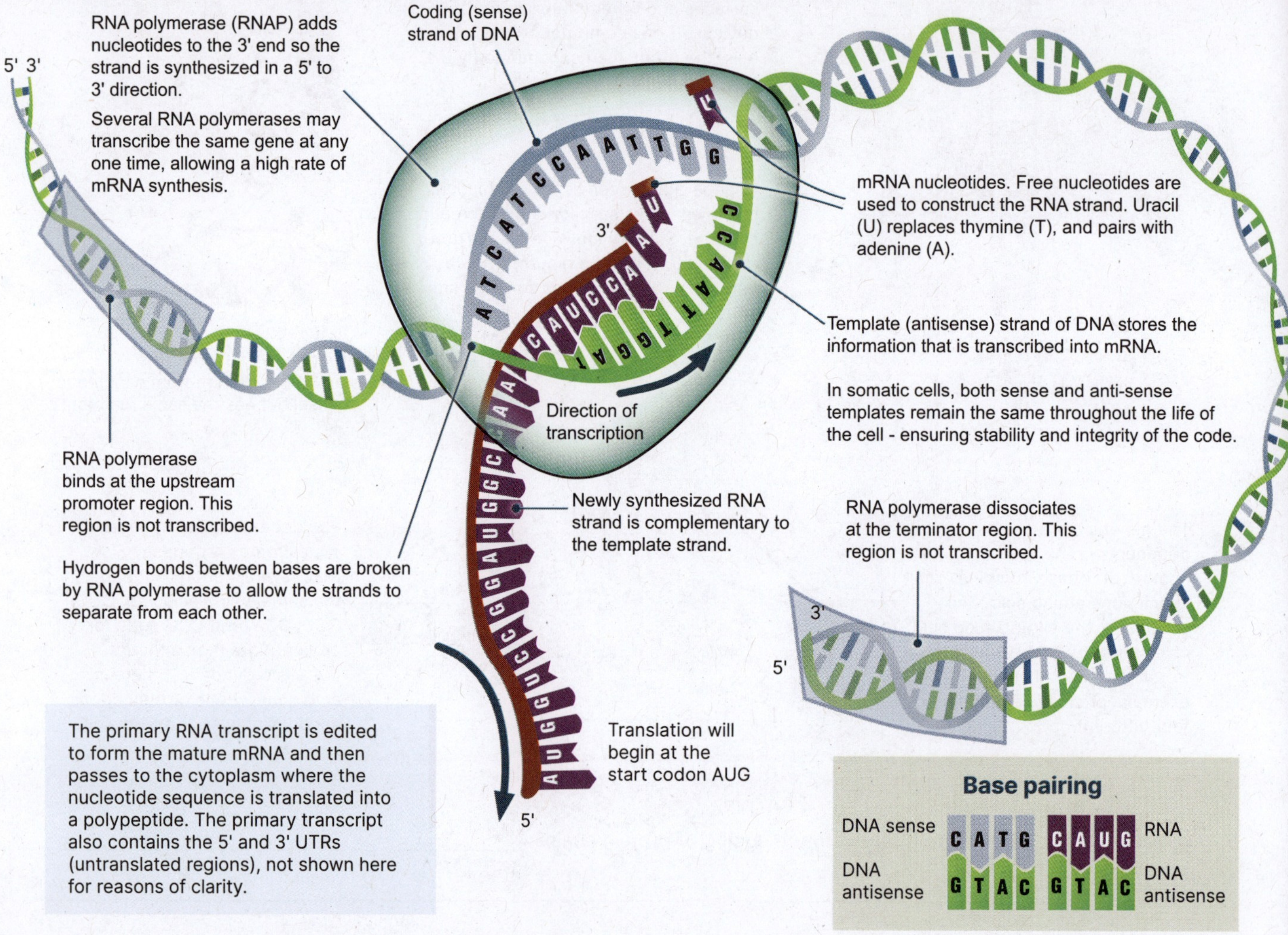

RNA polymerase (RNAP) adds nucleotides to the 3' end so the strand is synthesized in a 5' to 3' direction.

Several RNA polymerases may transcribe the same gene at any one time, allowing a high rate of mRNA synthesis.

Coding (sense) strand of DNA

mRNA nucleotides. Free nucleotides are used to construct the RNA strand. Uracil (U) replaces thymine (T), and pairs with adenine (A).

Template (antisense) strand of DNA stores the information that is transcribed into mRNA.

In somatic cells, both sense and anti-sense templates remain the same throughout the life of the cell - ensuring stability and integrity of the code.

RNA polymerase binds at the upstream promoter region. This region is not transcribed.

Direction of transcription

Newly synthesized RNA strand is complementary to the template strand.

RNA polymerase dissociates at the terminator region. This region is not transcribed.

Hydrogen bonds between bases are broken by RNA polymerase to allow the strands to separate from each other.

The primary RNA transcript is edited to form the mature mRNA and then passes to the cytoplasm where the nucleotide sequence is translated into a polypeptide. The primary transcript also contains the 5' and 3' UTRs (untranslated regions), not shown here for reasons of clarity.

Translation will begin at the start codon AUG

Base pairing

DNA sense	C A T G	C A U G	RNA
DNA antisense	G T A C	G T A C	DNA antisense

1. (a) Name the enzyme responsible for transcribing the DNA: _____

   (b) What strand of DNA does this enzyme use? _____

   (c) The code on this strand is the [same as / complementary to] the RNA being formed (circle correct answer):

   (d) Which nucleotide base replaces thymine in mRNA? _____

   (e) On the diagram, use a coloured pen to mark the beginning and end of the protein-coding region being transcribed.

2. (a) In which direction is the RNA strand synthesized? _____

   (b) Explain why this is the case: _____

   _____

   _____

3. (a) Why is AUG called the start codon? _____

   (b) What would the three letter code be on the DNA coding strand? _____

D1.2
1 - 4

©2024 **BIOZONE** International
ISBN: 978-1-99-101410-8
Photocopying prohibited

269 Translation

Key Idea: Translation converts RNA into amino acid polypeptides in the ribosomes.

In eukaryotes, translation occurs in the cytoplasm either at free ribosomes or ribosomes on the rough endoplasmic reticulum. Ribosomes translate the code carried in the mRNA molecules, providing the catalytic environment for the linkage of amino acids delivered by transfer RNA (tRNA) molecules. Protein synthesis begins at the start codon and, as the ribosome wobbles along the mRNA strand, the polypeptide chain elongates. On reaching a stop codon, the ribosome subunits dissociate from the mRNA, releasing the polypeptide chain.

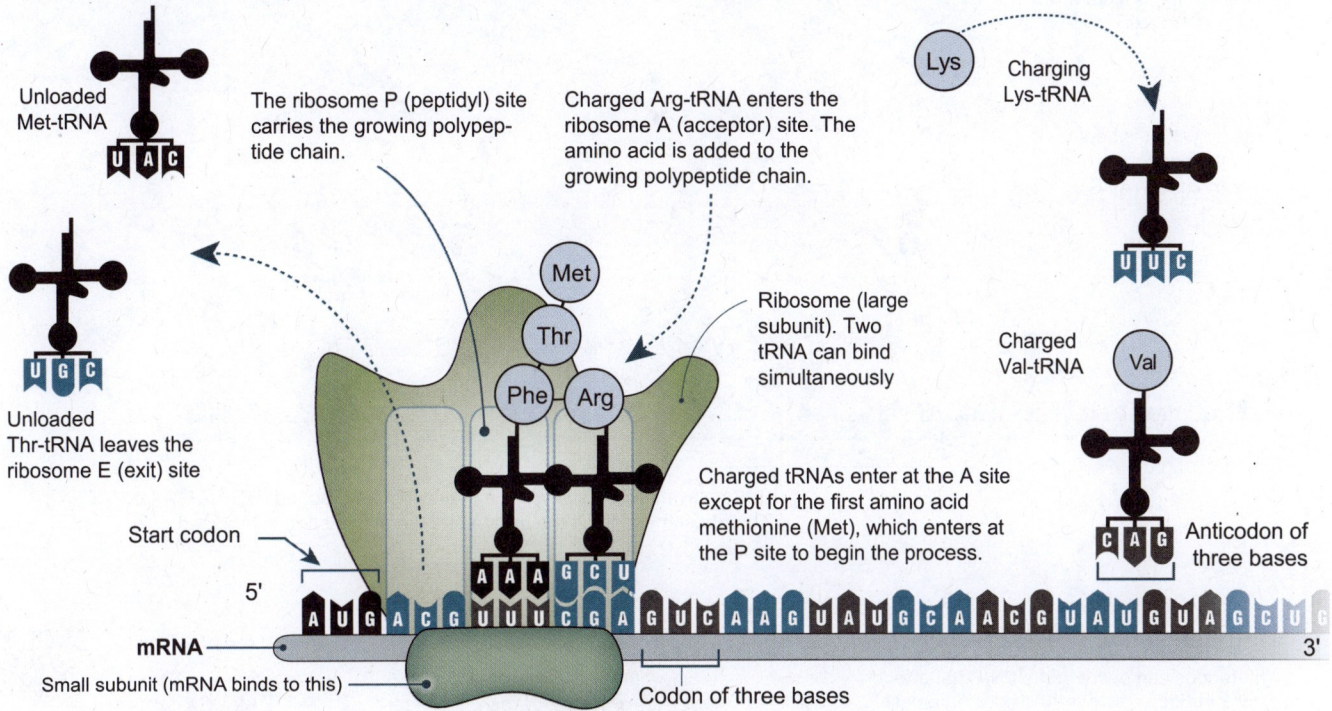

Unloaded Met-tRNA — UAC

The ribosome P (peptidyl) site carries the growing polypep-tide chain.

Charged Arg-tRNA enters the ribosome A (acceptor) site. The amino acid is added to the growing polypeptide chain.

Lys — Charging Lys-tRNA — UUC

Unloaded Thr-tRNA leaves the ribosome E (exit) site — UGC

Ribosome (large subunit). Two tRNA can bind simultaneously

Charged Val-tRNA — Val

Start codon

Charged tRNAs enter at the A site except for the first amino acid methionine (Met), which enters at the P site to begin the process.

CAG — Anticodon of three bases

5'

mRNA — AUGACGUUUCGAGUCAAGUAUGCAACGUAUGUAGCUG — 3'

Small subunit (mRNA binds to this)

Codon of three bases

tRNA molecules deliver amino acids to ribosomes, and then match amino acids with the appropriate codon on mRNA. As defined by the genetic code, the anticodon specifies which amino acid the tRNA carries. The tRNA delivers its amino acid to the ribosome, where enzymes join the amino acids to form a polypeptide chain. The ribosome 'wobbles' along the mRNA molecule joining amino acids together. Enzymes and energy are involved in charging the tRNA molecules (attaching them to their amino acid) and elongating the peptide chain. The polypeptide chain grows as more amino acids are added.

1. Describe the specific roles of the subunits of ribosomes, mRNA, and tRNA during the translation process:

_____

_____

_____

_____

_____

_____

2. Explain the process and function of tRNA and mRNA complementary base pairing, including the terms codon and anticodon:

_____

_____

_____

_____

3. Many ribosomes can work on one strand of mRNA at a time (a polyribosome system). What would this achieve?

_____

_____

273

D1.2

5 - 7,
10 - 11

Peptide bonding

As amino acids are delivered to the large subunit of the ribosome by tRNA, the two closest molecules undergo a hydrolysis reaction to remove a water molecule and form a peptide bond. This process continues and the peptide chain grows as more amino acids are linked.

H_2O removed to bond amine group (NH_2) to carboxyl group (COOH) forming a peptide bond between amino acids.

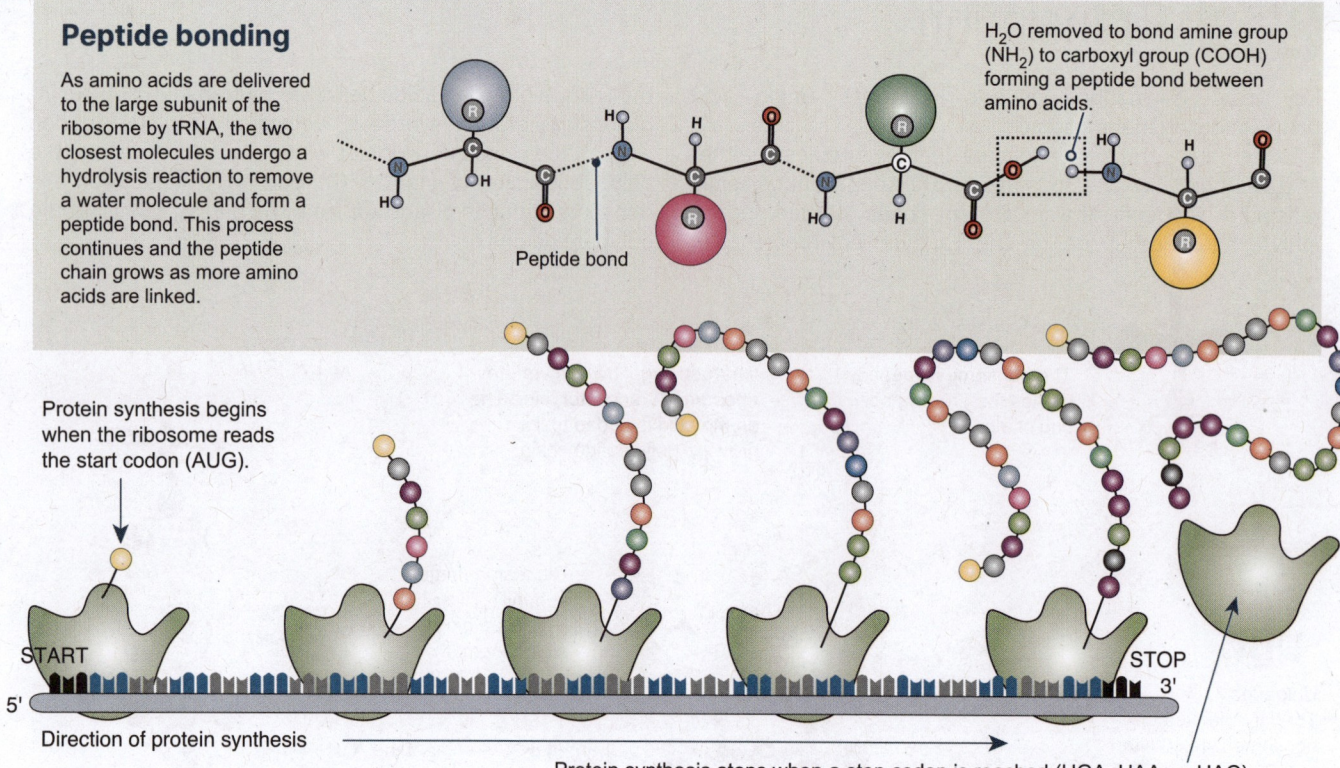

Peptide bond

Protein synthesis begins when the ribosome reads the start codon (AUG).

START
5'

Direction of protein synthesis

STOP
3'

Protein synthesis stops when a stop codon is reached (UGA, UAA, or UAC). The ribosome falls off the mRNA and the polypeptide is released.

Mutations and protein structure

Mutations can occur in the DNA that affect polypeptide chain synthesis. A single base change is known as a point mutation and is the substitution of a single nucleotide, e.g. A to G, or deletion/addition of a base. Deletion or addition mutations involve a frame shift of codon reading, and every codon will therefore be different in the remainder of the chain. Redundancy of genetic code allows for some substitution mutations to still code for the same amino acid. However, some substitutions cause an amino acid change which can lead to significant changes in the final protein form, and therefore its function. Sickle cell anaemia is an example of one base substitution mutation leading to significant overall changes in the protein structure.

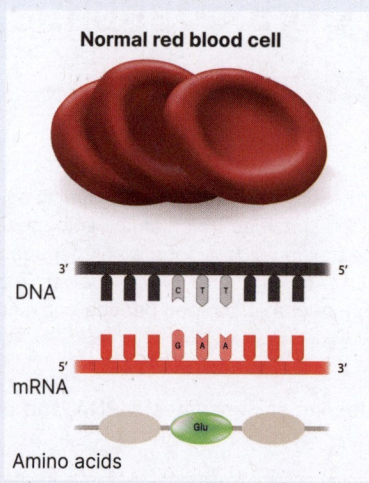

Normal red blood cell

DNA

mRNA

Amino acids

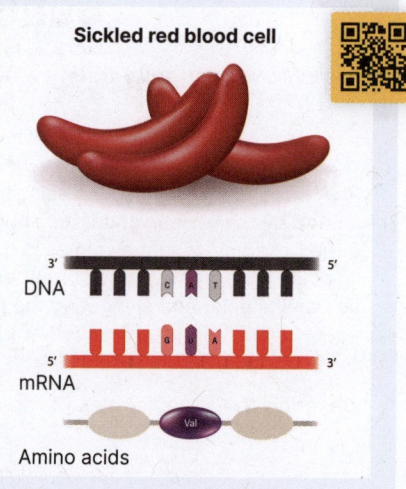

Sickled red blood cell

DNA

mRNA

Amino acids

4. List the sequence of events that occurs in the ribosomes from tRNA arrival to polypeptide chain formation:

_____

_____

_____

_____

5. Using a researched example other than sickle cells, discuss why a point mutation may cause a change in protein structure and function:

_____

_____

_____

_____

_____

©2024 **BIOZONE** International
ISBN: 978-1-99-101410-8
Photocopying prohibited

270 Regulating Transcription

Key Idea: Transcription factors initiate the formation of polypeptides from the coding sequences of DNA, which undergo further post-transcriptional modification.

The structural differences of the 3' and 5' ends of the DNA and RNA chains determine the placement of the bases during transcription. Newly added bases of the transcribed RNA strand bond with those on the existing DNA template strand, so the nucleotides are added from the 5' end of the template, with the RNA beginning from the 3' end. Transcription can only begin once transcription factors have prepared the promoter, allowing RNA polymerase II to facilitate assembly of the nucleotides. Not all of the DNA is transcribed into RNA. The segments used for the promoter and only a small percentage (around 2%) of the transcribed RNA are translated into protein. The remaining, non-coding sequences are used to regulate gene expression. Most of these are located in introns and removed before translation occurs. Other non-coding sequences are used as telomeres, to protect the ends of chromosomes, tRNA, which delivers nucleotides, or ribosomal RNA (rRNA), which assists in forming the peptide bonds between amino acids during translation. Several other processes occur in the pre-mRNA, known as post-transcriptional modifications, before it moves out of the nucleus to the ribosomes for translation.

Initiating transcription

▶ In eukaryotes, RNA polymerase II cannot initiate the transcription of structural genes alone. It requires the presence of transcription factors. Transcription factors are encoded by regulatory genes and have a role in creating an initiation complex for transcription.

▶ Transcription factors bind to distinct regions of the DNA, including the promoter and upstream enhancers. They will act as a guide to indicate to RNA polymerase where transcription should start.

▶ Once bound to the promoter sequence, the transcription factors capture RNA polymerase II, which can then begin transcription.

▶ Transcription is activated when a hairpin loop in the DNA brings the transcription factors (activators) attached to the enhancer sequence in contact with the transcription factors bound to RNA polymerase at the promoter (bottom).

Assembly of the transcription initiation complex

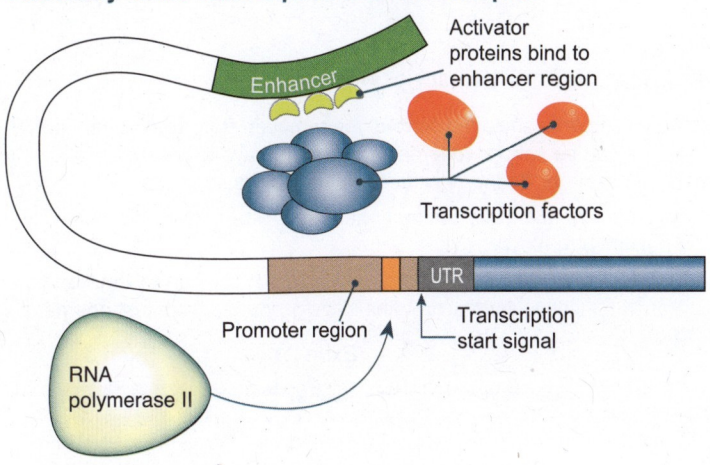

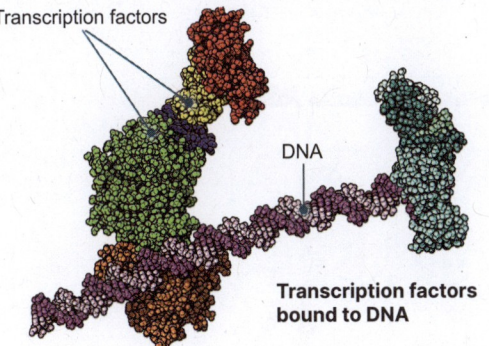

Transcription factors bound to DNA

RNA polymerase II binds and transcription begins

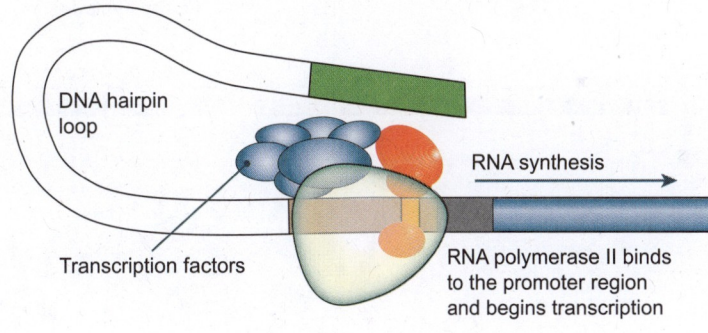

1. Why would a gene contain non-coding sequences in DNA be transcribed but not translated?

_____

_____

_____

2. (a) What is a transcription factor? _____

_____

(b) What type of genes encode transcription factors? _____

(c) Use the diagram to describe how transcription factors are involved in the regulation of gene expression:

_____

_____

_____

D1.2 12 - 16 AHL

Post-transcriptional modification

After transcription, the primary transcript is modified to produce the mRNA strand that will be translated in the cytoplasm. Only mRNA codes for polypeptides; tRNA and rRNA have other functions. Modifications to the 5' and 3' ends of the transcript enable the mRNA to exit the nucleus and remain stable long enough to be translated. Other post-transcriptional modifications remove non-protein coding intronic DNA and splice exons in different combinations to produce different protein end products.

Primary RNA is modified by the addition of caps and tails

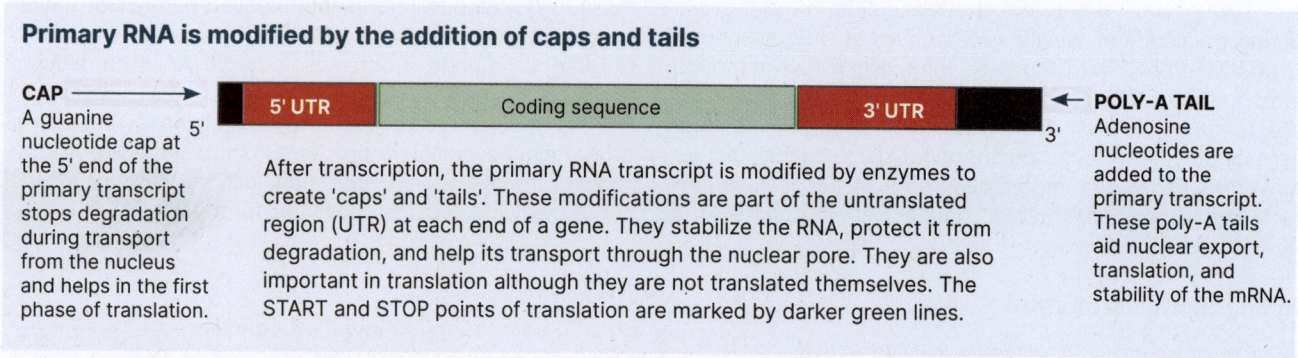

CAP
A guanine nucleotide cap at the 5' end of the primary transcript stops degradation during transport from the nucleus and helps in the first phase of translation.

After transcription, the primary RNA transcript is modified by enzymes to create 'caps' and 'tails'. These modifications are part of the untranslated region (UTR) at each end of a gene. They stabilize the RNA, protect it from degradation, and help its transport through the nuclear pore. They are also important in translation although they are not translated themselves. The START and STOP points of translation are marked by darker green lines.

POLY-A TAIL
Adenosine nucleotides are added to the primary transcript. These poly-A tails aid nuclear export, translation, and stability of the mRNA.

Modification after transcription

▶ As you have seen earlier, introns are removed from the primary mRNA transcript and the exons are spliced together. However, exons can be spliced together in different ways to create variations in the translated proteins. Exon splicing occurs in the nucleus, either during or immediately after transcription.

▶ In mammals, the most common method of alternative splicing involves exon skipping, in which not all exons are spliced into the final mRNA (below).

▶ Human DNA contains 25,000 genes, but produces up to 1 million different proteins. Modifications after transcription and translation allow several proteins to be produced from just one gene.

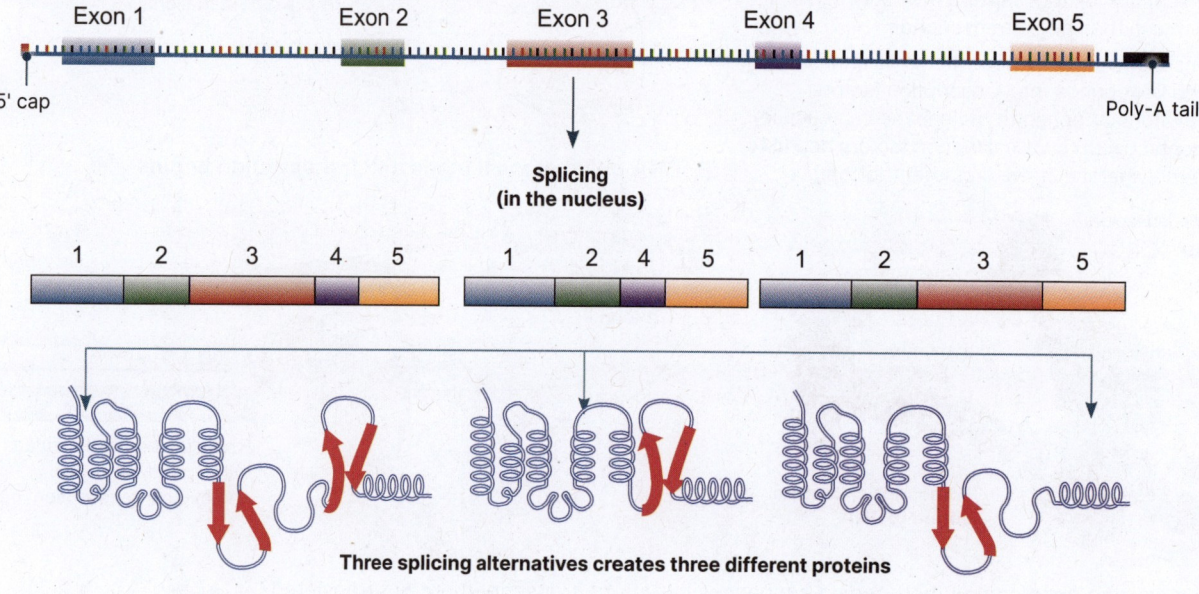

Three splicing alternatives creates three different proteins

3. Explain the purpose of the caps and tail on mRNA? _____
_____

4. (a) What happens to the intronic sequences in DNA after transcription? _____

   (b) What is one possible fate for these introns? _____

5. Explain how so many proteins can be produced from a much smaller number of genes: _____
_____
_____

6. If a human produces 1 million proteins, but human DNA codes for only 25,000 genes, on average, how many proteins are produced per gene?
_____

©2024 **BIOZONE** International
ISBN: 978-1-99-101410-8
Photocopying prohibited

271 Regulating Translation

Key Idea: Translation requires specific processes to occur before it is initiated and the polypeptides produced are further modified to synthesize functional substances.

Similar to transcription, translation requires an initiation process. Translation begins with an initiator tRNA attaching to the small ribosomal subunit, and only then does the large ribosomal subunit attach to begin the elongation of the polypeptide. Proteins may be modified in the endoplasmic reticulum (rER) and Golgi after they have been produced by post-translational modification, including adding carbohydrates or lipids to the protein. Glycoproteins are formed by adding carbohydrates to proteins as they pass through the rough endoplasmic reticulum and Golgi. Other proteins may have fatty acids added to them in the rER to form lipoproteins. Some modifications can be multi-stage and involve the removal of amino acids to form a final product, such as in the formation of insulin. Translation is continuous, so a readily available supply of amino acids is required. Proteins can be recycled by proteasomes to provide amino acids for further protein synthesis.

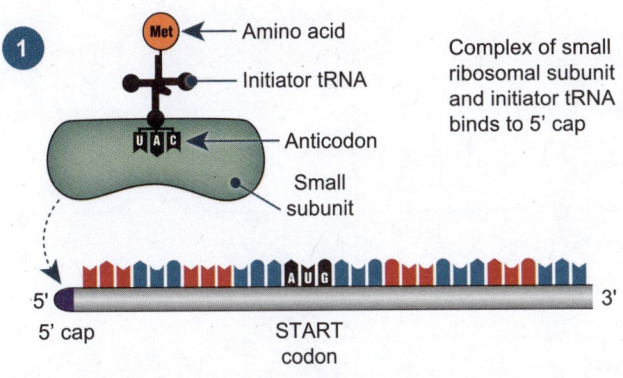

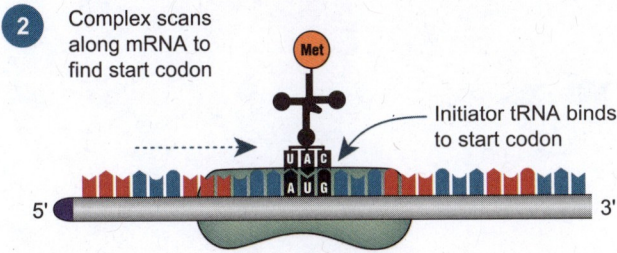

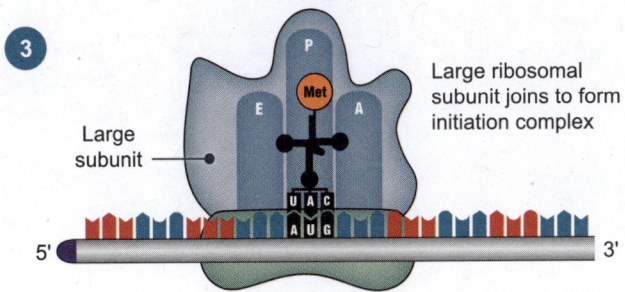

Initiation complex

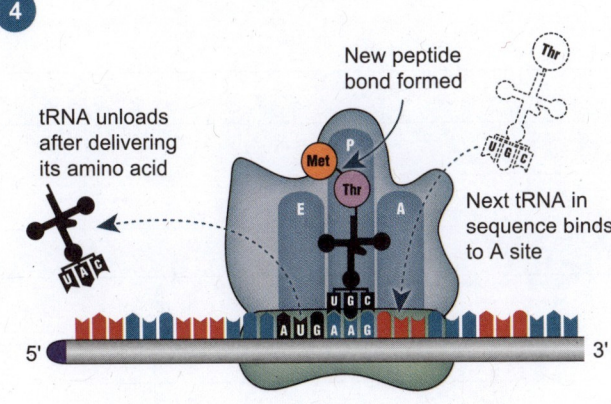

Step 1: initiator tRNA attaches to ribosomal subunit and binds to 5' cap

The tRNA, almost always carrying the amino acid methionine, combines with the small ribosomal subunit to form a complex, then binds to the 5' cap (guanine nucleotide).

Step 2: Complex moving to start codon

Once the complex is attached, it moves along the RNA from the 5' to the 3' end. When the start codon (AUG) on the RNA is located, the complex binds.

Step 3: Large subunit attaches

The large ribosomal subunit attaches to form the initiation complex. There are three binding sites: E, P, and A. The initiator tRNA locates itself in the middle site: P.

Step 4: Elongation at the binding sites

The next tRNA, carrying an amino acid (AA), arrives at the A site. A hydrolysis reaction binds the original MET amino acid to the second AA at the A site. The RNA then moves along through the ribosome so that the initial tRNA now sits in the E site. When it departs, the second tRNA now occupies the P site and the next tRNA now enters the vacant A site. This elongation process continues until a stop codon (UAA, UAG, or UGA) arrives.

1. What is the key function of the initiator complex?

    _____

    _____

    _____

    _____

2. What occurs in the 3 different binding sites?

    _____

    _____

    _____

    _____

    _____

    _____

D1.2
17 - 19

AHL

Modification of polypeptides

The modification of proteins allows the cell to specify their use and final destination. Modifications include glycoprotein formation, where the carbohydrates may help position and orient the glycoprotein in the membrane, guide a protein to its final destination, or help in cell-to-cell recognition and cell signalling. Lipoproteins are formed so they can transport lipids in the plasma between various organs in the body (e.g. gut, liver, and adipose tissue). Other common post-translational modifications include degradation, phosphorylation, and cleavage, a process used to form insulin through multi-steps (below).

Cleaving: Polypeptide chains may be cleaved to give smaller chains, which then fold or join to make the functional protein. An example is human insulin. This is transcribed as one, long polypeptide chain which is cleaved to form two shorter chains. These form the functional protein.

Glycosylation (adding carbohydrate groups): This is used to add an ID tag to the protein that will allow the cell to recognise its use and where it is to be transported (2a). The resulting glycoprotein may be used in the cell membrane or secreted. The carbohydrate tag may help position the glycoprotein within the membrane (2b).

Phosphorylation (the addition of phosphate groups) takes place in the Golgi. It may contribute to the protein's three dimensional structure or help with cell signalling.

Lipid attachment: Proteins may have lipids attached to them which anchor the protein to the plasma membrane.

Degradation: Some polypeptide chains may be tagged for degradation when they are no longer useful and their amino acids reused in the formation of other proteins (see more details on the next page).

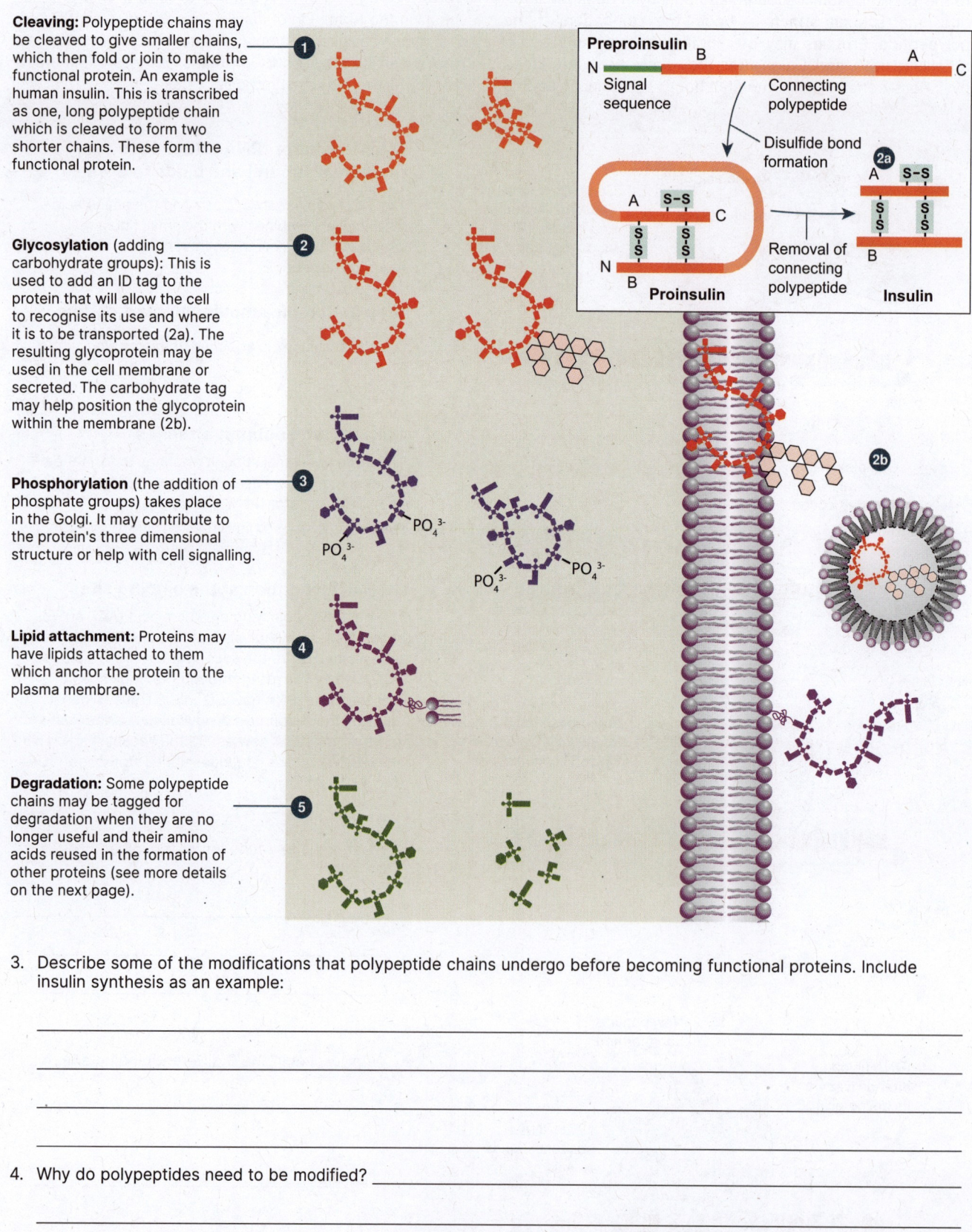

3. Describe some of the modifications that polypeptide chains undergo before becoming functional proteins. Include insulin synthesis as an example:

_____

_____

_____

_____

4. Why do polypeptides need to be modified? _____

_____

_____

_____

©2024 **BIOZONE** International
ISBN: 978-1-99-101410-8

Proteasomes and recycling of amino acids

The set of all proteins needed in the body is known as the proteome. Those proteins no longer required in cells are recycled back into amino acids so they can be re-used for translation. Ubiquitin is a small, regulatory protein that is involved in recycling unwanted proteins. The target proteins are bound to the ubiquitin in a series of reactions involving specific enzymes. These are then transported to a proteasome: a complex of protease enzymes. The protein is digested into amino acids and small peptides, while the ubiquitin undergoes a further reaction to make it available for the process once more.

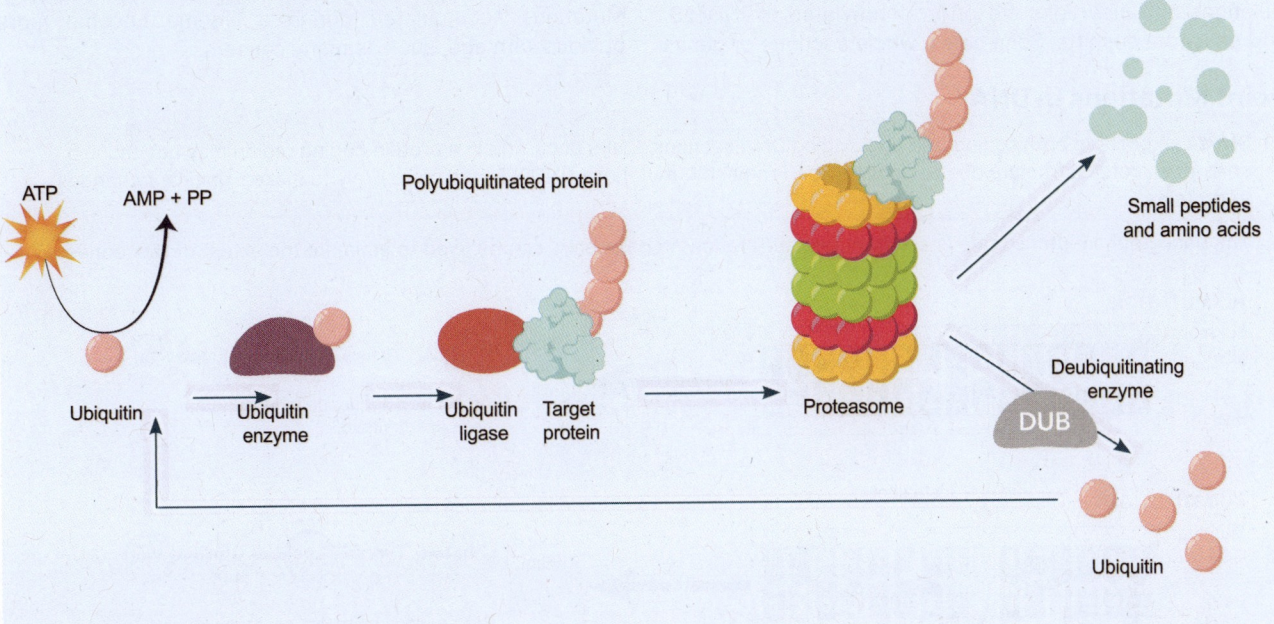

5. Draw a flowchart to represent the process of translation in eukaryotes. A starting point has been provided:

> mRNA enters the cytoplasm via a nuclear pore

6. Based on the diagram above, describe two requirements for translation to be initiated:

(a) _____

(b) _____

7. What is the difference between polypeptides and final function protein products? _____

_____

_____

8. What is the importance of recycling proteins? _____

_____

_____

272 Gene Mutations

Key Idea: Mutations change the DNA sequence.

Mutations can change the DNA sequence in many different ways. Point mutations, when just one base pair is involved, can be substitutions, deletions, and insertions of bases. They may involve just one nucleotide substitution, called a single-nucleotide polymorphism (SNP, pronounced snip). Deletions and insertions are often abbreviated to INDELS, and cause frameshifts. Sometimes, whole sections of genes

or chromosomes may be deleted or duplicated. Mutations more commonly involve just one or a few nucleotides and occur during replication. The enzymes involved in replication occasionally insert or leave out a nucleotide. Mistakes are normally repaired by proofreading enzymes. More mutations become more common in the presence of mutagens. Mutations accumulated during a lifetime become more obvious with age, such as many cancers.

Point Mutations in DNA

▶ SNPs can occur in both coding and non-coding DNA sections. SNPs occurring in a protein coding region may not cause a change in protein structure due to degeneracy. Frameshift mutations (INDELS) are more likely to make significant changes to proteins.

▶ The background mutation rate is constant, and differences in base pairs can be used to estimate the extent of relatedness.

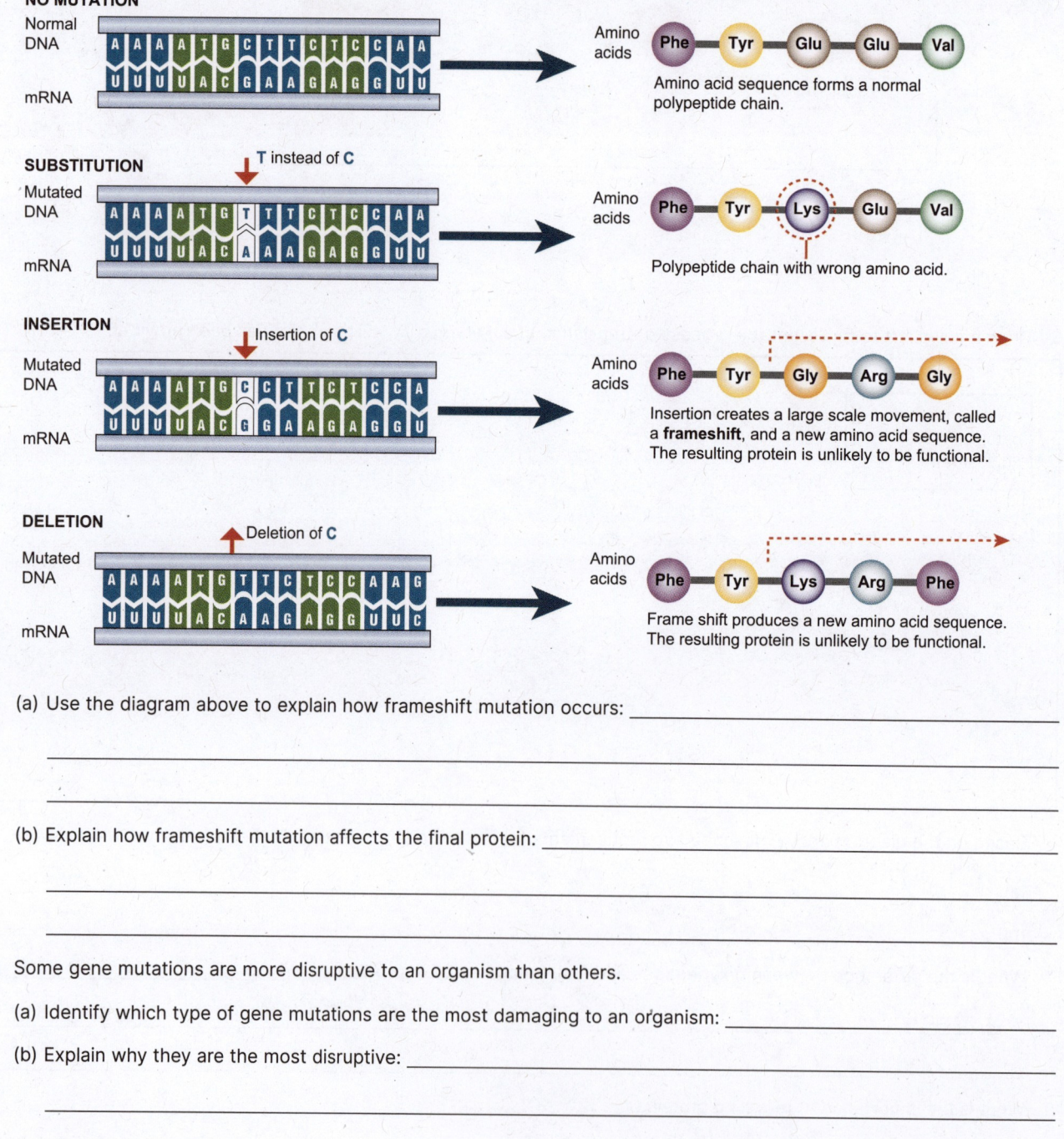

1. (a) Use the diagram above to explain how frameshift mutation occurs: _____

_____

   (b) Explain how frameshift mutation affects the final protein: _____

_____

_____

2. Some gene mutations are more disruptive to an organism than others.

   (a) Identify which type of gene mutations are the most damaging to an organism: _____

   (b) Explain why they are the most disruptive: _____

_____

_____

D1.3
1 - 3

©2024 **BIOZONE** International
ISBN: 978-1-99-101410-8
Photocopying prohibited

273 Causes and Consequences of Mutations

Key Idea: Mutations are random events that occur in both somatic and germ cells, where the rate of mutation can increase in frequency with exposure to mutagens.

A mutation is a permanent change to the DNA sequence of an organism. Mutations are the ultimate source of new alleles. Most mutations are harmful because they disrupt some important cellular process, often by causing a protein to fold incorrectly. Occasionally they may cause some beneficial change, such as making an enzyme more efficient. This phenomenon is the basis of natural selection. Although the DNA replication process is very accurate, it is estimated that, in humans, a mutation occurs once every 30 million base pairs copied during DNA replication prior to meiosis. This means that every person has about 200-300 new mutations that their parents did not have. Mutations occurring in germ cells can be passed down to offspring.

Some mutations are retained, others are eliminated

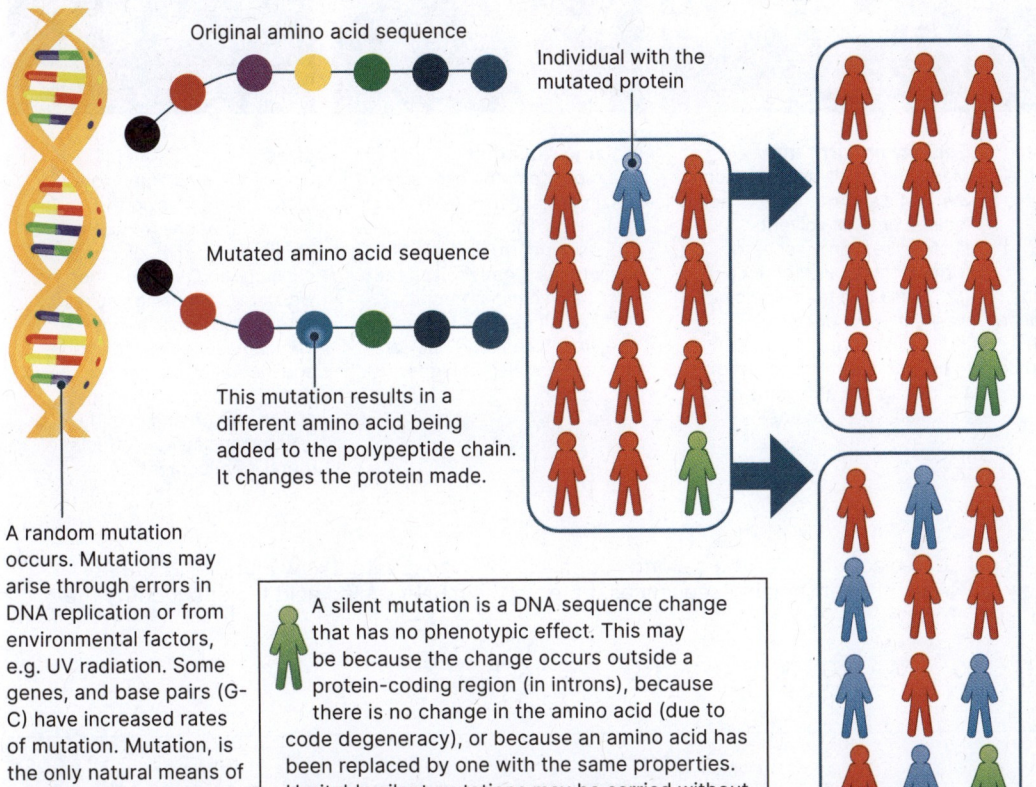

Original amino acid sequence

Individual with the mutated protein

Mutated amino acid sequence

This mutation results in a different amino acid being added to the polypeptide chain. It changes the protein made.

A random mutation occurs. Mutations may arise through errors in DNA replication or from environmental factors, e.g. UV radiation. Some genes, and base pairs (G-C) have increased rates of mutation. Mutation, is the only natural means of introducing new alleles.

A silent mutation is a DNA sequence change that has no phenotypic effect. This may be because the change occurs outside a protein-coding region (in introns), because there is no change in the amino acid (due to code degeneracy), or because an amino acid has been replaced by one with the same properties. Heritable silent mutations may be carried without effect and may only be subject to selection pressure when environmental conditions change.

If the mutation is harmful (reduces fitness) in the current environment, it is selected against and is usually eliminated from the population.

There is evidence that some 'silent' changes affect mRNA stability and transcription, even though they do not change codon information. In these cases, the mutations are not neutral.

If the mutation is beneficial (increases fitness) and it is heritable (occurs in the gametes), it is selected for and retained in the population. It may become more common over several generations.

1. What is a mutation? _____
_____

2. Why are some mutations retained within a population and others eliminated? _____
_____
_____
_____

3. (a) What is a silent mutation? _____
_____

(b) Explain the potential advantage of a silent mutation being retained within a population? _____
_____
_____
_____
_____

D1.3
4 - 7

Environmental factors can cause mutations

▶ Mutations occur naturally via mistakes in DNA replication or meiosis but the rate of mutations can be increased by mutagens. Mutagens may be chemicals or physical factors in the environment that increase the risk of errors in DNA replication. The longer a person is exposed to these factors, the greater their effect will be and the greater the risk of serious consequences.

▶ In most cases, the somatic cells will be affected and a possible outcome will be cancer and most cancers are caused by environmental factors. In some cases, the gametes or zygote will be affected and so, therefore, will the embryo.

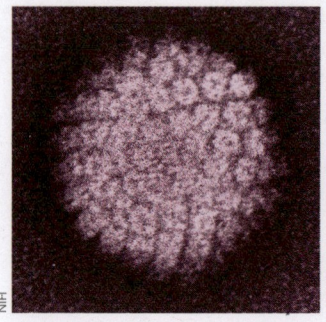   

Viruses and microorganisms
Some viruses integrate into the human chromosome, upsetting genes and triggering cancers. Examples include hepatitis B virus (liver cancer) and HPV (above) which is implicated in cervical cancer. Those at higher risk of viral infections include intravenous drug users and those with unsafe sex practices.

Poisons and irritants
Many chemicals are mutagenic. Synthetic and natural examples include organic solvents such as benzene, asbestos, formaldehyde and tobacco tar. Those most at risk include workers in the chemical industries. People involved in environmental clean-up of toxic spills are at high risk of exposure to mutagens.

Ionizing radiation
The most common exposure to ionizing radiation is from UV rays in sunlight. Too much exposure causes sunburn. Being sunburnt many times greatly raises the risk of skin cancer. People working with X-rays (from X-ray machines) and nuclear radiation are also at higher risk of cancers.

Lifestyle
One of the most important factors in increasing the risk of mutations is lifestyle. This includes diet, alcohol consumption, and whether we smoke. Smoking causes lung cancer, while heavy drinking causes liver damage and increases cancer risk. Heavy drinking at the time of conception and in early pregnancy severely affects foetal development.

4. In the following DNA sequence, replace the G of the second codon with an A to create a new mutant DNA, then determine both the new mRNA and the amino acid sequences. Refer to the mRNA-amino acid table to identify the amino acids coded for in each case.

   (a) Original DNA: CGT ATG AAA CTG GGG CTG TCA CCT AAT

   Mutated DNA: _____

   mRNA: _____

   Amino acids: _____

   (b) Identify the amino acid coded for by codon 2 (ATG) in the original DNA: _____

   (c) Explain the effect of the mutation: _____
   _____
   _____

5. Explain how mutagens cause mutations: _____
   _____
   _____

6. Mutation breeding is a biotechnological technique in which plant seeds are exposed to mutagens and then grown.

   (a) What would be the purpose of this technique? _____
   _____
   _____

   (b) Pollen is also treated with mutagens. What effect would this have on the plant produced by using this pollen?
   _____
   _____

©2024 **BIOZONE** International
ISBN: 978-1-99-101410-8
Photocopying prohibited

Somatic vs germ cell mutations

▶ Gametic cells are the reproductive (sex) cells of an organism (the egg and sperm). Mutations occurring in these cells are called germ-line mutations, or gametic mutations.

▶ Somatic cells (body cells) are all the remaining cells. Mutations to these cells are called somatic mutations.

▶ Only germ-line mutations will be inherited. Somatic mutations are not inherited but may affect an organism in its lifetime, e.g. a cancer.

▶ The red delicious apple (right) is a natural chimera (an organism with a mixture of two or more different genotypes). In the apple, a mutation occurred in the part of the flower that developed into the fleshy part of the apple. The seeds are unaffected by the mutation, so it is not inherited.

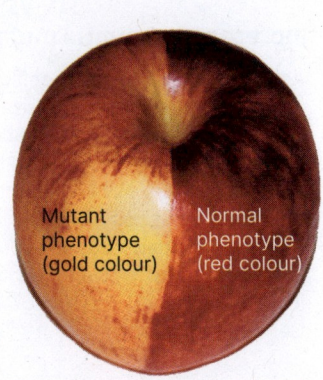

Mutant phenotype (gold colour)   Normal phenotype (red colour)

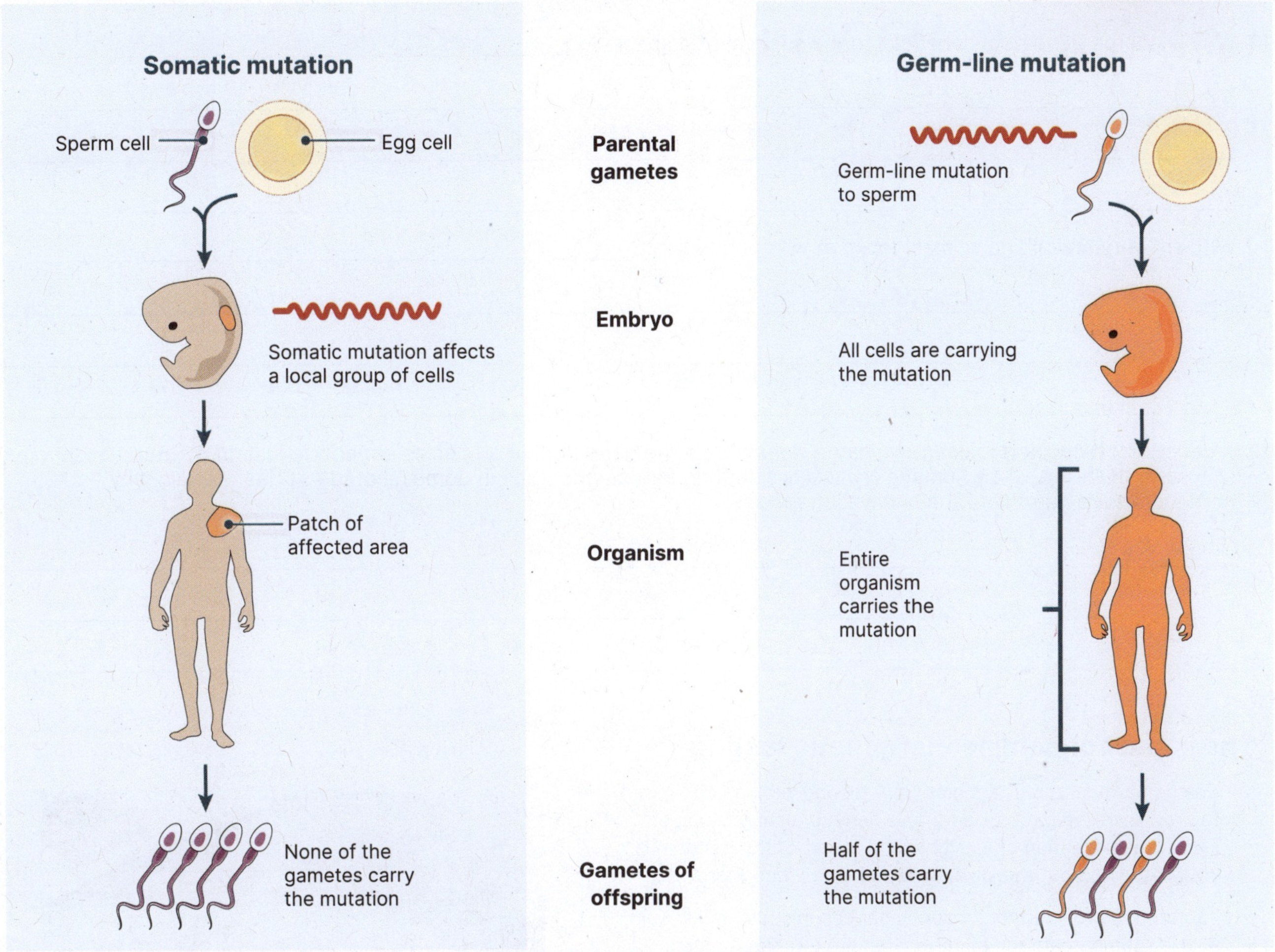

Somatic mutation

Sperm cell — Egg cell

Parental gametes

Somatic mutation affects a local group of cells

Embryo

Patch of affected area

Organism

None of the gametes carry the mutation

Gametes of offspring

Germ-line mutation

Germ-line mutation to sperm

All cells are carrying the mutation

Entire organism carries the mutation

Half of the gametes carry the mutation

7. Distinguish between somatic and germ-line mutations: _____

_____

_____

_____

8. Explain the consequences of these different mutation locations: _____

_____

_____

_____

_____

9. Chimeras can be produced artificially in both plants and animals. What kind of information could these organisms provide in studies of gene expression and gene regulation?

_____

_____

_____

Increases in harmful mutations

Artificially selecting for specific traits when breeding dogs has caused an accumulation of harmful mutations in certain breeds. This has led to specific, disease causing mutations becoming more common. High levels of inbreeding can lead to high levels of harmful traits, including deafness in dalmatians, and hip dysplasia in German Shepherds.

Straight backed German shepherds (left) are less likely to develop hip dysplasia than those bred to have extreme sloping backs (right).

10. Why is a large gene pool beneficial to the survival of a species? _____

_____

_____

_____

11. Why are very few albino animals found in wild populations? _____

_____

_____

_____

12. A genetic bottleneck is caused by a large reduction in the gene pool of a species either because of a natural event such as a bush-fire, or by humans artificially selecting for specific traits in domesticated species. Explain why bottlenecks are harmful to the survival of species:

_____

_____

_____

_____

The impact of commercial genetic testing

▶ The demand for direct to home DNA testing kits has continued to grow over the past decade. This genetic testing is unique in that the collection of a sample and report of the genome is completed without the involvement of a healthcare provider.

▶ Information provided to the consumer can include ancestry, such as the percentage of genes that are representative of different ethnic groups, but also the presence of genotypes linking to phenotypes of a range of traits, including the likelihood of developing various health related conditions or diseases.

▶ Some DNA testing companies offer an evaluation of children's future potential in sports, based on the presence of variant genes for 'fast-twitch' muscles.

▶ Ethical issues arise when medical information is shared without the support of medical professionals, leading to incorrect interpretations and assumptions. Consent from participants, especially children, can be missing, and companies can have access to personal data that consumers have no control over.

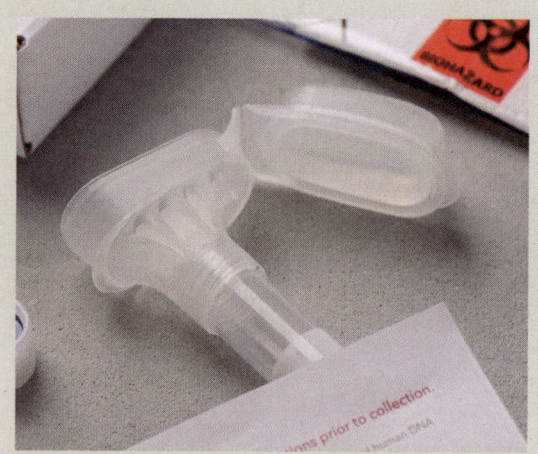

Commercial home DNA testing kits require a simple non-invasive saliva test, and consumers receive a report of the genome findings.

13. What might be some advantages and disadvantages of commercial genetic testing to potential carriers of genetic mutations?

_____

_____

_____

274 Genetic Technology

Key Idea: CRISPR is a complex made up of Cas9 enzymes and RNA. The CRISPR complex cuts DNA at very specific sequences and can be used to edit genes.

CRISPR-Cas9 (shortened to CRISPR and pronounced crisper) is an endonuclease complex occurring naturally in bacteria which use it to edit the DNA of invading viruses. CRISPR is able to target specific stretches of DNA and edit it at very precise locations. Two key components are required for CRISPR to work: an RNA guide that locates and binds to the target piece of DNA and the Cas9 endonuclease that unwinds and cuts the DNA. The technology has potential applications in correcting mutations responsible for disease, switching faulty genes off, adding new genes to an organism, or studying the effect of specific genes. It represents a major advance because it allows more precise and efficient gene editing at much lower cost than ever before.

How does CRISPR-Cas9 work?

CRISPR-CAS9 was adapted from a bacterial defence mechanism against viruses. IT is an enzyme used to guide RNA to cut target DNA sequence.

What can CRISPR-Cas9 be used for?

Virus resistance: immunity against turnip mosaic virus, Gemini virus, potyvirus.

Disease resistance: transgenic plants resistant to powdery mildew, rice blast.

Crop production enhancement: improving productivity, nutritional value, concentration.

Physical and chemical resistance: against herbicide, drought, post-harvesting processes.

Manipulating bioactive compounds: impacting flavonoids, hormones.

Research: create cell and animal models, screening studies for drug discovery, manufacturing stem cells.

Medical: cell and gene therapy, diagnostic tool (as used for Covid-19).

Biofuels: engineering algae to produce lipids.

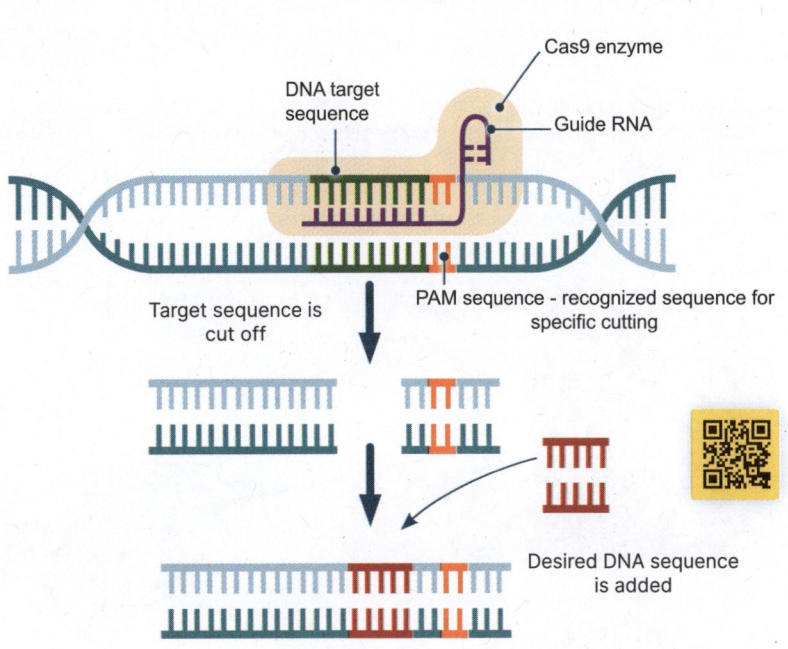

Gene knockout - 'gene silencing'

Frame-shift mutations can change the way the nucleotide sequence is read, either disabling gene function or producing a STOP signal. This technique can be used to silence a faulty gene. An existing gene may be deleted or deactivated (switched off) to prevent the expression of a trait, e.g. the deactivation of the ripening gene in tomatoes produced by the Flavr-Savr tomato.

Humans can not be used in knockout research for ethical reasons, so animal models such as mice are used. Many hundreds of strains of mice with different knockout genes are available in a living 'library' for genetic studies. The animal models can be genetically manipulated to more accurately represent a disease.

Manipulating gene action is one way to control processes such as ripening in fruit so that it stays fresher for longer.

1. Using examples from an online search, outline two ways in which CRISPR can be used to edit genes:

2. What benefits are offered by CRISPR technology for one of your chosen examples?

3. Using an example from an online search, outline one way gene knock out can 'repair' a genetic mutation:

©2024 **BIOZONE** International
ISBN: 978-1-99-101410-8
Photocopying prohibited

CASE STUDY: CRISPR used to alter starch in potatoes

▶ Potatoes are the number one most important vegetable and the third most important crop in the world after rice and wheat. Research into improving potato plants is therefore important for world food production.

▶ The potato tuber is mostly starch, a large polymer made up of two monomers: amylose and amylopectin. Starches high in amylopectin are more useful in the food industry because they produce better products such as stabilizers, thickeners and emulsifiers. They are also better for ethanol production.

▶ Researchers in Texas have recently used *Agrobacterium* to insert the CRISPR-Cas9 system into Yukon Gold potato plants to edit the gbss gene. This gene produces a protein that catalyzes the synthesis of amylose.

▶ The resulting potato plants grew normally and had high levels of amylopectin and no amylase in their starch (diagram below).

Alteration of amylase production in potatoes

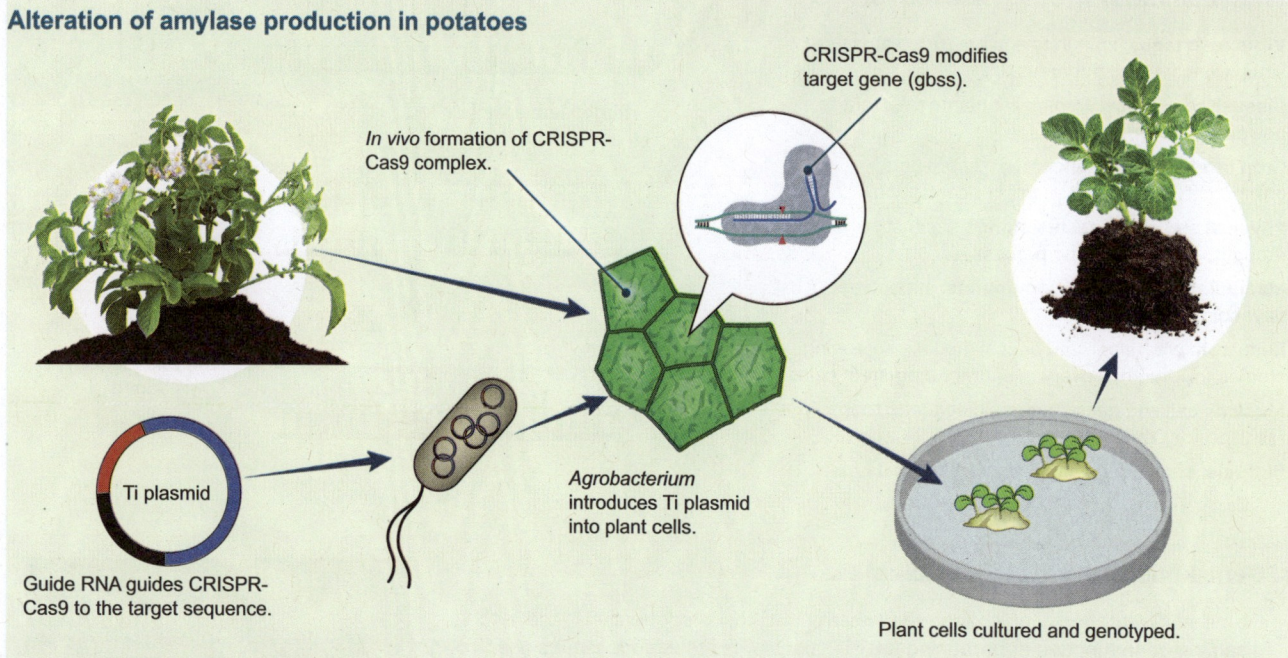

In vivo formation of CRISPR-Cas9 complex.

CRISPR-Cas9 modifies target gene (gbss).

Ti plasmid

Guide RNA guides CRISPR-Cas9 to the target sequence.

Agrobacterium introduces Ti plasmid into plant cells.

Plant cells cultured and genotyped.

An *in vivo* method (above) uses *Agrobacterium* to deliver a modified Ti plasmid into the plant cells. The Ti plasmid inserts the genes required to introduce CRISPR-Cas9 into the plant DNA. The cell's machinery then produces the CRISPR-cas9 protein and RNA, which then edits the cell's DNA.

The ethical implications of using CRISPR for genetic engineering in humans

▶ CRISPR has allowed genetic engineering to make huge and rapid progress. Placing genes from other species into organisms has been very useful, e.g. enabling the production of transgenic animal models to advance disease research.

▶ Nearly every government and organization has agreed that use of CRISPR for eugenics (altering the human gene pool) and human enhancement should be banned. International regulatory frameworks are still being developed and implemented for the use of this tool in humans, as the benefits of CRISPR need to be balanced with the ethical implications of permanently changing the genome, and potentially the future evolution of humans.

4. Work in groups and take part in a class debate. If your class is large, divide into smaller teams of around 3 students each. Debate the pros and cons of using CRISPR to genetically engineer genes in humans, either by adding sequences from another species or altering the bases.

(a) Which side of the debate are you arguing for? For (pro) or against (con)? _____

(b) Work together in your smaller group to decide upon at least three key points that you will use to support your side of the debate. Record those points down below.

_____

_____

_____

©2024 **BIOZONE** International
ISBN: 978-1-99-101410-8
Photocopying prohibited

275 Hypotheses for Conserved Genetic Sequences

Key Idea: Some gene sequences are highly conserved and so, therefore, are the gene products. Different hypotheses account for this conservation.

Some gene sequences show very little change and the conservation of the gene through many related species can be seen in a number of gene products, essentially unchanged from a past common ancestor. One hypothesis for this is that the gene products are essential and cannot be replaced by another product. Therefore, any gene mutations are likely to result in death of the organism and risk the vital genetic sequences not being passed on. A second hypothesis suggests that some gene sequences mutate less frequently than others. At a molecular level, the G-C base sequence in DNA has a higher mutation rate than A-T, and their position around a genetic sequence can be used to predict mutation rate frequency.

Essential requirements of protein function

The amino acid sequence of proteins can be used to establish molecular homologies (similarities) between organisms. Any change in the amino acid sequence reflects changes in the DNA sequence. As genetic relatedness increases, the number of amino acid differences due to mutation decreases.

Some proteins are common to many different species. These proteins are often highly conserved, meaning they mutate very little over time. For example, proteins involved in transcription and translation, histone proteins, or the proteins involved in glycolysis and the Krebs cycle.

One hypothesis for highly conserved gene sequences is that because the proteins have critical roles, e.g. in cellular respiration, any mutations are likely to be detrimental to their function, and the individual dies before reproducing.

Evidence indicates that these highly conserved proteins are homologous and have been derived from a common ancestor. Because they are highly conserved, changes in the amino acid sequence are likely to represent major divergences between groups during the course of evolution.

The Pax-6 protein is highly conserved

▶ The Pax-6 gene is a patterning gene. It produces a protein that regulates eye formation during embryonic development.

▶ The Pax-6 gene is so highly conserved that the gene from one species can be inserted into another species, both invertebrates and vertebrates, and still produce a normally functioning eye.

▶ This suggests the Pax-6 proteins are homologous, and the gene has been inherited, and retained from a common ancestor.

An experiment inserted mouse Pax-6 gene into fly DNA and turned it on in a fly's legs. The fly developed fly eyes on its legs!

Varying mutation rates of genetic sequences

The probability of mutations occurring in different genetic sequences is unequal. Genetic sequences within different species have shown variance in mutation rates. For example, recent research has hypothesized that the mutation rate of many genetic sequences in conifers, a cone bearing group of gymnosperms, is much slower than in angiosperms (flowering plants), leading to a slower rate of evolution. Positive correlations have been found between plant height (nearly all conifers are tall trees), and mutation rate. Conversely, the higher mutation rate of genetic sequences in angiosperms, a group of plants first emerging in the Cretaceous epoch, is correlated with rapid adaptive radiation. Consider that, currently, only 1000 species of gymnosperms exist compared to nearly 500, 000 species of angiosperms on Earth.

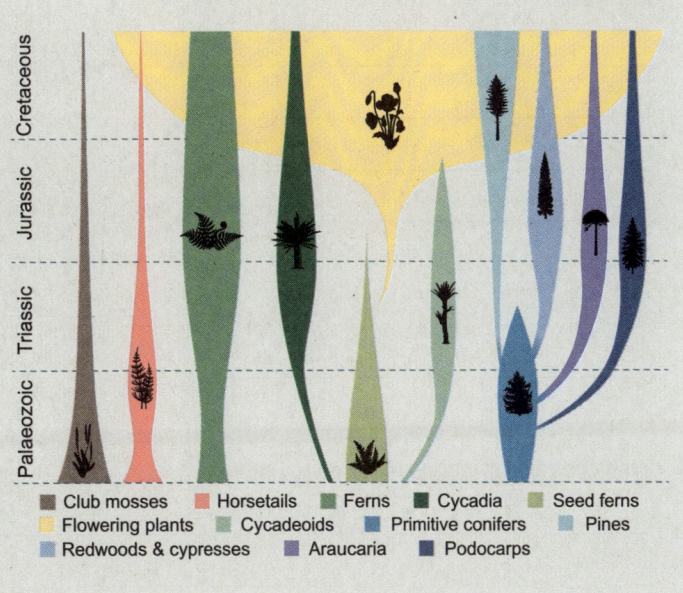

Club mosses ■ Horsetails ■ Ferns ■ Cycadia ■ Seed ferns ■ Flowering plants ■ Cycadeoids ■ Primitive conifers ■ Pines ■ Redwoods & cypresses ■ Araucaria ■ Podocarps

1. (a) What is a highly conserved genetic sequence? _____

_____

(b) Suggest why some genetic sequences are highly conserved: _____

_____

_____

2. Prepare a short research report on one example of a conserved genetic sequence across species and time, evident as a gene product such as haemoglobin. Use one of the above highlighted hypotheses to discuss a possible explanation for the conservation. Attach the report to the page.

276 Did You Get It?

1. (a) The schematic below shows the levels of control in gene expression. Choosing from the following word list, fill in the boxes, indicating the structures and processes. Word list: *5' cap, mRNA in the nucleus, polypeptide, mRNA in the cytoplasm, DNA packing, exon, intron, functional protein, folding and assembly, poly A tail, gene, cleavage or chemical modification, primary mRNA, protein degradation, translation, transcription, exon splicing, nuclear export.*

   (b) On the diagram, indicate processes with a P and structures with an S.

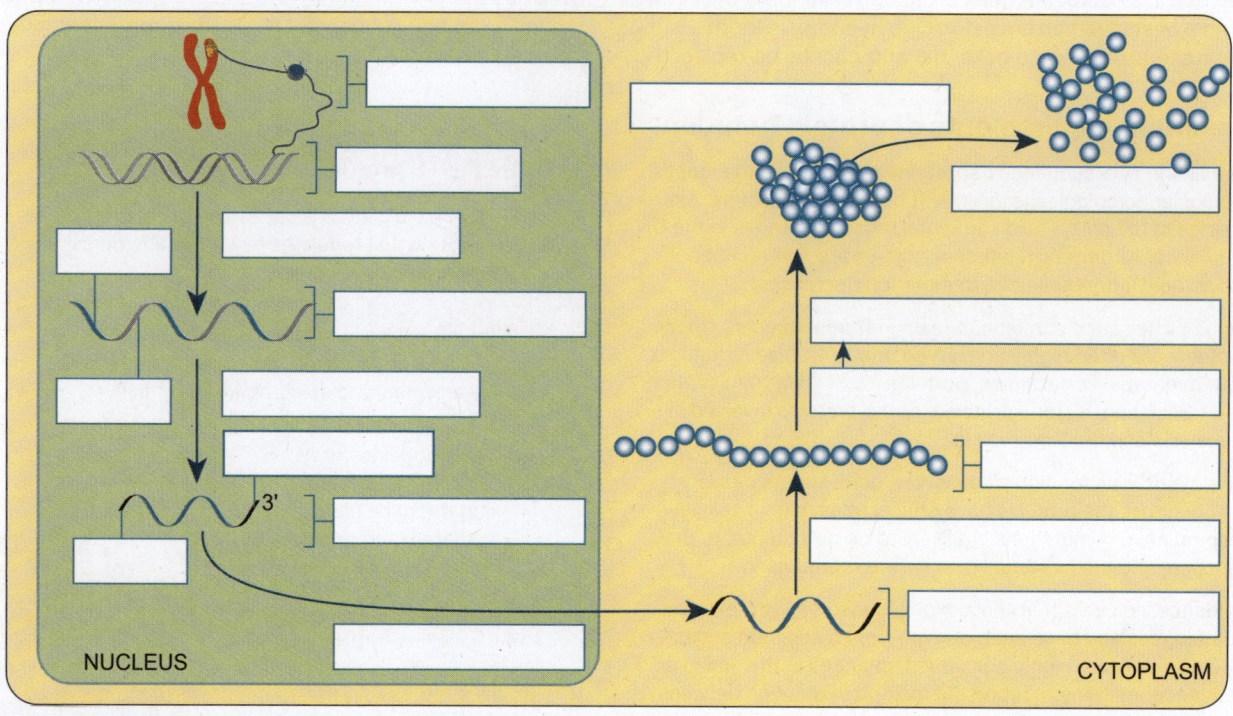

NUCLEUS   CYTOPLASM

2. Explain how accuracy in copying base sequences is enabled by semi-conservative DNA replication and complementary base pairing:

   _____
   _____
   _____
   _____

3. **AHL:** Explain how the different 5' and 3' ends of a DNA strand affect directionality of translation and transcription:

   _____
   _____
   _____

4. How does genetic code degeneracy reduce the impact of a SNP on the phenotype? _____

   _____
   _____

5. Apolipoprotein A1-Milano is a mutation that helps transport cholesterol through the blood. The mutation causes a change to one amino acid and increases the protein's effectiveness by ten times, dramatically reducing incidence of heart disease. The mutation can be traced back to its origin in Limone, Italy, in 1644. What factors could cause this mutation to spread further?

   _____
   _____
   _____
   _____

   The village of Limone, Italy

©2024 **BIOZONE** International
ISBN: 978-1-99-101410-8
Photocopying prohibited

 # Cells

D2.1 Cell and nuclear division

	Activity Number

Guiding Questions:
▶ What is the significance of mitosis to an organism?
▶ What is the significance of meiosis to an organism?

Learning Outcomes:

			Activity Number
☐	1	Summarize the process and function of cell division.	277
☐	2	Describe the steps of cytokinesis of a parent cell during cell division to form two daughter cells.	278
☐	3	Compare equal and unequal cytokinesis using the examples of cytokinesis that occurs during human oogenesis and yeast budding.	278
☐	4	Compare and contrast the roles of eukaryotic mitosis and meiosis.	280
☐	5	Explain why mitosis and meiosis require DNA replication as a prerequisite.	278, 280
☐	6	Elaborate on DNA condensation and the movement of chromosomes in both mitosis and meiosis.	278, 280
☐	7	Provide an overview of mitosis including identification of the names of each of the phases.	278
☐	8	**AOS:** Identify phases of mitosis in diagrams and micrographs.	279
☐	9	Explain why meiosis is considered a reduction division and the relevance of this to production of haploid gametes during reproduction.	280
☐	10	Link the occurrence of Down syndrome to a non-disjunction mutation occurring during meiosis.	281
☐	11	Discuss how meiosis functions as a source of variation with reference to the process of crossing over and bivalent orientation.	280
☐	12	**AHL:** Explain how growth, cell replacement, and tissue repair are possible due to cell proliferation; relate to plant meristems, early-stage animal embryos, skin cell replacement, and wound healing as examples.	282
☐	13	**AHL:** Identify and describe the phases and sequence of the cell cycle.	283
☐	14	**AHL:** Elaborate on the process of cell growth including biosynthesis of cell components during interphase.	282, 283
☐	15	**AHL:** Explain the role of cyclins in control of the cell cycle.	284
☐	16	**AHL:** Discuss the consequences of mutations in genes controlling the cell cycle including proto-oncogenes and tumour suppressor genes.	285
☐	17	**AHL:** Describe the difference in the rate of cell growth in different types of tumour and explain why some tumours become cancerous. **AOS:** Determine the mitotic index from observed cell populations and link to identification of possible tumours.	286

D2.2 Gene expression

	Activity Number

Guiding Questions:
▶ How can gene expression result in different outcomes in a cell?
▶ How are the results of gene expression conserved in subsequent generations?

Learning Outcomes:

			Activity Number
☐	1	**AHL:** Discuss how phenotype is the result of gene expression.	287
☐	2	**AHL:** Explain how transcription is regulated by promoters, enhancers, and transcription factors.	287
☐	3	**AHL:** Explain how translation can be regulated by controlling mRNA degradation.	287
☐	4	**AHL:** Link cell differentiation, and therefore changes in phenotype, to the process of epigenesis.	288
☐	5	**AHL:** Compare the differences between the genome, transcriptome, and proteome of individual cells.	287
☐	6	**AHL:** Explain how methylation of the promoter and histones repress transcription of DNA, acting as epigenetic tags.	288
☐	7	**AHL:** Explain how specific phenotypic changes can be passed from parent to offspring when epigenetic tags remain in place during mitosis or meiosis.	288

D2.3 Water potential

	Activity Number

Guiding Questions:
▸ What variables are involved with osmotic movement of water across cells?
▸ How is osmotic regulation different in plant and animal cells?

Learning Outcomes:

277 Cell Division

Key Idea: Mitosis and meiosis are two different types of cell division and are important parts of the cell cycle.

New cells are formed when existing cells divide. In eukaryotes, cell division begins with the replication of a cell's DNA followed by division of the nucleus. There are two forms of cell division. Mitosis produces two identical daughter cells from each parent cell. Mitosis is responsible for growth and repair processes in multicellular organisms, and asexual reproduction in some eukaryotes, e.g. yeasts. Meiosis is a special type of cell division concerned with producing haploid cells for sexual reproduction. It occurs in the sex organs of plants and animals.

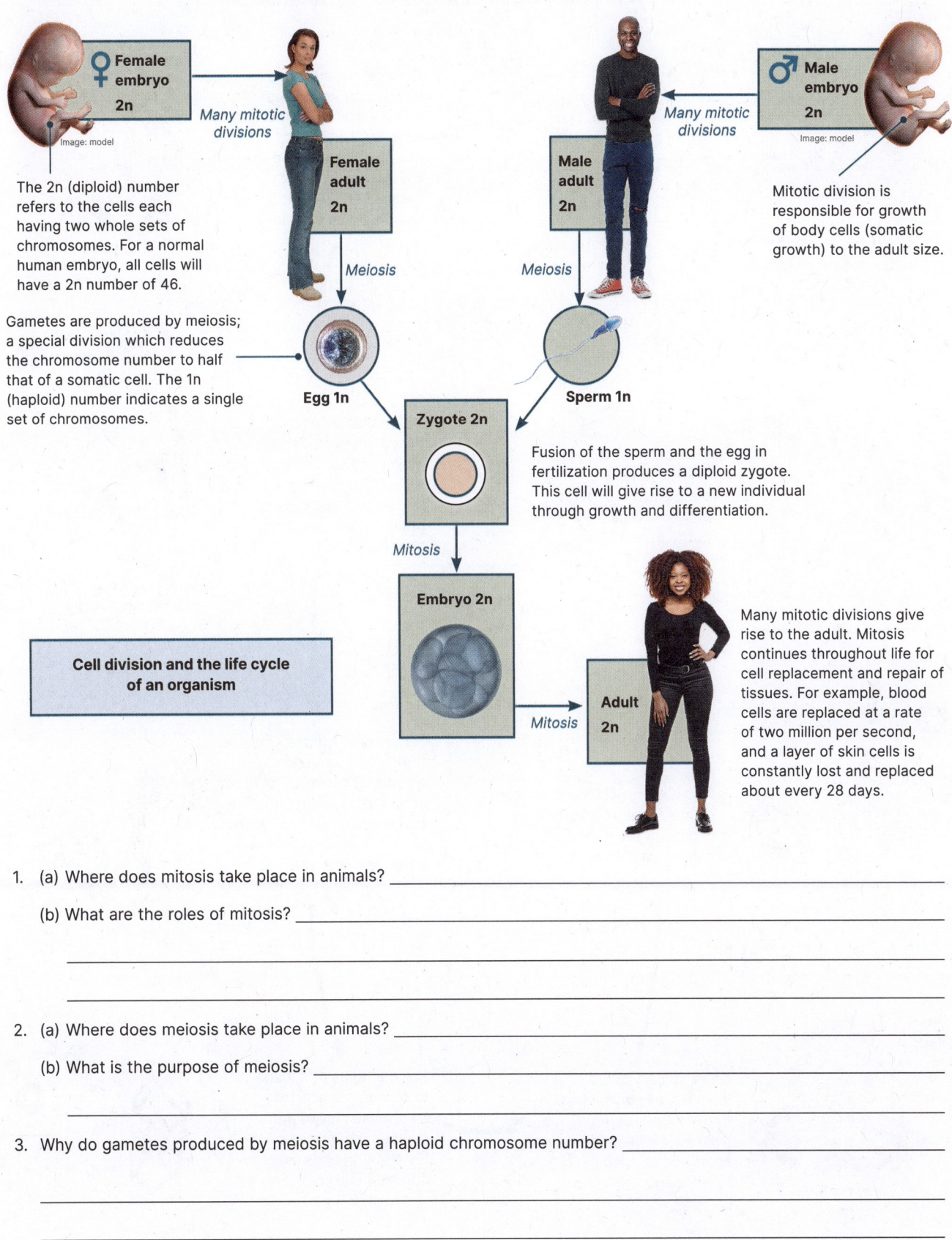

Female embryo 2n

Many mitotic divisions

Female adult 2n

Image: model

Meiosis

Male adult 2n

Meiosis

Many mitotic divisions

Male embryo 2n

Image: model

The 2n (diploid) number refers to the cells each having two whole sets of chromosomes. For a normal human embryo, all cells will have a 2n number of 46.

Gametes are produced by meiosis; a special division which reduces the chromosome number to half that of a somatic cell. The 1n (haploid) number indicates a single set of chromosomes.

Mitotic division is responsible for growth of body cells (somatic growth) to the adult size.

Egg 1n

Sperm 1n

Zygote 2n

Fusion of the sperm and the egg in fertilization produces a diploid zygote. This cell will give rise to a new individual through growth and differentiation.

Mitosis

Cell division and the life cycle of an organism

Embryo 2n

Adult 2n

Mitosis

Many mitotic divisions give rise to the adult. Mitosis continues throughout life for cell replacement and repair of tissues. For example, blood cells are replaced at a rate of two million per second, and a layer of skin cells is constantly lost and replaced about every 28 days.

1. (a) Where does mitosis take place in animals? _____

   (b) What are the roles of mitosis? _____
   _____
   _____

2. (a) Where does meiosis take place in animals? _____

   (b) What is the purpose of meiosis? _____
   _____

3. Why do gametes produced by meiosis have a haploid chromosome number? _____
   _____
   _____
   _____

D2.1
1

278 Mitosis and Cytokinesis

Key Idea: Mitosis and cytokinesis are key parts of the cell cycle, Mitosis can be divided into six main stages.

Mitosis refers to the division of the nuclear material and it is followed immediately by division of the cell. Although mitosis is part of a continuous cell cycle, it is divided into stages to help distinguish the processes occurring during its progression. Mitosis is one of the shortest stages of the cell cycle. Cytokinesis (the division of the newly formed cells) is part of M-phase but it is distinct from nuclear division. During cytokinesis the cell divides into two.

The cell cycle and stages of mitosis in an animal cell

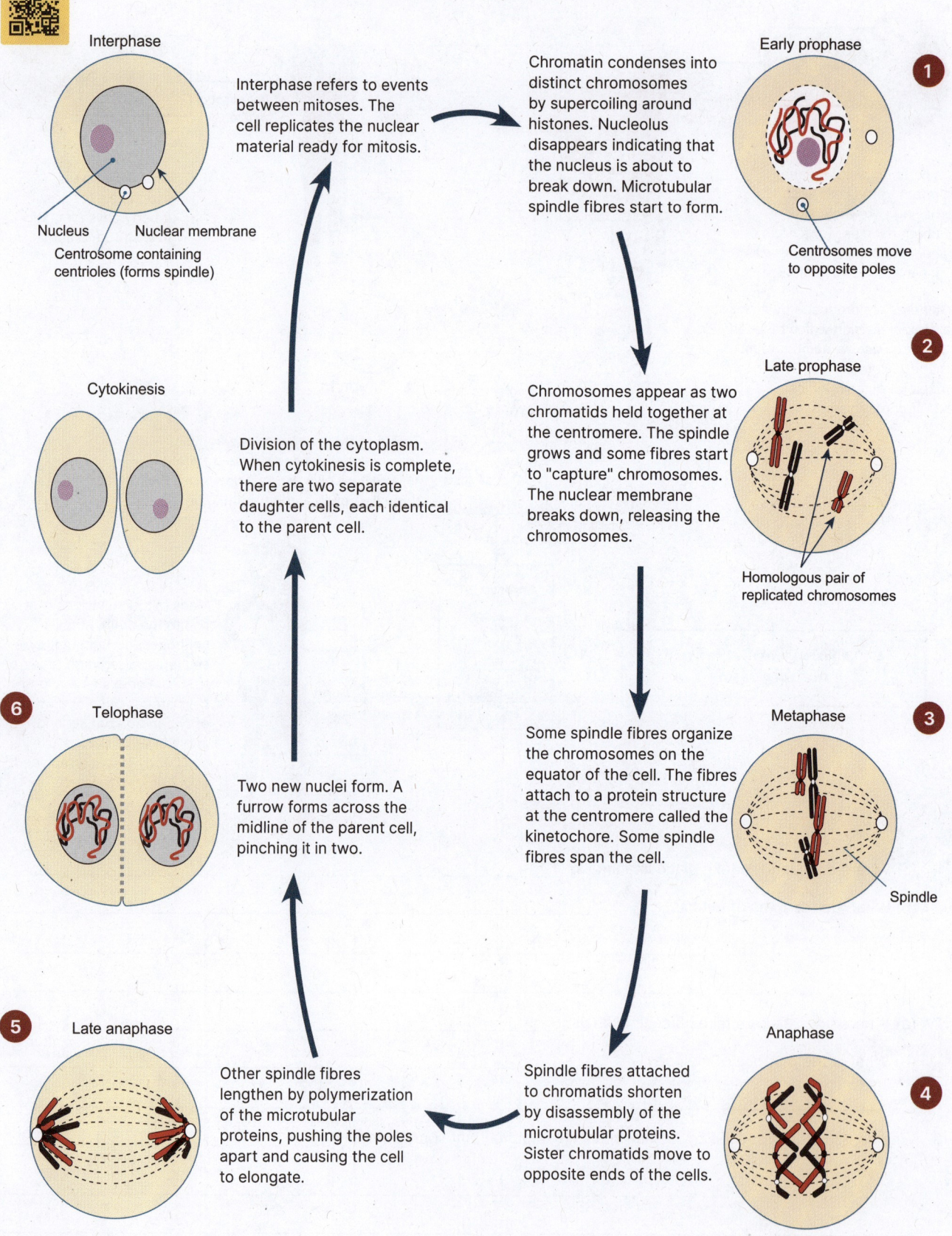

Interphase

Nucleus

Nuclear membrane

Centrosome containing centrioles (forms spindle)

Interphase refers to events between mitoses. The cell replicates the nuclear material ready for mitosis.

Early prophase 1

Chromatin condenses into distinct chromosomes by supercoiling around histones. Nucleolus disappears indicating that the nucleus is about to break down. Microtubular spindle fibres start to form.

Centrosomes move to opposite poles

Cytokinesis

Division of the cytoplasm. When cytokinesis is complete, there are two separate daughter cells, each identical to the parent cell.

Late prophase 2

Chromosomes appear as two chromatids held together at the centromere. The spindle grows and some fibres start to "capture" chromosomes. The nuclear membrane breaks down, releasing the chromosomes.

Homologous pair of replicated chromosomes

6 **Telophase**

Two new nuclei form. A furrow forms across the midline of the parent cell, pinching it in two.

Metaphase 3

Some spindle fibres organize the chromosomes on the equator of the cell. The fibres attach to a protein structure at the centromere called the kinetochore. Some spindle fibres span the cell.

Spindle

5 **Late anaphase**

Other spindle fibres lengthen by polymerization of the microtubular proteins, pushing the poles apart and causing the cell to elongate.

Spindle fibres attached to chromatids shorten by disassembly of the microtubular proteins. Sister chromatids move to opposite ends of the cells.

Anaphase 4

D2.1
2,3,5-7

©2024 **BIOZONE** International
ISBN: 978-1-99-101410-8
Photocopying prohibited

Cytokinesis (division of the cytoplasm)

Cytokinesis in an animal cell

Animal cells: Cytokinesis (below left) begins shortly after the sister chromatids have separated in anaphase of mitosis. A ring of microtubules assembles in the middle of the cell, next to the plasma membrane, constricting it to form a cleavage furrow. In an energy-using process, the cleavage furrow moves inwards, forming a region where the two cells will separate.

Cytokinesis in a plant cell

Plant cells: (below right): Cytokinesis involves construction of a cell plate (a precursor of the new cell wall) in the middle of the cell. The cell wall materials are delivered by vesicles derived from the Golgi. The vesicles join together to become the plasma membranes of the new cell surfaces.

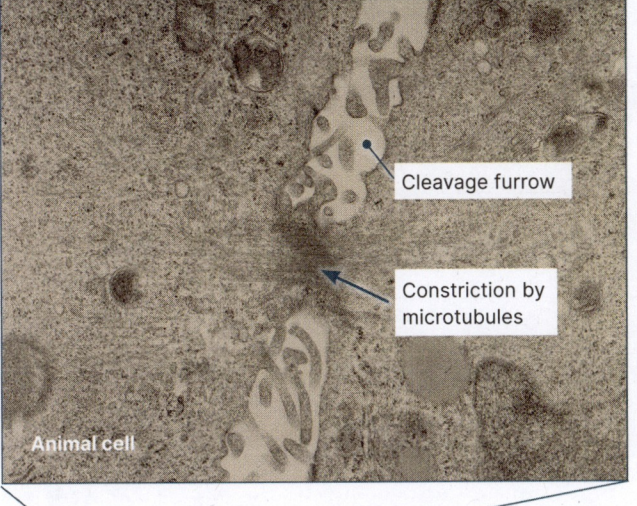

Cleavage furrow

Constriction by microtubules

Animal cell

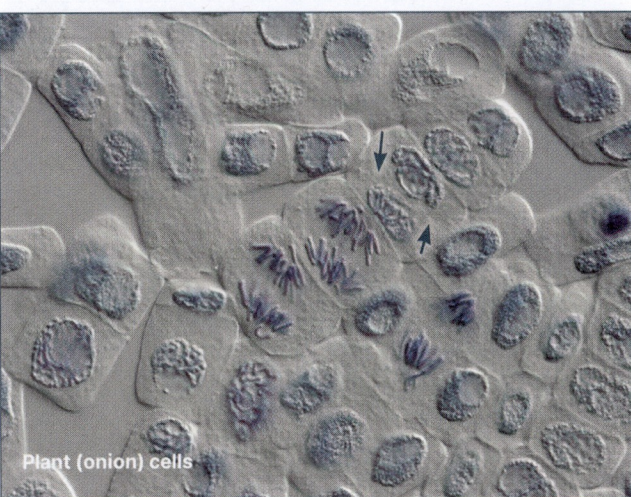

Plant (onion) cells

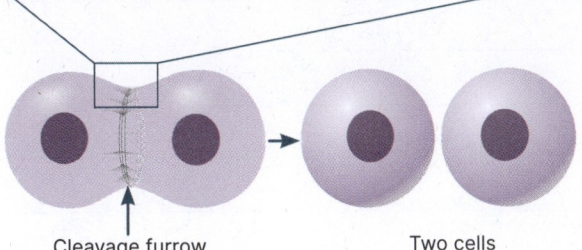

Cleavage furrow — Two cells

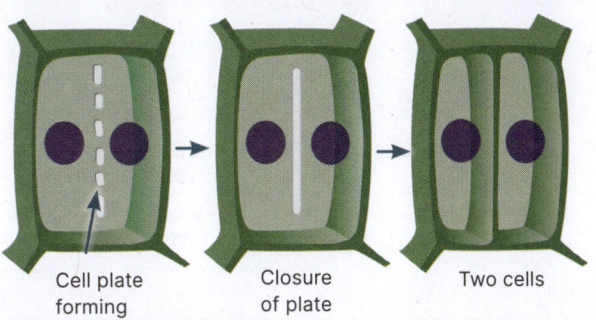

Cell plate forming — Closure of plate — Two cells

▶ During cell divisions each new cell takes a random assortment of organelles including mitochondria or chloroplasts (in plants) which reproduce independently of the cell during interphase. Cells need at least one mitochondria to produce ATP to drive cellular metabolism.

▶ In most cases of mitosis cell division ends in two cells that are the same size. However, that is not always the case. In the production of the oocyte (egg cell) (via meiosis) three of the four cells produced (called polar bodies (arrowed far right)) are tiny in comparison to the egg cell.

▶ Yeast cells reproduce via asexual reproduction, budding new cells (arrowed near right). The daughter cell is much smaller than the parent cell.

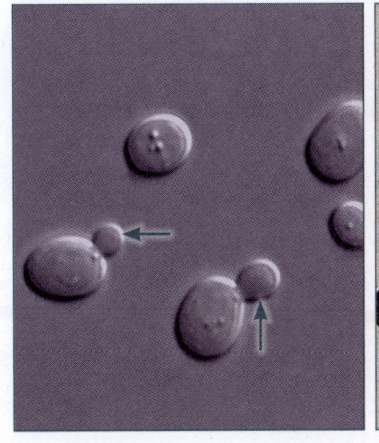
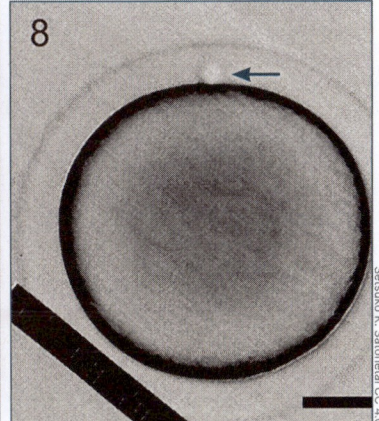

1. What must occur before mitosis takes place? _____

2. (a) What is the purpose of the spindle fibres? _____

(b) Where do the spindle fibres originate? _____

3. When do cell organelles replicate? _____

446

4. Suggest why mitosis and cytokinesis are energetically demanding processes: _____
_____
_____
_____

5. Summarize what happens in each of the following phases of mitosis:

   (a) Prophase: _____
   _____
   _____

   (b) Metaphase: _____
   _____

   (c) Anaphase: _____
   _____

   (d) Telophase: _____
   _____
   _____
   _____

6. (a) What is the purpose of cytokinesis? _____
   _____

   (b) Describe the differences between cytokinesis in an animal cell and a plant cell: _____
   _____
   _____
   _____
   _____
   _____

7. Use the information in the previous activity to identify which stage of mitosis is shown in each of the photographs below:

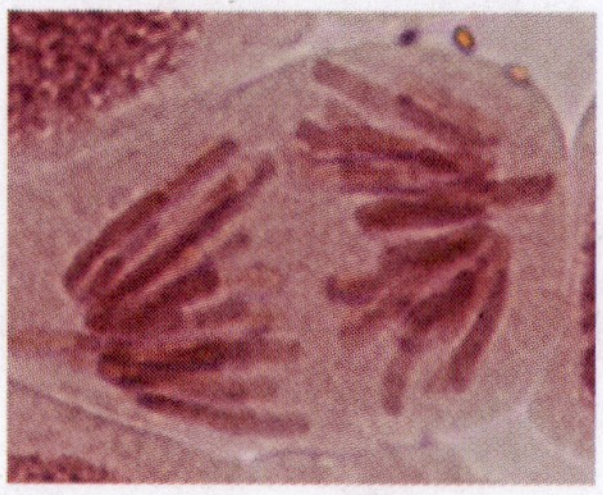

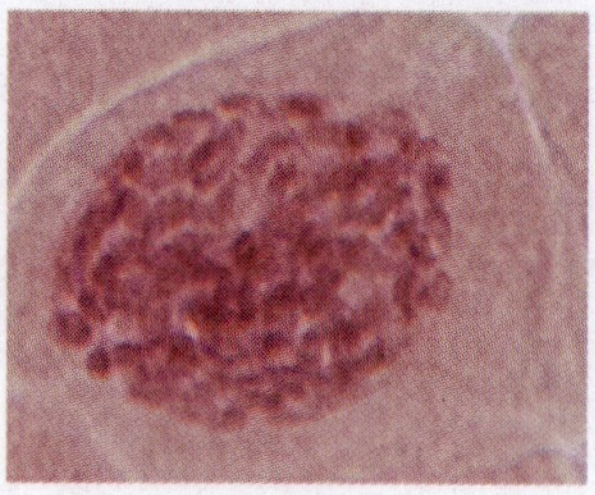

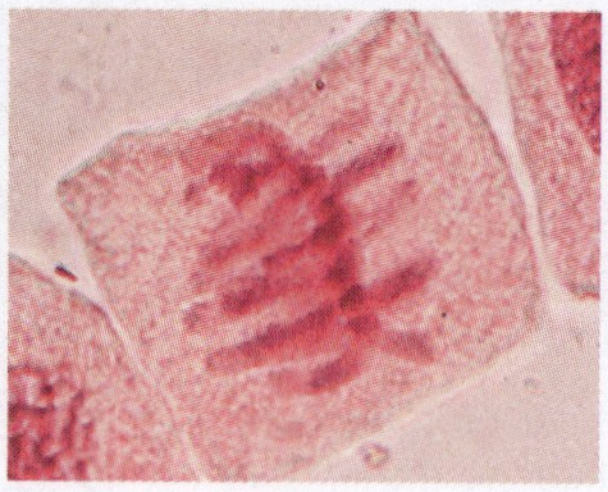

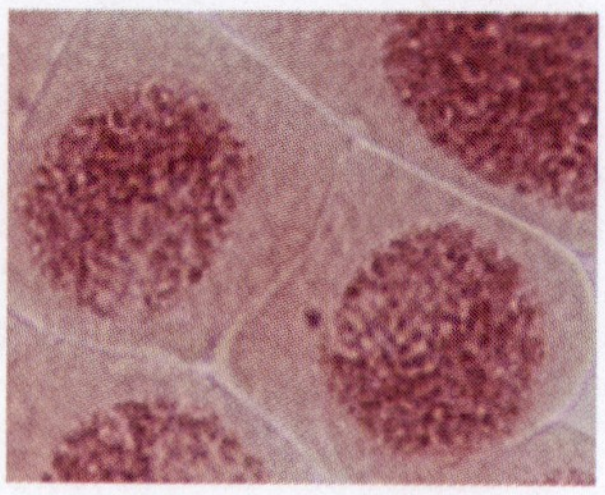

   (a) _____ (b) _____ (c) _____ (d) _____

8. Why might unequal cytokinesis be important in the production of oocytes?
   _____
   _____
   _____

9. How does a cell acquire its first mitochondria? _____
   _____
   _____

279 Modelling Mitosis

Key Idea: Modelling mitosis will help develop your understanding of the process.

Chenille stems (pipe cleaners) and yarn can be used to model stages of mitosis in an animal cell. For simplicity, it is easiest to start with 4 chromosomes (2n = 4).

Investigation 14.1 Modelling mitosis

See appendix for equipment list.

1. Use the information on the previous pages to model mitosis in an animal cell using chenille stems (pipe cleaners) and string. Work in pairs and use four chromosomes for simplicity (2n = 4). Photograph or film each stage.

2. Photo 1 (below) can be used as a starting point for your model. It represents a cell in interphase before mitosis begins. The circular structures are the replicated centrosomes.

3. Before you start, identify the structures A-C in photo

   A: _____ B: _____ C:_____

4. Remember to label your photos as you place them on the page.

Photo 1

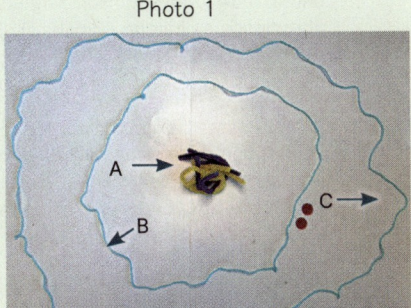

---

280 Meiosis and Variation

Key Idea: Meiosis produces haploid sex cells for the purpose of sexual reproduction.

Meiosis involves a single chromosomal duplication followed by two successive nuclear divisions, and results in a halving of the diploid chromosome number. Meiosis occurs in the sex organs of animals and the sporangia of plants. If genetic mistakes (mutations) occur here, they will be passed to the offspring (inherited). The first division in meiosis is a reduction division (the chromosome number is halved). The second meiotic division is similar to mitosis in that chromatids are pulled apart and the chromosome number remains the same. Meiosis creates genetic variation in the sex cells through two important processes: crossing over and independent assortment. These are described on the next page.

KEY EVENTS

MEIOSIS I (separation of homologous chromosomes)

Interphase
• DNA is copied and the cell prepares to divide.

REMEMBER! Homologous chromosomes (or homologs) are chromosome pairs with the same gene sequence. One comes from the mother (called maternal) and one comes from the father (called paternal).

Prophase I
• Chromosomes condense
• Homologs pair up
• Recombination of alleles occurs as homologous chromosomes exchange DNA in crossing over.

Metaphase I
• Random alignment of homologous chromosomes at the equator.
• Chromatids still held at centromere.

Anaphase I
• The pairs of chromosomes separate and move to opposing poles.

Telophase I
• Nuclear membranes reform, the cell divides and two cells are formed, each with n = 2 chromosomes.
• The spindle fibres disassemble.

MEIOSIS II (separation of chromatids)

Prophase II
• There are now two cells.
• DNA does not replicate again.
• Recombination of alleles occur

Metaphase II
• Individual chromosomes line up along the equator of the cell.

Anaphase II
• The chromatids split at the centromere and are moved by the spindle fibres to opposite poles of the cell.

Telophase II
• Nuclear membranes reform.
• There are 4 new haploid daughter cells.

2n = 4

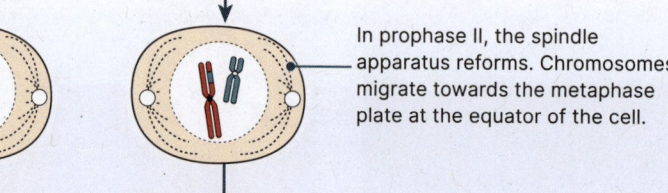

Maternal chromosome

Paternal chromosome

Spindle apparatus begins to form.

n = 2

n   n   n   n

When a cell is not dividing (interphase) the chromosomes are not visible, but the DNA is being replicated. The cell shown is 2n, where n is the number of copies of chromosomes in the nucleus. n = one copy of each chromosome (haploid). 2n = two copies of each chromosome (diploid).

═══ Meiosis starts here ═══

In prophase I, replicated chromosomes appear as two sister chromatids held together at the centromere (a specialized DNA sequence that holds the sister chromatids together). Homologous chromosomes pair up (synapsis). Crossing over may occur at this time making sister chromatids differ from one another.

In metaphase I, homologous pairs line up in the middle of the cell (equator) independently of each other. This results in paternal and maternal chromosomes assorting independently into the gametes.

In anaphase I, homologs separate, pulled apart by the disassembly of the spindle fibres that are attached to the kinetochores (protein discs) on the centromeres. Other spindle fibres grow longer to push the cell poles apart and elongate the cell.

In telophase I, two intermediate cells form.

In prophase II, the spindle apparatus reforms. Chromosomes migrate towards the metaphase plate at the equator of the cell.

In metaphase II, the chromosomes line up on the metaphase plate.

In anaphase II, the centromere divides and sister chromatids (now individual chromosomes) are separated, moving to opposite poles of the cell.

Telophase II produces four haploid gametes. Each one is n = 2.

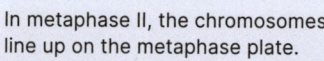

D2.1 4-6,9,11

©2024 **BIOZONE** International
ISBN: 978-1-99-101410-8
Photocopying prohibited

Crossing over and recombination

▸ Chromosomes replicate during interphase, before meiosis begins, to produce replicated chromosomes with sister chromatids held together at the centromere (below). The centromere is a specialized DNA sequence that links a pair of sister chromatids.

▸ When the replicated chromosomes are paired during the first stage of meiosis, non-sister chromatids may become entangled and segments may be exchanged in a process called crossing over. Crossing over results in the recombination of alleles (gene variants) producing greater variation in the offspring than would otherwise occur.

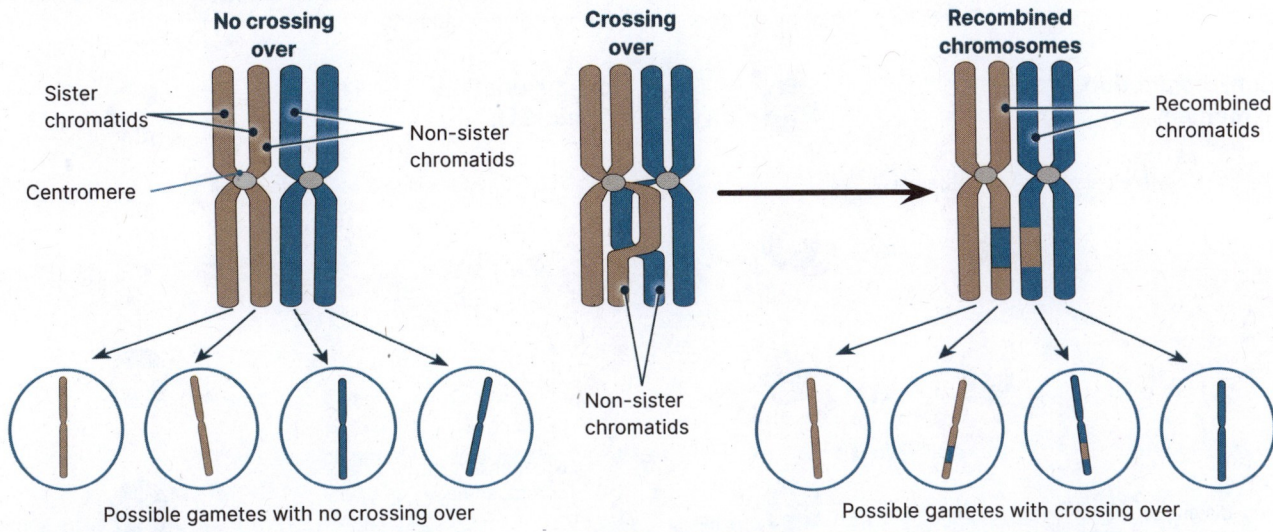

Independent assortment

▸ Independent assortment describes the random alignment and distribution of chromosomes during meiosis. Independent assortment is an important mechanism for producing variation in gametes.

▸ During the first stage of meiosis, replicated homologous chromosomes pair up along the middle of the cell. The way the chromosomes pair up is random. For the homologous chromosomes right, there are two possible ways in which they can line up resulting in four different combinations in the gametes. The intermediate steps of meiosis have been left out for simplicity.

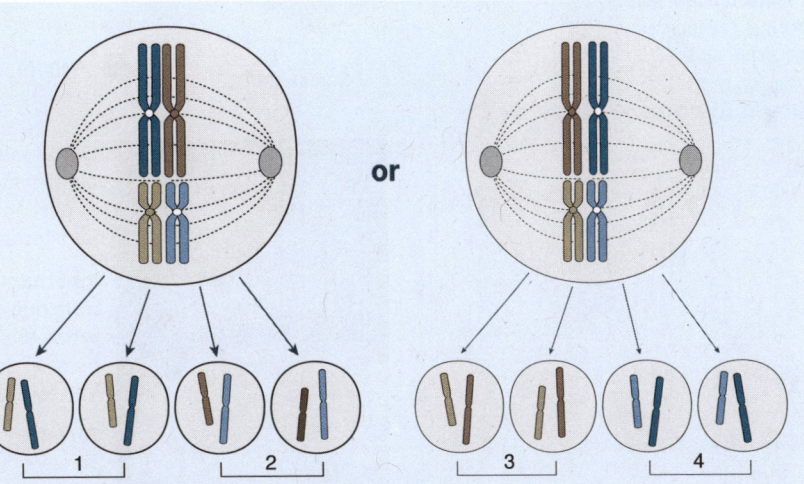

1. Describe the behaviour of the chromosomes in the first and then the second division in meiosis:

_____

2. Explain how independent assortment increases the variation in gametes: _____

3. (a) What is crossing over? _____

(b) Explain how crossing over increases the variation in the gametes (and hence the offspring):

©2024 **BIOZONE** International
ISBN: 978-1-99-101410-8

281 Nondisjunction in Meiosis

Key Idea: Nondisjunction during meiosis leads to gametes having more or less chromosomes than the haploid number. In meiosis, chromosomes are usually distributed to daughter cells without error. Occasionally, homologous chromosomes fail to separate properly in meiosis I, or sister chromatids fail to separate in meiosis II. In these cases, one gamete receives two of the same type of chromosome and the other gamete receives no copy. This error is known as non-disjunction and it results in abnormal numbers of chromosomes in the gametes. The union of an aberrant and a normal gamete at fertilization produces offspring with an abnormal chromosome number (known as aneuploidy).

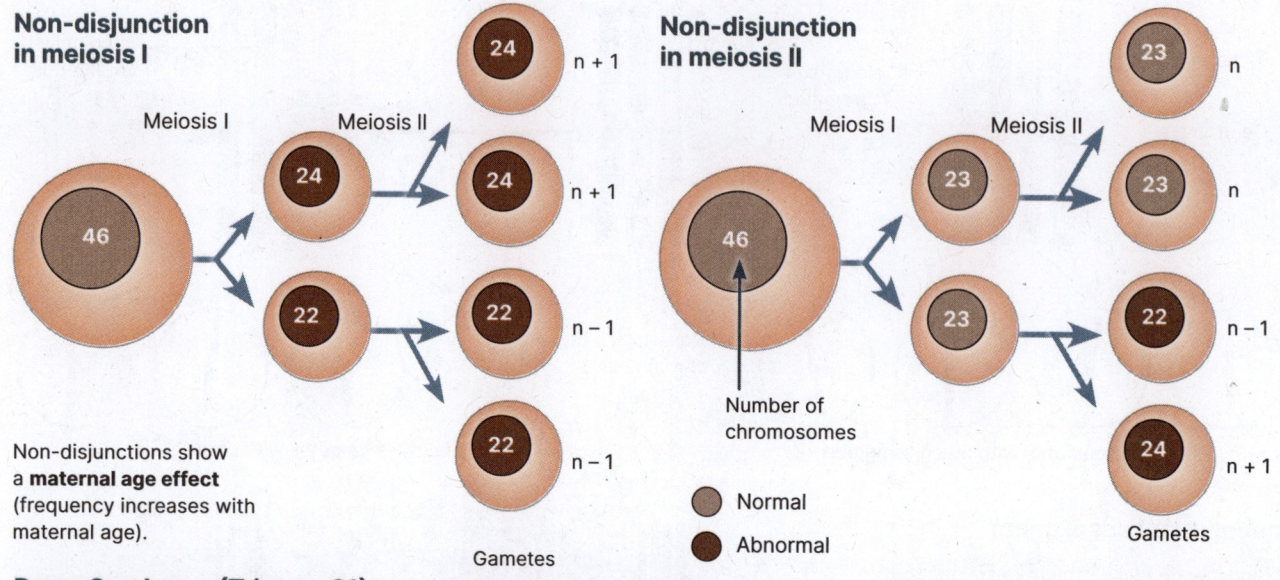

Non-disjunctions show a **maternal age effect** (frequency increases with maternal age).

Down Syndrome (Trisomy 21)

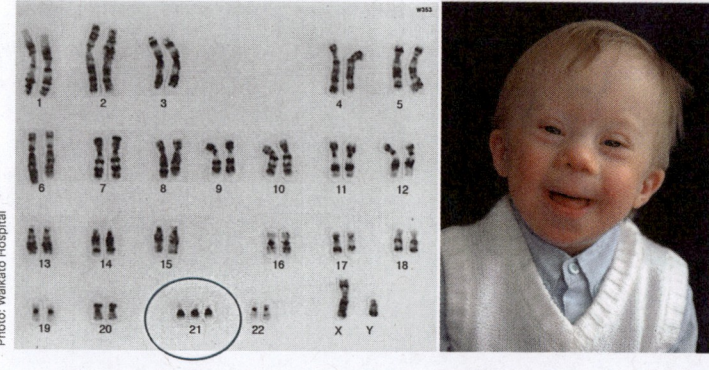

Incidence: Down syndrome is the most common of the human aneuploidies. The incidence rate in humans is about 1 in 800 births for women aged 30 to 31 years, with a maternal age effect.

Inheritance: Nearly all cases (approximately 95%) result from non-disjunction of chromosome 21 during meiosis. When this happens, a gamete (most commonly the oocyte) ends up with 24 rather than 23 chromosomes, and fertilization produces a trisomic offspring.

Far left: A karyogram for an individual with trisomy 21. The chromosomes are circled.

1. Describe the consequences of non-disjunction during meiosis:

2. Explain why non-disjunction in meiosis I results in a higher proportion of faulty gametes than non-disjunction in meiosis II:

3. What is the maternal age effect and what are its consequences?

4. Predict the outcome of non-disjunction occurring to all chromosomes during meiosis I:

©2024 **BIOZONE** International
ISBN: 978-1-99-101410-8
Photocopying prohibited

D2.1
9

282 Growth and Repair

Key Idea: Cell division is needed for growth and tissue repair. Mitotic cell division produces daughter cells identical to the original parent cell. This is important for growth, replacement of dead cells, or repair of tissues. In plants cell division takes place in specific regions called meristems. These are found in growing regions including the root and shoot tips.

In animals, cell division is carried out by stem cells found in all tissues of the body. In mammals, growth and repair is mainly carried out by adult stem cells specific to the different tissues. These cells can only differentiate into cells of the same type of tissue they came from (e.g. blood stem cells can only differentiate into the different blood cells).

Brocken Inaglory CC 3.0

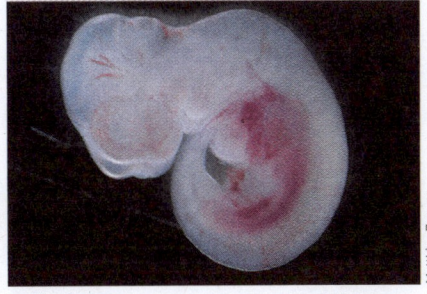

Matthias Zepper

Mitosis is vital in the repair and replacement of damaged cells. When you break a bone, or graze your skin, identical new cells are generated to repair the damage. Some organisms, e.g. sea stars, can generate new limbs if they are broken off (above).

Multicellular organisms develop from a single fertilized cell (zygote). Organisms, such as this 12 day old mouse embryo, grow by increasing their cell number. Cell growth is highly regulated and once the mouse reaches its adult size, growth stops.

The growing tip (meristem) is the site of mitosis in plants. The example above shows an onion root tip meristem. The root cap below the meristem protects the dividing cells.

1. The data below relates to the healing of the wound on the hand shown below. It shows the approximate area of the unhealed region of the wound over time:

Time (days)	Area unhealed (mm²)
0.02	350
0.66	350
1	340
2	252
12	171
13	135
17	72
18	64
21	16
30	0

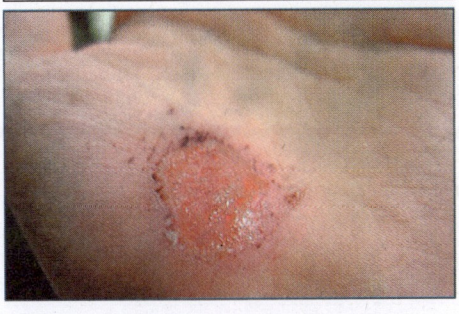

(a) Plot the data on the grid provided:

(b) Describe the trend in the wound size as shown by the data:

_____

_____

2. What feature of mitosis is central to its role in repairing and replacing damaged tissues in an adult organism?

_____

_____

D2.1 12 AHL

283 The Eukaryotic Cell Cycle

Key Idea: There are four major phases of the cell cycle characterized by growth, DNA replication, and mitosis.
The life cycle of a eukaryotic cell is called the cell cycle. It can be divided broadly into interphase, which is the time between cell divisions, and M phase, during which the cell divides. Aspects of the cell cycle can vary enormously between cells of the same organism. For example, intestinal cells divide around twice a day, while cells in the liver divide once a year, and those in muscle tissue do not divide at all. If any of these tissues is damaged, however, cell division increases rapidly until the damage is repaired. This variety of length in the cell cycle can be explained by the existence of regulatory mechanisms that are able to slow down or speed up the cell cycle in response to changing conditions.

Interphase

Cells spend most of their time in interphase. Interphase is divided into three stages:

▶ The first gap phase (G_1).

▶ The S-phase (S).

▶ The second gap phase (G_2).

During interphase the cell increases in size, carries out its normal activities, and replicates its DNA in preparation for cell division. Interphase is not a stage in mitosis.

Mitosis and cytokinesis (M-phase)

Mitosis and cytokinesis occur during M-phase. During mitosis, the cell nucleus (containing the replicated DNA) divides in two equal parts. Cytokinesis occurs at the end of M-phase. During cytokinesis the cell cytoplasm divides, and two new daughter cells are produced.

Exiting the cycle

Cells may exit the cycle in response to chemical cues and enter a resting phase, called G_0.

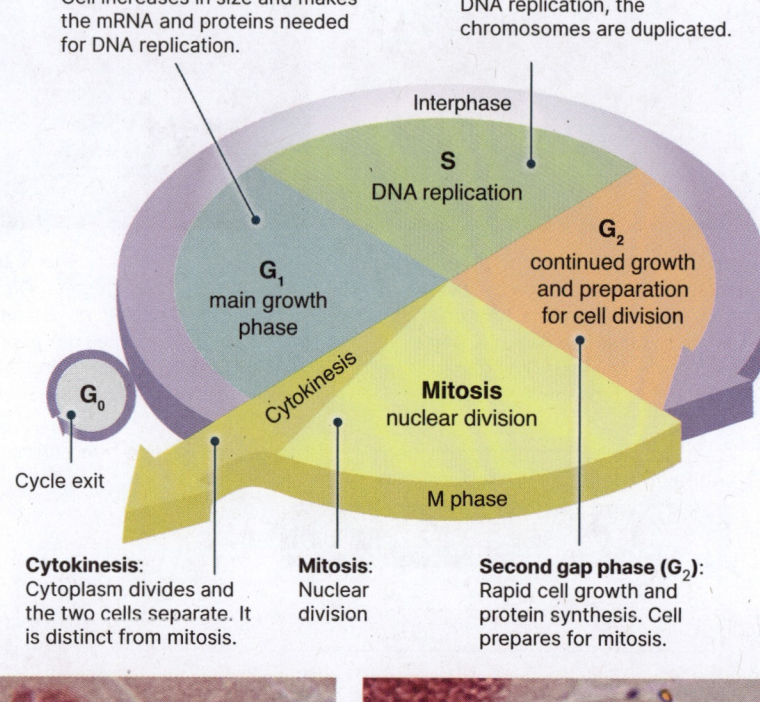

First gap phase (G_1): Cell increases in size and makes the mRNA and proteins needed for DNA replication.

S (synthesis) phase: DNA replication, the chromosomes are duplicated.

Interphase

S DNA replication

G_2 continued growth and preparation for cell division

G_1 main growth phase

G_0

Cycle exit

Cytokinesis

Mitosis nuclear division

M phase

Cytokinesis: Cytoplasm divides and the two cells separate. It is distinct from mitosis.

Mitosis: Nuclear division

Second gap phase (G_2): Rapid cell growth and protein synthesis. Cell prepares for mitosis.

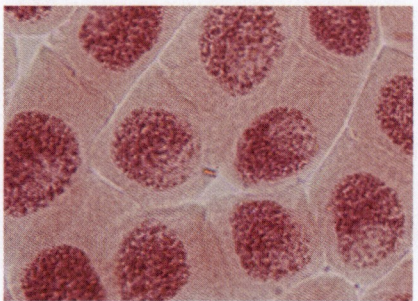

During the first stages of interphase, the cell grows and acquires the materials needed to undergo mitosis. It also prepares the nuclear material for separation by replicating it.

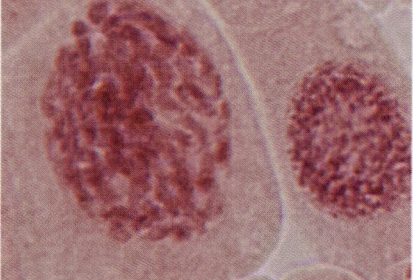

During the final, G_2 stage of interphase the cell grows more, duplicates its organelles and produces the proteins it needs to divide. The nuclear material begins to reorganize.

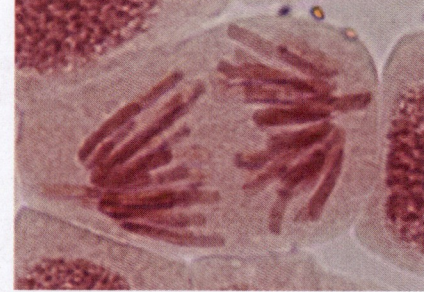

During mitosis the condensed chromosomes are separated. Mitosis is a highly organized process and the cell must pass checkpoints before it proceeds to the next phase.

1. In which stage of the cycle cell does the cell replicate the chloroplasts and/or mitochondria in the cell?

_____

2. What is the difference between S phase and M phases? _____

_____

_____

3. What is the difference between G_1 phase and G_2 phases? _____

_____

_____

4. What is the purpose of cytokinesis? _____

_____

AHL

D2.1
13, 14

©2024 **BIOZONE** International
ISBN: 978-1-99-101410-8
Photocopying prohibited

284 Regulating the Cell Cycle

Key Idea: The cell cycle is regulated to ensure cells only divide as and when required.

Mitosis is virtually the same for all eukaryotes but aspects of the cell cycle can vary enormously between species and even between cells of the same organism. For example, the length of the cell cycle varies between cells such as intestinal and liver cells. Intestinal cells divide around twice a day, while cells in the liver divide once a year. However, if these tissues are damaged, cell division increases rapidly until the damage is repaired. Variation in the length of the cell cycle are controlled by regulatory mechanisms that slow down or speed up the cell cycle in response to changing conditions.

Checkpoints during the cell cycle

▶ There are three checkpoints during the cell cycle. A checkpoint is a critical regulatory point in the cell cycle. At each checkpoint, a set of conditions determines whether or not the cell will continue into the next phase. For example, cell size is important in regulating whether or not the cell can pass through the G_1 checkpoint.

G_1 checkpoint
Pass this checkpoint if:
- Cell is large enough.
- Cell has enough nutrients.
- Signals from other cells have been received.

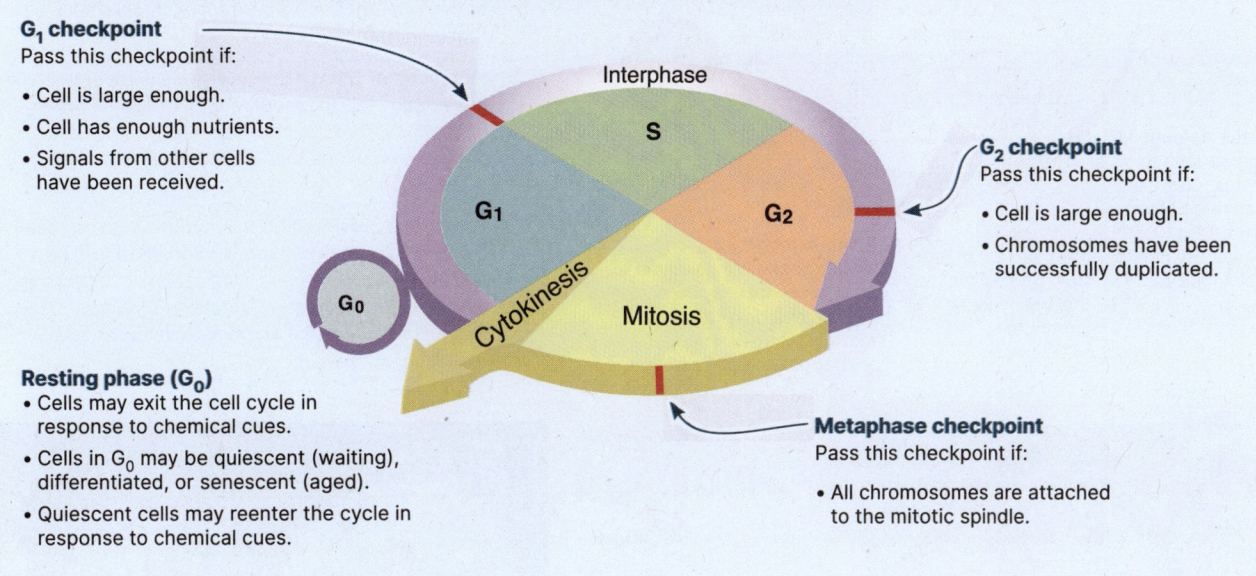

G_2 checkpoint
Pass this checkpoint if:
- Cell is large enough.
- Chromosomes have been successfully duplicated.

Resting phase (G_0)
- Cells may exit the cell cycle in response to chemical cues.
- Cells in G_0 may be quiescent (waiting), differentiated, or senescent (aged).
- Quiescent cells may reenter the cycle in response to chemical cues.

Metaphase checkpoint
Pass this checkpoint if:
- All chromosomes are attached to the mitotic spindle.

Cyclins help regulate the cell cycle

▶ Regulation of the cell cycle is important in detecting and repairing genetic damage, and preventing uncontrolled cell division. Tumours and cancers are the result of uncontrolled cell division.

▶ The cell cycle is regulated by proteins called cyclin dependent kinases (CDKs) which are always present in the cell, and cyclins, which vary in concentration (shown in the graph right).

▶ Cyclins control the progression of cells through the cell cycle by binding to and activating CDKs. CDKs phosphorylate other proteins to signal a cell is ready to proceed to the next stage in the cell cycle.

▶ The cyclin/CDK complexes are important at cell checkpoints, with different complexes playing different roles. For example, at the G_1 checkpoint the transcription factor E2F is required to produce cyclin E and so allow the cell to progress to S phase. However E2F is bound by the protein pRB. Cyclin D phosphorylates pRB, which releases E2F, allowing transcription and production of cyclin E. Cyclin D activity is suppressed by tumour suppressor proteins, e.g. p53, until other signalling molecules in the cell signal that conditions in the cell are favourable for DNA replication.

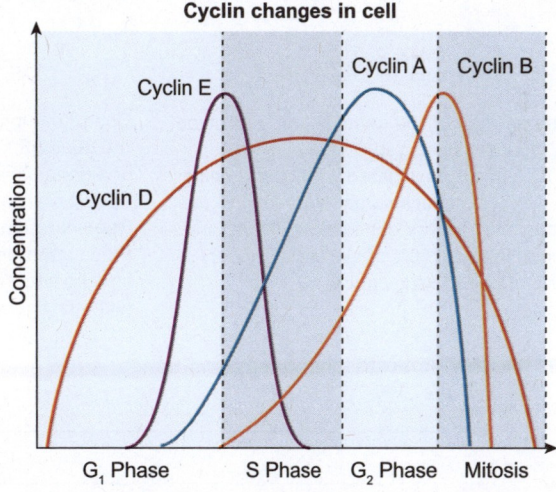

Cyclin changes in cell

1. Identify which cyclins are important in each of phases of the cell cycle: _____

_____

2. Why are checkpoints important in the cell cycle? _____

_____

_____

3. Suggest why the cell cycle is shorter in epithelial cells (such as intestinal cells) than in liver cells:

_____

_____

D2.1 1 - 15 AHL

285 Disruption to the Cell Cycle

Key Idea: Mutations in genes that control the cell cycle can result in uncontrolled cell division.

Cell division is a highly regulated process involving many genes. Some of these genes are specifically involved in suppressing tumour growth, e.g p53. However when these tumour suppressing genes themselves are damaged by mutations they are unable to regulate the cell cycle. Cell division becomes uncontrolled and cancerous tumours may develop. Carcinogens are agents that can produce mutations that can cause cancer.

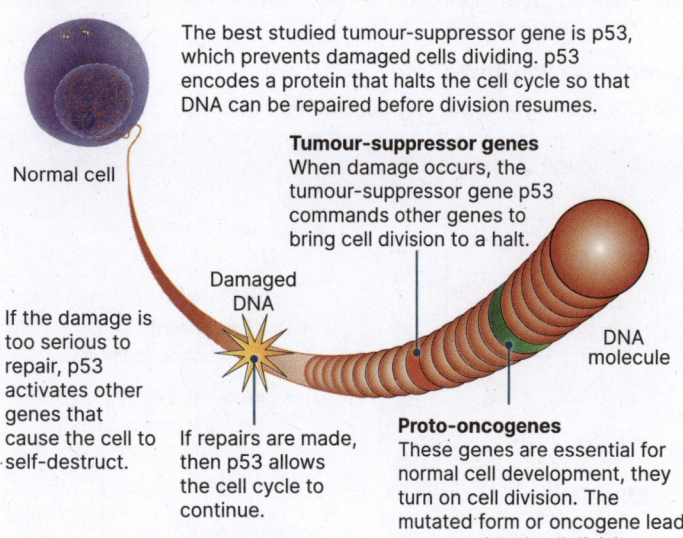

The best studied tumour-suppressor gene is p53, which prevents damaged cells dividing. p53 encodes a protein that halts the cell cycle so that DNA can be repaired before division resumes.

Normal cell

Tumour-suppressor genes
When damage occurs, the tumour-suppressor gene p53 commands other genes to bring cell division to a halt.

Damaged DNA

If the damage is too serious to repair, p53 activates other genes that cause the cell to self-destruct.

If repairs are made, then p53 allows the cell cycle to continue.

DNA molecule

Proto-oncogenes
These genes are essential for normal cell development, they turn on cell division. The mutated form or oncogene leads to unregulated cell division.

Cancer: cells out of control

Two types of gene are involved in controlling the cell cycle: **proto-oncogenes**, which start the cell division process and **tumour-suppressor genes**, which switch off cell division. In their normal form, both work together to perform vital tasks such as repairing defective cells and replacing dead ones.

Mutations (a change in the DNA sequence) in these genes can stop them operating normally. Proto-oncogenes, through mutation, can give rise to **oncogenes**; genes that lead to uncontrollable cell division.

Cancerous cells result from changes in the genes controlling normal cell growth and division. The resulting cells become immortal and no longer carry out their functional role. Mutations to tumour-suppressor genes initiate most human cancers.

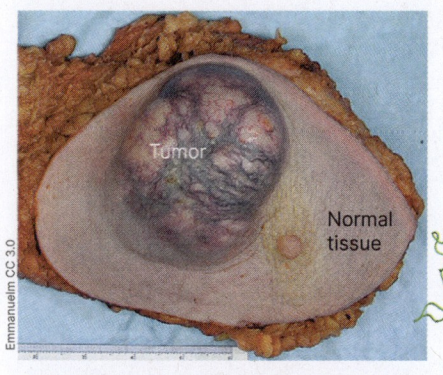

Emmanuelm CC 3.0

The product of the gene BRCA1 is involved in repairing damaged DNA and BRCA1 deficiency is associated with abnormalities in cell cycle checkpoints. Mutations to this gene and another gene called BRCA2 are found in about 10% of all breast cancers and 15% of ovarian cancers.

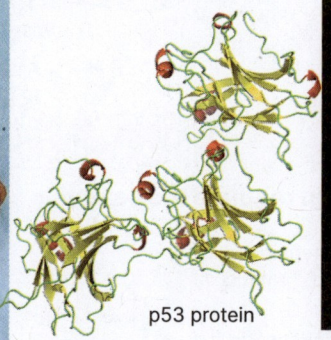

p53 protein

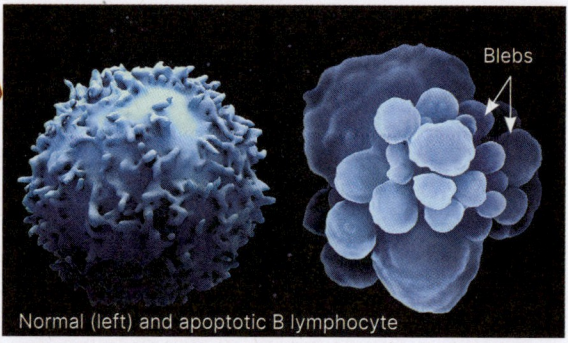

Blebs

Normal (left) and apoptotic B lymphocyte

Felix Hicks

One of the most important proteins in regulating the cell cycle is the protein produced by the gene p53. The p53 tumor-suppressor protein helps regulate the cell cycle, apoptosis, and genomic stability. Mutations to the p53 gene are found in about 50% of cancers. Apoptosis is a controlled process that involves cell shrinkage, blebbing (above), and DNA fragmentation. Apoptosis removes damaged or abnormal cells before they can multiply. When apoptosis malfunctions it can cause disease, including cancer. When cell cycle checkpoints fail, the normal rate of apoptosis falls. This allows a damaged cell to divide without regulation.

1. (a) How do proto-oncogenes and tumour suppresor genes normally regulate the cell cycle?

_____

_____

_____

(b) Explain how the normal controls over the cell cycle can be lost: _____

_____

_____

(c) How can these failures result in cancer? _____

_____

_____

_____

AHL

D2.1
15

©2024 **BIOZONE** International
ISBN: 978-1-99-101410-8
Photocopying prohibited

286 Determining the Rate of Growth

Key Idea: The stages of mitosis can be recognized by the organisation of the cell and chromosomes.

Cancerous tumours develop in stages, rather than just appearing. A tumour may first form, cause by cell proliferation. Cells in the tumour may then accumulate more mutations, becoming cancerous. Cell proliferation can be measured using the mitotic index, which measures the ratio of cells in mitosis to the number of cells counted. The mitotic index can be used to diagnose cancer because cancerous cells divide very quickly.

How tumours spread

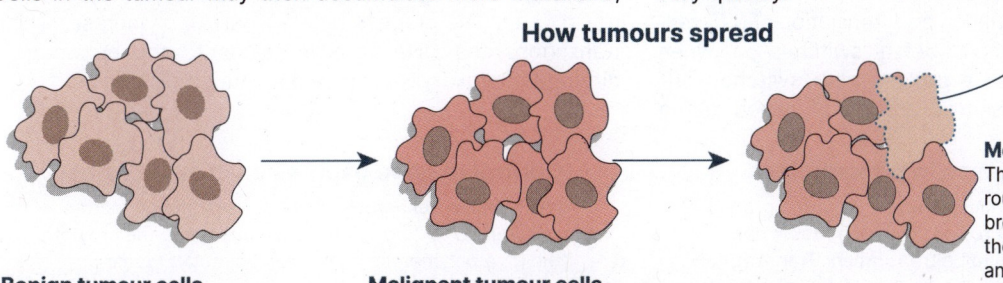

Benign tumour cells
Mutations cause the formation of a benign (harmless) tumour. The formation of new cells is matched by cell death. These cells do not spread.

Malignant tumour cells
More mutations may cause the cells to become malignant (harmful) forming a primary tumour. Changes to the cell chemistry encourage capillary formation. New capillaries grow into the tumour, providing it with nutrients so it can grow rapidly.

Metastasis
The new capillaries provide a route for the malignant cells to break away (metastasize) from the primary (original) tumour and travel to other parts of the body where they start new cancers (secondary tumours).

Measuring the mitotic index

Mitotic index = Number of cells in mitosis ÷ Total number of cells

Using a mitotic index it is possible to compare the rates of mitosis in different populations of cells, and also to compare these rate against an expected rate. The example below shows tissue from onion roots tips as these are simpler to analyse than cells from animal tissue. The mitotic index may be shown as a ratio e.g. 0.2, or as a percentage e.g. 20%.

1. The light micrograph (right) shows a section of cells in an onion root tip. Some of these cells are undergoing mitosis. Use the equation to calculate the mitotic index for the sample of cells shown.

    (a) Total number of cells: _____

    (b) Cells in mitosis: _____

    (c) Mitotic index: _____

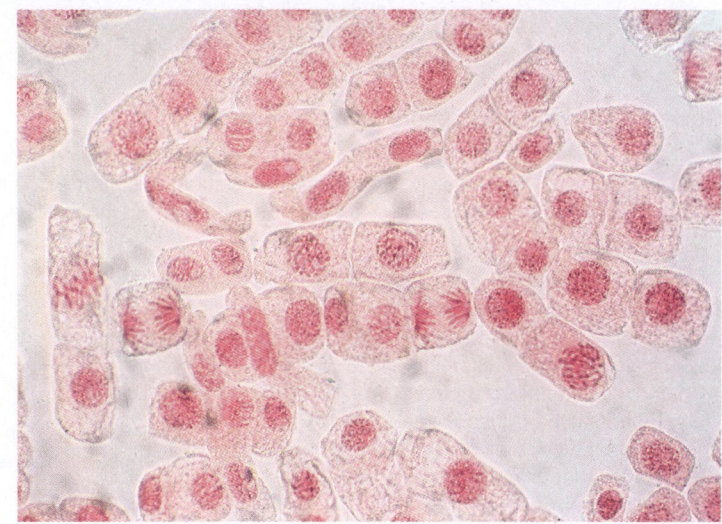

2. How would finding the mitotic index of a population of cells help identify possible cancerous tissue?

    _____

    _____

    _____

    _____

3. The graph on the right shows the effect of effluent from the dying industry on the mitotic index of onion root tip cells.

    (a) What is the effect of effluent concentration on the rate of mitosis in onion root tip cells?

    _____

    _____

    _____

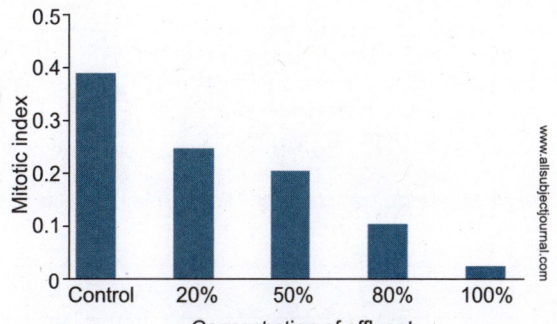

Concentration of effluent

www.allsubjectjournal.com

    (b) What would be the effect of this on the onion root tip cells and the onion as a whole?

    _____

    _____

287 Gene Expression and Regulation

Key Idea: Gene expression is regulated by transcription factors, which bind to promoters and enhancers on DNA. Recall that gene expression includes the process of transcription and translation, which interprets the genetic information in DNA to produce proteins. Different proteins are produced by the transcription and translation of different genes. A gene is referred to as being "switched on" when it is being used to produce a protein, and "switched off" when it is not. The switching on and off of genes during the development of a cell results in the cell's phenotype, i.e. what kind of cell it is. Transcription, the first stage in gene expression is, in eukaryotes, regulated by transcription factors, which bind to promoters, and enhancers found on the DNA. The promoter is upstream of the DNA to be transcribed. The process by which RNA polymerase, which transcribes the DNA to RNA, binds to the promoter is different in prokaryotes and eukaryotes (below).

Control of gene expression in prokaryotes

In prokaryotes, the RNA polymerase and associated proteins bind directly to the promoter region. The promoter is normally very close to the DNA region to be transcribed. Transcription of the structural gene is controlled by a regulator gene, which produces a repressor molecule that may bind to the operator and block transcription. The promoter, operator, and DNA to be transcribed are called an operon.

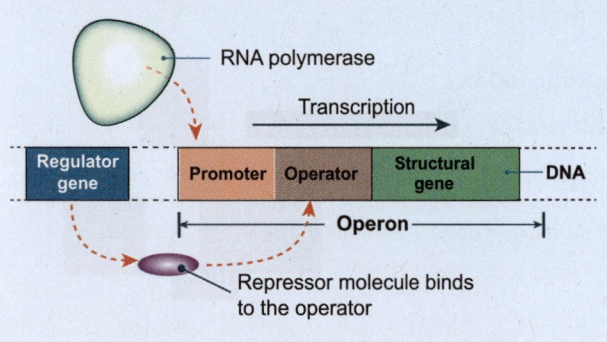

Control of gene expression in eukaryotes

Recall that in eukaryotes, several transcription factors are required to bind the RNA polymerase to the promoter (the RNA polymerase cannot bind directly). Some of these transcription factors may also bind to the enhancer region which may be quite distant from the promoter. The transcription factors cause the promoter and enhancer to come together, which initiates transcription.

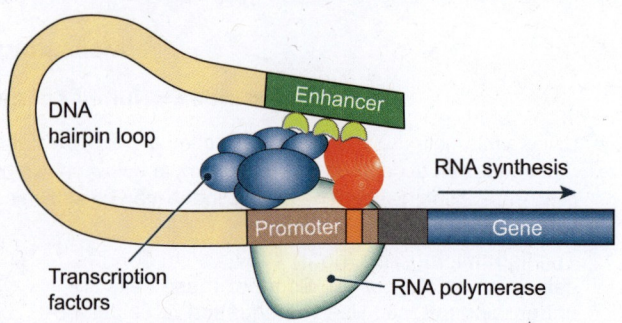

Gene expression and cellular differentiation

It is important for the cell to be able to regulate the mRNAs produced by transcription. Once an mRNA has been produced a cell must decide if it is translated, stored for later translation, or degraded. This must also occur after translation. mRNA is broken down by enzymes called nucleases. The differences in mRNA production between cells leads to cellular differentiation.

▸ Cell differentiation is an output of the transcription and translation of genes in the genome. The mRNAs produced from transcribed genes are known as the transcriptome. From the translated transcriptome comes the proteome, the proteins used by the cell that ultimately result in the cell's phenotype.

The zygote (fertilized egg) is able to form all the other cells in the body.

Transcription factors determine which genes will be transcribed and expressed, producing different changes in each generation of cells.

As genes are switched on and off the cell takes on a new phenotype.

As the cells differentiate their future pathways become more and more limited. By using different transcription factors at different times a large variety of different cells is produced.

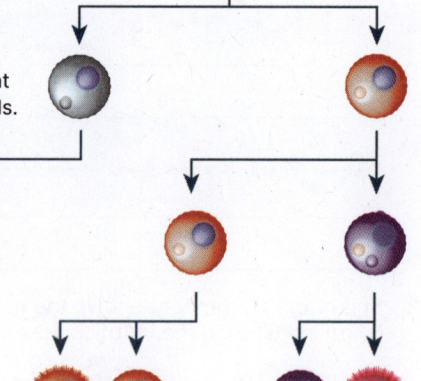

1. What role do transcription factors play in cellular differentiation? _____

_____

_____

2. Why would a cell need to degrade mRNA after it has been used? _____

_____

_____

©2024 **BIOZONE** International
ISBN: 978-1-99-101410-8
Photocopying prohibited

288 Epigenetics

Key Idea: The mechanisms by which the environment modifies the expression of genes are often epigenetic.

Cellular differentiation and phenotypes are the result of gene expression which is in part regulated by the genes themselves. However, phenotype and therefore gene expression, is also regulated by the environment. The environment can directly affect gene expression, such as colour-pointing in Siamese cats in which pigment production only occurs in cooler parts of the body (e.g. the ears). It can also affect gene expression in more subtle ways through epigenetic regulation. Epi- means 'on top of' or 'extra to'. Thus epigenetic factors are those external to the gene itself (e.g. chemical tags) that influence how that gene is expressed. Epigenetic regulation

is achieved by modifying the way the DNA is packaged and its availability to be transcribed. The packaging of DNA is affected by histone modification and DNA methylation. These modifications regulate gene expression either by making the chromatin (the DNA and proteins strands making up the chromatids) pack together tightly or more loosely. This affects whether or not RNA polymerase can attach to the DNA and transcribe it into messenger RNA and eventually translate it into proteins. Epigenetic tags help to regulate gene expression as the cell differentiates, and record a cell's history as it experiences different, changing environments. The DNA itself is not affected or changed.

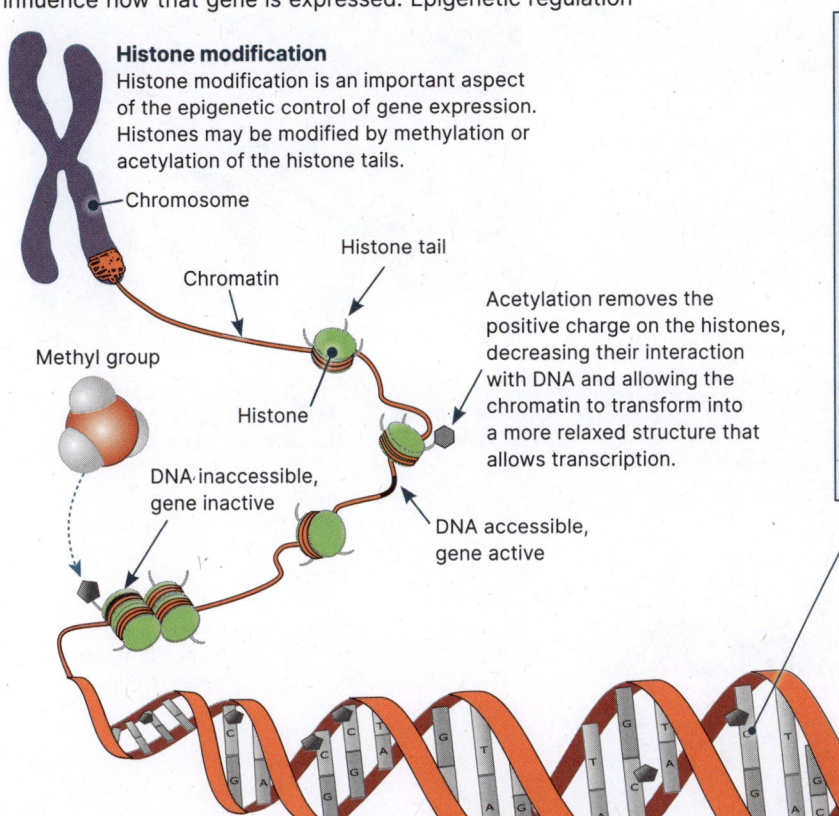

Histone modification

Histone modification is an important aspect of the epigenetic control of gene expression. Histones may be modified by methylation or acetylation of the histone tails.

- Chromosome
- Histone tail
- Chromatin
- Methyl group
- Histone
- DNA inaccessible, gene inactive
- DNA accessible, gene active

Acetylation removes the positive charge on the histones, decreasing their interaction with DNA and allowing the chromatin to transform into a more relaxed structure that allows transcription.

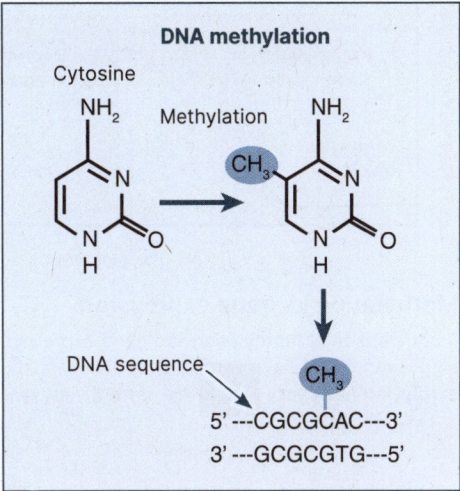

DNA methylation

Cytosine

NH_2

Methylation

CH_3

DNA sequence

5' ---CGCGCAC---3'
3' ---GCGCGTG---5'

Methyl groups bind only to cytosine. Cytosine methylation may either prevent the actual binding of transcription factors or cause the chromatin to bind tightly so that genes cannot be transcribed.

Environmental effects and epigentics

The environment can affect gene expression. Studies have shown stressful situations, starvation, and air pollution can all alter the way genes are expressed, normally through DNA methylation.

▶ Fresno county, California USA, has one of the highest levels of particulate air pollution in the United States. Not far from it, Palo Alto county has much less air pollution. Studies of asthma in children in the Fresno and Palo Alto counties revealed significant differences in the frequency of asthma between the two counties, with more asthma cases in Fresno. In both cases it was found that methylation of the gene Foxp3 was higher in asthma sufferers. However, the children with the most severe cases of asthma were from Fresno and also had the greatest amount of Foxp3 methylation. Studies in mice have shown that these epigenetic changes are heritable, with the offspring of mice exposed to air pollution developing asthma even when not exposed to air pollution themselves.

▶ The destruction of New York's Twin Towers on September 11, 2001, traumatized thousands of people. In those thousands were 1700 pregnant women. Some of them suffered severe, post-traumatic stress disorder (PTSD), others did not. Studies on the mothers who developed PTSD found very low levels of the stress-related hormone cortisol in their saliva. Low levels of cortisol can be caused by very high stress, as the body uses it faster than it can be produced. The children of these mothers also had much lower levels of cortisol than those whose mothers had not suffered PTSD, particularly those who had been in the third trimester of pregnancy, indicating a developmental response to severe stress.

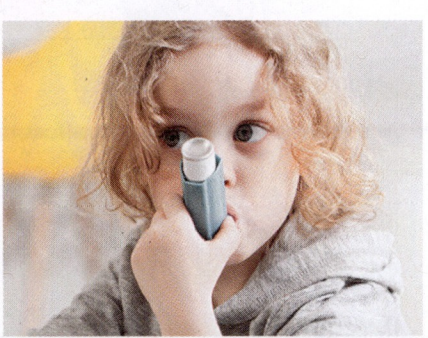

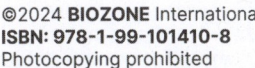

 **D2.2** 4, 6-10 **AHL**

Methylation and gene expression

▸ Gene expression changes as an organism develops. During the development of the embryo there are many genes that are switched on and off as development of tissues and organs proceeds. Some of this control is achieved by DNA methylation. Enzymes can add or remove methyl groups from cytosine bases and so activate or silence genes.

▸ Often methylation is affected by changes in the environment and this provides the developing organism with a rapid response mechanism.

▸ DNA methylation also differs in eggs and sperm (below). Genes received from the father have more methylation than those received from the mother. This results in quite different development in offspring (right).

Methylation in mammalian development

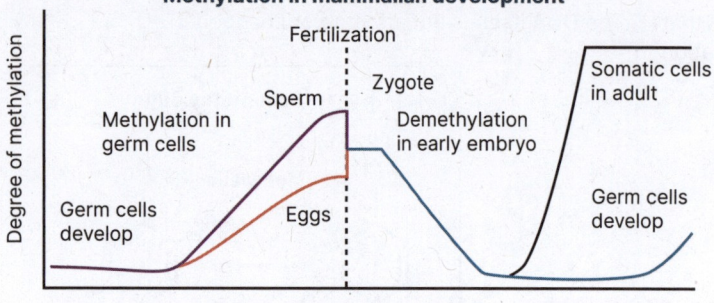

Methylation vs gene expression

A comparison of methylation and gene expression finds that highly expressed genes have very little methylation while genes that are not expressed have very high levels of methylation.

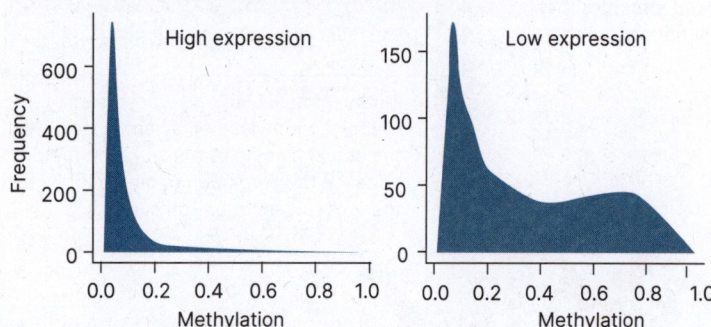

Methylation and imprinted genes

Genomic imprinting is a phenomenon in which the pattern of gene expression is different depending on whether the gene comes from the mother or the father. Imprinted genes are silenced by methylation and histone modification. A gene inherited from the father may be silenced while the gene inherited from the mother may be active or vice versa. Evidence of this is seen in two human genetic disorders, Angelman syndrome and Prader-Willi syndrome. Both are caused by the same mutation; a specific deletion on chromosome 15. Which syndrome is expressed depends on whether the mutation occurs on the maternal or paternal chromosome.

A liger

The effect of genomic imprinting can be seen in other mammals. Ligers (a cross between a male lion and a female tiger) although not occurring in the wild, are the biggest of the big cats. However, a tigon (a cross between a female lion and a male tiger) is no bigger than a normal lion. It is thought this difference in phenotype is due to the male lion carrying imprinted genes that result in larger offspring which are normally counteracted by genes from the female. Similarly the differences between a mule (male donkey + female horse) and a hinny (female donkey + male horse) may be to do with genomic imprinting.

1. (a) Describe the effect of histone modification and DNA methylation on DNA packaging: _____

_____

_____

_____

(b) How do these processes affect transcription of the DNA? _____

_____

_____

_____

2. Use examples to explain how genes are affected by the environment and how these effects can be heritable:

_____

_____

_____

_____

_____

_____

©2024 **BIOZONE** International
ISBN: 978-1-99-101410-8
Photocopying prohibited

3. When a zygote forms at fertilization, most of the epigenetic tags are erased so that cells return to a genetic "blank slate", ready for development to begin. However, some epigenetic tags are retained and inherited. Why do you think it might be advantageous to inherit some epigenetic tags from a parent?

_____

_____

_____

4. Prader-Willi syndrome is caused when a mutated gene on chromosome 15 is inherited from the father. How does this tell us that the mother must therefore have donated the imprinted gene?

_____

_____

_____

_____

Twin studies

▸ Monozygotic twin studies can provide a lot of information about how the environment affects gene expression. The studies are often done when identical twins have been separated at birth (usually because one or both of them are adopted out). Their similarities and differences can then be studied to assess how much the environment influenced their development.

Twins in space

▸ In 2015, NASA astronaut Scott Kelly blasted into space for a year long stay on the International Space Station. His identical twin brother Mark remained on Earth. This gave NASA a chance to study the real effects of space travel on the human body. Importantly, the gene expression of the men could be measured before and after Scott went to space.

▸ It was found that six months after Scott's return, 7% of his genes had not returned to their normal level of gene expression. Also, although there was no decrease in Scott's cognitive abilities, there was a decrease in his speed and accuracy until his readjustment to Earth gravity. The space environment had altered Scott Kelly's gene expression compared to his identical twin Mark Kelly.

Genes vs environment

▸ Just how much do genes and environment play a role in a person's development? There are many examples of twins showing both significant similarities and differences in their development. While some monozygotic twins develop in very similar ways, e.g. same physical skills or health, others do not. There are numerous examples of one of the twins developing cancer while the other never shows any symptoms. Twin studies have also shown that the relative chance of death from disease such as coronary heart disease (graphs right), is more likely to be influenced by genes at a young age, but become more influenced by the environment as age increases.

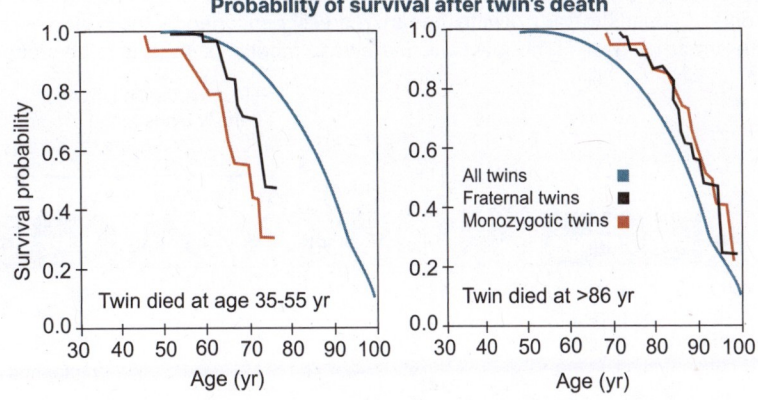

Probability of survival after twin's death

5. How might twin studies help the study of gene-environment responses? _____

_____

_____

_____

6. How does the effect of the environment on a person's development change over the lifetime of a person?

_____

_____

_____

_____

©2024 BIOZONE International
ISBN: 978-1-99-101410-8
Photocopying prohibited

289 Gene Regulation

Key Idea: External factors play an important role in the regulation of gene expression.

Tryptophan is an amino acid used by cells to synthesise proteins. Tryptophan (Trp) is an essential amino acid for mammals meaning it cannot be synthesised and must be obtained from food. However, free-living bacteria, such as *E. coli*, are able to synthesise tryptophan and regulate its production via the trp operon, a group of structural genes encoding enzymes and their regulatory sequences. When tryptophan is available genes responsible for its production are repressed. In mammals genes responsible for red blood cell (RBO) production are regulated by oxygen levels in the blood. Low oxygen levels promote RBC production. Higher oxygen levels reduce RBC production to replacement levels.

Prokaryotic genes occur as operons

A number of structural genes encoding the enzymes for a metabolic pathway are under the control of the same regulatory elements.

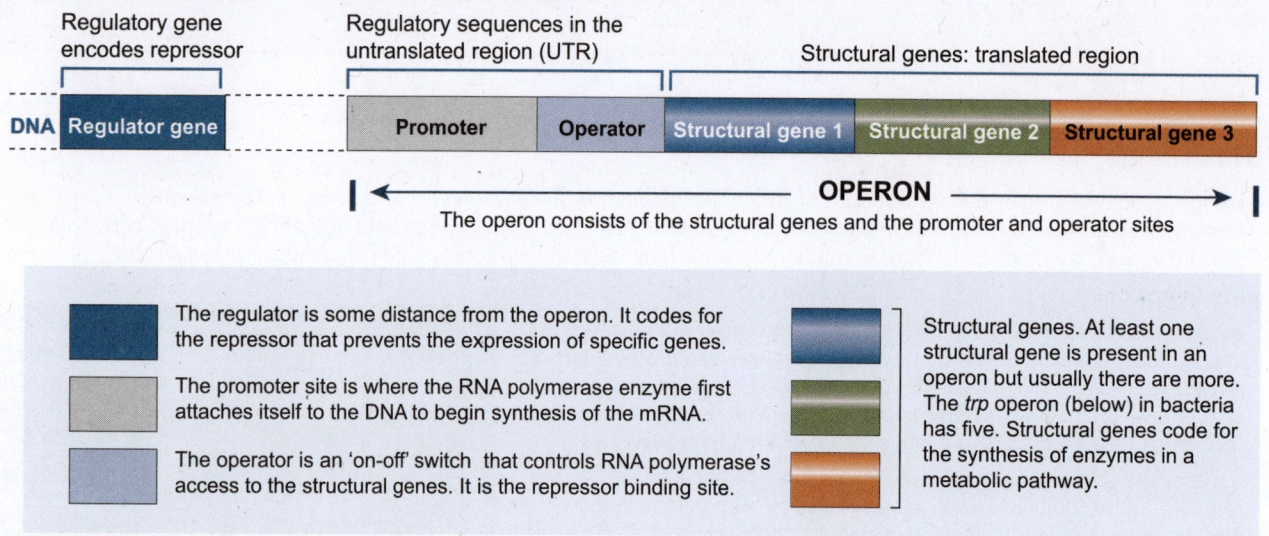

The regulator is some distance from the operon. It codes for the repressor that prevents the expression of specific genes.

The promoter site is where the RNA polymerase enzyme first attaches itself to the DNA to begin synthesis of the mRNA.

The operator is an 'on-off' switch that controls RNA polymerase's access to the structural genes. It is the repressor binding site.

Structural genes. At least one structural gene is present in an operon but usually there are more. The *trp* operon (below) in bacteria has five. Structural genes code for the synthesis of enzymes in a metabolic pathway.

The *trp* operon: a repressible operon

Transcription of the five structural genes in the trp operon is normally on. To stop transcription, the genes are switched off (repressed) by tryptophan itself, so this operon is called repressible. The genes trpA and trpB produce the enzyme tryptophan synthetase, which catalyses the formation of tryptophan from serine and trpC's product. When tryptophan is present in excess, some of it binds to and activates the trp repressor encoded by the regulator gene. The active repressor then blocks transcription of the structural genes. This mechanism allows tryptophan synthesis to be stopped when it is plentiful (in the cell or environment).

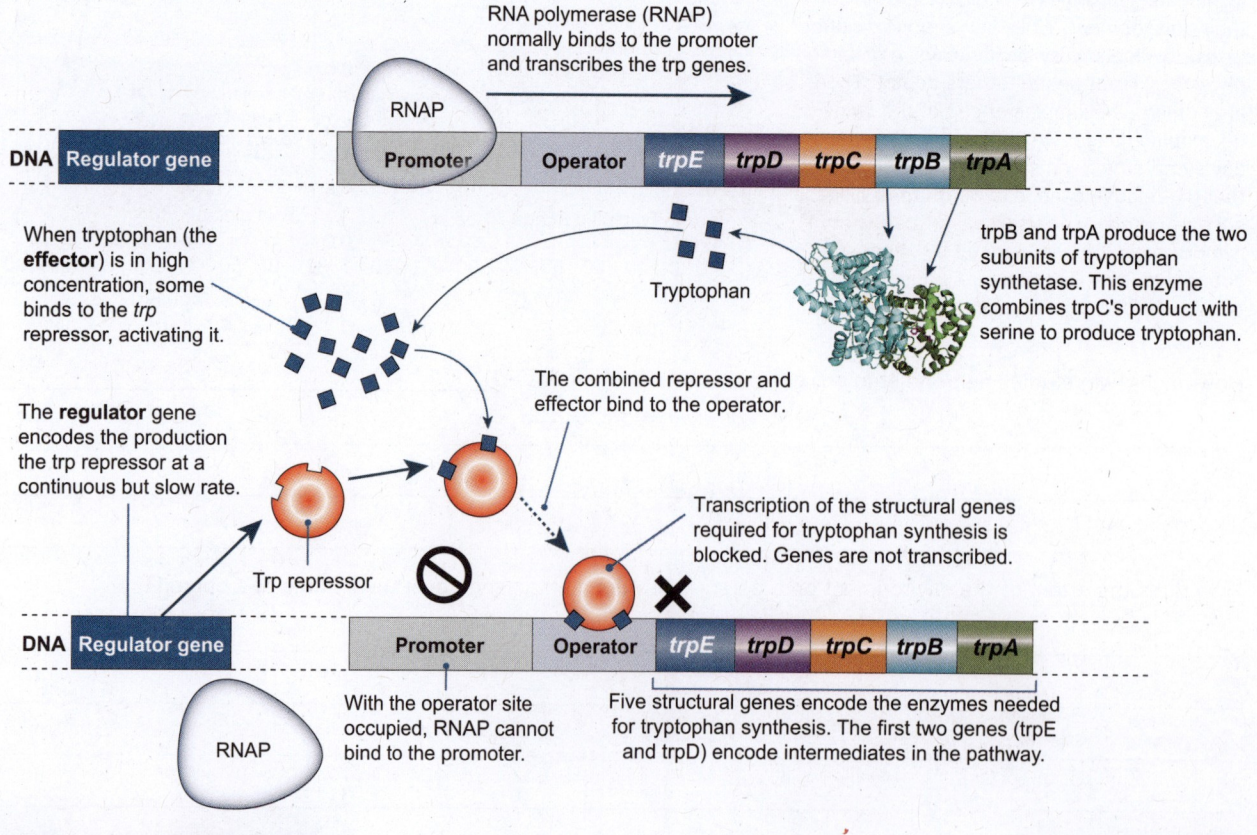

©2024 **BIOZONE** International
ISBN: 978-1-99-101410-8
Photocopying prohibited

Hormones and gene expression

▸ Recall that hormones play an important role in gene expression. Hormones bind to receptors (extracellular or intracellular) and initiate cellular mechanisms that lead to gene expression. For example, oestrogen binds to the oestrogen receptor, which acts as a transcription factor which binds to DNA to alter gene transcription.

▸ Erythropoietin (EPO) is a hormone involved in the production of red blood cells. Its production increases when oxygen is low. It then causes the increase in production of red blood cells in red bone marrow (below).

Erythropoietin and red blood cell production

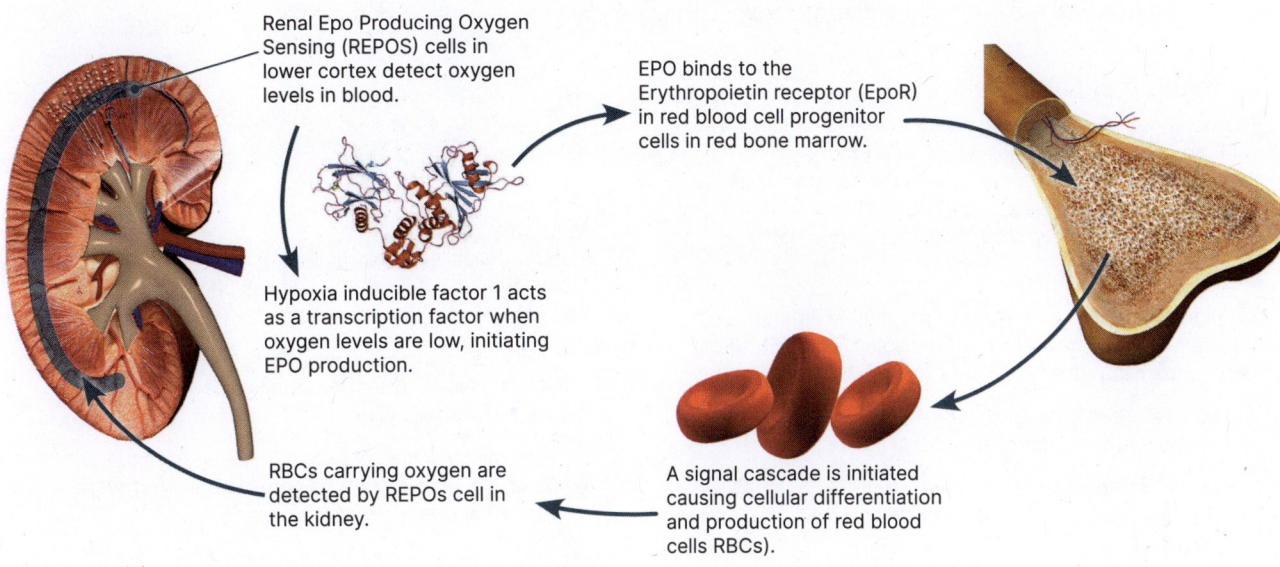

Renal Epo Producing Oxygen Sensing (REPOS) cells in lower cortex detect oxygen levels in blood.

EPO binds to the Erythropoietin receptor (EpoR) in red blood cell progenitor cells in red bone marrow.

Hypoxia inducible factor 1 acts as a transcription factor when oxygen levels are low, initiating EPO production.

RBCs carrying oxygen are detected by REPOs cell in the kidney.

A signal cascade is initiated causing cellular differentiation and production of red blood cells RBCs).

1. Outline the role of each of the following components of an operon:

   (a) Promoter: _____

   (b) Operator: _____

   (c) Structural genes: _____

2. What is the role of the repressor molecule in operon function? _____

3. (a) In the trp operon, is the repressor normally attached or not attached to the operator? _____

   (b) What is the role of the effector in the production of tryptophan? _____

4. Describe how the trp operon is a self regulating system (a negative feedback loop): _____

5. Explain how environmental factors regulate the production of red blood cells: _____

6. Why are environmental controls important in gene regulation? _____

290 Solutions

Key Idea: The concentration of solutes inside and outside of a cell affect the movement of water in these solutes.

Recall that water is a solvent, it is able to dissolve many substances, forming a solution. Water's polar nature allows it to form hydrogen bonds with or attract ions of soluble substances (e.g. most salts) and so dissolve them. Recall also that osmosis is the diffusion of water molecules from regions of lower solute concentration to regions of higher solute concentration across a partially permeable membrane. This movement of water is important in cells for maintaining their shape. Too much water entering a cell may result in it bursting. Too much water leaving a cell will see it shrink and lose shape. These two principles play important roles in the movement of water in and out of cells.

Osmotic potential

The presence of solutes (dissolved substances) in a solution increases the tendency of water to move into that solution. This tendency is called the osmotic potential or osmotic pressure. The more total dissolved solutes a solution contains, the greater its osmotic potential.

Describing solutions

Solutions separated by a partially permeable membrane are often described in terms of their solute concentrations relative to one another.

Isotonic solution: Having the same solute concentration relative to another solution (e.g. the cell's contents).

Hypotonic solution: Having a lower solute concentration relative to another solution.

Hypertonic solution: Having a higher solute concentration relative to another solution.

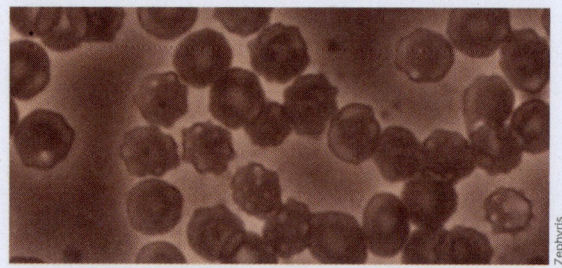

The red blood cells above were placed into a hypertonic solution. As a result, the cells have lost water and have begun to shrink, losing their usual discoid shape.

Aim
To demonstrate the effect of solute concentration on osmosis.

Method
Two model cells of dialysis tubing were filled with 5 cm^3 each of a 1 molL^{-1} sucrose solution and a 0.1 molL^{-1} sucrose solution.

The dialysis tubing cells were tied off and weighed to 2 decimal places. They were then placed in separate beakers of distilled water for 10 minutes.

After 10 minutes the cells were removed from the distilled water and blotted dry with a paper towel. They were reweighed and their masses recorded.

The experiment was carried out three times. Results are shown below.

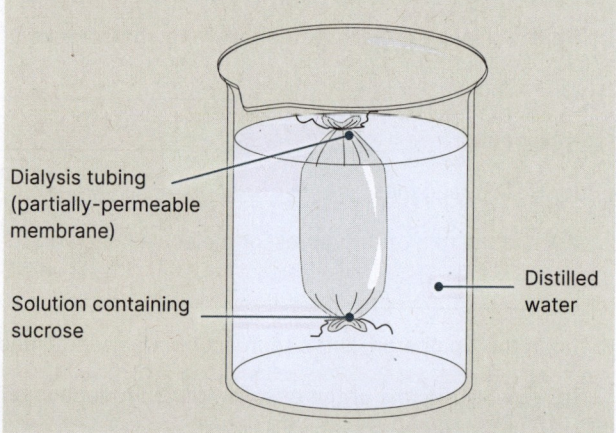

Dialysis tubing (partially-permeable membrane)

Solution containing sucrose

Distilled water

1 molL^{-1}				
Cell	Final mass (g)	Initial mass(g)	change (g)	% change
1	11.22	10.39		
2	11.23	10.33		
3	12.03	10.98		
Mean				

0.1 molL^{-1}				
Cell	Final mass (g)	Initial mass(g)	change (g)	% change
1	10.44	10.35		
2	10.56	10.47		
3	10.64	10.55		
Mean				

1. (a) Calculate the mean percentage change in mass for the two model cells in the table above:

   (b) Explain the results of the experiment: _____

   _____

   _____

   _____

   (c) What would you expect to happen if a third cell was used with a sucrose concentration of 0.01 molL^{-1}?

   _____

©2024 **BIOZONE** International
ISBN: 978-1-99-101410-8
Photocopying prohibited

 D2.3 1-3, 5-7  79

(d) What would you expect to happen if the model cell was to have distilled water in it and the beaker was to have a sucrose solution of 1 molL^{-1}?

_____

_____

(e) What would you expect to happen if the model cell had a sucrose concentration of 2 molL^{-1} and the beaker a sucrose solution of 1 molL^{-1}?

_____

_____

Tonicity and cells

Cells do not always find themselves in isotonic solutions and so need mechanisms to deal with water constantly entering or leaving the cell and to keep their shape. For cells without cell walls, too much water entering the cell can be catastrophic, bursting the cell.

Plants

When the watery contents of a plant cell push against the cell wall they create turgor (tightness) which helps to provide support for the plant body. When cells lose water, there is a loss of cell turgor and the plant will wilt. Complete loss of turgor from a cell is called plasmolysis and is irreversible.

Medical tissue

Osmosis is important when handling body tissues for medical transport or preparation. The tissue must be bathed in a solution with an osmolarity (a measure of solute concentration) equal to the tissue's to avoid a loss or gain of fluid in the tissue.

Unicellular organisms

Unicellular organisms regulate the water in their cells using a contractile vacuole (CV). Water moves from the cytoplasm into the CV which contracts and expels the water outside of the cell. The rate of expulsion depends on the osmolarity of the environment.

Wilted plant (cells have lost turgor)

Plant cells are turgid

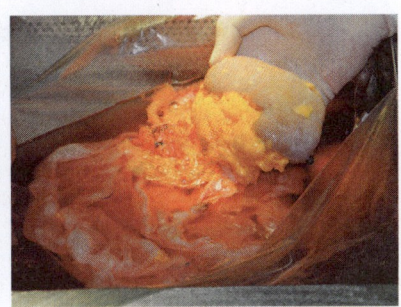

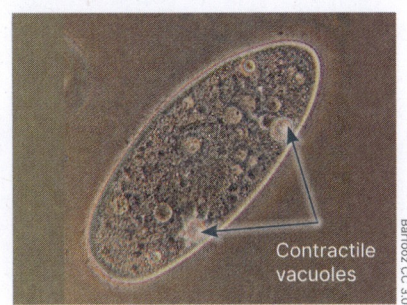
Contractile vacuoles

2. The diagrams below depict what happens when a red blood cell is placed into three solutions with differing concentrations of solutes. Describe the tonicity of the solution (in relation to the cell) and describe what is happening:

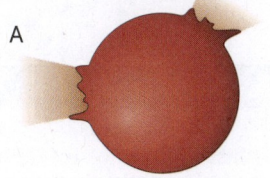
A

B

C

(a) _____

_____

_____

(b) _____

_____

_____

(c) _____

_____

_____

3. The rectangles below represent the cell walls of two plant cells. The cell on the left in placed in a hypotonic solution, while the cell on the right is placed in a hypertonic solution.

(a) Draw the cell membrane to show which cell would be in turgor and which would have undergone plasmolysis:

(b) Add arrows and labels to the cells to show the movement of water into or out of the cells:

Plant cell in hypotonic solution

Plant cell in hypertonic solution

4. Would you expect an intravenous solution administered to a patient to be hypertonic, isotonic, or hypotonic?

_____

291 Estimating Osmolarity of Cells

Key Idea: Calculating loss or gain of mass in tissues allows us to determine the osmolarity of the tissue's cells.

The osmolarity (a measure of solute concentration) of a cell or tissue can be estimated by placing part of the cell or tissue into a series of solutions of known concentration and observing if the tissue loses (hypertonic solution) or gains (hypotonic solution) water. The solution in which the tissue remains unchanged indicates the osmolarity of the tissue.

Investigation 14.2 Estimating osmolarity

See appendix for equipment list.

1. Prepare 6 beakers of sucrose ($C_{12}H_{22}O_{11}$, table sugar) solution with the concentrations of 0.0 (distilled water), 0.1, 0.2, 0.3, 0.4, and 0.5 mol/L of sucrose (0, 34.2 g , 68.5 g, 102.6 g, 136.9 g and 171.1 g per litre). Label the beakers so that they can be easily identified at the end of the experiment.

2. Peel a potato and cut it into 36 identical cubes 1 cm³ (1 cm x 1 cm x 1 cm) or use a cork borer to produce 36 identical cylinders of potato. Pat the potato cubes dry with a paper towel.

3. Weigh each cube and record their mass in the table below under initial mass. Place the cubes in the beaker of distilled water.

4. Repeat step 3 with the other 33 potato cubes and concentrations. Make sure you identify each beaker so the cubes can be weighed at the end of the experiment. Each beaker should have 6 potato cubes in it.

5. Leave the potato cubes in the solutions for at least 40 minutes (or up to 24 hours).

6. Remove the potato cubes from the distilled water and pat dry with a paper towel. Weigh each cube and record the mass in the table below under final mass.

7. Repeat for all the other concentrations of sucrose.

8. Calculate the change in mass (if any) for all the concentrations (final mass - initial mass) for each cube.

9. Calculate the % change for each cube. This removes any error based on the masses of the potato cubes not being identical (larger cubes have a larger change of mass). % change = change in mass ÷ initial mass x 100.

10. Calculate the mean % change for each sucrose concentration (mean mass change ÷ mean initial mass x 100).

11. Calculate the standard deviation and standard error for % change for each concentration. Review Activity 228 on standard deviation and standard error.

12. Plot a graph of % change vs concentration. Add error bars showing the standard error for the % change for each sucrose concentration.

Sucrose concentration: 0.0 molL⁻¹

Final mass (g)	Initial mass (g)	Mass change (g)	% Change
Mean			

Sucrose concentration: 0.1 molL⁻¹

Final mass (g)	Initial mass (g)	Mass change (g)	% Change
Mean			

Sucrose concentration: 0.2 molL⁻¹

Final mass (g)	Initial mass (g)	Mass change (g)	% Change
Mean			

Sucrose concentration: 0.3 molL⁻¹

Final mass (g)	Initial mass (g)	Mass change (g)	% Change
Mean			

Sucrose concentration: 0.4 molL⁻¹

Final mass (g)	Initial mass (g)	Mass change (g)	% Change
Mean			

Sucrose concentration: 0.5 molL⁻¹

Final mass (g)	Initial mass (g)	Mass change (g)	% Change
Mean			

AOS

D2.3
4

©2024 **BIOZONE** International
ISBN: 978-1-99-101410-8

1. Plot a graph of mean % change vs sucrose concentration on the grid right:

2. Calculate the standard deviation and standard error for the % change for each sucrose concentration. Using a spreadsheet will speed this up and reduce errors, see Activity 228). Use the space below for working. Add these to the graph:

3. Use the graph to estimate the concentration of the isotonic solution (the point where there is no change in mass):

_____

4. Which of the solutions are hypotonic? Which are hypertonic? _____

_____

_____

5. Describe the movement of water when a cell is placed in each of the following solutions:

(a) Hypertonic: _____

_____

(b) Isotonic: _____

_____

(c) Hypotonic: _____

_____

292 Water Relations in Plants

Key Idea: Solutions separated by a semi-permeable membrane create a potential for water movement (osmosis) which can be describe by the equation for water potential. The water potential of a solution, denoted by the Greek letter psi (ψ) is the term given to the tendency for water molecules to enter or leave a solution by osmosis. The tendency for water to move into or out of a living cell can be calculated on the basis of the water potential of the cell sap relative to its surrounding environment. The use of water potential to express water relations of plant cells is used in preference to osmotic potential and osmotic pressure, although these terms are often used in medicine and animal physiology.

Water potential (ψ) and water movement

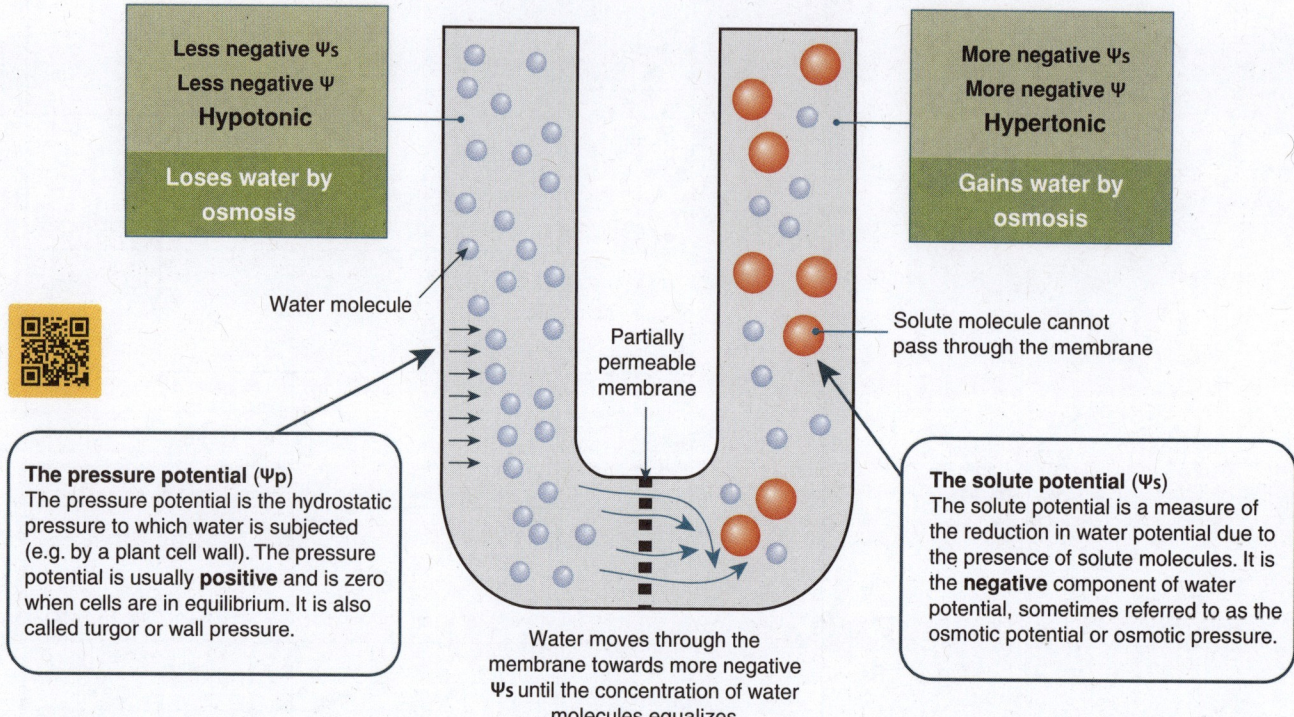

Less negative Ψ_s
Less negative Ψ
Hypotonic

Loses water by osmosis

More negative Ψ_s
More negative Ψ
Hypertonic

Gains water by osmosis

Water molecule

Solute molecule cannot pass through the membrane

Partially permeable membrane

The pressure potential (Ψ_p)
The pressure potential is the hydrostatic pressure to which water is subjected (e.g. by a plant cell wall). The pressure potential is usually **positive** and is zero when cells are in equilibrium. It is also called turgor or wall pressure.

The solute potential (Ψ_s)
The solute potential is a measure of the reduction in water potential due to the presence of solute molecules. It is the **negative** component of water potential, sometimes referred to as the osmotic potential or osmotic pressure.

Water moves through the membrane towards more negative Ψ_s until the concentration of water molecules equalizes

As water molecules move around, some collide with the plasma membrane and create pressure on the membrane called water potential (ψ).The greater the movement of water molecules, the higher their water potential. The presence of solutes (e.g. sucrose) lowers water potential because the solutes restrict the movement of water molecules. Pure water has the highest water potential (zero). Dissolving any solute in water lowers the water potential (makes it more negative).

Water always diffuses from regions of less negative to more negative water potential. Water potential is determined by two components: the solute potential, ψ_s (of the cell sap) and the pressure potential, ψ_p, expressed by:

$$\Psi_{cell} = \Psi_s + \Psi_p$$

The closer a value is to zero, the higher its water potential.

1. What is the water potential of pure water? _____

2. The diagrams below show three hypothetical situations where adjacent cells have different water potentials. For each pair (a)-(c) calculate ψ for each side and describe the net direction of water flow (A→B, B→A or no net movement):

(a)

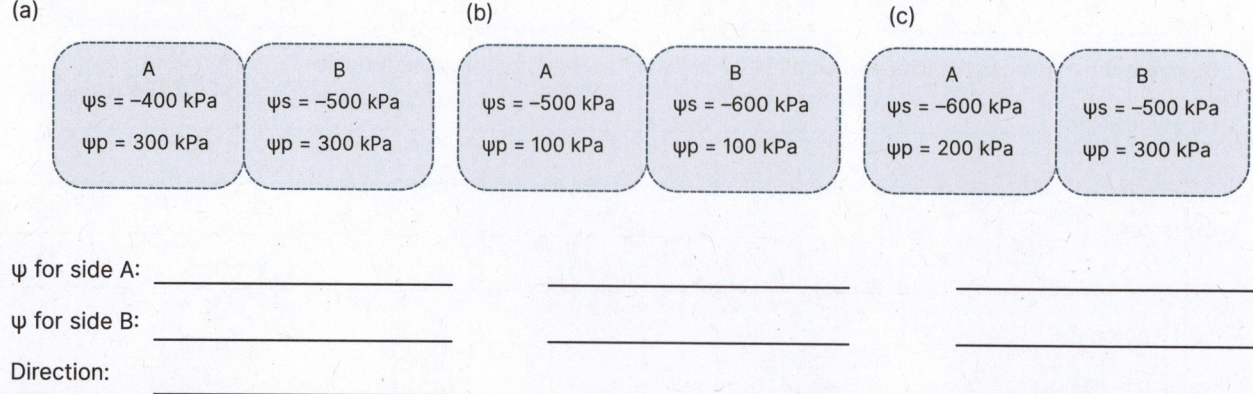

A	B
ψs = −400 kPa	ψs = −500 kPa
ψp = 300 kPa	ψp = 300 kPa

(b)

A	B
ψs = −500 kPa	ψs = −600 kPa
ψp = 100 kPa	ψp = 100 kPa

(c)

A	B
ψs = −600 kPa	ψs = −500 kPa
ψp = 200 kPa	ψp = 300 kPa

ψ for side A: _____   _____   _____

ψ for side B: _____   _____   _____

Direction: _____   _____   _____

AHL

D2.3
8-10

©2024 **BIOZONE** International
ISBN: 978-1-99-101410-8
Photocopying prohibited

▶ When the contents of a plant cell push against the cell wall they create turgor (tightness). Turgor provides support for the plant body. When cells lose water, there is a loss of turgor and the plant wilts. Complete loss of turgor from a cell is called plasmolysis and is irreversible.

▶ The diagram below shows the state of a cell when the external water potential is more negative than the cell's contents (left) and when it is less negative than the cell's contents (right). When the external water potential is the same as that of the cell, there is no net movement of water.

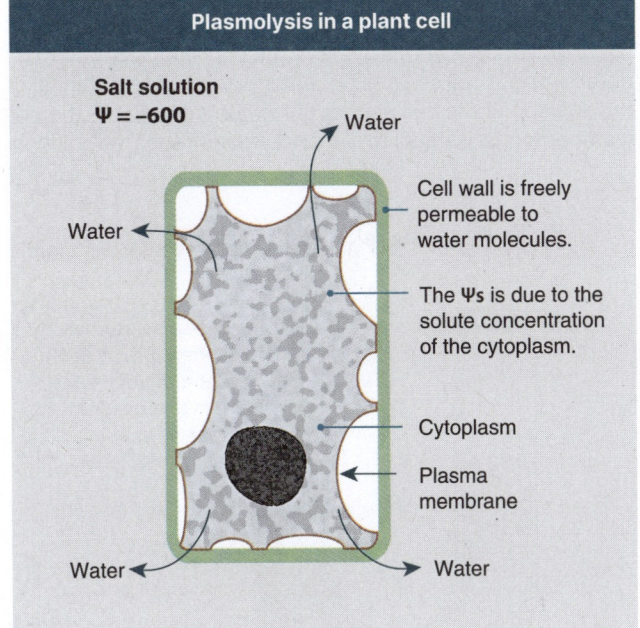

Plasmolysis in a plant cell

Salt solution
$\Psi = -600$

Water

Water

Cell wall is freely permeable to water molecules.

The Ψ_s is due to the solute concentration of the cytoplasm.

Cytoplasm

Plasma membrane

Water

Water

Turgor in a plant cell

Pure water
$\Psi = 0$

Water

Water

Cell wall bulges outward

Cytoplasm takes on water, putting pressure on the plasma membrane and cell wall. Ψ_p rises, offsetting Ψ_s at full turgor.

Water

Water

When external water potential is more negative than the water potential of the cell ($\Psi_{cell} = \Psi_s + \Psi_p$), water leaves the cell and, because the cell wall is rigid, the plasma membrane shrinks away from the cell wall. This process is termed plasmolysis and the cell becomes flaccid ($\Psi_p = 0$). Full plasmolysis is irreversible because the cell cannot recover by taking up water.

When the external water potential is less negative than the Ψ_{cell}, water enters the cell. A pressure potential is generated when sufficient water has been taken up to cause the cell contents to press against the cell wall. Ψ_p rises progressively until it offsets Ψ_s. Water uptake stops when the $\Psi_{cell} = 0$. The rigid cell wall prevents cell rupture. Cells in this state are turgid.

3. What is the effect of dissolved solutes on water potential? _____

_____

_____

4. Why don't plant cells burst when water enters them? _____

_____

_____

5. (a) Distinguish between plasmolysis and turgor: _____

_____

_____

_____

(b) Describe the state of the plant in the photo on the right and explain your reasoning:

_____

_____

6. (a) Explain the role of pressure potential in generating cell turgor in plants:

_____

_____

(b) Explain the purpose of cell turgor to plants: _____

_____

_____

293 Solute Potential and Cells

Key Idea: Solute potential is the pressure required to stop water flowing into a cell.

Recall the contributors to a cell's water potential and their relationship to the movement of water into and out of cells. The water potential of a plant cell can be calculated by measuring the gain or loss in mass when the cells are placed into solutions with a range of known concentrations. Cells placed into different concentrations will either gain or lose water depending on whether their internal water potential

(ψ) is higher (less negative) or lower (more negative) than the solution's. Cells with a water potential equal to the surrounding solution will neither gain or lose water (mass). A solution in an open beaker has no pressure acting on it ($\psi_p = 0$). For cells in this system, the only important factor in determining a cell's water potential is the solute potential of the solution. If the cell neither gains nor loses water, then its water potential is equal to the solute potential of the solution:

Solute potential

▸ Solute potential is the pressure needed to be applied to a solution to stop the inward flow of water across a partially permeable membrane due to solutes. It is always negative in a plant cell and zero in distilled water. It is measured in bars (100 Kpa = 1 bar = 1 atmosphere at sea level).

▸ Solute potential can be calculated using the formula:

$$\psi_s = -iCRT$$

i = ionization constant (this is the number of particles produced per unit when dissolved. For example the ionization constant for sucrose is 1, whereas for NaCl it is 2 (Na$^+$ and Cl$^-$).

C = molar concentration (molL^{-1})

R = pressure constant = 8.314 L kPa K^{-1} mol^{-1}

T = temperature (K) = 273 + °C of solution.

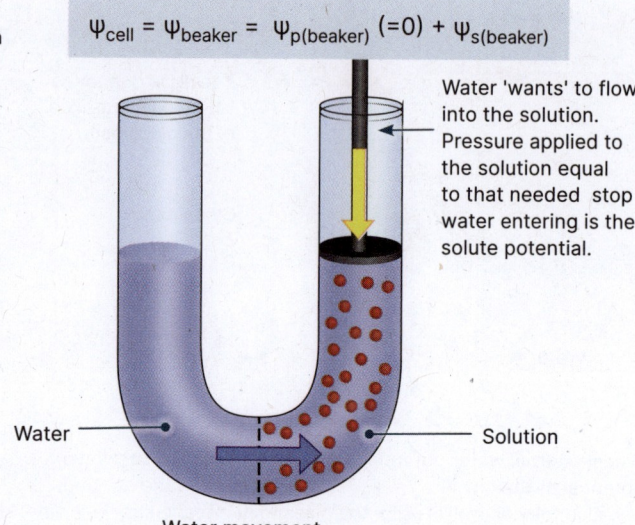

$\Psi_{cell} = \Psi_{beaker} = \Psi_{p(beaker)}$ (=0) + $\Psi_{s(beaker)}$

Water 'wants' to flow into the solution. Pressure applied to the solution equal to that needed stop water entering is the solute potential.

Water — — Solution

Water movement

1. What is the ionization constant for the following substances?

   (a) Glucose: _____ (b) KCl: _____ (c) MgCl$_2$: _____

2. A student made up three solutions of sucrose with the following concentrations: 1.00 molL^{-1}, 0.50 molL^{-1}, and 0.25 molL^{-1}. Calculate the solute potentials for each solution at 22°C:

   (a) 1.00 molL^{-1}: _____

   (b) 0.50 molL^{-1}: _____

   (c) 0.25 molL^{-1}: _____

3. Plot a graph of the of solute potential vs sucrose concentration on the grid below:

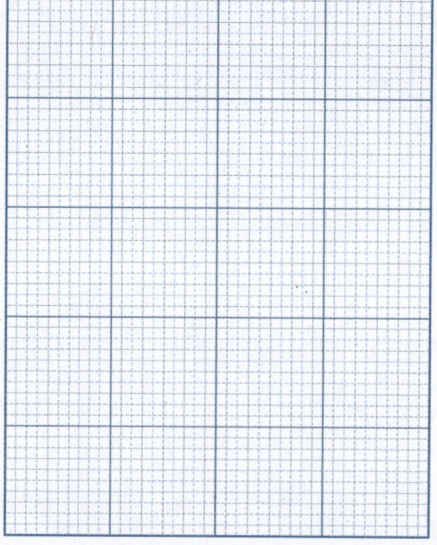

4. Use the plot to determine the solute potential of a 0.75 mol L^{-1} solution: _____

D2.3
10, 11

©2024 **BIOZONE** International
ISBN: 978-1-99-101410-8
Photocopying prohibited

294 Did You Get It?

1. The process right shows one of two types of cell division. Name the process shown:

   _____

2. What is the purpose of this process? _____

   _____

   _____

3. Describe the two events that have occurred at point A on the diagram:

   _____

   _____

   _____

   _____

4. (a) Name the second type of cell division (not shown here):

   _____

   (b) What is the purpose of this second type of cell division?

   _____

   _____

5. Contrast the location of mitosis in plants and animals: _____

   _____

6. A cell with 10 chromosomes undergoes mitosis.

   (a) How many daughter cells are created? _____

   (b) How many chromosomes does each daughter cell have? _____

   (c) Is the genetic material of the daughter cells the same as the parent cell? _____

7. Salt is often used to preserve meat such as salt fish and salt pork. Large amount of salt are stacked around the meat in containers. The meat may remain usable for many weeks or months. Use osmosis to explain this food preservation technique.

   _____

   _____

   _____

8. **AHL**: Cell A has a water potential (ψcell) of -300 kPa. Cell B has a water potential (ψcell) of -500 kPA. In which direction would the water move: from cell A to cell B, or from cell B to cell A?

   _____

9. (a) **AHL**: Label the cell cycle (right) with the following labels: G1, G2, M, S, cytokinesis:

   (b) **AHL**: Briefly describe what happens in each of the following phases:

   G_1: _____

   _____

   G_2: _____

   _____

   M phase: _____

   _____

INTERPHASE

M PHASE

©2024 **BIOZONE** International
ISBN: 978-1-99-101410-8
Photocopying prohibited

🦠 Organisms

Resource Hub
bit.ly/3taHvQZ

D3.1 Reproduction

Activity Number

Guiding Questions:
▶ How is change or continuity ensured by asexual and sexual reproduction?
▶ What are the requirements in organisms to ensure reproduction can occur??

Learning Outcomes:

☐ 1	Compare and contrast the advantages of asexual reproduction producing genetically identical offspring, and sexual reproduction producing offspring with variation, in adaptability to existing or new environments.	295
☐ 2	Describe how meiosis and the fusion of gametes through fertilization produce new allele combinations.	295
☐ 3	Compare and contrast the structure and function of male and female gametes, including relative number produced, and the different reproductive strategies of males and females.	295, 296, 298
☐ 4	Draw and annotate diagrams of male and female reproductive systems identifying key structures.	296, 298
☐ 5	Discuss the roles of hormones in regulating the menstrual cycle through positive and negative feedback.	302
☐ 6	Describe key stages in human fertilization.	300
☐ 7	Explain how artificial doses of hormones are utilized in the process of in vitro fertilization to induce superovulation.	304
☐ 8	Describe the process of sexual reproduction of flowering plants.	309
☐ 9	Draw and annotate key structures in a diagram of an insect-pollinated flower.	310
☐ 10	Discuss a range of cross-pollination methods in plants.	311
☐ 11	Evaluate a range of self-incompatibility mechanisms in flowering plants that prevent self-pollination.	311
☐ 12	Distinguish between seed dispersal and pollination processes and identify key stages of seed germination.	312 - 313
☐ 13	**AHL:** Discuss the role of changing hormone levels in triggering puberty onset.	303
☐ 14	**AHL:** Link the different processes of spermatogenesis and oogenesis to different gamete cytoplasm composition and the quantity of gametes produced.	348
☐ 15	**AHL:** Examine mechanisms to prevent polyspermy during fertilization including the acrosome and cortical reactions.	301
☐ 16	**AHL:** Identify the stages of blastocyst development and endometrium implantation during early embryo formation.	301
☐ 17	**AHL:** Discuss the use of monoclonal antibodies to identify the presence of human chorionic gonadotropin (hCG) in pregnancy testing.	305
☐ 18	**AHL:** Explain the role of the placenta in allowing exchange of substances between mother and baby and maintaining a lengthy gestation.	306
☐ 19	**AHL:** Describe the role of hormone levels in the process of pregnancy maintenance and the triggering of childbirth.	307
☐ 20	**AHL: NOS:** Evaluate evidence from epidemiological studies, including socioeconomic factors, exploring the type of causal relationship between hormone replacement therapy (HRT) and the incidence of coronary heart disease (CHD) in women.	308

D3.2 Inheritance

Guiding Questions:
- What patterns of inheritance can be observed in plants and animals?
- How do inheritance patterns result from molecular level genetic processes?

Learning Outcomes:

		Activity Number
☐ 1	Understand the connection between haploid gametes in the formation of a diploid zygote, leading to inheritance.	**314**
☐ 2	Use Punnett grids to perform F_1 and F_2 genetic crosses in various flowering plant species.	**316**
☐ 3	Distinguish between genes and alleles and define genotype using the terms 'homozygous' and 'heterozygous'.	**315**
☐ 4	Identify phenotypic traits influenced by either genotype or environmental factors and those influenced by both.	**317**
☐ 5	Explain how the same phenotype is the result of both homozygous-dominant genotype and a heterozygous genotype.	**316**
☐ 6	Explain how phenotypic plasticity can be reversible but not linked to changes in genotype.	**318**
☐ 7	Investigate phenylketonuria (PKU) as a recessive genetic condition caused by an autosomal gene mutation.	**319**
☐ 8	Understand the composition of gene pools including the contribution of single-nucleotide polymorphisms (SNPs) and multiple alleles.	**320**
☐ 9	Investigate ABO blood groups as an example of multiple alleles.	**321**
☐ 10	Distinguish between incomplete and codominance using AB blood type and four o'clock flower (*Mirabilis jalapa*).	**321 - 322**
☐ 11	Appraise the human sex determination mechanism, including X and Y chromosome structure, leading to the formation of male or female specific features.	**323**
☐ 12	Investigate haemophilia as an example of a sex-linked genetic disorder.	**323**
☐ 13	Utilize both pedigree charts and inheritance laws to identify genetic disorder patterns in a family tree of related individuals. **NOS:** Use both inductive and deductive reasoning to analyse pedigrees.	**324**
☐ 14	**AOS:** Investigate skin colour as an example of polygenic continuous variable, and ABO blood group as an example of a discrete variable, measuring the mean, median, and mode from collected data.	**325**
☐ 15	**AOS:** Construct a box-and-whisker plot from collected continuous variable data which displays outliers, minimum, first quartile, median, third quartile, and maximum.	**325**
☐ 16	**AHL:** Explain how genetic recombination during meiosis can be tracked by dihybrid cross outcomes in unlinked genes.	**326**
☐ 17	**AHL:** Use Punnett grids and apply Mendel's second law to predict genotypic and phenotypic ratios in dihybrid crosses. **NOS:** Explain that certain conditions must be present for this law to be applicable.	**327**
☐ 18	**AHL: AOS:** Investigate genetic databases to identify loci of gene pairs, and their polypeptide products.	**328**
☐ 19	**AHL:** Relate reduced rates of independent assortment to autosomal linked genes.	**329**
☐ 20	**AHL:** Determine the outcomes of individual heterozygous and homozygous recessive crosses identifying recombinants in gametes and genotypes and phenotypes of offspring.	**330**
☐ 21	**AHL:** Use a chi-squared test on dihybrid cross data, **NOS:** Apply this test as a means of statistical sampling (the F_2 generation) of a population.	**331**

D3.3 Homeostasis

Guiding Questions:
▶ How is homeostasis controlled in humans?
▶ Why is homeostasis regarded as an important life function in organisms?

Activity Number

Learning Outcomes:

295 How Do Organisms Reproduce?

Key Idea: Reproduction is the production of offspring in order to continue a genetic lineage.

Reproduction is the production of new life that will carry on the genetic lineage and ensure the continuity of a species. Without it, an organism's genetic lineage is lost. There are two types of reproduction, asexual and sexual. Asexual reproduction produces offspring that are genetically identical to the parent. Sexual reproduction combines half the genetic material from each of two parents to produce offspring that are genetically distinct. Both types of reproduction have their advantages and disadvantages. The gametes are formed during meiosis, a process that creates variation due to crossing over and independent assortment. The random nature of fertilization, i.e. the joining of the gametes, introduces even more variation when forming the zygote. The male and female gametes are different in form.

Sexual vs asexual reproduction

Snails mating

Sexual reproduction	Asexual reproduction
▸ Passes genetic information to the next generation.	▸ Ability to quickly take advantage of favourable environmental conditions.
▸ Production of offspring.	▸ One parent required.
▸ Produces variation in offspring.	▸ Offspring genetically identical to parent.
▸ Requires two parents.	▸ Passes genetic information to the next generation.
▸ Offspring genetically related to both parents, but different from either.	▸ Production of offspring.
▸ Relatively long time taken.	▸ Relatively short time taken.
▸ E.g. pollination, fertilization.	▸ E.g. budding, cloning.

Hydra budding

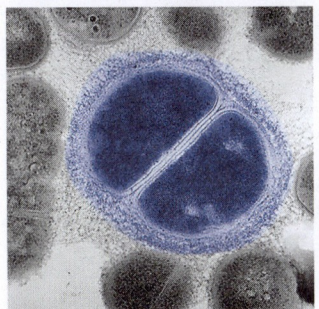

Prokaryotes, including bacteria, primarily reproduce by asexual reproduction via binary fission. This enables rapid population growth of identical individuals to utilize resources.

Most plants can reproduce asexually by bulbs (above), corms, tubers, or putting out rhizomes or stolons. Many plants have both a sexual and asexual phase in their life cycles.

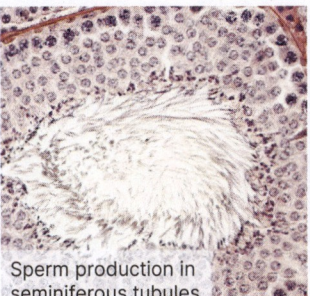

Sperm production in seminiferous tubules

Sexual reproduction requires the production of gametes: egg and sperm, formed via meiosis to create variation. Each gamete carries half the genetic complement.

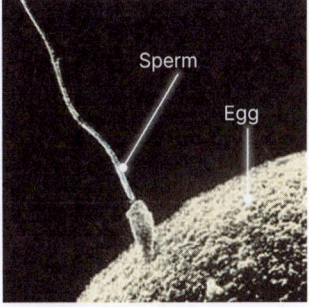

Sperm

Egg

Gametes are haploid (n). Unification of the gametes, called fertilization, restores the full genetic complement (2n) and produces the first cell of the new individual, called the zygote.

1. What is the purpose of reproduction? _____

_____

2. What is the difference in the genetic relationship of an offspring to its parent(s) in asexual reproduction, compared to sexual reproduction?

_____

_____

3. Populations in existing, stable environments are advantaged if offspring have the same adaptations as parents. However, in a changing environment, offspring that have variation to their parents may be advantaged if they have slightly different adaptations. Explain how asexual or sexual reproduction are the most beneficial in each scenario:

_____

_____

_____

_____

Sexual reproduction: Meiosis and fertilization

▶ All cells in the body, apart from germ cells, are somatic cells that undergo mitosis for growth and repair, producing identical cells. Germ cells in reproductive organs are the exception. They can produce non-identical gametes through meiosis.

▶ Meiosis enables gametes to be produced from one primary oocyte or spermatocyte, reducing from the 2n (diploid) set of 46 chromosomes in humans to the n (haploid) set of 23 chromosomes. Each of the gametes will have a set of chromosomes that is different from the other gametes due to the crossing over and independent assortment that occurs during meiosis.

▶ In meiosis in females, cell division is uneven. Only one mature ovum is produced (with three much smaller polar bodies produced) during meiosis, ensuring the ovum is n (by reduction division).

▶ Human females typically produce just one ovum per menstrual cycle at ovulation. Males continuously produce gametes, and average between 80 to 300 million gametes released per ejaculation.

▶ Fertilization occurs in the fallopian tube of the female. Here, the chromosomes of the ovum and sperm combine to restore the diploid set in the newly formed zygote.

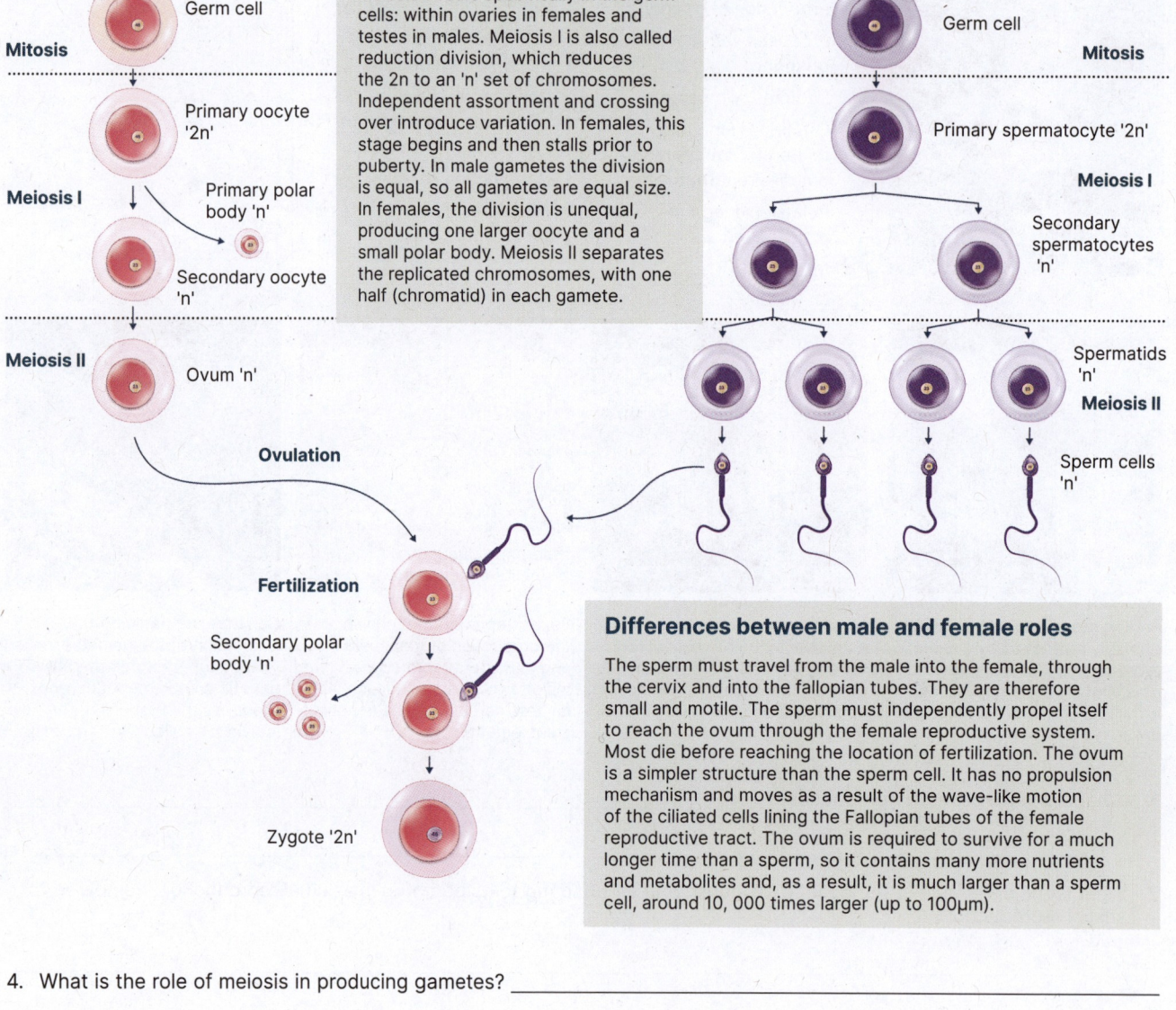

Meiosis and gamete formation

Meiosis occurs specifically in the germ cells: within ovaries in females and testes in males. Meiosis I is also called reduction division, which reduces the 2n to an 'n' set of chromosomes. Independent assortment and crossing over introduce variation. In females, this stage begins and then stalls prior to puberty. In male gametes the division is equal, so all gametes are equal size. In females, the division is unequal, producing one larger oocyte and a small polar body. Meiosis II separates the replicated chromosomes, with one half (chromatid) in each gamete.

Differences between male and female roles

The sperm must travel from the male into the female, through the cervix and into the fallopian tubes. They are therefore small and motile. The sperm must independently propel itself to reach the ovum through the female reproductive system. Most die before reaching the location of fertilization. The ovum is a simpler structure than the sperm cell. It has no propulsion mechanism and moves as a result of the wave-like motion of the ciliated cells lining the Fallopian tubes of the female reproductive tract. The ovum is required to survive for a much longer time than a sperm, so it contains many more nutrients and metabolites and, as a result, it is much larger than a sperm cell, around 10, 000 times larger (up to 100μm).

4. What is the role of meiosis in producing gametes? _____

_____

5. Compare and contrast the difference in the quantity and size of the ovum and sperm, linked to their roles in fertilization and reproduction:

_____

_____

_____

_____

_____

©2024 BIOZONE International
ISBN: 978-1-99-101410-8
Photocopying prohibited

296 The Male Reproductive System

Key Idea: The male reproductive system produces sperm cells for reproduction, and the hormone testosterone which drives the development of male characteristics.

The male reproductive system (below) is concerned with producing sperm and delivering them to the female urogenital tract. Mature sperm are ejaculated as semen with fluids from the seminal vesicles and prostate. When a sperm combines with an egg (ovum), it contributes half the genetic material of the offspring and, in humans and other mammals, determines its sex.

Male reproductive system: Front view

- Urinary bladder
- Seminal vesicle
- Ureter
- Ductus (or vas) deferens
- Prostate gland
- Lobules
- Testis
- Epididymis
- Urethra
- Sperm producing tubes
- Glans of penis
- Sperm

Male reproductive system: Side view

- Ureter
- Bladder
- Pubic bone
- Seminal vesicle
- Prostate gland
- Vas deferens
- Rectum
- Anus
- Urethra
- Testis
- Scrotum

Sperm structure

▸ Mature spermatozoa (sperm) are produced in the testes. Meiotic division of spermatocytes produces spermatids, which then differentiate into mature sperm.

▸ The sperm's structure reflects its purpose, which is to swim through the female reproductive tract to the ovum, penetrate the ovum's protective barrier, and donate its genetic material. A sperm cell comprises three regions: a headpiece, containing the nucleus and penetrative enzymes; an energy-producing mid-piece; and a tail for propulsion.

▸ Human sperm live for only about 48 hours, but they swim quickly and there are so many of them (millions per ejaculation) that usually some are able to reach the egg to fertilize it.

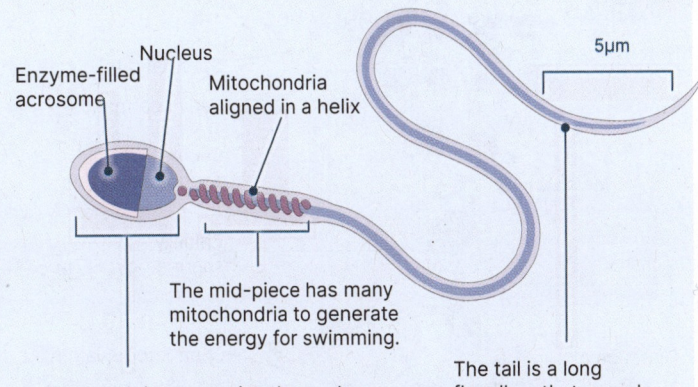

- Enzyme-filled acrosome
- Nucleus
- Mitochondria aligned in a helix
- 5µm

The mid-piece has many mitochondria to generate the energy for swimming.

The headpiece contains the nucleus and the acrosome, which contains the enzymes that help penetrate the egg.

The tail is a long flagellum that propels the sperm in its swim to the egg.

1. The male human reproductive system and associated structures are shown above. After studying the diagram above, on a separate sheet, draw a diagram of either the front or side view of the male reproductive system and label the following structures: *bladder, scrotal sac, sperm duct (vas deferens), epididymis, seminal vesicle, testis, urethra, prostate gland*. Attach the diagram to the page when completed.

2. Describe the path of the sperm from testis to ejaculation from the penis glans: _____

_____

_____

3. Link the specific structures of the sperm to their roles in sexual reproduction and fertilization: _____

_____

_____

_____

D3.1

3 - 4

297 Spermatogenesis

Key Idea: Sperm is produced in the seminiferous tubules of the testes.

Sperm is produced via the process of spermatogenesis which occurs in the seminiferous tubules. Sperm production is a continual process. It takes 60-70 days for a sperm cell to fully develop, but several million sperm cells will be produced each day. A sperm cell will survive in the epididymis for about 3 weeks before being reabsorbed.

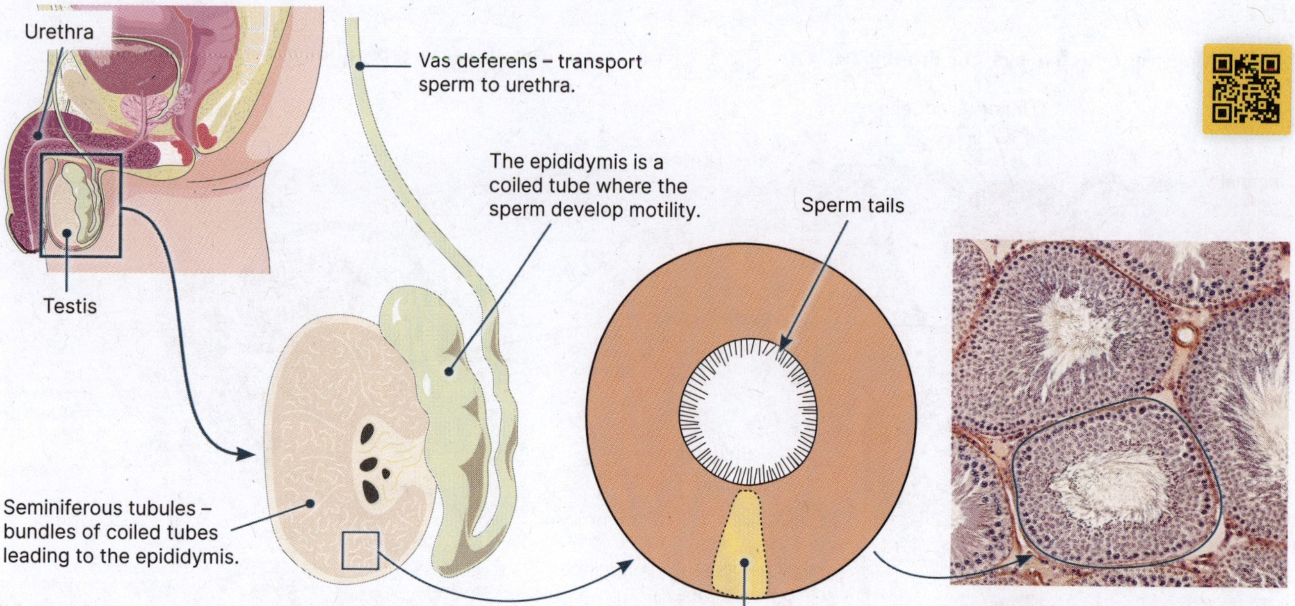

Urethra

Vas deferens – transport sperm to urethra.

The epididymis is a coiled tube where the sperm develop motility.

Sperm tails

Testis

Seminiferous tubules – bundles of coiled tubes leading to the epididymis.

Sertoli cell (enlarged below)

Section through testis showing seminiferous tubules. Each tubule is 50-60cm long. In total, there are about 600m of tubules.

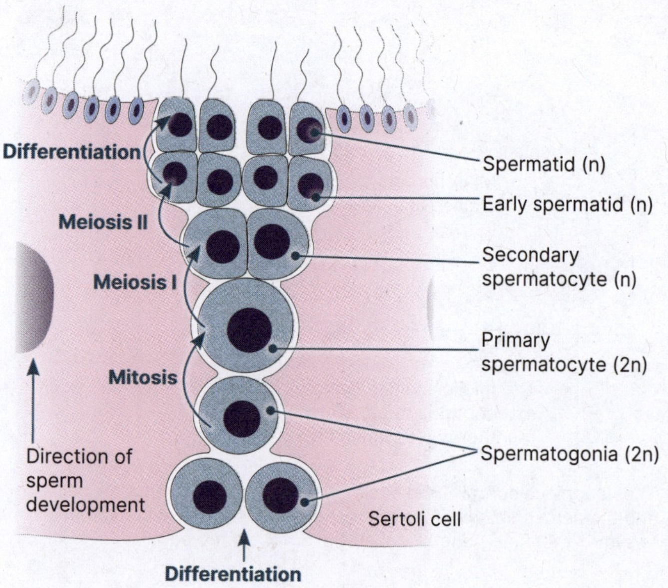

Differentiation

Meiosis II

Spermatid (n)

Early spermatid (n)

Secondary spermatocyte (n)

Meiosis I

Primary spermatocyte (2n)

Mitosis

Direction of sperm development

Spermatogonia (2n)

Sertoli cell

Differentiation

Spermatogenesis

▶ Sperm cell production is stimulated by follicle stimulating hormone (FSH), released from the anterior pituitary in response to gonadotropin releasing hormone (GnRH) released from the hypothalamus.

▶ Germ cells differentiate into spermatogonia that enter the gonads. Sperm (the male gametes) are produced by meiotic division of spermatogonia in the seminiferous tubules of the testes. The nucleus of the germ cell in the male divides twice to produce four similar-sized sperm cells. Spermatogenesis continues throughout an adult male's life.

▶ The developing sperm are nourished by supportive Sertoli cells. Sperm production occurs in response to the male steroid hormone, testosterone. Sperms formed in the testes enter the epididymis, where they mature and develop motility.

1. State the main role of the testes: _____

_____

_____

2. (a) Where does spermatogenesis occur? _____

(b) What hormone controls this process? _____

(c) From the diagrams above, what evidence do you have that spermatogenesis is most efficient at temperatures just below core body temperature?

_____

_____

3. Are sperm cells haploid or diploid? Explain: _____

_____

AHL

D3.1
14

298 The Female Reproductive System

Key Idea: The female reproductive system maintains female characteristics, produces egg cells for reproduction, and provides the environment for the growth and development of the fertilized egg.

The female reproductive system consists of the ovaries, Fallopian tubes, uterus, the vagina and external genitalia, and the breasts. Although both male and females have breasts, the female breasts (mammary glands) are modified so that they produce milk after childbirth. The female reproductive system produces eggs, receives the penis and sperm during sexual intercourse, protects and houses the developing foetus, and produces milk to nourish the young after birth.

Female reproductive system: Front view

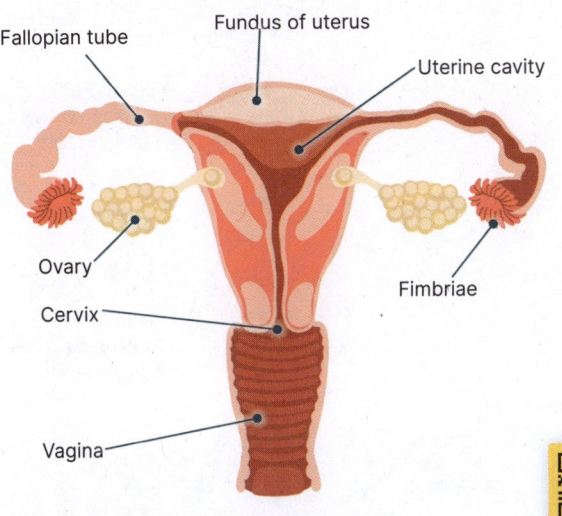

Female reproductive system: Side view

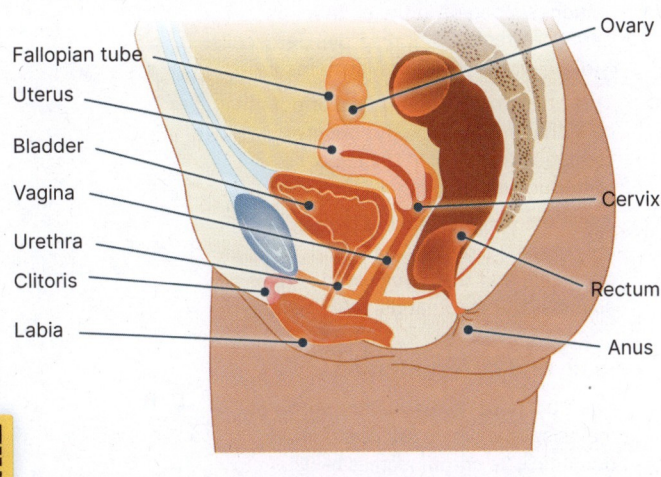

Ovum structure

▶ The contents of the ovum are similar to that of a typical mammalian cell, although it is externally surrounded by a jelly-like glycoprotein called the zona pellucida.

▶ A small polar body (the remnants of a sister cell) lies between the plasma membrane and zona pellucida. Cortical granules around the inner edge of the plasma membrane contain enzymes that are released once a sperm has penetrated the egg, forming a block to prevent further sperm entry (the cortical reaction).

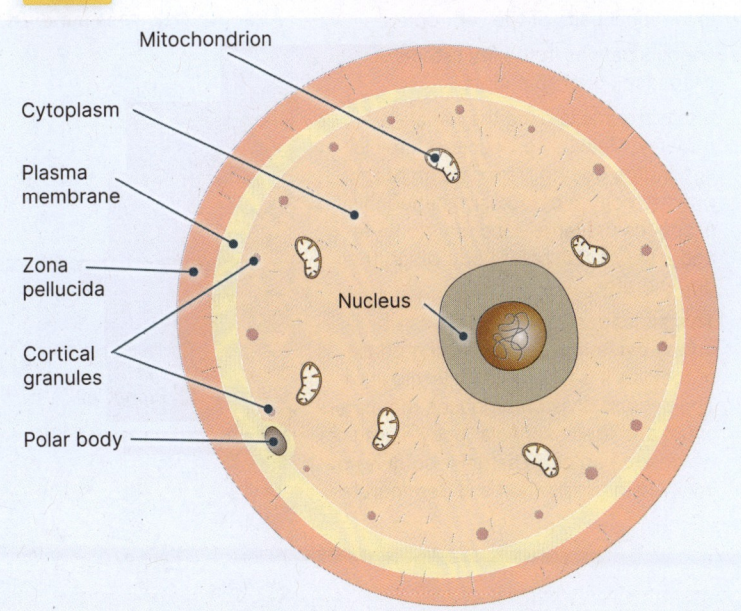

1. The female human reproductive system and associated structures are shown above. After studying the diagram above, on a separate sheet, draw a diagram of either the front or side view of the female reproductive system and label the following structures: ovary, uterus (womb), vagina, Fallopian tube (oviduct), cervix, clitoris. Attach the diagram to the page when completed.

2. Compare and contrast the position in the body of the gamete producing organs (ovary in females and testis in males) relating to their function:

_____

_____

_____

3. What monthly event releases an ovum from the ovary in females? _____

_____

D3.1
3 - 4

299 Oogenesis

Key Idea: The ova (eggs) are produced by a process called oogenesis, which is only completed if the egg is fertilized. Unlike males, who produce sperm throughout their lifetime, females are born with their full complement of eggs and do not produce more. Oogenesis is the process by which mature ova are produced in the ovary. Ova (egg) production initially begins in the embryo to produce oocytes from germ cells through mitosis; oogonia of the female foetus produces diploid (2n) oocytes. These then remain in a suspended state of the prophase stage of meiosis I until the female reaches puberty. At puberty, meiosis resumes. Each menstrual cycle, one or two ova resume development but, again, meiosis is suspended, this time in metaphase of meiosis II. The second meiotic division is only completed if the egg is fertilized. The final stages of meiosis II are completed when the polar bodies are formed after fertilization and the zygote develops.

Mitosis and cell growth

▸ Oogenesis begins in the embryo, as germline stem cells (primordial germ cells – PGC), which proliferate by mitosis.

▸ This phase builds up a 'bank' of potential cells that are available for oocyte formation throughout the female's reproductive life.

▸ While the mitosis phase is happening, the cells remain diploid (46 chromosomes in each cell) and the chromosomes remain as individual chromatids.

Meiosis and differentiation

▸ The PGC undergoes the first phase of meiosis I to develop into the primary oocytes. The chromosomes replicate to form double chromatids bound together in identical pairs at the centromere.

▸ Meiosis pauses in the final stage of prophase I prior to puberty.

▸ meiosis restarts once the female reaches puberty. Meiosis I produces a haploid secondary oocyte. The division of the primary oocyte cytoplasm is unequal and produces the large secondary oocyte and a very small polar body which then degenerates.

▸ This secondary oocyte is released during ovulation. The remaining phase of meiosis II is triggered by fertilization. The second division produces the mature ovum and the small polar body - both of which are haploid. The polar body is not involved in fertilization and degenerates.

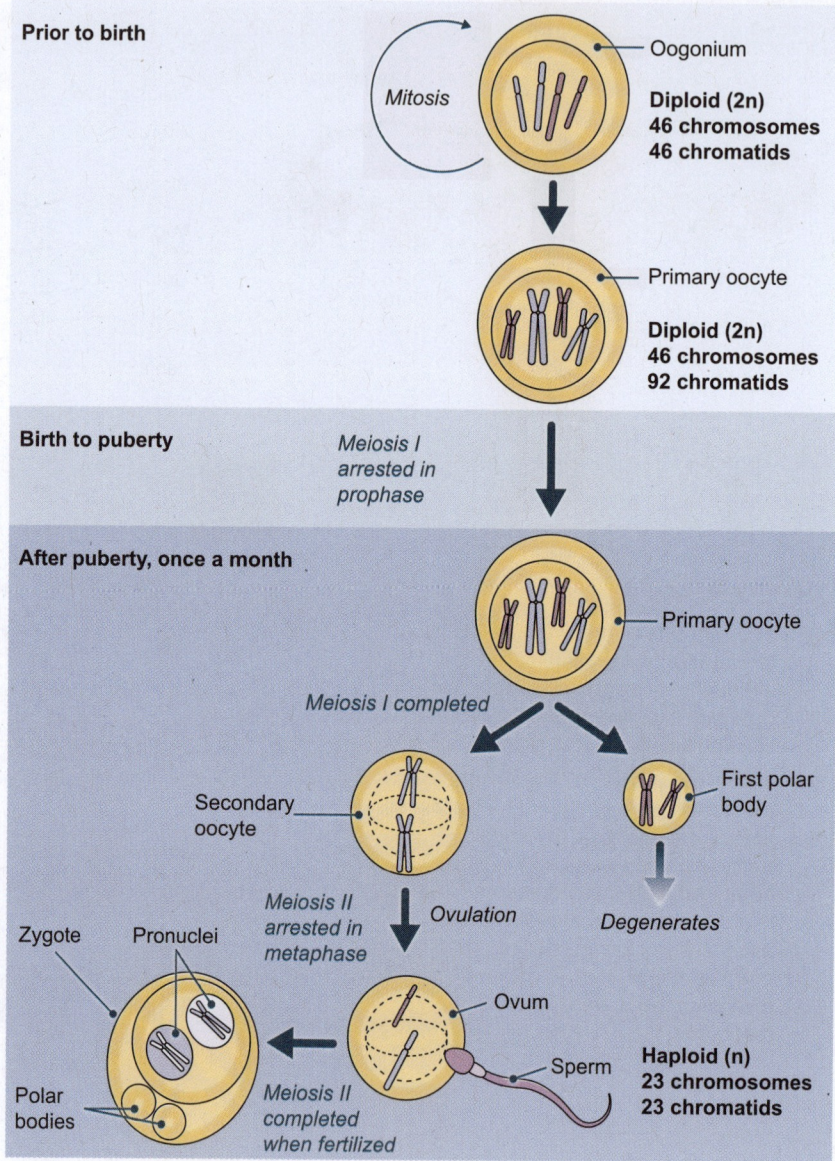

1. (a) Describe the difference between the lifetime production of gametes in males and females: _____

   _____

   _____

   (b) In humans, how might this affect reproductive decisions between males and females? _____

   _____

   _____

2. What is the trigger for the completion of oogenesis? _____

3. Where does fertilization occur? _____

   _____

©2024 **BIOZONE** International
ISBN: 978-1-99-101410-8
Photocopying prohibited

300 Fertilization in Humans

Key Idea: In order for fertilization to occur and a zygote to be formed, the nucleus containing chromosomes from both the ovum and sperm need to fuse.

Only one sperm cell can fuse with one egg cell to create a fertilized, one cell zygote. This process occurs in the fallopian tube of the female, although there are exceptions.

The fertilized zygote moves down towards the uterus for implantation in the wall. Once one sperm has gained access to the egg and the fusion of both membranes occur, other sperm are permanently blocked from entry; only one haploid set of sperm and egg can join. Cell division by mitosis occurs soon after the egg cell has been fertilized.

1. The sperm that gains first successful access to the egg initiates a process of fusing sperm and egg membranes. This is facilitated by enzymes in the sperm head. Any other sperm attempting to access the egg will be prevented from doing so by changes to the egg's surface.

2. The nucleus of the sperm enters into the cytoplasm of the egg, but the sperm tail and mitochondria are destroyed and play no further part in the fertilization process.

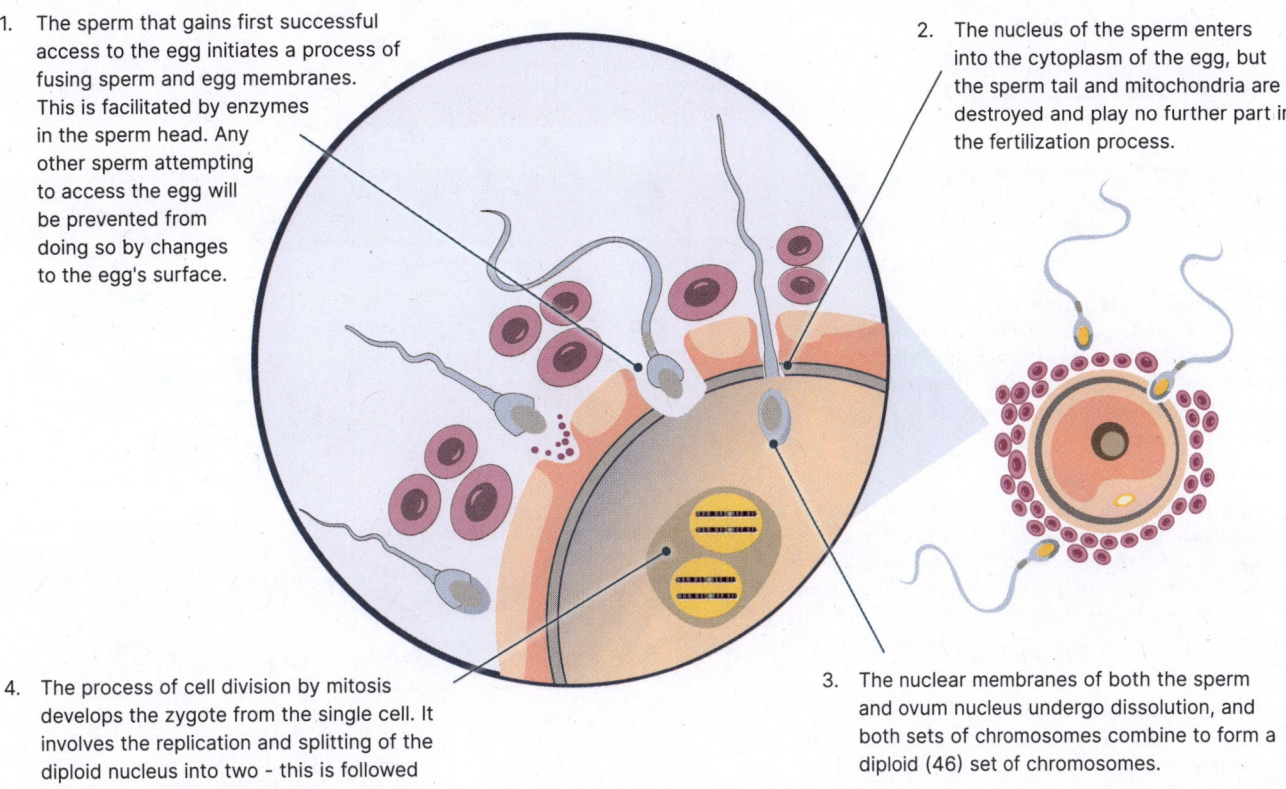

4. The process of cell division by mitosis develops the zygote from the single cell. It involves the replication and splitting of the diploid nucleus into two - this is followed by the first zygote division into two cells.

3. The nuclear membranes of both the sperm and ovum nucleus undergo dissolution, and both sets of chromosomes combine to form a diploid (46) set of chromosomes.

Ovulation and implantation

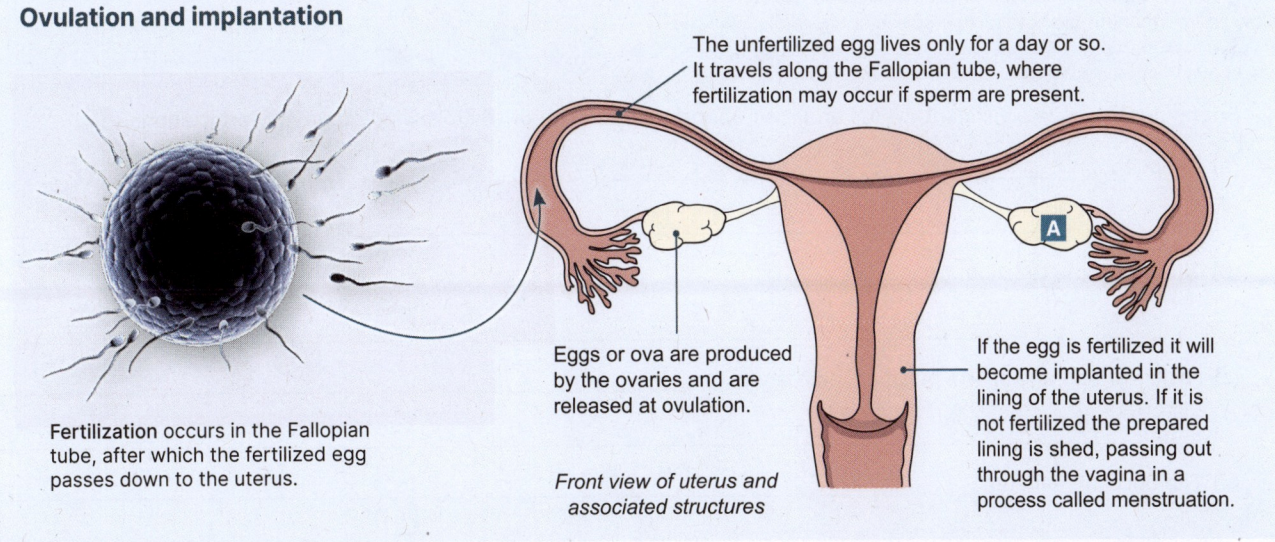

The unfertilized egg lives only for a day or so. It travels along the Fallopian tube, where fertilization may occur if sperm are present.

Eggs or ova are produced by the ovaries and are released at ovulation.

If the egg is fertilized it will become implanted in the lining of the uterus. If it is not fertilized the prepared lining is shed, passing out through the vagina in a process called menstruation.

Fertilization occurs in the Fallopian tube, after which the fertilized egg passes down to the uterus.

Front view of uterus and associated structures

1. Describe key stages in human fertilization, including the importance of only one sperm being able to fertilize one ovum:

_____

_____

_____

_____

©2024 **BIOZONE** International
ISBN: 978-1-99-101410-8
Photocopying prohibited

301 Fertilization and Early Growth

Key Idea: Fertilization involves several distinct stages and reactions, including the acrosome and cortical reactions. Fertilization occurs when a sperm penetrates an egg cell at the secondary oocyte stage. The sperm and egg nuclei then unite to form the zygote. In mammals, the entry of a sperm into the egg triggers specific mechanisms to prevent polyspermy (fertilization of the egg by more than one sperm). These include a change in membrane potential, and the cortical reaction (see below). A zygote resulting from polyspermy contains too many chromosomes and is not viable, i.e. it does not develop. Fertilization is seen as time 0 in a period of gestation (pregnancy) and has five stages (below). After fertilization, the zygote begins its development, i.e. its growth and differentiation into a multicellular organism.

Fertilization (time 0)

The stages in fertilization are represented below in a numbered sequence (1-5)

1. Capacitation
The surface of the sperm cell undergoes changes that are essential to enable the acrosome reaction and sperm entry.

2. The acrosome reaction
Enzymes from the acrosome (an enzyme-filled bag at the tip of the sperm) are released and digest a pathway through the follicle cells (not shown) and the jelly-like zona pellucida surrounding the egg cell (secondary oocyte).

3. Fusion of sperm head
The plasma membranes of the sperm and egg fuse, and the nucleus of the sperm enters the egg cytoplasm. Fusion causes a sudden membrane depolarisation that acts as a 'fast block' to further sperm entry. The fusion of the two plasma membranes also triggers the completion of meiosis II in the egg cell and induces the cortical reaction (below).

4. The cortical reaction
The fusion of the two plasma membranes induces a permanent change in the egg surface that prevents further sperm entry. Cortical granules in the egg cytoplasm release their contents into the space between the plasma membrane and the vitelline layer. Substances released from the granules raise and harden the vitelline layer to form a slow and permanent block to further sperm entry.

Zona pellucida (glycoprotein layer)

Egg plasma membrane

Egg nucleus (n)

Perivitelline space

Sperm nucleus (n)

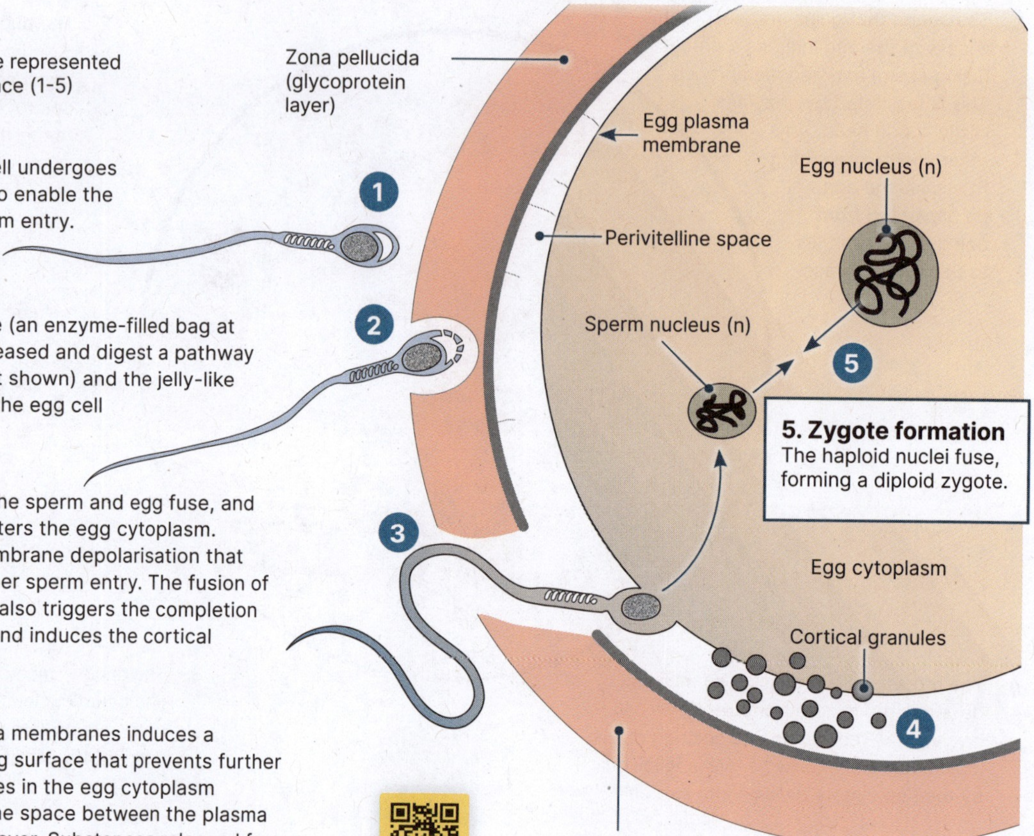

5. Zygote formation
The haploid nuclei fuse, forming a diploid zygote.

Egg cytoplasm

Cortical granules

Vitelline layer

1. Briefly describe the significant events and their importance at each of the following stages of fertilization:

(a) Capacitation: _____

_____

(b) The acrosome reaction: _____

_____

(c) Fusion of egg and sperm plasma membranes: _____

_____

(d) The cortical reaction: _____

_____

(e) Fusion of egg and sperm nuclei: _____

_____

2. Why is it important that fertilization of the egg by more than one sperm (polyspermy) does not occur? _____

_____

_____

©2024 **BIOZONE** International
ISBN: 978-1-99-101410-8
Photocopying prohibited

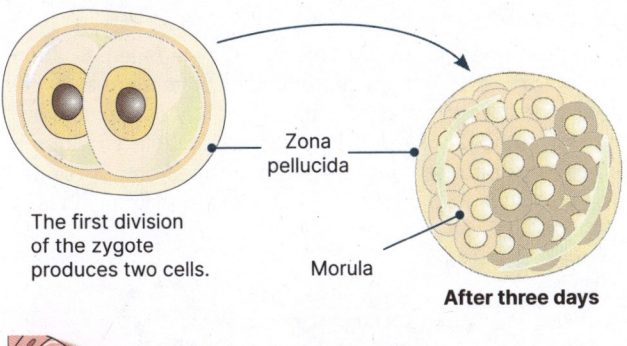

The first division of the zygote produces two cells.

Zona pellucida

Morula

After three days

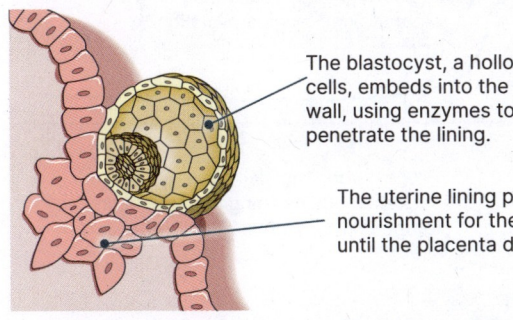

The blastocyst, a hollow ball of cells, embeds into the uterine wall, using enzymes to digest and penetrate the lining.

The uterine lining provides nourishment for the embryo until the placenta develops.

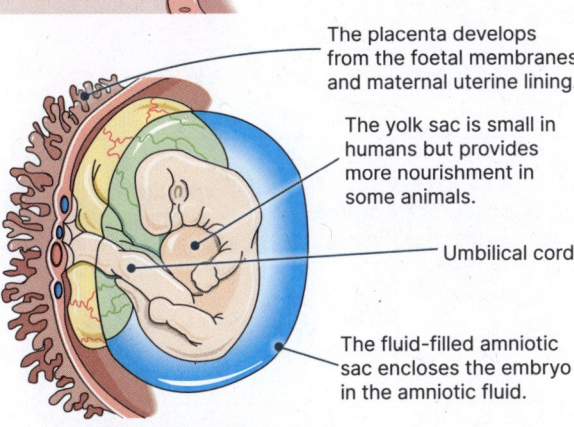

The placenta develops from the foetal membranes and maternal uterine lining.

The yolk sac is small in humans but provides more nourishment in some animals.

Umbilical cord

The fluid-filled amniotic sac encloses the embryo in the amniotic fluid.

Five week old embryo

Early growth and development

Cleavage, and the development of the morula

Immediately after fertilization, rapid cell division takes place. These early cell divisions are called cleavage and they increase the number of cells, but not the size of the zygote. The first cleavage is completed after 36 hours, and each succeeding division takes less time. After 3 days, successive cleavages have produced a solid mass of cells called the morula (left), which is still about the same size as the original zygote.

Implantation of the blastocyst (after 6-8 days)

After several days in the uterus, the morula develops into the blastocyst. It makes contact with the endometrium uterine lining and pushes deeply into it, ensuring a close maternal-foetal contact. Blood vessels provide early nourishment as they are opened up by enzymes secreted by the blastocyst. The embryo produces hCG (human chorionic gonadotropin), which prevents degeneration of the corpus luteum and signals that the woman is pregnant.

The embryo at 5-8 weeks

Five weeks after fertilization, the embryo is only 4-5 mm long, but already the central nervous system has developed and the heart is beating. The embryonic membranes have formed; the amnion encloses the embryo in a fluid-filled space, and the allanto-chorion forms the foetal portion of the placenta. From two months, the embryo is called a foetus. It is still small (30-40 mm long) but the limbs are well formed and the bones are beginning to harden. The face has a flat, rather featureless appearance with the eyes far apart. Foetal movements have begun and brain development proceeds rapidly. The placenta is well developed, although not fully functional until 12 weeks. The umbilical cord, containing the foetal umbilical arteries and vein, connects foetus and mother.

3. (a) Explain why the egg cell, when released from the ovary, is termed a secondary oocyte: _____

_____

(b) At which stage is its meiotic division completed? _____

4. What contribution do the sperm and egg cell make to each of the following:

(a) The nucleus of the zygote: Sperm contribution: _____ Egg contribution: _____

(b) The cytoplasm of the zygote: Sperm contribution: _____ Egg contribution: _____

5. What is meant by cleavage? Explain its significance to the early development of the embryo: _____

_____

_____

6. (a) What is the importance of implantation to the early nourishment of the embryo? _____

_____

(b) What is the purpose of hCG production by the embryo? _____

_____

_____

7. Why is the foetus particularly prone to damage from drugs towards the end of the first trimester (2-3 months)?

_____

302 The Menstrual Cycle

Key Idea: The menstrual cycle involves cyclical changes in the ovaries and uterus controlled by hormone concentration. In humans, fertilization of the ovum (egg) is most likely to occur around the time of ovulation. The uterine lining (endometrium) thickens in preparation for pregnancy, but is shed as a bloody discharge through the vagina if fertilization does not occur. This event, called menstruation, characterizes the human reproductive or menstrual cycle. The menstrual cycle starts from the first day of bleeding and lasts for about 28 days. It involves predictable changes in response to pituitary and ovarian hormones and is divided into three phases: the follicular, ovulatory, and luteal phases.

The menstrual cycle

Luteinising hormone (LH) and follicle stimulating hormone (FSH): FSH stimulates the development of the ovarian follicles, resulting in the release of oestrogen. Oestrogen levels peak, stimulating a surge in LH and triggering ovulation, in a positive feedback cycle.

Hormone levels: One of the follicles that begins developing in response to FSH (the Graafian follicle) becomes dominant. In the first half of the cycle, oestrogen is secreted by this developing Graafian follicle. Later, the Graafian follicle develops into the corpus luteum (below right) which secretes large amounts of progesterone (and smaller amounts of oestrogen).

The corpus luteum: The Graafian follicle continues to grow and, at around day 14, ruptures to release the egg (ovulation). LH causes the ruptured follicle to develop into a corpus luteum (yellow body). The corpus luteum secretes progesterone which promotes full development of the uterine lining, maintains the embryo in the first 12 weeks of pregnancy, and inhibits the development of more follicles.

Menstruation: Oestradiol acts to thicken the uterus wall. If fertilization does not occur, the corpus luteum breaks down. Progesterone secretion declines, causing the uterine lining to be shed (menstruation). If fertilization occurs, high progesterone levels maintain the thickened uterine lining. The placenta develops and nourishes the embryo completely by 12 weeks.

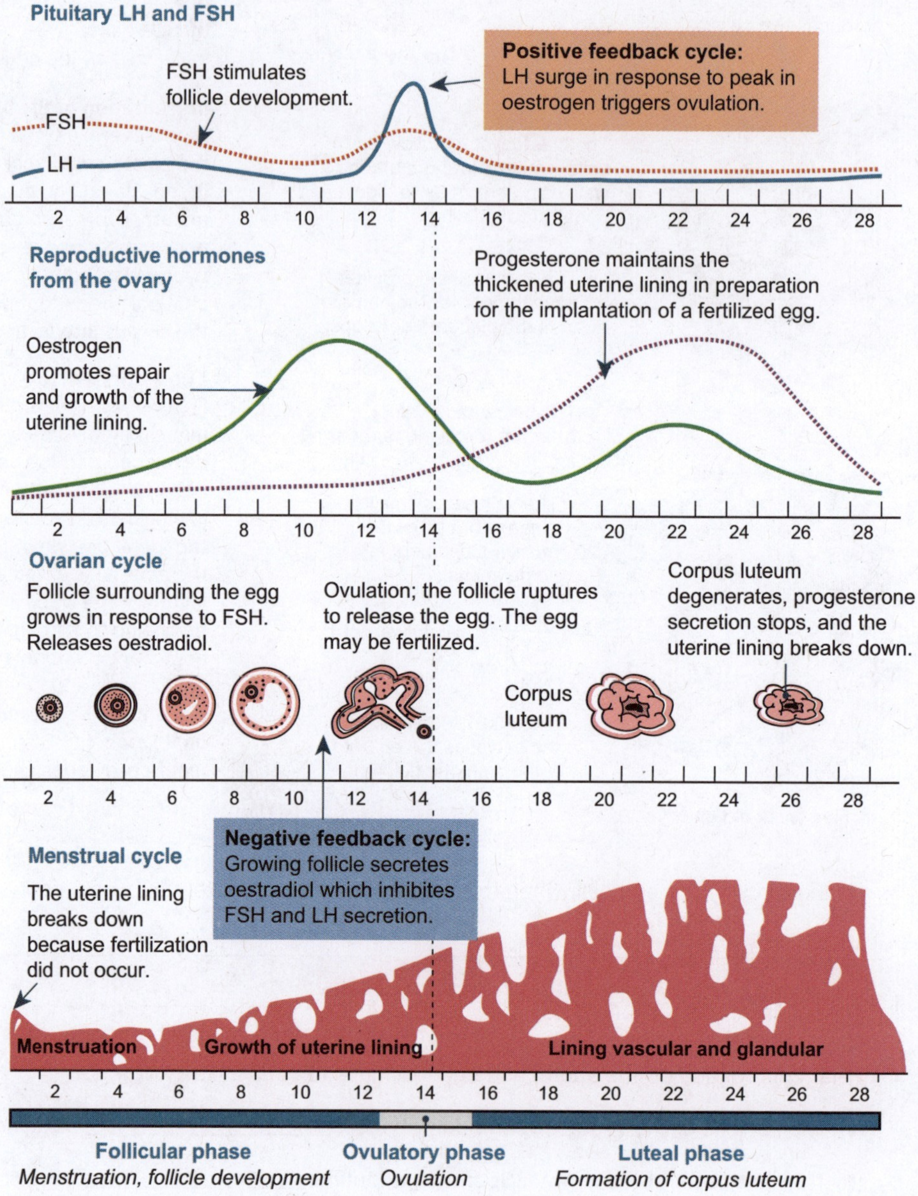

Pituitary LH and FSH

FSH stimulates follicle development.

FSH

LH

Positive feedback cycle: LH surge in response to peak in oestrogen triggers ovulation.

Reproductive hormones from the ovary

Oestrogen promotes repair and growth of the uterine lining.

Progesterone maintains the thickened uterine lining in preparation for the implantation of a fertilized egg.

Ovarian cycle

Follicle surrounding the egg grows in response to FSH. Releases oestradiol.

Ovulation; the follicle ruptures to release the egg. The egg may be fertilized.

Corpus luteum

Corpus luteum degenerates, progesterone secretion stops, and the uterine lining breaks down.

Menstrual cycle

The uterine lining breaks down because fertilization did not occur.

Negative feedback cycle: Growing follicle secretes oestradiol which inhibits FSH and LH secretion.

Menstruation | Growth of uterine lining | Lining vascular and glandular

Follicular phase
Menstruation, follicle development

Ovulatory phase
Ovulation

Luteal phase
Formation of corpus luteum

1. Identify the hormone responsible for:

   (a) Follicle growth: _____   (b) Ovulation: _____

2. Describe the role of oestrogen in positive feedback in the second half of the cycle: _____

   _____

   _____

   _____

3. Describe the role of oestradiol in negative feedback throughout the menstrual cycle: _____

   _____

   _____

   _____

D3.1

5

©2024 **BIOZONE** International
ISBN: 978-1-99-101410-8
Photocopying prohibited

303 Hormones and Puberty

Key Idea: Puberty onset and maintenance is triggered by hormones, including gonadotropin-releasing hormones (GnRH) and steroid sex hormones.

Humans differentiate into the male or female sex by the action of a combination of different hormones. The hypothalamus releases GnRH to trigger LH secretion which stimulates male gonads - producing testosterone, and FSH to stimulate female gonads - producing oestrogen. The hormones testosterone (in males), and oestrogen and progesterone (in females), are responsible for puberty (the onset of sexual maturity), the maintenance of sexual differences, and the production of gametes.

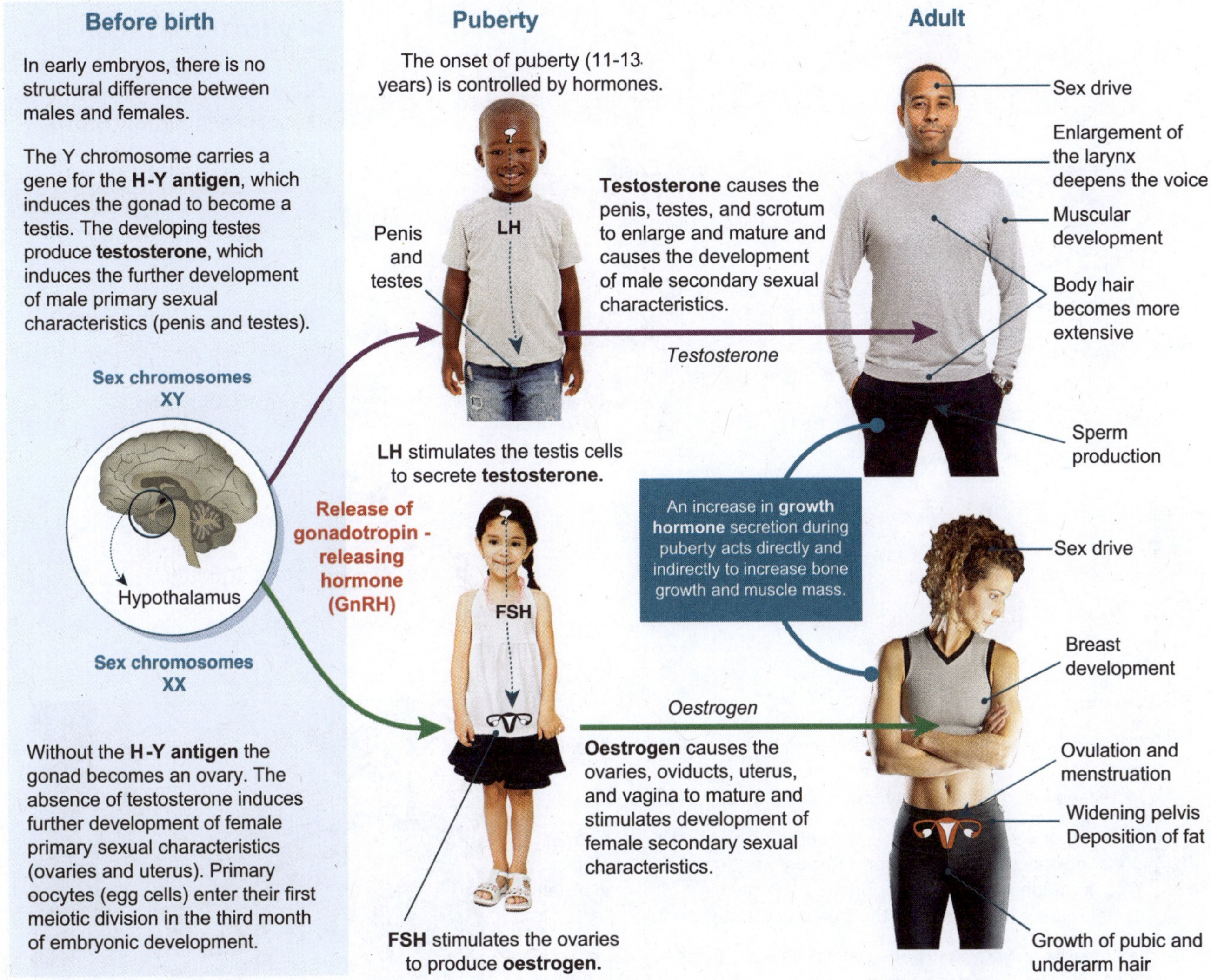

Before birth

In early embryos, there is no structural difference between males and females.

The Y chromosome carries a gene for the **H-Y antigen**, which induces the gonad to become a testis. The developing testes produce **testosterone**, which induces the further development of male primary sexual characteristics (penis and testes).

Sex chromosomes XY

Hypothalamus

Sex chromosomes XX

Without the **H-Y antigen** the gonad becomes an ovary. The absence of testosterone induces further development of female primary sexual characteristics (ovaries and uterus). Primary oocytes (egg cells) enter their first meiotic division in the third month of embryonic development.

Puberty

The onset of puberty (11-13 years) is controlled by hormones.

Penis and testes — **LH**

LH stimulates the testis cells to secrete **testosterone**.

Release of gonadotropin - releasing hormone (GnRH)

FSH

FSH stimulates the ovaries to produce **oestrogen**.

Testosterone causes the penis, testes, and scrotum to enlarge and mature and causes the development of male secondary sexual characteristics.

Testosterone

An increase in **growth hormone** secretion during puberty acts directly and indirectly to increase bone growth and muscle mass.

Oestrogen

Oestrogen causes the ovaries, oviducts, uterus, and vagina to mature and stimulates development of female secondary sexual characteristics.

Adult

- Sex drive
- Enlargement of the larynx deepens the voice
- Muscular development
- Body hair becomes more extensive
- Sperm production

- Sex drive
- Breast development
- Ovulation and menstruation
- Widening pelvis Deposition of fat
- Growth of pubic and underarm hair

---

1. What is the hormonal pathway to trigger oestrogen and testosterone at puberty? _____
   _____
   _____
   _____

2. What is the purpose of the widening of the pelvis during female puberty? _____
   _____

3. How does GnRH hormone affect the body during puberty? _____
   _____

4. What is the difference between primary and secondary sexual characteristics? _____
   _____
   _____
   _____

D3.1 13 | AHL |

304 Treating Infertility with In Vitro Fertilization

Key Idea: Female infertility may occur due to many factors and *in vitro* fertilization (IVF) is a possible treatment. Hormones can be used to induce super-ovulation.

Failure to ovulate is a common cause of female infertility. In most cases, the cause is hormonal although sometimes the ovaries may be damaged or not function normally. Female infertility may also arise through damage to the Fallopian tubes as a result of infection or scarring. These cases are usually treated with hormones, followed by IVF. Most treatments for female infertility involve the use of synthetic female hormones which stimulate ovulation, boost egg production, and induce egg release.

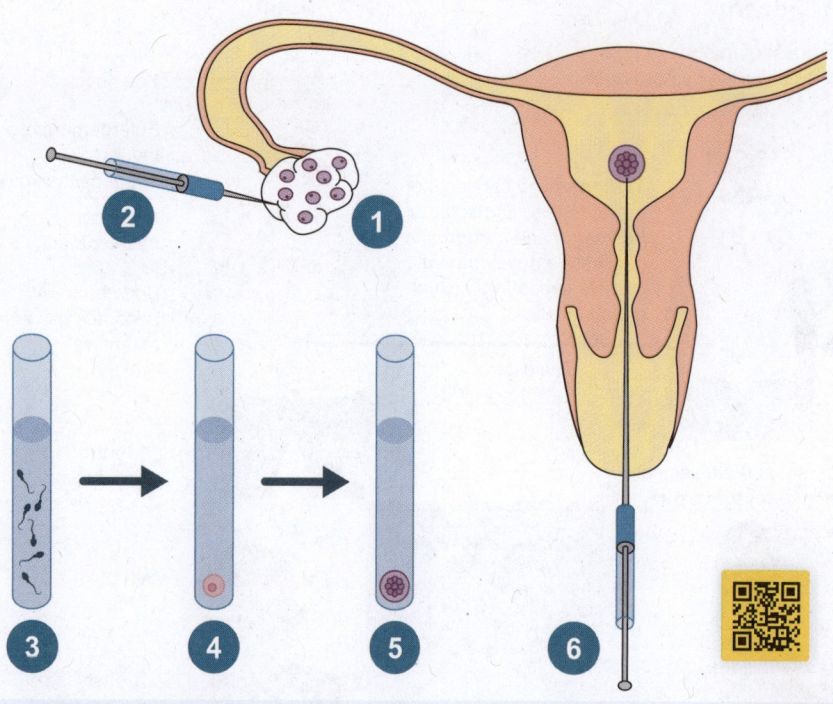

In vitro fertilization

1. Stimulate follicle growth (super-ovulation with FSH).

2. Aspiration of oocytes (assisted with HCG to release eggs).

3. Oocyte mixed with sperm.

4. *In vitro* fertilization.

5. Development of embryo.

6. Embryo transfer.

Superovulation

▶ When mature ova (eggs) are required to be harvested from the woman undergoing IVF treatment, the number of ova can be increased using a hormonal technique called super-ovulation.

▶ The woman's ovulation cycle is disrupted using an artificial form of gonadotropin releasing hormone (GnRH) as a nasal spray or injection. This hormone can induce menopause (the halting of the menstrual cycle) but is reversible after the treatment is stopped.

▶ In conjunction with GnRH, the patient self-administers a daily subcutaneous injection of follicle stimulating hormone (FSH), with the aim of developing 3 or more viable ova.

▶ Oestradiol measurements indicate when the oocytes have developed into mature ova; then human chorionic gonadotropin (HCG) is administered to induce release of ova ready for harvesting.

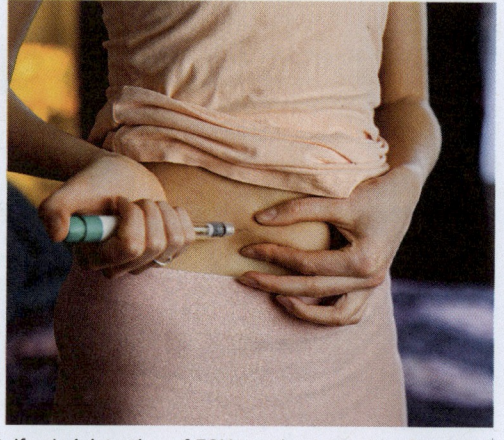

Self-administration of FSH to stimulate ovum formation.

1. Explain the role of each of the following hormones involved in IVF super-ovulation:

   (a) Gonadotropin releasing hormone: _____

   _____

   (b) Follicle stimulating hormone: _____

   _____

   (c) Oestradiol: _____

   _____

   (d) Human chorionic gonadotropin: _____

   _____

©2024 **BIOZONE** International
ISBN: 978-1-99-101410-8
Photocopying prohibited

D3.1
7

305 Detecting Pregnancy

Key Idea: Pregnancy can be detected by testing for hCG in the urine of women.

When a woman becomes pregnant, a hormone called human chorionic gonadotropin (hCG) is released by the chorion cells of the embryo. HCG accumulates in the bloodstream and is excreted in the urine. Pregnancy tests use antibodies to bind the hCG and produce a reaction that can be used to signal a positive test for pregnancy.

Pregnancy testing

▶ Human chorionic gonadotropin (hCG) is released by pregnant women soon after fertilization in rising concentrations, and is excreted in the urine. In pregnancy tests, hCG acts as an antigen and is acted on by monoclonal antibodies (Ab) in the test kit to determine if a woman is pregnant.

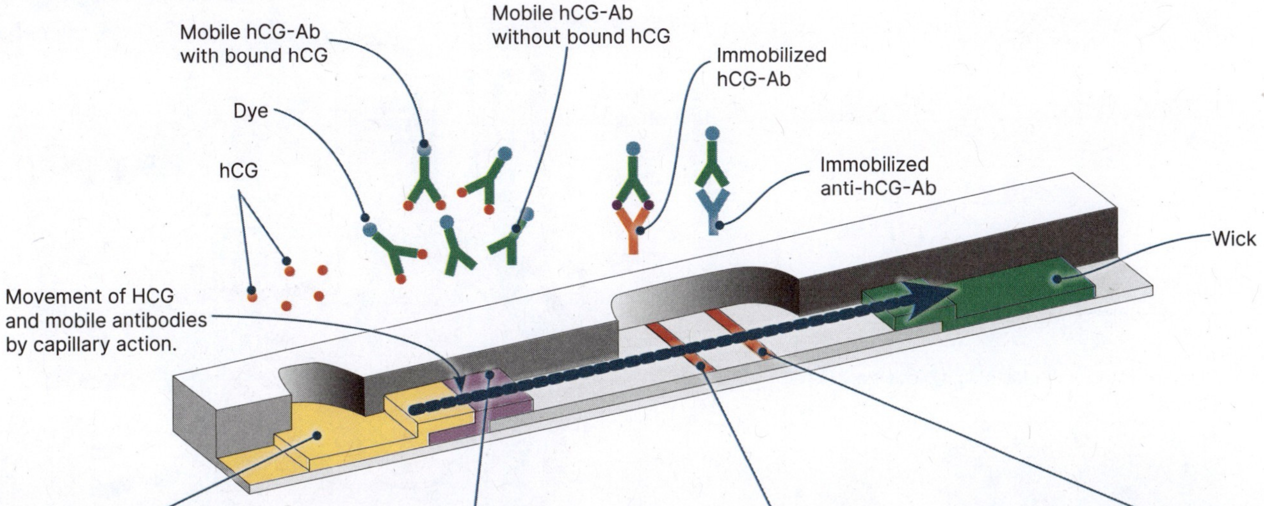

Mobile hCG-Ab with bound hCG

Mobile hCG-Ab without bound hCG

Immobilized hCG-Ab

Dye

Immobilized anti-hCG-Ab

hCG

Wick

Movement of HCG and mobile antibodies by capillary action.

1. The dipstick or sample pad absorbs urine and hCG, if present.

2. Reaction zone: hCG binds to mobile hCG antibodies (hCG-Ab) with dye attached.

3. Test line: If hCG is present hCG-Ab complexes will bind to a second immobilized hCG-ab, producing a line of dye.

4. Control line: hCG-ab that have not bound any hCG (it may or may not be present) continue on the anti-hCG-Ab and are bound in place, producing a second line.

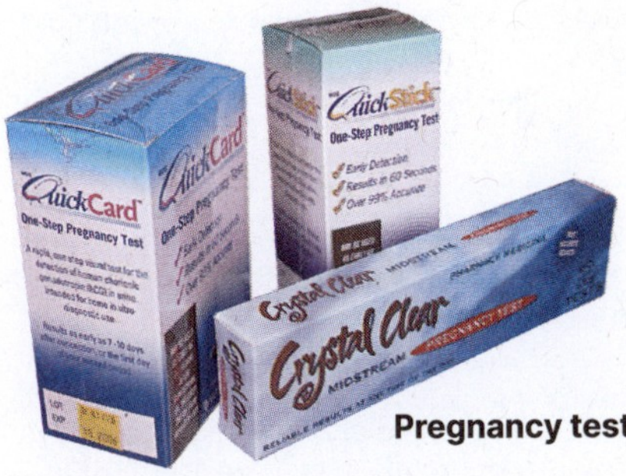

Pregnancy tests

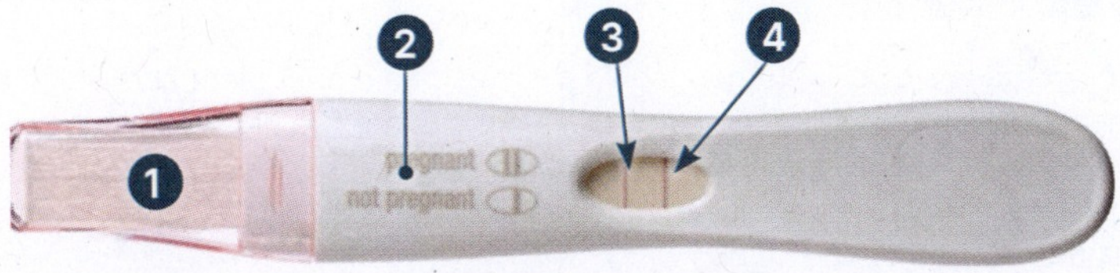

Pregnancy test device

1. Suggest why a monoclonal antibody-based test for hCG to detect pregnancy is valuable: _____

_____

_____

2. (a) Will a positive pregnancy test produce one or two lines in the result window? _____

(b) Explain your answer: _____

_____

_____

_____

_____

3. Progesterone is produced by the placenta during pregnancy. Why do pregnancy tests not test for this hormone?

_____

_____

_____

D3.1
17
AHL

306 The Placenta

Key Idea: The placenta allows materials to be exchanged between the foetus and its mother. It also acts as a temporary endocrine organ, secreting hormones to maintain pregnancy. The human foetus is dependent on its mother for nutrients, oxygen, and waste elimination. The placenta is the specialized organ that performs this role, enabling exchange between foetal and maternal tissues and allowing a prolonged period of foetal growth and development within the uterus. The placenta also has an endocrine role, producing progesterone and oestrogen to maintain the pregnancy.

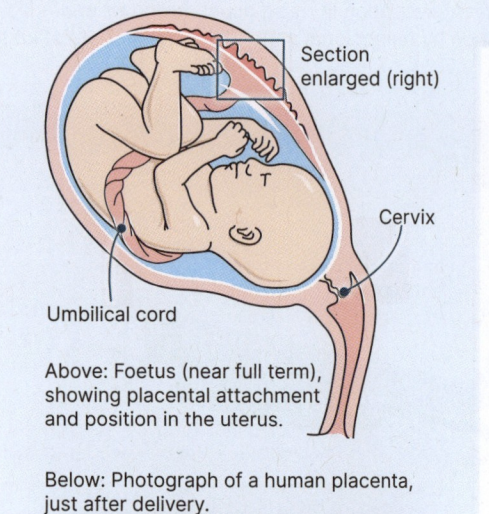

Section enlarged (right)

Cervix

Umbilical cord

Above: Foetus (near full term), showing placental attachment and position in the uterus.

Below: Photograph of a human placenta, just after delivery.

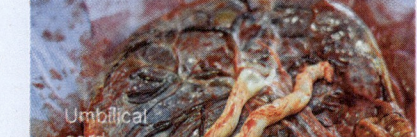

Umbilical cord

Schematic diagram showing part of the placenta in section

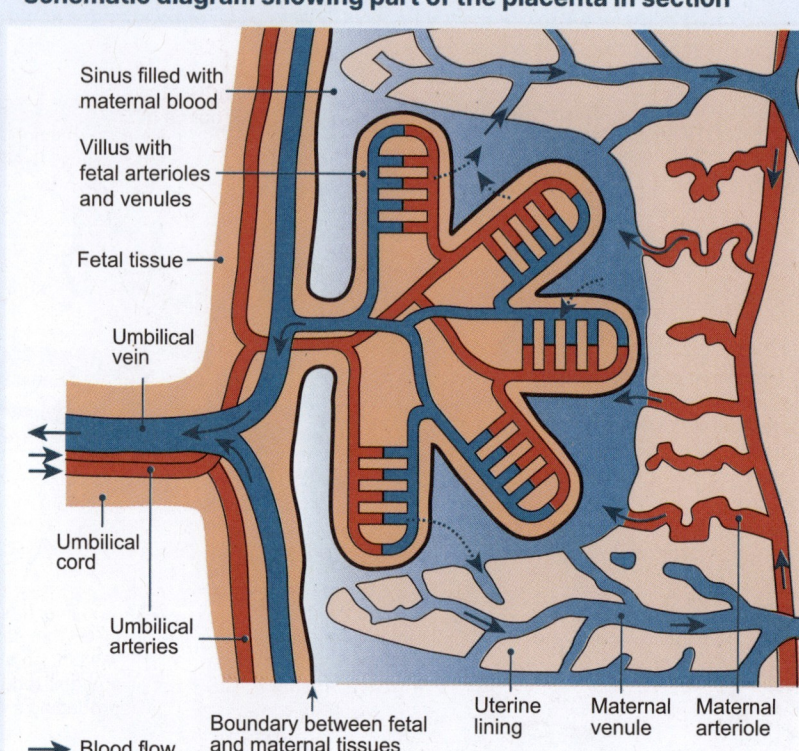

Sinus filled with maternal blood

Villus with fetal arterioles and venules

Fetal tissue

Umbilical vein

Umbilical cord

Umbilical arteries

Boundary between fetal and maternal tissues

Uterine lining

Maternal venule

Maternal arteriole

→ Blood flow

·····> Exchange of wastes and nutrients via diffusion

▸ The placenta is a disc-like organ, about the size of a dinner plate and weighing about 1kg. It develops when finger-like projections (villi) from the foetal membranes grow into the uterine lining. The villi contain the capillaries connecting the foetal arteries and vein. They continue invading the maternal tissue until they are bathed in the maternal blood sinuses. The maternal and foetal blood vessels are in such close proximity that oxygen and nutrients can diffuse from the maternal blood into the capillaries of the villi.

▸ From the villi, the nutrients circulate in the umbilical vein, returning to the foetal heart. Carbon dioxide and other wastes leave the foetus through the umbilical arteries, pass into the capillaries of the villi, and diffuse into the maternal blood. The foetal and maternal blood do not mix. The exchanges occur via diffusion through capillaries.

Placental mammals retain the foetus internally until a much later stage of maturity than mammals that do not develop a placenta (the marsupials and the monotremes). Marsupials, e.g. the kangaroo, give birth within weeks of fertilization. The young are very underdeveloped. Monotremes, e.g. the platypus, lay eggs. In both cases the young are nourished on milk and complete development outside the womb.

1. Describe the structure of the human placenta, including the placental villi, and explain its role: _____

_____

_____

_____

_____

2. The umbilical cord contains the foetal arteries and vein. Describe the status of the blood in each type of foetus vessel:

(a) Foetal arteries: Oxygenated and containing nutrients / Deoxygenated and containing nitrogenous wastes (delete one).

(b) Foetal vein: Oxygenated and containing nutrients / Deoxygenated and containing nitrogenous wastes (delete one).

AHL

D3.1
18

©2024 **BIOZONE** International
ISBN: 978-1-99-101410-8
Photocopying prohibited

307 The Hormones of Pregnancy

Key Idea: Hormones secreted during pregnancy maintain the pregnancy and prepare the body for birth.

Levels of oestrogen and progesterone regulate the secretion of the pituitary hormones that control the ovarian cycle in non-pregnant adult human women. Pregnancy interrupts this cycle and maintains the corpus luteum and the placenta as endocrine organs, with the specific role of maintaining the developing foetus during its development. During the last month of pregnancy, the hormone oxytocin induces the uterine contractions that will expel the baby from the uterus.

HCG (Human chorionic gonadotropin)
- Secreted by the chorion (foetal portion of the placenta).
- Maintains corpus luteum.

Progesterone
- Maintains endometrium.
- Inhibits uterine contraction.

Oestrogen
- Maintains endometrium.
- Prepare mammary glands for lactation
- High levels induce labour.

Human placental lactogen (HPL)
- Stimulates breast growth and development.

Relaxin
- Produced by the placenta towards the end of the pregnancy.
- Relaxes pubic symphysis at birth.
- Helps dilate cervix at birth.

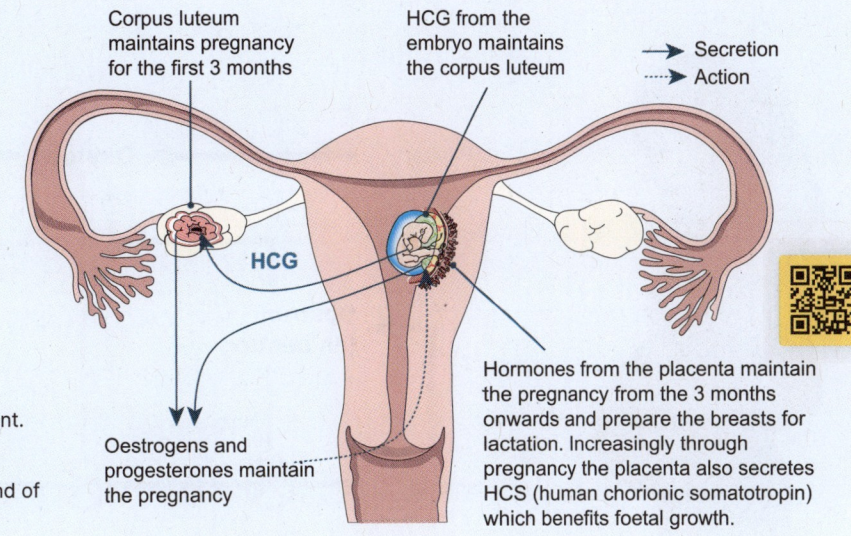

Corpus luteum maintains pregnancy for the first 3 months

HCG from the embryo maintains the corpus luteum

→ Secretion
⇢ Action

HCG

Oestrogens and progesterones maintain the pregnancy

Hormones from the placenta maintain the pregnancy from the 3 months onwards and prepare the breasts for lactation. Increasingly through pregnancy the placenta also secretes HCS (human chorionic somatotropin) which benefits foetal growth.

Hormonal changes during pregnancy, birth, and lactation

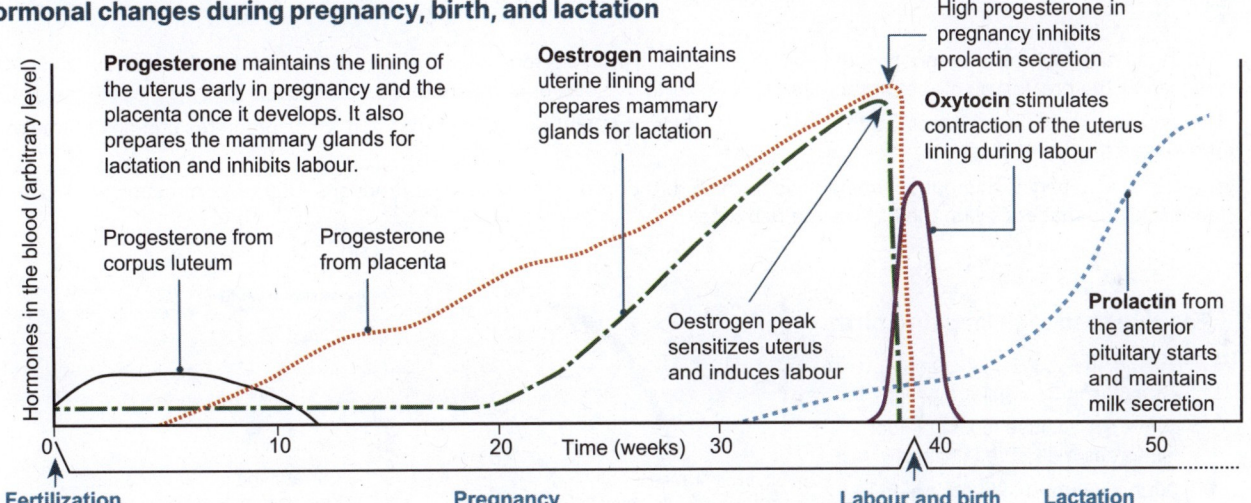

Progesterone maintains the lining of the uterus early in pregnancy and the placenta once it develops. It also prepares the mammary glands for lactation and inhibits labour.

Oestrogen maintains uterine lining and prepares mammary glands for lactation

High progesterone in pregnancy inhibits prolactin secretion

Oxytocin stimulates contraction of the uterus lining during labour

Progesterone from corpus luteum

Progesterone from placenta

Oestrogen peak sensitizes uterus and induces labour

Prolactin from the anterior pituitary starts and maintains milk secretion

Hormones in the blood (arbitrary level)

Time (weeks) — 0, 10, 20, 30, 40, 50

Fertilization — Pregnancy — Labour and birth — Lactation

- During the first 12-16 weeks of pregnancy, the corpus luteum secretes enough progesterone to maintain the uterine lining and sustain the developing embryo. After this, the placenta takes over as the primary endocrine organ of pregnancy. Progesterone and oestrogen from the placenta maintain the uterine lining, stop further ova (eggs) developing, and prepare the breast tissue for lactation (milk production).

- At the end of pregnancy, the placenta begins to break down and progesterone levels fall. The uterus becomes sensitive to the high oestrogen levels, triggering labour to start. The oestrogen peak coincides with an increase in oxytocin, which stimulates uterine contractions in a positive feedback loop: the contractions and the increasing pressure on the cervix from the infant stimulate release of more oxytocin and more contractions, until the infant exits the birth canal, which triggers the end of the feedback cycle.

- After birth, the secretion of prolactin increases. Prolactin maintains lactation during the period of infant nursing.

1. (a) Why is the corpus luteum the main source of progesterone in early pregnancy? _____

_____

(b) What hormones are responsible for maintaining pregnancy? _____

2. (a) Name two hormones involved in labour (onset of the birth process): _____

(b) Describe two physiological factors in initiating labour: _____

_____

_____

D3.1 19 · AHL

Hormonal regulation of birth

▶ Birth and lactation are both controlled by hormones and involve positive feedback. A human pregnancy lasts about 38 weeks after fertilization. It ends in labour, the birth of the baby, and expulsion of the placenta.

▶ During pregnancy, progesterone maintains the placenta and inhibits contraction of the uterus. At the end of a pregnancy, increasing oestrogen induces labour. Prostaglandins, an ageing placenta, and the state of the foetus itself also play a role. At the same time, the hormone oxytocin stimulates the contractions of the uterus that will expel the baby.

▶ After birth, the mother provides nutrition for the infant through lactation (milk production). Breast milk nourishes infants for the first 4-6 months of life. It also contains maternal antibodies which protect the infant against infection.

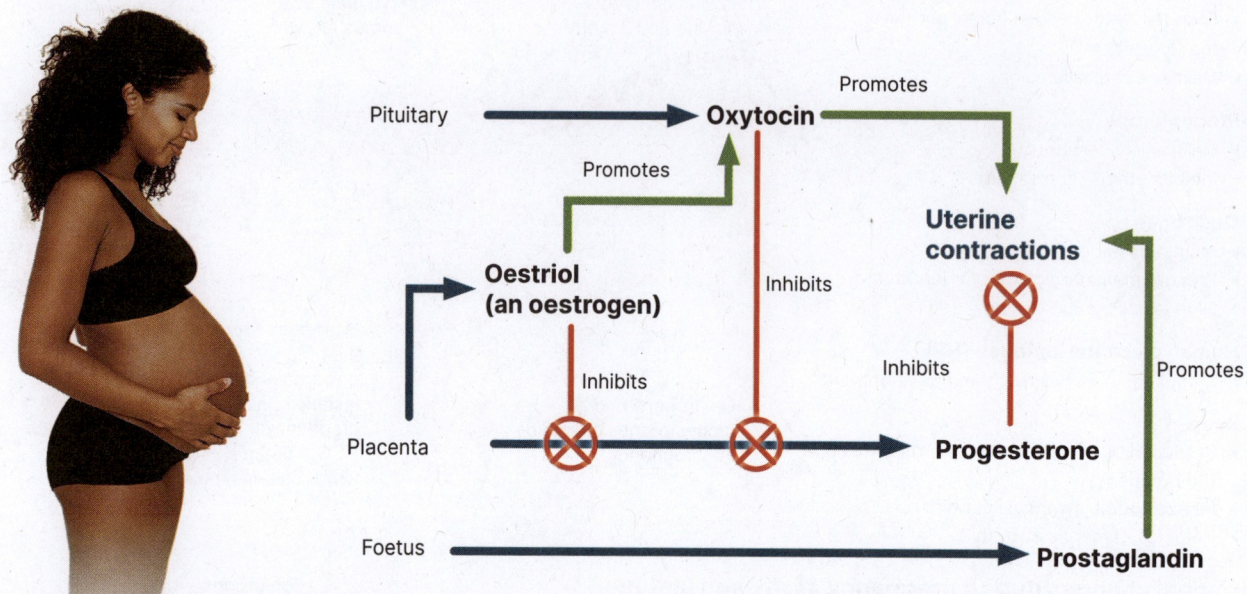

▶ During the nine months of pregnancy, the placenta produces progesterone which inhibits uterine contractions. At the end of the pregnancy, the breakdown of the placenta results in a fall in the level of progesterone.

▶ The oxytocin produced by the placenta is then able to promote the production of oxytocin which promotes uterine contractions and begins the birth process.

▶ The foetus also begins to produce prostaglandin which further promotes uterine contractions. Uterine contractions produce a feedback loop that ends with the birth of the baby (below).

Positive feedback loop and birth

▶ Childbirth involves a positive feedback loop under hormonal control. A positive feedback loop reinforces a detected change.

▶ During childbirth, stretching of the cervix by the foetus triggers the release of hormones that enhance the birthing process by causing the contraction of the muscles lining the uterus.

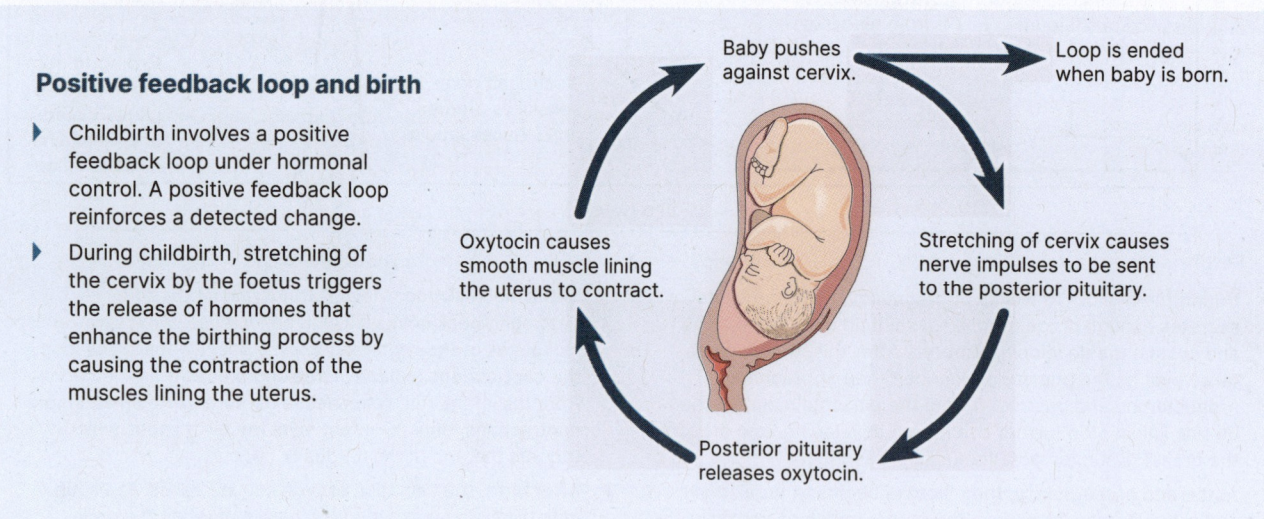

3. (a) What type of feedback loop is labour? _____

   (b) What two other factors might influence the timing of labour onset? _____

   _____

   _____

4. Which two hormones inhibit progesterone? _____

5. What causes the end of the birth feedback loop? _____

©2024 **BIOZONE** International
ISBN: 978-1-99-101410-8
Photocopying prohibited

308 Hormone Replacement Therapy

Key Idea: Hormone replacement therapy was originally linked to lowered risk of coronary heart disease. However, further research identified a causal relationship, rather than cause-and-effect.

Hormone replacement therapy (HRT) is a treatment available to women to counteract the loss of naturally produced hormones, such as oestrogen and progesterone, during menopause. Menopause can cause symptoms including anxiety, hot flushes, and discomfort. Epidemiological studies that were undertaken when HRT first became commonly used identified a correlation between women undergoing treatment and a lower incidence of coronary heart disease. Further research using more stringent randomized controlled trials revised the current thinking between the correlation of these two factors - shifting from a cause-and-effect relationship to a causal relationship.

Menopause and oestrogen levels

▶ Oestrogen levels in females, mostly produced from the ovaries, rise to a constant high level at puberty and start to fluctuate at the beginning of perimenopause. On average, perimenopause lasts for around 10 years, and completes when oestrogen levels fall to a low constant level. Women are recommended to begin taking HRT once oestrogen falls to a constant low level, as there is a risk of it inducing frequent and irregular bleeding if begun before this point.

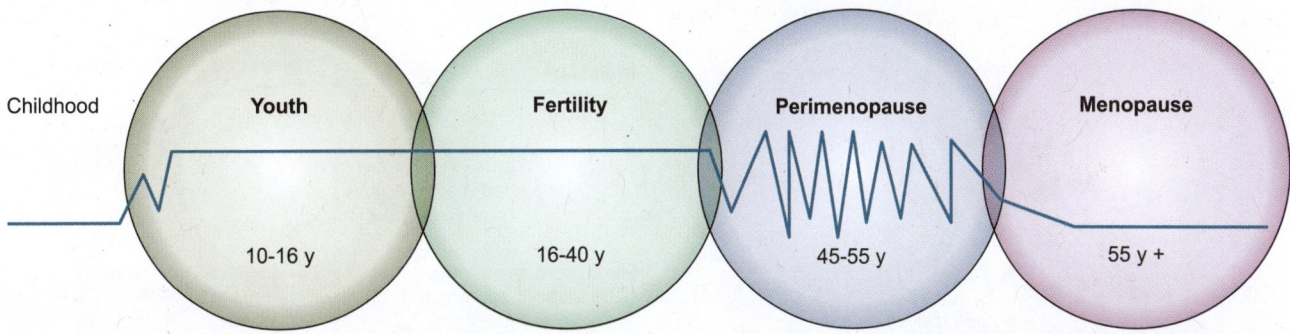

Childhood	Youth	Fertility	Perimenopause	Menopause
	10-16 y	16-40 y	45-55 y	55 y +

Cause-and-effect vs causal correlations

▶ Two decades of mainly observational studies of health profiles of women who were undergoing HRT suggested that their risk of coronary heart disease (CHD) was lower than for women not taking HRT.

▶ A cause-and-effect correlation was promoted, whereby women who used HRT (the cause) could benefit from reduced CHD risk (the effect).

▶ However, as HRT was almost always a medication that incurred a financial cost to the patient, the women who took HRT also tended to be from a higher socio-economic group, who already have better health outcomes due to their lifestyles.

▶ Further studies, including the Women's Health Initiative (WHI) and the Heart and oestrogen/progestin Replacement Study (HERS), had more robust protocols for their investigations. Scientists concluded that there was no cause-and-effect correlation between HRT and CHD, but instead a causal relationship, i.e. the same variable that increased the likelihood of women undergoing HRT (socioeconomic status) was also likely to influence CHD risk.

▶ Even more recent studies using randomized and controlled clinical trials, identified a small increase in CHD risk and HRT.

Possibilities if A is correlated with B

A causes B Cause-and-effect	B causes A Cause-and-effect
A ➔ B	A ⬅ B

A and B were caused by C Causal	A and B are unrelated and their correlation is a coincidence
A B ⬉⬈ C	A B ⬆⬆ C D

1. Why might the initial investigations have concluded that HRT and CHD risk were a cause-and-effect correlation?

_____

_____

_____

2. What is the benefit of controls and randomized clinical trials when conducting epidemiological studies?

_____

_____

_____

309 Sexual Reproduction in Flowering Plants

Key Idea: The life cycle of flowering plants includes alternation between a haploid (n) gametophyte generation and a diploid (2n) sporophyte generation.

The life cycles of all plants are characterized by a phenomenon known as alternation of generations. The haploid (n) gametophyte generation alternates with the diploid (2n) sporophyte generation. In vascular plants, the sporophyte generation is dominant, typically called the plant.

▶ The sporophyte and gametophyte generations are named for the type of reproductive cells they produce. Haploid gametophytes (one set of chromosomes, denoted N) produce gametes by mitosis, whereas diploid sporophytes (two sets of chromosomes, denoted 2n) produce spores by meiosis (production of cells for sexual reproduction).

▶ Spores develop directly into organisms. Gametes (egg and sperm) form inside the gametophyte: in the pollen grain contained in the anther, and inside the ovule in the ovary.

▶ Pollen carries the gametes from the male reproductive structures to the female reproductive structures during pollination.

▶ The gametes (n) unite during fertilization to form a zygote (2n), which gives rise to the sporophyte generation.

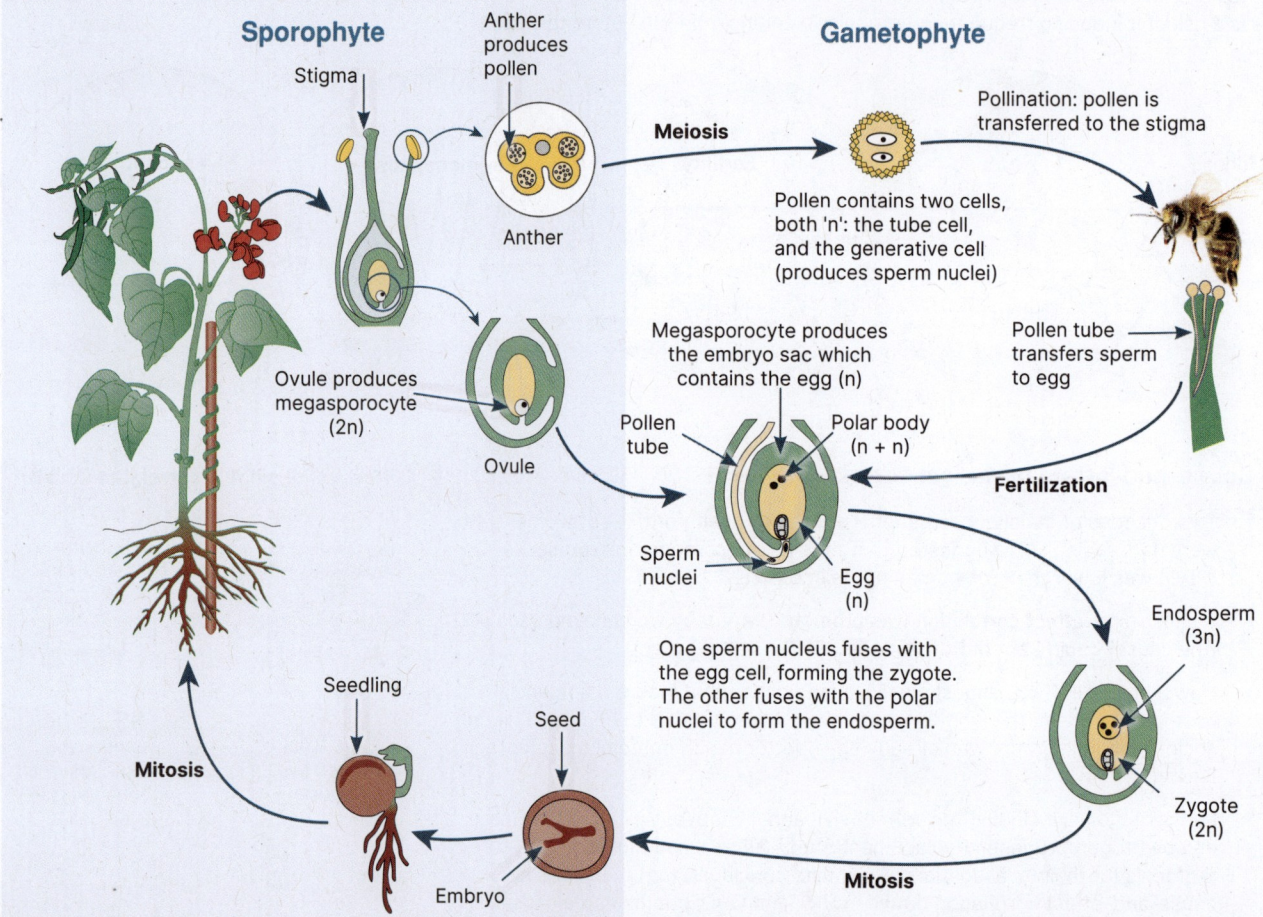

Most flowering plants have both male and female structures contained within the same flower, and therefore plant. Due to this, most flowering plants are hermaphroditic but most have mechanisms to avoid self-pollination. However, the flowering plants are still sexually reproducing, evidenced by the process of meiosis to form gametes which then combine in fertilization to form a zygote - and thus introduce variation in the species population.

1. Describe the alternation of generations in plants: _____

_____

2. (a) In which part of a flowering plant do the male gametophytes develop? _____

   (b) In which part of a flowering plant do the female gametophytes develop: _____

3. Complete the table below to show what produces each structure:

Structure	Spores	Gametes	Zygote
Produced by			
Process			

D3.1

8

©2024 **BIOZONE** International
ISBN: 978-1-99-101410-8
Photocopying prohibited

310 Insect-Pollinated Flower Structure

Key Idea: The flower of an insect-pollinated plant has typical adaptations to facilitate the pollination process.

Pollination is the transfer of pollen from the male anther to the female stigma. Pollination of flowers by insects is usually mutualistic. Mutualistic relationships involve exchanges between two species, where the insect benefits from the energy in the plant nectar or pollen it consumes. The plant

benefits by having its gametes transferred to another plant. Nearly 88% of all flowering plants are pollinated by animals, with the vast majority being insects. The petal colour, flower shape and scent, and location of the nectary are all adaptations to attract and position a specific insect species or insect group to most effectively collect the pollen and then deposit it on the stigma of a different flower.

Cross section of an insect pollinated flower

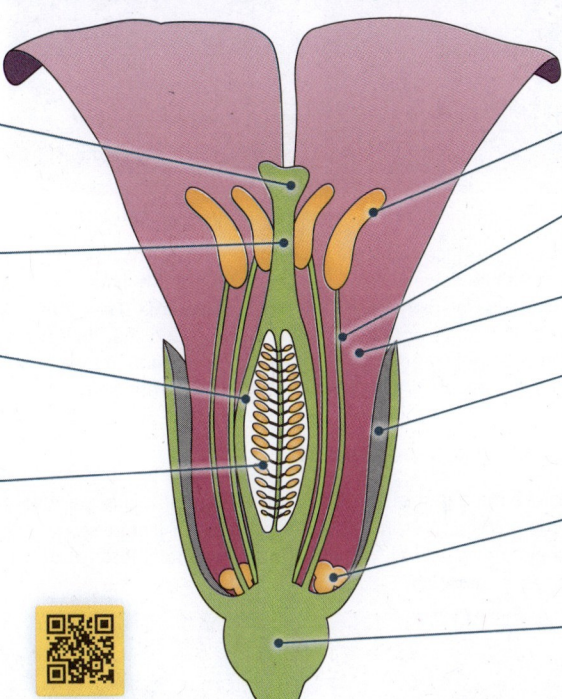

Female Reproductive Structures

Stigma: The receptive part of the carpel. Pollen grains will germinate only if they land here.

Style: The structure that supports the stigma.

Ovary: The base of the carpel where the ovules develop.

Ovules: These are eggs that, once fertilized, become the seeds. The ovule skin becomes the seed coat or testa.

An entire female part is the carpel. There may be one or more carpels per flower.

Male Reproductive Structures

Anther: Top portion of the stamen, the male organ of reproduction.

Filament: The slender stalk of the stamen that supports the anther.

Petal: Collectively, these form the corolla. Often brightly coloured.

Sepals: Together, form the calyx. Usually green, but sometimes the same colour as petals.

Nectary: Plants produce a sugary liquid called nectar to attract insects to the flower.

Receptacle: The swollen base of the flower. Sometimes, it forms the succulent tissue of the fruit.

The petals of flowers guide insects towards the pollen or nectar at the centre of the flower, using various colours and lines known as nectar guides. In this way, wandering insects are enticed into entering the flower and transfer pollen in the most efficient way. Insect pollinated flowers tend to be white, pink, or purple - within the range of light that insects can 'see'.

Bees, and many other insects, are able to detect ultraviolet light. Many flowers contain pigments that reflect UV, producing a specific pattern visible to insects but not to other animals. In this way, plants can use their flowers to specifically attract preferred insect pollinators.

1. The insect pollinated flower and associated structures are shown above. After studying the diagram above, on a separate sheet, draw a diagram of the cross-section of the flower and label the following structures: stigma, style, ovary, ovules, anther, filament, petal, sepals, nectary, receptacle, and identify male and female structures. Annotate the diagram with functions of the respective structures. Attach the diagram to the page when completed.

2. (a) Explain why flowering plants need to attract pollinators to their flowers: _____

   _____

   _____

   (b) Describe the adaptations in angiosperms (flowering plants) that attract specific pollinators to their flowers:

   _____

   _____

   _____

311 Cross-pollination and Fertilization Mechanisms

Key Idea: In plants, pollination is essential to ensure fertilization and production of seeds.

Pollination is the transfer of pollen grains from the male reproductive structures to the female reproductive structures of plants. This must happen before fertilization (the joining of the egg and sperm) can occur. Adaptations to ensure cross-pollination (pollination between different plants) include structural and physiological mechanisms associated with the flowers or cones themselves, and reliance on wind and animal pollinators.

Plant mechanisms to ensure cross-pollination

Male willow catkin

An effective way of ensuring cross-pollination is to have separate male and female plants. This occurs in about 6% of angiosperms, including willow, feijoa, and holly plants.

Male and female pine cones

Other plants produce separate male and female flowers or cones on the same plant. They may develop at different times to ensure that pollen does not fertilize the same plant.

Tulip anthers and stigma

Some angiosperms have flowers with both male and female structures. They can ensure cross-pollination if the anthers/pollen and stigma mature at different times.

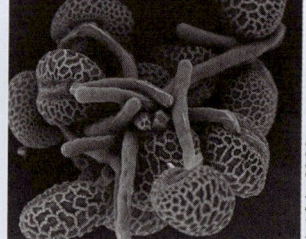

Germinating pollen grains

In many plants, pollen landing on the stigma of the same plant will not even germinate. This ensures that the egg cells are not fertilized by sperm from the same plant.

Animal and wind pollination to ensure cross-pollination

You will recall that pollen grains are immature male gametophytes, formed by mitosis of haploid microspores within the pollen sac. Pollination is the actual transfer of the pollen from the stamens to the stigma, or from the male cone to the female cone. Pollen grains cannot move independently and they are usually carried by wind or animals. Both methods distribute the pollen away from the parent plant, reducing the chance of self-pollination.

Relying on **animal pollination**, including insects, can be an energy-expensive mechanism for plants. Growing the flower, and synthesizing nectar from photosynthetic products direct resources away from plant growth. However, if the adaptations are targeted to specific pollinators who are likely to visit the flowers of the same species, less pollen needs to be produced and the chances of the pollen arriving at the correct location are increased.

Wind can be used to move pollen from one plant away to another. The plants invest less into reproductive flowers, as they do not need colourful petals or nectar to entice pollinators. However, the process is random and most of the pollen does not reach the desired location on another plant. This is countered by large scale production of small pollen grains and a high density of plants of the same species, such as grasses, growing close together.

1. Explain why plants have generally evolved to limit self-pollination: _____

   _____

   _____

   _____

2. Compare and contrast animal and wind pollination as methods to enable cross-pollination: _____

   _____

   _____

   _____

   _____

D3.1
10 – 11

©2024 **BIOZONE** International
ISBN: 978-1-99-101410-8
Photocopying prohibited

Fertilization in angiosperms

▶ In angiosperms, pollen lands on the sticky stigma and completes development, germinating and growing a pollen tube that extends down to the ovary (shown right).

▶ The pollen tube is directed by chemicals (usually calcium) to the ovule. It enters through the micropyle, a small gap in the ovule. A double fertilization takes place. One sperm nucleus fuses with the egg to form the zygote. A second sperm nucleus fuses with the two polar nuclei within the embryo sac to produce the endosperm tissue (3N).

▶ There are usually many ovules in an ovary, therefore many pollen grains (and fertilizations) are needed before the entire ovary can develop.

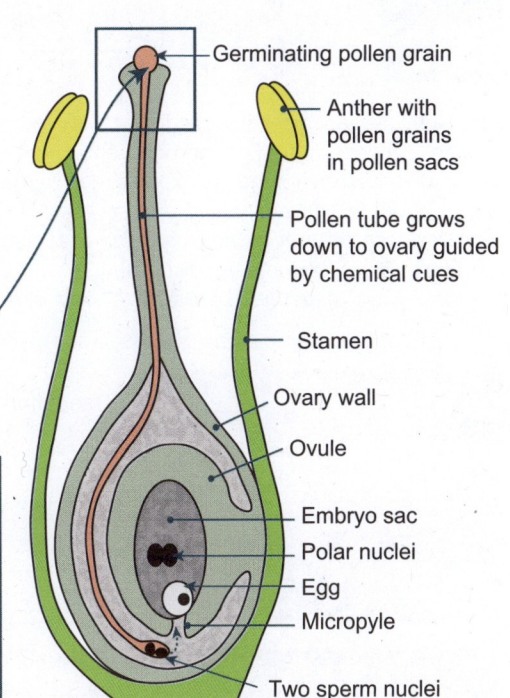

- Germinating pollen grain
- Anther with pollen grains in pollen sacs
- Pollen tube grows down to ovary guided by chemical cues
- Stamen
- Ovary wall
- Ovule
- Embryo sac
- Polar nuclei
- Egg
- Micropyle
- Two sperm nuclei

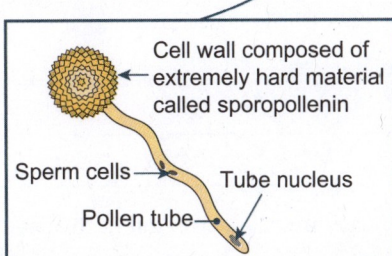

- Cell wall composed of extremely hard material called sporopollenin
- Sperm cells
- Tube nucleus
- Pollen tube

Self-incompatibility mechanisms

▶ At a genetic level, many plants have mechanisms to prevent self-pollination and avoid a loss of variation in offspring. Excessive self-pollination reduces the genetic diversity of a population of plants that occurs with inbreeding, and limits the extent at which a population can survive change in the environment.

▶ Self-incompatibility is the inability of a self-pollinated flower to produce a zygote, and involves cellular interactions between the pollen and the carpel of the same plant. The cell to cell recognition system is the result of one or more allele loci (allele pairs at the same location on homologous chromosomes), but operates with different mechanisms and different alleles, indicating that self-incompatibility has evolved more than once in different plant species.

▶ Brassicas have a signalling pathway that activates in the carpel when encountering same-plant pollen. This causes the pollen to be rejected.

▶ Plants in the nightshade family (*Solanaceae*) have an RNase (an enzyme that breaks down RNA) controlled by multiple alleles. This blocks pollen tube growth when a same-plant pollen is detected.

▶ Poppies (*Papaveraceae*) have a number of complex reactions, including programmed cell death of pollen tubes from incompatible pollen. California poppies, above right, form a ground level mat that is extensively covered in bright orange flowers that bloom at the same time for a lengthy period. Self-incompatibility limits inbreeding in closely grouped flowers.

3. Distinguish clearly between pollination and fertilization: _____

_____

_____

_____

4. Why is plant inbreeding detrimental for many wild plant populations? _____

_____

_____

5. Research another self-incompatibility mechanism that prevents inbreeding in plants and summarize below:

_____

_____

_____

312 Seed Dispersal

Key Idea: Seeds are dispersed from the parent plant to reduce competition for light and nutrients.

Plants disperse their seeds to expand their range and reduce competition. In some cases, the seed itself is the agent of dispersal, but often it is the fruit or an associated attached structure. Seeds are mainly dispersed by wind, water, and animals. Wind dispersed seeds have wing-like or feathery structures that catch the air currents and carry the seeds. Plants that rely on animals to spread their seeds may have hooks or barbs that catch the animal hair, sticky secretions that adhere to the skin or hair, or fleshy fruits that are eaten, leaving the seed to be deposited in faeces, away from the parent plant. Other dispersal mechanisms rely on explosive discharge or shaking from pods or capsules.

For each of the examples below, research the method of dispersal and the adaptive features associated with the method:

1. **Dandelion** seeds are held in a puff-like cluster:

   (a) Dispersal mechanism: _____

   (b) Adaptive features: _____

   _____

   _____

2. **Acorns** are heavy fruits in which the fleshy seeds are encased in a resistant husk:

   (a) Dispersal mechanism: _____

   (b) Adaptive features: _____

   _____

   _____

   _____

3. **Coconuts** are heavy, buoyant fruits with a thick husk:

   (a) Dispersal mechanism: _____

   (b) Adaptive features: _____

   _____

   _____

Seed pod

Flower

Changarra CC 2.5

4. **Banksia** seeds are attached to a light wing structure

   (a) Dispersal mechanism: _____

   (b) Adaptive features: _____

   _____

   _____

   _____

5. **Wattle** (*Acacia spp.*) seeds are enclosed in pods. A fleshy strip surrounds each seed:

   (a) Dispersal mechanism: _____

   (b) Adaptive features: _____

   _____

   _____

6. **LillyPilly** (*Acmena spp.*, *Syzgium spp.*) seeds are surrounded by fleshy fruits:

   (a) Dispersal mechanism: _____

   (b) Adaptive features: _____

   _____

D3.1
12

130

313 Seed Structure and Germination

Key Idea: The seed houses the dormant embryonic plant and its food store until conditions for germination are met.

After fertilization, the ovules within the ovary become the seeds. Recall the double fertilization in angiosperms: one fertilization produces the embryo and the other produces the endosperm. The development of the endosperm is important as it provides a nutrient store for the young plant. A seed is an entire reproductive unit, housing the embryonic plant in a state of dormancy. During the last stages of maturation, the seed dehydrates until its water content is only 5-15% of its weight. The embryo stops growing and remains dormant until the seed germinates. At germination, the seed takes up water and the food store is mobilized to provide the nutrients required for plant growth and development.

Dicot seeds: soy (above) cashew (below) There are two fleshy cotyledons. These store food absorbed from the endosperm.

In germination, the seed must rehydrate and reactivate its metabolism. The seed absorbs water through the seed coat (testa) and micropyle. As the seed tissue takes up water, the cells expand, metabolism is reactivated, and growth begins. Activation begins with the release of the hormone gibberellin (GA) from the embryo. GA promotes cell elongation so the root can penetrate the testa. It also stimulates the synthesis of enzymes that hydrolyze the starch to produce sugars. The mobilized food is delivered to the developing roots and shoots.

Seed structure and function

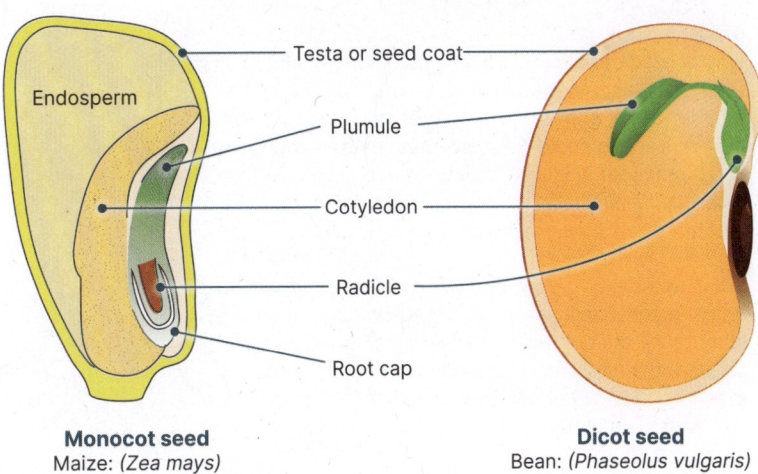

Testa or seed coat — Endosperm — Plumule — Cotyledon — Radicle — Root cap

Monocot seed
Maize: (*Zea mays*)

Dicot seed
Bean: (*Phaseolus vulgaris*)

Every seed contains an embryo comprising a rudimentary shoot (plumule), root (radicle), and one or two cotyledons (seed leaves). The embryo and its food supply are encased in a tough, protective seed coat or testa. In monocots, the endosperm provides the food supply, whereas in most dicot seeds, the nutrients from the endosperm are transferred to the large, fleshy cotyledons.

Germination in a dicot seed
Bean: (*Phaseolus vulgaris*)

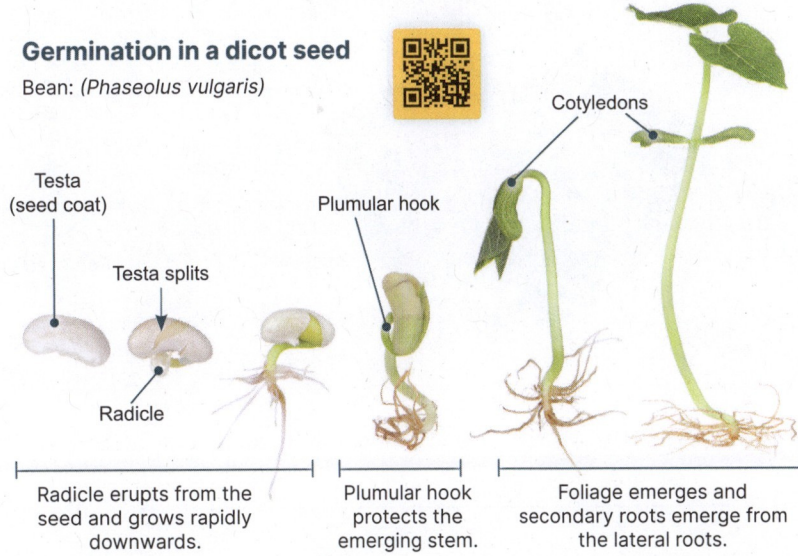

Testa (seed coat) — Testa splits — Radicle — Plumular hook — Cotyledons

Radicle erupts from the seed and grows rapidly downwards.

Plumular hook protects the emerging stem.

Foliage emerges and secondary roots emerge from the lateral roots.

1. Compare between pollination and seed dispersal: _____

_____

_____

_____

2. How are seeds able to develop an embryo prior to light being available for photosynthesis? _____

_____

_____

_____

314 The Genetic Basis of Inheritance

Key Idea: Inheritance involves the passing of genetic material held in chromosomes from one generation to the next.

All genetic material in eukaryotes is held in chromosomes located in the cell nucleus. Diploid cells contain two copies of each autosomal gene, as distinct from genes held on sex chromosomes. However, there may be some variation between the genes: variants are called alleles. Meiosis creates haploid gametes that have only one copy of each gene. Fertilization combines the male and female gametes to create a zygote with a restored diploid set of chromosomes. Gamete formation and fertilization to form a zygote is common to all eukaryotic processes of sexual reproduction.

Gregor Mendel and inheritance

Gregor Mendel (1822-1884), below, was an Austrian monk who carried out pioneering studies of inheritance. Mendel bred pea plants to study the inheritance patterns of a number of traits (specific characteristics). He showed that characters could be masked in one generation but could reappear in later generations and proposed that inheritance involved the transmission of discrete units of inheritance from one generation to the next.

When Mendel proposed his laws of inheritance, the actual mechanism was unknown. Knowledge of genetic inheritance did not nullify Mendel's inheritance laws. Instead, the new knowledge provided an accepted mechanism and we now know these units of inheritance as genes. The entire genetic makeup of an organism is its genotype.

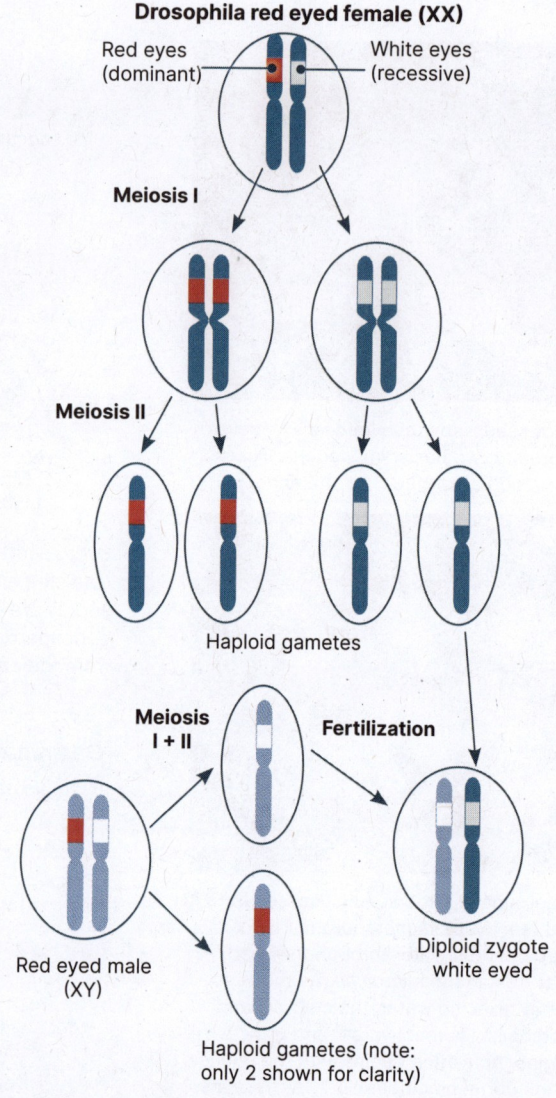

1. How did knowledge of genes and chromosomes support Mendel's laws of inheritance? _____

_____

_____

2. Compare and contrast diploid somatic cells and haploid gametes: _____

_____

_____

_____

3. What is the connection between haploid gametes and the formation of a diploid zygote and how does that relate to the concept of genetic inheritance?

_____

_____

_____

D3.2
1
©2024 **BIOZONE** International
ISBN: 978-1-99-101410-8
Photocopying prohibited

315 Genotype and Alleles

Key Idea: Eukaryotes generally have paired chromosomes. Each chromosome contains many genes and each gene may have a number of versions called alleles.

The combination of two alleles a person has of any gene is known as their genotype. The alleles can be the same, known as homozygous, or different, known as heterozygous. The physical characteristics that result from the genotype is known as the phenotype.

Homologous chromosomes

Most cells in sexually reproducing organisms have a homologous pair of chromosomes (one from each parent). Chromosomes are formed from DNA tightly wound around special proteins. This diagram shows the position of three different genes on the same chromosome that control three different traits (A, B and C).

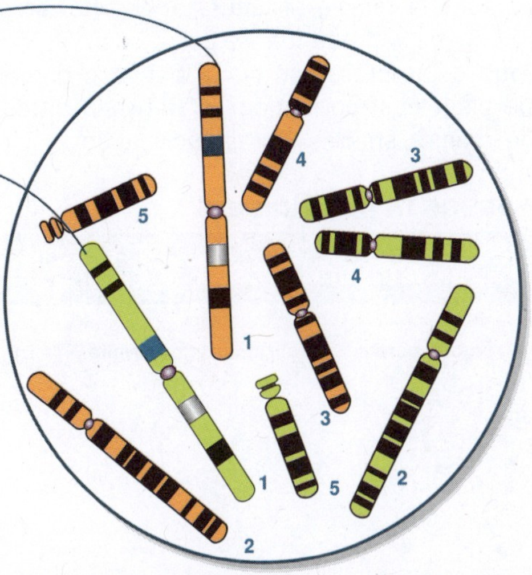

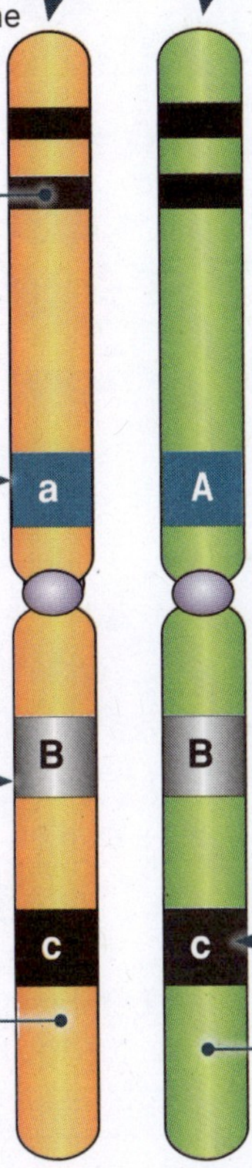

A gene is the unit of heredity. Genes occupying the same locus or position on a chromosome code for the same phenotypic character (e.g. eye colour).

Having two different versions (alleles) of gene A is called the heterozygous condition. Only the dominant allele (A) will be expressed. Alleles differ by only a few bases.

When both chromosomes have identical copies of the dominant allele for gene B the organism is homozygous dominant for that gene.

When both chromosomes have identical copies of the recessive allele for gene C, the organism is homozygous recessive for that gene.

Maternal chromosome originating from the egg of the female parent.

Paternal chromosome originating from the sperm of the male parent.

The diagram (above) shows the complete chromosome complement for a hypothetical diploid organism, with ten chromosomes as five nearly identical pairs. Each pair is numbered. Each parent contributes one chromosome to the pair. Sex chromosomes are not shown.

The pairs are called homologues or homologous pairs. Each homologue carries an identical assortment of genes, but the version of the gene (the allele) from each parent may differ.

Recall that mutations are often the result of copying errors during DNA replication. Mutations are the source of all new alleles.

1. Define the following terms used to describe the allele combinations in the genotype for a given gene:

   (a) Heterozygous: _____

   (b) Homozygous dominant: _____

   (c) Homozygous recessive: _____

2. For a gene given the symbol 'A', name the alleles present in an organism that is identified as:

   (a) Heterozygous: _____ (b) Homozygous dominant: _____ (c) Homozygous recessive: _____

3. What is a homologous pair of chromosomes? _____

   _____

   _____

   _____

4. Discuss the significance of genes existing as alleles: _____

   _____

   _____

   _____

©2024 **BIOZONE** International
ISBN: 978-1-99-101410-8

33 D3.2 3

316 Genetic Crosses in Flowering Plants

Key Idea: The outcome of a cross depends on the parental genotypes. A true breeding parent is homozygous for the gene involved.

A cross between plants occurs when male gametes in the pollen contact the female structure of the flower and subsequently combine with the female gametes in the ovule during fertilization. The cross can occur between gametes from the same plant (self-pollination) or between different plants of the same species (cross-pollination). These crossed plants are known as the parent generation, or 'P generation'. The offspring that result from the first cross are known as the 'F_1 generation' and if any two plants from that first generation breed, then their offspring are known as the 'F_2 generation'. This targeted crossing of plants is often used to develop new plant varieties for horticultural purposes and improve agricultural crop species. Breeders select plants to cross based on their phenotype, aiming to produce a pure-breeding (homozygous) plant with desired physical features.

Genetic crosses in pea flowers

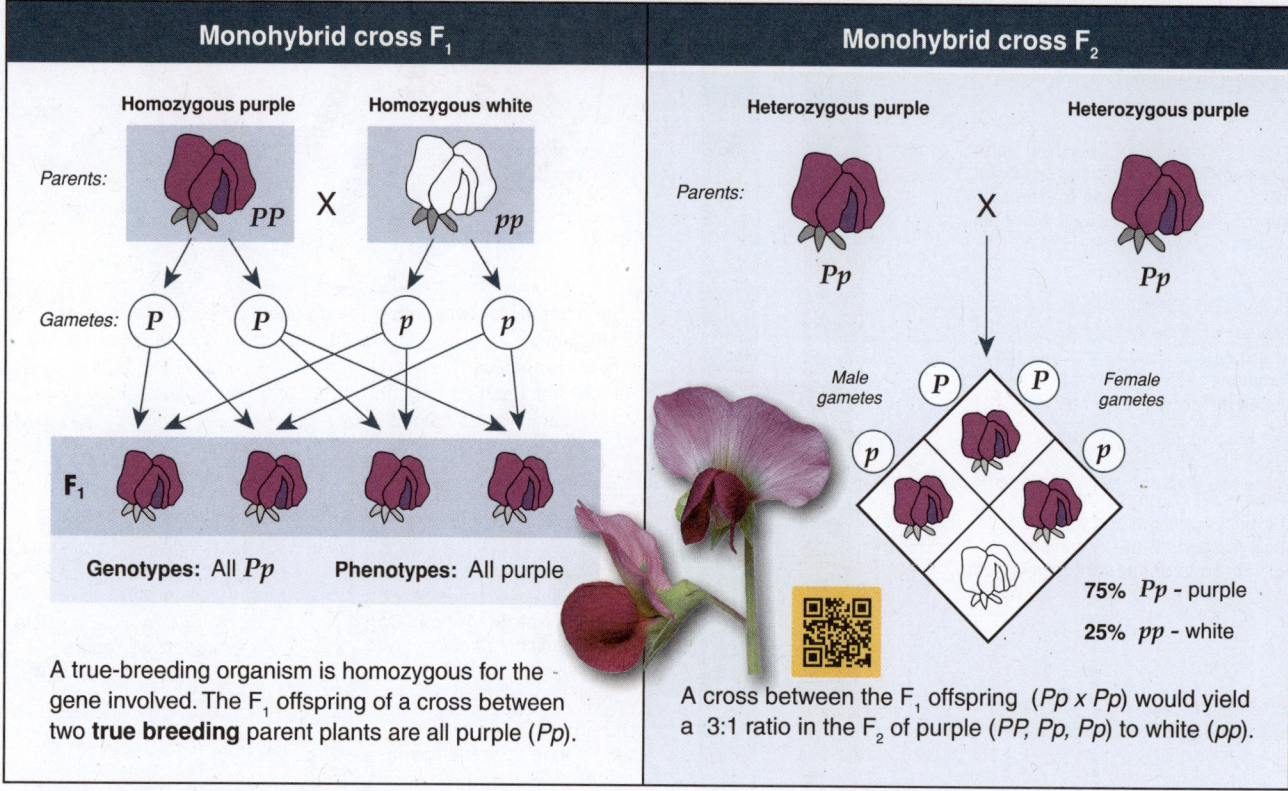

A true-breeding organism is homozygous for the gene involved. The F_1 offspring of a cross between two **true breeding** parent plants are all purple (*Pp*).

A cross between the F_1 offspring (*Pp* x *Pp*) would yield a 3:1 ratio in the F_2 of purple (*PP*, *Pp*, *Pp*) to white (*pp*).

The genotype of homozygous recessive organisms can be determined directly from the phenotype in a gene following simple Mendelian laws of inheritance. However, the genotype of a dominant phenotype can be either homozygous dominant or heterozygous, and can be ascertained by analysing the F_1 and F_2 generation crosses. Punnett squares are an efficient method of predicting ratios of expected genotypes and phenotypes in a cross, given the P generation's genotype.

1. Complete the crosses below:

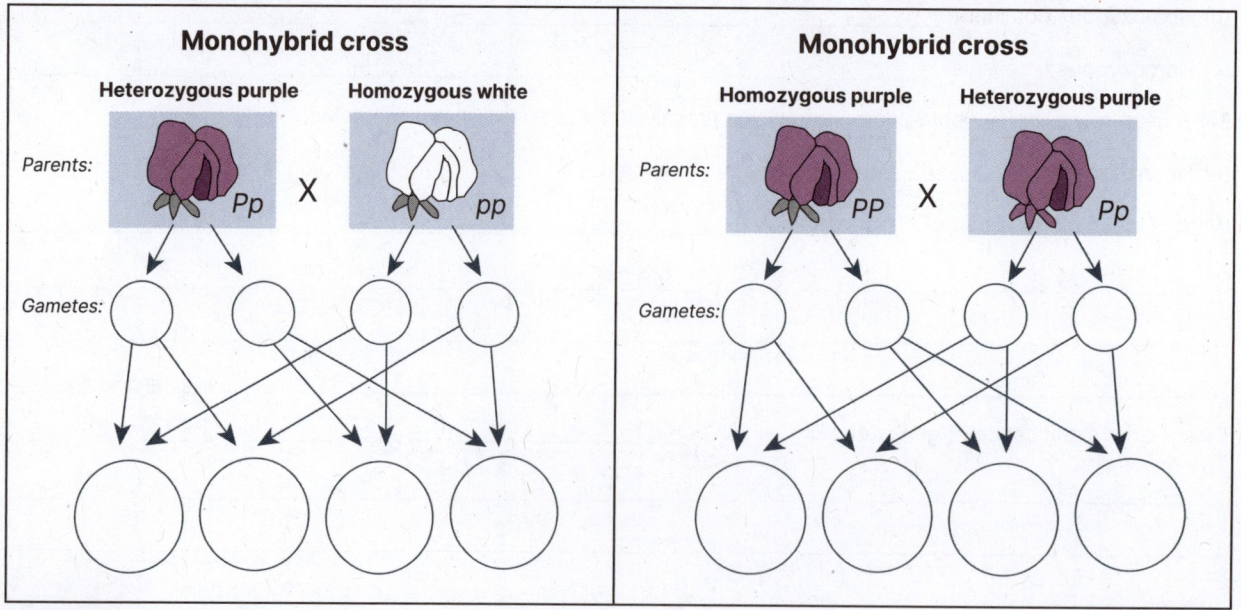

D3.2
2, 5

©2024 **BIOZONE** International
ISBN: 978-1-99-101410-8

Phenotypes and traits

▶ Traits are particular variants of a phenotypic (observed physical) characteristic. For example, a phenotype is eye colour, a trait is blue eye colour.

2. Describe the difference between the expression of dominant and recessive traits: _____

_____

_____

3. Study the top diagram on the previous pages and explain why the white flower colour trait does not appear in the F_1 generation but reappears in the F_2 generation:

_____

_____

_____

_____

4. Gregor Mendel carried out many inheritance crosses on pea plants (*Pisum sativum*). He examined seven phenotypic traits. Some of his results from crossing heterozygous plants are tabulated below. The numbers in the results column represent how many offspring had those phenotypic features.

   (a) Study the results for each of the six experiments below. Determine which of the two phenotypes is dominant and which is recessive. Place your answers in the spaces in the dominance column in the table below.

   (b) Calculate the ratio of dominant phenotypes to recessive phenotypes (to two decimal places). The first one has been done for you (5474 ÷ 1850 = 2.96). Place your answers in the spaces provided in the table below:

Trait	Possible phenotypes	Results		Dominance	Ratio
Seed shape	Wrinkled / Round	Wrinkled Round TOTAL	1850 5474 7324	Dominant: **Round** Recessive: **Wrinkled**	**2.96 : 1**
Seed colour	Green / Yellow	Green Yellow TOTAL	2001 6022 8023	Dominant: Recessive:	
Pod colour	Green / Yellow	Green Yellow TOTAL	428 152 580	Dominant: Recessive:	
Flower position	Axial / Terminal	Axial Terminal TOTAL	651 207 858	Dominant: Recessive:	
Pod shape	Constricted / Inflated	Constricted Inflated TOTAL	299 882 1181	Dominant: Recessive:	
Stem length	Tall / Dwarf	Tall Dwarf TOTAL	787 277 1064	Dominant: Recessive:	

5. Typical Mendelian crosses predict a 3:1 first cross. Why are the crosses above slightly different?

_____

_____

317 Phenotype and the Environment

Key Idea: Phenotype can be influenced by the environment, as well as the genotype in some genes.

The phenotype encoded by genes is a product not only of the genes themselves, but of their internal and external environment and the variations in the way those genes are controlled (epigenetics). For example, in colour-point cats, the darkened extremities of their coat colour is in response to temperature acting upon the genes. Humans have phenotypes that are controlled entirely by the genotype, such as earlobe attachment, the presence of a widow's peak, and the ability to curl the tongue. However, other traits are the result of both the genotype and environmental influences, including intelligence and behaviours, with evidence from identical twin 'nature-nurture' studies to support the claims.

Genotype is the primary influence on most qualitative traits, such as flower colour or comb shape in poultry; determined by one or two genes.

Single comb
rrpp

Epigenetic influences are important in providing the mechanism by which the environment influences phenotype, as in temperature dependent sex determination.

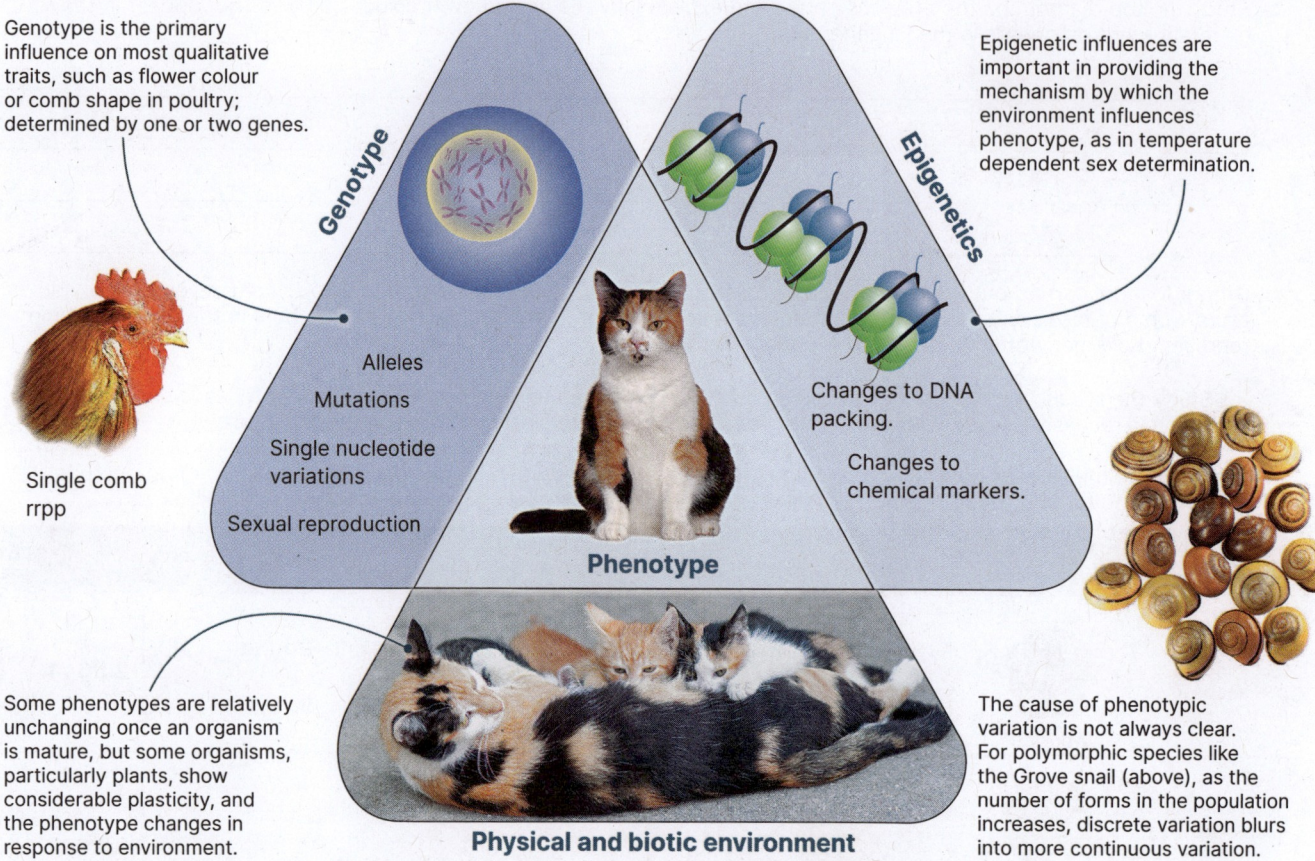

Genotype

Alleles
Mutations
Single nucleotide variations
Sexual reproduction

Phenotype

Changes to DNA packing.

Changes to chemical markers.

Epigenetics

Some phenotypes are relatively unchanging once an organism is mature, but some organisms, particularly plants, show considerable plasticity, and the phenotype changes in response to environment.

Physical and biotic environment

The cause of phenotypic variation is not always clear. For polymorphic species like the Grove snail (above), as the number of forms in the population increases, discrete variation blurs into more continuous variation.

The effect of temperature

▶ The sex of some animals is determined by the incubation temperature during their embryonic development. Examples include turtles, crocodiles, and the American alligator. In some species, high incubation temperatures produce males and low temperatures produce females. In other species, the opposite is true. Temperature regulated sex determination may provide an advantage by preventing inbreeding, since all siblings will tend to be of the same sex.

▶ Colour-pointing is a result of a temperature sensitive mutation to one of the melanin-producing enzymes. The dark pigment is only produced in the cooler areas of the body (face, ears, feet, and tail), while the rest of the body is a pale colour, or white. Colour-pointing is seen in some breeds of cats and rabbits, e.g. Siamese cats and Himalayan rabbits.

1. Investigate the incidence of one or more discrete traits (either an individual has the trait or does not have the trait) in your class. The traits can include those mentioned in the introduction, or hitch hiker's thumb (whether the thumb can bend back or not), or interlocking fingers - determining which thumb, left or right, is on top. Collate the class tally on a separate sheet, graph the data, and comment on results. Attach to this page.

2. Name one trait you have that has been influenced by your environment? _____

_____

_____

D3.2
4

©2024 **BIOZONE** International
ISBN: 978-1-99-101410-8
Photocopying prohibited

318 Variation and Phenotypic Plasticity

Key Idea: Some organisms are able to change the expression of their genes to develop traits that best suit a set of environmental conditions. This is called phenotypic plasticity. An organism's phenotype is not necessarily fixed. For many phenotypic traits, especially those controlled by multiple genes, organisms may show varying degrees of flexibility in their observable phenotype (phenotypic plasticity).

Plasticity can apply at both the individual level, e.g. aerial and submerged leaves in aquatic plants; and species level, e.g. winged and wingless morphs in aphids. and different morphs in peppered moths. The increase in average height in humans compared to that of previous centuries is an example of phenotypic plasticity, rather than evolutionary change. The onset age of puberty is another example.

Phenotypic plasticity in animals

▶ The presence of other individuals of the same species may control sex determination for some animals. Some fish species, including Sandager's wrasse (right), show this characteristic. The fish live in groups consisting of a single male with attendant females and juveniles. In the presence of a male, all juvenile fish of this species grow into females. When the male dies, the dominant female will undergo physiological changes to become a male. The male and female look very different.

▶ Some organisms respond to the presence of other, potentially harmful, organisms by changing their body shape. Invertebrates, such as some *Daphnia* species (right), grow a helmet when invertebrate predators are present. The helmet makes *Daphnia* more difficult to attack and handle. Such changes are usually in response to chemicals produced by the predator (or competitor) and are common in plants as well as animals.

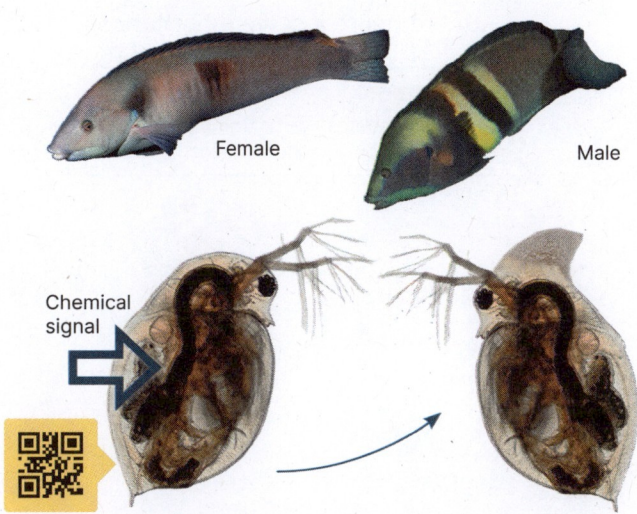

Female   Male

Chemical signal

Non-helmeted *Daphnia*   Helmeted *Daphnia*

1. What is the advantage of phenotypic plasticity in sex determination in Sandager's wrasse? _____

   _____

   _____

   _____

2. The helmet development on *Daphnia* is an example of a trait that is reversibly plastic and, after several moults in the absence of harmful predators, the *Daphnia* can revert back to the non-helmeted form. What is the advantage of this?

   _____

   _____

   _____

3. The squinting bush brown is a butterfly found in east Africa that feeds on fruit. The environment alternates between dry and wet seasons. In the dry season, the butterfly has subdued eye spots on its wings to blend in with the dry vegetation, helping it avoid predators. In the wet season, where food is green and plentiful, bold and striking spots appear which, instead of camouflage, now act as warning signs to predators.
   Additionally, the reproductive life cycle of the butterfly becomes much quicker in the wet season, with 2 or 3 cycles, compared to only one in the dry season. Explain how this difference in traits is an example of phenotypic plasticity and how this may advantage the butterfly:

   _____

   _____

   _____

   _____

   _____

   _____

   _____

Female squinting bush brown (*Bicyclus anynana*) in the dry season.

D3.2

6

319 Phenylketonuria

Key Idea: Autosomal recessive genetic conditions are only expressed when there are two copies of the allele.

An autosomal gene refers to a gene that is carried on an autosome, i.e. not a sex chromosome. A disease caused by a mutation on a recessive allele can be carried silently: the carrier is heterozygous for the disease but does not express it in the phenotype. Many rare diseases are autosomal recessive, and only appear in offspring when two adult carriers produce children who have a 1 in 4 chance of having the disease. Phenylketonuria (PKU) is one such example of a rare, recessive mutation. If the mutation is undetected without a newborn heel prick test and left untreated, the individual develops mental and health consequences.

Autosomal recessive inheritance

Chromosome with recessive abnormal gene

Chromosome with dominant gene

Carrier father

Carrier mother

Example: Autosomal Recessive PKU mutation to phenylalanine hydroxylase gene on chromosome 12.

Phenylketonuria (PKU)

Phenylketonuria (PKU) is an example of a metabolic disorder that occurs when an error exists in a metabolic pathway. Babies born with PKU are missing the enzyme needed to catalyze the first step in the pathway that metabolizes the essential amino acid phenylalanine. Without the enzyme, phenylalanine cannot be converted to the next substrate, tyrosine. Instead, it is metabolized to toxic derivatives which cause central nervous system damage. Children with PKU tend to have lighter skin and hair than people without the disorder.

Newborn babies are tested for a number of genetic disorders soon after birth, including PKU. Blood is collected from a heel prick onto a Guthrie card (below) and tested. The prognosis for PKU sufferers is good if the disease is detected early and a low phenylalanine diet is followed throughout life. People with PKU must take supplements to provide the amino acids that would otherwise be lacking in a low-phenylalanine diet, e.g. tyrosine, which is normally derived from phenylalanine and is needed for brain function.

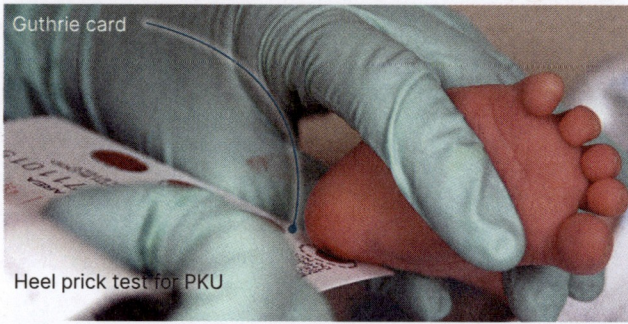

Guthrie card

Heel prick test for PKU

1. Explain the inheritance pattern of an autosomal recessive allele such as PKU: _____

_____

_____

2. Contrast the pattern of inheritance for an autosomal dominant and an autosomal recessive condition:

_____

_____

_____

3. If parents are closely related, i.e. first cousins, the chance of offspring having a rare autosomal disease is greatly increased. Explain why:

_____

_____

_____

4. Phenylketonuria affects only around 1 in ~23,900 people. Why is the heel-prick test still important even though PKU is so rare?

_____

_____

D3.2
7
272

©2024 **BIOZONE** International
ISBN: 978-1-99-101410-8
Photocopying prohibited

320 Alleles in Populations

Key Idea: Gene pool variation can be increased with SNPs and multiple alleles.

You will recall that a gene pool consists of all the alleles for each gene that are present in a population, and potentially available to be inherited by the next generation through the process of sexual reproduction. Genes that have single-nucleotide polymorphisms, where one base is swapped for another during a random gene mutation, adds further diversity in allele variation. Sometimes just one SNP can alter the phenotype of the gene, but in most cases the mutation is silent with no impact, although it can be used as a marker for common ancestry and evolutionary divergence. Multiple alleles indicate more than 2 allele variants exist in a gene pool, and this can lead to many more possible phenotypes.

Single-nucleotide polymorphisms (SNPs) and the gene pool

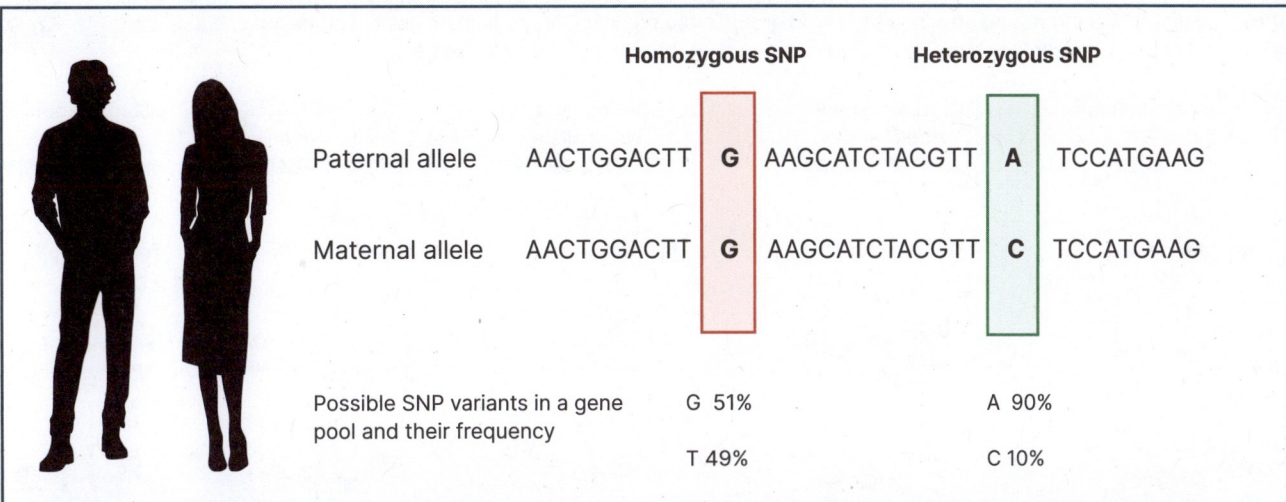

	Homozygous SNP	**Heterozygous SNP**
Paternal allele	AACTGGACTT **G** AAGCATCTACGTT	**A** TCCATGAAG
Maternal allele	AACTGGACTT **G** AAGCATCTACGTT	**C** TCCATGAAG
Possible SNP variants in a gene pool and their frequency	G 51% T 49%	A 90% C 10%

Single-nucleotide polymorphisms (SNPs) in genes occur due to mutation, and almost always during meiosis. Most nucleotide 'swaps' cause no noticeable change to the genotype, and therefore phenotype, due to degeneracy of amino acid coding. Therefore, if the SNP occurs in the non-coding DNA sections, there will be no impact on the genotype, and hence phenotype. However, all SNP variants add variety to the gene pool of a population, with an example above showing the variants of maternal and paternal alleles due to SNPs. You will recall that sexually reproducing organisms receive just one allele from each parent, regardless of the number of allele variants in the gene pool.

Multiple alleles and the gene pool

When there are more than two possible allele variants in a gene pool, these are known as multiple alleles. The alleles rank in dominance against each other and the final phenotype depends on which two alleles an individual receives from its parents.

For example, coat colour in rabbits is controlled by multiple alleles. C is the agouti or wildtype colour. The presence of this allele is dominant over all others. The chinchilla allele (c^{ch}) is dominant over all other alleles except agouti (C). The Himalayan allele (c^h) is only dominant over albino (c). Therefore, a rabbit can only have an albino colouration if no other alleles are in the genotype.

Himalayan colour: White fur with black extremities
Genotype: $c^h c^h$

Albino colour: White fur
Genotype: cc

Chinchilla colour: Black tipped white fur
Genotype: $c^{ch}c^{ch}$

Agouti colour: Brown tipped fur
Genotype: CC

1. How do SNPs and multiple alleles add variation into a gene pool despite an individual still only inheriting 2 alleles?

_____

_____

_____

_____

36 D3.2 8

321 Multiple Alleles and Codominance

Key Idea: Multiple alleles lead to non-Mendelian patterns of inheritance.

Not all genetic crosses follow Mendel's laws. In fact, Mendel was fortunate in that the traits he studied in pea plants were associated with genes on different chromosomes and had simple, single gene dominant/recessive relationships. Not all genes behave this way and, in fact, most do not.

Codominance is an inheritance pattern in which both alleles in a heterozygote contribute to the phenotype, and both alleles are independently and equally expressed. Examples include the human blood group ABO and roan coat colour in horses and cattle. Red coat colour is equally dominant with white. Animals with both alleles have coats that are roan: both red and white hairs are present.

ABO blood types and multiple alleles

Blood group type is controlled by three different alleles, I^A, I^B, and i , to produce four possible phenotypes. The presence of the recessive allele (i) has no effect on the blood group in the presence of a dominant allele. For example, If a person has the I^Ai or $I^A I^A$ allele combination then their blood group will be group A, the phenotype.

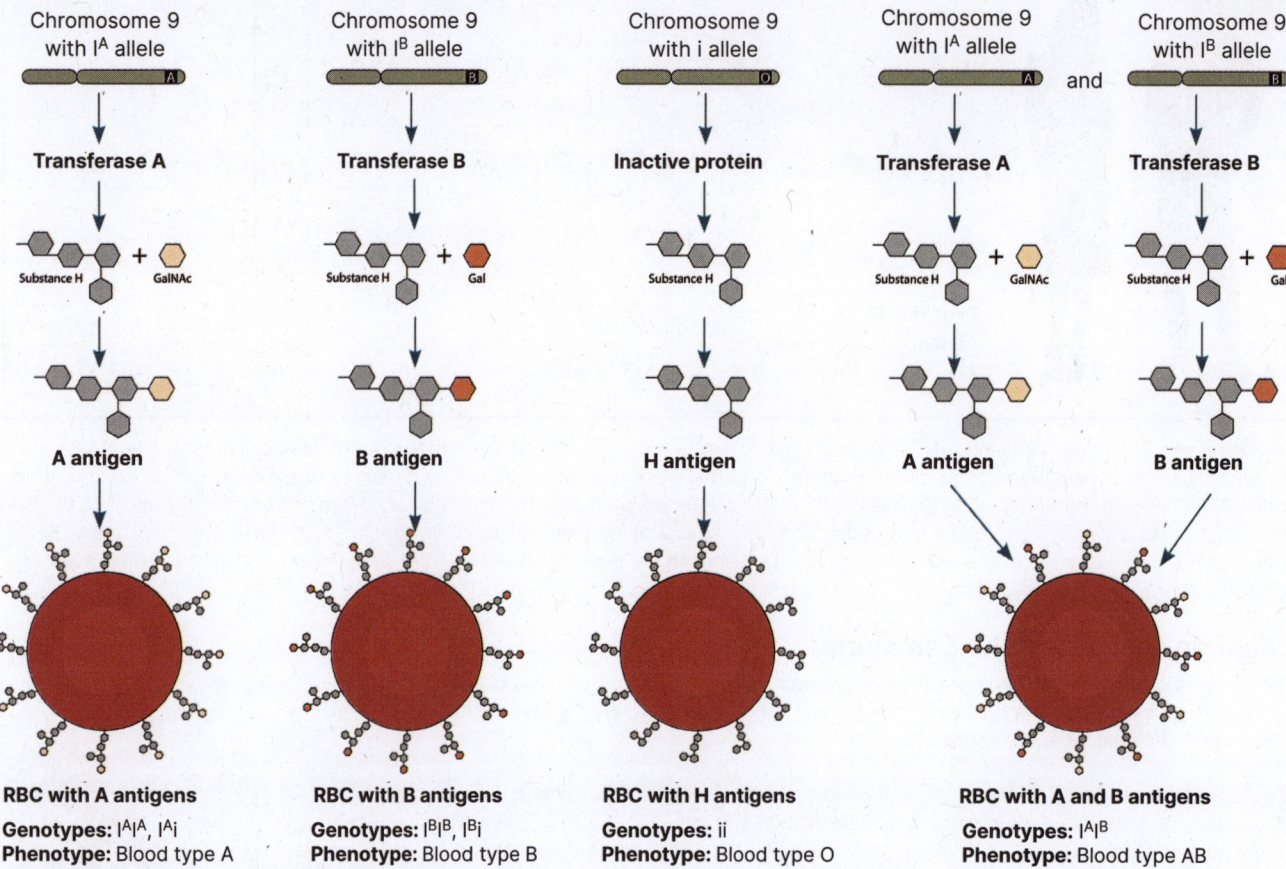

Genotypes: $I^A I^A$, I^Ai
Phenotype: Blood type A

Genotypes: $I^B I^B$, I^Bi
Phenotype: Blood type B

Genotypes: ii
Phenotype: Blood type O

Genotypes: $I^A I^B$
Phenotype: Blood type AB

ABO blood groups and codominance

▸ The human ABO blood group system also shows codominance. The ABO antigens consist of sugars attached to the surface of red blood cells. The alleles code for enzymes (proteins) that join these sugars together.

▸ The allele i is recessive. It produces a non-functioning enzyme that cannot make any changes to the basic sugar molecule. The other two alleles (I^A, I^B) are codominant and are expressed equally. They each produce a different functional enzyme that adds a different, specific sugar to the basic sugar molecule, and result in a different phenotype to just I^A and I^B expressed individually.

▸ The blood group A and B antigens are able to react with antibodies present in the blood of other people so blood must always be matched for transfusion.

1. In the shorthorn cattle breed, coat colour is inherited. White shorthorn parents ($R^w R^w$) always produce calves with white coats. Red parents ($R^r R^r$) always produce red calves. However, when a red parent mates with a white one, the calves have a coat colour that is different from either parent: a mixture of red and white hairs, called roan ($R^r R^w$). How do two alleles produce three phenotypes?

_____

_____

_____

_____

©2024 **BIOZONE** International
ISBN: 978-1-99-101410-8
Photocopying prohibited

2. Below are four crosses possible between couples of various blood group types. Complete the genotype and phenotype for the crosses below:

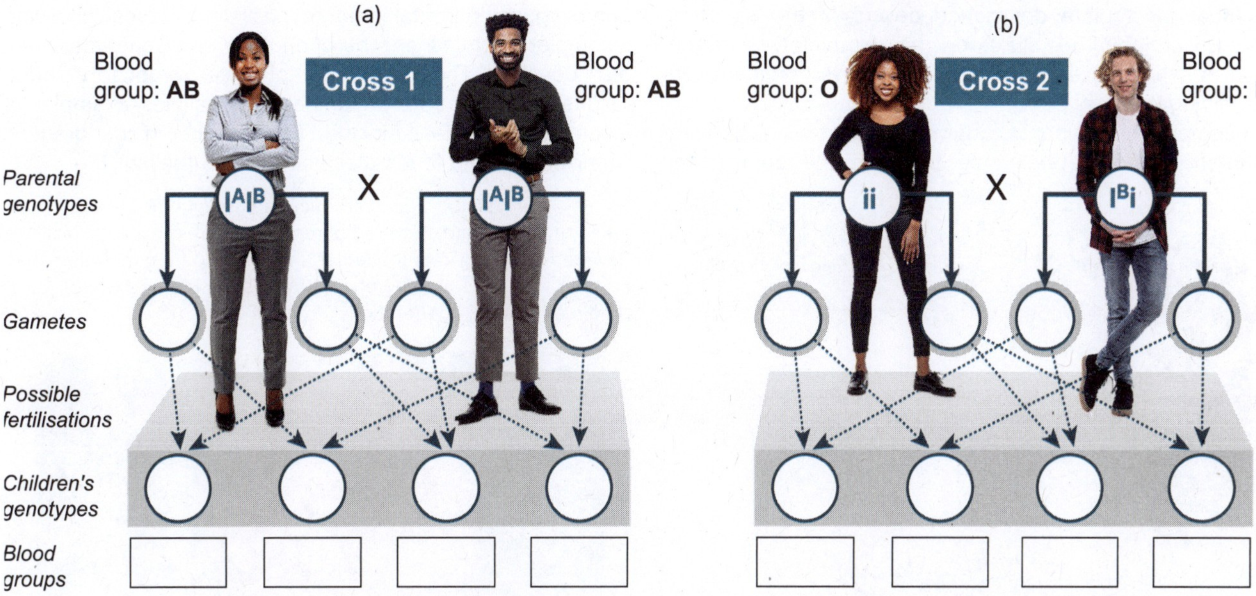

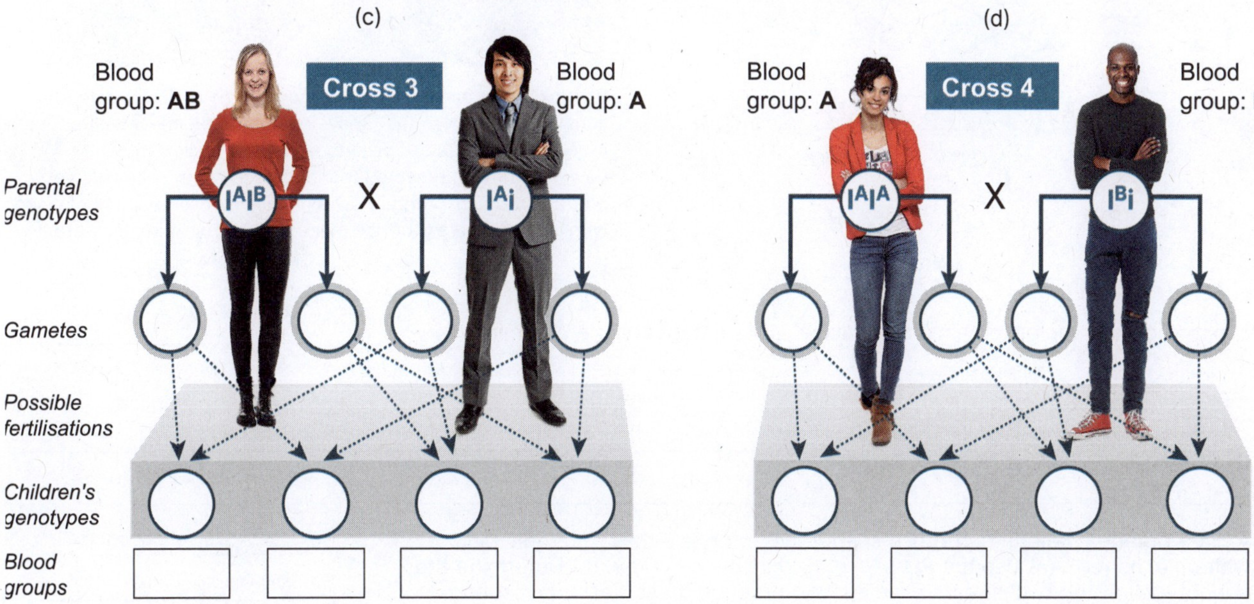

(a)

Blood group: **AB**　　Cross 1　　Blood group: **AB**

Parental genotypes　　$I^A I^B$　　X　　$I^A I^B$

Gametes ○ ○ ○ ○

Possible fertilisations

Children's genotypes ○ ○ ○ ○

Blood groups □ □ □ □

(b)

Blood group: **O**　　Cross 2　　Blood group: **B**

Parental genotypes　　ii　　X　　$I^B i$

Gametes ○ ○ ○ ○

Possible fertilisations

Children's genotypes ○ ○ ○ ○

Blood groups □ □ □ □

(c)

Blood group: **AB**　　Cross 3　　Blood group: **A**

Parental genotypes　　$I^A I^B$　　X　　$I^A i$

Gametes ○ ○ ○ ○

Possible fertilisations

Children's genotypes ○ ○ ○ ○

Blood groups □ □ □ □

(d)

Blood group: **A**　　Cross 4　　Blood group: **B**

Parental genotypes　　$I^A I^A$　　X　　$I^B i$

Gametes ○ ○ ○ ○

Possible fertilisations

Children's genotypes ○ ○ ○ ○

Blood groups □ □ □ □

3. A farmer has only roan cattle on his farm. He suspects that one of the neighbours' bulls may have jumped the fence to mate with his cows earlier in the year because half the calves born were red and half were roan. One neighbour has a red bull, the other has a roan.

(a) Fill in the spaces (right) to show the genotype and phenotype for parents and calves.

(b) Which bull serviced the cows? Red or roan (delete one)

4. Describe the classical phenotypic ratio for a codominant gene resulting from the cross of two heterozygous parents, e.g. a cross between two roan cattle:

_____

_____

_____

_____

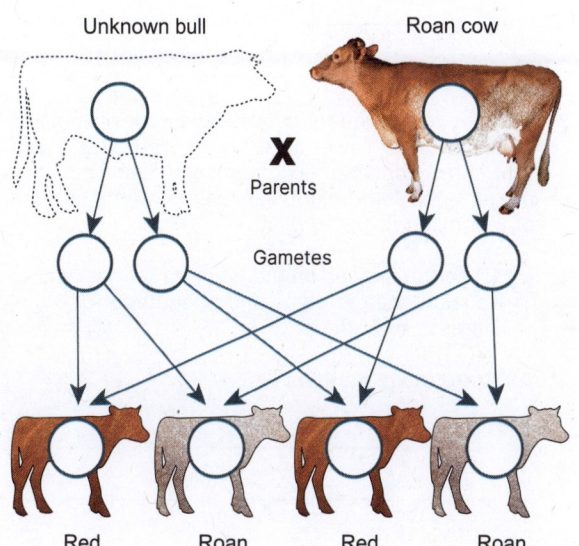

Unknown bull　　　　Roan cow

X
Parents

Gametes

Red　　Roan　　Red　　Roan

322 Incomplete Dominance

Key Idea: Incomplete dominance describes the situation where the action of one allele does not completely mask the action of the other and neither allele shows dominance in determining the trait.

In incomplete dominance, the heterozygous offspring are intermediate in phenotype between the contrasting homozygous parental phenotypes. In crosses involving incomplete dominance, the phenotype and genotype ratios are identical. The phenotype of heterozygous offspring results from the partial influence of both alleles. Examples of incomplete dominance include flower colour in snapdragons (*Antirrhinum*) and four o'clocks (*Mirabilis*) (below).

Pure breeding snapdragons produce red or white flowers (left). When red and white-flowered parent plants are crossed, a pink-flowered offspring is produced. If the offspring (F_1 generation) are crossed together, all three phenotypes (red, pink, and white) are produced in the F_2 generation.

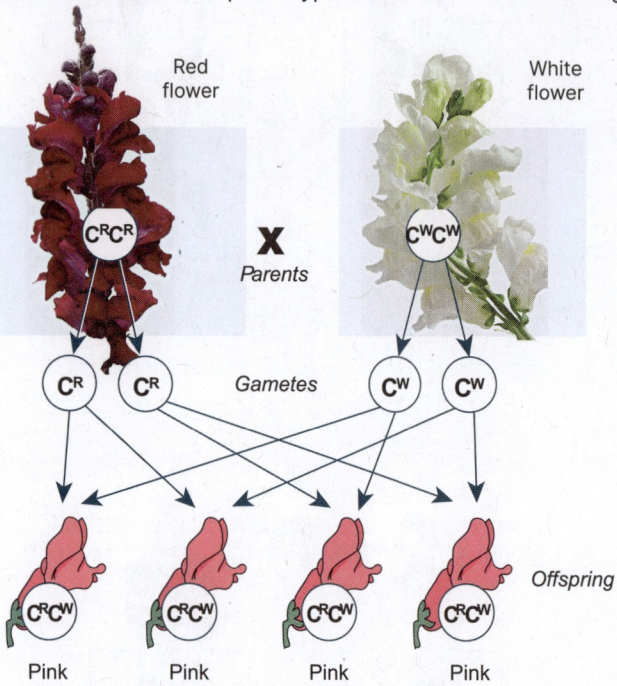

Four o'clocks (above) are also known to have flower colours controlled by alleles that show incomplete dominance. Pure breeding four o'clocks produces crimson, yellow, or white flowers. Crimson flowers (above) crossed with yellow flowers produce reddish-orange flowers, while crimson flowers crossed with white flowers produce magenta (reddish-pink) flowers.

1. Explain how incomplete dominance of alleles differs from complete dominance: _____
_____
_____
_____

2. A plant breeder wanted to produce snapdragons for sale that were only pink or white (i.e. no red). (a) Use the Punnett grid (right) to help you. (b) Determine the phenotypes of the two parents necessary to produce these desired offspring:

_____
_____
_____

	Gametes from male	
Gametes from female		

3. Another plant breeder crossed two four o'clocks, known to have flower colour controlled by alleles that show incomplete dominance. Pollen from a magenta flowered plant was placed on the stigma of a crimson flowered plant.

   (a) Fill in the spaces on the diagram on the right to show the genotype and phenotype for parents and offspring.

   (b) State the phenotype ratio:

   _____

   _____

   _____

   _____

Magenta flower X Crimson flower

Gametes

Possible fertilisations

Offspring

Phenotypes

D3.2
10
©2024 **BIOZONE** International
ISBN: **978-1-99-101410-8**

323 Sex Determination and Linkage

Key Idea: Many genes on the X chromosome do not have a match on the Y chromosome. In males, which are XY, a recessive allele on the X chromosome will be expressed.

Sex linkage refers to the way genes on the sex chromosomes are inherited and expressed. In humans, the sex chromosomes are X and Y, but sex linkage usually involves genes on the X chromosome, which has many more genes than the Y chromosome. X-linked recessive traits are usually seen only in males (XY) and occur rarely in the females (XX) because females may be heterozygous (carriers). X-linked dominant traits do not necessarily affect males more than females. In humans, recessive sex linked genes are responsible for a number of heritable disorders in males. Y-linked disorders are rare and usually associated with infertility.

Sex determination in humans

▶ Eggs can only contain an X chromosome, as females all have an XX genotype. Sperm can have either X or Y, as the males have an XY genotype. The sex chromosome contained within the fertilizing sperm, therefore determines the sex of the zygote.

▶ The SRY gene (sex-determining region Y gene) on the Y chromosome alters DNA properties and allows the testis to form in the developing embryo, leading onto male-specific characteristics. Without this gene, not carried on the X chromosome, the foetus will continue to develop female-specific characteristics.

X Sperm + X Egg = XX Female child

Y Sperm + X Egg = XY Male child

Sex-linked genetic disorder: haemophilia

Haemophilia is a recessive disorder linked to the X-chromosome that results in ineffective blood clotting when a blood vessel is damaged. The most common type, haemophilia A, occurs in 1 in 5000 male births. Any male who carries the gene will express the phenotype. Haemophilia is extremely rare in women.

1. A couple wishes to have children. The woman knows she is a carrier for haemophilia. The man is not a haemophiliac. (a) Use the notation X^h for haemophilia, X^H for the dominant allele, and Y for male to complete the diagram (right). Include the parent genotypes, gametes and possible fertilizations. (b) Write the genotypes and phenotypes in the table below.

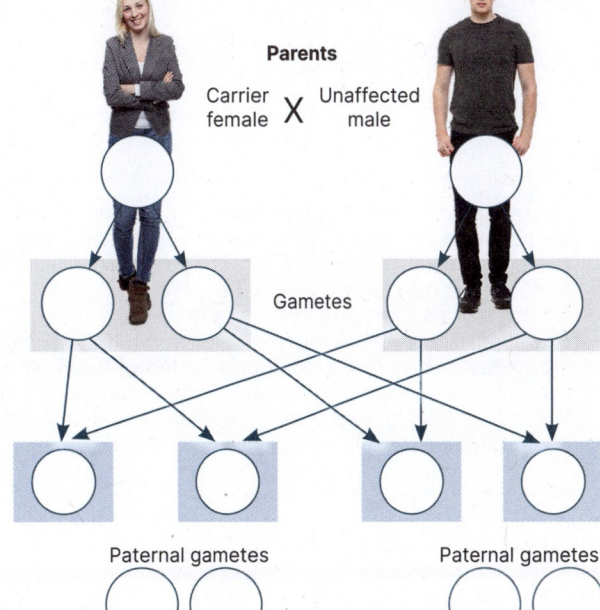

Parents: Carrier female X Unaffected male

Gametes

	Genotypes	Phenotypes
Male children		
Female children		

2. (a) A second couple also wishes to have children. The woman knows her maternal grandfather was a haemophiliac but neither her mother or father were. Determine the probability of her being a carrier ($X^H X^h$). Use the Punnett grids (right) to help you:

_____

_____

(b) The man is not a haemophiliac. Determine the probability that their first male child will have haemophilia. Use the Punnett grids to help you:

_____

_____

_____

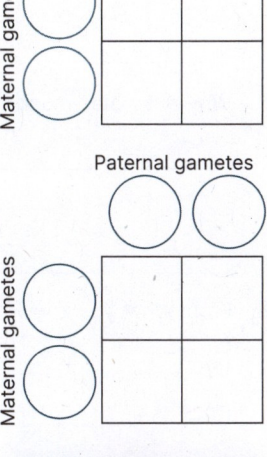

Sex linked rickets

A rare form of rickets in humans is determined by a dominant allele of a gene on the X chromosome. It is not found on the Y chromosome. This condition is not successfully treated with vitamin D therapy. The allele types, genotypes, and phenotypes are as follows:

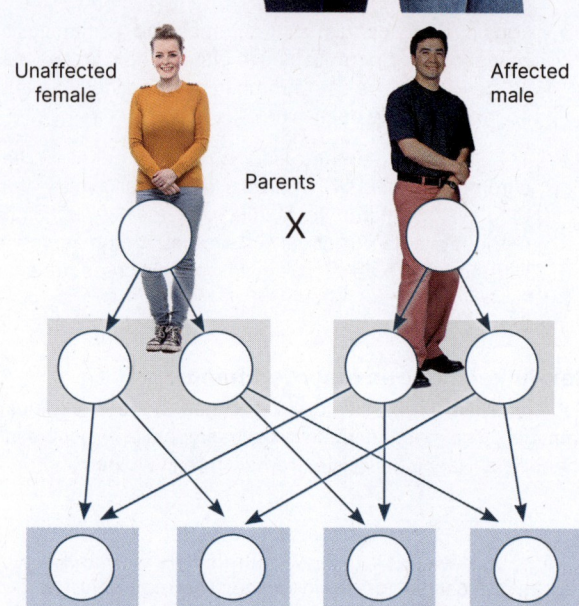

Allele types	Genotypes	Phenotypes
X^R = affected by rickets	$X^R X^R$, $X^R X$	= Affected female
X = unaffected	$X^R Y$	= Affected male
	XX, XY	= Unaffected female, unaffected male

3. As a genetic counsellor, you are presented with a couple, one of whose family has a history of this disease. The male is affected by this disease and the female is unaffected. The couple, who are thinking of starting a family, would like to know what their chances are of having a child born with this condition. They would also like to know what the probabilities are of having an affected boy or affected girl. (a) Use the symbols above to complete the diagram (right) and determine the probabilities stated below (expressed as a proportion or percentage).

Determine the probability of having:

(b) Affected children: _____

(c) An affected girl: _____

(d) An affected boy: _____

Unaffected female   Affected male

Parents X

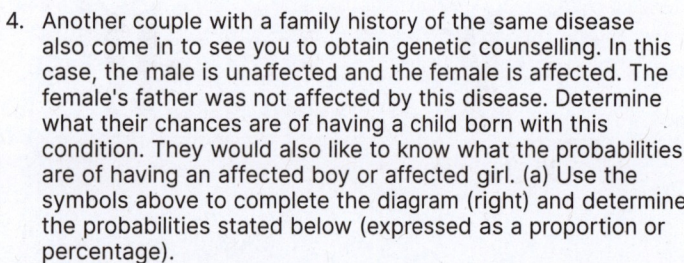

4. Another couple with a family history of the same disease also come in to see you to obtain genetic counselling. In this case, the male is unaffected and the female is affected. The female's father was not affected by this disease. Determine what their chances are of having a child born with this condition. They would also like to know what the probabilities are of having an affected boy or affected girl. (a) Use the symbols above to complete the diagram (right) and determine the probabilities stated below (expressed as a proportion or percentage).

Determine the probability of having:

(b) Affected children: _____

(c) An affected girl: _____

(d) An affected boy: _____

Affected female (unaffected father)   Unaffected male

Parents X

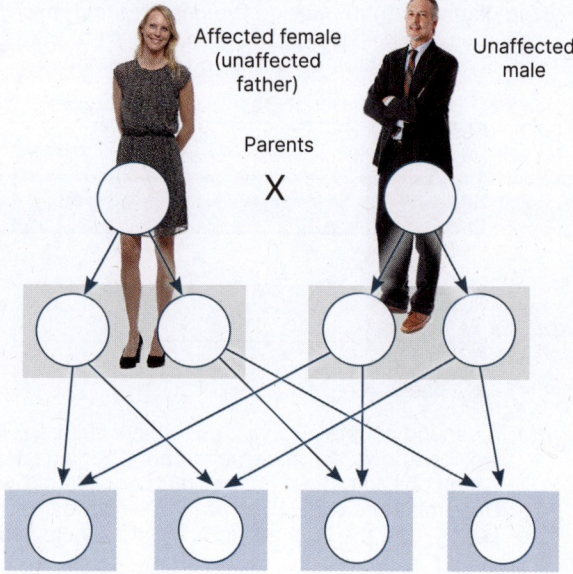

5. Why are males much more likely to inherit X-linked recessive disorders than females?

_____

_____

_____

6. From what you know about sex linkage, what two features could you use to detect a Y-linked disorder in a pedigree?

(a) _____

(b) _____

324 Pedigree Analysis

Key Idea: Pedigree charts are a way to graphically illustrate inheritance patterns over a number of generations.
One way in which to analyze the family history of an observable trait is to use a pedigree chart, which follows certain rules and uses particular symbols to indicate the sex and genotype of individuals across generations.

Pedigree charts

▶ A pedigree chart is a diagram that shows the occurrence of a particular gene or trait from one generation to the next. In humans, pedigree charts are often used to deduce the inheritance of heritable conditions. In domestic animals, pedigree charts are often used to trace the inheritance of traits in selective breeding programs.

▶ Pedigree charts use symbols to indicate an individual's particular traits. The key (right) explains the meaning of the symbols. Particular individuals are identified by their generation number and their order in that generational row. For example, II-6 is the sixth person in the second generation. The arrow indicates the person through whom the pedigree was discovered, i.e. reported the condition.

▶ The chart on the right represents three generations: grandparents (I-1 and I-2) with three sons and one daughter. Two of the sons (II-3 and II-4) are identical twins, but did not have any children. The other son (II-1) had a daughter and another child (sex unknown). The daughter (II-5) had two sons and two daughters, plus a child that died in infancy.

▶ Pedigrees can also indicate if a trait shows autosomal or sex linked inheritance. In autosomal patterns, both males and females are generally equally affected (more or less). For the particular trait being studied, the grandfather was expressing the phenotype (showing the trait) and the grandmother was a carrier. One of their sons and one of their daughters also show the trait, together with one of their granddaughters (arrow).

Key to symbols

◯ Normal female	◆ Sex unknown
☐ Normal male	● Died in infancy
● Affected female (expresses allele of interest)	Identical twins
■ Affected male (expresses allele of interest)	Non-identical twins
◉ Carrier (heterozygote)	I, II, III Generations
	1, 2, 3 Children (birth order)

1. (a) Brown eyes are the result of a dominant eye-colour allele and blue eyes are recessive. A brown-eyed man (A), whose mother had blue eyes and whose father had brown eyes, marries a blue-eyed woman (B) whose parents are both brown-eyed. They have a daughter who is blue-eyed. Draw a pedigree showing all four grandparents, the two parents, and the daughter. Indicate each individual's possible genotype. Use filled shapes to indicate the recessive trait.

(b) Identify the individuals that are definitely heterozygous (carriers): _____

(c) Identify the individual that could be heterozygous (a carrier): _____

(d) What is the probability of couple A and B having a blue-eyed boy as their next child? _____

(e) Explain your reasoning: _____

_____

_____

©2024 **BIOZONE** International
ISBN: 978-1-99-101410-8
Photocopying prohibited

510

Inbreeding and the Habsburg jaw

▶ The Habsburgs were a family of royal lineage who ruled for a period between 1282-1918 in Austria, and later in Hungary and Spain between the early 1500s and 1700. To keep the royal bloodline 'pure', marriages between close relatives were common and these unions often resulted in children, albeit not all surviving.

▶ Close relationship between parentage significantly increases the chances of rare mutations being present. Charles II of Spain (1661-1700) had generations of inbreeding occurring in his ancestors. In fact his parents, Philip IV and Mariana of Austria, were niece and uncle. Aside from an overly large tongue, epilepsy, and other illnesses, most likely genetically inherited, Charles II had prognathism, a recessive genetic condition where the jaw is pushed outward. The condition was common enough in the family to be called the 'Habsburg jaw'.

▶ Historians suspect that decreasing fertility from inbreeding led to the eventual demise of the family in the early 1900s.

Charles II and the Habsburg jaw

Inductive and deductive reasoning of inheritance.

▶ Inductive reasoning can be used to make generalized conclusions from specific evidence from the pedigree charts. For example, by examining the family tree of Charles II and indicating who had the Habsburg jaw, scientists were able to conclude that it was caused by a recessive allele.

▶ A complete understanding of inheritance also needs deductive reasoning, where generalized conclusions can be illustrated with specific examples when inheritance laws are applied. For the example below, the genetic condition is dominant but because 10 has a recessive phenotype, he must have received a recessive allele from both parents. We can therefore deduce that the father, 5, is heterozygous.

The pedigree of lactose intolerance

Lactose intolerance is the inability to digest the milk sugar lactose. It occurs because some people do not produce lactase, the enzyme needed to break down lactose. The pedigree chart (right) was one of the original studies to determine the inheritance pattern of lactose intolerance.

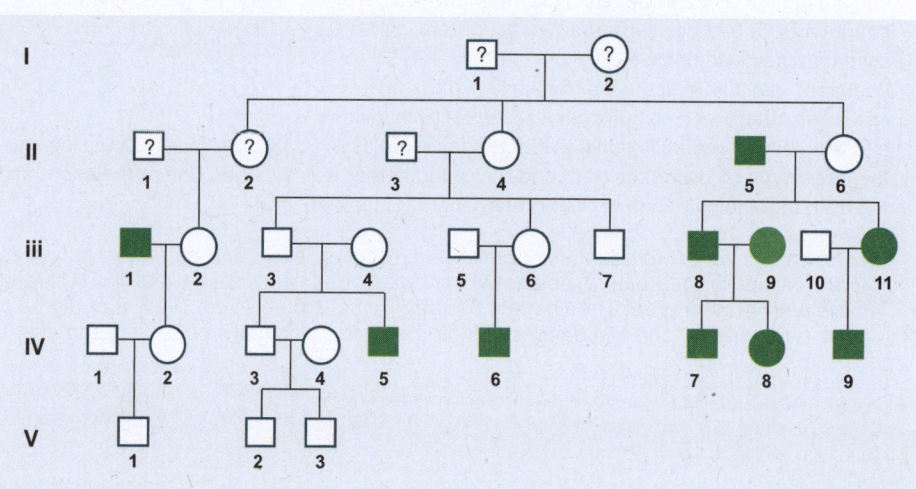

2. Use an analysis of the pedigree above to make a claim about the inheritance pattern of lactose intolerance. Support your claim with at least two pieces of evidence:

_____

_____

_____

_____

_____

3. (a) Draw a Punnett grid on paper (attach here) to show the cross between III-10 and III-11 in the pedigree chart above. Use the capital letter 'L' for the dominant allele and the lower case letter 'l' for the recessive allele.

(b) Explain how you can be certain about III-10's genotype: _____

_____

_____

_____

325 Polygenes

Key Idea: Many phenotypes are affected by multiple genes. Many phenotypes are controlled by more than one gene. This is called polygeny or polygenic inheritance. As phenotype is controlled by many genes and alleles, a large range of phenotypes is possible. Combined with environmental effects, this produces continuous variation within the population. Two examples in humans are skin colour and height.

Polygenic traits

Very pale	Light	Medium light	Medium	Medium dark	Dark	Black
0	1	2	3	4	5	6

Polygenic traits are usually identified by the following:

▸ Traits are usually quantified by measuring rather than counting.

▸ Two or more genes contribute to the phenotype.

▸ Phenotypic expression is over a wide range (often in a bell shaped curve).

▸ Polygenic phenotypes include skin colour, height, eye colour, and weight.

Note:
Multiple genes (many genes contributing to a phenotype) are quite different from multiple alleles (many alleles present in the population for one phenotypic characteristic).

It is estimated that skin colour is controlled by at least eight genes. There are various ways to compare skin colour. One is shown above, in which there are seven shades ranging from very pale to very dark. Most individuals are somewhat intermediate in skin colour.
The table (right) shows a cross between three genes involved in skin colour: A, B, and C, each with two alleles (AaBbCc x AaBbCc). This is sufficient to give the seven shades of skin colour above. The shaded boxes indicate their effect on skin colour when combined. No dominant allele results in a lack of dark pigment (aabbcc). Full pigmentation (black) requires six dominant alleles (AABBCC). Note that for three genes with two alleles each there are $2^3 \times 2^3 = 8 \times 8 = 64$ possible genotypes.

Gametes	ABC	ABc	AbC	Abc	aBC	aBc	abC	abc
ABC	AABBCC	AABBCc	AABbCC	AABbCc	AaBBCC	AaBBCc	AaBbCC	AaBbCc
ABc	AABBCc	AABBcc	AABbCc	AABbcc	AaBBCc	AaBBcc	AaBbCc	AaBbcc
AbC	AABbCC	AABbCc	AAbbCC	AAbbCc	AaBbCC	AaBbCc	AabbCC	AabbCc
Abc	AABbCc	AABbcc	AAbbCc	AAbbcc	AaBbCc	AaBbcc	AabbCc	Aabbcc
aBC	AaBBCC	AaBBCc	AaBbCC	AaBbCc	aaBBCC	aaBBCc	aaBbCC	aaBbCc
aBc	AaBBCc	AaBBcc	AaBbCc	AaBbcc	aaBBCc	aaBBcc	aaBbCc	aaBbcc
abC	AaBbCC	AaBbCc	AabbCC	AabbCc	aaBbCC	aaBbCc	aabbCC	aabbCc
abc	AaBbCc	AaBbcc	AabbCc	Aabbcc	aaBbCc	aaBbcc	aabbCc	aabbcc

1. (a) What is the difference between discrete and continuous variables? _____

_____

_____

(b) How does polygeny contribute to continuous variation? _____

_____

_____

_____

2. Study the cross between the A, B, and C genes above. Write down the frequencies of the seven phenotypes (0-6):

_____

_____

D3.2
14 - 15

Investigation 15.1 Measuring continuous variation

See appendix for equipment list.

1. Choose one variable which occurs as a result of continuous variation, e.g. height, weight, hand span, or foot length, and write the variable you will be investigating here:

   _____

2. Select 30-50 classmates to be your sample. Measure the variable of interest (to one decimal place) and record the results in the space for raw data below.

3. Decide on appropriate frequency for the data, then record it as a tally chart in the space below.

Raw data (e.g. weight in kg)

Tally chart

4. Apply measures of central tendency to your data in the table (right). You can enter your data on a spreadsheet to make it easier to calculate if you want.

   Mean = (sum of observations) ÷ (total number of observations)

   Median = place the numbers in value order and find the middle

   Mode = Place all numbers in order and then count how many times each number appears in the set. The one that appears the most is the mode.

   Number of entries:

   Mean:

   Median:

   Mode:

5. Plot the continuous data as a box-and-whisker plot on the grid using the 6 aspects of data:

    Median (M): from the previous page - line values up in order

    First quartile (Q1): the 'median' to the left of the central median (between median and lowest value)

    Third quartile (Q3): the 'median' to the right of the central median (between median and highest value):

    Minimum: smallest value   Maximum: highest value

    Outliers: more than 1.5 × IQR (interquartile range) above the third quartile or below the first quartile.

    Draw a box from Q1 to Q3 with a vertical line through M (or horizontal if variable is on x axis/frequency on y)

    Draw 'whiskers' from Q1 to minimum and from Q3 to maximum

    Draw a dot or cross to indicate outliers

3. (a) Describe the pattern of the distribution shown in your graph: _____

    _____

    _____

    _____

    _____

    (b) What is the genetic basis of this distribution? Give a brief explanation of what this means: _____

    _____

    _____

    _____

    _____

    _____

4. Compare the distributions of continuous and discontinuous variation giving examples to illustrate your answer:

    _____

    _____

    _____

    _____

326 Mendel's Laws of Inheritance

Key Idea: Genetic information is inherited from parents in units called genes.

Mendel's laws of inheritance govern how these genes are passed onto the offspring. The laws of segregation and independent assortment apply during the process of meiosis to unlinked genes. The effect of these laws on paired, but unlinked, genes can be demonstrated with dihybrid crosses using a Punnett grid as a tool.

Particulate inheritance

Characteristics of both parents are passed on to the next generation as discrete entities (genes). This model explained many observations that could not be explained by the idea of blending inheritance, which was universally accepted prior to Mendel's work. The trait for flower colour (right) appears to take on the appearance of only one parent plant in the first generation, but reappears in later generations.

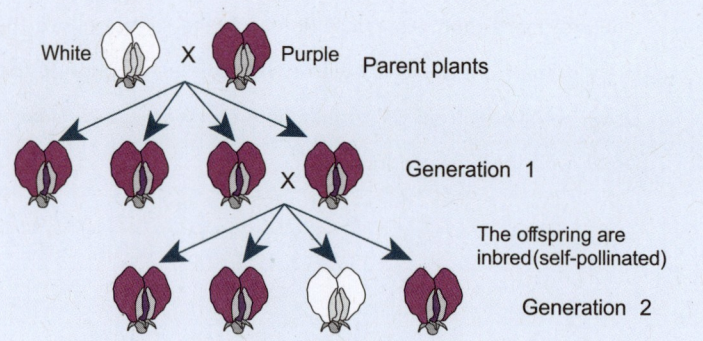

Law of segregation

During meiosis, the two members of any pair of alleles segregate unchanged and are passed into different gametes.

These gametes are eggs (ova) and sperm cells. The allele in the gamete will be passed onto the offspring.

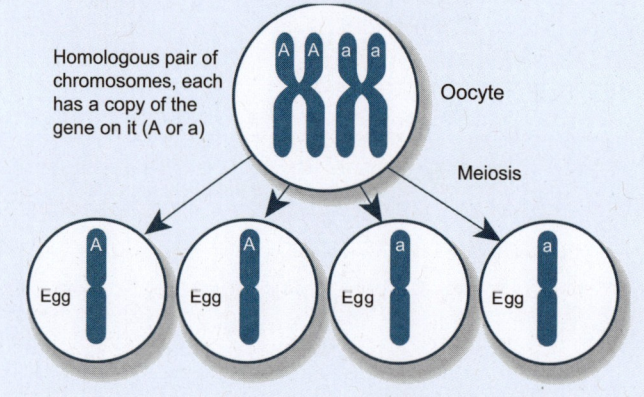

Law of independent assortment

Allele pairs separate independently during gamete formation, and traits are passed on to offspring independently of one another (this is only true for genes on separate chromosomes).

This diagram shows two genes (A and B) that code for different traits. Each of these genes is represented twice, one copy (allele) on each of two homologous chromosomes. The genes A and B are located on different chromosomes and, because of this, they will be inherited independently of each other, i.e. the gametes may contain any combination of the parental alleles.

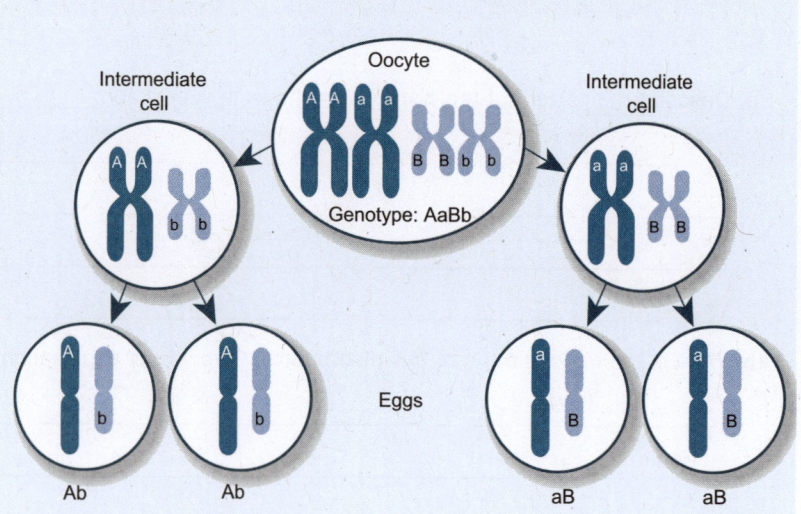

1. State the property of genetic inheritance which applies to the following scenario: the crossing of a tall and dwarf pea plant produces all tall pea plants in the first generation which, when crossed, produce both tall and dwarf pea plants in the second generation.

_____

2. An oocyte (egg producing cell) had the following allele combination: Tt Gg and produced the following combination of alleles in the egg cells: TG, Tg, tG, tg. Which law of inheritance applies to this scenario?

_____

327 Dihybrid Inheritance

Key Idea: A dihybrid cross studies the inheritance pattern of two genes. In crosses involving unlinked autosomal genes, the offspring occur in predictable ratios.

Four types of gamete are produced in a cross involving two genes on separate chromosomes and assorting independently. The two genes in the example of guinea pigs below are on separate chromosomes and control two unrelated characteristics, coat colour and length. Black (B) and short (L) are dominant to white (b) and long (l).

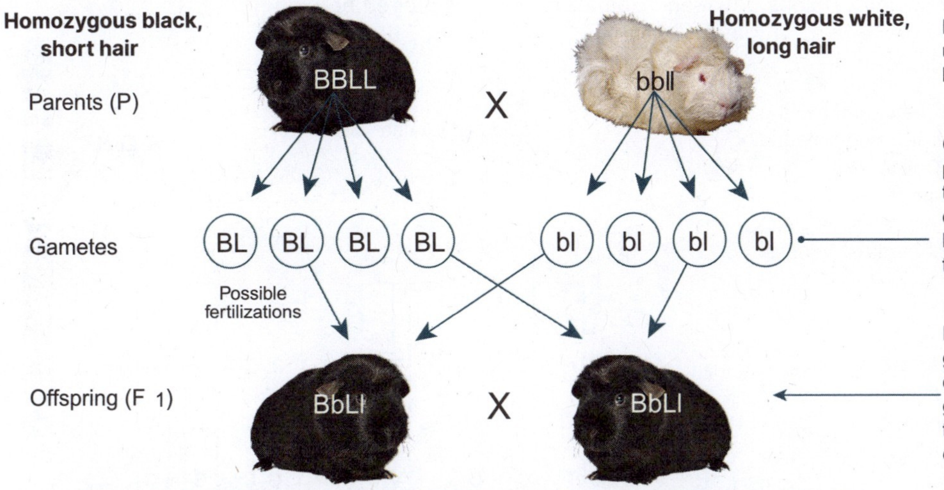

Homozygous black, short hair

Homozygous white, long hair

Parents (P)   BBLL   X   bbll

Parents: The notation P is only used for a cross between true breeding (homozygous) parents.

Gametes

BL BL BL BL   bl bl bl bl

Possible fertilizations

Gametes: Only one type of gamete is produced from each parent (although they will produce four gametes from each oocyte or spermatocyte). This is because each parent is homozygous for both traits.

Offspring (F₁)   BbLl   X   BbLl

F₁ offspring: There is only one kind of gamete from each parent, therefore only one kind of offspring produced in the first generation. The notation F₁ is only used to denote the heterozygous offspring of a cross between two true breeding parents.

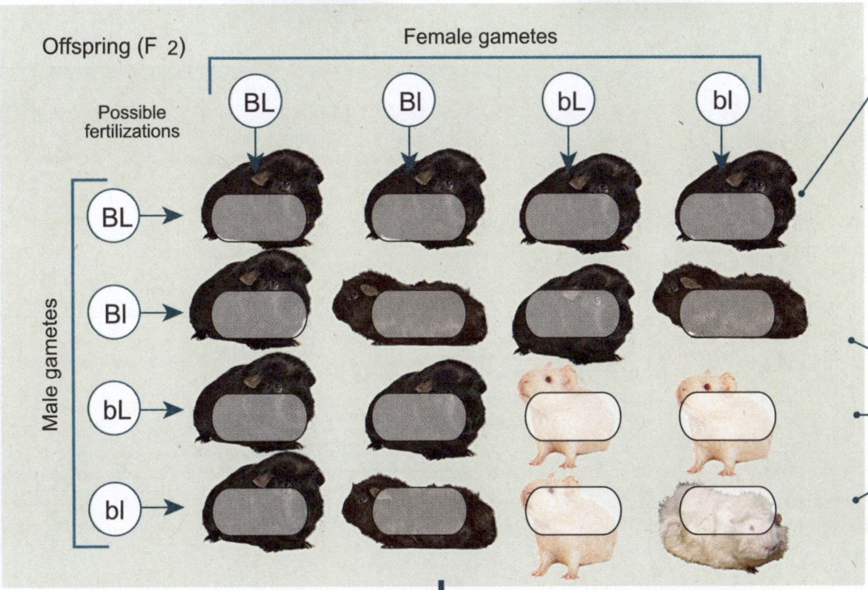

Offspring (F₂)

F₂ offspring: The F₁ were mated with each other (selfed). Each individual from the F₁ is able to produce four different kinds of gamete. Using a grid called a Punnett square (left), it is possible to determine the expected genotype and phenotype ratios in the F₂ offspring. The notation F₂ is only used to denote the offspring produced by crossing F₁ heterozygotes.

Female gametes

Possible fertilizations

BL   Bl   bL   bl

Male gametes

BL
Bl
bL
bl

Each of the 16 animals shown here represents the possible zygotes formed by different combinations of gametes coming together at fertilization.

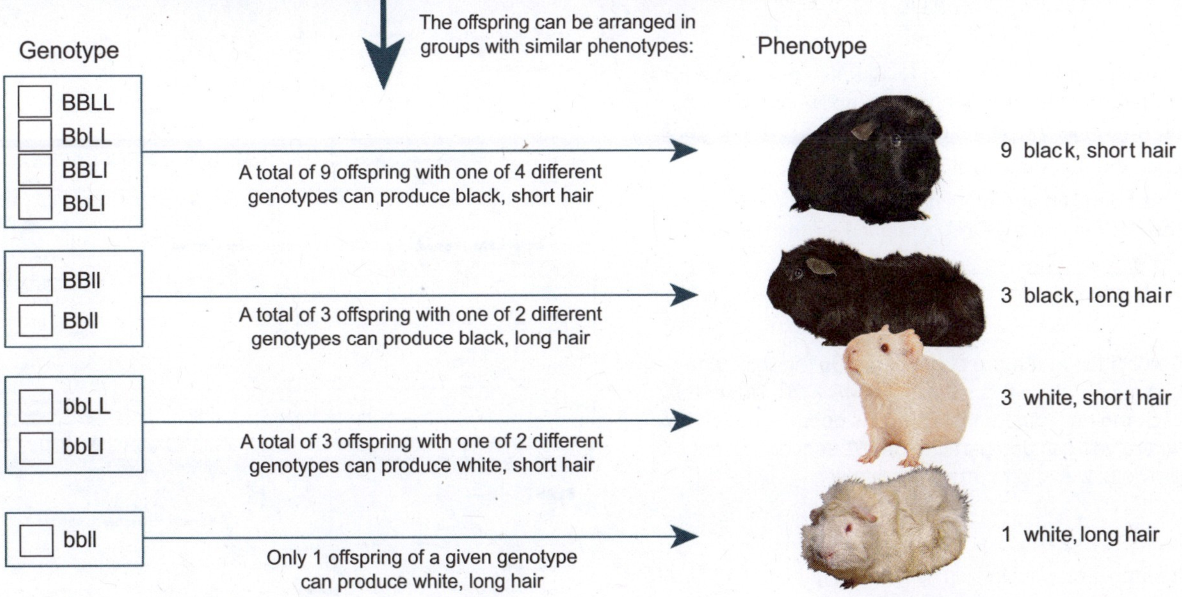

The offspring can be arranged in groups with similar phenotypes:

Genotype

BBLL
BbLL
BBLl
BbLl

A total of 9 offspring with one of 4 different genotypes can produce black, short hair

Phenotype

9 black, short hair

BBll
Bbll

A total of 3 offspring with one of 2 different genotypes can produce black, long hair

3 black, long hair

bbLL
bbLl

A total of 3 offspring with one of 2 different genotypes can produce white, short hair

3 white, short hair

bbll

Only 1 offspring of a given genotype can produce white, long hair

1 white, long hair

1. (a) Complete the Punnett grid above and (b) use it to fill in the number of each guinea pig genotype in the boxes (above left).

D3.2
17

328 Exploring Gene Databases

Key Idea: Genetic databases can be used to identify loci of gene pairs, and their polypeptide products.

Advancements in DNA sequencing have produced enormous amounts of information. The collection, storage, and analysis of this information by computers is called bioinformatics. Bioinformatics allows DNA sequence comparisons between species, a field called comparative genomics. Comparative genomics has provided the information to support (or overturn) established phylogenies. Genes and their polypeptide products can be located and compared at the same loci on different chromosomes, or between different genes on loci close together.

Using the sequence data

As genome sequencing has become faster and more genomes are sequenced, there has been a corresponding growth in the use of bioinformatics. Online DNA databases allows researchers to compare DNA being studied to known DNA sequences. This can allow quick identification of species, especially bacteria or viruses.

A gene of interest is selected for analysis.

Loci

... G A G A A C T G T T T A G A T G C A A A A...

Powerful computer software can quickly compare the DNA sequences of many organisms. Commonalities and differences in the DNA sequence can help to determine the organism and its evolutionary relationship to other organisms. The blue boxes indicate differences in the DNA sequences.

Organism 1 ...G A G A A C T G T T T A G A T G C A A A A...

Organism 2 ...G A G A T C T G T G T A G A T G C A G A A...

Organism 3 ...G A G T T C T G T G T C G A T G C A G A A...

Organism 4 ...G A G T T C T G T T T C G A T G C A G A G...

Online databases

DNA sequences can be studied using easily accessible online databases. The example below uses NCBI database at www.ncbi.nlm.nih.gov.

A search box at the top of the page has a drop down menu. Use the menu find and click Gene (IMAGE 1).
In the search box, type SATB2. This is a homeobox gene on the 2nd human chromosome. Homeobox genes control the early embryonic development of many body structures. Click search and a new window will appear showing the results.

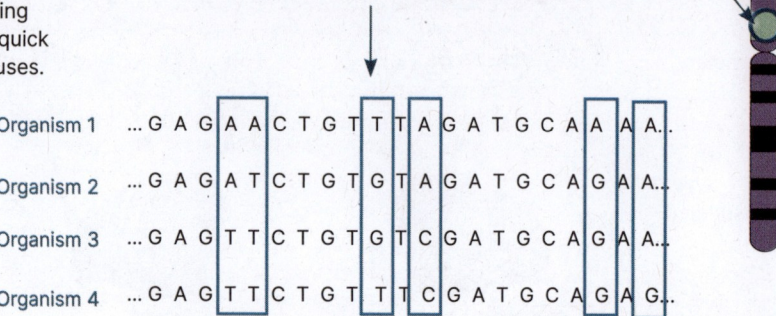

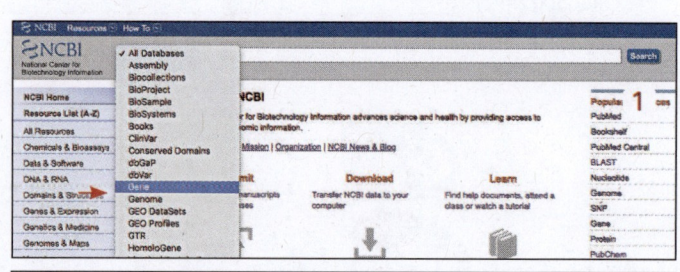

▸ Click the second SATB2 on the list (ID 23314) (IMAGE 2). A new window appears showing known information on the SATB2 gene. Scrolling down the screen provides a huge amount of information about this gene.

or go directly to: https://bit.ly/3QZx6zG

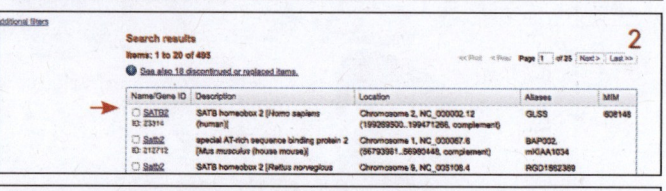

▸ Scroll down to the heading NCBI Reference Sequences. Under this is the smaller heading mRNA and proteins (IMAGE 3) listing all the mRNA transcripts of this gene.

▸ Click the first on the list (NM_001172509.2).

▸ This brings up information on the transcript. Scroll to the bottom and the DNA sequence for the mRNA is shown.

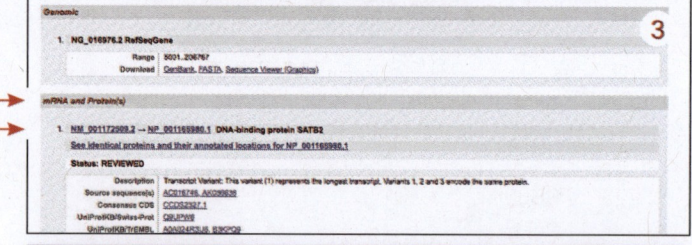

▸ Highlight and copy this sequence.

▸ Scroll back to the top of the page. On the right is the heading Analyze this sequence (IMAGE 4). Under this, click the heading Run BLAST. This opens a search box where you can compare the SATB2 sequence from the previous website to other sequences.

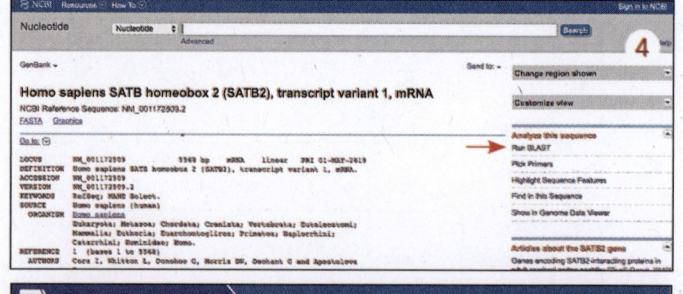

▸ Paste the SATB2 DNA sequence into the box headed 'Enter accession number(s), gi(s), or FASTA sequence(s)' or use the reference number shown in the box (IMAGE 5). Click on BLAST at the bottom of the page. The search may take a minute or so.

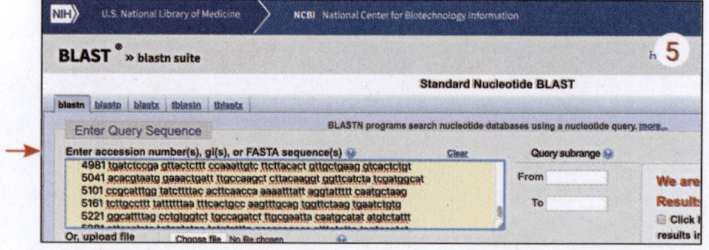

D3.2
18

- A new screen will appear showing the BLAST results. Scroll down to see a list of all the species with similar SATB2 sequences. Clicking on a species shows a match of the DNA in the sequences. Once you are familiar with this process go back to the opening page www.ncbi.nlm.nih.gov. Find HomoloGene in the drop down menu beside the search box and click it. This is the homologous gene database. Type SATB2 in the search box and click search.

- A new page comparing the SATB2 proteins in other species appears (IMAGE 6). Under the heading Protein alignments is a pair of drop down boxes. Here you can compare any two species' SATB2 proteins.

- Compare humans and dogs (*Canis lupus familiaris*).

- Click BLAST under the drop down boxes. After a minute, all the amino acids coded for by the SATB2 gene will appear compared as a query (what you are comparing to) and a subject (what is being compared). Any differences are shown in the middle line of text.

The steps above are a highly simplified step-through of the database. It is capable of many different searches and investigations and stores millions of DNA sequences. Even this simple process shows the power of technology when investigating species relationships.

1. Using the SATB2 gene and the NCBI gene database, answer the following:

    (a) What is the % similarity of the DNA between a human and a chimpanzee for the SATB2 gene? _____

    (b) What is the % similarity of the DNA between a human and a domestic dog for the SATB2 gene (variant X1)?

    _____

2. Computers scan the DNA sequences to find where the sequences most closely align. At what number DNA base pairs do the wolf and human DNA begin to align?

    _____

3. Using the Homologene search for the gene SATB2, answer the following:

    (a) What is the % similarity between the human and dog SATB2 amino acid sequence? _____

    (b) At what number amino acids do the differences occur? _____

4. Go to the NCBI web page shown in image 5 above. The search box is useful in that the DNA sequence can be typed directly into it. How might this be useful in identifying the origin of a DNA sequence?

    _____

    _____

    _____

    _____

5. Type a short DNA sequence of your choosing in the search box (about 20 to 30 letters). Click Blast and see if there are any matches. The search may take some time. Have you 'found' a unique DNA sequence or is it from a known species?

    _____

6. Use what you have learned in this activity to explain why bioinformatics is such a powerful tool:

    _____

    _____

    _____

    _____

    _____

329 Inheritance of Linked Genes

Key Idea: Linked genes tend to be inherited together. Linkage reduces the genetic variation in the offspring.

Genes are said to be linked when they are on the same chromosome. Linked genes tend to be inherited together. The likelihood of crossing over between linked genes decreases when genes are closer together. In genetic crosses, linkage is indicated when a greater proportion of the offspring are of the parental type than would be expected if the alleles were on separate chromosomes and assorting independently. Linkage reduces the genetic variation in the offspring.

Overview of linkage

Parent 1 (2N)
Normal body, normal eyes
B ‖ b
Pr ‖ pr

Parent 2 (2N)
Black body, purple eyes
b ‖ b
pr ‖ pr

Chromosomes before replication

Genes are linked when they are found on the same chromosomes. In this case, in *Drosophila*, B (body colour) and Pr (eye colour) are linked.

Linked gene notation uses vertical double links to show the chromosome pairs.

Each replicated chromosome produces two identical gametes. One of each is shown here.

Chromosomes after replication

X

Meiosis

Gametes

Offspring

Bb Prpr — Normal body, normal eyes

Bb Prpr — Normal body, normal eyes

bb prpr — Black body, purple eyes

bb prpr — Black body, purple eyes

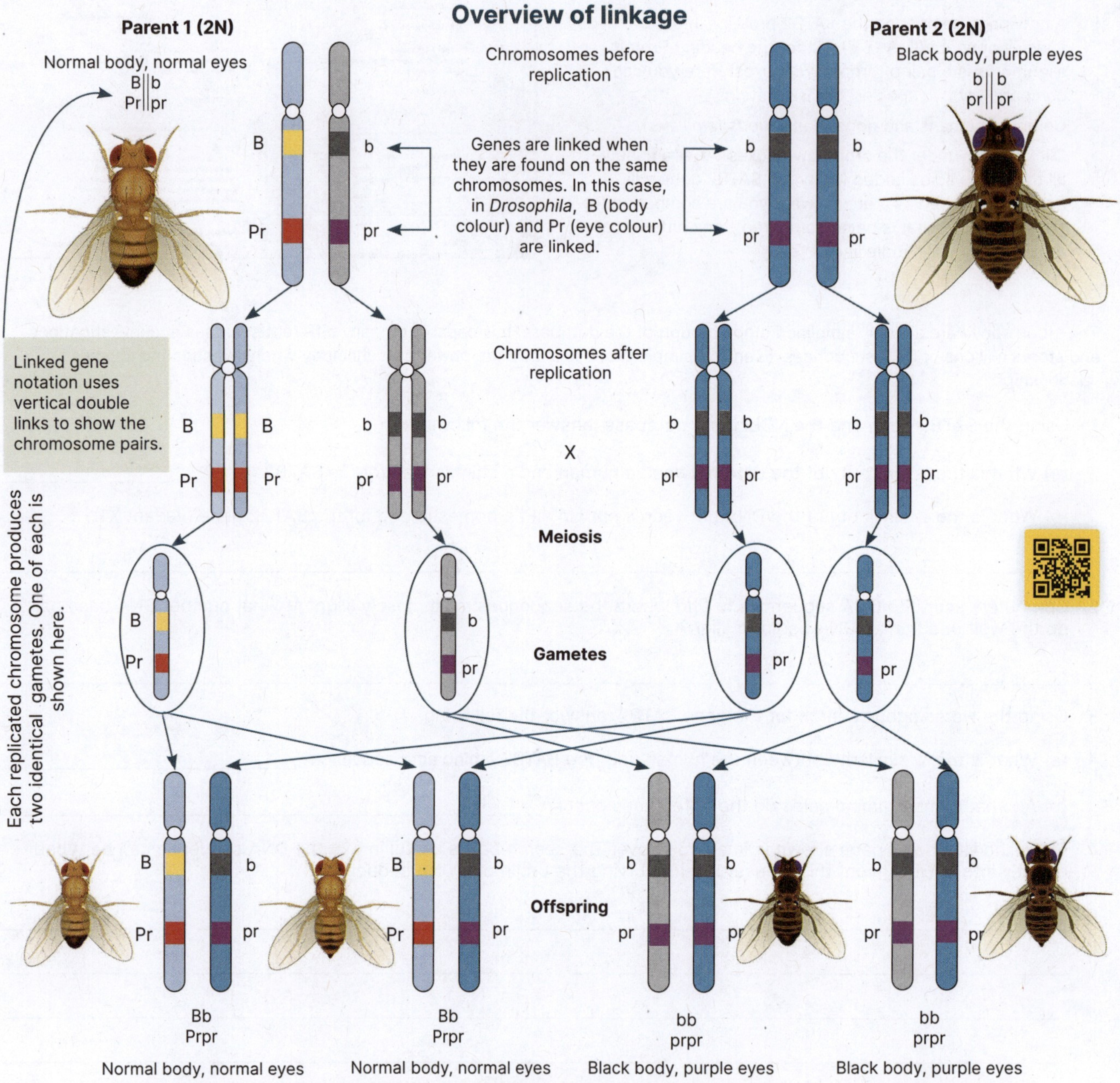

Possible offspring

Only two kinds of genotype combinations are possible. They are the same as the parent genotype.

1. What is the effect of linkage on the inheritance of genes? _____

_____

_____

2. Explain how linkage decreases the amount of genetic variation in the offspring: _____

_____

_____

_____

AHL

D3.2
19

©2024 **BIOZONE** International
ISBN: 978-1-99-101410-8
Photocopying prohibited

330 Recombination and Dihybrid Inheritance

Key Idea: Recombination is the exchange of alleles between homologous chromosomes as a result of crossing over. Recombination increases genetic variation.

The alleles of parental linkage groups can separate in crossing over so that new associations of alleles are formed in the gametes (alleles are reshuffled). Offspring formed from these gametes show combinations of characteristics not seen in the parents and are called recombinants. In contrast to linkage, recombination increases genetic variation in the offspring. Recombination between the alleles of parental linkage groups is indicated by the appearance of non-parental types in the offspring

Overview of recombination

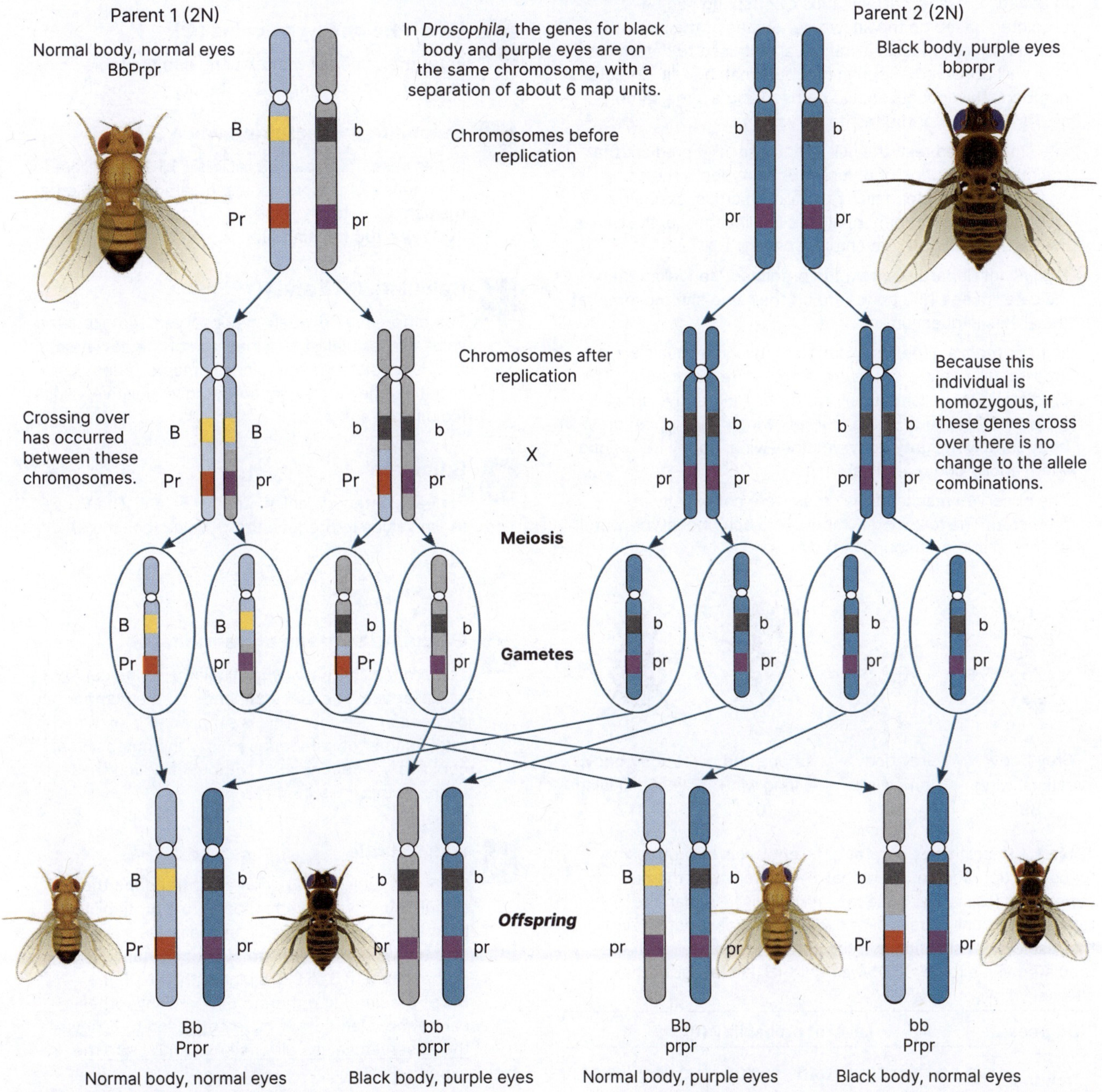

Parent 1 (2N)
Normal body, normal eyes
BbPrpr

In *Drosophila*, the genes for black body and purple eyes are on the same chromosome, with a separation of about 6 map units.

Chromosomes before replication

B
b

Pr
pr

Parent 2 (2N)
Black body, purple eyes
bbprpr

b
b

pr
pr

Chromosomes after replication

Crossing over has occurred between these chromosomes.

B   B
Pr pr

b   b
Pr pr

X

b   b
pr pr

b   b
pr pr

Because this individual is homozygous, if these genes cross over there is no change to the allele combinations.

Meiosis

Gametes

B   B   b   b
Pr pr Pr pr

b   b   b   b
pr pr pr pr

Offspring

B   b   b   b   B   b   b   b
Pr pr   pr pr   pr pr   Pr pr

Bb   bb   Bb   bb
Prpr   prpr   prpr   Prpr

Normal body, normal eyes | Black body, purple eyes | Normal body, purple eyes | Black body, normal eyes

Non-recombinant offspring
These two offspring show allele combinations that are expected as a result of independent assortment during meiosis. Also called parental types.

Recombinant offspring
These two offspring show unexpected allele combinations. They can only arise if one of the parent's chromosomes has undergone crossing over.

1. Describe the effect of recombination on the inheritance of genes: _____

D3.2 / 20 / AHL

331 Testing the Outcomes of Genetic Crosses

Key Idea: A chi-squared (χ^2) test can be used to determine whether the outcome of a genetic cross is significantly different from the expected outcome.

When using the chi-squared test, the null hypothesis predicts the ratio of offspring of different phenotypes according to the expected Mendelian ratio for the cross, assuming independent assortment of alleles (no linkage). Significant departures from the predicted Mendelian ratio indicate linkage of the alleles in question. Raw counts should be used and a large sample size is required for the test to be valid.

Using χ^2 in Mendelian genetics

▶ In genetic crosses, certain ratios of offspring can be predicted based on the known genotypes of the parents. The chi-squared test is a statistical test used to determine how well observed offspring numbers match or fit expected numbers. Raw counts should be used and a large sample size is required for the test to be valid.

▶ In a chi-squared test, the null hypothesis (H_0) predicts that the ratio of offspring of different phenotypes is the same as the expected Mendelian ratio for the cross, assuming independent assortment of alleles (no linkage, i.e. the genes involved are on different chromosomes).

▶ Significant departures from the predicted Mendelian ratio indicate linkage (the genes are on the same chromosome) of the alleles in question.

▶ In a *Drosophila* genetics experiment, two individuals were crossed (the details of the cross are not relevant here). The predicted Mendelian ratios for the offspring of this cross were 1:1:1:1 for each of the four following phenotypes: grey body-long wing; grey body-vestigial wing; ebony body-long wing; ebony body-vestigial wing.

▶ The observed results of the cross were not exactly as predicted. The following numbers for each phenotype were observed in the offspring of the cross:

| Grey body, vestigial wing **88** | Grey body, long wing **98** | Ebony body, long wing **102** | Ebony body, vestigial wing **112** |

Table 1: Critical values of χ^2 at different levels of probability. By convention, the critical probability for rejecting the null hypothesis (H_0) is 5%. If the test statistic is less than the tabulated critical value for $P = 0.05$ we cannot reject H_0 and the result is not significant. If the statistic is greater than the tabulated value for $P = 0.05$ we reject (H_0) in favour of the alternative hypothesis.

Degrees of freedom	Level of probability (P)					
	0.50	**0.20**	**0.10**	**0.05**	**0.02**	**0.01**
1	0.455	1.64	2.71	3.84	5.41	6.64
2	1.386	3.22	4.61	5.99	7.82	9.21
3	2.366	4.64	6.25	7.82	9.84	11.35
4	3.357	5.99	7.78	9.49	11.67	13.28
5	4.351	7.29	9.24	11.07	13.39	15.09

Do not reject H_0 ←   → Reject H_0

Steps in performing a χ^2 test for goodness of fit

1 **Enter the observed value (O).**

Enter the values of the offspring into the table (below) in the appropriate category (column 1).

2 **Calculate the expected value (E).**

In this case the expected ratio is 1:1:1:1. Therefore the number of offspring in each category should be the same (i.e. total offspring/ no. categories). 400 / 4 = 100 (column 2).

3 **Calculate O–E and $(O-E)^2$**

The difference between the observed and expected values is calculated as a measure of the deviation from a predicted result. Since some deviations are negative, they are all squared to give positive values (columns 3 and 4).

4 **Calculate χ^2**

For each category, calculate $(O-E)^2 / E$. Then sum these values to produce the χ^2 value (column 5).

$$\chi^2 = \sum \frac{(O-E)^2}{E}$$

5 **Calculate degrees of freedom**

The probability that any particular χ^2 value could be exceeded by chance depends on the number of degrees of freedom. This is simply one less than the total number of categories (this is the number that could vary independently without affecting the last value) In this case 4 – 1 = 3.

6 **Use χ^2 table**

On the χ^2 table with 3 degrees of freedom, the calculated χ^2 value corresponds to a probability between 0.2 and 0.5. By chance alone a χ^2 value of **2.96** will happen 20% to 50% of the time. The probability of 0.0 to 0.5 is higher than 0.05 (i.e 5% of the time) and therefore the null hypothesis cannot be rejected. We have no reason to believe the observed values differ significantly from the expected values.

	1	2	3	4	5
Category	O	E	O–E	$(O-E)^2$	$(O-E)^2/E$
GB, LW	98	100	-2	4	0.04
GB, VW	88	100	-12	144	1.44
EB, LW	102	100	2	4	0.04
EB, VW	112	100	12	144	1.44
				$\chi^2 \rightarrow$	2.96

   D3.2 21

©2024 **BIOZONE** International
ISBN: 978-1-99-101410-8
Photocopying prohibited

The following problems examine the use of the chi-squared (χ^2) test in genetics.

1. In a tomato plant experiment, two heterozygous individuals were crossed (the details of the cross are not relevant here). The predicted Mendelian ratios for the offspring of this cross were **9:3:3:1** for each of the **four following phenotypes**: purple stem-jagged leaf edge, purple stem-smooth leaf edge, green stem-jagged leaf edge, green stem-smooth leaf edge.

   The observed results of the cross were not exactly as predicted.
   The numbers of offspring with each phenotype are provided below:

Observed results of the tomato plant cross			
Purple stem-jagged leaf edge	12	Green stem-jagged leaf edge	8
Purple stem-smooth leaf edge	9	Green stem-smooth leaf edge	0

   (a) State your null hypothesis for this investigation (H0): _____

   _____

   (b) State the alternative hypothesis (HA): _____

   _____

2. Use the chi-squared (χ^2) test to determine if the differences between the observed and expected phenotypic ratios are significant. Use the table of critical values of χ^2 at different P values on the previous page.

   (a) Enter the observed values (number of individuals) and complete the table to calculate the χ^2 value:

Category	O	E	O − E	(O − E)²	$\frac{(O-E)^2}{E}$
Purple stem, jagged leaf					
Purple stem, smooth leaf					
Green stem, jagged leaf					
Green stem, smooth leaf					
Σ					Σ

   (b) Calculate χ^2 value using the equation:

   $$\chi^2 = \sum \frac{(O-E)^2}{E} \qquad \chi^2 = \underline{\qquad}$$

   (c) Calculate the degrees of freedom: _____

   (d) Using the χ^2 table, state the P value corresponding to your calculated χ^2 value:

   _____

   (e) State your decision: *(circle one)*

   reject H0 / do not reject H0

3. Students carried out a pea plant experiment, where two heterozygous individuals were crossed. The predicted Mendelian ratios for the offspring were **9:3:3:1** for each of the **four following phenotypes**: round-yellow seed, round-green seed, wrinkled-yellow seed, wrinkled-green seed.

   The observed results were as follows:

Round-yellow seed	**441**	Wrinkled-yellow seed	**143**
Round-green seed	**159**	Wrinkled-green seed	**57**

   Use a separate piece of paper to complete the following:

   (a) State the null and alternative hypotheses (H0 and HA).

   (b) Calculate the χ^2 value.

   (c) Calculate the degrees of freedom and state the P value corresponding to your calculated χ^2 value.

   (d) State whether or not you reject your null hypothesis: reject H0 / do not reject H0 (circle one)

   (e) Comment on whether the χ^2 values obtained above are similar. Suggest a reason for any difference:

Sampling and statistical testing

Often, it is impossible to gather data from every individual in a population to use in a statistical test so a sample is taken to represent the entire data set. The χ^2 test uses the F_2 generation as the sample (the parents are the two F_1 heterozygotes). Other sample data sets come from experiment repeats or replications. Accuracy in developing and then following the experimental method will ensure that each sample represents the entire population data set more accurately.

332 Principles of Homeostasis

Key Idea: Homeostasis is the process of sustaining a constant physiological state within the body, regardless of fluctuations in the external environment.

Organisms maintain a relatively constant physiological state, called homeostasis, despite changes in their external environment. Any change in the environment to which an organism responds is called a stimulus. Environmental stimuli are constantly changing, so organisms must adjust their behaviour and physiology constantly to maintain homeostasis. This requires the coordinated activity of the body's organ systems. Homeostatic mechanisms prevent potentially harmful deviations from the steady state and keep the body's internal conditions within strict preset limits.

Homeostasis is required to maintain constant body temperature, at about 37°C. Similarly, you must regulate blood sugar (glucose) levels and blood pH, water, and electrolyte balance (osmotic concentration), and blood pressure. Your body's organ systems coordinate to carry out these tasks.

How homeostasis is maintained: the stimulus-response model

To maintain homeostasis, the body must detect stimuli through receptors, process this sensory information in a control centre, and respond to it appropriately via an effector. The responses provide new feedback to the receptor. These three components are illustrated below.

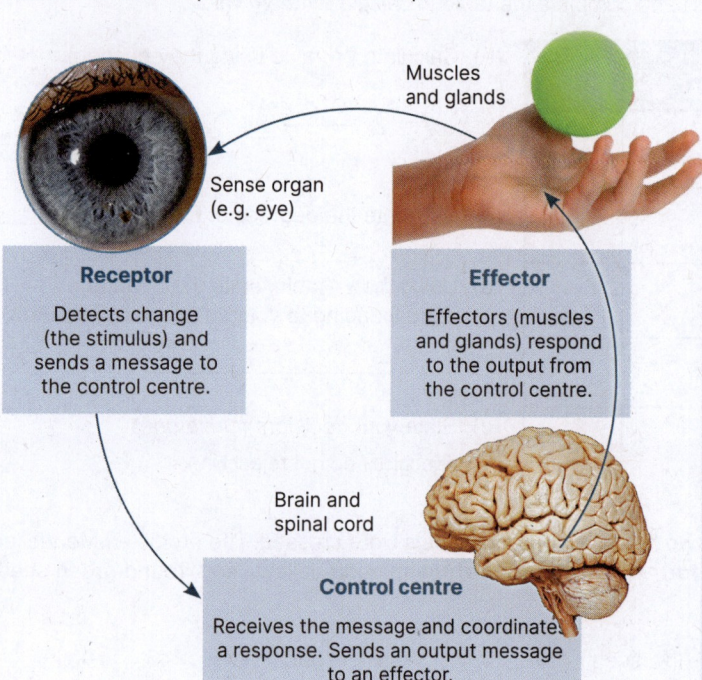

Sense organ (e.g. eye)

Muscles and glands

Receptor
Detects change (the stimulus) and sends a message to the control centre.

Effector
Effectors (muscles and glands) respond to the output from the control centre.

Brain and spinal cord

Control centre
Receives the message and coordinates a response. Sends an output message to an effector.

Homeostasis analogies

The analogy of a temperature setting on a heat pump is a good way to explain how homeostasis is maintained. A heat pump has sensors (a receptor) to monitor room temperature. It also has a control centre to receive and process the data from the sensors. Depending on the data it receives, the control centre activates the effector (heating/cooling unit), switching either on or off.

When the room is too cold, the heating unit switches on, and the cooling unit is off. When it is too hot, the heating unit switches off and the cooling unit is switched on. This system maintains a constant temperature, similar to homeostasis in the body.

The analogy of staying upright on a mountain bike, using body weight, arms, pedals, brakes, and steering, demonstrates that many homeostasis systems have multiple mechanisms to maintain a steady state.

1. Define the term 'homeostasis': _____

   _____

2. What is the role of the following components in maintaining homeostasis?

   (a) Receptor: _____

   _____

   (b) Control centre: _____

   _____

   (c) Effector: _____

   _____

©2024 **BIOZONE** International
ISBN: 978-1-99-101410-8

D3.3

1

333 Negative Feedback Loops

Key Idea: Feedback loops, driven by various mechanisms, can stabilize biological systems or exaggerate deviations from the median condition.

Two types of feedback loop are used in the body, each producing specific outcomes. Negative feedback loops

maintain homeostasis, e.g. regulation of body temperature. Positive feedback loops exaggerate any changes in the internal environment, moving the body away from a stable state by quickly amplifying changes in the internal environment e.g. blood clotting.

Negative feedback loops

▶ Negative feedback loops are control systems that maintain the body's internal environment at a relatively steady state.

▶ When variations from the norm are detected by the body's receptors, a response or output from the effectors that opposes the stimulus is classified as negative feedback.

▶ Negative feedback discourages variations from a set point and returns internal conditions to a steady state.

▶ Most physiological systems achieve homeostasis through negative feedback loops.

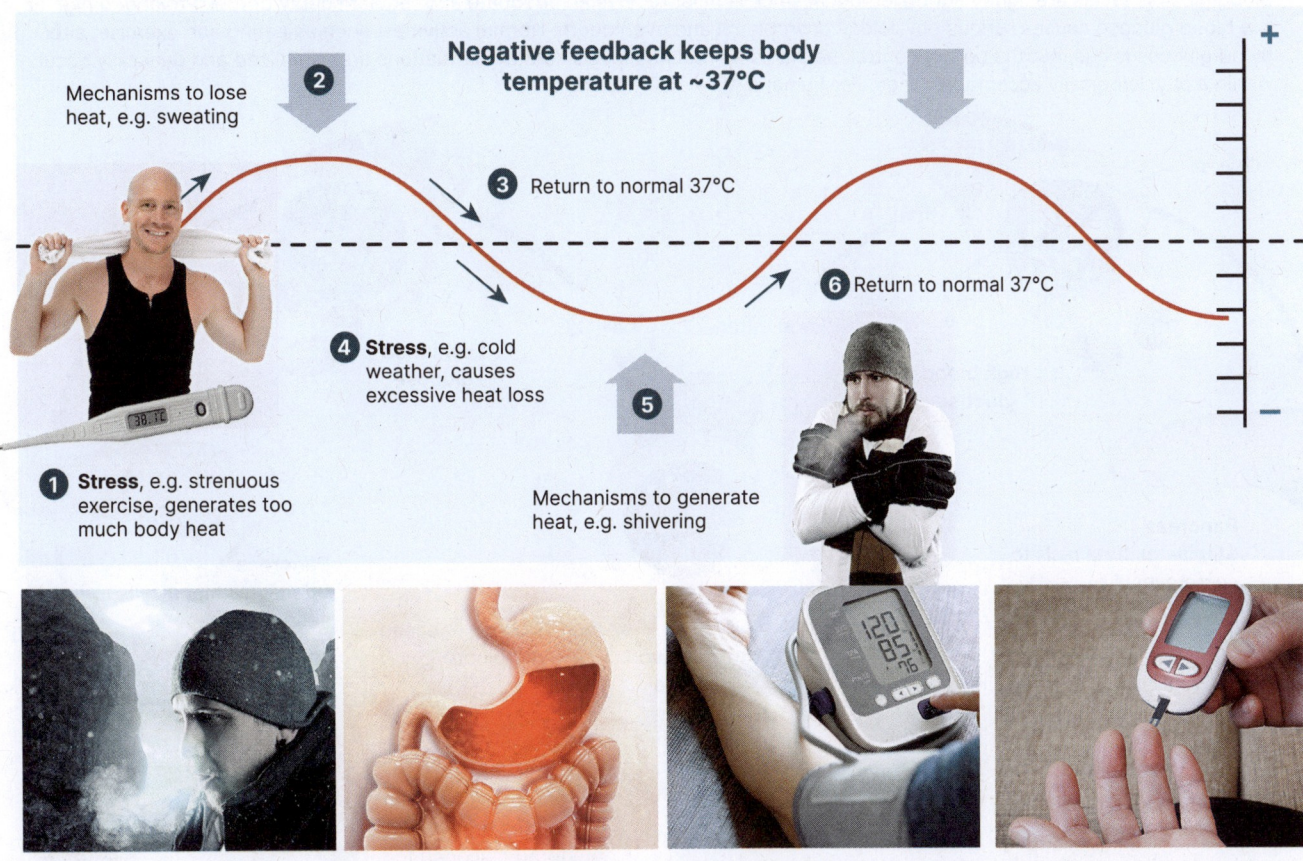

Negative feedback keeps body temperature at ~37°C

2 Mechanisms to lose heat, e.g. sweating

3 Return to normal 37°C

6 Return to normal 37°C

4 **Stress**, e.g. cold weather, causes excessive heat loss

5 Mechanisms to generate heat, e.g. shivering

1 **Stress**, e.g. strenuous exercise, generates too much body heat

We know when we are cold. Although we are unaware of most of the negative feedback loops operating in our bodies, they keeps our systems stable.

Food in the stomach activates stretch receptors, stimulating gastric secretion and motility. As the stomach empties, the stimulus for gastric activity declines.

Negative feedback loops control almost all the body's functioning processes, including heart rate, blood glucose, blood pressure, and pituitary secretions.

Maintaining a stable blood glucose level is an important homeostatic function regulated by negative feedback. It involves two antagonistic hormones.

1. Construct a diagram of a negative feedback loop that models thermoregulation (homeostasis of constant body temperature) using information and terms in this activity:

©2024 **BIOZONE** International
ISBN: 978-1-99-101410-8
Photocopying prohibited

D3.3
2

334 Control of Blood Glucose

Key Idea: Insulin and glucagon are the two hormones secreted by the pancreas that maintain blood glucose at a steady state via a negative feedback loop.

Insulin and glucagon are produced by the islet cells of the pancreas and control blood glucose levels. Insulin lowers blood glucose by promoting the uptake of glucose by the body's cells and the conversion of glucose into the storage molecule glycogen in the liver. Glucagon increases blood glucose by stimulating the breakdown of stored glycogen and the synthesis of glucose from amino acids. The liver has a central role in these carbohydrate conversions. Negative feedback stops hormone secretion when normal blood glucose levels are restored. Blood glucose homeostasis allows energy to be available to cells, as required.

The importance of blood glucose homeostasis

▶ Glucose is the body's main energy source. It is chemically broken down during cellular respiration to generate ATP, which is used to power metabolism. Glucose is the main sugar circulating in blood so it is often called blood sugar. Blood glucose levels are regulated by negative feedback involving two hormones: insulin and glucagon.

▶ Blood glucose levels are tightly controlled because cells must receive an adequate and regular supply of fuel. Prolonged high or low blood glucose causes serious physiological problems and even death. Normal activities, such as eating and exercise, alter blood glucose levels, but the body's control mechanisms regulate levels so that fluctuations are minimized and generally occur within a physiologically acceptable range. For humans, this is 60-110 mg dL^{-1}.

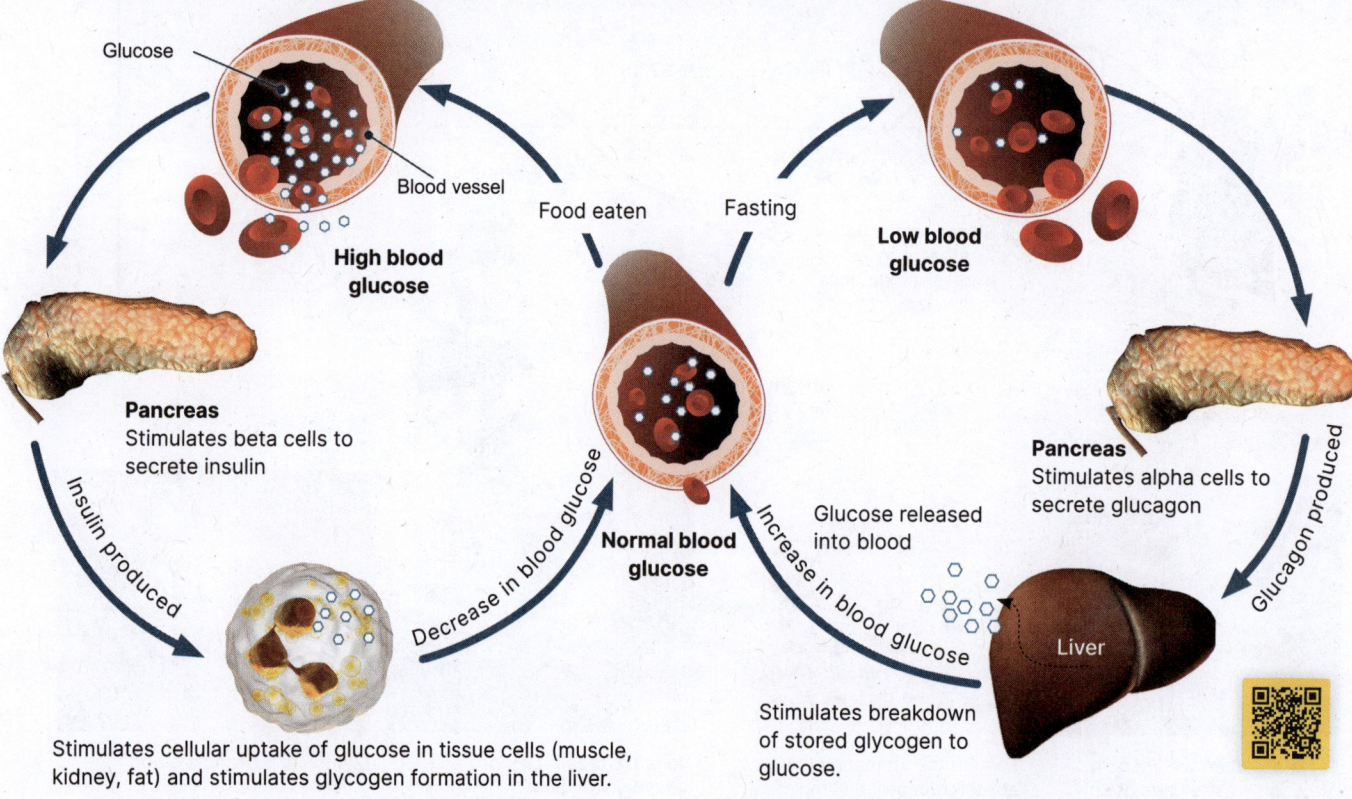

1. For the following two scenarios, describe how normal blood glucose level is restored:

   (a) Low blood glucose: _____

   _____

   (b) High blood glucose: _____

   _____

2. What is the role of the liver in blood glucose homeostasis? _____

   _____

3. In what way are the actions of the hormones insulin and glucagon 'antagonistic'? _____

   _____

   _____

   _____

D3.3
3

©2024 **BIOZONE** International
ISBN: 978-1-99-101410-8
Photocopying prohibited

335 Diabetes Mellitus

Key Idea: Diabetes mellitus is a condition in which blood glucose levels are elevated, either because of a lack of insulin (type 1) or because of resistance to insulin's effects (type 2). Diabetes mellitus (diabetes) is a condition in which blood glucose is too high because the body's cells cannot take up glucose in the normal way. It is usually detected by glucose appearing in the urine (glucose is normally reabsorbed and does not enter the urine). The two types of diabetes, type 1 and type 2, have different causes and treatments, but both are life threatening conditions if untreated (below).

Type 1 diabetes

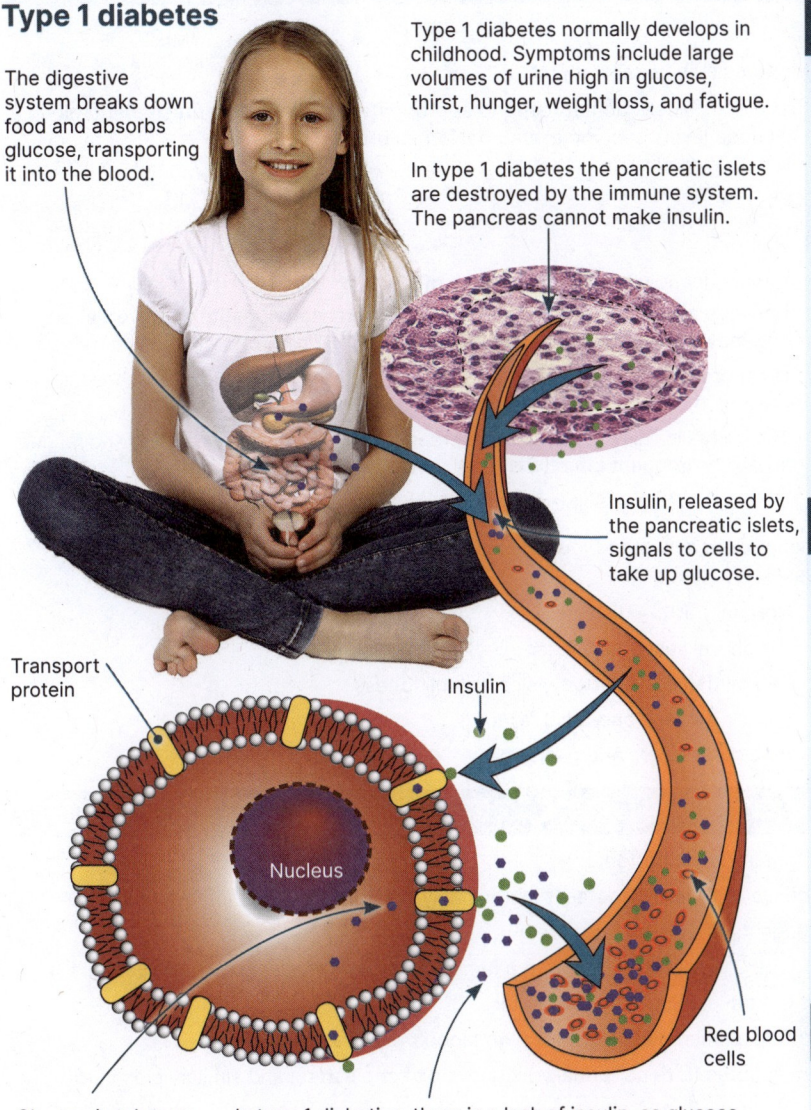

The digestive system breaks down food and absorbs glucose, transporting it into the blood.

Type 1 diabetes normally develops in childhood. Symptoms include large volumes of urine high in glucose, thirst, hunger, weight loss, and fatigue.

In type 1 diabetes the pancreatic islets are destroyed by the immune system. The pancreas cannot make insulin.

Insulin, released by the pancreatic islets, signals to cells to take up glucose.

Transport protein

Insulin

Nucleus

Red blood cells

Glucose is taken up by cells and used in cellular respiration.

In type 1 diabetics, there is a lack of insulin, so glucose builds up in the blood. The body tries to get rid of it by producing large volumes of urine. Regular injections of insulin allow diabetics to regulate blood glucose levels.

Type 1 diabetes summary

Age at onset: Early in life; often in childhood. Often called juvenile onset diabetes

Cause: Absolute deficiency of insulin due to lack of insulin production (pancreatic β cells are destroyed in an autoimmune reaction). There is a genetic component, but usually a childhood viral infection triggers the development of type 1 diabetes.

Treatment: Blood glucose is monitored regularly and insulin injections, combined with dietary management, are used to keep blood sugar levels stable. Therapies involving pancreatic transplants, or transplants of insulin-producing islet cells or stem cells are currently being investigated.

Short term effects

Low blood sugar (*hypoglycaemia*): Blood glucose levels below normal (4 mmol L^{-1}) can result in clumsiness, confusion, and seizures. It can result from too much insulin, usually after injection if glucose levels are already low.

High blood sugar (*hyperglycaemia*): High blood glucose levels occur when glucose fails to enter the cells. Effects include frequent urination, fatigue, thirst, and blurred vision

Ketoacidosis: A lack of insulin can result in a build up of molecules called ketones caused by metabolism of fats for fuel. Ketones are acidic and can lead to metabolic acidosis (fall in tissue pH), which can quite quickly be fatal.

Long term effects

General circulation: Over time, high blood glucose damages the lining of small blood vessels, making them prone to developing plaques and become narrow and clogged. The result of this is increased blood pressure.

Heart disease: Nearly 3 in 4 people with type 1 diabetes will suffer some form of heart disease. Causes may be from autoimmune responses and high blood glucose and blood pressure.

Kidney disease: Damage to the small blood vessels of the kidney causes kidney function to decline and produces many associated health problems. Glucose in the urine can also result in fungal infections in the bladder.

Eye problems: Damage to the blood vessels in the eyes leads to cataracts and retinal damage.

Nerve damage: High blood glucose levels cause nerve damage indirectly through blood vessel damage. Symptoms include tingling and weakness in the limbs. Numbness can lead to unnoticed, hard to treat infections and ulcers.

1. (a) What is type 1 diabetes? _____

_____

_____

(b) Explain how the usual mechanisms for blood glucose homeostasis are disrupted in a person with type 1 diabetes. How does this disruption result in the symptoms observed?

_____

_____

_____

_____

_____

_____

Type 2 diabetes

▶ Type 2 diabetes mellitus is characterized by a resistance to insulin's effects and relative (rather than absolute) insulin deficiency.

▶ The pancreas produces insulin, but the body's cells cease to respond to it and glucose levels in the blood remain high. Type 2 diabetes is a chronic, progressive disease, usually caused by lifestyle choices, and becomes worse with time if not managed. Its long-term effects include heart disease, strokes, blindness, and kidney failure. However, ketoacidosis (a feature of type 1 diabetes), is uncommon.

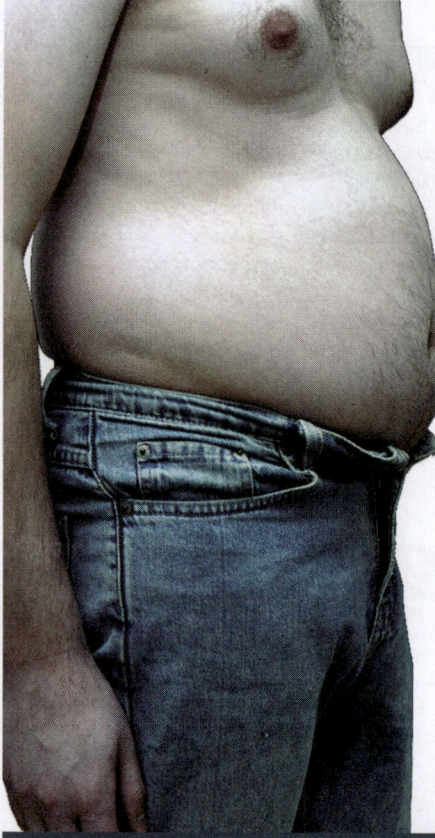

Symptoms of type 2 diabetes mellitus

Symptoms may be mild at first.

▶ The body's cells do not respond appropriately to the insulin present and blood glucose levels become elevated. Normal blood glucose level is 3.3-6.1 mmol L^{-1} (60-110 mg dL^{-1}). In diabetics, fasting blood glucose level is 7 mmol L^{-1} (126 mg dL^{-1}) or higher.

Symptoms occur with varying degrees of severity:

▶ Cells are starved of fuel. This can lead to increased appetite and overeating and may contribute to an existing obesity problem.

▶ Urine production increases to rid the body of the excess glucose. Glucose is present in the urine and patients are frequently very thirsty.

▶ The body's inability to use glucose properly leads to muscle weakness and fatigue, irritability, frequent infections, and poor wound healing.

c

Uncontrolled, elevated blood glucose eventually results in damage to the blood vessels and leads to:

▶ Coronary artery disease
▶ Peripheral vascular disease
▶ Retinal damage, blurred vision and blindness
▶ Kidney damage and renal failure
▶ Persistent ulcers and gangrene

Risk factors

▶ Obesity: BMI greater than 27. Distribution of weight is also important.
▶ Age: Risk increases with age, although the incidence of type 2 diabetes is increasingly being reported in obese children.
▶ Sedentary lifestyle: Inactivity increases risk through its effects on body weight.
▶ Family history: There is a strong genetic link for type 2 diabetes. Those with a family history of the disease are at greater risk.
▶ Ethnicity: Certain ethnic groups are at higher risk of developing type 2 diabetes.
▶ High blood pressure: Up to 60% of people with undiagnosed diabetes have high blood pressure.
▶ High blood lipids: More than 40% of people with diabetes have abnormally high levels of cholesterol and similar lipids in the blood.

Treatment: Increasing physical activity, losing weight (especially abdominal fat), and improving diet may be sufficient to control type 2 diabetes. The use of prescribed anti-diabetic drugs and insulin therapy (injections) may be required if lifestyle changes are insufficient on their own.

2. Type 2 diabetes is a non-infectious disease that interferes with the homeostatic control of blood glucose in the body.

(a) Describe the type 2 diabetes disease and common symptoms experienced by an individual:

(b) Describe the risk factors that increase the chance of developing type 2 diabetes:

(c) Explain how ONE current treatment of type 2 diabetes controls or eliminates the disease:

©2024 **BIOZONE** International
ISBN: 978-1-99-101410-8
Photocopying prohibited

336 Thermoregulation

Key Idea: Thermoregulation is controlled both by the hormonal and nervous systems of the body.

In humans, the temperature regulation centre is a region of the brain called the hypothalamus. The hypothalamus acts as the control centre in the thermoregulation negative feedback loop. It contains thermoreceptors that monitor core body temperature and has a 'set-point' temperature of 36.7°C. The hypothalamus acts like a thermostat: it registers changes in the core body temperature and also receives information about temperature changes from thermoreceptors in the skin. It then coordinates effector responses to counteract the changes and restore normal body temperature. Communication between components of the negative feedback loop uses both nervous and hormonal pathways. When normal body temperature is restored, the corrective mechanisms are switched off.

Regulating body temperature

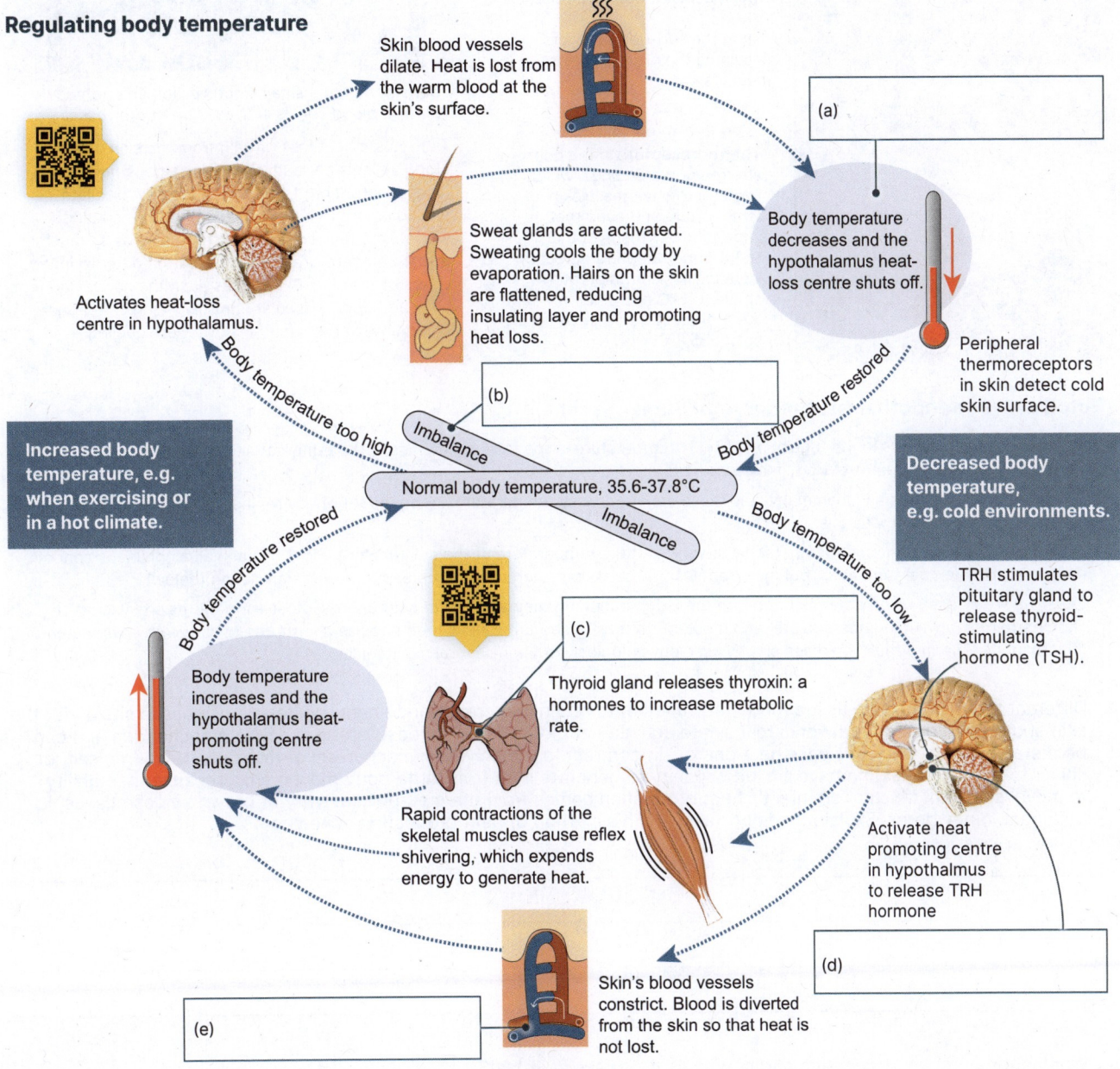

Skin blood vessels dilate. Heat is lost from the warm blood at the skin's surface.

(a)

Body temperature decreases and the hypothalamus heat-loss centre shuts off.

Peripheral thermoreceptors in skin detect cold skin surface.

Activates heat-loss centre in hypothalamus.

Sweat glands are activated. Sweating cools the body by evaporation. Hairs on the skin are flattened, reducing insulating layer and promoting heat loss.

(b)

Body temperature too high

Imbalance

Normal body temperature, 35.6-37.8°C

Imbalance

Body temperature restored

Increased body temperature, e.g. when exercising or in a hot climate.

Body temperature restored

Decreased body temperature, e.g. cold environments.

TRH stimulates pituitary gland to release thyroid-stimulating hormone (TSH).

(c)

Body temperature increases and the hypothalamus heat-promoting centre shuts off.

Thyroid gland releases thyroxin: a hormones to increase metabolic rate.

Rapid contractions of the skeletal muscles cause reflex shivering, which expends energy to generate heat.

Body temperature too low

Activate heat promoting centre in hypothalmus to release TRH hormone

(d)

Skin's blood vessels constrict. Blood is diverted from the skin so that heat is not lost.

(e)

1. In the diagram above showing the regulation of body temperature:

   (a) Identify the stimulus: _____

   (b) Identify the effectors: _____

   (c) What structure(s) would you add to represent the receptors? _____

2. Label the diagram above by appropriately adding the labels: stimulus, receptors, control centre, and effectors.

3. How do the effectors restore body temperature when it increases above the set point? _____

   _____

   _____

Skin cross-section

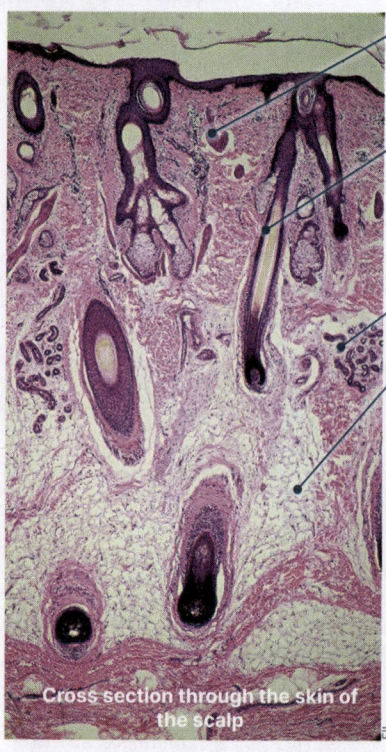

Blood vessels in the dermis dilate or constrict to promote or restrict heat loss.

The hair erector muscles allow hairs to be raised or lowered to increase or decrease the thickness of the insulating air layer between the skin and the environment.

Sweat glands produce sweat, which cools through evaporation.

Fat in the sub-dermal layers insulates the organs against heat loss.

Cross section through the skin of the scalp

Thermoreceptors in the dermis are free nerve endings, which respond to changes in skin temperature and send that information to the hypothalamus. Hot thermoreceptors detect an increase in skin temperature above 37.5°C and cold thermoreceptors detect a fall below 35.8°C.

Regulating blood flow to the skin

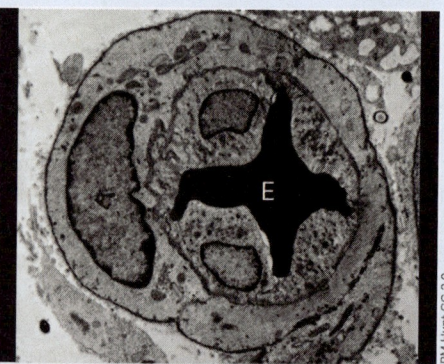

RM Hunt CC 3.0

Constriction of a small blood vessel. An erythrocyte (E) (red blood cell) is in the centre of the vessel.

To regulate heat loss or gain from the skin, the blood vessels beneath the surface constrict (vasoconstriction) to reduce blood flow or dilate (vasodilation) to increase blood flow. When blood vessels are fully constricted there may be as much as a 10°C temperature gradient from the outer to inner layers of the skin. Extremities such the hands and feet have additional vascular controls which can reduce blood flow to them in times of severe cooling.

Ectotherms or endotherms: thermoregulation in animals

▸ Thermoregulation refers to the regulation of body temperature in the face of changes in the temperature of the environment. Animals show two extremes of body temperature tolerance:

▸ Homeotherms (all birds and mammals) maintain a constant body temperature, independent of environmental variation, which represents a large energetic cost.

▸ Poikilotherms allow their body temperature to vary with the temperature of the environment. Most fish and all amphibians cannot regulate body temperature at all, but most reptiles use behaviour both to warm up and to avoid overheating (below).

▸ Animals are classed as ectotherms or endotherms, depending on their sources of heat energy. Most endotherms are also strict homeotherms (always thermoregulate) and most, but not all, ectotherms allow their body temperatures to vary with environmental fluctuations. Thermoregulation relies on physical, physiological, and behavioural mechanisms.

4. Different types of adipose tissue (body fat) function to retain and create body heat. White adipose tissue sits under the skin and around organs and works to insulate against heat loss. Brown adipose tissue is rich in mitochondria and can be metabolized to quickly release heat energy, in contrast to the shivering mechanism of muscles. Newborns cannot shiver to produce heat and have limited capacity to generate heat from large body movements because their ability to move is limited. Much of an infant's heat production comes from the metabolic activity of brown adipose tissue, abundant in newborns. Explain the importance of this method of heat production in newborns:

_____

_____

_____

_____

5. What is the purpose of sweating and how does it achieve its effect? _____

_____

_____

_____

6. How do the blood vessels help to regulate the amount of heat lost from the skin and body? _____

_____

_____

_____

©2024 **BIOZONE** International
ISBN: 978-1-99-101410-8
Photocopying prohibited

337 Kidneys and Osmoregulation

Key Idea: The kidneys filter and remove metabolic wastes from the blood, and have a role in osmoregulation.

The kidneys have a plentiful blood supply from the renal artery. The blood plasma is filtered by the kidneys to form urine, which is produced continuously, passing along the ureters to the bladder. Kidneys are very efficient, producing a urine that is concentrated to varying degrees, depending on fluid requirements at the time. The body's fluid and electrolyte balance is critical to metabolic function. Hormones are involved in osmoregulation, the control of water and salt in body fluids, where osmotic concentration is measured in osmoles per litre (OsMol L^{-1}). ADH promotes water reabsorption in the kidney collecting ducts, and aldosterone promotes sodium reabsorption in the kidney tubules.

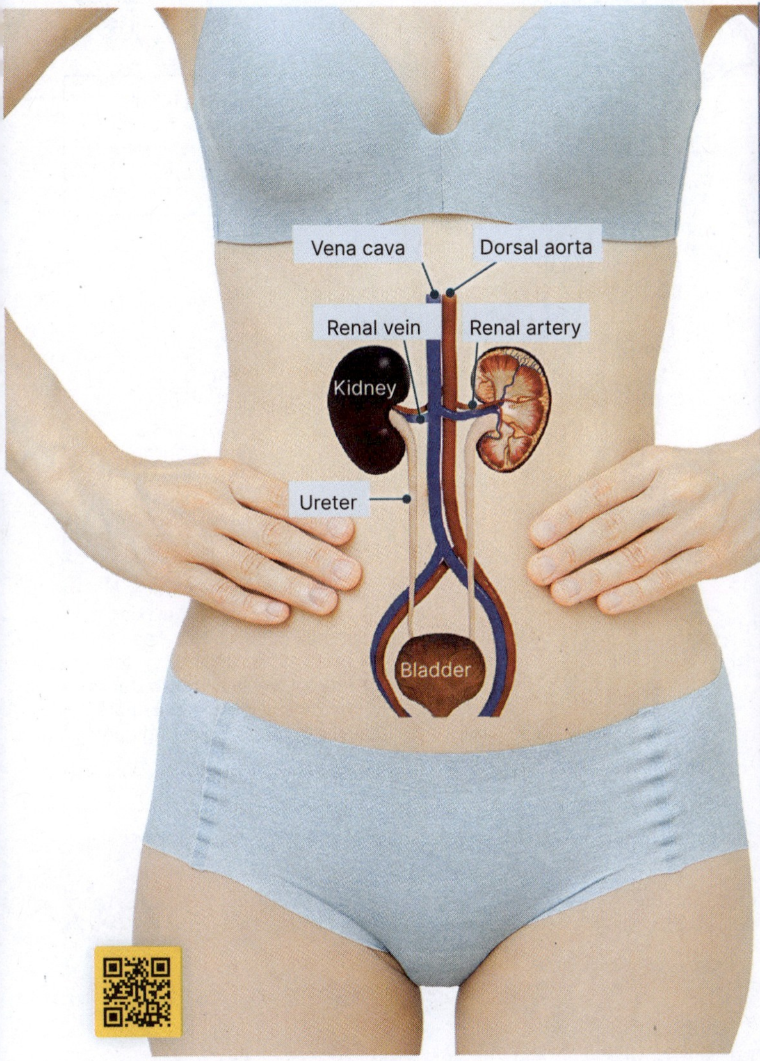

Vena cava · Dorsal aorta · Renal vein · Renal artery · Kidney · Ureter · Bladder

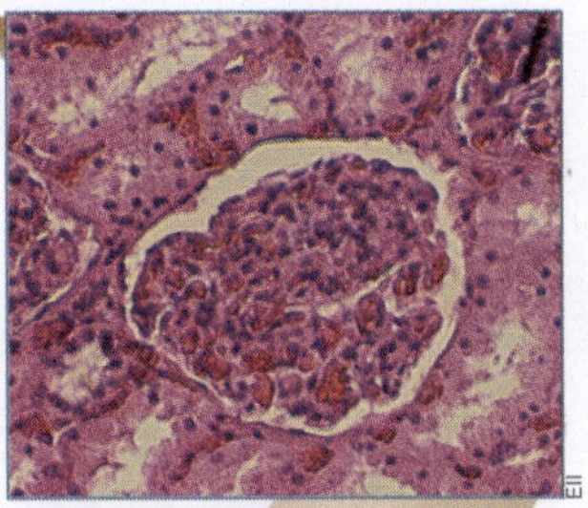

Blood is filtered in the kidneys by the glomerulus: a dense knot of capillaries. Blood pressure forces fluid through the capillary walls in a process called ultrafiltration. The filtrate is collected in the Bowman's capsule surrounding the glomerulus.

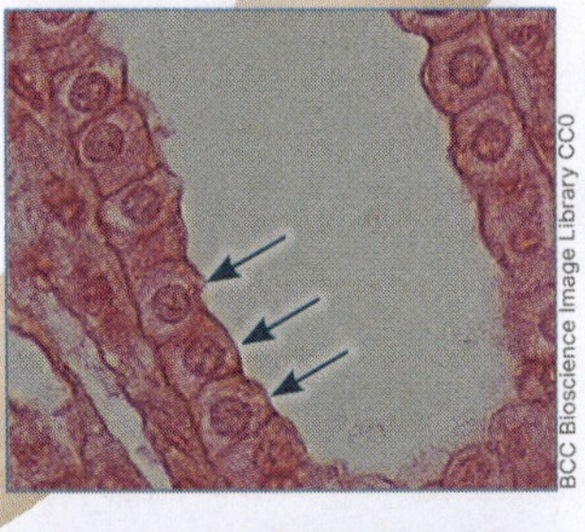

The filtrate moves from Bowman's capsule to the convoluted tubules. In the proximal tubule, the cuboidal epithelial cells (arrowed) have microvilli which increase the reabsorption of substances from the substrate. Most reabsorption occurs in the proximal tubule.

BCC Bioscience Image Library CC0

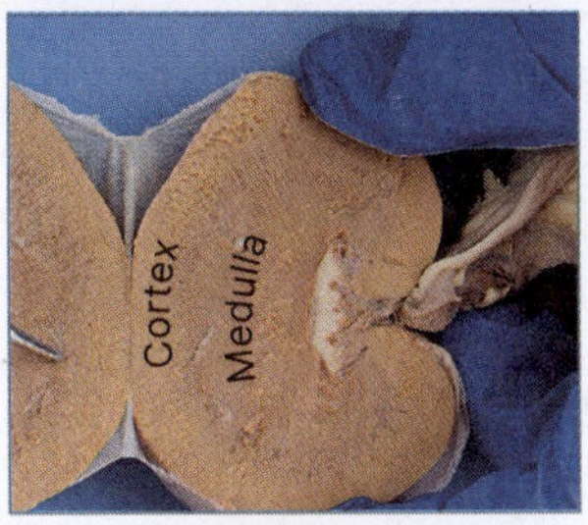

The glomerulus, capsule, and tubules form the nephron (the functional unit of the kidney). The thousands of nephrons are aligned and organized in an orderly way. The glomeruli and convoluted tubules are found in the outer cortex, while the 'loop of Henle' is found in the inner medulla region.

Cortex · Medulla

The filtrate passes to the renal ducts and then to the ureter and finally to the bladder. The kidney itself is bean shaped and is around 10 cm long in humans.

1. What is the purpose of the microvilli in the epithelial cells of the convoluted tubules? _____

_____

2. (a) How is the filtrate formed? _____

_____

_____

(b) How is the filtrate modified? _____

_____

3. The circulation rate of blood through the renal artery is about 1.2 L min^{-1}, about one quarter of the heart's total output. Why does so much blood need to pass through the kidneys every minute?

_____

_____

_____

D3.3 7 · AHL ·

338 Kidney Function

Key Idea: The functional unit of the kidney is the nephron. It is a selective filter element, comprising a renal corpuscle and its associated tubules and ducts.

Ultrafiltration, i.e. forcing fluid and dissolved substances through a membrane by pressure, occurs in the first part of the nephron, across the membranes of the capillaries and the glomerular capsule. The formation of the glomerular filtrate depends on the pressure of the blood entering the nephron (below). If it increases, filtration rate increases; if it falls, glomerular filtration rate also falls. This process is precisely regulated so that glomerular filtration rate per day stays constant. The initial filtrate, now called urine, is modified through secretion and tubular reabsorption, according to the body's needs at the time.

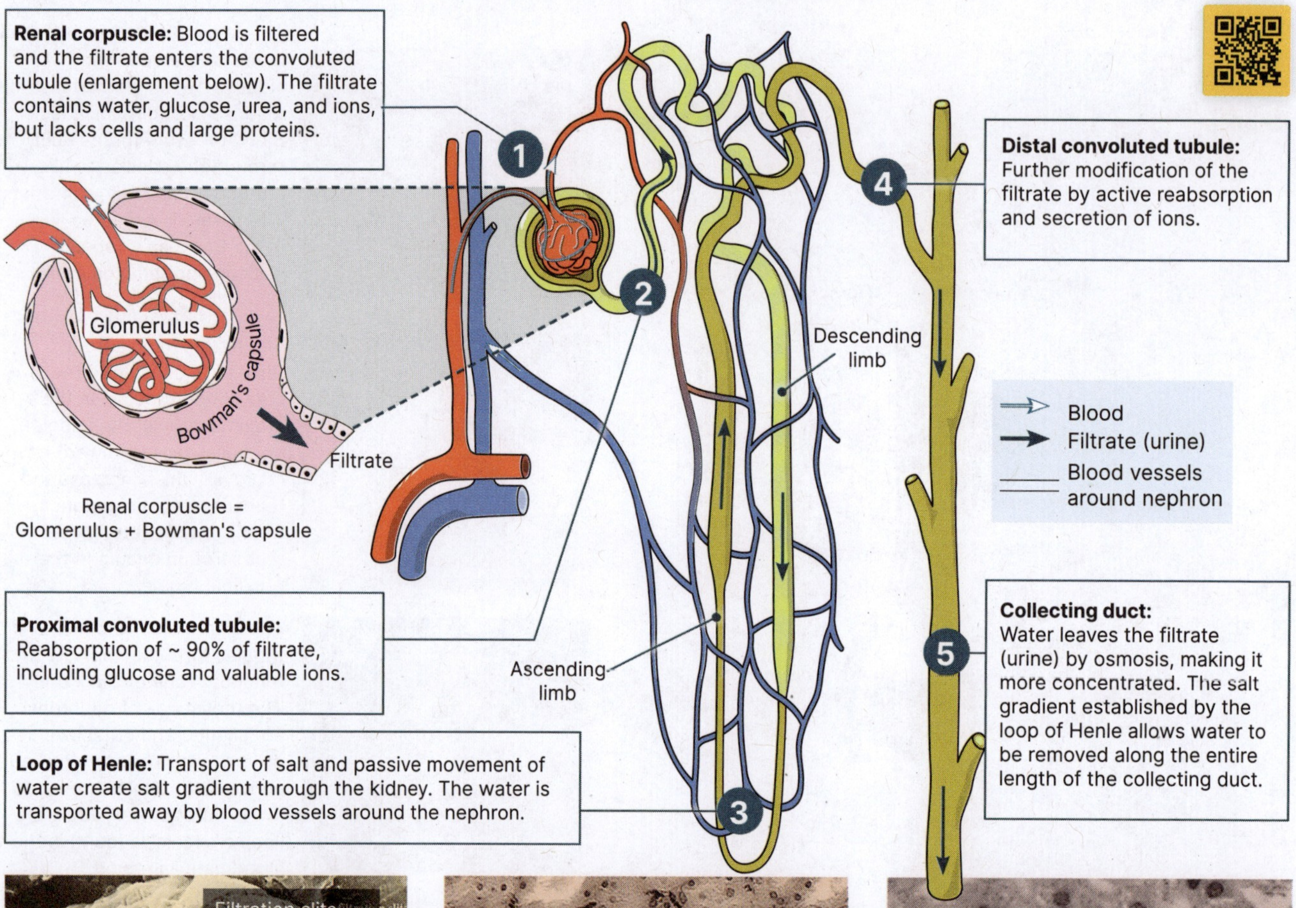

Renal corpuscle: Blood is filtered and the filtrate enters the convoluted tubule (enlargement below). The filtrate contains water, glucose, urea, and ions, but lacks cells and large proteins.

Glomerulus

Bowman's capsule

Filtrate

Renal corpuscle = Glomerulus + Bowman's capsule

1

2

Proximal convoluted tubule: Reabsorption of ~ 90% of filtrate, including glucose and valuable ions.

Loop of Henle: Transport of salt and passive movement of water create salt gradient through the kidney. The water is transported away by blood vessels around the nephron.

Descending limb

Ascending limb

3

4

Distal convoluted tubule: Further modification of the filtrate by active reabsorption and secretion of ions.

→ Blood
→ Filtrate (urine)
— Blood vessels around nephron

5

Collecting duct: Water leaves the filtrate (urine) by osmosis, making it more concentrated. The salt gradient established by the loop of Henle allows water to be removed along the entire length of the collecting duct.

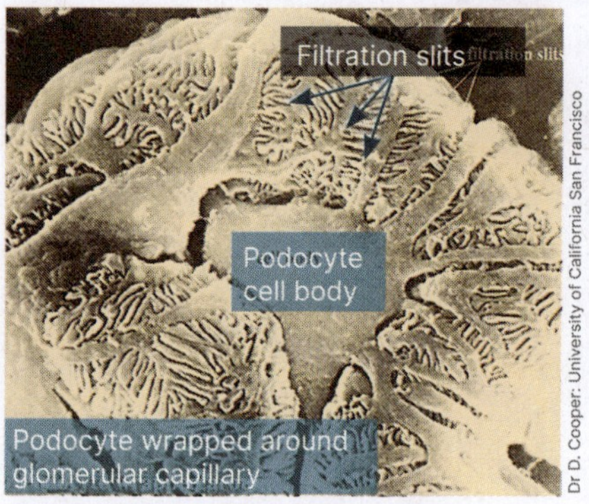

Filtration slits

Podocyte cell body

Podocyte wrapped around glomerular capillary

Dr D. Cooper: University of California San Francisco

The epithelium of Bowman's capsule is made up of specialized cells called podocytes. The finger-like cellular processes of the podocytes wrap around the capillaries of the glomerulus, and the plasma filtrate passes through the filtration slits between them.

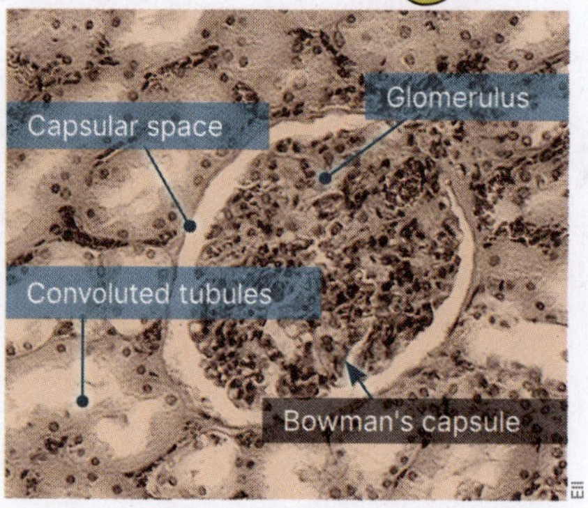

Capsular space

Glomerulus

Convoluted tubules

Bowman's capsule

Bowman's capsule is a double walled cup, lying in the cortex of the kidney. It encloses a dense capillary network called the glomerulus. The capsule and its enclosed glomerulus form a renal corpuscle. In this section, the convoluted tubules can be seen surrounding the renal corpuscle.

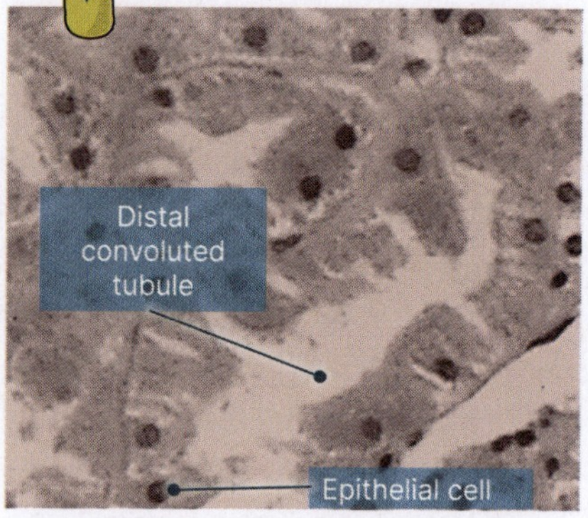

Distal convoluted tubule

Epithelial cell

There are around 16 different types of epithelial cells in the kidney, lining the surface of tubules, each with different functions. The kidney tissue also contains endothelial cells that line blood vessels, interstitial cells (in the space between functional cells), and immune cells.

1. Discuss the role of the glomerulus, Bowman's capsule, and proximal convoluted tubule in excretion, specifically the ultrafiltration process.

_____

_____

_____

_____

©2024 **BIOZONE** International
ISBN: 978-1-99-101410-8
Photocopying prohibited

Summary of activities in the kidney nephron

Urine formation begins by ultrafiltration of the blood, as fluid is forced through the capillaries of the glomerulus, forming a filtrate similar to blood but lacking cells and proteins. The filtrate is then modified by secretion and reabsorption to add or remove substances, e.g. ions. The processes involved in urine formation are summarized below for each region of the nephron (glomerulus, proximal convoluted tubule, loop of Henle, and distal convoluted tubule), and the collecting duct. The loop of Henle acts as a countercurrent multiplier, establishing and increasing the salt gradient through the medullary region. This is possible because the descending loop is freely permeable to water but the ascending loop is not.

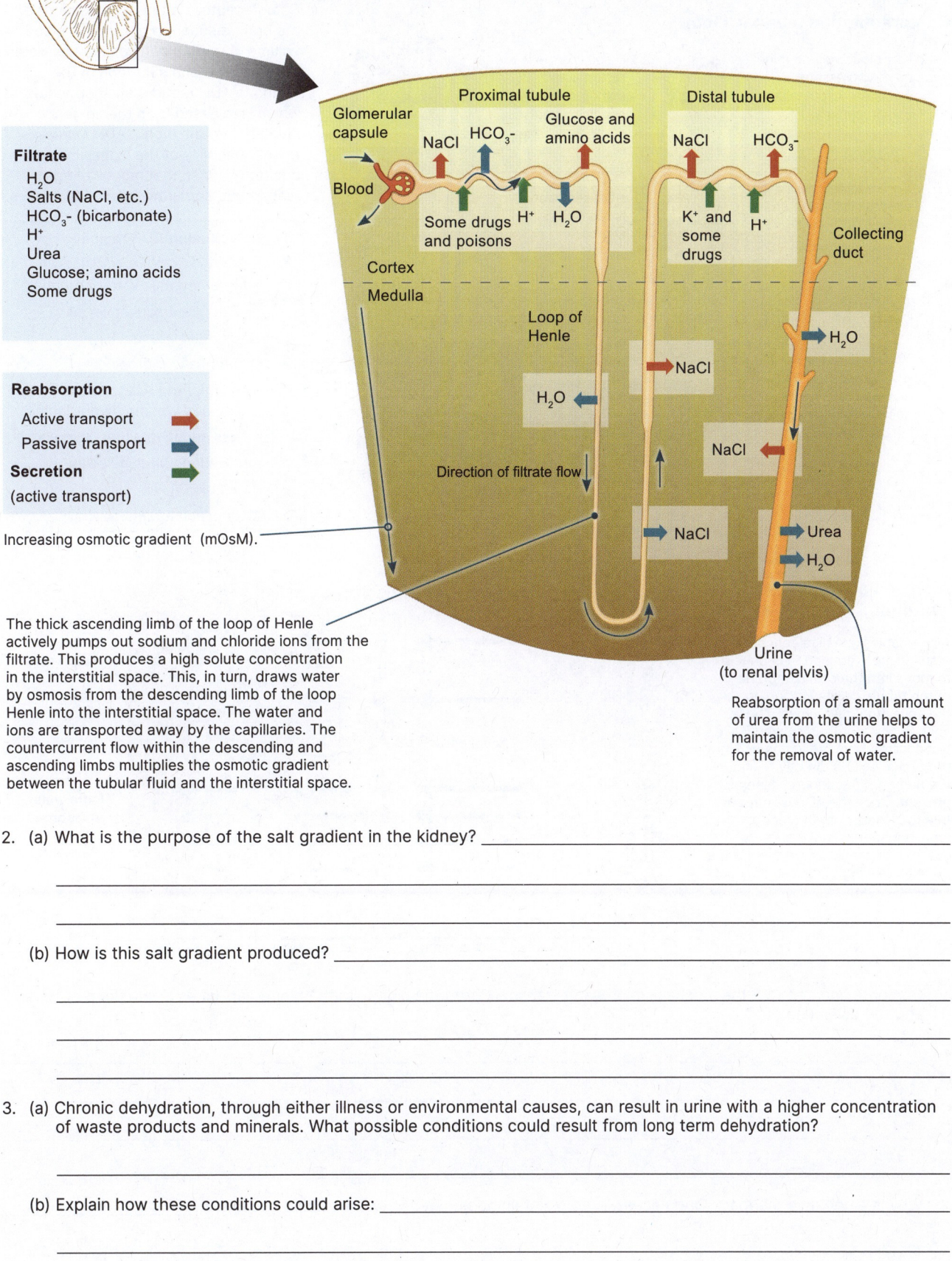

Filtrate

H_2O
Salts (NaCl, etc.)
HCO_3^- (bicarbonate)
H^+
Urea
Glucose; amino acids
Some drugs

Reabsorption

Active transport →
Passive transport →
Secretion →
(active transport)

Increasing osmotic gradient (mOsM).

The thick ascending limb of the loop of Henle actively pumps out sodium and chloride ions from the filtrate. This produces a high solute concentration in the interstitial space. This, in turn, draws water by osmosis from the descending limb of the loop Henle into the interstitial space. The water and ions are transported away by the capillaries. The countercurrent flow within the descending and ascending limbs multiplies the osmotic gradient between the tubular fluid and the interstitial space.

Reabsorption of a small amount of urea from the urine helps to maintain the osmotic gradient for the removal of water.

2. (a) What is the purpose of the salt gradient in the kidney? _____

   _____

   _____

   (b) How is this salt gradient produced? _____

   _____

   _____

   _____

3. (a) Chronic dehydration, through either illness or environmental causes, can result in urine with a higher concentration of waste products and minerals. What possible conditions could result from long term dehydration?

   _____

   (b) Explain how these conditions could arise: _____

   _____

   _____

339 Control of Urine Output in the Collecting Ducts

Key Idea: Mechanisms in the collecting ducts of the kidney can control osmoregulation.

Water is reabsorbed into blood vessels surrounding the collecting duct from the filtrate before it moves into the ureter, then bladder, and is finally excreted out of the body.

Aquaporins, specialized protein channels embedded in the collecting duct tubule cells, allow water to be passively transported across from collecting duct to blood vessels in the presence of ADH. An osmoregulation feedback loop controls ADH release from the pituitary gland.

Osmoregulation feedback loop

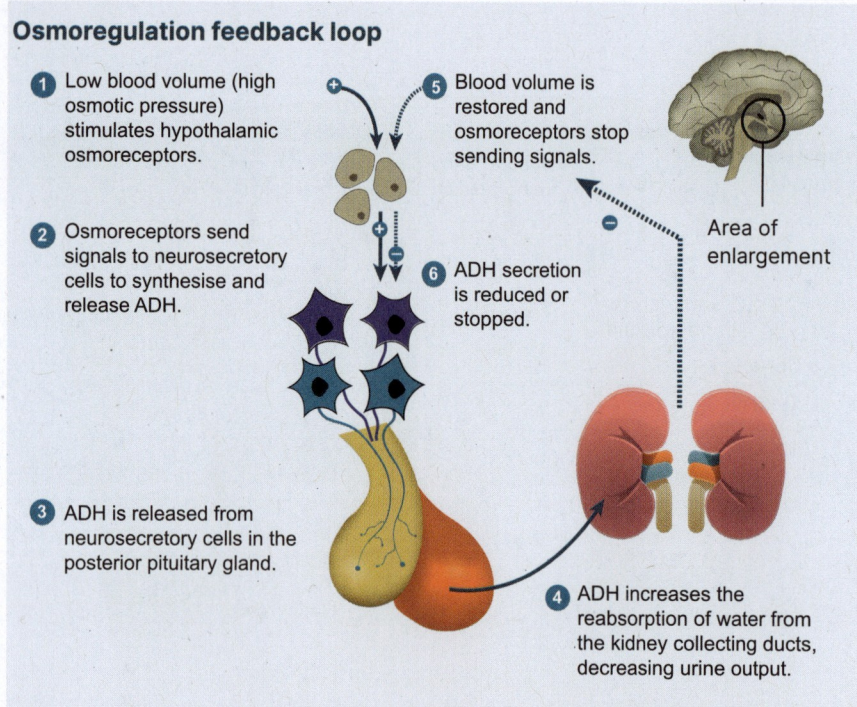

1. Low blood volume (high osmotic pressure) stimulates hypothalamic osmoreceptors.

2. Osmoreceptors send signals to neurosecretory cells to synthesise and release ADH.

3. ADH is released from neurosecretory cells in the posterior pituitary gland.

5. Blood volume is restored and osmoreceptors stop sending signals.

6. ADH secretion is reduced or stopped.

Area of enlargement

4. ADH increases the reabsorption of water from the kidney collecting ducts, decreasing urine output.

Osmoreceptors in the hypothalamus of the brain respond to changes in blood volume. A fall in blood volume stimulates the synthesis and secretion of the hormone ADH (antidiuretic hormone), which is released from the posterior pituitary into the blood. ADH increases the permeability of the kidney collecting duct to water so that more water is reabsorbed and urine volume decreases.

Factors causing ADH release
▸ Low blood volume (dehydration)
  = More negative water potential
  = High blood sodium levels
  = Low fluid intake
▸ Nicotine and morphine

Factors inhibiting ADH release
▸ High blood volume (hydration)
  = Less negative water potential
  = Low blood sodium levels
▸ High fluid intake
▸ Alcohol consumption

ADH and aquaporin switches

ADH enables chemical reactions that result in more aquaporin channels to move into the cell membrane of the tubule cells surrounding the collecting duct lumen. The volume of water that is osmotically transported from the lumen and into the tubule cells is directly related to the number of aquaporin channels present. Once the ADH is reduced due to signals from the feedback loop, aquaporins move back out of the membrane, reducing the volume of water leaving the collecting duct.

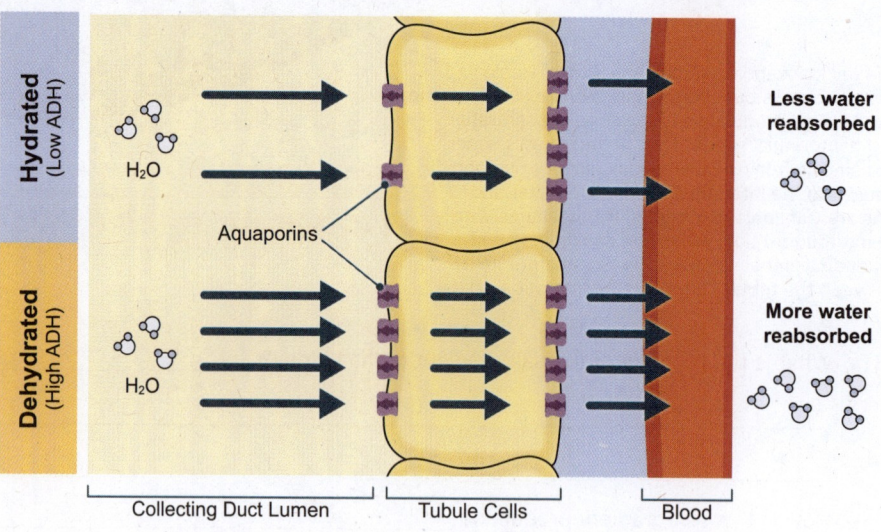

Hydrated (Low ADH)

H_2O

Aquaporins

Dehydrated (High ADH)

H_2O

Less water reabsorbed

More water reabsorbed

Collecting Duct Lumen   Tubule Cells   Blood

1. Compare and contrast the mechanism of water transport from the kidney to the blood capillaries in the Loop of Henle and the collecting duct:

_____

_____

_____

_____

2. Why is a negative feedback loop necessary to maintain homeostatic control of blood volume? _____

_____

_____

©2024 **BIOZONE** International
ISBN: 978-1-99-101410-8
Photocopying prohibited

D3.3
10

340 Changes to Blood Supply

Key Idea: Blood supply to organs changes during activity.
Vigorous physical activity places demands on the body's organs to function at a higher metabolic rate. This requires an increased level of energy. Both glucose and oxygen demands for cellular respiration increase, as well as mechanisms to remove wastes that accumulate at a faster rate.

Changes in blood supply during exercise

Brain: Blood flow remains constant at rest and in a vigorous exercise state. Oxygen and energy supply cannot be reduced without an impact to overall body functioning.

Muscles and tissues: Blood flow is increased dramatically (hyperemia) by increased heart pumping and vasodilation - observed as an increased heart rate.

Heart: Increase in blood flow to heart muscles to allow the organ to work faster and with more vigorous contractions.

Lungs: Increased blood flow to intercostal muscles to enable an increased ventilation rate, with active inhalation and expulsion of air, both to draw in more oxygen and expel carbon dioxide.

Liver: Reduction of blood flow to redirect blood to muscles.

Kidneys: Slight reduction of blood flow to redirect blood to muscles, as the organs need to be operational to remove built up wastes generated by increased cellular respiration and to retain water so the body does not become dehydrated.

Skin: Blood flow is slightly reduced to direct more blood towards the muscles. This can be observed as a paler appearance during extreme exercise. However, this needs to be balanced with releasing heat generated by the increased cellular respiration of muscles, both by vasodilation and sweating.

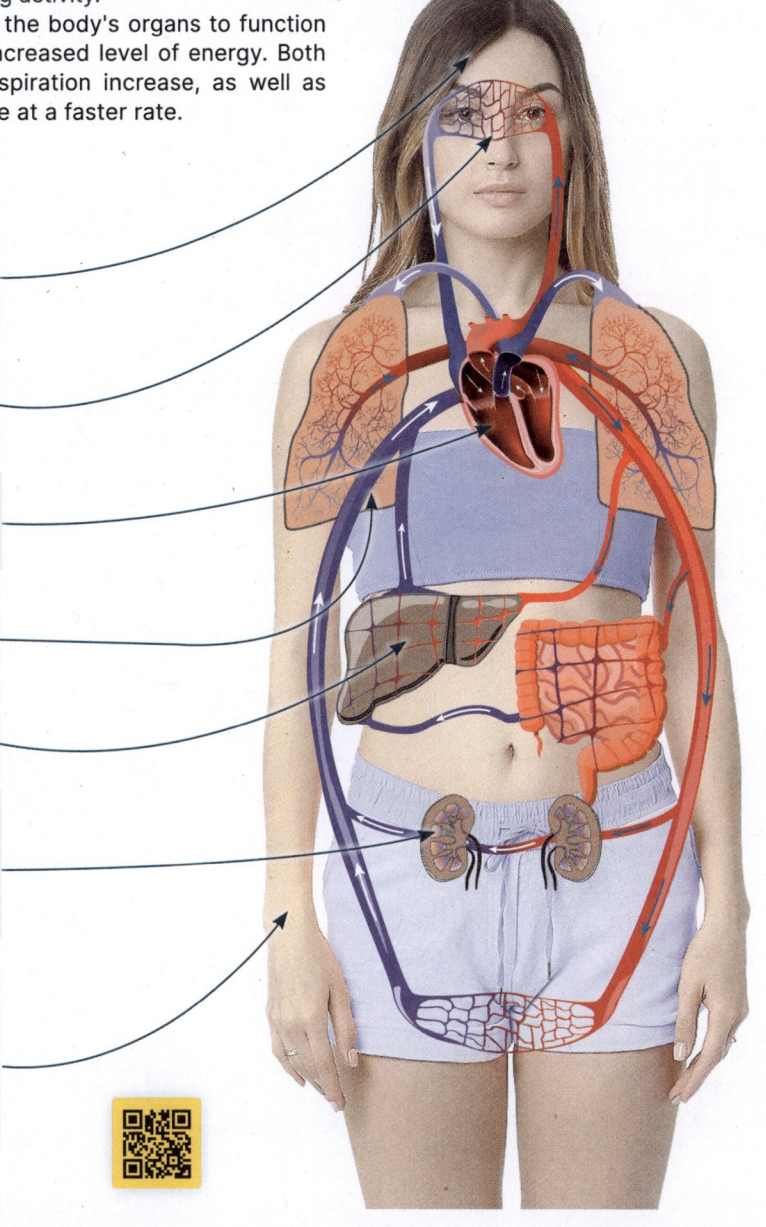

1. Complete the table below to summarize important changes of the blood supply to each system during exercise:

Organ	Blood flow change	Reason for change of blood flow
Brain		
Muscles and tissue		
Heart (muscles)		
Lungs (muscles)		
Liver		
Kidneys		
Skin		

©2024 **BIOZONE** International
ISBN: 978-1-99-101410-8
Photocopying prohibited

D3.3
11
AHL

341 Did You Get It?

1. Referring to the apple blossom on the right, what is the yellow 'dust' covering the bee, what does it contain, and from where, specifically, did it originate?

_____

_____

_____

2. Explain how insect pollination is a mechanism to reduce self-pollination:

_____

_____

_____

_____

3. A homozygous genotype for a gene in apples produces a much redder skin colour than a heterozyogous genotype. Explain this pattern of inheritance:

_____

_____

_____

4. **AHL:** The gene for apple flesh colour is linked to the gene for apple skin colour. Explain the term 'autosomal gene linkage' and how it would impact variation in gametes when developing a new apple variety:

_____

_____

_____

_____

_____

5. (a) What process does the diagram on the right show?

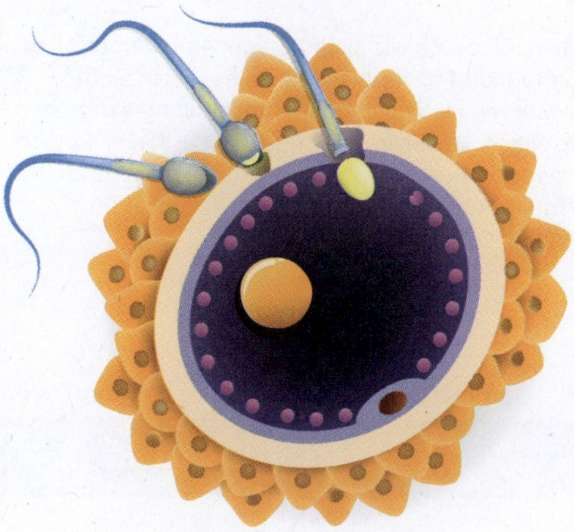

_____

   (b) How is the sex of the zygote determined?

_____

_____

_____

_____

6. **AHL:** Egg formation in humans is known as oogenesis. Link the processes which form each of the following: primordial germ cells - PGC, primary oocytes, secondary oocyte, ovum, zygote:

_____

_____

_____

_____

7. Describe how the levels of hormones FSH, LH, oestrogen, and progesterone change over a menstrual cycle:

_____

_____

_____

_____

8. **AHL:** Which hormone can be used to detect pregnancy and why? _____

_____

🌤 Ecosystems

Resource Hub
bit.ly/4a5bdaL

D4.1 Natural Selection

Activity Number

Guiding Questions:
▶ How does evolution cause allele frequency change within a population?
▶ Why is variation in populations a requirement for natural selection to occur?

Learning Outcomes:

☐ 1	Explain the link between cumulative scientific evidence for natural selection and the factors involved. **NOS:** Explain what is meant by a paradigm shift.	342
☐ 2	Evaluate how factors that generate variation such as mutation and sexual reproduction can cause a change of inherited characteristics in a population over time.	343
☐ 3	Link the overproduction of offspring and competition for resources such as food, as being factors that promote natural selection.	343
☐ 4	Describe how abiotic factors, including density-independent factors, act as selection pressures.	344
☐ 5	Define fitness as it relates to natural selection and explain its relationship to adaptation, survival, and reproduction in individuals.	345
☐ 6	Distinguish between heritable traits and those acquired due to environmental factors and relate each to the potential for evolutionary change.	345
☐ 7	Elaborate on sexual selection in animal species acting as a selection pressure illustrating with examples such as the plumage of birds of paradise.	346
☐ 8	**AOS:** Use data from John Endler's guppy experiments to model how selection pressure can influence sexual and natural selection.	347
☐ 9	**AHL:** Define a gene pool and explain how the concept is linked to the mechanism of natural selection.	348
☐ 10	**AHL: AOS:** Use allele frequency data from databases to investigate examples of geographically isolated populations, including human populations.	348
☐ 11	**AHL:** Explain how an understanding of allele frequency change in gene pools allows a distinction between the theory of evolution by natural selection and neo-Darwinism.	349
☐ 12	**AHL:** Compare and contrast directional, disruptive, and stabilizing selection.	350
☐ 13	**AHL:** Use the Hardy–Weinberg equation to calculate allele or genotype frequencies.	351
☐ 14	**AHL:** Explain how genetic equilibrium can be predicted by Hardy–Weinberg conditions being met.	351
☐ 15	**AHL:** Discuss the benefits and drawbacks of artificial selection by humans.	352

D4.2 Stability and Change

		Activity Number
Guiding Questions:	▶ How is ecosystem stability maintained in ecosystems over long time periods?	
	▶ How has human activity disrupted the stability of ecosystems?	

Learning Outcomes:

☐ 1	Discuss the phenomenon of ecosystem stability using examples demonstrating long-term stability.	353
☐ 2	Describe the requirements for stability in ecosystems including energy supply, nutrient recycling, genetic diversity, and tolerable climatic variance.	353
☐ 3	Explain the link between ecosystem stability and tipping points in the Amazon rainforest. **AOS:** Calculate percentage change data to demonstrate Amazon rainforest deforestation.	354
☐ 4	Use models to demonstrate the importance of functioning matter cycles to ecosystem stability. **NOS:** Take appropriate care of any organisms used in the investigation.	355
☐ 5	Link the loss of keystone species to ecosystem destabilization.	356
☐ 6	Evaluate how harvesting rates impact resource sustainability, using examples from both marine fish and terrestrial plant species.	357
☐ 7	Investigate how factors including soil erosion, nutrient leaching, fertilizers, pollution, and carbon footprint, can affect agriculture sustainability.	358
☐ 8	Explain how leaching can lead to eutrophication of aquatic and marine ecosystems.	359
☐ 9	Explain the process of pollutant biomagnification in ecosystems.	360
☐ 10	Discuss microplastic and macroplastic pollution of the oceans. **NOS:** Consider how media coverage of scientific research has contributed to addressing the above issue.	361
☐ 11	Evaluate a range of ecosystem restoration projects related to rewilding including examples.	362
☐ 12	**AHL:** Discuss the causes of ecological succession.	363
☐ 13	**AHL:** Analyze the changes that occur during primary succession illustrating with examples.	363
☐ 14	**AHL:** Describe the phenomenon of cyclical succession referring to examples in ecosystems.	364
☐ 15	**AHL:** Discuss the influence of human activity on climax communities and arrested succession using livestock grazing and wetlands drainage as examples.	364

D4.3 Climate Change

Guiding Questions:	▶ What forcings are the main drivers of the current climate change?	
	▶ How are ecosystems being impacted by climate change?	

Learning Outcomes:

☐ 1	Analyze increases in atmospheric concentrations of carbon dioxide and methane. **NOS:** Distinguish between causation and correlation to discuss anthropogenic causes of climate change.	365
☐ 2	Discuss a range of positive feedback cycles linked to global warming including deep ocean carbon dioxide release, the albedo effect, permafrost melting, and increased incidence of droughts and forest fires.	365
☐ 3	Explain the potential consequences of exceeding a tipping point due to net carbon loss in boreal forests.	365
☐ 4	Discuss the ecosystem impacts of land and sea ice melting including populations of emperor penguin in the Antarctic and walruses in the Arctic.	356
☐ 5	Discuss potential consequences resulting from timing and extent changes in nutrient upwelling due to warming surface temperatures affecting ocean currents.	367
☐ 6	Use evidence-based examples of temperate species to investigate poleward and upslope range shifts.	367
☐ 7	Investigate how ocean acidification and water temperature increases are linked to potential coral reef ecosystem collapse.	367
☐ 8	**NOS:** Evaluate the carbon sequestration effectiveness of afforestation and forest regeneration in both native and non-native tree species rewilding, and peat wetland restoration, in boreal or tropical peatlands.	368
☐ 9	**AHL:** Investigate how changes to photoperiod and temperature can influence the timing of biological events (phenological events) such as flowering, tree budding, bird migration, and nesting.	369
☐ 10	**AHL:** Discuss how climate change can lead to the disruption of phenological event synchrony disruption including Arctic mouse-ear chickweed spring growth, and arrival of migrating reindeer as an example.	369
☐ 11	**AHL:** Discuss how climate change can lead to increases to the number of insect life cycles within a year using the example of the spruce bark beetle.	369
☐ 12	**AHL:** Explain how climate change can lead to evolution using changes in frequency of tawny owl colour variants as an example.	370

342 Evolution and Change

Key Idea: Natural selection is the key mechanism that leads to evolutionary change over time.

Natural selection has operated since the appearance of the first living organism. It has led to the biodiversity of species present today, as well as those that have become extinct. Darwin's theory of evolution by natural selection was developed prior to any understanding about genetic mechanisms, but it understood that the process operated on the phenotypes of individuals and that it resulted in the differential survival of some phenotypes over others.

Individuals with phenotypes conferring greater fitness in the environment at the time will become relatively more numerous in the population. Over time, natural selection may lead to a permanent change in the population. Although other scientists, such as Lamarck, had proposed that species change over time, Darwin's theory created a paradigm shift in thinking. Darwin understood that evolutionary change was the result of traits that were inherent to an individual being passed on to successive generations, and not traits developed during a lifetime.

Why are there so many different types of bugs?

Charles Darwin

Charles Darwin (1809-1882) and Alfred Russel Wallace (1823-1913) jointly and independently proposed the theory of evolution by natural selection. Both amassed large amounts of supporting evidence, Darwin from his voyages aboard the Beagle and in the Galápagos Islands, and Wallace from his studies in the Amazon and the Malay archipelago. Wallace wrote to Darwin of his ideas, spurring Darwin to publish *On the Origin of Species by Means of Natural Selection*.

▶ Charles Darwin was a keen entomologist (a scientist who studies insects), and took every opportunity to observe and collect from those he encountered during his voyages.

▶ Darwin often used the sometimes extreme adaptations of insects as evidence for his natural selection theory. He explained that for each niche there would likely be an insect that had evolved to best exploit the resources in that environment, or had adapted to avoid predators found there.

▶ On receiving samples of an unusual Madagascan orchid with a 30cm nectary tube, Darwin predicted the presence of a yet undiscovered moth or butterfly with a equally long proboscis that could reach the nectar. 40 years later just such a moth was found.

1. A paradigm shift is caused by a complete change in scientific thought or the underlying assumptions that contradicted past knowledge. How was the shift from Lamarckism to Darwinism an example of a paradigm shift?

_____

_____

_____

2. How does the biodiversity of insects provide evidence for natural selection? _____

_____

_____

_____

3. Explain how Darwin was able to develop his theory of natural selection as a mechanism for evolution without understanding how genetic inheritance worked?

_____

_____

_____

©2024 **BIOZONE** International
ISBN: 978-1-99-101410-8
Photocopying prohibited

 47

 D4.1 1

 NOS

343 Variation and Natural Selection

Key Idea: Natural selection is the evolutionary mechanism by which organisms that are better adapted to their environment survive to produce a greater number of offspring.

Evolution is simply the change in inherited characteristics in a population over generations. Darwin recognized this as the consequence of four interacting factors: (1) the capacity of populations to increase in numbers, through overproduction of offspring (2) the phenotypic (physical characteristics) variation of individuals, due to sexual reproduction and mutation (3) that there is competition for resources, and (4) proliferation of individuals with better survival and reproduction (those having better fitness).

The factors of natural selection

▸ Natural selection needs to act upon populations that have variation: if there was no genetic variation, all individuals would have the same chance at survival and therefore survival would be random.

▸ Sexual reproduction ensures that offspring have a different combination of alleles from both their parents and siblings, while mutation may introduce completely new alleles.

▸ When better adapted organisms survive to produce a greater number of viable offspring, we call it natural selection. This has the effect of increasing their proportion in the population so that they become more common. It is the basis of Darwin's theory of evolution by natural selection.

We can demonstrate the basic principles of evolution using the model analogy of a 'population' of M&M's® candy.

#1

In a bag of M&M's®, there are many colours. This represents the variation in a population. As you and a friend eat through the bag of candy, you both leave the blue ones. Neither of you like this colour so you return them to the bag.

#2

The blue candy becomes more common...

#3

Eventually, you are left with a bag of blue M&M's®. Your selective preference for the other colours changed the make-up of the M&M's® population. This is the basic principle of selection that drives evolution in natural populations.

Darwin's theory of evolution by natural selection

Darwin's theory of evolution by natural selection is outlined below. It is widely accepted by the scientific community today and is one of the founding principles of modern science.

Overproduction

Populations produce too many young: many must die.

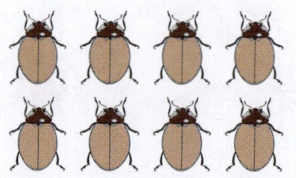

Populations generally produce more offspring than are needed to replace the parents. Habitats have a carrying capacity with limited resources, so natural populations normally maintain constant numbers. A certain number of individuals will die without reproducing.

Variation

Individuals show variation: some variations are more favourable than others.

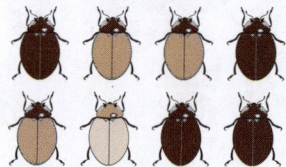

Individuals in a population have different phenotypes, and therefore genotypes, generated by mutation and sexual reproduction. Some traits are better suited to the environment, and individuals with these have better survival and reproductive success.

Natural selection

Natural selection favours the individuals best suited to the environment at the time.

Individuals in the population compete for limited resources. Those with favourable variations will be more likely to survive. Relatively more of those without favourable variations will die.

Inherited

Variations are inherited:

The best suited variants leave more offspring due to increased reproductive success.

Variation, selection, and population change

Natural populations, like the ladybird population above, show genetic variation and therefore different phenotypes. This is a result of mutations which alter and create new genetic material, and sexual reproduction which produces new combinations of genetic material. Some phenotypic variants are more suited to accessing the limited resources when competing with others. These variants will leave more offspring, as described for the hypothetical population (right).

1. Variation through mutation and sexual reproduction:
In a population of ladybirds, mutations independently produce red colouration and 2 spot marking on the wings. The individuals in the population compete for limited resources, such as small insects and pollen.

Red

Brown mottled

2. Selective predation:
Brown ladybirds are eaten by birds but red ones are avoided.

3. Change in the gene pool of the population:
Red ladybirds have better survival and fitness and become more numerous with each generation. Brown ladybirds have poor fitness and become rare.

1. Identify the four factors that interact to bring about evolution in populations: _____

2. What process creates new genetic material? _____

3. Darwin understood that populations exhibited variation in phenotypes but he did not understand the genetic mechanism that caused that variation. Explain how new alleles and new combinations of alleles are introduced into a population to create this variation:

4. Phenotypes such as better camouflage and speed are advantageous to predators. Predator species, such as big cats, often have more offspring than can survive to adulthood. Habitats have a carrying capacity that limits the number of predators in a given area. How do these features promote natural selection in a population of predators?

344 Abiotic Selection Pressures

Key Idea: Abiotic factors act as selection pressures.

The survival, growth, and reproduction of individual organisms and populations depend on the availability of essential environmental factors, such as specific nutrients, light, and temperatures. Most of these abiotic factors tend to be density independent and not influenced by population size. Environmental factors act as a selection pressure, favouring survivability of some traits over others. Individuals with phenotypes better suited to the environment have better reproductive success and there are more of them.

More of these successful alleles will exist in the population. In any environment, individuals will be able to function within a certain range of environmental conditions called the tolerance range. Individuals with phenotypes that allow them to survive and reproduce in areas outside the optimum range will have access to more resources and are more likely to be healthy and reproduce. They will pass on their alleles with a greater frequency, possibly leading to a population with a distinctly different gene pool over time in those habitats with a difference in abiotic factors.

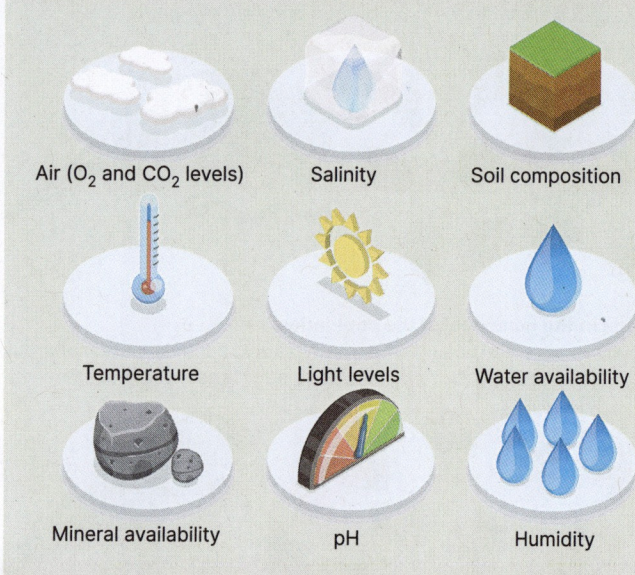

Air (O_2 and CO_2 levels) Salinity Soil composition

Temperature Light levels Water availability

Mineral availability pH Humidity

Abiotic factors

Changes in abiotic factors can act as selection pressures. Tolerance to abiotic factors is a phenotype, controlled by the alleles present in a genotype. Those individuals able to tolerate specific abiotic factors more effectively will have an increased chance of passing on their alleles to the next generation. At a population level, over time, there may be a shift in the gene frequency.

Changes in abiotic factors can occur over a long time period and be permanent, such as temperature rises due to climate change. This will reflect in a permanent change in genetic frequency, i.e. evolution.

Other events, such as natural disasters or extreme weather can cause temporary changes in abiotic factors. If alleles are lost due to the death of all individuals carrying them during an adverse event, the change to the allele frequency can be permanent.

Abiotic factors affect evolution

Abiotic factors also play a part in the direction and rate of at which species evolve. Rapid changes in the environment can increase the rate of evolution by creating new selection pressures.

A gradual increase in the environment's temperature, such as is currently occurring, may favour phenotypes able to tolerate the greater heat or reduce heat uptake. These phenotypes will proliferate.

Desert plants such as cacti have evolved to tolerate the extreme heat and solar radiation in the desert. Many plants in tropical rainforests have evolved leaves that quickly drain water from their surface.

1. Define abiotic factors and provide examples: _____

_____

_____

2. How does an abiotic factor act as a selection pressure? Use one of the examples above to explain how a change in a given abiotic factor, e.g. temperature or salinity, over time may lead to natural selection in a population:

_____

_____

_____

_____

_____

D4.1
4

239

345 Adaptation, Survival, and Reproduction

Key Idea: Differences between adaptations of individuals affect their fitness, i.e. their ability to survive and reproduce. In evolution, fitness is a mathematical measure of the genetic contribution an individual makes to the next generation. In simplest terms, it is a measure of reproductive success. The more offspring an individual (or genotype) contributes to the following generations, the greater its fitness. Fitness is linked to adaptation. The better adapted an individual is to its environment, the more likely it is to reproduce successfully and therefore the greater its fitness will be. Intraspecific competition (between individuals of the same species in a population) acts as a selection pressure. This occurs when there are fewer resources than can sustain a population and some individuals will gain an uneven share of the resources. Those individuals with better adaptations are more likely to outcompete other individuals and gain more or all of the resources, such as food, territory, or mates, and therefore have greater fitness. The alleles for their traits will be passed on with greater frequency. This differential reproductive success is the cornerstone of natural selection.

Intraspecific competition and the Malthusian dilemma

When resources are freely available for a given population, there is no intraspecific competition. No particular adaptation will have an advantage over another and there will be no selection pressure to drive natural selection.

Intraspecific competition for resources is created as a result of the same species in a population requiring more than is available. Thomas Malthus proposed his dilemma to explain how different growth rates in an ecosystem - linear for resources and exponential for populations - leads to a catastrophe point. After this point is reached, the population produces more offspring than the ecosystem can support (the carrying capacity) leading competition for resources to become a selection pressure.

The phrase, 'survival of the fittest' encapsulates the concept of differential adaptations leading to differential survival and reproductive rates which is created by different combinations of alleles in competing individuals who overproduce offspring beyond the number that can be supported by the available resources.

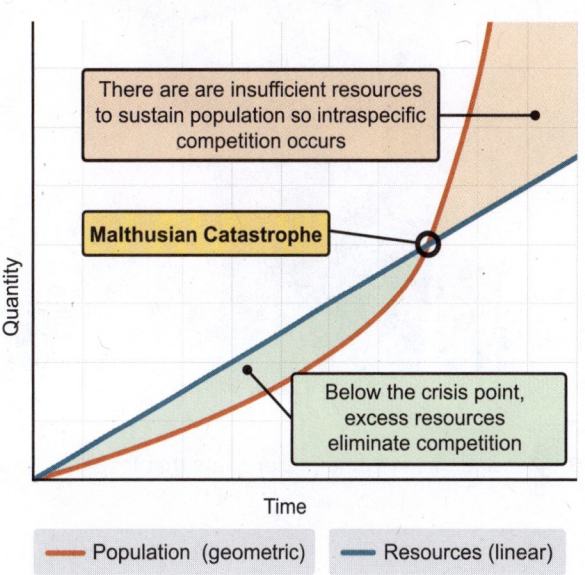

Legend:
- Population (geometric)
- Resources (linear)

Intraspecific competition driving natural selection in cichlids

T. temporalis are cichlid fish found in Lake Tanganyika in East Africa. The fish are polymorphic, in that they display a range of body sizes. Males are larger than females, but there is also variance between individuals. In low to no competition, the rock substrate of the lake is the preferred habitat for most fish, although smaller females move to the shelled areas to avoid predation. When the density of fish increases to drive intraspecific competition, the larger fish outcompete the smaller fish for food and territory. Smaller fish, both male and female, are then displaced to the shell substrate areas of the lake.

Note: The lighter colour is to distinguish size type for clarity - typical colour is dark grey.

1. Being adapted becomes more important as resources become more limited.

   (a) What adaptation in the cichlids determines fitness in rock environments with limited resources, and why?

   _____

   _____

   (b) What long term changes in the genotype might be seen in cichlid populations if competition remains a factor?

   _____

   _____

   _____

Tusks and African elephants

▶ African elephants can be found across 23 countries in Africa and have adapted to a wide range of habitats.

▶ Both male and female African elephants typically have tusks, which are adapted from extended teeth. The tusks can vary in length.

▶ The tusks are made of ivory, a material much sought after by humans. Poaching (illegal hunting) of elephants has greatly reduced the total population, with around 14,000 individuals being killed every year, mainly for their tusks.

▶ Tusklessness is a phenotype that is controlled by genes. Typically, around 2-4% of an African elephant population are born without tusks. Almost all are female.

▶ In some heavily poached areas, scientists have observed up to 60% of the elephants having the tuskless phenotype.

▶ Tusks can also be lost from injury and poaching, and the 'tuskless' elephants can survive and continue to breed.

Carved ivory tusk

2. Critique this statement: Elephants that lose their tusks are more likely to have tuskless offspring:

_____

_____

3. Tuskless females will pass the tuskless genes to 50% of their daughters. The tuskless gene is carried on the X (female) chromosome. Why is the increase in tuskless African elephants over time an example of natural selection?

_____

_____

_____

4. A long term study of a Mozambique population of African elephants showed changing proportions of 2,1, and no tusk elephants. Lack of poaching protection for elephants occurred during conflicts during the cival war period of 1977-1994. Use the data to discuss the role of tusklessness as a heritable trait and changes in population:

_____

_____

_____

_____

_____

_____

_____

_____

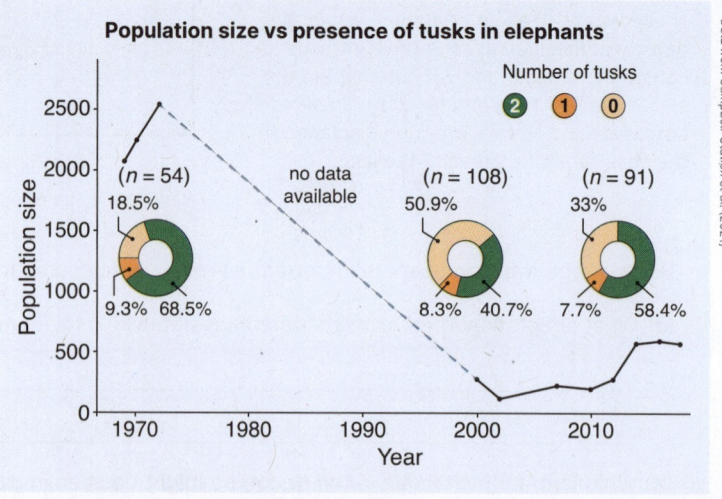

Data from Campbell-Station et al. (2021)

346 Sexual Selection

Key Idea: Sexual selection is a type of natural selection in which there is differential reproductive success or fitness among individuals of the same sex (and species) because of mate choice or competition within one sex.

The success of an individual is measured not only by the number of offspring it leaves, but also by the quality or likely reproductive success of those offspring. Therefore, its choice of mate is important. Darwin (1871) first introduced the concept of sexual selection: a type of natural selection that produces anatomical and behavioural traits affecting an individual's ability to acquire mates. Biologists recognize two types: intrasexual selection (usually male-male competition) and intersexual selection (also called mate choice). One result of either type is the evolution of sexual dimorphism (differences in appearance between males and females of the same species).

Intersexual selection

▶ In intersexual selection (or mate choice), individuals of one sex (usually the males) advertise themselves as potential mates and members of the other sex (usually the females) choose between them. This type of selection often results in dimorphism, where the female has a far more drab appearance compared to the male.

▶ Intersexual selection results in development of exaggerated ornamentation, such as elaborate colouration combined with behavioural display, as seen in the bird of paradise species.

▶ The more attractive a male is, the more fitness he will potentially have. The advantage for attracting and securing a mate is significant: many display features, such as long feathers, that can inhibit movement and the ability to secure food resources.

The Raggiana bird-of-paradise. The dull coloured female inspects the brightly coloured male during a courtship dance.

▶ Birds of paradise consist of 42 species located in the rain forests of New Guinea and surrounding areas. The birds have evolved from a Corvid (crow family) ancestor, which split from the group around 24 million years ago.

▶ The birds have no natural predators in their habitat: an important feature that reduces the impact of noticeable plumage that can inhibit escape from danger.

▶ Female preference for elaborate male ornaments is well supported by both anecdotal and experimental evidence. For example, in the Raggianna bird-of-paradise (*Paradisaea raggiana*), females prefer males with vibrant, long feathers that are displayed during a ritualized dance.

▶ The Raggianna bird-of-paradise, like many other brightly plumaged species, is polygynous and the male seeks as many females to breed with as possible.

▶ The most attractively plumaged male competes to occupy the highest position on a display tree, through intrasexual selection. These males secure most, if not all, of the females to breed with, ensuring the transfer of his genes to the next generation.

1. Explain how sexual selection can act as a selection pressure using birds of paradise as an example:

_____

_____

_____

_____

2. Suggest how sexual selection results in sexual dimorphism: _____

_____

_____

3. Why does sexual selection fix exaggerated characteristics in a population even though they might be harmful?

_____

_____

_____

D4.1

7

347 Modelling Selection

Key Idea: Sexual and natural selection can be investigated using models that alter selection pressures.

Darwin made use of observations in unaltered habitats to develop his theory of natural selection. However, experiments that occur in a lab, in closed outdoor settings, or even in computer models can be used to investigate how changes in selection pressures can influence the outcomes of sexual and natural selection. John Endler was an evolutionary biologist who began his research into guppy (*Poecilia reticulata*) evolution in the 1970s in the Caribbean island of Trinidad. After extensive observations, he began to alter different selection pressures in guppy populations, such as predation. His research became an important source of evidence to support the mechanism of natural and sexual selection.

John Endler's guppy experiments

The guppy is a small, tropical fish, often called the millionfish or rainbowfish, found in small ponds. It feeds on small, microscopic organisms. It is native to Trinidad, although now found all over the world as a popular aquarium fish. The guppy can be found in a variety of colourful forms which attract potential female mates. The bright colours also attract predatory fish, so in areas where predators are present, the guppies tend to be dull and camouflaged to enable them to hide against the pond floor.

Endler was able to alter a key selection pressure, the removal of a predator, or a swap with a less dangerous predator, by shifting guppy populations to different pools. He observed the results after 15 generations of guppies, which equated to around 2 years.

Guppy selection experiments in the wild

Region 1:
Dangerous predators, dull guppies

15 generations of selection

Region 2:
Dull guppies introduced to a new region with less dangerous predators
Introduced

15 generations of selection

Region 3:
Less dangerous predators, colorful guppies

15 generations of selection

1. What was the likely hypothesis of this experiment and did the results agree with it? _____

_____

_____

Guppy selection experiments in the laboratory

John Endler also conducted experiments in the laboratory. He observed that guppies had body spots that could be used to stand out from the pond base, therefore drawing attention to themselves and attracting mates. Large spots stood out from fine gravel and vice-versa. However, by being more visible, they also increased the chance of predation. Endler altered both the presence of predator and pond base material, both acting as selection pressures, and recorded his observations.

Initial Setup:
Coarse gravel, predator present | Fine gravel, predator present

Fewer than 15 generations of selection

Initial Setup:
Coarse gravel, no predator present | Fine gravel, no predator present

Fewer than 15 generations of selection

2. The initial populations of Endler's guppies showed variation in the size of spots. Why was this important and how is it linked to the final observations?

_____
_____
_____

3. Discuss the role that natural selection played in the experiments containing a predator fish (above, left):

_____
_____
_____

4. Why were the experimental results (large or small spots) 'reversed' when there was no predator (above, right)?

_____
_____
_____

5. Natural selection can be modelled using spreadsheets or computer programmes. Use a computer simulation to change a selection pressure and write a summary of your findings below. For example, you could use Connectedbio Multi-Level Simulation (MLS) to model change in phenotype (fur colour) over time in a population. You will model natural selection by adding a predator into a deer mouse population. Access the MLS programme here: **https://short.concord.org/lm3** (you can also access this simulation or one of the others offered via the **BIOZONE Resource Hub**).

_____
_____
_____
_____
_____

348 Gene Pools and Populations

Key Idea: A gene pool is the sum total of all the genes present in a population at any one time. Natural selection can alter the allele frequencies in gene pools.

Each individual carries part of the total genetic complement of a population but not all individuals will be breeding at a given time. Allele frequencies are more stable in large populations because they are less affected by changes involving only a few individuals. Populations may have distinct geographical boundaries which prevent or limit gene flow. Small, geographically isolated populations, such as found on islands, have fewer alleles in their gene pools to begin with, so the severity and speed of changes in allele frequencies can be greater. Databases can be used to analyze changes in gene pools, including in human populations.

Measuring gene pool change

1. The model below shows a hypothetical population of beetles undergoing changes as it is subjected to natural selection. The phases shown represent the same gene pool undergoing change over time. The beetles have two phenotypes, dark and pale, determined by the amount of pigment deposited in the cuticle. The gene controlling this character is represented by two alleles, A and a. Your task is to analyze the gene pool as it changes. For each phase in the gene pool below fill in the following tables (the first columns have been done for you):

   (a) Count the number of A and a alleles separately. Enter the count into the top row of the table (left hand columns).
   (b) Count the number of each type of allele combination (AA, Aa and aa) in the gene pool. Enter the count into the top row of the table (right hand columns).
   (c) For each of the above, work out the frequencies as percentages (bottom row of table):

Allele frequency = No. counted alleles ÷ Total no. of alleles x 100

Evolutionary mechanisms in gene pools

Four microevolutionary processes can contribute to genetic change in populations: 1. Mutation alters the genetic material and produces new genetic variations. 2. Recombination 'shuffles' the combinations of parent alleles in each offspring. Both mutation and recombination occur during sexual reproduction. 3. Migration creates gene flow as genetic material enters or leaves a population. 4. Genetic drift alters the frequency of genetic variants randomly; its effects are due to chance events. Increasingly, genetic drift is being recognized as an important agent of change, especially the **founder effect** seen in small, isolated populations, e.g. island colonizers.

Phase 1: Initial gene pool

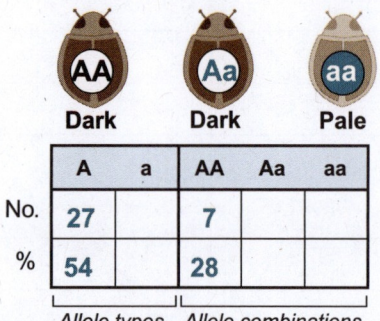

	A	a	AA	Aa	aa
No.	27		7		
%	54		28		

Allele types   *Allele combinations*

Phase 2: Natural selection

In the same gene pool at a later time, there was a change in allele frequencies. This was because of the loss of certain allele combinations due to natural selection. Some of those with a genotype of aa were eliminated (poor fitness).

These individuals (surrounded by small red arrows) are not counted for allele frequencies; they are dead!

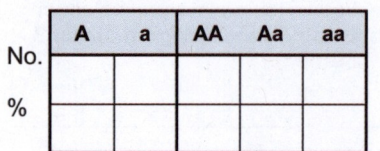

	A	a	AA	Aa	aa
No.					
%					

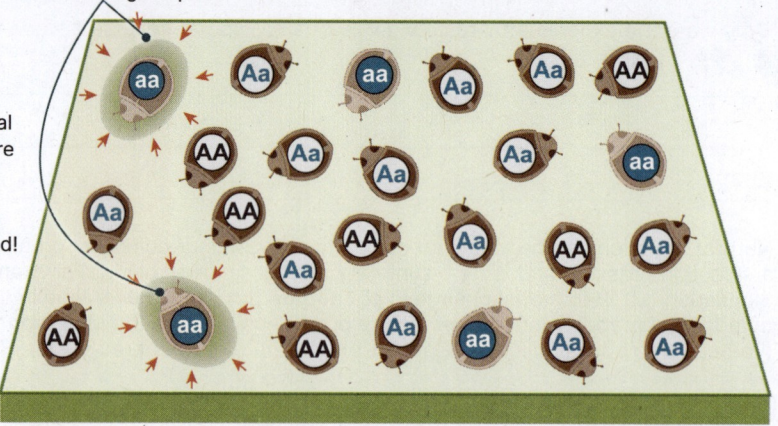

Two pale individuals died. Their alleles are removed from the gene pool.

2. What evidence demonstrates that evolution has occurred in the beetle population after phase 2 in the scenario above?

_____

_____

D4.1
9 - 10

©2024 **BIOZONE** International
ISBN: 978-1-99-101410-8
Photocopying prohibited

Island platypus and the perils of isolation

Platypus, Tasmanian mainland

Klaus cc 2.0

▶ A 2012 study of genetic diversity in platypus populations on mainland Australia, Tasmania, King Island, and Kangaroo Island has revealed very low immunological diversity in the island populations. The study (Lillie et al.) looked at the diversity of the MHC DZB gene and three MHC associated markers, all of which are involved in immune function. High allelic diversity in immune genes is important because it provides the variation necessary to resist different kinds of diseases. Without genetic variation, the animals are likely to have low resistance to new diseases and environmental change.

▶ The study found that populations on the Australian mainland and in Tasmania have high levels of genetic diversity within their populations, with 57 DZB alleles identified in 70 individuals. However, platypuses on King Island and Kangaroo Island (see maps), had very low levels of genetic diversity. For the isolated King Island populations, there was no variation at all (only one allele at the DZB locus).

▶ Why is the genetic diversity of these island populations so low compared to the mainland populations? The Kangaroo Island population was founded from an introduction of around 20 animals in the 1930s and 1940s. The population on King Island is endemic, separated for some 14,000 years since the last ice age.

▶ Inbreeding in a small population and genetic drift have resulted in the loss of alleles and a dangerously low diversity in immune genes. These island populations will now need careful management to protect them from disease risk.

Source: Diversity at the Major Histocompatibility Complex Class II in he Platypus, *Ornithorhynchus anatinus*
Mette Lillie *et. al.* Journal of Heredity 2012:103(4):467–478

Tentotwo cc 3.0
Nzeemin cc 3.0

Allele frequency change in brown anole lizards

▶ In 2004, Hurricane Francis wiped out the brown anole lizard (*Anolis sagrei*) populations on several cays (small sandy islands) around the Bahamas. Scientists used this as a chance to study the changes in allele frequencies. Limb length is a phenotype controlled by a genotype - which showed variation in alleles. They took pairs of lizards from the mainland and placed them on different cays.

▶ The vegetation on the cays is much smaller and scrub-like compared to the much larger trees of the mainland. On the mainland, scientists noted that the lizards use their long limbs to climb around the trees. They hypothesized that the populations isolated on the cays would eventually evolve shorter limbs to adapt to the scrub-like, less supportive vegetation. They measured the limb length over several years.

▶ It was found that limb length indeed became shorter over successive generations in all the populations. Importantly, populations founded by lizards with the longest legs still had the longest legs and populations founded by lizards with the shortest legs still had the shortest legs. The characteristics of the founder populations influenced the descendant populations.

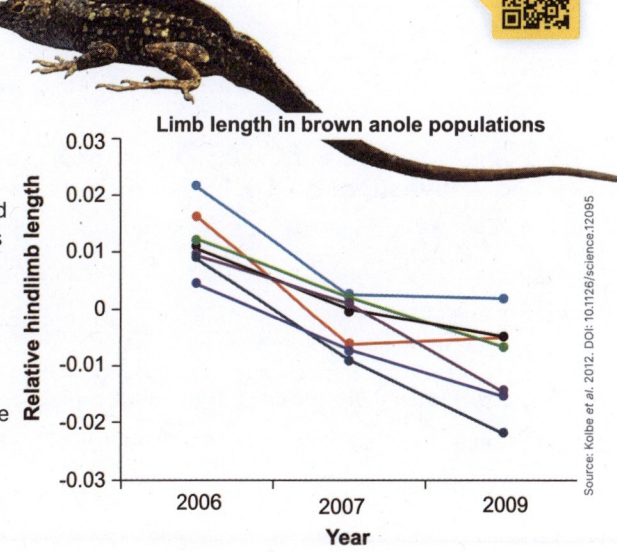

Source: Kolbe *et al.* 2012. DOI: 10.1126/science.12095

3. (a) Why were conditions good for setting up an experiment on the changing frequency of alleles over time on the cays around the Bahamas?

_____

_____

   (b) The different coloured lines in the graph above represent different small and isolated 'founder' populations of brown anole lizards. What impact did the allele frequency of tail length have on the final gene pool frequency three years after the observations started?

_____

_____

_____

_____

Allele frequency change in isolated human populations

Pitcairn Island

Norfolk Island

thinboyfatter CC 2.0

Tristan da Cunha

Brian Gratwicke CC 2.0

Due to the frequent episodes of human migration around the world, there are many instances of the founder effect in human populations. In 1790, nine mutineers from the ship HMS Bounty along with six Tahitian men, eleven Tahitian women, and a baby girl settled on Pitcairn island. The population eventually grew to 193 by 1856.

In 1856 the entire population of Pitcairn Island resettled on Norfolk Island after it was decided Pitcairn was over populated. The effect of this can still be seen in genetic studies of the Norfolk Island population. In 1859, 16 people returned to Pitcairn Island and founded a new population, that eventually reached 250 people by 1936. The population is now around 50.

Tristan da Cunha sits 2,400 km from Africa and more than 3,500 km from South America. The current settlement was founded by the English in 1817. In 1961, a genetic study traced 14% of all genes in the population of 300 to one founding couple. Around 47% of the population are now affected by asthma. From the 15 original settlers at least three had asthma.

4. (a) The rate of asthma in the UK is about 8%. Calculate the rate of asthma in the original Tristan da Cunha settlers:

_____

   (b) How has genetic drift affected the current population of Tristan da Cuhna? _____

_____

_____

_____

 Investigation 16.1 Using allele frequency databases

See appendix for equipment list.

A number of academic institutions, NGO organizations, and government departments develop and maintain human allele frequency databases. The data can be used in research to test hypotheses on allele frequency change, and hence evolution. In this investigation we will use ALFRED, but other options are provided in the Biozone Resource Hub.

1. Connect to the Allele Frequency Database (ALFRED) https://bit.ly/3MmUF3X (or QR, right)

Initial investigation

2. Select a gene you wish to investigate and place the name into search parameters, using a keyword or SNP. The gene could be any covered in previous activities, such as ABO blood groups, or one provided by your teacher.

3. Select a specific gene loci (a fixed position on a chromosome where a gene is located).

4. Select a polymorphism (a specific variant of allele).

5. Select a display format for allele frequency (graph or table).

Depth investigation

6. You will prepare a short report on the allele frequency of a select polymorphism, i.e. rs1265956. Provide information of the gene: what functional products it expresses or the phenotype it results in.

7. Once you have displayed your data, identify two human populations that have distinctly different allele frequency data. Research the demographics, location, history, and any other data of each population that may be relevant to defining its properties.

8. Develop a hypothesis to account for the allele frequency differences.

9. Attach your human allele frequency report below.

349 Neo-Darwinism and Modern Synthesis of Evolution

Key Idea: Darwin's theory of natural selection has now been modified to account for the underlying genetic mechanisms. The theory of natural selection was built upon preceding and contemporary scientific research. Although Charles Darwin is largely credited with the development of the theory of evolution by natural selection, his ideas did not develop in isolation but within the context of the work of others before him. Neo-Darwinism was coined by Romanes in 1883 to describe the modified Darwinian theory that also accounted for genetic variation in populations mostly attributed to gene mutation. From this, the Modern Synthesis of evolution arose (Huxley, 1942), adding in the principles of Mendel's inheritance laws. The diagram below summarizes just some of the important players in the story of evolutionary biology and the work of many has contributed to a deeper understanding of evolutionary processes. This understanding continues to develop with the use of molecular techniques and cooperation between scientists across many disciplines.

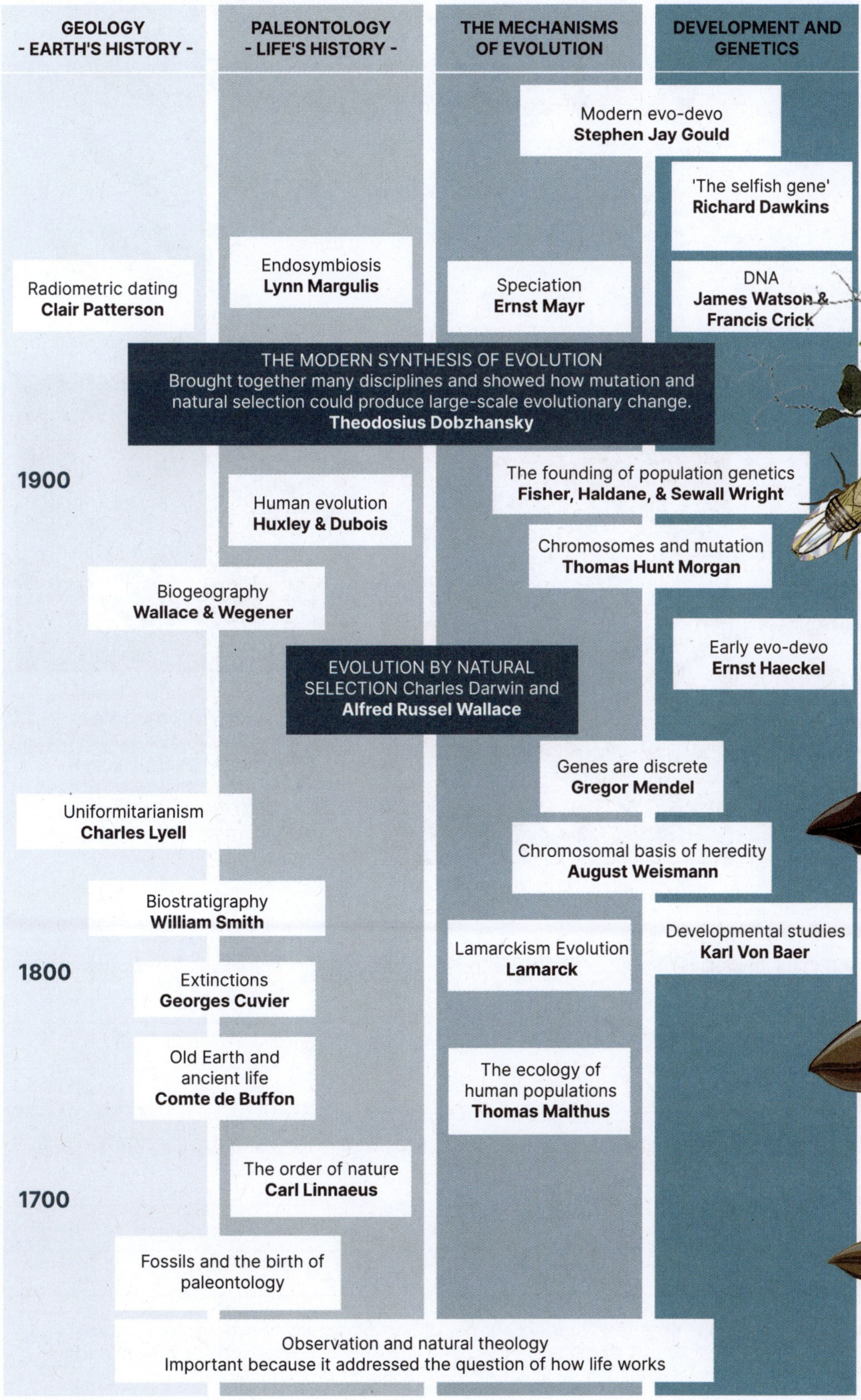

GEOLOGY - EARTH'S HISTORY -	PALEONTOLOGY - LIFE'S HISTORY -	THE MECHANISMS OF EVOLUTION	DEVELOPMENT AND GENETICS
			Modern evo-devo **Stephen Jay Gould**
			'The selfish gene' **Richard Dawkins**
Radiometric dating **Clair Patterson**	Endosymbiosis **Lynn Margulis**	Speciation **Ernst Mayr**	DNA **James Watson & Francis Crick**

THE MODERN SYNTHESIS OF EVOLUTION
Brought together many disciplines and showed how mutation and natural selection could produce large-scale evolutionary change.
Theodosius Dobzhansky

1900

	Human evolution **Huxley & Dubois**	The founding of population genetics **Fisher, Haldane, & Sewall Wright**	
		Chromosomes and mutation **Thomas Hunt Morgan**	
Biogeography **Wallace & Wegener**			Early evo-devo **Ernst Haeckel**

EVOLUTION BY NATURAL SELECTION Charles Darwin and **Alfred Russel Wallace**

Uniformitarianism **Charles Lyell**		Genes are discrete **Gregor Mendel**	
		Chromosomal basis of heredity **August Weismann**	
Biostratigraphy **William Smith**			Developmental studies **Karl Von Baer**

1800

| | Extinctions **Georges Cuvier** | Lamarckism Evolution **Lamarck** | |
| | Old Earth and ancient life **Comte de Buffon** | The ecology of human populations **Thomas Malthus** | |

1700

| | The order of nature **Carl Linnaeus** | | |
| | Fossils and the birth of paleontology | | |

Observation and natural theology
Important because it addressed the question of how life works

Mendelian inheritance

Mendel's inheritance laws were a key component of the Modern Synthesis along with understanding of genes and their contribution to evolution. Originally, peas and now more commonly fruit flies, are used to study and test inheritance.

Darwin's Finches

Observations of phenotypes were the key to Darwin developing his theory of evolution by natural selection.

Medium ground finch
(*Geospiza fortis*)

Large ground finch
(*Geospiza magnirostris*)

Small tree finch
(*Camarhynchus parvulus*)

D4.1
11

From: Wikipedia Commons CC 4.0

The development of Neo-Darwinism and the Modern Synthesis

Malthusian competition

(geometric population growth, limited resources)

Variation

(breeds, races, subspecies)

Mutation

(small changes in individual characteristics)

Natural selection

("survival of the fittest")

Genetic variation

(alleles of individual genes, combining to give continuous variation)

Mendelian inheritance

(2 copies of each gene, 1 from each parent)

Modern Synthesis

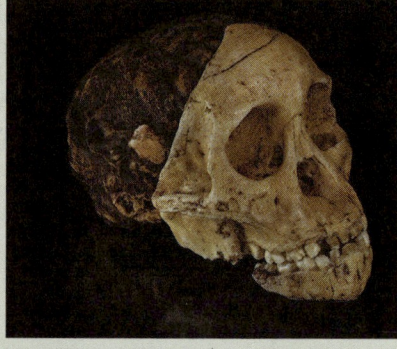

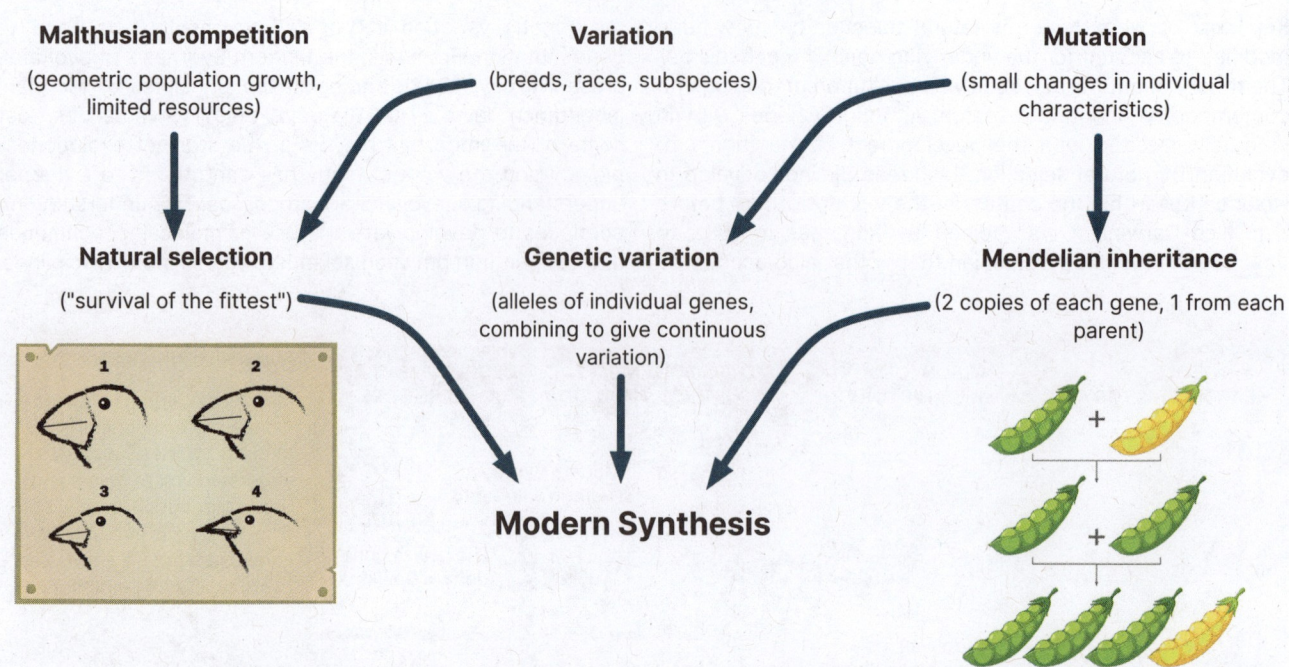

The Modern Synthesis today

James Watson and Francis Crick's discovery of DNA's structure in 1953 revolutionized evolutionary biology. The genetic code could be understood and deciphered, and the role of mutation as the source of new alleles was realized.

After Haeckel's flawed work on embryology, the evolutionary study of embryos was largely abandoned for decades. However, in the 1970s, Stephen Jay Gould's work on the genetic triggers for developmental change brought studies of embryological development back into the forefront. Today evo-devo provides strong evidence for how novel forms can rapidly arise.

Stephen Jay Gould (1941-2002)

Kathy Chapman online cc 3.0

In recent decades, DNA and protein analyses have revolutionized our understanding of phylogeny. Allan Wilson was one of a small group of pioneers in this field, using molecular approaches to understand evolutionary change and reconstruct phylogenies, including those of human ancestors.

1. (a) Did the discovery of the genetic mechanisms support Darwin's theory of natural selection? Explain your answer:

_____

_____

_____

   (b) How is Modern Synthesis a good example of the way that many foundational principles and theories are developed?

_____

_____

_____

2. Using a separate sheet, research and then write a 150 word account of the development of evolutionary thought and the importance of contributors from many scientific disciplines in shaping what became the Neo-Darwinism and/or Modern Synthesis. You should choose specific examples to illustrate your points of discussion.

350 Directional, Stabilizing, and Disruptive Evolution

Key Idea: Natural selection acts on phenotypes and results in the differential survival of some genotypes over others. It is an important cause of change in allele frequency in gene pools of populations.

Natural selection operates on the phenotypes of individuals, produced by their particular combinations of alleles. It results in the differential survival of some genotypes over others. As a result, organisms with phenotypes most suited to the prevailing environment are more likely to survive and breed than those with less suited phenotypes. Favourable

phenotypes will become relatively more numerous than unfavourable phenotypes. Over time, natural selection may lead to a permanent change in the genetic makeup of a population. Natural selection is always linked to phenotypic suitability in the prevailing environment so it is a dynamic process. It may favour existing phenotypes or shift the phenotypic median, as is shown in the diagrams below. The top row of diagrams below represents the population phenotypic spread before selection, and the bottom row the spread afterwards.

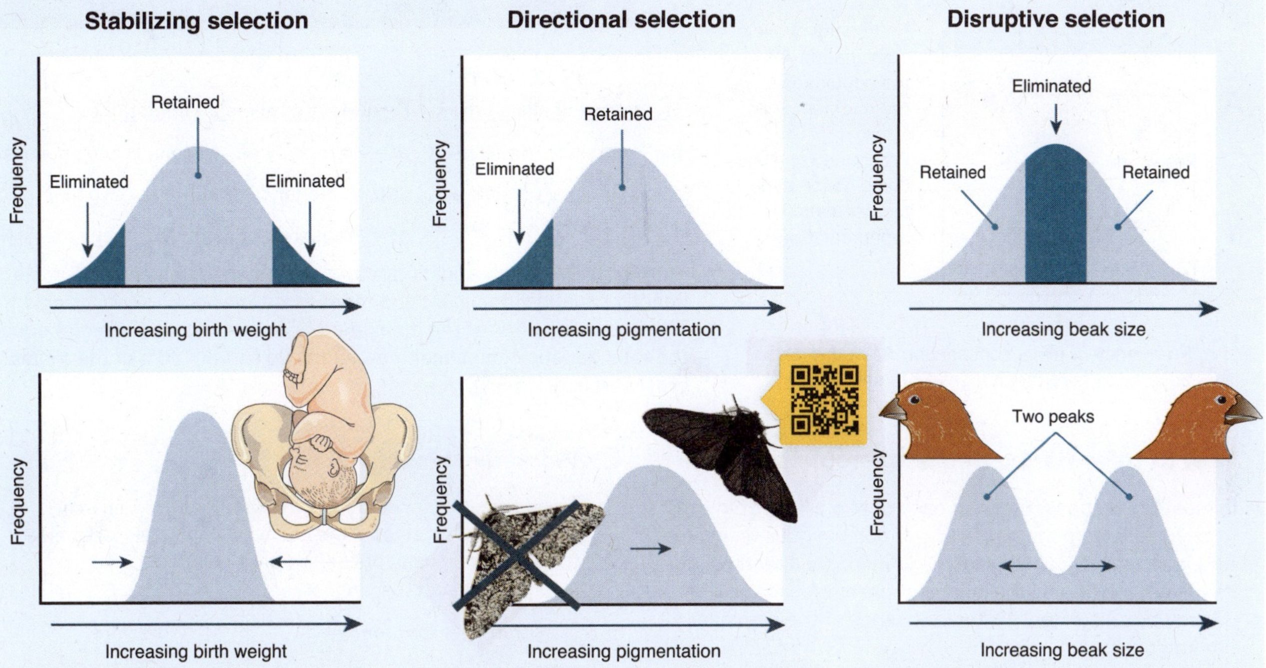

Stabilizing selection	Directional selection	Disruptive selection

Stabilizing selection — graphs of Frequency vs Increasing birth weight. Top graph: "Retained" (middle), "Eliminated" (both ends). Bottom graph shows narrowed distribution.

Directional selection — graphs of Frequency vs Increasing pigmentation. Top graph: "Retained" (peak), "Eliminated" (left tail). Bottom graph shows shifted distribution.

Disruptive selection — graphs of Frequency vs Increasing beak size. Top graph: "Eliminated" (middle), "Retained" (both sides). Bottom graph shows "Two peaks".

Extreme variations are selected against and the middle range (most common) phenotypes are retained in greater numbers. This type of selection operates most of the time in most populations, and prevents divergence of form and function, e.g. birth weight of human infants.

The adaptive phenotype is shifted in one direction and one phenotype is favoured over others. Directional selection was observed in peppered moths in England during the Industrial Revolution. Selection pressures are currently more balanced and the proportions of each form vary regionally.

Disruptive selection favours two phenotypic extremes at the expense of intermediate forms. During a drought on Santa Cruz Island in the Galápagos, disruptive selection resulted in a population of ground finches that was bimodal for beak size. Large or small beaks exploited the available seeds better than intermediate beak sizes.

1. Explain why fluctuating, as opposed to stable, environments favour disruptive selection: _____

_____

_____

2. Disruptive selection can be important in the formation of new species.

(a) Describe the evidence from the ground finches on Santa Cruz Island that provides support for this statement:

_____

_____

(b) The ground finches on Santa Cruz Island are one interbreeding population with a strong bimodal distribution for the character of beak size. Suggest what conditions could lead to the two phenotypic extremes diverging further:

_____

(c) Predict the consequences of the end of the drought and an increased abundance of medium size seeds as food:

_____

_____

©2024 **BIOZONE** International
ISBN: 978-1-99-101410-8
Photocopying prohibited

D4.1
12

351 Using the Hardy–Weinberg Equation

Key Idea: The Hardy-Weinberg equation is used to model the allele frequencies in a population.

One of the important theoretical concepts in population genetics is that of genetic equilibrium, which states that, 'for a large, randomly mating population, allele frequencies do not change from generation to generation'. If allele frequencies in a population are to remain unchanged, all of the following conditions must be met: the population must be large, there must be no mutation or gene flow, mating must be random, and there must be no natural selection. The Hardy-Weinberg equation provides a simple mathematical model of genetic equilibrium in a gene pool, but its main application in population genetics is in calculating allele and genotype frequencies in populations, particularly as a means of studying changes and measuring their rate, and if the genotype frequencies match Hardy-Weinberg conditions..

Punnett square

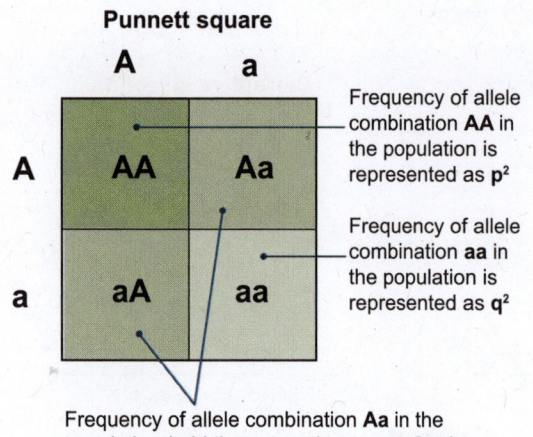

Frequency of allele combination **AA** in the population is represented as p^2

Frequency of allele combination **aa** in the population is represented as q^2

Frequency of allele combination **Aa** in the population (add these together to get **2pq**)

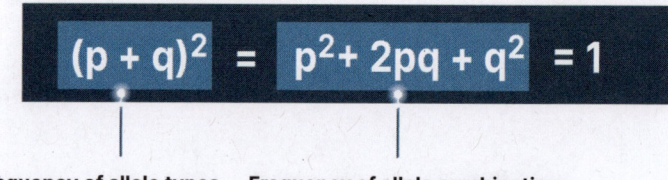

$$(p + q)^2 = p^2 + 2pq + q^2 = 1$$

Frequency of allele types

p = Frequency of allele A

q = Frequency of allele a

Frequency of allele combinations

p^2 = Frequency of AA (homozygous dominant)

2pq = Frequency of Aa (heterozygous)

q^2 = Frequency of aa (homozygous recessive)

The Hardy-Weinberg equation is applied to populations with a simple genetic situation: dominant and recessive alleles controlling a single trait. The frequency of all of the dominant (A) and recessive alleles (a) equals the total genetic complement, and adds up to 1 or 100% of the alleles present (i.e. p + q = 1).

How to solve Hardy-Weinberg problems

In most populations, the frequency of two alleles of interest is calculated from the proportion of homozygous recessives (q^2), as this is the only genotype identifiable directly from its phenotype. If only the dominant phenotype is known, q^2 may be calculated (1 – the frequency of the dominant phenotype).

The following steps outline the procedure for solving a Hardy-Weinberg problem:

Remember that all calculations must be carried out using proportions, NOT PERCENTAGES!

Examine the question to determine what piece of information you have been given about the population. In most cases, this is the percentage or frequency of the homozygous recessive phenotype q^2, or the dominant phenotype p^2 + 2pq (see note above).

The first objective is to find out the value of p or q, If this is achieved, then every other value in the equation can be determined by simple calculation.

Take the square root of q^2 to find q.

Determine p by subtracting q from 1 (i.e. p = 1 – q).

Determine p^2 by multiplying p by itself (i.e. p^2 = p x p).

Determine 2pq by multiplying p times q times 2.

Check that your calculations are correct by adding up the values for p^2 + q^2 + 2pq (the sum should equal 1 or 100%).

Worked example

Among white-skinned people in the USA, approximately 70% of people can taste the chemical phenylthiocarbamide (PTC) (the dominant phenotype), while 30% are non-tasters (the recessive phenotype).

Determine the frequency of:	**Answers**
(a) Homozygous recessive phenotype(q^2).	30% - provided
(b) The dominant allele (**p**).	45.2%
(c) Homozygous tasters (p^2).	20.5%
(d) Heterozygous tasters (**2pq**).	49.5%

Data: The frequency of the dominant phenotype (70% tasters) and recessive phenotype (30% non-tasters) are provided.

Working:
Recessive phenotype: q^2 = 30% *use 0.30 for calculation*

therefore: q = 0.5477 *square root of 0.30*

therefore: p = 0.4523 *1 – q = p*
                      1 – 0.5477 = 0.4523

Use p and q in the equation (top) to solve any unknown:
Homozygous dominant
        p^2 = 0.2046
        (p x p = 0.4523 × 0.4523)

Heterozygous: **2pq** = 0.4953

1. A population of hamsters has a gene consisting of 90% M alleles (black) and 10% m alleles (grey). Mating is random.
   Data: Frequency of recessive allele (10% m) and dominant allele (90% M).

   Determine the proportion of offspring that will be black and the proportion that will be grey (show your working):

   _____

   _____

Recessive allele:	q	= ☐
Dominant allele:	p	= ☐
Recessive phenotype:	q^2	= ☐
Homozygous dominant:	p^2	= ☐
Heterozygous:	2pq	= ☐

D4.1
13 - 14

2. You are working with pea plants and found 36 plants out of 400 were dwarf. Data: Frequency of recessive phenotype (36 out of 400 = 9%)

   (a) Calculate the frequency of the tall alele: _____

   (b) Determine the number of heterozygous pea plants:

   _____

   _____

(c)		
Recessive allele:	q	=
Dominant allele:	p	=
Recessive phenotype:	q^2	=
Homozygous dominant:	p^2	=
Heterozygous:	2pq	=

3. In humans, the ability to taste the chemical phenylthiocarbamide (PTC) is inherited as a simple dominant characteristic. You discovered that 360 out of 1000 college students could not taste the chemical.
Data: Frequency of recessive phenotype (360 out of 1000).

   (a) State the frequency of the allele for tasting PTC: _____

   _____

   (b) Determine the number of heterozygous students in this population:

   _____

   _____

(c)		
Recessive allele:	q	=
Dominant allele:	p	=
Recessive phenotype:	q^2	=
Homozygous dominant:	p^2	=
Heterozygous:	2pq	=

4. A type of deformity appears in 4% of a large herd of cattle. Assume the deformity was caused by a recessive allele.
Data: Frequency of recessive phenotype (4% deformity).

   (a) Calculate the percentage of the herd that carries the recessive allele:

   _____

   (b) Determine the frequency of the dominant allele in this case:

   _____

(c)		
Recessive allele:	q	=
Dominant allele:	p	=
Recessive phenotype:	q^2	=
Homozygous dominant:	p^2	=
Heterozygous:	2pq	=

5. Assume you placed 50 pure bred black guinea pigs (dominant allele) with 50 albino guinea pigs (recessive allele) and allowed the population to attain genetic equilibrium (several generations have passed).
Data: Frequency of recessive allele (50%) and dominant allele (50%).

   (a) Determine the proportion (%) of the population that becomes white:

   _____

   _____

(b)		
Recessive allele:	q	=
Dominant allele:	p	=
Recessive phenotype:	q^2	=
Homozygous dominant:	p^2	=
Heterozygous:	2pq	=

6. It is known that 64% of a large population exhibit the recessive trait of a characteristic controlled by two alleles (one is dominant over the other).
Data: Frequency of recessive phenotype (64%). Determine the following:

   (a) The frequency of the recessive allele: _____

   (b) The percentage that are heterozygous for this trait: _____

   (c) The percentage that exhibit the dominant trait: _____

   (d) The percentage that are homozygous for the dominant trait: _____

   (e) The percentage that has one or more recessive alleles: _____

(f)		
Recessive allele:	q	=
Dominant allele:	p	=
Recessive phenotype:	q^2	=
Homozygous dominant:	p^2	=
Heterozygous:	2pq	=

7. Explain how the Hardy-Weinberg equations allow the calculation of allele frequencies in a population while needing to know only the frequency of the recessive phenotype. Use the example that the recessive phenotype cystic fibrosis occurs at a frequency of 1 in 2500 births.

_____

_____

_____

_____

Analysis of a Squirrel Gene Pool

▶ In Olney, Illinois, there is a unique population of albino (white) and grey squirrels. Between 1977 and 1990, students at Olney Central College carried out a study of this population. They recorded the frequency of grey and albino squirrels.

▶ The albinos displayed a mutant allele expressed as an albino phenotype only in the homozygous recessive condition.

▶ The data they collected are provided in the table below. Using the Hardy-Weinberg equation, it was possible to estimate the frequency of the normal 'wild' allele (G) providing grey fur coloring, and the frequency of the mutant albino allele (g) producing white squirrels when homozygous.

Thanks to **Dr. John Stencel**, Olney Central College, Olney, Illinois, US, for providing the data for this exercise.

Grey squirrel, usual colour form   Albino form of grey squirrel

Population of grey and white squirrels in Olney, Illinois (1977-1990)

Year	Grey	White	Total	GG	Gg	gg	Freq. of g	Freq. of G
1977	602	182	784	26.85	49.93	23.21	48.18	51.82
1978	511	172	683	24.82	50.00	25.18	50.18	49.82
1979	482	134	616	28.47	49.77	21.75	46.64	53.36
1980	489	133	622	28.90	49.72	21.38	46.24	53.76
1981	536	163	699	26.74	49.94	23.32	48.29	51.71
1982	618	151	769	31.01	49.35	19.64	44.31	55.69
1983	419	141	560	24.82	50.00	25.18	50.18	49.82
1984	378	106	484	28.30	49.79	21.90	46.80	53.20
1985	448	125	573	28.40	49.78	21.82	46.71	53.29
1986	536	155	691	27.71	49.86	22.43	47.36	52.64
1987	No data collected this year							
1988	652	122	774	36.36	47.88	15.76	39.70	60.30
1989	552	146	698	29.45	49.64	20.92	45.74	54.26
1990	603	111	714	36.69	47.76	15.55	39.43	60.57

8. Graph population changes: (a) Use the data in the first 3 columns of the table above to plot a line graph. This will show changes in the phenotypes: numbers of grey and white (albino) squirrels, as well as changes in the total population. Plot grey, white, and total for each year:

(b) Determine by how much (as a %) the total population numbers have fluctuated over the sampling period:

_____

(c) Describe the overall trend in total population numbers and any pattern that may exist:

_____

_____

_____

_____

_____

_____

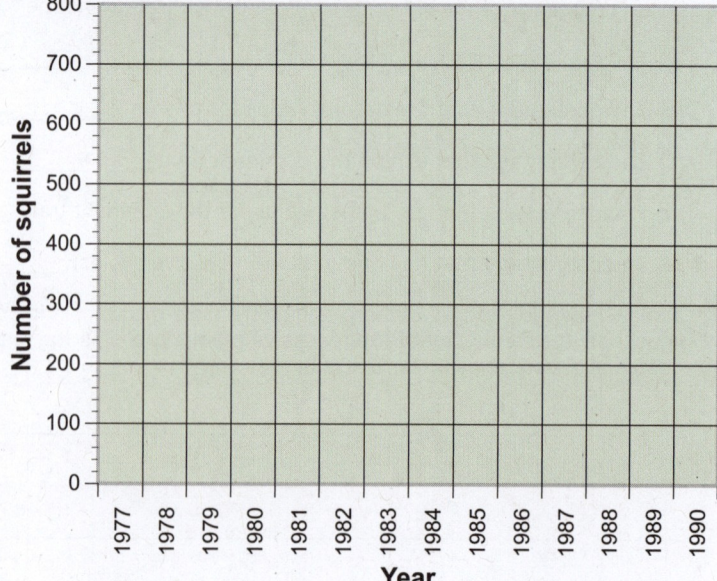

9. Graph genotype changes: (a) Use the data in the genotype columns of the table on the previous page to plot a line graph. This will show changes in the allele combinations (GG, Gg, gg). Plot: GG, Gg, and gg for each year:

Describe the overall trend in the frequency of:

(b) Homozygous dominant (GG) genotype:

_____

_____

Heterozygous (Gg) genotype:

_____

_____

Homozygous recessive (gg) genotype:

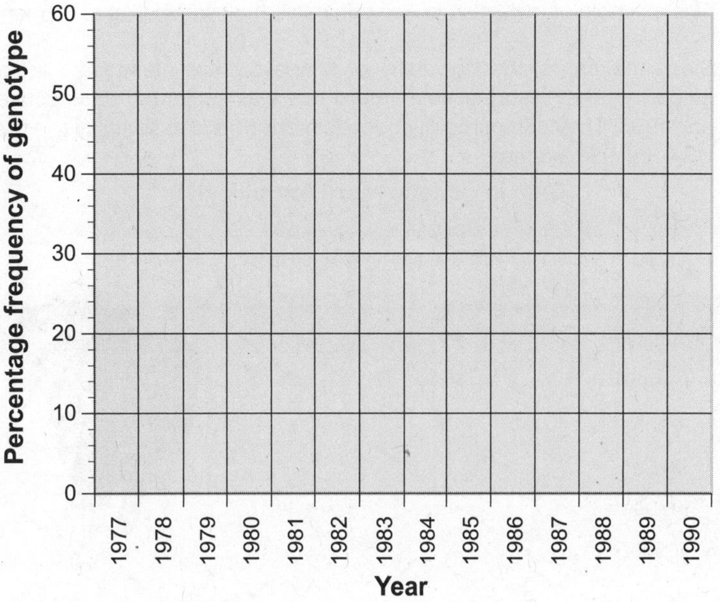

_____

_____

_____

[Graph: Percentage frequency of genotype (y-axis, 0–60) vs Year (x-axis, 1977–1990)]

10. Graph allele changes: (a) Use the data in the last two columns of the table on the previous page to plot a line graph. This will show changes in the allele frequencies for each of the dominant (G) and recessive (g) alleles. Plot: the frequency of G and the frequency of g:

(b) Describe the overall trend in the frequency of the dominant allele (**G**):

_____

_____

(c) Describe the overall trend in the frequency of the recessive allele (g):

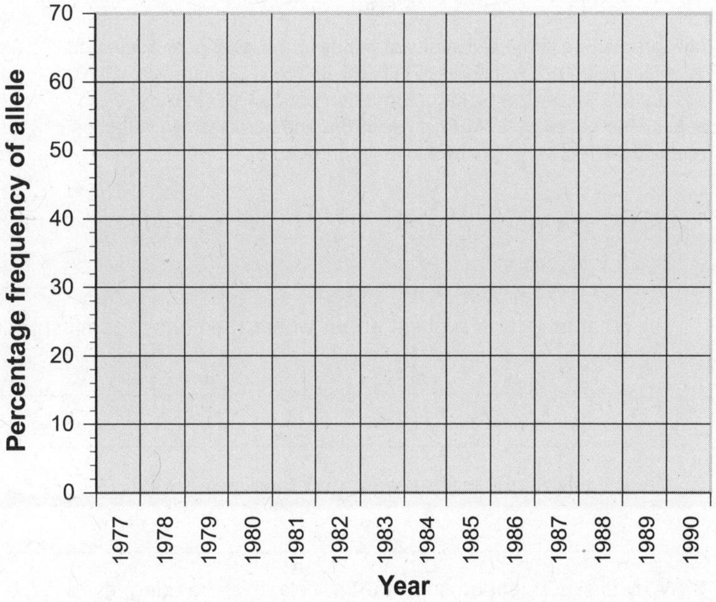

_____

_____

_____

11. (a) State which of the three graphs best indicates that a significant change may be taking place in the gene pool of this population of squirrels:

_____

(b) Give a reason for your answer: _____

_____

_____

_____

[Graph: Percentage frequency of allele (y-axis, 0–70) vs Year (x-axis, 1977–1990)]

12. Describe a possible cause of the changes in allele frequencies over the sampling period: _____

_____

_____

_____

_____

13. Visit the simulation websites listed on the **Biozone Resource Hub** for this activity to model how gene pools change from one generation to the next when variables such as selection, mutation, and migration are manipulated.

352 Artificial Selection

Key Idea: Selective breeding can result in rapid change in the phenotypic characteristics of a population.

Humans may create selection pressure for evolutionary change by choosing to breed individuals with particular traits. The example of milk yield in Holstein cows (below) illustrates how humans have directly influenced the genetic makeup of Holstein cattle with respect to milk production and fertility. Other examples include domesticated dogs and crops. Due to the nature of genes, artificial selection may favour one phenotype at the expense of another.

Since the 1960s, the University of Minnesota has maintained a Holstein cattle herd that has not been subjected to any selection. They also maintain a herd that was subjected to selective breeding for increased milk production between 1965 and 1985. They compared the genetic merit (this is essentially the breeding value) for milk yield in these groups to that of the USA Holstein average.

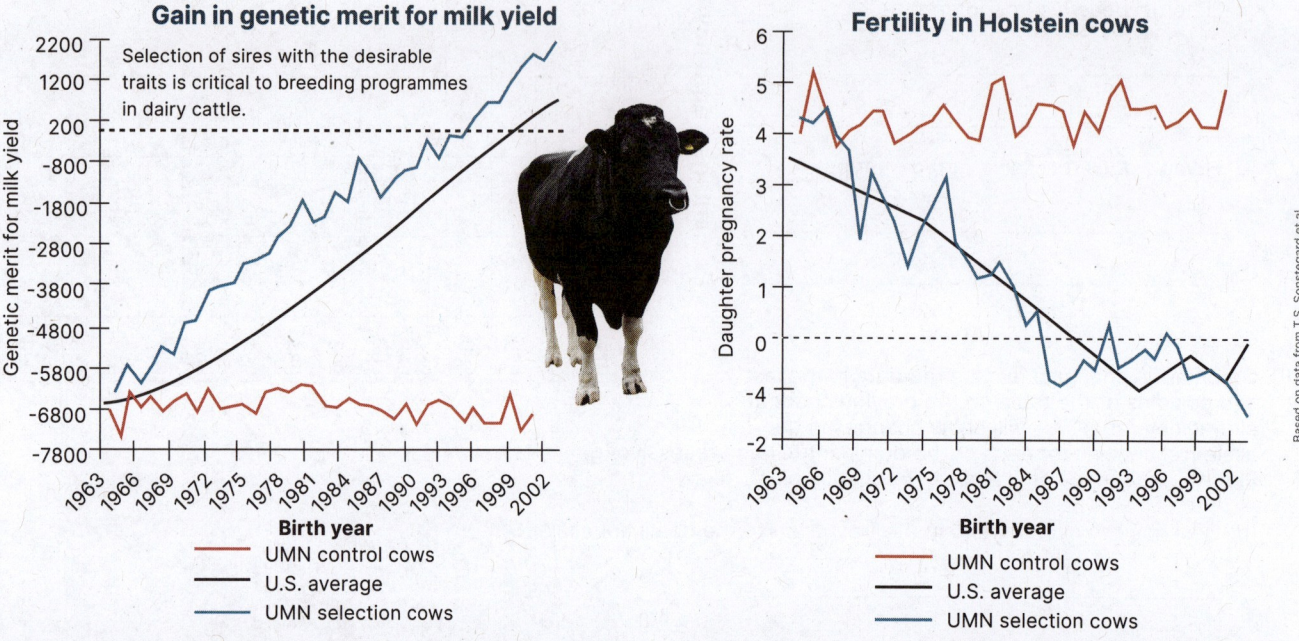

Selection of sires with the desirable traits is critical to breeding programmes in dairy cattle.

Based on data from T.S. Sonstegard et al

Milk production in the University of Minnesota (UMN) herd subjected to selective breeding increased in line with the U.S. average production. In real terms, milk production per cow per milking season increased by 3740 kg since 1964. The herd with no selection remained effectively constant for milk production.

Along with increased milk production there has been a distinct decrease in fertility. The fertility of the University of Minnesota (UMN) herd that was not subjected to selection remained constant while the fertility of the herd selected for milk production decreased with the U.S. fertility average.

1. (a) Describe the relationship between milk yield and fertility on Holstein cows: _____

_____

(b) What does this suggest about where the genes for milk production and fertility are carried? _____

_____

_____

2. What limits might this place on maximum milk yield? _____

_____

3. Why is sire selection important in selective breeding, even if the characters involved are expressed only in the female?

_____

4. Natural selection is the mechanism by which organisms with favourable traits become proportionally more common in the population. How does selective breeding mimic natural selection? How does the example of the Holstein cattle show that reproductive success is a compromise between many competing traits?

_____

_____

_____

_____

AHL

D4.1

15

©2024 **BIOZONE** International
ISBN: 978-1-99-101410-8
Photocopying prohibited

The origins of domestic dogs

▶ All breeds of dog are members of the same species, *Canis familiaris* and provide an excellent example of artificial selection. The dog was likely the first domesticated species and, over centuries, humans have selected for desirable traits, so extensively that there are now more than 400 breeds of dogs.

▶ Until very recently, the grey wolf was considered to be the ancestor of the domestic dog. However, 2015 genetic studies provide strong evidence that domestic dogs and grey wolves are sister groups and shared a now extinct wolf-like common ancestor, which gave rise to the dog before the agricultural revolution 12,000 years ago. Based on genetic analysis, four major clusters of ancient dog breeds are recognized. Through artificial selection, all other breeds are thought to have descended from these clusters.

1: Older lineages
The oldest lineages, including Chinese breeds, basenji, huskies, and malamutes.

2: Mastiff-type
An older lineage that includes the mastiffs, bull terriers, boxers, and rottweilers.

3: Herding
Includes German shepherd, St Bernard, borzoi, collie, corgi, pug, and greyhound.

4: Hunting
Most arose in Europe. Includes terriers, spaniels, poodles, and modern hounds.

Problems with artificial selection

▶ Selection for a desirable phenotype can result in undesirable traits being emphasized, often because genes for particular characteristics are linked and selection for one inadvertently selects for the other. For example, the German shepherd is a working dog, originally bred for its athleticism and ability to track targets.

▶ In German shepherds bred to meet the specific appearance criteria of show dogs, some traits have been exaggerated so much that it causes health issues. The body shape of the show German shepherd has been selected for a flowing trot and it has a pronounced slope in the back. This has resulted in leg, hip, and spinal problems. In addition, artificial selection has increased the incidence of some genetic diseases such as epilepsy and blood disorders.

Sloped-backed German shepherd

Straight-backed German shepherd

5. What are some ethical considerations of using artificial selection to 'improve' dog breeds and what would it take to change breed standards to avoid health issues?

_____

_____

_____

_____

_____

6. Some consequences of artificial selection include low fertility or immunity. Technology and/or medicine may be needed to compensate for this. Antibiotics may be required and their use can lead to antibiotic resistance. Explain why antibiotic resistance is considered natural selection, and compare with artificial selection:

_____

_____

_____

_____

353 Ecosystem Stability

Key Idea: A number of factors contribute to ecosystem stability, which exists as a natural property.

Ecosystems are dynamic and constantly changing. Many ecosystem components including the seasons, predator-prey cycles, and disease cycles, are cyclical. Some cycles may be short term, such as the change of seasons. Others are long term, such as the growth and retreat of deserts.

Although ecosystems may change constantly over the short term, they may be relatively stable over longer periods. For example, some tropical areas have wet and dry seasons, but over hundreds of years the ecosystem as a whole remains unchanged.

Changes in stability

▶ Some ecosystems have been stable for an extremely long time, such as deep oceans, frozen tundra, or rain forests.

▶ Change can be introduced by a shift of any biotic or abiotic factors away from the seasonal normal.

▶ The four key requirements for ecosystem stability (right) can be altered by natural cycles, natural disasters, such as volcanoes, or by human activity.

▶ Human impact has been responsible for many ecosystem changes, both on an ecosystem specific scale and a global scale. Examples are over exploitation of organisms, or human-induced climate change. These changes affect numerous abiotic factors across multiple ecosystems, creating instability.

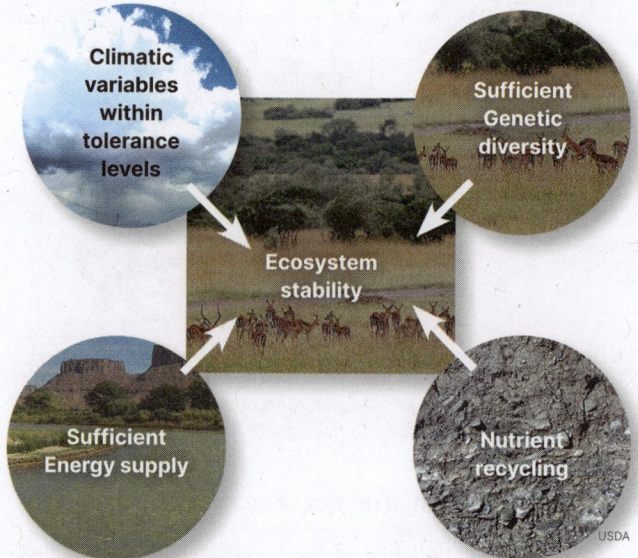

Ecosystem stability depends upon a number of requirements being maintained - reduction of any one can impact stability.

An ecosystem may remain stable for many hundreds or thousands of years, provided that the components interacting within it remain stable.

Small scale changes usually have little effect on an ecosystem. Fire or flood may destroy some parts, but enough is left for the ecosystem to return to its original state relatively quickly.

Large scale disturbances such as volcanic eruptions or large scale open cast mining can destabilize an ecosystem and change it forever.

1. What is meant by the term dynamic ecosystem? _____

_____

_____

_____

2. (a) Describe a small scale event occurring in your local area that an ecosystem recovered from: _____

_____

_____

(b) Describe two large scale events that an ecosystem may not recover from: _____

_____

_____

D4.2

1 - 2

©2024 **BIOZONE** International
ISBN: 978-1-99-101410-8
Photocopying prohibited

Ecosystem stability and biodiversity

▶ Ecosystem stability is affected by its inertia (the ability to resist disturbance) and resilience (the ability to recover from external disturbances).

▶ Ecosystem stability is closely linked to biodiversity. More biodiverse systems are more stable. Researchers hypothesize that greater diversity causes a greater number of biotic interactions, and few, if any, vacant niches. The system is resistant to invasions and enough species are present to protect ecosystem functions if one is lost. This hypothesis is supported by experimental evidence. However, there is uncertainty over what level of biodiversity provides stability or what factors will stress a system beyond its tolerance.

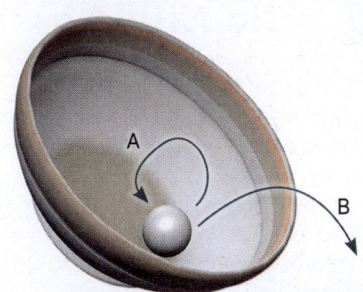

The stability of an ecosystem can be illustrated by a ball in a tilted bowl. Given a slight disturbance the ball will eventually return to its original state (line A). However, given a large disturbance the ball will roll out of the bowl and the original state with never be restored (line B).

Response to environmental change

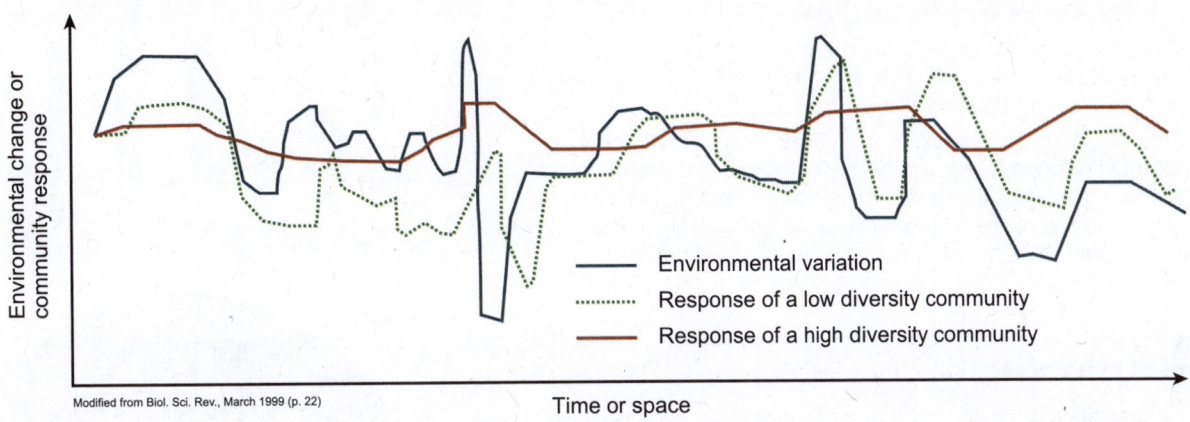

Modified from Biol. Sci. Rev., March 1999 (p. 22)

Time or space

- Environmental variation
- Response of a low diversity community
- Response of a high diversity community

▶ In models of ecosystem function, higher species diversity increases the stability of ecosystem functions, such as productivity and nutrient cycling. In the graph above, note how the low diversity system varies more consistently with variations in the environment, whereas changes in the high diversity system are more gradual.

▶ In any one ecosystem, some species have a disproportionate effect on ecosystem stability due to their key role in an ecosystem function, e.g. nutrient recycling. These species are called keystone species.

3. Why is ecosystem stability higher in ecosystems with high biodiversity than in those with low biodiversity?

_____

_____

4. The effect of changes in ecosystems can be difficult to measure in the field so researchers often build small scale simulations. The graph (right) shows the effect of adding nutrients to a marine ecosystem e.g. nutrient runoff from land into the sea. Algal growth-promoting medium was added at 2, 10, or 20% to seawater, together with 0.1 mL of an algal mix. Two days after adding the growth medium and algae, six copepods were added to each chamber. The chambers were sealed and the population size in each chamber was measured over time.

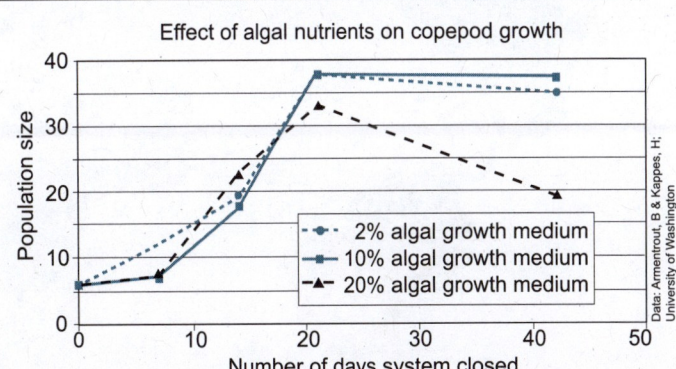

Effect of algal nutrients on copepod growth

- 2% algal growth medium
- 10% algal growth medium
- 20% algal growth medium

Population size

Number of days system closed

Data: Armentrout, B & Kappes, H; University of Washington

(a) Which chamber had the greatest environmental disturbance?

_____

(b) Which chamber(s) were able to withstand the environmental disturbance? _____

(c) What does this tell us about the stability and resilience of the system being studied? _____

_____

_____

354 The Amazon Rainforest and Deforestation

Key Idea: Deforestation of the Amazon rainforest could push the climate into a tipping point.

Tipping points are reached when feedback mechanisms, such as transpiration, air flow, and rainfall, that enable ecosystem stability are irrevocably damaged, causing an ecosystem to undergo rapid change. Scientists estimate that when the Amazon rainforest reaches around 20-25% deforestation then the tipping point will be reached. Currently, in 2023, nearly 17% of the rainforest has been lost. In equatorial regions, the pace of deforestation is accelerating. This is of global concern as species biodiversity is highest in the tropics and habitat loss puts a great number of species at risk. Deforestation fronts (below) are large areas of forests that are under threat, with around 10% of forest cover lost between 2004 - 2017, and the remaining areas at significant risk of further future losses.

Deforestation

▶ At the end of the last glacial period, about 10,000 years ago, forests covered around 45% of the Earth's land surface. Forests now cover about 31% of Earth's surface. These include the cooler temperate forests of North and South America, Europe, China, and Australasia, and the tropical forests of equatorial regions.

▶ Over the last 5000 years, the loss of forest cover is estimated at 1.8 billion hectares. A net loss of around 5.2 million hectares has occurred the last 10 years alone.

▶ Temperate regions where human civilizations have historically existed the longest, e.g. Europe, have suffered the most, but now the vast majority of deforestation is occurring in the tropics. Intensive clearance of forests during settlement of the most recently discovered lands has extensively altered their landscapes and permanently changed or decreased biodiversity.

Deforestation by type and period

(Bar chart: Million hectares, y-axis 0-450. Categories: Pre 1700, 1700-1849, 1850-1919, 1920-1949, 1950-1979, 1980-1995, 1995-2010. Legend: Temperate, Tropical)

Causes of deforestation

▶ Deforestation is the end result of many interrelated factors which often centre around socioeconomic drivers. In many tropical regions, the vast majority of deforestation is the result of subsistence farming.

▶ Poverty and a lack of secure land can be partially solved by clearing small areas of forest and producing family plots. However, huge areas of forests have been cleared for agriculture, including ranching and production of palm oil plantations. These produce revenue for governments through taxes and permits, producing an incentive to clear more forest.

▶ Just 14% of deforestation is attributable to commercial logging although, combined with illegal logging, it may be much higher.

Causes of deforestation

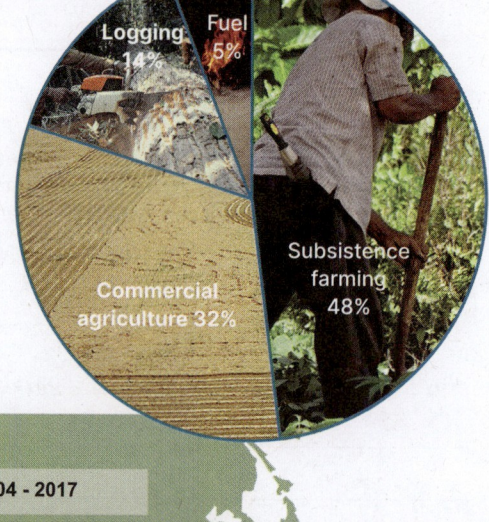

Logging 14%
Fuel 5%
Subsistence farming 48%
Commercial agriculture 32%

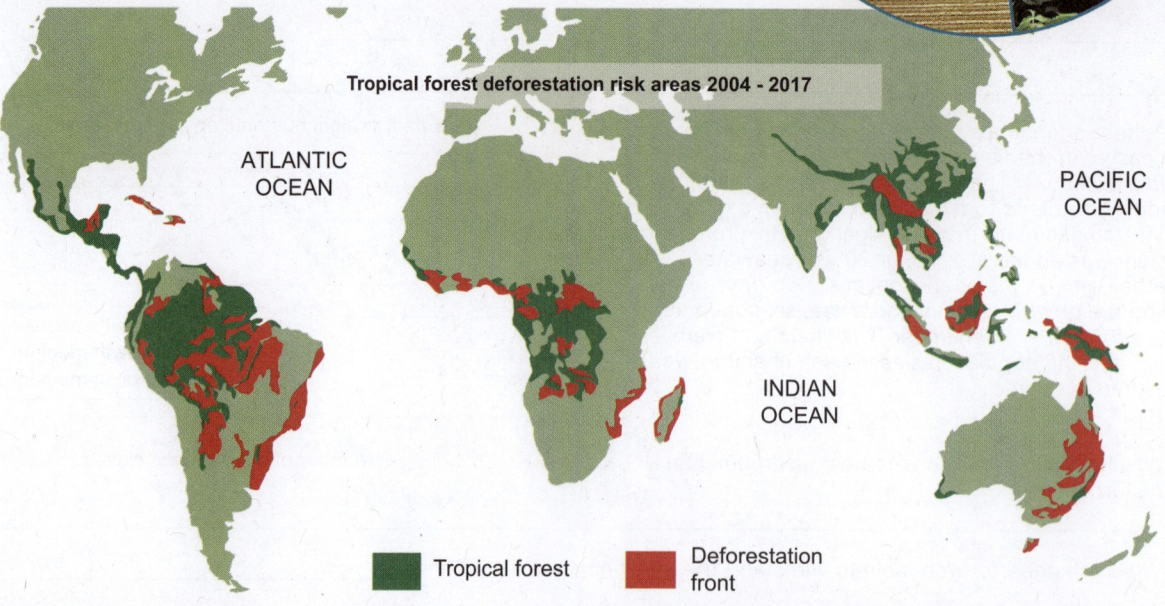

Tropical forest deforestation risk areas 2004 - 2017

ATLANTIC OCEAN
PACIFIC OCEAN
INDIAN OCEAN

■ Tropical forest   ■ Deforestation front

▶ It is important to distinguish between deforestation involving primary (old growth) forest and deforestation in plantation forests. Plantations are regularly cut down and replaced, and can artificially inflate a country's apparent forest cover or rate of deforestation. The loss of primary forests is far more important as these are refuges of high biodiversity, including rare species, many of which are not found anywhere else in the world.

AOS   D4.2 3

©2024 **BIOZONE** International
ISBN: 978-1-99-101410-8
Photocopying prohibited

Feedback cycles and deforestation tipping points in the Amazon rainforest

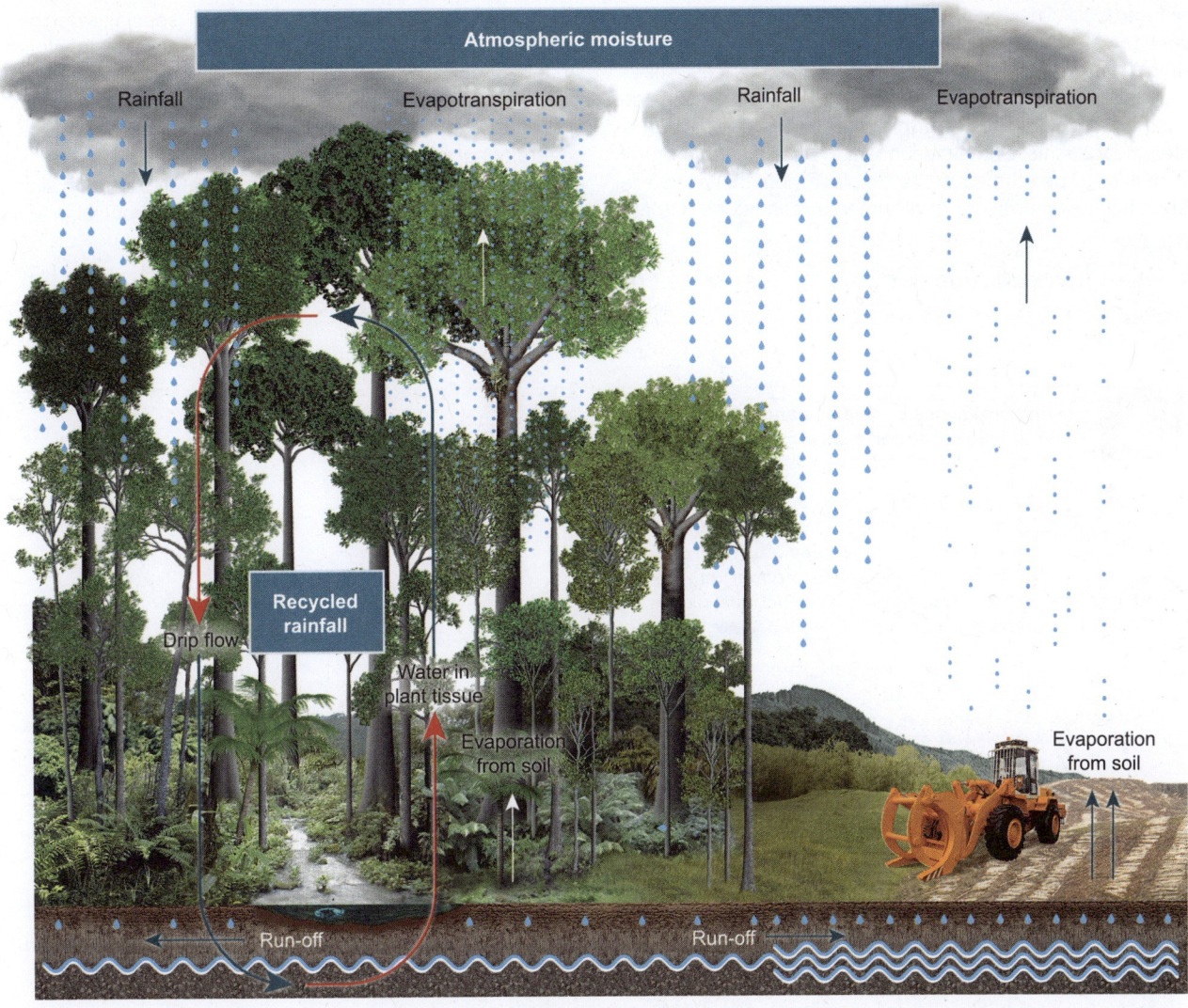

A healthy forest draws water from the soil. The water circulates through the plants and returns to the atmosphere via transpiration, causing humidity. This humid air continues to circulate further through the forest in airflows. Deforested land does not have enough plants to return the water to the atmosphere and most of the rain is lost in runoff. The resulting lower humidity impacts forested areas downwind such that less rainfall occurs, leading to degradation. Once the tipping point is reached, even areas that have not been deforested will lose ecosystem stability, damaging the whole region in a domino effect.

1. (a) Define a tipping point: _____

_____

(b) Why is reaching a tipping point so damaging to the Amazon rainforest? _____

_____

_____

2. Percentage change can be used as a statistical tool to analyze deforestation rates.
   (a) Go to https://www.globalforestwatch.org/map locate Brazil (Amazon rainforest area) and click on the analysis to show forest loss from 2001 to 2022.
   (b) Use the data to calculate percentage change of forest loss (deforestation).

   _____   _____

   Percentage change (negative)

   % change = $\dfrac{(\text{initial value} - \text{final value})}{\text{initial value}} \times 100$

   For a percentage increase, final value - initial value

(c) Does the data confirm that the Amazon forest is close to tipping point?

_____

355 Modelling Ecosystems

Key Idea: Models such as a sealed mesocosm can be used to investigate stability in a closed system.

Aspects of ecosystem function, including responses to changes in inputs and long term stability, can be investigated using physical representations of ecosystems called mesocosms. Some mesocosm studies allow a natural community to be studied in situ (in place), but still allow the researcher to control the environmental conditions. Others are carried out at research facilities in specially designed containers. Mesocosms can be open or sealed (enclosed) systems. Sealed mesocosms allow the researcher to fully control the experimental conditions, including the entry and exit of matter. Mesocosms, especially small ones, are generally not stable in the long term, and change over time as a result of their smaller scale and isolated nature. A simple mesocosm model can be developed in the school laboratory.

Investigation 16.2 A mesocosm model of an ecosystem

See appendix for equipment list.

> ⚠ **Living organisms should be handled with care and respect: See IB experimental guidelines.**

1. In this investigation, you will make a simple ecobottle to model stability in a small, closed ecosystem. Your group will be provided with the following equipment: a large, clear soda bottle with a lid; filtered pond water; aquarium gravel; a source of detritus, e.g. dead leaves; aquatic plants (such as Cabomba); small pond snails.

2. Use the equipment to set up two bottle ecosystems (a and b). You will need to think about how long you wait before you close the system off, how much air gap you will have, how much organic material you will add, and where you will put your ecosystem (light/dark). Alter ONE variable (the same) in each of the bottles.

3. Draw a scientific drawing of your bottle ecosystems. Label the picture to include important design features and a key with the type and total numbers of each organism. You may need additional paper to attach drawings.

4. Leave your bottle ecosystems for a week. Observe it carefully at various times during the week. After a week, note down any changes since you set it up.

Cabomba

5. Return any living organisms back to the aquarium and dispose of any waste materials.

1. (a) What variable did you change in your bottle ecosystems? _____

   (b) Write a hypothesis for how the change will affect the stability of your ecosystem: _____

   _____

   (c) Did your observations allow you to see any differences between your two bottle ecosystems and, if so, did they support your hypothesis?

   _____

   _____

2. The pond water contains small microorganisms. What is their role in this system? _____

   _____

3. Were your systems stable? Explain why (or why not): _____

   _____

   _____

   _____

©2024 **BIOZONE** International
ISBN: 978-1-99-101410-8
Photocopying prohibited

356 Keystone Species and Stability

Key Idea: Keystone species have a disproportionate effect on ecosystem stability.

Every species has a functional role in an ecosystem (its niche), but some have a much bigger effect on ecosystem processes and stability than their abundance would suggest because their activities are crucial to the way the ecosystem as a whole functions. These species are called keystone species. They are often top (apex) predators, or have a critical role in seed dispersal or nutrient cycling. The loss of a keystone species can have a large and rapid impact on the structure and function of an ecosystem, changing the balance of relationships and leading to instability. This has important implications for the management of threatened ecosystems because many keystone species are endangered.

Why are keystone species important?

The term 'keystone species' comes from the analogy of the keystone in a true arch. An archway is supported by a series of stones, the central one being the keystone. If the keystone is removed, the arch collapses.

The idea of the keystone species was first hypothesized in 1969 by Robert Paine. He determined through experimentation that the ochre starfish (*Pisaster*), a predator in rocky shore communities, had a role in maintaining community diversity. When the starfish were removed, their prey species increased, crowding out algae and reducing species richness in the area from 15 to 8.

Trophic cascades following the return of grey wolves to Yellowstone

Yellowstone wolves run down a bull elk

▶ Grey wolves are a top predator in North American ecosystems, yet federal extermination programs in the late 1800s-early 1900s reduced them to near extinction in the US, including in Yellowstone National Park (YNP) where National Park status did not protect them.

▶ Once the wolves were gone, numbers of elk (their primary prey) increased and the deciduous vegetation became severely overgrazed. This had a number of consequences. Without wolves, coyotes also increased, and the numbers of pronghorn antelope (coyote prey) then declined. Beavers became largely absent.

▶ In 1974, the grey wolf was declared endangered under the Endangered Species Act (1973) and in 1995 its reintroduction to the park began. Since that time, wolf numbers in the park have grown, elk have declined and shifted into less favourable habitats, deciduous vegetation has recovered, beavers have returned, and the coyote population has stabilized at a lower level.

Figures, right, show some of the ecological changes recorded in YNP following the 1995/1996 reintroduction of grey wolves. Individual plots show numbers of wolves (A), elk (B), and beaver (C), together with vegetation changes. Aspen heights are recorded in areas with downed logs, which regenerate somewhat faster than areas without downed logs.

Dashed lines represent time periods with at least 1 year of missing data.

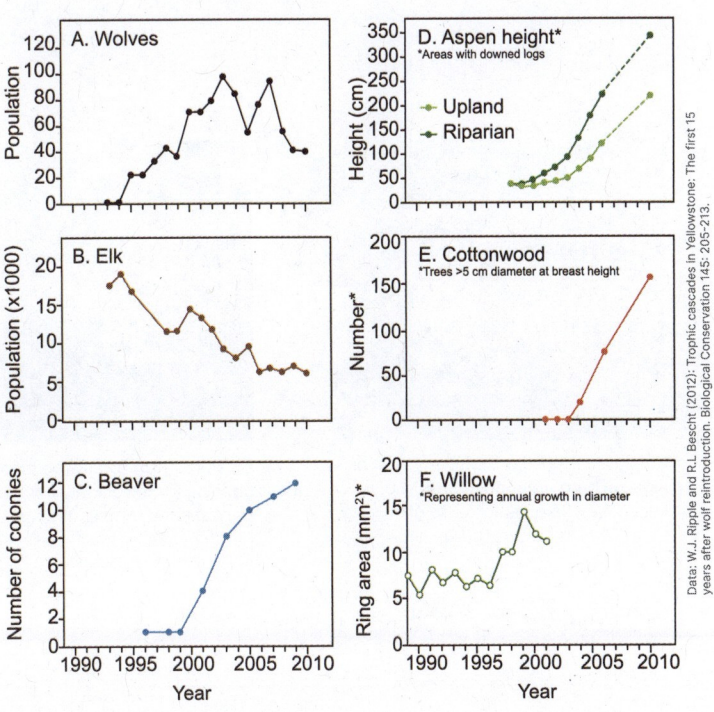

Data: W.J. Ripple and R.L. Bescht. (2012): Trophic cascades in Yellowstone: The first 15 years after wolf reintroduction. Biological Conservation 145: 205-213.

1. Use the figures (A-F) above to discuss the evidence for the role of grey wolves as a keystone species:

_____

_____

_____

_____

_____

_____

_____

D4.2
5

Sea otters as keystone species

▶ Sometimes the significant keystone effects of a species are evident when a species declines rapidly to the point of near extinction. This is illustrated by the sea otter example described below.

▶ Sea otters live along the Northern Pacific coasts and were hunted for hundreds of years for their fur. Extensive commercial hunting between 1741 and 1911 reduced the global population to fewer than 2000 animals.

▶ The drop in sea otter numbers had a significant effect on local marine environments. Sea otters feed on shellfish, particularly sea urchins, and keep their populations in check. Sea urchins eat kelp, on which many marine species depend for food and habitat. Without the sea otters, sea urchin numbers increased and the kelp forests were destroyed or severely reduced.

▶ Sea otters have been protected since 1911 and reintroduced throughout much of their original range. The most secure populations are now in Russia.

Sea otter feeding on a sea urchin
Matt knoth cc BY 2.0

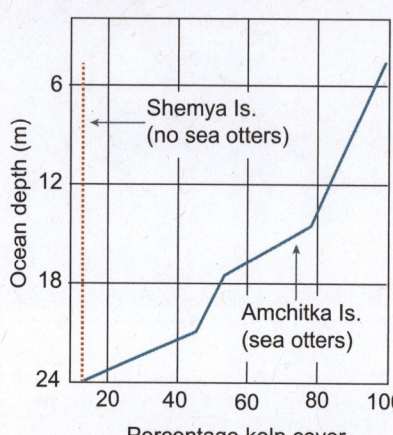

Sea urchin density (individuals per m²)

Ocean depth (m)

Shemya Is. (no sea otters)

Amchitka Is. (sea otters)

Percentage kelp cover

Sea otters are critical to the functioning of North Pacific coastal ecosystems. Their widespread decline, including many local extinctions, was associated with sea urchin increases and widespread disappearance of the kelp forests (left). In the Aleutian Island group, most islands experienced local extinctions of sea otters. This provided the opportunity to record the ecology of coastal systems with and without sea otters (Palmisano and Estes 1976).

The effect can be seen on Shemya and Amchitka Islands. Where sea otters are absent, there are large numbers of sea urchins and almost no kelp. Sea otters began recolonizing Shemya in the 1990s and the kelp has since recovered.

Kelp are large seaweeds, a type of brown algae. There are many forms and species of kelp. Giant kelp can grow to 45 m long.

Similar to how forests on land provide diverse habitats for terrestrial species, kelp provides habitat, food, and shelter for a variety of marine animals.

Sea urchins kill kelp by eating the holdfast that secures the kelp to the seabed. Unchecked urchin populations can quickly turn a kelp forest into an 'urchin barren'.
NPS: Dan Richards

2. (a) What evidence is there that the sea otter is a keystone species in these Northern Pacific coastal ecosystems?

_____

_____

_____

_____

_____

_____

(b) Predict the effect on the Amchitka Island ecosystem if sea otters became locally extinct: _____

_____

_____

©2024 **BIOZONE** International
ISBN: 978-1-99-101410-8
Photocopying prohibited

357 Sustainable Harvesting

Key Idea: Allowing resources to recover and renew themselves in between harvests assures their sustainability. In general, sustainability means the ability to use resources without using them up faster than they are renewed. The vast majority of the resources used by humans are either non-renewable (such as fossil fuels) or are being used faster than they are being renewed, i.e. unsustainably. The sustainable harvesting of any resource requires that its rate of harvest is no more than its replacement rate. If the harvest rate is higher than the replacement rate, then the resource will reduce at ever increasing percentages (assuming a constant harvest rate) and eventually be lost. This applies to many renewable resources such as hardwood tree species logged in South East Asian rainforests, and many ocean fisheries.

Sustainable methods of forestry

Selective logging

A mature forest is examined, and trees are selected for removal based on height, girth, or species. These trees are felled individually and directed to fall in such a way as to minimize the damage to the surrounding younger trees. The forest is managed to ensure continual regeneration of young seedlings (reforestation) and provide a balance of tree ages that mirrors the natural age structure of the forest.

Foresters can use monitoring platforms (such as EOSDA) to assess tree health and density so the most sustainable approach to tree selection is used.

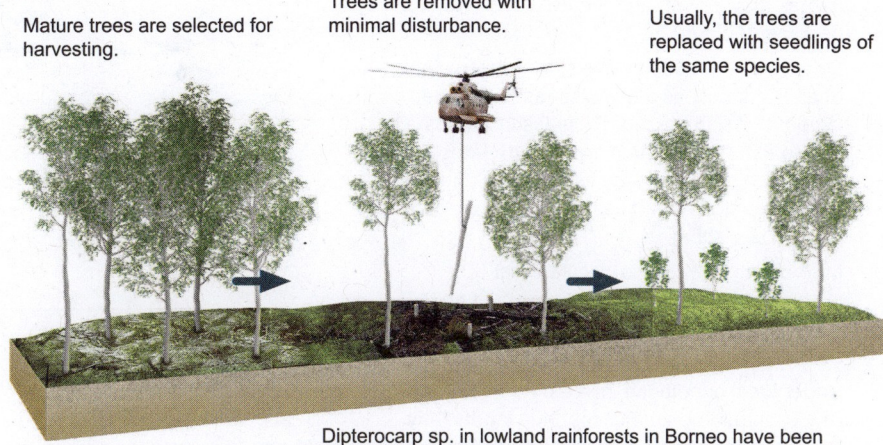

Mature trees are selected for harvesting.

Trees are removed with minimal disturbance.

Usually, the trees are replaced with seedlings of the same species.

Dipterocarp sp. in lowland rainforests in Borneo have been selectively harvested and researched for 20 years. Minimal disturbance to remaining trees, and ensuring there is no over harvesting of any one species can lead to sustainable harvesting.

Strip cutting

Strip cutting is a variation of clearcutting. Trees are clear cut out of a forest in strips. The strip is narrow enough that the forest on either side is able to reclaim the cleared land. As the cleared forest re-establishes (3-5 years) the next strip is cut.

Strip cutting allows the forest to be logged with minimal effort and damage to forest on either side of the cutting zone, while at the same time allowing natural re-establishment of the original forest. Each strip is not cut again for around 30 years, depending on regeneration time.

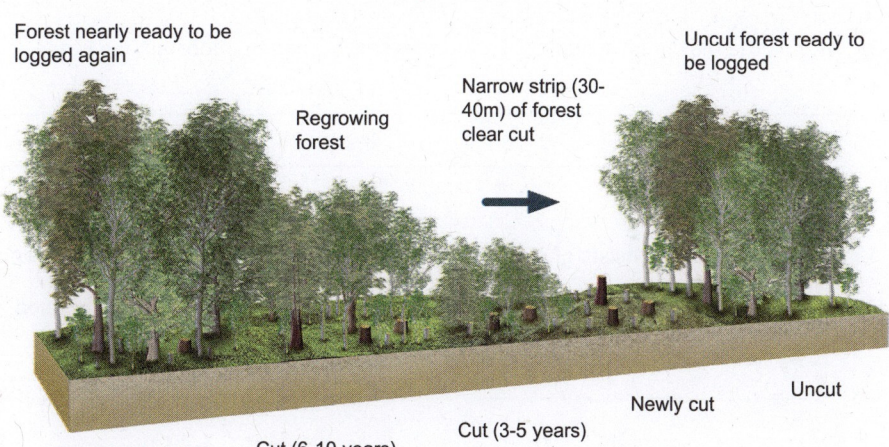

Forest nearly ready to be logged again

Regrowing forest

Narrow strip (30-40m) of forest clear cut

Uncut forest ready to be logged

Reestablished forest

Cut (6-10 years)

Cut (3-5 years)

Newly cut

Uncut

Mixed tree species in the Peruvian Amazon have been harvested in 40 year cycles. Although sustainable in an ecological sense, the slow growing trees makes this approach less economically viable, and after the first cut, there are less valuable pioneer species

1. What is the importance of sustainable harvesting from natural resources? _____

_____

_____

2. Identify advantages and disadvantages of each of the logging methods above: _____

_____

_____

_____

_____

_____

D4.2

6

How do scientists use data to measure sustainable fishing?

▸ The sustainable harvesting of any food resource requires that its rate of harvest is no more than its replacement rate. If the harvest rate is higher than the replacement rate, then it follows that the food resource will continually reduce at ever increasing percentages (assuming a constant harvest rate), and eventually be lost. Scientists can collect data and use mathematical calculations, such as maximum sustainable yield, to establish how many fish can be harvested without affecting future populations.

▸ The maximum sustainable yield (MSY) is the maximum amount of fish that can be taken without affecting the stock biomass and replacement rate. Calculating an MSY relies on obtaining precise data about a population's age structure, size, and growth rate. If the MSY is incorrectly established, unsustainable quotas may be set, and the fish stock may become depleted.

▸ Scientists often use biomass, the total weight of all fish stock, as a useful measure for the 'amount' of fish present. Biomass increases due to fish reproduction and growth rates, which can differ for each fish species. Biomass decreases due to fish death, both natural and caused by fishing.

▸ MSY biomass (B_{MSY}) is calculated at 50% of the maximum (unfished) biomass of an ecosystem and identifies the most effective amount of fish harvesting / fishing.

▸ Under ideal conditions, harvesting at this rate (B_{MSY}) should be able to continue indefinitely. However, the growth rate of a population is likely to fluctuate from year to year.

▸ If a population has below-average growth for several years while the take remains the same, there is a high risk of population collapse because an ever-increasing proportion of the population will be taken with each harvest.

1. Fishing below B_{MSY}

Less available fish. Reduced catch rates and average fish size due to fewer fish in the water. Fish size profile is altered.

2. Fishing above B_{MSY}

Larger and older fish dominate, and therefore result in less productive fish stock.

3. What is the maximum sustainable yield and why is that an important indicator of sustainable harvesting? _____

_____

_____

_____

4. A fish population consists of about 3.5 million individuals. A study shows that about 1.8 million are of breeding age.

(a) Researchers want to know the maximum sustainable yield for the population so that it can be fished sustainably. What factors will they need to know to accurately determine the MSY?

_____

_____

_____

(b) Should these smaller, non-breeding individuals be included in the catch? Explain your reasoning: _____

_____

_____

(c) It is found that the larger a breeding individual is, the more fertile it is. What implications might this have on the harvesting method for these fish and the viability of the fishery?

_____

_____

_____

_____

_____

_____

©2024 **BIOZONE** International
ISBN: 978-1-99-101410-8
Photocopying prohibited

358 Sustainable Agriculture

Key Idea: Changes to agricultural practices can help to provide a more sustainable way of farming.

Sustainable agriculture refers to farming practices that maximize the net benefit to society by meeting current and future food and material demands while maintaining ecosystem health and services. Two key issues in sustainable agriculture are biophysical and socio-economic. Biophysical issues centre on soil health and the biological processes essential to crop productivity. Socio-economic issues centre on the long-term ability of farmers to manage resources, such as labour, and obtain inputs, such as seed. Sustainable agricultural practices aim to maintain yields and improve environmental health. Crops are often grown as polycultures (more than one crop type per area), which reduces pest damage by providing a trap crop or pest confuser, e.g. planting onions in a carrot crop masks the carrots' odour and reduces damage by carrot sawfly. However, yields obtained using sustainable practices can be up to 25% lower than those obtained using intensive practices. Food needs are projected to be 50% greater by 2050 than today, so this is a major disadvantage that must be overcome, either in the management of agriculture or by society as a whole.

Factors affecting sustainable agriculture

Fertilizers
Adding artificial fertilizer into an agricultural system can increase crop production. However, excess amounts can kill soil bacteria and reduce the amount of nitrogen naturally produced.

Soil erosion
Terracing along the slope, instead of downwards reduces soil erosion by breaking long slopes into a series of shorter ones. Terraces protect water quality by intercepting agricultural runoff.

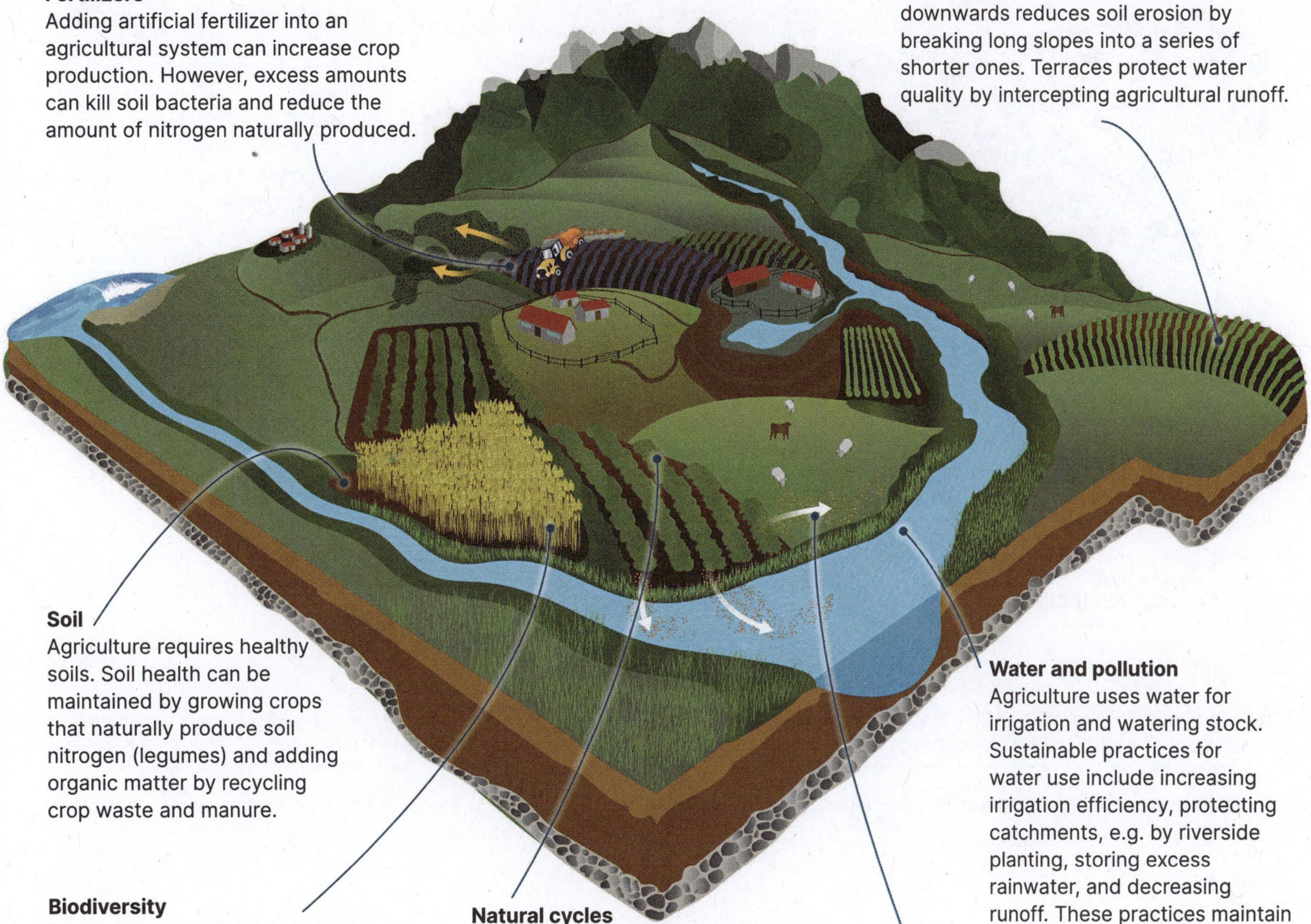

Soil
Agriculture requires healthy soils. Soil health can be maintained by growing crops that naturally produce soil nitrogen (legumes) and adding organic matter by recycling crop waste and manure.

Water and pollution
Agriculture uses water for irrigation and watering stock. Sustainable practices for water use include increasing irrigation efficiency, protecting catchments, e.g. by riverside planting, storing excess rainwater, and decreasing runoff. These practices maintain and improve water quality.

Biodiversity
Biodiversity in agriculture is important for soil, plant, and animal health. Using many different agricultural crops (rotation) or grasses in a paddock decreases the risks of pests and diseases spreading in the soil and affecting crop yield. It also reduces the need for pesticides.

Natural cycles
Sustainable agriculture matches crops with natural cycles and systems. Legumes fix nitrogen and reduce the need for applied fertilizer. Crops are grown in suitable climates, reducing the need for irrigation or pest management. Materials are recycled as much as possible to promote environmental health.

Nutrient leaching
Excess nutrients from fertilizer and plant waste can leach into the waterways and result in eutrophication. Excess algae growth can reduce dissolved oxygen and damage the ecosystem.

1. Why are sustainable agricultural practices important and what factors should be considered by farmers using this system?

_____

_____

_____

_____

©2024 **BIOZONE** International
ISBN: 978-1-99-101410-8
Photocopying prohibited

The carbon footprint and sustainable agriculture

▸ Recall that areas where carbon is stored are known as sinks or carbon pools. In agriculture, soil, plant biomass, and microbial biomass can act as sinks.

▸ Photosynthesis removes carbon from the atmosphere and converts it into organic molecules, stored in crops.

▸ This organic carbon may eventually be returned to the atmosphere as CO_2 through plant respiration, or respiration of stock or humans who eat the crops. The microbial biomass also respires. Additionally, carbon can be released from frozen soil when it thaws. Agricultural machinery is nearly always run on fossil fuels which, when combusted, adds CO_2 to the atmosphere.

▸ Stable ecosystems rely on a balanced carbon cycle. When carbon outputs in an agricultural system equal the inputs, the system has a NET zero carbon footprint. However, most unsustainable agricultural systems add to the atmospheric carbon pool.

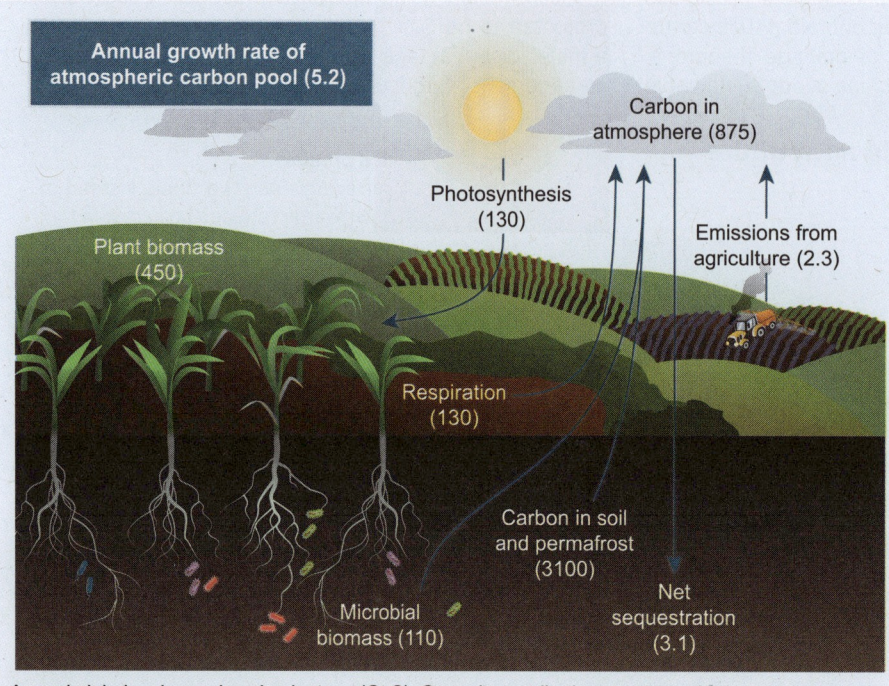

Annual global carbon values in gigatons (Gt C). Oceanic contributions to atmospheric pool not shown. Information from Trends in plant science (Hirt et al.)

How can the carbon footprint be reduced?

A reduction in the carbon footprint equates to reducing greenhouse gas emissions, specifically CO_2. Agricultural practices that decrease the use of fossil fuel powered machinery can be utilized. These include low tillage methods that keep the soil structure intact, reduction of bare soil between crops by use of a cover crop, and accurate application of fertilizer that can be reduced with rotated crops and nitrogen fixing plants. New technology is leading to greater use of electric farm vehicles and renewable energy to reduce the reliance on fossil fuelled machinery.

2. What is a carbon footprint and why is the reduction of it relevant to agriculture? _____

_____

_____

_____

3. Explain how sustainable agriculture manages each of the following resources to meet its goals of long term sustainability:

(a) Biodiversity: _____

_____

_____

_____

(b) Water: _____

_____

_____

_____

(c) Soil: _____

_____

_____

_____

©2024 **BIOZONE** International
ISBN: 978-1-99-101410-8
Photocopying prohibited

359 Eutrophication

Key Idea: Eutrophication affects water quality, damaging aquatic and marine ecosystems.

Excess nitrogen from fertilizer use can leach into groundwater and run off into surface waters. This extra nitrogen load is one of the causes of increased enrichment (eutrophication) of lakes and coastal waters. An increase in algal growth increases decomposer activity, depleting oxygen and leading to the death of fish and other aquatic organisms.

At the same time, eutrophic conditions allow undesirable species, tolerant of low oxygen, to increase in numbers. This can permanently disrupt ecosystem stability, removing food sources for many other animals. Many aquatic microorganisms also produce toxins. These can accumulate in the water, fish, and shellfish. The rate at which nitrates are added has increased faster than the rate at which nitrates are returned to the atmosphere as N_2 gas.

Eutrophication causes

▶ Eutrophication can occur naturally, but is usually the result of human activity. Discharge or runoff of leached nitrate or phosphate-containing detergents, fertilizers, or sewage into a waterway are the main causes of eutrophication. Phosphorus enrichment is contributory to freshwater eutrophication.

▶ The high nutrient levels cause excessive algal growth (an algal bloom).The algal bloom prevents sunlight penetrating far beneath the water's surface and aquatic plants (macrophytes) begin to die because they cannot photosynthesize. Oxygen levels begin to fall.

▶ Eventually, the algae die and are decomposed along with dead plants by microbes. The decomposition process uses up oxygen and the oxygen levels become low (hypoxia).

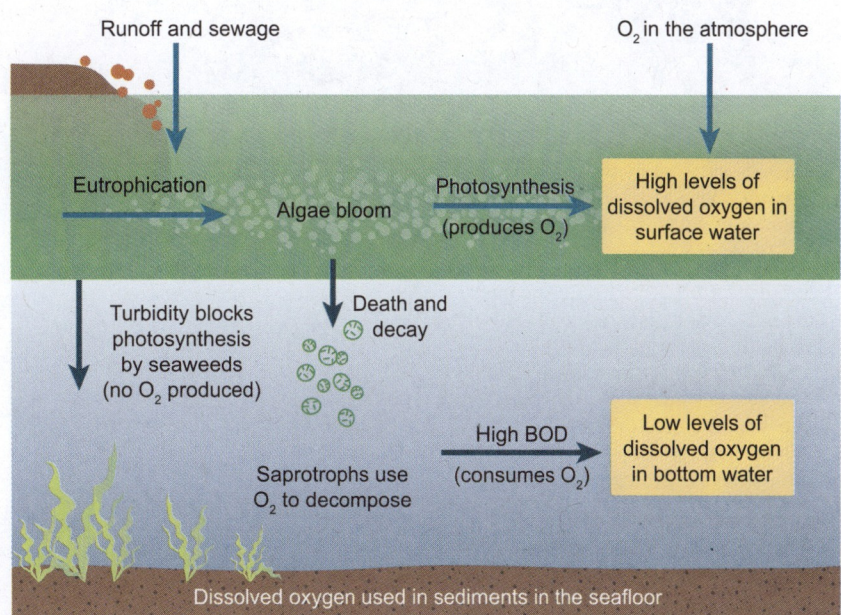

▶ The use of oxygen by decomposing microorganisms is called biochemical oxygen demand (BOD), and a high BOD deprives other organisms from oxygen, leading to the habitat becoming beyond tolerance levels, so organisms die or migrate out.

Oligotrophic waterway
(nutrient poor, high oxygen, few organisms)

Hypereutrophic waterway
(nutrient rich, low oxygen, high density of organisms)

Alexandr Trubetskoy CC 3.0

Waterways with very low primary production due to correspondingly low nutrient levels are known as oligotrophic. They are often found near the top of watershed areas, such as in mountains, and are clear, often deep, with little evidence of aquatic organisms. Nutrient rich waterways have high primary production, often murky or green with algae, and in an extreme state; hypereutropic. They are found in areas with run-off, especially from agriculture.

1. Describe the causes and consequences of eutrophication: _____

_____

_____

_____

_____

2. Describe the differences between an oligotrophic waterway and an eutrophic waterway: _____

_____

_____

_____

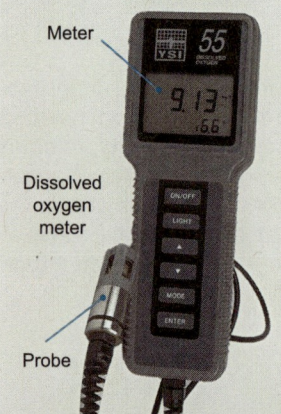

Meter

Dissolved oxygen meter

Probe

Dissolved oxygen (DO) varies with depth in a lake and from place to place in a stream or river, e.g. DO is lower above a waterfall than below it. At 20°C, the maximum DO is 9.07 mg L-1.

In order for a BOD to be significant, it needs to be compared to BODs taken from other parts of the stream or lake, and to a control. Two samples are taken at each site: one to be tested for DO immediately, the other to be tested after 5 days of storage.

If organic matter is discharged into a stream, the BOD increases markedly immediately after the point of discharge (right).

Biological Oxygen Demand

The amount of polluting organic material in a water body can be inferred by the biological oxygen demand (BOD). This is a measure of how much oxygen the organic material is using for its decomposition. BOD is measured as 'the weight (mg) of oxygen used by one litre of sample effluent stored in darkness at 20°C for 5 days'. The more oxygen that is used, the greater the bacterial activity, and (therefore) the greater the pollution. A high BOD results in less oxygen being available for aquatic organisms such as fish and invertebrates, which will die, or shift away from the polluted area.

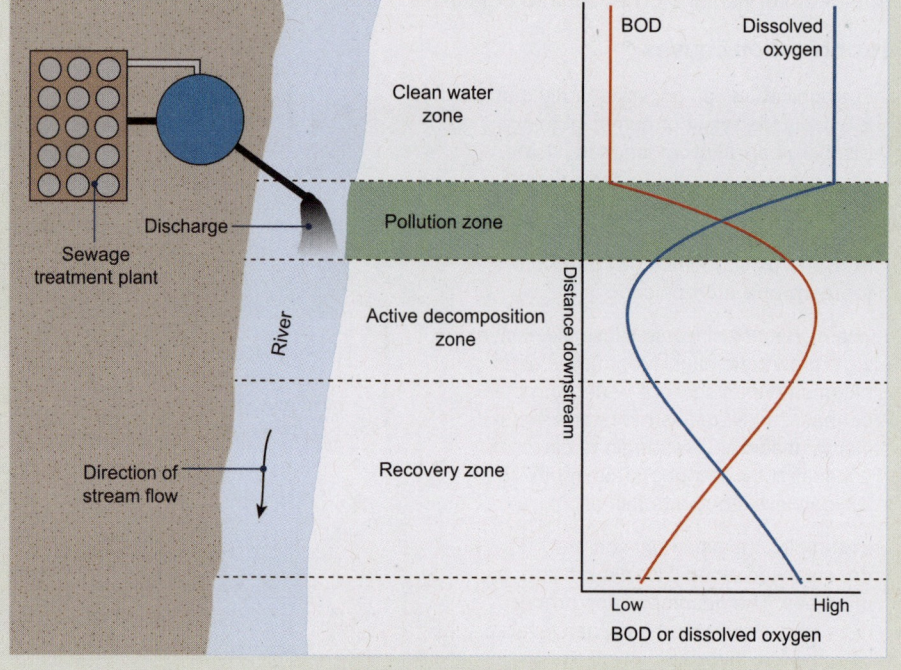

3. Discuss the link between water quality and land use. Relate to use of nitrogen and phosphate fertilizers:

_____

_____

_____

4. Explain why an increased BOD leads to disrupted ecosystems and a loss of stability: _____

_____

_____

_____

_____

5. Indicator species, such as some insects, snails, or crustaceans, have low tolerance to an increase in BOD and are often the first to die or disappear. When they are present, it 'indicates' that the water is still sufficiently oxygenated. How could an indicator species to detect pollution in a stream?

_____

_____

_____

6. How could a BOD test identify the source of a pollutant in a water system when there is no visible sign of the source?

_____

_____

_____

_____

©2024 **BIOZONE** International
ISBN: 978-1-99-101410-8
Photocopying prohibited

360 Biomagnification

Key Idea: Persistent toxins in ecosystems can lead to harmful biomagnification.

Persistent toxins such as heavy metals, e.g. mercury, and organic pesticides and industrial chemicals, e.g. DDT, resist degradation and stay in the environment for a long time. They can leach into the surrounding waterways and soil due to unsafe practices. These persistent toxins can be taken up by organisms in their food or absorbed from the surrounding environment and accumulate in their tissues. This is called bioaccumulation. Once within an organism, toxins can be passed through a food chain, becoming more concentrated at each trophic level. This is called biomagnification.

The biomagnification of a persistent organic pollutant in a food chain

▶ Persistent organic pollutants (POPs) are organic compounds that are highly resistant to being broken down through chemical, biological, or photolytic processes.

▶ DDT is a man-made (synthetic) insecticide and was first made in the 1940s. In the past, it was widely used to control insect vectors carrying diseases such as malaria and typhus, and to control agricultural insect pests.

▶ However, it soon became obvious that DDT was harming non-target organisms too. For example, in the US, DDT sprayed onto agricultural crops washed into waterways and accumulated in the fatty tissues of fish. Bald eagles became poisoned when they ate the fish and their reproduction was affected. Their eggs had thin shells and broke during incubation.

▶ Agricultural use of DDT is banned in most countries and its use is restricted to controlling disease vectors, e.g. the mosquitoes that carry the malaria parasite.

▶ Biomagnification of POPs occurs within food chains (right). The DDT accumulates in the tissues of organisms. Higher order consumers may ingest toxic levels of a chemical because they eat a large number of lower order consumers. This bioaccumulation can prove fatal to apex predators, such as eagles and polar bears.

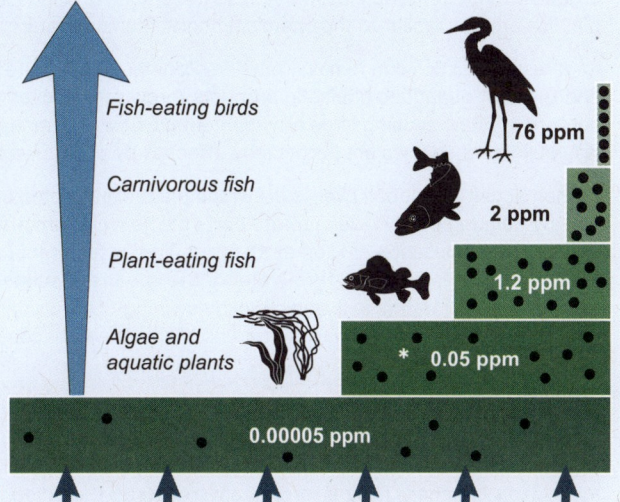

Fish-eating birds — 76 ppm
Carnivorous fish — 2 ppm
Plant-eating fish — 1.2 ppm
Algae and aquatic plants — * 0.05 ppm
0.00005 ppm

DDT enters the water as runoff from farmland sprayed with the insecticide. * ppm = parts per million

Mercury in food chains

▶ Just as POPs accumulate in tissues and increase in concentration in food chains, so too do heavy metals such as mercury. During the California gold rush of the 1800s, mercury was used to extract gold. Some mining sites still leach mercury into the Sacramento Delta and the San Francisco Bay, highlighting just how persistent mercury is.

▶ Recall that methylmercury is the most toxic form of mercury. Like POPs, methylmercury accumulates in the tissues of organisms. This occurs because it is taken in at a faster rate than it can be excreted (removed by metabolic processes) from the body.

▶ As methylmercury passes through successive trophic levels in a food chain it becomes more concentrated (biomagnification). When a predatory fish consumes smaller fish, it acquires the mercury in all those smaller fish, which themselves have acquired mercury from the organisms they ate. In this way, mercury becomes more concentrated at higher trophic levels. This also applies to humans. Eating contaminated fish and shellfish is the most common source of methylmercury for humans.

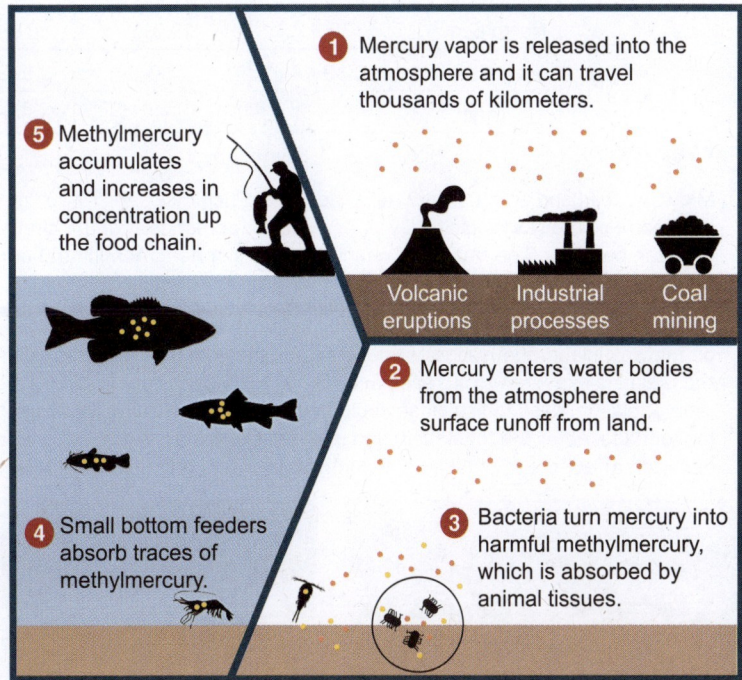

1. Mercury vapor is released into the atmosphere and it can travel thousands of kilometers.

Volcanic eruptions | Industrial processes | Coal mining

2. Mercury enters water bodies from the atmosphere and surface runoff from land.

3. Bacteria turn mercury into harmful methylmercury, which is absorbed by animal tissues.

4. Small bottom feeders absorb traces of methylmercury.

5. Methylmercury accumulates and increases in concentration up the food chain.

1. Research an example of biomagnification in an ecosystem. Prepare a short report to explain the process of pollutant biomagnification and how it has impacted the species in that ecosystem. Attach the report to this page.

 253 D4.2 9

361 Plastics in the Ocean

Key Idea: Plastic waste from human activity has detrimental impacts on the ocean ecosystems.

Plastic is a substance used in a wide range of products and much of it is discarded after it has served its purpose. Plastic makes its way to the oceans in the form of macroplastics, larger pieces; or microplastics, tiny particles broken apart by water movement. The plastic in many cases is non-biodegradable and persists in the environment permanently. Ocean currents have distributed the plastic to all coastal areas on Earth. Microplastics are ingested and enter the food chain, while macroplastics can entangle marine animals. Publicity, through media-presented scientific findings, is driving the remediation efforts to slow, and in some cases even reverse, the plastic pollution problem.

Plastic is a problem

▸ The problem with plastic is its stability. In nature, organic material is broken down by enzymes and microbes that have evolved over billions of years to deal with the chemical bonds found in nature. Very few organisms can degrade plastic because the chemical bonds in most plastics are not similar to the chemical bonds found in nature.

▸ As a result, plastics can remain in the environment for hundreds of years, and the vast quantities of plastic products thrown away over the last half century are now causing large environmental and waste management issues. Marine species are particularly affected by plastic waste.

Marine species, such as the turtle above, can mistake plastic for jellyfish and be harmed when eating it. Additionally, plastic can become entangled.

▸ Human activity is global. Every part of the planet is affected in some way by human activity. Even remote parts of the world are affected by activities thousands of kilometres away. The circulation of the oceans tends to concentrate plastic waste into certain areas of the ocean. The surface water of the oceans circulates in giant whirlpools called gyres. In the same way that you can concentrate debris in a small pool by swirling the water around, these gyres also concentrate floating debris. When this happens with floating plastic, giant areas of the ocean become plastic 'garbage patches'.

Henderson Island

▸ Henderson Island sits in the South Pacific Ocean. It is part of the Pitcairn group and is about 5000 km from the nearest significant land mass. The island is small, at only 37.3 km2, and uninhabited. A study in 2017 measured the amount of plastic on the island's beaches. The research team measured the amount of plastic already on the beach, and then cleared a control area to measure the rate at which plastic accumulated (below).

Site	Mean density on beach (items per m^2)	Rate of accumulation (items km^{-1} d^{-1})	Estimated total debris on beach (items)		Estimated island total including buried items and back beach (items)	
			Number	Mass (kg)	Number	Mass (kg)
North Beach	30.3	13,316	812,116	2985	7,634,052	4,744
East Beach	239.4	–	3,053,901	12,611	30,027, 343	12,857
Total			3,866,017	15,597	37,661,395	17,601

Lavers, JL. 2017.

Midway Atoll

▸ Midway Atoll (land area 6.2 km2) is in the North Pacific Ocean, 2,400 km west of Hawaii (the nearest significant land), and near the centre of the North Pacific Gyre. As with Henderson Island, the circulation of the ocean washes vast quantities of plastic onto the beaches. The National Oceanic and Atmospheric Administration (NOAA) regularly removes plastic debris from the beaches. Since 1999, they have removed 125 tonnes of plastic.

▸ On both these islands, and many others, plastic is mistaken for food by seabirds who eat it or feed it to their young. Every year on Midway Island, thousands of young albatrosses die from ingesting plastic products. The proportion of plastic deposited in the ocean is expected to increase at a rate of 4.8% each year until 2025, and at a rate of 3% from 2025 to 2050. At this rate, the ratio of plastic mass to fish mass could be roughly 1:1 by 2050. Microplastics and nanoplastics, tiny pieces of plastic, including microbeads, enter the food chain and concentrate toxins, which accumulate in the fish that eat them. Nanoplastics can enter cells and affect their functioning. Scientists have recently discovered nanoplastics suspended in clouds.

Plastic in dead albatross chick, Midway Atoll.

Plastic debris on beach, Henderson Island.

Discarded nets can trap and drown marine mammals, reptiles, and birds (above).

NOS

D4.2
10

©2024 **BIOZONE** International
ISBN: 978-1-99-101410-8
Photocopying prohibited

Plastic pollution and the media

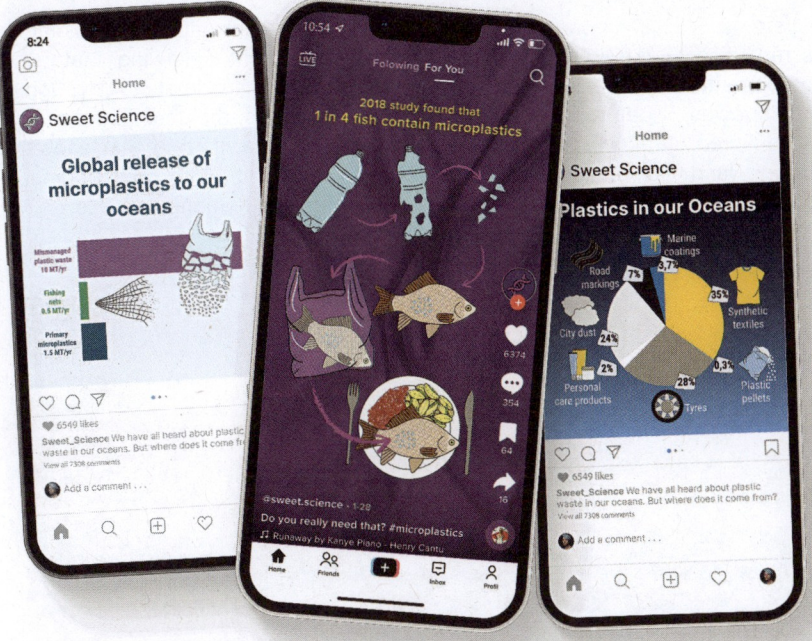

▶ In 2023, over 55% of the world's population has access to social media platforms. Environmental agencies (NGOs) have turned towards social media as a means to spread awareness about plastic waste and pollution in the hopes of spurring action from citizens.

▶ Research has shown that, if claims are linked to a confirmed study demonstrating that plastic pollution can be attributed to lifestyle choices, and possible manageable solutions are provided, then social media posts can be an effective method to elicit behavioural changes in people towards reducing plastic use and recycling.

▶ Social media posts have the advantage of delivering targeted messages, sharing the media between people and groups, and allowing a personal contribution with the chance to comment.

1. Why do plastics persist in the environment? _____

_____

_____

2. Some islands in the middle of the North Pacific Ocean are very isolated, yet they have a massive problem with plastic washing up on the beaches. Explain how this happens:

_____

_____

_____

_____

3. Describe some potential problems created by plastic waste in the oceans: _____

_____

_____

_____

_____

4. What effect do micro and nanoplastics have on the animals that ingest them? _____

_____

_____

5. Scientists and conservation organizations are using social media to inform citizens about the problems created by plastic and microplastic pollution. What is your perspective on the effectiveness of these messages and how might that encourage you to take action against this problem?

_____

_____

_____

_____

_____

_____

362 Rewilding

Key Idea: Rewilding is a method of restoring ecosystems.
An important tool in ecosystem conservation is rewilding. Damaged ecosystems can be restored if biotic components, such as apex predators, keystone species, and plant species are re-introduced to areas, and pest species are removed. Disjointed ecosystems can be reconnected through land bridges or restoration of the land between them. Once the components of the ecosystem are introduced, rewilding requires minimal human involvement, allowing natural processes to return stability. Human impact, through deforestation, hunting, or farming, is the main cause of ecosystem damage, but rewilding is a method that shows even the most depleted areas can return to their original state. Numerous successful examples demonstrate this.

Reintroduction of blue wildebeest into the Serengeti

Blue wildebeest in the Serengeti, Tanzania, were decimated by viruses introduced by domestic cattle and were reduced to fewer than 300,000 individuals by the mid-20th century. Overgrown grasslands became a wildfire hazard, with up to 80% of them destroyed by fire each year. The small numbers of wildebeest could not support sufficient apex predators, and the food web collapsed. Rewilding of the ecosystem began with disease control in the wildebeest, restoring their numbers up to 1.5 million in less than a decade. Grasslands were once more utilized by the wildebeest, sometimes coined the 'lawnmowers' of the Serengeti for their ability to maintain the vegetation in a healthy condition, allowing other herbivore species to increase in number as well. With apex predators, such as lions, returning, as well as scavengers like hyenas, the Serengeti regained ecosystem stability.

Hinewai reserve in New Zealand

▸ Hinewai reserve is situated on Banks Peninsula, near Christchurch, New Zealand. What was once a small, 100 hectare block of native forest surrounded by bare farmland is now a 1250 hectare rewilded reserve, due to reconnecting smaller blocks of land into one.

▸ Although some pest species, such as deer, pigs, and goats, are under removal management, the reserve has effectively been left to regenerate without human interference. Gorse species, as seen in the lower left of the photo, are not native plant species, but have been left in place to act as a nurse canopy for growing tree saplings, as natural succession occurs.

▸ As the forest is regenerating, the native bird, reptile, and fish species returning. (You may recall that New Zealand has no native mammal species, aside from two species of bat, so all mammals are considered conservation pests).

▸ Hinewai reserve is an example of rewilding in that, if sufficient land is set aside and native species can access the area, and it remains undisturbed by human impacts, then natural processes can allow a healthy ecosystem to be restored.

1. What are the key features of rewilding conservation? _____

_____

2. There are many examples of successful rewilding around the world, including: bison reintroduction in the USA Midwest plains and in the Southern Carpathian mountains in Romania; Grey wolves returned to Yellowstone (see activity 356), trout reintroduced to South London Wandle river; Eurasian beavers being returned to UK, from where they had previous become extinct; Arctic muskoxen; and Tasmanian devils in Australia. Select one of the listed examples or one of choice as the focus of a short report. Detail the rewilding methods used and the subsequent impacts on the ecosystem.

D4.2

11

363 Primary Succession

Key Idea: Succession is the community changes that occur in an ecosystem as diversity returns to a natural state. Ecological succession is a natural process of continuous, sequential change in an ecological community. It usually occurs in response to a disturbance and is the result of the dynamic interactions between biotic and abiotic factors over time. Earlier communities modify the physical environment, making it more favourable for the species that will make up later communities. Over time, a succession may result in a stable, mature, or climax, community, although this is not always the case. Succession occurring where there is no pre-existing vegetation or soil is called primary succession.

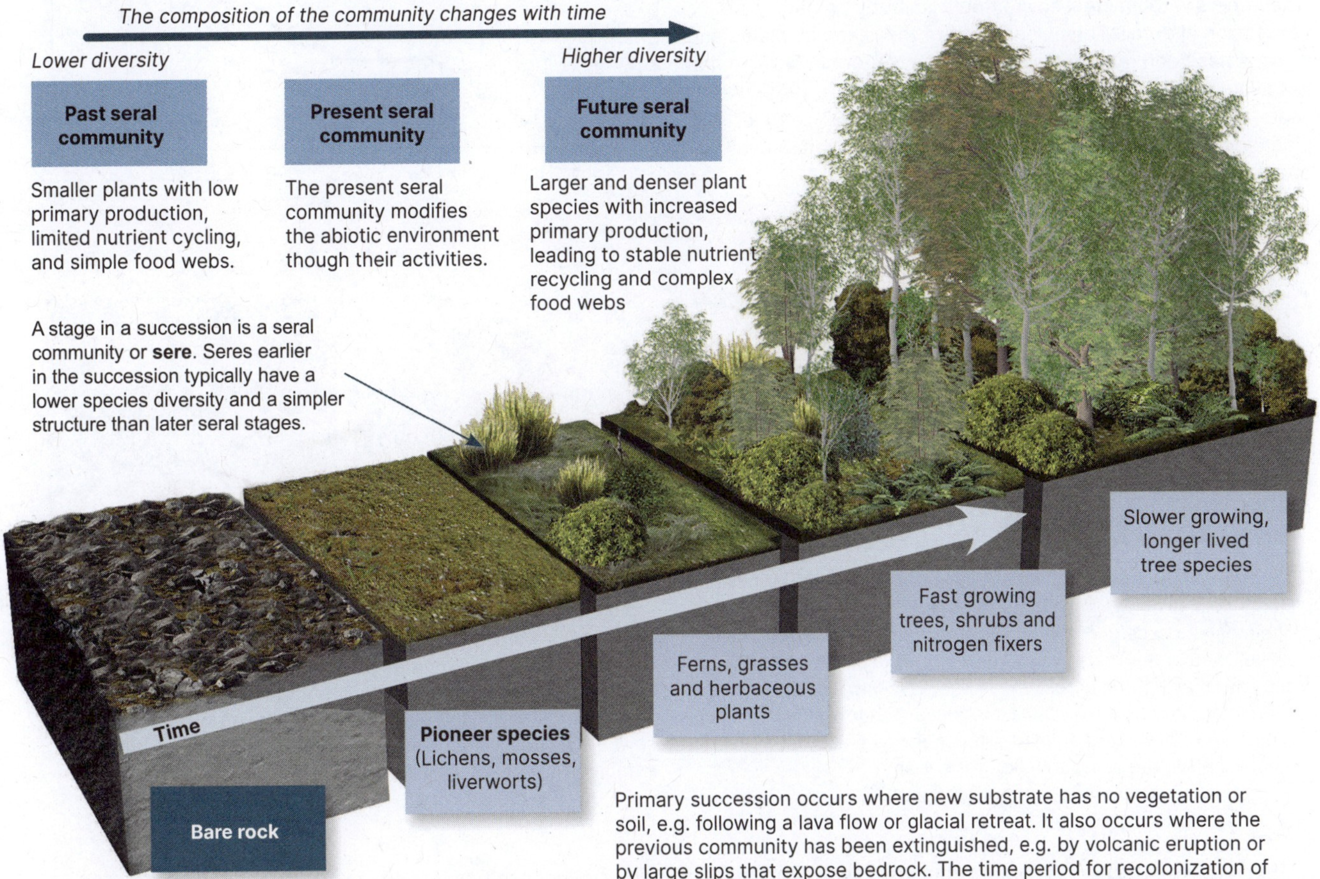

The composition of the community changes with time

Lower diversity → *Higher diversity*

Past seral community

Smaller plants with low primary production, limited nutrient cycling, and simple food webs.

Present seral community

The present seral community modifies the abiotic environment though their activities.

Future seral community

Larger and denser plant species with increased primary production, leading to stable nutrient recycling and complex food webs

A stage in a succession is a seral community or **sere**. Seres earlier in the succession typically have a lower species diversity and a simpler structure than later seral stages.

Slower growing, longer lived tree species

Fast growing trees, shrubs and nitrogen fixers

Ferns, grasses and herbaceous plants

Pioneer species (Lichens, mosses, liverworts)

Time

Bare rock

Primary succession occurs where new substrate has no vegetation or soil, e.g. following a lava flow or glacial retreat. It also occurs where the previous community has been extinguished, e.g. by volcanic eruption or by large slips that expose bedrock. The time period for recolonization of the area and the composition of the final community depends on the local environment. Recovery is quicker when vegetation is nearby.

Features of pioneer species

The earliest pioneer species are microorganisms, e.g. cyanobacteria, and simple photosynthetic plants and algae. They are able to survive on exposed substrates lacking in nutrients and make their own food using sunlight energy. Even at this level, ecological associations are important. Lichens, which are important pioneers, are a symbiosis between fungi and algae. Associations between mosses and cyanobacteria (which can fix nitrogen) are also important. Pioneers begin the process of soil formation by breaking down the substrate and adding organic matter through their own death and decay. Their growth therefore creates a more favourable environment for vascular plant growth.

Lichen on bare rock

Moss on bare rock

Bob Blaylock CC 4.0

Note the vascular plants establishing in the crevices where soil is forming.

Associations between mosses and cyano-bacteria provides mosses with nitrogen.

1. Describe situations in which a primary succession is likely to occur: _____

_____

_____

2. (a) Identify pioneers during the colonization of bare rock: _____

(b) Describe two important roles of species that are early colonizers of bare slopes: _____

_____

_____

D4.2
12 - 13

AHL

Surtsey: A case study in primary succession

Surtsey Island is a volcanic island lying 33 km off the southern coast of Iceland. The island was formed over four years from 1963 to 1967 when a submarine volcano 130 m below the ocean surface built up an island that initially reached 174 m above sea level and covered 2.7 km2. Erosion has since reduced the island to around 150 m above sea level and 1.4 km^2.

As an entirely new island, Surtsey was able to provide researchers with an ideal environment to study primary succession in detail. The colonization of the island by plants and animals has been recorded since the island's formation. The first vascular plant there (sea rocket) was discovered in 1965, two years before the eruptions on the island ended. Since then, 69 plant species have colonized the island and there are a number of established seabird colonies.

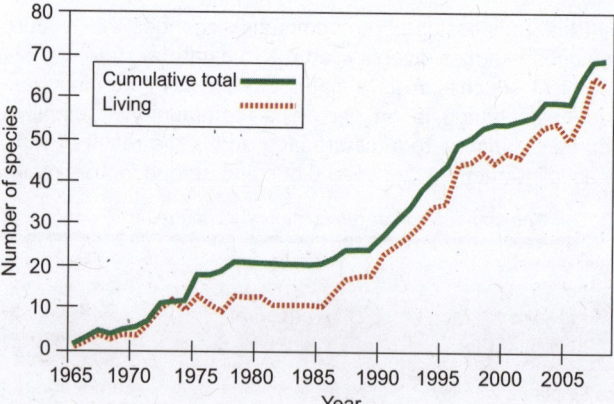

Number of vascular plant species found on Surtsey

Sea rocket

H. peploides

P. annua

S. phylicifolia

The first stage of colonization on Surtsey was dominated by shore plants colonizing the northern shores, brought by ocean currents. The most successful of these was *Honckenya peploides*, which established on tephra sand and gravel flats. It set seed in 1971 and subsequently spread across the island. This initial colonization by shore plants was followed by a lag phase with few new colonizers. A number of new plant species arrived after a gull colony became established at the southern end of the island.

Populations of plants within or near the gull colony expanded rapidly to about 3 ha, while populations outside the colony remained low but stable. Grasses such as *Poa annua* formed extensive patches of vegetation. After this rapid increase in plant diversity, the arrival of new colonizers again slowed. A third wave of colonizers began to establish following this slower phase and soil organic matter increased markedly. The first bushy plants established in 1998, with the arrival of willow, *Salix phylicifolia*.

3. Explain why Surtsey provided ideal conditions for studying primary succession: _____

_____

_____

4. Why did the first colonizing plants establish in the north of the island but later colonizers establish in the south?

_____

_____

_____

5. There are three distinct phases on Surtsey where species richness increased rapidly.

   (a) Label the graph to indicate the three phases of rapid increase in species richness on Surtsey.

   (b) Label the two lag phases where species richness increased slowly.

6. A gull colony established on the island in 1985. How did this affect the number of plant species on the island?

_____

_____

7. Why is the living number of plant species on the island less than the cumulative number colonizing the island?

_____

_____

_____

©2024 **BIOZONE** International
ISBN: 978-1-99-101410-8
Photocopying prohibited

364 Cyclical and Arrested Succession

Key Idea: Some ecosystems experience continual cyclical change, while human impact can permanently disrupt others. Some communities are constantly changing and experience cyclical succession, where the repeated removal or appearance of species, due to natural cycles, such as fires or changes in lifestyle stages in animals, initiates repeated succession to a climax community. Climax communities are ecosystems that have reached a steady and stable equilibrium state, and are seen at the terminal or end stage of succession. Some ecosystems reach a climax community, but due to human impacts, the final state is quite different from the original state, as seen when farmed grasslands replace deforested areas, or drainage of wetlands takes place. This is known as arrested succession.

Cyclical succession

▶ Some events, such as natural wildfires, or death of old trees in forests, can change the composition of a community and may occur in repeated cycles. This cyclical succession can be in varying lengths depending on the community.

▶ Examples of cyclic succession include regular wildfires in the chaparral ecosystem in coastal California which occur naturally every 30-150 years. After each fire, a succession of different vegetation species replace each other in the same order. In the Sonoran desert in Southwestern US and Mexico, two plants, cholla and creosote, grow together, but cyclically replace each other with assistance from animals dispersing seeds. Regular wildfires are also common in parts of Australia.

The rate of succession depends on the type of forest and the circumstances that cleared the land. For example, full recover from forest fire in Australian eucalypt woodland can be extremely rapid (within a few years) whereas recovery from fire in tropical rainforest may take decades

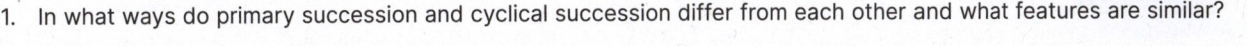

Time

Fire removes plants above ground

Primarily bare earth

Grasses and low growing perennials

Seedlings regrow and grasses recover

Scrub and small trees. Surviving trees regrow

Mature woodland

1. In what ways do primary succession and cyclical succession differ from each other and what features are similar?

_____

_____

2. Research another example of cyclic succession and provide a summary of the cause and species involved:

_____

_____

_____

_____

©2024 **BIOZONE** International
ISBN: 978-1-99-101410-8

 368 →
 D4.2 | 14 - 15
 AHL

575

Arrested succession

Human impact (and sometimes nature) may deflect or arrest the natural course of succession, e.g. by mowing or fire, and the climax community that results will differ from the community that would occur if there had been no disturbance.

Time

Mature or developing forest land

Forest is felled. Low scrub and grasses begin to regenerate.

Livestock eat saplings and scrub, so grasses are not over-topped.

Grasses spread and become the dominant vegetation.

A climax community arising from arrested succession is called a plagioclimax.

Mammoths and arrested succession

▸ Woolly mammoths belonged to the same family as modern Asian and African elephants. They lived on Earth from about 300,000 years ago to around 10,000 years ago. Alongside other large, grazing herbivores, they occupied an ecosystem of treeless grasslands. In winter, they scraped off snow with their tusks, grazing and trampling the grassland. This maintained the landscape, keeping the ground compacted and frozen and preventing trees and shrubs from establishing.

▸ Evidence suggests human hunting activity, in conjunction with climatic warming, may have contributed to the extinction of the large grazers, including mammoths, in these grasslands. Without the trampling effect of the grazers, the ground grew softer and other plants were able to establish themselves. What was formerly grassland changed such that small shrubs and trees grew. The ground began to thaw, melting the permafrost cover, and changing the ecosystem permanently in arrested succession.

3. Wetland areas present a special case of ecological succession. They are constantly changing as plant invasion of open water leads to silting and infilling. In well drained areas, pasture or heath may develop as a result of succession from freshwater to dry land. When the soil conditions remain non-acidic and poorly drained, a swamp will eventually develop into a seasonally dry fen. In special circumstances, an acid peat bog may develop. The domes of peat produce a hummocky landscape with a unique biota. Wetland peat ecosystems may take more than 5000 years to form but are easily destroyed by excavation and lowering of the water table.

(a) Why does drainage of wetland peat systems nearly always result in arrested succession? _____

_____

_____

(b) Rewilding is a conservation method discussed previous that can return damaged ecosystems to their original state. Why is this method unlikely to be successful in wetland peat ecosystems?

_____

_____

_____

©2024 BIOZONE International
ISBN: 978-1-99-101410-8
Photocopying prohibited

365 Climate Change: Causes and Tipping Points

Key Idea: Climate change is caused by many factors including some human activities.

Climate scientists can measure trapped CO_2 in polar ice cores to determine concentrations prior to modern recording. They have found that the concentration of CO_2 has cycled relatively consistently over the last several hundred thousand years. The steep rise in CO_2 concentration over the last 60 years has been attributed to the burning of fossil fuels. Methane, like CO_2, is a greenhouse gas and has shown similar rises in atmospheric concentration although it has a lesser effect on temperature rise. Greenhouse gas levels are increasingly close to pushing physical systems to their tipping points, where 'runaway' positive feedback cycles begin to accelerate the effects caused by climate change.

Changes in atmospheric CO_2 and global temperature

Earth's climate has varied considerably when viewed over the long term. Frequent ice ages consisting of glacials and interglacials have occurred. Carbon dioxide and other greenhouse gases play a part in climate variation, helping to either trigger changes or enhance an effect already underway. Studies of gas trapped in ice cores taken from polar ice caps have helped reveal these climatic changes. For the last 50 years or so, a network of observatories around the world has constantly measured the concentration of CO_2 in the atmosphere.

In November 2023, the CO_2 levels were at 416 ppm. These concentration levels have not occurred since the Pliocene, about 3-5 million years ago, when global temperatures were at least 4 C° higher than today.

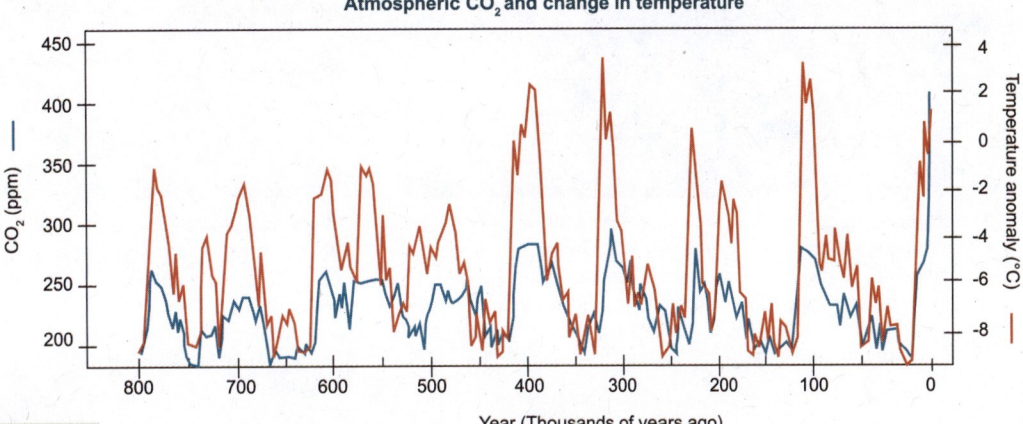

Atmospheric CO_2 and change in temperature

Correlation or causation

Greenhouse gas (GG) levels in the atmosphere, primarily CO_2, have risen rapidly since the middle of the 20th century. The average temperature on Earth has also risen almost 1.5 C° since that time. This is an example of a positive correlation.

Correlation does not always mean causation so, in order to establish a causation claim between these two factors, further evidence was required. This included the identification of the increased atmospheric CO_2 molecules as being those from fossil fuel combustion and the scientific principle of GG creating radiative forcing. Currently, with multiple lines of evidence from different fields, 97% of climate scientists link causation of climate change to anthropomorphic GG emissions.

The warming influence of CO_2 and methane acting as greenhouse gases

Greenhouse gases in the atmosphere cause radiative forcing, leading to heating. A balanced energy budget on Earth requires energy outgoing to equal the energy incoming (0 watts). The increase in greenhouse gases from anthropogenic causes results in 'extra' energy being retained - held by the greenhouse gas molecules - and making Earth's energy budget unbalanced. The result is incremental increases in temperature, leading to climate change.

Energy retained is measured in watts per square metre (vertical axis). [net energy gain]

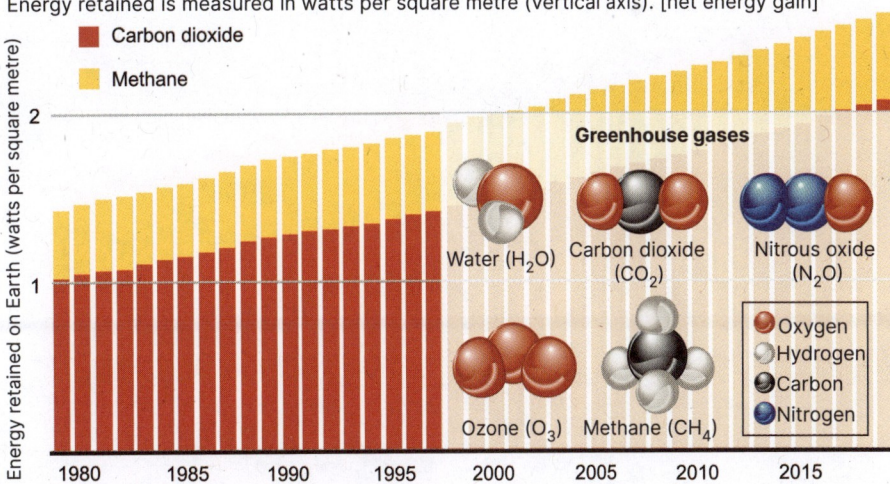

1. How can scientists be confident that a rise in greenhouse gases, leading to climate change, is due to anthropomorphic activity and is the causation of global warming and how is this different from a simple positive correlation?

_____

_____

_____

_____

_____

Positive feedback cycles are accelerating global warming

▶ Typically, feedback cycles allow physical systems to adjust to change to maintain stability. However, positive feedback cycles in global warming accelerate the change, leading to increasing non-linear temperature rise. Many positive feedback loops contributing to the increase in CO_2 and CH_4 gases in the atmosphere, and hence climate change, have been identified.

▶ Some positive feedback loops impact others, that is they are interconnected, and an increase in effects in one loop will increase impacts in the loop of another. For example, the albedo effect increases heat absorbed, both on land, increasing methane release from permafrost, and also under the oceans, increasing methane and carbon dioxide release from previously frozen clathrates and hydrates, respectively.

Icebreaker in melting sea ice

The Albedo effect: The high albedo (reflectivity) of sea-ice helps to maintain its presence. As sea-ice retreats, more non-reflective surfaces are exposed. Heat is absorbed instead of reflected, warming the air and water and causing sea-ice to form later in the fall than usual. Thinner and less reflective ice forms, continuing the cycle.

Water evaporation from lake surface

Water vapour release: Water vapour adds to the greenhouse effect in the atmosphere, trapping heat energy, although it has a shorter lifespan than other gases. Warmer temperatures result in higher rates of evaporation from water on Earth's surface, therefore increasing the amount of atmospheric water vapour. This energized water cycle can lead to intensified weather events.

The link between ocean temperature and CO_2 release

▶ Colder, polar oceans can absorb more CO_2 than warmer oceans. Since pre-industrial times, the oceans have absorbed up to 30% of total anthropogenic CO_2 emissions, reducing the impacts of climate change but resulting in their increased acidification. Warming oceans hold less CO_2 and become net CO_2 emitters rather than net carbon sinks. Additionally, the increasing temperature reduces the mixing of ocean waters, so acidified water remains trapped under a warmer band of water, reducing nutrient and oxygen mixing.

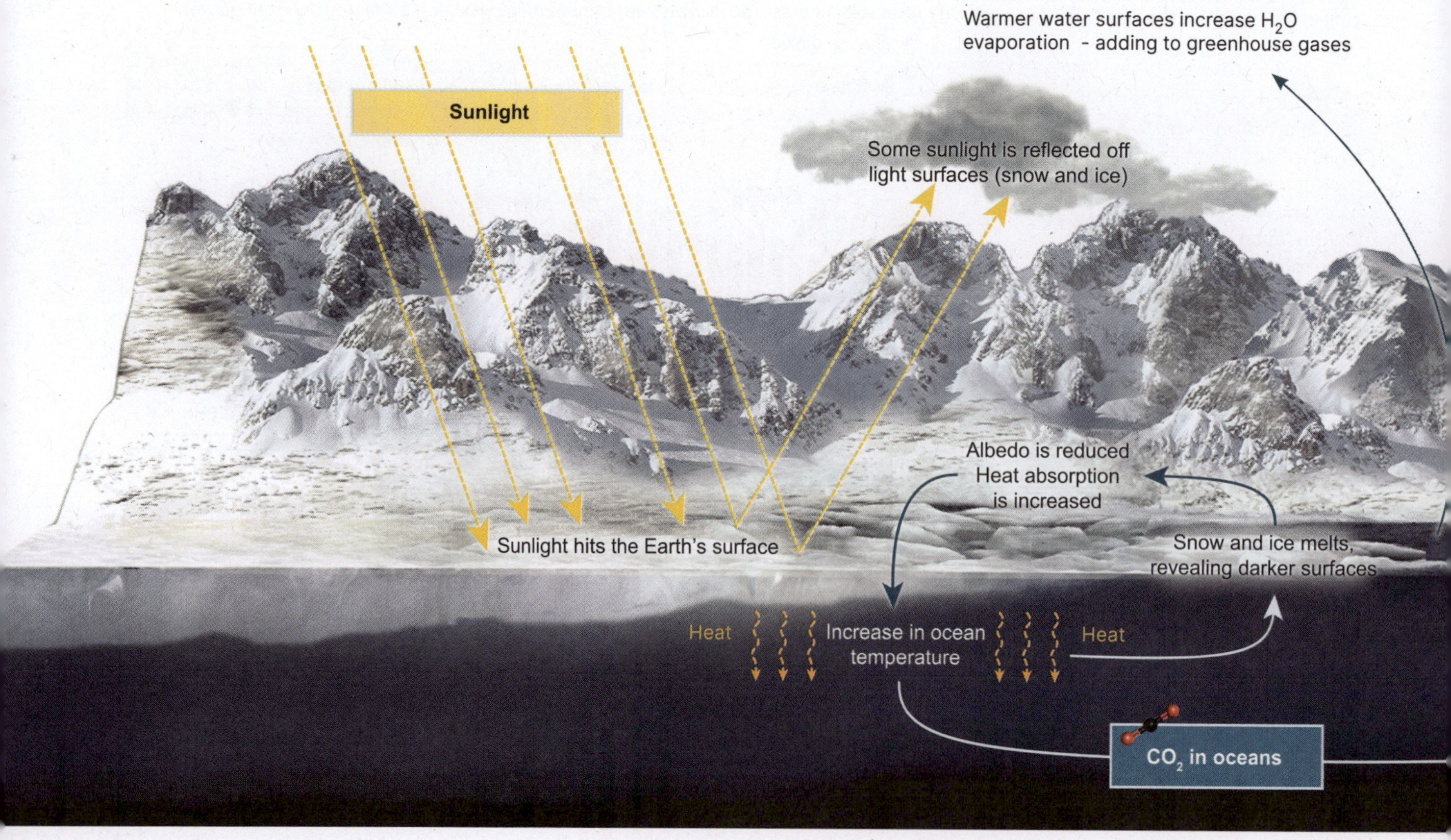

Warmer water surfaces increase H_2O evaporation - adding to greenhouse gases

Sunlight

Some sunlight is reflected off light surfaces (snow and ice)

Sunlight hits the Earth's surface

Albedo is reduced Heat absorption is increased

Snow and ice melts, revealing darker surfaces

Heat

Increase in ocean temperature

Heat

CO_2 in oceans

Gas hydrate under carbonate rock on ocean floor

Methane ice bubbles under Spray Lake, Canada

Carbon dioxide release from deep oceans: Carbon is transported to ocean floor sediments via dead microorganisms. It is stored as CO_2 in hydrates, under pressure and in a near freezing state. Warming oceans create instability in the hydrates and release the CO_2 into the ocean, eventually releasing it from the ocean surface.

Methane release from permafrost: Large amounts of methane are stored in frozen ground in polar regions and as clathrates under the ocean. Increased warming is thawing long-frozen permafrost, enabling a chemical process called methogenesis, which ultimately releases methane gas to the atmosphere. Thawing clathrates directly release methane gas.

The greenhouse effect

The Earth's atmosphere comprises a mix of gases including nitrogen, oxygen, and water vapour. Small quantities of carbon dioxide (CO_2), methane, and a number of other trace gases are also present. Together, water, CO_2, and methane produce a greenhouse effect that moderates the surface temperature of the Earth.

The term greenhouse effect describes the natural process by which heat is retained within the atmosphere by these greenhouse gases letting in sunlight, but trapping the heat that would normally radiate back into space. The greenhouse effect results in the Earth having a mean surface temperature of about 15°C, 33°C warmer than it would have without an atmosphere. About 75% of the natural greenhouse effect is due to water vapour. The next most significant agent is CO_2.

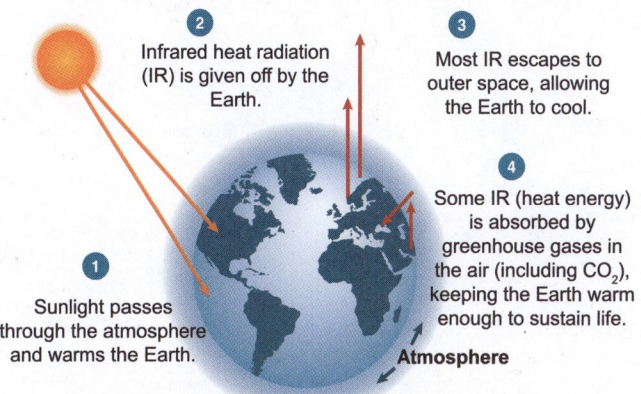

2 Infrared heat radiation (IR) is given off by the Earth.

3 Most IR escapes to outer space, allowing the Earth to cool.

1 Sunlight passes through the atmosphere and warms the Earth.

4 Some IR (heat energy) is absorbed by greenhouse gases in the air (including CO_2), keeping the Earth warm enough to sustain life.

Atmosphere

5 **Enhanced greenhouse effect:** Increasing levels of CO_2 increase the amount of heat retained, causing the atmosphere and Earth's surface to heat up.

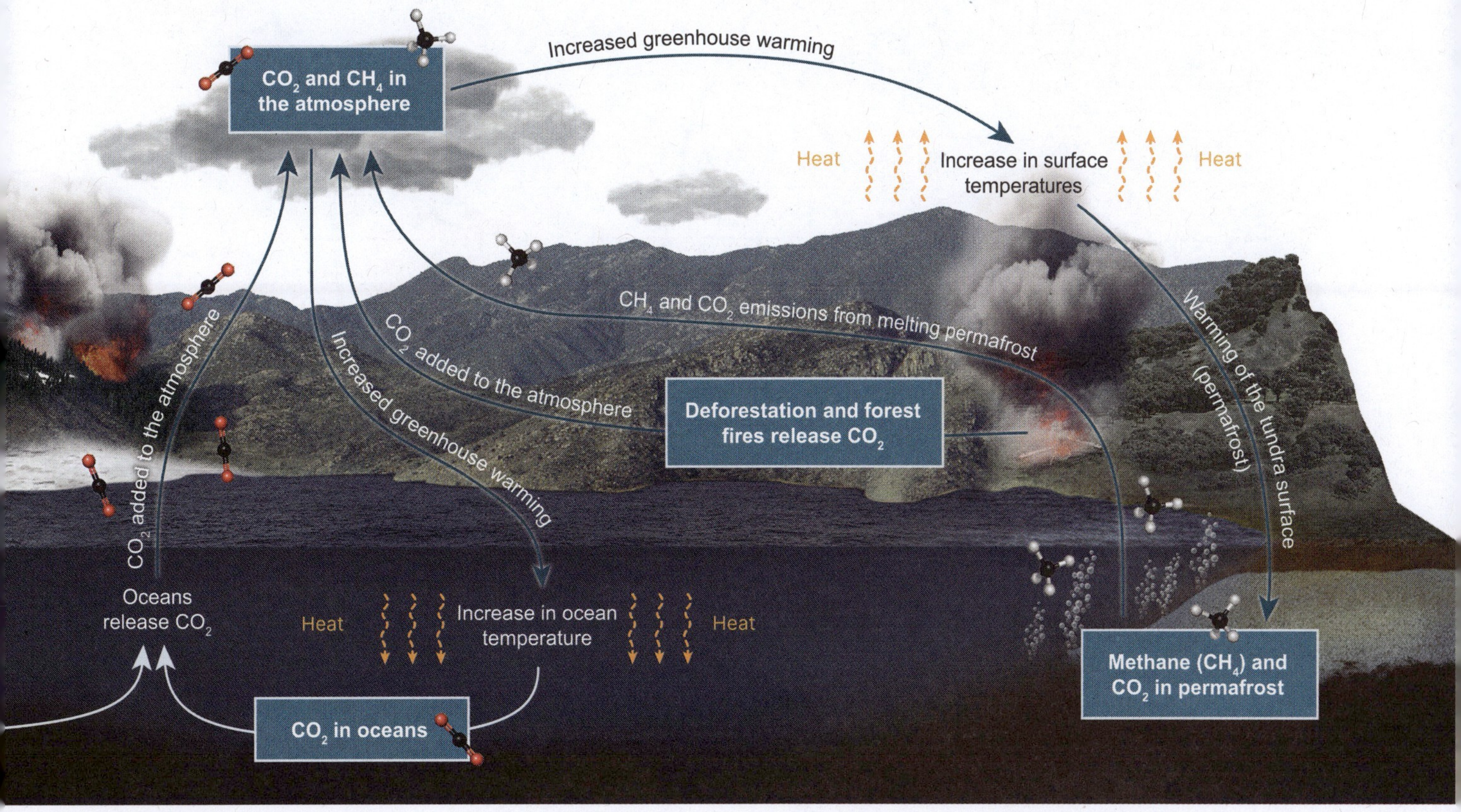

CO_2 and CH_4 in the atmosphere

Increased greenhouse warming

Heat — Increase in surface temperatures — Heat

CO_2 added to the atmosphere

Increased greenhouse warming

CH_4 and CO_2 emissions from melting permafrost

Warming of the tundra surface (permafrost)

Deforestation and forest fires release CO_2

Oceans release CO_2

Heat — Increase in ocean temperature — Heat

CO_2 in oceans

Methane (CH_4) and CO_2 in permafrost

Tipping points and climate change

▶ Ecological systems have mechanisms to adjust to change to maintain stability. However, when the changes become too great, the system can rapidly shift from one state to another. This phenomenon is known as a tipping point, and the changes become self-reinforcing. Many scientists, groups, and governments are working to avoid reaching the point of no return.

▶ Climate change effects are already noticeable, with a worldwide average global temperature rise of about 1.2 °C since pre-industrial times. The Intergovernmental Panel on Climate Change (IPCC) publishes regular, in-depth reports on the causes, mitigation, and adaptation responses to climate change, and recommends that the worst of the effects can be prevented by limiting global warming to 1.5 °C. Governments meet at yearly UN climate change conferences to discuss and sign accords (agreements or letters of intent) for how they can reduce emissions to contribute to global reductions, and meet the Paris Agreement - an international climate change treaty.

1. Greenland ice sheet

The albedo effect, exposing dark rock of Greenland's bare surface, and the loss of glaciers around the coast are causing positive feedback loops. Scientists estimate a rise of 2.7 °C will shift Greenland into an irreversible tipping point of ice melt. Total lost of ice from Greenland will cause a sea rise of over 7m over the entire Earth.

10. Thermoline current

The Atlantic current (AMOC) circulates warm water north up the Eastern coast of USA as the gulf stream and maintains the temperate climate of Great Britain, and Western European countries. An increase in fresh water from melting Greenland icesheets slows this ocean current. A collapsing Gulf Stream current will result in a colder climate in Northern Europe.

11. Sahel

The West Africa monsoon (WAM) brings regular heavy rain to areas in a band south of the Sahara. Warm sea surface temperatures (SSTS) influence the variability of timing of the occurrence of the WAM and any disruption can result in widespread drought, as rainfall shifts southward. The WAM system is sensitive and a change in the Atlantic current (AMOC) can also add to destabilization.

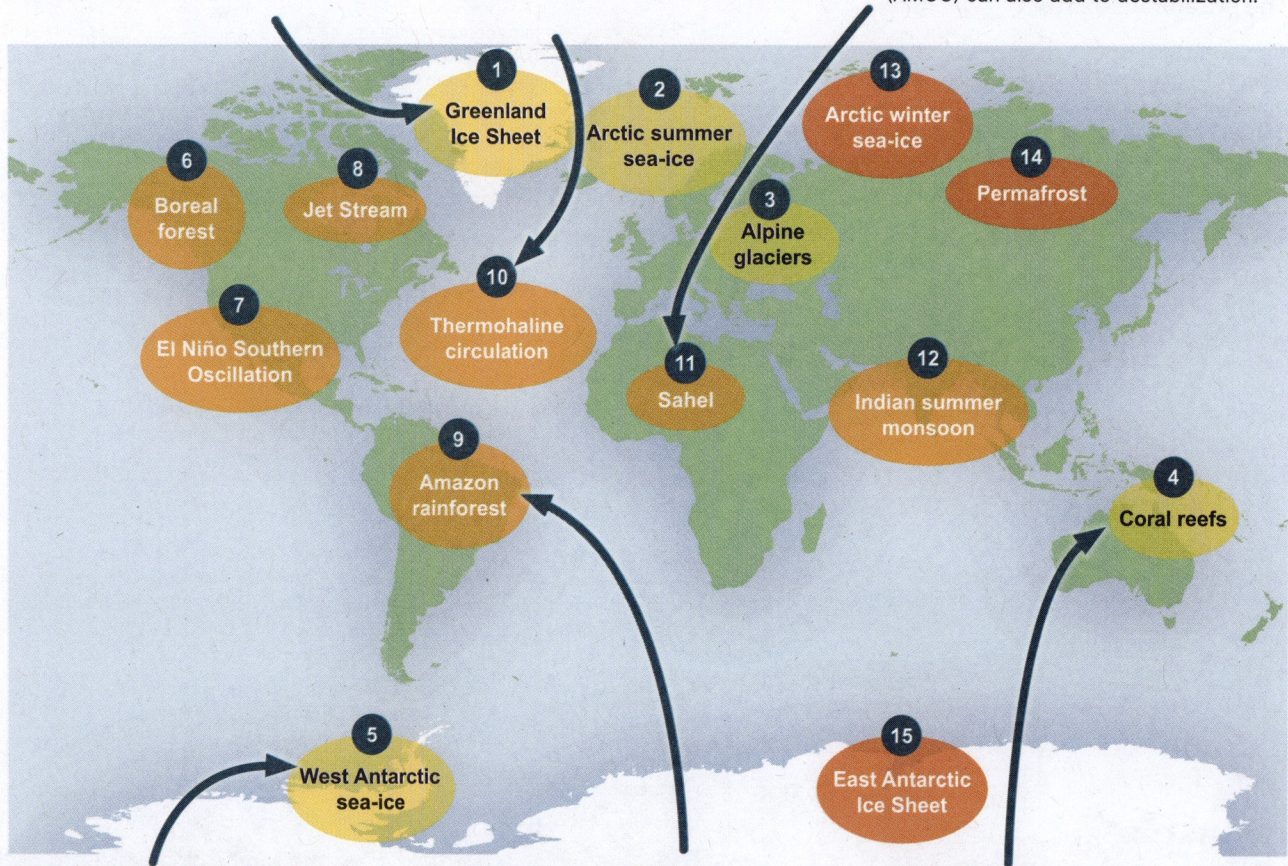

5. West Antarctic icesheet

This icesheet holds enough water to raise sea level by around 3.3 metres if it is completely lost. The bedrock of this icesheet sits below sealevel, making it especially vulnerable to rising ocean temperatures and warming sea breezes. The thinning ice sheet allows a retreating grounding line (where the ice attaches to bedrock) and then accelerates loss due to ice sheet flow breaking off. (see Activity 366).

9. Amazon rainforest

Scientists predict that a deforestation of around 25% of the Amazon rainforest will lead to a tipping point. By 2023, around 15-17% had been lost. Removal of timber for agriculture and export is the main cause of deforestation currently, but rising global temperatures are leading to less rainfall and death of parts of the forest due to drought. This results in lower rainfall in other areas of the forest (see Activity 354).

4. Coral reefs

Many coral reefs depend on a very narrow temperature range in order to sustain their mutualistic relationship with residing photosynthetic protists. Scientists have estimated that coral reefs take on average 10 years to recover from an ocean heat wave, but heat wave events are occurring more than once every 8 years. A 2 °C rise would cause up to 99% of coral reefs to become extinct. (see Activity 367).

2. Create a small report by selecting one of the numbered tipping points above. Research the feedback cycle involved in the tipping point, the predicted warming (in °C rise) that scientists estimate the tipping point will be reached, and the consequences of the tipping point being reached. You may wish to share your findings, either as a presentation, or digitally, so everyone in class has access to information on all tipping points.

Increases in droughts and boreal forest fires

Evidence shows that increasing temperatures leading to climate change are creating more prolonged and severe droughts in areas around the world. Forests act as a carbon sink, storing huge amounts within their plant material. When under drought conditions, vegetation and dead matter on the forest floor become drier which increases the risk of wildfires. When wildfires occur, combustion of the plant material releases large amounts of CO_2 into the atmosphere.

Soot and dust from the wildfire can also reduce albedo if near ice and snow.

These phenomena are also examples of positive feedback loops driving climate change.

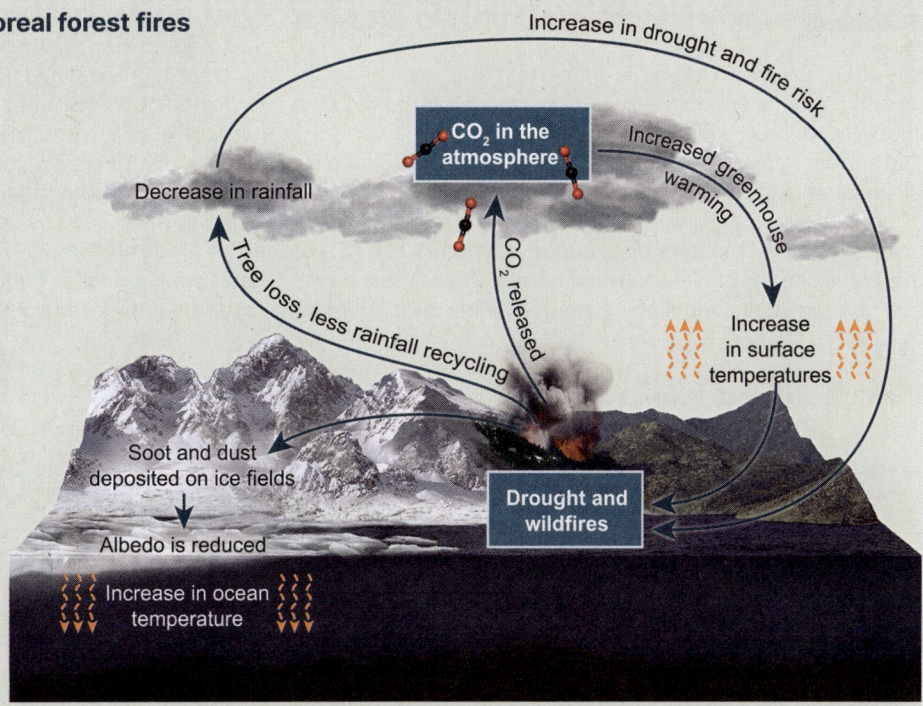

Boreal forests, or taiga, make up 30% of all global forest cover and surround polar regions. The forests consist mainly of cold-tolerant conifer species and typically act as carbon sinks, containing 20% of the global carbon emissions. Boreal forests are at risk of reaching a tipping point due to global warming leading to climate change. Warmer temperatures and subsequent loss of snowfall are creating drought stress, browning of forests, increased vulnerability to insect pests, and more fire risk, increasing in both intensity and frequency, all of which are decreasing primary production.

Carbon enters the boreal sink through photosynthesis, so lower primary production leads to less carbon being stored. Additionally, more CO_2 is released into the atmosphere as trees burn (see above). The irreversible tipping point is reached when the boreal forest shifts from a net carbon sink to a net carbon emitter. Models have predicted tipping points in some regions will begin when global warming reaches 1.5 C° (above pre-industrial levels), and become biome-wide at around 4 C°.

3. Explain what a positive feedback cycle is and why these are so significant for climate change and global warming:

_____

_____

_____

4. What are the consequences of interconnected positive feedback cycles? _____

_____

_____

_____

5. Lowering carbon dioxide and other greenhouse gas emissions is called mitigation and can help us avoid some of the worse predicted consequences of climate change. Why does this approach have less impact once a climate change tipping point is reached?

_____

_____

_____

6. Work in small groups to discuss how climate change has already impacted you or your community. Record some actions that you have been involved in, either individually or part of a community, to reduce emissions:

_____

_____

_____

366 Climate Change and Polar Regions

Key Idea: Climate change is increasing the loss of land-fast and sea ice in polar regions, reducing polar habitats for specially adapted animals.

The surface temperature of the Earth is, in part, regulated by the amount of ice on its surface which reflects a large amount of heat into space. However, the area and thickness of the polar sea-ice is trending downwards. From 1980 to 2023, the Arctic sea-ice minimum almost halved. The 2012 summer saw the greatest reduction in sea-ice since the beginning of satellite recordings. In 2023, the Antarctic sea winter ice extent was so low it was outside any possible natural variability, with evidence supporting global warming as the causation. Melting sea ice and land-fast ice has little impact on sea level rise but a lack of it can hugely impact the animals that use polar habitats to live, breed, and feed. Both walrus in the Arctic and emperor penguins in the Antarctic are facing catastrophic habitat damage due to changes that are much faster than they can adapt to.

Walrus and sea ice

▶ Walrus are large carnivorous marine mammals, distributed in the oceans that surround the Arctic and sub-Arctic polar regions.

▶ On average, adults weigh around 1000 kg, and require an extensive and continuous intake of molluscs, a wide variety of ocean invertebrates, and the occasional polar cod fish, if available.

▶ The walrus dive for their food, but need to rest between dives on floating sea ice, usually in depths less than 80m. The sea ice is essential for their young calves which tire easily. From April-May, female walrus give birth and nurse their young on sea ice. The floating sea ice allows the walrus to reduce density of populations, and therefore competition for food.

▶ Rising ocean temperatures and other impacts of climate change are reducing sea ice, and in some regions, such as the Chukchi Sea along the Russian coast, it completely disappears through summer. Current trends have led to models predicting total sea ice melt during summer by 2040.

▶ Walrus can migrate up to 3000 km a year but when the sea ice has disappeared they are forced to occupy coastal rocky shorelines, far away from their ocean feeding grounds. This is called haulout. These areas tend to be crowded, with colonies of up to 35, 000 walrus, as they look for space that is close enough to feeding areas. Walrus have to make round trips of up to 250 km in order to feed, as food availability decreases when competition rises. Young walrus are easily squashed to death without the space afforded by sea ice.

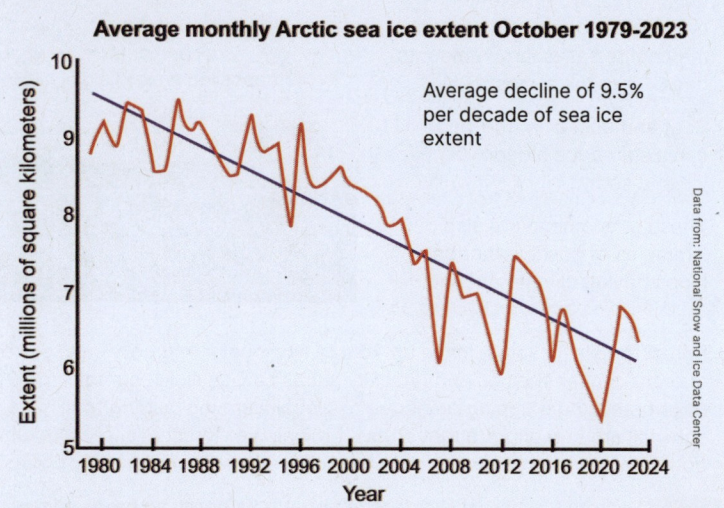

Average monthly Arctic sea ice extent October 1979-2023

Average decline of 9.5% per decade of sea ice extent

Data from: National Snow and Ice Data Center

Walrus resting on sea ice between feeding trips

1. What processes are contributing to the destabilization of Arctic habitats? _____

_____

_____

2. Why is the floating sea ice so important to the walrus? _____

_____

_____

3. What might be some long term consequences to walrus populations due to sea ice melt? _____

_____

_____

D4.3

4

Antarctica land-fast ice melt

▶ Land-fast ice (or sometimes called just fast ice) are shelves of frozen sea water attached to the edge of land, or anchored to the sea floor. They surround the coastline of Antarctica, controlling the speed of ice shelf and glacial flow into the ocean.

▶ The size of the land-fast ice ebbs and flows seasonally, typically reducing from mid-September onwards, but never quite disappearing, even in mid-summer in the southern hemisphere. In June 2022 (winter) a record was reached. Satellite data showed a minimum area of just 1.23 thousand square kilometres, compared to a typical average of 1.68 - 2.95 thousand square kilometres for June.

▶ The Antarctic ice sheet covers 98% of the continent at 14 million square kilometres. It contains 26.5 million cubic kilometres of ice, enough to raise sea levels by 58 meters if all of it was to melt.

▶ Recent studies of the ice sheet show rapid melting in some areas, with the 2023 season showing the greatest melt in modern times. Not only did a greater amount of last-fast sea ice disappear, but the melt began much earlier in the spring. This had huge consequences for Antarctic species living and breeding in the region, specifically the emperor penguin.

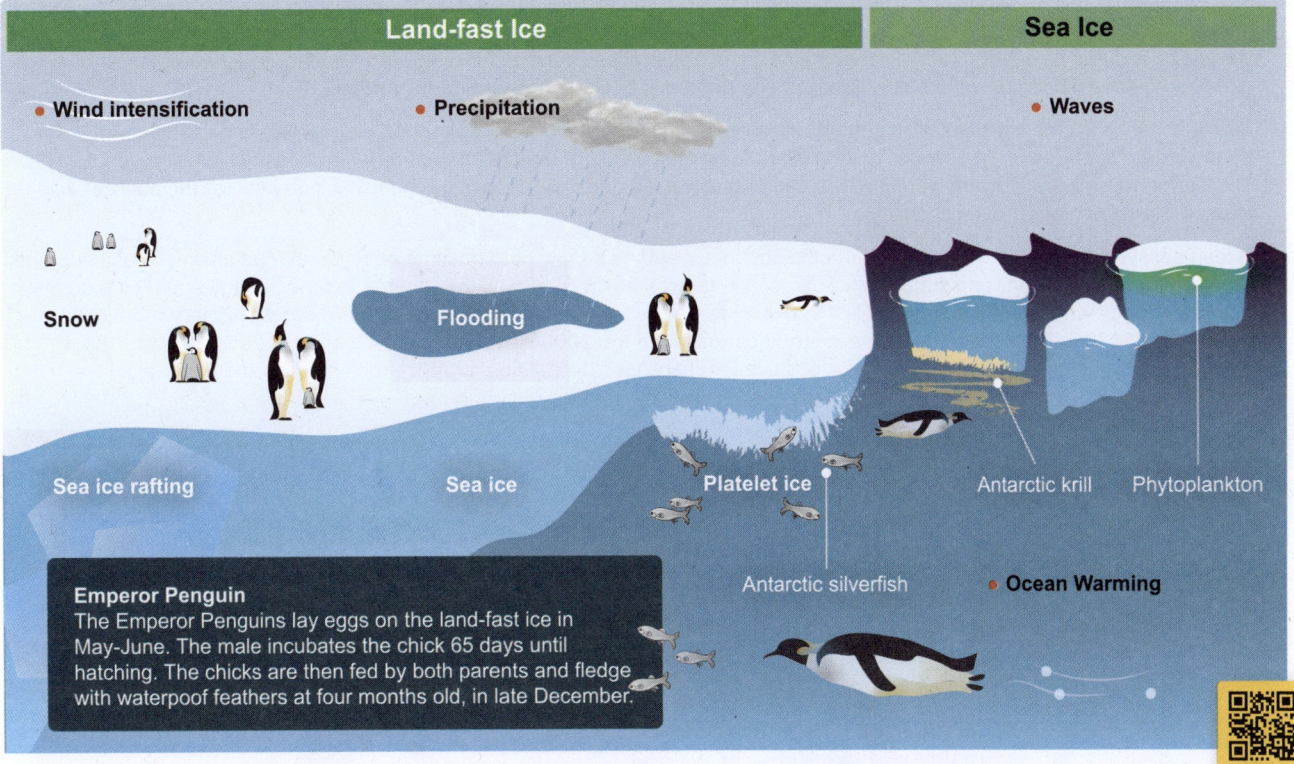

The decline of the Emperor penguin

▶ Emperor penguins are the only animals to remain in Antarctica over winter. They need to be close enough to the open sea to access silver fish and krill, however, only the female feeds during winter. They therefore lay their eggs in colonies on the fast ice areas, such as the Bellingshausen sea (right).

▶ In 2022, four out of the five emperor penguin colonies lost 100% of their chicks, who drowned due to the land-fast ice melting before they could fledge. In 2023, the ice did not form until late July, delaying the laying of eggs by a month and resulting in delayed fledging.

▶ Nearly 30% of all Antarctic penguin colonies were impacted by early land-fast ice melt from 2018-2022. Climate change models are projecting an increasingly greater loss of this ice, from forming later, a smaller area forming, and melting earlier.

4. Why can the emperor penguin not be relocated to other areas to breed and live? _____

_____

_____

5. Create a short report on Antarctic habitats and the emperor penguin by addressing the following points: the location of the breeding site and the link to the food supply/penguin mobility, the period required for breeding to fledging of chicks, the link to land-fast ice formation, and the current and projected fate of both the land-fast sea ice and the emperor penguin populations. Attach the report to this page.

367 The Effects of Climate Change

Key Idea: Climate change, due to global warming, is causing wide ranging impacts on physical systems and ecosystems. Warmer global temperatures are impacting long established, and stable systems. In some regions, such as the poles, the temperature increase is even more pronounced, and the changes occurring are impacting the whole globe. Ocean circulation and current changes are affecting the quantity and timing of nutrient laden upwellings. Various mechanisms are increasing dissolved CO_2 in the ocean, leading to ocean acidification, which combined with warmer waters are decimating coral ecosystems - areas that are rich in diversity and support around 25% of marine species. On land, warmer temperatures are forcing the migration, for those species that are mobile, into cooler or higher altitude regions.

Ocean circulation

▶ The oceans transport heat, moisture, and nutrients around the globe through a system of interconnected currents. Surface currents move from cold to warmer water; a consequence of differential heating.

▶ The deep-water ocean currents (the thermohaline circulation) are driven by the cooling and sinking of water masses in polar and subpolar regions, due to their difference in temperature and salinity. Cold water circulates through the Atlantic, penetrating the Indian and Pacific oceans, before returning as warm upper ocean currents to the South Atlantic. Deep water currents move slowly and, once a body of water sinks, it may spend hundreds of years away from the surface.

▶ Climate change is creating weakened currents. Greater than normal melting of ice at the poles is releasing vast amounts of freshwater to the oceans. Freshwater is less dense than seawater, and stratification of different density layers prevents mixing. This slows the sinking of ocean waters which carry dissolved O_2 and CO_2, decaying organisms, and heat, at the poles and so alters patterns of global ocean circulation.

▶ Ocean ecosystems rely on the upwelling of cold ocean currents to supply nutrients into ecosystem food chains. These nutrients 'fertilize' the upper oceans and support microorganism growth which, in turn, allows energy to flow through a marine food chain. Warmer surface waters are leading to stratification, slowing the vertical mixing of waters, and reducing primary productivity. Research in 2020 indicated a 5% increase in stratification in the top 2000m, with an intensification of 20% in the top 150m alone. This warming is another example of a positive feedback loop, increasing stratification and reducing upwelling.

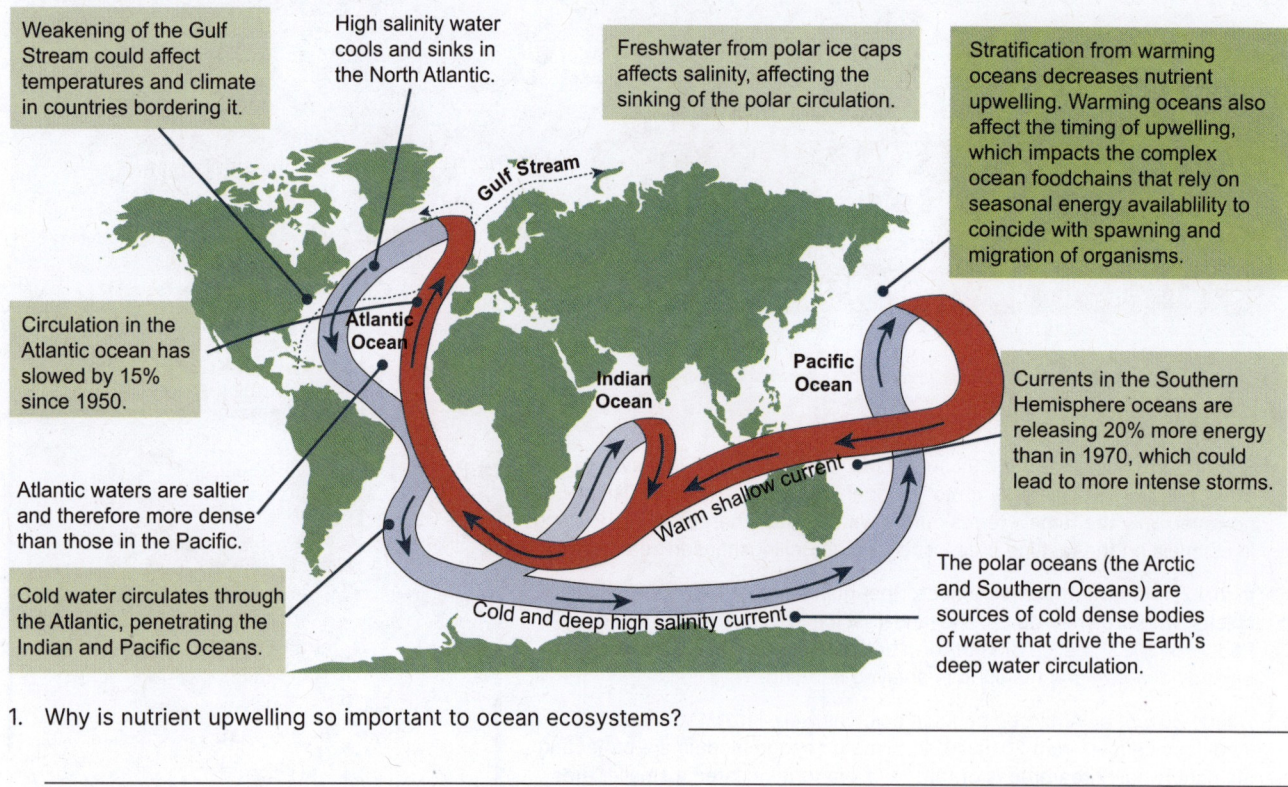

Weakening of the Gulf Stream could affect temperatures and climate in countries bordering it.

High salinity water cools and sinks in the North Atlantic.

Freshwater from polar ice caps affects salinity, affecting the sinking of the polar circulation.

Stratification from warming oceans decreases nutrient upwelling. Warming oceans also affect the timing of upwelling, which impacts the complex ocean foodchains that rely on seasonal energy availablility to coincide with spawning and migration of organisms.

Circulation in the Atlantic ocean has slowed by 15% since 1950.

Atlantic waters are saltier and therefore more dense than those in the Pacific.

Cold water circulates through the Atlantic, penetrating the Indian and Pacific Oceans.

Currents in the Southern Hemisphere oceans are releasing 20% more energy than in 1970, which could lead to more intense storms.

The polar oceans (the Arctic and Southern Oceans) are sources of cold dense bodies of water that drive the Earth's deep water circulation.

Gulf Stream

Atlantic Ocean

Indian Ocean

Pacific Ocean

Warm shallow current

Cold and deep high salinity current

1. Why is nutrient upwelling so important to ocean ecosystems? _____

_____

_____

_____

2. Humpback whales migrate from warm and shallow, but food-poor, breeding grounds to colder ocean waters in winter where they take advantage of ocean upwelling to provide them with rich feeding grounds of krill and small fish. How would a disruption in the timing of the ocean upwelling impact the humpback whales?

_____

_____

_____

D4.3

5 - 7

©2024 **BIOZONE** International
ISBN: 978-1-99-101410-8
Photocopying prohibited

Ocean acidification and warming - impacts on coral reef ecosystems

▶ The rise in the global atmospheric temperature due to increasing greenhouse gas emissions must ultimately affect the oceans. Ocean temperatures have always fluctuated, but measurements show that the average ocean temperature is rising, and have been higher in the past thirty years than in the past 150 years, since scientific observations have been recorded.

▶ Climate change is a concern to ocean ecosystems for a number of reasons. The change in ocean upwelling has been previously discussed, but rising ocean surface temperatures will also affect marine communities adapted to live at certain temperatures, such as coral reefs. The increase of CO_2 in the atmosphere is leading to increased amounts dissolved in the ocean, lowering the pH through the formation of carbonic acid and H^+ ions - resulting in ocean acidification. Changes in this abiotic factor reduce the calcification rate of many species, including coral reefs and shelled molluscs, limiting their ability to form and maintain skeletal structures.

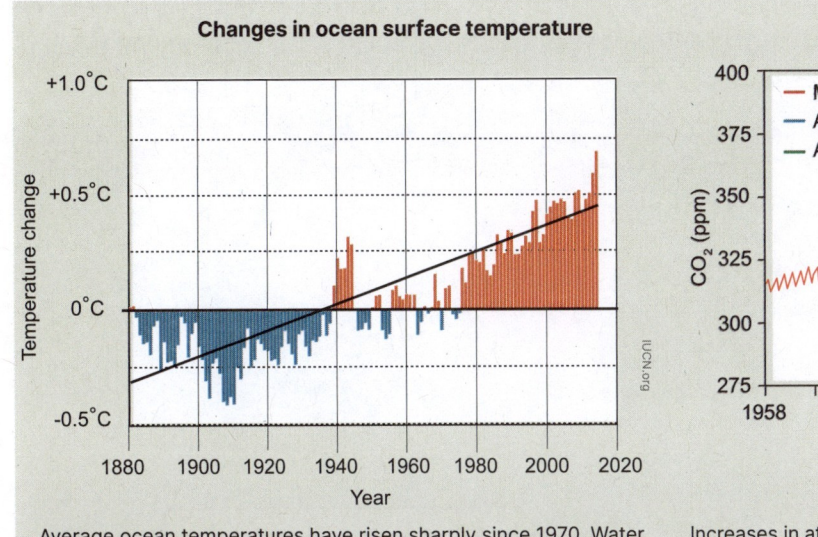

Changes in ocean surface temperature

Average ocean temperatures have risen sharply since 1970. Water absorbs a large amount of energy for every degree Celsius it rises (4.2 joules per millilitre or gram). Thus, even a small rise in sea temperature equates to the absorption of an enormous amount of energy when considering the entire oceans.

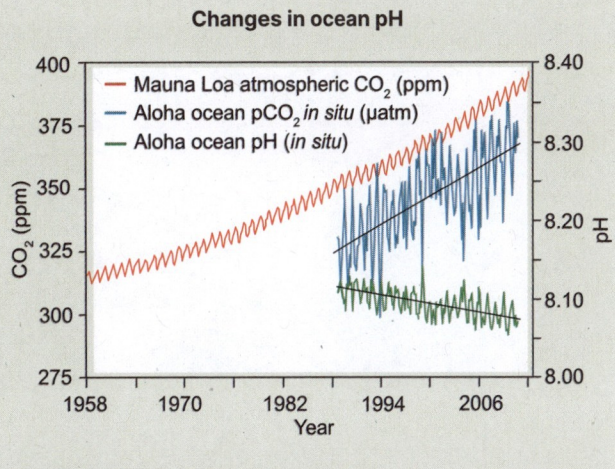

Changes in ocean pH

- Mauna Loa atmospheric CO_2 (ppm)
- Aloha ocean pCO_2 *in situ* (µatm)
- Aloha ocean pH (*in situ*)

Increases in atmospheric CO_2 levels cause more CO_2 to dissolve into ocean waters. This reacts with water to form carbonic acid and lowers the pH of the water.

Healthy coral
Zooxanthellae in coral tissue

Bleached coral
Zooxanthellae expelled from coral tissue

Dead coral
Skeleton covered in filamentous algae

Water temperature increases — Prolonged temperature stress

Days - Weeks   Days - Weeks

Healthy coral   Bleached coral   Dead coral

Weeks - Months

Water temperature returns to normal

- ∿ Zooxanthellae
- ∿∿∿ Coral polyps

Coral bleaching process

An increase in sea temperatures could mean the death of coral reefs. Healthy coral reefs depend on the symbiotic relationship between a coral polyp that builds the reef and photosynthetic protists called zooxanthellae. Zooxanthellae live within the polyp tissues and provide coral with most of its energy. A 1-2°C temperature increase is enough to disrupt the photosynthetic enzymes. The zooxanthellae either die, or are expelled from the coral due to stress. The result is coral bleaching. Some coral bleaching is reversible, if water temperature cools once more to tolerance levels, but this process is much slower than the original bleaching event. Once the coral is dead, this state is irreversible, and coral cannot recover.

3. Coral bleaching events can impact more than just the organisms themselves; whole ecosystems rely on the coral to transfer energy through the food chain and provide habitats for other species to live, breed, and be protected.
Research one recent coral bleaching event, and present the consequences in the form of a flowchart or mind map. Show details of the factors leading to the coral bleaching, such as temperature rise of ocean waters, and observed and potential consequences impacting the ecosystem. Attach your presentation to this page.

The chemistry of ocean acidification

▶ The pH of the oceans has fluctuated throughout geologic history but has always remained at around pH 8.1 - 8.2. Recent studies have measured current ocean pH at around 8.0.

▶ The oceans act as a carbon sink, absorbing much of the CO_2 produced from burning fossil fuels. When CO_2 reacts with water, it forms carbonic acid (H_2CO_3) which decreases the pH of the oceans.

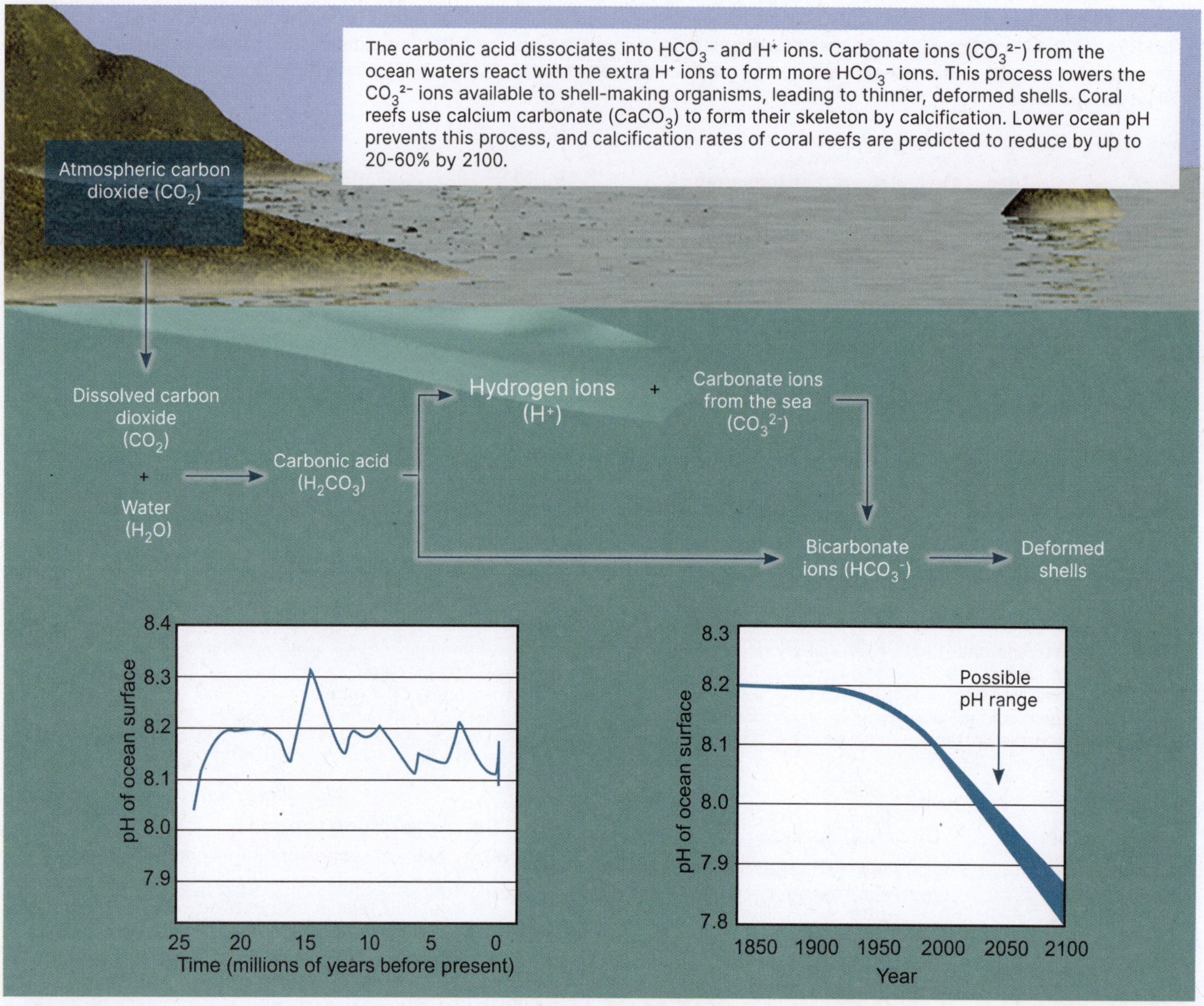

The carbonic acid dissociates into HCO_3^- and H^+ ions. Carbonate ions (CO_3^{2-}) from the ocean waters react with the extra H^+ ions to form more HCO_3^- ions. This process lowers the CO_3^{2-} ions available to shell-making organisms, leading to thinner, deformed shells. Coral reefs use calcium carbonate ($CaCO_3$) to form their skeleton by calcification. Lower ocean pH prevents this process, and calcification rates of coral reefs are predicted to reduce by up to 20-60% by 2100.

Atmospheric carbon dioxide (CO_2)

Dissolved carbon dioxide (CO_2)
+
Water (H_2O)

→ Carbonic acid (H_2CO_3)

→ Hydrogen ions (H^+) + Carbonate ions from the sea (CO_3^{2-})

→ Bicarbonate ions (HCO_3^-) → Deformed shells

Left graph: y-axis "pH of ocean surface" (7.9–8.4), x-axis "Time (millions of years before present)" (25–0)

Right graph: y-axis "pH of ocean surface" (7.8–8.3), x-axis "Year" (1850–2100), labelled "Possible pH range"

▶ pH is a logarithmic scale so even a small change in pH represents a large change in H^+ concentration. Some areas of the ocean, e.g. areas of increased human activity or underwater volcanic eruptions are more affected by pH change than others.

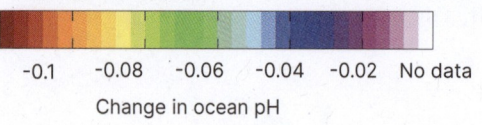

-0.1   -0.08   -0.06   -0.04   -0.02   No data
Change in ocean pH

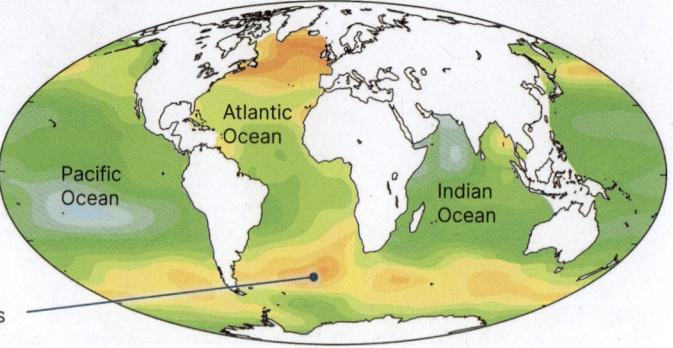

Atlantic Ocean

Pacific Ocean

Indian Ocean

Change of -0.09 pH units

4. (a) Define the term ocean acidification: _____

_____

(b) Describe the trend in ocean pH since the 1850s: _____

_____

5. Suggest what might be causing this: _____

_____

Climate change and range shift

▶ Climate change is changing the habitats of organisms. This may have profound effects on the biodiversity of specific regions, as well as on the planet overall. As temperatures rise, organisms may be forced to move to areas better suited to their temperature tolerances. Those that cannot move or tolerate the temperature change may face extinction.

▶ Changes in precipitation as a result of climate change will also affect where organisms can live. Long term changes in climate will ultimately result in a shift in vegetation zones as some habitats contract and others expand.

Effects of increases in temperature on populations

A number of studies indicate that animals are beginning to be affected by increases in global temperatures. Data sets from around the world show that birds are migrating up to two weeks earlier to summer feeding grounds that are further poleward, to compensate for temperature increase. This is also reflected in their wintering range, which is proportionally more poleward as well. Collated research on a wide range of North American birds demonstrates the movement poleward.

Animals living at altitude are also affected by warming climates and are being forced to shift their normal range. As temperatures increase, the snow line increases in altitude, pushing alpine animals to higher altitudes. In some areas of North America, this has resulted in the local extinction of the North American pika (*Ochotona princeps*).

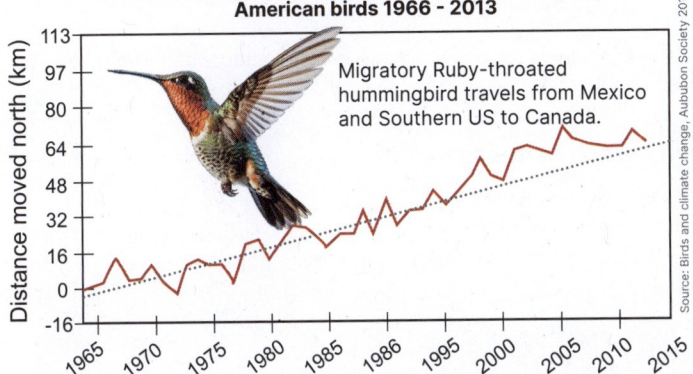

Change in centre of abundance in 305 widespread North American birds 1966 - 2013

Migratory Ruby-throated hummingbird travels from Mexico and Southern US to Canada.

Source: Birds and climate change, Audubon Society 2014

Wiki Commons PD

Upslope range shifts in tropical-zone montane bird species in New Guinea

The montane (mountain) regions of New Guinea form a group of unique, but unconnected, habitats. Due to their altitude-induced, cooler microclimates within a tropical region, the montane forests contain endemic species that are specifically adapted, and restricted to altitudes that correspond with their tolerance of temperature.

Bird species in the montane regions of New Guinea appear to be particularly sensitive to long-term climate change and temperature increase in their habitats. Research has measured a shift upslope of 147m on Karkar Island, and 107m on Mt Karimui over the past 42 years, corresponding to around a 0.4 °C rise over the same period. Scientists project that just a further 1 °C rise will likely result in the extinction of at least 4 endemic bird species, but they also predict another 2.5 °C rise is likely by 2100 in these habitats.

PomFoto

The Eastern crested berrypecker is a New Guinean montane species at risk of extinction from climate change.

6. Why are montane forest species particularly at risk from temperature increases, attributable to climate change?

_____

_____

_____

7. North American tree species range movement has been documented over the past 30 years. The average shift for trees is around 32km north and 40km west. However, deciduous trees tend to move westward and precipitation patterns are demonstrating less rainfall in the east. Conifer species are moving poleward as temperatures rise beyond the plants' tolerance levels. What might be some barriers to current and future movement of habitats for tree species?

_____

_____

_____

368 Carbon Sequestration

Key Idea: Ecosystem development and restoration can be used as a method of carbon sequestration.

If greenhouse gas production was to completely stop today, Earth's temperature would continue to rise slightly. To prevent continued climate change, climate scientists agree that we need to reduce our level of greenhouse gas emissions to slow the most damaging effects. Carbon sequestration (carbon capture and storage) is a viable means of reducing atmospheric concentrations of CO_2. Technology can be used to develop a wide range of possible sequestration solutions, but natural ecosystem development approaches can be effective: reforesting bare areas (afforestation), regenerating depleted forest, and restoring peat wet-lands. Scientific research is still ongoing to assess the effectiveness of both natural rewilding methods, and native and non-native plantations.

Reforestation and afforestation carbon sequestration

▶ Before carbon storage can occur, there needs to be a mechanism to capture atmospheric CO_2 from the air. Plants naturally capture carbon through photosynthesis, converting CO_2 into glucose, which is then further processed into other, carbon-based organic substances. Some carbon is stored in the soil when the plant decomposes, other carbon passes through the food chain when eaten, and of course carbon is released through respiration as CO_2. The longer a plant lives, the longer it sequesters carbon. Plants grown in cooler areas, such as boreal forests, tend to grow much more slowly and are therefore more effective at sequestering carbon, compared with faster growing tropical forests.

▶ Damaged and depleted forest can be restored to tree covered ecosystems, or bare land can be afforested with either native or plantation (non-native) crops as a method of increasing carbon sequestration.

What trees should be planted for afforestation-driven carbon sequestration?

When afforestation is used as a carbon sequestration method, there are a number of planting options. These are: planting and maintaining native tree species; allowing the land to rewild with no intervention; growing a non-native short-rotation plantation; or growing a slower growing, non-native (using conifer) plantation. The four different methods are represented below:

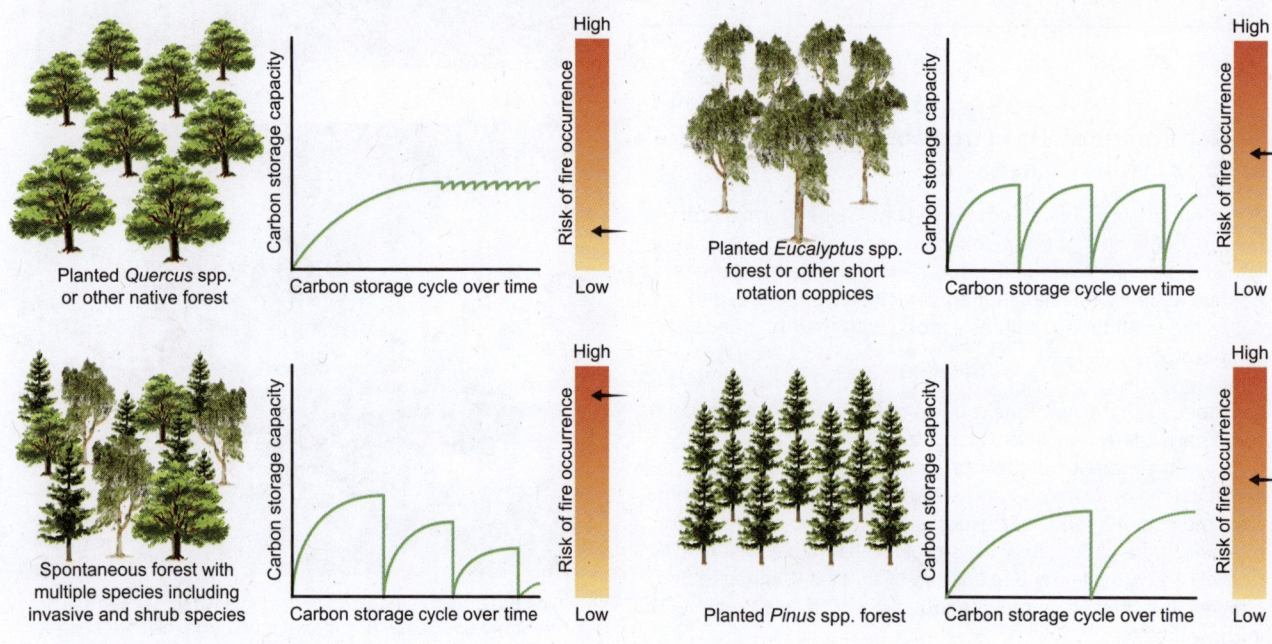

Planted *Quercus* spp. or other native forest

Planted *Eucalyptus* spp. forest or other short rotation coppices

Spontaneous forest with multiple species including invasive and shrub species

Planted *Pinus* spp. forest

1. What does carbon sequestration mean? _____

_____

_____

2. Evaluate the effectiveness of both native and non-native tree species afforestation as carbon sequestration methods:

_____

_____

_____

_____

_____

_____

©2024 **BIOZONE** International
ISBN: 978-1-99-101410-8
Photocopying prohibited

Restoration of peat forming wetlands

▶ Most natural peat lands are found in the Northern hemisphere, mainly in Canada, Siberia, and Alaska. They develop in areas with a high water table and permanent water saturation, low in both nutrients and pH, and a cool and humid climate. Although only covering about 3% of the land (200-400 million hectares), they store up to 30% of all terrestrial carbon.

▶ Wetland plants initially take up carbon through photosynthesis and build up biomass. The low oxygen and pH environment inhibits decomposition so the biomass remains in the soil and peat forms over time. Regardless of peat formation being a slow process, it is still up to 10-20 time faster than boreal forest growth as a method of carbon sequestration.

▶ Many peat wetlands in more temperate regions have been drained for farmland, and/or the peat is stripped off and used as heating fuel. The initial burning of peat releases huge CO_2 emissions, and the switch to agricultural land significantly reduces future carbon storage. These drained wetlands switch from being a net carbon sink to being a net carbon emissions contributor.

▶ Restoring suitable areas back into peat wetlands has the potential to act as a valuable method of long-term carbon sequestration as carbon and dissolved organic carbon (DOC). Methods of 're-wetting' include backfilling in drainage canals, damming waterways, and replanting of wetland plants or removal of non-wetland species, i.e. pasture grass. Rewilding is one of the simplest approaches: once artificial drainage and pest plants are removed, the ecosystem is allowed to restore without human intervention.

Peatlands carbon cycle

LIFE Peatcarbon projects are taking place in Europe as part of Climate Change Mitigation (CCM). The peatland restoration (above) is in Loch nan Con-donna, Scotland. Invasive lodgepole pines have been felled as they inhibit the peat rewilding process.

3. Why is peatland able to store so much more carbon, and for a longer-term, than just grassland or temperate forests?

_____

_____

_____

4. What might be some advantages for using restoration of peatlands as a carbon sequestration tool rather than using technology alone?

_____

_____

_____

5. Peatlands can form in boreal, temperate, and tropical ecosystems. What might some considerations be for using each of those areas for peatland restoration when considering other land use options?

_____

_____

_____

6. What is the difference between forest regeneration and afforestation as carbon sequestration methods?

_____

_____

_____

369 Climate Change and Phenology

Key Idea: Phenology studies how the timing of biological events are influenced by environmental cues.

A number of biological events in organisms are synchronized to climate events, such as the changes in length of day to night (response: photoperiodism) and temperature shifts. Reproductive life stages in deciduous plants include budburst and budset, flowering, and dormancy. Each event must coincide with other abotic and biotic ecosystem events, to align with seasonal light and warmth for growth, and presence of pollinators and seed dispersers. Likewise, animal and bird migration, and bird nesting must be correctly timed with resource availability. Climate change is causing disruption to the environmental cues in some species, although other species are adapting with behaviour changes

Examples of behavioural response to environmental changes

Plant dormancy: is a condition of arrested growth. The plant, or its seeds or buds, do not resume growth until increasing day length (photoperiod) and temperatures provide favourable growing conditions in spring.

Plant bud burst and flowering: follow exposure to a cold period in many plants, including bulbs and many perennials. This process is called vernalization and it ensures that reproduction occurs in spring and summer, not autumn.

Bird migration: Many birds move great distances at different times of the year or at certain stages in their life cycle. The behaviour is triggered by an environmental cue, e.g. a change of season. Bar-tailed godwits migrate over 22,000km each season in a round trip.

Bird nesting: This reproductive process needs to coincide with food availability, including budburst, that impacts the entire foodweb in spring. Environmental cues, such as mean night time temperature, initiate the egg laying process.

The impact of climate change on phenological events

▶ Climate change, specifically the global rise in temperature, is disrupting the timing of biological events. In an ecosystem, the timing of events for one species is often crucial and interlinked to the timing of another. This is particularly important when the budding and growth of a plant species needs to occur when an arriving or breeding animal species relies on the species for food at an important time.

▶ When warmer temperatures bring forward biological events in one species, but another species uses a day-night length - an immutable photoperiod phenomena not influenced by climate change, then timed events can become unsynchronized.

Reindeer (*Rangifer tarandus*) in the Northern tundra begin to migrate to their spring feeding grounds in late spring. The timing of this event is cued by day length so is consistent year-by-year. They are herbivores who consume energy-rich herbs and vegetation. Arctic mouse-ear chickweed (*Cerastium arcticum*) is an important food source, but its growth is cued by temperature. Increasing temperatures are pushing forward peak plant productivity so when reindeer arrive at the feeding grounds this point has already been passed, reducing the quantity of food available to them.

The great tit (*Parus major*) is found in north European forests. It requires energy rich food, such as caterpillars, to successfully raise chicks. Warming temperature thresholds induce peak caterpillar growth and population, and climate change is attributed to pushing forward this date. The breeding initiation of great tits is less influenced by temperature, and the mistiming of hatching chicks and peak food supply is leading to unsuccessful breeding. Scientists predict that a gap of just 24 days of mis-timing between these two factors can result in local extinction of the great tit.

1. Discuss how climate change can lead to the disruption of phenological event synchrony disruption using a provided or researched example (you may continue your response on paper and attach to the page):

_____

_____

_____

©2024 **BIOZONE** International
ISBN: 978-1-99-101410-8
Photocopying prohibited

Phenotypic plasticity and climate change

▶ You will recall that most species have what is termed phenotypic plasticity. This means they are able to change their behaviour, physiology, or morphology as their environment changes (within limits). This includes aspects such as learning a new behaviour or changing breeding strategies (timing and frequency). If this plasticity is extensive enough, individuals can keep up with environmental changes, e.g. rising temperatures. However, phenotypic plasticity is not adaptation in the genetic sense. It involves changes to the phenotype (behaviour, morphology, etc.) without a change in genotype. It may allow species to track environmental changes, such as rising temperatures generated by climate change, and give them time to adapt genetically.

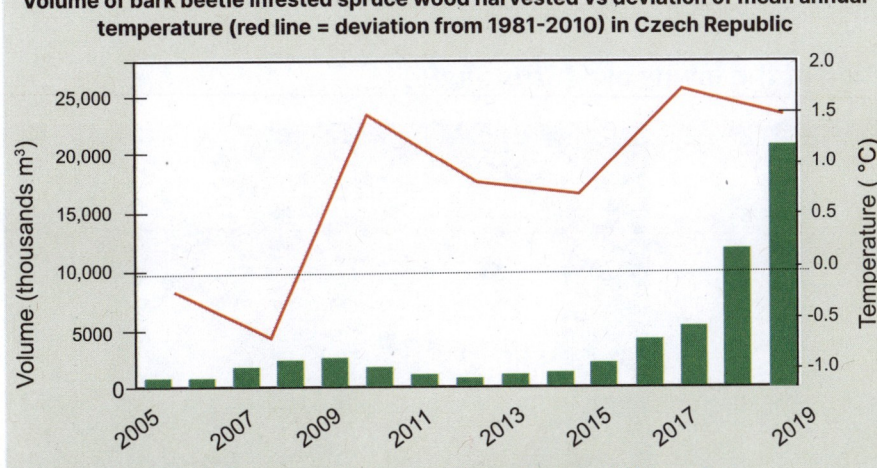

Volume of bark beetle infested spruce wood harvested vs deviation of mean annual temperature (red line = deviation from 1981-2010) in Czech Republic

Spruce bark beetle (*Ibs typopraphus*) emerging from bark in a spruce tree.

Spruce bark beetle - increasing number of insect life cycles

▶ Spruce bark beetles typically have a two-year life cycle. This begins with egg laying in bark in June; the tree eating larvae start consuming the spruces a month later; they then pupate and emerge as mature adults the following spring when ambient temperatures reach around 16 °C. If summers are warm, bark beetles can compress their life cycle to one year.

▶ Climate change is increasing the amount of damage inflicted by destructive bark beetles. Warmer temperatures are resulting in shortened life cycles of the beetles and therefore more beetles are reproducing, leading to increased population numbers. Typical predator and prey systems, where trees regenerate between population bursts, are being disrupted, leading to the death of spruce tree strands.

2. How is phenotypic plasticity helping the spruce bark beetle adapt to climate change? _____

_____

_____

_____

_____

_____

3. Explain why some organisms are more likely to become extinct as a result of climate change than others:

_____

_____

_____

_____

4. (a) Why is climate change leading to a phenology mismatch in the great tit and caterpillar peak food needs and mass?

_____

_____

_____

(b) What are some possible consequences for the stability of the ecosystem due to this phenology mismatch?

_____

_____

_____

370 Climate Change and Evolution

Key Idea: Climate change is creating new selection pressures, which is driving evolution in some species.

Climate change is creating changes to ecosystems. Increase in temperature, increase or decrease in precipitation, changes or range shift in vegetation or food sources, and a decrease in ice cover all contribute. For many species, these changes act as selection pressures. Those individuals in a population with variation that allows them to compete effectively for the available resources in the changed environment are more likely to survive and produce more offspring. Eventually, this is likely to result in a gene pool change, and therefore drive evolution as successful alleles increase in frequency. These evolutionary changes can affect both behavioural or physical traits.

Climate change induced evolution of the tawny owl (*Strix aluco*)

▶ The tawny owl lives in woodlands across most of Europe and into Western Siberia and are non-migratory. The species is polymorphic in feather colour on their backs, ranging from greyish brown (left) to rusty brown (right). This is a genetic trait.

▶ The feather colours correlate to success in hunting, mostly for small mammals, which they almost exclusively undertake at night. Grey-brown colouration provides a hunting advantage in snow-covered northern regions, whereas rusty-brown is more common in humid, western areas.

▶ Females have no sexual preference for male colour; therefore sexual selection is not a selection pressure.

▶ Data collected from over 28 years in Finland show a strong correlation between reduced snow cover due to climate change and a frequency increase (from 30% to 50% of the population) in rusty brown colour morphs, as evidence of evolution. Scientists are still unclear how much the rusty brown colour adds to the fitness of the owl, but hypothesize that the mechanism may be similar to London's peppered moths that enabled them to be better camouflaged on 19th century pollution covered trees.

1. Using changes in frequency of tawny owl colour variants as an example, explain how climate change can act as a selection pressure, and how evolutionary changes could be irreversible after a certain period:

_____

_____

_____

_____

2. New Zealand red-billed gulls, a type of sea bird, are becoming more slender. Data from over 46 years indicates the gulls are being born at a normal size, but with less body fat correlating to temperature rise due to climate change. Discuss how this might be an example of an evolutionary advantage as a consequence of climate change:

_____

_____

_____

AHL

D4.3
12

©2024 **BIOZONE** International
ISBN: 978-1-99-101410-8
Photocopying prohibited

371 Did You Get It?

1. What factors were required for natural selection to act on the fennec and Arctic ancestral fox population?

_____

_____

2. What are selection pressures and how were they different between the two fox species on the right?

_____

_____

_____

_____

_____

The fennec fox illustrates the adaptations for dessert survival: a small body size and lightweight fur, and long ears, legs, and nose.

The Arctic fox illustrates adaptations for Arctic survival: a stocky, compact body shape with small ears, short legs and nose, and dense fur.

Drew Avery cc2.0

3. **AHL:** Does global warming in the Arctic act as directional, disruptive, or stabilizing selection to the Arctic fox population? Explain the difference between the three types of selection:

_____

_____

4. **AHL:** Silver foxes can be domesticated over generations, and aside from docility (tameness), other unintended genotypes are also selected for, such as colouration. Compare between natural selection and artificial section and describe some of the issues that might arise from artificial selection:

_____

_____

_____

Frozen tundra, much as it would have appeared for mammoths

Modern, thawing tundra

5. Mammoth were a keystone species. Explain what that term means using the frozen tundra ecosystem as an example:

_____

_____

_____

6. How does climate change contribute to thawing permafrost and why is this considered an example of a feedback loop?

_____

_____

_____

7. Rewilding is a method to increase carbon sequestration. Some scientists have suggested genetically engineered mammoths could be returned to the tundra. Explain how both statements are linked:

_____

_____

_____

372 Summary Assessment

1. Where do replication, transcription, translation, and post-transcriptional modification occur, in order, in a eukaryote cell?

   (a) Nucleus, cytoplasm, ribosome, organelles

   (b) Nucleus, nucleus, ribosome, ribosome

   (c) Nucleus, nucleus, ribosome, organelles

   (d) Nucleus, ribosome, ribosome, cytoplasm

2. What is changed in a SNP mutation?

   (a) A gene

   (b) A base

   (c) A chromosome

   (d) A codon

3. What is the correct order of stages in mitosis?

   (a) Prophase, metaphase, anaphase, telophase

   (b) Prophase, anaphase, metaphase, telophase

   (c) Anaphase prophase, metaphase, telophase

   (d) Prophase, metaphase, telophase, anaphase

4. What role does oestradiol play in pregnancy?

   (a) Initiates development of placenta

   (b) A growth factor for the embryo

   (c) Thickens the uterus lining

   (d) Initiates contractions for birth

5. Which of the following is an example of co-dominance?

   (a) AB blood type

   (b) four o'clock flower

   (c) ABO blood groups

   (d) Haemophilia

Question 6 and 7 refers to the graph below

Changes in atmospheric CO_2 since 1960

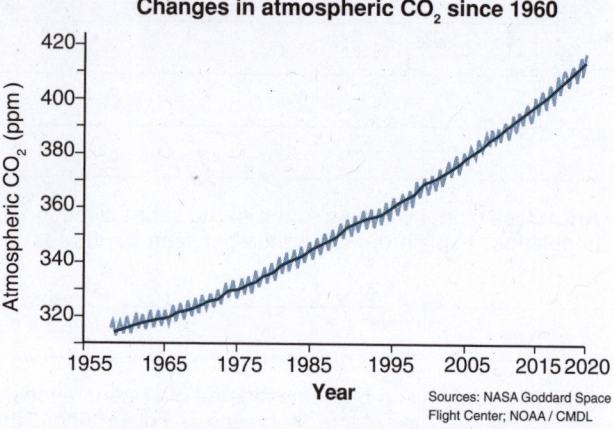

Sources: NASA Goddard Space Flight Center; NOAA / CMDL

6. Which of the following is the likely cause of the trend in the graph?

   (a) Increased combustion of fossil fuels

   (b) Increased levels of solar activity

   (c) A constant annual increase in atmospheric temperature

   (d) An increase in evaporation from the oceans

7. An increase in atmospheric CO_2 is leading to increased acidification of the ocean. Explain the link between these two phenomena and describe some consequences to marine species and ecosystems:

   _____

   _____

   _____

   _____

Cas9 unwinds the DNA and cuts both strands at a specific point.

3'
5'
Target DNA sequence

8. (a) **AHL:** The CRISPR-Cas9 system is used in gene editing. Explain a successful use of this technology:

   _____

   _____

   _____

   _____

   (b) **AHL:** What is the difference between the 3' and 5' strands of DNA and how does this affect the directionality of DNA polymerase action during replication?

   _____

   _____

   _____

   _____

9. (a) Define the term fitness:

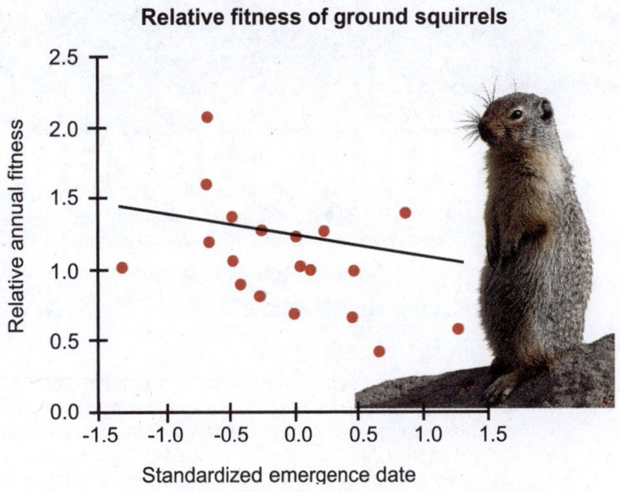

_____

_____

_____

(b) How would the emergence date of ground squirrels influence intraspecific competition?

_____

_____

_____

_____

Relative fitness of ground squirrels

(c) How might continued intraspecific competition lead to natural selection in the ground squirrel population?

_____

_____

_____

(d) How does meiosis lead to genetic variation in the ground squirrel population and why is variation important?

_____

_____

_____

10. Below, left, is an example of a positive feedback loop. Why are negative feedback loops required for homeostasis and explain how a negative feedback loop is used to control blood glucose levels in humans:

_____

_____

_____

_____

11. (a) Why do positive feedback loops have the potential to lead to climate change tipping points? _____

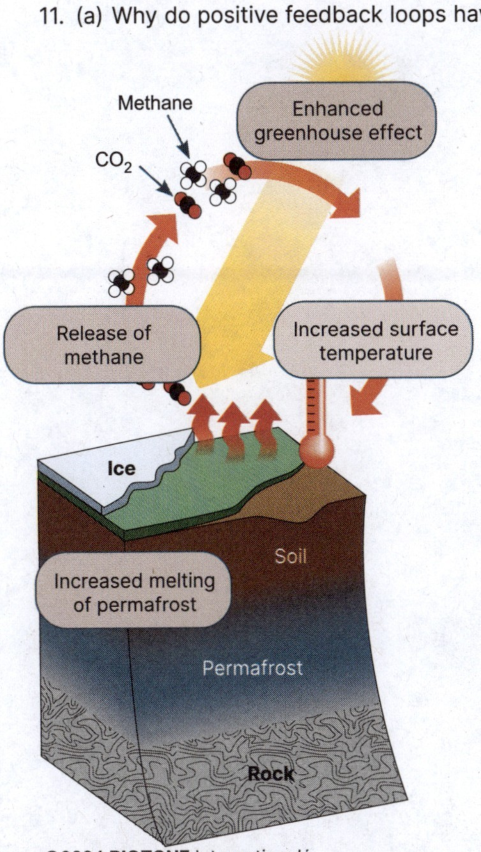

_____

_____

_____

_____

(b) What phenomenon is the diagram to the left showing and how do anthropogenic causes contribute to it?

_____

_____

_____

_____

_____

(c) **AHL:** Why might the thawing of permafrost ecosystems be considered an example of arrested succession?

_____

_____

_____

Chapter 17

Scientific Investigation

Scientific Investigation	Page Number

☐ 1   Design a unique research question for investigation. This will be investigated in the laboratory or by fieldwork.
    • Select methods for measuring dependent and independent variables.
    • Describe area of research, its relevance and state the research question within the context of this area of research.
    • Carefully and clearly describe the method for data gathering, associating relevant steps with the data being gathered.
    • Clearly write in a way that would allow the method and results to be replicated.
	599

☐ 2   Analyse data in ways that are relevant to the investigation.
    • Record data clearly to allow for interpretation using the correct units where relevant.
    • Present tables and graphs following conventions such as units and numbering e.g. significant figures.
    • Process data in a way that is relevant to the investigation using methods and presentation that can be clearly understood by a reader.
	600

☐ 3   Draw conclusions from the data that will answer the questions set up in the investigation.
    • Draw conclusions consistent with the data gathered and analysed in the investigation.
    • Acknowledge other published material or relevant information.
	602

☐ 4   Evaluate the investigation including methods and suggested improvements.
    • Identify methodological weaknesses or limitations and their possible effects on the investigation.
    • Identify improvements that might produce more accurate or precise data if the investigation was to be repeated.
    • Cite all sources used in a reference list.
	602

373 Scientific Investigation

Key Idea: Four components of a scientific investigation are research design, data analysis, drawing a conclusion and evaluating the experiment.

In a practical investigation, you will directly collect data for analysis (called primary data). For example, measuring the pH of water samples or recording the perching times of birds in an aviary. The data you gather will have an independent and dependent variable, e.g. the rate of bubbles produced by *Elodea* leaves(dependent variable) vs the intensity of the light (independent variable). It is important to consider what data you will collect and how you will collect it when you plan your investigation so that your analysis is meaningful. For the purposes of easy analysis, it is best to collect quantitative data where possible.

Planning your scientific investigation

▶ Once you have decided on a research question or hypothesis for investigation you will need to decide what data is needed to support your investigation and how you will gather it.

▶ Gathering evidence to test a hypothesis is central to a scientific investigation. For a practical investigation, you need to ensure that the methods used to gather and analyse data are fair, i.e. without deliberate or unknowing bias, or the data may produce results that supports hypotheses that are flawed.

▶ For example, it is very easy to gather data that supports the idea that light objects fall more slowly that heavy objects. Dropping a feather and a hammer from head height in a closed room will undoubtedly result in the hammer hitting the ground first. However, that is not a valid result because of the biased nature of the test.

▶ You need to plan how you will organize the data you gather. What statistical tests can be carried out? You may need to carry out preliminary investigations and modify your method based on the outcome. This could include a literature review to gather background information.

How good was the investigation's design? Was it a fair test?

Identifying variables

A variable is any characteristic or property able to take any one of a range of values. Investigations often look at the effect of changing one variable on another. It is important to identify all variables in an investigation: independent, dependent, and controlled, although there may be nuisance factors of which you are unaware. In all fair tests, only one variable is changed by the investigator.

Dependent variable
- Measured during the investigation.
- Recorded on the y axis of the graph.

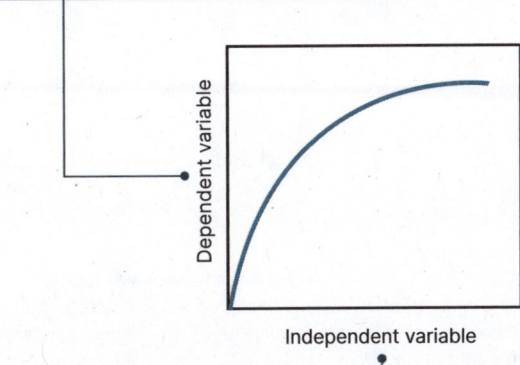

Dependent variable (y-axis)
Independent variable (x-axis)

Controlled variables
- Factors that are kept the same or controlled.
- List these in the method, as appropriate to your own investigation.

Independent variable
- Set by the experimenter.
- Recorded on the graph's x axis.

Experimental controls

A control refers to a standard or reference treatment or group in an experiment. It is the same as the experimental (test) group, except that it lacks the one variable being manipulated by the experimenter. Controls are used to demonstrate that the response in the test group is due a specific variable, e.g. temperature. The control undergoes the same preparation, controlled conditions, observations, measurements, and analysis as the test group. This helps to ensure that the responses observed in the treatment groups can be reliably interpreted.

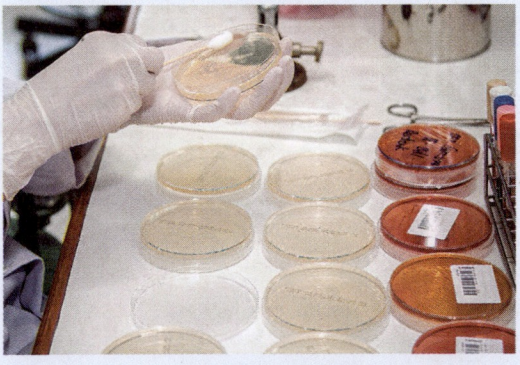

▶ The experiment above tests the effect of a certain nutrient on microbial growth. All the agar plates are prepared in the same way, but the control plate does not have the test nutrient applied.

▶ Each plate is inoculated from the same stock solution, incubated under the same conditions, and examined at the same set periods. The control plate sets the standard; any growth above that seen on the control plate is attributed to the nutrient.

Review your initial data

▶ After you have collected your first set of data (your preliminary data) it is a good idea to spend a short period of time analysing it.

▶ You may discover that you need to collect your data differently from how you first planned, e.g. taking more measurements or changing the way you collect it, such as automation for rapidly occurring changes or prolonged time series data.

▶ Take some time to plot the data or calculate summary statistics as these will allow you to see trends and patterns more easily than when the data is recorded in a logbook. Once you are satisfied that your methods of data collection are adequate, you can continue with your investigation.

How do I analyse my data?

▶ Check your data to see that it makes sense. Do the results seem logical? Are there any outliers? If so, you must decided whether to include them in your analysis.

▶ Raw data may need to be transformed to see trends and patterns. These transformations are often quite simple, e.g. percentages, rates, ratios. Other transformations are used to normalize the data so that it can undergo further analysis, e.g. log transformations when working with large numbers.

▶ Descriptive statistics, e.g. mean and standard deviation, provide a way to summarize your data and provide results that can easily be presented and compared across groups. Summary statistics are also useful in identifying trends and patterns in the data.

▶ Sometimes, an appropriate statistical analysis is required to test the significance of results. However, with simple experiments, if the design is sound, the results are often clearly shown in a plot of the data.

Presenting data in graphs

▶ Graphs are a good way to show trends, patterns, and relationships visually without taking up too much space. Complex data sets tend to be presented as graphs rather than tables, although the raw data can sometimes be included as an appendix.

▶ Presenting graphs properly requires attention to a few basic details, including correct orientation and labelling of the axes, accurate plotting of points, and a descriptive, accurate title.

▶ Before representing data graphically, it is important to identify the kind of data you have. Common graphs include scatter plots and line graphs (for continuous data), and bar charts (for categorical data). For continuous data with calculated means, points can be connected. On scatter plots, a line of best fit is often drawn. If fitting by eye, 50% of the points should fall above the line and 50% below. A line of best fit is also easily fitted using a spreadsheet program such as Microsoft Excel.

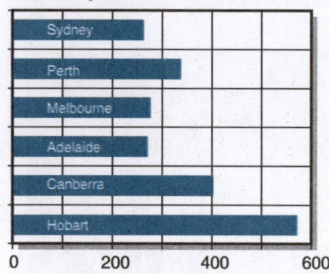

Average household water consumption in Australian cities
Household consumption (L per year x 1000)

Temperature vs metabolic rate in a rat
Line connecting points
Metabolic rate vs Temperature (°C)

Body length vs brood size in Daphnia
Line of best fit
Number of eggs in brood vs Body length (mm)

Guidelines for bar/column graphs

• Column graphs (above) are appropriate for data that are non-numerical and categorical for one variable. Data is discontinuous so the bars do not touch.

• Multiple data sets can be displayed side by side for comparison using a key, e.g. males and females.

• A histogram is superficially similar to a column graph but is used when one variable is continuous and the other is a frequency (counts). These plots produce a frequency distribution.

Guidelines for line graphs

• Line graphs are used when one variable (the independent variable or treatment) affects another, the dependent variable (the response variable).

• The data must be continuous for both variables. The relationship between two variables can be represented as a continuum and the plotted data points are connected directly (point to point).

• A double axis allows two independent variables with different measurement scales to be plotted on the same graph.

Guidelines for scatter graphs

• A scatter graph is used to plot continuous data where the two variables are interdependent.

• There is no independent (manipulated) variable, but the variables are often correlated, i.e. they vary together in a predictable way.

• The points on the graph are not connected, but a line of best fit, either by eye or computer, is often drawn through the points to show the relationship between the variables.

©2024 **BIOZONE** International
ISBN: 978-1-99-101410-8
Photocopying prohibited

Your report should include:

When writing your report, it is usual to write the methods or the results first, followed by the discussion, conclusion, and evaluation. The introduction should be one of the last sections that you write. Writing the other sections first gives you a better understanding of your investigation within the context of other work in the same area.

Important points to communicate

▶ All reports or presentations are based around a similar reporting format even though the information may be presented in different ways. In general, the format for reporting or responding to information you have researched and collated is:

1. Introduction: Introduce your investigation. This may include background information, your aim and your hypothesis.

2. Methods: Describe how you carried out your study including any equipment you used.

3. Results or analysis: Present your data in an appropriate format along with an explanation of it.

4. Conclusions and evaluation: Write a brief conclusion based on the results of the investigation. Identify limitations to the method and suggest improvements based on how the investigation proceeded.

5. Acknowledgements/references: Include a section listing your sources of information. This helps validate your arguments.

Methods

The methods section of a report should include enough detail to enable the study to be repeated, but should omit the details of standard procedures, e.g. how to use a balance. Details of statistical analyses can be included as well as the rationale for the choice of these. The diagram below describes information that should be included in the methods of a field study. It is not exhaustive but indicates the type of information that should be presented.

Study site & organisms

- Site location and features
- Why that site was chosen
- Species involved

Specialized equipment

- pH and oxygen meters
- Thermometers
- Nets and traps

Data collection

- Number and timing of observations/collections
- Time of day or year
- Sample sizes and size of the sampling unit
- Temperature at time of sampling
- Methods of sample preservation or staining
- Weather conditions on the day(s) of sampling
- Methods of measurement/sampling
- Methods of recording

Results

The results section is arguably the most important part of any research report; it is the place where you can bring together and present your findings. When properly constructed, this section will present your results clearly and in a way that shows you have organized your data and carefully considered the appropriate analysis. A portion of the results section from a scientific paper on the habitat preference of New Zealand black mudfish is presented below (Hicks, B. and Barrier, R. (1996), NZJMFR. 30, 135-151). It highlights some important features of the results section and shows how you can present information concisely, even if your results are relatively lengthy. Use it as a guide for content when you write up this section.

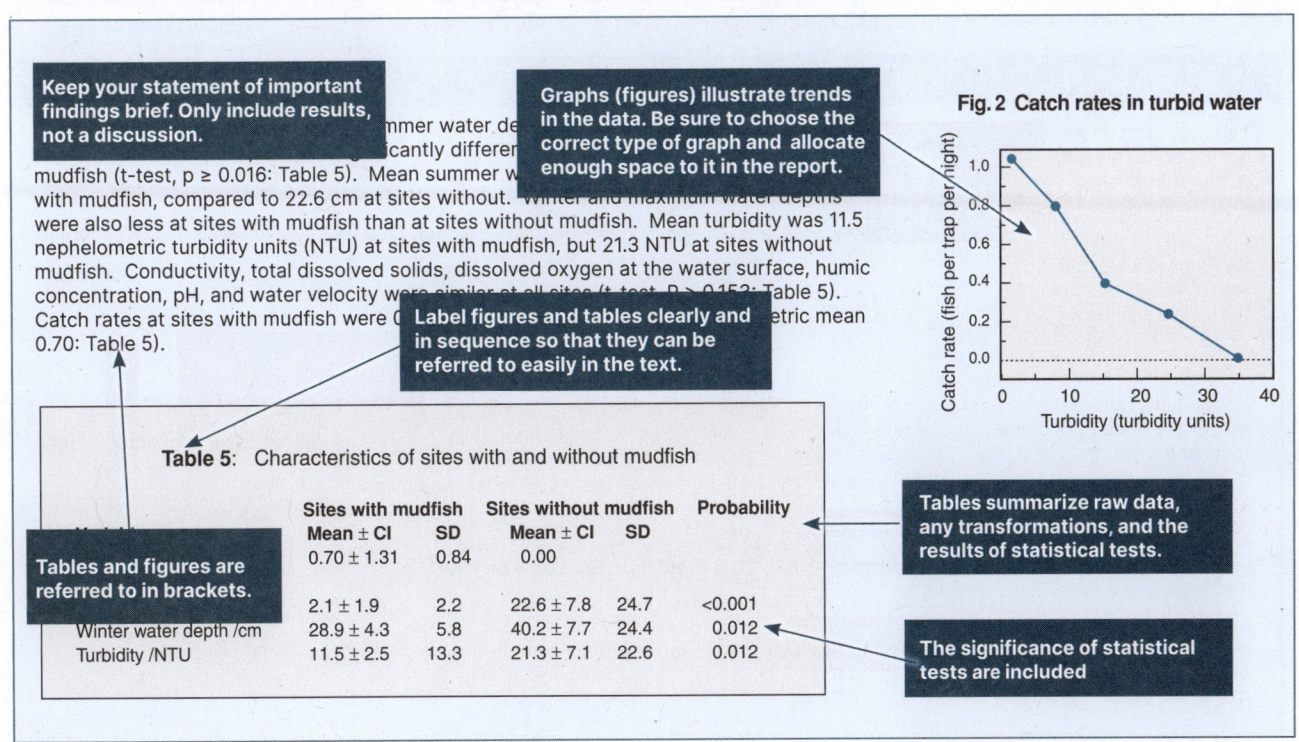

Keep your statement of important findings brief. Only include results, not a discussion.

Graphs (figures) illustrate trends in the data. Be sure to choose the correct type of graph and allocate enough space to it in the report.

Fig. 2 Catch rates in turbid water

...mmer water de... ...cantly differen... mudfish (t-test, p ≥ 0.016: Table 5). Mean summer w... with mudfish, compared to 22.6 cm at sites without. ...inter and maximum water depths were also less at sites with mudfish than at sites without mudfish. Mean turbidity was 11.5 nephelometric turbidity units (NTU) at sites with mudfish, but 21.3 NTU at sites without mudfish. Conductivity, total dissolved solids, dissolved oxygen at the water surface, humic concentration, pH, and water velocity we... ...milar at all sites (t-test, P > 0.152: Table 5). Catch rates at sites with mudfish were ... ...etric mean 0.70: Table 5).

Label figures and tables clearly and in sequence so that they can be referred to easily in the text.

Tables and figures are referred to in brackets.

Table 5: Characteristics of sites with and without mudfish

	Sites with mudfish		Sites without mudfish		Probability
	Mean ± CI	SD	Mean ± CI	SD	
	0.70 ± 1.31	0.84	0.00		
	2.1 ± 1.9	2.2	22.6 ± 7.8	24.7	<0.001
Winter water depth /cm	28.9 ± 4.3	5.8	40.2 ± 7.7	24.4	0.012
Turbidity /NTU	11.5 ± 2.5	13.3	21.3 ± 7.1	22.6	0.012

Tables summarize raw data, any transformations, and the results of statistical tests.

The significance of statistical tests are included

Conclusions and evaluation

In this section of your report, you must interpret your results in the context of the specific questions / hypothesis you set out to answer or test in the investigation. You should also place your conclusions in the context of any broader, relevant issues. If your results coincide exactly with what you expected, then your conclusions and evaluation will be relatively brief. However, your evaluation should discuss any unexpected or conflicting results and critically evaluate any problems with your study design.

Black mud[...] [Support your statements with reference to Tables and Figures from the Results section.] n be adequately described for predictive purposes by four variables that are easy to measur[...] r depth, extent of disturbance (as indicated by vegetation), and turbidity. Catch rates of bl[...] In the present study, catch rates ranged from 0.2 to 8.4 mudfish per trap per night (mean 0.70) between May and October 1992, and were similar to those of Dean (1995[...] 1994 in the Whangamarino Wetland comple[...] (0.0-2.0 mudfish per trap per night). The highe[...] [The discussion describes the relevance of the results of the investigation.] .4 mudfish per trap per night, was at Site 24 (Table 1, Figure 1). The second highest (6.4 mudfis[...] 2, in a drain about 4 km east of Hamilton. Black mudfish in the Waikato region were most commo[...] absence of water in summer, moderate depth of water in winter, limited modification of the vegetation (low DSR), and low turbidity (Fig. 2). There are similarities between the habitat requirements of black mudfish and those of brown mudfish and the common river [...] [State any limitations of your approach in carrying out the investigation and what further studies might be appropriate.] nudfish inhabited shallow water, sometimes at the edges of deeper water bodies, but were [...] an about 30-50 cm (Eldon 1978). The common river galaxias also has a preference for [...] ns < 20 cm deep (Jowett and Richardson 1995).

Sites where black mudfish were found were not just shallow or dry in summer, but also had substantial seasonal variation in water depth. A weakness of this study is the fact that sites were trapped only once; however, [Reference is made to the work of others that you may used to help design your experiment or gather data from.] portant for black mudfish, in the form of emergent or overhanging vegetation, or t[...] ning the presence or absence of black mudfish is predictable, considering the sh[...] gh nocturnally active as adults, are likely to require cover during the to protect them f[...] predators, such as bitterns (*Botaurus stellaris poiciloptilus*) and kingfishers (*Halcyon sancta vagans*). Predation of black mudfish by a swamp bittern has been recorded (Ogle & Cheyne 1981). Cover is also [...] [Further research is suggested] bitats may be a key to the successful coexistence of mudfish with their predators. Mosquitofish are known predators of mudfish fry (Barrier & Hicks 1994), and eels would presumably also prey on black mudfish, as they do on Canterbury mudfish (Eldon 1979b). If, however, black mudfish are relatively uncompetitive and vulnerable to predation, the question remains as to how they manage to coexist with juvenile eels and mosquitofish. The habitat variables measured in this study can be used to classify the su[...] [A clear conclusion is made based on the data gathered.] n future. The adaptability of black mudfish allows them to survive in some altered habitats, su[...] n this study, we can conclude that the continued existence of suitable habitats appears to be m[...] n the presence of predators and competitors. This study has also improved methods of identifying suitable mudfish habitats in the Waikato region.

References

Proper referencing of sources of information is an important aspect of report writing. It shows that you have explored the topic and recognize and respect the work of others. There are two aspects to consider: citing sources within the text (making reference to other work to support a statement or compare results) and compiling a reference list at the end of the report. A bibliography lists all sources of information, but these may not necessarily appear as citations in the report. In contrast, a reference list should contain only those texts cited in the report. Citations in the main body of the report should include only the authors' surnames, publication date, and page numbers (or internet site) and the citation should be relevant to the statement it claims to support (see above).

Example of a reference list

Lab notes can be listed according to title if the author is unknown. → Advanced biology laboratory manual (2000). Cell membranes. pp. 16-18. Sunhigh College.

Cooper, G.M. (1997). The cell: A molecular approach (2nd ed.). Washington D.C.: ASM Press
[Book titles in italics] [Publisher]

References are listed alphabetically according to the author's surname.

Davis, P. (1996). Cellular factories. New Scientist 2057: Inside science supplement.
[Journal titles in italics]

Publications from a single author are listed from oldest to most recent. Indge, B. (2001). Diarrhea, digestion and dehydration. Biological Sciences Review 14(1), 7-9.
[Article title] [Volume (Issue number), Pages]

Last name and initials of authors. Indge, B. (2002). Experiments. Biological Sciences Review, 14(3), 11-13.

Kingsland, J. (2000) Border control. New Scientist 2247: Inside science supplement.
[Publication date]

Internet sites change often so the date accessed is included. The person or organization in charge of the site is also included. → http://www.cbc.umn.edu/~mwd/cell_intro.html (Dalton, M. "Introduction to cell biology" 12.02.03)

©2024 **BIOZONE** International
ISBN: 978-1-99-101410-8
Photocopying prohibited

Appendix: Equipment List

2: Unity and diversity: Cells

INVESTIGATION 2.1
Modelling protein structure

Per group:

- [ ] Light microscope
- [ ] Onion/onion leaf
- [ ] Glass microscope slides
- [ ] Coverslips
- [ ] Scalpel or razor
- [ ] Iodine stain
- [ ] Filter paper/tissue paper

3: Unity and diversity: Organisms

INVESTIGATION 3.1
Develop a dichotomous key

Per student:

- [ ] Leaves of various local plants or photos of local animals.

7: Form and function: Organisms

INVESTIGATION 7.1
Measuring lung volumes

Per student:

- [ ] Balloon
- [ ] Large measuring container

INVESTIGATION 7.2
Comparing stomatal density

Per group:

- [ ] Light microscope
- [ ] Dicot plants (e,g, buttercup sunflowers)
- [ ] Monocot plant (e.g. maize or corn)
- [ ] Glass microscope slides
- [ ] Coverslips
- [ ] Scalpel or razor
- [ ] Access to a computer or device with internet connection

INVESTIGATION 7.3
Investigating effect of exercise on heart rate

Per group:

- [ ] Stopwatch

INVESTIGATION 7.4
Investigating vascular tissue

Per group:

- [ ] Light microscope
- [ ] Dicot plants (e,g, buttercup sunflowers)
- [ ] Monocot plant (e.g. maize or corn)
- [ ] Glass microscope slides
- [ ] Coverslips
- [ ] Scalpel or razor
- [ ] Access to a computer or device with internet connection

INVESTIGATION 7.5
Investigating motion in a shoulder joint

Per group:

- [ ] Goniometer or a similar device to measure angles.

8: Form and function: Ecosystems

INVESTIGATION 8.1
Correlating abiotic factors and population distribution

Per student:

- [ ] Quadrats
- [ ] Tape measure
- [ ] Devices for measuring abiotic factors (e.g. thermometer, lux meter, aneomometer)

9: Interaction and interdependence: Molecules

INVESTIGATION 9.1
Investigating peroxidase activity

Per group:

- [ ] 13 x boiling tubes
- [ ] 42 mL distilled water
- [ ] 1.8 mL 0.1% H_2O_2 solution
- [ ] 1.2 mL prepared guaiacol solution
- [ ] Parafilm
- [ ] 6 mL of each pH buffered solution (pH 3, 5, 6, 7, 8, 10)
- [ ] 9 mL turnip peroxidase solution
- [ ] Test tube rack
- [ ] Timer

INVESTIGATION 9.2
Measuring respiration in germinating seeds

Per group:

- [ ] 3 x boiling tubes
- [ ] Marker pen
- [ ] 6 x cotton balls
- [ ] 15% KOH solution
- [ ] 2 x eye dropper or plastic pipette
- [ ] 3 x gauze pieces
- [ ] Germinated bean seeds (enough to fill one quarter of the boiling tube)
- [ ] Ungerminated bean seeds (enough to fill one quarter of the boiling tube)
- [ ] Glass beads (enough to fill one quarter of the boiling tube)
- [ ] 3 × 2-hole tube stoppers
- [ ] 3 x bent glass tubes or pipettes
- [ ] 3 x tubes (must be able to be clamped shut)
- [ ] 3 x screw clips
- [ ] A few drops of colored liquid
- [ ] 3 x syringes (must fit tube with screw clamp attached)
- [ ] 3 x clamp stands or rack
- [ ] Water bath (25°C)
- [ ] Ruler
- [ ] Timer

INVESTIGATION 9.3
Investigating yeast fermentation

Per group:

- [ ] 1 × 100 mL beaker
- [ ] 10 g of active yeast
- [ ] 50 mL tap water at 24°C
- [ ] 25 g of substrate (glucose, maltose, sucrose, or lactose)
- [ ] 1 x glass stirring rod
- [ ] 1 x conical flask (to hold 275 mL)
- [ ] Paraffin oil
- [ ] Single hole stopper
- [ ] Tubing
- [ ] 1 × 100 mL measuring cylinder
- [ ] 1 x small basin to hold inverted cylinder
- [ ] Stopwatch

INVESTIGATION 9.4
Separating photosynthetic pigments

Per group:

- [ ] Leaves of silverbeet or spinich
- [ ] Toothpick
- [ ] Boiling tube or test tube
- [ ] Filter paper or chromatography paper
- [ ] Pencil
- [ ] Ethanol
- [ ] Clingwrap or parafilm
- [ ] Motar and pestle
- [ ] Sand
- [ ] Scissors

INVESTIGATION 9.5
Investigating the effect of light intensity on photosynthetic rate

Per group:

- [ ] 1.0 g Cabomba aquatica
- [ ] Balance
- [ ] Scissors
- [ ] Water
- [ ] 1 x large beaker (large enough to hold the glass funnel)
- [ ] 1 x glass funnel
- [ ] 0.2 molL^{-1} sodium hydrogen carbonate solution (enough to cover the plant)
- [ ] 1 x test tube
- [ ] 1 x lamp with a 60W bulb
- [ ] Lux meter
- [ ] Timer
- [ ] 1 x ruler or tape measure

©2024 **BIOZONE** International
ISBN: 978-1-99-101410-8
Photocopying prohibited

11: Interaction and interdependence: Organisms

INVESTIGATION 11.1
Investigating phototropism response to light in seedlings

Per group:

- [ ] 4 petri dishes
- [ ] Cotton wool
- [ ] Mustard or cress seeds
- [ ] Light source
- [ ] Cardboard boxes to place over petridishes

12: Interaction and interdependence: Ecosystems

INVESTIGATION 12.1
Carry out a sampling survey using quadrats to estimate the population of a plant or sessile animal

Per group:

- [ ] Quadrats
- [ ] Tape measure
- [ ] Maps
- [ ] Identification notes for plants or animals.

INVESTIGATION 12.2
Modelling yeast growth

Per group:

- [ ] 500 mL beaker
- [ ] 10 g sugar (sucrose)
- [ ] 2.5 g yeast
- [ ] Light microscope
- [ ] Microscope slides
- [ ] Coverslips

14: Continuity and change: Cells

INVESTIGATION 14.1
Modelling mitosis

Per group:

- [ ] String / yarn
- [ ] Pipe cleaner / chenille stems
- [ ] Scissors

INVESTIGATION 14.2
Estimating osmolarity

Per group:

- [ ] 6 × 500 mL beakers
- [ ] Balance and equipment to weigh sugar
- [ ] Table sugar or lab sucrose
- [ ] Potato
- [ ] Cork borer or scalpel
- [ ] Paper towels
- [ ] Marker pen

15: Continuity and change: Organisms

INVESTIGATION 15.1
Measuring continuous variation

Per student:

- [ ] Measuring equipment as required, e.g. tape measure

16: Continuity and change: Ecosystems

INVESTIGATION 16.1
Using allele frequency databases

Per student/group:

- [ ] Computer or device with internet access

INVESTIGATION 16.2
A mesocosm model of an ecosystem

Per student/group:

- [ ] 1× 2L clear soda bottle with lid
- [ ] 1 scoop aquarium gravel
- [ ] Several dead leaves
- [ ] 1x aquatic plant (e.g. Cabomba)
- [ ] 3-4 small pond snails
- [ ] 2 L filtered pond water.

©2024 **BIOZONE** International
ISBN: 978-1-99-101410-8

Image Credits

We acknowledge the generosity of those who have provided photographs for this edition:

• PASCO for photographs of probeware • Berkshire Community College Bioscience Image Library • Dr Lucille K. Georg • Cell Image Library • Cytogenetics Dept, Waikato Hospital for the karyotypes • Louisa Howard, Katherine Connollly Dartmouth College • John Green and Stephen Moore • Jonathan Wright Bernard Field Station, Claremont College • UC Berkley • Louisa Howard Dartmouth Electron Microscopy • Andrea Braakhius, Wintec • Rhys Barrier, University of Waikato • Waikato Hospital

We also acknowledge the photographers that have made their images available through Wikimedia Commons under Creative Commons Licences 1.0, 2.0, 2.5, 3.0, or 4.0:

Graham Beards • Michael Rygel • NIAID-RML • Deuterosome • Sbj1976 • Tom Bruns • JM Garg • Ryan Somm • asonhgraham3 • PLOS • Changqing Liu • Argyris JM • Giorgi D • Dr. Raju Kasambe • Jo Naylor • Bruce Marlin • Lorax • DinosaursRoar • Dinosaurs20 • Raul654 • DAVID ILIFF.• Michael J Benton • Rufino Uribe • J. Miquel *et al* • Zephyris • Christian Schmelze • A2-33 • Vossman • Itayba • Kristian Peters • Massimini *et al* • Atudu•Brett Taylor • NJR ZA • Deuterostome • Eunice Laurent • Mogana Das Murtey and Patchamuthu Ramasamy • Stan Shebs• Ryna Somma • Didier Descouens • Bernard DUPONT • Vinayaraj • Professor Dr. habil. Uwe Kils • Margaret McFall-Ngai • Roadnottaken • Kronoman • James Heilman • Mike Baird • Lilly M • Marco vinc • Luc Viatour • JJ Harrison • Marc Tarlock • Aris Gerakis • Paul Whippey • BirdPhotos.com • Artem Topchiy • Stefan Goen, Uckermarck • kdee64 (Keith Williams • Eric Guinther • Crulina 98 • Jan Kronsell • Wojsyl • Alex Proimos • it:Utente:Cits • Setsuko K. Satohetal • Brocken Inaglory • Matthias Zeppe • dsworth Center: New York State Department of Health • Emmanuelm • Barfooz • University of Wisconsin–Milwaukee • Zephyris • Neutr0nics • Ernie • Cnangarra • Cindy McCravey • RM Hunt • Klaus • Nzeemi • Tentotwo • Brian Gratwicke • Oxford University Museum of Natural History.• Kathy Chapman online • Matt knoth • Alexandr Trubetskoy • Duncan Wright • Bob Blaylock • Janke • Daderot • Rasbak • Mikrolit • Pomfoto • JJ Harrison • Drew Avery• JM Garg • Jasonhgraham3 • Ryan Somma • Changqing Liu,*et al* • Argyris JM *et al* • Giorgi D *et al* • Luc Viatour • Roger Griffith • Дмитро Леонтьєв • Usman Bashir

Contributors identified by coded credits are:

BCC: Berkshire Community College Image Library • NASA: National Aeronautics and Space Administration • NOAA: National Ocean and Atmosphere Administration • KP: Kent Pryor • PLOS: Public Library of Science • CDC: Centers for Disease Control and Prevention • EII: Education Interactive Imaging • USDE: United States Department of Energy • NIH: National Institute of Health • WMU: Waikato Microscope Unit • PEIR: Pathology Education Informational Resource Digital Library • BF: Brian Finerran (University of Canterbury) •WBS: Warwick Silvester • BH: Brendan Hicks • PDB: Protein Database • USFWS: United States Fish and Wildlife Service • CSIRO Commonwealth Scientific and Industrial Research Organisation • NPS: National Park Service

Royalty free images, purchased by BIOZONE International Ltd, are used throughout this workbook and have been obtained from the following sources:

Corel Corporation from their Professional Photos CD-ROM collection; IMSI (Intl Microcomputer Software Inc.) images from IMSI's MasterClips® and MasterPhotos™ Collection, 1895 Francisco Blvd. East, San Rafael, CA 94901-5506, USA; ©1996 Digital Stock, Medicine and Health Care collection; © 2005 JupiterImages Corporation www.clipart.com; ©Hemera Technologies Inc, 1997-2001; ©Click Art, ©T/Maker Company; ©1994., ©Digital Vision; Gazelle Technologies Inc.; PhotoDisc®, Inc. USA, www.photodisc.com. • TechPool Studios, for their clipart collection of human anatomy: Copyright ©1994, TechPool Studios Corp. USA (some of these images were modified by Biozone) • Totem Graphics, for their clipart collection • Corel Corporation, for use of their clipart from the Corel MEGAGALLERY collection • 3D images created using Poser and Pymol • iStock images • Art Today • Adobestock • Image stills from Sketchfab

Index

©2024 **BIOZONE** International
ISBN: 978-1-99-101410-8
Photocopying prohibited

©2024 **BIOZONE** International
ISBN: 978-1-99-101410-8
Photocopying prohibited

©2024 **BIOZONE** International
ISBN: 978-1-99-101410-8
Photocopying prohibited